Coral Reefs of the USA

Coral Reefs of the World

Volume 1

Coral Reefs of the USA

Bernhard M. Riegl and Richard E. Dodge

Editors

Editors:
Bernhard M. Riegl
National Coral Reef Institute
Nova Southeastern University
Dania Beach, Florida
USA

Richard E. Dodge
National Coral Reef Institute
Nova Southeastern University
Dania Beach, Florida
USA

ISBN 978-1-4020-6846-1   e-ISBN 978-1-4020-6847-8
DOI: 10.1007/978-1-4020-6847-8

Library of Congress Control Number: 2007939564

Cover illustration:
Small figure top: Photomicrograph of a Miocene *Porites* coral fossil in a dolostone from Navassa (Photo: R. Halley, Chapter 10).
Small figure bottom: Gag Grouper (*Mycteroperca microlepis*) in Broward County, Florida (Photo: L. Jordan, Chapter 5).
Large figure: A view across Tanapag Lagoon, Saipan, Mariana Islands (Photo: B. Riegl, Chapter 18).

Printed on acid-free paper

9 8 7 6 5 4 3 2 1

springer.com

This book is dedicated to our own and the children of the USA in the hope that they
will still be able to experience healthy coral reef ecosystems

Preface

Coral reefs are certainly one of the crown jewels of the USA's natural heritage and the nation is aware of that. Coral reefs receive much attention by a broad cross section of society – tourists, scientists, politicians, fishermen, conservationists are but some of those who cannot escape their charm. But there is no love-affair without some responsibility for the object of affection and the US has been very actively engaged in living up to this responsibility through a long history of research and efficient management. Being prone to boast about what it has, the USA can rightly show off its coral reefs. It possesses some of the most visited coral reefs (in Hawaii and the Florida Keys) and some of the world's biggest marine reserves (the Northwestern Hawaiian Islands= Papahanaumokuakea Marine National Monument and the Pacific Remote Islands National Wildlife Refuge Complex). It is also home to some of the world's most degraded as well as the most pristine coral reefs. Not surprisingly, its coral reef research has a long history and the coral reef research community is as vibrant and active today as it has been from the early beginnings.

This book was produced with the goal of providing an overview of coral reefs in the USA and to provide a uniform entry point for the study of any specific region. It is thus a scholarly review as opposed to a status report, such as those produced by the National Oceanographic and Atmospheric Administration (NOAA) every 4 years, which should be consulted for updates regarding biological monitoring status and trajectories of the living resources under federal jurisdiction and surveillance. In contrast, the various authors who contributed to this present book provide a sampling of published and sometimes yet unpublished knowledge of the geology, biology and oceanography of the various reef types and their inhabitants. The book is designed to provide a big picture overview of what makes each region unique (or not) and to provide a scholarly review of some key facts together with key literature. It should thus be of use both to the casual visitor seeking to generally inform him/herself of an area to be visited, and the serious scholar who wishes to receive some guidance where to begin his/her readings regarding an area of specific interest. Although most reef areas within US territorial waters are covered, it is also obvious that any such task will never be fully exhaustive and a large amount of existing literature and some information that a reader might look for will not be included. There is therefore no claim that this book is complete, it never will be, but we hope that it will be useful as an overview and entry point for each individually treated area. For any shortcomings we are to blame.

We have not included in this book the coral reefs of the Freely Associated States (the Republic of Palau, the Republic of the Marshall Islands, The Federated States of Micronesia) since we respect their political independence and therefore could not subsum them under the title "...of the USA". However, US management agencies such as NOAA and the USGS are, at least partially, involved in research and management. Information about the reef status and management issues in the Freely Associated States is provided by the various national agencies and can also be found in the NOAA Status of the Coral Reefs reports that appear every 4 years.

The organization of chapters in this present book is alphabetically by ocean and does not represent an order of listing according to any preference: first Atlantic (Florida–US Virgin

Islands–Puerto Rico–Gulf of Mexico–Navassa), then Pacific (Main Hawaiian Islands–Northwest Hawaiian Islands–Line and Phoenix Islands–Wake and Johnston Atolls–Guam and the Commonwealth of the Northern Mariana Islands–American Samoa) and finally the deep reefs of all oceans.

The editors wish to express their deep appreciation and gratitude to the authors of the various chapters for their hard work despite the exceedingly short deadline that was imposed on them in a busy time of the year. We would like to thank our publishers at Springer for believing in the project and our reviewers for providing us with excellent feedback. Reviewers for this volume were: Russel Brainard, Francine Fioust, Peter Glynn, Richard Grigg, Bob Halley, Peter Houk, Jochen Halfar, Ken Deslarzes, Greg Foster, Kristy Foster, Greg Jacoski, Jack Kindinger, Judith Lang, Barbara Lidz, Joyce Miller, Ryan Moyer, David Gulko, Christy Pattengill-Simmens, Chris Perry, Werner Piller, Sam Purkis, John Rooney, Gwilym Rowlands, Eugene Shinn, Michael Trianni, Bernardo Vargas-Angel, Andrew Wheeler, Wendy Wood and the many co-authors of the chapters who cross-checked and reviewed the other sections of their respective chapters. Other reviewers within the government agencies saw and approved the chapters by their employees and we thank them for their efforts – not all names are known to us. Maureen Trnka deserves an extra thank-you for finding literature, tirelessly digitizing graphs and formatting and reformatting manuscripts. Greg Jacoski proofread many of the chapters. We apologize if we have forgotten to thank anyone.

Bernhard Riegl
Richard Dodge
Dania, Florida
December 2007

Contents

Contributors

Richard S. Appeldoorn
Department of Marine Sciences,
University of Puerto Rico, Mayagüez, PR 00681,
USA
rappeldo@uprm.edu

Roy Armstrong
Department of Marine Sciences,
University of Puerto Rico, Mayagüez, PR 00681,
USA
roy@cacique.uprm.edu

Jerald S. Ault
Rosenstiel School of Marine
and Atmospheric Science,
University of Miami,
4600 Rickenbacker Causeway, Miami, FL 33149,
USA
jault@rsmas.miami.edu

Paula Ayotte
Marine Sciences Department,
University of Hawaii, 200 W. Kawili St., Hilo,
HI 96720, USA
payotte@hawaii.edu

David L. Ballantine
Department of Marine Sciences,
University of Puerto Rico, Mayagüez,
PR 00681, USA
dballant@uprm.edu

Jason Baker
NOAA Pacific Island Fisheries Science Center,
2570 Dole St., Honolulu, HI 96822, USA
jason.baker@noaa.gov

William Bane Schill
US Geological Survey Leetown Science Center,
National Fish Health Research Laboratory,
11649 Leetown Rd., Kearneysville,
WV 25430, USA
bane_schill@usgs.gov

Kenneth W. Banks
Broward County Environmental Protection
Department, 1 University Drive, Plantation,
FL 33322, USA
kbanks@broward.org

James P. Beets
Department of Marine Science,
University of Hawaii at Hilo, 200 W. Kawili St.,
Hilo, HI 9672, USA
beets@hawaii.edu

Charles Birkeland
Hawaii Cooperative Fishery Research Unit,
United States Geological Survey,
University of Hawaii, Honolulu, HI 96822, USA
charlesb@hawaii.edu

Chris Bochicchio
University of Hawaii,
Department of Geology and Geophysics,
1680 East West Rd., Honolulu, HI 96822, USA
bochicch@hawaii.edu

James A. Bohnsack
Southeast Fisheries Science Center,
75 Virginia Beach Drive, Miami, FL 33149,
USA
jim.bohnsack@noaa.gov

Victor Bonito
Reef Explorer Fiji, PO Box 183 Korolevu,
Fiji Islands
reefexplorerfiji@yahoo.com

Rafe Boulon
National Park Service,
Virgin Islands National Park/Virgin Islands Coral
Reef National Monument, 1300 Cruz Bay Creek,
St. John, VI 00830, USA
rafe_boulon@nps.gov

Russell Brainard
NOAA Pacific Island Fisheries Science Center,
2570 Dole St., Honolulu, HI 96822, USA
rusty.brainard@noaa.gov

Sandra D. Brooke
Ocean Research & Conservation Association Inc.,
Duerr Laboratory for Marine Conservation,
1420 Seaway Drive, 2nd Floor, Fort Pierce,
FL 34949, USA
sbrooke@oceanrecon.org

Andrew Bruckner
NOAA Fisheries Office of Habitat Conservation,
NOAA Coral Reef Conservation Program,
1315 East West Highway, Silver Spring,
MD 20910, USA
andy.bruckner@noaa.gov

Randolph B. Burke
Formerly: North Dakota Geological Survey;
600E Blvd. Ave., Bismarck, ND 58505
Present address: Burke Consulting, Research
Collaborator, Smithsonian Institution,
E 117 NHB, Washington, DC 20560, USA
burke.randy@gmail.com

Genevieve Cabrera
Department of Community and Cultural Affairs,
Historic Preservation Office, PO Box 10007
Saipan, MP 96950, USA
gscabrera@cnmihpo.com

Chris Caldow
National Oceanic & Atmospheric Administration
National Ocean Service, 1305 East West Highway
NSCI1 SSMC4 Silver Spring, MD 20910, USA
chris.caldow@noaa.gov

Joseph Chojnacki
Joint Institute for Marine and Atmospheric
Research, University of Hawai'i and NOAA

Pacific Island Fisheries Science Center, Kewalo
Research Facility, 1125B Ala Moana Blvd.,
Honolulu, HI 96814, USA
jchojnacki@mail.nmfs.hawaii.edu

Kate T. Ciembronowicz
U.S. Geological Survey Center for Coastal and
Wetland Studies, 140 7th Ave. South,
St. Petersburg, FL 3370, USA
kciembro@usgs.gov

Robert Clarke
Plymouth Marine Laboratory, Prospect Place,
West Hoe, Plymouth PL1 3 DH, UK
cac@hedge.u net.com

Chris L. Conger
State of Hawaii, Department of Land and Natural
Resources, Office of Conservation and Coastal
Lands, 1151 Punchbowl St., Honolulu, HI 96813,
USA
conger@hawaii.edu

Peter Craig
National Park of American Samoa, Pago Pago,
American Samoa 96799, USA 011
peter_craig@nps.gov

Gerry Davis
NOAA/National Marine Fisheries Service,
Pacific Islands Regional Office,
1601 Kapiolani Blvd, Honolulu,
HI 96814, USA
Gerry.Davis@noaa.gov

Edward DeMartini
NOAA Fisheries Service,
Pacific Islands Fisheries Science Center,
Honolulu, HI 96822 2396, USA
edward.demartini@fws.gov

Barry E. Devine
Eastern Caribbean Center,
University of the Virgin Islands,
#2 John Brewers Bay, St. Thomas,
VI 00802, USA
bdevine@uvi.edu

Richard E. Dodge
National Coral Reef Institute,
Nova Southeastern University
Oceanographic Center,
8000 N. Ocean Drive, Dania, FL 33004, USA
dodge@nova.edu

Brian T. Donahue
College of Marine Science,
University of South Florida, St. Petersburg,
FL 33701, USA
donahue@marine.usf.edu

Jennifer Dupont
University of South Florida,
College of Marine Science,
140 7th Ave. South, St. Petersburg,
FL 33701 USA
jdupont@seas.marine.usf.edu

Peter Edmunds
Department of Biology,
California State University 18111 Nordhoff St.,
Northridge, CA 91330 8303, USA
peter.edmunds@csun.edu

Mary S. Engels
Sea Education Association, PO Box 6,
Woods Hole, MA 02543, USA
mengels@sea.edu

Eden J. Feirstein
University of Hawaii, Department of Geology
and Geophysics, 1680 East West Rd., Honolulu,
HI 96822, USA
eden@hwr.arizona.edu

Douglas Fenner
Department of Marine and Wildlife Resources,
Government of American Samoa, PO Box 3730,
Pago Pago, AS 96799, USA
douglasfenner@yahoo.com

Scott Ferguson
Coral Reef Ecosystem Division, NOAA Fisheries
Service, Pacific Islands Fisheries Science Center, 1125
B. Ala Moana, Blvd., Honolulu, HI 96814, USA
scott.ferguson@noaa.edu

Charles H. Fletcher
University of Hawaii, Department of Geology
and Geophysics, 1680 East West Rd., Honolulu,
HI 96822, USA
fletcher@soest.hawaii.edu

Elizabeth Flint
Pacific Remote Islands, National Wildlife
Refuge Complex, U.S. Fish and Wildlife Service,
300 Ala Moana Blvd., Rm. 5231, PO Box 50167,
Honolulu, HI 96850, USA
beth_flint@fws.gov

L. Neil Frazer
Hawaii Institute of Geophysics,
University of Hawaii at Manoa,
Honolulu,
Hawaii 96822, USA
neil@soest.hawaii.edu

Alan M. Friedlander
Fisheries Ecologist NOAA/NOS/CCMA
Biogeography Team & The Oceanic Institute;
Makapu'u Point/41 202 Kalanianaole Highway,
Waimanalo, HI 96795, USA
afriedlander@oceanicinstitute.org

Jorge R. Garcia
Department of Marine Sciences,
University of Puerto Rico, Mayagüez,
PR 00681, USA
renigar@caribe.net

Michael Garcia
Department of Geology and Geophysics,
University of Hawaii, 1680 East West Rd,
Honolulu, HI 96822, USA
mgarcia@soest.hawaii.edu

Ivan P. Gill
Department of Geology and Geophysics,
University of New Orleans, New Orleans, LA
70146, USA and Ben Franklin Senior High
School, 2001 Leon C. Simon Ave., New Orleans,
LA 70148, USA
giloslon@bellsouth.net

Elizabeth H. Gladfelter
Woods Hole Oceanographic Institution,
Marine Policy Center Woods Hole,
MA 02543, USA
egladfelter@whoi.edu

Arthur C.R. Gleason
Division of Marine Geology and
Geophysics, Rosenstiel School of
Marine and Atmospheric Sciences,
University of Miami, 75 Virginia Beach Drive,
Miami, FL 33149, USA
agleason@rsmas.miami.edu

Craig R. Glenn
Department of Geology & Geophysics,
University of Hawaii, 1680 East-West Rd.,
Honolulu, HI 96822,
glenn@soest.hawaii.edu

Scott Godwin
Hawaii Institute of Marine Biology,
University of Hawaii, PO Box 1346, Kaneohe,
HI 96744 1346, USA
lgodwin@hawaii.edu

Jamison Gove
University of Hawaii, Joint Institute for Marine and
Atmospheric Research,
c/o Coral Reef Ecosystem Division,
NOAA Pacific Islands Fisheries Science Center, 1125
B. Ala Moana Blvd., Honolulu, HI 96814, USA
jamison.gove@noaa.gov

Richard W. Grigg
University of Hawaii,
Department of Oceanography, Honolulu,
HI 96822, USA
rgrigg@soest.hawaii.edu

Eric E. Grossman
USGS Pacific Science Center,
400 Natural Bridges Drive, Santa Cruz,
CA 95060, USA
egrossman@usgs.gov

Robert B. Halley
US Geological Survey, 600 4th St. South,
St. Petersburg, FL 33701, USA
Previous Address: US Geological Survey,
Fisher Island Station, Miami Beach,
FL 33139, USA
rhalley@usgs.gov

Jodi N. Harney
Coastal & Ocean Resources Inc., 214 9865
West Saanich Rd., Sidney BC V8L 5Y8, Canada
jodi@coastalandoceans.com

Kevin P. Helmle
National Coral Reef Institute,
Nova Southeastern University Oceanographic
Center, 8000 N. Ocean Drive,
Dania, FL 33004, USA
kevinh@nova.edu

Edwin A. Hernandez Delgado
University of Puerto Rico,
Department of Biology, Coral Reef Research
Group, PO Box 23360,
San Juan, PR 00931, USA
coral_giac@yahoo.com

Emma L. Hickerson
National Oceanic and Atmospheric
Administration Flower Garden Banks National
Marine Sanctuary 4700 Ave. U, Building 216
Galveston, TX 77551, USA
Emma_Hickerson@noaa.gov

Zandy Hillis-Starr
National Park Service Buck Island Reef National
Monument/SARI/CHRI, 2100 Church St. #100,
Christiansted, VI 00820, USA
zandy_hillis starr@nps.gov

Albert C. Hine
College of Marine Science, University of South
Florida, St. Petersburg, FL 33701, USA
hine@marine.usf.edu

Ronald Hoeke
Joint Institute for Marine and Atmospheric
Research, University of Hawaii and
NOAA Pacific Island Fisheries Science Center,
Kewalo Research Facility, 1125B
Ala Moana Blvd.,
Honolulu, HI 96814, USA
ronald.hoeke@noaa.gov

Peter Houk
CNMI Division of Environmental Quality,
PO Box 1304, Saipan, MP 96950, USA
peterhouk@deq.gov.mp

Dennis K. Hubbard
Department of Geology, Oberlin College,
Oberlin, OH 44074, USA
dennis.hubbard@oberlin.edu

J. Harold Hudson
National Oceanic and Atmospheric
Administration, Florida Keys National Marine
Sanctuary, 95230 Overseas Highway, Key Largo,
FL 33037, USA
Previous Address: US Geological Survey,
Fisher Island Station, Miami Beach,
FL 33139, USA
harold.hudson@noaa.gov

Ebitari Isoun
Department of Geology and Geophysics,
School of Ocean and Earth Science and Technology,
University of Hawaii, 1680 E-W Road,
PO Box 721, Honolulu, Hawaii 96822, USA

Nasseer Idrisi
MacLean Marine Science Center, University of the
Virgin Islands, 2 John Brewers Bay, St. Thomas,
VI 00802, USA
nidrisi@uvi.edu

Karilyn Jaap
Lithophyte Research, 273 Catalan Blvd NE,
St. Petersburg, FL 33704, USA
jaap@nelson.usf.edu

Walter C. Jaap
University of South Florida, College of Marine
Science, 140 7th Ave. South, St. Petersburg,
FL 33701 USA and Lithophyte Research, 273
Catalan Blvd. NE, St. Petersburg,
FL 33704 USA
wjaap@tampabay.rr.com

Bret D. Jarrett
College of Marine Science,
University of South Florida, St. Petersburg,
FL 33701, USA
Present address: N.S. Nettles & Associates Inc.,
201 ALT 19, Palm Harbor,
FL 34683, USA
bretjarrett@tampabay.rr.com

Christopher F.G. Jeffrey
National Oceanic & Atmospheric
Administration National Ocean Service,
1305 East West Highway, NSCI1 SSMC4
Silver Spring,
MD 20910, USA
chris.jeffrey@noaa.gov

Paul L. Jokiel
Hawaii Coral Reef Assessment and Monitoring
Program, Hawaii Institute of Marine Biology,
PO Box 1346, Kaneohe, HI 96744, USA
jokiel@hawaii.edu

Lance K.B. Jordan
Nova Southeastern University Oceanographic
Center, 8000 N. Ocean Drive,
Dania, FL 33004, USA
jordanl@nova.edu

G. Todd Kellison
Southeast Fisheries Science Center,
75 Virginia Beach Drive, Miami,
FL 33149, USA
todd.kellison@noaa.gov

Steven G. Kellison
Southeast Fisheries Science Center,
75 Virginia Beach Drive, Miami,
FL 33149, USA
todd.kellison@noaa.gov

Lisa Kerr Lobel
Boston University, Biology Department,
5 Cummington St., Boston,
MA 02215, USA
plobel@bu.edu

William E. Kiene
Fagatele Bay National Marine Sanctuary,
PO Box 4318, Pago Pago,
American Samoa 96799, USA
William.Kiene@noaa.gov

Barbara Kojis
Tropical Discoveries, PO Box 305731,
St. Thomas, VI 00803, USA
bkojis@vipowernet.net

Ilsa B. Kuffner
US Geological Survey, 600 4th St. South,
St. Petersburg, FL 33701, USA
ikuffner@usgs.gov

Barbara H. Lidz
US Geological Survey, 600 4th St. South,
St. Petersburg, FL 33701, USA
Previous Address: US Geological Survey,
Fisher Island Station, Miami Beach,
FL 33139, USA
blidz@usgs.gov

Craig Lilyestrom
Marine Resources Division,
Department of Natural and Environmental
Resources, PO Box 366147, San Juan,
PR 00936 6147, USA
craig_02@mac.co

Phillip S. Lobel
Boston University, Biology Department,
5 Cummington St., Boston,
MA 02215, USA
plobel@bu.edu

Stanley D. Locker
College of Marine Science, University of South
Florida, St. Petersburg, FL 33701, USA
stan@marine.usf.edu

Emily Lundblad
University of Hawaii, Joint Institute for Marine
and Atmospheric Research, c/o Coral Reef
Ecosystem Division, NOAA Pacific Islands
Fisheries Science Center, 1125 B. Ala Moana
Blvd., Honolulu, HI 96814, USA
Emily.Lundblad@noaa.gov

Ian Lundgren
National Park Service Buck Island Reef National
Monument/SARI/CHRI 2100, Church St. #100,
Christiansted, VI 00820, USA
ian_lundgren@nps.gov

David J. Mallinson
Department of Geology, East Carolina University,
Greenville, NC 27858, USA
mallinsond@ecu.edu

James Maragos
Pacific Remote Islands National Wildlife Refuge
Complex, U.S. Fish and Wildlife Service,
300 Ala Moana Blvd., Rm. 5231, PO Box 50167
Honolulu, HI 96850, USA
jim_maragos@fws.gov

Sarah McTee
Department of Zoology, University of Hawaii,
2538 McCarthy Mall, Honolulu,
HI 96822, USA
mctee@hawaii.edu

Charles Menza
National Oceanic & Atmospheric
Administration National Ocean Service,
1305 East West Highway, NSCI1
SSMC4 Silver Spring,
MD 20910, USA
charles.menza@noaa.gov

Charles G. Messing
Oceanographic Center,
Nova Southeastern University,
8000 N Ocean Drive, Dania Beach,
FL 33004, USA
messingc@nova.edu

Jeff Miller
National Park Service South Florida/Caribbean
Inventory and Monitoring Network,
1300 Cruz Bay Creek, St. John,
VI 00830, USA
william_j_miller@nps.gov

Joyce Miller
University of Hawaii, Joint Institute for Marine
and Atmospheric Research, c/o Coral Reef
Ecosystem Division, NOAA Pacific Islands
Fisheries Science Center, 1125 B. Ala Moana
Blvd., Honolulu, HI 96814, USA
joyce.miller@noaa.gov

Margaret W. Miller
NOAA Fisheries, Southeast Fisheries Science
Center, 75 Virginia Beach Drive, Miami,
FL 33149, USA
margaret.w.miller@noaa.gov

Mark E. Monaco
National Oceanic & Atmospheric
Administration, National Ocean Service,
1305 East West Highway,
NSCI1 SSMC4 Silver Spring, MD 20910, USA
mark.monaco@noaa.gov

Erinn M. Muller
Florida Institute of Technology,
Department of Biological Science,
150 West University Blvd., Melbourne,
FL 32901, USA
emuller@fit.edu

H. Gray Multer
Professor Emeritus, Fairleigh Dickinson
University, 3005 Watkins Rd., Horseheads,
NY 14845, USA
multerg@infoblvd.net

Bruce Mundy
NOAA Fisheries Service, Pacific Islands
Fisheries Science Center, Honolulu,
HI 96822 2396, USA
bruce.mundy@noaa.gov

Colin V. Murray-Wallace
University of Wollongong, School of Geosciences,
New South Wales, 2522, Australia
cwallace@uow.edu.au

Craig Musburger
Department of Zoology, University of Hawaii,
2538 McCarthy Mall, Honolulu, HI 96822, USA
musburge@hawaii.edu

David F. Naar
College of Marine Science, University of South
Florida, St. Petersburg, FL 33701, USA
naar@marine.usf.edu

Richard S. Nemeth
Center for Marine and Environmental Studies,
University of the Virgin Islands,
2 John Brewer's Bay, St. Thomas,
VI 00802, USA
rnemeth@uvi.edu

David Obura
CORDIO East Africa, 8/9 Kibaki Flats,
Kenyatta Public Beach, PO Box 10135,
Mombasa 80101, Kenya
dobura@cordioea.org

John C. Ogden
Florida Institute of Oceanography,
830 1st St. South, St. Petersburg,
FL 33701, USA
jogden@marine.usf.edu

Nancy B. Ogden
Florida Institute of Oceanography,
830 1st St. South, St. Petersburg,
FL 33701, USA
nogden@marine.usf.edu

Ernesto Otero
Department of Marine Sciences,
University of Puerto Rico, Mayagüez,
PR 00681, USA
eotero@uprm.edu

Francisco Pagan
Department of Marine Sciences,
University of Puerto Rico, Mayagüez,
PR 00681, USA
fpagan@cima.uprm.edu

Olga Pantos
Centre for Marine Studies,
University of Queensland, St. Lucia,
OLD 4042, Australia
o.pantos@uq.edu.au

Frank Parrish
NOAA Pacific Island Fisheries Science Center,
2570 Dole St., Honolulu, HI 96822, USA
frank.parrish@noaa.gov

Valerie J. Paul
Smithsonian Marine Station at Fort Pierce, 701
Seaway Drive, Fort Pierce, FL 34949, USA
paul@sms.si.edu

Janet Phipps
Environmental Resources Management,
2300 N. Jog Rd., 4th floor, West Palm Beach,
FL 33411 2743, USA
jphipps@co.palm beach.fl.us

Jeffrey Polovina
National Marine Fisheries Service/NOAA, 2570
Dole St., Honolulu, HI 96822, USA
Jeffrey.Polovina@noaa.gov

William F. Precht
Battelle Inc., West Palm Beach, Florida, USA
prechtw@battelle.org

Samuel J. Purkis
National Coral reef Institute,
Nova Southeastern University
Oceanographic Center, 8000 N. Ocean Drive,
Dania, FL 33004, USA
purkis@nova.edu

Wilson R. Ramirez
Department of Geology,
University of Puerto Rico,
Mayagues, PR 00681, USA
ramirezwilson@aol.com

John K. Reed
Harbor Branch Oceanographic Institution,
5600 U.S. 1, North, Fort Pierce,
FL 34946, USA
jreed@hboi.edu

Vincent P. Richards
National Coral Reef Institute,
Nova Southeastern University Oceanographic
Center, 8000 N. Ocean Drive, Dania,
FL 33004, USA
ardsrich@nova.edu

Laurie L. Richardson
Department of Biological Sciences,
Florida International University, Miami,
FL 33199, USA
richardl@fiu.edu

Robert H. Richmond
Kewalo Marine Laboratory, University of Hawaii
at Manoa, 41 Ahui St., Honolulu, HI 96813,
USA
richmond@hawaii.edu

Bernhard M. Riegl
National Coral Reef Institute,
Nova Southeastern University
Oceanographic Center, 8000 N. Ocean Drive,
Dania, FL 33004, USA
rieglb@nova.edu

Daniel M. Robbin
Veterans Administration, Department of Veterans
Affairs, 1492 West Flagler St., Miami, FL 33135,
USA Previous Address: US Geological Survey,
Fisher Island Station, Miami Beach,
FL 33139, USA
daniel.robbin@med.va.gov

Caroline S. Rogers
US Geological Survey Caribbean Field Station,
1300 Cruz Bay Creek, St. John, VI 00830, USA
caroline_rogers@usgs.gov

Steven O. Rohmann
NOS/NOAA, 1305 East West Highway #9653,
Silver Spring, MD 20910, USA
steve.rohmann@noaa.gov

John J. Rooney
NOAA Pacific Island Science Center,
Kewalo Research Facility, 1125B Ala Moana
Blvd., Honolulu, HI 96814, USA
john.rooney@noaa.gov

Steve W. Ross
Center for Marine Sciences, University of North
Carolina at Wilmington, 5600 Marvin Moss Ln,
Wilmington, NC 28409, USA
rosss@uncw.edu

Ken H. Rubin
University of Hawaii, Department of Geology
and Geophysics, 1680 East West Rd.,
Honolulu, HI 96822, USA
krubin@hawaii.edu

Enric Sala
Center for Marine Biodiversity
and Conservation, Scripps Institution of Oceanography, La Jolla, CA 92093, USA
esala@coast.ucsd.edu

Stuart Sandin
Center for Marine Biodiversity
and Conservation, Scripps Institution
of Oceanography, La Jolla, CA 92093, USA
ssandin@ucsd.edu

George P. Schmahl
National Oceanic and Atmospheric
Administration Flower Garden Banks National
Marine Sanctuary 4700 Ave. U, Building 216
Galveston,
TX 77551, USA
George_schmahl@noaa.gov

Clark E. Sherman
Department of Marine Sciences,
University of Puerto Rico, Mayagues,
PR 00681, USA
c_sherman@cima.uprm.edu

Eugene A. Shinn
College of Marine Science, University of South
Florida, 140 7th Ave. South, St. Petersburg,
FL 33701, USA
Previous Address: US Geological Survey,
Fisher Island Station, Miami Beach,
FL 33139, USA
eshinn@marine.usf.edu

Mahmood S. Shivji
Nova Southeastern University
Oceanographic Center, 8000 N. Ocean Drive,
Dania, FL 33004, USA
mahmood@nova.edu

Daria Siciliano
Remote Sensing Center, Physics Department,
Naval Postgraduate School, Monterey,
CA 93943, USA
siciliano@biology.acsc.edu

Lance Smith
Department of Zoology,
University of Hawaii, Honolulu,
HIi 96822, USA
lancesmi@hawaii.edu

Tyler B. Smith
Center for Marine and Environmental Studies,
University of the Virgin Islands,
2 John Brewers Bay, St. Thomas,
VI 00802, USA
tsmith@uvi.edu

Paul Somerfield
Plymouth Marine Laboratory, Prospect Place,
West Hoe, Plymouth PL1 3 DH, UK
cac@hedge.u net.com

Richard E. Spieler
National Coral Reef Institute, Nova Southeastern
University Oceanographic Center, 8000 N. Ocean
Drive, Dania, FL 33004, USA
spielerr@nova.edu

Anthony Spitzack
US Geological Survey Caribbean Field Station,
1300 Cruz Bay Creek, St. John,
VI 00830, USA
t_spitzack@hotmail.com

Alina Szmant
UNCW Center for Marine Science,
5600 Marvin K. Moss Ln,
Wilmington NC 28409, USA
szmanta@uncw.edu

Molly Timmers
University of Hawaii, Joint Institute for Marine
and Atmospheric Research,
c/o Coral Reef Ecosystem Division, NOAA
Pacific Islands Fisheries Science Center, 1125 B.
Ala Moana Blvd., Honolulu, HI 96814, USA
molly.timmers@noaa.gov

Michael Trianni
Division of Fish and Wildlife, Commonwealth
of the Northern Mariana Islands, PO Box 10007,
Lower Base, Saipan, MP 96950, CNMI
mstdfw@gmail.com

Roy Tsuda
Natural Sciences, Bernice P. Bishop Museum,
1525 Bernice St., Honolulu,
HI 96817 2704, USA
roy.tsuda@bishopmuseum.org

Sean Vitousek
University of Hawaii, Department
of Geology and Geophysics, 1680 East West Rd.,
Honolulu, HI 96822, USA
seanfkv@hawaii.edu

Joshua D. Voss
Harbor Branch Oceanographic Institution,
5600 US 1 North, Ft. Pierce, FL 34946, USA
vossjd@eckerd.edu

Peter Vroom
Joint Institute for Marine and Atmospheric
Research, University of Hawai'i and NOAA

Pacific Island Fisheries Science Center, Kewalo
Research Facility, 1125B Ala Moana Blvd.,
Honolulu, HI 96814, USA
pvroom@mail.nmfs.hawaii.edu

Brian K. Walker
National Coral Reef Institute,
Nova Southeastern University Oceanographic
Center, 8000 N. Ocean Drive, Dania,
FL 33004, USA
walkerb@nova.edu

Todd Wass
University of Hawaii, Joint Institute for Marine
and Atmospheric Research,
c/o Coral Reef Ecosystem Division,
NOAA Pacific Islands Fisheries Science Center,
1125 B. Ala Moana Blvd., Honolulu,
HI 96814, USA
wasstodd@hotmail.com

Ernesto Weil
Department of Marine Sciences, University of
Puerto Rico, Mayagüez, PR 00681, USA
eweil@caribe.net

Jonathan Weiss
Joint Institute for Marine and Atmospheric
Research, University of Hawai'i and NOAA
Pacific Island Fisheries Science Center,
Kewalo Research Facility, 1125B Ala Moana
Blvd., Honolulu, HI 96814, USA
jonathan.weiss@noaa.gov

Pal Wessel
Department of Geology and Geophysics,
University of Hawaii, 1680 East West Rd,
Honolulu, HI 96822, USA
pwessel@soest.hawaii.edu

Eric Wolanski
Australian Institute of Marine Science,
PMB No. 3, Townsville MC, Qld. 4810,
Australia
e.wolanksi@aims.gov.au

Paul Yoshioka
Department of Marine Sciences,
University of Puerto Rico, Mayagüez,
PR 00681, USA
p_yoshioka@cima.uprm.edu

1

Introduction: A Diversity of Oceans, Reefs, People, and Ideas: A Perspective of US Coral Reef Research

Bernhard Riegl and Richard Dodge

1.1 Historical Perspective

By virtue of its geographical extent and the size and wealth of its population, US surveyors and academics entered the scientific coral reef world soon after the study of the latter became of interest. But even earlier, the coral reefs of what are today territories of the USA have been noted and, at least cursorily, studied out of necessity since they were threats to vessels along the trade routes. Also the fossil coral reefs of the USA, of which the country has many famous examples, have received much early study and maybe even more attention than the living coral reefs. They hold a special place in sedimentology and economic geology since some of them are associated with the important oil finds that set off the early twentieth-century oil-boom in places like Texas and Utah. We will not treat these in the present volume, but restrict ourselves to the living coral reefs that can be observed in the ocean today. Recent reviews and entry points to the study of the fossil system can be found, among many others, in Stanley (2001) and Kiessling et al. (2002).

Early knowledge of, and comment on, US coral reefs dates back to the Spanish who were of course well aware of the Florida reef tract which was situated along their trade routes to and from their Caribbean possessions. Already Ponce de Leon, who explored Florida in 1513 while purportedly searching for the fountain of youth, sighted and remarked upon the reefs. Very soon, these reef-strewn shallow and treacherous waters would be the preferred haunt of pirates and buccaneers (like Edward Teach aka. the fabled "Blackbeard")

attacking trade vessels while running the gauntlet in the Straits of Florida between the reefs and shallows of the Florida shelf and the Bahamas banks (whose name possibly derives from the Spanish "islas del bajamar" – isles of the low tide, or shallow sea. Or it is a derivation of a Lucayan Indian name). In the Pacific Ocean, the reefs around Hawaii were noted by Captain Cook and the place where he found his untimely death at the hands of unfriendly Polynesians is now protecting both the historical site and an adjacent coral reef (Kealakua Bay Marine Protected Area on the Big Island). Also the Mariana Islands and their reefs were known to early explorers, traders and pirates, who repeatedly called on Tinian to use the good port made by the one small section of barrier reef (that in 1944/45 would be dredged and changed into a port which allowed the delivery and offloading of the first nuclear bombs).

When Darwin published his "Structure and Distribution of Coral Reefs" in May 1842 (after having presented his theories at the Geographical Society on 31 May 1837 in a talk titled "On certain areas of Elevation and Subsidence in the Pacific and Indian Oceans as deduced from the Study of Coral Formations" which was published as an abstract in the Society's Proceedings) his ideas were rapidly taken up and vividly discussed also among scientists in the USA. J.D. Dana was at the time a scientist aboard the US Exploring Expedition headed by Captain Wilkes (an efficient but stern and not much-loved character. Captain Ahab of "Moby Dick" fame was possibly modelled on him since Melville knew Wilkes–Keating

1992). Dana heard of Darwin's theories (reading about them in a newspaper while in Australia) and became an ardent supporter. The US Exploring Expedition went to the most forlorn corners of the territories, such as Rose Atoll near Samoa (next to be visited 80 years later by Mayor in 1920). This expedition also started the tradition of collaboration between US government agencies, then the Navy, and civilian research, now embodied by agencies like NOAA (the National Oceanographic and Atmospheric Administration). Dana (1843) was the first to show that cold currents prevented reef growth preferentially on the eastern sides of oceans and he charted many of the Hawaiian Islands for the first time (Grigg 2006). Consurrently, A.E. Verrill studied the taxonomy of corals at Harvard and also jointly published with Dana (Verrill 1864, 1866, 1872). In the meanwhile, the British Challenger expedition (1873–1876) visited the reefs in what was to become American Guam and the Hawaiian Islands. Dana remained loyal to Darwin's views of subsidence as the driver of atoll formation and defended them against competing ideas that pre-existing platforms are the foundation of atolls (Dana 1885). Another titan of early US reef science, Alexander Agassiz, who had been a member of the Challenger Expedition, was in correspondence with Darwin and keenly interested in the discussion regarding reef formations. Having made a fortune from Michigan copper, he himself funded an almost decade-long series of exploring expeditions to all major coral reef areas of the world. He sailed on the *Blake* and the *Albatross* from 1893 to 1902. Agassiz was the first to study the Florida Keys and Marquesas reefs in detail and came to the conclusion that they had grown on a bank until reaching the surface and found no evidence of atoll-like subsidence. Wanting to know more about reefs, he then proceeded to explore the reefs of the Pacific, among them the Great Barrier Reef and the US territories in the Mariana Islands and Samoa. At the same time as Agassiz worked in the Florida Keys, and produced one of the first habitat maps of a coral reef at the Dry Tortugas (Agassiz 1883), the eminent Harvard zoologist and geologist T.W. Vaughan also entered the field and subsequently published a series of works. Vaughan's work in Florida, the Bahamas, Cuba and Australia led him to propose antecedent control of reef formation (i.e. reefs can only form on pre-existing structures and

are thus pre-defined in their position and shape) in the form of flooded platforms (Vaughan 1914a, b). He also published the first exhaustive monograph on the corals of Hawaii and Laysan (Vaughan 1907). Verrill (1868) had earlier produced a review of corals on the American west coast, as well as of the North Pacific (Verrill 1866). Also at Harvard, R.A. Daly developed his Glacial Control Theory along similar lines of thinking and pioneered a better understanding, and eventually uniform acceptance, of glacio-eustatic processes (i.e. that sea level is driven primarily by water retention/mobilisation in ice during cold/warm-ages) and their control on shelf morphology and coral reefs (Daly 1910, 1915, 1916, 1917, 1934, 1948). Daly also studied beachrocks in the Tortugas and the reefs of American Samoa (Daly 1924). The idea that all coral formation was uniquely defined by sea level and erosional forces was not accepted by W.M. Davis, Daly's predecessor at Harvard who stayed scientifically active after his early retirement and strongly argued for cycles of uplift and subsidence for which he provided an eloquent platform in his "Coral Reef Problem" (Davis 1928). In 1904, A.G. Mayer (name changed to Mayor in 1918) had received permission from the Carnegie Institution in Washington to establish the Tortugas Marine Laboratory on Loggerhead Key in the Dry Tortugas (Mayer 1903). After its opening in 1905 it rapidly turned into an important center of tropical marine research, attracting many renowned American and international scientists and from which a resultant wealth of marine biological research poured forth (Shinn and Jaap 2005). At the laboratory R.C. Wells (1922) was among the first to study CO_2 saturation state, a hot topic in today's greenhouse world. Mayer studied the thermal tolerances of reef animals and showed how closely they live to their upper tolerance limits (Mayer 1914, 1918). Daly studied beachrock (Daly 1920), and T.W. Vaughan and J.W. Wells studied the corals (Vaughan 1910, 1914a, b, 1915, 1916a, b; Wells 1932). Mayor himself studied reefs all over the Caribbean and the Pacific, among others those of American Samoa (Tutuila, the Manu'a Islands and Rose atoll) (Mayer 1914, 1918, 1924). The transects established by Mayor are still being used for reef monitoring in Pago Pago harbour (Chapter 20, Birkeland et al.). While in Samoa, Mayor undertook research that sounds surprisingly modern – among many other things he studied the survival of corals in seawater of changed pH and how acidification would act on coral death by bleaching. Inspired by Mayor, C.H. Edmondson pioneered ecological and physiological studies in Hawaii (Edmondson 1928, 1933, 1946; Chapter 12, Jokiel). During these early days of the century, the Tanager Expedition of 1923/24 explored the Hawaiian and NW Hawaiian Islands and started Hawaii as a hub of coral reef science which it has remained to this day. Also, some of the first established large-scale marine reserves were declared in the NW Hawaiian Islands when President Roosevelt in 1909 declared them (with the exception of Midway) protected as the National Wildlife Bird Reservation (more to protect birds, which were being flagrantly over-exploited at the period, than coral reefs). In 1940, this became the Hawaiian Islands National Wildlife Refuge and in 2006 the Northwest Hawaiian Islands (=Papahnaumokuakea) National Monument, now the largest coherent coral reef reserve in the world.

In the 1930s, J.E. Hoffmeister and H.S. Ladd contributed to the discussion how coral reefs form their findings that extensive shoals can be developed by volcanic activity leading to emergence and erosion (Hoffmeister and Ladd 1935, 1944, 1945). Hoffmeister then went on to study the Florida Keys reefs and, together with H.G. Multer, the Key Largo Limestone which is the rock the upper Florida Keys are made from (Hoffmeister and Multer 1964, Hoffmeister et al. 1967, Multer and Hoffmeister 1969; Chapter 2, Lidz et al.). In the Pacific, a large contribution to the study of coral reefs and island geology was made by H. Stearns who, prior to World War II, had investigated sea level by studying the shore benches of Hawaii (Stearns 1935) and Guam (Stearns 1940, 1941) and had a detailed look at coral reefs in his treatise of Samoan geology (Stearns 1944). He also regarded the antecedent platform as essential for the formation of a reef (Stearns 1946).

During the Japanese presence in what were later to become, for a time, US territories, namely the Mariana Islands and Palau, their very active coral reef scientists established a research station in Palau, which led to a wealth of research in the region that influenced later US research (Burke 1951). Yabe and Sugiyama (1935) provided full taxonomic accounts of the corals. Yabe (1942) and Asano (1942) showed that the shapes of many

Pacific reefs were defined by subaerial solution (i.e. the exposed parts of islands being weathered away). This idea was later developed by F.S. McNeil (1954) who stressed the importance of organismic growth during submergence but the equal importance of subaerial erosion for the formation of lagoons and the shape of the reefs. This original model was further developed by E. Purdy (1974) to interpret reef shape in general by the "antecedent karst model" (karst being the rugose and gnarly form limestones take due to dissolution by freshwater when lifted out of the sea) suggesting preferential reef accretion on topographic highs. This has now been generally accepted (Hopley 1981). McNeil's paper marks a watershed in reef science inasmuch it clearly shows that reefs do not all have a similar evolution and that structure (its stratigraphic and sedimentologic make-up) and morphology (its surficial aspect) must not be confused. General, ubiquitously applicable theories subsequently lost their appeal and a wide array of empirical studies of form and process began (Hopley 1981).

The post-war period in the Pacific led to important government-funded work in the US territories. Groundbreaking geological investigations took place in the US and occupied territories, such as the Mariana Islands, under the Pacific Islands Mapping Program of the US Geological Survey (Whitmore 2001). H.S. Ladd conceived the idea of a long-term geologic mapping program in the Pacific Islands, which achieved many landmark studies on fossil and modern reefs (Cloud et al. 1956, Tracey et al. 1964, Emery et al. 1954, Emery 1962, Schlanger 1963, 1964). A further government-sponsored boost to US coral reef science was provided by the desire for a Pacific nuclear test ground that led to deep drilling at Bikini (775 m core) and Enewetak (2,307 and 2,530 m cores) which reached the Eocene basalt base of the atoll. These investigations also included studies of the coral fauna (Wells 1954a, b). Further drilling and seismic explorations at Midway, Enewetak, Funafuti, Kwajalein and Nukufetau furthered the understanding of the structure of US Pacific reefs and were of great marine geological importance in general (Ladd et al. 1953, 1970; Ladd and Schlanger 1960, Menard 1986, Tracey 2001). The Office of Naval Research also sponsored biological investigations in the Pacifc, which allowed F. Bayer to conduct extensive research on Pacific octocorallia, research he also conducted in the Atlantic. The eminent botanist F.R. Fosberg had been involved in the USGS's Pacific Islands mapping initiative, and, worried that much information would only remain in unpublished reports, founded the Atoll Research Bulletin in 1951, which has remained a mainstay of US coral reef literature (now edited by the eminent sedimentologist I. Macintyre).

Durham (1947) evaluated the coral fauna on the American West coast and the E. Pacific, which was later continued by P. Glynn in many papers (p. ex. 1997, 371). In the 1950s, coral reef studies came to full bloom in Florida and R.N. Ginsburg developed the principles of comparative sedimentology in the Florida Keys (Ginsburg 1953, 1956, 1957; Ginsburg and Lowenstam 1958, Ginsburg and Shinn 1964). Many US American scientists began working throughout the nearby Caribbean and a number of marine laboratories were established. In Jamaica T.F. Goreau produced seminal papers regarding the ecology and physiology of corals (Goreau and Goreau 1959, 1960). Connell (1973, 1978), in Australia, followed in Jamaica by Jackson and Buss (1975), applied aspects of competition and disturbance theories as community-shaping processes to coral reefs.

In 1964, H.G. Multer used a portable trailer-mounted rig to drill fossil and modern reefs in the Florida Keys, and I. Macintyre introduced submersible drilling in 1974, which revolutionized reef framework studies. Saturation, at least in a diving sense, was achieved in 1970, when the Tectite II program allowed 12 missions of 5 scientists each to spend 2 weeks in an underwater habitat in the US Virgin Islands (Miller et al. 1971) and became a precursor to today's Aquarius habitat in the Florida Keys, maintained by the University of North Carolina. Enos and Purkins (1977) provided the first exhaustive overview of facies and habitats throughout the Florida Keys, and Enos (1974) provided the first complete sediment facies map of the Florida–Bahamas plateau. Marszalek (1977) mapped the habitats of the entire Florida Keys. Shinn and co-workers published much about reefs in Florida and the Bahamas (and elsewhere) while Wells (1932, 1954a, b, 1956) worked up the taxonomy of living and fossil corals in US waters and worldwide. The Third International Coral Reef Conference in Miami in 1977, the Seventh ICRS

in Guam in 1992, the eighth ICRS 1996 in Panama (hosted by the Smithsonian Institution) allowed the US to welcome the international reef science community and present its own progress. Over the past 30 years, such a plethora of coral reef work was produced by US scientists that a full review here is impossible. It is also not necessary, since key aspects are mentioned in the following chapters.

1.2 Diverse Country, Diverse Oceans, Diverse Reefs

As diverse as its inhabitants and geography, the coral reefs of the USA show almost all the variability of which these systems are capable. The reader of this book can therefore vicariously travel through almost all types of carbonate depositional systems within which coral reefs can be embedded, as well as experience much of the biological differentiation experienced by coral reefs as a result of geographic position (Pacific versus Atlantic), latitude or different oceanographic and climatic control.

Coral reefs occur in US territories on one of the most stable passive margins (i.e. one where no plate subduction or collision occurs, which keeps tectonic deformation relatively minor) which, in Florida, has created one of the largest and thickest carbonate platforms found in the ocean today. That platform shows all transitions from a rimmed platform (where a reef at the edge encloses a carbonate platform with very gentle bathymetry), to an unrimmed platform (where the carbonate platform is unprotected by shelf-edge reefs), to a homoclinous ramp (where the seafloor slopes uniformly towards the deep – unlike in the platforms, which have an abrupt change in topography at the shelf-edge), to a distally steepened ramp (where the uniform slope is distally accentuated), and shows many of the responses coral reefs are capable of producing in response to shelf morphology (Chapter 2, Lidz et al.; Chapter 4, Hine et al.; Chapter 11, Fletcher et al.; Chapter 13, Rooney et al.).

On the other extreme, coral reefs also exist in one of the world's most active ocean margins, where the Pacific plate gets pulled underneath the overriding West Philippine Plate, in the Commonwealth of the Northern Mariana Islands and Guam. There, we clearly see the effects of tectonic activity on the establishment of coral reefs, and how they can be used as indicators of isostatic sealevel variation (when sealevel changes in an entire ocean basin), variably caused by uplift or the ocean reacting to plate adjustments due to loading/unloading of iceshields (Chapter 18, Riegl et al.).

In the Hawaiian Islands and Samoa, we see the effect of oceanic hotspots (melt-anomalies in earth's mantle that break the ocean's crust to form oceanic islands) on reef development, and clear illustrations of some of Darwin 's own coral reef theories in action (Chapter 11, Fletcher et al.; Chapter 13, Rooney et al.; Chapter 14, Grigg et al.; Chapter 20, Birkeland et al.). Samoa demonstrates the effects of volcanism on modern, Holocene, reef building, which has been influenced by eruptions as well as emergence and subsidence of the islands. Both in Hawaii and Samoa, we observe carbonate sedimentation in a mid-oceanic island setting.

True atolls and submerged carbonate banks are found in the northwestern Hawaiian Islands and Rose atoll in American Samoa, and the Pacific Remote Islands (like the Line Islands, Johnston atoll, etc.; Chapter 15, Maragos et al.; Chapter 17, Lobel and Lobel).

In the US Virgin Islands, and parts of the territory of Puerto Rico, we experience reef building in the context of what used to be a large carbonate shelf in the Oligocene to Pliocene (from about 28 to 5 million years ago; Van Gestel et al. 1999), but since has acquired a strong tectonic overprint with rifting, faulting and volcanism that has generated a variety of landforms made up by different rock types that all have different influences on reef building (Chapter 7, Hubbard et al.; Chapter 8, Rogers et al.; Chapter 9, Ballantine et al.).

US territories stretch from the tropics to beyond the latitudinal limits of coral reef distribution, which has provided for much biological interest since early on and the opportunity to study latitudinal attenuation of reef building and biodiversity (Vaughan 1914; Chapter 4, Hine et al.; Chapter 5, Banks et al.; Chapter 14, Grigg et al.). Zonation and within-reef differentiation has been an important subject of US coral reef science since the 1950s and continues to be so. Physiological studies regarding the environmental tolerances of corals also have a long history and some key advances regarding upper and lower limits of thermal tolerances were obtained on US coral reefs (Mayer 1914, Coles and Jokiel 1977). US coral reefs are far flung, and connectivity between

them is a big issue. Johnston atoll is arguably one of the most remote coral reefs in the world (Maragos and Jokiel 1986). The entire Hawaiian island chain exists in a relatively isolated setting and much has been hypothesized regarding where its tropical fauna originates from and how it is maintained (Chapter 14, Grigg et al.; Chapter 15, Maragos et al.; Chapter 17, Lobel and Lobel). Thus a host of connectivity studies, using variable techniques, have been conducted. Naturally, a fair amount of endemism can be observed in these isolated settings, which has proven a fruitful subject of study (Chapter 12, Jokiel; Chapter 14, Grigg et al.; Chapter 15, Maragos et al.; Chapter 17, Lobel and Lobel). The coral reefs in the US Caribbean, on the other hand, all exist in relative close proximity to each other in an ocean where almost ubiquitous connectivity has been postulated – but is increasingly being questioned (Chapter 6, Banks et al.; Baums et al. 2005, Vollmer and Palumbi 2007).

Also, the health trajectories of US coral reefs differ among oceans. The Caribbean has seen spectacular die-back of its dominant reefbuilder *Acropora palmata* (Chapter 3, Jaap et al.; Chapter 8, Rogers et al.; Chapter 9, Ballantine et al.). The cumulative effects of diseases possibly as a result of, or at least following, the die-off of the long-spined sea urchin *Diadema antillarum* (Lessios et al. 1984) and problematic levels of algal growth have decimated previously flourishing Caribbean reefs. Although reefs in the Pacific have also been badly affected by plagues of the coral-eating starfish *Acanthaster planci,* for example in Guam and the CNMI, no similar species-specific mortality of a dominant reef-builder has been observed (Chapter 19, Richmond et al.) and some reefs seem to exhibit significant resilience (Chapter 20, Birkeland et al.). It is interesting to note that early Holocene coral communities in the Mariana Islands exhibit a very similar community to what is found on their reefs today, while the comparable coral community composition in the Caribbean has been dramatically altered (Chapter 5, Banks et al.; Chapter 7, Hubbard et al.; Chapter 18, Riegl et al.). But the coral reefs of the Pacific face other threats. The Hawaiian Islands currently face major problems with introduced noxious species that may eventually turn out to threaten the very existence of these coral reefs as we know them today (Chapter 12, Jokiel). Of course Caribbean reefs are also threatened by introduced species, however, impacts there are yet less obvious than in the Pacific (Chapter 8, Rogers et al.; Chapter 9, Ballantine et al.; Chapter 10, Miller et al.). Unique impacts have been created on many US coral reefs in the Caribbean and the Pacific through their use as military facilities and bombing ranges. Most of these lands have been handed to civilian, mostly conservation, authorities, but interesting "case studies" remain (Chapter 9, Ballantine et al.; Chapter 16, Maragos et al.; Chapter 17, Lobel and Lobel). While much management effort is expended throughout the US territories, examples of overfishing are unfortunately easy to find (Chapter 8, Rogers et al.; Chapter 10, Miller et al.; Chapter 12, Jokiel; Chapter 14, Grigg et al.; Chapter 16; Maragos et al.; Chapter 20, Birkeland et al.). However, the US also possesses some of the world's most pristine reefs and is making strong efforts to protect them (Chapter 14, Grigg et al.; Chapter 16, Maragos et al.).

Some US coral reef scientists have been at the forefront of decrying the negative effects of anthropogenic impacts (Jackson 2001, Jackson et al. 2001, Pandolfi et al. 2005) and key scientific advances now allow a better understanding of the negative effects of nutrient enrichment, overfishing, rising temperatures, and ocean acidification, disease epizootics, to name but a few. The study and forecasting of such impacts is and increasingly important theme in US and international coral reef science.

In short, the coral reefs of the USA are interesting, well-studied and therefore a deserving subject for a synoptic scientific review. Alternatively, a closer look at them is justified merely by their biological wealth and beauty. They are certainly one of this country's most cherished and most valuable natural treasures.

References

Agassiz A (1883) Exploration of the surface fauna of the Gulf Stream, under the auspices of the United States coast survey, II. The Tortugas and Florida reefs, Mem Am Acad Arts Sci, Centennial II:107–132

Asano D (1942) Coral reefs of the South Sea islands. Tok Imp Univ Geol Paleo Inst Rep 39:1–19

Baums IB, Miller MW, Hellberg ME (2005) Regionally isolated populations of an imperiled Caribbean coral, *Acropora palmata*. Molec Ecol 14:1377–1390.

Burke HW (1951) Contributions by the Japanese to the study of coral reefs. US Geol Surv Military Branch Memo, 1–43

Cloud PE, Schmidt RG, Burke HW (1956) Geology of Saipan, Mariana Islands, Part 1. General Geology. US Geol Surv Prof Pap 280-A:1–126

Coles SL, Jokiel P (1977) Effects of temperature on photosynthesis and respiration in hermatypic corals. Mar Biol 43:209–216

Connell JH (1973) Population ecology of reef building corals. In: Jones OA, Endean R (eds) Biology and geology of coral reefs, vol 1. Academic, pp 205–223

Connell JH (1978) Diversity in tropical rain forests and coral reefs. Science 199:1302–1310

Daly RA (1910) Pleistocene glaciation and the coral reef problem. Amer J Sci 30:297–308

Daly RA (1915) The glacial control theory of coral reefs. Proc Nat Acad Arts Sci 51:155–251

Daly RA (1916) A new test of the subsidence of coral reefs. Proc Nat Acad Sci 2:664–670

Daly RA (1917) Origin of the living coral reefs. Scientia 22:188–199

Daly RA (1920) Origin of beach rock. Carnegie Inst Wash Yearbook 18:192

Daly RA (1924) The geology of American Samoa. Carnegie Inst Wash Pub 340:93–143

Daly RA (1934) The changing world of the ice age. Yale University Press, 271 pp

Daly RA (1948) Coral reefs – a review. Amer J Sci 10:281–313

Dana JD (1843) On the temperature limiting the distribution of corals. Am J Sci 45:130–131

Dana JD (1885) Origin of coral reefs and islands. Amer J Sci 30:89–105, 169–191

Darwin C (1842) The structure and distribution of coral reefs. Smith, Elder and Co, London

Davis WM (1928) The coral reef problem. Am Geogr Soc Spec Pub 9:1–596

Durham JW (1947) Corals from the Gulf of California and the North Pacific Coast of America. Mem Geol Soc Am 20:1–68

Edmondson CH (1928) The ecology of a Hawaiian coral reef. Bull B P Bishop Museus 45:1–64

Edmondson CH (1933) Reef and shore fauna of Hawaii. Spec Pub Berenice P Bisho Mus 22:1–295

Edmondson CH (1946) Behavior of coral planulae under altered saline and thermal conditions. Occas Pap Berenice P Bishop Mus 18:283–304

Emery KO (1962) Marine geology of Guam. US Geol Surv Prof Pap 403-B:1–76

Emery KO, Tracey JI, Ladd HS (1954) Geology of Bikini and nearby atolls. US Geol Surv Prof Pap 260-A:1–265.

Enos P, Purkins RD (1977) Quaternary sedimentation in South Florida. Geol Soc Am Mem 147:1–198

Enos P (1974) Surface sediment facies map of the Florida -Bahamas Plateau. Geol Soc Am Map 5(4):1–5

Ginsburg RN (1953) Intertidal erosion in the Florida Keys. Bull Mar Sci Gulf Caribbean 3:55–69

Ginsburg RN (1956) Environmental relationships of grain size and constituent particles in some south Florida carbonate sediments. AAPG Bull 40:2384–2427

Ginsburg RN (1957) Early diagenesis and lithification of shallow-water carbonate sediments in South Florida. Soc Econ Petrol Mineral Spec Pub 5:80–100

Ginsburg RN, Lowenstam HA (1958) The influence of marine bottom communities on the depositional environment of sediments. J Geol 66:310–318

Ginsburg RN, Shinn EA (1964) Distribution of reef-building community in Florida and the Bahamas. AAPG Bull 48:527

Glynn PW (1997) Eastern Pacific reef coral biogeography and faunal flux: Durham's dilemma revisted. Proc 8th Int Coral Reef Sym Panama:371–378

Goreau TF, Goreau NI (1959) The physiology of skeleton formation in corals, 2. Calcium deposition by hermatypic corals under various conditions in the reef. Biol Bull (Woods Hole) 117:239–250

Goreau TF (1960) The physiology of skeleton formation in corals, 3. Calcification rate as a function of colony weight and total nitrogen content in the reef coral Manicina areaolata (Linn.). Biol Bull (Woods Hole) 118:419–429

Grigg RW (2006) The history of marine research in the northwestern Hawaiian Islands : lessons from the past and hopes for the future. Atoll Res Bull 543:13–22

Hoffmeister JE, Ladd HS (1935) The foundation of atolls. J Geol 43:653–665

Hoffmeister JE, Ladd HS (1944) The antecedent-platform theory. J Geol 52:388–502

Hoffmeister JE, Ladd HS (1945) Solution effects on elevated limestone terraces. Geol Soc Am Bull 56:809–818

Hoffmeister JE, Multer HG (1964) Pleistocene limestones of the Florida Keys. In: Ginsburg HG (compiler) South Florida carbonate sediments. Geol Soc Am Field Trip No 1:57–61

Hoffmeister JE, Stockman KW, Multer HG (1967) Miami limestone of Florida and its recent Bahamian counterpart. Geol Soc Am Bull 78:175–190

Hopley D (1981) The geomorphology of the Great Barrier Reef. Quaternary development of coral reefs. Wiley, 453 pp

Keating BH (1992) Contributions of the 1838–1842 U.S. exploring expedition. In: Keating BH, Bolton BR (eds) Geology and offshore mineral resources of the central Pacific basin, Circum-Pacific Council for Energy and Mineral Resources Earth Science Series 14, Springer, New York, pp 1–9

Kiessling W, Fluegel E, Golonka J (2002) Phanerozoic reef patterns. Soc Econ Petrol Mineral Spec Pub 72:1–775 pp

Ladd HS, Ingerson E, Townsend RC, Russell M, Stephenson HK (1953) Drilling on Eniwetok Atoll, Marshall Islands. AAPG Bull 37:2257–2280

Ladd HS, Schlanger SO (1960) Drilling operations on Eniwetok Atoll. US Geol Surv Prof Pap 206-Y:863–899

Ladd HS, Tracey JI, Gross MG (1970) Drilling on Midway Atoll. US Geol Surv Prof Pap 680-A:1–22

Lessios HA, Robertson DR, Cubit JD (1984) Spread of diadema mass mortality throughout the Caribbean. Science 226:335–337

Maragos JE, Jokiel PL (1986) Reef corals of Johnston Atoll : one of the world's most isolated reefs. Coral Reefs 4:141–150

Marszalek DS (1977) Florida reef tract marine habitats and ecosystems: maps published in cooperation with State of Florida Department of Natural Resources; U.S. Department of Interior Bureau of Land Management, New Orleans Outer Continental Shelf Office; and University of MiamiRosenstiel School of Marine and Atmospheric Science, scale 1:30,000, 9 sheets.

Mayer AG (1903) The Tortugas, Florida, as a station for research in biology. Science 17:190–192

Mayer AG (1914) Effects of water temperature on tropical marine animals. Carnegie Inst Wash Pub 183:1–24

Mayer AG (1918) Toxic effects due to high temperature. Carnegie Inst Wash Pub 252:175–178

Mayer (1918) Ecology of the Murray Island coral reef. Carnegie Inst Wash Pub 213:1–48

Mayor AG (1924) Structure and ecology of Samoan reefs. Carnegie Inst Wash Pub 340:1–25

MacNeil FS (1954) The shape of atolls: an inheritance from subaerial erosion forms. Amer J Sci 252:402–427

Menard HW (1986) Islands. HW Freeman, New York, 205 pp

Miller JW, VanDerwalker JG, Waller RA (1971) Tektite 2: scientists in the sea. US Department of Interior, Washington, DC, 582 pp

Multer HG, Hoffmeister JE (1969) Petrology and significance of the Key Largo Limestone, Florida Keys (abs.). Geol Soc Am Spec Paper 121:211–212

Jackson JBC, Buss L (1975) Allelopathy and Spatial Competition among Coral Reef Invertebrates. Proc Nat Acad Sci 72:5160–5163

Jackson JBC (2001) What was natural in the coastal ocean? Proc Nat Acad Sci 98:5411–5418

Jackson JBC, Kirby MX, Berger WH, Bjorndal KA, Botsford LW, Bourque BJ, Bradbury RH, Cooke R, Erlandson J, Estes JA, Hughes TP, Kidwell S, Lange CD, Lenihan HS, Pandolfi JM, Peterson CH, Steneck RS, Tegner MJ, Warner RR (2001) Historical overfishing and the recent collapse of coastal ecosystems. Science 293:629–638

Pandolfi JM, Jackson JBC, Baron N, Bradbury RH, Guzman HM, Hughes TP, Kappel CV, Micheli F, Ogden JC, Possingham HP, Sala E (2005) Are U.S. coral reefs on the slippery slope to slime? Science 307:1725–1726

Purdy EG (1974) Karst-determined facies patterns in British honduras: holocene carbonate sedimentation model. Amer Assoc Petrol Geol Bull 58:825–855

Schlanger SO (1963) Subsurface geology of Eniwetok Atoll. US Geol Surv Prof Pap 260-BB:991–1066

Schlanger SO (1964) Petrology of limestones of Guam. US Geol Surv Prof Pap 403-D:1–52

Shinn EA, Jaap WC (2005) Field guide to the major organisms and processes building reefs and islands of the Dry Tortugas : the Carnegie Dry Tortugas Laboratory centennial celebration (1905–2005). US Geological Survey and Florida Fish and Wildlife Research Institute, St. Petersburg, FL, 40 pp

Stanley Jr., GD (2001) The history and sedimentology of ancient reef systems. Kluwer/Plenum, 458 pp

Stearns HT (1935) Shore benches on the island of Oahu, Hawaii. Bull Geol Soc Am 46:1467–1482

Stearns HT (1940) Geologic history of Guam (abstr). Geol Soc Am Bull 52:1948

Stearns HT (1941) Shore benches on north Pacific Islands. Geol Soc Am Bull 52:773–780

Stearns HT (1944) Geology of the Samoan Islands. Bull Geol Soc Am 55:1279–1332

Stearns HT (1946) An integration of coral reef hypotheses. Amer J Sci 244:245–262

Tracey Jr., JI, Schlanger SO, Stark JT, Doan DB, May HD (1964) General geology of Guam. US Geol Surv Prof Pap 403-A:1–104

Tracey JI (2001) Working in the Pacific. Atoll Res Bull 494:11–22

Van Gestel J-P, Mann P, Grindlay NR, Dolan JF (1999) Three-phase tectonic evolution of the northern margin of Puerto Rico as inferred from an integration of seismic reflection, well, and outcrop data. Mar Geol 161:257–286

Vaughan TW (1907) Recent madreporaria of the Hawaiian Islands and Laysan. US Nat Mus Bull 59:1–427

Vaughan TW (1910) A contribution to the geologic history of the Floridian plateau. Carnegie Inst Wash Pub 4:99–185

Vaughan TW (1914a) Sketch of the geologic history of the Florida reefs tract and comparisons with other coral reef areas. J Wash Acad Sci IV:26–34

Vaughan TW (1914b) Building of the Marquesas and Tortugas atolls and a sketch of the geologic history of the Florida reef tract. Carnegie Inst Wash Pub 182:55–67

Vaughan TW (1915) Growth-rate of the Floridian and Bahamian shoal-water corals. Carnegie Inst Wash Yearbook 13:221–231

Vaughan TW (1916a) Results of investigations of the ecology of the Floridian and Bahaman shoal-water corals. Proc Nat Acad Sci 2:95–100

Vaughan TW (1916b) The temperature of the Florida coral reef tract. Carnegie Inst Was Pub 213:321–339

Vaughan TW (1907) Recent Madreporaria of the Hawaiian Islands and Laysan. Smithsonian Institution US Nat Hist Mus Bull 59:1–420

Verrill AE (1864) A list of the corals and polyps sent by the Museum of Comparative Zoology to other institutions in exchange, with annotations. Bull Mus Comp Zool Harvard 1:29–60

Verrill AE (1866) Synopsis of the polyps and corals of the North Pacific exploring expedition, 1853–1856 with descriptions of some additional new species from the west coast of North America, part 3: Madreporania. Proc Commun Essex Institute 5:17–32, 33–50, 315–333

Verrill AE (1868) Review of the corals and polyps of the west coast of America. Trans Conn Acad Arts Sci 1:351–372

Verrill AE (1872) Names of the species of corals. In: Dana JD (ed) Corals and coral islands. New York, Dodd and Mead, pp 379–388

Vollmer SV, Palumbi SR (2007) Restricted gene flow in the Caribbean staghorn coral *Acropora cervicornis*: implications for the recovery of endangered reefs. J Heredity 98(1):40–50

Wells JW (1932) Study of the reef corals of the Dry Tortugas. Carnegie Inst Wash Yearbook 31:290–291

Wells JW (1954a) Recent corals of the Marshall Islands. Bikini and nearby atolls. US Geol Surv Prof Pap 260-I: 285–486

Wells JW (1954b) Fossil corals from Bikini Atoll. US Geol Surv Prof Pap 260:609–617

Wells JW (1956). Scleractinia. In: Moore RC (ed) Treatise on invertebrate paleontology. Geological Society of America/University of Kansas Press, Part F, pp 328–440

Wells RC (1922) Carbon-dioxide content of sea water at Tortugas. Pap Tortugas Lab 18:87–93

Whitmore Jr., FC (2001) The Pacific Island mapping program of the U.S. geological survey. Atoll Res Bull 494:1–7

Yabe H, Sugiyama T (1935) Revised lists of the reef corals from the Japanese seas and of the fossil reef corals of the raised reefs and the Ryukyu Limestone of Japan. J Geol Soc Japan 42:379–403

Yabe H (1942) Problems of the coral reefs. Toh Imp Univ Geol and Paleo Inst Rep 39:1–6

2

Controls on Late Quaternary Coral Reefs of the Florida Keys

Barbara H. Lidz, Eugene A. Shinn, J. Harold Hudson, H. Gray Multer, Robert B. Halley, and Daniel M. Robbin

2.1 Regional Setting and Early Cultural History

The Florida Keys is an arcuate, densely populated, westward-trending island chain at the south end of a karstic peninsular Florida Platform (Enos and Perkins 1977; Shinn et al. 1996; Kindinger et al. 1999, 2000). The "keys" mark the southernmost segment of the Atlantic continental margin of the United States. The islands are bordered by Florida Bay to the north and west, the Atlantic Ocean to the east and southeast, Gulf of Mexico to the west, and Straits of Florida to the south. Prevailing southeasterly trade winds impinge on the keys, creating a windward margin. The largest coral reef ecosystem in the continental United States rims this margin at a distance of ~5–7 km seaward of the keys and occupies a shallow (generally <12 m), uneven, westward-sloping shelf (Parker and Cooke 1944; Parker et al. 1955; Enos and Perkins 1977). The platform is tectonically stable at present (Davis et al. 1992; Ludwig et al. 1996; Toscano and Lundberg 1999). The reefs and 240-km-long island chain parallel the submerged shelf margin, corresponding roughly to the 30-m depth contour that marks the base of a fossil shelf-edge reef (studies cited use the same criterion). The modern reef tract extends west-southwest from Soldier Key southeast of Miami (25°60′ N, 80°20′ W) to the Dry Tortugas in the Gulf of Mexico (24°40′ N, 83°10′ W). Reef-tract habitats lie within the protective domain of the Florida Keys National Marine Sanctuary (Fig. 2.1a–c; Multer 1996).

Prehistoric Paleoindians inhabited the Floridan Peninsula around 12 ka (Zeiller 2005). The Archaic Period of human progress followed (from ~7 to 2 ka) as aboriginal tool making became more sophisticated. The Formative or Ceramic Period (from ~2 ka to AD 1513) was next as the creation of pottery for transportation and storage of food and water became important. The Historic Period began in 1513. By the mid-1500s, Florida had become part of a Spanish monopoly in the Americas. Conquistadors first settled in La Florida in St. Augustine on the East Coast in 1567. In 1763, England took Canada from France, and Spain ceded all of La Florida to England. Spain again took possession of La Florida in the 1783 Treaty of Paris (Zeiller 2005).

The United States acquired Florida from Spain by treaty in 1821 largely for the potential military advantage that the Florida Keys offered (see articles in Gallagher et al. 1997, and selected human-interest notes in Appendix 2.A). The government recognized a need to protect shipping between the Atlantic and Gulf Coasts, and the keys were natural sites for military bases for this purpose. The US Army and US Navy established bases on several islands, and upon admission to the Union as the 27th State in 1845, forts were built at Key West (Fort Zachary Taylor) and the Dry Tortugas (Fort Jefferson). The Florida Keys played major roles in the Second Seminole War (1835–1842), the Spanish-American War (April–August 1898), World War I (1916–1918, when Key West first became a major naval training base), World War II (1941–1945), the Cuban Missile Crisis (1962), the war on drugs

B.M. Riegl and R.E. Dodge (eds.), *Coral Reefs of the USA*,
© Springer Science + Business Media B.V. 2008

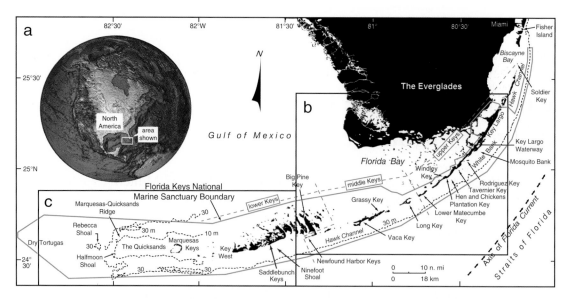

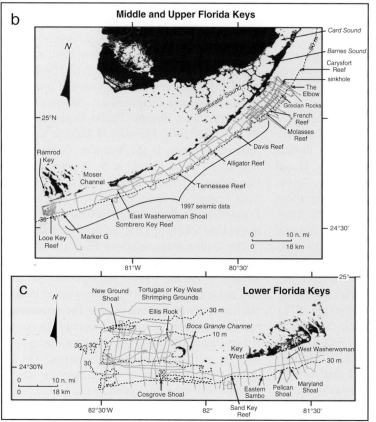

Fig. 2.1. **a** Index map of South Florida and the Florida Keys. Dashed red dogleg line separates areas of Pleistocene ooid bank (Miami Limestone) of the lower Keys and coral reef (Key Largo Limestone) of the westernmost middle Keys. An ooid bank also formed at the east end of the reef and today underlies the city of Miami (Halley and Evans 1983). Note major tidal passes in middle Keys. Dotted line (30-m-depth contour) marks the shelf margin, which lies within the Florida Keys National Marine Sanctuary boundary (blue line). **b** Index map shows locations of major Holocene coral reefs and USGS geophysical surveys (gray lines) for a portion of the upper and all of the middle Keys. Sinkhole at the northeast end is discussed in Shinn et al. (1996). **c** Index map shows survey lines for the lower Keys, Marquesas Keys, and The Quicksands areas. Contours are in meters

(1970s), and the Mariel Boatlift (1980). Financier Henry Flagler's Overseas Railroad transported tourists south to Key West and agricultural produce north from Cuba to Miami for 23 years before the Labor Day hurricane of 1935 destroyed both train and railway tracks (Parks 1968). The keys and other areas of South Florida today remain favored destinations for Caribbean immigrants seeking asylum in the US. But for the past three decades, the coral reefs have fueled the economy of the keys, providing lucrative commercial fisheries and colorful easily reachable habitats that draw tourists from around the world.

Accessibility of the shallow and emergent late Quaternary sequences to scientists makes the Florida windward margin one of the best-studied modern carbonate platforms. In the early years, Florida reefs intrigued researchers interested in the tropical marine-carbonate environments. Shinn and Jaap (2005) recount some of the classic carbonate studies that were carried out in the Dry Tortugas. Louis Agassiz mapped benthic communities in the Tortugas (Agassiz 1880). His son Alexander published the map (Agassiz 1883). In an effort to protect shipping, Louis also examined reefs for the Lighthouse Service (the US Coast Survey, predecessor of the US Coast Guard) with the intent of determining how to prevent the reefs from growing. Reefs took a heavy toll on shipping and in those days were considered a costly nuisance. Failing to discover how to halt reef growth, Louis decided the logical solution was installation of lighthouses. A 46-m-high structure was completed on Loggerhead Key in the Tortugas in 1858 and still functions today, though with updated illumination. In 1905, Alfred Goldsborough Mayer, a student of Alexander Agassiz, built and directed the Carnegie Institution's Dry Tortugas Laboratory on Loggerhead Key. To help justify the laboratory, he documented the so-called black-water event (a red tide) of 1879 that killed fish and essentially all acroporids at the Tortugas (Mayer 1903). He published his landmark treatise on medusae (Mayer 1910) and contributed to research on temperature tolerance of corals and other marine organisms (Mayer 1914, 1918). Without the aid of drilling, T. Wayland Vaughan, a close friend of Mayer, correctly deduced that the Tortugas was an elliptical atoll-like structure built primarily of Pleistocene coral, which spurred his interest in reef geology, ecology, and coral growth rates (Vaughan 1914a, b, 1915a, b, 1916). After Mayor's death in 1922 (Mayer changed the spelling of his name to Mayor in 1918), William H. Longley (who with Hildebrand 1941, pioneered the first underwater color photography of tropical Atlantic fishes), then David Tennent (sea-urchin embryology) directed the Carnegie Laboratory until its closure in 1939 for economic reasons. Today, little is left of the facility. A memorial plaque designed by Mayor's artist wife was erected near the site a year after Mayor died. The monument stands in lone testimony to the benchmark tropical marine-biology research that Mayor had envisioned and that he and his colleagues had achieved (Stephens and Calder 2006).

Prior to being designated a National Marine Sanctuary in 1990, reefs in the vicinity of the Florida Keys were drilled in the search for oil. Hydrocarbons are being produced from Lower Cretaceous limestone, anhydrite, and dolomite that compose the Sunniland Formation of Florida (Winston 1969, 1972). Seventeen exploratory wells were drilled in south and central Florida and in the keys beginning at about the time oil was discovered at the Sunniland Field in 1943 (Fig. 2.2; Dustan et al. 1991). All wells had oil shows, but no show was economically viable. All wells left magnetic signatures due to borehole casing. Most offshore well sites evolved into 'artificial reefs' as sessile organisms colonized discarded wires and casings, and great numbers of fish congregated in borehole cavities that formed havens in otherwise featureless seafloor sites (Shinn et al. 1989a, 1993). Conclusions drawn from the well-site studies were that none of the environments sustained permanent biological damage during the one-time perturbations of drilling, even to depths of several thousand meters, and that the biological impact was negligible. Conclusions could not be drawn from those studies for wells that would become producing wells with longer-term on-site perturbations.

2.2   Overview of Large-scale Geologic Parameters

South Florida is built of thousands of meters of Cenozoic limestone deposited on top of an igneous Mesozoic basement (e.g., Applin and Applin 1965;

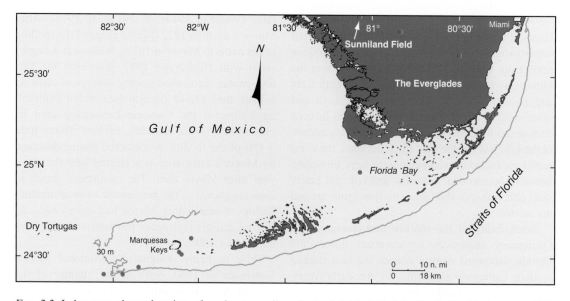

FIG. 2.2. Index map shows location of exploratory oil wells (red dots) drilled in South Florida between 1943 and 1962. The well at Sunniland Field is an oil producer

Milton and Grasty 1969; Winston 1969). The uppermost sections are stratigraphically successive late Pleistocene non-coralline marine sequences, later Pleistocene coral reefs and cemented-ooid tidal bars, and Holocene corals. The exposed Florida Keys consist of an emergent coral reef (Key Largo Limestone; Multer et al. 2002) and tidal-bar oolite (Miami Limestone; Halley and Evans 1983) that accumulated during the eustatic highstand of ~125 ka. The climate today is subtropical, and the reef tract is regarded as being marginal for coral growth (Jaap 1984; Shinn 1988; Shinn et al. 1989b). Marginality is due in part to a high-latitude flooded shelf that allows tidally induced coastal, bay, and gulf waters to flow over many of the reefs. Variable salinity, nutrient content, temperature extremes, and turbidity unfavorable for coral growth characterize these waters.

The reef-rimmed Florida margin differs in morphology and depositional processes from the classic windward carbonate-margin model (James and Ginsburg 1979). The shelf is shallow, topographically asymmetrical at the shelf edge (reefs and reentrants), several meters deeper to the southwest than northeast, and has an average present gradient of ~0.8 m/km from the keys to the break in slope. Pleistocene gradients were generally less pronounced (Perkins 1977; Multer et al. 2002). Other typical

windward margins are steeply inclined (James and Ginsburg 1979; Hine et al. 1981; Hine and Mullins 1983). Though the modern shallow setting is marginal for coral growth, Pleistocene corals produced imposing structures in paleoenvironments that were optimal for cumulative coral accretion (Lidz et al. 1991; Multer et al. 2002; Lidz 2004). Processes associated with the present shelf involve complex, multidimensional, tidally induced exchanges of inimical inshore waters with the reefs. In comparison, processes associated with the late Pleistocene shelf were simple and were topographically controlled. The result was vigorous coral growth at the edge of an extensive, broad, protective landmass (Lidz 2006) vs. diminished vigor of later Holocene reef growth along a shallow, coastal, shelf.

The primary controls on late Quaternary shelf and reef-framework buildup were position of sealevel maxima relative to topographic elevation and effects of those maxima and antecedent topography on coral growth and reef location (Lidz 2006; Lidz et al. 2006). Secondary were surf-zone duration, oceanographic influences, and karstification during periods of subaerial exposure. The geologic record shows that reef growth varied locally through time – in location, dimension, structure, geometry, depth, age distribution, and accretionary direction. Yet the recurring regional theme is the replication of antecedent facies and geomorphic landforms, and thus repetition of stratigraphic shelf-edge asymmetry.

2.2.1 The Pleistocene Key Largo Limestone Coral Reef

All of South Florida during the Pleistocene was submerged many times as a wide, shallow platform whose curving southernmost edge lay isolated, often 160 km from any mainland to the north. Sea level fluctuated periodically (Fig. 2.3a, b), and discontinuous arcuate chains of bank-barrier and patch reefs evolved, flourished, and died. These coral reefs often provided protection for the production of carbonate sediments that became widespread leeward mudstones, packstones, grainstones (Appendix 2.B, Plates 2.1, 2.2), and finally bryozoan- and ooid-rich sediments.

Sanford (1909, p. 214) first applied the name Key Largo Limestone to "the elevated reef that forms the backbone of the main chain of the Florida Keys from Soldier Key to Bahia Honda." During Key Largo time, repeated fluctuations of sea level allowed karst features to form during periods of subaerial exposure, including precipitation of dark-brown laminated subaerial crusts or calcrete (Fig. 2.4a–e; Multer and Hoffmeister 1968). Positions of these crusts in cores (Appendix 2.B, Plate 2.3) serve as markers, indicating times of land emergence above the sea due to a lowered sea level. Presence of the crusts allowed division of the Pleistocene into distinctive marine sequences, termed the Q1–Q5 Units for Quaternary (Perkins 1977). Multer et al. (2002) compiled a schematic summary overview of the five units in the Key Largo Limestone (Appendix 2.B, Fig. 2.B-1).

Thick siliciclastic carbonates and the presence of topographic highs strongly influenced formation of Q1-Unit sediments at the south end of the platform (Ginsburg et al. 1989; Guertin et al. 1996; Cunningham et al. 1998). Quartz grainstones and mollusc-quartz wackestones characterize Q1 rocks.

Quartz became less common during the succeeding Q2-Unit period with red coralline algae more abundant. Deposition of an initial deep-water arenaceous facies was followed by accumulation of successive skeletal-grain beds, reducing water depths, and increasing water temperature.

Such factors encouraged skeletal-carbonate production by shallow-water organisms and corals (*Montastraea annularis, Porites* sp., and *Acropora cervicornis)*, especially on the higher submarine topography.

Pleistocene carbonate production at the Florida Platform margins may have begun during the interval between 420 and 360 ka (marine oxygen-isotope Stage 11), which is regarded as the longest and warmest period during the past 500 ka (Droxler and Farrell 2000; Kroon et al. 2000). Maximum coral reef accretion, however, occurred during Q3 time with rigid and fused pillars and thickets forming bank-barrier and patch reefs. Non-rigid and stabilized thickets as well as rubble were also common (Appendix 2.B, Fig. 2.B-2). Perkins (1977) described the Q3 reef framework beneath Big Pine Key as being 29 m thick. Multer et al. (2002) found the same coralline framework in nine successive core holes from Grassy Key to north Key Largo (Fig. 2.1a). Lateral growth of patch reefs may have been responsible for the formation of many intermittent bank-barrier reefs along the shelf edge during Q3 and subsequent times. Coring occasionally revealed successive "vertical mimicking" in Q3–Q5 coral-rich frameworks superimposed on ancient topographic highs under both the emergent Florida Keys and Florida Bay (Multer et al. 2002).

Following Q3 time, prolonged subaerial exposure (Fig. 2.3a) of the Florida shelf produced widespread laminar subaerial crusts (Appendix 2.B, Plates 2.3-1, 2, 3, 7; 2.4-1, 3; 2.5-2). During Q4 time, abundances of coral and coralline algae decreased, and quartz appeared locally as a basal constituent in Q4 limestones (Harrison and Halley 1979; Harrison et al. 1984). Cores (Appendix 2.B, Plate 2.4-3) show wide distribution of quartzose grainstones that occasionally display inclined bedding. Field studies and SEM photomicrographs of the quartz point to possible former regolith and/ or beachrock environments (Multer et al. 2002). Molluscan-peloid-bryozoan packstones and grainstones with patch reefs dominate the upper Q4-Unit sections with rapid facies changes and channels to backreef areas.

Rocks assigned to Q5e time (terminology of Multer et al. 2002) are exposed today along the Florida Keys and represent the high stand (~125 ka) of isotope-substage 5e (Fig. 2.3a). In the lower Keys, ooids precipitated from tidal currents passing

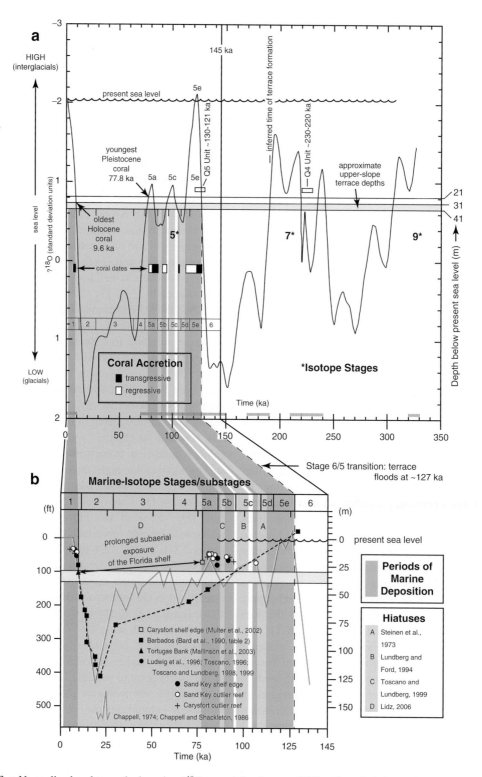

FIG. 2.3. **a** Normalized and smoothed marine $\partial^{18}O$ record for the past 350 ka (from Imbrie et al. 1984) with select data from South Florida added. Note periods of transgressive (black boxes representing coral dates) and regressive (white boxes) coral accretion. Gray horizontal bar represents depths (~30–40 m) of present upper-slope terrace beneath the outlier reefs as based on actual depths (21 and 31 m) of hiatal end-member corals. Those corals have radiometric ages of 77.8 (Multer et al. 2002) and 9.6 ka (Mallinson et al. 2003) and bracket the period of

around reefs into a shallow backreef platform (Hoffmeister et al. 1967; Coniglio 1981; Harrison et al. 1982; Kindinger 1986; Shinn et al. 1989b). In the upper Keys, the Windley Key Fossil Coral Reef State Park and the Key Largo Cross Canal (also called Adams Cut and the Key Largo Waterway, Fig. 2.1a) provide excellent examples of Q5e coral boundstone, grainstone, and packstone containing red coralline algae, molluscs, foraminifera, and fecal pellets. Some 17% of the rock volume in some of the limestone on the island of Key Largo consists of *Halimeda* (Stanley 1966). A shallow-water mix of sediment with random rigid and non-rigid patch reefs borders the leeward side of the Q5e rocks (Appendix 2.B, Fig. 2.B-2). A submerged, gently sloping ramp extends 5–7 km offshore (Fig. 2.5; Multer et al. 2002).

During Q5e time, high nutrient loading, abnormal water temperatures, and/or excessive water turbidity resulting from strong Gulf and Atlantic tidal currents flowing over the accreting proto-Florida Keys (Ginsburg and Shinn 1964; Shinn 1966; Jaap 1979; Lidz and Hallock 2000) may have affected growth rates of organisms. Periodic influx of airborne African dust (Shinn et al. 2000a) may have influenced growth rates. Some of these events may also have impacted ooid production.

After the maximum sea level of Q5e time, lower highstands occurred, providing settings for cumulative accretion of later Pleistocene coralline-rich units on the shelf-edge reef and the buildup of offshore outlier reefs. The outlier reefs (terminology of Lidz et al. 1991) are illustrative case-study examples of multiple Pleistocene geological processes in the area. Outlier characteristics and processes in the keys are referred to throughout this chapter.

2.2.2 Geochronology, Unconformities, Stratigraphic Sequences, and Dominant Corals

Lidz et al. (2007) consolidated the regional spectrum of reef history during the past 350 ka into a detailed database that covers the record in ~3,140 km² of the Florida Keys National Marine Sanctuary. The geographic database area extends from the keys seaward to a margin-wide upper-slope terrace (30–40 m deep) and from a shelf-edge reef known as The Elbow (central Key Largo) across the Marquesas Keys and The Quicksands to Halfmoon Shoal (Gulf of Mexico; all named reefs are Holocene; Fig. 2.1a–c). Six types of sea-level-controlled events took place: non-coralline marine deposition, coral accretion/ooid precipitation, prolonged platform exposure, mangrove-peat/calcrete formation, substantial nearshore erosion, and platform progradation.

Correlation of shelf-wide coral-age and facies data with interglacial maxima (Lidz 2004) shows three recurring trends in the late Quaternary: (a) 13 marine transgressive/regressive cycles influenced deposition (Table 2.1, #1–13); (b) fluctuations in sea level produced seven successive coralline stratigraphic sections (Table 2.1, arrows); (c) six major Pleistocene tracts of discontinuous reefs accumulated – the inland Key Largo Limestone (and contemporaneous oolitic Miami Limestone), the shelf-edge reef, and four tracts of outlier reefs (Lidz et al. 1991, 2003). The shelf-edge and outlier reefs contain corals that date to the same six highstands (Table 2.1, letters A–F). The discontinuous reef groupings at the margin occur in an en-echelon pattern common to transverse and sinuous barchanoid sand dunes (McKee and Ward 1983) and are suggestive of their substrates.

Fɪɢ. 2.3. (continued) prolonged subaerial exposure. Closely spaced tick marks at bottom represent 1-ka increments. Variation in normalized $\partial^{18}O$ mainly reflects changes in sea level due to waxing and waning of continental ice sheets during glacial/interglacial cycles. Approximate durations of substages within Stage 5 are: 5e (127–116 ka), 5d (116–108 ka), 5c (108–96 ka), 5b (96–86 ka), and 5a (86–75 ka). **b** Detailed record of sea-level change during the last glacial/interglacial cycle (modified from Ludwig et al. 1996). Gray horizontal bar represents depths (~30–40 m) of present upper-slope terrace. Jagged gray curve was derived from dating of tectonically uplifted coral reefs on the Huon Peninsula, New Guinea, and is shown for comparison with coral-age data from Barbados (dashed line) and tectonically stable Florida (triangles, circles, and plus signs). All datasets and coral ages point to subaerial exposure of the outer Florida shelf (for ~68 ka) and the paleoplatform interior (for >100 ka) between isotope-substage 5a and the Holocene

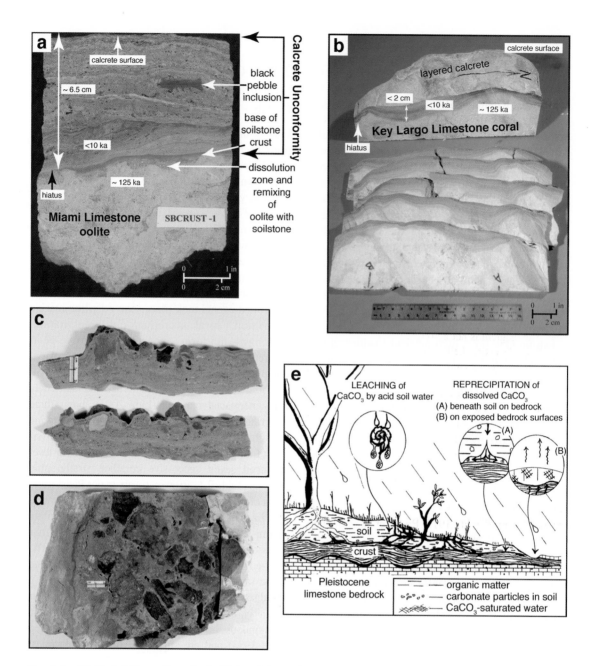

Fig. 2.4. **a** Slabbed drill core from the Saddlebunch Keys (SBCRUST-1, lower Keys, Fig. 2.1a) shows thick, reddish-brown laminar calcrete that precipitated as an unconformity on top of Miami Limestone oolite. **b** Rock sample from Key Largo Waterway (also called Adams Cut and Cross Key Canal, upper Keys, Fig. 2.1a) shows the same calcrete unconformity on top of Key Largo Limestone coral. Note difference in crust thickness between the two substrates, thought to be due to a difference in limestone porosity. Less porous than the coral, oolite retained rainfall moisture longer, allowing longer periods of layered-calcrete buildup. Reddish and brownish laminae in both calcrete samples represent periods of influx of non-carbonate minerals on African dust. The hiatus or gap in these rock records represents an interval of more than 115 ka during which no marine deposition occurred. **c** Large, angular, blackened pebbles embedded in naturally brown, layered, fine-grained calcrete form a breccia. Breccias are consolidated coarse-grained clastic rocks whose fragments are angular and consist of any type of pre-existing rock. This sample is from Ramrod Key (lower Keys, Fig. 2.1b). **d** A multicolored, well-cemented, artificially blackened breccia was collected at a depth of ~6 m below sea level from quarry tailings in a solution pit on Big Pine Key (lower Keys, Fig. 2.1a). Note layered crust on left side of specimen. Blackened fragments include pre-existing calcrete and fossiliferous Key

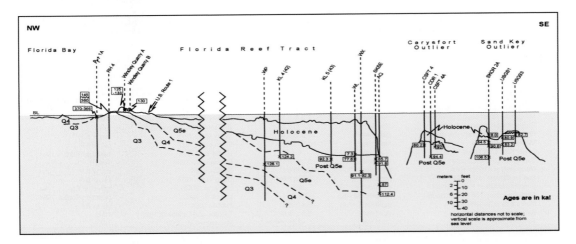

FIG. 2.5. Generalized schematic of Florida Keys shelf area (not to horizontal scale; vertical scale approximate) shows dated cores (Fig. 7 of Multer et al. 2002) (Reprinted with kind permission of Springer Science and Business Media) Note: The diagram is not a normal cross section in that cores involved come from distant (Carysfort to Sand Key) areas along the entire shelf, and subsurface facies change radically along this trend. The diagram represents only one possible subsurface picture based on very limited control existing along ~193 km of shelf. Cross sections with core facies and references for core driller and core dates are given in Multer et al. (2002). General core locations (Fig. 2.1a, b): Ppt 1A: Point Pleasant in Florida Bay at edge of Blackwater Sound. RH 4: Key Largo Canal West from Hawk Channel into Blackwater Sound. Windley Key Quarry A, B: Windley Key. WP: Basin Hill Shoals patch reef on inner shelf midway between Barnes Sound and Carysfort Reef. KL 4: ~3.2 km southeast of Rodriquez Key. KL 5: ~3.6 km southeast of KL 4. WL: beneath Carysfort Reef Light. WX: Looe Key Reef. SKSE: Sand Key Reef. AQ: adjacent to *Aquarius* (18-m-deep underwater research habitat) at Conch Reef. CSFT 4, CDR 1, CSFT 4A: Carysfort outlier reef. SKOR 2A: Sand Key outlier reef. USGS 1, USGS 3: Sand Key outlier reefs

Cemented beach-dune ridges underlie reefs in areas north of Miami (Lighty 1977; Banks et al. 2007) and are believed to underlie all Pleistocene reefs along the South Florida shelf (Shinn et al. 1977, 1991; Lidz et al. 1991, 1997a; Lidz 2004). Wind-blown sand dunes, now rock ridges, are well known throughout the Bahamas and Caribbean (e.g., Ball 1967; McKee and Ward 1983).

Calcrete or soilstone crust is found nearly everywhere offshore, along with mangrove peat indicative of paleoshorelines (Figs. 2.4a–e, 2.6a–d). Calcrete unconformities separate the five stratigraphic Q Units and commonly contain breccias of blackened limestone pebbles (Fig. 2.4c, d). The Q1 and Q2 Units are essentially non-coralline. Shelf corals become dominant in the Q3 Unit (Perkins 1977; Multer et al.

2002). Corals in the Q3 and Q4 Units are tentatively correlated to isotope Stages 9 and 7 (Fig. 2.3a, b; Multer et al. 2002). The shelf-edge and outlier reefs (Lidz et al. 1991, 2003) are inferred to have initiated growth during the Stage-6 transgression that culminated in the substage-5e highstand at ~125 ka (Lidz 2004). The top of the Q5 Unit (the Florida Keys) consists of the ~125-ka Key Largo Limestone coral reef and Miami Limestone oolite (Hoffmeister et al. 1967; Hoffmeister and Multer 1968; Halley and Evans 1983). Substage-5e corals are also found offshore (Perkins 1977) and are inferred to be present near the bases of the outlier reefs in 30+ m of water (Lidz et al. 2003). Later offshore corals accreted during isotope substages 5c, 5b, and 5a (~106.5, ~94–90, and ~86–77.8 ka; Fig. 2.3a, b). Mortality of

---

FIG. 2.4. (continued) Largo Limestone. Left part of cut specimen was heated in the laboratory to reproduce darkening similar to colors of blackened pebbles widely found in the Pleistocene record (Shinn and Lidz 1988). Original rock colors are visible in unheated section at right. **e** Schematic shows some processes responsible for calcrete forming subaerial crusts. Graphic modified from Multer and Hoffmeister (1968, their Fig. 2) (Reprinted with permission of the Geological Society of America Bulletin)

TABLE 2.1. Summary of late Quaternary geochronology and sea-level status along the Florida reef tract.

Event/Evidence	~Lower sea level (m) relative to present	Oxygen-isotope (ka)	stages	Authors	Dating methodologies
Holocene					
Florida and Biscayne Bays flood	slowing rise (at 2)	~0.5		Lidz and Shinn, 1991	based on Robbin, 1984
Carysfort Reef grows	slowing rise	4.8 to 4.8[†]		Multer et al., 2002	TIMS
Long Reef grows	slowing rise	6.3 to 5.8[†]		Shinn et al., 1977	cal. yr B.P.
Upper Keys shelf floods	slowing rise	~7.1		Lidz et al., 1997a	based on Robbin, 1984
Sand Key outlier crest ridge dies	slowing rise	7.1 to 6.8[†]		Toscano and Lundberg, 1998	TIMS
Mangroves SW of Marquesas Keys	slowing rise (> 6.7)	7.2 to 6.8[†]		Robbin, 1984	cal. yr B.P.
Carysfort Reef grows	slowing rise	7.7 to 7.3[†]		Toscano and Lundberg, 1998	TIMS
Fort Lauderdale barrier reefs die	rapid rise	7.8 to 7.5[†]		Lighty et al., 1978	cal. yr B.P.
Marker G reef grows	rapid rise	7.8 to 7.5[†]		Shinn et al., 1977	cal. yr B.P.
Lower Keys shelf floods	rapid rise	~8.1		Lidz et al., 1997a	based on Robbin, 1984
Sand Key outlier crest grows	rapid rise	8.3 to 8.1[†]		Ludwig et al., 1996	TIMS
Mangroves at Alligator Reef	rapid rise (> 7.2)	8.6 to 8.2[†]		Robbin, 1984	cal. yr B.P.
Fort Lauderdale reef grows	rapid rise	9.0 to 8.6[†]		Lighty et al., 1978	cal. yr B.P.
Shoreline facies off Pelican Shoal	stillstand (at ~25)	~9.1		Lidz et al., 2003	based on Robbin, 1984
Shoreline ridge(?) off Pelican Shoal	stillstand (at ~30)	~9.2		Lidz et al., 2003	based on Robbin, 1984
Mangroves at Alligator Reef	rapid rise (> 7.4)	9.4 to 8.4[†]		Robbin, 1984	cal. yr B.P.
(F) Sand Key outliers grow	stillstand	9.8 ± 8.0[†]		Toscano and Lundberg, 1998	TIMS
Tortugas Bank reefs grow	(13) rapid rise	~9.6[†] ← 1		Mallinson et al., 2003	TIMS
Pleistocene					
Marquesas S4 drowns	rapid rise			Locker et al., 1996	
Marquesas shoreline S4 forms	stillstand (at 65)	~13.8[‡]		Locker et al., 1996	AMS ¹⁴C - CRA
Marquesas S3 drowns	rapid rise			Locker et al., 1996	
Marquesas shoreline S3 forms	stillstand (at 71)	~14.0[‡]		Locker et al., 1996	AMS ¹⁴C - CRA
Marquesas S2 drowns	rapid rise			Locker et al., 1996	
Marquesas shoreline S2 forms	stillstand (at 80)	~14.5[‡]		Locker et al., 1996	AMS ¹⁴C - CRA
Marquesas S1 drowns	rapid rise			Locker et al., 1996	
Marquesas shoreline S1 forms	stillstand (at 124)	~18.9[‡]		Locker et al., 1996	AMS ¹⁴C - CRA
	rapid rise	~20.0		Milliman and Emery, 1968	
Sea level at major lowstand	lowstand (> 130)			Chappell and Shackleton, 1986	
	(12) highstand (~41)	30.0[§]	2	Bloom et al., 1974; Chappell, 1974	
	lowstand	?			
	(11) highstand (~38)	50-40[§]	3	Bloom et al., 1974; Chappell, 1974	
	lowstand	?			
	(10) highstand (~28)	60.0[§]	4	Bloom et al., 1974; Chappell, 1974	
Prolonged platform exposure	rapid fall	?		Toscano and Lundberg, 1999	
Carysfort Reef grows	highstand	77.8[§]	5a	Multer et al., 2002	MC-ICP-MS
Sand Key outliers grow	highstand (~9)	~83-80[§]	5a	Ludwig et al., 1996; Toscano and Lundberg, 1999	TIMS
Carysfort Reef grows	highstand (~9)	85.3[§]	5a	Toscano and Lundberg, 1999	TIMS
(E) Sand Key outliers grow	(9) highstand (~9)	86.6[§] ← 5a		Ludwig et al., 1996	TIMS
	lowstand				
(D) Sand Key outliers grow	(8) highstand (14-10)	94-90[§] ← 5b		Toscano and Lundberg, 1999	TIMS
	lowstand				
	highstand	~105.0[§]	5c	Bloom et al., 1974; Chappell, 1974	
(C) Sand Key outliers grow	(7) highstand (~15)	~106.5[§] ← 5c		Toscano and Lundberg, 1999	TIMS
Hiatus in records; Q5 Unit unconf. forms	lowstand (70-80)	~110.0	5d	Steinen et al., 1973; Perkins, 1977	
Q5 Key Largo Ls. & marine unit form	highstand (at ~+10.6)	~125.0[§] ← 5e		Hoffmeister and Multer, 1968	
Miami Limestone oolite facies forms	(6) highstand	~125.0	5e	Hoffmeister and Multer, 1968	
(B) Sand Key outliers grow(?)	rising	~125.0	5e	Lidz et al., 2003	Event/timing inferred
	lowstand(?)				
(A) Sand Key dune ridges colonized(?)	(5) highstand	~127.0	6	Chappell, 1974; Lidz et al., 2003	Event/timing inferred
Dune ridges form on terrace(?)	lowstand	~190.0		Lidz et al., 2007	Event/timing inferred
Q4 Unit unconformity forms	lowstand	~190.0[§]		Perkins, 1977	
Upper-slope terrace forms(?)	falling	~190.0		Lidz et al., 2007	Event/timing inferred
Q4 Key Largo Ls. & marine unit form	highstand	~230.0 ← 7		Muhs, 2002; Muhs et al., 2004	U-series, ∂¹⁸O
Miami Ls. bryozoan facies forms	(4) highstand	~230.0		Hoffmeister and Multer, 1968	
Q3 unconformity forms	lowstand	<366.8[§]		Perkins, 1977	
Q3 Key Largo Ls. & marine unit form	(3) highstand	~366.8[§] ← 9		Multer et al., 2002	TIMS
Q2 unconformity forms	lowstand	?		Multer et al., 2002	
Q2 Key Largo Ls. & marine unit form	(2) highstand	?	11(?)	Perkins, 1977	
Q1 unconformity forms	lowstand	?		Multer et al., 2002	
Q1 Key Largo Ls. & marine unit form	(1) highstand	?	11(?)	Perkins, 1977	

Notes: Table is read chronostratigraphically from bottom to top. Fluctuating sea level was the primary control on all events.

(A-F): Dated corals document four periods of outlier-reef growth off Sand Key between 106.5 ka and 6.9 ka (C-F). Two older growth periods are inferred at the Stage-6/ substage-5e transition (A-B), during which corals are believed to have initiated outlier-reef development at ~127 ka and continued to accrue through substage-5e time.

Marquesas S1-S4, left column: Four subtidal, shoal, dune-ridge complexes formed a series of elevated, shore-parallel upslope shorelines during a pulsating rise in sea level at the end of the Pleistocene. A fifth pulsation produced a shoreline-associated change in sediment stratification off Pelican Shoal (Fig. 2.26a; Lidz et al., 2003).

(1-13): At least 13 periods of marine highstands occurred. Alteration of original mineralogy within the Q3 and Q4 Units has rendered radiometric coral dates for those stratigraphic sections uncertain (Szabo and Halley, 1988; Muhs et al., 1992). Assignments of isotope stages to the Q1-Q4 Units are tentative (Multer et al., 2002). Three highstands (#10-12) remained below elevation of the shelf.

Abbreviations: unconf.—unconformity; Ls.—Limestone; AMS—accelerator mass spectrometry; CRA—conventional radiocarbon age; MC-ICP-MS—multicollector inductively coupled plasma mass spectrometry; TIMS—thermal ionization mass spectrometry; U-series—uranium series; ∂¹⁸O—variations in marine oxygen isotope.

[†]Radiometric dates in ka based on cal. yr B.P. 2-sigma ranges for conventional radiocarbon ages.

[‡]Conventional radiocarbon ages based on accelerator mass spectrometer (AMS) ¹⁴C dating of ooids corrected for an assumed air-sea reservoir effect of -400 years.

[§]Abbreviated radiometric dates: no indications of precision were reported for dates obtained by Mitterer (1974) and cited by Perkins (1977; Q2 at 324 ka, Q3 at 236 ka, Q4 at 180 ka, Q5 at 134 ka). The amino-acid age-dating method used in the 1970s and the more recent U-series age-dating method are sensitive to environmental factors and subject to error (Schroeder and Bada, 1976; Muhs et al., 2004). Using TIMS, Multer et al. (2002) obtained actual Q3-Unit dates of 366.8 and 370.2 ka.

Arrows: Dated lithofacies in Florida represent seven periods of Quaternary coral growth—during accumulation of (a) stratigraphic units Q3, Q4, and Q5 of Perkins (1977), the Q5 Unit correlating with deep-sea oxygen-isotope substage 5e, and (b) younger corals that date to substages 5c, 5b, 5a, and Stage 1 (the Holocene).

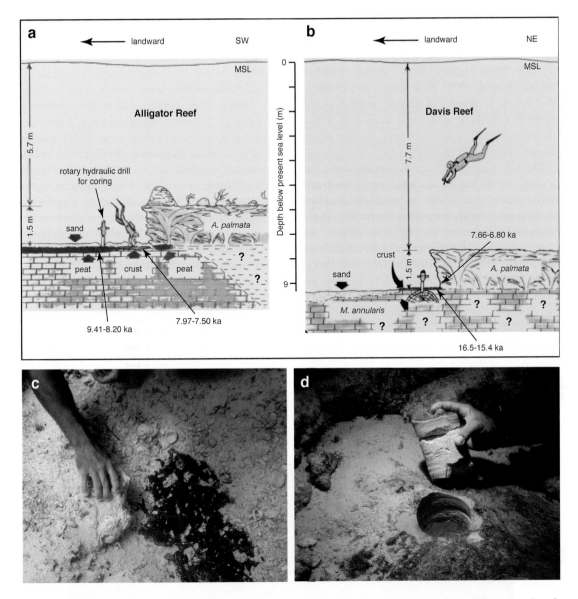

FIG. 2.6. **a** and **b** Sketch shows locations of soilstone crust relative to present sea level and proximity to coral reefs at Alligator and Davis Reefs in the upper Keys (modified from Robbin 1981). Davis Reef is ~11 km northeast of Alligator Reef (Fig. 2.1b). Though relatively close, Davis sample site is deeper than Alligator site due to local irregularities in bedrock surface. Ages on figure have been converted from cal. year BP to ka for brevity. Corrected ^{14}C age range (7.97–7.50 ka) for crust at Alligator Reef indicates reef growth before ~7.5 ka would not have been possible due to subaerial conditions. An older age range (9.41–8.20 ka) for adjacent mangrove peat indicates the peat was forming in a moisture-laden depression at Alligator Reef site before crust began to accumulate on higher adjoining topography. Corrected ^{14}C age range (16.5–15.4 ka) for top 2 cm of crust at deeper Davis Reef site is consistent with submergence of the deeper site during crust formation at Alligator Reef. The corrected ^{14}C age range for coral at the Davis site (7.66–6.80 ka) suggests that corals at the Alligator site are probably younger than 6.8 ka due to later flooding of higher-elevation bedrock at Alligator Reef. MSL = mean sea level. **c** Photo of mangrove peat over soilstone crust under ~30 cm of coarse sand at a site within ~5 m of shoreward margin of Alligator Reef. Note higher-elevation crust ledge under diver's hand. **d** Photo of cored crust on top of *Montastraea annularis* (under diver's thumb) at Davis Reef. Base of dead *Acropora palmata* reef is at upper left

the six Pleistocene systems was ultimately due to a naturally lowered sea level, a point in contrast to the cause of reef decline observed today (Hallock and Schlager 1986; Hallock 2005; Shifting Baselines 2005; Lidz et al. 2007). Now, waters of a rising sea level link an assortment of adverse, natural, inshore elements with the seventh coral ecosystem. Effects of anthropogenic activities compound the natural stresses (Jameson et al. 2002).

The Key Largo and later Pleistocene reefs vary in coral species but are characterized by the framework-building head corals *Montastraea annularis*, *M. cavernosa*, *Porites astreoides*, *Siderastrea siderea*, *Solenastrea bournoni*, *Colpophyllia natans*, and *Diploria strigosa* (Hoffmeister et al. 1967; Hoffmeister and Multer 1964, 1968; Hoffmeister 1974; Perkins 1977; Harrison and Coniglio 1985; Toscano and Lundberg 1999). *Montastraea annularis* is by far the dominant species. Head corals require calmwater conditions (deeper and/or sheltered), as does the staghorn coral *Acropora cervicornis*. Presence of head corals implies that sea level was sufficiently high to preclude violent wave action in their vicinity or that they were protected from wave energy in some other manner. During deposition of the Key Largo Limestone ~125 ka, sea level may have been as much as ~10.7 m higher than at present (Stanley 1966; Hoffmeister 1974; Halley and Evans 1983).

Platy, branched, acroporid corals require well-circulated water. The reef-framework builder *A. palmata* grows where wave energy is high (Shinn 1963, 1980). These corals act as baffles, dispersing wave energy. The depth threshold for *A. palmata* is at 5 m below sea level (Shinn et al. 1977). Except for a few small fragments (Hoffmeister 1974), Pleistocene *A. palmata* is not found on the Florida shelf. Its absence is consistent with highstand water depth greater than 5 m or with lack of a surf zone where the reefs grew. Acroporids are temperature-sensitive species. A theory has also been proposed that periodic infusion of hot platform-interior water to the outer shelf during the 125-ka highstand may have precluded their growth (Harrison and Coniglio 1985). Exposure to cold winter temperatures from Gulf of Mexico waters cannot be ruled out. However, *A. palmata* is found in the Pleistocene outlier reefs, albeit in sparse frequency (Toscano and Lundberg 1999). In contrast to its general absence in Pleistocene facies, *A. palmata* was prolific and was the primary builder of outer-shelf Holocene reefs as recently as ~7–2 ka (Fig. 2.7; Shinn et al. 1977; Dustan 1985).

Fɪɢ. 2.7. Photo shows a magnificent stand of dense *Acropora palmata* at Carysfort Reef prior to the 1970s (Photo courtesy of Phil Dustan)

2.2.3 Morphologies and Limestone Formations

Natural composition and orientation divide the Florida Keys into the upper, middle, and lower Keys (Fig. 2.1a). The Key Largo Limestone, a linear fossil coral reef that parallels the shelf margin, forms the upper and middle Keys. The largest tidal passes demarcate the middle Keys. The Miami Limestone, oolitic tidal bars normal to the margin, forms the lower Keys. Both formations are well dated to the highstand of ~125 ka (Multer et al. 2002). The Key Largo reef extends from beneath Miami Beach, where its elevation is below sea level, to Soldier Key, where it makes its first appearance as an island, to the Newfound Harbor Keys, where the reef last appears as islands, to the Dry Tortugas, where it is again submerged (Shinn et al. 1977). Core borings around the Marquesas Keys and scattered rock samples collected west of those islands show the Miami Limestone oolite extends westward into the Gulf of Mexico and may reach as far as Halfmoon Shoal (Shinn et al. 1990). The Marquesas Keys and Halfmoon Shoal mark the respective east and west ends of a rectangular limestone ridge beneath The Quicksands, a large-volume, westward-migrating belt of non-oolitic carbonate sand (Fig. 2.8a–c; Shinn et al. 1990). The sand consists primarily of fragmented *Halimeda* plates (Hudson 1985). On average 1–2 m thick and molded by the currents, the sand-belt surface is ornamented with large (5-m-high) north-trending tidal bars topped by smaller (1-m-high) west-trending sand waves (Shinn et al. 1990). The low-amplitude sand waves overlie bedrock near the Marquesas Keys. The high-amplitude waves at the west end of the ridge overlie westward-dipping sand beds as thick as 12 m. Coring has shown the broad ridge beneath the active sand bars is Pleistocene oolite (Shinn et al. 1990). Judging from the (undated) oolite and from isolated coral facies along a narrow ridge on its northern edge that are similar to the Key Largo Limestone (Shinn et al. 1990), the ridge is inferred to be the same age (~125 ka) as the two limestone formations of the keys (Lidz 2006).

Major inshore morphologies consist of landmasses and tidal passes of the keys, backed by the inland lagoon of Florida Bay (Fig. 2.1a; Multer

1977). Seaward of the keys, the shelf is informally divided into inner and outer areas (Fig. 2.9; Enos 1977). A prominent nearshore rock ledge that slopes ~2.5 km seaward from the keys shoreline and a margin-parallel bedrock depression under Hawk Channel mark the inner shelf (Enos 1977; Marszalek 1977; Lidz et al. 2003). Three distinct types of Pleistocene reefs characterize the outer shelf and margin: coral ridges, a shelf-edge reef, and outlier reefs. All are discontinuous. Eight narrow coral ridges and intervening sediment-filled swales form a ridge-and-swale geometry that is most prominent behind the shelf-edge reef off the lower Keys (Fig. 2.10a, b; Shinn et al. 1977; Lidz et al. 2003). A ridge-and-swale couplet is several tens of meters across. The ridges are below seismic resolution (<1 m). A core transect across two ridges behind Marker G in the lower Keys (Figs. 2.1c, 2.10a) demonstrated that, where cored, the ridges consist of Holocene corals overlying Pleistocene corals and are separated by a sediment-filled swale (Shinn et al. 1977). The swale core penetrated loose Holocene sands and recovered cemented, chalky, calcrete-coated Pleistocene grainstone (Fig. 2.11). The cores and photos combined show the ridges clearly form a distinctive outer-shelf ridge-and-swale architecture keys-wide (Shinn et al. 1991; Lidz et al. 2003).

In contrast to the narrow ridge-and-swale geometry, a broad reef-and-trough structure consisting of a shelf-edge reef and four outlier reefs many hundreds of meters across marks the margin (Fig. 2.12a, b; Enos and Perkins 1977; Lidz 2004). Seaward of the shelf-edge tract, a margin-wide (in most places) upper-slope terrace supports tracts of imposing outlier reefs, each backed by a trough (Lidz et al. 1991, 2003). Four parallel outlier tracts occur together off Sand Key Reef southwest of Key West (Fig. 2.1a, c), but the tracts extend primarily and discontinuously as a single or double tract seaward of the shelf-edge reef for >200 km along the margin. Shelf-wide, the sections of the shelf-edge and outlier reefs that are in contact with Holocene overgrowth are well dated to the isotope substage-5a highstand of ~80 ka (Fig. 2.3b). In turn and in stratigraphic order, reefs dated to 5b and 5c time underlie the 5a reefs. Substage-5e corals are presumed to compose the base of the shelf-edge and outlier reefs. Similar in size, width, relief, water depth, offshore location, and later ages (Tables 2.1, 2.2), the five tracts attest

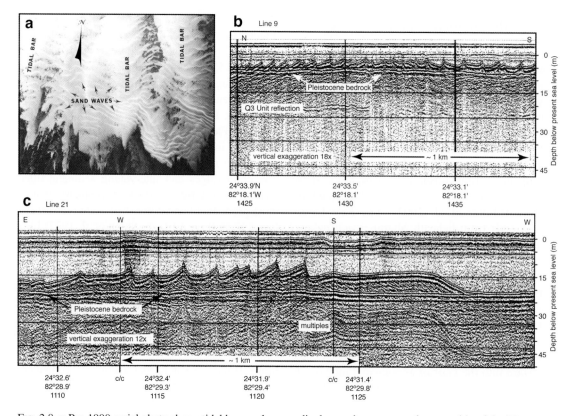

Fig. 2.8. **a** Pre-1990 aerial photo shows tidal bars and perpendicular sand waves near the west side of the Marquesas Keys in the Gulf of Mexico. Many sand waves are awash at spring low tide. Strong, reversing, north/south currents produce the sandbar orientations. *Thalassia testudinum* and species of *Halimeda* populate the dark areas. Corals are absent. Distance across photo is ~2.5 km. **b** Seismic profile in area of **a** shows 1-m-high sand waves lying directly on Pleistocene bedrock. **c** Profile across deep channel bottom west of Halfmoon Shoal shows large (5-m-high) sand waves. Seismic profiles show that lower-elevation topography surrounds the Marquesas-Quicksands ridge (Shinn et al. 1990; Lidz et al. 2003). Sands on the ridge are thus formed *in situ* and migrate westward off the ridge. c/c = course change

to optimal platform-margin sites and settings for reef growth during substage-5c and later time.

2.3 A Fluctuating Sea Level and Pleistocene Settings

The high-precision radiometric dates on bedrock corals recovered in drill cores have established multiple periods of coral accrual during the late Pleistocene (Fig. 2.3a, b, Tables 2.1, 2.2). The coral ages correlate with eustatic high stands of sea level on the global $\partial^{18}O$ marine-isotope paleotemperature curve (Imbrie et al. 1984). Sea-level maxima derived from dates and depths below present sea

level of recovered local corals indicate that in Florida sea level during those times was at or near its apex at minus ~15 m (substage 5c), ~12 m (range 14–10 m; 5b), and ~9 m (5a), relative to present sea level (Table 2.2). The youngest Pleistocene and oldest Holocene coral dates (77.8 ka of Multer et al. 2002, and 9.6 ka of Mallinson et al. 2003) bracket a period (~68 ka) when the shelf was subaerially exposed (Lidz et al. 1985, 1991, 1997a, 1997b, 2003, 2007; Shinn et al. 1990). However, the $\partial^{18}O$ curve shows the interval of prolonged exposure of the higher-elevation paleoplatform interior was longer (>100 ka) and continued from the time the substage-5e sea abandoned the shelf (~112 ka) to the time the Holocene sea again submerged the

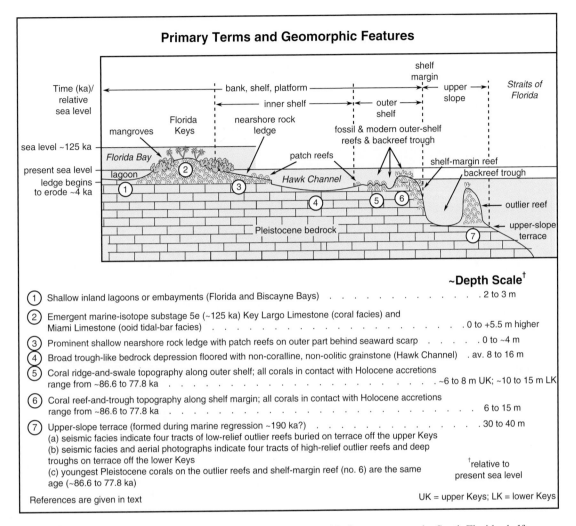

Primary Terms and Geomorphic Features

FIG. 2.9. Schematic cross section (not to scale) illustrates geomorphic features across the South Florida shelf

area (~8 ka). At a sea-level position of 9 m below present, bedrock topography and bathymetry show the existence of an emergent curved promontory that was connected to the mainland and extended as far west as Halfmoon Shoal (Lidz et al. 2003; Lidz 2006). Relative to deep-sea stratigraphic sequences that are often regionally widespread, Holocene accretions capping the submerged Florida shelf are patchy (Lidz et al. 2003). Thus, for regional paleo-topographic purposes, hydrographic bathymetry becomes a useful tool to supplement seismic data and help delineate landmass extent during periods of lower late Pleistocene sea level.

Successive periods of sea-level zeniths that failed to breach a continuously emergent platform promontory constituted the primary control on evolution of large Pleistocene reefs off the Florida Keys (Lidz 2006). The resultant geomorphic settings provided pristine conditions for offshore coral growth – accommodation space and protection. The landmass separated the Gulf of Mexico and Straits of Florida. Card, Barnes, and Blackwater Sounds, the Everglades, and Florida Bay were dry land, as was most of the shelf seaward of the upper and middle Keys (Fig. 2.1c; Lidz and Shinn 1991). Shallow coastal waters with harmful temperatures, nutrients, salinity, and turbidity that might have reached the reefs were distant and minimal. Corals, unencumbered by inimical elements, were able to thrive, nourished by the clear, warm, low-nutrient,

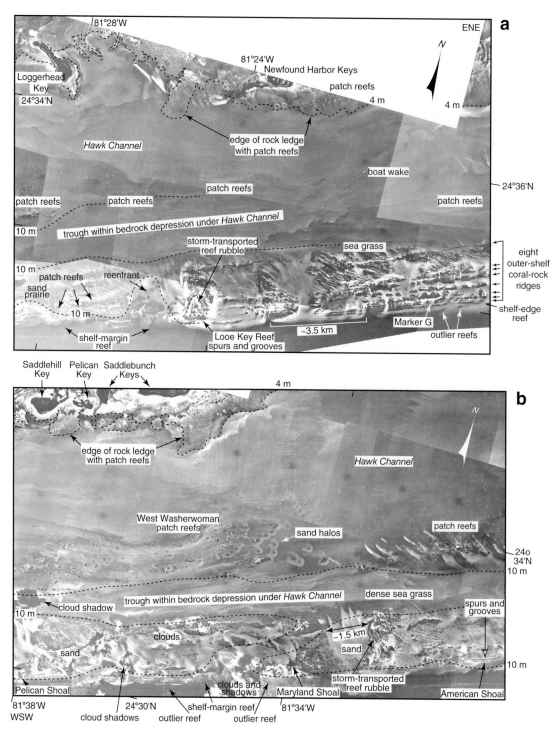

FIG. 2.10. Contiguous aerial photos (1975) show seabed features and habitats seaward of **a** the Newfound Harbor Keys and **b** the Saddlebunch Keys (Fig. 2.1a). Shelf-edge reefs include Looe Key Reef and reefs at American, Maryland, and Pelican Shoals, and at Eastern Sambo. Eastern Sambo lies just out of photo **b** at left edge. Visible onshore habitats are mangroves. Note patch reefs on nearshore rock ledge. Water depth over the ledge ranges from 0 at shoreline to ~4 m. Also note eight linear outer-shelf rock lines (arrows), believed to be narrow coral reefs separated by sediment-filled swales such as have been cored near Marker G (photo **b**)

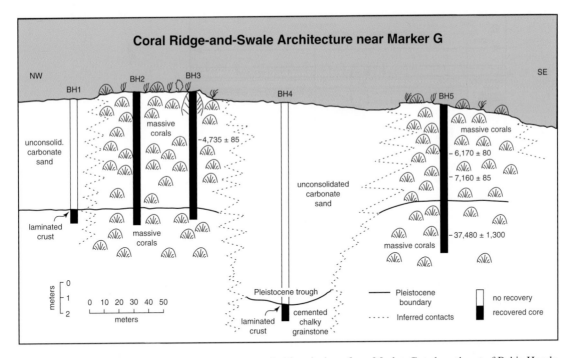

FIG. 2.11. Cross section of core transect across two rock ridges inshore from Marker G and southeast of Bahia Honda (BH) in the lower Keys. Cores show the ridges are coral reefs, are growing vertically, and are separated by a sediment-filled swale. Note lack of a Pleistocene trough on landward (NW) side of left coral ridge. Also note uncorrected radiocarbon ages of recovered corals. The 37,480±1,300 ^{14}C date, obtained on a recrystallized coral, was believed to be much too young due to contamination of younger carbon. Inferred age of bedrock coral sections is between ~86.2 and 77.8 ka (substage 5a), which is the age range for the youngest Pleistocene corals on the shelf and for those Pleistocene corals in direct contact with Holocene accretions (Toscano 1996; Multer et al. 2002; Lidz et al. 2003)

normal-salinity oceanic waters of the Gulf Stream. Reefs at the shelf edge and on the upper-slope terrace reached imposing seismic relief of 30 m and widths of ~1 km (Fig. 2.12a, b; Lidz et al. 1991, 1997a, 2003, 2006), attesting to optimal region-wide conditions. Relative to the long landmass duration of >100 ka, collective time of reef accrual was comparatively short, perhaps only about 20 ka (Fig. 2.3a; Lidz 2006). Buildup occurred during both transgressive and regressive conditions, which has implications for rates of late Pleistocene coral growth and rise and fall of sea level (not addressed in this chapter).

Comparison of bathymetry on a National Geophysical Data Center (NGDC) map (Divins 2003) with contoured bedrock topography (Lidz et al. 2003) shows a correlative emergent-landmass shape. However, the seismic data and oolitic

nature of bedrock beneath The Quicksands indicate that dry land extended farther west to include Halfmoon Shoal. Both bedrock and bathymetric datasets verify subaerial exposure of bedrock in the Boca Grande Channel and on the oolite ridge when sea level was 9 m lower than present. Thus, both types of data corroborate integrity of the emergent paleoplatform interior during substage-5a time. Elevation of the ridge beneath The Quicksands (6–8 m deep) relative to the substage-5a apex at 9 m below present supports an inferred ridge-bedrock age of ~125 ka (Lidz 2006), the same age as the emergent Key Largo/Miami Limestone and older than the submerged 5a shelf off the keys. Skeletal Holocene reef facies at the west end of the reef tract essentially terminates south of Halfmoon Shoal, where the emergent paleoplatform ended.

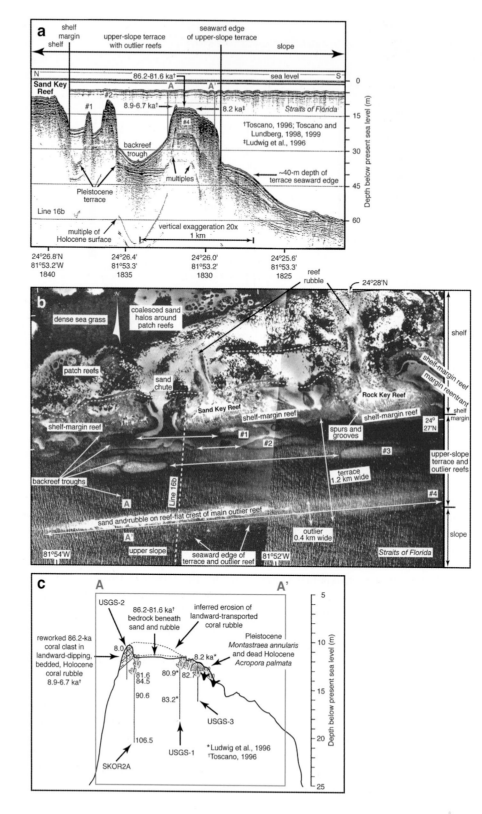

2.4   A Rising Sea Level and Holocene Settings

The same geomorphic setting and pristine conditions for coral growth that existed during the late Pleistocene also existed during the early and middle Holocene. Rate of Holocene sea-level rise, at first rapid, had slowed by the time the sea began to cross the Pleistocene reef crests and creep landward (Fig. 2.13a, b). The slowed rate permitted a protracted movement of the surf zone across the crests and the broad low-gradient outer-shelf expanse. *Acropora palmata* initially flourished, constructing lush stands of coral spurs along the outer shelf and on the shelf-edge and outlier reefs (Figs. 2.7, 2.14a, b; Dustan 1985). Later and with continued sea-level rise, corals responded by changing species and growth habits, even within reefs, and by backstepping over higher-elevation topography (Shinn 1980; Lidz et al. 1985; Toscano and Lundberg 1998). Similar conditions prevailed north of the Florida Keys (Lighty 1977; Banks et al. 2007). With the continuous rise, however, came eventual flooding of lower-elevation areas of the emergent promontory. Landmass connectivity soon vanished (Shinn et al. 1989b; Lidz and Shinn 1991).

As extent of shallow water expanded in the later Holocene and islands formed offshore, new sea-level-induced controls came into effect and influenced circulation, current and wave energy, light attenuation, and distribution of present reefs and benthic habitats. Contrary to most growth habits and patterns that existed during the Pleistocene, Holocene reefs responded to the new controls by changing:

- Location – from the shelf edge and outer shelf to the inner shelf (landward margin of White Bank in the upper Keys, middle of Hawk Channel in the middle and lower Keys, and isolated areas on the nearshore rock ledge; Figs. 2.1a, 2.10a, b).
- Dimension – from broad reef-and-trough shelf-edge structures of high relief to narrow ridge-and-swale outer-shelf ridges of low relief (Figs. 2.10a, b, 2.11).
- Structure – from species of deeper, quiet-water head corals (primarily *Montastraea annularis*) to those of shallow-water branching corals (primarily *Acropora palmata*), resulting in a coral zonation on some reefs; Grecian Rocks is a prime, well-studied example that includes *M. annularis* followed by oriented *A. palmata* on the seaward side and unoriented *A. palmata* on the landward side (Fig. 2.15a; Shinn 1980).
- Geometry – from primarily dense, linear, margin-parallel reefs of massive head corals to spectacular, open-fabric spur-and-groove systems of platy branching corals growing normal to the margin; spurs and grooves formed on nearly every named reef at the shelf edge (Figs. 2.10a, b, 2.16, 2.17) and on some landward outer-shelf reefs; spurs and grooves faced incoming high-energy waves (Shinn 1963, 1980; Shinn et al. 1991); whereas most vertically accreting reefs initiate growth on topographic

FIG. 2.12. **a** Original seismic documentation of three tracts of outlier reefs off Sand Key Reef, located southwest of Key West (from Lidz et al. 1991). Numbered reefs correspond to numbered tracts in **b**. Largest outlier, cored and dated, grew at the seaward edge of a broad upper-slope terrace. Note discontinuity of Holocene section on reef crest. Also note reflection representing nearly flat terrace surface between the largest outliers and lack of reflections beneath the reefs. Coral reefs and reef rubble typically obscure sound-wave penetration to underlying rock surfaces. Latitude and longitude in degrees and decimal minutes based on GPS coordinates. Hours (military time) below coordinates serve as navigational correlation points along seismic line. **b** Aerial photo (1975) of Sand Key Reef area shows four tracts of outlier reefs (#1–4) and their sandy backreef troughs (from Lidz et al. 2003). Seismic line 16b in **a** just missed tract #3, as shown on photo **b**. Note discontinuous nature of Pleistocene shelf-margin reef, and discontinuous and hummocky outlines of outliers. Also note landward patch reefs, linear Holocene spurs and grooves at seaward edge of Sand Key and Rock Key Reefs, and ovate zones of storm-transported reef rubble (red dotted lines) behind the two reefs. White dotted lines mark trends of outer-shelf coral-rock ridges. Approximate location of cross section A-A' across the largest most seaward outlier is shown west of seismic line 16b. This key photo, diagnostic of the actual number of outlier-reef tracts present off Sand Key Reef, is courtesy of Jim Pitts, Department of Transportation, Tallahassee, Florida. **c** Sketch shows approximate locations of three USGS cores and core SKOR2A of Toscano (1966) on largest outlier reef. Landward pinnacle on outlier, visible in **a**, consists of landward-dipping Holocene coral rubble and is separated from an *in situ* seaward Holocene reef patch by substage-5a bedrock. These features are interpreted to represent erosion of a storm-transported reef-crest accumulation of reef rubble derived from the seaward reefs

TABLE 2.2. Data and assumptions used to infer youngest possible age of upper-slope terrace.

Material dated	Dates (ka)	Geologic age	Coral depths (m)[†]	Thickness in outlier reef (m)	Maximum highstand (m)[†]	Authors
A. palmata, C. natans	8.9- 6.9	Holocene	-12.3 to -8.9	3.4	0	Toscano and Lundberg, 1998
1 *A. palmata*, 1 *C. natans*, 8 *M. annularis*	84.5- 80.9	5a[‡]	-24.0 to -12.3	11.7	-9	Ludwig et al., 1996; Toscano and Lundberg, 1999
3 *M. annularis*, 1 *A. palmata*	94.4- 90.6	5b[‡]	-19.8 to -15.5	4.3	-14 to -10	Toscano and Lundberg, 1999
M. annularis	106.5	5c[‡]	-21.7	?	-15	Toscano and Lundberg, 1999
M. annularis	128.1-112.4	5e[‡]	-36.6 to -15.9		~ +10.6	Multer et al., 2002
M. annularis, top of Q5 Unit	~125	5e[‡]	~ +5.5		~ +10.6	Hoffmeister and Multer, 1968; Perkins, 1977; Halley and Evans, 1983
(highstand)	140		not identified in Florida			Chappell, 1974
Marine Q4 Unit top	mid-Pleistocene		-5.0			Muhs, 2002; Muhs et al., 2004
(highstand)	185		not identified in Florida			Chappell, 1974
(highstand)	220		?			Chappell, 1974
Marine Q3 Unit top	~366.8		-8.0			Multer et al., 2002; Muhs et al., 2004
Marine Q2 Unit top	>324		-37.5			Perkins, 1977

Notes: Coral dates and depths of Toscano (1996) and Toscano and Lundberg (1998, 1999) are from the largest Sand Key outlier reef. Coral dates of other authors are from elsewhere in the Florida Keys.

[†]Depths relative to present sea level. Depths of Q2-Q4 unconformities from Big Pine Key core 56 of Perkins (1977). Elevation for top of Q5 Unit is the highest elevation in the keys (at Windley Key; Stanley, 1966; Lidz and Shinn, 1991).

[‡]Substages of marine oxygen-isotope Stage 5.

Data:
- Upper-slope terrace on which the outlier reefs grew is 30-40 m below present sea level.
- Largest Sand Key outlier reef is ~28 to 30 m (mean 29 m) in seismic relief.
- The oldest corals dated (106.5 ka) are also the deepest (21.7 m below sea level) cored from the outlier reef, which leaves ~7.3 m of corals of unknown age at the base of the reef.
- The nearest subsurface data to the Sand Key Reef area come from a long core (core 56 at ~58.5 m) recovered from Big Pine Key (Perkins, 1977). Unconformities on top of the Q1 and Q2 Units in core 56 are found at respective depths of ~48.0 and 37.5 m below sea level. These depths are below elevation of the terrace.
- Six stands of Pleistocene sea level during the past 350 ka were high enough to have flooded the terrace and shelf (Imbrie et al., 1984).

Assumptions:
- The terrace under the outlier reefs pre-dates the reefs.
- The terrace is erosional and was formed during a falling sea level.
- Age of the bottom 7.3 m of the outlier reef is older than 106.5 ka.
- The closest possible time to 106.5 ka for older coral growth on the terrace is during the marine-isotope substage-5e highstand or during a slightly older time at ~127 ka as a rising sea level flooded the terrace during the isotope Stage-6/5 transition.
- Prior to 127 ka, the apex of the last stand of sea level to flood the terrace and shelf occurred at ~190 ka (Imbrie et al., 1984). The marine regression from that highstand reached terrace depths at ~190 ka, the most recent time possible that the terrace could have been formed. If this assumption is true, then the terrace is a regressional erosional feature.

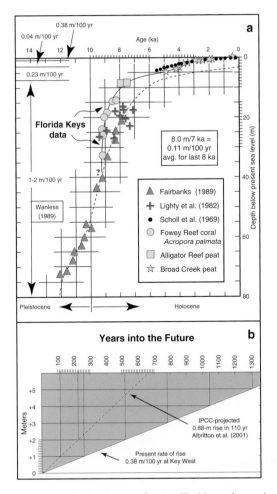

highs, spur-and-groove systems do the same but then backstep over sand and rubble (Fig. 2.15b).

- Depth – from relatively deep (~20 m) spurs and coalesced outer-shelf patch-reef colonies to coalesced patch-reef colonies on the landward margin of White Bank and in the middle of Hawk Channel, and individual patch reefs on the shallow (<4 m) nearshore rock ledge (Fig. 2.10a, b).
- Age distribution – in the keys, from oldest (first to colonize Pleistocene bedrock) in the southwest to youngest in the northeast (Lidz and Shinn 1991); duration of coral growth also lasted ~2 ka longer in the northeast due to a more protected setting than the open-ocean environment in the southwest (Ginsburg and Shinn 1964; Lighty 1977; Lighty et al. 1978, 1982).
- Accretionary direction – from vertical accretion to backstepped growth, as at Grecian Rocks and Looe Key Reef (Fig. 2.15b; Looe Key Reef was named after the *HMS Looe Key* went aground on the reef in 1744).

The elongate island of Key Largo in the upper Keys has no natural tidal passes and offers offshore habitats protection from the tidal exchange of turbid nutrient-rich Florida Bay and cold Gulf of Mexico waters (Roberts et al. 1982; Hallock and Schlager 1986; Shinn et al. 1989b; Lidz and Shinn 1991). Shallower water depths off Key Largo generate a restricted setting, or one in which circulation and current energy are low, vs. an open-marine setting with wider tidal passes and deeper waters off the lower Keys (Ginsburg and Shinn 1964; Ginsburg 1956; Turmel and Swanson 1976; Enos 1977; Smith 1994). The open-marine setting is subject to harsher weather and temperature and greater water-mass exchanges than the restricted setting. A lower Keys Gulf-to-Atlantic gradient in sea level assures net seaward water flow and associated offbank sediment transport (Lidz et al. 1985; Shinn et al. 1990; Smith 1994; Locker and Hine 1995). Periodic confluences of Gulf of Mexico, Straits of Florida, and meandering Loop Current and Gulf Stream eddies contribute to sediment movement (Lidz et al. 2003). Hurricanes and winter storms enhance these endemic elements. These differences in the modern setting are important to understanding the geologic record, because conditions in Florida are thought to have been similar (Ball 1967; Harrison and Coniglio 1985) at the times sea level twice attained its present

FIG. 2.13. **a** Sea-level curve for the Florida reef tract is well constrained by local proxy data (in conventional ^{14}C ages) as modified (Lidz and Shinn 1991) from the curve of Robbin (1984). Data from 8 ka to the present are considered reliable. Question mark denotes lack of dates older than ~8 ka due to lack of *local* material >8 ka (excluding a 9.6-ka date from Southeast Reef at Dry Tortugas; Mallinson et al. 2003). Closely spaced tick marks at top represent 20-ka increments; those at right represent 1-m increments. Rates of sea-level rise are shown at left. Upper part of figure shows rise as measured since 1932 by tide gauges at Key West. **b** Diagram shows time that it will take for sea level to inundate the Florida Keys at its present measured rate of rise (solid diagonal). Dashed line represents predicted rise in mean global sea level from 1990 to 2100 (Albritton et al. 2001) extrapolated to +1-m mark (~110 years) and +2-m mark (~225 years) and beyond. Most dry land in the keys will disappear with a rise in sea level of 1–2 m (Lidz and Shinn 1991). Closely spaced tick marks at top represent 10-year increments; those at left represent 0.1-m increments. IPCC = Intergovernmental Panel on Climate Change.

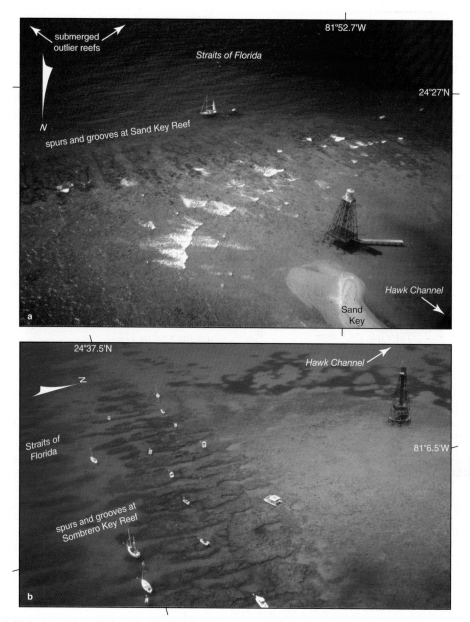

FIG. 2.14. Holocene spurs and grooves off reefs in **a** the lower Keys and **b** middle Keys (Figs. 2.1b, c, 2.17). Spur-and -groove systems are also found off the upper Keys

elevation during the 125-ka transgressive/regressive cycle (Fig. 2.3a). The highstand at +10.7 m relative to present would have altered interactions among endemic elements.

Contours on a map of Pleistocene bedrock topography are a tool to infer positions and shapes of paleoshorelines as the Holocene sea submerged the Florida shelf (Lidz et al. 2003). The paleoshorelines migrated from southwest to northeast with rising sea level. Radioisotope dates on Holocene corals con-

firm the flooding history (Shinn et al. 1977, 1989b; Multer et al. 2002; see Fig. 1 in Lidz 2004). Time-series sketches showing 2-ka increments of flooding over the past 8 ka (Lidz and Shinn 1991) show that numerous ephemeral islands of Pleistocene bedrock formed on the outer shelf, and then disappeared.

At ~4 ka, the shallow bedrock depression that would become Florida Bay began to flood through lower-elevation breaks between the keys, creating tidal passes. With rising water and time, the changes have

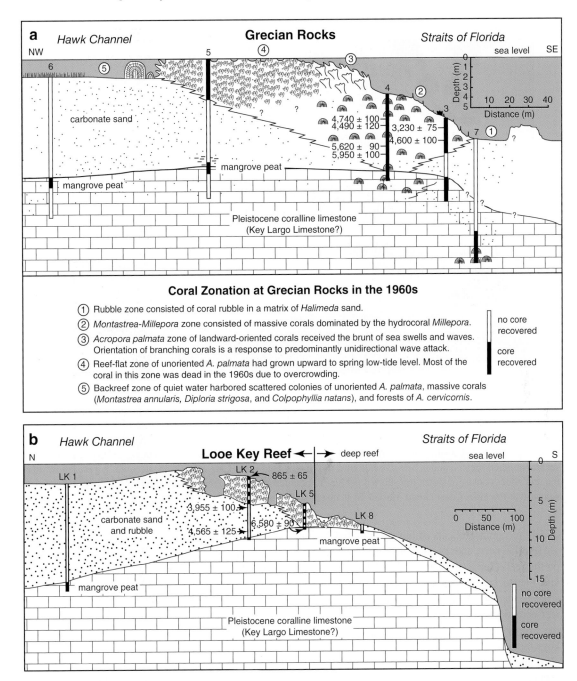

FIG. 2.15. **a** Cross section of Grecian Rocks shows reef components and uncorrected radiocarbon ages in years before present (YBP) of corals recovered in cores. Coral zonation at Grecian Rocks in the 1960s showed five distinct zones. Field observations in 2002 found that most corals including hydrocorals were dead. **b** Cross section of Looe Key Reef (LK) shows the Holocene reef began accretion on the crest of a Pleistocene topographic high

resulted today in a complete shift in offshore benthic communities as non-framework-building organisms characteristic of hardbottoms have replaced reef-framework-building corals (Fig. 2.18a–h; Lidz and Hallock 2000; Shinn et al. 2003; Shifting Baselines 2005; Lidz et al. 2007). With tidally induced influx of deleterious coastal, bay, and gulf waters, the stressed reefs have succumbed. Despite the now-present and

24°32'50"N   81°24'37"W   storm-transported coral rubble 'horns'   81°24'9"W

direction of day-to-day sand movement

sea grass

reef flat   reef flat

Marker 24

shallow reef

intermediate reef

sand-transport directions
during hurricanes (N)
and winter fronts (S and W)

deep shelf-margin reef

~30 m

0   meters   200

Straits of Florida

24°32'40"N

FIG. 2.16. Close-up aerial view (1975) of Looe Key Reef Sanctuary 'core' area (black dotted lines) shows different spur-and-groove zones. Sand is burying each zone. Sand coverage extends seaward and westward, indicating southward and westward transport directions. Intermediate reef is actually two reef trends. Deep reef is a section of the shelf-edge reef at the south edge of the Looe Key Sanctuary. Deep reef here is continuous, extends east and west several hundred meters, and is being buried by a prominent sand lobe (Lidz et al. 1985). Rubble 'horns' landward of the reef zones consist of storm-transported pebble- to boulder-size coral debris (4–256 mm). Area of reef flat at right is awash at low tide. Coast Guard Marker 24 is inside southeast corner of core boundary

widely recognized effects of human activities, rising sea level has been, and continues to be, the main stressor on the reefs (Jameson et al. 2002). Conditions in the late Pleistocene (lower sea levels, accommodation space, protective emergent promontory) and resultant reef accretions of that time clearly demonstrate that the primary cause of reef decline observed in the keys today is natural – a rising sea has flooded the shelf and fragmented the landmass promontory (Lidz 2006).

2.4.1 Rates of Coral Growth

Apart from radical climate change, the markedly uneven coral accretions reflect some of many subtle local limiting factors that affect coral growth. Coral reefs grow at different rates (Korniker and Boyd 1962; Shinn et al. 1977; Aronson and Precht 1997). Not all reefs keep up with rising sea level (Neumann

and Macintyre 1985), and not all topographic highs are colonized (Shinn et al. 1989b; Lidz et al. 2003). Temperature extremes and excess nutrients stress corals (Roberts et al. 1982; Glynn 1984; Hallock and Schlager 1986; Hallock 2005). Non-endemic microbes and minerals (natural, and today anthropogenic) imported on African dust may inhibit growth and have been shown in some instances to be toxic (D'Almeida 1986; Shinn et al. 2000a; Griffin et al. 2001; Walsh and Steidinger 2001; Garrison et al. 2003). Shifting sands interfere with rigorous or continuous growth (Lidz et al. 1997b). These factors notwithstanding and lacking anthropogenic influence, late Pleistocene corals thrived offshore during lowered stands of sea level.

Many investigators, beginning in Florida and the Tortugas with Mayer (1914) and Vaughan (1915a, b), have studied growth rates of some coral species. Shinn (1966) measured the growth rate of *Acropora*

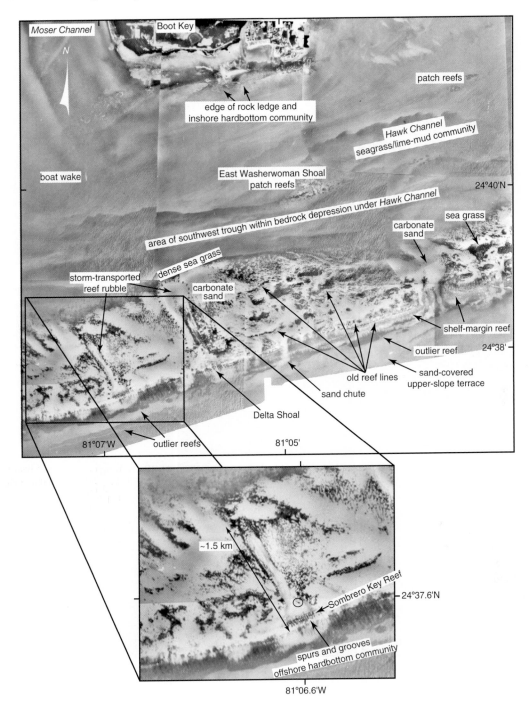

FIG. 2.17. Aerial photomosaic shows Sombrero Key Reef and Delta Shoal off the middle Keys (Fig. 2.1c). Note elongate areas of storm-transported coral rubble (red dotted lines) and irregularity of skeletal spurs and grooves at Sombrero Key Reef (inset). Spurs and grooves dominate the seaward side of most Holocene reefs and are popular diving sites in the keys. Spurs consist of skeletal Holocene *Acropora palmata*. Though no longer growing, they nonetheless harbor colorful hard-bottom communities. Also note irregular topographic relief of shelf-edge reef (numerous areas of sand spillovers or chutes) and discontinuous upper-slope outlier reefs. Sediment analyses show that highest percentages (>60%) of coral grains along the reef tract occur in sands at Sombrero Key Reef (Fig. 2.22a, d). High coral-grain percentages correlate with high abundances of bioerodable (i.e., skeletal) coral. Sombrero Key Reef Light and its shadow are circled in inset

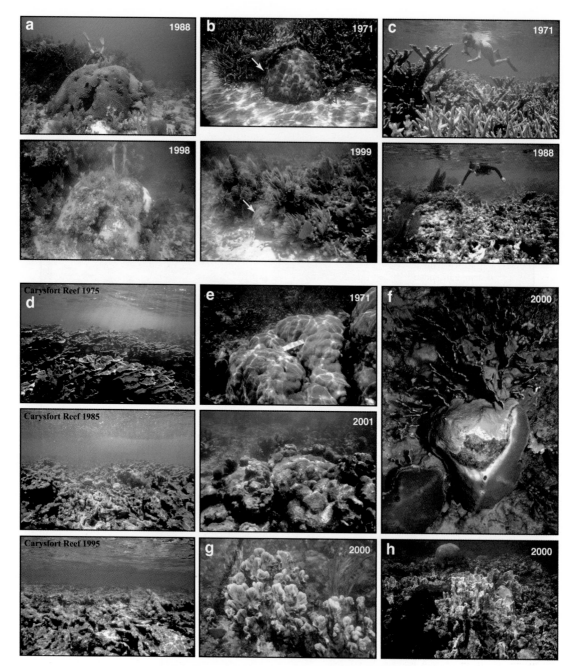

FIG. 2.18. **a–e** Photo pairs (three for **d**) show vitality of head and branching corals in the recent past and their recent state of decline in the Florida Keys. Each set is of the same corals at different points in time. Note coral rubble, macroalgae, sea whips, and sea fans in the later photos. Grecian Rocks: **a** *Colpophyllia natans*, **b** *Montastraea franksi* (arrow), **c** *M. franksi* under diver, *Acropora palmata* at left and back right, *A. cervicornis* in foreground (photos in **c** are from Shinn et al. 2003). Carysfort Reef: **d** *Acropora palmata* (compare with healthy staghorn coral in Fig. 2.7; photos courtesy of Phil Dustan), **e** *M. franksi*. Elsewhere on the reef tract: **f** *Millepora complanata* overgrowing a dying *M. annularis*. **g** *M. complanata*. **h** A field of *M. complanata* surrounds a brain coral (*Diploria* or *Colpophyllia* species) visible in background

cervicornis (~10 cm/year), and Mayer (1914) and Shinn (1966) both demonstrated that changes in seasonal water temperature affect growth rate. Few growth-rate measurements of healthy Florida *A. palmata* have been determined. However, Lewis et al. (1968) obtained linear rates of growth on the order of 10–12 cm/year in Caribbean *A. palmata*. Such rates are consistent with observations of the authors in Florida over the past 40 years. *Acropora palmata* is robust, surf resistant, and the major reef-framework builder and spur former in the Florida-Caribbean region (Shinn 1963).

Sclerochronology, the counting and measuring of annual growth density-band couplets and bandwidths in stony corals, is used to detect paleoecologic patterns in coral-bearing accretions (Hudson et al. 1976; Hudson 1977, 1981a). X-radiography makes bands visible. Sclerochronology was employed in the most comprehensive growth-rate study of the massive star coral *Montastraea annularis* conducted to date (Hudson 1981b). Measurement of >7,000 annual bands in cores from more than 100 large living *M. annularis* off Key Largo led to delineation of growth rates well back into the nineteenth century. Rates in Florida varied depending upon location. Average rate was lowest (6.3 mm/year) in offshore areas deeper than 6 m. Average rate was slightly higher (8.2 mm/year) in nearshore patch reefs and was highest (11.2 mm/year) in shallow (<3 m) offshore areas. A similar study conducted in Biscayne National Park (Hudson et al. 1994) revealed that growth rates in *M. annularis* varied from 5.0 mm/year in the northernmost sector to 11.3 mm/year in the southernmost sector of the park. A reasonable conclusion from such rates is that a 1-m-high *M. annularis* is ~100-years old. Using similar banding measurements, Hudson (1979) reported that Pleistocene *M. annularis* in the Key Largo Limestone grew only 5 mm/year. Apparently, growth conditions during the higher position of sea level at 125 ka were not as favorable as in the earlier Holocene. Like the possible exclusive effect on temperature-sensitive *Acropora palmata* in Pleistocene shelf facies, periodic influx of waters with extremely high summer (platform interior) and low winter (Gulf of Mexico) temperatures may have stressed corals forming the proto-Florida Keys, slowing their growth rates.

Sclerochronologic analyses of *Diploria strigosa* and *D. labyrinthiformis* cores from Bermuda showed growth rates in those corals from that location are lower (3.3 mm/year) than rates measured in *M. annularis* from the Florida Keys (Dodge and Vaisnys 1975). A rate in a *D. labyrinthiformis* core collected off Key Largo was comparable at 3.5 mm/year (Ghiold and Enos 1982).

Alizarin Red-S is a bone stain used to create an internal marker in corals that have indistinct annual bands. Landon (1975) stained corals collected off Key West to ascertain growth rates for *Porites porites* (16 mm/year) and *Siderastrea siderea* (2.4 mm/year). *Siderastrea siderea* is a massive species, resistant to cold and sedimentation. Sclerochronology is imprecise in this coral, because bands are not distinct. Preliminary results of a new Sr/Ca-ratio method have essentially verified the short-term rates of Landon (1975; EAS).

A specimen of *Solenastrea bournoni*, a massive coral from the keys, grew at a rate of 8.9 mm/year (Hudson et al. 1989). *Dendrogyra cylindrus*, a pillar coral collected off Key Largo, grew at ~18 mm/year (Hudson and Goodwin 1997). The pillar coral is rare and is not a significant reef builder, although individual colonies can reach 2 m or more in height.

Individual *Acropora cervicornis* sticks extracted from a few centimeters below the sand surface in 39 widespread areas from Key Largo to Key West were dated with ^{14}C (Shinn et al. 2003). Dates ranged from 6 ka to the present. The distribution of ^{14}C ages was pretty much random except for a significant absence of dates (i.e., absence of *A. cervicornis* sticks of that age) during a 500-year interval centered on 4.5 ka. A less significant break in its occurrence was centered on 3 ka.

The thickness of reefs and lime sediment produced by these major coral species should be significant. Consider that at 10 cm/year, *A. palmata* or *A. cervicornis* have the potential to grow 100 m in 100 years! The potential production rate of *M. annularis* at just 10 mm/year (=1 m/100 years, or 10 m/1 ka) is considerable. What do these data mean in the context of geologic time and reef building in the Florida Keys? The hypothetical upward linear growth rates of certain species over the past 6 ka are graphically summarized (Fig. 2.19). At the known growth rates for the corals depicted, reefs should have easily kept pace with the well-documented rate of sea-level rise over the past 6 ka. Surprisingly, the number of reefs that have actually kept pace, such as those presently at sea level, i.e.,

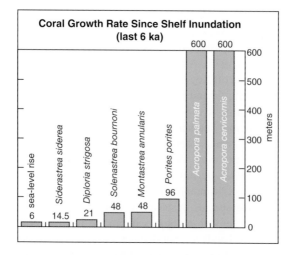

FIG. 2.19. Graph shows theoretical rates of growth based on documented annual growth rates of selected coral species (from Shinn et al. 2000b). Compare growth rates with the degree of sea-level rise over the past 6 ka, the approximate duration of time during which the shelf off the keys has been flooded as shown by all data

Grecian Rocks, Key Largo Dry Rocks, Carysfort Reef, Looe Key Reef, and several others, is low, on the order of 0.01% of the total accommodation space (USGS, unpublished data). Furthermore, none of these reefs kept pace by purely vertical growth. Cores have shown that, for the most part, the shallowest parts of these reefs, the parts presently at sea level, are the result of either backstepped growth or landward-transported coral debris. The shallow reefs are mostly the well-known named reefs with lighthouses. The only reef area shown by drilling to have maintained continuous upward growth is Southeast Reef at Dry Tortugas (Shinn et al. 1977; Shinn and Jaap 2005). True Holocene reef framework in the Florida Keys exists only on the seaward side of named reefs, and even at those locations framework is mainly restricted to corals of the spur-and-groove system. Cores and exposures created by occasional ship groundings verify the skeletal nature of the spurs and show that most spur systems ceased accreting more than 1 ka. Even more perplexing is that the vast majority of Holocene platform-margin reefs, the "senile reefs" of Lidz et al. (1997a, 2006), accumulated less than 2 m of reef framework during the past 6 ka. Senile reefs, also termed senescent reefs (Lidz and Hallock 2000), are those that are not actively accreting reef framework. Today, the primary area

of senile reefs is at the shelf edge (Lidz et al. 2006), where the discontinuous Pleistocene (substage-5a) reef supports a thin Holocene hardbottom coral community rather than a geologically construed coral reef. Though biologically constructed, skeletal spurs and other coral reef framework are solid limestone (geologic) features.

Although much remains to be learned, coral growth has apparently not lived up to its potential, literally. Considering the time interval of 6 ka and high growth rates of some framework-building species, continuous widespread growth should have filled the vast majority of accommodation space. Clearly, hurricanes, temperature extremes, and other natural phenomena have influenced growth during the 6 ka of platform flooding. The well-documented coral demise experienced in the Florida Keys, and throughout the Caribbean, over the past 30 years may serve as an analog to periodic declines in coral vigor over the past 6 ka. The ultimate sources of pathogens, anthropogenic or natural, causing the recent and ongoing demise are arguable, but there is general consensus that microbial diseases are the root cause. Beginning in 1983, diseases lethal to corals, including the unknown reason for demise in populations of *Diadema* spp. (Lessios et al. 1984), have simultaneously affected corals and reefs throughout the Caribbean. In conjunction with periodic hurricanes, diseases can set reef accretion back for many dozens of years.

2.4.2 The Sedimentary Signature as Geoindicator and Its Implications

Sediments began forming on the modern Florida reef tract about 6 ka when sea level was ~5.7 m lower than at present and inner-shelf bedrock lows began to flood (Lidz and Shinn 1991). Reef sands are the product of *in situ* breakdown of endemic calcareous organisms. General similarity between the distribution of living sediment-producing organisms in the Florida reef tract and their fragmentary remains (Ginsburg 1956; Enos 1977) implies that little major transportation of sand-size sediment occurs from one area to another. Even intense wind and wave actions of winter or tropical storms and hurricanes simply mix but do not remove sands from the *general* area of their origin (Ball et al. 1967).

Human activities have substantially increased the natural flux of fixed nitrogen to coastal systems worldwide (Lidz and Hallock 2000; Hallock 2005, and many others). Coastal waters of Florida are not immune to biological consequences of increasing anthropogenic-nutrient influx. Spatial and temporal trends in sediment constituents have been adapted to a previously published model that predicts changes in benthic-community structure with increasing nutrient flux into the environment (Fig. 2.20a, b; Hallock et al. 1988). Whether nutrification is a factor in the decline of Florida reefs is unproven, but the effects of nutrification on corals in general are recognized (Table 2.3). The natural variety of nutrient sources to Florida's reefs ranges from the Florida Current and African dust (Fig. 2.21) to the Everglades, Florida and Biscayne Bays, and groundwater flow (e.g., Hallock et al. 1993). *Aspergillus sydoweii*, a soil fungus exogenous to a marine-carbonate environment, has been identified in soil samples from Mali, Africa (Walsh and Steidinger 2001; Garrison et al. 2003), in atmospheric-dust samples collected from above the Atlantic Ocean (Griffin et al. 2001), and in cultures from a disease (*Aspergillosis*) that is killing sea fans in Florida and the Caribbean (Smith et al. 1996; Shinn et al. 2000a). Human activities contribute nutrients to these natural sources and also add purely anthropogenic sources, such as air pollution and farm fertilizers (e.g., Fanning 1989; Bryant et al. 1998), offshore sewage outfalls, and direct delivery that injects detergents and both human and boat effluents into the system (Smith et al. 1981; Shinn et al. 1994; Multer 1996). Over time, the ecosystem responds, and pollution- or stress-tolerant species replace other, once-dominant organisms. These shifts in community structure eventually become part of the sedimentary record with breakdown of calcareous organisms upon death.

Few historical data exist with which to make comparisons of the only keys-wide petrologic database for the Florida reef tract (Lidz and Hallock 2000). The limited historical data are also based on sparse samples from local transects (Ginsburg 1956; Swinchatt 1965; Lidz et al. 1985). Three principal grain types have remained dominant over the 37 years of sampling periods (1952, 1963, 1989): *Halimeda*, mollusc, and coral (Fig. 2.22a–e). Grain percentages have changed, however, in response to environmental and anthropogenic parameters. The most important percentage with the most significant implications is that of particulate coral. Coral grains have decreased in the upper Keys and increased in the middle and lower Keys over the 37 years (Fig. 2.20a; Lidz and Hallock 2000). Mechanical breakage due to turbulence and current and wave action fragments large coral heads or breaks large chunks from the reef framework, producing smaller grains in the process, but recovery is rapid (Ball et al. 1967; Shinn 1976). Such breakage is limited to the outer reefs and the area immediately behind the reefs, however (Swinchatt 1965). The only other alternative for coral breakdown elsewhere on the shelf is through biological action. Skeletal coral is fodder for bioeroders. Herbivores such as *Diadema antillarum* remove scraps of dead coral as they feed upon attached algae (Glynn 1988). Certain species of fish bioerode reefs by biting off bits of either

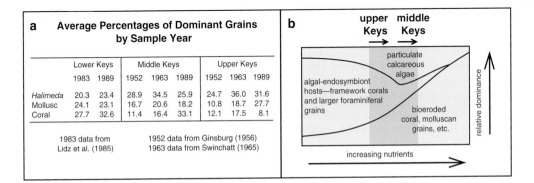

a	Average Percentages of Dominant Grains by Sample Year									
	Lower Keys		Middle Keys			Upper Keys				
	1983	1989	1952	1963	1989	1952	1963	1989		
Halimeda	20.3	23.4	28.9	34.5	25.9	24.7	36.0	31.6		
Mollusc	24.1	23.1	16.7	20.6	18.2	10.8	18.7	27.7		
Coral	27.7	32.6	11.4	16.4	33.1	12.1	17.5	8.1		

1983 data from Lidz et al. (1985)

1952 data from Ginsburg (1956)
1963 data from Swinchatt (1965)

FIG. 2.20. **a** Summary of petrologic data averaged by sample year for upper, middle, and lower Keys. Note data sources. **b** Conceptual model predicting effects of nutrification on biogenic sedimentary components in a reef ecosystem (adapted from Hallock et al. 1988). Gray area represents present conditions in the Florida Keys. Arrows show postulated shifts in upper and middle Keys bioerosional components (i.e., changes in benthic communities) between 1963 and 1989

TABLE 2.3. Examples of nutrient sources and effects on corals in general.

Environment	Effects	Authors
Nutrient-poor water	Reef ecosystems flourish	Margalef, 1968; Falkowski et al., 1993
Nutrient-enriched water	Reef ecosystems succumb	Smith et al., 1981; Hallock and Schlager, 1986
Increased nutrient supply	Subtropical/tropical benthic communities change from coral/algae-dominant to algae/sponge-dominant	Birkeland, 1987, 1997; Triffleman et al., 1992
Annual terrestrial nitrogen flux into aquatic systems has increased		Barber, 1988
Annual fixed-nitrogen flux to terrestrial ecosystems is double pre-anthropogenic levels		Vitousek et al., 1997a, 1997b
Saharan dust, the source of iron- and clay-rich soils on isolated Caribbean islands, is hypothesized to be the common forcing factor	Episodic benthic algal blooms and cyanobacterial infestations	Shinn, 1988; Rawls et al., 1988; Hayes et al., 1988
Oceanic N-fixation is iron-regulated, whether iron source is eolian, tectonic, sedimentary, or anthropogenic		Barber, 1988
Saharan dust is believed to be the effective agent for infiltration of *Aspergillus* (soil fungus) spores in Caribbean and Florida reefs	Sea fans infected with *Aspergillus*	Prospero and Nees, 1986; Muhs et al., 1990; Smith et al., 1996
One billion tons of air-borne Saharan dust are carried into the Caribbean each year		D'Almeida, 1986

live coral or attached algae and excreting coral sand (e.g., Scoffin et al. 1980; Glynn 1987). Numerous boring organisms, such as endolithic algae, fungi, barnacles, clionid sponges, lithophagid bivalves, and other molluscs bore into coral skeletons (Neumann 1966; Edwards and Perkins 1974; Rützler 1975; Hudson 1977; Glynn 1987).

Reef recovery from weather phenomena, other than storms, is less rapid, allowing time for breakdown of freshly killed coral. Recovery from thermal stress, for example, requires settlement of imported planktic coral larvae and a substrate suitable for larval attachment (Shinn 1976). Unusually severe winter cold fronts impacted coral communities keys-wide in 1970 (Hudson et al. 1976) and again in 1977 (Porter et al. 1982; Roberts et al. 1982), when snow fell in Miami. Both events killed not only the more temperature-sensitive branching corals (*Acropora* spp.), but also the much hardier head corals including *Montastraea annularis* as far inshore as at Hen and Chickens (Fig. 2.1a; Hudson et al. 1976; Hudson 1981b). Unusually high water temperatures also stress coral by causing the coral to bleach (Glynn 1984; Ogden and Wicklund 1988; Porter et al. 1989; International Society for Reef Studies 1998; Hoegh-Guldberg 1999; Anderson et al. 2001; Fitt et al. 2001; Rowan 2004). A coral is said to bleach when it loses its symbiotic zooxanthellae. If temperatures remain high and corals do not regain their expelled algae, coral mortality may result (Jaap 1985).

Halimeda and molluscan grains are produced in sandy and hardbottom environments through-

out the keys. Coral debris is produced mostly in hardbottom settings, where bioerodable substrate includes fossil coral, recently killed coral, and the exposed skeleton underlying live coral (Lidz and Hallock 2000). Historically and as recently as in the 1960s, species of *Acropora* were the principal reef-building and spur-forming corals in Florida (e.g., Shinn 1963; Shinn et al. 1981) and in the Caribbean (Aronson and Precht 1997). In the late 1970s and 1980s, however, white-band and black-band diseases decimated acroporids and massive species such as *Montastraea annularis* and *Colpophylia natans* throughout these regions (Antonius 1973; Gladfelter 1982; Rützler and Santavy 1983; Dustan and Halas 1987) and in Bermuda (Garrett and Ducklow 1975). In 1983, an unknown plague destroyed about 90% of *Diadema antillarum* populations in Florida and the Caribbean (Lessios et al. 1984; Lessios 1988). The sea-urchin communities have not recovered. Without this important check to reduce algal competition, the balance has swung to favor algae, grazing molluscs, and hydrocorals (Fig. 2.18a–h; Carpenter 1985; Hughes 1994; Lapointe et al. 1990). Similar extreme changes have occurred in hardbottom communities in the upper Keys, where percentages of live-coral cover had dropped into the single digits in quadrats surveyed between 1996 and 1997 (EPA 1998). In cases where algal overgrowth has occurred but bioerosional rates of reefs have not increased significantly, the proportions of particulate *Halimeda* and grazing molluscs should increase relative to those of coral

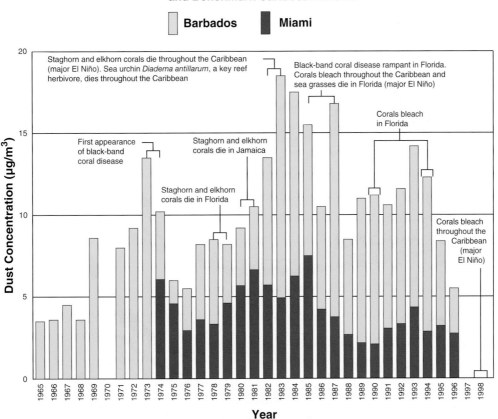

FIG. 2.21. Average African dust concentrations measured in Miami and Barbados (Windward Islands) and major events that occurred in reef organisms in the Florida-Caribbean region. Data for 1965–1986 from Prospero and Nees (1986). Data from 1987 to 1996 courtesy of J.M. Prospero. African dust events are readily observed on the Internet from NASA SeaWiFS and TOMS satellite data (http://coastal.er.usgs.gov/african-dust and http://seawifs.gsfc.nasa. gov/cqibrs/levvel3.pl?DAY = = = tcir). A short video can be viewed at http://coastal.er.usgs.gov/african dust/dust-documentary.html

grains (Fig. 2.20b). Numerous researchers have concluded that rates of bioerosion are a function of abundances of bioeroders and of their food supplies (e.g., Highsmith 1980; Glynn 1987, 1988; Hallock 1988).

An increase in coral-grain debris in the middle Keys may be related to Hurricane Donna in 1960 but is also consistent with increased dead-coral substrate and the model prediction of accelerated bioerosion by boring bivalves and sponges in response to increased plankton productivity. Nutrients from Florida Bay and well-documented eutrophication of nearshore environments stimulate plankton productivity (Lapointe et al. 1990;

Shinn et al. 1993, 1994). Plumes of brown bay water have been reported flowing seaward to offshore reefs through passes on either side of Long Key (Fig. 2.1a; Szmant and Forrester 1996). The discolored water contains elevated levels of nutrients and chlorophyll. Aerial photographs show similar plumes drifting seaward from the offshore Key West sewage outfall (B.E. Lapointe, 1999, personal communication). Total nitrogen and total phosphorous have been measured in sediments off the keys (Szmant 2002).

In Florida reef sands, coral-grain percentages are related to geographic location, tidal-pass exposure to nutrient-enriched waters, and rates of bioero-

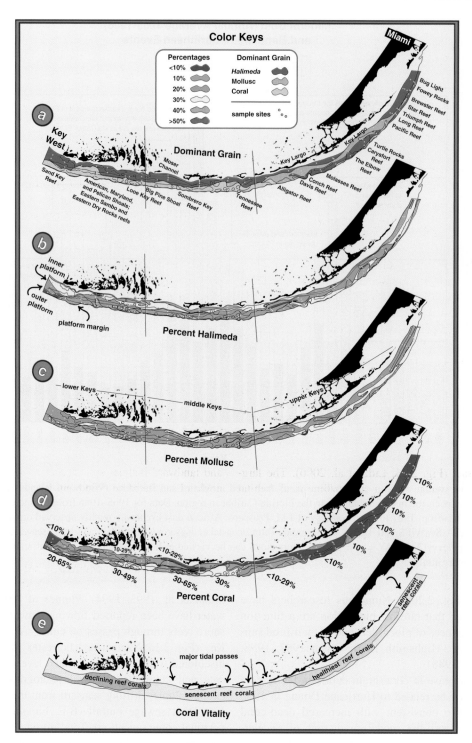

FIG. 2.22. Contour maps of dominant grains in sands of the Florida Keys reef tract in 1989 (Lidz and Hallock 2000)

sion. In general, high coral-grain percentages and high frequency of declining or senescent reefs occur together and are found opposite tidal passes, where rates of bioerosion by plankton-feeding boring organisms are also high (Lidz and Hallock 2000). Percentages of coral grains are highest at reefs along the lower and lower-middle Keys outer-shelf margin and lowest off the upper Keys, where reefs are somewhat removed from bay and near-shore influences (Fig. 2.22d). A relative decrease in coral debris off the upper Keys over the 37 years may reflect suppressed rates of bioerosion in the absence of *Diadema*. However, an obvious prolif-eration of algae and herbivorous molluscs may be a sign that the ecosystem there is changing. These patterns imply that the sedimentary signature in the Florida Keys involves interplay between physi-cal and biological processes and landforms that influence water quality. The results are evidence that the composition of reef sediments is a reliable geoindicator of reef health and thus is a useful tool to assess and monitor past, present, and possibly future status and trends.

When interpreted in detail from aerial photo-mosaics, large- and small-scale benthic habitats and environments (Marszalek 1977; Multer 1996; Lidz et al. 1997a) allow classification of three ter-restrial and 19 marine substrates throughout the reef tract (Fig. 2.23; Lidz et al. 2006). The larg-est ecosystems are a seagrass/lime-mud habitat (map area: 27.5%) found in Hawk Channel, and seagrass/carbonate-sand (18.7%) and bare carbon-ate-sand (17.3%) zones found on the outer shelf and on the Marquesas-Quicksands ridge. A lime-mud/seagrass-covered muddy carbonate-sand zone (9.6%) abuts the keys. Hardbottom communities (13.2%) consist of inshore bare Pleistocene coral-line and oolitic limestone and offshore coral rubble and senile coral reefs. Combined, the three onshore (4.0%) and remaining 11 offshore habitats classi-fied (9.7%), including live outer-shelf patch reefs (0.7%), are small. Each habitat has its own associ-ated organisms (Enos 1977; Lidz et al. 2006).

2.4.3 An Erosional Nearshore Rock Ledge and Its Implications

Whereas a late Pleistocene sea level higher than at present gave rise to the Key Largo and Miami Limestones that now form the Florida Keys, the Holocene transgression also gave rise to a keys-specific event: island erosion. Few sand beaches are found in the keys, but a rocky shoreline with a 0.5- to 1-m-high scarp is present on many islands, especially on Key Largo. Presence of a nearshore rock ledge on the seaward side of the keys has long been known (e.g., 'limestone bedrock' of Marszalek 1977), but its significance has only recently been established (Lidz et al. 2006).

The ledge as an entity, *per se*, has not been stud-ied (Fig. 2.24a, b). Ledge composition is inferred to consist of the 125-ka Key Largo Limestone reef and Miami Limestone oolite for numerous reasons. The ledge is found keys-wide (Lidz et al. 2003). Its landward edge is cut into the two limestone forma-tions and is the present seaward shoreline of the keys. The ledge extends seaward as much as 2.5 km with an offshore gradient. Its seaward edge is a 30-cm-high scarp elevated above adjacent grain-stone underlying Hawk Channel, indicating the ledge may not be the same type of substrate as the channel limestone. Its seaward edge is also jagged, consistent with the irregular nature of a fore reef. A core drilled through the ledge for installation of a water-monitoring well off north Key Largo recovered coral (undated) similar to that of the Key Largo Limestone (Shinn et al. 1994). The sloped ledge surface is flat but pockmarked with potholes, particularly in the lower Keys oolite. The width and landward incline are consistent with an initial angle of repose of an ooid bank and with a broad, lower-elevation fore reef in front of the narrower emergent reef crest. Such structure and dimension are found in most reefs, such as at Grecian Rocks (Shinn 1980).

The most compelling argument that the ledge may represent the seaward extent of the Key Largo and Miami Limestone islands, as Lidz et al. (2003) have proposed, is that its landward edge cuts into the islands (both oolite and reef) and forms the seaward shoreline keys-wide. Color hues on the NGDC bathymetric map (Lidz et al. 2006) clearly delineate extent of the Miami Limestone tidal-bar field and seaward edge of the nearshore rock ledge, supporting the interpretation. The landward edge/shoreline, elevated seaward scarp, landward slope, smooth surface, and potholes are each consistent with surface erosion. By definition, the *time* of formation of an erosional surface is not datable. However, the ledge was clearly carved into the island chain. The time of erosion is thus post-125-ka deposi-tion of the Key Largo/Miami Limestone. The question is, did the erosion occur on the 125-ka regression or the

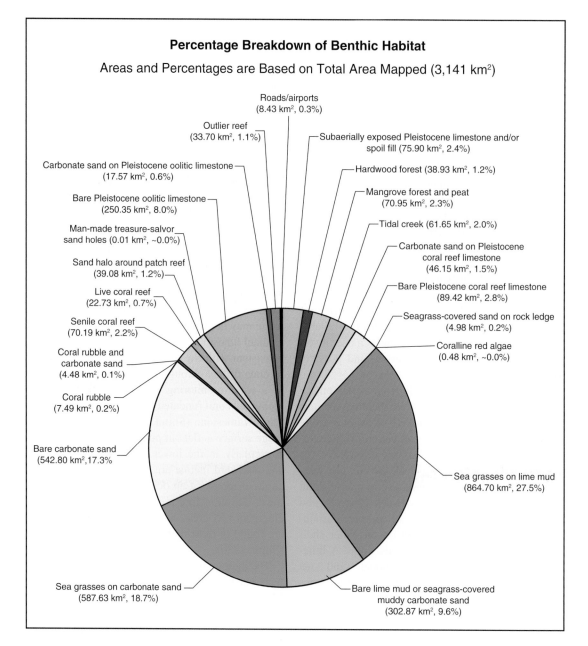

FIG. 2.23. Pie chart shows primary benthic habitats mapped in the Florida Keys National Marine Sanctuary (Lidz et al. 2006). Three are terrestrial. All others are marine units. Area (in km^2) given for each map unit represents spatial area of that unit within the total area mapped (3,140.5 km^2). The km^2 values are converted to percentages for each unit within the total area. Examples of features/habitats mapped are discernible in the aerial photomosaics from which the interpreted map was primarily derived (see aerial photos in this chapter and in Lidz et al. 2003; Lidz 2004)

Holocene transgression? If the ledge had been carved during the highstand regression, as may be argued, correlation of the present shoreline with the landward ledge edge would likely not be precise. Proxy sea-level data point to the Holocene transgression as being the only other possibility for time of erosion.

The highest post-Key Largo Limestone Pleistocene sea level, that of isotope substage 5a, reached its maximum at 9 m below present sea level (Table 2.3; Toscano and Lundberg 1999), 5 m below the ledge scarp. Ledge width (~2.5 km) and correlation of maximum depth of the ledge surface

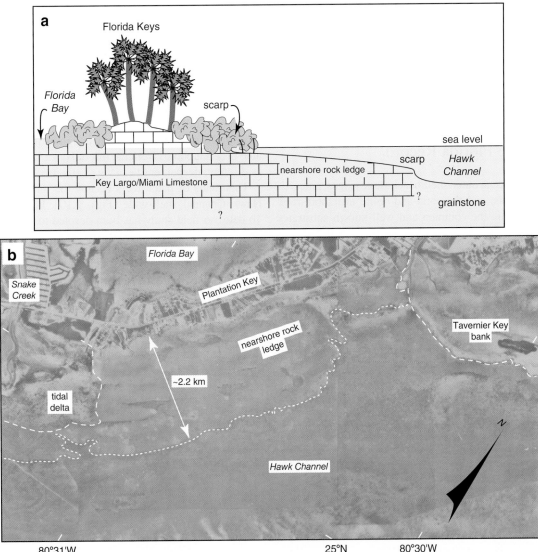

FIG. 2.24. **a** Sketch (not to scale) shows the nearshore rock ledge. **b** Photomosaic (1975) shows the clearly visible ledge (outlined in short dashes) that lines the seaward side of the Florida Keys. Area shown is in the upper Keys. A tidal delta at Snake Creek and a muddy bank at Tavernier Key (long dashes) cover parts of the ledge. Note ledge is much wider than adjacent island, Plantation Key (Fig. 2.1a). Plantation Key consists of the emergent Key Largo Limestone coral reef. Total width including island is consistent with a sloping forereef and a landward reef crest. Thin sands cover inner part of ledge. Outer part is generally bare rock with scattered patch reefs, especially in lower Keys

(~4 m) with the local sea-level curve (Fig. 2.13a) indicate that island erosion has occurred over a distance of ~2.5 km in the past 4 ka. This equates to 6.25 m/100 years. Given that sea level is rising at a present rate of 0.38 m/100 years (Wanless 1989) and is projected to rise as much as 0.88 m in the next 110 years (Albritton et al. 2001), an erosional interpretation has significant implications for past and future rate of land loss in the densely populated

Florida Keys. A large extent of the keys is uninhabitable mangrove forest. The highest elevation is ~5.5 m (Stanley 1966). Most inhabitable areas will disappear in a sea-level rise of 1 m (Lidz and Shinn 1991). Under the present 0.38-m-rise scenario, the +2-m mark would be reached in ~520 years (Fig. 2.13b). Under the projected 0.88-m-rise scenario, the +2-m mark would be reached in ~225 years. To compound the issue, the island chain is subject to hurricanes 6 months of the year. Effects of hurricane-induced wind damage and storm surge increase with higher sea levels (Fig. 2.25a, b) and are heightened even more with greater storm intensity, as is also widely predicted to occur for the next 10–20 years.

2.5 Coral Reef Nuclei in the Florida Record

Pleistocene outer-shelf beach-dune ridges, antecedent reefs, and the coral-and-oolite nearshore rock ledge are the common and well-known substrates for corals along the Florida reef tract. A recent combination of select datasets (NGDC bathymetry of Divins 2003, and benthic habitats of Lidz et al. 2006) revealed a fourth type of antecedent topography for reef substrate: landward edges of two inner-shelf bedrock troughs within the shallower depression beneath Hawk Channel (Lidz et al. 2006). The troughs are landward of outer-shelf ridge-and-swale structures and are indistinguishable in aerial photos from the rest of Hawk Channel as discrete, deeper, entities (Fig. 2.10a, b).

Thousands of patch reefs occupy the middle of Hawk Channel shelf-wide (Figs. 2.10a, b, 2.17). To date, no field study has been devoted to assess-

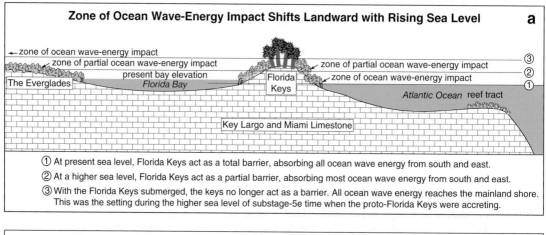

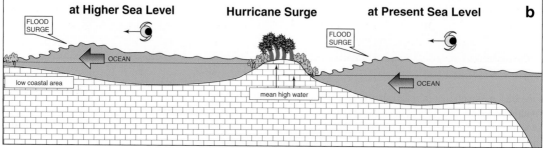

FIG. 2.25. Schematic diagram (not to scale) shows generalized hurricane-associated processes and impact of rising sea level as the eye of the storm approaches a low-lying coastline. Note the dome or flood surge of water that is pushed ahead of the eye. Once run up on land, the dome of water will return to the sea as ebb surge. In effect, surge forces potentially impact shoreline settings twice. **a** Landward movement of ocean wave-energy impact zones occurs with rising sea level. Note that present water level in Florida Bay is higher than the water level in the Atlantic. **b** Hurricane-flood surge at present sea level and at a higher sea level. Right side of figure: dome of water would engulf the keys and raise water level in Florida Bay as the hurricane moves inland. At present sea level, the keys receive the most damage. Left side of figure: under conditions of higher sea level, the flood-surge dome would cross the keys and would come ashore in the low-lying coastal region of the Everglades. The coastal region would receive greater damage and would also experience an ebb surge. Higher sea levels in **a** are arbitrary, but the sketch implies that sea level at position 2 is not as high as highest elevation in the keys (~ +5.5 m, Lidz and Shinn 1991), and at position 3 the keys are inundated

ing their underlying substrate or the basis for their location and alignment. The reefs are large, live head corals that occur individually and in clusters. Sand haloes encircle the reefs, coalescing in the clusters. The combined benthic-habitat and bathymetric datasets clearly establish that nearly all the patch reefs align along the landward margins of two topographic troughs bordering the seaward edge of the bedrock depression beneath Hawk Channel. Channel bedrock is believed to be grainstone/packstone, as has been cored off north Key Largo (upper Keys) and Ninefoot Shoal (lower Keys, Fig. 2.1a; Shinn et al. 1994). Consistent with a westward platform slope, the deeper trough (edge 10 m below sea level; depth >12 m) lies off the lower and middle Keys. The shallower trough (edge 6 m below sea level; depth <8 m) lies off the upper Keys.

Correlation of trough-edge depths with time on a sea-level curve derived from local data (Fig. 2.13a) indicates the lower and middle Keys mid-channel patch reefs should be younger than 7.8 ka and those off the upper Keys should be ~6.2 ka. Both inferred ages are consistent with ages of dated outer-shelf corals landward of Marker G and at Looe Key Reef (Figs. 2.11a, 2.15b; Shinn et al. 1977; Lidz et al. 1985) and at Grecian Rocks (Fig. 2.15a; Shinn 1980). Bathymetric elevation of the ridge-and-swale system seaward of both troughs is generally 2–6 m higher relative to trough-edge depths (Lidz et al. 2006). If sea level today were 10 m lower, broad Pleistocene islands would intermittently line the outer shelf. Though ~2 m (Shinn et al. 1977) of that higher elevation today is presumed Holocene overgrowth (except for thicker rubble infill in the trough behind the shelf-edge reef), the outer-shelf corals would have recruited to bedrock highs, likely beach-dune ridges or coral ridges of the ridge-and-swale architecture. Where present, those ridges would have blocked incoming waves from reaching the troughs when sea level was lower. Patch-reef corals colonized bare trough-edge substrate prior to sedimentation and vegetation of the channel floor. Today, both troughs are located within, and harbor, the seagrass/lime-mud habitat characteristic of Hawk Channel (Figs. 2.10a, b, 2.17, 2.24; Lidz et al. 2006). The troughs are the first such nuclei for corals—non-coralline, non-oolitic, non-beach-dune, inner-shelf trough edge, likely Pleistocene grainstone/packstone substrate — known in the Florida reef record.

2.6   Models of Margin Evolution

Geomorphogenic models were generated for four sites along the South Florida margin on the basis of presence of coral reef complexes, clarity of seismic reflections, stratigraphic differences, and quantity of coral dates at those sites (Lidz 2004). Geomorphogeny of a feature usually addresses shape. These databased models focus on origin, timing, accretion, and changes in surface landform as derived from seismic profiles and coral ages. The sites modeled are seaward of Carysfort Reef, The Elbow, Pelican Shoal, and Sand Key Reef (Fig. 2.1a). Due to lack of seismic penetration below the Pleistocene terrace, partial assumptions were used in basal parts of the models. The models indicate that: (1) a simple morphology consisting of a horizontal upper-slope terrace fronted by an elevated, cemented beach dune likely existed marginwide during the mid-Pleistocene, as is observed in seismic profiles off Pelican Shoal (Fig. 2.26a, b). (2) The corals initially colonized beach-dune ridges along the outer shelf, at the shelf edge, and on the upper-slope terrace (Fig. 2.26c, d), likely during the substage-5e transgression (Lidz 2004). (3) Age of the upper-slope terrace is no younger than ~190 ka (Table 2.2; Lidz 2006; Lidz et al. 2007), because (a) the marine regression at about that time is the most recent event during which an erosional terrace at a present water depth of 40 m could have formed (Fig. 2.3a), (b) pre-Holocene corals in outlier reefs overlying the terrace have been dated to younger Pleistocene ages (Fig. 2.3b, Table 2.2; Ludwig et al. 1996; Toscano 1996; Toscano and Lundberg 1998), and (c) seismic data and direct observation by diving show that sediments on the terrace are unconsolidated, i.e., they have not been subaerially exposed and are therefore Holocene (Figs. 2.12a, 2.26a, c).

The models were derived from tracings of seismic reflections marking the Pleistocene and Holocene surfaces. Reflections representing the upper-slope terrace surface (inferred to be Pleistocene) and the higher-elevation, dated (substage-5a) outlier-reef surface clearly demarcate thickness of the upper Pleistocene sequence. Inferred stratigraphic 'layers' (not visible on the seismic profiles) were incrementally inserted into the Pleistocene section based on the inferred initial period of coral accretion (during substage-5e time) and the known (dated)

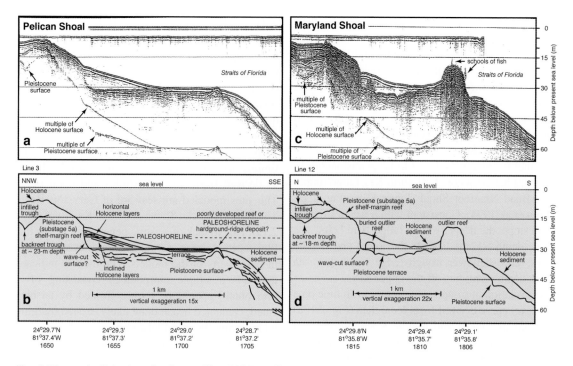

FIG. 2.26. **a** and **c** Seismic-reflection profiles (1989) and **b** and **d** interpretations show upper-slope terrace off Pelican Shoal and terrace with single outlier reef off nearby Maryland Shoal (Fig. 2.10**b**). Area is ~20 km east of multiple outlier tracts off Sand Key Reef. Note facies change in sediment wedge on terrace in **a**. Horizontal layers are interpreted to result from a still sea-level stand that eroded tops of inclined strata and redeposited the sands in a paleointertidal zone (Lidz et al. 2003). Note smooth reflection marking mounded bedrock surface buried at the seaward terrace edge in **a**. The mound is interpreted to represent a cemented dune ridge that was not colonized by corals. For comparison, note jagged nature of reflection across known Holocene shelf-edge corals at crest of Maryland Shoal in **c**. Also note small outlier reef beneath Holocene sediments near base of shelf-edge reef. Latitude and longitude in degrees and decimal minutes based on GPS coordinates. Hours (military time) below coordinates serve as navigational correlation points along seismic line

accretionary periods of substages 5c, 5b, and 5a. Seismic reflections and stratigraphic layers in the models show that margin progradation potential had previously existed to the extent that backstepping Pleistocene reefs were slowly filling landward troughs, but the time needed (i.e., duration of transgressive intervals), or the rate of coral growth, was insufficient and the trough-infill process at those sites was incomplete.

Carysfort Reef and The Elbow are Holocene buildups located ~9 km apart off the upper Keys (Figs. 2.1a, 2.27a, b, 2.28a, b). Both overlie areas of the Pleistocene shelf-edge reef. Two main geomorphic differences noted in seismic profiles distinguish the sites: (1) A prominent outlier reef exists seaward of Carysfort Reef but not off The Elbow. Based on differences in scale, this outlier is considered to represent a tract different from those off Sand Key Reef (Fig. 2.12a, b; Lidz et al. 1997b;

Toscano and Lundberg 1999; Lidz et al. 2003). (2) The trough behind the Carysfort outlier is only partially filled with Holocene sediments, whereas the trough behind the shelf-edge reef at The Elbow is completely filled. An elevated bedrock ridge is observed at the seaward edge of the upper-slope terrace in both reef areas. Seismic facies in both areas display buried low-relief accretions on the terrace interpreted to represent a series of parallel coralcapped ridges (Lidz et al. 1997b, 2003). Pleistocene sections on the Carysfort outlier and at The Elbow have accreted upward and landward, apparently indicating reef backstepping through time. Coral dates from a landward-oriented core transect on the Carysfort outlier verify at least one instance where Pleistocene corals backstepped ~0.25 km between 85.3 and 80.2 ka (Toscano and Lundberg 1998). Several other dates on the outlier indicate that Holocene coral growth advanced seaward over more than 0.5 km before backstepping toward the end of accrual (Toscano and Lundberg 1998). Coral dates from three Sand Key outlier-reef cores confirm that periods of backstepping took place after ~82.7 ka: at 81.6 and 80.9 ka (Toscano and Lundberg 1998). The Carysfort model shows that the Pleistocene shelf-edge reef also has seaward-growth components.

Holocene sediment is thinner at Carysfort Reef and on the Carysfort outlier than at The Elbow. The Holocene reef buildup is ~6–7 m thick on the outlier-reef crest and thins along the forereef slope to ~2–3 m (Toscano and Lundberg 1998). Holocene sediments are ~27 m thick in the trough behind The Elbow (Lidz et al. 2003). Seismic stratigraphic thickness of the cumulative Pleistocene 'layers' modeled is about the same in both areas.

Pelican Shoal and Sand Key Reef are Holocene buildups on the shelf-edge reef ~26 km apart off the lower Keys (Figs. 2.1a, 2.29a, b, 2.30a, b). These sites show different geomorphic and stratigraphic characteristics than at the Carysfort and Elbow areas. Whereas the terrace surface off Pelican Shoal has remained essentially unchanged from when it was formed (at ~190 ka; Figs. 2.3a, 2.26a), four massive tracts of high-relief outlier reefs accumulated on the terrace off Sand Key Reef (Fig. 2.12a, b). A concave, possibly wave-cut surface is present at the base of the shelf margin in both areas. A sharp seismic stratigraphic contact at a water depth of ~25 m divides inclined and horizontal bedding in a Holocene sediment wedge off Pelican Shoal (Fig. 2.26a, b). No such contact is detected in sediments between the Sand Key outlier reefs. A low-relief reef or beach-dune ridge caps the seaward terrace edge in ~30 m of water off Pelican Shoal but is inferred to have served as the nucleus for a distinct outlier reef off Maryland Shoal (Fig. 2.26c, d) and for the largest outlier reef off Sand Key. The seismic dune-ridge reflection is analogous to those of elevated ridge deposits on the deeper (50–124 m) slope southwest of the Marquesas Keys (Fig. 2.1a; Locker et al. 1996, and this volume). The trough behind Pelican Shoal is V-shaped but is U-shaped behind Sand Key. Despite an alongshore direction of sand transport in the lower Keys, both troughs behind the shelf-margin reef sections are filled with ~16 m of Holocene sands, indicating some cross-shelf transport, likely during storms. Field observations reveal landward-dipping infill bedding (e.g., Ball et al. 1967; Enos 1977). Troughs behind the Sand Key outlier reefs

are not yet filled, an observation consistent with increased current intensity and sediment transport due to the open-marine and off-shelf setting (e.g., Ginsburg 1956; Enos 1977).

Seismic stratigraphic thickness of late Quaternary shelf-edge sections is comparable at the lower Keys sites, but seaward Holocene reefs on the largest Sand Key outlier are sporadic. As elsewhere along the shelf, most Holocene corals off the lower Keys have died since the 1970s, especially the acroporids. Two Holocene reef accumulations are present on the outlier, a 3.4-m-thick landward pinnacle of secondarily emplaced coral rubble (EAS, personal observation) and a thinner seaward *in situ* reef-crest patch (Figs. 2.12a, c, 2.30b, c; Ludwig et al. 1996; Toscano and Lundberg 1998).

Comparison of the four models shows that Pleistocene accretion of the shelf-edge reef was the same (upward and landward, partly filling a backreef trough) at The Elbow (northeast area of reef tract) and Pelican Shoal (southwest) and the same (upward, landward, and seaward) at Carysfort Reef (northeast) and Sand Key Reef (southwest). The Pleistocene sections in the freestanding outlier reefs off Carysfort and Sand Key Reefs and in the shelf-edge reef margin-wide display keep-up profiles, but the buried reef-like seismic facies on the terrace off the Carysfort outlier and The Elbow (Lidz 2004; Lidz et al. 1997a, 2003) show give-up profiles (terminology of Neumann and Macintyre 1985). Interestingly, the opposite is true for Holocene accretions. Holocene sections at Pelican Shoal and Sand Key Reef, including those on the Sand Key outliers, are very thin and lie many meters below sea level, i.e., they represent give-up reefs. At Carysfort Reef in the northeast, however, and to some degree at The Elbow, Holocene corals have grown to or near sea level (see Figs. 2.18d, 2.28a) and are regarded as having been keep-up reefs before their recent demise.

Little vertical accretion at the shelf edge has occurred during the Holocene. For the most part, Holocene reefs off the keys are less than 2 m thick. Holocene accretions are thickest at named reefs and in the shelf-edge backreef trough. The thickest area of Holocene corals cored (17 m) on the reef tract is at Southeast Reef in the Dry Tortugas (Shinn et al. 1977).

In general, carbonate platforms throughout geologic history have built seaward by aggrading (coral accretion) or prograding (off-bank accretion of shelf-derived sediments; Hine and Neumann

1977; Hine et al. 1981; Ginsburg 2001). Classic ancient reefs that formed during periods of stable sea levels had steeply inclined margins that are able only to aggrade seaward. The South Florida reefs are atypical because they formed during periods of rapidly changing sea level. The windward margin has accreted through periods of retreat and advance, as depicted in the models. The Pleistocene record shows two variations on the standard progradational theme: backstepping and lateral stacking of reefs/infilled troughs (Shinn et al. 1991). The respective stratigraphic results have been termed *a backstepped reef-complex margin* and *a coalesced reef-complex margin* (Lidz 2004). Coral growth and sediment production will eventually fill the troughs.

The outlier reefs off Carysfort and Sand Key Reefs forecast new episodes of rapid seaward-stepping progradation when their troughs become filled (Lidz et al. 1991). Stratigraphy resulting from such lateral accretion would represent a coalesced reef-complex margin (Lidz 2004) that would be manifest as parallel coral facies spanning distances of several kilometers in cross-sectional seismic profile (Figs. 2.27b, 2.30b). Interpreting this type of margin at an ancient platform might be difficult. Landward- and seaward-dipping bedding would be present, and contacts between coral facies may dip landward.

In addition to episodes of backstepping and coalescing, a third variation on the standard progradational theme is visible in the seismic records. Though long known that Holocene sediments have filled many areas of the Pleistocene trough behind the present shelf-edge reef (Hoffmeister 1974; Perkins and Enos 1977), the stratigraphic result was only formally termed (Lidz 2004) *a backfilled prograded margin* upon model completion. The infill is mostly storm-transported coral rubble derived mainly from the shelf-edge reef with an admixture of seaward-moving sediments from the outer-shelf reefs (Shinn et al. 1991). Reef-rubble zones are clearly visible in aerial photographs as elongated debris fields that extend behind most major shelf-edge reefs and that overlie and conceal margin-parallel seabed features (Figs. 2.10a,b, 2.12b, 2.17; Lidz 2004; Lidz et al. 2003, 2007). Field observations show landward-dipping bedding in the rubble (Ball et al. 1967; Enos 1977). Infilling was rapid, having occurred solely in the time since the shelf was submerged – during the last 7–6 ka. Infilling was asymmetric, erasing surface expressions of some, but not all, troughs, and leaving

reentrants essentially unchanged. Elevation of the filled troughs was raised to or above shelf elevation, thus prograding those parts of the shelf seaward to the new margin, the bare face of the shelf-edge reef.

The discontinuous shelf-edge reef has produced a 'notched' margin. The presence of outlier reefs off the shelf-edge reef but not off large reentrants such as at Alligator and Tennessee Reefs (Figs. 2.1c, 2.31; Lidz et al. 2003) implies increasing shelf-edge intricacy or irregularity through time (Lidz et al. 2007). Where outlier reefs and backreef troughs exist seaward of the present shelf-edge reef, those parts of the shelf and margin will build seaward in sea-level-controlled jumps through infilling of troughs. Each jump will be as great as the width of the trough, and in each jump the outlier reef whose trough was filled will become the new shelf margin. Under the right conditions as is evident in the Holocene record, infilling will require only a few thousand years. These laterally accreting areas may form promontories of coalesced reef complexes that may flank reentrants (Lidz 2004).

Core transects across Looe Key Reef in the lower Keys (Fig. 2.15b) show that Holocene corals initiated growth on a Pleistocene high composed of massive corals (Lidz et al. 1985; Shinn et al. 1991). The high has the elevation and dimensions of an outlier reef. Seismic profiles and drilling confirm a sediment-filled trough exists behind the high and beneath the modern reef. The modern reef has backstepped to a position landward of the high crest on which its Holocene counterpart first began growth. Other Holocene outer-shelf reefs have also backstepped in response to rising sea level (Shinn 1980; Shinn et al. 1989). Using the stratigraphy at Looe Key Reef as an example, Shinn et al. (1991) proposed that during periods of lowered sea level, berms, dunes, and possibly fringing reefs formed along the ancient off-bank shorelines and became subaerially cemented. The hardened features provided elevated nuclei for outlier-reef growth as sea level rose. As reefs flourished and grew upward, sediment-lined troughs formed on their landward side. When sea level fell, freshwater diagenesis cemented the reefs and trough sediments, and new berms, dunes, or fringing reefs initiated accretion on a new seaward shoreline and became cemented, creating new topographic highs for outlier-reef growth during the next transgression. Storm waves periodically removed and transported sediment and coral debris into the landward troughs, eventu-

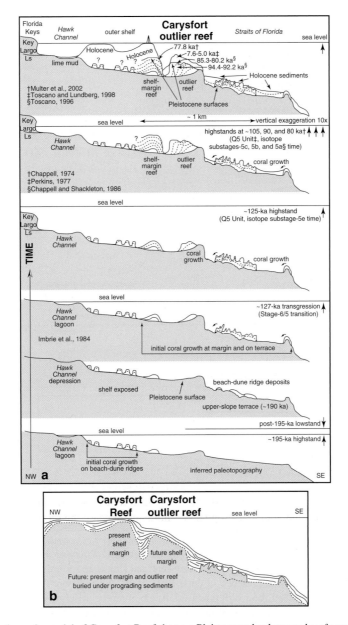

FIG. 2.27. **a** Geomorphogenic model of Carysfort Reef shows a Pleistocene *backstepped reef complex* in the outlier reef, and Holocene *backfilled progradation* of the outer shelf. Up arrows represent the seven highstands that produced coral reef complexes (bold arrows, Table 2.1). Dotted lines signify Pleistocene stratigraphic 'layers.' Overlying Holocene accretions have no dotted lines. The Holocene surface at and behind the shelf edge in this model is drawn from field knowledge. Outer-shelf and shelf-edge reefs in the Carysfort area are too shallow to obtain seismic data. Note 'tripod' representing Carysfort Light on Carysfort Reef behind the outlier. Though beach-dune ridges are suspected to underlie all coral reefs shelf-wide, they are discontinuous; their presence behind Carysfort Reef has not been confirmed. Note shapes of the outlier reef and its landward trough are very different from those of the Sand Key Reef outliers (Fig. 2.12**a**). Also note buried reef-like features on upper-slope terrace. In all four models (Figs. 2.27–2.30), horizontal scale and vertical exaggeration, shown for present stratigraphies in the top panels in part **a**, apply to features at and seaward of the shelf margin as based on seismic profiles in Lidz et al. (2003). Shelf features are based on cores behind Marker G (Fig. 2.11) and off Bal Harbor (Shinn et al. 1977) and are not drawn to scale. **b** Filling of the existing trough behind the outlier reef would extend the margin seaward. However, if corals resume growth on the outlier, elevation of the new margin may be as high as the present shelf edge, unless corals also recolonize Carysfort Reef. If corals do not colonize either reef, sediment would eventually bury both. This part of the shelf margin would become a pronounced slope. Seismic profiles would show multiple lateral sections of landward- and seaward-dipping beds (accretionary foresets) on either side of two buried parallel facies of *coalesced reef complexes*

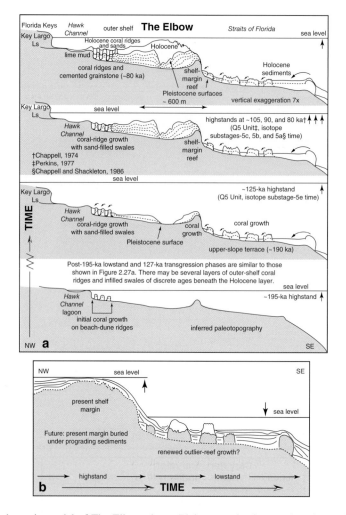

FIG. 2.28. **a** Geomorphogenic model of The Elbow shows Pleistocene *backstepped reef complex* at the margin and Holocene *backfilled progradation* of the outer shelf. Pleistocene and Holocene surfaces (in Figs. 2.28a, 2.29a, 2.30a) are traced from seismic profiles. Note presence of coral-ridge bands and sediment-filled swales. Storm-generated bedding in Holocene fill behind The Elbow dips landward. Angle of repose at debris surface is ~45°. Buried terrace outlier reefs are ~5–12 m high with crests ~35–38 m below sea level. Compare with dimensions of high-relief outlier reefs at Sand Key Reef (Fig. 2.12a). **b** The Elbow presently exhibits a progradational-margin profile. If corals renew growth on the linear terrace features, new outlier reefs and backreef troughs may form. If there is no coral growth on these features, sediments should build up on the terrace, raising elevation of its horizontal surface. Renewed growth of outlier reefs on terrace features would produce multiple reef complexes. When buried, the complexes would form a *coalesced reef-complex margin.* Unlike at Sand Key Reef, the future reef complexes on the terrace and the present shelf-edge reef would be of very different geologic ages

ally filling them. Asymmetric margin progradation resulted from lateral stacking of reefs and sediment-filled troughs where the reef-and-trough geometry was present. Where this geometry was absent, reentrants became enlarged as the shelf-edge reefs on either side prograded seaward (Shinn et al. 1991; Lidz et al. 2007).

2.7 Reef Responses to Geomorphic Settings

Geologic imprint in the late Quaternary reef record in the Florida Keys is one of consistencies and contrasts. Offshore evolutionary themes are repetitive: successive coral reef ecosystems and architectures

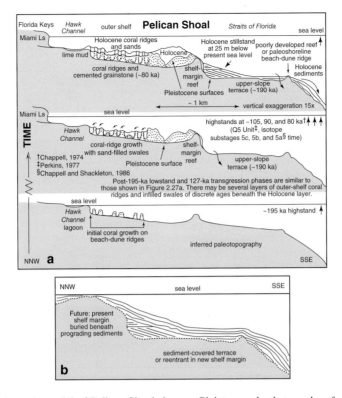

FIG. 2.29. **a** Geomorphogenic model of Pelican Shoal shows a Pleistocene *backstepped reef complex* at the margin, Holocene *backfilled progradation* at the margin, and a sediment-covered beach-dune or hardground ridge at the seaward terrace edge. Note presence of coral-ridge bands and sediment-filled swales on the outer shelf. **b** On the terrace, no hard, sediment-free surface for coral colonization presently exists. Sands will continue to accumulate, extending the sediment-wedge dip angle upward and seaward across the terrace. The beach-dune or hardground ridge at this site lies between, and in the same position on the terrace as, two outlier-reef segments at neighboring Maryland Shoal and Eastern Sambo. The ridge is believed to underlie those and all segments of all outlier-reef tracts. The Pelican Shoal site may become a reentrant in a new shelf margin with trough infilling behind the outlier segments at Maryland Shoal and Eastern Sambo. With continued sediment accrual, a future seismic record at Pelican Shoal would show a classic prograded margin with landward-dipping beds behind a buried shelf-edge reef on either side of a reentrant. If sediments were to fill space on the terrace to sea level (e.g., Tucker 1985), a new shelf margin would be formed above the point at which the present upper-slope gradient slopes below terrace level

at a consistently irregular margin. The Pleistocene reef systems formed two basic structures: a narrow bank-top coral ridge-and-swale architecture and a broad shelf-edge reef-and-trough architecture. Corals of the Holocene ecosystem built magnificent spurs facing incoming waves of a slowly rising sea. Based on geomorphic elements, vertical and lateral stacking of narrow coral reefs and cemented sediments in intervening swales was the major process that built the outer shelf upward and seaward. Lateral stacking of shelf-edge and outlier reefs and infilling of backreef troughs built alternate parts of the margin upward, landward, and seaward. Massive head corals dominated the

Pleistocene. Branching acroporids dominated the Holocene. The recurring evolutionary theme has been replication of antecedent facies and geomorphic landforms, and thus repetition of stratigraphic shelf-edge asymmetry.

Inshore morphologies are consistent: a Pleistocene (substage-5a) bedrock depression beneath Hawk Channel and a prominent Pleistocene (substage-5e) nearshore rock ledge on the seaward side of each Florida Key. Inferred age of the offshore Marquesas-Quicksands ridge is Pleistocene (substage 5e). Contrasts are evident in the physical environments along the modern shelf and are due to margin curvature, westward platform slope, varied inshore

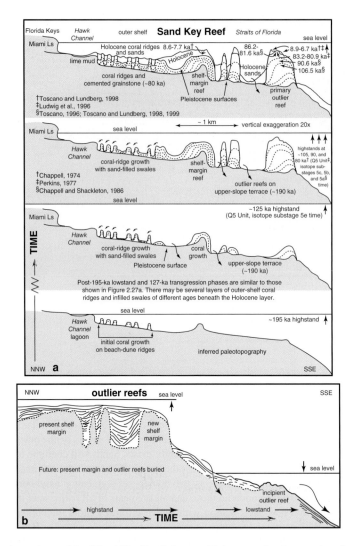

FIG. 2.30. **a** Geomorphogenic model of Sand Key Reef shows a Pleistocene *backstepped reef complex* at the margin and Holocene *backfilled progradation*. Note bands of outer-shelf coral ridges. Outlier-reef rendition (three reef tracts) is based on seismic data. **b** When the troughs behind the outlier reefs become filled with sediment and/or coalescing fringing reefs, the shelf margin would instantaneously "step" seaward, relative to geologic time. Considering present status of The Elbow and its infilled backreef trough (Fig. 2.28**a**), some parts of the South Florida shelf may have built seaward in this fashion in the past. The seismic geologic record would show a progradational margin with multiple lateral sections of landward- and seaward-dipping beds between multiple buried parallel facies of *coalesced reef complexes*. New outlier reefs may form in front of present reef complexes. These scenarios are dependent on the position of sea level and its duration at that position

topographies of islands and tidal passes, and prevailing southeasterly winds and waves impinging on the curved margin. Pleistocene accommodation space is several meters greater to the southwest than northeast, yet Holocene accretions average 3 to 4 m thick shelf-wide. This anomaly demonstrates offbank sediment transport in the southwest.

Contrasts between Pleistocene and Holocene records are also detected in the types of framework-building corals that were dominant, the sites of reef distribution relative to antecedent topography, the types of topographic nuclei, and reef size. Positions of lower sea level restricted nuclei common to Pleistocene corals to higher-elevation outer-shelf

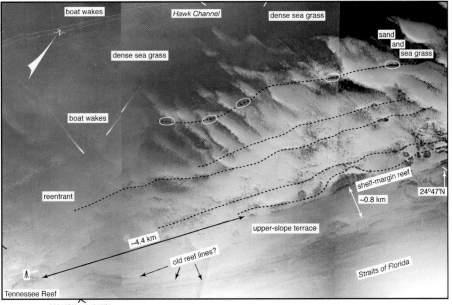

FIG. 2.31. Photograph (1975) shows reentrant seafloor morphology in vicinity of Tennessee Reef. 'Tripod' represents location of Tennessee Reef Light. Lower bedrock topography similar in elevation to that under Hawk Channel surrounds Tennessee Reef. Rock ridges are visible along shelf margin (dotted lines) and at seaward edge of the upper-slope terrace (arrows). Ovals mark areas of patch reefs on landward ridge. Note terrace expanse and absence of outlier reefs opposite reentrant

features. Four substrate types focused reef initiation: antecedent reefs and beach-dune ridges along the outer shelf and upper-slope terrace, and grainstone/packstone topographic-trough margins and the reef/oolite rock ledge on the inner shelf. Pleistocene reefs grew on the first two. Holocene reefs grew on all four.

Overall, vigorous coral growth during several Pleistocene highstands formed substantial tracts of cumulative reef framework. Early Holocene corals also thrived, but later corals were increasingly less able to do so. Recorded within biologic and geologic components of the reef ecosystems and their relict structures are responses of corals and associated organisms to the geomorphic settings and ambient processes in which they lived (Table 2.4). In South Florida, the premier dynamic was position of highstand apices relative to paleotopography and effects of those zeniths and antecedent landforms on coral growth and reef distribution. A changing sea level induced very different late Pleistocene and late Holocene geomorphic settings. Within those settings, very different ambient conditions generated very different responses in reef evolution.

Geologic imprint of the Holocene transgression is twofold. Parts of the shelf surface have stepped seaward in less than 7 ka through infilling of Pleistocene backreef troughs, a process termed backfilled progradation. The seaward side of the Florida Keys has eroded landward at a rate of ~6.25 m/100 years over the past ~4 ka, forming a 2.5-km-wide landward-sloping nearshore rock ledge. Under a present rise in sea level and a projected increase in rate of rise, the erosion has real-time relevancy and implications for residents of the low-lying keys.

The Quaternary windward margin of Florida has a spatial geomorphic complexity unlike other windward margins in the Caribbean-Bahamas. Physically and biologically predisposed, the late Quaternary depositional system along the South Florida shelf has replicated the structure and form of antecedent shelf-edge morphologies that are distinctly different from what would be generated at a classic steeply inclined windward margin. Present irregularity of shelf-edge morphologies and future differential accretion of reefs and sediments will increase lateral shelf-edge asymmetry through time. Yet

TABLE 2.4. Examples of biologic and geologic indicators and the events they record in the Florida coral reef ecosystem. References to indicators are cited in the text.

Indicator	Event, process, setting, or status recorded
Stratigraphic sequence	**Sediment deposition, coral reef accretion**
(a) marine—	high stand of sea level, warm global climate, expansive oceans
(b) non-marine—	subaerial exposure, low stand of sea level, cool global climate with oceanic waters locked up in glaciers
(c) in chronostratigraphic order—	lack of bioturbation or structural deformation; tectonic stability
Sedimentary grains	**Ecosystem vitality, dominant organisms**
(a) coral—	
• high percentages	poor coral health, high bioerosion; algal proliferation; lack of *Diadema* to control algae; reduced substrate for colonization
• low percentages	good coral health; low-nutrient, clear, warm oceanic waters of normal salinity favorable for coral growth
(b) mollusc—	
• high percentages	likely settings are seaward edges of carbonate sands or areas of coral rubble
(c) *Halimeda*—	sandy inner shelf in the Florida Keys; combined sandy and hardbottom settings in The Quicksands
(d) blackened limestone pebbles—	lightning-sparked fire at Pleistocene subaerial unconformities
(e) rounded *vs.* angular grains—	mechanical transport by water, wind, gravity (and ice in higher latitudes); secondary deposition *vs.* autochthonous *in-situ* accumulation
Sedimentation	**Process of sediment accumulating in layers**
(a) thickness—	quantity produced, quantity transported, or non-deposition; bedrock high or low; current strength and direction; offbank sediment transport
(b) bedforms, orientation, change in bedding—	current strength and direction; direction of sand movement; erosion and redeposition of sediments at a paleoshoreline or by marine or fluvial currents
(c) size fraction—	
• fine grained	calm or restricted-circulation setting
• coarse grained	high circulation or winnowing of fine grains by currents
Stony coral	**Stratigraphic sequences of fossiliferous reef-framework limestone**
(a) rubble zones—	storm transport, *in-situ* coral mortality, settings suitable for abundant clionid sponges/boring molluscs
(b) backstepped coral or zonation—	rising sea level, landward-moving surf zone, backstepped reef complex
• head corals	low-energy, quiet, or protected waters
• branching corals	high-energy or surf zone
• oriented stands	unidirectional wave attack
• unoriented stands	possible reef flat behind surf zone
(c) skeletal growth bands—	
• normal high- and low-density couplet (thin, narrow bands)	'normal' conditions suitable for healthy coral growth and accretion
• abnormal couplet (wide high-density band)	environmental stress (extreme high or low temperatures)
• fluorescent band	environmental change (influx fresh/acidic water)
(d) high-relief outlier reefs—	optimal paleotopographic settings, paleoenvironments, time, and accommodation space for cumulative, accretionary coral growth
(e) substrates—	antecedent coral reefs, beach-dune ridges, edges of grainstone/packstone bedrock troughs, reef/oolite nearshore rock ledge
Offshore subaerial features	**Past low stands of sea level**
(a) sinkhole—	limestone karstification; dissolution; sediment catchment
(b) calcrete—	subaerial exposure; leaching and cementation of soluble calcium salts; secondary deposition by capillary action and evaporation; reprecipitation of dissolved $CaCO_3$
• red laminae	periods of influx of exogenous minerals on African dust
(c) mangrove peat—	locations and positions of paleoshorelines
Bedrock topography (contours of)	**Holocene paleoshorelines, thus flooding history**
(a) bedrock highs—	sediment-free areas suitable for coral accretion; marine (ooid) precipitation and tidal-bar formation; beach-dune formation; first to undergo subaerial cementation and (calcrete) precipitation; sediments accumulate in front of topographic highs at west end of The Quicksands
(b) bedrock lows—	sediment catchments unsuitable for coral accretion; erosion or karstification; last to undergo non-marine subaerial cementation and (calcrete) precipitation; first to accumulate mangrove peat
Other	
Commensurate decrease in live coral, increase in algae	Ecosystem shift from coral reef to hardbottom communities, observed first in extant organisms and later in the sedimentary record
Infilled backreef trough at shelf margin	Backfilled prograded margin
Laterally stacked coral reefs	Coalesced reef-complex margin
Margin-wide upper-slope terrace at 30 to 40 m below present sea level	Regressive(?) erosion, likely during the later Pleistocene (~190 ka?)
Nearshore rock ledge on seaward side of every Florida Key	Transgressive erosion, during the late Holocene (over the last ~4 ka)

geomorphogenic definition of seismic stratigraphic sequences provides perspectives on characterizing reef-complex buildup in general. The Florida reef models show a regular, consistent stratigraphic simplicity in mode of evolution. Fluctuating sea level and paleotopography were the principal controls on the processes and resultant geomorphic architectures. The cause, at least in part, of the uniqueness of this margin among windward margins is probably its curvature and low-gradient slope. Most carbonate platforms consist of many thousands of meters of accreted carbonates, or volcanoes, both of which have precipitous slopes.

Acknowledgments This chapter is dedicated to the memory of modern grandfathers of Florida-Caribbean coral reefs and carbonate geology – Mahlon M. Ball, Robert F. Dill, and J. Edward Hoffmeister – and to the memory of Captain Roy Gaensslen who made so much of what we know possible through scores of USGS cruises. Sincere appreciation is given to colleagues from USGS Fisher Island Field Station days (Fig. 2.1a), associates too many to name who generated significant contributions to our research of those years (1974–1989) and to the history of the Florida Keys coral reefs. Sincere appreciation is also given to colleagues of the USGS St. Petersburg Field Center for their equally significant later contributions: Russell L. Peterson, Christopher D. Reich, and Lance Thornton. Gratitude is extended to Jim Pitts of the Department of Transportation in Tallahassee, Florida, who kindly supplied the 1975 aerial photographs displayed in this chapter, and to Joan Rikon and Patricia Kibler of the NOAA/National Geodetic Survey in Silver Spring, Maryland, who provided the 1991 photographs used in previous related publications. All photographs were instrumental in constructing the benthic habitat map that allowed classification of habitats and computation of areal extents in the Florida Keys National Marine Sanctuary. In two cases (i.e., number, four, of outlier-reef tracts and extent, keys-wide, of the nearshore rock ledge), the photos yielded the only concrete evidence for parts of the geologic record that would otherwise not have been known from other datasets. The authors thank Jack L. Kindinger, Associate Director of the USGS St. Petersburg Field Center, for constructive comments on the manuscript.

The USGS Coastal and Marine Geology Program funded the data-synthesis project, completed in 2005 (published in Lidz et al. 2007), and two later studies (Lidz 2006; Lidz et al. 2006) from which this summary paper is derived. The 1997 seismic geophysical dataset was acquired under Florida Keys National Marine Sanctuary Permit FKNMS-27–97. All products using or referencing any portion of the 1997 dataset are attributable to and should acknowledge that permit number.

Appendix 2.A: Human-interest Notes

Good compilations of many short articles on the natural and anthropogenic history of the Florida Keys are found in Gallagher et al. (1997) and Zeiller (2005). Examples of human-interest notes are:

- **Fossils and archeological finds** show that prehistoric peoples shared the land south of Miami with Ice Age mammals, including dire wolves, the huge American lion, saber-toothed tigers, mastodons, mammoths, and giant ground sloths.
- **The sea was much lower** in prehistoric times, exposing more landmass than today, with the coastline on the Gulf Coast being 160 km farther offshore than at present. The climate was much drier, more closely resembling that of modern African savannas.
- **With each rise and fall of sea level**, prehistoric peoples migrated north or south in response to fluctuations in landmass size and extent, which, in turn, altered availability of fresh water, animals, and botanicals to hunt for food. Freshwater and food sources also ebbed and flowed with the sea.
- **More than 1,000 shipwrecks** dating to the 1500s, 1600s, and 1700s lie off the Florida Keys. Most are Spanish, but some are Dutch, British, and American. They wrecked as a result of storms or illness that disabled the crew. The best known is the Spanish galleon *Nuestra Señora de Atocha*, which ran aground on a calcrete-coated bedrock high on the south side of the Marquesas-Quicksands ridge in a 1622 hurricane. The bottom of the vessel remained on the sea floor at the site of impact, but the entire upper structure and treasure were scattered several kilometers to the northwest. Much of her gold, silver, and gemstone treasure is on exhibit at the Mel Fisher Maritime Heritage Society in Key West.

- **The first non-native Americans** settled in Key West around 1819. Key West was in the early years more closely related to Cuba than to the U.S.
- **Key West** has been home to famous writers, the most familiar being Tennessee Williams and Ernest Hemingway. Today, the Hemingway house, replete with its many well-cared-for multi-toed cats, is a historic residence, and Earnest Hemingway look-alike contests are annual events on the island.
- **The Key West naval base** often hosted visits from President Harry Truman. Truman Avenue and the Truman Annex remain part of the base today.
- **Industries of early settlers** in the keys included cigar and salt manufacturing, rum smuggling, pirating, sailing-craft construction, sponging, agriculture, and fishing.

- **"Wrecking"** was also a legal profession (as well as a racket). Wreckers boated out to a stranded vessel to rescue survivors and recover cargo. The cargo was taken to a wreckers' court where a Federal judge sold the goods at auction. The captain and crew got a cut, but the U.S. Government got its share as duty on imported goods. Cargo salvage included rich French silk fabrics, Spanish shawls, European clothing, jewels, sewing machines, lumber, coal, scrap iron, bales of cotton, and casks of wine. Modern-day booty includes bales of pot.

Appendix 2.B: Figures and Plates from Multer et al. (2002)

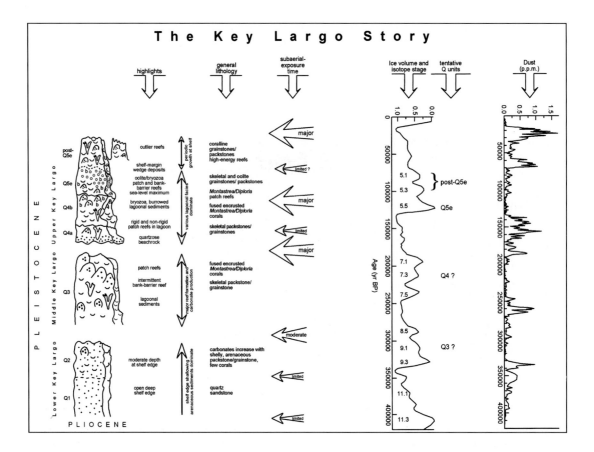

Fig. 2.B-1. Summary overview from Multer et al. (2002, their Fig. 9) shows various aspects of lower, middle, and upper Key Largo Limestone intervals. Ice volume, isotope stage, and dust data from Petit et al. (1999) (Reprinted with kind permission of Springer Science and Business Media)

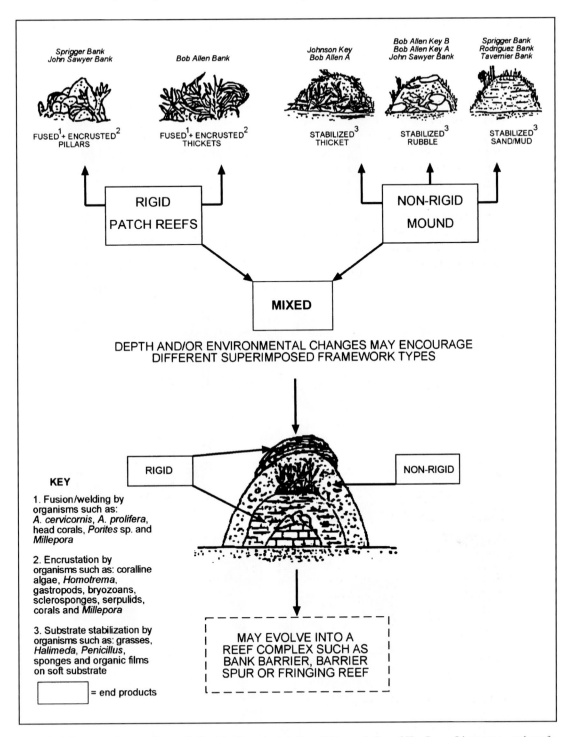

FIG. 2.B-2. Framework types that are believed to have been influential in evolution of Key Largo Limestone patch reefs. Possible examples of specific reefs are noted. Diagram from Multer et al. (2002, Fig. 11) as modified from Multer and Zankl (1988) (Reprinted with kind permission of Springer Science and Business Media)

Color Plates 2.1–2.5 reproduced unaltered from Multer et al. (2002, their Plates 41, 42, 44, 45, 46). All photos depict typical Key Largo Limestone facies, a Pleistocene shelf-edge facies in South Florida.

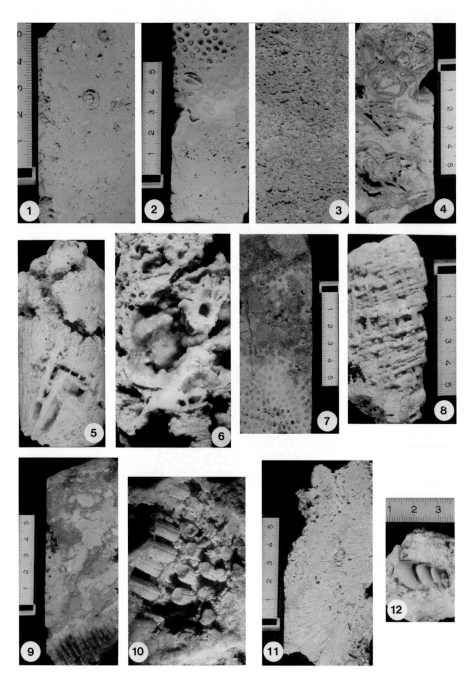

PLATE 2.1

Plate 2.1   Key Largo Limestone. All depths are below core tops. Core locations are given in Multer et al. (2002) [Reprinted from Multer et al. (2002, their Plate 41) with kind permission of Springer Science and Business Media]

1. Molluscan-foraminiferal wackestone with moderate moldic porosity, core W5 (Key West), depth 57.3 m, Q1 Unit of Perkins (1977). Scale in centimeters
2. Molluscan wackestone to packstone, moldic porosity, boring bivalve at base of recrystallized *Montastraea* coral, core W5 (Key West), depth 45.5 m, Q2 Unit (Perkins 1977). Scale in centimeters
3. Quartzose molluscan grainstone with moldic porosity; rock displays rare angular bedding, core W3 (Winston Waterway, Key Largo), depth 14.2 m, Q2 Unit. Width of photo 5 cm
4. Quartzose molluscan packstone with chaotic bedding and moldic porosity; spar coats open vugs, fragment of soilstone at base of sample, core 9 (Sandy Key #2, well #9, Florida Bay), depth 9.7 m, Q2 Unit. Scale in centimeters
5. Diagenetically altered coral and quartzose molluscan packstone displaying intense dissolution and spar coating, core 16 (Bob Allen Key-A), depth 4.2 m, Q3 Unit (Perkins 1977). Width of photo 5 cm
6. Bioturbated molluscan packstone-grainstone with solution voids coated with thick layers of drusy calcite, core WB (Florida Bay), depth 23.9 m, Q3 Unit. Width of photo 6 cm
7. *Montastraea* coral overlain by grainstone with fragments of coral (*Porites*?) and unidentified (plant?) tubes, core 6 (Old Dan Bank, northwest off Long Key, Florida Bay), depth 3.0 m, Q4b Unit (terminology of Harrison et al. 1984). Scale in centimeters
8. *Montastraea* coral displaying dissolution and recrystallization, vugs coated with chalky wackestone, core W9 (Jewfish Creek), depth 6.0 m, Q4b Unit. Scale in centimeters
9. Vuggy coral (*Porites* and *Montastraea*) boundstone with brownish skeletal wackestone infill, core W9 (Jewfish Creek), depth 3.0 m, Q4b Unit. Scale in centimeters
10. Cast of a *Montastraea* coral, core W5 (Key West), depth 11.2 m, Q4 Unit (Perkins 1977). Width of picture 7.5 cm
11. *Montastraea* coral encrusted by crustose coralline algae, core WT (Dry Tortugas), depth 17.0 m, Q Unit unidentified. Scale in centimeters
12. Mold of gastropod in foraminiferal-peloid-molluscan packstone facies, core 5 (Key West), depth 39.4 m, Q3 Unit. Scale in centimeters

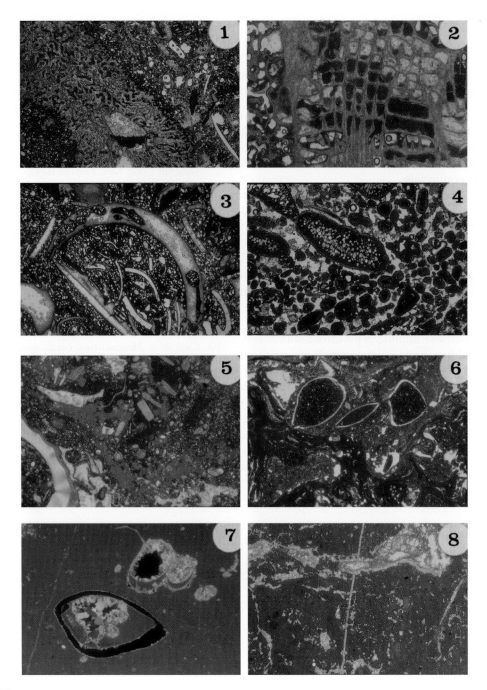

PLATE 2.2

Plate 2.2 Key Largo Limestone microfacies. All depths are below core tops. Core locations are given in Multer et al. (2002) [Reprinted from Multer et al. (2002, their Plate 42) with kind permission of Springer Science and Business Media]

1. Grainstone with coral (with geopetal fabric), molluscs, foraminifera, echinoids, quartz, moldic porosity, core W4 (North Key Largo), depth 30.3–31.8 m, Q3 Unit. Width of picture 15 mm
2. Recrystallized *Montastraea* coral, partial infill of skeletal framework with fine sediment and calcite spar, note geopetal infill, core 13 (Boundary Marker, Florida Bay), depth 1.5 m, Q4 Unit. Width of picture 7 mm
3. Quartz-rich molluscan packstone with fragments of peneroplid foraminifera, moldic porosity, voids partially filled with blocky calcite, micrite envelopes preserved; note bored shell; echinoid fragment shows syntaxial cement, core WB (Florida Bay), depth 44.2 m, Q2 Unit. Width of picture 15 mm

4. Grainstone with ooids, peloids, *Halimeda*, molluscs, foraminifera, and rare quartz grains, meniscus cement, core 10 (John Sawyer Bank, 4.8 km north of Marathon, Florida Bay), top of core, Q4 Unit. Width of picture 6.5 mm
5. Mollusc-foraminifera packstone with two generations of fine-grained quartzose-skeletal mud infill, core W5 (Key West), depth 55.2 m, Q1 Unit? Width of picture 12 mm
6. Packstone to wackestone with coralline algae, molluscs, foraminifera, core W2 (Grassy Key), depth 18.2 m, Q3 Unit. Width of picture 15 mm
7. Molluscan wackestone with recrystallized shells and moldic porosity, molds have geopetal fabric and are partially filled by sparry calcite, core 4 (Lake Key), depth 1.8 m, Q4 Unit. Width of picture 7 mm
8. Peloidal packstone that was subaerially exposed; note subsequent quartz-grain infill and vertical shrinkage crack, core WB (Florida Bay), depth 9.1 m, top of Q4b. Width of picture 14 mm

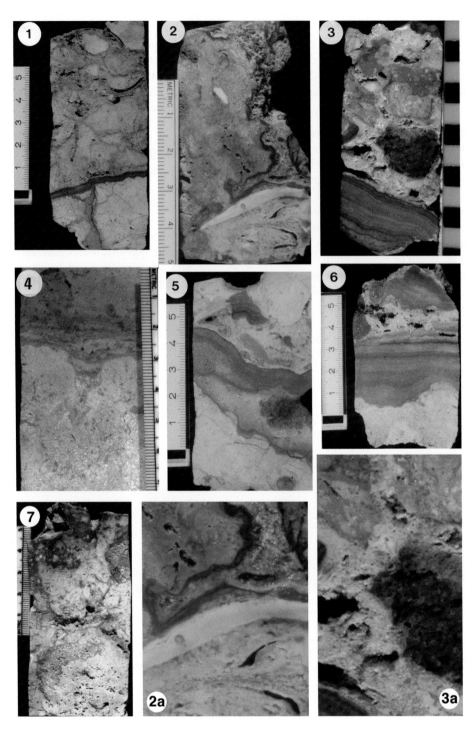

PLATE 2.3

Plate 2.3 Boundaries between Q Units in the Key Largo Limestone. All depths are below core tops. Core locations are given in Multer et al. (2002) [Reprinted from Multer et al. (2002, their Plate 44) with kind permission of Springer Science and Business Media]

1. A dark-brown laminated subaerial crust and vein-filling microcrystalline rind define contact between a Q3-Unit packstone with molluscs, foraminifera, peloids, and moldic porosity and overlying Q4a-Unit rubble of grainstone containing molluscs, foraminifera, peloids, quartz, and rounded reworked encrusted fragments of rock from below crust, core 3 (Cotton Key, Florida Bay), depth 3.3 m, boundary between Q3 and Q4 Units (Q4a of Harrison et al. 1984). Scale in centimeters

2. Example of subaerial diagenesis including (1) encrusted solution cavity filled with soilstone crust (calcrete) displaying moldic porosity, and (2) sheltering by pelecypod shell from further meteoric alteration, core 10 (John Sawyer Bank, Florida Bay), depth 1.2 m, boundary between Q3 and Q4 Units. Scale in centimeters. (2a). Detail

3. Thick subaerial laminated crust at top of Q3 Unit with rare rhizoids and quartz grains underlying base of a Q4a-Unit quartz-grainstone rubble; rubble consists of fragments of coral, crust, and a variety of packstone pebbles containing red coralline algae, molluscs, peloids, and foraminifera, core 30 (Bayside Well Cluster #2, Florida Bay), depth 5.8 m. Scale in centimeters. (3a). Detail

4. Subaerial contact between Q2 and Q1(?) Units seen here (and also found between Q5 and Q4 Units elsewhere) are well known to be thin, discontinuous and subtle with the lithology above and below the contact very similar in nature. Here a thin brown crust with rhizoids coats and fills fractures at the top of a tan quartz-rich packstone with molluscs and rare foraminifera. Above the crust is a similar darker-brown packstone containing only an increase in molluscan debris, core 9 (Sandy Key #2), depth 13.0 m. Scale in centimeters

5. Molluscan-foraminiferal-peloid packstone overlain by laminated crust containing thick interbeds of peloidal packstone and random quartz, in turn overlain by a pebbly grainstone rubble with fragments of skeletal wackestone and grainstone, laminated pebbles, and rare mud infilling, core 23 (Lignumvitae Channel, Florida Bay), depth 3.2 m, boundary between Q4b(?) and Q4a(?) Units. Scale in centimeters

6. Molluscan-peloidal-foraminiferal packstone overlain by thick laminated crust, which in turn is overlain by foraminiferal-peloidal grainstone rubble with few quartz grains and displaying a large laminated pebble; black soilstone pebble with "floating" molluscs, foraminifera, and quartz grains, core 23 (Lignumvitae Channel, Florida Bay), depth 4.5 m, boundary between Q4a(?) and Q3(?) Units

7. Thin discontinuous laminated crust coats deeply weathered surfaces, solution cavities, and fractures to depth of 0.3 m. This skeletal (molluscs, foraminifera) quartzose packstone contains unidentified tubes, skeletal debris, and common open voids partially filled with quartz, core 4 (Lake Key, Florida Bay), depth 3 m, top of Q3(?) Unit. Scale in centimeters

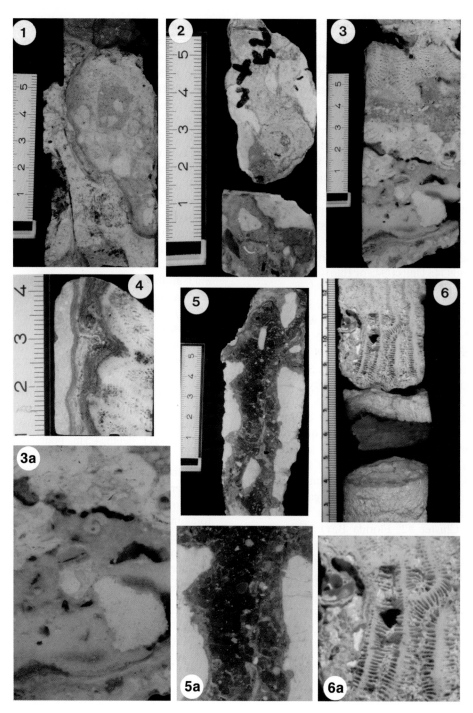

PLATE 2.4

Plate 2.4   Boundaries between Q Units in the Key Largo Limestone. All depths are below core tops. Core locations are given in Multer et al. (2002) [Reprinted from Multer et al. (2002, their Plate 45) with kind permission of Springer Science and Business Media]

1. Thin laminated crust often coats solution cavities shown here filled with skeletal and peloidal pebbles and capped with thick laminated crust, core W2 (Grassy Key), depth 5.5 m, boundary between Q3 and Q4 Units

2. Different thicknesses of dark-brown to gray soilstone (diagenetic mudstones) locally containing "floating" shells; brown and black pebbles are at the top surface of Q Units, core WX (Looe Key), depth 15.8 m, top of post-Q5e Unit (terminology of Multer et al. 2002). Scale in centimeters

3. At the boundary between Q3 and Q4a Units, quartz sandstone with pebbles of packstone overlie laminated crust; above the quartz layer are skeletal packstone and coral, core 93 (Bayside Cluster D, Florida Bay), depth 12.1 m. Scale in centimeters. (3a). Detail

4. Coral with vertical cavities coated with thick laminated crust containing black pebbles, core W2 (Grassy Key), depth 0.3 m, top of Q5 Unit. Scale in centimeters

5. Vertical cavity in a molluscan-foraminiferal wackestone is filled with dark-gray soilstone and black pebbles, core WP (Basin Hill Shoals), depth 10.6 m, top of post-Q5e Unit. Scale in centimeters. (5a). Detail

6. A thin microcrystalline rind (sometimes all that is present at contacts) is illustrated by three consecutive core pieces in descending order displaying (1) *Diploria* coral, (2) encrusted grainstone composed of molluscs, foraminifera, *Halimeda*, and peloids with thin (1–2 mm) black and dark-brown microcrystalline rind coating surfaces and voids; cavities in fossils near crust show vadose whisker-calcite needles, and (3) a second encrusted-grainstone core piece. Core MQ-4 (southwest Marquesas area), sample depth is 3.9 m from top of rock and 16.1 m below sea level. Q Units involved are unknown. Scale in centimeters. (6a). Detail

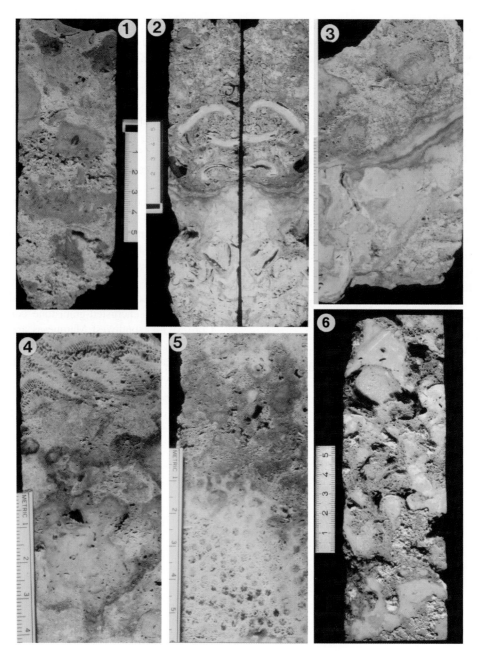

PLATE 2.5

Plate 2.5 Examples of possible framework types capable of producing rigid patch reefs and non-rigid mounds or beds within the Key Largo Limestone. Explanations of terminologies given in Fig. 2.B-2). All depths are below core tops. Core locations are given in Multer et al. (2002) [Reprinted from Multer et al. (2002, their Plate 46) with kind permission of Springer Science and Business Media]

1. Non-rigid stabilized thicket or bed of branching coral, molluscs, serpulid tubes, and red coralline algae in a matrix of subangular quartzose wackestone, core 25 (Johnson Key), depth 5.8 m, Q2 Unit. Scale in centimeters

2. Rigid, fused, encrusted thicket or bed with coral, red coralline algae, molluscs, peloids, and mud infill overlain by subaerial laminated crust incorporating small mollusc shells and black round pebbles. Above crust is a non-rigid stabilized rubble mound or bed of *Porites* coral, molluscs, black pebbles, fragments of laminated crust, and reworked grainstone pebbles (containing foraminifera and peloids), interspersed with serpulid tubes and mud, core 17 (Bob Allen Key-B, near Bob Allen Key-A core, Florida Bay), depth 0.6 m, Q3/Q4 Unit contact. Scale in centimeters

3. Non-rigid stabilized rubble of coral (*Montastraea?*), molluscs, foraminifera, bryozoans, peloids, and trace of quartz overlain by laminated crust. Above crust is a stabilized coral-thicket mound or bed of *Porites* coral with molluscs, foraminifera, bryozoans, peloids, and a few round black pebbles, core 16 (Bob Allen Key-A, Florida Bay), depth 1.2 m, Q4b/Q4a-Unit contact. Scale in centimeters

4. Possible stabilized mud mound or bed with common mollusc and foraminiferal fragments, small rhizoid(?) holes, and larger former root or bore holes filled with skeletal debris, all floating in a wackestone matrix; overlying this stabilized skeletal mud mound is a rigid coral (*Diploria*) framework, core 15 (Sprigger Bank, Florida Bay). Approximate depth 4.5 m, Q4 Unit. Scale in centimeters

5. Rigid, fused, pillar framework of *Montastraea* coral with overlying non-rigid stabilized rubble, including coral, mollusc, coralline algae, foraminifera, pellets, random quartz grains, and unidentified plant(?) tubes filling interstices; micritic replacement common, core 10 (John Sawyer Bank, north of Marathon, Florida Bay), depth 1.8 m, Q3 Unit. Scale in centimeters

6. Fused encrusted thicket of stick corals lying above *Montastraea* (not shown) with skeletal packstone and unidentified tubes partially filling voids between corals that include *Porites* sp., core W9 (Jewfish Creek), depth 4.3 m, Q4b Unit. Scale in centimeters

References

Agassiz A (1883) Exploration of the surface fauna of the Gulf Stream, under the auspices of the United States Coast Survey, II. The Tortugas and Florida reefs. Mem Am Acad Arts Sci, Centennial 2:107–132

Agassiz L (1880) Report on the Florida Reefs. Mem Mus Comp Zool at Harvard College 7:1–61, 23 pls

Albritton DL, Allen MR, Baede APM et al. (2001) Summary for Policymakers. In: Houghton JT, Ding Y, Griggs DJ, Noguer M, van der Linden PJ, Dai X, Maskell K, Johnson CA (eds) Climate Change 2001: The Scientific Basis, Contribution of Working Group I to the Third Assessment Report of the Intergovernmental Panel on Climate Change. Cambridge University Press, Cambridge and New York, pp 1–20

Anderson S, Zepp R, Machula J, Santavy D, Hansen L, Mueller E (2001) Indicators of UV exposure in corals and their relevance to global climate change and coral bleaching. Hum Ecol Risk Assess 7:1271–1282

Antonius A (1973) New observations on coral destruction in reefs (abs). Assoc Isl Mar Lab Carib 10: p.3

Applin PL, Applin ER (1965) The Comanche Series and associated rocks in the subsurface in Central and South Florida. US Geol Surv Prof Pap 447:1–84

Aronson RB, Precht WF (1997) Stasis, biological disturbance, and community structure of a Holocene coral reef. Paleobiology 23:326–346

Ball MM (1967) Carbonate sand bodies of Florida and the Bahamas. J Sed Petrol 37:556–591

Ball MM, Shinn EA, Stockman KW (1967) The geologic effects of Hurricane Donna in South Florida. J Geol 75:583–597

Banks KB, Riegl BM, Piller WE, Dodge RE, Shinn EA (2007) Geomorphology of the southeast Florida continental reef tract (Dade, Broward, and Palm Beach Counties, USA). Coral Reefs 26(3):617–633

Barber RT (1988) Iron, geologic-scale nutrient variations plus regulation of new production and carbon export can explain glacial-interglacial changes in atmospheric carbon dioxide (abs). Oceanography Society 5th Scientific Meeting, Seattle, Washington, DC, p 45

Bard E, Hamelin B, Fairbanks RG, Zindler A, Mathieu G, Arnold M (1990) U/Th and ^{14}C ages of corals from Barbados and their use for calibrating the ^{14}C time scale beyond 9000 years BP. Nuc Instrum Meth Phys Res B52: 461–468

Birkeland C (1987) Nutrient availability as a major determinant of differences among coastal hard-substratum communities in different regions of the tropics. In: Birkeland C (ed) Differences Between Atlantic and Pacific Tropical Marine Coastal Systems: Community Structure, Ecological Processes, and Productivity. UNESCO, Paris, France, pp 45–90

Birkeland C (1997) Chapter 12: Geographic differences in ecological processes on coral reefs. In: Birkeland C (ed) Life and Death of Coral Reefs. Chapman & Hall, New York, pp 273–287

Bloom AL, Broecker WS, Chappell JMA, Matthews RK, Mesolella KJ (1974) Quaternary sea level fluctuations on a tectonic coast: New ^{230}Th/^{234}U dates from the Huon Peninsula, New Guinea. Quat Res 4:185–205

Bryant D, Burke L, McManus J, Spalding M (1998) Reefs at Risk: A Map-Based Indicator of Potential Threats to the World's Coral Reefs. World Resources Institute, Washington, DC, 56 pp

Carpenter RC (1985) Sea urchin mass mortality: Effects on reef algal abundance, species composition and metabolism and other coral reef herbivores. Proc 5th Int Coral Reef Symp 4:53–59

Chappell J (1974) Geology of coral terraces, Huon Peninsula, New Guinea: A study of Quaternary tectonic movements and sea-level changes. Geol Soc Am Bull 85:553–570

Chappell J, Shackleton NJ (1986) Oxygen isotopes and sea level. Nature 324:137–140

Coniglio M (1981) Sedimentology of Pleistocene carbonates from Big Pine Key, Florida. M.S. thesis, University of Manitoba, 261 pp

Cunningham KJ, McNeill DF, Guertin LA, Scott TM, Ciesielski PF, de Verteuil L (1998) A new Tertiary stratigraphy for the Florida Keys and southern peninsula of Florida. Geol Soc Am Bull 110:231–258

D'Almeida GA (1986) A model for Saharan dust transport. J Clim Appl Meteor 24:903–916

Davis RA, Hine AC, Shinn EA (1992) Holocene coastal development on the Florida Peninsula. In: Fletcher C, Wehmiller J (eds) Quaternary Coasts of the United States: Marine and Lacustrine Systems. SEPM (Society for Sedimentary Geology) Spec Publ 48:193–212

Divins DL (2003) Coastal Relief Model. National Geophysical Data Center, Boulder, CO http://www.ngdc.noaa.gov/mgg/coastal/coastal.html

Dodge RE, Vaisnys JR (1975) Hermatypic coral growth banding as environmental recorder. Nature 258:706–708

Droxler AW, Farrell JW (2000) Marine isotope Stage 11 (MIS 11): New insights for a warm future. Global Planet Change 24:1–5

Dustan P (1985) Community structure of reef-building corals in the Florida Keys: Carysfort Reef, Key Largo and Long Key Reef, Dry Tortugas. Atoll Res Bull 288:1–27

Dustan P, Halas J (1987) Changes in the reef-coral population of Carysfort Reef, Key Largo, Florida, 1975–1982. Coral Reefs 6:91–106

Dustan P, Lidz BH, Shinn EA (1991) Impact of exploratory wells, offshore Florida: A biological assessment. Bull Mar Sci 48:94–124

Edwards BD, Perkins RD (1974) Distribution of micro-borings within continental margin sediments of the southeastern United Sates. J Sed Petrol 44:1122–1135

Enos P (1977) Holocene sediment accumulations of the South Florida shelf margin, Pt. I. In: Enos P, Perkins RD (eds) Quaternary Sedimentation in South Florida. Geol Soc Am Mem 147:1–130

Enos P, Perkins RD (1977) Quaternary Sedimentation in South Florida. Geol Soc Am Mem 147:1–198

EPA (Environmental Protection Agency) (1998) Florida Keys National Marine Sanctuary Water Quality Protection Program, Coral Reef and Hardbottom Monitoring Project, 1997. Annual and 4th Quarterly Report, 39 pp

Fairbanks RG (1989) A 17,000-year glacio-eustatic sea level record. Influence of glacial melting rates on the Younger Dryas event and deep ocean circulation. Nature 342:637–642

Falkowski PG, Dubinsky Z, Muscatine L, McCloskey L (1993) Population control in symbiotic corals. BioScience 43:606–611

Fanning KA (1989) Influence of atmospheric pollution on nutrient limitation in the ocean. Nature 339:460–463

Fitt W, Brown B, Warner M, Dunne R (2001) Coral bleaching: Interpretation of thermal tolerance limits and thermal thresholds in tropical corals. Coral Reefs 20:51–65

Gallagher D, Causey BD, Gato J, Viele J (eds) (1997) The Florida Keys Environmental Story, A Panorama of the Environment, Culture and History of Monroe County, Florida. Seacamp Association Inc, Big Pine Key, FL, 371 pp

Garrett P, Ducklow H (1975) Coral diseases in Bermuda. Nature 235:349–350

Garrison VH, Shinn EA, Foreman WT, Griffin DW, Holmes CW, Kellogg CA, Majewski MS, Richardson LL, Ritchie KB, Smith GW (2003) African and Asian dust: from desert soils to coral reefs. BioScience 53:469–480

Ghiold J, Enos P (1982) Carbonate production of the coral *Diploria labyrinthiformis* in South Florida patch reefs. Mar Geol 45:281–296

Ginsburg RN (1956) Environmental relationships of grain size and constituent particles in some South Florida carbonate sediments. Am Assoc Petrol Geol Mem 47:1–130

Ginsburg RN (ed) (2001) Subsurface Geology of a Prograding Carbonate Platform Margin, Great Bahama Bank. Results of the Bahamas Drilling Project. SEPM (Society for Sedimentary Geology) Spec Publ 70, Tulsa, OK, 271 pp

Ginsburg RN, Shinn EA (1964) Distribution of the reef-building community in Florida and the Bahamas (abs.). Am Assoc Petrol Geol Bull 48: p.527

Ginsburg RN, Browne KM, Chung GS (1989) Siliciclastic foundation of South Florida's Quaternary carbonates. Geol Soc Am, Abstracts with Programs 21:A-290

Gladfelter WG (1982) Whiteband disease in *Acropora palmata*: Implications for the structure and growth of shallow reefs. Bull Mar Sci 32:639–643

Glynn PW (1984) Widespread coral mortality and the 1982–1983 El Niño warming event. In: Ogden J, Wicklund R (eds) Mass Bleaching of Coral Reefs: A Research Strategy. Natl Undersea Res Prog Res Rep 88:42–45

Glynn PW (1987) Chapter 14: Bioerosion and coral-reef growth: A dynamic balance. In: Birkeland C (ed) Life and Death of Coral Reefs. New York, Chapman & Hall, pp 68–95

Glynn PW (1988) El Niño warming, coral mortality and reef framework destruction by echinoid bioerosion in the eastern Pacific. Galaxea 7:129–160

Griffin DW, Kellogg CA, Shinn EA (2001) Dust in the wind: Long range transport of dust in the atmosphere and its implications for global public and ecosystem health. Glob Change Hum Health 2:20–33

Guertin LA, McNeill DF, Cunningham KJ (1996) Geologic history of the Florida Keys: A model for carbonate-siliciclastic mixing, reef foundation, and current-controlled deposition. Geol Soc Am, Abstracts with Programs 28:A-64

Halley RB, Evans CE (1983) The Miami Limestone, A Guide to Selected Outcrops and their Interpretation (with a Discussion of Diagenesis in the Formation). Miami Geological Society, Miami, FL, 65 pp

Hallock P (1988) The role of nutrient availability in bioerosion: Consequences to carbonate buildups. Palaeogeogr, Palaeoclim, Palaeoecol 51:467–474

Hallock P (2005) Global change and modern coral reefs: New opportunities to understand shallow-water carbonate depositional processes. Sed Geol 175:19–33

Hallock P, Schlager W (1986) Nutrient excess and the demise of coral reefs and carbonate platforms. PALAIOS 1:389–398

Hallock P, Hine AC, Vargo GA, Elrod JA, Jaap WC (1988) Platforms of the Nicaraguan Rise: Examples of the sensitivity of carbonate sedimentation to excess trophic resources. Geology 16:1104–1107

Hallock P, Muller-Karger FE, Halas JC (1993) Coral reef decline. Natl Geogr Res Exp 9:358–378

Harrison RS, Coniglio M (1985) Origin of the Pleistocene Key Largo Limestone, Florida Keys. Bull Can Petrol Geol 33:350–358

Harrison RS, Halley RB (1979) Key Largo-subsurface core study (abs). Am Assoc Petrol Geol Bull 63:p.463

Harrison RS, Coniglio M, Halley RB (1982) Late Pleistocene deposits of the Florida Keys (abs). Am Assoc Petrol Geol Bull 66:578–579

Harrison RS, Cooper LD, Coniglio M (1984) Late Pleistocene carbonates of the Florida Keys. In: Carbonates in Subsurface and Outcrop. Canadian Society of Petroleum Geologists Core Conference, pp 291–306

Hayes M, Bunce L, Howd KB, Paerl H, Barber R, Shinn E (1988) A test of the hypothesis that dust stimulation of cyanobacteria and epizootic pathogens is causing widespread decline of coral reefs (abs). Oceanography Society, 5th Scientific Meeting, Seattle, Washington, DC, p 45

Highsmith RC (1980) Geographic patterns of coral bioerosion: A productivity hypothesis. J Exp Mar Biol Ecol 46:177–196

Hine AC, Mullins HT (1983) Modern carbonate shelf-slope breaks. In: Stanley DJ, Moore GT (eds) The Shelf Break: Critical Interface on Continental Margins. Soc Econ Paleontol Mineral Spec Publ 33:169–188

Hine AC, Neumann AC (1977) Shallow carbonate bank margin growth and structure, Little Bahama Bank, Bahamas. Am Assoc Petrol Geol Bull 61:376–406

Hine AC, Wilber RJ, Neumann AC (1981) Carbonate sand bodies along contrasting shallow bank margins facing open seaways in northern Bahamas. Am Assoc Petrol Geol Bull 65:261–290

Hoegh-Guldberg O (1999) Climate change, coral bleaching and the future of the world's coral reefs. Mar Freshwater Res 50:839–866

Hoffmeister JE (1974) Land from the Sea, the Geologic Story of South Florida. The University of Miami Press, Coral Gables, FL, 143 pp

Hoffmeister JE, Multer HG (1964) Pleistocene limestone of the Florida Keys. In: Ginsburg RN (compiler) South Florida Carbonate Sediments. Geol Soc Am Field Trip 1:57–61

Hoffmeister JE, Multer HG (1968) Geology and origin of the Florida Keys. Geol Soc Am Bull 79:1487–1502

Hoffmeister JE, Stockman KW, Multer HG (1967) Miami Limestone of Florida and its Recent Bahamian counterpart. Geo Soc Am Bull 78:175–190

Hudson JH (1977) Long-term bioerosion rates on a Florida reef – A new method. Proc 3rd Int Coral Reef Symp, Geology 2:491–497

Hudson JH (1979) Absolute growth rates and environmental implications of Pleistocene *Montastrea annularis* in southeast Florida. Geol Soc Am Bull, Abstracts with Program 11:p.85

Hudson JH (1981a) Response of *Montastraea annularis* to environmental change in the Florida Keys. Proc 4th Int Coral Reef Symp 2:233–240

Hudson JH (1981b) Growth rates in *Montastraea annularis*: A record of environmental change in Key Largo Coral Reef Marine Sanctuary, Florida. Bull Mar Sci 31:444–459

Hudson JH (1985) Growth rate and carbonate production in *Halimeda opuntia*, Marquesas Keys, Florida. In: Toomey DF, Nitecki MH (eds) Paleoalgology: Contemporary Research and Applications. Berlin, Heidelberg, Springer, pp 257–263

Hudson JH, Goodwin WB (1997) Restoration and growth rate of hurricane damaged pillar coral (*Dendrogyra cylindrus*) in the Key Largo National Marine Sanctuary, Florida. Proc 8th Int Coral Reef Symp 1:567–570

Hudson JH, Hanson KJ, Halley RB, Kindinger JL (1994) Environmental implications of growth rate changes in *Montastraea annularis*. Biscayne National Park, Florida. Bull Mar Sci 54:647–669

Hudson JH, Powell GVN, Robblee MB, Smith III TJ (1989) A 107-year-old coral from Florida Bay: Barometer of natural and man-induced catastrophes? Bull Mar Sci 44:283–291

Hudson JH, Shinn EA, Halley RB, Lidz BH (1976) Sclerochronology – A tool for interpreting past environments. Geology 4:361–364

Hughes TP (1994) Catastrophes, phase shifts, and large-scale degradation of a Caribbean coral reef. Science 265:1547–1551

Imbrie J, Hays JD, Martinson DG, McIntyre A, Mix AC, Morley JJ, Pisias NG, Prell WL, Shackleton NJ (1984) The orbital theory of Pleistocene climate: Support from a revised chronology of the marine $\partial^{18}O$ record. In: Berger AL, Imbrie J, Hays J, Kukla G, Saltzman B (eds) Milankovitch and Climate. Understanding the Response to Astronomical Forcing. Proc NATO Adv Res Workshop, Palisades, NY, NATO Sci Series C 126:269–305

International Society for Reef Studies (1998) ISRS Statement on Bleaching. Reef Encounter 24:19–20

Jaap WC (1979) Observations on zooxanthellae expulsion at Middle Sambo Reef, Florida Keys. Bull Mar Sci 29:414–422

Jaap WC (1984) The ecology of the South Florida coral reefs. A community profile. U.S. Department of the Interior, Fish and Wildlife Service Report FWS/OBS-82/08 and Minerals Management Service Report MMS 84-0038, pp 1–138

Jaap WC (1985) An epidemic zooxanthellae expulsion during 1983 in the lower Florida Keys coral reefs: Hyperthermic etiology. Proc 5th Int Coral Reef Symp 6:143–148

James NP, Ginsburg RN (1979) The seaward margin of Belize barrier and atoll reefs: Morphology, sedimentology, organism distribution and late Quaternary history. Int Assoc Sedimentol Spec Publ 3, Blackwell, Oxford, Edinburgh, UK, Melbourne, Australia, 191 pp

Jameson SC, Tupper MH, Ridley JM (2002) The three screen doors: *Can* marine "protected" areas be effective? Mar Pol Bull 44:1177–1183

Kindinger JL (1986) Geomorphology and tidal-belt depositional model of lower Florida Keys (abs). Am Assoc Petrol Geol Bull 70:607

Kindinger JL, Davis JB, Flocks JG (1999) Geologic controls on the formation of lakes in north-central

Florida. In: Pitman JK, Carroll AR (eds) Modern and Ancient Lake Systems. Utah Geological Association Guidebook 26, Salt Lake City, UT, pp 9–30

Kindinger JL, Davis JB, Flocks JG (2000) Geologic controls on the formation of Florida sinkhole lakes. US Geological Survey Open-File Report 00–294, 4 pp

Korniker LA, Boyd DW (1962) Shallow water geology and environment of Alacran reef complex, Campeche Bank, Mexico. Am Assoc Petrol Geol Bull 46:640–673

Kroon D, Reijmer JJG, Rendle R (2000) Mid- to late-Quaternary variations in the oxygen-isotope signature of *Globigerinoides ruber* at Site 1006 in the western subtropical Atlantic. Proc ODP Sci Results 166:13–22

Landon SM (1975) Environmental Controls on Growth Rates in Hermatypic Corals from the Lower Florida Keys. M.S. thesis, SUNY, 81 pp

Lapointe BE, O'Connell JE, Garrett GS (1990) Nutrient couplings between on-site sewage disposal systems, groundwaters, and nearshore surface waters of the Florida Keys. Biogeochemistry 10:289–307

Lessios HA (1988) Mass mortality of *Diadema antillarum* in the Caribbean: What have we learned? Ann Rev Ecol Syst 19:371–393

Lessios HA, Robertson DR, Cubit JE (1984) Spread of *Diadema* mass mortalities through the Caribbean. Science 226:335–337

Lewis JB, Axelson F, Goodbody I, Page C, Chislett G (1968) Comparative growth rates of some reef corals in the Caribbean. Marine Science Manual Report no. 10, McGill University, Montreal, Canada, pp 1–10

Lidz BH (2004) Coral reef complexes at an atypical windward platform margin: late Quaternary, southeast Florida. Geol Soc Am Bull 116:974–988

Lidz BH (2006) Pleistocene corals of the Florida Keys: Architects of imposing reefs – Why? J Coast Res 22:750–759

Lidz BH, Hallock P (2000) Sedimentary petrology of a declining reef ecosystem, Florida reef tract (USA). J Coast Res 16:675–697

Lidz BH, Shinn EA (1991) Paleoshorelines, reefs, and a rising sea: South Florida, USA. J Coast Res 7:203–229

Lidz BH, Hine AC, Shinn EA, Kindinger JL (1991) Multiple outer-reef tracts along the South Florida bank margin: Outlier reefs, a new windward-margin model. Geology 19:115–118

Lidz BH, Reich CD, Peterson RL, Shinn EA (2006) New maps, new information: Coral reefs of the Florida Keys. J Coast Res 22:61–83

Lidz BH, Reich CD, Shinn EA (2003) Regional Quaternary submarine geomorphology in the Florida Keys. Geol Soc Am Bull 115:845–866, plus oversize color plate of Pleistocene topographic and Holocene isopach maps

Lidz BH, Reich CD, Shinn EA (2007) Systematic mapping of bedrock and habitats along the Florida reef tract: Central Key Largo to Halfmoon Shoal (Gulf of Mexico). US Geological Survey Prof Paper 1751 (CD-ROM)

Lidz BH, Robbin DM, Shinn EA (1985) Holocene carbonate sedimentary petrology and facies accumulation, Looe Key National Marine Sanctuary, Florida. Bull Mar Sci 36:672–700

Lidz BH, Shinn EA, Hansen ME, Halley RB, Harris MW, Locker SD, Hine AC (1997a) USGS Investigations Series Map #I-2505. Maps Showing Sedimentary and Biological Environments, Depth to Pleistocene Bedrock, and Holocene Sediment and Reef Thickness from Molasses Reef to Elbow Reef, Key Largo, South Florida. Three descriptive and interpretive sheets with aerial photomosaic

Lidz BH, Shinn EA, Hine AC, Locker SD (1997b) Contrasts within an outlier-reef system: Evidence for differential Quaternary evolution, South Florida windward margin, USA. J Coast Res 13:711–731

Lighty RG (1977) Relict shelf-edge Holocene coral reef, southeast coast of Florida. Proc 3rd Int Coral Reef Symp, Geology 2:215–221

Lighty RG, Macintyre IG, Stuckenrath R (1978) Submerged early Holocene barrier reef, southeast Florida shelf. Nature 276:59–60

Lighty RG, Macintyre IG, Stuckenrath R (1982) *Acropora palmata* reef framework: a reliable indicator of sea level in the western Atlantic for the past 10,000 years. Coral Reefs 1:125–130

Locker SD, Hine AC (1995) Late Quaternary sequence stratigraphy, South Florida margin. Proc 27th Ann Ocean Technology Conf pp 319–327

Locker SD, Hine AC, Tedesco LP, Shinn EA (1996) Magnitude and timing of episodic sea-level rise during the last deglaciation. Geology 24:827–830

Longley WH, Hildebrand SF (1941) Systematic catalogue of the fishes of Tortugas, Florida, with observations on color, habits, and local distribution. Papers Tortugas Laboratory 34:1–331

Ludwig KR, Muhs DR, Simmons KR, Halley RB, Shinn EA (1996) Sea-level records at ~80 ka from tectonically stable platforms: Florida and Bermuda. Geology 24:211–214

Lundberg J, Ford DC (1994) Late Pleistocene sea-level change in the Bahamas from mass spectrometric U-series dating of submerged speleothem. Quat Sci Rev 13:1–14

Mallinson D, Hine A, Hallock P, Locker S, Shinn E, Naar D, Donahue B, Weaver D (2003) Development of small carbonate banks on the South Florida platform margin: Response to sea level and climate change. Mar Geol 199:45–63

Margalef R (1968) The pelagic ecosystem of the Caribbean Sea. In: Symposium on Investigations and Resources of the Caribbean Sea and Adjacent Regions. UNESCO, Paris, France, pp 484–498

Marszalek DS (1977) Florida Reef Tract Marine Habitats and Ecosystems. Nine map sheets published in cooperation with State of Florida Department of Natural Resources; U.S. Department of Interior Bureau of Land Management, New Orleans Outer Continental Shelf Office; and University of Miami Rosenstiel School of Marine and Atmospheric Science

Mayer AG (1903) The Tortugas, Florida, as a station for research in biology. Science 17:190–192

Mayer AG (1910) Medusae of the World. Carnegie Institution of Washington Publication 109, Quarto, 3 vols, 735 pp

Mayer AG (1914) The effects of temperature on tropical marine organisms. Carnegie Institution of Washington Publication 183, pp 3–24

Mayer AG (1918) Toxic effects due to high temperature. Carnegie Institution of Washington Publication 252, pp 175–178

McKee ED, Ward WC (1983) Eolian environment. In: Scholle PA, Bebout DG, Moore CH (eds) Carbonate Depositional Environments. Am Assoc Petrol Geol Mem 33:131–170

Milliman JD, Emery KO (1968) Sea levels during the past 35,000 years. Science 162:1121–1123

Milton C, Grasty R (1969) Basement rocks of Florida and Georgia. Am Assoc Petrol Geol Bull 53:2482–2493

Mitterer RM (1974) Pleistocene stratigraphy in southern Florida based on amino acid diagenesis in fossil *Mercenaria*. Geology 2:425–428

Muhs DR (2002) Evidence for the timing and duration of the last interglacial period from high-precision uranium-series ages of corals on tectonically stable coastlines. Quat Res 58:36–40

Muhs DR, Bush CA, Stewart KC (1990) Geochemical evidence of Saharan dust parent material for soils developed on Quaternary limestones of Caribbean and Western Atlantic islands. Quat Res 33:157–177

Muhs DR, Szabo BJ, McCartan L, Maat PB, Bush CA, Halley RB (1992) Uranium-series age estimates of corals from Quaternary marine sediments of southern Florida. In: Scott TM, Allmon WD (eds) Plio-Pleistocene Stratigraphy and Paleontology of Southern Florida. FL Geol Surv Spec Publ 36:41–49

Muhs DR, Wehmiller JF, Simmons KR, York LL (2004) Quaternary sea-level history of the United States. Dev Quat Sci 1 [doi: 10.1016/S1571-0866(03)01008-X]

Multer HG (1977) Field Guide to Some Carbonate Rock Environments: Florida Keys and Western Bahamas. 6th edition, Kendall/Hunt, Dubuque, IA, 425 pp

Multer HG (1996) Seakeys Habitat Guide to the Florida Keys National Marine Sanctuary: Past, Present, Future. Double-sided colored descriptive and interpretive map, 1 sheet, scale 1.400,000. Copies available from The Florida Keys National Marine Sanctuary, P.O. Box 500368, Marathon, FL 33050, www.fknms.nos.noaa.gov

Multer HG, Hoffmeister JE (1968) Subaerial laminated crusts of the Florida Keys. Geol Soc Am Bull 79:183–192

Multer HG, Zankl H (1988) Holocene reef initiation and framework development, Antigua, WI. Proc 6th Int Coral Reef Symp 3:413–418

Multer HG, Gischler E, Lundberg J, Simmons KR, Shinn EA (2002) Key Largo Limestone revisited: Pleistocene shelf-edge facies, Florida Keys, USA. Facies 46:229–272

Neumann AC (1966) Observations on coastal erosion in Bermuda and measurements of the boring rate of the sponge, *Cliona lampa*. Limnol Oceanog 11:92–108

Neumann AC, Macintyre IG (1985) Response to sea level rise: keep-up, catch-up, or give-up. Proc 5th Int Coral Reef Symp 3:105–110

Ogden J, Wicklund R (eds) (1988) Mass Bleaching of Coral Reefs in the Caribbean: A research strategy. National Undersea Research Program, Res Rep 88:1–9

Parker GG, Cooke CW (1944) Late Cenozoic resources of southeastern Florida, with a discussion of the ground water. FL Geol Survey Bull 27:1–119

Parker GG, Hoy ND, Schroeder MC (1955) Geology. In: Parker GG, Ferguson GE, Love SK et al. (eds) Water resources of southeastern Florida. US Geol Surv Water-Supply Paper 1255:57–125

Parks P (1968) The Railroad that Died at Sea, Florida East Coast Railway Flagler System, the Florida East Coast's Key West Extension. Langley, Key West, FL, 48 pp

Perkins RD (1977) Depositional framework of Pleistocene rocks in South Florida, Pt. II. In: Enos P, Perkins RD (eds) Quaternary Sedimentation in South Florida. Geol Soc Am Mem 147:131–198

Petit JR, Jouzel J, Raynaud D, Barkov NI, Barnola J-M, Basile I, Bender M, Chappellaz J, Davis M, Delaygue G, Delmotte M, Kotlyakov VM, Legrand M, Lipenkov VY, Lorius C, Pepin L, Ritz C, Saltzman E, Stievenard M (1999) Climate and atmospheric history of the past 420,000 years from the Vostok ice core, Antarctica. Nature 399:429–436

Porter JW, Battey JF, Smith JG (1982) Perturbation and change in coral reef communities. Proc Natl Acad Sci 79:1678–1681

Porter JW, Fitt WK, Spero HJ, Rogers CS, White MW (1989) Bleaching in reef corals: Physiological and stable isotopic responses. Proc Natl Acad Sci 86:9342–9346

Prospero JM, Nees RT (1986) Impact of the North African drought and El Niño on mineral dust in the Barbados trade winds. Nature 320:735–738

Rawls M, Paerl H, Barber RT (1988) Varying eolian iron flux to marine systems and the putative increase in harmful algal blooms (abs). Oceanography Society 5th Scientific Meeting, Seattle, Washington, DC, p 45

Robbin DM (1981) Subaerial CaCO₃ crust: A tool for timing reef initiation and defining sea level changes. Proc 4th Int Coral Reef Symp 1:575–579

Robbin DM (1984) A new Holocene sea level curve for the upper Florida Keys and Florida reef tract. In: Gleason PJ (ed) Environments of South Florida: Present and Past. Miami Geol Soc Mem 2:437–458

Roberts HH, Rouse Jr., LJ, Walker ND, Hudson JH (1982) Cold-water stress in Florida Bay and northern Bahamas – A product of winter cold-air outbreaks. J Sed Petrol 52:145–155

Rowan R (2004) Coral bleaching – Thermal adaptation in reef coral symbionts. Nature 430:742

Rützler K (1975) The role of burrowing sponges in bioerosion. Oecologia, Berlin 19:203–216

Rützler K, Santavy DL (1983) The black band disease of Atlantic reef corals, Pt. I: Description of the cyanophyte pathogen PSZNI. Mar Ecol 4:301–319

Sanford S (1909) The topography and geology of southern Florida. FL Geol Surv Ann Rep 2:175–231

Scholl DW, Craighead Sr FC, Stuiver M (1969) Florida submergence curve revised: Its relation to coastal sedimentation rates. Science, New Series 163:562–564

Schroeder RA, Bada JL (1976) A review of the geochemical applications of the amino acid racemization reaction. Earth-Sci Rev 12:347–391

Scoffin TP, Stearn CW, Boucher D, Frydll P, Hawkins CM, Hunter IG, MacGeachy JK (1980) Calcium carbonate budget of a fringing reef on the west coast of Barbados, Pt. II. Erosion, sediments and internal structure. Bull Mar Sci 30:475–508

Shifting Baselines (2005) The truth about ocean decline. http://www.shiftingbaselines.org/news/photocont. html (accessed 5 March 2007).

Shinn EA (1963) Spur and groove formation on the Florida reef tract. J Sed Petrol 33:291–303

Shinn EA (1966) Coral growth rate, an environmental indicator. J Paleont 40:233–240

Shinn EA (1976) Coral reef recovery in Florida and the Persian Gulf. Env Geol 1:241–254

Shinn EA (1980) Geologic history of Grecian Rocks, Key Largo Coral Reef Marine Sanctuary. Bull Mar Sci 30:646–656

Shinn EA (1988) The geology of the Florida Keys. Oceanus 31:46–53

Shinn EA, Jaap WC (2005) Field guide to the major organisms and processes building reefs and islands of the Dry Tortugas: The Carnegie Dry Tortugas Laboratory centennial celebration (1905–2005). US Geol Survey Open-File Report 2005–1357, 72 pp

Shinn EA, Lidz BH (1988) Blackened limestone pebbles: fire at subaerial unconformities. In: James NP, Choquette PW (eds) Paleokarst, Springer. IO, pp 117–131

Shinn EA, Halley RB, Hine AC (2000b) SEPM Field Guide to the Florida Reef Tract, Key Largo Area. SEPM (Society for Sedimentary Geology), Tulsa, OK, 47 pp

Shinn EA, Hudson JH, Halley RB, Lidz BH (1977) Topographic control and accumulation rate of some Holocene coral reefs, South Florida and Dry Tortugas. Proc 3rd Int Coral Reef Symp, Geology 2:1–7

Shinn EA, Hudson JH, Robbin DM, Lidz BH (1981) Spurs and grooves revisited, construction versus erosion, Looe Key Reef, Florida. Proc 4th Int Coral Reef Symp, Biology 1:475–483

Shinn EA, Lidz BH, Dustan P (1989a) Impact assessment of exploratory wells offshore South Florida. OCS Study MMS 89–0022. US Department of the Interior, Minerals Management Service, Gulf of Mexico OCS Region, New Orleans, LA, 111 pp

Shinn EA, Lidz BH, Hine AC (1991) Coastal evolution and sea-level history: Florida Keys. In: Coastal Depositional Systems in the Gulf of Mexico, Quaternary Framework and Environmental Issues. Gulf Coast Section/SEPM Foundation 12th Annual Research Conference Program with Extended and Illustrated Abstracts, pp 237–239

Shinn EA, Lidz BH, Holmes CW (1990), High-energy carbonate sand accumulation, The Quicksands, southwest Florida Keys. J Sed Petrol 60:952–967

Shinn EA, Lidz BH, Kindinger JL, Hudson JH, Halley RB (1989b) Reefs of Florida and the Dry Tortugas: A guide to the modern carbonate environments of the Florida Keys and the Dry Tortugas. International Geological Congress Field Trip Guidebook T176, Am Geophys Union, Washington, DC, 55 pp

Shinn EA, Lidz BH, Reich CD (1993) Habitat impacts of offshore drilling: Eastern Gulf of Mexico. OCS Study MMS 93–0021, US Department of the Interior, Minerals Management Service, Gulf of Mexico OCS Region, New Orleans, LA, 73 pp

Shinn EA, Reese RS, Reich CD (1994) Fate and pathways of injection-well effluent in the Florida Keys. US Geol Survey Open-File Report 94–276, 116 pp

Shinn EA, Reich CD, Hickey DT, Lidz BH (2003) Staghorn tempestites in the Florida Keys. Coral Reefs 22:91–97

Shinn EA, Reich CD, Locker SD, Hine AC (1996) A giant sediment trap in the Florida Keys. J Coast Res 12:953–959

Shinn EA, Smith GW, Prospero JM, Betzer P, Hayes ML, Garrison V, Barber RT (2000a) African dust and the demise of Caribbean coral reefs. Geophys Res Lett 27:3029–3032

Smith NP (1994) Long-term Gulf-to-Atlantic transport through tidal channels in the Florida Keys. Bull Mar Sci 54:602–609

Smith GW, Ives LD, Nagelkerken IA, Ritchie KB (1996) Caribbean sea-fan mortalities. Nature 383:p 487

Smith SV, Kimmerer WJ, Laws EA, Brock RE, Walsh TW (1981) Kaneohe Bay sewage diversion experiment: Perspectives on ecosystem responses to nutrient perturbation. Pac Sci 35:279–402

Stanley SM (1966) Paleoecology and diagenesis of Key Largo Limestone, Florida. Am Assoc Petrol Geol Bull 50:1927–1947

Steinen RP, Harrison RS, Matthews RK (1973) Eustatic low stand of sea level between 125,000 and 105,000 BP: Evidence from the subsurface of Barbados, West Indies. Geol Soc Am Bull 84:63–70

Stephens LD, Calder DR (2006) Seafaring Scientist – Alfred Goldsborough Mayor, Pioneer in Marine Biology. The University of South Carolina Press, Columbia, SC, 221 pp

Swinchatt JP (1965) Significance of constituent composition, texture, and skeletal breakdown in some Recent carbonate sediments. J Sed Petrol 35:71–90

Szabo BJ, Halley RB (1988) ^{230}Th/^{234}U ages of aragonitic corals from the Key Largo Limestone of South Florida. American Quaternary Association, Program and Abstracts of the 10th Biennial Meeting, University of Massachusetts, p 154

Szmant AM (2002), Nutrient enrichment on coral reefs: Is it a major cause of coral reef decline? Estuaries 25:743–766

Szmant AM, Forrester A (1996) Water column and sediment nitrogen and phosphorus distribution patterns in the Florida Keys, USA. Coral Reefs 15:21–42

Toscano MA (1996) Late Quaternary Stratigraphy, Sea-Level History, and Paleoclimatology of the Southeast Florida Outer Continental Shelf. Unpublished Ph.D. thesis, University of South Florida, St. Petersburg, FL, 280 pp

Toscano MA, Lundberg J (1998) Early Holocene sea-level record from submerged fossil reefs on the southeast Florida margin. Geology 26:255–258

Toscano MA, Lundberg J (1999) Submerged late Pleistocene reefs on the tectonically stable SE Florida margin: High-precision geochronology, stratigraphy, resolution of substage 5a sea-level elevation, and orbital forcing. Quat Sci Rev 18:753–767

Triffleman NJ, Hallock P, Hine AC, Peebles MW (1992) Morphology, sediments, and depositional environments of a small carbonate platform: Serranilla Bank, Nicaraguan Rise, Southwest Caribbean Sea. J Sed Petrol 62:591–606

Tucker ME (1985) Shallow-marine carbonate facies and facies models. In: Brenchley PJ, Williams BPJ (eds) Sedimentology – Recent Developments and Applied Aspects. The Geological Society, Blackwell, pp 147–169

Turmel RJ, Swanson RG (1976) The development of Rodriguez Bank, a Holocene mudbank in the Florida reef tract. J Sed Petrol 46:497–518

Vaughan TW (1914a) Sketch of the geologic history of the Florida coral reef tract and comparisons with other coral reef areas. J Wash Acad Sci IV:26–34

Vaughan TW (1914b) Building of the Marquesas and Tortugas atolls and a sketch of the geologic history of the Florida reef tract. Carnegie Institution of Washington Publication 182, Papers of the Department of Marine Biology, 5:55–67

Vaughan TW (1915a) Growth rate of the Floridian and Bahamian shoal-water corals. Carnegie Institution of Washington Yearbook no. 13, pp 221–231

Vaughan TW (1915b) The geological significance of the growth rate of the Floridan and Bahamian shoal water corals. J Wash Acad Sci 5:591–600

Vaughan TW (1916) On the Recent *Madreporaria* of Florida, the Bahamas, and the West Indies, and on collections from Murray Island, Australia. Carnegie Institution of Washington Year Book No. 14, pp 220–231

Vitousek PM, Aber JD, Howarth RW, Likens GE, Matson PA, Schindler DW, Schlesinger WH, Tilman GD (1997a) Human alteration of the global nitrogen cycle: Sources and consequence. Ecol Appl 7:737–750

Vitousek PM, Mooney HA, Lubchenco J, Mellilo JM (1997b) Human domination of Earth's ecosystems. Science 277:494–499

Walsh JJ, Steidinger KA (2001) Saharan dust and Florida red tides: The cyanophyte connection. J Geophys Res 106:11597–11612

Wanless HR (1989) The inundation of our coastlines: Past, present and future with a focus on South Florida. Sea Frontiers 35:264–271

Winston GO (1969) A deep glimpse of West Florida's Platform. Oil Gas J

Winston GO (1972) Oil occurrence and the Lower Cretaceous carbonate-evaporite cycle in South Florida. Am Assoc Petrol Geol Bull 56:158–160

Zeiller W (2005) A Prehistory of South Florida: Jefferson, North Carolina and London. McFarland & Co, 226 pp

3

A Perspective on the Biology of Florida Keys Coral Reefs

Walter C. Jaap, Alina Szmant, Karilyn Jaap, Jennifer Dupont,
Robert Clarke, Paul Somerfield, Jerald S. Ault, James A. Bohnsack,
Steven G. Kellison, and G. Todd Kellison

Dedication   This chapter is dedicated to Dr. Carl Robert Beaver; a coral reef scientist employed by the Florida Fish and Wildlife Research Institute and formerly by Texas A&M University, Corpus Christi. In the short time he had with us, his heart and soul were passionately devoted to coral reef science and conservation.

3.1   Introduction

South Florida is a unique enclave of the Caribbean thanks to the nexus of geography and environmental factors. Tropical mangrove, sea grass, coral reef epifaunal and infaunal sedimentary communities are common from Stuart on the east coast to Tampa Bay on the west coast. Florida is the only state in the continental United States to have such an ecosystem in its coastal waters. Climate and hydrodynamic features support a variety of plants and animals. The Florida Keys are the most Caribbean-like region in Florida. These "islands in the sun" have attracted millions of visitors and residents; some of the more famous include: George Meade (Union General in the Civil War), James Audubon (artist), President Harry Truman (built the Little White House in Key West), Humphrey Bogart and Lauren Bacall (who made the movie Key Largo in Key Largo), Ted Williams (baseball player and avid fisherman), Tennessee Williams (playwright), Ernest Hemingway (writer), and Jimmy Johnson (football coach). Many respected scientists worked their crafts in the coral reefs of the Florida Keys including Louis and Alexander Agassiz, Louis Pourtalès, Alfred G. Mayer, Thomas W. Vaughan, William Longley, Reginald Daly, Lawrence Cary, Walter Stark, Robert Ginsburg, and Eugene Shinn. A significant portion of the foundation of coral reef science is the result of research conducted in

the Florida Keys. The first underwater photographs (some in color) of coral reef fish were taken in the Keys. The first coral reef underwater park (John Pennekamp) and marine protected area (Dry Tortugas) were created in the Keys. Coral reefs are important economic, ascetic, and natural assets to Florida and the United States (Table 3.1).

Florida Keys reefs are equivalent to many Caribbean reefs in the biodiversity and richness of coral cover that is present, in the proliferation of benthic algae, and in the scarcity of *Diadema antillarum* (following the post-1983 disease epidemic) and the apex predator fish species, e.g., grouper and snapper. Unfortunately, also like the Caribbean, pressures associated with coastal development, the heavy extraction of fish and invertebrates, and climate -related incidents have degraded the reefs. Since 2004, the reefs from the Lower Keys to Dry Tortugas have endured five hurricanes (the impact of these hurricanes will be discussed in detail later in this chapter). In the twenty-first century, the coral reefs in the Florida Keys continue on; however, their fate is in our hands. Can society control the urge to use more carbon and extract more protein and ornamental species? Are there ways to protect the corals from the virulent diseases that are a relatively recent development? Parks, sanctuaries, wildlife refuges, and marine protected areas have been created in the Florida Keys (Shinn 1979; NOAA 2007); management has reduced physical destruction

TABLE 3.1. Coral reef habitat estimates, Florida, east coast (Johns et al. 2001).

Region (county)	Habitat area (ha)	Capitalized value (Billions of 2001 $)	Annual usage (person days, millions)
Palm Beach	12,000	1.4	2.83
Broward	8,300	2.8	5.46
Miami -Dade	7,200	1.6	6.22
Monroe	115,290	1.8	3.64

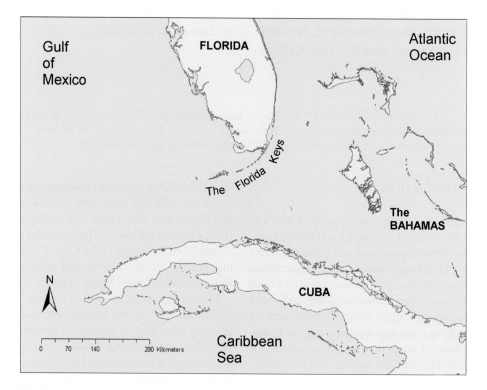

FIG. 3.1. Map displaying the geographic setting of the Florida Keys

from anchoring and emphasized education and public awareness. The goal in this chapter is to provide an overview with moderate details and sufficient reference documentation for those wishing to delve deeper into the subjects.

3.2 Geography

The Florida peninsula (151,670 km²) is a large carbonate plateau and projects south from the continental land mass of North America into the Atlantic Ocean and Gulf of Mexico (Chapter 2, Lidz et al.). The Peninsula's axis of projection is about 160° and extends approximately 724 km (391 miles) from the northern border to Florida Bay. The distance from the St. Lucie Inlet (Stuart) to the Tortugas Banks is 667 km (360 nmi). The Florida Keys archipelago (Fig. 3.1) arcs southwest from Biscayne Bay over a distance of 253 km (137 nmi) to Key West and 378 km (204 nmi) to the Dry Tortugas.

The islands (keys or cays) from Soldier (25°31.4′ N, 80°10.5′ W) to Boca Grande (24°32.5′ N, 82°00.4′ W) are built of Pleistocene marine limestone. The Florida Reef Tract (Vaughan 1914) parallels the Florida Keys archipelago from Fowey Rock (25°35.4′ N, 80°05.8′ W) to the Dry Tortugas (24°36.9′ N, 82°53.6′ W), and shallow coral reef communities continue north to the St. Lucie Inlet near Stuart (27°10′ N, 80°09′ W). Details on temperate-deep reef communities are covered in Chapter 21 by Messing et al.

The well-developed coral reefs of the Florida Keys are found on the eastern–southern sides of the archipelago (locals say, "ocean side"), parallel to the Florida Current. Biscayne Bay, Card Sound, Little Card Sound, Barnes Sound, Blackwater Sound, and Florida Bay are situated between the Florida Keys and the peninsula. These are shallow bodies of water; average depth in Florida Bay is 1–2 m. Net transport of water is from Florida Bay into the Atlantic Ocean (Smith 1994). Water masses in these bays, particularly Florida Bay, are rapidly altered by environmental events: cold fronts chill the waters, rainstorms reduce salinity, windy weather produces turbid water, hot and dry periods of weather elevate the salinity and temperature, and algal blooms can result in toxic water masses. The Upper Keys, defined here as from the north end of Elliott Key (25°31.4′ N, 80°10.6′ W) to Upper Matecumbe Key (24°54.1′ N, 80°39.6′ W), have three narrow channels leading from the bays to the ocean. From Upper Matecumbe to Grassy Key (24°46.5′ N, 80°55.9′ W) there are six major channels, some as wide as 4.3 km (2.2 nmi). From Hog Key (24°42.4′ N, 81°07.5′) to Cudjoe Key (24°39.8′, 81°28.1′) there are nine major channels, including a 13 km wide (7 nmi) opening into the Atlantic from Florida Bay. This middle portion of the Florida Keys has small islands and wide channels. Reef distribution is influenced by the "inimical waters" generated in Florida Bay (Ginsburg and Shinn 1964). In general, there is poor reef development in those areas where there is unrestricted water movement from Florida Bay into the Atlantic. Wide passes between keys that allow large volumes of Florida Bay water to flow onto the reef tract impede coral reef formation (the part of the reef tract known as the Middle Keys). Areas isolated from Florida Bay (Key Largo and the area of Key West) have well-developed offshore bank reefs (Ginsburg and Shinn 1964; Jaap 1984; Jaap and Hallock 1990).

3.3   History

Shallow-water coral reefs occurred intermittently during the glacial periods along the east coast of Florida and the Florida Keys (Lighty 1977). Sea level change resulted in a hiatus of coral reefs in some areas and development in others (Lighty 1977; Shinn et al. 1977). The oldest Holocene coral reef (8,000 YBP) is located on the Tortugas Banks (Mallinson et al. 2003). Interestingly, the Tortugas Banks reefs did not keep pace with rising sea levels while Bird Key Reef, a Dry Tortugas National Park reef did (Shinn et al. 1977). Mallinson et al. (2003) hypothesized that the Wisconsin glacial melt-water flowed across the Tortugas Banks, so that lowered salinity and increased turbidity inhibited reef growth.

Florida and the Florida Keys show evidence of human habitation dating back at least 12,000 years. Florida was discovered by Europeans in 1513 when Spanish conquistador Juan Ponce de Leon came ashore at several locations along the coast he named "La Florida" (from the Spanish *Pascua Florida*, "feast of flowers") because his first landing occurred during Easter. Among his stops were Biscayne Bay (the northern end of the Florida Keys), Key West, and Dry Tortugas. Florida was in the hands of the Spanish, French, and British until the Spanish sold Florida to the United States in 1819; it subsequently became the 27th state in 1845.

The Florida Keys are situated in Monroe County, established in 1823, with its county seat in Key West. This town has grown from a remote port and fishing (and ship wrecking) settlement in the mid 1500s to a resident population of about 25,000 at present. The Florida Keys are connected to the mainland by the Overseas Highway (US 1), first built in 1912 as a railroad by Henry Flagler. After a severe hurricane destroyed much of the railroad in 1935, bridges and rail bed were converted to a toll road. This is the only land-based transportation corridor into and out of the Keys; it facilitates a lively economy based largely on tourism and commercial and recreational fishing centered in and around coral reefs (Johns et al. 2001; Andrews et al. 2005). The three counties (Miami-Dade, Broward, Palm Beach) to the north of the Keys represent a mass of urban development from Homestead-Florida City to Palm Beach; the 2000 census documented 2.2 million private dwellings and 5.1 million permanent residents in these counties; population increases substantially during the winter tourist season (December–March). The residents are active users of the marine environment; in 2004–2005 fiscal year there were over 173,870 boat registrations and 74,454 fishing licenses (Florida Statistical Abstract 2006; FWC licenses sales data, personal communication) issued in Martin, Palm Beach, Broward, Miami-Dade, and Monroe counties alone. Commonly, from spring through early fall, the residents of the counties north of the Keys will take vacations and extended weekend visits in the Keys, often taking their boats with them.

Miami-Dade and Monroe counties include southern components of the greater Everglades ecosystem. Water resources in South Florida are managed by the south Florida Water Management District and the US Army Corps of Engineers. An extensive canal and dam system is in the process of being redesigned with the goal of reversing the former policy of diverting water resources from the Everglades. Originally, water flowed south from the Lake Okeechobee–Kissimmee River watershed, through the Everglades and Big Cypress Swamp and into Florida Bay and the Ten Thousand Islands. Drainage projects forced the water east (St. Lucie Waterway) and west (Caloosahatchee River), starving the Everglades system of water (Douglas 1947; Sklar et al. 2002; Steinman et al. 2002).

Because of their impact on the economic and social well-being of Floridians and humanity in general, the natural environments of south Florida, including the Everglades and the Florida Reef Tract, have been protected by the establishment of a number of managed areas (parks, wildlife refuges, reserves, and marine sanctuaries). By executive order in 1908, President Theodore Roosevelt designated the Dry Tortugas a wildlife refuge, with the main purpose of protecting bird rookeries. Dry Tortugas National Monument was established in 1935; bird rookeries and cultural resources (Fort Jefferson) were still the principal concerns. In 1980, the enacting language was amended to include coral reefs and resident marine life; in 1992, the area was designated Dry Tortugas National Park (DTNP). The park occupies approximately 25,900 ha (100^2 miles). Dry Tortugas and Everglades National Parks were designated World Heritage Sites in 1979. Current regulations at DTNP prohibit commercial fishing, although recreational fishing based on Florida and federal regulations is allowed, prohibits taking of lobster, and seasonally closes bird rookeries and turtle nesting islands. Everglades National Park, created in 1947, includes large portions of Florida Bay. John Pennekamp Coral Reef State Park (JPCRSP) was designated in 1963; it was the first park to emphasize underwater resources (coral reefs) in the United States. Pennekamp includes the mangrove and upland hammocks, sea grass beds, and patch reefs that are within the 3 nmi (5.6 km) limit of state jurisdiction. The rules and regulations in JPCRSP protect coral, allow for recreational and commercial fishing, and close shallow areas to powered boats. Biscayne National Park (previously a National Monument) was created in 1980. It includes portions of Biscayne Bay, Card Sound, the Ragged Keys, Elliott Key, and a dense concentration of patch reefs seaward of Elliott Key (Voss et al. 1969). Regulations protect the benthic resources but allow for recreational and commercial fishing.

The enabling legislation for the Florida Keys National Marine Sanctuary (FKNMS) mandates management of a vast marine area (NOAA 1996): 2,800 nmi^2 from south of Miami (25°17.683′ N, 80°13.145′ W) to the Tortugas Banks (24°36.703′ N, 82°52.212′ W). The motivation to designate the entire region as FKNMS was a series of ship groundings in 1989: *Alec Owen Maitland* near Carysfort Reef, *Elpsis* near Elbow Reef, and the *Mavro Vetranic* at Pulaski Shoal. The process included a series of workshops and meetings to establish goals and receive public input. The FKNMS encompasses the water column and seafloor from the coast to 91 m depth. The FKNMS area includes mangroves, sea grasses, sedimentary habitat, hardbottom, and coral reefs (Jaap 1984; Jaap and Hallock 1990; Chiappone 1996). Partial habitat mapping of the region shows that coral reef habitat constitutes 7.79% of the area (Florida Marine Research Institute 1998). The Tortugas Ecological Reserve (an addition to the FKNMS) was established in 2001 and, at that time, was the largest marine protected area within US waters, a total of 514.5 km^2 (150 nmi^2). The FKNMS is divided into use zones. Some zones prohibit taking of anything, others are open to recreational harvest, and others are open to commercial and recreational fishers.

The designation of parks and sanctuaries has efficacy in on-site use. One particular success is the reduction of coral injuries at popular reefs through the introduction of mooring systems (Halas 1985). However, the more subtle and difficult issues (water quality, coastal development, limiting access to certain areas, and region-wide problems such as coral diseases) are not easily managed (fishing issues are discussed in Section 3.16 and water quality in Section 3.17).

3.4 Heritage of Past Research

There is a rich legacy of marine research that focused on Florida Keys and south Florida coral reefs. Louis Agassiz (1852, 1869, 1880), Alexander Agassiz (1885, 1888, 1890) and Louis Pourtalès (1863, 1869, 1871, 1878, 1880a, b) were pioneers in studying the reefs and reef organisms. Early studies were, for the most part, exploratory surveys, taxonomic collections, and efforts to see if there were ways to reduce shipping losses along the Florida Reef Tract.

Andrew Carnegie was a wealthy industrialist and wanted to accelerate the United States into the forefront of science. To do this, in 1902, he set up the Carnegie Institution and appointed a board of trustees to guide and fund scientific research. Alfred G. Mayer, who changed his name to Mayor in 1918, lobbied the trustees that they should establish a tropical marine laboratory at Dry Tortugas. In December 1903, the trustees awarded Mayer funds to establish a marine laboratory (Colin 1980; Stephens and Calder 2006). In 1904, the Carnegie Institution obtained a lease on the north end of Loggerhead Key, Dry Tortugas, from the US Lighthouse Service for the construction of a research laboratory. Construction began in 1904 under Mayer's direction (Mayer 1902; Stephens and Calder 2006); he had been a student/technician of Alexander Agassiz and was well qualified to pick this site for a tropical research station. The first researchers arrived in the summer of 1905. In spite of remoteness, lack of many amenities, and near destruction of the laboratory by hurricanes in 1910 and 1919, approximately 146 different researchers worked at the laboratory between 1905 and 1939 (Colin 1980; Schmidt and Pikula 1997). The world's first underwater photographs of coral reef fishes were taken by Longley and an assistant at Dry Tortugas; these were published in the *National Geographic* in January 1927. Two well-equipped research vessels were available to Carnegie researchers: the *Physilia*, a 39 ft auxiliary

powered ketch served from 1905 until 1910. In 1910, she was replaced by a powered vessel, the *Anton Dohrn*, built in Miami; the *Dohrn* served until the lab closed in 1939. The *Dohrn* supported Vaughan's coral growth rate studies at Golding Cay, Bahamas and other Caribbean cruises (Stephens and Calder 2006). Vaughan's coral growth rates are quite similar to rates reported in modern studies (Jaap 1984). Additionally, the use of surface supplied diving gear (a helmet and hand powered air pump) began in 1920s, another example of the pioneering fieldwork that went on at Dry Tortugas. Even today, Mayer and Vaughan are recognized for their groundbreaking work on corals and reefs. Mayor studied the physiology of corals, especially their temperature and salinity tolerances. Vaughan was a leading researcher in the origin and formation of coral reefs, a follow-up to the argument between Darwin, Agassiz, and Daley (Dobbs 2005). Articles on the overview of south Florida coral reefs include Jaap (1984), Jones et al. (1985), Jaap and Hallock (1990), Chiappone (1996), NOAA (2002, 2005) and Porter and Porter (2002).

3.5 Climate

"The maritime influence of the Caribbean Sea and the Gulf of Mexico transforms Florida's climate" (Chen and Gerber 1990). Climate is also influenced by the Bermuda high pressure system that reduces precipitation and by northern frontal systems that lower the air temperature (November to April). The "jet stream," a narrow band of strong winds in the upper atmosphere, steers the cold air masses; it meanders and is influenced by the El Niño Southern Oscillation (ENSO) system. Florida's climate is superimposed on the ENSO cycles. El Niño results in warmer and wetter winters, fewer hurricanes, and doldrums in late summer that often lead to coral bleaching events. La Niña results in drier and cooler winters, and more frequent hurricanes. Tropical cyclones (hurricanes) force radical changes in the coral reef, sea grass, and mangrove communities; hurricanes are most common in August and September. Climatic data for Key West and Stuart provide a synoptic summary (Fig. 3.2). Variation in seasonal air temperature is greater in Florida than in the more tropical Grand Cayman or Mayaguez, Puerto Rico; and rainfall is less in Florida than in Mayaguez.

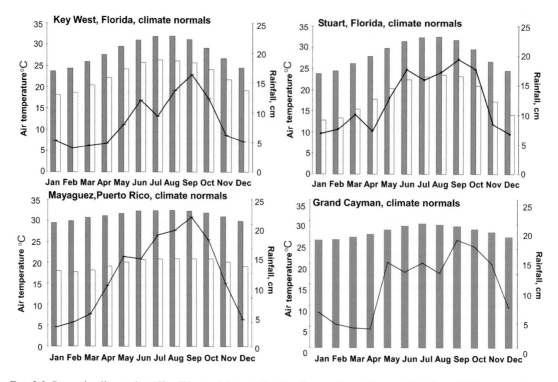

FIG. 3.2. Synoptic climate data: Key West and Stuart, Florida, George Town Cayman Islands, and Mayaguez, Puerto Rico. grey bars = monthly temperature maxima, white bars = monthly temperature minima, line = precipitation (National Weather Service, NOAA, and Grand Cayman Government)

Southeast Florida climate ranges from subtropical in Stuart to tropical maritime in Key West (Fig. 3.2). Low air temperatures around 17.6°C occur in January with an average maximum of 32.5°C in August. The temperature in Key West is slightly cooler than Stuart in the summer and warmer in the winter. The dry season is typically November to April/May and the wet season is May/June to October. Total rainfall is around 100 cm a year in Key West and 148 cm in Stuart. Climate data has been summarized from the National Weather Service web page, as well as Schomer and Drew (1982), Chen and Gerber (1990), and McPherson and Halley (1996).

3.6 Ocean Temperatures

Environmental recordkeeping for coastal Florida goes back 125 years (Vaughan 1918); light-house keepers [Loggerhead Key, Dry Tortugas (1879–1907); Sand Key, Key West (1878–1890); Carysfort Reef, Key Largo (1878–1899); and Fowey Rock, near Cape Florida (1879–1912)] recorded the water temperature using a bucket and a thermometer (Fig. 3.3).

The overall range for these data is 15.6–32.2°C (a change of 16.6°C annually); lowest temperatures occurred at Fowey Rock in February and the highest in August at Sand Key. Sand Key water temperature was warmest most of the year. Dry Tortugas was cooler January to May; Fowey Rocks was cooler June to November. Vaughan averaged these temperatures over 10-day periods so the extreme range of temperatures is not in the data set.

The National Oceanographic Atmospheric Administration's (NOAA), National Data Buoy Center operates meteorological–oceanographic data collecting systems off the Florida coast (Table 3.2). Figs. 3.4–3.8 provide synoptic data on the monthly means, standard deviations, and ranges for air and water temperatures and wind speed for a 10-year period. Data are acquired hourly for each of these parameters. These data confirm the lighthouse

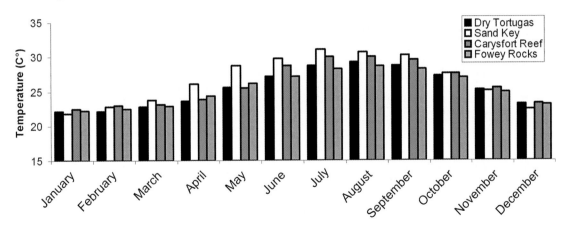

FIG. 3.3. Seawater temperature (1878–1912, C°), Florida Reef Tract Lighthouses (Vaughan, 1918)

TABLE 3.2. Selected meteorological–oceanographic data buoys off Florida.

Buoy code	Location	Latitude	Longitude
42036 W Tampa	Near the Florida Middle Grounds	28°30′00″	84°31′00″
DRYF1	Dry Tortugas	24°38′18″	82°51′42″
SANF1	Sand Key	24°27′25″	81°52′42″
SMKF1	Sombrero Key	25°37′36″	81°06′36″
FWYF1	Fowey Rocks	25°35′25″	80°05′48″
LKWF1	Lake Worth	26°36′42″	80°05′48″

keepers' records in the context of general trends. Extreme high temperatures were recorded on occasions in tide pools (Mayer 1914, 1918) and 14°C (a stressful low water temperature) was recorded in 1978 (Davis 1982).

3.7 Caribbean –West Indian Connections

Southern Florida [the region south of an arbitrary line drawn between the St. Lucie inlet (27°12.9′ N, 80°12.5′ W) and the southern tip of Sanibel Island (26°27.1′ N, 82°01.7′ W)] displays a greater similarity to the Caribbean and West Indian flora and fauna than to the flora and fauna of northern Florida. The West Indian zoogeographic area is defined as a sub-region of the Neotropical Province. As used here, it includes the Bahamas, Greater and Lesser Antilles, the northern coast of South America, the eastern coast of Central America, and southern Florida. On land, iconic vegetation providing evidence of Caribbean–West Indian affinity includes the coconut palm (*Cocos nucifera*), sea grape (*Coccoloba uvifera*), red mangrove (*Rhizophora mangle*), Gumbo Limbo (*Bursera simaruba*), and the mahogany tree (*Swietenia mahogani*). In the sea, the alga *Halimeda opuntia*, turtle grass (*Thalassia testudium*), invertebrates such as the horse conch (*Pleuropoca gigantia*), the long-spine, black urchin (*Diadema antillarum)*, fire coral (*Millepora complanata*), purple sea fan (*Gorgonia ventalina*), and Elkhorn coral (*Acropora palmata*) are typical of this particular ecosystem.

3.8 Hydrodynamic Connectivity

The coral reef fauna found in the Bahamas, Cuba, and southern Florida are remarkably similar. The lack of land barriers, connectivity of the water masses, and ocean currents facilitate larval transport of progeny among these areas.

Southern Florida's climate and marine systems are influenced by the large western boundary current (often referred to as the Caribbean Current, Florida Current, or the Gulf Stream). This current enters Lesser Antilles as the Guiana

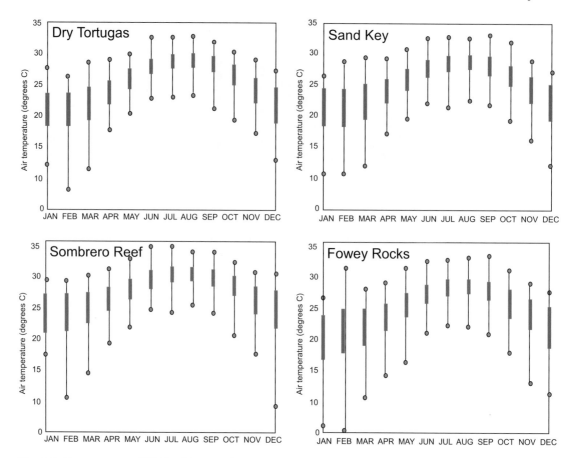

FIG. 3.4. Air temperature (C°), Dry Tortugas, Sand Key, Sombrero Reef, Fowey Rock 10-year average, range, standard deviation. .Graphs present the maximum (circle at upper end of a line representing the range), minimum (circle at the lower end of a line representing the range), mean (circle with a dot in the middle), ± 1 standard deviation of the mean (red vertical column) (National Data Buoy Center, NOAA)

Current flowing northwest of the Venezuelan coast. These waters are forced into the Caribbean Sea north and south of St. Lucia Island; the axis of the current is westward, passing the Aruba Gap and the Columbia Basin before turning north into the Cayman Basin, west to 85–86° W, north through the Yucatan Strait and into the Gulf of Mexico as the "Loop Current" (Fairbridge 1966). The Loop Current is variable in volume and penetration into the Gulf of Mexico (Lee et al. 1992, 1994, 2002); it reverses course in the northern Gulf, returns to the area around the Dry Tortugas, and enters the Straits of Florida, a deep channel between Florida, Cuba, and the Bahamas. The current at this point is typically called the "Florida Current." It meanders offshore–onshore, depending on a multitude of environmental influences. One of the facets of

biological significance is that the Florida Current sets up a series of gyres (rotating water masses) that persist for 60–100 days at a time (Lee et al. 1994, 2002). The largest of these is the Tortugas Gyre. The Pourtalès Gyre is formed east and north of the Tortugas Gyre; it can extend eastward to Vaca Key. Gyres retain and distribute invertebrate and fish larvae in the Florida reefs (Ingle et al. 1963; Sims and Ingle 1966; Lee et al. 1994; McGowan et al. 1994a, b, c). Upwelling, the other condition resulting from the current flow, occurs when the boundary current is near the Florida coast. This phenomenon results in cool, nutrient-rich water entering the reef system (Leichter et al. 2003). Short-term pulses of nutrients result in benthic algal blooms (*Dictyota* and *Cladophora*) in the reef system. Beyond the Keys, the Florida Current parallels the coast. Typically it

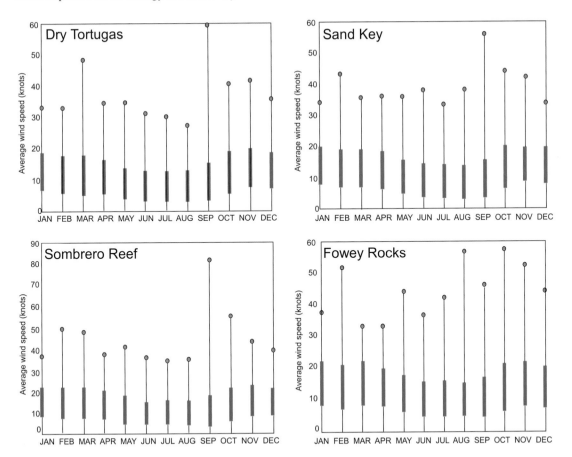

Fig. 3.5. Wind speed (knots), Dry Tortugas, Sand Key, Sombrero Reef, Fowey Rock ten year average, range, standard deviation. .Graphs present the maximum (circle at upper end of a line representing the range), minimum (circle at the lower end of a line representing the range), mean (circle with a dot in the middle), ± 1 standard deviation of the mean (red vertical column) (National Data Buoy Center, NOAA)

comes closest to the coast in Palm Beach County and tends to stray away from the coast after passing Palm Beach.

The Caribbean Sea and Gulf of Mexico are quite different. The Caribbean is totally tropical in nature and continental land masses enclose it to the south and west. Large and high islands (Hispaniola and Puerto Rico) and smaller islands of the Lesser Antilles form an incomplete barrier to the east. Two large islands, Cuba and Jamaica, are situated in the middle, and volcanic activity is common throughout the area. In contrast, the Gulf of Mexico is a large lake, with no large islands, open to the south, and surrounded by continental land masses. Narrow, sand-barrier islands parallel the west coast of Florida to the Texas–Mexico border. Coral cays are a feature off the Yucatan Peninsula, east coast

of Mexico, and the Florida Keys. The climate is temperate in the north and tropical to the south. Tectonic activity is uncommon.

The Bahamas is an extensive area of low islands, carbonate banks, and coral reefs. The Bahamas archipelago is 260,000 km² and extends 800 km from SE Florida to northern Hispaniola. The majority of the Bahamas is located on two shallow banks, ideal for coral reef development. Land (2,750 islands and shoals extending above the highest tides) occupies 11,400 km² (4.4% of the Bahamas area). The islands are low with few cases of greater than 30 m elevation (Gerace 1988). The northern Bahamas climate is similar to Florida; Little Bahama Bank (Grand Bahama, Walker's Cay, and Abaco) is occasionally impacted by winter cold fronts.

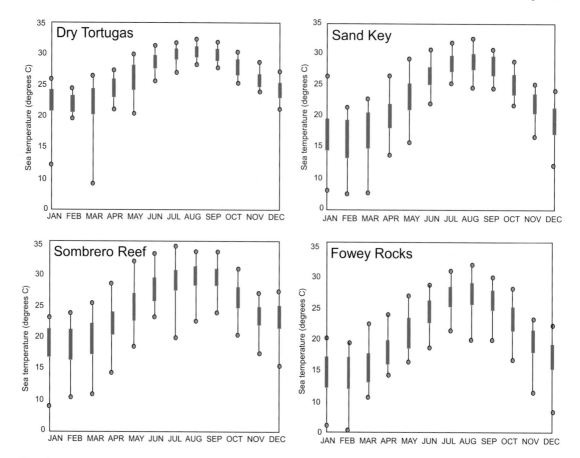

Fɪɢ. 3.6. Seawater temperature (C°), Dry Tortugas, Sand Key, Sombrero Reef, Fowey Rock 10-year average, range, standard deviation. graphs present the maximum (circle at upper end of a line representing the range), minimum (circle at the lower end of a line representing the range), mean (circle with a dot in the middle), ± 1 standard deviation of the mean (red vertical column) (National Data Buoy Center, NOAA)

Cuba is the closest high Caribbean island to Florida (167 km [90 miles] south of Key West) with a surface area of 110,860 km². The island has a 5,700 km long estimated coastline; the continental shelf is 100–140 km wide; and there are approximately 4,200 islands in four groupings: Los Colorados and Jardines del Ray (northern coast); Los Canarreos and Jardines de la Reina (southern coast) (Jiménez 1982). Zlatarski and Estalella (1982) and Kühlman (1974a) describe multiple reef types (bank, fringing, cluster, patch, coral canyons, and horst). Jiménez (1984) estimated that the total length of coral reefs on Cuba's north coast was 2,150 km and 1,816 km on the south coast.

3.9 Biodiversity and Taxonomic Distinctness of the Scleractinia (Stony Corals)

The zooxanthellate Scleractinia are representative of the biodiversity and taxonomic character of the region. Order Scleractinia: Phylum Cnidaria: Class Anthozoa is exclusively marine, occurs from the subarctic to Antarctica, and from near sea level to below 6300 m (Cairns 2001). Physiologically, ecologically (and roughly phylogenetically), the Scleractinia can be divided into two groups: those containing zooxanthellae (dinoflagellate algae of the genus *Symbiodinium*) in their tissues (the

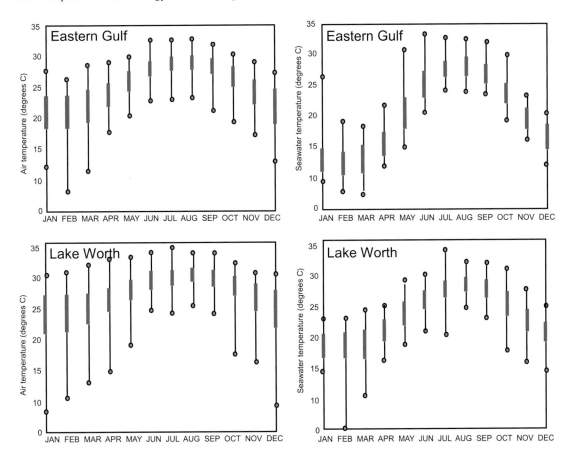

FIG. 3.7. Air temperature (C°), Seawater temperature (C°), Eastern Gulf of Mexico and Lake Worth (Palm Beach). Graphs present the maximum (circle at upper end of a line representing the range), minimum (circle at the lower end of a line representing the range), mean (circle with a dot in the middle), ∀1 standard deviation of the mean (red vertical column) (National Data Buoy Center, NOAA)

zooxanthellate Scleractinia) and those that do not (the azooxanthellate Scleractinia). Zooxanthellate species are restricted to the photic zone and are typically found in tropical–subtropical regions in depths that rarely exceed 70–80 m. The zooxanthellate species include species attaining sizes exceeding 3 m in diameter and height (e.g., *Colpophyllia natans* and *Montastraea annularis* complex), as well as species that rarely exceed 10 cm in diameter (such as *Cladocora arbuscula* and *Favia fragum*). Most zooxanthellate species are colonial (with multiple polyps), and their morphology includes branching, columnar, encrusting, foliaceous, and massive skeletal structures. Within a zooxanthellate species, the morphology has great variability, reflecting local environmental conditions, many of which vary with depth, e.g., ambient light, water

movement, sedimentation, and temperature. The zooxanthellate species manifest reticulate evolution, a mechanism that facilitates hybridization and sibling species but confuses and complicates taxonomy (Veron 1995; Medina et al. 1999; Willis et al. 2006). Tissue color (plant pigments) displays large ranges of variability, contributing to the challenge of in situ species determination.

Zooxanthellate corals are sometimes called "hermatypic " corals, because they construct shallow-water reef communities, whereas azooxanthellate corals, sometimes called "ahermatypic" corals, are usually solitary in habit and thus do not form reefs. But there are many exceptions to these generalizations, one being that deep-water azooxanthellate colonial corals, such as *Lophelia pertusa* and *Madrepora carolina*, may form reefal structures

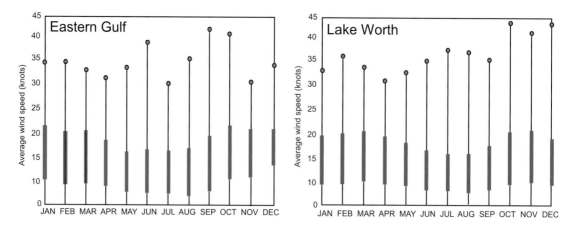

Fɪɢ. 3.8. Wind speed (knots), Eastern Gulf of Mexico and Lake Worth (Palm Beach). Graphs present the maximum (circle at upper end of a line representing the range), minimum (circle at the lower end of a line representing the range), mean (circle with a dot in the middle), ± 1 standard deviation of the mean (red vertical column) (National Data Buoy Center, NOAA)

at continental slope depths (Cairns 2001; Cairns and Stanley 1982; Williams et al. 2006). There are also a few species, e.g., *Astrangia poculata* and *Madracis pharensis*, that occasionally have zooxanthellae in their endodermic tissues (Wells 1973; Peters et al. 1988). *Solenastrea hyades* is reported to expel and regain zooxanthellae on a seasonal basis in Onslow Bay, North Carolina (Wells, personal communication 1975).

The zooxanthellate species found in Florida were first reported and described in pioneering studies of Louis Agassiz (1852, 1869, 1880), Alexander Agassiz (1885), Pourtalès (1880a, b), Vaughan (1901, 1911) and Verrill (1902). Following Smith's 1948 and 1954 publications, understanding of the zooxanthellate species in the Florida area has benefited greatly from studies conducted in the Bahamas, Caribbean, and Mexico using underwater photographs, compiling information on habitat, bathymetric ranges, ecological relationships, and/ or paleontological ranges. Relevant publications include: Voss and Voss (1955) Soldier Key; Voss et al. (1969) Elliott Key (Upper Keys, Florida); Squires (1958), Jaap and Olson (2000) Bahamas; Goreau (1959), Goreau and Wells (1967), Wells (1973), Wells and Lang (1973) Jamaica; Lewis (1960), Ott (1975), James et al. (1977) Barbados; Almy and Carrión-Torres (1963) Puerto Rico; Duarte-Bello (1961), Kühlmann (1971, 1974a, b), Zlatarski and Estalella (1982), González-Ferrer (2004) Cuba; Roos (1964, 1971), Bak (1975)

the Netherlands Antilles; Pfaff (1969), Antonius (1972), Geister (1973), Erhardt and Werding (1975), Werding and Erhardt (1976) Colombia; Porter (1972) Panama; Goldberg (1973) Boca Raton, SE Florida; Scatterday (1974) Bonaire; Adey et al. (1976) Martinique; Ogden (1974), Adey et al. (1977) St. Croix; US Virgin Islands; Roberts (1977) Grand Cayman; Cairns (1982) Belize; Horta-Punga and Carricart-Ganivet (1993), Jordán-Dahlgren and Rodríguez-Martínez (2003) Mexico; Grimm and Hopkins (1977), Jaap et al. (1989), Coleman et al. (2005) Florida Middle Grounds; Halley et al. (2005) Pulley Ridge; Brooks (1963), Davis (1979, 1982), Jaap et al. (1989) Dry Tortugas; Hoffmeister (1974), Jaap (1984), Wheaton and Jaap (1988), Jaap and Hallock (1990) Florida Keys; Zlatarski and Estalella (1982), González-Ferrer (2004) Cuba.

Field guides to shallow-water stony corals (*Millepora* and Scleractinia) of Florida and the Greater Caribbean include Smith (1948, 1971), Zeiller (1974), Voss (1976), Greenberg (1977), Colin (1978), Kaplan (1982) and Humann and DeLoach (2002). Additionally, Littler and Littler's (2000) field guide of Caribbean algae has numerous plates with zooxanthellate species identified.

Smith (1954) reported 40 zooxanthellate species of Scleractinia (Madreporaria) from unidentified Florida records. He misrepresented the following: *Acropora prolifera* is now recognized to be a hybrid and *Astrangia solitaria* is azooxanthelate. *Agarica nobilis* is not recognized; *Porites divaricata* and *P. furcata* are considered forms of *P. porites*; *Colpophyllia amaranthus* is a junior synonym of *C. natans. Manicina mayori* is a form of *M. areolata, Meandrina brasilensis* a form of *M. meandrites,* and *Isophyllyia multifora* a form of *I. sinuosa.* Taking these reductions/revisions into account, Smith's reporting in 1954 includes 31 species from Florida. Subsequently, Jaap et al. (1989) listed 43 species for the Dry Tortugas. Smith (1954) missed species that are relatively common in Florida reefs.

The zooxanthellate species requirements for optimal success were reported by Wells (1956): temperatures greater than 18°C, low turbidity, high solar illumination, stable levels of oceanic salinity, and a solid substratum. Southeastern Florida, Cuba, Bahamas, and southwestern portions of the Gulf of Mexico have more of these conditions in time and space than do the northeastern and northwestern areas of the Gulf. Endemic species are few: *Oculina robusta* is limited to the eastern Gulf of Mexico (Florida Middle Grounds to Dry Tortugas).

The "Florida Reef Tract," as defined by Vaughan (1914), includes the region from south of Soldier Key (25°31.4′ N, 80°10.5′ W) to Dry Tortugas (24°38.4′ N, 82°51.8′ W) but excludes the reefs and hardbottom habitats in Miami -Dade (north of Fowey Rock), Broward, Palm Beach, and Martin counties that continue north to the St. Lucie Inlet at a latitude of 27°10.06′ N. There are 47 zooxanthellate species at Dry Tortugas, 38 at Looe Key, 28 in Biscayne National Park, 24 in the area north of the Miami harbor channel in Miami-Dade County, 36 off Broward County, 24 off Palm Beach County, and 8 in Martin County on the reefs south of the St. Lucie Inlet (Herren 2004). North of the St. Lucie Inlet to Cape Hatteras, the zooxanthellate species fauna is sparse: *Stephanocenia intersepta, Leptoseris cucullata, Solenastrea hyades,* and *Cladocora arbuscula* are reported. Mexican reefs range from 25 species in the Tuxpan region to 33 at Campeche Bank (Beltrán-Torres and Carricart-Ganivet 1999). For Cuba (Sancho Pardo to Rio Camarioca) zooxanthellate species richness ranges from 19 at Sancho Pardo and Punta Seboruco to 29 offshore of the Oceanographic Institute (Zlataski and Estalella 1982). Banks in the northwestern Gulf are most depauperate in zooxanthellate species, e.g., Geyer two, McGrail five, Sonnier seven, and Bright nine (Flower Garden Banks National Marine

Sanctuary faunal records). The Flower Garden Banks reports 23 zooxanthellate species (Flower Garden Banks National Marine Sanctuary faunal records), and the Florida Middle Grounds is known to have 19 zooxanthellate species (Grimm and Hopkins 1977; Jaap et al. 1989; Coleman et al. 2005). Areas in temperate latitudes and deeper depths also have fewer zooxanthellate species. The eastern Gulf region from Tampa Bay to Sanibel Island (the "Hourglass" region) includes 14 zooxanthellate species. The Florida Bay –Gulf of Mexico side of the Keys is relatively impoverished; in 2000, the FKNMS coral reef monitoring program in the Gulf of Mexico found 12 zooxanthellate species at Content Key and 24 at Smith Shoal.

Ubiquitous zooxanthellate species include *Stephanocoenia intersepta, Siderastrea radians, Agaricia agaricites, Porites astreoides, P. porites, Montastraea annularis* complex, and *M. cavernosa.* Rarer species include *Madracis formosa, Oculina robusta,* and *O. valenciennesi.* Species with restricted distribution include *Cladocora arbuscula, Dendrogyra cylindrus,* and *Solenastrea* spp.

The Atlantic zooxanthellate species display a gradient of decreasing similarity based on spatial separation (Fig. 3.9). Areas such as the west coast of Africa and the Azores are most dissimilar to the Gulf and Caribbean, whereas Bermuda and the south Atlantic (Brazil) present a case of intermediate similarity. Brazil has more endemic zooxanthellate species than any region (Neves et al. 2006). This can largely be explained by current patterns within the Gulf and Caribbean, which provide reasonable connectivity for Bahamas, Cuba, Florida, Mexico, and the Caribbean; however, they are not thought to ordinarily connect the eastern, southern, and western Atlantic in the time frame necessary to transport viable larvae. Interestingly, there are a few species that have managed ampi-Atlantic transit (*Madracis decactis, Siderastrea radians, Porites astreoides, P. porites,* and *Montastraea cavernosa*) and are found from Brazil to Bermuda.

Average taxonomic distinctness, $\Delta +$, defined as: $\Delta + = \{\Sigma\Sigma_{i < j}\omega\}/\{S(S - 1)/2\}$; S is the number of species present, i and j are the range of species observed (Clarke and Warwick 1998; Warwick and Clarke 2001); and variation in taxonomic distinctness, $\lambda +$, defined as $\lambda + = 2(s - m)[m(m - 1) (s - 2)(s - 3)]^{-1}$ (Clarke and Warwick 1998; Warwick and Clarke 2001) were evaluated using

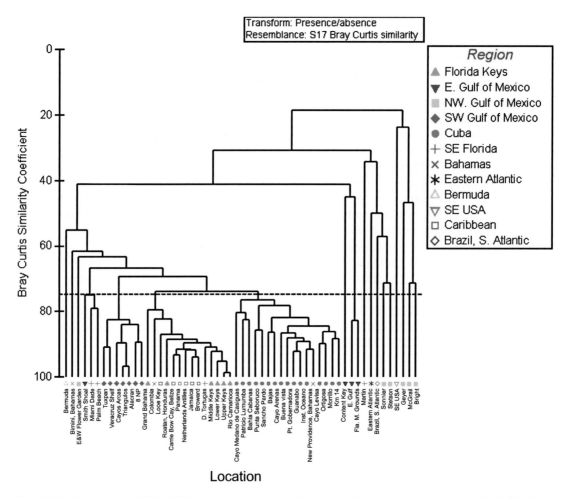

FIG. 3.9. Dendrogram, classifying 56 locations and presence of zooxanthellate species using the Bray Curtis Similarity Coefficient and the dendrogram formed by group average sorting; 75% boundary line inserted on the y-axis

regional distribution of zooxanthellate species, including the regions of the Caribbean, Gulf of Mexico, Florida (Keys, SE Florida), eastern Atlantic, Bermuda, and Brazil (Fig. 3.10). A comprehensive species list of zooxanthellate species for the region was used to compare the individual locality to an expected taxonomic distinctness based on the comprehensive zooxanthellate species list (Warwick and Clarke 2001); the null expectation is that any species present at a specific location behaves like a random selection from the regional species pool. If the average taxonomic distinctness for a specific location is outside the confidence limits, the null hypothesis is rejected. Variation in the taxonomic distinctness idea is that, due to

habitat heterogeneity, anthropogenic or natural disturbance, some species are over-represented and others are under-represented by comparison with the regional zooxanthellate species list. Warwick and Clarke proposed that, under disturbance pressure, species with high taxonomic distinctness (species-poor higher taxa; e.g., a genus with one species or a family with two genera) are typically extirpated before species with lower taxonomic distinctness (Warwick and Clarke 1998).

Clustering (Fig. 3.11) showed that zooxanthellate species distribution at Florida Keys locations was similar to Cuban reefs (75% similarity); Florida Keys reefs also were quite similar to the Mexican Reefs (69% similarity). Ranking by decreasing

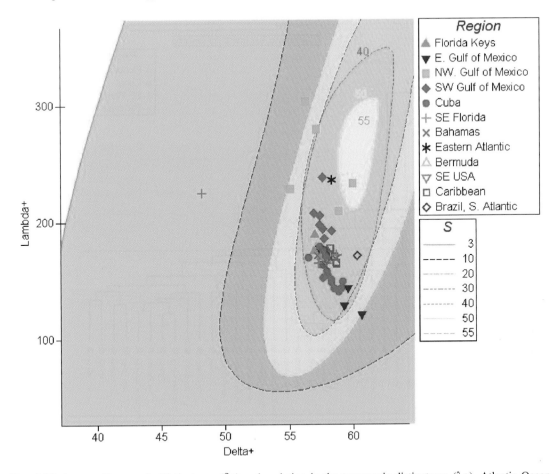

Fig. 3.10. Average Taxonomic Distinctness (δ+) and variation in the taxonomic distinctness (λ+), Atlantic Ocean zooxanthellate species, for 56 locations using 5,000 permutations of these data and a master list of 59 zooxanthellate species for the Atlantic

similarity order is: Cuba, Mexico, Bahamas, SE Florida, E Gulf of Mexico, NW Gulf of Mexico (Fig. 3.11). Martin County was outside the confidence limits of the taxonomic distinctness (Fig. 3.12). The Tuxpan region of Mexico, Dry Tortugas, Looe Key, and Upper Florida Keys in Florida were either on or just outside the confidence limits (Fig. 3.12). The implication is that species assemblages in these areas are slightly reduced in taxonomic distinctness. We suspect that the causes include thermal extremes, habitat degradation and, perhaps, these areas have less hydro-connectivity to the main body of the Caribbean.

Regional analysis reports that zooxanthellate species distribution/diversity is highly similar (82% for Colombia, Belize, Panama, the Netherlands Antilles, Florida Keys, Jamaica, Cuba, and the Bahamas). The eastern Atlantic, Martin County, Brazil, NW and NE Gulf of Mexico show greater difference, whereas Miami -Dade, Broward, and Palm Beach counties are 70% similar to the Florida Keys and Caribbean (Fig. 3.9). The taxonomic distinctness analysis reported that, regionally, most localities are within the confidence limits (Fig. 3.12); however, the southeastern USA and Martin County fell outside the confidence limits. They are both marginal areas for zooxanthellate species prospering. Bahamas, Florida, Mexico, and Caribbean localities are much alike in taxonomic distinctness (Fig. 3.12). Distinctions for the eastern Atlantic, Brazil, NW Gulf of Mexico are clearly quite different. Brazil has the greatest number of endemic species and has the greatest δ + value, 191.75; Bermuda's zooxanthellate species fauna

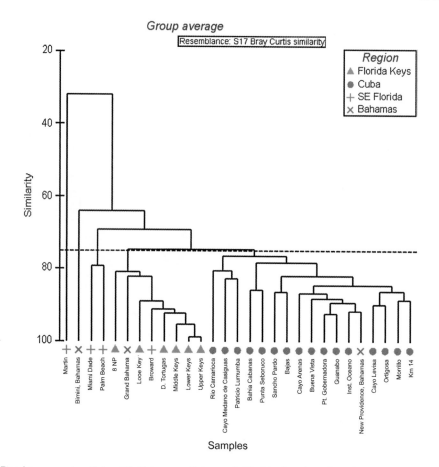

FIG. 3.11. Dendrogram, classifying 29 Bahamas, Cuba, and Florida locations based presence of zooxanthellate species using the Bray Curtis Similarity Coefficient and the dendrogram formed by group average sorting, 75% boundary line inserted on the *y*-axis

is impoverished and has the lowest δ + value, 179.05.

3.10   Florida Keys Coral Reefs

The seaward-most reefs in the Florida Keys are a series of bank reefs separated by expanses that lack coral reef development. In the Holocene reef development, *Acropora palmata* was a principal builder of Florida reefs (Shinn et al. 1977). In the early to mid-1970s, *A. palmata* was common or abundant on most of the outer reefs; however, due to hurricanes and disease, it has become non-extant on many offshore bank reefs.

A prominent feature on many outer reefs (Grecian Rocks, Molasses, Sombrero, Looe Key, western Sambo, and Rock Key) is a distinct spur and groove system (Shinn 1963; Shinn et al. 1981). These ridges are constructional (Shinn et al. 1981) from coral growth and deposition, while the grooves (surge channels) facilitate water and sediment transport. The seaward margin of the spurs is typically around 10 m below sea level; the spur may be continuous to the reef flat or it may be interrupted by a gap connecting adjacent surge channels. Zonation along the spur follows the depth (light) and wave energy gradients. Seaward, the spur horizontal surface is typically a series of massive corals (*Montastraea* spp., *Diploria* spp., and *Colpophyllia natans*); the vertical spur walls often have plates of *Agaricia* spp., *Leptoseris cucullata*, and *Mycetophyllia* spp.; shallow caverns and shaded overhangs are colonized by *Scolymia* spp., *Agaricia fragilis*, *Mussa angulosa*, and *Madracis pharensis* (Wheaton and Jaap 1988). As the depth decreases, faunal elements change; at

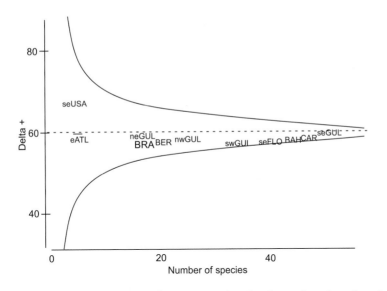

Fig. 3.12. Average taxonomic distinctness values, δ+, plotted against the observed number of species for the Atlantic zooxanthellate species at 56 locations. Dashed lines indicate the simulated mean δ+ from 5000 random selections from the master list of 59 zooxanthellate species. Intervals (solid lines) within which 95% of the simulated δ+ lie (the expected range of δ+ for a given number of species) are constructed for each of the 56 locations and represented as a probability funnel (Warwick and Clarke 2001). seUSA = southeast USA, eATL = eastern Atlantic, neGUL = NE Gulf of Mexico, BRA = Brasil, BER = Bermuda, swGUL = SW Gulf of Mexico, seFLO = SE Florida, BAH = Bahamas, CAR = Caribbean, seGUL = SE Gulf of Mexico

approximately 3 m depth, the surface of the spur tops are in a surge zone with the waves breaking and retreating seaward, resulting in turbulent water. Tops of the spurs may be veneered with *Acropora palmata*, *Millepora complanata*, and *Palythoa caribea*; this was described by Geister (1977) as a shallow spur and groove community, common in the Caribbean –western Atlantic. As the depth decreases to less than 1 m, the surge of water makes it impossible to swim during windy periods. The organisms that characterize this shallowest portion of the spur are *Palythoa caribea*, *Millepora complanata*, *Zoanthus pulchelus*, and *Porites astreoides*. These organisms are very tolerant of heavy wave energy, intense sunlight, and exposure-desiccation during spring low tides. Inshore of the spur and groove system, the spurs coalesce to form a reef flat – a field of rubble material that is loosely held together by calcareous algae and naturally forming marine cements.

The grooves-surge channels usually have a carpet of sediment composed of broken bits of carbonate material and conspicuous plates of *Halimeda*. Following a severe storm, such as a hurricane,

the sediment may be completely washed out of the grooves exposing large, sub-recent corals, the foundation of the reef.

The spur and groove system is absent on many outer reefs (Carysfort, Conch, Alligator, Tennessee, Maryland Shoal, Pelican Shoal). These reefs may be too young to have developed such a system or they may have failed to keep pace with sea level rise. A prominent shoal at or near sea level, and a variety of morphological attributes, lacking in a common pattern, characterizes these reefs.

The deeper spur and groove is a prominent feature of some reefs. A good example is Bird Key Reef, Dry Tortugas National Park (Jaap et al. 1989); it is 2 km long, 400 km wide, and the spur and groove system is restricted to 6–21 m depth; shallower areas (0–5 m) are low-irregular relief, populated with numerous octocorals. Deep spur and groove formations are interpreted as a situation where the reef growth failed relative to rising sea level. Octocorals are a prominent element in deep spur and groove habitats.

The most common reef type in the Florida Keys is the patch reef. These are widely distributed inshore of the bank reefs. Some are very close to shore; for example, there is a cluster of high relief patch reefs 0.9 km (0.5 nmi) off Big Munson Island, New Found Harbor. Although most of the patch reefs are unnamed, named patch reefs of some notoriety include a cluster of patch reefs at Bache Shoal and Margot Fish Shoal off Elliott Key; two clusters of reefs off Upper Key Largo in John Pennekamp Coral Reef State Park (Mosquito Banks and Basin Hill Shoals); the Rocks and Hen and Chickens (off Plantation Key); the Garden (off Lower Matecumbe Key); Coffins Patch, East and West Turtle Shoal, East Washerwoman Shoal (off Vaca Key); West Washerwoman (off the Saddlebunch Keys); Western Head (off Key West); and Texas Rock, Iowa Rock, Middle Ground (in Dry Tortugas).

Patch reefs have considerable variation in morphology and size. High-relief patch reefs are constructed of massive-framework corals (*Colpophyllia natans*, *Diploria* spp., *Montastraea* spp., and *Siderastrea siderea*). In some patch reefs, the massive corals are on the periphery of the reef and the center of the reef is a dense forest of octocorals and smaller Scleractinian species (*Porites porites*, *Meadrina meandrites*, *Dichocoenia stokesi*). Corals may be in close proximity or widely dispersed. The epitomized patch reef is dome shaped, circular in outline, surrounded by sedimentary-sea grass habitat, and 30–700 m in diameter (Jaap 1984). A good example is the aptly named Dome Reef in Biscayne National Park (Jaap et al. 1989). Here the octocorals tend to be more specious and abundant than the Scleractinian corals (Table 3.3).

TABLE 3.3. Species abundance, Dome Reef, Biscayne National Park, 1977 (From Jaap 1989).

Species	Category	Number of colonies	Percent abundance	Cumulative abundance
Plexaura homomalla	Octocoral	49	17.63	17.63
Plexaura flexuosa	Octocoral	32	11.51	29.14
Pseduplexaura porosa	Octocoral	30	10.79	39.93
Pseudopterogorgia americana	Octocoral	26	9.35	49.28
Pseudopterogorgia acerosa	Octocoral	25	8.99	58.27
Pseudoplexaura flagellosa	Octocoral	12	4.32	62.59
Eunicea caliculata	Octocoral	12	4.32	66.91
Briareum asbestinum	Octocoral	12	4.32	71.22
Siderastrea siderea	Scleractinian	10	3.60	74.82
Gorgonia ventalina	Octocoral	10	3.60	78.42
Eunicea tourneforti	Octocoral	9	3.24	81.65
Muriceopsis flavida	Octocoral	5	1.80	83.45
Plexaurella grisea	Octocoral	4	1.44	84.89
Plexaurella fusifera	Octocoral	4	1.44	86.33
Muricea elongata	Octocoral	4	1.44	87.77
Eunicea succinea	Octocoral	4	1.44	89.21
Dichocoenia stokesi	Scleractinian	4	1.44	90.65
Muricea atlantica	Octocoral	3	1.08	91.73
Millepora alcicornis	Hydrozoan	3	1.08	92.81
Eunicea fusca	Octocoral	3	1.08	93.88
Porites porites	Scleractinian	2	0.72	94.60
Porites astreoides	Scleractinian	2	0.72	95.32
Montastraea cavernosa	Scleractinian	2	0.72	96.04
Eunicea laciniata	Octocoral	2	0.72	96.76
Diploria labyrinthiformis	Scleractinian	2	0.72	97.48
Agaricia agaricites	Scleractinian	2	0.72	98.20
Plexaurella nutans	Octocoral	1	0.36	98.56
Montastraea annularis	Scleractinian	1	0.36	98.92
Favia fragum	Scleractinian	1	0.36	99.28
Eusmilia fastigiata	Scleractinian	1	0.36	99.64
Eunicea clavigera	Octocoral	1	0.36	100.00
	Total	278		

Above data, from two, 25 m long transects

3.11   Tortugas Banks

A massive complex of reefs (the Tortugas Banks) in 70–90 ft (21–27 m) is situated west of Loggerhead Key, Dry Tortugas (Jaap et al. 2001). These banks were created during a lower sea level, and it is thought that they were unable to keep pace with rising sea level following the Wisconsin glacial epoch (Mallinson et al. 2003). Growth was retarded after the water became too deep for light penetration to stimulate vigorous coral activity. A region referred to as Sherwood Forest is characterized by a low relief foundation of Swiss cheese-like karst limestone and moderate sized platy and mushroom-like *Montastraea cavernosa* corals overlying the foundation. Black Coral Rock is a high relief structure with peaks protruding from the seafloor. Multiple peaks and valleys provide a mosaic of horizontal, vertical, and angled surfaces, offering niches to a wide variety of plants and animals, including colonies of *Antipathes* spp. (black coral) that is listed as locally extinct in Florida (Deyrup and Franz 1994). Black corals are moderately abundant in the Tortugas Banks.

3.12   Epibenthic Hardbottom Communities

Hardbottom habitat was identified by Davis (1982) as major bottom type. He reported 39.65 km² of octocoral-covered hardbottom within Dry Tortugas National Park (4.08% of the seafloor in the park). Throughout the Keys on both coasts, this is a very abundant and conspicuous habitat type. It is characterized by a great number of sponges and octocorals (sea whips, sea plumes, sea fans), and the topography is rather flat. Octocoral species density at a monitoring station at Pulaski Shoal was 15.50±3.50 species and 92.60 ± 31.74 colonies/m². The area resembles a jungle with the sea floor totally obscured by the octocoral canopy. Octocoral hard grounds have a rich diversity in other species that use the canopy for refuge, to seek prey, and to breed.

3.13   Sedimentary Habitats

Sedimentary habitat is important to coral reefs; a large portion of the Dry Tortugas sea floor is composed of sediments (silt, sand, gravel); Davis (1982) estimated that sediments made up the largest component (108.92 km² or 47.80%) of the benthic habitat in the Dry Tortugas National Park. If sea grasses are included (because they grow in sediments), the sediment benthic contribution in the Tortugas is 78%. Sedimentary habitats provide niches for virtually every marine phyla, thus the biodiversity of these habitats is relatively high. Because the organisms live (for the most part) under the surface of the sediments, there is a misconception that this area is barren (Cahoon et al. 1990; Snelgrove 1999). Diatoms, protozoa, molluscs, crustaceans, echinoderms, ploychaetes, gobies, and blennies are examples of taxonomic categories that are found in the sediments. The sediments are a forage area for larger predators (Cox et al. 1996) and are a pool of recyclable calcium carbonate.

3.14   Status and Trends in Florida Keys Reefs

The FKNMS Water Quality Protection Plan (initiated by EPA and FKNMS in 1996) monitoring program utilizes repeated sampling; random, stratified protocols were used to select monitoring sites with replicate sampling stations. Data from 1996 though 2004 documents that disturbance events (bleaching, black water, and hurricanes) are a principal cause of change in these coral reef communities. Sampling protocols (quality assurance/quality control plans) were approved by US Environmental Protection Agency in 1996 to generate four annual station level data products:

1. A species list of all stony corals (*Millepora* and Scleractinia)
2. A qualitative list of conditions affecting the vitality of the scleractinian corals (bleaching and diseases by coral species)
3. Benthic cover based on video imagery
4. A census of *Diadema antillarum*.

For detailed sampling protocols see Porter et al. (2002) and www.floridamarine.org.

From 1996 to 2004 stony coral species numbers declined in virtually all of the spatial strata. Losses outnumbered the gains and unchanged status. The taxonomic losses for habitat types were 72–73%, except for hardbottom (55%). Hardbottom sites generally were taxonomically depressed in species, and those species that are found (e.g., *Stephanocenia intersepta*, *Siderastrea radians*, *Porites astreoides*) in these habitats are often less sensitive to environmental perturbations (temperature and turbid-

ity). Upper Keys stations had a higher percentage loss than Middle and Lower Keys stations. Dry Tortugas taxonomic losses were similar to Upper Keys stations. Gains and unchanged status were greatest for the hardbottom stations.

The confidence interval predicts that the temporal flux of station species richness is described by plus or minus four to five taxa; however, losses were greater than gains and some stations lost as many as 13 taxa from 1996 to 2003 (Fig. 3.13); very few stations gained species.

Thirty-four species declined in occurrence, seven increased, and four remained unchanged (Fig. 3.14). Decline in taxonomic richness in stony corals was consistent for all habitats (Fig. 3.15).

Sanctuary-wide, the number of stations where *Acropora cervicornis* and *Scolymia lacera* were present decreased significantly ($\alpha = 0.05$) while *Copolphyllia natans*, *Madracis mirabilis*, *Porites porites*, *Siderastrea radians*, *Mycetophyllia ferox* and *M. lamarkiana* showed decreases at the $\alpha = 0.1$ level. Only *Siderastrea siderea* was observed at

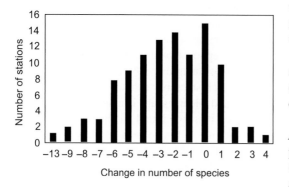

FIG. 3.13. Change in number of species observed at CREMP monitoring stations, contrasting 1996 and 2003

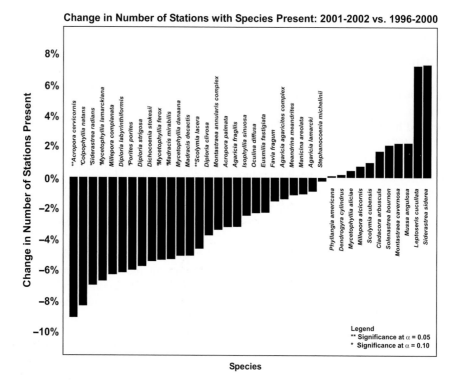

FIG. 3.14. Number of CREMP stations that lost or gained particular species of corals between 1996 and 2003. A negative number, such as −5, reports that the species occurred at five fewer stations in 2003 than in 1996

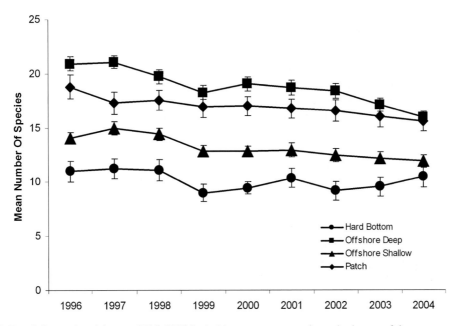

FIG. 3.15. Trends in species richness, 1996–2003 by habitat type, mean and standard error of the mean

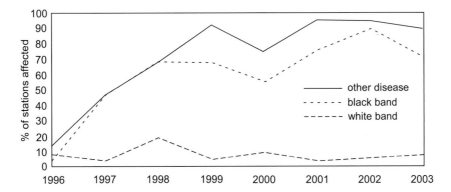

FIG. 3.16. Trends in coral disease, 1996–2003, compiled by the occurrence of the disease at a station

a significantly greater number of stations in 2001 and 2002 than in previous years.

Black band disease (Rützler and Santavy 1983) was the least common of the conditions monitored; the incidence of black band was slightly higher in 1998 and has wavered at low levels in subsequent years (Fig. 3.16). *Colpophyllia natans*, *Montastraea annularis*, *M. cavernosa*, and *Siderastrea siderea* were infected by black band disease. This disease is more active in the summer and is positively correlated with nutrient enrichment (Voss and Richardson 2006).

The category "white disease" includes white plague and white pox which manifest, not surprisingly, a white appearance (Patterson et al. 2002). White plague was first identified in the Florida Keys in 1977. The disease is characterized by an abrupt band of white that separates living coral tissue from exposed skeleton. The disease usually initiates at the base of the colony and travels upward. Three types of white plague have been reported. Type I plague affects ten species of corals, causing tissue mortality at a rate of about 3 mm/day. Type II plague affects 32 species of corals and can cause up to 2 cm of tissue mortality per day. Type III plague affects the large reef building corals (*Colpophyllia natans* and *Montastraea* spp.) and can cause tissue loss at a rate much greater than either the Type I or Type II (Richardson 1998). White pox disease affects Elkhorn coral (*Acropora palmata*) in the Florida Keys. It was first found in 1996 and is characterized by white circular lesions on the surface of infected colonies. Tissue loss and mortality may be very rapid; the cause of the disease is still unknown.

In 1996, white disease was recorded at five stations; in 2002 it was present at 90 stations. *Agaricia agaricites* complex was not infected with white disease in 1996 but was observed at 33 stations in 2002. *Montastraea annularis* complex followed the previous pattern very closely; no reports in 1996 and 32 stations reporting infection in 2002. Other conditions include several suspected diseases; an example is the purple spot on *Siderastrea siderea*. Fourteen species exhibited increased infection by other diseases/conditions: *Agaricia agaricites*, *Colpophyllia natans*, *Dichocoenia stokesii*, *Eusmila fastigiata*, *Favia fragum*, *Meandrina meandrites*, *Millepora alcicornis*, *M. complanata*, *Montastraea cavernosa*, *M. annularis* complex, *Porites astreoides*, *P. porites*, *Siderastrea sidera*, and *Stephanocenia michelinii*.

A *Diadema antillarum* census recorded a total of 188 urchins between 1996 and 2003, with the greatest number in 2001 and the fewest in 1996. Western Sambo is the site with the greatest cumulative number of *Diadema*: 47 were observed over the 7-year period. Since the *D. antillarum* demise in 1983, recovery has been exasperatingly retarded. Sites in Dry Tortugas, Middle, and Lower Keys had more *Diadema* than the other sites (White Shoal, Western Sambo, Jaap, West Turtle Shoal). *Diadema* numbers testify that recovery is slow with a very slight increase in numbers from 1996 through 2003; densities range from 0.02 to 0.13 m².

Between 1996 and 2002, sampling stations exhibited a 38% decline in stony coral cover. Declines were most noteworthy between 1997 and 1998 (11.3–9.6%) and 1998 and 1999 (9.6–7.4%). The decline from 1996 to 1999 was significant *p*-value of 0.03, Wilcoxon rank-sum test. Since 1999 the mean percentage cover for all stations has wavered between 7.4% and 7.5% and is statistically insignificant. Hypothesis testing by geographic

area (Upper, Middle, and Lower Keys) and habitat type (patch reef, shallow and deep reefs) exhibit the same trends (declining coral cover) as observed in the sanctuary-wide analysis.

In 1996, the most cover was contributed by *Montastraea annularis*, *M. cavernosa*, *Acropora palmata*, *Siderastrea siderea*, *Millepora complanata*, and *Porites astreoides*. The average cover per station for *M. annularis* was 4.6% in 1996 and 2.7% in 2002. *Acropora palmata* average cover declined from 1.1% in 1996 to 0.1% in 2002 (91% reduction). The Staghorn coral, *A. cervicornis,* was rare in 1996 (0.2% cover) and had nearly disappeared by 2002 (0.01%).

3.14.1 Episodic Disturbances

Bleaching episodes occurred in summer and fall 1997 and 1998. The areas most influenced were the shallow offshore sites. The water temperature at the Sea Keys C-Man Station at Sand Key peaked near 32°C on 10 August 1997 (unfortunately, the temperature recorder failed on 11 August). Temperatures were high enough to cause zooxanthellae expulsion, discoloring many of the zooanthids, fire coral, stony corals, and some octocorals such as *Briareum* spp. The organisms that were stressed the most by bleaching were *Millepora complanata* and *Palythoa caribea*. These species tend to expel their zooxanthellae at a slightly lower threshold than the other zooxanthellate organisms (Jaap 1979). *Millepora complanata* cover decline was greatest between 1998 and 1999 (Fig. 3.17). There was a slight improvement in percentage cover and frequency of occurrence after 2001. The 1997 bleaching event is suspected to have stressed *M. complanata*, and a second exposure to hyperthermia in 1998 also reduced the population.

The golden sea mat, *Palythoa caribea*, is abundant and conspicuous in shallow reef, high energy communities (Geister 1977). The Coral Reef Environmental Monitoring Program (CREMP) analysis pools all zooanthids (*Zoanthus* spp., *Palythoa* spp., *Ricordia* spp.) as a single category; however, point count experience is that virtually all zoanthids observed in these images are *P. caribea*. Unlike the fire coral, *Millepora complanata*, *P. caribea* showed little change in cover after the bleaching disturbance (Fig. 3.18). There was a slight reduction in the mean percentage cover

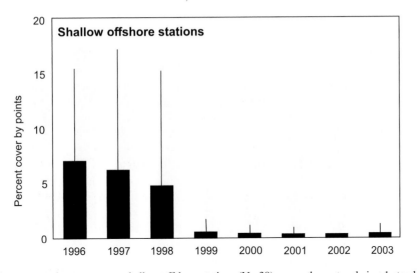

FIG. 3.17. *Millepora complanata* cover at shallow offshore stations (N=39) mean, the rectangle is ±1 standard deviation, and vertical line is the range

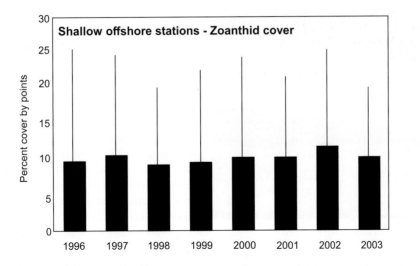

FIG. 3.18. Zooanthid cover (*Palythoa mammillosa*) at shallow offshore stations (N=39). The mean, the rectangle is ±1 standard deviation, and vertical line is the range

between 1997 and 1998; however, 2000 and subsequent years, the coverage levels were equal to or exceeded the pre-bleaching period.

Mustard hill coral (*Porites astreoides*) is also common in the shallow reef communities and creates small mounds and encrustations. This coral also was resilient and did not suffer population reduction from the bleaching episode (Fig. 3.19). We are cognizant that *P. astreoides* will expel zooxanthellae, but it has the capacity to recover after hyperthermic conditions dissipate.

Hurricane Georges crossed the Straits of Florida near Key West on 25 September 1998. The Sombrero Key C-MAN buoy (SMKF1) recorded a maximum sustained wind of 82 knots with a peak gust to 92 knots at 1500 UTC 25 September. Hurricane Georges's greatest influence on coral reef communities was between Sombrero Key and Dry Tortugas. The storm's impact was evidenced by the change in *Acropora palmata* cover (Figs. 3.20 and 3.21). All of the descriptive statistics (range, mean, and frequency of occurrence) declined after

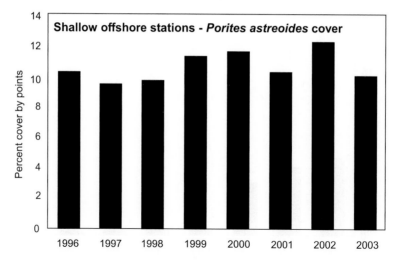

FIG. 3.19. *Porites astreoides* cover at shallow reef stations (N = 39). The mean, the rectangle is ± 1 standard deviation, and the vertical line is the range

FIG. 3.20. Mosaic image of Western Sambo Shallow Station 2 (approximately 4 m of middle video transect, 4) 1996 and 2002 (Image product of Ravenview application software)

Hurricane Georges. We sampled before the hurricane struck in 1998, thus the major decline is most noticeable in 1999 and subsequent years.

Western Sambo station two exhibited the greatest *Acropora palmata* cover: 15.28% in 1996 and 16.34% in 1997. The mosaic image (Fig. 3.20) created from video at this station highlights the severity of the loss. From 1996 to 1998, there was a large thicket of *A. palmata* on station two; it disappeared after 1999, and recruitment into the area had not occurred through 2004.

The National Hurricane Center reported that at 2300 UTC on 14 October 1999, Irene reached hurricane status over the Florida Straits. The center moved over Key West at 1300 UTC the next day. The hurricane force winds were concentrated east of Irene's center over the Lower to Middle Florida Keys. Irene made its fourth landfall near Cape Sable and then moved across southeast Florida bringing sustained 39–73 mph winds and 10–20 in. of rainfall. Irene's sustained and peak wind gusts were less than those of Georges. This second hurricane in a year disturbed the offshore shallow reefs, but, since Hurricane Georges had already reduced the *Acropora palmata* populations, Hurricane Irene's influence was muted.

In early 2002, a body of dark-colored water was reported between Marco Island and Key West (Hu et al. 2003). Hu and colleagues obtained satellite imagery data from the Advanced Very High Resolution Radiometer and the Sea Viewing Wide Field-of-view Sensors. Water-leveling radiance and ocean color index were used to create a daily image time series, which documented the water mass anomaly from January through April 2002 (Hu et al. 2003). High concentrations of Rhizosoliniaceae diatoms and the toxic dinoflagellate *Karenia brevis* (red tide) were found in water samples. This phytoplankton bloom extended over two CREMP sampling sites: Content Keys and Smith Shoal. The 2002 and 2003 data for these sites document

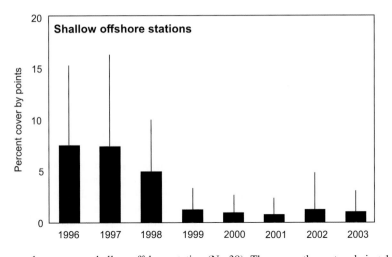

FIG. 3.21. *Acropora palmata* cover, shallow offshore station (N = 39). The mean, the rectangle is ± 1 standard deviation, and vertical line is the range

a reduction in the coral cover at both sites. The Content Key site is a hardbottom site that has had consistently low species richness and cover since 1996. The Smith Shoal site is a high profile patch reef ; it exhibited cover losses in all but one species after the black water event, making it probable that the black water anomaly caused the coral decline. Red tide has previously resulted in severe losses in the eastern Gulf of Mexico sponge-octocoral–stony coral communities (Smith 1971, 1976).

3.14.2 Multivariate Analyses

The 1996–2003 Florida Keys National Marine Sanctuary coral reef monitoring database includes files for 1,360 stations, 4,080 transects, and records for 2,400,000 points related to 200,000 images. The multivariate statistical approach (Clarke 1993; Clarke and Green 1988; Clarke and Gorley 2005; Clarke et al. 2005) exhibits and contrasts the different sources of variation in assemblage structure by strata: hardbottom, shallow, deep and patch reefs; geographical region: Dry Tortugas, Lower, Middle and Upper Keys; site within geographic regions; and across years within each site.

There is a general trend of similarity in assemblage structure from patch reefs through deep reefs to shallow reefs. Hardbottom assemblages tend not to overlap with those on reefs. While some deep reefs, such as Looe Key, eastern Sambo, Tennessee,

and Rock Key, are relatively similar in terms of the community of stony corals inhabiting them, others such as Molasses and Alligator differ from them substantially. Shallow reefs tend to be less similar to each other than deep reefs and exhibit the greatest degree of spatial variability. The major pattern is one of difference between the strata. Analysis of Similarities (ANOSIM) confirms this observation, as differences between strata are highly significant (Global R = 0.458, $p < 0.001$, for all pair-wise comparisons $p < 0.003$).

Within shallow reef assemblages (Fig. 3.22a), there is little evidence of a trend in community structure related to each reef's position along the Keys, as assemblages from shallow reefs in the Upper and Lower Keys tend to be more similar to each other than to those from the Middle Keys. A similar lack of geographic pattern is shown for deep reefs (Fig. 3.22b) and patch reefs (Fig. 3.22c). No significant differences were found using ANOSIM tests for differences in average assemblage structure between regions within reef types ($p > 0.167$).

Changes through time at each site (Fig. 3.23) reveal the effects of man-made and natural changes in the Florida Reef Tract. It is readily apparent that the changes in some reefs are greater than the changes in others. For example, among shallow reefs in the Lower Keys (Fig. 3.23a), changes over time at Looe Key and western Sambo are similar

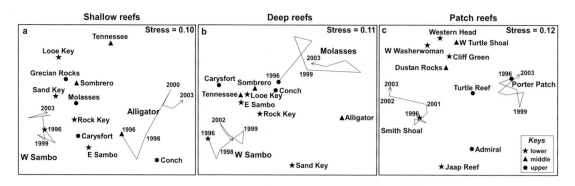

FIG. 3.22. Ordination by MDS of Bray-Curtis similarities between fourth-root transformed percentage cover data for stony corals from each site within: (a) shallow reefs; (b) deep reefs; (c) patch reefs. Sites are grouped according to their location along the reef tract (Lower, Middle and Upper Keys). Data from all sampling occasions are averaged for most sites, except for two within each plot for which samples from different times are kept separate in order to illustrate the relative magnitude of changes in time compared to differences between locations

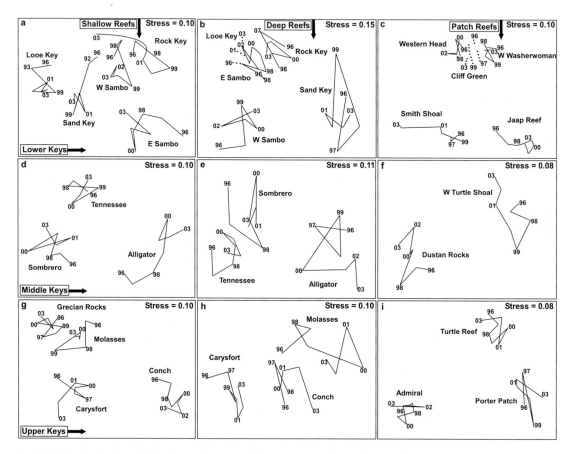

FIG. 3.23. Ordination by MDS of Bray-Curtis similarities between fourth-root transformed percentage cover data for stony corals from each site within: (a) shallow reefs in the Lower Keys; (b) deep reefs in the Lower Keys; (c) patch reefs in the Lower Keys; (d) shallow reefs in the Middle Keys; (e) deep reefs in the Middle Keys; (f) patch reefs in the Middle Keys; (g) shallow reefs in the Upper Keys; (h) deep reefs in the Upper Keys; (i) patch reefs in the Upper Keys. Sites are grouped according to their location along the reef tract (Lower, Middle and Upper Keys). Data from each sampling occasion are the average percent cover across all samples at each site. Lines join consecutive samples from each site, and are intended as an aid to interpreting changes through time (time-trajectories) at different sites

in magnitude and direction, but less than changes at Sand Key, Rock Key and eastern Sambo. Sand Key shows a marked change between 1998 and 1999, which may be related to the passage of Hurricane Georges or the bleaching event, whereas at Rock Key these events did not produce marked changes, but something happened between 2002 and 2003 that changed assemblages at the site substantially. Changes through time at eastern Sambo are generally larger than changes at Looe Key and western Sambo, but there is no evidence of a major shift in community structure in any particular year. Among deep reefs in the same areas (Fig. 3.23b), eastern Sambo, Looe Key, Rock Key and western Sambo show variations through time, which are similar in magnitude, though essentially random in direction, while changes at Sand Key tend to be greater. Large changes in community structure at Sand Key occurred between 1996 and 1997, between 1998 and 1999, and between 1999 and 2000, as the community returned to a state similar to that observed in most other years. Among patch reefs in the Lower Keys, small, essentially random, temporal variation is seen at the western Head, Cliff Green and west Washerwoman sites (Fig. 3.23c). In contrast, at Smith Shoal and Jaap Reef relatively large changes are seen in some years, especially between 1999 and 2000 and between 2001 and 2002 at Smith Shoal, and between 1996 and 1997, and between 1998 and 1999 at Jaap Reef. Although changes in some years are greater than in others, assemblages on both these reefs show diverging linear trends such that differences between the two reefs at the start of sampling, in 1996, are very much less than differences between them in 2003.

3.15   A Case Study: Dry Tortugas

The Dry Tortugas is the western terminus of the Florida Reef Tract and lies 100–112 km west of Key West, Florida; the geographic coordinates are 24°33'''–24°44''' N latitude, 82°46'''–83°15' W longitude (Fig. 3.24). At the convergence of the Gulf of Mexico, Caribbean Sea, and the Atlantic Ocean, the area is characterized as a mosaic of coral reefs, sedimentary shoals, sea grass meadows, and small islands. A 100 mile^2 Dry Tortugas National Park is the prominent feature (Fig. 3.24).

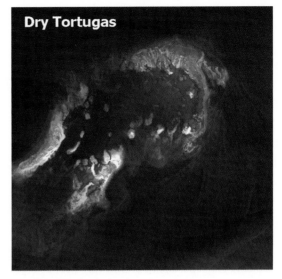

FIG. 3.24. The Dry Tortugas, southernmost Florida

3.15.1   Environmental Setting

Water temperatures at the Loggerhead Key Lighthouse, recorded between 1879 and 1907, ranged from 17.9°C to 31.1°C (Vaughan 1918). The most extreme cold temperature at Dry Tortugas occurred in the winter of 1977–1978, when a cold front chilled the waters to 14°C and caused a massive die off of *Acropora cervicornis* (Davis 1982). Extremely warm temperatures (>32°C) are uncommon but do occur during doldrums (Jaap, unpublished thermograph data, 1988–1996). The National Park Service sponsored a reef-monitoring project at DTNP from 1989 to 1997. At Pulaski Shoal, in a 30 ft (10 m) depth, from 28 August 1993 to 30 August 1994, the temperature ranged from 20.13°C to 30.35°C (hourly thermograph data). During a coral bleaching event spanning 31 August to 13 September 1991, the extreme high temperatures ranged from 30.4°C to 31.5°C. From December through April, the extreme low temperatures should range from 17.9 to 21.3 based on Vaughan's lighthouse data.

Salinity off the Dry Tortugas typically ranges from 37 to 32 parts/1,000. Lower salinity values result from entrained Mississippi River freshwater runoff (Ortner et al. 1995; Gilbert et al. 1996).

Waves are a function of wind speed, duration, and fetch. During the period of frontal passages, wave heights can exceed 3.1 m. Currents that influence this area are tidal and episodic gyres from the Loop Current. Tides are semidiurnal or mixed; two tide current stations are listed for Southwest and Southeast Channels, and maximum current speed listed is 51.44 cm s^{-1} (Tide Current Tables 2002).

Two major currents influence the Tortugas area: the Florida Current in the Straits of Florida and the Loop Current that sweeps into the eastern Gulf of Mexico from the Yucatan Straits and reenters the southern Straits of Florida to join the Florida Current. Lee et al. (1994) report that gyres propagate and move off the Tortugas as the Loop Current reenters the southern Straits of Florida. The gyre is important because of larval transport as well as influencing the thermal regime because of cold water in the gyres. The cooler water can be detected at 20 m; a thermograph positioned at Bird Key Reef detected a cold-water gyre from 7 to 11 August 1988. Water temperature pulsed from 22.5°C to 29°C (Jaap et al. 1991). Loop Current gyres or eddies routinely occur on the Tortugas Banks. In 2002 (May 9–15), while the FMRI Coral Reef Monitoring Project group was sampling at Black Coral Rock, the current speed was so strong that our research team could not work underwater. During the March 1993 "storm of the century," current speeds ranged from 42.2 to 50.2 cm s^{-1} at an oceanographic buoy in 30 m of water between Southwest and Southeast Channels. This is similar to maximum tide speeds in SW Channel (51.44 cm s^{-1}).

3.15.2 Early Research

As previously mentioned, natural history expeditions to the area in the nineteenth century include Louis and Alexander Agassiz and Louis Pourtalès. The greatest contribution in documenting marine benthic resources during this era is a map of submerged habitats published by Alexander Agassiz (1882). In 1905, when the Carnegie Institution established a laboratory on Loggerhead Key, Dry Tortugas (Colin 1980; Stephens and Calder 2006) this facility became a leader in early twentieth-century marine research, studying the biology, geology, and environmental conditions of the Dry Tortugas and adjacent area (Davenport 1926; Colin 1980; Stephens and Calder 2006). Papers of the Tortugas Laboratory published by the Carnegie Institution, Washington, DC, contain a complete set of the publications resulting from this research.

Seminal coral reef work includes: Vaughan (1911, 1914–1916), Mayer (1914, 1918), and Wells (1932). Vaughan studied the growth rate of corals and the geologic history of coral reefs; Mayer did extensive studies on the physiology of medusa and temperature tolerance of corals. Subsequent publications on Tortugas coral reefs include Shinn et al. (1977), Thompson and Schmidt (1977), Davis (1979, 1982), Halley (1979), Dustan (1985), Jaap et al. (1989), and Jaap and Sargent (1993). Schmidt and Pikula (1997) published an annotated bibliography of scientific studies within Dry Tortugas National Park. An excellent history of the Dry Tortugas island dynamics is found in Robertson (1964). Robertson reported that Bird Key was a major island with a large rookery of terns (documented by Audubon in 1832). Severe hurricanes in 1910 and 1919 destroyed the vegetation (bay cedars, many reaching 8 ft high). This was followed by chronic erosion of the island, and, in 1929, the Audubon warden abandoned his house on Bird Key and moved to Garden Key. It is a possibility that the hurricane and subsequent erosion of Bird Key was partially responsible for the loss of coral in the shallow portions of Bird Key Reef.

3.15.3 Habitat Description

The major reef types at Dry Tortugas include bank reefs, patch reefs, and thickets of Elkhorn and Staghorn corals. The Elkhorn corals today are situated in the middle of 5 ft channel, the sea floor is limestone with virtually no sediment. The Staghorn coral thickets are situated on sedimentary rubble at White Shoal and west of Loggerhead Key. Benthic habitat mapping began over a century ago and continues today (Agassiz 1882; Davis 1982; Florida Marine Research Institute 1998; Franklin et al. 2003). In a recent mapping effort, Franklin and his team classified approximately 200 km^2 of previously unmapped habitat. Habitats were classified into nine reef and hardbottom habitats ranging from low-relief hardbottom habitats to high-relief coral pinnacles/reef knolls.

Over the past 30 years, coral communities in the Dry Tortugas have trended downward in coral cover and coral diversity (Jaap et al. 2002). The once abundant Elkhorn coral (*Acropora palmata*) assemblages have virtually disappeared from the area (Davis 1982; Jaap and Sargent 1993). Staghorn

(*A. cervicornis*, *A. prolifera*) populations have also declined due to hypothermic stress (Roberts et al. 1982) and a virulent disease of unknown etiology (Peters et al. 1983). More recently the Coral Reef Environmental Monitoring Program (CREMP) has documented a decrease of mean percentage stony coral cover at several Tortugas reefs between 2001 and 2003. This decrease was largely attributed to losses in two stony coral species, *Montastraea annularis* complex and *Colpophyilla natans*, and is most likely the result of an unknown disease infecting these two corals.

Bird Key Reef, in the southern portion of the park, was intensely studied during the 1975/76 TRACTS Program (Shinn et al. 1977; Jaap et al. 1989), including the stony coral fauna community structure. *Montastraea annularis*, *M. cavernosa*, and *Siderastrea siderea* were the principal framework builders on this reef. Coral diversity, cover, and habitat complexity increased with depth. Coral cover (as determined by linear measurement) was highest in depths between 9 and 13 m. If this is used as the baseline for temporal trends, the patterns of change are remarkable. Average coral cover was 47% in 1975, declined to 21–28% from 1989 to 2001, and in 2004 ranged from 12% to 14% (Fig. 3.25). The decline was caused by winter cold fronts, bleaching, coral diseases, and hurricanes.

The trends in Florida Keys reefs are disturbing, but they are no different from most other Caribbean and Bahamian reefs. Declines began in the late 1970s to early 1980s. Early on, *Acropora*

cervicornis and *A. palmata*, particularly, exhibited significant population losses. One hundred years earlier, in 1882, Alexander Agassiz published a map of the marine communities. This map included the spatial coverage of the principal marine community components (Table 3.4). At that time, the Staghorn and Elkhorn reefs contributed most of the reef habitat at Dry Tortugas shallow reefs.

These are the earliest qualitative data for any western Atlantic reef system and are significant comparative aids in documenting the downward trend in reef diversity and vitality. Although the Carnegie workers added greatly to the knowledge of corals and reefs, they did not map reefs. John W. Wells studied Tortugas reefs in 1932; his field notes (Wells 1932, unpublished) reflect that the *Acropora palmata* reefs were similar to what Agassiz had reported. From the closing of the Carnegie laboratory in 1939 until 1973, there was a hiatus in reef research at Dry Tortugas. In 1982, Davis published a map of Dry Tortugas marine communities (Table 3.5).

The principal differences between Agassiz's and Davis's evaluations are:

Davis included sea grasses and algae, which were not detailed in Agassiz's evaluation.

A great increase in the octocorals hardbottom habitat from 1882 to 1982. (Note the actual surveys occurred in 1881 for Agassiz and in 1975–1977 for Davis).

The Staghorn (*Acropora cervicornis*, *A. prolifera*) reefs had increased slightly while the Elkhorn reefs (*A. palmata*) virtually disappeared.

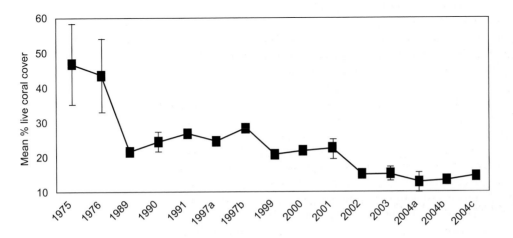

FIG. 3.25. Coral cover trends, Bird Key Reef, Dry Tortugas : 1975–2004 (Jaap et al. 1989, Jaap unpublished, J. Miller unpublished)

TABLE 3.4. Terrestrial and marine habitats, Dry Tortugas. (From Agassiz 1882)

Habitat	Acres	Hectares	Percent
Land	108.7	44	0.20
Astrea and *Meandrina* Reefs	380.5	154	2.80
Staghorn reefs	1030.4	417	1.90
Elkhorn reefs	108.7	44	0.20
Broken coral heads	163.1	66	0.30
Total coral reef	1682.8	681	3.09
Octocoral -hardbottom	2607.0	1,055	4.80
Sediments	49,952.3	20,215	91.90
Total	54,350.8	21,995	

TABLE 3.5. Terrestrial and marine habitats, Dry Tortugas. (From Davis 1982)

Habitat	Acres	Hectares	Percent
Land	113.7	46	0.20
Bank reefs	338.5	137	0.60
Coral head buttresses	620.2	251	1.10
Staghorn reefs	1181.2	478	2.10
Elkhorn corals	1.5	0.6	(.0026)
Total coral reef	2140.7	867	3.8
Octocoral hardbottom	9,797.7	3,965	17.40
Benthic algae	281.7	114	0.50
Sea grasses	17,060.1	6,904	30.29
Sediments	26,914.7	10,892	47.80
Total	56,309.3	22,788	

In January 1978, an extreme cold-water disturbance resulted in a 95% or greater loss of *Acropora cervicornis* from the Dry Tortugas (Davis 1982; Porter et al. 1982; Roberts et. al. 1982). Projects at Dry Tortugas have monitored coral abundance and cover at Bird Key Reef, west of Loggerhead Key (*A. cervicornis* thickets prior to the cold water disturbance), Pulaski Shoal, Texas Rock, and White Shoal. Neither *A. cervicornis* nor *A. palmata* populations have recovered (Jaap and Wheaton 1995). In 2002 and 2004, the Dry Tortugas sites were surveyed again and some areas with a few colonies of *A. cervicornis* were found but no general recovery (Jaap et al. 2002).

Jaap and Sargent (1993) published details on mapping of the *Acropora palmata* community in Five Foot Channel. They identified the community as occupying 1,400 m^2 (containing both sparse and dense concentrations of *A. palmata*; however, the dense concentration was confined to within 728 m^2). The area is slightly larger than what Davis (1982) reported.

In the 2002 expedition to Five Foot Channel, an inspection of the *Acropora palmata* and the *A. prolifera* communities revealed that they had suffered from disease and/or environmental stresses. In 2000, a channel that separated Garden and Bush Keys filled in with sediment thus changing water circulation. The circulation at flood tide is from the southeast through the Five Foot Channel gap between the Long Key and Bird Key Reef rampart. At ebb tide, the flow reverses and the source of water changes; now water from and around the Garden Key anchorage (possibly including overflow from the Garden Key septic tank field) flows out Five Foot Channel (previously, water entering the channel between Garden and Bush Keys formed the principal volume).

In October 2004, approximately a month after the eye of Hurricane Charlie passed over the Dry Tortugas, another visit was made to Five Foot Channel. It was observed that the storm had fragmented many corals and the pieces had been scattered inshore as far as 100 m from the *Acropora palmata* and *A. prolifera* communities concentrated area. Some fragments had healthy-looking tissue. The patches were reduced in upward relief but did not seem to have suffered catastrophic destruction.

A site off the northeast side of Loggerhead Key (3 m deep) was also impacted by the storm. D. Williams (personal communication 2002) observed in 2002 a moderate population of *Acropora cervicornis*. Following the hurricane, this site exhibited severe disturbance. There were very few multi-branched colonies; mostly it was small branch fragments, many of which had washed inshore (west) and ended up in a sparse *Thalassia* bed. For the most part, the vitality of the fragments in the *Thalassia* appeared satisfactory (color, few signs of disease, or predation).

Recent disturbances at Dry Tortugas have radically changed the area's shallow coral reef communities, although deeper reefs on the Tortugas Banks were not impacted. Based on the past history of coral reef recovery in Florida (Shinn 1975) after Hurricane Donna, the expectation is that recovery will be decades or longer for the Tortugas shallow reef communities.

3.16 Fisheries Issues in the Florida Keys Coral Reef Ecosystem

3.16.1 Florida Keys Coral Reef Fisheries

Coral reefs in southeastern Florida and the Florida Keys provide the ecological foundation for vital fisheries and a tourism-based economy that generated an estimated 71,000 jobs and $6 billion of economic activity in 2001 (Johns et al. 2001). They also contributed to the designation of Florida as the "fishing capital of the world" by the state legislature (FWC 2007). Coral reef ecosystem goods and services, however, extend beyond fishing to include a range of educational, scientific, aesthetic, and other recreational uses, such as snorkeling, scuba diving, and tourism.

Fisheries in southern Florida are complex. Adult reef fishes are caught for food and sport around bridges and on offshore patch and barrier reefs. Commercial and sport fisheries also target spiny lobster, marine aquarium fishes and invertebrates, inshore and offshore. Pink shrimp (*Penaeus duorarum*), a principal prey item of the snapper –grouper complex, are intensively exploited. Offshore, a substantial commercial food fishery targets adult pink shrimp inhabiting sedimentary areas near coral reefs. In coastal bays and near barrier islands, juvenile pink shrimp are commercially targeted as live bait for the recreational fishery. Both food and sport fisheries target pre-spawning subadult pink shrimp as they emigrate from coastal bay nursery grounds to offshore spawning grounds. Inshore, sport fisheries pursue highly prized game fishes, including spotted sea trout (*Cynosscion nebulosus*), sheepshead (*Archosargus probatocephalus*), black and red drum (*Sciaenops* spp.), snook (*Centropomus undecimalis*), tarpon (*Megalops atlanticus*), bonefish (*Albula vulpes*), and permit (*Trachinotus falcatus*), while commercial fisheries primarily target sponges and crabs (blue and stone). Offshore of the deep margin of the bank reefs, commercial and sport fisheries capture an assortment of species including amberjack (*Seriola dumerili*), king (*Scomberomorus cavalla*) and Spanish (*S. maculatus*) mackerel, barracuda (*Sphyraena barracuda*), sharks (Class Chondrichthyes), and small bait fishes (e.g., Exocoetidae, Mullidae, Carangidae, Clupeidae, and Engraulidae). Further offshore (seaward of the 40 m isobath), commercial and sport fisheries catch dolphinfish (*Corypaena hippurus*), tunas (*Thunnus* spp.), and swordfish (*Xiphias gladius*), and sport fishers target sailfish (*Istiophorus* spp.), wahoo (*Acanthocybuim solandri*), and white (*Tetrapterus albidus*) and blue (*Makaira nigricans*) marlin.

3.16.2 Florida Keys Fishery Management

Fisheries are managed along the Florida Keys ecosystem (Fig. 3.26) by the Florida Fish and Wildlife Conservation Commission and two federal fishery management councils (south Atlantic and Gulf of Mexico). Special regulations can also apply in the Florida Keys National Marine Sanctuary (FKNMS) under NOAA (Department of Commerce); in three national parks (Biscayne, Everglades, Dry Tortugas) and four National Fish and Wildlife Refuges (Department of Interior); and in John Pennekamp Coral Reef State Park (Florida Department of Environmental Protection).

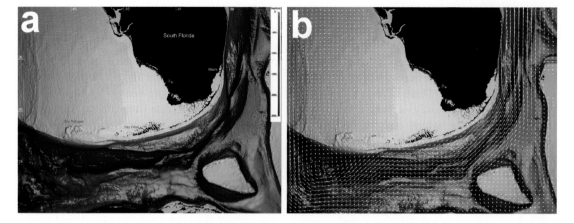

Fɪɢ. 3.26. (**a**) Topographic and (**b**) oceanographic complexity along the Florida Keys coral reef ecosystem from Miami to the Dry Tortugas. Circulation gyre exhibited in the proximity of the Dry Tortugas in the lower panel

3.16.2.1 Biodiversity of the Major Reef Fauna Groups

The Florida Keys has more than 500 fish species, including 389 that are reef associated (Stark 1968), and thousands of invertebrates, including corals, sponges, shrimps, crabs, and lobsters. Snapper–grouper utilize a mosaic of cross-shelf habitats and oceanographic features over their life spans (Ault and Luo 1998; Lindeman et al. 2000). Most adults spawn on the bank reefs and sometimes form large spawning aggregations (Domeier and Colin 1997). The Dry Tortugas region in particular contains numerous known spawning aggregation sites (Schmidt et al. 1999). Pelagic eggs and developing larvae are transported from spawning sites along the barrier reef tract by a combination of seasonal wind-driven currents and unique animal behaviors to eventually settle as early juveniles in a variety of inshore benthic habitats (Lee et al. 1994; Ault et al. 1999). Some of the most important nursery habitats are located in the coastal bays and near barrier islands (Lindeman et al. 2000; Ault et al. 2001). As individuals develop from juveniles to adults, ontogenetic habitat utilization patterns generally shift from coastal bays to offshore reef environments. The frequency of occurrence of the more common fish species (Table 3.6) notes that the most common species have similar ranking over time, the less common species are more dynamic in temporal rank.

3.16.2.2 Human Impacts and Conservation Issues

Coral reefs in the Florida Keys are impacted by fishing and by habitat degradation from other human activities including coastal development, altered freshwater flow, and changes in water quality from pollution, sedimentation, and excess nutrients (CERP 1999; Cowie-Haskell and Delaney 2003). Human impacts have escalated as a result of Florida's tenfold population growth from 1.5 million people in 1930 to 16 million in 2000. In 2000, over five million residents, nearly a third of Florida's population, lived in the five southern counties adjacent to coral reefs (Palm Beach, Broward, Miami -Dade, Monroe, and Collier). In addition, over three million tourists visit the Keys annually (Leeworthy and Vanasse 1999).

The Florida Keys reef ecosystem is considered one of the nation's most significant, yet most stressed, marine resources and is managed by Florida, NOAA, and the National Park Service. Reef fisheries target the "snapper –grouper complex", which consists of 73 species of mostly groupers and snappers, but also grunts (Pomadasyidae), jacks (Carangidae), porgies (Sparidae), and hogfish (Labridae). The fishery has been intensively exploited over the past 75 years, during which time the local human population has grown exponentially, generating concerns over sustainable fishery productivity. Many reef species are extremely sensitive to exploitation, and coastal development subjects coral reefs to a suite of other stressors that can cumulatively impact reef fish

TABLE 3.6. Ranked top 50 reef fish species in terms of percent occurrence for 2004 compared to 1999–2000. Common names in bold denote species in the exploited snapper –grouper complex. (From Ault et al. 2006).

Common name	Scientific name	Family	Occurrence rank 2004	1999–2000	Change.
Bluehead	*Thalassoma bifasciatum*	Labridae	1	1	=
Striped parrotfish	*Scarus iseri*	Scaridae	2	2	=
Cocoa damselfish	*Stegastes variabilis*	Pomacentridae	3	3	=
Redband parrotfish	*Sparisoma aurofrenatum*	Scaridae	4	5	+
Yellowhead wrasse	*Halichoeres garnoti*	Labridae	5	10	+
Blue tang	*Acanthurus coeruleus*	Acanthuridae	6	8	+
Bicolor damselfish	*Stegastes partitus*	Pomacentridae	7	11	+
Spotted goatfish	*Pseudupeneus maculatus*	Mullidae	8	20	+
White grunt	*Haemulon plumieri*	Haemulidae	9	4	–
Slippery dick	*Halichoeres bivittatus*	Labridae	10	7	–
Yellowtail snapper	*Ocyurus chrysurus*	Lutjanidae	11	9	–
Saucereye porgy	*Calamus calamus*	Sparidae	12	6	–
Stoplight parrotfish	*Sparisoma viride*	Scaridae	13	15	+
Bridled goby	*Coryphopterus glaucofraenum*	Gobiidae	14	12	–
Red grouper	*Epinephelus morio*	Serranidae	15	13	–
Purple reef fish	*Chromis scotti*	Pomacentridae	16	28	+
Ocean surgeon	*Acanthurus bahianus*	Acanthuridae	17	18	+
Blue angelfish	*Holacanthus bermudensis*	Pomacanthidae	18	16	–
Spotfin butterfly fish	*Chaetodon ocellatus*	Chaetodontidae	19	17	–
Butter hamlet	*Hypoplectrus unicolor*	Serranidae	20	29	–
Greenblotch parrotfish	*Sparisoma atomarium*	Scaridae	21	24	+
Masked goby	*Coryphopterus personatus*	Gobiidae	22	25	+
Blue hamlet	*Hypoplectrus gemma*	Serranidae	23	38	+
Yellowhead jawfish	*Opistognathus aurifrons*	Opistognathidae	24	21	–
Gray angelfish	*Pomacanthus arcuatus*	Pomacanthidae	25	23	–
Hogfish	*Lachnolaimus maximus*	Labridae	26	19	–
Foureye butterfly fish	*Chaetodon capistratus*	Chaetodontidae	27	32	+
Clown wrasse	*Halichoeres maculipinna*	Labridae	28	26	–
Threespot damselfish	*Stegastes planifrons*	Pomacentridae	29	30	+
Beaugregory	*Stegastes leucostictus*	Pomacentridae	30	34	+
Harlequin bass	*Serranus tigrinus*	Serranidae	31	27	–
Saddled blenny	*Malacoctenus triangulatus*	Labrisomidae	32	14	–
Barred hamlet	*Hypoplectrus puella*	Serranidae	33	31	–
Neon goby	*Elacatinus oceanops*	Gobiidae	34	22	–
Graysby	*Cephalophilis cruentatus*	Serranidae	35	37	+
Black grouper	*Mycteroperca bonaci*	Serranidae	36	44	+
Blue chromis	*Chromis cyanea*	Pomacentridae	37	45	+
Mutton snapper	*Lutjanus analis*	Lutjanidae	38	52	+
Tobaccofish	*Seranus tabacarius*	Serranidae	39	46	+
Bar jack	*Caranx ruber*	Carangidae	40	40	=
Queen angelfish	*Holacanthus ciliaris*	Pomacanthidae	41	43	+
Great barracuda	*Sphyraena barracuda*	Sphyraenidae	42	49	+
Sharpnose puffer	*Canthigaster rostrata*	Tetraodontidae	43	39	–
Spanish hogfish	*Bodianus rufus*	Labridae	44	47	+
Tomtate	*Haemulon aurolineatum*	Haemulidae	45	41	–
Princess parrotfish	*Scarus taeniopterus*	Scaridae	46	61	+
Reef butterfly fish	*Chaetodon sedentarius*	Chaetodontidae	47	36	–
French grunt	*Haemulon flavolineatum*	Haemulidae	48	50	+
Cero	*Scomberomorus regalis*	Scombridae	49	117	+
Bucktooth parrotfish	*Sparisoma radians*	Scaridae	50	101	+

populations by degrading water quality and damaging nursery and adult habitats.

3.16.3   Fishing and Fishing Intensity

Extensive recreational and commercial fishing occurs in Florida Keys waters. Recreational fishers emanate either locally (from SE Florida or the Keys) or from more distant venues desiring to experience "The Fishing Capital of the World," as the state of Florida promotes itself (Ault et al. 2005a, b; FWC 2007).

In addition to SE Florida's expanding human population, recreational vessel registrations in south Florida increased more than 100% from 1964 to 2006. Commercial vessel registrations increased by about 100% from 1964 to 1998 but have since decreased by 37% (Fig. 3.27). Commercial fisheries target reef and pelagic fish species, spiny lobster, stone crabs, blue crabs, shrimp, and ballyhoo. Headboat fisheries, in which customers pay "by the head" to fish from vessels with a typical capacity of from 10 to 20 people, predominantly target reef species. Precise data on fishing effort on coral reefs do not exist but are reflected by statewide and regional fishing statistics. In the five most recent years for which recreational fishery estimates are available (2001–2005) for Florida, more than 6.4 million anglers averaged 27.2 million marine fishing trips

annually. An estimated 173.3 million fish were caught annually, of which a little more than 50% were released (86.9 million) (NMFS 2007). Two recent (2000–2001, 2003) non-concurrent studies showed that 3.64 million person/days were spent fishing on natural reefs annually in the Florida Keys (Johns et al. 2001, 2004). Concomitant with increasing fishing pressure associated with increasing population, average fishing power (the proportion of stock removed per unit of fishing effort) may have quadrupled in recent decades because of technological advances in fishing tackle, hydroacoustics (depth sounders and fish finders), navigation (charts and global positioning systems), communications, and vessel propulsion (Bohnsack and Ault 1996; Mace 1997).

Fishing can impact coral reefs by removing targeted species and by killing non-target species as bycatch, both of which may result in cascading ecological effects (Frank et al. 2005). Because fishing is size-selective, concerns exist about ecosystem disruption by removal of ecologically important keystone species, top predators (groupers, snappers, sharks, and jacks), and prey (e.g., shrimps and baitfish).

Fishing can also negatively impact reef ecosystems via fishing-related habitat damage. Commercial fisheries in the Keys for lobsters and stone crabs utilize traps that are deployed in habitats adjacent to reefs. Currents associated with

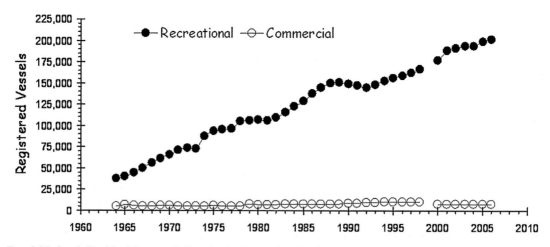

FIG. 3.27. South Florida (Monroe, Collier, Dade, Broward and Palm Beach Counties commercial and recreational vessel registrations by year (Florida Statistical Abstract 2006, Table 19.45 – State of Florida, Department of Highway Safety and Motor Vehicles, Bureau of Vessel Titles and Registrations; http://www.hsmv.state.fl.us/dmv/vslfacts.html)

strong storms can move traps onto reefs, where corals and other benthic organisms are damaged or killed (e.g., Sheridan et al. 2005). In 2005, it is estimated that approximately 300,000 lobster traps were lost during a series of hurricanes and strong storms (Clark 2006). Many reefs throughout the Keys are littered with lost traps and with monofilament line lost by recreational anglers. Reef damage may also occur from anglers anchoring on reefs (Davis 1977). Stress associated with fishing-related removal of species and habitat damage may be compounded when combined with other stressors such as pollution and climate change (Wilkinson 1996).

In response to declining trends in reef fishery catches, a series of regional federal and state management regulations were imposed including recreational bag limits, minimum size limits, commercial quotas and trip limits, seasonal closures, gear restrictions, limited commercial entry, closed fisheries, species moratoria, game fish status, and restrictions on sale and possession. These regulations were implemented to stabilize catches, protect spawning stock biomass, and reduce fishing mortality rates. In general, the history of regional regulations for reef fishes has been complex and has tended to be more restrictive over time. Nonetheless, despite the bevy of regulations imposed in the Florida Keys, recent fishery assessments indicate that, for example, black grouper spawning stock biomass was less than 10% of its historical size (Ault et al. 2005b).

In recent years, new ecosystem-based management measures have been enacted in the Florida Keys, including the 1997 implementation of a network of 23 No Take Marine Reserves (NTMRs) by the Florida Keys National Marine Sanctuary (FKNMS, NOAA, www.fknms.noaa.gov). These are relatively small (mean area $2\,km^2$, range 0.16–$31\,km^2$), comprising $46\,km^2$ in total area and have varying levels of protection: four allow catch-and-release surface trolling and four can only be accessed by special permit. In July 2001, the Florida Keys network was expanded to become the largest in North America with the implementation of two NTMRs in the Dry Tortugas region covering about $566\,km^2$. The Tortugas region is believed to be extremely important for coral reefs and fisheries as a source of recruitment because of its upstream location in the Florida Current that facilitates advective dispersion and transport of eggs and larvae to the rest of the Keys.

3.16.4 Fisheries History

Native Americans fished for reef fishes on Florida reefs long before the arrival of European settlers (Oppel and Meisel 1871). Reef fishing accelerated in the 1920s. Following growing public conflicts and sharp declines in catches, monitoring programs at the species level began in the early 1980s (Bohnsack et al. 1994; Bohnsack and Ault 1996; Harper et al. 2000; Ault et al. 2005a, b).

Reef fish landings trends for the period 1981–1992 were reported for the Florida Keys by Bohnsack et al. (1994). Depending on year, recreational landings comprised between 40% and 66% of total landings. Reef fishes accounted for 58% of total fish landings, 69% of recreational landings, and 16% commercial landings. Commercial landings were dominated by invertebrates (spiny lobster, shrimp, and stone crabs), which comprised 63% of total landings.

In a report to the US Congress, NMFS classified 11 species that are landed in the Keys as overfished (i.e., depleted below minimum standards), and 11 as subject to overfishing (i.e., being fished at a rate that would lead to being overfished), with some overlap between the two categories (NMFS 2005). Included in these totals are reef-associated species such as gag (*Mycteroperca microlepis*), black (*M. bonaci*), red (*Epinephelus morio*), snowy (*E. niveatus*), Warsaw (*E. nigritus*), goliath (*E. itajara*) and Nassau (*E. striatus*) groupers, speckled hind (*E. drummondhayi*), and red (*Lutjanus campechanus*) and vermilion (*Rhomboplites aurorubens*) snappers. Fisheries for goliath and Nassau groupers and for queen conch (*Strombus gigas*) were closed in the 1990s and remain closed today, although the goliath grouper stock continues to indicate signs of recovery (Porch et al. 2003, 2006) to the extent that considerable debate occurs regarding reopening the fishery.

Ault et al. (1998) assessed the status of multiple reef fish stocks and determined that 13 of 16 groupers (Epinephinlae), 7 of 13 snappers (Lutjanidae), 1 wrasse (hogfish; Labridae), and 2 of 5 grunts (Haemulidae) were overfished according to federal (NMFS) standards. It was suggested that some stocks appeared to have been chronically overfished since the 1970s, and that the Florida keys fishery exhibits classic "serial overfishing" in which the largest, most desirable species are depleted by fishing (Ault et al. 1998). Ault et al. (2001) found that the average size of adult black grouper in the Upper Keys was close to 40% of its 1940 value, and that the spawning stock for this species is now less than 5% of its historical unfished maximum. In subsequent analyses, Ault et al. (2005a, b) determined that, of 34 species within the snapper –grouper complex for which sufficient data were available for analysis, 25 were experiencing overfishing (Fig. 3.28).

Partly in response to concerns about fisheries exploitation, the Florida Keys National Marine Sanctuary (FKNMS) established a series of "Sanctuary Preservation Areas" (SPAs) in 1997. Comparison of fish and benthic communities within versus outside of the SPAs is underway. The FKNMS also created the Tortugas Ecological Reserve in 2001 to protect reef resources and sup-

port sustainable reef fisheries. The reserve covered 151 miles[2]; prohibited all anchoring, fishing, and other extractive activities; and, at the time, was the largest marine reserve in North America. Scientists at the University of Miami and NOAA Fisheries Service have studied and reported on responses of coral reef fish populations to this reserve. Based on data collected during over 4,000 research dives, they compared changes in the Dry Tortugas region between 1999 and 2000 before the reserve was established and in 2004, 3 years after the reserve was established (Ault et al. 2006). As predicted by marine reserve theory, significant regional increases in abundance for several exploited and non-exploited species were detected. Significantly greater abundance and larger fish sizes were found in the Tortugas Ecological Reserve for black grouper (Fig. 3.29), red grouper (Fig. 3.30), and mutton snapper compared to the baseline period. No significant declines were detected for any exploited species in the reserve, while non-exploited species

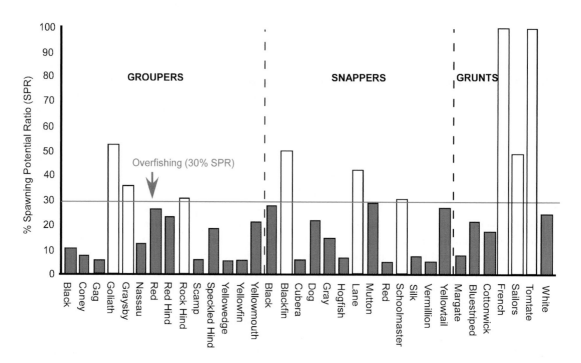

Fig. 3.28. Spawning potential ratio (SPR) analysis for 34 exploited species in the snapper –grouper complex from the Florida Keys for the period 2000–2002. Dark bars indicate overfished stocks and open bars indicate stocks above the 30% SPR standard (Ault et al. 2005. ICES Journal of Marine Science 62: 417–423. Reproduced by permission of Elsevier)

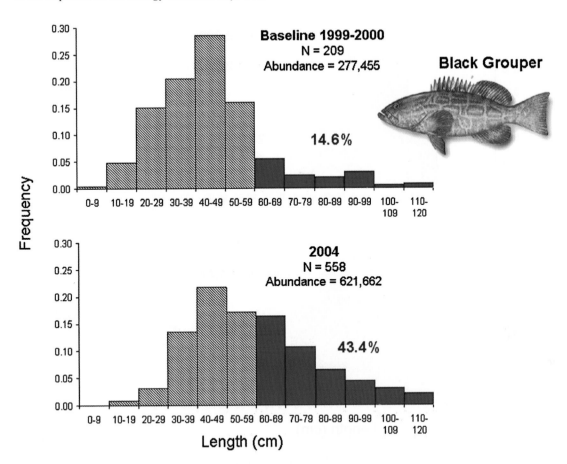

FIG. 3.29. Black grouper (*Mycteroperca bonaci*) size distributions from the Tortugas region in 1999–2000 (Top) and 2004 (Bottom), before and after the establishment of the Tortugas Ecological Reserve in 2001. Blue bars represent size classes larger than the minimum legal minimum size. Percentages show the proportion of the population larger that the legal minimum size of capture

showed both increases and declines. Abundance of exploited species in fished areas on the Tortugas Bank either declined or did not change. A comparison of black grouper size distributions as a function of management zone is given in Fig. 3.31.

On 19 January 2007, the National Park Service established a 46 miles[2] Research Natural Area within Dry Tortugas National Park. This area was contiguous to the FKNMS Ecological Reserve and effectively expanded the marine reserve network since it also prohibited all anchoring and extraction. Ongoing research and monitoring are planned to ascertain whether patterns observed in protected areas in the Tortugas are due to influences of marine reserves, confounding effects of recent changes in fishing regulations, hurricane disturbances, or random oceanographic and chance recruitment events.

3.17 Human Influences and Water Quality

The US Environmental Protection Agency Water Quality Protection Plan for the FKNMS includes a quarterly sampling dating from 1995 (Boyer and Briceño 2005). They report trends implying anthropogenic input to coastal waters that include:

- Elevated concentration of dissolved inorganic nitrogen
- Elevated concentration of total organic carbon
- The "back country" (Florida Bay) exhibits trends of elevated concentrations of dissolved inorganic nitrogen, total organic carbon, chlorophyll A, greater turbidity, and greater nitrite concentrations

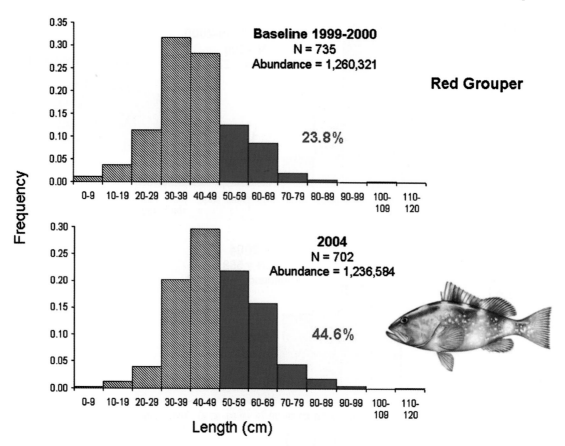

Fig. 3.30. Red grouper (*Epinephelus morio*) size distributions from the Tortugas region in 1999–2000 (Top) and 2004 (Bottom), before and after the establishment of the Tortugas Ecological Reserve in 2001. Blue bars represent size classes larger than the minimum legal minimum size. Percentages show the proportion of the population larger that the legal minimum size of capture

• Water quality measurements are consistent from year to year with seasonal fluctuations

Boyer and Briceño (2005) reported that the water quality in FKNMS is mostly driven by hydrological and meteorological forces that extend beyond the Florida Keys boundaries.

Episodic disturbance events such as hurricanes, bleaching episodes, and harmful algal blooms are most responsible for structuring the coral reef communities. Vessel groundings are a growing concern (Jaap 2000; Jaap et al. 2006) as recovery is frequently impeded by repeated groundings on the same reef sites. Anthropogenic activities add stress to a system that is already under pressure from nat-

ural forces, especially the extensive urbanization of the coast and resource utilization that is perceived by many to exceed a reasonable carrying capacity. Florida's reefs are in crisis. Although the faunal population has shown resilience over the thousands of years of natural change, the speed and variety of man-made pressures on the community are overwhelming the habitat's ability to recover. Figures 3.32–3.34 document a cross section of Octocorallia and Scleractinia from Florida Keys reefs.

Transport of risky materials into the region has marine, terrestrial, and airborne components. The hydrodynamic system brings both freshwater and pollutants from as far away as Montana and Wisconsin via the Mississippi, Missouri, and Ohio

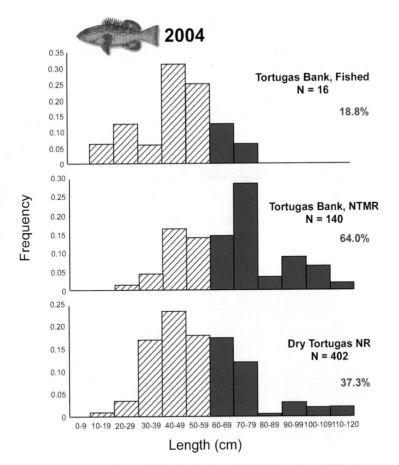

FIG. 3.31. Black grouper (*Mycteroperca bonaci*) size distributions in 2004 from three different managed zones: Fished areas on the Tortugas Bank (Top), unfished areas on the Tortugas Bank in the Ecological Reserve (Middle), and recreational angling only areas in Dry Tortugas National Park (Bottom). Blue bars represent size classes larger than the minimum legal minimum size. Percentages show the proportion of the population larger that the legal minimum size of capture

drainage basin (Ortner et al. 1995; Gilbert et al. 1996). Runoff from the urban areas brings nutrients (pet feces and fertilizer), petroleum, trace metals, and chemicals into the marine system. The upper level winds transport dust from Africa (Darwin 1846; Gillette 1981; Prospero et al. 1987; Goudie and Middleton 2001; Garrison et al. 2003) including metals and pathogens that enter directly or indirectly into the marine system (Duce and Tindale 1991; Shinn et al. 2000, 2003; Jickells 2005).

Anthropogenic alterations of the environment began when the Flagler railroad (1912) built causeways and bridges linking Key West with the mainland. Urbanization and development in the last century radically altered coastlines that were dominated by protective buffers: mangroves and wetlands that filter upland runoff, stabilize sediments, absorb nutrients, and maintain water quality, benefiting sea grass and coral reef communities. Natural coastal habitats have been replaced with concrete seawalls and cul-de-sac canals that have increased water turbidity and nutrient enrichment while decreasing sea grass cover and degrading juvenile fish habitat.

Estimates indicate that approximately 1,000 new residents move to Florida each day, many with

Fig. 3.32. (**a**) *Millepora complanata* (Lamarck 1816) [Bladed Fire Coral], Western Sambo Reef, 3–5 m, July 26, 2004 (**b**) *Gorgonia ventalina* Linnaeus, 1758 [Common sea fan], Western Sambo Reef, 3–5 m, July 26, 2004 (**c**) *Briareum asbestinum* (Pallas 1766) [Deadman's fingers], Western Sambo Reef, 3–5 m, July 26, 2004, (Photo credit W. Jaap). (**d**) *Siderastrea radians* (Pallas 1766) [Lesser Starlet Coral], Smith Shoal, 6–7 m, July 24, 2004 (Photo credits W. Jaap)

dreams of waterfront property and ocean views. The pressure on coastal development is intense; investors purchase developed properties such as shipyards, marinas, and trailer parks and convert them into high-rise condominiums, increasing the population. Historically, Keys homesteads relied on rainwater because there was very limited fresh water; today, homes and businesses import their fresh water from the mainland. Wastewater treatment is a challenge (Kruczynski 1999), since building advance water treatment plants on multiple islands is very expensive. Experiments funded by US EPA and FDEP (Florida Department of Environmental Protection) indicate that advanced wastewater treatment systems for each dwelling

are not economical or, from a practical standpoint, feasible. The enigma that must be solved is how to deal with millions of gallons of wastewater. The porous nature of the Florida Keys geologic strata and extensive canalization enhances wastewater migration to the marine system (Paul et al. 2000; Griffin et al. 2003) and has generated near shore nutrient loading. Chemicals entering the near shore marine system are the soup of society: human waste (Griffin et al. 2003), household chemicals, residues from medications and health products, and yard and garden chemicals. The biological results include algal blooms, displacement of species that do not tolerate high loads of nutrients, and reproductive failure in some species. The best example of

FIG. 3.33. (**a**) *Madracis auretenra* (Locke, Weil, and Coates 2007) Yellow pencil coral], Western Sambo Reef, 3–5 m, July 26, 2004 (**b**) *Acropora palmata* (Lamarck 1816) [Elkhorn coral], Western Sambo Reef, 3–5 m, July 26, 2004 (**c**) *Acropora cervicornis* (Lamarck 1816) [Staghorn coral], Western Sambo Reef, 3–5 m, July 26, 2004 (**d**) *Agarica agaricites* (Linnaeus 1758) [Lettuce coral], Smith Shoal, 6–7 m, July 24, 2004 (Photo credits W. Jaap)

reproductive failure is the queen conch (*Strombus gigas*) an icon of the Florida Keys. In nearshore waters, queen conch no longer produce progeny (Glazer and Quintero 1998; Delgado et al. 2004). If you move the individuals offshore, they resume a reproductive cycle and produce progeny. Individuals from offshore that were sexually functional lose that ability when moved to nearshore locations. Experts speculate that birth control chemicals flushed in the waste water may be involved.

Scientific debate on the transport of nutrient-enriched waters beyond the nearshore zones in the Florida Keys corresponds to the complexity of the hydrodynamic processes and the distance from shore (Szmant and Forrester 1996; Fourqurean and Zieman 2002). Net flow of water is from the Gulf side to the Atlantic (Smith 1994); however, the distance the water moves in a tide cycle is limited. Wind events will drive water far to sea and bring it into contact with patch and bank reefs. Reef corals are typically considered to thrive in oligotrophic conditions; however, the precise tolerances for nutrients are unclear (Szmant 2002). Mosquito spraying is chronic in the Florida Keys, and there is evidence of dibrom residue resulting in impacts to coral physiology (Morgan and Snell 2002).

A Working Group of the Intergovernmental Panel on Climate Change (IPCC) recently released their fourth assessment report titled "Climate Change 2007: The Physical Science Basis" (Intergovernmental Panel on Climate Change 2007). They state, unequivocally, that "global atmospheric

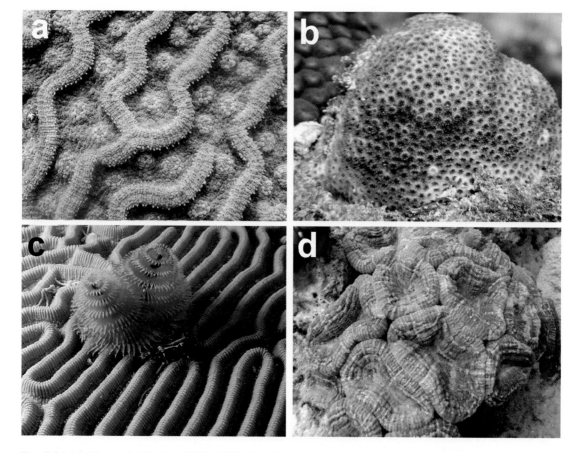

Fig. 3.34. (**a**) *Mycetophyllia ferox* Wells, 1973 [Rough cactus coral], Western Sambo Reef, 3–5 m, July 26, 2004
(**b**) *Solenastrea bournoni* Milne Edwards and Haime, 1849 [Smooth star coral], Smith Shoal, 6–7 m, July 24, 2004
(**c**) *Diploria strigosa* (Dana 1846) [Symmetrical brain coral] and *Spirobranchus giganteus* (Pallas, 1766) [Christmas
tree worm], Western Sambo Reef, 3–5 m, July 26, 2004 (**d**) *Mussa angulosa* (Pallas 1766) [Large flower coral],
Western Sambo Reef, 3–5 m, July 26, 2004 (Photo credits W. Jaap)

concentration of carbon dioxide, methane and nitrous oxide have increased markedly as a result of human activities since 1750 and now far exceed pre-industrial values." Fossil fuel burning, land use change, and agriculture are the primary anthropogenic forces in these changing atmospheric conditions. Ocean conditions have also been altered as a result of the intrinsic link that exists between the atmosphere and the ocean. Observations since 1961 show that the average temperature of the global ocean has increased to depths of at least 3,000 m and the ocean has been absorbing more than 80% of the heat added to the climate system (Intergovernmental Panel on Climate Change 2007).

Bleaching is strongly associated with elevated seawater temperatures as stressed, overheated organisms expel their zooxanthellae and become pale or white. If thermal stress is severe (greater than 31°C) and prolonged (weeks), most zooxanthellate organisms (fire corals, octocorals, zooanthids, and scleractinian corals) will bleach and many may die. A model by Hoegh-Guldberg (1999) shows an invariant bleaching "threshold" at approximately 1°C above mean summer temperatures. This threshold will be exceeded over the next 50 years as temperatures continue to rise, leading to predictions of massive coral losses. Other abiotic factors, such as increased ultraviolet radiation, salinity extremes, high turbidity, sedimentation, and bacterial infection, have also been implicated in coral bleaching. However, the seven major episodes of bleaching that have occurred since 1979 have been primarily attributed to increased sea water temperatures associated with global climate change and El Niño/La Niña events, with a possible synergistic effect of elevated ultraviolet and visible light (Hoegh-Guldberg 1999). Detrimental physiological and morphological effects on the corals as a result of bleaching include reduced skeletal growth and reproductive activity, lowered mucus production and subsequent decrease in a capacity to shed sediments and resist bacterial/viral invasion (Glynn 1996). Affected colonies can regain their zooxanthellae within weeks to months only if stressful conditions abate and the bleaching is not too severe (Jaap 1979; Glynn 1996).

Thermal expansion of seawater in response to elevated temperatures, combined with melting ice sheets, glaciers, and ice caps, has contributed to a global average rise in sea level of approximately 100 mm (data from 1961 to 2003) with models showing continued rise. If rates of sea level rise exceed the rates of coral growth, then reefs could essentially be "drowned" as they sink beneath their optimal photic zone. Another consequence of rising sea level is the impact on aquatic and terrestrial habitats. The saltwater will kill the plants, and the detritus will be washed to sea and add nutrients to system. This is referred to as the reef being shot in the back by the flood of coastal effluents (Neumann and Moore 1975).

Climatic shifts over decades or less have occurred in the past; in the Pleistocene and Holocene, extant coral species underwent dramatic shifts in geographic range in response to periods of warming and cooling (Paulay 1996) and sea level change (Veron 1995). Coral species display a range of adaptive capabilities with certain species able to survive rapid changes much better than others through migration, morphological changes, and expulsion/assimilation of certain zooxanthellae clades. The major difference between today's climate driven changes in reefs and the recent past is that current reef dwelling species are profoundly affected by people. Human impacts and increased fragmentation of the coral reef habitat has undermined reef resilience, rendering them increasingly susceptible to climate change. Persistently elevated rates of mortality, reduced recruitment of larvae,

and slow recovery rates have been observed on a more frequent basis (Hughes et al. 2003).

Ocean acidification accompanied by increased atmospheric CO_2 concentration may have profound effects on calcifying organisms in general, with particular negative implications for coral reefs. In the past 200 years, oceans have absorbed approximately half of the CO_2 produced by fossil fuel burning and cement production (Royal Society 2005). This is a reduction of surface water pH of 0.1 units, or a 30% increase in the concentration of hydrogen ions. If global emissions continue to rise at the current rate, the Royal Society predicts a 0.5 decrease in pH by the year 2100. This hydrogen ion concentration is probably higher than what has been experienced in hundreds of millennia and, more critically, the rate at which it will be achieved will far exceed that of any past records. This is important to corals and other plants and animals that remove calcium from seawater and combine it with the bicarbonate ions to build a limestone shell or skeleton (mollusks and corals).

The carbon cycle provides the needed bicarbonate and carbonate ions for calcium carbonate ($CaCO_3$) construction. Increased CO_2 concentrations and the resultant lowered pH reduces the saturation state of $CaCO_3$ and raises that saturation horizon in the ocean (the depth at which temperature and pressure effects cause the rate of $CaCO_3$ dissolution to exceed $CaCO_3$ formation). Calcium carbonate is precipitated by organisms in two forms: calcite (foraminifera and coccolithophores) and aragonite (corals and pteropods); mollusks may have both aragonite and calcite in their shells. Aragonite is the more soluble form and it is postulated that corals and pteropods will be most susceptible to dissolution as oceans continue to acidify.

Increased CO_2 concentration is lowering the aragonite saturation state of the oceans, further compounding the effects of elevated ocean temperatures, bleaching, disease, and phase shifts. Kleypas et al. (1999) have proposed that the calcification rates of corals would decrease by 10–30% under a doubling of atmospheric CO_2 concentrations. The synergistic interactions of anthropogenic activities will most likely cause a major shift in the distribution, diversity, and abundance of corals. Over longer timescales, reef frameworks will begin to erode and dissolve, leading to a possible reduction in biodiversity and carrying with it severe implications for coastlines, which depend on the presence of reefs for protection from hurricanes, cyclones, and tsunamis.

The long term health and longevity of Florida Keys reefs are analogous to a patient that has suffered major trauma and medical intervention. Unfortunately, the prognosis is fair to poor (Hallock et al. 2003). Wilkinson (2004) ranked the Atlantic coral reefs, including Florida, as a "low C" based on 18 criteria and noted that climate change may inhibit recovery after bleaching events and hurricanes in heavily populated areas (added stress from use and pollution will impact the resilience of the organisms). Local, regional, and global threats are stressing the system. The interactions of the stressors and the unpredictability of the episodic disturbances are challenging. Employing an intermediate disturbance model (Connell 1978) is simplistic but relevant. When the stressing factors are chronic in time and space, long-term longevity of the coral reefs is doubtful. A prognosis that might be reasonable is complete loss of the shallowest (less than 3 m depth) coral reef communities and marginal survival of corals in depths that exceed 3 m. In special locations, such as Tortugas Banks and Pully Ridge, the buffer of depth and remoteness will provide some protection for the corals. We can only hope that Florida's reefs are up to the challenge.

Acknowledgements We thank the Florida Fish and Wildlife Research Institute for the support of data mining used in the section dealing with coral reef monitoring and Dry Tortugas. Florida FWC, US EPA, NOAA, and National Park Service funded this work in the recent period. In the past funding sources included the John D. and Katherine T. MacArthur Foundation. We appreciate the work of the multitude of people who worked and collaborated on these projects. We gratefully appreciate their contributions to the effort of expanding our understanding of Florida's natural treasure: her coral reefs.

References

Adey WH, Adey PJ, Burke R, Kaufman L (1976) Holocene reef systems of eastern Martinique, French West Indies. Atoll Res Bull 211, p 40

Adey WH, Gladfelter W, Ogden J, Dill R (1977) Field guidebook to the reefs and reef communities of St. Croix, U. S. Virgin Islands. Atlantic Reef Committee, University of Miami, Miami, FL, p 52

Agassiz A (1882) Explorations of the surface fauna of the Gulf Stream under the auspices of the US Coast Survey II. The Tortugas and Florida Reefs. Mem Am Acad Arts Sci Cent 2:107–134

Agassiz A (1885) The Tortugas and Florida reefs. Mem Am Acad Arts Sci 11:107–134

Agassiz A (1888) Three cruises of the US Coast Survey steamer Blake in the Gulf of Mexico, Caribbean, and along the Atlantic coast of the US from 1877 to 1880. Houghton-Mifflin, Boston, MA, 2 vols

Agassiz A (1890) On the growth rate of corals. Bull Mus Comp Zool 2:61–64

Agassiz L (1852) Florida reefs, keys, and coast. Ann Rep Superintendent US Coast Survey 1851:107–134

Agassiz L (1869) Florida reefs, keys, and coast. Ann Rep Superintendent US Coast Survey 1866:120–130

Agassiz L (1880) Report on the Florida reefs. Mem Mus Comp Zool 7:1–61, pls. 1–23 [plate text by LF Pourtalès]

Almy CC Jr, Carrión-Torres C (1963) Shallow-water stony corals of Puerto Rico. Carib J Sci 3:133–162

Andrews KA, Nall L, Jeffret C, Pittman S (eds) (2005) The state of coral reef ecosystems of Florida. NOAA, Silver Springs, MD, pp 150–200

Antonius A (1972) Occurrence and distribution of stony corals (Anthozoa and Hydrozoa) in the vicinity of Santa Marta, Colombia. Mitteilungen aus dem Instituto Colombo-Alemán de Investigaciones Científicas "Punta De Betín" 6:89–103

Ault JS, Luo J (1998) Coastal bays to coral reefs: systems use of scientific data visualization in reef fishery management. International Council for the Exploration of the Seas. ICES C. M. 1998/S:3

Ault JS, Bohnsack JA, Meester G (1998) A retrospective (1979–1995) multispecies assessment of coral reef fish stocks in the Florida Keys. Fish Bull 96(3):395–414

Ault JS, Luo J, Smith SG, Serafy JE, Wang JD, Humston R, Diaz GA (1999) A spatial dynamic multistock production model. Can J Fish Aquat Sci 56(S1):4–25

Ault JS, Smith SG, Meester GA, Luo J, Bohnsack JA (2001) Site characterization for Biscayne National Park : assessment of fisheries resources and habitats. NOAA Technical Memorandum NMFS-SEFSC-468, 185 p

Ault JS, Smith SG, Bohnsack JA (2005a) Evaluation of average length as an indicator of exploitation status for the Florida coral-reef fish community. ICES J Mar Sci 62:417–423

Ault JS, Bohnsack JA, Smith SG, Luo J (2005b) Towards sustainable multispecies fisheries in the Florida, USA, coral reef ecosystem. Bull Mar Sci 76(2):595–622

Ault JS, Smith SG, Bohnsack JA, Luo J, Harper DE, McClellan DB (2006) Building sustainable fisheries in Florida 's coral reef ecosystem: positive signs in the Dry Tortugas. Bull Mar Sci 78(3):633–654

Bak RPM (1975) Ecological aspects of the distribution of reef corals in the Netherlands Antilles. Bijdragen tot de Dierkunde 45:181–190

Beltrán-Torres A, Carricart-Ganivet JP (1999) Lista revisada y clave de determinación de los corales pétreos zooxantelados (Hydrozoa: Milleporina; Anthozoa: Scleractinia) del Atlántico mexicano. Revista de Biologia Tropical 47:813–829

Bohnsack JA, Ault JS (1996) Management strategies to conserve marine biodiversity. Oceanography 9:73–82

Bohnsack JA, Harper DE, McClellan DB (1994) Fisheries trends from Monroe County, Florida. Bull Mar Sci 54:982–1018

Boyer JN, Briceño HO (2005) FY 2005 report for the water quality monitoring project, for the water quality protection program of the Florida Keys National Marine Sanctuary. Southeast Environmental Research Center, Florida International University, Miami, FL, p 91

Brooks HK (1963) Reefs and bioclastic sediments of the Dry Tortugas. Geol Soc Amer, Special Paper 73 (abstract)

Cahoon LB, Llindquist DG, Clavijo IE (1990) Live Bottoms in the continual shelf ecosystem: a misconception. Proc Amer Acad Underwater Sci 10th Symp, pp 39–47

Cairns SD (1982) Stony corals: (Cnidaria: Hydrozoa, Scleractinia) of Carrie Bow Cay, Belize. In: Rützler K, Macintyre IG (eds) The Atlantic barrier reef ecosystem at Carrie Bow Cay, Belize, I Structure and Communities. Smithsonian Contribution to the Mar Sci 12, Smithsonian Institution Press, Washington, DC, pp 271–302

Cairns SD (2001) A brief history of taxonomic research on azooxanthellate Scleractinia (Cnidaria: Anthozoa). Bull Biol Soc Washington, DC 10:191–203

Cairns SD, Stanley GD (1982) Ahermatypic coral banks: living and fossil counterparts. Proc 4th Int Coral Reef Symp 1:611–618

CERP (Comprehensive Everglades Restoration Plan) (1999) Central and southern Florida project, comprehensive review study, final integrated feasibility report and programmatic environmental impact statement. April 1999. U.S. Army Corps of Engineers and South Florida Water Management District. http://www.evergladesplan.org/about//rest_plan.cfm

Chen E, Gerber JF (1990) Climate. In: Myers RL, Ewel JJ (eds) Ecosystems of Florida. University of Central Florida Press, Orlando, pp 11–34

Chiappone M (1996) Marine benthic communities and habitat maps of the Florida Keys. The Nature Conservancy, Miami, FL

Clark C (2006) Lobster fishermen stake it all on 2006 season. Miami Herald, Aug 6, 2006

Clarke KR (1993) Non-parametric multivariate analyses of change in community structure. Aust J Ecol 18:117–143

Clarke KR, Gorley RN (2005) PRIMER v6: User manual/tutorial. PRIMER-E, Plymouth, UK

Clarke KR, Green RH (1988) Statistical design and analysis for a 'biological effects' study. Mar Ecol Prog Ser 46:213–226

Clarke KR, Warwick RM (1998) A taxonomic distinctness index and its statistical properties. J Appl Ecol 35:523–531

Clarke KR, Warwick RM, Somerfield PJ, Gorley RN (2005) Change in marine communities: an approach to statistical analysis and interpretation, 3rd edition. PRIMER-E, Plymouth, UK

Coleman C, Dennis G, Jaap WC, Schmahl GP, Koenig C, Reed S, Beaver CR (2005) Status and trends of the Florida Middle Grounds. Tech Rep Gulf of Mexico Fisheries Management Council, Tampa, FL, p 135

Colin PL (1978) Caribbean Reef Invertebrates and Plants. A field guide to the invertebrates and plants occurring on coral reefs of the Caribbean, the Bahamas, and Florida. Tropical Fish Hobbyist, Neptune City, NJ

Colin PL (1980) A brief history of the Tortugas marine laboratory and the Department of Marine Biology, Carnegie Institution of Washington. In: Sears M, Merriman D (eds) Oceanography: the past. Springer, Berlin, pp 138–147

Connell JH (1978) Diversity in tropical rain forests and coral reefs. Science 199:1302–1310

Cowie-Haskell BD, Delaney JM (2003) Integrating science into the design of the Tortugas ecological reserve. MTS Journal 37:1–14

Cox C, Hunt JH, Lyons WG, Davis GE (1996) Nocturnal foraging in the Caribbean spiny lobster, *Panulirus argus*. Proc 24th Benthic Ecol Meeting:30 (abstract)

Darwin C (1846) An account of fine dust which often falls on vessels in the Atlantic Ocean. Q J Geol Soc London 2:26–30

Davenport CB (1926) Alfred Goldsborough Mayor. Biogr Mem Nat Acad Sci 8:1–10

Davis GE (1977) Anchor damage to a coral reef on the coast of Florida. Biol Conserv 11(1):29–34

Davis GE (1979) Outer continental shelf resource management map, coral distribution, Fort Jefferson National Monument, the Dry Tortugas. Bureau of Land Management, New Orleans, LA

Davis GE (1982) A century of natural change in coral distribution at the Dry Tortugas : a comparison of reef maps from 1881–1976. Bull Mar Sci 32:608–623

Delgado GA, Bartels CT, Glazer RA, Brown-Peterson NJ, McCarthy KJ (2004) Translocation as a strategy to rehabilitate the queen conch (*Strobus gigas*) population in the Florida Keys. Fish Bull 102:278–288

Deyrup M, Franz R (eds) (1994) Rare and endangered biota of Florida, Vol IV Invertebrates. University Press of Florida, Gainesville, FL

Dobbs D (2005) Reef madness: Charles Darwin and Alexander Agassiz, and the meaning of coral. Pantheon Books, New York

Domeier ML, Colin PL (1997) Tropical reef fish spawning aggregations: defined and reviewed. Bull Mar Sci 60:698–726

Douglas MS (1947) Everglades : river of grass. Rinehart, New York

Duarte-Bello PP (1961) Corales de los Arrecifes Cubanos. Acuario Nacional Series Educacional 2, Marianao, Cuba, p 84, 74 plates

Duce RA, Tindale NW (1991) Atmospheric transport of iron and its deposition in the ocean. Limnol Oceanogr 36:1715–1726

Dustan P (1985) Community structure of reef building corals in the Florida Keys: Carysfort Reef, Key Largo and Long Key Reef, Dry Tortugas. Atoll Res Bull 228: 27

Erhardt H, Werding B (1975) Los corales (Anthozoa e Hidrozoa) de la ensenada de Granate Pequeña Bahia al este de Santa Marta, Colombia. Caldasia 11:107–138

Fairbridge RW (ed) (1966) The encyclopedia of oceanography. Van Norstrand Reinhold, New York

Florida Marine Research Institute (1998) Benthic habitats of the Florida Keys. Technical Report TR-4, p 53

Fourqurean JW, Zieman JC (2002) Nutrient content of the sea grass Thalassia testudinum reveals regional patterns of relatively available nitrogen and phosphorus in the Florida Keys, USA. Biogeochemistry 61:229–245

Frank KT, Petrie B, Choi JS, Leggett WC (2005) Trophic cascades in a formerly cod-dominated ecosystem. Science 308(5728):1621–1623

Franklin EC, Ault JS, Smith SG, Luo J, Meester GA, Diaz GA, Chiappone M, Swanson DW, Miller SL, Bohnsack JA (2003) Benthic habitat mapping in the Tortugas region, Florida, Marine Geodesy 26:19–24

FWC (Florida Fish and Wildlife Conservation Commission) (2007) Fishing capital of the world. http://fishingcapital.com. Accessed 29 Jan 2007

Garrison VH, Shinn EA, Foreman WT, Griffin DW, Holmes CW, Kellogg CA, Majewski MS, Richardson LL, Ritchie KB, Smith GW (2003) African and Asian dust: from desert soils to coral reefs. BioScience 53:469–480

Geister J (1973) Los arrecifes de la Isla de San Andrés (Mar Caribe, Colombia). Mitteilungen aus dem Instituto Colombo-Alemán de Investigaciones Científicas "Punta De Betín" 7:211–228

Geister J (1977) The influence of wave exposure on the ecological zonation of Caribbean coral reefs. Proc 3rd Int Coral Reef Symp 1:23–29

Gerace DR (1988) Bahamas. In: Wells S (ed) Coral reefs of the World, I, Atlantic and Eastern Pacific. UNEP and IUCN. Geland, Switzerland, pp 13–28

Gilbert PS, Lee TN, Podesta G (1996) Transport of anomalous low-salinity waters from the Mississippi river flood of 1993 to the Straits of Florida. Cont Shelf Res 16:1065–1085

Gillette DA (1981) Production of dust that may be carried great distances. Geol Soc Am Spec Pap 186: 11–26

Ginsburg RN, Shinn EA (1964) Distribution of the reef-building community in Florida and the Bahamas. Amer Assn Petrol Geol 48: 527

Glazer RA, Quintero L (1998) Observations on the sensitivity of the queen conch to water quality : implications for coastal development. Proc Gulf and Caribb Fish Inst 50:78–93

Glynn PW (1996) Coral reef bleaching : facts, hypotheses, and implications. Glob Chang Biol 2:495–509

Goldberg W (1973) The ecology of the coral-octocoral communities off the southeast Florida coast: geomorphology, species composition, and zonation. Bull Mar Sci 23:465–488

González-Ferrer S (2004) Corales Pétreos de Cuba. In: González-Ferrer S (ed) Corales Pétreos, Jardines sumergidos de Cuba, Instituto de Oceanología, Editorial Academica, pp 79–195

Goreau TF (1959) The ecology of Jamaican coral reefs I. Species composition and zonation. Ecology 40:67–90

Goreau TF, Wells JW (1967) The shallow-water Scleractinia of Jamaica : revised list of species and their vertical distribution range. Bull Mar Sci 17:442–453

Goudie AS, Middleton NJ (2001) Saharan dust storms: nature and consequences, Earth-Sci Rev 56:179–201

Greenberg I (1977) Guide to the corals and fishes of Florida, the Bahamas, and the Caribbean. Seahawk, Miami, FL

Griffin DW, Donaldson KA, Paul JH, Rose JB (2003) Pathogenic human viruses in coastal waters. Clin Microbiol Rev 16:129–143

Grimm D, Hopkins T (1977) Preliminary characterization of the octocorallian and scleractinian diversity at the Florida Middle Grounds. Proc 3rd Int Coral Reef Symp 1:135–142

Halas J (1985) A unique mooring system for reef management in the Key Largo National Marine Sanctuary. Proc 5th Int Coral Reef Symp 4:237–242

Halley RB (1979) Guide to sedimentation for the Dry Tortugas. Southeast Geol Publ 21, p 98

Halley RB, Dennis GP, Weaver D, Coleman F (2005) Characterization of Pulley Ridge Coral and Fish Fauna. Technical Report Gulf of Mexico Fisheries Management Council, Tampa, FL, p 72

Hallock P, Lidtz BH, Cockey-Burkhard EM, Donnelly KB (2003) Foraminifera as bioindicators in coral reef assessment and monitoring. Environ Monit Assess 81:221–238

Harper DE, Bohnsack JA, Lockwood B (2000) Recreational fisheries in Biscayne National Park, Florida, 1976–1991. Mar Fish Review 62:8–26

Herren L (2004) St. Lucie Inlet State Park reef monitoring program. Florida Dept Env Protection, Progress Rep 2, p 20 and appendices

Hoegh-Guldberg O (1999) Climate change, coral bleaching, and the future of the world's coral reefs. Mar Freshwater Res 50:839–866

Hoffmeister JE (1974) Land from the sea. The geological story of south Florida. University of Miami Press, Coral Gables, FL

Horta-Puga G, Carricart-Ganivet JP (1993) Corales pétreos recientes (Milleporina, Stylasterina y Scleractinia) de México. In: Salazar-Vallejo SI Gonzales NE (eds) Biodiversidad Marina y Costera de México. Comisión Nacional para la Biodiversidad y Centro de Investigaciones de Quintana Roo, México, pp 66–79

Hu C, Hackett K, Callahan MK, Andréfouë S, Wheaton JW, Porter JW, Muller-Karger FE (2003) The 2002 ocean color anomaly in the Florida Bight: a cause of local coral reef decline. Geophys Res Lett 30:1151–1154

Hughes TP, Baird AH, Bellwood DR, Card M, Connolly SR, Folke C, Grosberg R, Hoegh-Guldberg O, Jackson JBC, Kleypas J, Lough JM, Marshall P, Nyström M, Palumbi SR, Pandolfi JM, Rosen B, Roughgarden J (2003) Climate change, human impacts, and the resilience of coral reefs. Science 301:929–933

Humann P, DeLoach N (2002) Reef coral identification – Florida, Caribbean, Bahamas, including marine plants, Second Edition. New World, Jacksonville, FL

Ingle RM, Eldred B, Sims HW Jr, Eldred EA (1963) On the possible Caribbean origin of Florida 's spiny lobster populations. Florida State Board of Conser Mar Lab Tech Ser 40: p 12

Intergovernmental Panel on Climate Change (2007) Climate change 2007: the physical science basis. WGI Fourth Assessment Report, Paris, p 18

Jaap WC (1979) Observation on zooxanthellae expulsion at Middle Sambo Reef, Florida Keys. Bull Mar Sci 29:414–422

Jaap WC (1984) The ecology of south Florida coral reefs: a community profile. US Fish and Wildlife Service. FWS/OBS-82/08, Washington, DC, p 138

Jaap WC (2000) Coral reef restoration. Ecol Eng 15:345–364

Jaap WC, Hallock P (1990) Coral reefs. In: Myers RL, Ewell JJ (eds) Ecosystems of Florida. University of Central Florida Press, Orlando, FL, pp 574–616

Jaap WC, Olson D (2000) Scleractinia Coral Diversity and Community Structure: Lucaya, Grand Bahama Island, Bahamas. Proc 20th AAUS Diving for Sci Symp, St. Petersburg, FL, pp 18–25

Jaap WC, Sargent FJ (1993) The status of the remnant population of Acropora palmata (Lamarck, 1816) at Dry Tortugas National Park, Florida, with a discussion of possible causes of changes since 1881. Proc Colloquium on global aspects of coral reefs: hazards and history. University of Miami, Florida, pp 101–105

Jaap WC, Wheaton JW (1995) Annual report for amendment 5, sub-agreement CA-5000–09027/1, Cooperative agreement CA-5000–8–8014. U.S. National Park Service, US Dept Interior: Benthic coral reef monitoring at Dry Tortugas National Park: 1 October 1994 to 30 September 1995, St Petersburg, p 80

.Jaap WC, Lyons WG, Dustan P, Halas JC (1989) Stony coral (Scleractinia and Milleporina) community structure at Bird Key Reef, Ft. Jefferson National Monument, Dry Tortugas, Florida. Florida Mar Res Pub 46, p 31

Jaap WC, Wheaton JW, Donnelly KB (1991) A three-year evaluation of community dynamics of coral reefs at Fort Jefferson National Monument, Dry Tortugas, Florida, USA. Florida Mar Res Inst. St. Petersburg, FL, p 22 + figures and tables

Jaap WC, Mallinson D, Hine A, Muller P, Jarrett B, Wheaton JL (2001) Deep reef communities : Tortugas Banks and Pulley Ridge. In: Martin Willison JH, Hall J, Gass SE, Kentchington ELR, Butler M, Doherty P (eds) Proc 1st Int Symp on deep sea corals. Ecol Action Centre, Halifax, Nova Scotia, p 209 (abstract)

Jaap WC, Wheaton JL, Hackett KE, Callahan MK, Kupfner S, Kidney J, Lybolt M (2002) Long-term (1989–2002) monitoring of selected coral reef sites at Dry Tortugas National Park. Florida Fish and Wildlife Conservation Commission FMRI, p 43

Jaap WC, Hudson JH, Dodge RE, Gilliam D, Shaul R (2006) Coral reef restoration with case studies from Florida. In: Cote IM, Reynolds JD (eds) Coral reef conservation. University of Cambridge Press, Cambridge, pp 478–514

James NP, Stearn C, Harrison R (1977) Field guidebook to modern and Pleistocene reef Carbonates, Barbados, West Indies. Atlantic Reef Committee, University of Miami, Florida, p 30

Jickells TD (2005) Global iron connections between desert dust, ocean biogeochemistry, and climate. Science 308:67–71

Jiménez AN (1982) Cuba: La Naturaleza y el Hombre El Archipiélago 1: Editorial letras Cubanas. Habana, Cuba

Jiménez AN (1984) Cuba: La Naturaleza y el Hombre El Archipiélago 2: Bojeo Editorial letras Cubanas. Habana, Cuba

Johns GM, Leeworthy VR, Bell FW, Bonn MA (2001) Socioeconomic study of reefs in southeast Florida.

Hazen and Sawyer, NOAA, and Florida Fish and Wildlife Conser Comm, p 235 and appendices

Johns GM, Milon JW, Sayers D (2004) Socioeconomic study of reefs in Martin County, FL. Final Report. Martin County, Technical Report

Jones AC, Berkeley SA, Bohnsack JA, Bortone SA, Camp DK, Darcy GH, Davis JC, Haddad KD, Hedgepth MY, Irby EW Jr, Jaap WC, Kennedy FS Jr, Lyons WG, Nakamura EL, Perkins TH, Reed JK, Steidinger KA, Tilmant JT, Williams RO (1985) Ocean habitat and fishery resources of Florida. In: Seaman Jr W (ed) Florida aquatic habitat and fishery resources. Florida Chapter Am Fish Soc, Kissimmee, FL, pp 437–542

Jordán-Dahlgren E, Rodríguez-Martínez RE (2003) The Atlantic coral reefs of México. In: Cortés J (ed) Latin American coral reefs. Elsevier, Amsterdam, The Netherlands, pp 131–158

Kaplan EH (1982) A field guide to the coral reefs of the Caribbean and Florida. Peterson Field Guide Series, number 27. Houghton, Boston, MA

Kleypas JA, Buddemeier RW, Archer D, Gattuso JP, Langdon C, Opdyke BN (1999) Geochemical consequences of increased atmospheric CO2? on coral reefs. Science 284:118–120

Kruczynski WL (1999) Water quality concerns in the Florida Keys: sources, effects, and solutions. US EPA and FKNMS Rep, p 65

Kühlmann DHH (1971) Die Korallen riffe Kubas II. Zur Ökologie der Bankriffe und ihrer Korallen. Int Revue der gesamten Hydrobiologie 2:145–199

Kühlmann DHH (1974a) Die Korallen riffe Kubas III. Riegel riff und Korallenterrasse, zwei verwandte Erscheinungen des Bankriffes. Int Revue der gesamten Hydrobiologie 159:305–325

Kühlmann DHH (1974b) The coral reefs of Cuba. Proc 2nd Int Coral Reef Symp:69–83

Lee TN, Williams RE, McGowan M, Szmant AF, Clarke ME (1992) Influence of gyres and wind-driven circulation on transport of larvae and recruitment in the Florida Keys coral reefs. Continental Shelf Res 12:97–1002

Lee TN, Clarke ME, Williams E, Szmant AF, Berger T (1994) Evolution of the Tortugas gyre and its influence on recruitment in the Florida Keys. Bull Mar Sci 54:621–646

Lee TN, Williams E, Johns E, Wilson D, Smith NP (2002) Transport processes linking south Florida coastal ecosystems. In: Porter J, Porter K (eds) The Everglades, Florida Bay, and coral reefs of the Florida Keys : an Ecosystem Sourcebook, CRC, Boca Raton, FL, pp 309–342

Leeworthy VR, Vanasse P (1999) Economic contribution of recreating visitors to the Florida Keys/Key West : updates for years 1996–1997 and 1997–1998. NOAA, Silver Springs, MD, 20 p

Leichter J, Stewert J, Miller SL (2003) Episodic nutrient transport to Florida coral reefs. Limnol Oceanogr 48:1394–1407

Lewis JB (1960) The coral reefs and coral communities of Barbados, West Indies. Can J Zool 38:1133–1145

Lighty RG (1977) Relict shelf-edge Holocene coral reef: southeast coast of Florida. Proc 3rd Int Coral Reef Symp 2:215–222

Lindeman KC, Pugliese R, Waugh GT, Ault JS (2000) Developmental pathways within a multispecies reef fishery: management applications for essential fish habitats and protected areas. Bull Mar Sci 66:929–956

Mace P (1997) Developing and sustaining world fishery resources: state of science and management. In: Hancock DA, Smith DC, Grant A, Beumer JP (eds) Proc Second World Fishery Congress:1–20

Littler DS, Littler MM (2000) Caribbean reef plants. Offshore Graphics, Washington, DC

Mallinson D, Hine A, Hallock P, Locker S, Shinn E, Naar D, Donahue B, Weaver D (2003) Development of small carbonate banks on the south Florida Platform margin: response to sea level and climate change. Mar Geol 199:45–63

Mayer AG (1902) The Tortugas, Florida as a station for research in biology. Science 17:190–192

Mayer AG (1914) The effects of temperature on tropical marine organisms. Carnegie Inst Wash Publ 183:3–24

Mayer AG (1918) Toxic effects due to high temperature. Carnegie Inst Wash Publ 252:175–178

McGowan MF, Cha SS, Richards WJ (1994a) Effects of the Pourtalès Gyre and coastal oceanography on the recruitment of larval fishes to the Florida Keys: a multivariate approach. In: Florida Coastal Ocean Symp, University of Miami, Florida, p 67

McGowan MF, Limouzy-Paris CB, Richards WJ, Cha SS (1994b) Reef fish larval transport in the Florida Keys. SERCAR results. In: Florida Coastal Ocean Symp, University of Miami, Florida, p 77

McGowan MF, Yeung C, Richards WJ (1994c) South Florida 's lobster recruitment from the Caribbean : additional evidence from plankton distribution and satellite -tracked drifters. In: Florida Coastal Ocean Symp, University of Miami, Florida, p 66

McPherson BF, Halley RB (1996) The south Florida environment: a region under stress. Washington, DC. US Geol Survey circular 1134: 61

Medina M, Weil E, Szmant AM (1999) Examination of the *Montastraea annularis* Species Complex (Cnidaria: Scleractinia) Using ITS and COI Sequences. Mar Biotechnol 1:89–97

Morgan MB, Snell TW (2002) Characterizing stress gene expression in reef building corals exposed to the mosquitoside dibrom. Mar Poll Bull 44:1206–1218

Neves N, Johnsson R, Sampaio C, Pichon M (2006) The occurrence of Scolymia cubensis in Brazil: revising the problem of the Caribbean solitary mussids. Zooraxa 1366:45–54

Neumann AC, Moore WS (1975) Sea level events and Pleistocene coral ages in the northern Bahamas. Quant Res 5:215–224

NMFS (National Marine Fisheries Service) (2005) Status of fisheries of the United States – Report to Congress. http://www.nmfs.noaa.gov/sfa/statusoffisheries/SOSmain.htm

NMFS (National Marine Fisheries Service), Fisheries Statistics Division (2007) Marine recreational fisheries statistics survey (MRFSS). http://www.st.nmfs.gov/st1/recreational/queries/index.html.

NOAA (1996) Florida Keys National Marine Sanctuary final management plan/environmental impact statement. Vol 1 The management plan, Washington, DC

NOAA (2002) The state of the coral reef ecosystems of the United States and Pacific Freely Associated States, Washington, DC

NOAA (2005) The state of the coral reef ecosystems of the United States and Pacific Freely Associated States, Washington, DC

NOAA (2007) Report on the status of marine protected areas in coral reef ecosystems of the United States 1: marine protected areas managed by states, territories, and commonwealths. NOAA Tech Mem CRCP 2, Silver Springs, MD

Ogden J C (1974) The major marine environments of St. Croix. In: Multer HG, Gerhard LC (eds) Guide book to the geology and ecology of some marine and terrestrial environments of St. Croix, U. S. Virgin Islands. Spec Publ 5 West Indies Lab, Farleigh Dickinson University, Christiansted, St. Croix, USVI, pp 5–32

Oppel F, Meisel T (1871) Along the Florida reef. In: Tales of old Florida. Castle, Seacaucus, NJ, pp 265–309

Ortner P, Lee T, Milne P, Zika R, Clarke M, Podesta G, Stewert P, Tester P, Atkinson L, Johnson W (1995) Mississippi River flood waters that reached the Gulf Stream. J Geophys Res 100:13595–13601

Ott B (1975) Community patterns on a submerged barrier reef at Barbados, West Indies. International Revue der gesamten Hydrobiologie 60:719–736

Patterson KL, Porter JW, Ritchie KB, Polson SW, Mueller E, Peters EC, Santavy DL, Smith GW (2002) The etiology of white pox, a lethal disease of the Caribbean Elkhorn coral, Acropora palmata. Proc Natl Acad Sci 99:8725–8730

Paul JH, McLaughlin MR, Griffin DW, Lipp EK, Stokes R, Rose JB (2000) Rapid movement of wastewater from on-site disposal systems into surface waters in the lower Florida Keys. Estuaries 23:662–668

Paulay G (1996) Dynamic clams: changes in the bivalve fauna of Pacific islands as a result of sea-level fluctuations. Am Malacol Bull 12:45–57

Peters EC, Oprandy JJ, Yevich PP (1983) Possible causal agent of "White Band Disease" in Caribbean Acroporid corals. J Invert Path 41:394–396

Peters EC, Cairns SD, Pilson MEQ, Wells JW, Jaap WC, Lang JC, Visclosky CEC, Gollahon L St. P (1988) Nomenclature and biology of Astrangia poculata (=A. danae = A. astreiformis (Cnidaria: Anthozoa). Proc Biol Soc Wash 101:234–250

Pfaff R (1969) Las Scleractinia y Millepora de las Islas del Rosario. Mitteilungen aus dem Instituto Colombo-Alemán de Investigaciones Científicas "Punta De Betín" 3:17–24

Porch CE, Eklund AM, Scott GP (2003) An assessment of rebuilding times for goliath grouper. SEDAR6-RW-3, 25 pp

Porch CE, Eklund AM, Scott GP (2006) A catch-free stock assessment model with application to goliath grouper (Epinephelus itajara) off southern Florida. Fish Bull 104: 89–106

Porter JW (1972) Ecology and species diversity of coral reefs on opposite sides of the Isthmus of Panama. Bull Biol Soc Wash 2:89–116

Porter JW, Porter KG (eds) (2002) The Everglades, Florida Bay, and coral reefs of the Florida Keys. CRC, Boca Raton, FL

Porter J, Battey J, Smith G (1982) Perturbation and change in coral reef communities. Proc Natl Acad Sci 79:1678–1681

Porter JW, Kosmynin V, Patterson KL, Porter KG, Jaap WC, Wheaton JL, Hackett K, Lybolt L, Tsokos CP, Yanev G, Marcinek DM, Dotten J, Eaken D, Patterson M, Meier OW, Brill M, Dustan P (2002) Detection of coral reef change by the coral reef monitoring project. In: Porter JW, Porter KG (eds) The Everglades, Florida Bay, and coral reefs of the Florida Keys. CRC, Boca Raton, FL, pp 749–769

Pourtalès LF (1863) Contributions to the fauna of the Gulf Stream at great depths. Bull Mus Comp Zool 1:103–120

Pourtalès LF (1869) Contributions to the fauna of the Gulf Stream at great depths. Bull Mus Comp Zool 1:121–142

Pourtalès LF (1871) Deep sea corals. Illustrated Cat. Bull Mus Comp Zool 4:1–93

Pourtalès LF (1878) Reports on the results of dredging in the Gulf of Mexico by the US Coast Survey Steamer Blake II. Crinoids and corals. Bull Mus Comp Zool 4:63–87

Pourtalès LF (1880a) Reports on the results of dredging in the Gulf of Mexico by the US Coast Survey Steamer Blake VI. Corals and Antipatharia. Bull Mus Comp Zool 6:95–120

Pourtalès LF (1880b) Figure explanations, pls 1–22. In: Agassiz L (ed) Report on the Florida reefs. Mem Mus Com Zool 7

Prospero JM, Nees RT, Uematsu M (1987) Deposition rate of particulate and dissolved aluminum derived from Saharan dust in precipitation at Miami, Florida, J Geophys Res 92:14723–14731

Richardson LL (1998) Coral diseases: what is really known? Trends in Ecol and Evol 13:438–443

Roberts HH (1977) Field guidebook to the reefs and geology of Grand Cayman Island, British West Indies. Atlantic Reef Committee, University of Miami, Florida, p 41

Roberts HH, Rouse LJ Jr, Walker ND, Hudson H (1982) Cold water stress in Florida Bay and northern Bahamas : a product of winter frontal passages. J Sed Petrol 52:145–155

Robertson WB Jr (1964) The terns of the Dry Tortugas. Bull Florida State Mus 8: 93

Roos PJ (1964) Distribution of reef corals in Curaçao. Studies fauna Curaçao and other Caribbean Islands 20, p 57, 12 pls

Roos PJ (1971) The shallow-water stony corals of the Netherlands Antilles. Studies fauna Curaçao and other Caribbean Islands 37, p 108, 53 pls

Royal Society (2005) Ocean acidification due to increasing atmospheric carbon dioxide. Clyvedon, Cardiff, UK

Rützler K, Santavy DL (1983) The black band disease of Atlantic reef corals. I. Description of a cyanophyte pathogen. PSZN I: Mar Ecol 4:329–358

Scatterday JW (1974) Reefs and associated coral assemblages off Bonaire, Netherlands Antilles and their bearing on Pleistocene and Recent models. Proc 2nd Int Coral Reef Symp 2:85–106

Schmidt TW, Pikula L (1997) Scientific studies on Dry Tortugas National Park: an annotated bibliography. US Dept. of Commerce, NOAA and U.S. Department of the Interior, NPS, p 108

Schmidt TW, Ault JS, Bohnsack JA, Luo J, Smith SG, Harper DE, Meester GA, Zurcher N (1999) Site characterization for the Dry Tortugas region: fisheries and essential habitats. NOAA Tech Mem NMFS-SEFSC-000, 115 pp

Schomer NS, Drew RD (1982) An ecological characterization of the lower Everglades, Florida Bay, and the Florida Keys. US Fish and Wildlife Service, Washington, DC, FWS/OBS-82/58.1, Washington, DC, p 246

Sheridan P, Hill R, Matthews G, Appeldoorn R, Kojis B, Matthews T (2005) Does trap fishing impact coral reef ecosystems? An update. Proc Gulf Caribb Fish Inst 56:511–519

Shinn EA (1963) Spur and groove formation on the Florida Reef Tract. J Sed Petrol 33:291–303

Shinn EA (1975) Coral reef recovery in Florida and the Persian Gulf. Environ Geol 1:241–254

Shinn EA (1979) Collecting biological and geological specimens in south Florida. Atlantic Reef Committee Mem 1: 4

Shinn EA, Hudson JH, Halley RB, Lidz B (1977) Topographic control and accumulation rates of some Holocene coral reefs: south Florida and Dry Tortugas. Proc 3rd Int Coral Reef Symp 2:1–7

Shinn EA, Hudson JH, Halley RB, Lidz B (1981) Spur and grooves revisited: construction versus erosion. Proc 4th Int Coral Reef Symp 1:475–484

Shinn EA, Smith GW, Prospero JM, Betzer P, Hayes ML, Garrison V, Barber RT (2000) African dust and the demise of Caribbean coral reefs. Geophys Res Lett 27:3029–3032

Shinn EA, Griffin DW, Seba DB (2003) Atmospheric transport of mold spores in clouds of desert dust. Arch Envir Health 58:498–504

Sims HW Jr, Ingle RM (1966) Caribbean recruitment of Florida 's spiny lobster population. Quart J Florida Acad Sci 29:207–242

Sklar F, McVoy C, VanZee R, Gawlik DE, Tarboton K, Rudnick D, Miao S, Armentano T (2002) The effects of altered hydrology on the ecology of the Everglades. In: Porter J, Porter K (eds) The Everglades, Florida Bay, and coral reefs of the Florida Keys : an ecosystem sourcebook. CRC, Boca Raton, FL, pp 39–82

Smith FGW (1948) Atlantic Reef Corals, a Handbook of the common reef and shallow water Corals of Bermuda, Florida, and the West Indies, and Brazil. University of Miami Press, Coral Gables, FL

Smith FGW (1954) Gulf of Mexico Madreporaria. Fish Bull 89:291–295

Smith FGW (1971) Atlantic reef corals, a handbook of the common reef and shallow water Corals of Bermuda, The Bahamas, Florida, the West Indies, and Brazil, revised edition. University of Miami Press, Coral Gables, FL

Smith GB (1971) The 1971 red tide and its impact certain communities in the mid-eastern Gulf of Mexico. Environ Lett 9:141–152

Smith GB (1976) Ecology and distribution of eastern Gulf of Mexico reef fishes. Florida Mar Res Publ 19, p 78

Smith NP (1994) Long-term Gulf to Atlantic transport through tidal channels in the Florida Keys. Bull Mar Sci 54:602–609

Snelgrove PVR (1999) Getting to the bottom of marine biodiversity: sedimentary habitats. BioSci 49:129–138

Squires DF (1958) Stony corals from the vicinity of Bimini, Bahamas, British West Indies. Bull Mus Nat Hist 115:215–262

Stark WA II (1968) A list of fish of Alligator Reef, Florida, with comments on the nature of the Florida reef fish fauna. Undersea Biol 1:4–40

Steinman AD, Havens KE, Carrick HJ, VanZee R (2002) The past present, and future hydrology and ecology of Lake Okeechobee and its watersheds. In: Porter J, Porter K (eds) The Everglades, Florida Bay, and coral reefs of the Florida Keys : an ecosystem sourcebook. CRC, Boca Raton, FL, pp 19–37

Stephens LD, Calder DR (2006) Seafaring scientist: Alfred Goldsborough Mayor, pioneer in marine biology. University of South Carolina Press, Columbia, South Carolina,

Szmant AL (2002) Nutrient enrichment on coral reefs: is it a major cause of coral reef decline? Estuaries 25:743–766

Szmant AL, Forrester A (1996) Water column and sediment nitrogen and phosphorus distribution patterns in the Florida Keys, USA. Coral Reefs 15:21–41

Thompson MJ, Schmidt TW (1977) Validation of the species/time random count technique sampling fish assemblages at Dry Tortugas. Proc 3rd Int Coral Reef Symp 1:283–288

Vaughan TW (1901) The stony corals of Porto Rican waters. Bull US Fish Comm 2:291–320, 38 pls

Vaughan TW (1911) Recent Madreporaria of southern Florida. Carnegie Inst Wash Yearbook 8:135–144

Vaughan TW (1914) Building of the Marquesas and Tortugas Atolls and a sketch of the geologic history of the Florida reef tract. Carnegie Inst Wash Publ 182 Dept of Marine Biology 5:55–67

Vaughan TW (1915) Growth-rate of the Floridian and Bahamian shoal-water corals Carnegie Inst Wash Year book 13:221–231

Vaughan TW (1916) The results of investigations of the ecology of the Floridian and Bahamian shoal-water corals. Proc Natl Acad Sci 2:95–100

Vaughan TW (1918) The temperature of the Florida coral reef tract. Carnegie Inst Wash Publ 213:321–339

Veron JEN (1995) Corals in Space and Time. The Biogeography and Evolution of the Scleractinia. University of New South Wales Press, Sydney

Verrill AE (1902) Variations and nomenclature of Bermudian, West Indian, and Brazilian reef corals, with notes on various Indo-Pacific corals. Trans Connecticut Acad Arts and Sci 1:63–167

Voss GL (1976) Seashore life of Florida and the Caribbean. EA Seamann, Miami, FL

Voss GL, Voss NA (1955) An ecological survey of Soldier Key, Biscayne Bay, Florida. Bull Mar Sci, Gulf and Caribbean 5:201–229

Voss GL, Bayer FM, Robins CM, Gomon M, Laroe ET (1969) The marine ecology of Biscayne National Monument. Inst Mar Atmospheric Sci, University of Miami. Rep to the National Park Service, p1 28, 40 figs

Voss JD, Richardson LL (2006) Nutrient enrichment enhances black band disease progression in corals. Coral Reefs 25:569–576

Warwick RM, Clarke KR (1998) Taxonomic distinctness and environmental assessment. J Applied Ecol 35:532–543

Warwick RM, Clarke KR (2001) Practical measures of marine biodiversity based on relatedness of species. Oceanog Mar Biol Ann Rev 2001 39:207–231

Wells JW (1932) A Study of the reef Madreporaria of the Dry Tortugas and sediments of coral reefs. Cornell University Of Ithaca, NY, p 138 (unpublished manuscript)

Wells JW (1956) Coral reefs. In: Hedgpeth J (ed) Treatise on marine ecology and paleoecology, Vol 1, Ecology. Geol Soc Am Mem 67, New York Lithographic Corp, New York, pp 609–631

Wells JW (1973) New and old scleractinian corals from Jamaica. Bull Mar Sci 23:16–55

Wells JW, Lang JC (1973) Appendix, systematic list of Jamaican shallow-water Scleractinia. Bull Mar Sci 23:55–58

Werding B, Erhardt H (1976) Los corales (Anthozoa e Hidrozoa) de la Bahía Chenque en al Parque Nacional "Tairona" (Colombia). Mitteilungen aus dem Instituto Colombo-Alemán de Investigaciones Científicas "Punta De Betín" 8:45–57

Wheaton JW, Jaap WC (1988) Corals and other prominent benthic Cnidaria of Looe Key National Marine Sanctuary, Florida. Florida Mar Res Publ 43, p 25

Wilkinson CR (1996) Global change and coral reefs: impacts on reefs, economies and human cultures. Global Change Biol 2(6):547–558

Wilkinson CR (ed) (2004) Status of coral reefs of the world: 2004. Aus Inst Mar Sci, vol 1, 301 pp, vol 2, 255 pp

Williams T, Kano A, Ferdelman T, Henriet J-P, Abe K, Andres MS, Bjerager M, Browning EL, Cragg BA, DeMol B, Dorschel B, Foubert A, Frannk TD, Fuya Y, Gaillot P, Gharib JJ, Gregg JM, Huvenne VAI, Léonide P, Li X, Mangelsdorf K, Tanaka A, Monteys X, Novosel I, Sakai S, Samarkin VA, Sasaki K, Spivack AJ, Takashima C, Titshack J (2006) Cold-water coral mounds revealed. EOS Trans Am Geophysical Union 87:525–526

Willis BL, van Oppen MJH, Miller DJ, Vollmer SV, Ayre DJ (2006) The role of hybridization in the evolution of reef corals. Ann Rev Ecol Evol and Systematics 37:489–517

Zeiller RW (1974) Tropical marine invertebrates of southern Florida and the Bahama Islands. Wiley, New York

Zlatarski VN, Estalella NM (1982) Les Scléractiniaires de Cuba. Academy of Sciences Bulgare, Sofia

4

Coral Reefs, Present and Past, on the West Florida Shelf and Platform Margin

Albert C. Hine, Robert B. Halley, Stanley D. Locker, Bret D. Jarrett,
Walter C. Jaap, David J. Mallinson, Kate T. Ciembronowicz,
Nancy B. Ogden, Brian T. Donahue, and David F. Naar

4.1 Introduction

In spite of the subtle, low-relief contours seen on bathymetric maps of Florida's Gulf of Mexico (west Florida) shelf and slope (Fig. 4.1), this rim-to-ramp carbonate platform has and continues to support a surprisingly wide variety of coral reefs as compared to much better-known morphologically complex areas such the Great Barrier Reef. From the mid-shelf to the upper slope, light-dependent, hermatypic coral reefs have formed as a result of hard substrate availability, ideal oceanographic conditions, and sea-level fluctuations. Indeed, the west Florida slope even supports living light-independent, ahermatypic coral reefs in ~550 m water depth (Newton et al. 1987).

This paper summarizes the geomorphic variability of these different reef types, their geologic setting, and the present coral-reef biological community. The paper is organized along a virtual depth transect by presenting different reef settings and types starting from the shallower mid-shelf or mid-ramp setting, moving to the shelf edge, and then to the deeper upper slope.

4.2 Background

The west Florida shelf/slope is an excellent example of a distally-steepened, carbonate ramp setting, that is fundamentally different than the well-known, rimmed carbonate platforms such as the Bahamas (Ahr 1973; Hine and Mullins 1983; Read 1982, 1985; Mullins et al. 1987, 1988, 1989; Hine 1997). Distally-

steepened ramps are accretionary shelves that have a break-in-slope that occurs in deeper water. The only other well-developed carbonate ramps in the modern warm-temperate/tropical ocean are the Campeche Bank located just across the Yucatan Straits from west Florida (Logan et al. 1969) and the south coast of the Persian Gulf (more of a homoclinal ramp where slope gradient is relatively uniform; Purser 1973; Tucker and Wright 1990). A rimmed carbonate platform has a wave or current-dominated shelf margin supporting shallow reefs or sand shoals. The Bahamas, the Great Barrier Reef, the Belize Shelf, and the Florida Keys are examples of rimmed carbonate margins.

The west Florida shelf/slope rests on the Florida Platform, a huge, shallow-water carbonate depositional system that began accumulating sediments in the late Jurassic on top of a mostly crystalline bedrock basement (Klitgord et al. 1984; Sheridan et al. 1988; Poag 1991). The south-southwestern portion of this platform remains to this day a carbonate-dominated province. The southern portion supports reefs along the margin, which define the shelf-slope break. This area is a classic rimmed platform margin (Hine and Neumann 1977; Tucker and Wright 1990). To the west, this reef rim disappears and the Florida Platform becomes an open, non-rimmed margin as the margin curves around to the north. Here, the Florida Platform becomes a classic ramp. The distally-steepened ramp terminates on top of the West Florida Escarpment, a huge erosional, submarine sea-cliff having ~1,800 m of relief (Fig. 4.1).

Prior to the mid-Cretaceous, the western portion of the Florida Platform was dominantly a shallow-water system that supported rudist reefs at the margin.

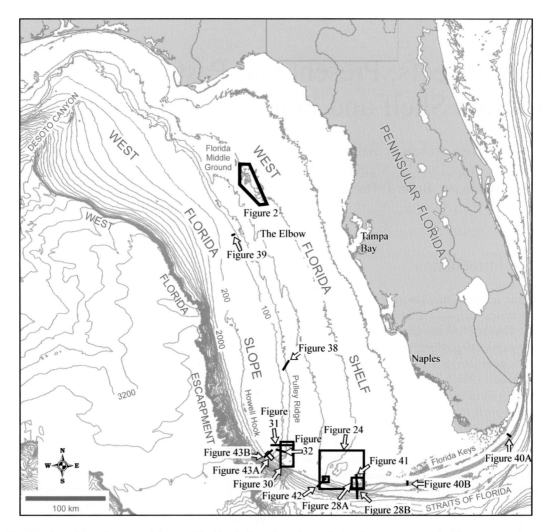

FIG. 4.1. Overall location map of the west Florida platform. Locations of other figures presented in this paper are shown. In the captions, figures are numbered as 4.1–4.43. All depths in meters.

However, most likely due to a succession of oceanic anoxic events (OAEs) coupled with sea-level rise in the mid-Cretaceous, this margin drowned resulting in a widespread and prominent unconformity known as the mid-Cretaceous sequence boundary (Arthur and Schlanger 1979; Corso et al. 1989; Vogt 1989; Buffler 1991). Eventually, the outer portion of the west Florida Platform subsided below the photic zone becoming dominated by pelagic sediments. A foraminifera-pteropod ooze, mixed with siliciclastic muds from the Mississippi River plume, is now the predominant sediment beyond ~600 m water depth (Mullins et al. 1987, 1988, 1989). The modern shelf/slope boundary is located roughly at 75 m water depth, and the slope

extends down to about 1,800 m water depth where the West Florida Escarpment drops off to ~3,200 m water depth.

The end result of this geologic history is one of the largest continental shelf/slope systems in the world that extends 900 km along the 75 m bathymetric line, passes through 6.5° latitude (700 km), and is up to 250 km wide. The low shelf gradient (0.2 –4 m/km) provides a template over which numerous sea-level driven transgressions and regressions have passed. In addition, its limestone foundation provides unique surficial and subterranean karst features. Finally, the siliciclastic influx from the eastern, exposed portion of the platform and from the modern fluvial systems to the north have produced a regional, carbonate-siliciclastic system (Hine 1997; Hine et al. 2003; Hine et al., in press). As a result, the west Florida shelf/slope, situated on top of the Florida Platform, presents pronounced significant depth, substrate, oceanographic, and climatic transitions that convolved to produce and maintain the coral reefs we see today.

4.3 Mid Shelf Reefs

4.3.1 Florida Middle Ground

The Florida Middle Ground (FMG) is a complex cluster of small carbonate banks that supports a diverse benthic environment with variable relief in a mid-to-outer shelf setting. (Figs. 4.1, 4.2). In general, the FMG trends north-northwest, parallel to the platform margin and is ~60 km long by ~15 km wide. Individual banks are approximately 12–15 m in height, and as much as 2–3 km in width. Water depths range from 45 m in surrounding basinal areas to a minimum depth of 24 m on bank-tops (Fig. 4.2). The FMG represents a relict or "give-up" reef (Neumann and Macintyre 1985), with a diverse and complex geomorphology, which is the product of the convoluted interplay of carbonate production, climate and sea-level change, and physical oceanographic processes. This reefal complex is unique, as it occurs in the middle of a carbonate ramp and is the northernmost large reef structure (albeit no longer framework building) in the Gulf of Mexico. Although a number of geologists and biologists have examined this unique reef, many critical questions remain regarding the age of reef growth, the paleoceanographic setting (which was certainly different from today), and the foundation for initial coral recruitment.

The entire west Florida shelf lies within the global "chlorozoan zone" (skeletal sediments dominated by hermatypic corals and calcareous green algae) predicted by Lees (1975) in his model, which is based on sea-surface temperature and salinity. Temperatures in the FMG area range from ~16°C to 30°C and are amenable to reef growth. However, reef growth is likely limited by excess nutrients and the associated increase in bioerosion rates (Hallock and Schlager 1986; Hallock 1988). At present, the northern half of the west Florida shelf is affected by periodic upwelling and

seasonal chlorophyll plumes indicating relatively high-nutrient water conditions (Gilbes et al. 1996). These plumes have been recognized each spring using Coastal Zone Color Scanner imagery. Plumes begin north of the FMG area near Apalachicola Bay and migrate south-southeast directly over the FMG. Mechanisms proposed for the origin of the plume include Loop Current interactions with the platform margin producing upwelling and entrainment of high-nutrient water masses from fluvial discharge.

The FMG was initially investigated for its biological significance. The first mapping effort by fathometer was performed by Jordan (1952) and Ludwick and Walton (1957). Gould and Stewart (1956) suggested the FMG relief was related to reef growth, based upon analysis of bottom samples. Brooks (1962) first investigated the area using SCUBA and defined the occurrence of reef communities. Austin and Jones (1974) described the physical oceanographic setting and the relationship to productivity. Grimm and Hopkins (1977) described the occurrence of zooxanthellate corals on the FMG. Geological investigations began in the 1970s with Back (1972), who defined the sediment textures and distributions and related them to carbonate production on the reef banks. Hilde et al. (1981) first performed seismic investigations in the area, and defined three prominent seismic stratigraphic units including a Miocene "basement". More detailed sediment work was performed by Doyle et al. (1980) and Brooks (1981). Even more recent seismic investigations include the work of Brooks and Doyle (1991), and Mallinson et al. (1996, 1999, 2006).

4.3.1.1 Geology

Geophysical data reveal a wide variety of geomorphic and acoustic facies that represent a complex geologic framework of relict Pleistocene limestone mantled with Holocene corals and sediment. Diver and ROV observations reveal highly bioeroded areas with abundant cryptic environments, including arches, overhangs, and shallow caverns. High-relief hardbottoms occur in the northern study area and are characterized by vertical reef growth exhibiting relief of ~12–20 m above the surrounding shelf environment. This general hardbottom morphology includes three varieties: (1) semi-continuous

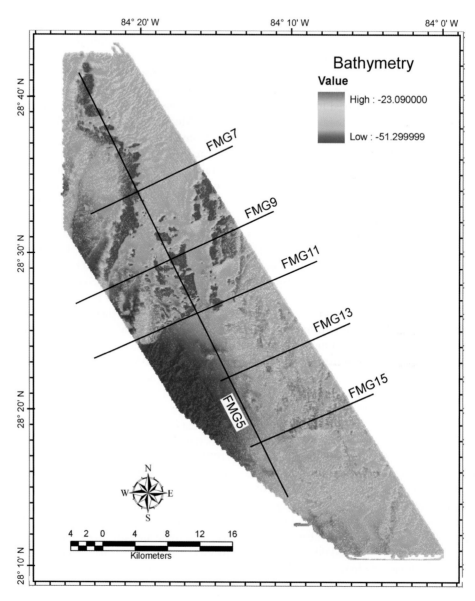

FIG. 4.2. Multibeam bathymetry data reveal a complex geomorphology of small carbonate banks constituting FMG. Location of seismic profiles in Figs. 4.9 and 4.10 are shown

(>8 km) margin-parallel banks 2–3 km in width, (2) isolated flat-topped banks with no particular orientation or maximum dimension, and (3) isolated patch reefs generally <1 km in diameter (Fig. 4.3).

Low-relief hardbottom environments occur in the southern study area and are characterized by relief of ~2–8 m above the surrounding shelf environment. This general hardbottom morphology includes five varieties (Fig. 4.4): (1) undulating/crenulated surfaces exhibiting no discernable patterns, (2) elon-

gate, parallel, overlapping biohermal structures, (3) rounded depressions of unknown origin with central patch reefs ~200 m in diameter, (4) isolated patch reefs, and (5) parallel ridges trending NW–SE.

Soft-bottom environments occur in the surrounding areas and between carbonate banks. Soft-bottom morphologies include: (1) featureless flats, (2) scour depressions surrounding hardbottom environments, (3) scour depressions with active sediment transport, and (4) sand waves

High-Relief Hardbottom Morphologies

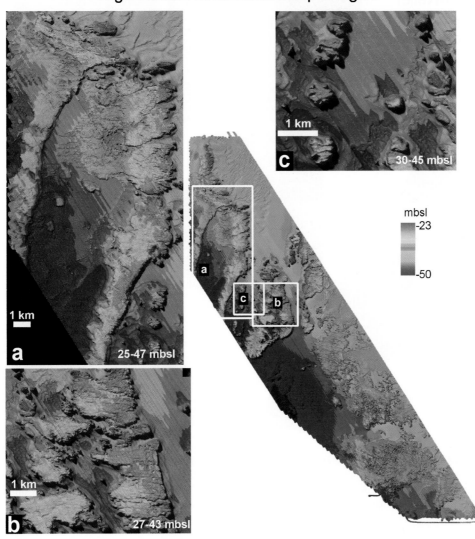

FIG. 4.3. FMG high-relief hardbottoms occur in the northern study area, stand approximately 12–15 m above the surrounding shelf environment, and include three varieties: (**a**) semi-continuous (>8 km) margin-parallel banks 2–3 km in width; (**b**) isolated flat-topped banks with no particular orientation or maximum dimension; (**c**) isolated patch reefs generally <1 km in diameter

(submarine dunes) ~1–2 m in height and 200 m in spacing (Fig. 4.5).

Bedforms, sediment distribution patterns, and scour patterns indicate an influence of southward-flowing and off-shelf directed currents (Fig. 4.5). Typical bottom scenery is shown in Fig. 4.6. Acoustic backscatter data from side-scan and multibeam sonar data reveal a distinctive contrast between coarse carbonate sediments shed from the reefs,

and fine siliciclastic sediments, which are prevalent across much of the surrounding shelf. Low-back-scatter areas (light blue in Fig. 4.7) consist of fine quartz and carbonate sand with a median grain size of ~2.5–3.2 phi. High-backscatter areas (dark blue in Fig. 4.7) consist of very coarse carbonate shell material with a median grain size of ~0 phi. Hardbottom /livebottom areas also yield high backscatter and may be distinguished based on texture (Fig. 4.8).

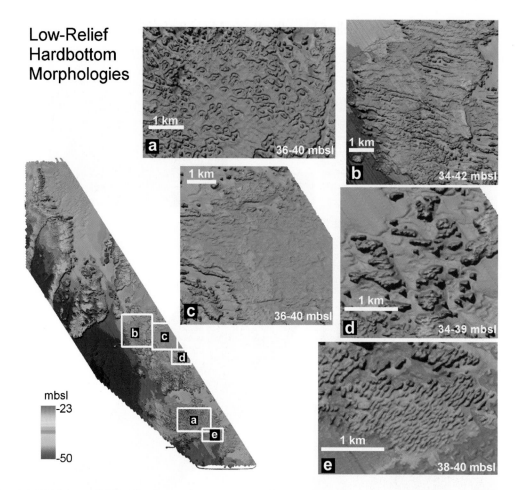

Fig. 4.4. FMG low-relief hardbottom environments occur in the southern study area, stand approximately 2–8 m above the surrounding shelf environment, and include five varieties: (**a**) rounded depressions of unknown origin with central patch reefs, ~200 m in diameter; (**b**) parallel ridges trending NW–SE; (**c**) crenulated surfaces exhibiting no discernable patterns; (**d**) isolated patch reefs; (**e**) elongate, parallel, overlapping biohermal structures

Several high-amplitude, continuous reflections are apparent in seismic data from the FMG (Figs. 4.9, 4.10). The basal reflection occurs at ~65 mbsl (meters below sea level) beneath the western portion of the FMG and rises to ~45 mbsl beneath the eastern portion of the FMG (Fig. 4.10). This surface exhibits up to 4 m of relief and is interpreted to be a karst surface. Based upon previous studies in this region, rocks beneath this surface are likely to be Miocene limestone (Hilde et al. 1981).

Above the karst surface, several seaward-dipping Plio-Pleistocene sequences are recognized. These sequences define a wedge that pinches out beneath the easternmost hardbottoms of the FMG. Sequence boundaries are irregular and indicate dissolution features within a carbonate substrate (Fig. 4.9).

A high-amplitude reflection truncates the Plio-Pleistocene sequences, and separates the dipping sequences from the younger aggradational FMG reefs and surrounding sediments (Figs. 4.9, 4.10). The sediments above this prominent unconformity range in thickness from ~4 to 15 m (under reefs). Soft sediment between reefs is nearly acoustically transparent, whereas the hardbottoms are acoustically opaque. Seismic data are somewhat ambiguous in that a prominent reflection underlies an acoustically transparent bed within the lows between the banks.

The prominent reflection could be interpreted as the Last Glacial Maximum (LGM ~20–18 ka) unconformity, making the FMG very early Holocene. However, there is a faint, discontinuous,

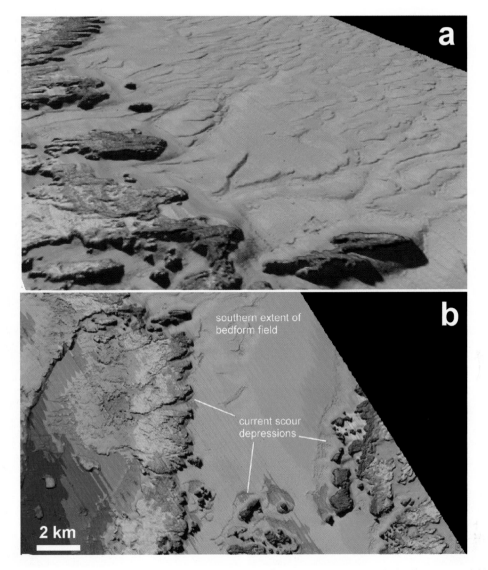

FIG. 4.5. South-flowing shelf currents promote active bedform fields and interact with the bathymetry to yield complex scour patterns surrounding the banks at FMG. (**a**) Bedform field consisting of sand waves of ~1–2 m height and ~500 m spacing (oblique view). (**b**) Current scour depressions surrounding the banks

low-amplitude reflection intermittently observable above the high-amplitude reflection, which has the characteristics of several broad, mound-like structures. One interpretation is that these are lithified, paleoshoreline features upon which the FMG corals were recruited.

Based on the depth, it is possible that the predominance of FMG reef growth occurred rapidly during the early Holocene, well before ~4 ka, when Mississippi River discharge was low due to arid climate conditions over the North American

continental interior (Knox 2000; Forman et al. 2001). In this scenario, these reefs would be contemporaneous with the reefs of Tortugas Bank (near the Dry Tortugas). The reefs at Tortugas Bank grew ~8 m in relief from 8.3 to 4.2 ka (Mallinson et al. 2003). Possible enhanced Mississippi River discharge ~4.2 ka may have terminated reef development (Mallinson et al. 2003) as a result of enhanced turbidity, nutrient loading, and decreased salinity. Alternatively, the reef framework may be much older corresponding to Marine Isotope Stage

FIG. 4.6. Selected photographs of various dive sites at FMG illustrating the predominance of hydrozoans (H), soft corals (G), and sponges (S)

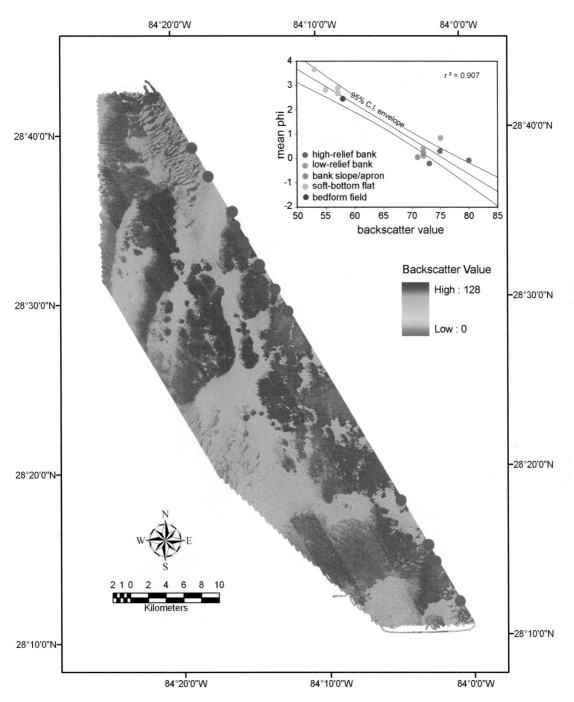

FIG. 4.7. Acoustic backscatter intensity measured with the multibeam system at FMG. Red dots indicate the location of bottom grab samples. A plot of backscatter intensity versus mean phi is shown at upper right. Low backscatter areas (light blue) consist of fine quartz and carbonate sand. High backscatter areas (dark blue) consist of very coarse carbonate shell material. Hardbottom /livebottom areas also yield high backscatter and may be distinguished based on texture

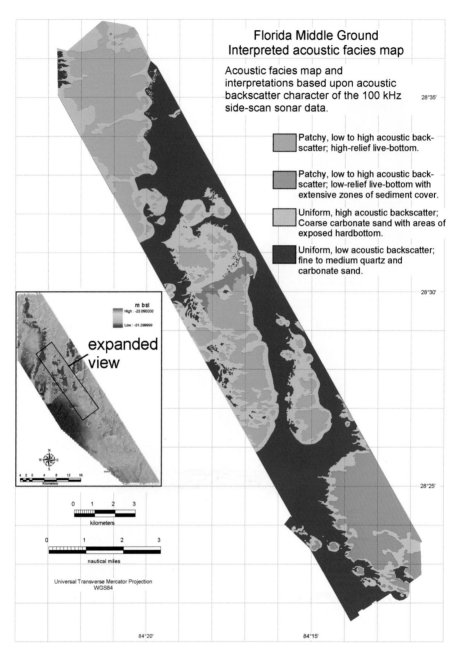

Fɪɢ. 4.8. Generalized acoustic facies map based upon backscatter intensity from side-scan sonar data at FMG

(MIS) 5 or 3 (~80 or ~40 ka), and provided recruitment sites for Holocene coral-reef development.

4.3.1.2 Biology

In 1982 a portion of the FMG was designated a Habitat Area of Particular Concern (HAPC) (Mallinson 2000) in the Coral-Coral Reef Fishery Management Plan under the Magnuson Act (Gulf of Mexico and South Atlantic Fishery Management Councils 1982). In spite of its remote location and the HAPC designation, the FMG faces environmental circumstances that reduce the biological richness. Tropical species are unable to tolerate the cool, winter temperatures while temperate organisms are excluded by the warm, summer temperatures.

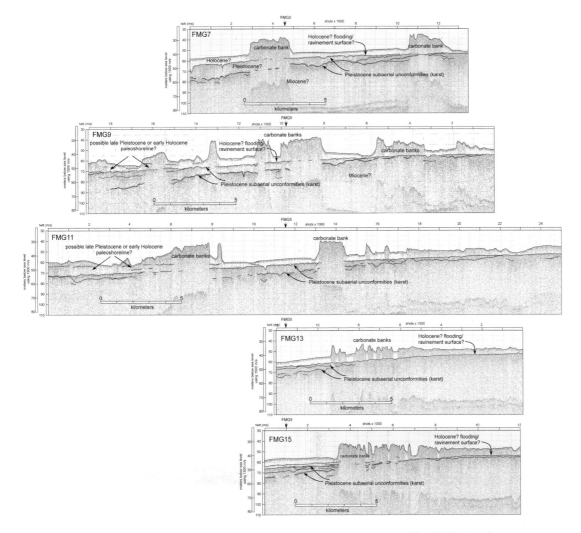

FIG. 4.9. FMG seismic profiles FMG 7, 9, 11, 13, and 15 with interpretations (see Fig. 4.2 for location)

A data buoy near the FMG reports that seawater temperature extreme ranges were 15–33°C from 1994 to 2001. The lowest extreme temperatures occurred in March and the highest in June (Fig. 4.11). Additionally, Mississippi River runoff reaches the FMG on occasion. In May 2003, a plume of Mississippi runoff covered much of the area. This water was cool to cold, reduced in salinity, silt-laden (very turbid), and polluted.

The FMG flora and fauna are eurythermic species. Algae, in particular, exhibit a seasonal pattern (Cheney and Dyer 1974) with the greatest abundance and biomass existing in the late summer and early fall. Late winter and early spring is the period of minimum algal biomass. Common tropical species include *Diadema antillarum* (black, spiny sea urchin), *Spondlus americana* (thorny oyster), *Millepora alcicornis* (fire coral), and *Hermodice carunculata* (fire worm). The area is an enclave of tropical species that co-exist with temperate species. Suspected controlling factors include the complex topographic reef structures providing a diversity of habitat or niches for these plants and animals. Eastern Gulf of Mexico rock-ledge, epibenthic communities are ephemeral due to Red Tides and thermal disturbances (Collard and D'asaro 1973; Godcharles and Jaap 1973; Lyons and Collard 1974; Lyons 1980; Jones et al. 1985). Recruitment of tropical organisms is from local sources as well as propagules that are carried from

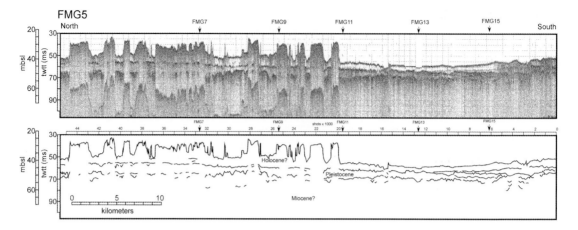

FIG. 4.10. FMG seismic profile FMG5, extending north to south over much of the high-relief hardbottom area (see Fig. 4.2 for the survey location)

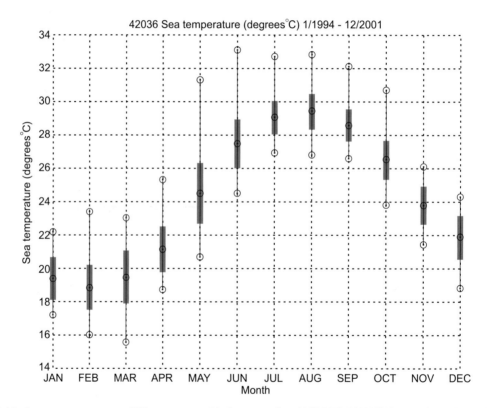

FIG. 4.11. Seawater temperature (C°), oceanographic buoy number 42036 W, 28°30'00"; N, 84°31'00", (NOAA, NDBC ,website; data inclusive 1994–2001near FMG)

the tropics to the FMG by intrusions of the Loop Current (Lee et al. 2002).

Studies in the 1970s documented 103 species of algae, 40 sponge species, 75 mollusk species, 56 decapod crustacean species, 41 polycheate species, 23 echinoderm species (Hopkins et al. 1977) and 170 species of fish (Smith et al. 1975) in the FMG. Most of the massive Caribbean reef-building Scleractinia genera (*Acropora*, *Montastraea*, *Diploria*, and *Colpophyllia*) do not occur at the

FMG (Grimm and Hopkins 1977; Grimm 1978), although it is the northern-most reef in the Gulf of Mexico with significant numbers of zooxanthellate Scleractinia. Grimm and Hopkins (1977) described faunal zones based on octocoral and stony coral (*Milleporina* and *Scleractinia*) distribution and abundance (Table 4.1) and a recent study has confirmed these associations (Coleman et al. 2005). The uppermost levels of the FMG reefs are typically horizontal with some indentations or depressions (depth to 26–28 m) and *Muricea* spp. *Dichocoenia stokesi* and *Porites porites* forma *divaricata* are the characteristic species in this habitat (Figs. 4.12, 4.13).

The slope descends from the horizontal platform to a depth of 30–36 m. Characteristic organisms in this zone include *Madracis decactis* and *Millepora alcicornis* (Figs. 4.14, 4.15). The slope margin is complex with fissures and caves supporting invertebrates, including *Stenorhyinchus seticornis* (arrow crab) and *Diadema antillarum* (Fig. 4.16), the most visible mobile invertebrate animals. The *Millepora alcicornis* occurs in dense clusters at the slope margin, and the ubiquitous purple reef fish (*Chromis scotti*) utilizes *Millepora* colonies as refuge habitat (Fig. 4.15).

Cheney and Dyer (1974) reported a marked seasonal difference in algal species richness and abundance in the FMG. In spring 2003, Rhodophyta were the most diverse and abundant phyla, comprising 62% of the species present, Chlorophyta contributed 20%, and Phaeophyta 17% (Coleman et al. 2005). Algal cover in 2003 was remarkably similar to information reported in Hopkins et al. (1977). He reported 61% of the species present were Rhodophyta, 28% Chlorophyta, and 11%

TABLE 4.1. Faunal zonation, Florida Middle Grounds (Grimm and Hopkins 1977).

Zone depth (m)	Characteristic coral assemblage	Relative position on the reef
26–28	*Muricea-Dichocoenia -Porites*	Horizontal platform
28–30	*Dichocoenia -Madracis*	Margin of the slope and horizontal platform
30–31	*Millepora*	Upper levels of the slope
31–36	*Millepora -Madracis*	Middle and lower levels of the slope

FIG. 4.12. *Muricea* spp., Octocorallia, typical of the upper surfaces of the FMG structures (Photo credit G.P. Schmahl)

Fig. 4.13. *Dichocoenia stokesi*, Scleractinia, two colonies displaying different pigmentation at FMG. This species often is found in clusters of multiple colonies (Photo credit G.P. Schmahl)

Fig. 4.14. *Madracis decactis*, Scleractinia, FMG (Photo credit G.P. Schmahl)

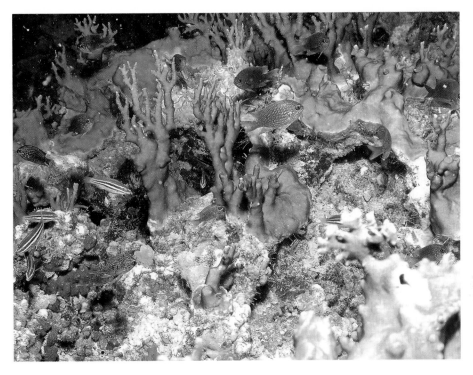

FIG. 4.15. *Millepora alcicornis*, Hydrozoa: Milleporina, with a school of purple reef fish, *Chromis scotti*, The purple reef fish is one of the most abundant fish species in the FMG (Photo credit G.P. Schmahl)

FIG. 4.16. *Diadema antillarum* (long-spine, black sea urchin), and several arrow crabs (*Stenorhyinchus seticornis*) in the back ground at FMG (Photo credit G.P. Schmahl)

were Phaeophyta. He also found that Chlorophyta often dominated in terms of biomass. Gelatinous red algae were the most conspicuous and abundant in late May 2003 (Table 4.2).

Sponges are the second most important component based on percentage cover of the FMG benthic community (Fig. 4.17). In a recent study, 32 sponge species (Table 4.3) were observed at the FMG historical sites (Coleman et al. 2005). *Pseudoceratina crassa*, *Niphates erecta*, *Amphimedon compressa*, and *Cribochalina vasculum* were the most abundant species in quadrat surveys. Sponge density ranged from 9 to12 colonies/m^2 (Table 4.4).

Octocorals are the most ubiquitous epibenthic organism on FMG reef structures, but there are only a few genera and species (Table 4.5, Figs. 4.18, 4.19; Coleman et al. 2005).

Twenty-two species of shallow-water zooxanthellate and azooxanthellate Scleractinia are known to inhabit the FMG (Tables 4.6, 4.7). *Dichocoenia stokesi* and *Meandrina meandrites* only occur in the FMG. They are not found in epibenthic communities between Naples, FL and the FMG. The other species are found throughout the eastern Gulf of Mexico in depths between 10 and 40 m (Jaap et al. 1989; Jaap and Hallock 1990). Both *D. stokesi* and *M. meandrites* are common in the Florida Keys and Dry Tortugas reefs. The FMG Scleractinian community is dissimilar to the Flower Garden Banks, the Florida Keys, and Southeast Florida (Fig. 4.20). The FMG has a greater taxonomic distinctness (Delta plus value; Fig. 4.21) than any of these other locations because the species found in the FMG come from higher-order taxonomic categories (genera, families, suborders) that have fewer species (Warwick and Clarke 2001). For example, the genera *Stephanocenia*, *Meandrina*, and *Dichocoenia* have a single species each. In contrast, the genera *Porites* and *Agaricia* have multiple species.

The FMG is the prime fishing ground in the eastern Gulf of Mexico (Steidinger et al. 1985); groupers are the principal target species (Figs. 4.22, 4.23).

TABLE 4.2. FMG benthic algae at diving stations (Coleman et al. 2005).

Species station	047	247	151	251	GGR	FR	Frequency
Kallymenia westii	XX		XX				0.33
Kallymenia sp.				XX			0.17
Halmenia floridana	XX		XX	XX			0.50
Champia salicornides	XX	XX	XX	XX			0.67
Trichogloea herveyi	XX	XX					0.33
Dictota menstrualis	XX	XX		XX		XX	0.67
Dictyota pulchella			XX				0.17
Sporolithon episporum	XX						0.17
Porlithon pachydermum	XX						0.17
Lithoththamnion reptile				XX			0.17
Lithothamnion sp.	XX						0.17
Gelidium sp.	XX						0.17
Udotea dixonii		XX		XX	XX	XX	0.67
Udotea verticulosa			XX				0.17
Codium coralinanum		XX					0.17
Codium ithmocladum		XX					0.17
Codium intertextum		XX					0.17
Halimeda discoidea		XX					0.17
Halimeda dixonii				XX			0.17
Mesohyllum mesomorphum		XX					0.17
Jania adhaerens		XX					0.17
Lithophyllum congestum		XX					0.17
Botrycladia pyriformis			XX	XX			0.33
Champia parvula				XX			0.17
Chrysymenia enteromorpha			XX	XX		XX	0.50
Chrysymenia sp.						XX	0.17
Colopermenia sinuosa			XX		XX		0.33
Rhiiplia sp.			XX				0.17
Number of taxa	9	11	9	10	2	4	

Fig. 4.17. Sponges are a dominant feature of the FMG, seen here (*Erylus formous*, grey; *Agelas clathrodes*, red) with a blue angelfish, *Holocanthus bermudensis* (Photo credit G.P. Schmahl)

Table 4.3. Sponge species identified from Florida Middle Grounds surveys, G. Schmahl, 2003 (Coleman et al. 2005).

Phylum Porifera		
Class Demospongiae		
Subclass Homoscleromorpha		
Order Homosclerophorida		
	Family Plakinidae	
		1. *Plakortis angulospiculatus*[*]
Subclass Tetractinomorpha		
Order Spirophorida		
	Family Tetillidae	
		2. *Cinachyra alloclada*
Order Astrophorida		
	Family Geodiidae	
		3. *Erylus formosus*
		4. *Geodia neptuni*
Order Hadromerida		
Family Clionaidae		
		5. *Cliona delitrix*[*]
		6. *Anthosigmella varians*
		7. *Spheciospongia vesparium*[*]
	Family Placospongiidae	
		8. *Placospongia melobesioides*
	Family Spirastrellidae	
		9. *Spirastrella coccinea*
Order Chondrosida		
	Family Chondrillidae	
		10. *Chondrilla nucula*

(continued)

TABLE 4.3. (continued)

Subclass Ceractinomorpha		
Order Agelasida		
	Family Agelasidae	
		11. Agelas clathrodes
Order Poecilosclerida		
'	Suborder Microcionina	
	Family Microcionidae	
		12. Clathria (Thalysias) juni-perina
	Suborder Myxillina	
Family Crambeidae		
		13. *Monanchora unguiferus*
		14. *Monanchora barbadensis*
	Suborder Mycalina	
Family Desmacellidae		
		15. *Neofibularia nolitangere**
	Family Mycalidae	
		16. *Mycale laxissima*
Order Halichondrida		
	Family Axinellidae	
		17. *Axinella corrugata*
		18. *Dragmacidon lunaecharta*
	Family Dictyonellidae	
		19. *Scopalina ruetzleri*
Order Haplosclerida		
	Family Callyspongiidae	
		20. *Callyspongia vaginalis*
		21. *Callyspongia armigera*
		22. *Callyspongia fallax**
	Family Niphatidae	
		23. *Amphimedon compressa*
		24. *Cribrochalina vasculum*
		25. *Niphates erecta*
Order Dictyoceratida		
	Family Ircinidae	
		26. *Ircinia felix*
		27. *Ircinia campana*
		28. *Ircinia strobilina*
	Family Thorectidae	
		29. *Smenospongia aurea**
Order Dendroceratida		
	Family Darwinellidae	
		30. *Aplysilla sulfurea*
Order Halisarcida		
	Family Halisarcidae	
		31. *Halisarca* sp.
Order Verongida		
	Family Aplysinidae	
		32. *Aplysina lacunosa*
		33. *Aplysina cauliformis**
	Family Pseudoceratinidae	
		34. *Pseudoceratina crassa*

Smith et al. (1975) compiled a list of 170 species of fish from the area. In the 2003 study, 95 species were identified (Coleman et al. 2005) with the five most abundant species being *Chromis scotti* (purple reef fish), *Chromis rysura* (yellow reef fish), *Halichoeres bivittatus* (slippery dick), *Scarus iserti* (stripped parrot fish), and *Stegastes variabilis* (coco damself-ish). The more abundant commercial food fishes

TABLE 4.4. Quadrat sponge abundance, George Schmahl, 2003 (Coleman et al 2005).

Sponge species	Site 251	Site 247	Site 151	Total	Frequency
Cinachyra alloclata	1	3	7	11	1
Erylus formosus	0	1	2	3	0.67
Geodia neptuni	2	3	0	5	0.67
Anthosigmilla varians	3	1	0	4	0.67
Placospongia melobesioides	2	0	2	4	0.67
Spirastrella coccinea	0	0	1	1	0.33
Agles clathrodes	0	2	2	4	0.67
Clathria (Thalysias) juniperina	0	0	1	1	0.33
Monanchora unguiferus	0	4	2	6	0.67
Monanchora barbadensis	2	0	0	2	0.33
Mycale laxissima	0	0	2	2	0.33
Axinella corrugata	0	0	2	2	0.33
Dragmacidon lunaecharta	1	1	0	2	0.67
Scopalina reutzleri	5	3	2	10	1
Callyspongia vaginalis	4	1	4	9	1
Callyspongia armigera	0	1	0	1	0.33
Niphates erecta	4	10	3	17	1
Amphimedon compressa	8	1	3	12	1
Cribochalina vasculum	6	1	5	12	1
Ircina felix	2	4	0	6	0.67
Ircina campana	1	0	3	4	0.67
Ircina strobilina	2	0	1	3	0.67
Aplysilla sulfurea	1	1	0	2	0.67
Halisarca sp.	0	1	0	1	0.33
Aplysina laculosa	1	0	0	1	0.33
Pseudoceratina crassa	5	10	7	22	1
Gray, encrusting on Geodia	1	1	0	2	0.67
Black ball	1	0	1	2	0.67
Hard, orange	1	4	7	12	1
Orange, ball	1	0	2	3	0.67
Brown, encrusting	1	0	0	1	0.33
Orange, Tethya	0	1	0	1	0.33
Geodia ?	0	0	2	2	0.33
Total					
Species	22	20	21	33	
Colonies	55	54	61	170	
Diversity, H'n, log base e	2.83	2.62	2.85	3.09	
Eveness, J'n	0.94	0.92	0.95	0.88	
No. of 1 m^2 quadrats	5	6	5	16	
Density, colonies m^2	11	9	12.2	10.63	

were *Lutjanus greiseus* (grey snapper), *Mycetoperca phenax* (scamp grouper), and *Epinephelus morio* (red grouper). The area is popular with commercial and recreational fishers. Compared to 30 years ago, there is a marked decline in food fish populations based on visual observations.

Scleractinian corals are a minor component, and the species are hardy, possessing the ability to survive in harsh conditions in contrast to the Florida Keys or the Caribbean. Fish communities are dominated by planktivorous species, such as *Chromis scotti*. In the Florida Keys, reef-fish com-

munities are dominated by *Haemulidae* (grunts), *Scaridae* (parrott fish), and *Pomacentridae* (damsel fish) (Tilmant 1984). While harmful algal blooms, such as Red Tide, are uncommon in the FMG, there are unpublished reports that cold-water upwelling extirpated the benthic flora and fauna in 1977 (T. Hopkins, personal communication 2004). Coleman et al. (2005) reported that the benthic community in the same areas that Hopkins et al. (1977) studied is very similar to the pre-hypo-thermic disturbance community. The abundant epifaunal elements include the same species of algae,

TABLE 4.5. Octocoral observations, FMG, 2003 (Coleman et al. 2005).

Species site	151	247	251	FR	GGR
Eunicea calculata	XX	XX	XX	XX	
Muricea spp.	XX	XX	XX	XX	XX
Plexaurella spp.		XX	XX		
Pseudoplexaura spp.			XX		
Pseudopterogorgia spp.			XX		XX
Eunicea spp.			XX	XX	XX
Lophogorgia cardinalis				XX	XX
Lophogorgia hebes				XX	XX
Diodigorgia nodinifera				XX	
Pseudoplexaura wagneri				XX	

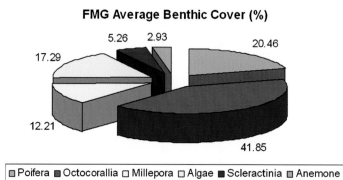

FMG Average Benthic Cover (%)

5.26 2.93 20.46 17.29 12.21 41.85

☐ Poifera ■ Octocorallia ☐ Millepora ☐ Algae ■ Scleractinia ☐ Anemone

FIG. 4.18. Benthic cover contribution of algae, Porifera, Octocoralia, anemones, Millepora, and Scleractiniia, FMG, 2003, N=7 sampling sites (Coleman et al. 2005)

FIG. 4.19. Octocorals and *Madracis decactis*, FMG. (Photo credit T. Hopkins 1976)

TABLE 4.6. Stony corals: milleporina and scleractinia, FMG.

Order Milleporina (Hickson, 1901)		
	Family Milleporidae (Fleming, 1928)	
		1. *Millepora alcicornis* (Linné, 1758)
Order Scleractinia (Bourne, 1900)		
	Family Astrocoeniidae (Koby, 1890)	
		2. *Stephanocenia intersepta* (Lamarck, 1816)
	Family Pocilloporidae (Gray, 1847)	
		3. *Madracis decactis* (Lyman, 1857)
		4. *Madracis pharensis* (Heller, 1868)
	Family Agariciidae (Gray, 1847)	
		5. *Agaricia fragilis* (Dana, 1846)
	Family Siderastreidae (Vaughan and Wells, 1943)	
		6. *Siderastrea radians* (Pallas, 1766)
		7. *Siderastrea siderea* (Ellis and Solander, 1786)
	Family Poritidae (Gray, 1842)	
		8. *Porites porites divaricata* (LeSueur, 1821)
		9. *Porites branneri* (Rathbun, 1888)
	Family Faviidae (Gregory, 1900)	
		10. *Manicina areolata* (Linné, 1758)
		11. *Solenastrea hyades* (Dana, 1846)
	Family Rhizangidae (D'Orbigny, 1851)	
		12. *Astrangia poculata* (Ellis and Solander, 1786)
	Family Oculinidae	
		13. *Oculina diffusa* (Lamarck, 1816)
		14. *Oculina robusta* (Pourtalès, 1871)
	Family Meandrinidae (Gray, 1847)	
		15. *Dichocoenia stokesii* (Milne Edwards and Haime, 1848)
		16. *Meandrina meandrites* (Linné, 1758)
	Family Mussidae (Ortman, 1890)	
		17. *Scolymia lacera* (Pallas, 1766)
		18. *Scolymia cubensis* (Milne Edwards and Haime, 1849)
		19. *Isophyllia sinuosa* (Ellis and Solander, 1786)
	Family Carophylliidae (Vaughan and Wells, 1943)	
		20. *Cladocora arbuscula* (LeSueur, 1821)
		21 *Phyllangia americana* (Milne Edwards and Haime, 1849)
	Family Dendrophylliidae (Gray, 1847)	
		22 *Balanophyllia floridana* (Pourtalès, 1868)

sponges, octocorals, and stony corals (*Millepora* and Scleractinia) that were reported in 1977.

4.4 Small Bank Reefs

Situated along the south Florida margin are three small carbonate banks, Dry Tortugas, Tortugas Bank, and Riley's Hump located ~120 km west of Key West (Figs. 4.1, 4.24–4.28). They have ~40 m of relief, extend ~25 km across, and reside in ~20–30 m water depth. The Dry Tortugas, the largest of the three, does support small sandy cays and shallow living coral reefs (Vaughan 1914; Shinn et al. 1989; Mallinson et al. 1999, 2003). These banks constitute the western extent of the rimmed margin which defines the Florida Keys reef tract. Further to the west, the south Florida margin transitions from a rimmed platform to a carbonate ramp (Jarrett 2003; Jarrett et al. 2005). The morphologic, seismic, and core data all indicate that these banks are reef edifices and have formed by smaller patch

TABLE 4.7. Abundance of stony Corals, FMG, May 2003 (Coleman et al. 2005).

Species site	147	247	151	251	GGR	Total	Frequency
Madracis decactis	9	0	15	14	25	63	0.80
Millepora alcicornis	15	9	13	5	1	43	1
Dichocoenia stokesii	2	1	14	3	0	20	0.80
Scolymia cubensis	0	0	1	6	0	7	0.40
Porites porites divaricata	0	0	3	0	0	3	0.20
Oculina diffusa	0	1	1	0	0	2	0.40
Siderastrea radians	0	0	0	2	0	2	0.20
Porites branneri	0	0	1	1	0	2	0.40
Scolymia lacera	0	0	1	0	0	1	0.20
Madracis pharensis	0	0	0	1	0	1	0.20
Stephanocenia intersepta	0	0	0	0	1	1	0.20
Oculina robusta	0	0	0	0	1	1	0.20
Total							
Number of species	3	3	8	6	4	12	
Number of colonies	26	11	49	32	28	146	
Diversity H'n, log base e	0.88	0.60	1.56	1.49	0.41	1.53	
Eveness, J'n	0.80	0.55	0.75	0.83	0.59	0.61	
No. of 1 m^2 quadrats	9	5	6	8	5	33	
Density, colonies m^2	2.88	2.20	8.17	4.00	5.60	4.42	

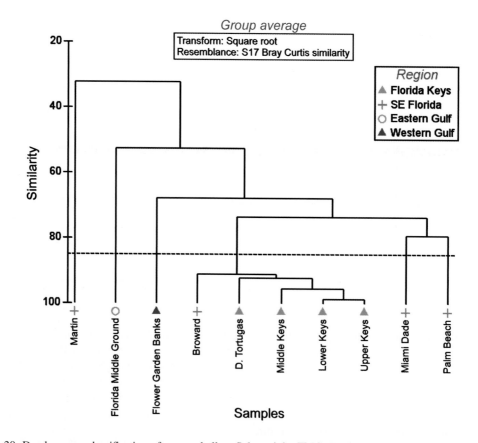

FIG. 4.20. Dendrogram, classification of zooxanthellate Scleractinia, FMG, Florida Keys, Dry Tortugas, Southeast Florida, and the Flower Garden Banks using Bray Curtis Similarity Coefficient and group average sorting to generate the clusters

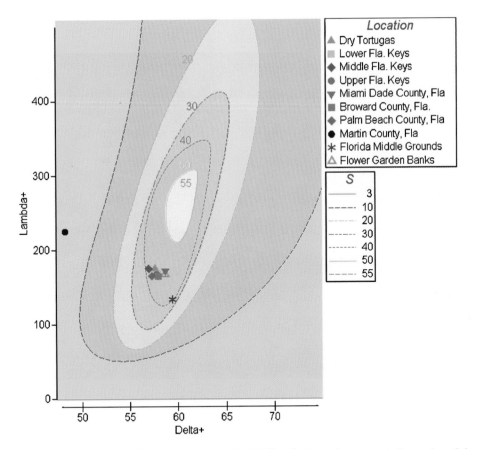

FIG. 4.21. Taxonomic Distinctness (δ +) and Taxonomic Variability (λ +) for the zooxanthellate scleractinia: FMG, Florida Keys, Dry Tortugas, Southeast Florida, and Flower Garden Banks

FIG. 4.22. Scamp grouper, *Mycetoperca phenax*, FMG (Photo credit G.P. Schmahl)

FIG. 4.23. Red grouper, *Epinephelus morio*, FMG (Photo credit G.P. Schmahl)

reefs coalescing and interior basins filling with reef-generated sediment. Radiocarbon age dating further indicates that a ~10 m thick Holocene reef, commencing as early as 9.6 ka, rests unconformably on top of a Pleistocene reef that maybe as much as 20 m in thickness (Fig. 4.25). The 9.6 ka ^{14}C date from Tortugas Bank represents the oldest Holocene date obtained from corals in the Florida Keys.

Mallinson et al. (2003) presented a keep-up, catch-up, give-up scenario (Neumann and MacIntyre 1985) whereby the topographically lower banks were stressed by rapid rates of sea-level rise (Fig. 4.29). As a result, coral-reef growth "steps back" to the next-higher paleo-topographic high. They also proposed that reef-growth demise on Tortugas Bank at 4.2 ka might have been climate-related as a result of decreased salinity, increased nutrients, and increasingly turbid waters emanating from enhanced Mississippi River discharge starting about that time due to a more humid mid-continent climate.

4.5   Reefs and Paleoshorelines

Previous workers have shown that past sea-level stillstands have produced important paleoshore-

lines along the west Florida shelf (Locker et al. 1996; Jarrett 2003; Jarrett et al. 2005). These shorelines have been preserved because they have been constructed from carbonate sand-sized sediments, which cement rapidly in the shallow marine and freshwater/salt-water groundwater transition. Indeed, bottles and beer cans are seen tightly embedded in newly formed limestone indicating the speed of this process (E.A. Shinn, personal communication 2007). So, when a beach/dune complex is generated during a brief sea-level stillstand, it may be largely preserved during the ensuing sea-level rise as a result of pervasive cementation. This drowned, intact shoreline then becomes a rocky substrate surrounded by finer-grained, uncemented sediments (Locker et al. 1996) and provides a surface upon which a coral reef could form.

4.5.1   The Deep, Light-dependent Reef at Pulley Ridge

4.5.1.1   Geology

Southern Pulley Ridge (Fig. 4.30) presents an ideal example of a coral reef that formed on

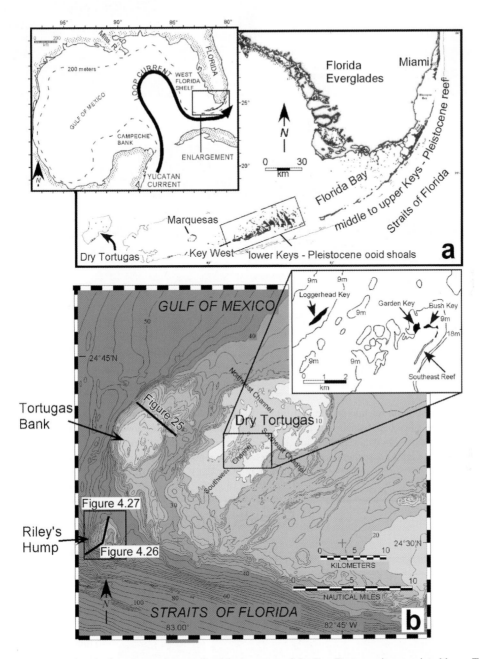

FIG. 4.24. Maps showing the location and generalized bathymetry of the Dry Tortugas, its associated keys, Tortugas Bank, and Riley's Hump. Locations of seismic lines are shown (After Fig. 1 of Mallinson et al. 2003, reproduced by permission of Elsevier)

cemented, coastal carbonate sedimentary deposits (Jarrett 2003; Jarrett et al. 2005). Pulley Ridge is a ~300 km long multiple-ridge complex that lies between 60–90 m water depth and extends N–S along the west Florida outer shelf (Fig. 4.1).

Along its southernmost 30 km portion, multibeam bathymetry reveals an intact barrier island featuring multiple beach ridges, recurved spits, closed-off tidal inlets, cat's eye ponds, and a cuspate foreland – all features found on modern barrier

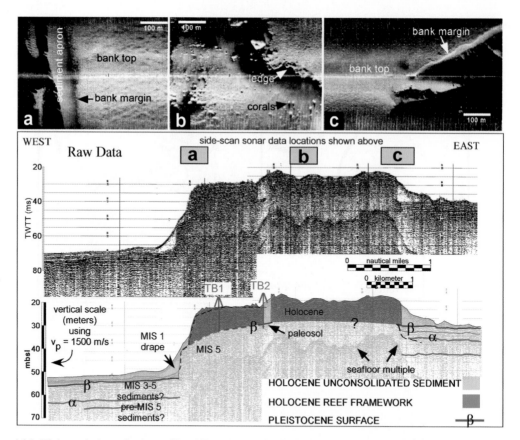

FIG. 4.25. High-resolution seismic profile of Tortugas Bank with interpretation and locations of drill sites TB1 and TB2 and side-scan sonar data (panels a, b, and c). Locations of side-scan data shown in boxes above reef bank. Dark shades of gray in side-scan data represent low back-scatter corresponding to muddy sands. Lighter shades in side-scan data represent high back-scatter corresponding to live-bottom communities, sand, and exposed rock (After Fig 4. of Mallinson et al. 2003; reproduced by permission of Elsevier Ltd.)

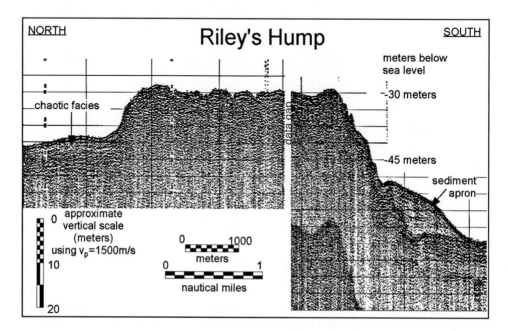

FIG. 4.26. High-resolution seismic profile across Riley's Hump revealing steep reef front and sediment -debris apron at base (After Fig. 6 of Mallinson et al. 2003; reproduced by permission of Elsevier)

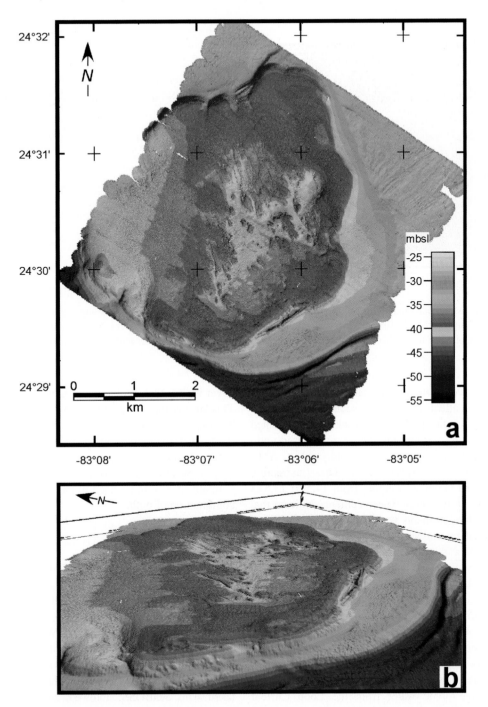

FIG. 4.27. (a) Plan view of 300 kHz multibeam bathymetry of Riley's Hump with shading from artificial light source from SE. (b) 3D perspective of same data looking towards N55E with 10× vertical exaggeration (after Fig. 7 of Mallinson et al. 2003; reproduced by permission of Elsevier)

islands (Figs. 4.30–4.32). Observations from submersibles and video imagery from ROV-mounted cameras show bedded, rocky outcroppings (Fig. 4.32). These outcroppings are identical to those studied by

Locker et al. (1996) at the similar depths located approximately 150 km to the east seaward of the south-facing Florida Keys reef tract just SW of Key West. Rock samples retrieved from this eastern area

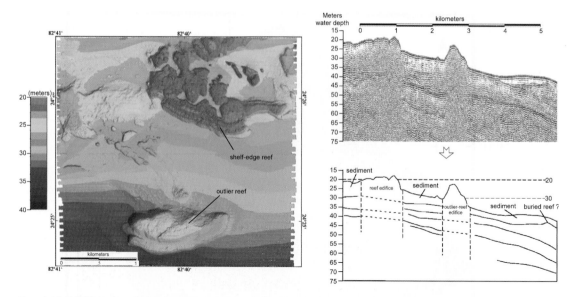

Fig. 4.28. Multibeam and high-resolution seismic images of shelf-edge and outlier reef along southern margin of the Florida Platform. Multibeam image reveals morphologic complexity of this bank reef. Seismic data show sediment bodies in between the shelf-edge reef and the outlier reef. Seaward of the outlier reef is a buried reef

are very-well sorted, finely-laminated oolitic grain-stones identical to eolianites seen in the Bahama Islands (Kindler 1992). These rocks were cemented in the fresh groundwater vadose zone based upon cement mineralogy and texture and yield ^{14}C AMS (accelerator mass spectrometer) dates (just dating outer ooid lamellae) ranging from 14.5 to 13.8 ka (Locker et al. 1996). These cements indicate that the marine sediments were ultimately deposited as subaerial, coastal sand dunes formed during multiple, temporary stillstands between Marine Isotopic Stage (MIS) 2 and 1 as the Laurentide Ice Sheet was melting in a non-linear manner. We conclude that the drowned, but intact, barrier-island forming southern Pulley Ridge is a contemporary western extension of these shorelines found off Key West.

At some point after the deposition, cementation, and flooding of the barrier island, when warmer water temperatures due to climatic amelioration and oligotrophic water from the Loop Current intrusions bathed this barrier-island substrate, a coral reef formed (Fig. 4.33a–c) (Jarrett et al. 2005). Since there are no rock cores from Pulley Ridge we do not know when this reef initiation occurred. However, because the underlying barrier-island geomorphology is so easily seen in the multibeam imagery, we conclude that the reef

cover is very thin, probably no thicker than 1–2 m. This means that, in geologic terms, this reef is a biostrome (laterally extensive) and not a bioherm (vertical framework constructed) (Jarrett et al. 2005). Additionally, this suggests that the reef is very slow growing (~14 cm/ka over past 14 ka), is relatively young (formed in the past ~6 ka) or a combination of both.

4.5.1.2 Biology

The depth of southern Pulley Ridge (~65–70 m) precludes reef characterization by the conventional methods for shallow reefs using SCUBA. Even tech divers using mixed gasses have only 20 min bottom time for sampling and observations. We characterized the seafloor using a combination of towed camera photos and analytical methods used on shallow reefs. More than 1,000 vertical bottom photographs were taken with SeaBOSS, a towed, still-photography and sampling system developed by the USGS (Valentine et al. 2000; Blackwood et al. 2000). The photographs were taken along 14 transects, each a kilometer or longer in length (Fig. 4.30). The scale of Pulley Ridge reef, approximately 5 × 15 km, requires much longer transects (10–100 times) than those used for shallow-water reef surveys. At this

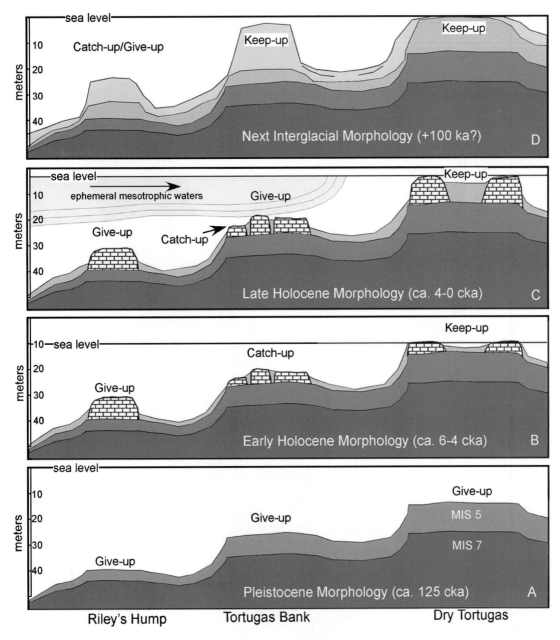

FIG. 4.29. Cross-section of Tortugas Bank, Dry Tortugas, and Riley's Hump with chronostratigraphic framework illustrated. Series of diagrams through time illustrate successive phases of development in response to sea-level changes. Each bank aggrades incrementally during successive sea-level events. Upon drowning of the lower banks (Riley's Hump and Tortugas Bank) corresponding to high rates of sea-level rise, the reef growth back-stepped to an available edifice occurring at an elevation commensurate with slower sea-level rise (Dry Tortugas). Tortugas Bank may have been influenced by the incursion of mesotrophic water from Mississippi River discharge (After Fig. 11 of Mallinson et al. 2003; reproduced by permission of Elsevier)

scale, southern Pulley Ridge is equivalent to a sector of the Florida Keys reefs, and a transect measures a site in the terminology of Murdoch and Aronson (1999). Figures 4.34 and 4.35 illustrate the common stony corals photographed by SeaBOSS. The entire data set is available in Cross et al. (2004, 2005).

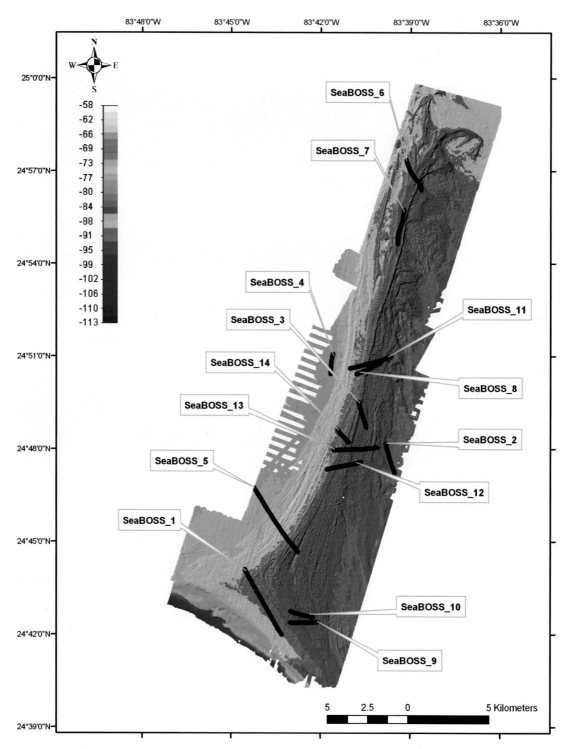

FIG. 4.30. Multibeam image of southern Pulley Ridge clearly showing preserved barrier-island morphology. Location of the 14 SeaBOSS transects on southern Pulley Ridge. Transect lines (black) are superimposed on the seafloor bathymetry (From Jarrett et al. 2005; reproduced by permission of Elsevier)

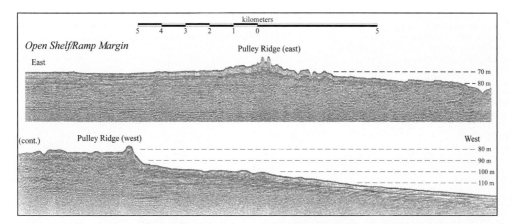

FIG. 4.31. One long E-W seismic line across the two ridges (~−65 m and ~−90 m) that defines Pulley Ridge. The living deep, light-dependent reef covers the ~−65 m deep ridge. Both ridges are paleo-shorelines

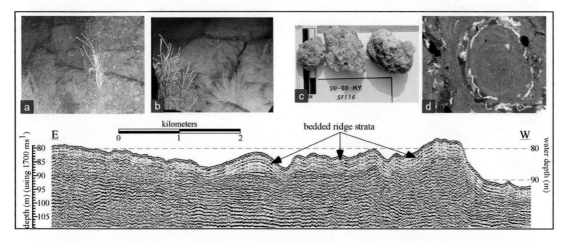

FIG. 4.32. Seismic line across the deeper paleoshoreline at Pulley Ridge. Video images (A, B) illustrate bedded, rectilinear nature of limestone consistent with lithified paleoshoreline facies. This deep ridge is covered with rhodoliths (C) coralline algal nodules. Thin-section photomicrograph (D) confirms rhodolith origin of these nodules (From Fig. 2 of Jarrett et al. 2005; reproduced by permission of Elsevier)

Point-count data were collected in 32 classes and combined into 7 major categories (Fig. 4.36) that account for 87.5% of the observations. Of the remaining 12.5% collectively termed "other", 5% is macroalgae other than coralline algae or *Anadyomene menziesii* (Tables 4.8, 4.9). Rubble, very coarse gravel, accounts for about 4% of the seafloor, 2% are unknown and 1.5% are identified miscellaneous (octocorals, echinoderms, bryozoans, tunicates, fish, etc.). *Anadyomene menziesii* is in its own category because of the abundance of this otherwise rare alga and because it might be identified as a keystone

species for this reef in future studies. Limestone is hardground not covered by previously mentioned organisms. It includes recently dead corals and coralline algae that have lost their color but may be colonized by micro flora and fauna (Fig. 4.35e).

Based on these data, Pulley Ridge reef is a coralline algae-coral dominated reef with these two components comprising 45–56% of the substrate along 11 of the 14 transects. The amount of coralline algae documented here is unusual for a Florida reef, but is typical of reefs in general. Adey (1976) and Adey et al. (1976) found coralline algae to be more abundant

FIG. 4.33. A, B, C Bottom photographs (~65 m depth) taken by Tim Taylor (breathing mixed gas) revealing the coral reef seascape superimposed on top of the cemented barrier island along southern Pulley Ridge.

in the deep fore-reef (30–80 m) of Indo-Pacific reefs than Caribbean reefs. Wray (1977) estimated 20–50% of the mass of modern reef accumulations to be composed of coralline algae. At the depth and low light levels of Pulley Ridge, these algae may be less cryptic than on the shallower Florida Keys reef because there is less light reflected into shaded areas. Adding limestone to the categories of coral and coralline algae provides an estimate of total hardbottom that ranges from 61% to 76% within the 11 transects indicated above. Two of the transects extended off

the reef: (1) transect 2 (SeaBoss, Fig. 4.30) on the gravel and sand plain to the east and (2) transect 4 on the sand and gravel plain to the west. Transect 6 is the northern limit for Agaricid corals. Three kilometers north of transect 6, two ROV surveys did not encounter platy corals. This northern portion of the ridge lacks photosynthetic scleractina and many of the large macro algae found further south. The northern ridge supports a heterotrophic octocoral-dominated community that does not contribute to a reefal accumulation like that in the south.

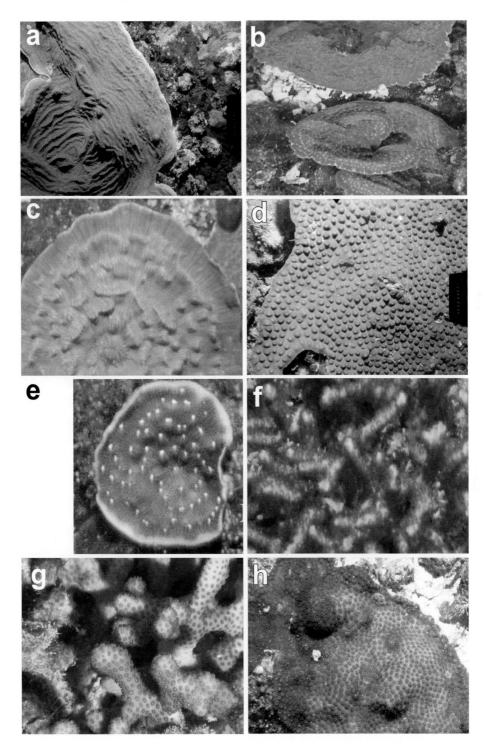

FIG. 4.34. Common stony corals of southern Pulley Ridge. (**a**) *Agaricia undata*; (**b**) *A. lamarki*; (**c**) *Leptoceris cucul-lata*; (**d**) *Montastraea cavernosa*; (**e**) *A. fragilis*; (**f**) *Oculina sp.* (**g**) *Madracis* sp.; (**h**) *Madracis decactis*

FIG. 4.35. Common benthic cover types of southern Pulley Ridge : (**a**) stony corals, (**b**) coralline algae, (**c**) *Anadyomene menziesii*, (**d**) sponges, (**e**) limestone (with algal turf cover), (**f**) gravel, and (**g**) sand. Bottom types not in these seven types are classified as "other" in Table 4.8

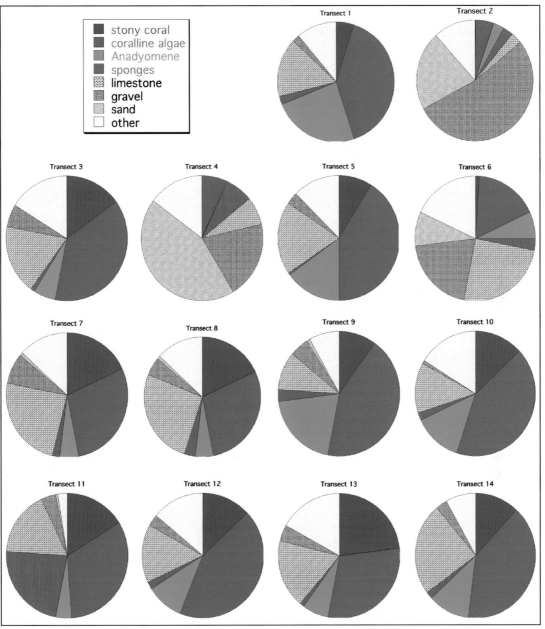

FIG. 4.36. Distribution of benthic cover compiled from 14 SeaBoss transects on Pulley Ridge (see Fig. 4.30): transects 1–7, and transects 8–14

Both coralline algae and the platy Agaricid corals result in foliate framework elements (Fig. 4.37a, b) capable of coating and preserving underlying topography. As mentioned above, the relatively thin coating of reef framework (probably less than a meter) and the platy nature of the framework account, in large part, for barrier-island morphology being so mark-edly apparent in the present-day bathymetry (Fig. 4.30; Jarrett et al. 2005).

In his now-classic study on the zonation of Jamaican reefs, Goreau (1959) described a reef zonation to depths of about 15 m that became a standard frame of reference for later reef surveys throughout the Caribbean, Florida and the

TABLE 4.8. Percentage distribution of major benthic cover categories at southern Pulley Ridge reef.

Dive #	Stony coral	Coralline algae	Anadyomene menziesii	Sponges	Limestone	Gravel	Sand	Other
1	5	40	24	2	16	2	0	11
2	0.2	5	3	3	3	53	21	11.8
3	15	38	6	1	18	6	0	16
4	0	7	0.3	7	7	20	44	14.7
5	9	41	15	0.5	19	3	0	12.5
6	1	17	7	3	25	20	9	18
7	18	29	5	2	24	8	1	13
8	18	29	5	3	26	5	1	13
9	10	43	20	3	10	5	1	8
10	13	42	13	2	13	1	0	16
11	16	33	4	23	17	4	0.4	2.6
12	13	43	10	2	15	3	0	14
13	23	30	7	1	18	4	0	17
14	12	40	11	2	24	3	0	8

TABLE 4.9. Species list of algae identified from Pulley Ridge (June 2005).

Rhodophyta
    Antithamnion sp.
    Agardhinula browneae (J. Agardh De Toni, 1897)
    Cladhymenia lanceifolia (Taylor, 1942)
  Ceramium sp.
  Chrysymenia planifrons (Melville)
    Chrysymenia new sp.?
    Dasya bailouviana (Gmelin) (Montagne, 1841)
  Halymenia integra (Howe et Taylor or n.s.?)
    Halymenia vinacea (Howe et Taylor, 1931)
    Herposiphonia sp.
  Hypoglossum anomalum (Wynne et Ballantine, 1986)
    Ochtodes secundiramea (Montagne) (Howe, 1920)
    Peysonnelia conchicola (Piccone et Grunow)
    Rhododictyon bermudense (Taylor)
Phaeophyta
    Dictyota bartayresiana Lamouroux (sensu Taylor, 1960)
    Dictyota cervicornis Kutzing w/reservations
  Lobophora variegata (Lamouroux) Womersley
    ex E.C. Oliveira – erect and encrusting forms
Chlorophyta
    Anadyomene menziesii Harvey
    Cladophora ?
    Codium isthmocladum Vickers vs Pseudocodium
    Codium spongiosum Harvey
    Halimeda discoidea Decaisne V. platyloba Boergesen
    Halimeda gracilis Harvey
    Ostreobium sp.
  Ventricaria ventricosa (J. Agardh) (J.L. Olsen et J.A. West)
    Verdigelas peltata (D.L. Ballantine et J.N. Norris)

Bahamas. At that time, the deepest part of the fore-reef explored (15–20 m) was characterized by large multi-lobed colonies of *Montastraea annularis*. Deeper diving during the 1960s revealed a different community on the fore-reef slope and deep fore-reef (40–75 m) described by Goreau and Goreau (1973). These zones are characterized by the platy corals *Agaricia undata, A. grahamae* and *A. lamarcki* and *Helioseris* (now *Leptoseris*) *cucullata*. During the 1970s, this zone proved to be widespread in Belize. Using a submersible, James and Ginsburg (1979) identified a zone of platy corals (*A. grahamae*, *A. lamarcki* and *Leptoceris cucullata*) between 40 and 70 m at five dive sites along the Belize barrier reef and Glovers Reef atoll. A very similar zone termed the "algal sponge" zone, is described between 48 and 88 m from Flower Gardens Reef (northwest Gulf of Mexico) by Rezak et al. (1985). They described the zone as being dominated by coralline algal nodules but with colonies of *Helioseris* (*Leptoseris*) cucullata and *Agaricia* and *Madracis*

"abundant enough among the algal nodule to be major sediment producers" (Rezak et al. 1985).

It seems likely that the coral community found at southern Pulley Ridge is widespread throughout the Caribbean, Bahamas and Gulf of Mexico and awaits further exploration in a depth range that is not easily accessible by SCUBA, but considered shallow for many submersibles. Pulley Ridge reef might be considered a stand-alone deep fore-reef, detached from any shallower reef zones by the combined effects of antecedent topography and sea-level rise (Jarrett et al. 2005). The deep fore reef, because

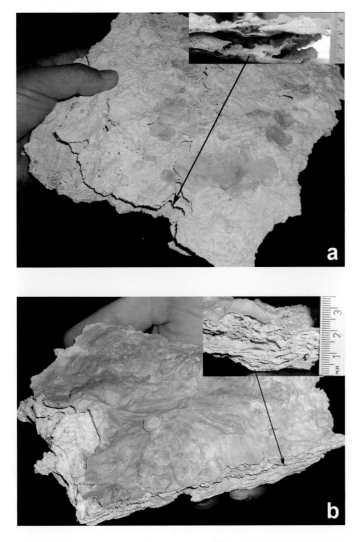

Fig. 4.37. Framework elements of Pulley Ridge reef: (**A**) Agaricid coral plates overgrowing and encrusting one another form a framework with centimeter-scale spaces (inset) that may be wholly or partially filled with sediment in the reef, and (**B**) coralline algae form platy masses of encrusting skeletons with millimeter-scale spaces (inset) partially filled with sediment

of its depth, is not included in reef-monitoring programs and, to the authors' knowledge, has not been re-visited in Jamaica or Belize in 30 years. The apparent health of the Pulley Ridge reef bodes well for this coral zone during decades when shallow reefs of the region have been devastated by disease and coral bleaching. Nevertheless, the coral-reef community covering Pulley Ridge, in ~65 m water, is the deepest, light-dependent coral reef on the US continental shelf known at this point in time (Halley et al. 2004).

Satellite SST and chlorophyll *a* data confirm the influence of the Loop Current on Pulley Ridge deep reef. This oceanographic current separates low-nutrient, outer-shelf waters from cooler, higher-nutrient, interior-shelf waters that are exported seaward across the shallow Florida reef tract. However, we do not understand the complex physical oceanographic interactions of the Loop Current and the west Florida shelf, particularly at the seafloor and how the reef is directly affected by the water column processes (He and Weisberg 2003; Weisberg and He 2003; Weisberg et al. 2005; Law 2003; Jarrett et al. 2005). Even with this very clear, oligotrophic water bathing the deep reef, measurements to date indicate that light at the seafloor is only about 1% of that at the surface.

Finally, the deep reef at southern Pulley Ridge: (1) provides an up-current source of larvae for downstream reefs – a potential refuge for shallow-water species, (2) supports a commercially-viable fishery industry, (3) should eventually provide understanding on how light-dependent corals can grow at such depths, (4) should eventually provide understanding on how this deep reef seems to be absent of disease or other stressors; a new type of natural laboratory, (5) provides clues as to where other similar deep reefs might exist and may not be such an apparent exception, (6) should eventually provide clues concerning timing of coral-reef development in the eastern GOM during the last 15 ka and provide insight to paleoceanographic and paleoclimatic changes, (7) provides an understanding of how the western boundary current (Loop Current component) interacts with the continental shelf and controls benthic communities, and (8) should provide a sanctuary for human study and enjoyment.

4.5.2 Northern Extension of Pulley Ridge

The 30 km long, 5 km wide barrier island forming southern Pulley Ridge seems to be unique. Here-to-fore unpublished side-scan sonar and high-resolution seismic reflection data indicate multiple, less pronounced ridges and paleoshorelines, and in some places a distinct 2–3 m high scarp at ~85 m water depth to the north (Fig. 4.38). This suggests that the modern relief of some ridges may have been produced by erosional processes during sea-level

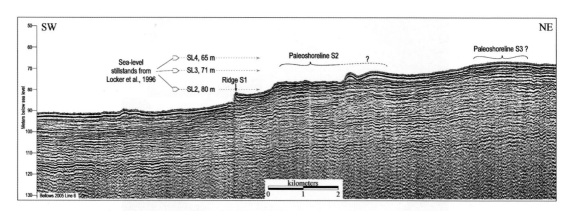

FIG. 4.38. Seismic profile along Pulley Ridge significantly north of the deep reef. Side-scan sonar and high-resolution seismic data reveal morphologies strikingly similar to paleoshoreline and ridge structures identified by Locker et al. (1996) on the south Florida margin facing the Straits of Florida. These alongslope structures are much narrower than the paleo-barrier island of southern Pulley Ridge. Much of this northern shoreline of Pulley Ridge was probably sediment starved and produced minimal coastal sedimentary facies. The depth of sea-level stillstands and structures S1–S3 are identified corresponding to Locker et al. (1996)

rise, or as with the case for ridge S1 in Fig. 4.38, the deeper (>80 m) ridges may reflect late relic Pleistocene reef buildups – yet to be identified. This further suggests that the carbonate sediment production, probably oolitic in nature, was much greater along the extreme SW corner of the Florida Platform than anywhere else. Widely-spaced dredge hauls from the paleoshoreline did not yield any coral-reef material. Additionally, coral reefs have a distinctive seismic facies (chaotic and discontinuous), which was not seen along this paleoshoreline. So, we tentatively conclude that the coral-reef cover is limited to the very southern portion of Pulley Ridge – the area that is influenced by oligotrophic Loop Current intrusions, which are not known to occur further to the north along the west Florida shelf.

4.5.3 Last Glacial Maximum Lowstand Ramp Reefs

Seaward of a prominent shelf ridge called The Elbow (Fig. 4.1) in ~120 m of water is a 1–2 km wide belt of "seismic patch reefs", that are ~7 m in relief and 50 m across (Fig. 4.39). They are evenly spaced ~200 m apart. In the absence of ground-truthing data, we interpret these to be shallow-water patch reefs that formed when sea-level was lower during the LGM. Instead of an obvious paleoshoreline, this sea-level lowstand resulted in the formation

of this belt of isolated mounds (believed to be patch reefs) making it the first feature of its kind found on the west Florida shelf/upper slope. The features are at the same depth as those found on southern Howell Hook (see below) suggesting that they are contemporaries.

4.6 Shelf-edge Reefs

4.6.1 Outlier Reefs

Defining the western extent of the south Florida Platform margin is a primary reef tract that is the extension of the well-known and well-studied main, massive reef tract located seaward of and parallel to the Florida Keys (Fig. 4.1). This nearly continuous shelf-edge reef defines the boundary of the platform top and the upper slope. Lidz et al. (1991, 1997, 2006) and Lidz (2006) have recognized outlier reefs, a reef-and-trough system, that occur along this main reef tract forming what appears to be a double reef edifice separated by a ~30 m deep, 1 km wide basin (Fig. 4.40a, b). In places, this basin has been completely filled with detritus shed from the main reef as well as the outlier reef. These discontinuous outlier reefs extend westward along the south-facing margin of the Florida Platform where they eventually disappear as this margin begins to transition from a rim to a ramp system west of Riley's Hump (Fig. 4.41).

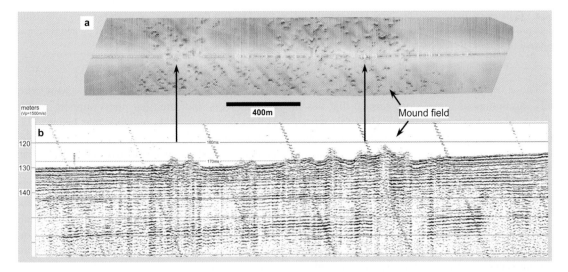

FIG. 4.39. Side-scan sonar image and seismic profile indicating presence of a possible sea-level lowstand patch reef belt that probably formed during the Last Glacial Maximum (~20 ka) when sea-level was ~120 m lower than present

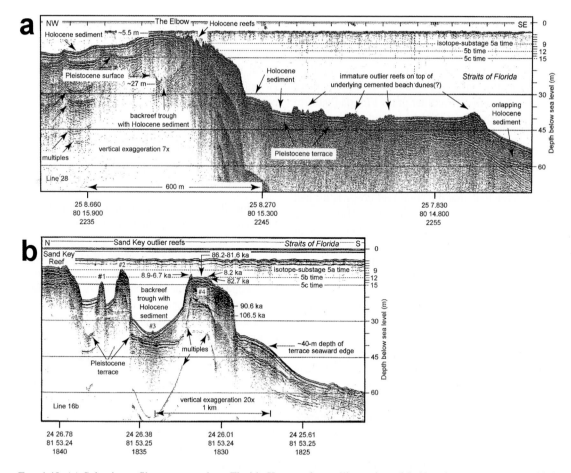

FIG. 4.40. (a) Seismic profile across northern Florida Keys reef tract illustrating ~35–40 m deep terrace upon which outlier reefs have been developed. (b) Seismic profile across Florida Keys reef tract SW of Key West (near Sand Key) illustrating multiple outlier reefs situated seaward of main reef tract (far left or to north). These late Pleistocene (see ^{14}C dates on figure) outliers have been developed on this 35–40 m deep terrace of unknown origin (Seismic data provided to B. Lidz by A.C. Hine and S.D. Locker; Figures from Lidz et al. 2006; reproduced by permission of Journal of Coastal Research)

These outlier reefs are situated upon a laterally discontinuous upper-slope terrace that lies in ~35 m water depth (Fig. 4.40b). How this terrace was formed and under what sea-level and other environmental conditions are unknown due to the absence of core data. Additionally, its laterally discontinuous nature is enigmatic. It is clearly a constructional feature, not an erosional one. Additionally, all outlier reefs have formed on this terrace. However, there are sections of the margin where the terrace exists but only incipient or no outlier reefs are present. So, the presence of the terrace did not guarantee the growth and development of outliers.

Diamond core drilling into these ~30 m relief outlier reefs indicate that they support a thin Holocene veneer of A. palmata that began ~9 ka and essentially ceased ~5 ka. A poorly preserved paleo-subaerial exposure horizon generally exists at ~16–12 mbsl. Below this unconformity lies a dominantly M. annularis reef whose TIMS U-Th dates range from 80.9 to 106.5 ka (MIS 5c, b, and a) with most between 80.9 and 85 ka (Toscano 1996). The longest core extends to 21.29 mbsl and is 10 m in length. No cores reached the terrace upon which the outlier reefs rest. So, MIS 5c, b, and a must have been a period of extensive shelf-margin reef building seaward of the Florida Keys and westward along the south Florida margin from ~12 to ~24 m below present sea level.

4.7   Upper Slope Reefs

4.7.1   Miller's Ledge

Located seaward of Riley's Hump along the upper slope is a 30 m high prominent escarpment called Miller's Ledge (Figs. 4.1, 4.42). From high-resolution seismic reflection data, this appears to be the seaward scarp of a reef that now lies in ~80 m water depth. Although this water depth is technically at shelf depth, it lies seaward of the reef-dominated rimmed margin of the Florida Platform. So, from a morphological viewpoint, it lies along the very upper slope. It probably is contemporary with the deeper component constituting Pulley Ridge (~90 m). It has not yet become buried by sediments being shed from the shallower reefs lying to the north, although that may yet become its fate. This escarpment may be the result of submarine erosion by strong, slope-parallel flows associated with the Loop Current.

4.7.2   Howell Hook

Extending along the upper slope of the SW margin of the Florida Platform in ~120–160 m water depth is a discontinuous ridge consisting of a series of "seismic reefs". This is the southernmost component of Howell Hook – a prominent bathymetric feature that runs ~300 km along the west Florida upper slope roughly parallel to Pulley Ridge, which lies in shallower water (60–90 m; Fig. 4.1). Earlier work by Holmes (1985) revealed that Howell Hook is encrusted with coralline algae where dredged.

Bathymetric maps and multibeam data clearly show this lack of linearity (Fig. 4.43a). Additionally, seismic reflection data reveal that rather than a single ridge line, there are multiple, stair-stepped ridges that occur along this upper slope (Fig. 4.43b). The acoustically chaotic seismic facies strongly suggest that these features were coral reefs. Figure 4.43b shows a 10 m high escarpment at ~125–135 m water depth, which could be an erosional feature probably formed during the sea-level lowstand occurring during the LGM. However, seaward of this scarp are three mound-like structures reaching 140, 155, and 185 m. The

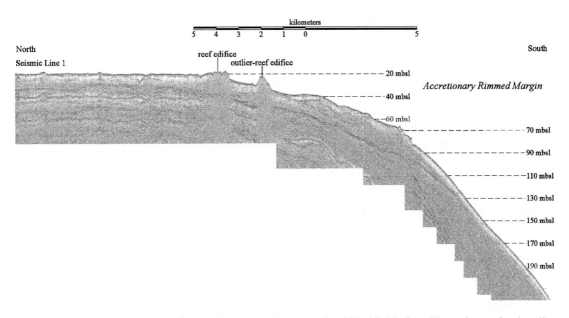

FIG. 4.41. High-resolution seismic line N–S across southern margin of Florida Platform illustrating reef and outlier reef structures. This is near the western-most extent of the Florida Keys reef tract and the end of the rimmed type of platform margin. Note buried reef at ~−40 m depth. Also, note paleoshoreline feature at ~−70 m

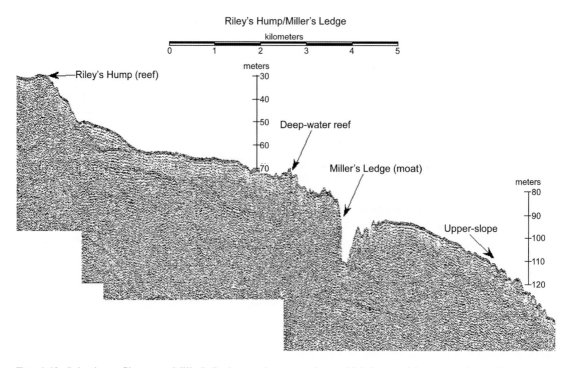

FIG. 4.42. Seismic profile across Miller's Ledge on the upper slope which is a reefal structure located at approximately the same depth (~90 m) as the deeper ridge associated with Pulley Ridge.

mound reaching 140 m is enigmatic as the basin behind it was filled by layered strata and the seaward side has been eroded perhaps by biological activity stimulated by currents associated with the Loop Current. Increased water flow may enhance bio-erosion activity (A.C. Neumann, personal communication 1975). In fact, this entire upper slope appears to have been significantly influenced by Loop Current erosion as the seismic stratigraphy reveals multiple, buried erosional surfaces.

To date there have been no ROV deployments, dredging, or drilling of these features. As a result, we simply do not know what type of reefs these features constitute. So, they remain a mystery, but based on their depth, may have been formed some 20 ka indicating that environmental conditions were appropriate for coral-reef growth at that time.

4.8   Discussion and Summary

The morphologic diversity of coral reefs, present and past, on the south, south-west and west Florida platform results from the interplay of antecedent topography, substrate type, sea-level fluctuations and water circulation. The broad nature of a ramp allows for extensive, lateral movement of the shoreline during sea-level cycles thus allowing for a diverse distribution of shorelines and shallow-water coral reefs. Where margins are steep, reefs become laterally compressed and vertically stacked. The rim-to-ramp transition along the south margin of the Florida platform adds to this complexity.

The small circular reef banks (Dry Tortugas, Tortugas Bank, and Riley's Hump, Florida Middle Ground) seem anomalous as compared to expected reef linearity imposed by laterally extensive shelf margins and paleoshorelines. Along the south Florida margin, perhaps the reef banks were originally sited on cemented, tide-dominated sand shoals that comprise the southern Florida Keys? In the Florida Middle Ground area, perhaps karst pinnacles provided the original antecedent topography on which these numerous small reef banks became established? Additionally, all reefs seem

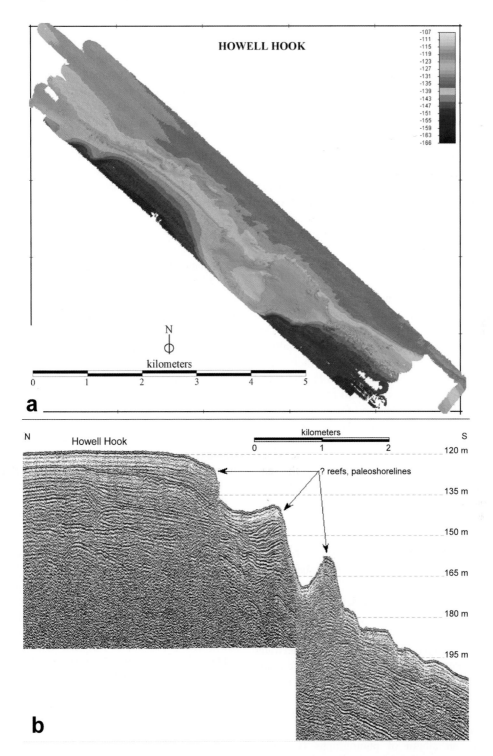

FIG. 4.43. (**a**) Multibeam data of SW margin of Florida Platform revealing multiple ridges and large mound structure that may be a sea-level lowstand reef. This is the southern extension of Howell Hook. (**b**) Seismic profile across this area revealing multiple "seismic reefs", one of which appears to have been significantly eroded. These are the deepest "seismic reefs" along the entire south, southwest, and west Florida margin

to be dominated by aggradation, backstepping, or drowning rather than progradation suggesting that rapid sea-level changes have been critically important in their development.

Sea-level fluctuations result not only in depth changes but also significant changes in shelf/adjacent deepwater circulation, water temperature, water quality as well as concomitant changes in continental/marine interaction. Healthy coral-reef development occurs when ideal conditions for substrate availability, water quality, and stable or slowly rising sea-level simultaneously converge. The diversity of reefs dispersed on the Florida platform indicate that such convergence is not simultaneous nor ubiquitous.

Acknowledgments For the south Florida and SW Florida margin work (outlier reefs, Dry Tortugas area, Pulley Ridge, Howell Hook) we thank Drs. A.C. Neumann, Sylvia Earle, Pamela Hallock, David Twichell, and Chuanmin Hu for their expertise. We thank Beau Suthard, David Palandro and many other graduate students and research associates (Chris Reich, Don Hickey) for their assistance at sea and in the laboratory. We thank the US Geological Survey, the National Oceanographic and Atmospheric Administration – National Underwater Research Center (particularly Mr. Lance Horn), Office of Naval Research, the Naval Oceanographic Office, the Florida Institute of Oceanography, the Harte Institute, the National Fish and Wildlife Foundation, and the Sustainable Seas Expedition for financial support. We thank Ben Haskell of the Florida Keys National Marine Sanctuary. We thank the crews of the Florida Institute of Oceanography's research vessels; *R/V Suncoaster* and *R/V Bellows*. We also thank the crew of the *M/V Spree*. We thank Mr. Tim Taylor of the *R/V Tiburon*, Inc. for use of the underwater photos of Pulley Ridge taken by divers. We thank G.P. Schmahl for photos taken in the FMG area.

For the Florida Middle Ground work, we are grateful for a 2003 cruise – the findings from which are provided in a report with appendices, including an album of photographs (Coleman et al. 2005). We thank all of the participants: Felicia Coleman, Mike Dardeau, George Dennis, Tom Hopkins, George Schmahl, Chris Koenig, Sherry Reed, Carl Beaver, Lance Horn, Selena Kupfner, Mike Callahan, Matt Lybolt, Anne McCarthy, and Jim Kidney for their persistence under difficult circumstances. We thank the Gulf of Mexico Fishery Management Council for funding support for the expedition. Tom Hopkins, Mike Dardeau, Paul Johnson, and others from the 1970s MAFLA expeditions provided photographs and unpublished data from their work. Tom Hopkins and Mike Dardeau confirmed the station locations using the ROV video which was a great help in locating the Hopkins et al. (1977) sampling stations.

References

Adey WH (1976) Crustose coralline algae as micro-environmental indicators for the Tertiary, in Gray J, Boucot AJ (eds) Historical Biogeography. Plate Tectonics, and the Changing Environment, Oregon State University Press, Corvallis, OR, pp 459–464

Adey WH, Adey PJ, Burke R, Kaufman L (1976) Holocene reef systems of eastern Martinique, French West Indies. Atoll Res Bull 211:1–40

Ahr WM (1973) The carbonate ramp : an alternative to the shelf model. Gulf Coast Assoc of Geol Soc 23:221–225

Arthur MA, Schlanger SO (1979) Cretaceous "oceanic anoxic events" as causal factors in reef-reservoired oil fields. Geol Soc Am Bull 69:870–885

Austin H, Jones J (1974) Seasonal variation of physical oceanographic parameters on the Florida Middle Ground and their relation to zooplankton biomass on the West Florida Shelf. Florida Scientist 37:16–32

Back RM (1972) Recent Depositional Environment of the Florida Middle Ground. Florida State University, Tallahassee, FL, 237 pp

Blackwood D, Parolski K, Valentine P (2000) Seabed observation and sampling system. U.S. Geological Survey Fact Sheet FS-142–00 (http://woodshole.er.usgs.gov/operations/sfmapping/seaboss.htm)

Brooks HK (1962) Observations on the Florida Middle Ground (abstract). Geol Soc Am Spec Publ 68:65–66

Brooks GR (1981) Recent Carbonate Sediments of the Florida Middle Ground Reef System; Northeastern Gulf of Mexico. University of South Florida, St. Petersburg, FL, 137 pp

Brooks GR, Doyle LJ (1991) Geologic development and depositional history of the Florida Middle Ground: a mid-shelf, temperate-zone reef system in the northeastern Gulf of Mexico, in Osborne RH (ed) From Shoreline to Abyss – Shepard Commemorative Volume:189–203

Buffler RT (1991) Seismic stratigraphy of the deep Gulf of Mexico and adjacent margins, in Salvador A (ed)

The Gulf of Mexico basin: the geology of North America, Geol Soc Am:353–387

Cheney DP, Dyer JP, III (1974) Deep-water benthic algae of the Florida Middle Ground. Mar Biol 27:185–190

Coleman C, Dennis G, Jaap WC, Schmahl GP, Koenig C, Reed S, Beaver CR (2005) Status and trends of the Florida Middle Ground. Technical Report of Gulf of Mexico Fisheries Management Council, Tampa, FL, 135 pp

Collard SB, D'asaro CN (1973) Benthic invertebrates of the eastern Gulf of Mexico., in Jones JI, Ring RE, Rinkel MO, Smith RE (eds) A summary of knowledge of the eastern Gulf of Mexico, Coordinated by the State Univ Sys Florida Inst Oceanogr, St. Petersburg, FL IIIG1–IIIG28

Corso W, Austin JA, Buffler RT (1989) The early Cretaceous platform off northwest Florida: controls on morphologic development of carbonate margins. Mar Geol 86:1–14

Cross VA, Blackwood DS, Halley RB, Twichell DC (2004) Bottom photographs from the Pulley Ridge deep coral reef. DVD-ROM US Geological Survey Open-file Report 2004–1228

Cross VA, Twichell DC, Halley RB, Ciembronowicz KT, Jarrett BD, Hammar-Klose ES, Hine AC, Locker SD, Naar DF (2005) GIS Compilation of data collected from the Pulley Ridge deep coral reef region. DVD-ROM US Geological Survey Open-file Report 2005–1089

Doyle LJ, Steinmetz JC, Brooks GR, Parker DM (1980) Gulf of Mexico topographic features study, Florida Middle Ground. Final Report, BLM, New Orleans, LA (BLM Contract AA5512-CT8–35)

Forman SL, Oglesby R, Webb RS (2001) Temporal and spatial patterns of Holocene dune activity on the Great Plains of North America: megadroughts and climate links. Global Planet Change 29:1–29

Gilbes F, Tomas C, Walsh J, Muller-Karger F (1996) An episodic chlorophyll plume on the West Florida Shelf. Cont Shelf Res 16:1201–1224

Godcharles MF, Jaap WC (1973) Fauna and flora in hydraulic clam dredge collections from Florida west and southeast coasts. Fla Dep Nat Resour Mar Res Lab Spec Sci Rep 40:89

Goreau TF (1959) The ecology of Jamaican reefs. I. Species composition and zonation. Ecology 40:67–90.

Goreau TF, Goreau NI (1973) The ecology of Jamaican coral reefs. II. Geomorphology, zonation, and sedimentary phases. Bull Mar Sci 23:399–464

Gould HR, Stewart RH (1956) Continental terrace sediments in the northeastern Gulf of Mexico., in Finding ancient shorelines, SEPM Spec Publ 3:2–19

Grimm D, Hopkins T (1977) Preliminary characterization of the octocorallian and scleractinian diversity at the Florida Middle Ground. Proc 3rd Int Coral Reef Symp, Miami, 1:135–141

Grimm DE (1978) The occurrence of the Octocorallia (Colenterata: Anthozoa) on the Florida Middle Ground. M.S. thesis, University of Alabama, 85 pp

Gulf of Mexico and South Atlantic Fishery Management Councils (1982) Fishery Management Plan Final Environmental Impact Statement for Coral and Coral Reefs. Gulf of Mexico Fishery Management Council, Tampa, FL, 178 pp

Halley RB, Hine AC, Jarrett BD, Twichell DC, Naar DF, and Dennis GD (2004) Pulley Ridge : The US's Deepest Hermatypic Coral Reef. Geological Society of America Annual Meeting, Denver, CO, Abstracts with Programs, 302 pp

Hallock P (1988) The role of nutrient availability in bioerosion: consequences to carbonate buildups. Palaeogeogr Palaeoclim Palaeoecol 63:275–291

Hallock P, Schlager W (1986) Nutrient excess and the demise of coral reefs and carbonate platforms. Palaios 1:389–398

He R, Weisberg RH (2003) A Loop Current intrusion case study on the West Florida shelf. J Phys Oceanogr 33:465–477

Hilde WC, Sharman G, Waris W, Lee CS, Feely M, Meyer M (1981) Mapping and Sub-bottom Profiling, Texas A&M University, College Station, TX.

Hine AC (1997) Structural and paleoceanographic evolution of the margins of the Florida Platform, in Randazzo AF, Jones DS (eds) The Geology of Florida, University Press of Florida:169–194

Hine AC, Mullins HT (1983) Modern carbonate shelf-slope breaks. Soc Econ Paleontol Mineral Spec Publ 33:169–188

Hine AC, Neumann AC (1977) Shallow carbonate bank margin growth and structure, Little Bahama Bank. Bull Am Assoc Petrol Geol 61:376–406

Hine AC, Brooks GR, Davis Jr. RA, Duncan DS, Locker SD, Twichell DC, Gelfenbaum G (2003) The west-central Florida inner shelf and coastal system; a geologic conceptual overview and introduction to the special issue. Mar Geol 200:1–17

Hine AC, Buthard B, Locker SD, Cunningham KJ, Duncan DS, Evans M, Morton RA (in press) Karst subbasins and their relationship to Florida transport of Tertiary siliciclastics, in Swart PK, Eberli, GP McKenzie J (eds) Perspectives in Sedimentary Geology: A Tribute to the Career of Robert Nathan Ginsburg, IAS Special Publication, Sedimentology, v.41

Holmes CW (1985) Accretion of the south Florida platform, late Quaternary development. AAPG Bull 69:149–160

Hopkins TS, Blizzard DR, Brawley SA, Earle SA, Grimm DE, Gilbert DK, Johnson PG, Livingston EH,

Lutz CH, Shaw JK, Shaw BB (1977) A preliminary characterization of the biotic components of composite strip transects on the Florida Middle Ground, northeastern Gulf of Mexico. Proc 3rd Int Coral Reef Symp 1:31–37

Jaap WC, Hallock P (1990) Coral reefs, in Myers RL, Ewell JJ (eds) Ecosytems of Florida. University of Central Florida Press, Orlando, FL, pp 574–616

Jaap WC, Lyons WG, Dustan P, Halas JC (1989) Stony coral (Scleractinia and Milleporina) community structure at Bird Key Reef, Ft. Jefferson National Monument, Dry Tortugas, Florida. Fla Mar Res Publ 46:1–31

James NP, Ginsburg RN (1979) The Seaward Margin of Belize Barrier and Atoll Reefs. Special Publication 3, International Association of Sedimentology, Blackwell, Oxford, 191 pp

Jarrett BD (2003) Late Quaternary carbonate sediments and facies distribution patterns across a ramp to rim transition: a new conceptual model for the southwest Florida platform. Ph.D. dissertation, College of Marine Science, University of South Florida, 135 pp

Jarrett BD, Hine AC, Halley RB, Naar DF, Locker SD, Neumann AC, Twichell D, Hu C, Donahue BT, Jaap WC, Palandro D, Ciembronowicz K (2005) Strange bedfellows – a deep hermatypic coral reef superimposed on a drowned barrier island; southern Pulley Ridge, SW Florida platform margin. Mar Geol 214:295–307

Jones AC, Berkeley SA, Bohnsack JA, Bortone SA, Camp DK, Darcy GH, Davis JC, Haddad KD, Hedgepth MY, Irby EW, Jaap Jr. WC, Kennedy FS, Lyons Jr. WG, Nakamura EL, Perkins TH, Reed JK, Steidinger KA, Tilmant JT, Williams RO (1985) Ocean habitat and fishery resources of Florida., in Seaman Jr., W (ed) Florida Aquatic Habitat and Fishery Resources. Florida Chapter American Fisheries Society, Kissimmee, Fl, pp 437–542

Jordan GF (1952) Reef formation in the Gulf of Mexico off Apalachicola Bay, Florida. Geol Soc Am Bull 63:741–744

Kindler P (1992) Coastal response to the Holocene transgression in the Bahamas : episodic sedimentation vs continuous sea-level rise. Sediment Geol 80: 319–329

Klitgord KD, Popenoe P, Schouten H (1984) Florida: a Jurassic transform plate boundary. J Geophys Res 89:7753–7772

Knox JC (2000) Sensitivity of modern and Holocene floods to climate change. Quat Sci Re 19:439–458

Law J (2003) Deep ocean effects on outer continental shelf flow: a descriptive study in the Loop Current, Florida Current, and Gulf Stream systems. M.S. thesis, University of South Florida, St. Petersburg, FL, 102 pp

Lee TN, Williams E, Johns E, Wilson D, Smith NP (2002) Transport processes linking south Florida coastal ecosystems, in Porter J, Porter K (eds) The Everglades, Florida Bay, and Coral Reefs of the Florida Keys : An Ecosystem Sourcebook, CRC, Boca Raton, FL, pp 309–342

Lees A (1975) Possible influence of salinity and temperature on modern shelf carbonate sedimentation. Mar Geol 19:159–198

Lidz BH (2006) Pleistocene corals of the Florida Keys: architects of imposing reefs – why? J Coastal Res 22:750–759

Lidz BH, Hine AC, Shinn EA, Kindinger JL (1991) Multiple outer-reef tracts along the south Florida bank margin: outlier reefs, a new windward model. Geology 19:115–118

Lidz BH, Shinn EA, Hine AC, Locker SD (1997) Contrasts within an outlier-reef system: evidence for differential quaternary evolution, south Florida windward margin, USA. J Coastal Res 13:711–731

Lidz BH, Reich CD, Peterson RL, Shinn EA (2006) New maps, new information: coral reefs on the Florida Keys. J Coastal Res 22:260–282

Logan BW, Harding JL, Ahr WM, Williams JD, Snead RG (1969) Late Quaternary carbonate sediments of Yucatan Shelf, Mexico. Am Assoc Petrol Geol Mem 11:5–128

Locker SD, Hine AC, Tedesco LP, Shinn EA (1996) Magnitude and timing of episodic sea-level rise during the last deglaciation. Geology 24:827–830

Ludwick JC, Walton WR (1957) Shelf-edge calcareous prominences in the northeastern Gulf of Mexico. Am Assoc Petrol Geol Bull 41:2054–2101

Lyons WG (1980) Molluscan communities of the west Florida shelf. Bull Am Malac Union 1979:37–40

Lyons WG, Collard S (1974) Benthic invertebrate communities of the eastern Gulf of Mexico., in Smith RE (ed) Proceedings of Marine Environmental Implications of Offshore Drilling in the Eastern Gulf of Mexico, State University System of Florida Institute of Oceanography, St. Petersburg, FL, pp 157–165

Mallinson D (2000) A Biological and Geological Survey of the Florida Middle Ground – HAPC: Assessing Seafloor Impacts of Fishing-related Activities, National Marine Fisheries Service, St. Petersburg

Mallinson D, Hine AC, Muller PH (1999) Structure and origin of relict reefs on the West Florida Margin. Eos transactions, American Geophysics Union Fall Meeting

Mallinson D, Hine A, Hallock P, Locker S, Shinn E, Naar D, Donahue B, Weaver D (2003) Development of small carbonate banks on the south Florida Platform margin: response to sea level and climate change. Mar Geol 199:45–63

Mallinson DJ, Hine AC, Locker SD (1996) Paleocirculation and paleoclimatic implications of the Florida Middle Ground reefs. Geological Society of America Abstracts with Programs 1996 National Meeting, Denver, CO

Mallinson DJ, Donahue B, Naar DF, Hine AC, Locker SD (2006) Pleistocene and Holocene geologic controls on the Florida Middle Ground relict reef complex; a diverse seafloor environment on the west Florida shelf, AGU Ocean Science Meeting OS261–04

Murdoch TJT, Aronson RB (1999) Scale-dependent spatial variability of coral assemblages along the Florida Reef tract. Coral Reefs 18:341–351

Mullins HT, Gardulski AF, Wise SW, Applegate J (1987) Middle Miocene oceanographic event in eastern Gulf of Mexico: implications for seismic stratigraphic succession and Loop Current/Gulf Stream circulation. Geol Soc Am Bull 98:702–713

Mullins HT, Gardulski AF, Hinchey EJ, Hine AC (1988) The modern carbonate ramp slope of central west Florida: J Sediment Petrol 58:273–290

Mullins HT, Gardulski AF, Hine AC, Melillo AJ, Wise SW, Applegate J (1989) Three-dimensional sedimentary framework of the carbonate ramp slope of central west Florida: a sequential stratigraphic perspective. Geol Soc Am Bull 100:514–533

Newton CR, Mullins HT, Gardulski AF, Hine AC, Dix GR (1987) Coral mounds on the west Florida slope: unanswered questions regarding the development of deep-water banks. Palaios 2:356–367.

Neumann AC, MacIntyre I (1985) Reef response to sea level rise: keep-up, catch-up, or give-up. Proc 5th Int Coral Reef Congr, Tahiti 2:105–110

Poag W (1991) Rise and demise of the Bahama – Grand Banks gigaplatform, northern margin of the Jurassic proto-Atlantic seaway. Mar Geol 102:63–150

Purser BH (1973) The Persian Gulf: Holocene Carbonate Sedimentation and Diagenesis in a Shallow Epicontinental Sea. Berlin, Springer, 471 pp

Read JF (1982) Carbonate platforms of passive (extensional) continental margins: types, characteristics, and evolution. Tectonophysics 81:195–212

Read JF (1985) Carbonate platform facies models. Am Assoc Petrol Geol Bull 64:1575–1612

Rezak R, Bright TJ, McGrail DW (1985) Reefs and Banks of the Northwestern Gulf of Mexico, Wiley, New York, 259 pp

Sheridan RE, Mullins HT, Austin Jr. JA, Ball MM, Ladd JW (1988) Geology and geophysics of the Bahamas., in Sheridan RE, Grow JA (eds) The geology of North America, V. I-2, The Atlantic continental margin, U.S. Geol Soc Am :329–364

Shinn EA, Lidz BH, Kindinger JL, Hudson JH, Halley RB (1989) Reefs of Florida and the Dry Tortugas. American Geophysics Union Field Guide T176, p 53

Smith GB, Austin HM, Bertone SA, Hasting RW, Ogren LH (1975) Fishes of the Florida Middle Ground with comments on ecological zoogeography. Fla Mar Res Publ 9:14

Steidinger KA, Tilmant JT, Williams RO (1985) Ocean habitat and fishery resources of Florida., in Seaman Jr., W (ed) Florida Aquatic Habitat and Fishery Resources. Florida Chapter American Fisheries Society, Kissimmee, Fl, a: pp 437–542

Tilmant JT (1984) Reef fish., in Jaap WC (ed) The Ecology of the South Florida Coral Reefs: A Community Profile, US Fish and Wildlife Service FWS/OBS-82/08:52–62

Toscano MA (1996) Late Quaternary stratigraphy, sea-level history, and paleoclimatology of the southeast Florida outer continental shelf: unpublished Ph.D. dissertation, College of Marine Science, University of South Florida, 280 pp

Tucker ME, Wright VP (1990) Carbonate Sedimentology, Blackwell, p 482

Valentine P, Blackwood D, Parolski K (2000) Seabed observation and sampling system. U.S. Geological Survey Fact Sheet FS-142–00 http://woodshole.er.usgs.gov/operations/sfmapping/seaboss.htm)

Vaughan TW (1914) A contribution to the geologic history of the Florida plateau, in papers from the Tortugas Laboratory, v. IV. Carnegie Inst Wash Spec Publ 133:99–185

Vogt PR (1989) Volcanogenic upwelling of anoxic, nutrient-rich water: a possible factor in carbonate bank/reef demise and benthic faunal extinctions? Geol Soc Am Bull 101:1225–1245

Warwick RM, Clarke KR (2001) Practical measurers of marine biodiversity based on relatedness of species. Oceanog Mar Biol Ann Rev 39:207–231

Weisberg RH, He R (2003) Local and deep-ocean forcing contributions to anomalous water properties on the West Florida Shelf. J Geophys Res 108:C6, 15, doi:10.1029/2002JC001407

Weisberg RH, He R, Liu Y, Virmani JR (2005) West Florida shelf circulation on synoptic, seasonal, and inter-annual time scales, in Sturges W, Lugo-Fernandez, A (eds) Circulation in the Gulf of Mexico. Amer Geophys Union Monogr Ser 161:325–347

Wray JL (1977) Calcareous Algae, Amsterdam, Elsevier, 185 pp

5

The Reef Tract of Continental Southeast Florida (Miami-Dade, Broward and Palm Beach Counties, USA)

Kenneth W. Banks, Bernhard M. Riegl, Vincent P. Richards, Brian K. Walker, Kevin P. Helmle, Lance K.B. Jordan, Janet Phipps, Mahmood.S. Shivji, Richard E. Spieler, and Richard E. Dodge

5.1 Introduction

Although South Florida coral reefs are frequently considered to be confined to the Florida Keys, a complex of relict early Holocene shelf-edge and mid-shelf reefs as well as limestone ridges extends along the continental coast of Southeast Florida (Fig. 5.1) from offshore south Miami (N25°34′) northward to offshore West Palm Beach (N26°43′). This extends the distance spanned overall by reefs in SE Florida by 125 km (Fig. 5.2). The nomenclature proposed by Moyer et al. (2003) and Banks et al. (2007) identifying these structures as ridge complex and inner, middle, and outer reef will be used herein. The reefs are arranged linearly and parallel to the trend of the shoreline. They are separated by sandy sedimentary deposits of varying thicknesses that overly erosional hardground surfaces (Duane and Meisburger 1969a, b; Raymond 1972; Shinn et al. 1977; Banks et al. 2007). The reefs themselves are presently not framebuilding but are colonized by a rich tropical fauna otherwise characteristic of the West Atlantic/Caribbean reef systems.

The continental shelf along the SE Florida coast is narrow and bathed by the relatively warm waters of the Florida Current, a branch of the Gulf Stream flowing northward between SE Florida and the Bahamas banks. Although SE Florida is located at the convergence of the subtropical and temperate climate zones (Chen and Gerber 1990), the influence of the Florida Current and the absence of any major rivers in the early and mid Holocene provided conditions suitable for reef building and,

after the demise of framebuilding, the maintenance of extensive coral reef associated communities.

Rohmann et al. (2005) estimated that 30,801 km^2 of inshore areas are situated in less than 18.3 m depth around South Florida and could potentially support shallow-water coral reef ecosystems. An area of 19,653 km^2 remains outside the Florida Keys and Dry Tortugas and is discussed here with regard to SE Florida and in Chapter 4 by Hine et al. with regard to the West Florida shelf. In comparison, estimates for other areas capable of providing habitat for reefs and reef-associated fauna in the United States are 108 km^2 in Guam, 1,231 km^2 in the Main Hawaiian Islands and 2,302 km^2 in Puerto Rico.

Goldberg (1973) provided the first description of the reef communities north of the Florida Keys based on studies of the reef community offshore Boca Raton, Florida (N26°20.8′). This work was limited to a small area in the northern part of the reef complex. Infrequent and regionally isolated reef community monitoring carried out in association with dredging for beach nourishment projects followed (Courtney et al. 1972, 1975, 1980; Continental Shelf Associates 1980, 1984; Goldberg 1981; Blair and Flynn 1989; Dodge et al. 1995). These projects, however, did not provide a continuous record of biotic dynamics nor were the study sites chosen to describe regional patterns of reef community structure.

Beginning in 1997, the Broward County Environmental Protection Department instituted a long term status and trends monitoring program at 18 fixed sites distributed across the shelf from N26°00.26′ to N26°20.80′ latitudes (Gilliam et al.

B.M. Riegl and R.E. Dodge (eds.), *Coral Reefs of the USA*,
© Springer Science + Business Media B.V. 2008

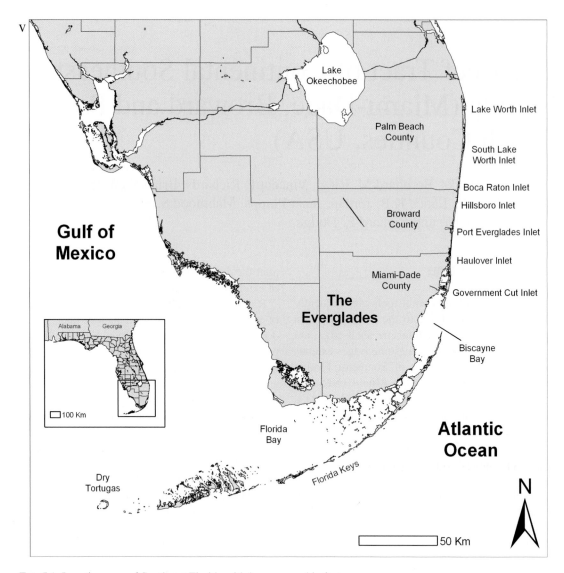

FIG. 5.1. Location map of Southeast Florida with key geographic features

2007). Coral cover, as well as octocoral and sponge density, have been measured annually using a single, fixed, 30 m² belt-transect (1.5×20 m) at each site. In 2004 the program was expanded to 25 sites.

In 2003, the Florida Fish and Wildlife Conservation Commission extended the Florida Keys Coral Reef Evaluation and Monitoring Program (CREMP) into the SE Florida region (Southeast Florida Coral Reef Evaluation and Monitoring Program [SECREMP]) by installing 10 permanent stations. Relative bottom cover is determined annually from video transects at each station (total transect area at each station = 528 m²). In addition, stony coral species inventory, clionid sponge cover (at 66 m² transect), *Diadema antillarum* abundance, and stony coral condition are measured. This activity covers the area from Palm Beach County (N26°42.63′) to Miami-Dade County (N25°50.53′) with 3 sites in each county and one additional site in Broward County at a site of unusually high *Acropora cervicornis* cover (Gilliam 2007).

An extensive investigation of spatial patterns in community structure among reef tracts was carried out in Broward County by Moyer et al.

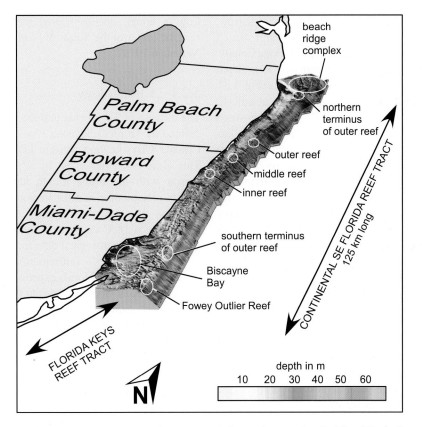

FIG. 5.2. The continental Southeast Florida reef tract extends from Biscayne Bay in Miami-Dade County (N25°34′) northward to West Palm Beach in northern Palm Beach County (N26°43′). It is composed of a complex of limestone ridges and shelf-edge and mid-shelf reefs

(2003) who measured relative bottom cover using six replicate, 50 m point intercept transects at 31 sites within three cross-shelf corridors (Broward County north, central and south), each containing all three reef tracts. While limited to the central part of SE Florida, the study was extensive enough to identify some latitudinal and cross-shelf patterns. Foster et al. (2006) expanded this study to include community patterns within reef tracts at the three corridors, the effects of sampling scale on the analysis of community structure, and the influence of certain environmental factors on community structure. Their measurements were based on 32 photoquadrats (0.75×0.50 m each), replicated six times (72 m²) at each of 99 sites (33 in each corridor; north, central, south).

A SE Florida regional reef habitat mapping project was begun in 2004 by the National Coral Reef Institute (NCRI) for the Florida Fish and Wildlife Research Institute (FWCC). Data from laser air-borne depth sounder (LADS) bathymetry, multi- and single-beam bathymetry, acoustic seafloor discrimination, ecological assessments, and ground-truthing were integrated into maps created with Geographic Information System (GIS) (Walker et al. 2008). Habitat categories were based on the NOAA biogeography program. Field verification of habitat types was used for quality assurance. Maps were completed for Broward County and Palm Beach and Miami-Dade counties are planned or underway (NCRI 2004).

5.2 Regional Setting

The reef tracts are believed to be founded on shore-parallel lithified Late Pleistocene beach ridges (Shinn et al. 1977; Banks et al. 2007). These ridges are likely an offshore extension of the shore-parallel ridge complex (Fig. 5.2).

The continental shelf of SE Florida is narrow, approximately 3 km wide offshore Palm Beach County and 4 km wide offshore Miami-Dade County. Bottom slope steepens at approximately 80 m depth where the East Florida Escarpment begins (Fig. 5.3). The East Florida Escarpment terminates at a depth of 200–375 m at the Miami Terrace, a drowned early to middle Tertiary carbonate platform (Mullins and Neumann 1979). The Terrace extends from N25°20′ northward to N26°30′ and resembles a long, low obtuse triangle. It covers approximately 740 m^2 with a maximum width of 22 km and two levels separated by a discontinuous ridge. The upper terrace is a drowned limestone formation which was subaerially exposed in the middle to late Miocene and is presently marked by karst features. The ridge is probably a drowned Miocene or post-Miocene bank margin complex. The lower terrace (600–700 m depth) is erosional and discontinuous. Its formation is believed to be related to increased flow of the Florida Current and subsequent bioerosion at the time of tectonic uplift and closure of the Isthmus of Panama in mid-Miocene time. Seaward of the lower terrace a large linear depression parallel to the Florida-Hatteras slope separates the Miami Terrace

from a broad, unconsolidated sedimentary ridge near the center of the Straits of Florida (Mullins and Neumann 1979).

5.2.1 Climatology

The climate of SE Florida and the Florida Keys is defined as Tropical Savanna (Aw) in the Köppen Climate Classification System (Trewartha 1968). This class is characterized by a pronounced dry season with the driest month having less than 60 mm precipitation and the total annual precipitation less than 100 mm. Average temperature for all months is 18°C or greater. Table 5.1 presents a summary of climatological data for SE Florida.

During the dry fall/winter/spring months (November–March), Florida experiences the passage of mid-latitude synoptic-scale cold fronts (Hodanish et al. 1997) which bring strong winds from the northeast. These "northeasters" usually last for 2–3 days. From late spring to early fall (the wet season, June–September), differential heating generates mesoscale fronts, creating sea breezes. Convergence of these moisture-laden sea breezes, developing from the different water bodies (Atlantic Ocean, Gulf of

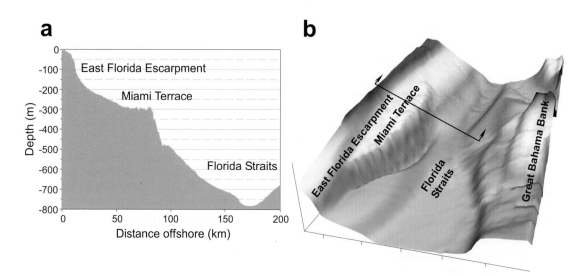

Fig. 5.3. The continental shelf of Southeast Florida is narrow and bordered by deep waters (d ~800–900 m) of the Florida Straits. Warm water of the Florida Current moderates sub-tropical conditions and allows for the existence of tropical coral reef associated communities

TABLE 5.1. Climate data for West Palm Beach and Miami, Florida (West Palm Beach, upper number; Miami, lower number). Temperatures (C°) are based on means from 1971 to 2000. West Palm Beach and Miami wind data are based on means from 1942 to 2005 and 1949 to 2005, respectively (T_{ave} = average monthly temperature, T_{max} = maximum monthly temperature, T_{min} = minimum monthly temperature, W_{ave} = average monthly wind speed (m/s), W_{dir} = average monthly wind direction) (NOAA 2005).

	Jan.	Feb.	Mar.	Apr.	May	June	July	Aug.	Sept.	Oct.	Nov.	Dec.	Average
T_{ave}	19.0	19.6	21.4	23.2	25.7	27.3	28.1	28.2	27.6	25.6	22.8	20.2	24.1
	20.1	20.6	22.4	24.3	26.4	28.0	28.7	28.7	28.0	26.0	23.6	21.1	24.8
T_{max}	23.9	24.6	26.2	27.8	29.9	31.4	32.3	32.3	31.5	29.4	26.9	24.7	28.3
	24.7	25.4	27.1	28.8	30.7	31.9	32.7	32.6	31.7	29.7	27.3	25.3	29.0
T_{min}	14.1	14.6	16.6	18.6	21.4	23.2	23.9	24.1	23.7	21.8	18.8	15.6	19.7
	15.3	15.8	17.8	19.8	22.2	24.0	24.7	24.7	24.3	22.3	19.7	16.8	20.6
W_{ave}	4.5	4.7	4.9	4.9	4.4	3.7	3.4	3.4	3.9	4.5	4.6	4.5	4.3
	4.2	4.5	4.6	4.7	4.2	3.7	3.5	3.5	3.7	4.1	4.3	4.1	4.1
W_{dir}	NW	NW	NW	NW	SE	SE	SE	SE	ESE	ESE	ESE	ESE	ESE
	N	N	ESE	ESE	ESE	ESE	ESE	ESE	E	ENE	E	N	ESE

Mexico, and Lake Okeechobee), coupled with high humidity in the Everglades, result in a low pressure trough developing across the Florida peninsula. This leads to intense thunderstorm activity, which moves from inland to the coasts, delivering large amounts of freshwater to the coastal shelf. South Florida receives 70% of its annual rainfall during these months. Trewartha (1968) refered to the daily sea breeze circulation as a "diurnal monsoon". The mean wind direction during most of the SE Florida wet season is from southeast (tropical)

5.2.2 Hurricanes

From June through November, Florida is a prime landfall target for tropical cyclones, although storms have been documented as early as March and as late as December. The numbers of direct hits of hurricanes (strength based on the Saffir-Simpson scale) affecting SE Florida in the 100 years from 1899–1998 (Neumann et al. 1999) are: 5, Category 1 (winds of 119–153 km/h); 10, Category 2 (winds of 154–177 km/h); 7, Category 3 (winds of 178–209 km/h); 4, Category 4 (winds of 210–249 km/h); and 1, Category 5 (winds >249 km/h). Table 5.2 presents the number of hurricanes and tropical storms affecting Palm Beach, Broward, and Miami-Dade counties from 1871–2006.

Hurricanes Floyd, 1987; Andrew, 1992; Irene, 1999; Frances, 2004; Katrina and Wilma, 2005

affected SE Florida in the last decade. Hurricane Andrew was the most severe with winds at landfall in SE Florida (Elliot Key in Biscayne Bay) of 269 km/h (Landsea et al. 2004). Tropical storm winds extended 225 km from the center. Wave hindcast predicted significant wave heights of 7 m at water depths of 8–9 m. This prediction accounted for the attenuation of the Bahamas Banks (Neumann et al. 1999; Grymes and Stone 1995) which, coupled with the high forward speed of the storm, substantially weakened Hurricane Andrew (Boss and Neumann 1995).

Hurricanes that form in June and July are spawned entirely in lower latitudes on the western side of the Atlantic and in the Western Caribbean. Storms at this time of year are usually weak. Hurricanes occurring in August and September usually form in the Atlantic Ocean and often become mature, severe storms. Hurricanes late in September through October and into November form mainly in the western Caribbean and Gulf of Mexico (USACE 1996).

Hurricanes can have significant influence on the reefal biota of Florida (Tilmant et al. 1994), which can range from wholesale destruction of biota with subsequent regeneration (Shinn 1976) to facilitation of asexual recruitment in some coral species (Fong and Lirman 1995; Lirman 2003) and by cleaning macroalgal and bacterial overgrowth from corals (K. Banks, personal communication, 2007).

TABLE 5.2. Storm frequencies for Southeast Florida (USACE 1996). Number of tropical storms or hurricanes passing within a 50-mi radius of Palm Beach, Broward, and Miami-Dade Counties (a single storm may affect more than one county).

Period	Palm Beach County		Broward County		Miami-Dade County	
	Hurricanes	Tropical storms	Hurricanes	Tropical storms	Hurricanes	Tropical storms
1871–1880	3	0	1	0	0	0
1881–1890	2	2	1	2	2	2
1891–1900	0	2	0	1	1	1
1901–1910	2	4	2	3	3	2
1911–1920	0	0	0	0	0	1
1921–1930	3	1	4	0	3	0
1931–1940	3	0	2	1	1	2
1941–1950	5	1	4	1	5	1
1951–1960	0	2	0	2	0	2
1961–1970	2	0	2	0	2	0
1971–1980	1	1	1	1	0	1
1981–1990	0	2	0	2	1	1
1991–2000	0	2	1	0	1	1
2001–2006	3	0	1	0	0	1

5.2.3   Regional Physical Oceanographic Processes

5.2.3.1   Water Temperature

Thermograph data from July 2001 to December 2003 are presented in Fig. 5.4 (monthly mean, minimum and maximum temperatures for the 3-year period). Highest monthly temperature observed was 30.5°C in August, 2000, on the ridge complex (Fig. 5.4). The lowest minimum temperature was 18.3°C, also on the ridge complex, was recorded in January 2001. Temperatures on the inner, middle and outer reefs were generally similar to one another but the ridge complex was warmer in the summer and cooler in the winter than the other tracts. This reflects the more rapid heat gain or loss of the shallower water over the ridge complex due to air temperature and/or solar insolation. For the 3-year period, the minimum air temperature also occurred in 2001 (11.9°C, Miami WSCMO Airport).

Long-term temperature records will be available in the future from the FWRI (Fish and Wildlife Research Institute) Southeast Coral Reef Ecological Monitoring Program (Gilliam 2007). Temperature loggers have recently been installed at each of the monitoring sites and continuous temperature data are being collected every 2 h. These data will be publicly available from 2007 onward from FWRI.

5.2.3.2   Circulation

The Florida Current flows north (with intermittent reversals) and is the dominant ocean current affecting the SE Florida shelf. It is a portion of the Gulf Stream that intrudes into the Gulf of Mexico as the Loop Current and reverses flow to return to the Straits of Florida before moving in a northeasterly direction towards Europe (Jaap and Hallock 1990). The average monthly low velocity is 1.0 m/s in November and the average monthly high is 2.3 m/s in July. Approximately 2 km offshore, the current velocity is 2.5 m/s (USACE 1996). The western edge of the current meanders from far offshore on to mid-shelf.

The coastal circulation along the SE Florida shelf is strongly related to the dynamics of the Florida Current. The Florida Current follows the steep bottom terrain along the shelf break separating the deep ocean (Florida Straits) from the coastal zone. Mixing between the shelf and deeper ocean waters is affected by transient features created at the western edge of the current. Sub-mesoscale spin-off eddies (Lee and Mayer 1977;

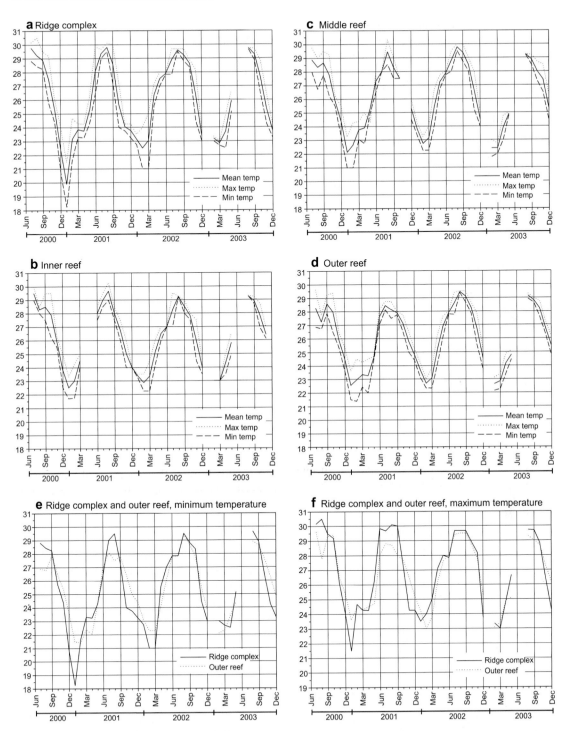

FIG. 5.4. Monthly average, minimum, and maximum water temperatures (C°) from data collected hourly on the seafloor of the: (a) ridge complex; (b) inner reef; (c) middle reef; (d) outer reef offshore central and south Broward County from July, 2000, to December, 2003. Cross-shelf variations in water temperature are illustrated by comparing ridge complex and outer reef water temperatures: (e) minimum water temperature on ridge complex are lower than the outer reef; (f) maximum temperatures are higher on the ridge complex

Shay et al. 2002) are important to local coastal circulation because they affect the continental shelf and largely determine the water properties on the shelf (Soloviev et al. 2003).

Tides in the region are semi-diurnal with amplitudes of approximately 0.8 m and tidal plumes influence coastal circulation near navigation inlets. Seven navigational inlets, approximately 16 km apart, are maintained in SE Florida. At the southern extent of the region, tidal passes allow exchange of water from Biscayne Bay onto the coastal shelf. The relative contribution of the inlets to coastal circulation can be estimated by comparing inlet tidal prisms (volume of water exchanged in the estuary between high and low tide) (Table 5.3). Other factors also affect coastal circulation, such as inlet dimensions, shelf width at the inlets, offshore distance of the Florida Current, tidal plume constituents and salinity. A. Soloviev (personal communication 2007) believes that salinity stratification in the plumes substantially influences local circulation. The salinity of the plumes from the inlets is significantly different in the wet season (June–September) than the dry season (October–May).

5.2.3.3   Ocean Waves

In winter, low pressure systems form on the Atlantic coast of the USA. Short-period, wind-driven waves develop near the center of these lows. As these seas

move away from the center of low pressure they can develop into long period swells, locally known as "ground swells", and affect SE Florida. The wave climate of SE Florida is influenced by the shadowing effect of the Bahamas and, to a lesser extent, Cuba. In the northern part of the SE Florida region, swells from the north are of relatively high energy since they are not influenced by the shallow Bahamas Banks. Broward and Miami-Dade Counties are less affected by this wave energy because of the shadowing effect of the Bahamas Banks.

Long period swells result in increased sediment suspension and turbidity, particularly in shallow water. If northeasters occur when the moon is in perigee, abnormally high tides can result in greater ebb flows at the inlets, increasing the delivery of inlet waters to the reef system. This combination of events can cause more sediment suspension and turbidity than an average hurricane due to relatively short time duration of hurricanes (USACE 1996). Hanes and Dompe (1995) measured turbidity concurrently with waves and currents *in situ* at depths of 5 m and 10 m offshore Hollywood, Florida (Broward County) from January, 1990 to April, 1992. They found a significant correlation between wave height and turbidity. In addition, there was a threshold wave height (0.6–0.6 m) below which waves do not influence turbidity.

The primary sources of regional wave data are the US Army Corps of Engineers Wave Information Study (WIS, frf.usace.army.mil/cgi-bin/wis/atl_main.html) hindcast and the Summary of Synoptic Meteorological Observations (SSMO, NOAA National Data Center). SSMO is a large dataset of observations and was used to calibrate WIS which is hindcast at 3-h time steps for 1956–1975. Figure 5.5 presents wave information for the SE Florida region from WIS. It is important to note that wave energy flux decreases in a southerly direction due to the increasing wave shadow cast by the Bahamas.

5.2.4   Environmental Records in Coral Skeletons

Massive reef building corals, *Montastraea* sp., *Siderastrea* spp., and *Diploria* spp. which possess annual density bands, are present in SE Florida. The size and concomitant age within SE Florida

TABLE 5.3. Southeast Florida tidal inlet characteristics (Stauble 1993; Powell et al. 2006).

Tidal inlet/pass	Latitude	Flow area (m²)	Tidal prism (m³)	Tidal range (m)
Lake Worth Inlet	N26°46.35′	1,400	24,000,000	1.0
South Lake Worth Inlet	N26°32.72′	100	3,000,000	0.9
Boca Raton Inlet	N26°20.16′	180	4,900,000	0.9
Hillsboro Inlet	N26°15.44′	300	8,100,000	0.9
Port Everglades Inlet	N26°05.63′	2,900	18,000,000	0.9
Bakers Haulover Inlet	N25°54.00′	520	10,194,666	0.8
Government Cut	N25°45.63′	1,400	2,700,000	0.8

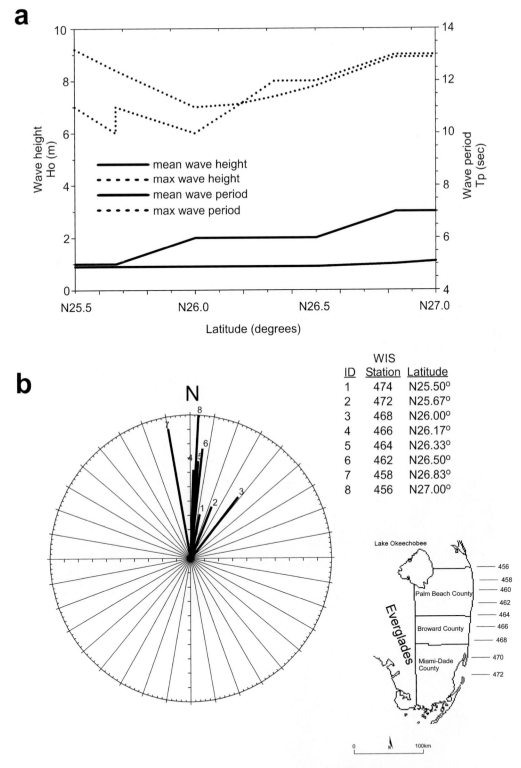

FIG. 5.5. Wave conditions throughout the Southeast Florida region show increasing northerly component of wave energy flux in the northern part of the region. Information based on US Army Corps of Engineers Wave Information System (WIS) hindcast data (http: frf.usace.army.mil/cgi-bin/wis/atl_main.html). WIS data is hindcast at 3-h time steps for 1956–1975: **(a)** mean and maximum wave height (Ho) and period (Tp) increase with latitude (large maximum wave heights and periods at N25.5° are due to the passage of Hurricane Andrew in 1992); **(b)** vector plot of average wave direction and relative magnitude shows that average wave direction trends more northerly with increasing latitude and wave magnitude increases with latitude

is generally limited to small colonies (<1 m) on the order of a few decades in age. Recently, large *Montastraea faveolata* corals with ages of 200–300 years have been discovered and cored (Fig. 5.6). The longest of these records has been dated by the annual density band chronology back to 1694 (K.P. Helmle, personal communication 2007). These multi-century coral records are long enough to identify a range of conditions from natural growth rates to anthropogenically influenced growth and include climatic conditions

from the Little Ice Age up to the recent period of rapid climate change.

Measurements of extension, bulk-density, and calcification from coral slab X-radiographs provide a metric for assessing the influences of an ever-changing environment on coral growth rates. For example, a three decade period of stress ca. 1940–1970 (Fig. 5.7, K.P. Helmle, personal communication 2007), defined by significantly decreased extension rates and increased bulk-densities, coincides with dramatically increased fresh-

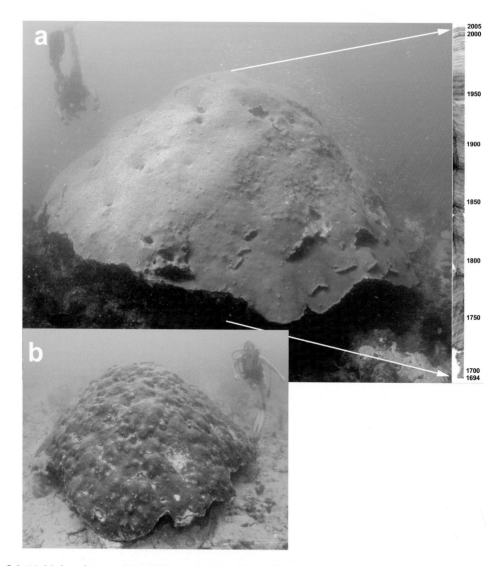

FIG. 5.6. (**a**) *M. faveolata* coral head 2.5 m in height with core X-radiograph on right dating this coral back to 1694. (**b**) *Montastraea faveolata* coral head 2 m in height and dating back to the early 1800s

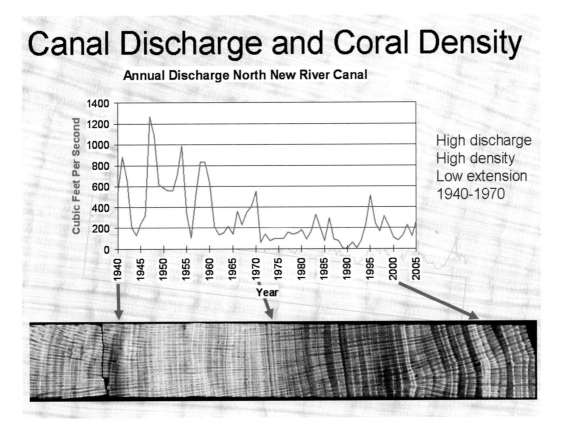

FIG. 5.7. Canal discharge data from North New River Canal illustrating high levels of discharge from 1940 to 1970. Arrows indicate 1940, 1970, and 2000 on coral core X-radiograph below. Note high density skeleton (dark) and low extension rate from 1940 to 1970 compared with 1970–2000

water discharge rates from construction of major canal systems linking Lake Okeechobee to the SE Florida coast. These consecutive stress bands primarily reflect local-scale anthropogenic influences. Conversely, stress bands in 1988 and 1998–99 are present and correspond with two of the strongest El Niños of the past 50 years in 1987 and 1997–98 which illustrates connections between coral growth and global-scale climate patterns.

Linear extension rates for *M. faveolata* in Broward County Florida exhibit strong correlation within and between mid (9 m) and deep water (18 m) from Hollywood, Fort Lauderdale, Pompano Beach, and Deerfield Beach (Table 5.4, provided by Kevin P. Helmle and Richard E. Dodge) The strong correlation within and between depths demonstrates that linear extension rates in *M. faveolata* of SE Florida generally respond com-

TABLE 5.4. Growth correlations for *Montastraea faveolata* (*M.f.*) for mid (9 m) and deep (18 m) depths.

Correlation coefficients between master chronologies over 1985–1970 n = 16, d.f. = 14 for p < 0.01, r_{crit} > 0.624

	MID *M.f.*	DEEP *M.f.*	ALL *M.f.*
MID *M.f.*	–	0.82	0.97
DEEP *M.f.*	0.82	–	0.93
ALL *M.f.*	0.97	0.93	–
Average internal correlation			
	Mean	N	
MID *M.f.*	0.84	6	
DEEP *M.f.*	0.73	3	
ALL *M.f.*	0.75	21	

monly to environmental influences and thus provide robust records of historical growth response to anthropogenic impacts and climate change over hundreds of years.

5.3   Characterization of Southeast Florida Reefs

5.3.1   Geomorphology

Reef growth in Florida is frequently said to be unique to the modern Florida Keys reef tract and considered to terminate at Fowey Rocks (Vaughan 1914; Jaap and Adams 1984; Shinn et al. 1989), but nonetheless reefs and reef-like ridges persist further north (Fig. 5.2). The reef-like ridges are a relict (no active accretion due to exceedingly low cover of reef builders; Moyer et al. 2003). The location of these reefs identifies them as a distinct and also presently non-accreting reef tract (Macintyre 1988), and their geomorphology has recently been described in detail by Finkl et al. (2005) and Banks et al. (2007) and will only be briefly reiterated here.

Between southern Miami-Dade County and Palm Beach County, beginning at the northern end of Biscayne Bay and terminating off the city of Palm Beach at N26°43.1′, up to three shore-parallel, ridge-systems occur in increasing depth. In the early to mid Holocene these were the locus of reef framework development and are called, in increasing distance from shore, the inner reef, middle reef and outer reef (Fig. 5.8). Their morphology is more complicated in the south (Fowey Rocks to Port Everglades) than north of Hillsboro Inlet, where the inner and the middle reefs eventually disappear. The northern termination of the outer reef is off Palm Beach County, where it is replaced by a series of beach ridges that probably represent a drowned headland (Fig. 5.2). The southern termination of the SE Florida reef tract is off Biscayne Bay. In southern Dade County, the middle reef disappears and only the inner and outer reefs remain which then both disappear in a sandy environment seaward of Biscayne Bay. Banks et al. (2007) demonstrated that the SE Florida outer reef is located on a ridge that bends landward near Fowey rocks, and continues behind the Fowey rocks outlier reef. Therefore, the SE Florida outer reef would be most likely equivalent with the Florida Keys shelf-edge-reefs, while the outlier reefs would constitute a separate, more seaward trend initiated on a deeper terrace.

The outer reef (Fig. 5.8b) is a relic acroporid-framework reef (Macintyre and Milliman 1970; Lighty 1977; Lighty et al. 1978) that crests at ~16 m below sea level. It extends more or less uninterrupted (with the exception of reef gaps) from Biscayne Bay northward to its distinct terminus at latitude N26°43′. At a lower sea level

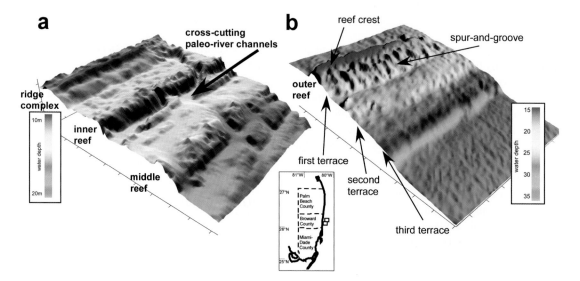

FIG. 5.8. Bathymetric block diagrams showing representative samples of morphology of (**a**) the ridge complex, inner and middle reef, (**b**) the outer reef. Inset map shows location of bathymetry blocks (modified from Banks et al. 2007 by permission of Springer)

stand this distance of about 125 km would have made it one of the best-developed fringing reefs in the western Atlantic, especially when combined with the Florida Keys reefs. Geomorphologic zonation as well as that of major constructors within the framework of the SE Florida outer reef is similar to that of modern Florida Keys reefs (Shinn 1963; Enos and Perkins 1977; Shinn et al. 1981; Lidz et al. 2006). Rubble aprons (talus), back reef, reef crest, spur-and-groove zones and two seaward terraces (17 and 23 m deep) (Fig. 5.6). Reef gaps and collapse features are striking and repetitive features (Banks et al. 2007). The reef gaps are erosional structures with bases deeper than the reef framework and correspond to similar erosional features in the middle and inner reefs. They, thus likely, represent the courses of paleo-river channels. The outer reef initiated before 10.2 cal BP (calibrated ^{14}C-dated years before present) and grew until 8 cal BP (Lighty 1977).

The middle reef is also a mostly continuous feature where it exists and crests at ~15 m below sea level. It extends from South Miami-Dade County northward to Boca Raton Inlet. Unlike the outer reef, it does not display a detectable zonation and reef framework is apparently not continuous throughout or at least variable in development. Any frameworks that do exist are developed on, or drape, a well-defined antecedent slope that is interpreted as the shoreline of the time when the outer reef initiated and began to accrete. Frameworks are mostly dominated by massive corals (*Montastraea* spp., *Diploria* spp.) and only few isolated *A. palmata* frameworks have been found to date. Also the middle reef is dissected by erosional features that connect to reef gaps in outer and inner reefs. The growth history of the middle reef is incompletely understood, but surficial ages obtained so far range from 4.2 to 3.7 cal BP.

The inner reef is the most variable and discontinuous of the three reef tracts and is, in most areas, a complicated amalgamation of patch reefs that can be fused to form longer structures, with individual patch reefs frequently remaining identifiable. It crests at ~8 m below sea level and generally consists of *A. palmata* framework. The inner reef begins south of the middle reef off North Dade County at N25°40′ and extends northward to Hillsboro Inlet at N26°15′ in Broward County where it disappears under the shoreline that

in this region changes trend. It is between 2 and 3 m thick and rests either on coquina (Broward County) or laminated soilstone crusts (South Broward and Miami-Dade Counties). Ages obtained from the inner reef range from 5.9 to 6.2 cal BP.

The nearshore ridge complex extends from N25°51′ in Miami-Dade County to N26°15′ (Hillsboro Inlet) and consists of shoreline deposits with visible karst features. Sediment in cores varied in coarseness from shell hash to coarse sand and had variable siliclastic content. The sediment was interpreted as cemented beach, or immediately nearshore deposits, consisting of a mixture of reworked Pleistocene Anastasia Formation and Holocene deposits. A possibly wave-cut feature exists at 6 m below sea level with a relief of 1.5 m on the outer ridge. The sea-level curve of Toscano and Macintyre (2003) would put erosion of that cliff at 3.5–6.5 cal BP, which would make it a possible shoreline for the period when the inner reef was alive and accreting. Lidz et al. (2003, 2006) also describe a nearshore rock ledge and scarp that could be an equivalent structure in the Florida Keys. In Palm Beach County, the substrate of the nearshore ridges is covered by colonies of tube-building polychaete (*Phragmatopoma*) worms (commonly known as "worm rock"), which considerably expand the complexity of this habitat and supports a diversity of macroalga, small stony corals, boring sponges, worm rock, and tunicates.

Banks et al. (2007) proposed a conceptual model of development of this reef system that included stepwise aggradation and backstepping in response to sea-level history. The basis of any reefal accretion is believed to be submarine sand dunes that originally formed during the oxygen isotope substage-5e highstand (approximately 125 BP). At that time, the SE Florida shelf experienced higher wave-energy due to the submergence of the Bahamas Banks with consequently less buffering of waves. These sand shoals fell dry during subsequent lowstands and became indurated. In the early Holocene, as sea level rose, the sand shoals near the shelf became the locus of accretion of the outer reef. Lighty et al. (1978) noted that it initiated as a fringing reef and transitioned to an extensive shelf-edge barrier reef as rising sea level submerged the back reef shelf margin during its growth from >10.2 cal BP and its demise at 8 cal BP (Lighty 1977). The middle reef is situated possibly on the shoreline that might

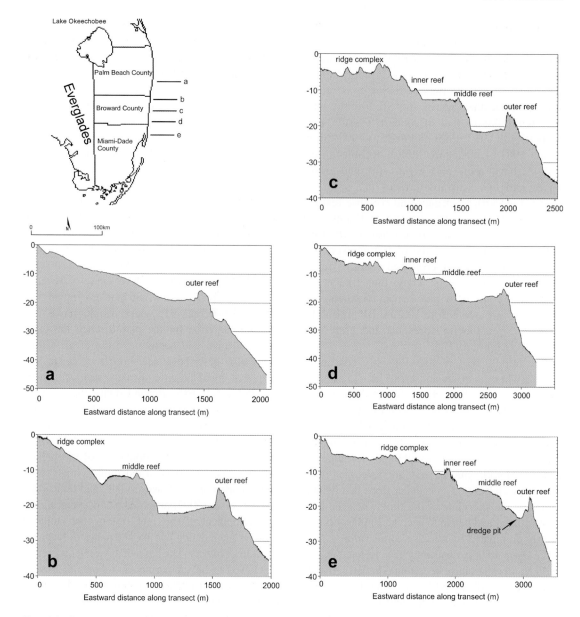

FIG. 5.9. Cross-sections of the ridge complex and reef tracts offshore of Southeast Florida. Bathymetry is extracted from a Lidar dataset of Broward County (Banks et al. 2007)

have been coeval with formation of the outer reef. A further rise in sea level led to the initiation of the inner reef. The available inner reef ages (~6 cal BP) are almost 2 KY younger than the uppermost outer reef (Lighty 1977). At least from the outer to the inner reef, true reef backstepping had occurred. The surface of the middle reef is at 4.3–3.7 cal BP another 2 KY younger than the inner reef, at least

on its surface and may thus have initiated sooner and persisted longer. Banks et al. (2007) propose the following backstepping sequence: growth of the outer reef from ~12–8 cal BP at which time a beach and beach-ridge system existed at the later locus of the middle reef; backstepping to the locations of the present middle and inner reefs during transgression. The initiation sequence of middle and inner reefs is still unresolved, but one would assume that the deeper middle reef initiated first. An ecological differentiation seems to have existed: *Acropora palmata* framework on the shallower inner reef, and massive, less environmentally sensitive corals (*Montastraea* spp. *Diploria* spp., *Siderastrea* spp.) on the middle reef. Causes for termination of inner reef accretion are unclear but cannot be linked to the rate of sea level rise since at the time of demise it was 2–3 mm year^{-1} (Toscano and Macintyre 2003), far less than the maximum reef-accretion rate of 14 mm year^{-1} (Buddemeier and Smith 1988).

Despite being largely unexplained, the demise of SE Florida *Acropora* is interesting insofar as it was apparently not unique to South Florida. Hubbard et al. (2005) found a Caribbean-wide gap in *A. palmata* reef building at around the same time (~6–5.2 cal BP). Parallels to the modern *Acropora* crisis in the Florida Keys and Caribbean become obvious. In each case, the cause or causes remain elusive.

5.3.2 Benthic Habitat Mapping

Walker et al. (2008) created maps of the nearshore benthic habitats from 0 to 35 m depth by employing a combined-technique approach which incorporated high resolution laser bathymetry, aerial photography, acoustic ground discrimination (AGD), video groundtruthing, and limited subbottom profiling (Fig. 5.10a–c). Features were classified based on their geomorphology and benthic fauna using similar criteria to NOAA Caribbean biogeography mapping including a similar classification scheme. *In situ* data, video camera groundtruthing, and AGD were used to help substantiate the classification of the habitat polygons.

Acoustic ground differentiation, which evaluates the shape of sound waves bounced off the seafloor from which different categories of wave shapes are classified that correspond to different habitats, was also used to further discriminate the sea floor based on the density of organisms (Moyer et al. 2005; Riegl et al. 2005). These data supplemented the geomorphology-based layer to include not only mapping between features (inner, middle, and outer reefs), but also the variability of habitat within these features (Walker et al. 2008).

This combined technique approach ensured high accuracy by utilizing the data with highest resolution (LADS bathymetry) as the base and supplementing it with lower resolution data of different information content. The maps yielded user and producer accuracies comparable to the photo-interpreted NOAA Caribbean maps (near 90%).

5.4 Biogeography of Southeast Florida Reefs

The biogeographic setting of the SE Florida reef system is a complex of potentially interconnected habitats and the confluence of oceanic (clear, oligotrophic water; stable temperature) and continental (higher temperature variability, land run-off, human impacts, submarine groundwater discharge) influences. The Florida current delivers planktonic larvae from the upstream sources (Yeung and Lee 2002) in the Florida Keys and Tortugas (tropical, insular) and the estuaries and lagoons of Biscayne Bay and Florida Bay (subtropical, continental). Estuaries within the southeast continental region with tidal inlet connectivity to the reef system include northern Biscayne Bay with seagrass and mangroves, mangrove habitat in Broward County, and mangrove and sea grass habitats in the Lake Worth Lagoon of Palm Beach County. The western Bahamas Banks lie only approximately 80 km eastward of SE Florida but are separated by the deep water and rapid currents of the Florida Straits.

Studies on the extent and direction of gene flow or connectivity among reefs in the SE Florida biogeographic region are limited to work on amphipods and ophiuroids by Richards et al. (2007). Genetic connectivity was studied along 355 km of the Florida Keys and SE Florida reef tract (Fig. 5.11) from assessment of gene flow in three commensal invertebrate species displaying contrasting reproductive strategies. The three species, two amphipods (*Leucothoe kensleyi* and *Leucothoe ashleyae*) and a brittle star (*Ophiothrix lineata*), are all commensal within the branching vase sponge *Callyspongia vaginalis* (Fig. 5.12). The brittle star is a broadcast spawner, whereas the amphipods brood their young and lack planktonic larvae. Although *O. lineata* is a broadcast spawner, it possesses an apparently rare form of development in ophiuroids where embryos develop inside a fertilization membrane for 6–8 days before emerging

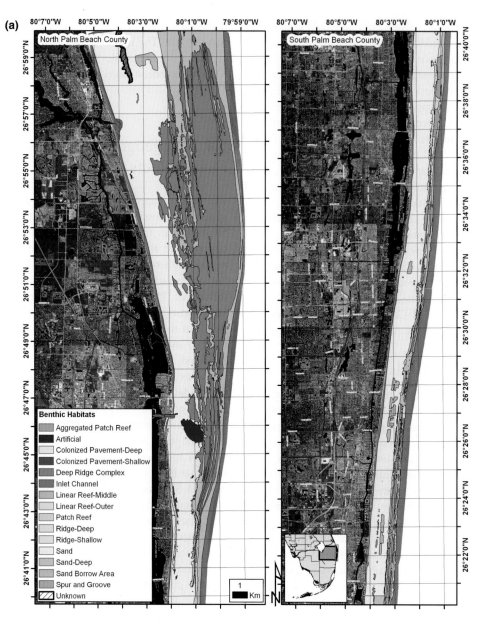

Fig. 5.10. Habitat maps for the Continental SE Florida reef tract based on a combined method approach using laser bathymetry, acoustic ground discrimination, visual groundtruthing (Walker et al. 2008). (**a**) Palm Beach County

as miniature crawl-away juveniles (V.P. Richards, unpublished data). This mode of development suggests that its embryos are passive propagules and exposure to currents for up to 8 days should give *O. lineata* enhanced dispersal abilities. These contrasting reproductive strategies led to expectations of strong gene flow (high connectivity) among reefs for the broadcasting brittle star, but low connectivity for the brooding amphipods.

Findings suggest that reproductive life history is not always a reliable predictor of genetic connectivity. Commonplace within reefs are positive species interactions such as commensalism and mutualism (i.e., facilitation), which have substantial influence

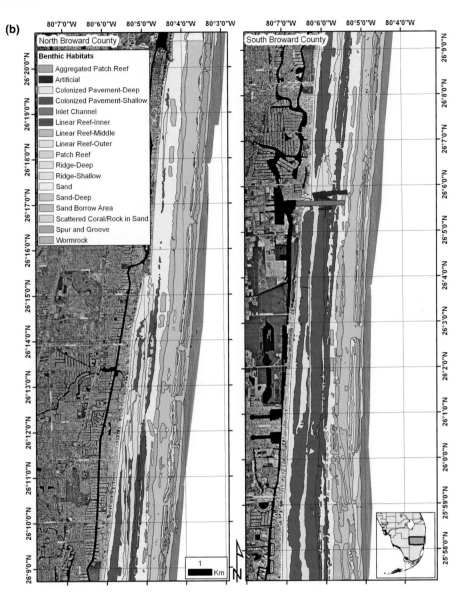

FIG. 5.10. (continued) (**b**) Broward County

on the structure and function of these communities (Bruno et al. 2003). The dynamics of connectivity among reef populations is complex, and factors such as shallow coastlines, deep expanses of open water, and interaction among species involved in facilitation also need to be considered.

Measures of genetic differentiation among populations showed high levels of connectivity overall along the Florida reef tract for all three species (i.e., there was no significant difference in the distribution of haplotypes among populations; Fig. 5.13). The Ft Lauderdale population of *L. ashleyae* did not follow this general pattern and represented the only instance of restricted gene flow along the SE Florida coastline (see Richards et al. 2007 for further discussion on possible reasons for this exception). Paradoxically, only the brittle star showed a statistically significant pattern of genetic isolation by geographic distance along the Florida reef tract (i.e., individuals from

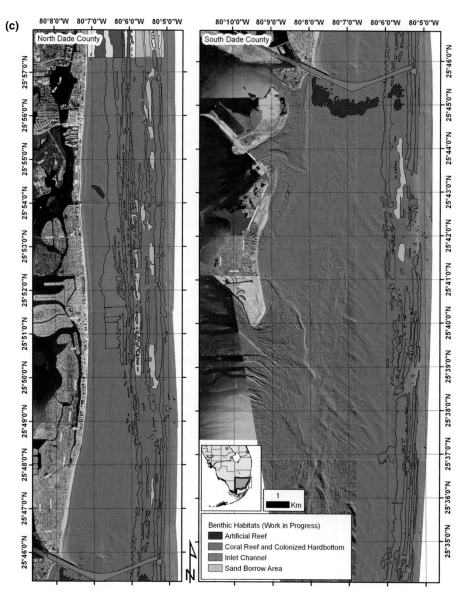

FIG. 5.10. (continued) (c) Miami-Dade County (modified from Walker et al. 2008 by permission of Journal of Coastal Research)

neighboring populations were more closely related to each other than individuals from more distant populations).

The high levels of connectivity detected for both amphipod species along the Florida reef tract were unexpected given their lack of planktonic larvae and raises the question of how they are able to disperse so effectively along the SE Florida coastline. If the amphipods (which usually only attain an adult size of <5.0 mm) were dispersing via crawling or occasional short-range swimming bursts, a pattern of genetic isolation by distance would exist. The lack of this signal suggests that another dispersal mechanism is operating. A possible

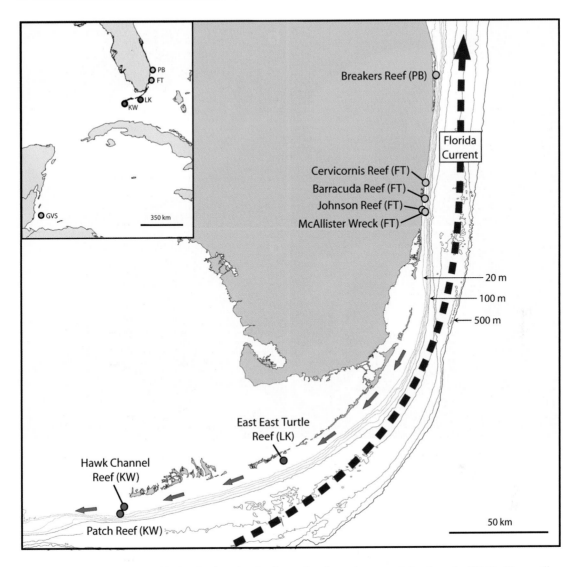

FIG. 5.11. Map showing individual collection sites for the study of genetic connectivity along the SE Florida coastline (color scheme corresponds to Fig. 5.13.). Blue arrows depict a counter current flowing through Hawk Channel (after Yeung and Lee 2002). Inset shows the four major collection locations relative to the collection site in Belize: PB, Palm Beach; FT, Ft Lauderdale; LK, Long Key; KW, Key West; GVS, Glover's Reef. Depth contour data from http://www.ngdc.noaa.gov/mgg/ibcca

explanation is one where the amphipods are being dispersed along the reef tract inside sponge fragments generated during strong storms and hurricanes. Asexual fragmentation is an important dispersal mechanism for many branching sponge species (Wulff 1991), and the type of severe storms and hurricanes often experienced along the SE Florida coastline are capable of detaching and transporting numerous sponge species con-

siderable distances as demonstrated in other reef environments by Wulff (1985, 1995a, b). Thus, at least a portion of SE Florida's benthos may have recruited by simply tumbling along the linear drowned reef systems.

An estimate of migration indicated that the amphipods were migrating up and down the SE Florida coastline with approximately equal frequency, a result consistent with random, bi-directional

Fɪɢ. 5.12. Richards et al. (2007) used the amphipods: (**a**) *Leucothoe kensleyi*; (**b**) *L. ashleyae*; (**c**) brittle star, *Ophiothrix lineata*, which are all commensals with: (**d**) the sponge, *Callyspongia vaginalis,* to study gene flow among reefs in the Southeast Florida biogeographic region. (Photos by: **a**, **b** Vince Richards; **c** scanned with permission from Hendler et al. 1995)

transport mediated by storms. In notable contrast, the brittle star showed a strong southerly migration bias in the Florida Keys. Although the strong northerly flow of the Florida Current is assumed to be an important dispersal agent along the Florida coastline; instead, the well characterized counter current (Fig. 5.11) that runs southwest through Hawks Channel in the Florida Keys (Lee and Williams 1999; Yeung and Lee 2002) may be the dominant dispersal agent for *O. lineata* embryonic propagules in this region. The southerly bias to *O. lineata* migration was not evident along the Broward and Palm Beach county coastlines, where migration occurred with high frequency in both directions. This complex pattern may result from the dynamic

counter currents and eddies created as the Florida current intrudes over the shelf break (Lee and Mayer 1977; Shay et al. 2002; Soloviev et al. 2003).

Richards et al. (2007) also tested the hypothesis that deep water acts as a barrier between populations by examining the connectivity between Florida and Belize. Because *O. lineata* embryo propagules are non-swimming, their dispersal can be assumed to be passive and, therefore more likely to be influenced by physical oceanographic factors. Consequently, the deep water between Florida and Belize and entrapment in eddy currents over the Meso-American Barrier Reef System (Sheng and Tang 2004) are factors that could affect dispersal. Furthermore,

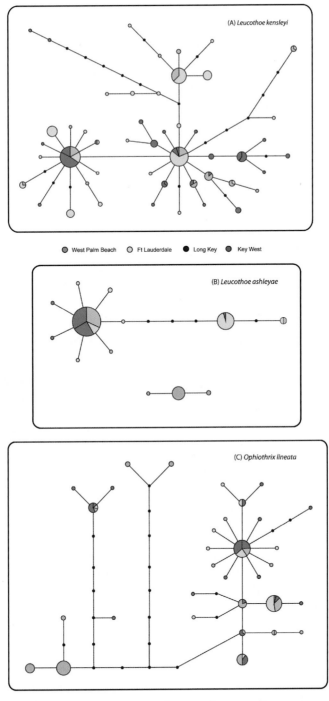

○ West Palm Beach   ○ Ft Lauderdale   ● Long Key   ● Key West

○ West Palm Beach   ○ Ft Lauderdale   ● Long Key   ● Key West   ○ Glovers Reef

Fɪɢ. 5.13. Statistical Parsimony networks depicting the relationship among different mitochondrial COI haplotypes for (A) *Leucothoe kensleyi*, (B) *Leucothoe ashleyae*, and (C) *Ophiothrix lineata*. Colored circles = individual haplotypes; small black circles = haplotypes that hypothetically should exist in the population, but were not sampled; connecting lines = one base pair mutation. Circle size for each haplotype is proportional to its frequency of occurrence and all three networks have the same scale. Different colors correspond to the five major geographic sampling regions (see Fig. 5.11). Due to the large genetic distance between the Florida and Belize *L. ashleyae* haplotypes (79 base pair mutations), there is no statistical support for any connection point between them. Consequently, connection of these haplotypes in the same network is precluded. Networks were created using the software package TCS version 1.13 (Clement et al. 2000)

the strong genetic isolation by distance signal detected across all locations suggests that geographic distance is also an important factor influencing connectivity patterns in this species. The results for *L. ashleyae* represent a dramatic contrast in dispersal ability dependant on the physical environment: high gene flow along a shallow coastline despite a brooding reproductive strategy, with absence of gene flow between populations separated by 'deep water resulting in potential cryptic speciation.

Richards also sampled the amphipods, *Leucothoe kensleyi* and *L. ashleyae* in Bimini, Bahamas which is 95 km from the Florida coast, yet separated by deep, high current velocity water. Although the geographic distance from Fort Lauderdale to Bimini is less than a third the distance from Palm Beach to Key West, the genetic distance was much higher, so these amphipods are strongly connected along shallow coastlines, yet show no connectivity across very short distances of deep water.

The finding of high levels of genetic connectivity for three species within the Florida reef tract (Figs. 5.10 and 5.12) has important implications for the management and conservation of Florida reefs. Northern reefs receive considerably less management attention than reefs in the south (Causey et al. 2002), and are likely being adversely impacted by extensive urban development in this region (Lapointe 1997; Finkl and Charlier 2003). The continued decline of the northern reefs could impact southern reefs, as a reduction in gene flow from the north could reduce genetic diversity in the southern reefs rendering them less adaptable to environmental perturbation.

5.5   Patterns in Reef Community Structure

Although the continental SE Florida reefs have a fauna similar to the Florida Keys, Bahamas and Caribbean, the community structure is different (Moyer et al. 2003). The only major reef building coral missing from SE Florida is *Acropora palmata*, although isolated colonies do exist (Banks, personal observation). Recent claims that *Acropora* occurrences in Broward county are indicative of ocean warming (Precht and Aronson 2004) require further verification, since widespread rubble of the species in question indicates that these corals have already previously had a repeated, but ephemereal presence in the area over the last centuries. Aside from a dominance of bare substratum, relative cover is dominated by macroalgae or octocorals, while scleractinian cover is low. Isolated patches of higher coral cover can be found on the ridge complex offshore central Broward County where one site, dominated by massive corals, has approximately 16% cover and another area is covered by large colonies of *Acropora cervicornis* with 34% cover.

5.5.1   Region-Wide Benthic Community Structure

A summary of relative bottom cover data for SE Florida is given in Table 5.5. Methodological and sample size differences likely contribute to variations in cover among studies. Based on the averages of data collected by Blair and Flynn (1989), Thanner et al. (2006), and Gilliam et al. (2007) overall faunal density is dominated by porifera ($15.7/m^2$), followed by octocorals ($7.7/m^2$) and scleractinia ($2.5/m^2$). A list of all reported species of macroalgae, sponges, octocorals and stony corals reported is provided in Table 5.13.

Macroalgal cover is generally lower in SE Florida than in the Florida Keys and Tortugas (Beaver et al. 2005; FWCC 2005). Foster et al. (2006) found that *Dictyota* spp. and *Halimeda* spp. were the dominant algal species. Abundance of macroalgae can be seasonal or vary over different time scales.

Paul et al. (2005) reported that the cyanobacteria *Lyngbya confervoides* and *L. polychroa* (Fig. 5.14) covered extensive areas of reef offshore Broward County in recent years, particularly 2003. Extensive blooms subsided in 2005 and 2006, although shorter time scale boom/bust cycles are apparent. Tichenor (2005) reported a persistent bloom on the outer reef offshore of Palm Beach County. Causes of these blooms are unknown although he attributed them to treated wastewater discharge from an outfall pipe up-current of his

TABLE 5.5. Average relative bottom cover for ridge complex and reef tracts of Southeast Florida. FWCC (2006) data averaged over 2003–2004 and based on three sites per county (fourth site in Broward is large stand of *Acropora cervicornis*). Two Broward sites have unusually high stony coral cover (one, 40% cover *A. cervicornis* and one, 12% cover of massive corals on ridge complex). Values in parentheses do not include these high cover sites. Data from other studies are averaged over all sites for each county.

	Palm Beach County	Broward County			Miami-Dade County
	(3)	(1)	(3)	(2)	(3)
Bare substrate	70%	10%	73% (80%)	54%	73%
Macroalgae	1%	66%	4% (4%)	15%	9%
Octocoral	20%	12%	8% (12%)	16%	12%
Porifera	7%	8%	2% (4%)	8%	3%
Scleractinia	1%	2%	13% (0%)	5%	1%
other	1%	2%	1% (0%)	3%	2%

(1) Foster et al. (2006)
(2) Moyer et al. (2003)
(3) FWCC (2006)

study site. Foster et al. (2006) included presence/absence data of *Lyngbya* spp. during their studies and found significant occurrences on all reef tracts but not on the ridge complex. The fine filamentous morphology of this genus probably contributes to an underestimated cover based on point count of images, therefore not contributing significantly to macroalgal cover estimates.

In spring, 2007, a macroalgae bloom comprised of *Cladophora liniformis*, *Enteromorpha prolifera*, *Centroceras clavulatum* (Fig. 5.7) (identification by Brad Bedford, Harbor Branch Oceanographic Institute) occurred. Other species may have been present but were not identified. These macroalgae formed a thick mat on sand bottom and reef in northern Broward and southern Palm Beach Counties. The cause has not been identified.

All datasets (Goldberg 1973; Moyer et al. 2003; Gilliam et al. 2007; Foster et al. 2006) consistently showed that octocorals dominate faunal cover on SE Florida reefs (Table 5.6), although porifera dominated in abundance. Forty-eight species were reported by Foster et al. (2006) and the dominant groups were *Eunicea/Muricea* spp. and *Briareum asbestinum*. Thanner et al. (2006) reported similar dominant taxa in Miami-Dade

County. Octocoral cover on SE Florida reefs is similar to that in the Florida Keys (Beaver et al. 2005; Gilliam et al. 2007).

The dominant sponges are the basket sponge, *Xestospongia muta*, *Anthosigmella varians* and *Spheciospongia vespariuum*. 43 species of scleractinian have been reported for SE Florida (Moyer et al. 2003; Foster et al. 2006; FWCC 2006; Goldberg 1973), including the Indo-Pacific azooxanthellate coral, *Tubastrea coccinea* (Fenner and Banks 2004). Cover is low and most colonies are of small size, typically less than 50 cm. SE Florida has lower species richness and cover than Florida Keys and Tortugas (1.5% versus >5% coverage; Beaver et al. 2005; FWCC 2006). *Montastraea cavernosa* is dominant in terms of cover, although *Siderastrea siderea* is numerically the most abundant. This species recruits frequently but does not usually reach large sizes. Similar dominance of *M. cavernosa* has been reported for other high latitude reefs (Bermuda and Brazil; Laborel 1966; Castro and Pires 2001) and reefs in turbid settings like windward Barbados (Lewis 1960). Loya (1976) found a correlation between the abundance of *M. cavernosa* and heavy turbidity and sedimentation, and the SE Florida reefs can certainly be called a usually turbid environment. Species of the *Montastraea annularis* complex are all present in SE Florida but, in contrast to the Caribbean, are not dominant (Knowlton 2001; Moyer et al. 2003).

Moyer et al. (2003) reported significant north/south variation in benthic community structure on the ridge complex and outer reefs. In general, diversity increased from north to south in Broward County. A number of factors could account for this, including temperature variations, planktonic larval supply, substrate characteristics, and wave energy. Planktonic larval supply is transported northward by the Florida Current from the Florida Keys and Caribbean provinces, thus the observed S-N gradient may indicate attenuation of larval supply towards the north. Studies that could verify such a hypothesis are lacking. Substrate differences may also affect community structure. Vertical relief of the ridge complex declines towards the north, eventually reaching <1 m. This seems to be accompanied by decreasing live benthic cover and increased sponge dominance. In the south, where relief is greater than 1 m, the benthic fauna is dominated by octocorals and the zoanthid, *Palythoa caribbaeorum*. Combined with

FIG. 5.14. (**a**, **b**) The cyanobacteria, *Lyngbya confervoides* and *L. polychroa* formed persistent blooms on the reefs in 2003 and smothered many reef organisms, particularly branching octocorals, such as *Pseudopterogorgia* spp. (**c**) Extensive blooms of *Cladophora liniformis*, *Enteromorpha prolifera*, *Centroceras clavulatum*, and others bloomed on inter-reef sand plains and reef tracts of northern Broward County and southern Palm Beach County in the spring of 2007. (Photos by: **a**, **b** Karen Lane; **c** Jeff Torode)

lower relief, the northern areas also receive higher wave energy, since the protection of the Bahamas banks decreases less towards the north. Wind waves from the north and northeast are generally higher in the northern portions of SE Florida which may results in more sediment suspension and turbidity in the northern nearshore.

The cover by scleractinians and octocorals showed no significant temporal changes in the 1997–2005 monitoring dataset obtained by Gilliam et al. (2007), but sponges declined significantly in 2000/2001 and stabilized after that. Concomitant sedimentation monitoring showed high rates at all sites in the winter of 2001 which may have contributed to sponge mortality.

5.5.2   Cross-Shelf Patterns in Benthic Community Structure

The reef cross sections shown in Fig. 5.9 illustrate the depths of the reef tracts and ridge complex. In addition to depth, variation in substrate composition and morphology among reef tracts can impact benthic composition. Among other factors, the porosity of the limestone on the narrow shelf may allow groundwater to seep up onto the reef (Finkl and Charlier 2003) potentially exposing the community to groundwater-borne pollutants. In general, environmental conditions on the ridge complex and possibly the inner reef are more variable than on the middle and outer reefs due to less depth and proximity to shore. Moyer et al. (2003) found that benthic communities on the middle and outer reefs were similar but both differed from the inner reef communities (Figs. 5.15 and 5.16).

Foster et al. (2006) investigated the effect of depth, bottom slope, rugosity and sediment thickness on the structure of benthic communities. Similar to Moyer et al. (2003), they found the community on the ridges and inner reefs to differ from the middle and outer reefs (Tables 5.7 and 5.8). Living benthic substrate cover increased from onshore to offshore. Table 5.9 shows lowest scleractinian density on the inner reef when sites with high *Acropora cervicornis* and anomalously high coral cover were removed (Gilliam et al. 2007). Overall, octocoral and sponge densities are lowest on the ridge complex and inner reef, which might be a direct response to increased wave energy. Kinzie (1973) and Yoshioka and Yoshioka (1991)

suggested that wave energy may influence octocoral populations by detachment of colonies. While living cover clearly differed, species richness of hard and soft corals and sponges among reef tracts was relatively constant.

Vargas-Angel et al. (2003) reported extensive patches of flourishing *Acropora cervicornis* approximately 400–800 m offshore on the ridge complex in Broward County in 3–7 m depth. The area of the patches ranged from ~0.1 to 0.8 ha. Coral cover was 5–28% within reef patches with 87–97% of the cover by *A. cervicornis*. In 2002 they reported Type I White Band Disease (WBD) in all thickets but no bleaching. The WBD was more common at the center of the patches than the periphery where cover was lower. These populations were found to be fertile and to spawn each summer (Vargas-Angel et al. 2003; Vargas-Angel et al. 2006).

5.5.3   Fish Community Structure

5.5.3.1   Miami-Dade and Broward Counties

In total, over 350 fish species have been recorded from Broward and Miami-Dade Counties. However, this total is certainly an underestimate and could substantially increase if piscicide (e.g., rotenone) collections of cryptic fishes were conducted (Ackerman and Bellwood 2000; Willis 2001; Collette et al. 2003).

Generally, the fish assemblages associated with hardbottom reef communities in SE Florida resemble those found throughout the Florida Keys, Greater Caribbean, and Gulf of Mexico. All the common families are represented (Labridae, Pomacentridae, Haemulidae, Gobiidae, etc.; Fig. 5.17). Nonetheless, given the northern latitude at which these reef communities exist, temperate fish (*Orthopristis chrysoptera* [Haemulidae]) can be commonly observed in winter months. The Broward County reef fish community exhibits cross-shelf variation in fish assemblage structure correlated with changes in depth. The deeper, outer reef sites harbor higher fish densities and more species than the shallower inner reef sites (Ferro et al. 2005).

Nearshore hardbottom (<300 m from the shoreline; colonized pavement) fish assemblages show considerable differences in assemblage structure

FIG. 5.15. Photos of ridge complex communities illustrate: (**a**) and (**b**) typical flat pavement-like substrate and abundance of octocorals; (**c**) the encrusting zoanthid, *Palythoa caribaeorum*, is common on the ridge complex and inner reef; (**d**) patches of relatively high stony coral cover are occasionally found on back- and foreslopes; (**e**) and (**f**) relatively large, monotypic patches of *Acropora cervicornis* are found offshore central Broward County (Photos by: **a**, **e**, **f**, Kenneth Banks; **b**, **c**, **d**, David Gilliam, Susan Devictor)

when compared to the more offshore sites on the inner, middle, and outer reefs. Typically, the inner reef is dominated by juvenile grunts (Haemulidae) (Jordan et al. 2004), whereas wrasses (Labridae) and damselfishes (Pomacentridae) dominate the middle and outer reefs. Due to its close proximity to shore and shallow depths (<7 m), the nearshore hardbottom habitat is ephemeral with movements

Fig. 5.16. Photos of the outer reef illustrate: (**a**) typical diversity of sponges, octocorals and scleractinians; (**b**) the massive sponge, *Xestospongia muta*, is common on all reef tracts in Southeast Florida; (**c**) the alcyonacean, *Icilligorgia schrammi* is common on the outer reef and areas of high rugosity on the middle reef; (**d**) reef-perpendicular sand channels commonly incise the outer reef (Photos by David Gilliam, Susan Devictor)

TABLE 5.6. Average relative faunal cover for ridge complex and reef tracts of Southeast Florida. FWCC (2006) data averaged over 2003–2004 and based on three sites per county (fourth site in Broward is large stand of *Acropora cervicornis*). Two Broward sites have unusually high stony coral cover (one, 40% cover *A. cervicornis* and one, 12% cover of massive corals on ridge complex sites). Values in parentheses do not include these high cover sites. Data from other studies are averaged over all sites for each county.

	Palm Beach County	Broward County			Miami-Dade County
	(3)	(1)	(3)	(2)	(3)
Octocoral	68%	48%	34% (75%)	50%	65%
Porifera	23%	34%	10% (22%)	25%	16%
Scleractinia	4%	8%	53% (2%)	16%	6%
other	5%	10%	3% (1%)	9%	13%

(1) Foster et al. (2006)
(2) Moyer et al. (2003)
(3) FWCC (2006)

of large amounts of sediments during major storm events (Walker 2007). Nevertheless, when compared to other natural reef habitats in the area, this low-relief hardbottom contains disproportionately high densities of juvenile fishes (Lindeman and Snyder 1999; Baron et al. 2004; Jordan and Spieler 2006). The ephemeral nature of the nearshore hardbottom and its high proportion of juveniles (with their intrinsic recruitment variability) are likely responsible for the large annual population fluctuations recorded for this dynamic habitat (Jordan and Spieler 2006). Additionally, anthropogenic impact appears to be intense in this habitat. For example, recent (2005–2006) beach renourishment activities (deposition of sand to widen beaches) buried approximately 30,000 m^2 of nearshore hardbottom along a 9-km segment of coastline. Although the habitat loss was mitigated by deploying large boulder artificial reefs (total area ~8.9 acres),

TABLE 5.7. Average relative bottom cover for ridge complex and reef tracts of Broward County. FWCC (2006) data averaged over 2003–2004 and based on three sites per county (fourth site in Broward is large stand of *Acropora cervicornis*). Two Broward sites have unusually high stony coral cover (one, 40% cover *A. cervicornis* and one, 12% cover of massive corals on ridge complex sites) and are not included in the average values.

	Ridge complex d < 6 m			Inner reef d = 6–10 m			Middle reef d = 10–20 m			Outer reef d = 15–30 m		
	(1)	(2)	(3)	(1)	(2)	(3)	(1)	(2)	(3)	(1)	(2)	(3)
Bare substrate	15%	72%	–	12%	53%	–	8%	46%	–	4%	43%	–
Macroalgae	68%	4%	–	73%	17%	–	70%	22%	–	60%	18%	–
Octocoral	7%	11%	–	6%	12%	–	10%	17%	–	23%	25%	–
Porifera	3%	3%	–	4%	7%	–	10%	10%	–	12%	10%	–
Scleractinia	3%	8%	4%	2%	3%	1%	1%	4%	2%	1%	4%	2%
Other	4%	2%	–	3%	8%	–	1%	1%	–	0	1%	–

(1) Foster et al. (2006)
(2) Moyer et al. (2003)
(3) FWCC (2006)

TABLE 5.8. Average relative faunal cover for ridge complex and reef tracts of Broward County. FWCC (2006) data averaged over 2003–2004 and based on three sites per county (fourth site in Broward is large stand of *Acropora cervicornis*). Two Broward sites have unusually high stony coral cover (one, 40% cover *A. cervicornis* and one, 12% cover of massive corals on ridge complex sites) and are not included in the average values.

	Ridge complex d < 6 m			Inner reef d = 6–10 m			Middle reef d = 10–20 m			Outer reef d = 15–30 m		
	(1)	(2)	(3)	(1)	(2)	(3)	(1)	(2)	(3)	(1)	(2)	(3)
Octocoral	41%	46%	–	40%	40%	–	45%	53%	–	64%	63%	–
Porifera	18%	13%	–	27%	23%	–	45%	31%	–	33%	25%	–
Scleractinia	18%	25%	4%	13%	10%	–	5%	13%	2%	3%	10%	2%
Other	24%	8%	–	20%	27%	–	5%	3%	–	0	3%	–

(1) Foster et al. (2006)
(2) Moyer et al. (2003)
(3) FWCC (2006)

TABLE 5.9. Average faunal density for ridge complex and reef tracts of Broward County. FWCC (2006) data averaged over 2003–2004 and based on three sites per county (fourth site in Broward is large stand of *Acropora cervicornis*). Two Broward sites have unusually high stony coral cover (one, 40% cover *A. cervicornis* and one, 12% cover of massive corals on ridge complex sites) and are not included in the average values.

	Ridge complex d < 6 m	Inner reef d = 6–10 m			Middle reef d = 10–20 m				Outer reef d = 15–30 m			
	(1)	(1)	(2)	Average	(1)	(2)	(3)	Average	(1)	(2)	(3)	ave
Octocoral	5.9	9.8	3.0	6.5	4.9	4.0	9.5	8.5	8.9	11.5	12.9	11.1
Porifera	7.7	7.5	4.9	6.2	12.6	3.3	50.8	22.2	8.6	3.3	43.3	14.0
Scleractinia	2.4	1.0	4.0	2.5	2.6	3.5	2.2	2.8	2.7	1.4	2.8	2.3

(1) Gilliam et al. (2007)
(2) Blair and Flynn (1989)
(3) Thanner et al. (2006)

preliminary results show that, while these mitigation boulders harbor juveniles at densities similar to the nearshore hardbottom, the percent contribution of juveniles to the total fish assemblage was substantially lower. That is, more adult-stage fishes and a higher piscivore density were found on the boulders, possibly affecting the natural predation rate on neighboring natural substrate (S. Freeman, personal communication 2007).

Reef fish community studies in Miami-Dade County are limited to work by Thanner et al. (2006) who compared benthic and fish communities on artificial and natural reefs offshore northern Miami-Dade County. They found that the middle reef was strongly dominated by Gobiidae, while the outer reef was dominated by Pomacentridae with Scaridae and Labridae closely following. They reported 44 and 48 fish species on the middle and outer reefs, respectively.

As previously mentioned, Broward County has sites with unusually high densities of *Acropora cervicornis* for SE Florida (Section 5.1). The corresponding fish assemblages are also unique to the area. High densities of both juveniles (mainly *Haemulon flavolineatum* [Haemulidae]) and piscivores have been recorded at the *A. cervicornis* thickets (Gilliam et al. 2007). Given the size of the juveniles present, it is likely that this habitat acts as refuge for early juvenile and subadult haemulids. Despite the attraction of juvenile fishes, the natural meshwork created by the branching coral appears to limit larger piscivores from entering.

Also characteristic of Broward County reef fish assemblages are its conspicuously low densities of legal-size groupers and snappers (variable for different species). Ferro et al. (2005) analyzed data from 667 point-counts in Broward County and found only two grouper (Serranidae) of legal size. When compared to reefs in the Upper Florida Keys (Dixie Shoals), Broward County fish assemblages have a substantially lower density of groupers. Point-counts from Dixie Shoals exhibited an abundance of groupers that was four times greater than surveys from Broward County. At Dixie Shoals, on average, two commercially and recreationally important groupers were seen per point-count survey; while in Broward County even single grouper was infrequently seen (Table 5.10). Although legal-sized grouper were recorded during only two of 667 surveys, large grouper exist in Broward County waters albeit at a lower frequency than in the Florida Keys, as evidenced by low commercial landings (Johnson et al. 2007).

Remotely operated vehicle (ROV) surveys along deep habitat at 50–120 m depth (Bryan 2006) revealed little hardbottom. What existed had low vertical relief (<1 m). Of the 27 species recorded from ROV surveys on the 50–120 m depth natural substrate, nine were absent from the more extensive, shallow reef survey.

TABLE 5.10. Grouper (Serranidae) abundances from point-count surveys in Broward County, Florida and Dixie Shoals (Florida Keys).

Species	Broward County abundance	Dixie Shoals abundance
Cephalopholis cruentata (Graysby)	127	40
Cephalopholis fulva (Coney)	2	20
Epinephelus adscensionis (Rock hind)	4	2
Epinephelus guttatus (Red hind)	8	2
Epinephelus morio (Red grouper)	232	5
Mycteroperca bonaci (Black grouper)	0	16
Mycteroperca interstitialis (Yellowmouth grouper)	1	1
Mycteroperca phenax (Scamp)	8	1
Mycteroperca venenosa (Yellowfin grouper)	1	8
Number of point-count surveys	**667**	**47**
Grouper per point-count survey	**0.57**	**2.02**

In addition to its natural hardbottom, Broward County has an abundant and diverse range of artificial reefs deployed with the intention of enhancing fishing and scuba diving activities. Vessel-reefs consisting of ship hulls intentionally scuttled for recreational fishing and diving use at ~21 m depth (Arena et al. 2007) have significantly higher fish abundance and species richness than natural hardbottom, as well as different species composition and trophic structure. Planktivores dominated the vessel-reef fish assemblages (53% versus 27% of total abundance on the nearby natural hardbottom). The higher planktivore densities recorded on vessel-reefs may have been due to entrainment of planktonic food resources as a result of laminar flow disruptions from the tall vertical profiles of the structures (Arena et al. 2007). Vessel-reefs also harbored several species not seen during natural hardbottom surveys (*Lutjanus buccanella* [Lutjanidae] and *Epinephelus niveatus* [Serranidae]) (Arena et al. 2004). Bryan (2006) conducted ROV surveys on three vessel-reefs located between depths of 50–120 m. Anthiinae (mostly *Pronotogrammus martinicensis* [Serranidae]) numerically dominated vessel-reefs and were likely the forage base of the numerous, large, piscivorous fishes inhabiting these same vessel-reefs. Unlike at shallow vessel-reefs, herbivorous fishes were absent from the deep vessel-reefs (Bryan 2006).

Besides the many vessel-reefs in SE Florida, several hundred artificial reef modules (~1 m³ in size) have been deployed for use as: replicate units for scientific studies examining reef fish colonization and predation, habitat mitigation, and restoration tools. Gilliam (1999) compared fish recruitment on caged and uncaged artificial reef modules at 8 m depth and showed that seasonal density-dependent predation is one of several factors affecting the structure of the associated fish assemblages. Nearshore (8 m) sites had a lower disparity

FIG. 5.17. (**a**) Gag grouper (*Mycteroperca microlepis*) at staghorn coral thicket. (**b**) Juvenile lane snapper (*Lutjanus synagris*) on nearshore hardbottom. (**c**) Longspine squirrelfish (*Holocentrus rufus*) on outer reef. (**d**) Red grouper (*Epinephelus morio*) at staghorn coral thicket. (**e**) Foureye butterflyfish (*Chaetodon capistratus*) on outer reef. (**f**) Goldspot goby (*Gnatholepis thompsoni*). (Photos by: **a–e** L. Jordan, **f** by K. Kilfoyle)

in newly settled Haemulidae abundance between caged and uncaged modules suggesting predation pressure to be lower in the nearshore environment. This supports the notion that nearshore hardbottom habitats are nurseries for juvenile fishes (L. Jordan, personal communication 2007).

5.5.3.2 Palm Beach County

A total of 2,440 fish surveys conducted at 109 sites in Palm Beach County (PBC) between 1993 and 2007 documented 400 species of fish. Of these, seven were sightings of exotic species found only on offshore reefs. Comparison of species reported for north versus south Palm Beach County found 300 species in common. 43 species were recorded in the north and not in the south compared to 56 additional species recorded in the south but not the north. The most frequently sighted species differed markedly. Table 5.11 lists the top 20 species documented for both north and south PBC. Markedly different assemblages reflect the differing reef habitats between north and south Palm Beach County. Only five of the most common species are shared between the north and the south (porkfish, sergeant major, bluehead, French grunt, and bluestriped grunt).

Comparison of species reported from nearshore reefs for north versus south PBC (Table 5.12) found

TABLE 5.11. The 20 most frequently sighted fish species in North and South Palm Beach County (courtesy REEF database).

Family	Family	Common name	Scientific name	N freq.	S freq.
Surgeonfish	Acanthuridae	Ocean Surgeon	Acanthurus bahianus	33.20%	70.80%
Surgeonfish	Acanthuridae	Doctorfish	Acanthurus chirurgus	87.50%	59.20%
Surgeonfish	Acanthuridae	Blue Tang	Acanthurus coeruleus	47.30%	86.70%
Butterflyfish	Chaetodontidae	Foureye butterflyfish	Chaetodon capistratus	19.50%	69.70%
Butterflyfish	Chaetodontidae	Spotfin Butterflyfish	Chaetodon ocellatus	24.50%	70.20%
Butterflyfish	Chaetodontidae	Reef Butterflyfish	Chaetodon sedentarius	32.50%	70.60%
Grunt	Haemulidae	Black Margate	Anisotremus surinamensis	61.70%	62.50%
Grunt	Haemulidae	Porkfish	Anisotremus virginicus	91.60%	89%
Grunt	Haemulidae	Tomtate	Haemulon aurollneatum	54.20%	66.10%
Grunt	Haemulidae	French Grunt	Haemulon flavolineatum	78.30%	77.80%
Grunt	Haemulidae	Sailor's Choice	Haemulon parra	66.80%	48.90%
Grunt	Haemulidae	Bluestriped Grunt	Hemulon sciurus	66.40%	77.60%
Spadefish	Ephippidae	Atlantic Spadefish	Chaetodipterus faber	73.30%	36.50%
Chub	Kyphosidae	Bermuda/Yellow chub	Kyphosus sp.	69.50%	36.50%
Wrasse	Labridae	Spanish Hogfish	Bodianus rufus	38%	75%
Wrasse	Labridae	Slippery Dick	Halichoeres bivittatus	67.50%	33.30%
Wrasse	Labridae	Yellowhead Wrasse	Halichoeres garnoti	30%	67.30%
Wrasse	Labridae	Bluehead Wrasse	Thalassoma bifasciatum	79.60%	58%
Snapper	Lutjanidae	Schoolmaster	Lutjanus apodus	68.50%	31.60%
Snapper	Lutjanidae	Gray Snapper	Lutjanus griseus	66.80%	50.70%
Snapper	Lutjanidae	Lane Snapper	Lutjanus synagris	63.30%	23.60%
Goatfish	Mullidae	Spotted goatfish	Pseudupeneus maculatus	33.60%	78.90%
Angelfish	Pomacanthidae	Queen Angelfish	Holacanthus ciliaris	64.60%	48.70%
Angelfish	Pomacanthidae	Rock Beauty	Holacanthus tricolor	25.80%	74.50%
Angelfish	Pomacanthidae	French Angelfish	Pomacanthus paru	85%	88.10%
Damselfish	Pomacentridae	Sergeant Major	Abudefduf saxatilis	85.90%	86.20%
Damselfish	Pomacentridae	Beaugregory	Pomacentrus leucostictus	73.90%	32.40%
Damselfish	Pomacentridae	Bicolor Damselfish	Pomacentrus partitus	52.40%	82.30%
Parrotfish	Scaridae	Redband Parrotfish	Sparisoma aurofrenatum	29.50%	67.70%
Parrotfish	Scaridae	Yellowtail Parrotfish	Sparisoma rubripinne	64.20%	21.70%
Parrotfish	Scaridae	Stoplight Parrotfish	Sparisoma viride	28.50%	79.30%
Drum	Sciaenidae	High-Hat	Equetus acuminatus	71%	63.40%
Barracuda	Sphyraenidae	Great Barracuda	Sphyraena barracuda	73.80%	26%

"N freq." is the sighting frequency for north Palm Beach County
"S freq." is the sighting frequency for south Palm Beach County

TABLE 5.12. Top 20 nearshore species listed for both North and South Palm Beach County with sighting frequencies.

Family	Family	Common name	Scientific name	N freq.	S freq.
Surgeonfish	Acanthuridae	Ocean Surgeon	*Acanthurus bahianus*	10.90%	47.30%
Surgeonfish	Acanthuridae	Doctorfish	*Acanthurus chirurgus*	96.70%	73.60%
Surgeonfish	Acanthuridae	Blue Tang	*Acanthurus coeruleus*	28%	47.30%
Jack	Carangidae	Bar Jack	*Caranx ruber*	50.20%	73.60%
Snook	Centropomidae	Snook	*Centropomus undecimalis*	95%	5.20%
Anchovies	Engraulidae			89%	26.30%
Spadefish	Ephippidae	Atlantic Spadefish	*Chaetodipterus faber*	94.90%	10.50%
Mojarra	Gerreidae	Yellowfin mojarra	*Gerres cinereus*	76.70%	57.80%
Grunt	Haemulidae	Black Margate	*Anisotremus surinamensis*	62.60%	71%
Grunt	Haemulidae	Porkfish	*Anisotremus virginicus*	94.30%	60.50%
Grunt	Haemulidae	French Grunt	*Haemulon flavolineatum*	86%	50%
Grunt	Haemulidae	Sailor's Choice	*Haemulon parra*	80.30%	52.60%
Chub	Kyphosidae	Bermuda/Yellow chub	*Kyphosus sp.*	88.40%	39.40%
Wrasse	Labridae	Slippery Dick	*Halichoeres bivittatus*	81.10%	73.60%
Wrasse	Labridae	Bluehead Wrasse	*Thalassoma bifasciatum*	83%	60.50%
Snapper	Lutjanidae	Schoolmaster	*Lutjanus apodus*	91.90%	5.20%
Snapper	Lutjanidae	Gray Snapper	*Lutjanus griseus*	76.70%	50%
Snapper	Lutjanidae	Lane Snapper	*Lutjanus synagris*	86.60%	52.60%
Filefish	Monacanthidae	Orangespotted Filefish	*Cantherhines pullus*	10%	55.20%
Angelfish	Pomacanthidae	French Angelfish	*Pomacanthus paru*	96.10%	42.10%
Damselfish	Pomacentridae	Sergeant Major	*Abudefduf saxatilis*	97.60%	94.70%
Damselfish	Pomacentridae	Beaugregory	*Pomacentrus leucostictus*	92.50%	52.60%
Damselfish	Pomacentridae	Cocoa Damselfish	*Pomacentrus variabilis*	50.90%	68.40%
Parrotfish	Scaridae	Yellowtail Parrotfish	*Sparisoma rubripinne*	97.70%	60.50%
Drum	Sciaenidae	High-Hat	*Equetus acuminatus*	87.80%	50%
Porgy	Sparidae	Silver Porgy	*Diplodus argenteus*	95.30%	65.70%
Barracuda	Sphyraenidae	Great Barracuda	*Sphyraena barracuda*	92.90%	42.10%

"N freq." is the sighting frequency for north Palm Beach County
"S freq." is the sighting frequency for south Palm Beach County

163 species in common. 92 species were recorded in the north and not in the south compared to only 26 additional species recorded in the south but not the north. The most frequently sighted species recorded for nearshore reefs shared 11 species in common in the top 20 species recorded. In Fig. 5.3 the top 20 species documented for both north and south PBC nearshore reefs are listed with sighting frequencies.

In north Palm Beach County, 343 species of fish were recorded. Of these, 255 were recorded at nearshore sites and 285 were recorded offshore. In south PBC, 356 species were recorded, and of these, 189 species were recorded in the nearshore, 171 recorded at the second reef, and 351 recorded offshore. As would be expected, there is an increase in species richness on the more rugose offshore tracts compared to the inshore tracts (Ettinger et al. 2001).

5.6 Environmental Factors Influencing Reef Biology

In addition to water depth and topographic variations, cooling of surface waters during severe winter cold fronts can be a major environmental control on the distribution of corals with depth. Offshore southern Miami-Dade County, shallow reefs nearest to tidal passes (points of cool water discharge) have the least developed reef communities (Burns 1985). The relatively high latitude of the SE Florida reef system (N25°34′ northward to N26°43′, a distance of 125 km) exposes the SE Florida reefal biota to

TABLE 5.13. List of species reported for Southeast Florida ridge complex and reefs.

Macroalgae Cyanobacteria	Scleractinia/Milliporina/ Zoanthidia	Octocorallia	Porifera
Dictyota bartayressi	*Acropora cervicornis*	*Briareum asbestinum*	*Agelas clathrodes*
Dictyota spp.	*A. palmata*	*Diodogorgia nodulifera*	*A. conifera*
Galaxaura obtusata	*Agaricia agaricites*	*Ellisella barbadensis*	*A. wiedermyeri*
Halimeda discoidea	*A. fragilis*	*Erythropodium caribaeorum*	*Amphimedon compressa*
H. opuntia	*A. humilis*	*Eunicea calyculata*	*Anthosigmella varians*
Jania adherens	*A. lamarcki*	*E. clavigera*	*Aplysina cauliformis*
Lyngbya confervoides	*Astrangia solitaria*	*E. fusca*	*A. fistularis*
L. polychroa	*Cladocora arbuscula*	*E. laciniata*	*A. fulva*
Padina spp.	*Colpophyllia natans*	*E. laxispica*	*A. lacunosa*
	Dendrogyra cylindrus	*E. palmeri*	*Callyspongia plicifera*
	Dichocoenia stokesii	*E. pinta*	*C. vaginalis*
	Diploria clivosa	*E. succinea*	*Chondrilla nucula*
	D. labyrinthiformis	*E. tourneforti*	*Cinachyra* spp.
	D. strigosa	*Gorgonia ventalina*	*Cliona celata*
	Eusmilia fastigiata	*Iciligorgia schrammi*	*C. delitrix*
	Favia fragum	*Lophogorgia cardinalis*	*Diplastrella megastellata*
	Isophyllia sinuosa	*Muricea laxa*	*Dysidea* spp.
	Leptoseris cucullata	*M. muricata*	*Ectyoplasia ferox*
	Madracis decactis	*M. pendula*	*Haliclona* spp.
	M. mirabilis	*Muriceopsis petila*	*Holopsamma helwigi*
	M. pharensis	*Nicella schmitti*	*Iotrochota birotulata*
	Manicia areolata	*Plexaura flexuosa*	*Ircinia campana*
	Meandrina meandrites	*Plexaurella dichotoma*	*I. felix*
	Millepora alcicornis	*P. fusifera*	*I. strobilina*
	Montastraea annularis f. annularis	*P. grisea*	*I. variablis*
	M. annularis f faveolata	*P. pumila*	*Microciona juniperina*
	M. annularis f franksii	*Pseudoplexaura crucis*	*Monachora barbadensis*
	M. cavernosa	*Pseudopterogorgia acerosa*	*M. unguifera*
	Mussa angulosa	*P. americana*	*Mycale laevis*
	Mycetophyllia danaana	*P. elisabethae*	*Myrmekioderma styx*
	M. lamarckiana	*P. navia*	*Niphates digitalis*
	M. aliciae	*P. rigida*	*N. erecta*
	Oculina diffusa	*Pterogorgia citrina*	*Ophiactis* spp.
	Palythoa caribaeorum	*P. guadalupensis*	*Pellina carbonaria*
	Phyllangia americana	*P. citrina*	*Pseudaxinella lunaecharta*
	Porites astreoides	*P. guadalupensis*	*Pseudoceratina crassa*
	P. porites	*Swiftia exserta*	*Ptilocaulis* spp.
	Scolymia cubensis		*Spheciospongia vesparium*
	S. lacera		*Strongylacidon* spp.
	Siderastrea radians		*Tedania ignis*
	S. siderea		*Ulosa ruetzleri*
	Solenastrea bournoni		*Verongula rigida*
	S. hyades		*Xestospongia muta*
	Stephanocoenia intersepta		
	Stylaster rosea		
	Tubastrea coccinea		

Based on data from Blair and Flynn (1989), Foster et al. (2006), Gilliam et al. (2007), Gilliam (personal communication 2006), Goldberg (1973), Kosmynin (personal communication 2006)

low air temperatures during the passage of winter cold fronts. At the northern and southern end of the reef system air temperatures of 14°C and 15°C, respectively, have been reported. The durations of temperature minima are usually short (hours) and water temperatures are moderated by the large mass of warm water of the Florida Current. In a regional sense, the influences of cold air are confounded by

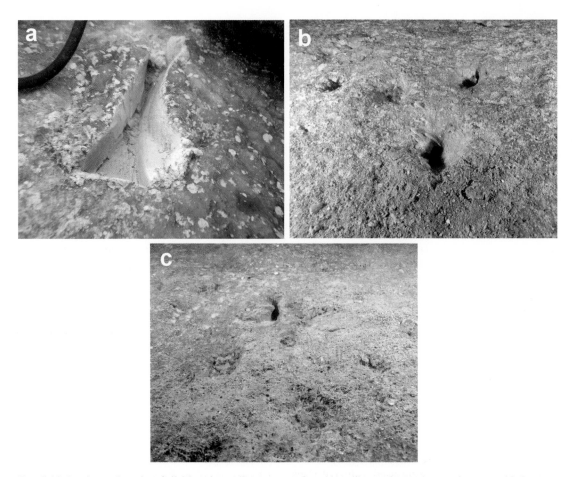

Fig. 5.18. Persistent deposits of silt/clay size sediments were found in sediment depressions and topographic lows on the reef tract and **a, b** and **c** inter-reef sand plains following the passages of hurricanes, Charley, Frances, Ivan, and Jeanne, in 2004 (Photos by: **a** Vladimir Kosmynin; **b, c** Kenneth Banks)

the proximity of the Florida Current to the shelf, i.e., closer to shore in the north and farther in the south. Such air temperature minima are near the lower tolerance limit for many reef-associated biota with scleractinia in Florida having a reported lethal limit at 14°C (Porter et al. 1982). Persistence of cold air can have deleterious consequences for many reef associated organisms. In the Florida reef tract, even further south, Roberts et al. (1982) documented water temperatures of 12.6–16.0°C at a depth of 4.3 m for 8 days in January 1977 (24°56′ N) and Lee (in Burns 1985) reported surface temperatures of 13.3°C (25°13′ N) and 15.2°C (25°01′ N) in January 1981. During summer 2003 water tempera-

tures were 5–7°C lower than air temperatures at the northern end of the region (Aretxabaleta et al. 2006) due to upwelling caused by a complex interaction between local and remote atmospheric forcing and open ocean effects which apparently caused mortality of fish and sea-turtles (local media reports).

Lowest average monthly temperatures were in January 2001 with values of 18.3°C (Fig. 5.4e). Average maximum monthly water temperatures in this time period exceeded 29°C in the summers of 2001 and 2002 (Fig. 5.4f). Elevated water temperatures can induce coral bleaching and while bleaching of some stony corals, octocorals, and *Palythoa caribbeorum* colonies has been observed, mass bleaching events have not occurred since the El Nino of 1997–98. Roberts et al. (1982) reported cold-water disturbances in Florida Bay in January 1977 with temperatures below 16°C sustained for 8 days (minimum recorded water temperature was 12.6°C). Burns (1985) attributed the lack of acroporids on the shallow fore-reef, the increase in total coral cover with depth and the greater abundance of *Montastraea annularis* in the deepest zones in the upper Florida Keys as a result of cooling of surface waters during severe winter cold fronts. Water temperatures in SE Florida were likely similar or lower than the low temperatures recorded in the Florida Keys in 1977 (Roberts et al. 1982). Thus, aperiodic chilling processes have a limiting influence on reef community development throughout the Florida Keys Reef Tract but probably even more so at the higher latitude reefs of SE Florida.

Sedimentation and turbidity can affect reef organisms by shading, burial of epibiota, and burial of substrate necessary for recruitment and survival of epibiota. The responses of organisms to shading and temporary burial are complicated by adaptation. Telesnicki and Goldberg (1995) documented depressed photosynthesis:respiration ratios and mucous production in the corals *Dichocoenia stokesii* and *Meandrina meandrites* due to elevated turbidity.

Gilliam et al. (2007) have used sediment traps to measure sedimentation on all reef tracts in Broward County since 1997. Their data indicate that the inner reef typically has the highest rate of sedimentation, as well as largest grain size, followed by the middle reef and then the outer reef. This trend is a result of increasing depth and therefore decreasing wave energy. This was illustrated in 2005 when severe sea conditions during hurricane Katrina (25 August 2005) resulted in the highest sedimentation rates measured in that year. FWCC (2006) reported burial of fixed monitoring sites on the ridge complex offshore Palm Beach County in 2005. The cause was unknown, although hurricanes Jeanne and Frances in 2004 may have contributed to substantial sand movement in the shallow ridge complex area.

In 2004, after the Gulf of Mexico and Atlantic coasts of Florida were impacted by four hurricanes (Charlie, Frances, Ivan, and Jeanne), widespread accumulation of silt/clay size sediment was observed on nearshore hardbottom from Cape Canaveral to Fort Lauderdale (Kosmynin and Miller 2007). This material filled sand depressions on the inter-reef sand plains and buried patches of nearshore hardbottom and persisted through at least the winter of 2007 (Fig. 5.18). The origin is unknown but Kosmynin and Miller (2007) hypothesized that hurricane generated waves and currents transported silt/clays shoreward from deeper parts of the continental shelf.

5.7 Human Impacts and Conservation Issues

The proximity of the SE Florida reef tract to a highly urbanized coastal zone contributes a number of human-related stressors to the reef communities. Water pollution, over-fishing, coastal construction activities, vessel anchoring and grounding, as well as ballast water discharge impact the region's reefs. While balancing economic growth with environmental protection is challenging, the economic value of the coral reefs is considerable. Johns et al. (2003) determined the economic contribution by recreational users of artificial and natural reefs (fishers, divers, snorkelers, visitors viewing reefs from glass-bottomed boats) over the period June 2000 to May 2001 to have been US$2.3 billion in sales and US$1.1 billion in income. 36,500 full and part-time jobs were related to the recreational use of the reefs.

5.7.1 Water Quality

Water quality monitoring in SE Florida is limited to inland waters (Trnka and Logan 2006; Caccia et al. 2005; Torres et al. 2003; Carter 2001). Long-term data does not exist for ocean waters, however the Broward County Environmental Protection Department began a coastal water quality monitoring program in 2005 (Craig 2004). Three study sites were established around Port Everglades Inlet where nutrients, chlorophyll, salinity, dissolved oxygen, and pH are measured monthly.

Lapointe (1997) provided evidence that macroalgal blooms on the reefs offshore Palm Beach County were caused by nitrogen from land-based sewage. Finkl and Charlier (2003) and Finkl and Krupa (2003) estimated that nutrient loading of nitrogen and phosphorus from inland agriculture to the

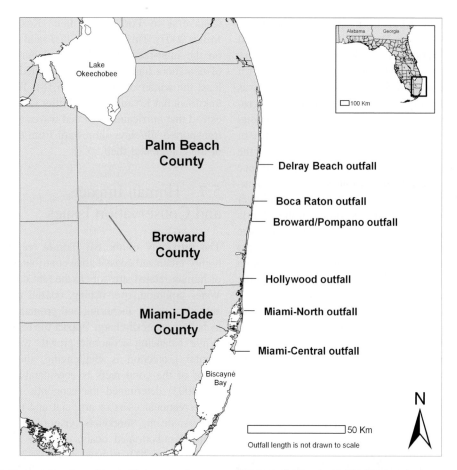

FIG. 5.19. General location of treated human wastewater outfall pipes in Southeast Florida that discharge effluent on lower foreslope of the outer reef

coastal waters offshore of Palm Beach County via surface water discharge are 2,473 and 197 MT/year, respectively, and via submarine groundwater discharge (SGD) 5,727 and 414 MT/year, respectively. They projected that even if nutrient loading to groundwater was to be stopped immediately, effects would still persist 5–8 decades into the future due to slow groundwater flow. Fauth et al. (2006) used cellular diagnostics to detect signs of nutrient-related stress in *Porites astreoides* offshore Broward County when compared to samples from the Bahamas. Stress responses of corals adjacent to treated (secondary treatment) human wastewater discharges as well as corals from the Florida Keys National Marine Sanctuary were consistent with sewage exposure while responses of offshore colonies were consistent with xenobiotic detoxification.

5.7.2 Coastal Construction

Before 2000, coastal construction activities in SE Florida were primarily related to installation of cross-shelf wastewater outfall pipes and dredging projects for inlet channel maintenance and beach restoration. The proximity of the reef tracts and hard bottom communities to the coast increases the potential for impacts from any dredging or other coastal construction activities.

Six wastewater pipes cut through the reef tracts in SE Florida (Fig. 5.19), which were constructed at a time when there was little awareness of reef resources or concern for the environment in general. The pipes were placed in deep trenches cutting through the reef. An overburden of boulders or articulated concrete block mats was used for protection where pipes traversed sand. The discharge points for the pipes are on the lower foreslope of the outer reef.

Seven navigational inlets exist in SE Florida (Fig. 5.1). North Lake Worth Inlet (locally called Palm Beach Inlet), Port Everglades Inlet and Government Cut service seaports and are maintained for large vessel traffic. South Lake Worth Inlet (also called Boynton Inlet), Boca Raton Inlet, Hillsboro Inlet, and Haulover Inlet are limited to use by smaller vessels due to depth constraints or bridge height clearances. Although all of the inlets are jettied, littoral transport of sand from the north results in in-filling of the inlet channels with ebb or flood shoals. Maintenance dredging usually involves removing this sand to down-drift beaches or to deepwater Offshore Dredge Material Disposal Sites (ODMDS) (EPA 2004). A number of harbor deepening projects have been proposed in recent years as a result of a trend toward larger, deeper draft commercial ships. Since the inlet channels cross the reef tracts, large areas of reef will be removed. Physical damage to reef and hardbottom in the vicinity of dredging activity can

occur from slack tow cables scraping the bottom, sedimentation resulting from tug boat propeller thrust in shallow water, and vessel anchors. Potential impacts to reefs from these dredging projects are primarily related to suspended sediments and turbidity. While the larger grain size sediments settle out rather quickly, the finer grain material may remain suspended for long periods of time. Subsequent advection by ebb tidal flows can lead to exposure of the surrounding reefs to turbidity and sediment deposition.

Another common type of dredging project in SE Florida is for the restoration of eroded beaches. Beach erosion is a problem because inlet jetties interrupt the transport of quartzose sands from the north causing sand starvation downdrift. In addition, coastal development has narrowed the beach zone, preventing the beach from absorbing wave energy. Regular nourishing of the beaches began in the early 1970s and continues to the present day. Eroded beaches are filled with sand dredged from between reef tracts and are generally re-nourished on 10–15-year intervals. Cutter-head and hopper

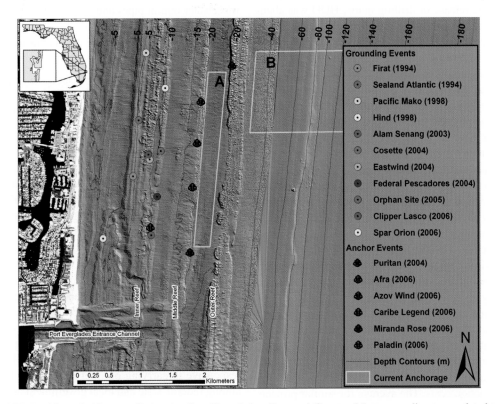

FIG. 5.20. Position of shipgroundings near Port Everglades, Broward County. Most groundings are related to the proximity of the middle reef to the shallow anchorage A

dredges used for beach nourishment are large vessels and must maneuver in small areas when dredging around reefs. The result is an increased risk of physical impacts from cutter-heads and drag arms hitting the reef and tow cables of support vessels dragging over the reef. As in the case for inlet dredging, sedimentation and turbidity are also problematic. Nearshore hardbottom is often directly buried during beach filling or during the adjustment of the beach profile after filling.

Rapid population growth in the SE Florida region has resulted in an increased demand for energy and communications infrastructure. A number of linear, cross-shelf natural gas pipelines and communications cables are planned or have been constructed.

Three natural gas pipelines have been proposed for SE Florida. Initial plans were to use horizontal directional drilling (HDD) to create pilot holes under the reef tracts which would subsequently be enlarged to the necessary diameters for pipeline installation. Because of the risk of frac-outs (leakage of drilling muds through the porous substrate underlying the reefs) and other potential impacts from the release of drilling muds, HDD was abandoned in favor of tunneling under the coastal shelf. This technology is well developed and minimizes risks to the reefs.

Communications cables are laid directly on the sea floor and although their diameters are small (24 cm), their lengths and numbers can result in significant primary physical reef impacts. Since cable-laying vessels have to come into shallow waters for the installation there is a high risk of impact from propeller wash and tow cable scrapes. ATT's installation of two fiberoptic telecommunications cables resulted in dislocated scleractinian and octocoral colonies, as well as physical damage to many remaining colonies. Re-cementing of stony coral colonies and mitigation were required (PBS&J 1999).

Environmental protection measures used during coastal construction projects has progressed greatly in recent years because of a greater conservation ethic by the public and increased awareness of the resources present. In a recent beach restoration project completed by Broward County, environmental protection and monitoring costs for the project were approximately 20% of the total construction costs. The average for similar projects is approximately 10% (Chris Creed, personal communication). The availability of high resolution bathymetry and advances in positioning technology and remote, real-time monitoring of vessels' position allow the establishment of transit corridors for vessels to minimize vessel-related impacts.

The increased extent and duration of reef impact monitoring associated with coastal construction projects has resulted in an increased knowledge of the reef-associated communities, however, understanding of long-term impacts of these projects remains unclear. Highly urbanized SE Florida presents a number of stressors to reef communities so it is difficult to find suitable experimental control for monitoring. The recent use of molecular and organismal level techniques to determine stress before impacts occur at the community level may be a portent of the future monitoring of project impacts. Vargas-Angel et al. (2006) used histological techniques to determine sedimentation induced stress on corals and calibrated a visual method of determining organismal stress. Fauth et al. (2006) used enzymatic biomarkers in the stony coral, *Porites astreoides*, coupled with analysis of community structure and healing of lesions to look for possible impacts of wastewater outfall pipes and inlet discharges. These techniques using multiple levels of monitoring could be expanded to examine impacts from coastal construction.

5.7.3 Ship Groundings and Anchor Damage on Reefs

Commercial shipping into the ports at Palm Beach, Port Everglades, and Miami is an important part of the economy of SE Florida and increased 150% between 1964 and 2002 (Andrews et al. 2005). The proximity of reefs to the navigational inlets and commercial ship anchorages leads to a high risk for ship groundings and anchor damage with subsequent reef damage. This reaches an extreme around Port Everglades Inlet where a relatively shallow (d = 20 m) anchorage lies in sand offshore of the middle reef tract. Between 1993 and 2007, eleven ships have grounded on reefs inshore of this anchorage, impacting over 40,000 m^2 (Fig. 5.20) and, fortunately, vessel owners have been relatively responsive in carrying out reef restoration. Efforts among federal, state and local government agencies to eliminate the shallow anchorage are underway in order to reduce impacts and a study of alternative anchorages has been completed (Moffatt and Nichol 2006).

Anchoring of ships outside the designated anchorage and even small boat anchors pose other problems. The number of recreational boats in SE Florida increased by 500% from 1964 to 2002 (Andrews et al. 2005). There is no documentation of the extent of resulting reef damage. To lessen anchor damage by small boat, over 100 moorings were installed in Broward County. Reef impacts due to concentration of reef users around moorings are currently being investigated by Klink et al. (2006).

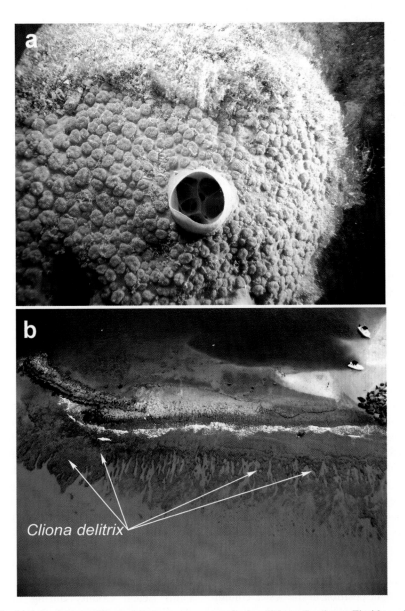

FIG. 5.21. **a** The boring sponge, *Cliona delitrix*, occurs extensively offshore Southeast Florida and may indicate human sewage contamination of the coastal waters (Ward-Paige et al. 2005). **b** Aerial photo of the hardbottom adjacent to the north jetty at Hillsboro Inlet shows high coverage of *C. delitrix* that may correlate with tidal plume contamination (Photos by: **a** K. Banks; **b** D. Behringer)

5.7.4 Climate Change

Increases in sea temperature, sea level rise and, possibly, increasing levels of ultraviolet radiation due to global climatic change may affect coral reefs in SE Florida. Locally or regionally, changes in tropical cyclone patterns may directly impact the coral communities, and changes in rainfall patterns may affect sedimentation, salinity, nutrient and pollutant inputs (Edwards 1995). Rainfall data for 1890–2000 show that there has been a decline in rainfall since the 1960s for unknown reasons, and global climate models predict a reduction of precipitation for South Florida ultimately resulting in decreased runoff (SFWMD 1996). This may, in itself, be beneficial to reef biota. For example, Dodge and Helmle (2003) found that lower salinities (relative to normal seawater) slowed coral growth rates (see also Section 5.2.4). However, a region's landscape (urbanization) can influence rainfall (Pielke et al. 1999) so the prediction of future rainfall levels is complicated by other factors, including water use patterns by an ever increasing local population.

While many local climatic changes occur on relatively smaller time scales, global events such as eustatic sea level rise and atmospheric warming occur on much larger time scales. Sea level rise is of great concern for low elevation coastal regions, such as SE Florida. The Intergovernmental Panel on Climate Change (IPCC 2007) reported that global sea level rise for the period 1961–2003 averaged 1.8 mm/year (1.3–2.3) and increased to 3.1 (2.4–3.8) mm/year over 1993–2003. Wanless (1989) reported that since 1932 tide gauge records from Key West and Miami show relative sea level rise in South Florida has accelerated and more recent rates are 3–4 mm/year. Titus (1995) estimated that by the year 2050 eustatic sea level will likely rise at least 15 cm (2.7 mm/year) and there is a 10% probability that it will rise 30 cm (5.4 mm/year). Others (Buddemeier and Smith 1988) estimate future rates of 15 ± 3 mm/year as probable over the next century. Such high rates could impact corals directly by shifting them to a deeper, lower light position in the water column. Acroporid reefs would drown under these conditions since their sustained reef accretion rates are only about 10 mm/year. Since SE Florida's reefs are already non-framebuilding relics in which reef-building biota are small and the reef-associated biota dominate space, one would assume them not to be as sensitive to climatic changes in the shorter term. However, secondary impacts, such as increased sedimentation and turbidity from coastal flooding and erosion, would degrade water quality and could affect reef growth.

5.8 Present Status of Reef Health

It is difficult to compare the health of the reefs of SE Florida with extant Acroporid coral dominated reefs in other areas of the western Atlantic and Caribbean. The Acroporids ceased dominating cover in SE Florida 5–7 cal BP (Lighty 1978; Banks et al. 2007). Most of the declines reported in other areas have been a result of loss of *Acropora* spp. to white band disease (Gardner et al. 2003). White band disease has been reported in 1.8% of the cover of the *Acropora cervicornis* thickets offshore of Broward County described by Vargas-Angel et al. (2003).

The Caribbean-wide decrease of *Diadema antillarum* was also experienced in SE Florida where Goldberg (1973) reported this sea-urchin to have been abundant offshore Boca Raton. In contrast, FWCC (2006) reported none at 10 SECREMP sites from Palm Beach to Miami-Dade Counties. Six were seen at for sites in 2004 and 15 at six sites in 2005. Recovery therefore seems to be lagging.

Ward-Paige et al. (2005) surveyed clionid sponges on the Florida Keys reef tract and found a relationship with sewage contamination. In SE Florida FWCC (2006) reported *Cliona delitrix* at all sites, except at the *A. cervicornis* thickets. *Montastraea cavernosa*, the regionally most abundant scleractinian, was most affected by this sponge. Diver observations by one of us (KB) indicate that *C. delitrix* is abundant throughout Broward County, particularly on the ridge complex and inner and middle reefs (Fig. 5.21).

In SE Florida harmful algal blooms of *Caulerpa brachypus* have occurred extensively offshore Palm Beach County during the past decade (Lapointe et al. 2006). In February 2007, *Caulerpa brachypus* spread into northern Broward County. Paul et al. (2005) reported extensive blooms of the cyanobacteria, *Lyngbya confervoides* and *L. poly-chroa*, on the reefs offshore of Broward County. These blooms have had a significant impact on reef-associated organisms by smothering and out-competing recruits of sessile benthos (Lapointe 1997). For example, at a study site of Gilliam et al. (2007) on the inner reef, significant coverage of *Lyngbya* spp. in 2003 had affected most erect octocorals and their densities declined steadily from 2003–2005. Also sponge density dropped from ~13/m^2 in 2002 to ~6/m^2 in 2003. Some subsequent recovery increased numbers in 2004 to ~8/m^2. Scleractinian cover did not appear to be impacted by *Lyngbya* spp.

The incidence of coral bleaching and disease has been relatively low in SE Florida since 2004, when data were first collected. In 2004, 19 diseased coral colonies were identified in the 10 study sites. In 2005, 21 diseased colonies were identified, 10 of which had apparently been infected in 2004. Nine of those were *Siderastrea siderea* with dark spot syndrome and had recovered by 2005. White complex disease was more prevalent in 2005 (FWCC 2006). No totally bleached coral colonies were observed although partial bleaching was more common than disease.

5.9 Conclusions

The SE Florida reef system consist of relict, early Holocene *Acropora palmata* framework reefs and indurated sand ridges that still maintain a rich, typically Caribbean, but non-framebuilding fauna today. The dominant hard corals are *Montastrea* spp. but living space cover by hard corals is overall low (<6%). Rich alcyonacean communities of typically Caribbean composition cover the majority of benthic space, allowing high benthic space cover. Three shore-parallel reefs (inner, middle, outer) are separated by sandy plains. The middle and outer reefs generally harbor denser benthic cover, dominated by sponges and alcyonacean soft corals. The inner reef is generally more sparsely settled, but has some large patches of dense *Acropora cervicornis* growth, which represents the northern latitudinal distribution limit for these corals The fish communities are typically Caribbean and similar in composition

to the Florida Keys, but changes in community composition are observed in Palm Beach County. Heavy recreational fishing pressure has reduced size classes and population densities of groupers and snappers. Threats to the area's reefs are pollution, coastal construction and dredging projects. Recently, benthic cyanobacteria and algae blooms have caused heavy mortality among alcyonacean and hard corals. The local economic value of these reefs is in the range of $2.3 billion in sales and $1.1 billion in income per year. 36.500 jobs rely on the use of the reefs.

Acknowledgements Special thanks to Alexander Soloviev of NSU, for information on SE Florida coastal circulation; Chris Creed, coastal engineer at Olsen Associates, Inc., for comments on relative costs of beach nourishment projects; and Brettany Cook (NSU) and Guynette Alexandre (Broward County Environmental Monitoring Division) for organizing ocean temperature data. Many of the components in this paper were produced under funding by NOAA, such as NOAA grant NA16OA1443 to NCRI.

References

Ackerman JL, Bellwood DR (2000) Reef fish assemblages: a re-evaluation using enclosed rotenone stations. Mar Ecol Progr Ser 206:227–237

Andrews K, Nall L, Jeffrey C, Pittman S (eds) (2005) The State of the Coral Reef Ecosystems in Florida. In The State of Coral Reef Ecosystems of the United States and Freely Associated States. National Oceanic and Atmospheric Administration, Tech Mem NOS NCCOS 11, 150–200

Arena PT, Quinn TP, Jordan LKB, Sherman RL, Gilliam DS, Harttung F, Spieler RE (2004) Presence of juvenile blackfin snapper, *Lutjanus buccanella*, and snowy grouper, *Epinephelus niveatus*, on shallow-water artificial reefs. Proc 55th Ann Gulf and Caribbean Fish Inst Meeting, Xel-Ha, Mexico, pp.700–712

Arena PT, Jordan LKB, Spieler RE (2007) Fish assemblages on sunken vessels and natural reefs in southeast Florida, USA Hydrobiologia. 580:157–171

Aretxabaleta A, Nelson JR, Blanton JO, Seim HE, Werner FE, Bane JM, Weisberg R (2006) Cold event in the South Atlantic Bight during summer of 2003: Anomalous hydrographic and atmospheric conditions. J Geophys Res 111:C06007, doi:10.1029/2005JC003105

Banks KW, Riegl BM, Shinn EA, Piller WE, Dodge RE (2007) Geomorphology of the Southeast Florida continental reef tract (Dade, Broward, and Palm Beach Counties, USA). Coral Reefs 26:

Baron RM, Jordan LKB, Spieler RE (2004) Characterization of the marine fish assemblage associated with the nearshore hardbottom of Broward County Florida, USA. Estuar Coast Shelf Sci 60:431–433

Beaver CR, Jaap WC, Porter JW, Wheaton J, Callahan M, Kidney J, Kupfner S, Torres C, Wade S, Johnson D (2005) Coral Reef Evaluation and Monitoring Project (CREMP), 2004 Executive Summary. Prepared by: Florida Fish and Wildlife Conservation Commission and the University of Georgia, 12 pp

Blair SM and Flynn BS (1989) Biological monitoring of hard bottom reef communities off Dade County Florida: community description. In Land M, Jaap W (eds), Proc Am Ac Underw Sci 9th Ann Sci Diving Symp, Woods Hole, MA, 9–24

Boss SK, Neumann AC (1995) Hurricane Andrew on Northern Great Bahamas Bank: insights into storm behavior on shallow seas. J Coastal Res SI 21:24–48

Bruno JF, Stachowicz JJ, Bertness MD (2003) Inclusion of facilitation into ecological theory. Trends Ecol Evol 18:119–125

Bryan DR (2006) Reef fish communities on natural substrate and vessel-reefs along the continental shelf of southeastern Florida between 50 and 120 m depth. MSC Thesis. Nova Southeastern University, 60 pp

Buddemeier RW, Smith SV (1988) Coral reef growth in an era of rapidly rising sea level: predictions and suggestions for long-term research. Coral Reefs 7:51–56

Burns TP (1985) Hard-coral distribution and cold-water disturbances in South Florida: variation with depth and location. Coral Reefs 4:117–124

Caccia VG, Boyer JN (2005) Spatial patterning of water quality in Biscayne Bay, Florida as a function of land use and water management. Mar Poll Bull 50:1416–1429

Carter K (2001) Broward County, Florida Historical Water Quality Atlas: 1972–1997, Technical Report Series, TR: 01–03. Broward County Department of Planning and Environmental Protection, February 28, 2001, Fort Lauderdale, FL, 480 pp

Castro CB, Pires DO (2001) Brazilian coral reefs: what we already know and what is still missing. Bull Mar Sci 69(2):357–371

Causey B, Delaney J, Diaz E et al. (2002) Status of coral reefs in the US Caribbean and Gulf of Mexico. In: Wilkinson C (ed), Status of Coral Reefs of the World. Aust Inst Mar Sci, Townsville, pp. 251–276

Chen E, Gerber JF (1990) Climate. In Myers RL and Ewel JJ (eds), Ecosystems of Florida. University of Central Florida Press, Orlando, pp. 11–34

Collette BB, Williams JT, Thacker CE, Smith ML (2003) Shore fishes of Navassa Island, West Indies:

a case study on the need for rotenone sampling in reef fish biodiversity studies. J Ichthyol Aquat Biol 6:89–131

Continental Shelf Associates (1980) Biological survey of proposed offshore borrow areas north of Haulover Beach Park, Dade County, Florida. Report to US Army Corps of Engineers, Jacksonville, FL, 21 pp

Continental Shelf Associates (1984) Biological analysis of macro-epibiotal and macro-infaunal assemblages following beach nourishment north Broward County, Florida. Report to Broward County Environmental Quality Control Board. 56 pp

Courtney WR Jr, Herrema DJ, Thompson MJ, Assinaro WP, Van Montfrans J (1972) Ecological monitoring of two beach renourishment projects in Broward County, Florida. Shore Beach 40(2):8–13

Courtney WR Jr, Blakesley HL, Reed JR, Waldner RE (1975) Environmental assessment of offshore reefs Miami Beach, Dade County, Florida. Research Report to US Army Crops of Engineers, Jacksonville District, Jacksonville, FL, 22 pp

Courtney WR Jr, Hartig BC, Loisel GR (1980) Ecological evaluation of a beach nourishment project at Hallandale (Broward County) Florida. US Army Corps of Engineers Coastal Engineering Research Center Misc. Report 80–1, Vol I:1–25

Craig N (2004) A long term vision for Broward County's coastal monitoring plan with a proposed pilot study. Broward County Environmental Protection Department, Environmental Monitoring Division, 20 pp

Dodge RE, Goldberg W, Messing CG, Hess S (1995) Final Report Biological Monitoring of the Hollywood-Hallandale Beach Renourishment. Prepared for the Broward County Board of County Commissioners, Broward County Department of Natural Resources Protection, Biological Resources Division. 103 pp

Dodge RE, Helmle KP (2003) Past stony coral growth (extension) rates on reefs of Broward County, Florida: possible relationships with Everglades drainage. Presented Poster, Joint Conference on the Science and Restoration of the Greater Everglades and Florida Bay Ecosystem. Palm Harbor, FL, April pp. 13–18

Duane DB, Meisburger EP (1969a) Geomorphology and sediments of the inner continental shelf, Palm Beach to Cape Kennedy, Florida. USACE Coastal Eng Res Cent Tech Mem 34, 82 pp

Duane DB, Meisburger EP (1969b) Geomorphology and sediments of the inner continental shelf, Miami to Palm Beach. USACE Coastal Eng Res Cent Tech Mem 29, 47 pp

Edwards AJ (1995) Impact of climatic change on coral reefs, mangroves, and tropical seagrass ecosystems. In Eisma D (ed), Climate Change-Impact on Coastal Habitation. Lewis Publishers, Boca Raton, FL, pp. 209–234

Enos P, Perkins RD (1977) Quaternary sedimentation in South Florida. Geol Soc Am Mem 147, 198 pp

EPA (2004) Final Environmental Impact Statement for designation of the Palm Beach Harbor Ocean Dredged Material Disposal Site and the Port Everglades Ocean Dredged Material Disposal Site, July 2004. US Environmental Protection Agency, Region 4, Atlanta, Georgia, 51 pp

Ettinger BD, Gilliam DS, Jordan LKB, Sherman RL, Spieler RE (2001) The coral reef fishes of Broward County Florida, species and abundance: a work in progress. Proc. 52nd Ann Gulf Caribb Fish Insti. 748–756

Fauth JE, Dustan P, Ponte E, Banks K, Vargas-Angel B, Downs C (2006) Final Report: Southeast Florida Coral Biomarker Local Action Study. Florida Department of Environmental Protection, Southeast Florida Coral Reef Initiative, 69 pp

Fenner D, Banks K (2004) Orange cup coral *Tubastrea coccinea* invades Florida and the Flower Garden Banks, Northwest Gulf of Mexico. Coral Reefs 23:505–507

Ferro F, Jordan LKB, Spieler RE (2005) The marine fishes of Broward County, Florida: final report of 1998–2002 survey results. NOAA Technical Memorandum NMFS SEFSC-532

Finkl CW (2005) Nearshore geomorphological mapping. In Schwartz ML (ed), The Encyclopedia of Coastal Science. Kluwer Academic, Dordrecht, The Netherlands, pp. 849–865

Finkl, CW, Benedet L, Andrews JL (2005) Interpretation of seabed geomorphology based on spatial analysis of high-density airborne laser bathymetry. J Coastal Res 21:510–514

Finkl CW, Charlier RH (2003) Sustainability of subtropical coastal zones in southeastern Florida: challenges for urbanized coastal environments threatened by development, pollution, water supply, and storm hazards. J Coastal Res 19:934–943

Finkl CW, Krupa SL (2003) Environmental impacts of coastal-plain activities on sandy beach systems: hazards, perception and mitigation. J Coastal Res SI35:132–150

Fong P, Lirman D (1995) Hurricanes cause population expansion of the branching coral Acropora palmata (scleractinia): wound healing and growth patterns of asexual recruits. PSZNI Mar Ecol 16:317–335

Foster K, Foster G, Riegl B (2006) Reef Mapping Final Report. National Coral Reef Institute for Broward County Environmental Protection Department, Biological Resources Division, 117 pp

FWCC (2006) Southeast Florida Coral Reef Evaluation and Monitoring Project: 2005 Year 3 Final Report. A report of the Florida Fish and Wildlife Conservation Commission, Fish and Wildlife Research Institute and the National Coral Reef Institute, Nova Southeastern University Oceanographich Center pursuant to DEP contract number #G0099 for Florida Department of Environmental Protection, Office of Coastal and Aquatic Managed Areas, Miami, FL, 25 pp

Gardner TA, Côte IM, Gill JA, Grant A, Watkinson AR (2003) Long-term region-wide declines in Caribbean corals. Science 301:958–960

Gilliam DS (1999) Juvenile reef fish recruitment processes in South Florida: a multifactorial field experiment. Ph.D. Dissertation. Nova Southeastern University, 150 pp

Gilliam DS, Dodge RE, Spieler RE, Jordan LKB, Walczak JC (2007) Marine Biological Monitoring in Broward County, Florida: Year 6 Annual Report. Prepared for the Broward County Environmental Protection Department, Biological Resources Division, 93 pp

Gilliam DS (2007) Southeast Florida Coral Reef Evaluation and Monitoring Project (SECREMP) 2006 Year 4 Final Report. Prepared for: Florida Fish and Wildlife Conservation Commission/Fish & Wildlife Research Institute, Florida Department of Environmental Protection. 31 pp.

Goldberg WM (1973) The ecology of the coral-octocoral communities off the southeast Florida coast: geomorphology, species composition, and zonation. Bull Mar Sci 23:465–488

Goldberg W (1981) A resurvey of coral communities adjacent to a beach restoration area in South Broward County, Florida. Report to the Broward County Environmental Quality Control Board. 23 pp

Grymes JM, Stone GW (1995) A review of key meteorological and hydrological aspects of Hurricane Andrew. J Coastal Res SI 21:6–23

Hanes DM, Dompe PE (1995) Field observations of fluctuations in coastal turbidity. J Mar Environ Eng 1:279–294

Hendler G, Miller JE, Pawson DL, Porter MK (1995) Sea Stars, Sea Urchins, and Allies: Echinoderms of Florida and the Caribbean. Smithsonian Press, Washington DC, 390 pp

Hodanish, S., Sharp D, Collins W, Paxton C, Orville RE (1997) A 10-yr monthly lightning climatology of Florida: 1986–95. Weather Forecasting, 12:439–448

Hubbard DK, Zankl H, Van Heerden I, Gill IP (2005) Holocene reef development along the northeastern St. Croix shelf, Buck Island, US Virgin Islands. J Sedim Res 75(1):97–113

IPCC (2007) Climate Change 2007: the Physical Science Basis. Intergovernmental Panel on Climate Change, Summary for Policymakers. WMO, UNEP, Geneva, Switzerland, 21 pp

Jaap WC, Adams JK (1984) The Ecology of the South Florida Coral Reefs: A Community Profile. US Fish and Wildlife Service, FWS/OBS-82/08, 138p

Jaap WC, Hallock P (1990) Coral Reefs. In Myers RL, Ewel JJ (eds), Ecosystems of Florida. University of Central Florida Press, Orlando, 574–616

Johns GM, Leeworthy VR, Bell FW, Bonn MA (2003) Socioeconomic Study of Reefs in Southeast Florida, April 18, 2003. Silver Spring, Maryland: Special Projects NOS, 255p

Johnson DR, Harper DE, Kellison GT, Bohnsack JA (2007) Description and discussion of southeast Florida fishery landings, 1990–2000. NOAA Tech Mem NMFS SEFSC-550

Jordan LKB, Gilliam DS, Sherman RL, Arena PT, Harttung FM, Baron RM, Spieler RE (2004) Spatial and temporal recruitment patterns of juvenile grunts (*Haemulon* spp.) in South Florida. Proc 55th Ann Gulf and Caribb Fish Inst Meeting, Xel-Ha, Mexico pp. 322–336

Jordan LKB, Spieler RE (2006) Implications of natural variation of fish assemblages to coral reef management. Proc 10th Int Coral Reef Sym:1391–1395

Kinzie III, RA (1973) The zonation of West Indian gorgonians. Bull Mar Sci 23:93–155

Klink LH, Gilliam DS, Stout D, Dodge RE (2006) An overview of international mooring buoy programs and the effect of mooring buoys on coral reefs offshore Broward County, FL. Poster, Int Soc Reef Stud European Coral Reef Meeting, Bremen, Germany

Knowlton N (2001) Who are the players on coral reefs and does it matter? The importance of taxonomy for coral reef management. Bull Mar Sci 69(2):305–308

Kosmynin VN, Miller CL (2007) Mud accumulation on nearshore hardbottom communities in Florida as a result of hurricane season of 2004. Abstract for 2007 Benthic Ecology Meeting, Georgia Institute of Technology, March 21–24, 2007

Laborel J (1966) Contibution a l'etudes des madreporaires des Bermudes (systematique et repartition). Bull Mus Nat Hist Nat Paris, Ser 2, 38:281–230

Landsea CW, Franklin JL, McAdie CJ, Beven JL, Gross JM, Jarvinen BR, Pasch RJ, Rappaport EN, Dunion JP, Dodge PP (2004) A reanalysis of Hurricane Andrew's intensity. Bull Am Meteorol Soc May 2004:1699–1712

Lapointe BE (1997) Nutrient thresholds for bottom-up control of macroalgal blooms on coral reefs in Jamaica and southeast Florida. Limnol Oceanogr 42:1119–1131

Lapointe, BE, Bedford BJ, Baumberger R (2006) Hurricanes Frances and Jeanne remove blooms of the invasive green alga *Caulerpa brachypus* forma *parvifolia* (Harvey) Cribb from coral reefs off northern Palm Beach County, Florida. Estuar Coasts 29(6A):966–971

Lee TN, Mayer DA (1977) Low-frequency current variability and spin-off eddies along the shelf off southeast Florida. M Mar Res 35:193–220

Lee TN, Williams E (1999) Mean distribution and seasonal variability of coastal currents and temperature in the Florida Keys. Bull Mar Sci 64:35–56

Lewis JB (1960) The coral reefs and coral communities of Barbados, West Indies. Can J Zool 38:1133–1145

Lidz BH, Reich CD, Shinn EA (2003) Regional Quaternary submarine geomorphology in the Florida Keys. Geol Soc Am Bull 115:845–866

Lidz BH, Reich CD, Peterson RL, Shinn EA (2006) New maps, new information: coral reefs of the Florida Keys. J Coastal Res 22:260–282

Lighty RG (1977) Relict shelf-edge Holocene coral reef: southeast coast of Florida. Proc 3rd Int Coral Reef Symp 2:215–221

Lighty RG, Macintyre IG, Stuckenrath R (1978) Submerged early Holocene barrier reef south-east Florida shelf. Nature 275:59–60

Lindeman KC and Snyder DB (1999) Nearshore hardbottom fishes of southeast Florida and effects of habitat burial caused by dredging. Fish Bull 97:508–525

Lirman D (2003) A simulation model of the population dynamics of the branching coral *Acropora palmata* Effects of storm intensity and frequency. Ecol Mod 161:169–182

Loya Y (1976) Effets of water turbidity and sedimentation on the community structure of Puerto Rican corals. Bull Mar Sci 26:450–456

Macintyre IG (1988) Modern coral reefs of western Atlantic: new geologic perspectives. Am Assoc Petrol Geol Bull 72:1360–1369

Macintyre IG, Milliman JD (1970) Physiographic features on the outer shelf and upper slope, Atlantic continental margin, southeastern United States. Geol Soc Am Bull 81:2577–2598

Moffatt and Nichol Intl (2006) Port Everglades, Florida, Offshore Anchorage Feasibility Study Final Report, MNI project no. 5905, New York, 119p

Moyer RP, Riegl B, Banks K, Dodge RE (2003) Spatial patterns and ecology of benthic communities on a high-latitude South Florida (Broward County, USA) reef system. Coral Reefs 22:447–464

Moyer RP, Riegl B, Banks K, Dodge RE (2005) Assessing the accuracy of acoustic seabed classification for mapping coral reef environments in South Florida. Rev Biol Trop 53:175–184

Mullins HT, Neumann C (1979) Geology of the Miami Terrace and its paleo-oceanographic implications. Mar Geol 30:205–232

NCRI – National Coral Reef Institute (2004) Final Report: Development of GIS Maps for Southeast Florida Coral Reefs. National Coral Reef Institute, Nova Southeastern University Oceanographic Center, Dania Beach, FL, for Florida Fish and Wildlife Research Institute, 71 pp

Neumann CJ, Jarvinen BR, McAdie CJ, Hammer GR (1999) Tropical Cyclones of the North Atlantic Ocean, 1871–1998, 5th Revision. National Climatic Data Center,

National Oceanic and Atmospheric Administration (NOAA), Asheville, North Carolina, USA 206 pp

NOAA (2005) Comparative Climate Data for the US through 2005. National Environmental Satellite Data and Information Service, National Climatic Data Center, Asheville, North Carolina, USA, 152 pp

Paul VJ, Thacker RW, Banks K, Golubic S (2005) Benthic cyanobacterial bloom impacts the reefs of South Florida (Broward County, USA). Coral Reefs, 24:693–697

PBS&J (1999) Mitigation Plan for the Deployment of Telecommunication Cables in the Nearshore Wates off Hollywood, Broward County, Florida. PBS&J, Inc, Jacksonville, FL, 44 pp

Pielke RA, Walko RL, Steyaert LT, Vidale PR, Liston GE, Lyons WA, Chase TN (1999) The influence of anthropogenic landscape changes on weather in South Florida. Am Meteor Soc 127:1663–1672

Porter JW, Battey J, Smith GJ (1982) Perturbation and change in coral reef communities. Proc Nat Acad Sci USA 79:1678–1681

Powell MA, Thieke RJ, Mehta AJ (2006) Morphodynamic relationships for ebb and flood delta volumes at Florida's tidal entrances. Ocean Dynamics 56: 295–307

Precht WF, Aronson RB (2004) Climate flickers and range shifts of reef corals. Front Ecol Environ 2:307–314

Raymond WF (1972) A Geologic investigation of the offshore sands and reefs of Broward County, Florida. MS Thesis, Florida State University, Tallahassee, FL, 95 pp

Richards VP, Thomas JD, Stanhope MJ, Shivji MS (2007) Genetic connectivity in the Florida reef system: comparative phylogeography of commensal invertebrates with contrasting reproductive strategies. Mol Ecol 16:139–157

Riegl B, Moyer RP, Morris LJ, Virnstein RW, Purkis SJ (2005) Distribution and seasonal biomass of drift macroalgae in the Indian River Lagoon (Florida, USA) estimated with acoustic seafloor classification (QTCView, Echoplus) J Exp Mar Biol Ecol 326:89–104

Roberts HH, Rouse LJ, Walker ND, Hudson JH (1982) Cold-water stress in Florida Bay and Northern Bahamas. A product of winter cold-air outbreaks. J Sedim Petrol 52:145–155

Rohmann SO, Hayes JJ, Newhall RC, Monaco ME, Grigg RW (2005) The area of potential shallow-water tropical and subtropical coral ecosystems in the United States. Coral Reefs 24:370–383

SFWMD (1996) Climate Change and Variability: How Should the District Respond? South Florida Water Management District, Planning Department, West Palm Beach, FL

Shay LK, Cook TM, Peters H, Mariano AJ, Weisberg RH, An PE, Soloviev, Luther MRE (2002) Very high-frequency radar mapping of surface currents. IEEE J Oceanic Eng 27:155–169

Sheng J, Tang L (2004) A two-way nested-grid ocean-circulation model for the Meso-American Barrier Reef System. Ocean Dynamics 54: 232–242

Shinn, EA (1963) Spur and groove formation on the Florida reef tract. J Sedim Petrol 33:291–303

Shinn EA (1976) Coral reef recovery in Florida and the Persian Gulf. Environ Geol 1:241–254

Shinn EA, Hudson JH, Halley RB, Lidz BH (1977) Topographic control and accumulation rate of some Holocene coral reefs, south Florida and Dry Tortugas. Proc Int Coral Reef Sym 2:1–7

Shinn EA, Hudson JH, Robbin DM, Lidz B (1981) Spurs and grooves revisited – construction versus erosion, Looey Key Reef, Florida. Proc 4th Int Coral Reef Sym 1:475–483

Shinn EA, Lidz BH, Kindinger JL, Hudson JH, Halley RB (1989) Reefs of Florida and the Dry Tortugas: A guide to the modern carbonate environments of the Florida Keys and Dry Tortugas. International Geological Congress Field Trip Guidebook T176, AGU, Washington DC, p. 55

Soloviev AV, Luther ME, Weisberg RH (2003) Energetic baroclinic super-tidal oscillations on the southeast Florida shelf, Geophys Res Lett 30(9):1463

Stauble DK (1993) An overview of Southeast Florida inlet morphodynamics. J Coastal Res SI18:1–28

Telesnicki GJ, Goldberg WM (1995) Effects of turbidity on the photosynthesis and respiration of two South Florida reef coral species. Bull Mar Sci 57:527–539

Thanner SE, McIntosh TL, Blair SM (2006) Development of benthic and fish assemblages on artificial reef materials compared to adjacent natural reef assemblages in Miami-Dade County, Florida. Bull Mar Sci 78:57–70

Tichenor E (2005) Environmental Conditions Status Report Cyanobacteria Proliferation Gulf Stream Reef, Boynton Beach, Florida, May 2005. Palm Beach County Reef Rescue, Boynton Beach, FL, 83 pp

Tilmant JT, Curry RW, Jones R, Szmant A, Zieman JC, Flora M, Robblee MB, Smith D, Snow RW, Wanless H (1994) Hurricane Andrew's effects on marine resources. BioScience 44:230–237

Titus JG (1995) The probability of sea level rise. Environmental Protection Agency, Rockville, MD

Torres AE, Higer AL, Henkel HS, Mixon PR, Eggleston JR, Embry TL, Clement G (2003) US Geological Survey Greater Everglades Science Program: 2002 Biennial Report. USGS Open File Report 03–54, Tallahassee, FL

Toscano MA, Macintyre IG (2003) Corrected western Atlantic sea-level curve for the last 11,000 years based on calibrated ^{14}C dates from *Acropora palmata* framework and intertidal mangrove peat. Coral Reefs 22:257–270

Trewartha GT (1968) An Introduction to Climate. McGraw-Hill, 408 pp

Trnka M, Logan K (2006) Land-based Sources of Pollution Local Action Strategy, Combined Projects 1 and 2. Florida Department of Environmental Protection, Southeast Florida Coral Reef Initiative, 196 pp

USACE (1996) Coast of Florida Erosion and Storm Effects Study-Region III, appendix D-Engineering Design and Cost Estimates (draft). US Army Corps of Engineers, Jacksonville, Florida, District, 233 pp

Vargas-Angel B, Thomas JD, Hoke SM (2003) High-latitude *Acropora cervicornis* thickets off South Lauderdale, FL. Coral Reefs 22:465–473

Vargas-Angel B, Riegl B, Gilliam D, Dodge R (2006) An experimental histopathological rating scale of sedimentation stress in the Caribbean coral *Montastraea cavernosa*. Proc 10th Int Coral Reef Sym:1168–1173

Vargas-Angel B, Colley SB, Hoke SM, Thomas JD (2006) The reproductive seasonality and gametogenic cycle of *Acropora cervicornis* off Broward County, FL. Coral Reefs 25:110–122

Vaughan TW (1914) Investigations of the geology and geologic processes of the reef tracts and adjacent areas in the Bahamas and Florida. Carnegie Inst Wash Yb 12:183

Walker BK, Riegl B, Dodge RE (2008) Mapping coral reef habitats in SE Florida using a combined-technique approach. J Coastal Res

Walker BK (2007) Seascape approach to predicting reef fish distribution. PhD dissertation, Nova Southeastern University, 226pp

Ward-Paige CA, Risk MJ, Sherwood OA (2005) Clionid sponge surveys on the Florida Reef Tract suggest land-based nutrient inputs. Mar Poll Bull 51:570–579

Wanless HR (1989) The inundation of our coastlines: past, present, and future with a focus on South Florida. Sea Frontiers 35(5):264–271

Willis TJ (2001) Visual census methods underestimate density and diversity of cryptic reef fishes. J Fish Biol 59:1408–1411

Wulff JL (1985) Dispersal and survival of fragments of coral reef sponges. Proc 5th Int Coral Reef Congr 5:119–124

Wulff JL (1991) Asexual fragmentation, genotype success, and population dynamics of erect branching sponges. J Exp Mar Biol Ecol 149:227–247

Wulff JL (1995a) Effects of a hurricane on survival and orientation of large erect coral reef sponges. Coral Reefs 14:55–61

Wulff JL (1995b) Sponge-feeding by the Caribbean starfish *Oreaster reticulates*. Mar Biol 123:313–325

Yeung C, Lee TN (2002) Larval transport and retention of the spiny lobster, *Panulirus argus*, in the coastal zone of the Florida Keys, USA. Fish Oceanogr 11:286–309

Yoshioka PM, Yoshioka BB (1991) A comparison of the survivorship and growth of shallow-water gorgonian species of Puerto Rico. Mar Ecol Progr Ser 69:253–260

6
Biology and Ecology of Coral Reefs and Coral Communities in the Flower Garden Banks Region, Northwestern Gulf of Mexico

George P. Schmahl, Emma L. Hickerson, and William F. Precht

6.1 Introduction

In the northwestern Gulf of Mexico, 50–100 miles off the coasts of Texas and Lousiana, dozens of underwater features rise from the seafloor near the edge of the continental shelf to form a complex of reefs and banks. While the crests of most of these features lie more than 50 m deep, a small number of them are shallow enough for coral reefs and coral communities to have become established. Two of these features, the East and West Flower Garden Banks, reach within 18 m of the surface and contain well-developed coral reefs (Bright 1977). In addition, a handful of other features in this region, including Stetson, Bright, Geyer, McGrail and Sonnier Banks, contain a mix of coral reefs and coral communities (Rezak et al. 1985). This chapter summarizes information about these high latitude coral ecosystems, and proposes a revision to a habitat classification system to describe the biological communities of this region.

The Flower Garden Banks are located approximately 185 km (115 miles) south of the Texas /Louisiana border in the Gulf of Mexico (Fig. 6.1). They are among the northernmost coral reefs in federal waters on the continental shelf of the United States. The US Department of Commerce, National Oceanic and Atmospheric Administration (NOAA) designated the East and West Flower Garden Banks as the Flower Garden Banks National Marine Sanctuary (FGBNMS) in 1992. Stetson Bank, located 48 km (30 miles) west/northwest of the West Flower Garden Bank, was added to

the FGBNMS in 1996. The East and West Flower Garden Banks contain some of the healthiest coral reefs in the Caribbean and western Atlantic region (Lang et al. 2001). The reefs exhibit an average of over 50% living coral cover and are dominated by massive brain and star corals (Fig. 6.2). Many common indicators of coral reef vitality, such as low macroalgal cover, low incidence of coral disease, and vigorous growth rates demonstrate the health of this coral system (Hickerson and Schmahl 2005). Stetson Bank is an uplifted claystone feature, that while not containing significant hermatypic coral reef structure, does harbor coral and coral-associated communities.

The coral reefs of the FGBNMS are the best known and most studied of the reefs and banks of the northwestern Gulf of Mexico. Stetson (1953) first reported the presence of coral at the Flower Garden Banks, and suggested that these banks were bioherms built on top of salt domes. Subsequent to this initial report, there ensued a spirited debate on whether living coral reefs could exist at such high latitudes in the Gulf of Mexico. The fact that the Flower Garden Banks did contain actively growing coral communities was substantiated by Pulley (1963) in a series of diving expeditions utilizing newly emerging SCUBA technology. Ever since, the coral reefs of the FGBNMS have been extensively studied and monitored. The earliest quantitative data on coral and other sessile invertebrate cover at the East and West Flower Garden Banks was collected in 1972 (Bright and Pequegnat 1974). What became the first of a series of regularly collected data on benthic communities occurred in 1978 (Viada 1980) and continued through 1983 (Continental Shelf

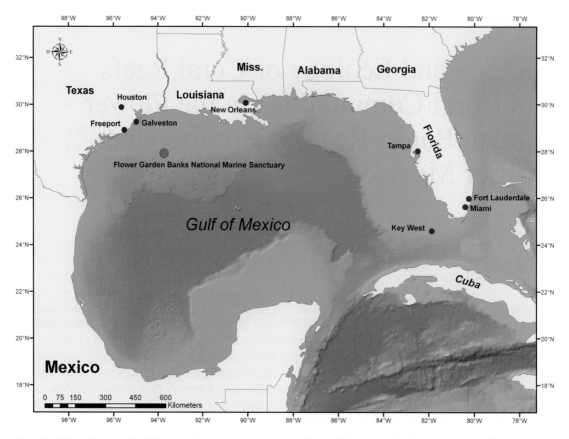

FIG. 6.1. View of the Gulf of Mexico showing location of the Flower Garden Banks National Marine Sanctuary.

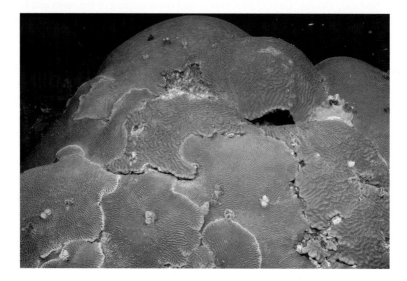

FIG. 6.2. Massive brain and star corals dominate the landscape at the East and West Flower Garden Banks. (Photo credit: FGBNMS/G.P. Schmahl)

Associates, Inc. 1985). This work was prompted by proposed exploratory drilling activities for oil and gas resources within 1.7 km (1 mile) of the coral reefs of the East Flower Garden Bank.

In the late 1980s, with the expansion of production operations within this region, the US Department of Interior, Minerals Management Service (MMS), in consultation with academia and industry, developed a comprehensive long-term benthic monitoring program for the coral reefs of the Flower Garden Banks. This program was initiated in 1988 and has remained in place, with some modifications, ever since (Gittings 1998).

The biological communities of Stetson Bank were first described by Bright et al. (1974) in a study conducted for Signal Oil and Gas Company. Additional surveys of Stetson Bank were conducted as part of the "Northern Gulf of Mexico Topographic Features Study" (Rezak et al. 1982), an investigation funded by the Bureau of Land Management (BLM – now MMS) in response to the increasing interest in offshore oil and gas development in the region. A monitoring program of benthic communities of Stetson Bank was first established in 1993. The Gulf Reef Environmental Action Team (GREAT), a volunteer organization of divers and scientists, initiated and installed a number of repetitive photostations, and conducted the first data collection cruises. After Stetson Bank was added to the FGBNMS in 1996, annual monitoring has continued by sanctuary staff and volunteers (Fig. 6.3) (Bernhardt 2000).

Although designated as a National Marine Sanctuary, the East and West Flower Garden and Stetson Banks are but three of approximately 130 reefs, banks and topographic features that occur in this region. The topographic features study, conducted during the 1970s and 1980s by Texas A&M University for BLM, investigated dozens of these features, with the objective to determine the level of regulation necessary to protect their associated biological communities. These studies produced a series of BLM publications and culminated in the publication of a comprehensive monograph on the reefs and banks of the northwestern Gulf of Mexico (Rezak et al. 1985). In addition to a diverse array of biological resources associated with these topographic features, this study documented the presence of other coral communities in the region. These included Bright, Geyer, McGrail (then known as 18 fathoms) and Sonnier Banks (Fig. 6.4).

FIG. 6.3. Diver collecting repetitive photostation image at Stetson Bank during the annual monitoring cruise. (Photo credit: Joyce and Frank Burek)

6.2 Regional Setting

The East and West Flower Garden Banks, located near coordinates 27°54′ N, 93°35′ W and 27°53′ N, 93°49′ W respectively, contain well known, highly developed coral reefs that are in excellent health (Lang et al. 2001). The outer continental shelf and continental slope off the coast of Texas and Louisiana is geologically complex. The continental shelf in this region slopes gradually from the shoreline to depths between 100 and 200 m, and is characterized primarily by sediments of terrigenous origin. However, this region is punctuated by a series of topographic features, scattered along an area parallel to the edge of the shelf. Most of these features were formed as the result of the movement of underlying salt deposits (also called salt domes or diapirs), which created topographic

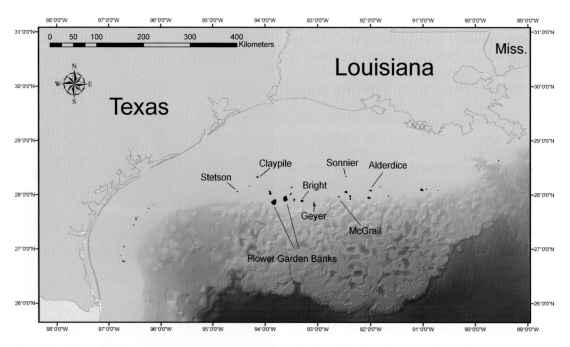

FIG. 6.4. Selected reefs and banks in the northwestern Gulf of Mexico that contain coral reefs and coral communities

complexity and exposed hard substrate (Rezak et al. 1990). Many of these features display relief of 50 m or more, extending well into the photic zone (Fig. 6.5). Baseline investigations documented that these features contain significant marine resources, including a number that harbor important coral reef communities (Rezak et al. 1985).

Environmental conditions in this portion of the northwestern Gulf of Mexico are favorable for coral reef and coral community development on hard bottom areas shallow enough to allow sufficient light penetration. Warm tropical water is transported from the Caribbean into the eastern Gulf of Mexico via the Loop Current and travels to the western Gulf through the action of spin off eddies (Fig. 6.6). Water temperatures in the region of the Flower Garden Banks typically range from 18°C (February) to 30°C (August), and salinity ranges from 34 to 36 parts per thousand. These values are well within the range of that necessary for coral reef growth, although winter temperatures approach the lower limit of some coral species tolerance. The reefs and banks of the northwestern Gulf of Mexico are located from 90 to over 160 km (60–100 miles) from the shore, so they are positioned well away from the normal influence of coastal runoff and nearshore eutrophication. Chlorophyll and nutrient levels are typically low, and indicative of oligotrophic oceanic conditions.

Biological surveys conducted since 2000 have revealed a broad array of exceptional marine communities beyond that previously known (Schmahl et al. 2003). The upper reaches of many of the features support tropical reef communities that provide significant habitat for a variety of fish species of commercial and recreational importance. Detailed investigations have re-confirmed the existence of a significant coral reef at McGrail Bank (Schmahl and Hickerson 2006), as well as lesser coral communities at Stetson, Bright, Geyer and Sonnier Banks. The deeper areas of all the banks provide habitat for non-reef building corals and associated communities. A wide variety of alcyonarian corals (sea whips and fans), antipatharians (black coral), sponges and other benthic invertebrates have been documented through surveys conducted by FGBNMS. Deep water coral communities have recently received much deserved attention as areas in need of research and management. Preliminary observations from the FGBNMS surveys also

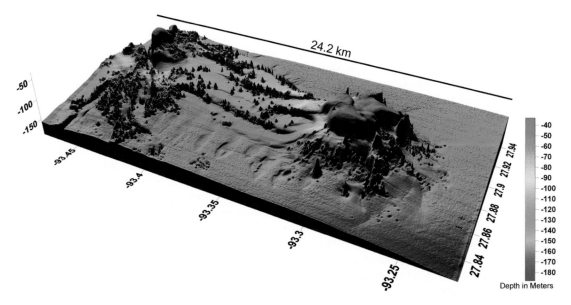

Fig. 6.5. High resolution bathymetry of Bright, Rankin, and 28 Fathom Banks showing high relief and habitat complexity (Bathymetry credit: USGS, NOAA's Office of Ocean Exploration, MMS, NOAA/FGBNMS, image rendition by FGBNMS/D.C. Weaver)

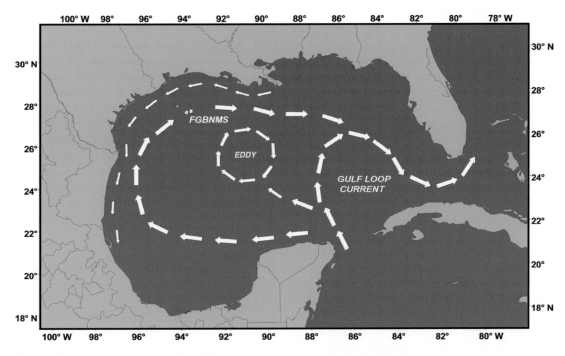

Fig. 6.6. Diagrammatic representation of the prevailing currents in the Gulf of Mexico

indicate that some of these areas are important aggregation and possible spawning sites for prominent fish species, such as scamp (*Mycteroperca phenax*) and marbled grouper (*Dermatolepis inermis*).

Results of the bathymetric and biological surveys indicate that many of the reefs and banks in the northwestern Gulf of Mexico are interconnected, both physically and ecologically. It is now clear that in many cases what was once thought to be individual, isolated banks are in actuality a system of features associated with the same geological structure (Fig. 6.7) (Gardner and Beaudoin 2005). This structure can provide a physical connection between two or more banks in the form of an exposed hard-bottom ridge, or as a series of rocky outcroppings. Such physical connections may provide for continuity of habitat between features that are utilized by marine organisms. For hard-bottom associated fish species, the features may provide "habitat highways" that allow for direct movement between banks. For example, a recent effort to track movements of manta rays (*Manta birostris*) utilizing acoustic tags has documented the use of multiple banks by individual rays (R. Graham, Wildlife Conservation Society, personal communication 2007). For non-mobile invertebrate species, ridges and outcrops provide suitable substrate "stepping stones" for colonization and development of benthic communities along a geographic continuum. For these reasons, it is important to consider the reefs and banks of the northwestern Gulf of Mexico as an ecological network of marine communities. Like shallow water coral reef ecosystems that are enhanced by the presence of adjacent seagrass and mangrove communities, the deeper coral reefs of the northwestern Gulf rely on adjacent algal dominated areas, soft bottom features, deep coral communities and rocky outcrops to provide feeding areas, spawning sites and habitat for critical life history stages for a variety of reef organisms. These areas, consisting of patch reefs, scarps and ridges, also provide physical habitat connections that facilitate organism movement. Thus, it is important to protect not only the coral reef and coral communities themselves, but also the habitat that connects these communities (Schmahl and Hickerson 2005).

Oceanographic processes in the vicinity of the Flower Garden Banks are relatively complex. The circulation within the Gulf of Mexico is dominated by the Loop Current and its associated "spin-off" eddies and gyres (Sturges et al. 2005). The Loop Current enters the Gulf of Mexico through the Yucatan Channel between Cuba and Mexico as a massive river of warm water, reaching speeds up to 2 m/s (almost 4 knots) (Badan et al. 2005). The current flows northward to a variable extent, at times reaching as far as 28° N, before looping clockwise along the west Florida shelf to exit through the Florida Straits. Here, the waters of the Loop Current flow northward along the southeastern US coast and become the Gulf Stream. As the Loop Current reaches its maximum northern position in the Gulf of Mexico, it often becomes unstable, shedding large eddies (or gyres) that spin clockwise as they drift westward at speeds of 1–8 km/day. These eddies can have a diameter of 200–400 km (125–250 miles), and last for intervals of 0.5–18.5 months (Schmitz et al. 2005) before they eventually spin down in the western Gulf. Figure 6.8 shows a satellite image of the sea surface in the Gulf of Mexico in February 2006. The warm water of the loop current, along with the formation of a spin-off eddy, is easily visible.

The presence of the Loop Current has had a major influence on the biology and ecology of the reefs and banks of the northwestern Gulf of Mexico. Biggs (1992) demonstrated that spin-off eddies from the Loop Current can transport biological materials from the Caribbean into the western Gulf. Based on drifter analysis, Lugo-Fernandez (2006) determined that it would take 55–135 days for larvae of coral reef organisms to be transported from the Yucatan peninsula to the Flower Garden Banks. These relatively long dispersal times may account for the lack of certain coral reef organisms (such as most shallow water octocorals) in the northern Gulf. Conversely, Lugo-Fernandez et al. (2001), using satellite and drifter buoy data, established that coral larvae produced at the Flower Garden Banks may provide a source of coral recruits to Florida and the Mexican Caribbean.

6.3 Biogeographical Setting of the Area

Rocky outcrops and hard banks are common on the continental shelf throughout the Gulf of Mexico. The biological communities associated with these features range from tropical coral reefs to more temperate assemblages. The development of coral

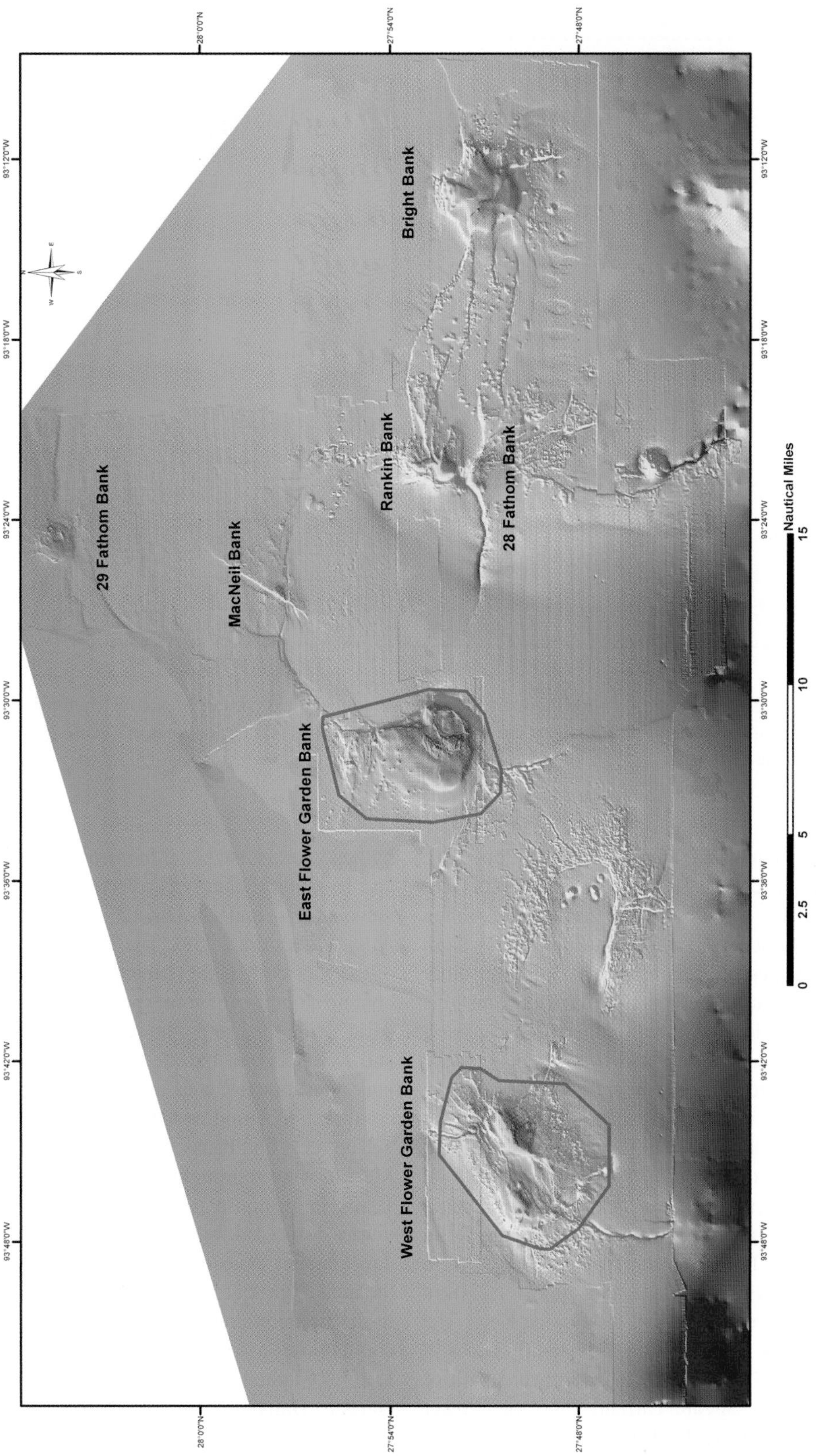

FIG. 6.7. High resolution bathymetry showing the structural connectivity between the reefs and bank features in the vicinity of the Flower Garden Banks (Bathymetry credit: USGS, NOAA's Office of Ocean Exploration, MMS, NOAA/FGBNMS, image rendition by FGBNMS/D.C. Weaver)

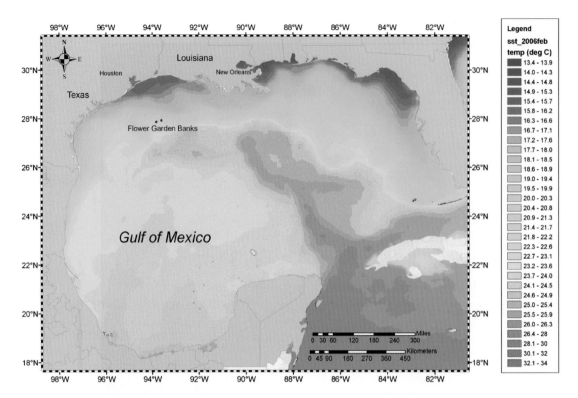

Fig. 6.8. Remote sensing imagery showing sea surface temperatures in the Gulf of Mexico. Data for this image was collected in February, 2006, and illustrates the spin-off eddies coming off the loop current that enters the Gulf of Mexico from the Caribbean

reefs can occur throughout the Gulf as long as a relatively narrow range of ecological factors are satisfied. The well-developed coral reefs of the Flower Garden Banks and vicinity in the northern Gulf of Mexico are relatively isolated from other reef systems in the region (Fig. 6.9). The nearest coral reefs are located in the southern Gulf, approximately 690 km (430 miles) away on the Campeche Bank of Mexico. These reefs, including Alacran reef, Cayo Arenas, Nuevo reef and Triangulos, lie on the outer Yucatan shelf adjacent to islands and on the crests of submerged banks. The reefs of the northwestern Gulf are located approximately 725 km (450 miles) from the reefs of Cabo Rojo, Mexico (near Tampico), 980 km (610 miles) from the coral reefs of Veracruz, and almost 1,270 km (790 miles) from the reefs of the Florida Keys. The most northerly coral reefs in the Atlantic occur in Bermuda (32.3° N) due to the northward path of the warm Gulf Stream current. Like the Flower Garden Banks, the reefs of Bermuda are dominated

by head corals, such as *Montastraea* and *Diploria*, and lack significant occurrences of shallow water alcyonarians and the branching coral *Acropora palmata* (Moore 1969). The benthic communities of the Florida Middle Grounds (28.5° N, 84.4° W) contain a number of coral species, primarily the fire coral *Millepora* and the scleractinians *Madracis* and *Dichocoenia*, but are dominated by alcyonarians and sponges, and are not actively accreting coral reefs (Grimm and Hopkins 1977).

The reefs and banks of the northwestern Gulf of Mexico were thoroughly studied in the 1970s and 1980s in response to the concern over potential impacts associated with the impending exploration and development of extensive oil and gas resources in the area (Rezak et al. 1985). These studies resulted in the first comprehensive classification of benthic communities associated with these features. Of the features studied, only the East and West Flower Garden Banks were considered to be "true" (actively accreting) coral reefs. Three coral reef

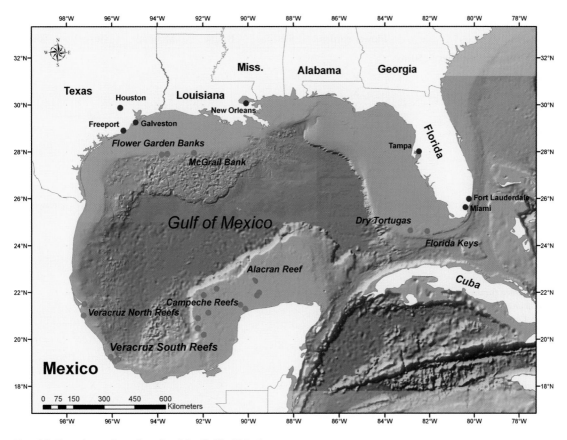

FIG. 6.9. Locations of coral reefs of the Gulf of Mexico

community types were observed (Rezak et al. 1990). The first two were called the *Diploria-Montastraea-Porites* zone and the "*Madracis* and Leafy Algae" zone. Both of these types were only observed on the East and West Flower Garden Banks. The third type was called the *Stephanocoenia-Millepora* zone. This zone was observed in the deeper areas of the Flower Garden Banks (36–50 m) and at Bright and McGrail (then known as 18 Fathom) Banks. In addition, a fourth reef type was classified within an area of "minor reef-building activity" as the "*Millepora*-Sponge" zone. This zone was observed primarily on mid-shelf reefs such as Stetson, Sonnier and Claypile Banks. Geyer Bank, a shelf-edge reef, was also classified as a "*Millepora*-Sponge" community.

Recent investigations by the FGBNMS have provided additional observations of coral communities in the northwestern Gulf of Mexico. Bright Bank, located 12 miles east of the East

Flower Garden Bank, is a hardbottom feature that contains only scattered occurrences of large living scleractinian coral heads. Although the coral species *Stephanocoenia* is found on Bright Bank, surveys indicate that it is not comprised of significant communities of this coral type. Inspection of the underlying substrate, however, demonstrates that portions of Bright Bank have been created by extensive vertical accumulations of the finger coral *Madracis* and other coral species. While living coral cover is low, it is obvious that this area can support coral reef development and exhibits minor reef communities at this time. Geyer Bank is similar in many ways to Bright Bank, and while it does not exhibit significant reef development, contains a variety of coral species scattered throughout the feature. In contrast, McGrail Bank (previously known as 18 Fathom Bank) supports an active hermatypic coral reef community that is much different than those observed at Bright, Geyer or the

TABLE 6.1. Scleractinians of the coral reefs and coral communities of the northwestern Gulf of Mexico.

	Flower garden	Stetson	Bright	McGrail	Geyer	Sonnier
"Shallow-water" Species (<50 m)						
Acropora palmata	X					
Agaricia agaricites	X					
Agaricia fragilis	X					X
Agaricia undata	X	X	X	X	X	
Colpophyllia amaranthus	X		X			
Colpophyllia natans[a]	X	X	X	X		
Dichocoenia stokesi[a]	X		X	X		
Diploria strigosa	X	X	X			
Leptoseris cucullata	X					
Madracis decactis	X	X				X
Madracis mirabilis	X	X				X
Madracis pharensis		X				X
Millepora alcicornis[a]	X	X	X	X	X	X
Montastraea annularis	X	X				
Montastraea cavernosa[a]	X	X	X	X		X
Montastraea faveolata	X					
Montastraea franksi	X		X			
Mussa angulosa	X					
Porites astreoides	X					
Porites furcata	X					
Scolymia cubensis[a]	X	X		X		
Siderastrea radians	X					X
Siderastrea siderea	X					
Stephanocoenia intersepta[a]	X	X	X	X	X	X
Tubastraea coccinea	X				X	X
Total	24	12	9	7	4	9
"Deeper-water" Species (>50 m)						
Polycyathus senegalensis chevalier	X					
Agaricia lamarcki			X	X		
Madracis asperula[b]	X	X	X			
Madracis brueggemanni	X					
Madracis cf. formosa	X					
Madracis myriaster	X					
Madrepora carolina	X					
Oxysmilia rotundifolia	X					
Paracyathus pulchellus	X	X			X	
Total	8	2	2	1	1	0

[a]Also documented below 50 m

[b]Also documented above 50 m

Flower Garden Banks. McGrail Bank contains a unique coral reef that is dominated by large numbers of *Stephanocoenia* colonies at densities covering up to 24% of the benthic substrate (Schmahl and Hickerson 2006). The reef communities at Stetson and Sonnier Banks are similar to each other. The dominant coral species is the hydrozoan *Millepora,* which occurs in densities of up to 30% cover on the shallowest (20–25 m) pinnacles. A small number of other coral species are also found at these locations. Table 6.1 lists the known species of scleractinian corals from various banks in the northwestern Gulf of Mexico.

We suggest that all of these features exhibit various gradations of coral reef development and should be considered coral reefs and coral communities. The Flower Garden Banks and McGrail Bank are "true" hermatypic coral reefs, surrounded in deeper depths by algal reef communities and deep reef assemblages. The reefs of Stetson, Sonnier, Bright and Geyer Banks are more accurately described as coral communities (following definitions of Geister 1983; Wainwright 1965; Stoddart 1969). While corals are a contributing component of the benthos at these locations, they are not the dominant organismal group. However, they do provide significant potential habitat for future coral reef development.

6.3.1 Biodiversity of Major Organismal Groups

The reefs and banks of the northwestern Gulf of Mexico support a diverse assemblage of marine organisms. For example, within the boundaries of the FGBNMS, the following numbers of species within major organismal groups have been recorded through published records and direct observations: 1 moneran, 113 algae, 30 protozoans, 36 bryozoans, 30 sponges, 68 cnidarians, 62 echinoderms, 632 mollusks, 170 crustaceans, 21 annelids, 280 fish, 2 sea turtles, and 3 marine mammals. This list includes organisms that occur not only on the coral reef areas, but from the deeper areas (below 50 m), as well. A complete listing of these species can be downloaded from: http://www. flowergarden.noaa.gov.

The coral reef caps of the Flower Garden Banks extend from approximately 17 to 46 m depth, and are dominated by large boulders of brain and star coral: *Montastraea franski*, *M. faveolata*, and *Diploria strigosa*. A total of 24 species of coral have been documented with in the coral reef of the East and West Flower Garden Banks. Coral cover is high, ranging from around 50% on the crest of the coral cap, to over 70% on its deeper margins (Precht et al. 2005). The reefs of the Flower Garden Banks vary from typical Caribbean reefs in that there is a striking lack of large branching corals, and no shallow water alcyonarians are found on the reef crest. The number of species of coral is about one third less than the most well developed Caribbean reefs. However, the reefs of the FGBNMS have some of the highest percentage of coral cover for the region (Kramer 2003).

Table 6.1 lists the scleractinian coral species reported from coral reefs and coral communities of the northwestern Gulf of Mexico. A total of 25 shallow water (<50 m depth) and 9 deepwater (>50 m depth) scleractinian coral species have

been recorded to date. It should be noted that while this list includes *Acropora palmata*, only two colonies have been documented – one each at the East and West Flower Garden Banks (Zimmer et al. 2006). *Dichocoenia stokesi*, typically a shallow water species, has only been documented in deeper coral reef zones on the East Flower Garden Bank, and at Bright and Geyer Banks. The orange cup coral, *Tubastraea coccinea*, is an invasive Indo-Pacific species that has become established on numerous oil and gas platforms in the Gulf of Mexico. This species has also been recorded on natural substrates at the East Flower Garden Bank reef crest (Fenner and Banks 2004), Sonnier and Geyer Banks.

Below the coral reef (>50 m) occurs a diverse assemblage of deeper water habitats, including deep coral communities. Table 6.2 lists the antipatharian and alcyonarian coral species recorded from these areas within the FGBNMS. Since 2000, the FGBNMS has conducted over 160 remotely operated vehicle (ROV) surveys within the sanctuary, and collected over 200 biological samples for documentation and identification. A rich variety of deep reef biota has been recorded, collected, and identified, including species of sponges, antipatharians, alcyonarians, fish, algae, and other invertebrates. Regional catalogs are under development by the FGBNMS to assist future survey efforts. Deepwater surveys (50–150 m) have also been conducted at several of the northwestern reefs and banks outside of the sanctuary boundaries, and have shown similar deepwater biological communities. Figure 6.10 illustrates a typical deepwater community from 93.5 m (307′) depth during ROV surveys at McGrail Bank.

The tropical reef fish communities, like the coral assemblages, are relatively less diverse compared with other areas in the tropical western Atlantic and Caribbean. Dennis and Bright (1998) cite three major factors that limit the number of reef fishes occurring in the northwestern Gulf of Mexico: limited habitat diversity, limited habitat area, and distance from source populations. In spite of these limiting factors, 280 species of reef fish have been recorded from the FGBNMS. Groups that are typically observed on coral reefs of the Caribbean but are either rare or absent from the FGBNMS include most species of grunts and hamlets.

TABLE 6.2. Antipatharians and alcyonarians of the Flower Garden Banks National Marine Sanctuary (FGBNMS). Samples of all species of antipatharians were collected by ROV between 48 and 131 m (158′ and 431′) within the FGBNMS, and identified by Dr. Dennis Opresko, Smithsonian Institute. Samples of all species of alcyonarians were collected by ROV between 46 and 110 m (150′ and 362′) within the FGBNMS and identified by Dr. Gary Williams (California Academy of Sciences), Drs. Stephen Cairns and Ted Bayer (Smithsonian Institute), and Peter Etnoyer (Texas A&M University Corpus Christi).

Antipatharians:		
Phylum Cnidaria		
Class Anthozoa		
Order Antipatharia		
Family Antipathidae		
	Antipathes atlantica	
	Antipathes furcata	
	Stichopathes lutkeni	
Family Myriopathidae		
	Tanacetipathes barbadensis	
	Tanacetipathes tanacetum	
	Plumapathes pennacea	
Family Aphanipathidae		
	Aphanipathes pedata	
	Elatopathes abientina	
	Phanopathes expansa	
	Acanthopathes thyoides	
Octocorals:		
Order Alcyonacea		
Family Nidaliidae		
	Subfamily Siphonogorgiinae	
		Chironephthya caribaea
	Subfamily Spongiodermatinae	
		Diodogorgia nodulifera
Family Keroeididae		
	Thelogorgia stellata	
Family Plexauridae		
	Bebryce cinerea	
	Caliacis nutans	
	Hypnogorgia spp.	
	Muricea pendula	
	Muriceides sp. cf. *furta*	
	Placogorgia mirabilis	
	Placogorgia tenuis	
	Scleracis sp. cf.	
	Guadalapenses	
	Swiftia sp.	
	Thesea rubra	
Family Gorgoniidae		
	Leptogorgia sp.	
Family Ellisellidae		
	Ellisella spp.	
	Nicella deichmanni	
	Nicella goreaui	
	Nicella guadalupensis	
Family Primnoidae		
	Callogorgia gracilis	

Fig. 6.10. Deepwater community at McGrail Bank. Similar species composition has been evident at many of the northwestern Gulf of Mexico reefs and banks. Genera and species present in this photograph are: crinoid (*Comactinia meridionalis*), a soft coral (*Chironephthya caribaea*), and several ocotocorals; *Nicella* sp., *Callogorgia gracilis*, *Scleracis* sp., *Ellisella* sp. *and Bebryce* sp. Deepwater fish species are bank butterflyfish (*Prognathodes aya*) and roughtongue bass (*Pronotogrammus martinicensis*) (Photo credit: FGBNMS/NURC-UNCW)

Schooling planktivores constitute the most abundant trophic guild, and accounted for approximately 50% of fishes during stationary diver surveys published by Pattengill (1998). These were dominated by six species: blue chromis (*Chromis cyanea*), brown chromis (*C. multilineata*), creole wrasse (*Clepticus parrae*), creolefish (*Paranthias furcifer*), bonnetmouth (*Emelichthyops atlanticus*) and boga (*Inermia vittata*). The dominance of planktivores is confirmed by recent surveys conducted for the long-term coral reef monitoring program (Precht et al. 2006).

Sessile invertebrate feeders, including butterflyfish, angelfish, boxfish, and filefish, accounted for approximately 20% of the reef fish community surveyed by Pattengill (1998). Herbivores, including parrotfish, damselfish, filefish, surgeonfish, and several species of gobies and blennies, accounted for between 8.9% and 13.6% of the fish surveyed (Pattengill 1998). Mobile invertebrate feeders and piscivores, together, including jacks, snapper, porcupinefish, squirrelfish, hogfish, sharks, moray eels, cobia, mackerel, wrasse, grouper, scorpionfish, barracuda, and lizardfish accounted for only

4–6.3% of the fish surveyed. This group includes the more conspicuous fish inhabitants of the reef, as well as the fish most targeted by fishers.

Interestingly, planktivores were also the dominant assemblage in deepwater (19–90 m) communities at Sonnier, McGrail, and Alderdice Banks (Weaver et al. 2006). Creole fish and brown chromis were among the dominant planktivorous species listed at those locations, a feeding guild that constituted over 87% by number in the deepwater surveys. Epibenthic browsers, including sessile invertebrate feeders, accounted for just 1.9% of the fish surveyed, and herbivores accounted for just 0.2% in the deeper surveys – both considerably less than those reported on the coral reef by Pattengill (1998). Benthic carnivores, general carnivores and piscivores together accounted for 11% of the fish surveyed in the deepwater habitats. This is double the percentage reflected in the shallower coral reef data.

The FGBNMS and surrounding reefs and banks support a wealth of diversity, and provide researchers opportunities to study marine ecosystems in relatively pristine condition. Regardless of

FIG. 6.11. Mardi Gras wrasse, currently being described, is found only at the FGBNMS and Vera Cruz reefs. This is a photograph of a terminal phase male at Stetson Bank. (Photo credit: FGBNMS/G.P. Schmahl)

the many hundreds of hours spent to date exploring the region, new species continue to be described as technology allows scientists to stay longer, and explore deeper. Several species new to science have been reported from the region recently. Numerous species of algae have been described, including a red algae, *Martensia hickersonii* (Gavio et al. 2005). A new species of snapping shrimp, *Alpheus hortensis*, was reported and described by Wicksten and McClure (2003) from Stetson Bank. Of particular note is the discovery of an undescribed species of wrasse of the genus *Halichoeres* (Weaver and Rocha, in press). This wrasse is proposed to be given the common name of "mardi gras wrasse", because of the vibrant colorings displayed by the terminal male phase of the animal (Fig. 6.11). This fish species is only known to occur at the FGBNMS (primarily at Stetson Bank), and from coral reefs near Vera Cruz, Mexico. Also of note is the occurrence of a "golden" phase of the smooth trunkfish (*Lactophrys triqueter*) (Pattengill 1999). Although not a new species, this color morph is only known to occur at the FGBNMS, and in the Gulf of Honduras. It is likely that as investigations continue, undescribed species of alcyonarians, sponges and other invertebrates may also be found.

6.3.2   Zonation and Community Patterns

Rezak et al. (1985) developed a classification and characterization system for biological communities associated with the reefs and banks of the northwestern Gulf of Mexico. This classification structure was the culmination of a large body of work on the Flower Garden Banks (Bright and Pequegnat 1974; Bright et al. 1985) and other reefs and banks in the area. Subsequent investigations have demonstrated the accuracy and usefulness of this classification framework. However, based on recent information, we propose a modification and reorganization of the classification scheme proposed by Rezak et al. (1985).

Coral reefs and coral communities form a mosaic of biological habitats at the Flower Garden Banks and other reefs and banks of the northwestern Gulf of Mexico. We propose a classification hierarchy characterized by broad biological zones. Within each zone there are multiple habitat types. Although the focus of this chapter is shallow water coral reefs, this classification includes associated reef building assemblages and deep coral communities. A summary of the biological zone classification is given in Table 6.3. These biological zones are described below.

TABLE 6.3. Summary of biological zones and habitat classification categories of reefs and banks of the northwestern Gulf of Mexico.

Zone	Habitats	Geomorphology	Depth
Coral reef	*Montastraea*	Bank reef	16–52 m
	Madracis	Patch reef	
	Stephanocoenia	Carbonate sand	
	Coral sand		
Coral community	*Millepora*-sponge	Hardbottom	16–52 m
	Low density coral	Patch reef	
	Leafy algae	Pinnacle	
	Sponge		
	Mixed coral		
Coralline algal reef	Algal nodules	Algal nodules	45–98 m
	Coralline algal reef	Pavement	
	Leafy algae	Patch reef	
		Low relief outcrop	
		Molluscan reef	
Deep coral	Antipatharian	Rock outcrops	50–200+ m
	Octocoral	Drowned reef	
	Crinoid	Pavement	
	Stony coral	Rubble	
		Eroded outcrop	
Soft bottom	*Amphistegina* sand	Sand	16–200+ m
	Quartz	Mud	
	Molluscan hash		

6.3.2.1 Coral Reef Zone

This zone includes the actively accreting hermatypic coral assemblages. The coral reef crest (or "cap") of the East and West Flower Garden Banks exemplifies this zone. The Coral Reef zone proposed here includes the *Diploria-Montastraea-Porites* zone, the "*Madracis* and Leafy Algae" zone and the *Stephanocoenia -Millepora* zones as described by Rezak et al. (1985, 1990). We propose that these classifications are sub-components of the coral reef zone. Major habitats within this zone are described by the dominant coral species that characterize the assemblage. The primary habitat of the coral reef zone of the Flower Garden Banks is the *Montastraea* habitat (Fig. 6.12a and b). Rezak et al. (1985) called this community the *Diploria-Montastraea-Porites* zone, but this is somewhat misleading in that the brain coral *Diploria* is not the dominant component of the species assemblage. Members of the genus *Montastraea* account for over 65% of the coral species encountered, while *Diploria* accounts for about 11%. Other habitats within the coral reef zone include those typified by *Madracis* (*Madracis* and Leafy Algae zone of Rezak et al. 1985), *Stephanocoenia* (*Stephanocoenia-Millepora* of Rezak et al. 1985), and coral sand. The *Montastraea* habitat of the Flower Garden Banks includes at least 24 species of stony corals. This habitat is interspersed by sand channels comprised of coral sand (coral debris with molluscan and algal components). The *Madracis* habitat occurs on the peripheral parts of the primary reef structure in depths ranging from 28 to 44 m, where large knolls characterized by almost monospecific stands of the small branching coral *Madracis mirabilis* can occur (Fig. 6.13). The *Stephanocoenia* habitat is a lower diversity coral community occurring in water depths primarily below 36 m. While dominated by the blushing star coral *Stephanocoenia intersepta*, other species such as *Millepora alcicornis*, *Colpophyllia natans*, *Agaricia* spp., *Mussa angulosa* and *Scolymia* sp. are also encountered. This habitat occurs in areas surrounding the Flower Garden Banks, and is the primary coral reef habitat at McGrail Bank (Fig. 6.14).

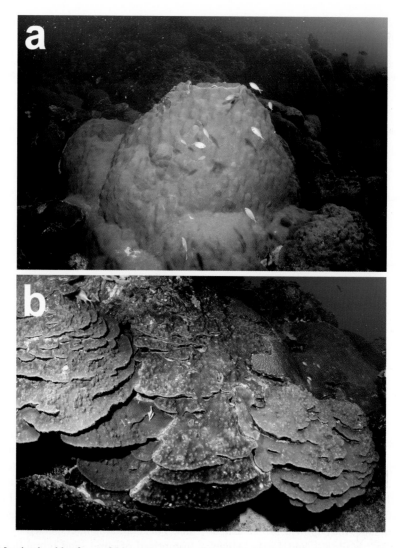

Fig. 6.12. (**a**) Massive boulder form of *Montastraea faveolata*, the typical morphology on the coral reef cap; (**b**) At depths below around 30 m, many coral species exhibit a plating morphology in order to maximize the surface exposed to sunlight (Photo credit: G.P. Schmahl)

6.3.2.2   Coral Community Zone

This zone is comprised of areas that, while not considered to be "true" coral reefs (where the primary reef structure is composed of reef building corals), do contain hermatypic coral species at low densities, or are characterized by other coral reef associated organisms, such as the hydrozoan *Millepora* spp. (fire coral), sponges and tropical macroalgae. Coral communities are found in depth ranges similar to those that contain coral reefs (18–50 m), where other environmental factors have not allowed full development of reef building species to occur. The "Coral Community" includes the "*Millepora*-Sponge" zone described by Rezak et al. (1985), and also includes some other coral associated assemblages. The most distinctive habitat type in this zone is the *Millepora*-sponge community that characterizes the shallowest peaks of the mid-shelf reefs at Stetson and Sonnier Banks (Fig. 6.15). The fire coral, *Millepora*, can account for up to 30% of the benthic cover on the pinnacles of Stetson Bank (Bernhardt 2000). In addition to fire coral, sponges comprise up to an additional 30% of the substrate.

FIG. 6.13. Extensive fields of *Madracis mirabilis*, the yellow pencil cover portions of coral, several flanks of the East and West Flower Garden Bank (Photo credit: FGBNMS/G.P. Schmahl)

FIG. 6.14. McGrail Bank is dominated by colonies of the blushing star coral, *Stephanocoenia intersepta*, reaching 2–4 m in size. This photograph was taken at approximately 43 m depth by a ROV mounted camera system (Photo credit: FGBNMS/NURC-UNCW)

The Coral Community zone also includes habitats that are characterized by scattered occurrences of stony corals or fire coral at relatively low densities. This habitat, called the low-density coral habitat, also includes a mix of other components including leafy algae, coralline algae, and sponges (Fig. 6.16). Habitats within the Coral Community can also be characterized by algae or sponges when they

Fɪɢ. 6.15. Multiple reefs and banks in the northwestern Gulf of Mexico harbor significant coral communities, including Stetson Bank, pictured here. Large pinnacles dominated by *Madracis decactis* provide habitat for thousands of fish, including, pictured, creole fish (*Paranthias furcifer*), blackbar soldierfish (*Myripristis jacobus*), sergeant majors (*Abudefduf saxatilis*), and graysby (*Cephalopholis cruentatus*). Spiny lobster (*Panulirus argus*) also find refuge amongst the coral formations (Photo credit: FGBNMS/E.L. Hickerson)

Fɪɢ. 6.16. Geyer Bank is an example of a coral community dominated by the *Millepora*/sponge habitat type. *Millepora alcicornis* (firecoral), sponges (*Xestospongia muta* and an unidentified species) and leafy algae (*Sargassum* sp. and *Lobophora*) dominate the landscape in this image. This type of habitat supports high numbers of an assortment of reef fish, including reef butterflyfish (*Chaetodon sedentarius*), spanish hogfish (*Bodianus rufus*), bluehead wrasse (*Thalassoma bifasciatum*), sunshinefish (*Chromis insolata*), bicolor damselfish (*Stegastes partitus*), rock beauty (*Holacanthus tricolor*), yellowtail reeffish (*Chromis enchrysura*), and cherubfish (*Centropyge argi*) (Photo credit: FGBNMS/G.P. Schmahl)

dominate a particular area, although small percentages of coral species also occur.

6.3.3.3   Coralline Algae Zone

This zone is characterized by crustose coralline algae that actively produce carbonate substrate, including rhodoliths, or algal nodules. The Coralline Algae zone is consistent with that designated as the "Algal-Sponge zone" by Rezak et al. (1985), but includes additional habitat. This zone extends from 45 m to over 90 m in depth and includes both the algal nodule habitat and rocky outcrops where coralline algal crusts cover a substantial percentage of the hard substrate. This is the largest reef-building zone by area in the Flower Garden Banks. Leafy algae are abundant in this zone to depths of at least 70 m. Algal nodules, or rhodoliths, are formed by species of coralline algae that lay down successive, concentric layers of carbonate around an initial "nucleus" (such as a rock fragment) to form irregular spheres 1 cm to over 20 cm in size. Between 50 m and 75 m, the nodules can cover 60–90% of the bottom (Minnery 1984) and can often occupy 100% of the sea floor in some areas (Fig. 6.17). Primary species include the coralline algae *Lithothamnium* sp., the squamariacean *Peyssonnelia* sp. and the encrusting foraminiferan *Gypsina plana*. Several species of hermatypic corals are scattered throughout the algal nodule zone, and can be locally abundant, including saucer shaped specimens of *Agaricia* spp. and *Leptoseris cucullata*. Leafy algae and sponges, most notably the toxic sponge *Neofibularia nolitangere*, are also common in this habitat. The Coralline Algae zone also includes deepwater coralline algal reefs, which are typically low-relief (1–2 m high), flat-topped rocky outcrops, ridges and patch reefs. While coralline algae is the dominant benthic group on these reefs, the rocky outcrops provide habitat for a variety of gorgonians, antipatharians, sponges and other organisms (Fig. 6.18). This zone corresponds with the area called "partly drowned reefs" by Bright and Pequegnat (1974) and Bright et al. (1985). Since the concept of "drowned reef" implies certain geological origins and temporal history, this terminology is not used here in relation to present-day biological communities. In fact, Bright et al. (1985) defined "partly drowned reefs " as reef structures below the depths of hermatypic corals, but within a depth range favoring crustose coralline

FIG. 6.17. A field of algal nodules within the coralline algae zone. Coralline algae and encrusting sponges are noticeable components of these formations. Photograph taken by ROV mounted camera at 67 m (220′) depth (Photo credit: FGBNMS/NURC-UNCW)

FIG. 6.18. Coralline algae patch reef habitat within the coralline algae zone, photographed at 81 m (266'). The ledge is encrusted with coralline algae, gorgonians (*Hypnogorgia* sp., *Ellisella* sp.), antipatharians (*Stichopathes* sp.), and a mosaic of sponges (Photo credit: FGBNMS/NURC-UNCW)

algae. This is consistent with the concept as used in the present classification.

6.3.3.4 Deep Coral Zone

The deep coral zone is consistent with what Bright et al. (1985) called the "drowned reef zone", and includes the "Antipatharian transitional zone" of Rezak et. al (1985, 1990). This zone occurs in water depths below that which support active photosynthesis by coralline algae (90 m and greater). Solitary corals and deepwater branching corals, such as *Madrepora* and *Oculina* are also found in this zone. The deep coral zone is characterized by a diverse assemblage of antipatharian and alcyonarian corals, crinoids, bryozoans, sponges, azooxanthellate branching corals and small, solitary hard corals (Fig. 6.19). It includes both low and high relief rock outcroppings of various origins. Rock outcrops are often highly eroded, and lack coralline algal growth. Reef outcrops may be covered with a thin layer of silt in areas subject to frequent resuspension of sediments. This area of high sediment resuspension and turbid water was identified as the "Nepheloid" zone by Bright et al. (1985) and Rezak et al. (1985). Since this terminology refers

to a physical oceanographic condition and not a biological classification, it is not used here.

Biological habitat maps utilizing the classification system described in this section are presented for East Flower Garden (Fig. 6.20a), West Flower Garden (Fig. 6.20b) and Stetson Banks (Fig. 6.21c).

6.4 Environmental Factors Influencing Reef Biology

The coral reefs and coral communities of the Flower Garden Banks region are bathed in clear, warm water entering into the Gulf of Mexico via the Loop Current. Winter water temperatures fall to around 18°C (65°F). The summer season occasionally brings hurricanes that sweep across the Gulf of Mexico, usually from the south or east. In 2005, Hurricane Rita's track brought the storm to within 50 miles of the East Flower Garden Bank, bringing waves reaching almost 10 m (30') in height. Hurricane Rita's landfall on the Texas /Louisiana border resulted in a large body of discolored water moving south from the coast over the FGBNMS

FIG. 6.19. Photograph at 108 m (355 ft) depicting a deep coral community. In this image is a soft coral, *Chironephthya caribaea*, octocorals: *Nicella deichmani*, *Ellisella* sp., *Caliacis nutans*, *Scleracis* sp., and antipatharians: *Antipathes furcata*, *Elatopathes abientina*, *Phanopathes expansa*, and *Tanacetipathes tanacetum* (Photo credit: FGBNMS/ NURC-UNCW)

(Fig. 6.21). It has not been determined if contaminants were present in this water mass, or what effect it may have had on the coral reefs. Hurricane Rita also caused some physical damage to the reefs of the Flower Garden Banks. The elevated sea state resulted in mass movement of sand that blanketed some coral heads or exposed previously covered reef structure (Fig. 6.22). Additionally, the storm resulted in injury of large specimens of *Xestospongia muta* (barrel sponges), and the toppling and movement of some coral colonies up to 3–4 m in diameter (Fig. 6.23). Large areas of the delicate finger coral *Madracis mirabilis* sustained considerable impacts.

Typical maximum summer water temperatures reach around 30°C (86°F). Historically, severe coral bleaching resulting in significant mortality was considered a rare occurrence at the FGBNMS. However, in the summer and fall of 2005, prolonged elevated water temperatures caused the most pronounced coral bleaching event on record at the Flower Garden Banks. By October 2005, FGBNMS surveys documented that approximately 45% of individual coral colonies to 29 m (95′)

depth were affected by bleaching to some degree (Fig. 6.24), ranging from partially to completely bleached. When considered from a percent cover perspective, up to 7% of the living coral tissue was categorized as bleached in November 2005 (Precht et al. 2006). Most severely affected species were *Montastraea cavernosa* and *Millepora alcicornis*. Fortunately, by March 2006, only 4–5% of coral colonies were still affected by the bleaching event. Although most species seemed to recover from the bleaching event, it appears that the fire coral, *Millepora alcicornis*, experienced significant mortality.

The reefs of the Flower Garden Banks have historically been relatively unaffected by coral disease. Although coral anomolies have been occasionally reported (Borneman and Wellington 2005), there have been no documented incidences of significant mortality associated with disease. During the early months of 2005 however, researchers were alarmed to find the widespread occurrence of symptoms that were consistent with that described as "plague-like" coral disease (Fig. 6.25). This phenomenon affected multiple colonies and multiple

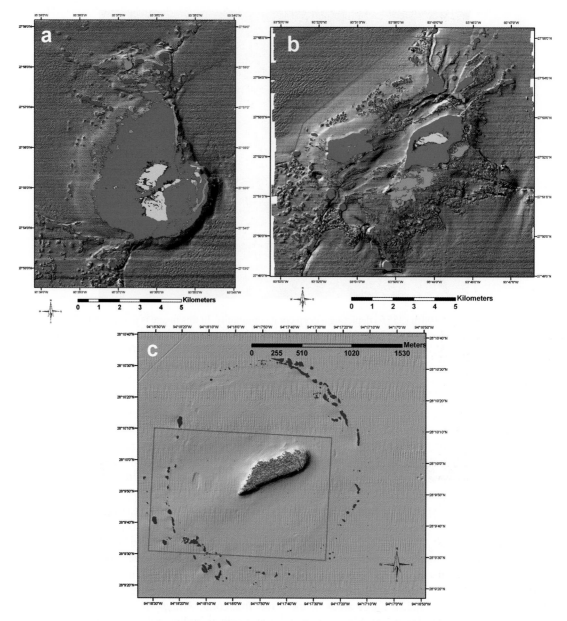

Fɪɢ. 6.20. Biological Zone and habitat classification charts of (**a**) the East Flower Garden Bank, (**b**) West Flower Garden Bank, (**c**) Stetson Bank. Sanctuary boundaries (red line); Coral reef zone (yellow); Coral community zone (pink); Coralline algae zone, containing two habitat types: algal nodules (red) and coralline algae reefs (green); and Deep coral zone (blue). These zones and habitats are persistent throughout the northwestern Gulf of Mexico reefs and banks (Image credit: FGBNMS/D.C. Weaver)

species, and occurred throughout the reef. This was the first time that such an occurrence of this magnitude had been documented at the FGBNMS. It was an unusual event, as typically, many types of coral disease are more pronounced during the warmer, summer months, whereas this outbreak occurred during the winter months. The outbreak appeared to slow down as water temperatures increased, but was once again active in the winters of 2006 and 2007. The specifics on the type of disease and impact on the reef are still under investigation.

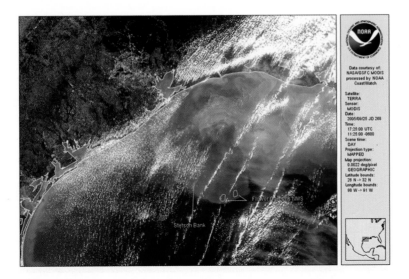

FIG. 6.21. Plume of discolored water resulting from landfall of Hurricanes Katrina and Rita in 2005 (Photo credit: NASA/GSFC MODIS/NOAA Coast Watch)

FIG. 6.22. Hurricane Rita was the cause of this major scouring around the base of a coral outcrop at the East Flower Garden Bank (Photo credit: Joyce and Frank Burek)

The high coral cover, large colonies, and healthy condition of the coral reefs of the Flower Garden Banks result in a spectacular mass spawning event, 7–10 nights after the full moon in August or September (Fig. 6.26). This mass spawning event is considered by many recreational divers to be the most visually impressive coral spawning event in the Caribbean. Other coral reef inhabitants, such as certain species of sponges, fish, and brittle stars, also spawn during this time. More detail on this event is included as a case study at the end of this chapter.

Fɪɢ. 6.23. In 2005, Hurricane Rita caused physical damage to the coral reef. Colonies up to 4 m across were shifted during the disturbance (Photo credit: Joyce and Frank Burek)

Fɪɢ. 6.24. A major bleaching event was documented during the fall of 2005. Up to 45% of the coral colonies were affected by coral bleaching to some degree (Photo credit: Joyce and Frank Burek)

6.5   Present Status of Reef Health

6.5.1   Monitoring Results 2004–2005

Long-term monitoring of the coral reefs of the FGBNMS has occurred annually since 1988, with intermittent monitoring events dating to 1972. Gittings (1998) reviewed the results of the monitoring data up until that time and found remarkable stability of the coral reef community. Subsequent monitoring studies support the continuation of this trend (Dokken et al. 2003; Precht et al. 2006). In spite of the fact that coral reefs in the Caribbean and western Atlantic have declined significantly in the last 3 decades (Gardner et al. 2003), the coral reefs of the Flower Garden Banks

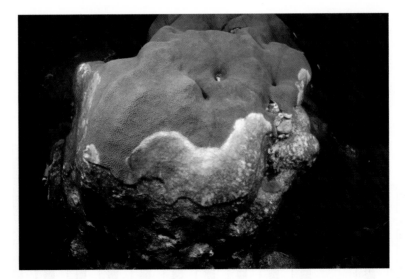

FIG. 6.25. Possible "plague-like" coral disease affecting a massive *Montastraea faveolata* colony at the East Flower Garden Bank in 2005 (Photo credit: FGBNMS/G.P. Schmahl)

FIG. 6.26. *Montastraea franksi* releasing gametes into the water during a mass spawning event at the Flower Garden Banks. A ruby brittle star, *Ophioderma rubicundum* takes advantage of the opportunity to collect bundles for consumption (Photo credit: FGBNMS/E.L Hickerson)

have shown no appreciable reduction in coral cover or in certain aspects of coral vitality during the monitoring period (Fig. 6.27) (Hickerson and Schmahl 2005).

Monitoring results for 2004–2005 continue to highlight the relative "health" of these reefs, expressed as consistently high coral cover, with a mean of 57.14% for both banks and both years, as well as the continuing trend of coral growth seen in repetitive photographic quadrats and lateral growth of individual colonies of the brain coral *Diploria strigosa*. Robust fish populations and oligotrophic water

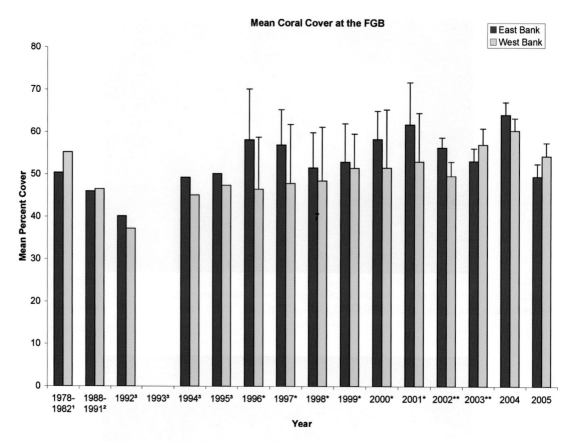

FIG. 6.27. Time series graph of mean coral cover at the East and West Flower Garden Banks as reflected through long term monitoring photographic data collection and analysis, 1978–2005 (Image credit: PBS&J)

conditions persisted while occurrences of disease and bleaching were low, ranging from 0–0.50% for both banks in both years. Sea urchins continued to occur at low densities, averaging 0.033/m^2 (both banks, both years). Herbivorous fishes continue to be present in large numbers and appear to keep algal cover under control as they represented the largest fish guild on both banks for both years.

Random video transect results revealed high coral cover at both banks, consistent with previous monitoring results, with 61.95% coral cover in 2004 and 51.4% coral cover in 2005. Macroalgal cover was also stable at ~15%, but higher at East Bank in 2005, and showed a significant site by year interaction ($P<0.001$). Crustose corallines, turf and bare (CTB), the largest cover category after coral cover, averaged ~15% for 2004 and 2005, and also showed a significant site by year interaction ($P = 0.012$).

The *Montastraea annularis* complex was the predominant component of coral cover at both banks in both years. Due to difficulty in differentiating the three species of the complex in the videographic images, *M. annularis*, *M. faveolata* and *M. franksi* were combined. In 2004, cover for this component group at the East Bank was 30.15±8.06% and 26.8±5.15% in 2005. At the West Bank, cover was reported at 31.67±6.1% in 2004 and 36.21±5.21% in 2005. *Diploria strigosa* was the next most abundant species, ranging from 12.13±2.82% at the West Bank in 2004 to 5.87 ±1.28% in 2005. The East Bank estimates were 13.41±1.74% and 6.68±1.29% in 2004 and 2005, respectively.

Repetitive quadrats were photographed to monitor 8 m^2 areas and their coral communities over time. Repetitive quadrats showed changes in coral species cover and coral condition (disease, paling, bleaching, and fish biting) from 2004 to 2005. The incidence of disease, paling and bleaching were low at both banks in both years; none of these metrics was above 0.50%, and there was no evidence of disease in any of the repetitive quadrats analyzed. Planimetry results of select colonies within repetitive quadrats showed an increase from 2004–2005 at both banks. Nine deep repetitive quadrats (32–40 m depth) were established on the East Bank in April 2003 and photographed in June 2005. Coral cover was high, at 76% overall at these deeper stations. The *Montastraea annularis* complex and *M. cavernosa* were the dominant species at these sites. Algae cover was quite low and CTB was high.

Lateral growth stations were monitored to measure changes in *Diploria strigosa* colonies. *Diploria strigosa* is important at the FGB because it is the largest contributor to coral cover after the *M. annularis* complex within the 100 m by 100 m monitoring sites. Overall there was a 35% increase in *Diploria strigosa* margins from 2004 to 2005. Sclerochronology was used to measure the accretionary growth rates of *Montastraea faveolata*. Cores taken at both banks revealed annual growth bands spanning 1992–2005. Yearly growth rates ranged from 2.75 mm to 14.54 mm between banks. Interestingly, a disruption in accretion was seen in three quarters of the samples from both banks centered in and around the 1997–1998 growth band (the year of widespread bleaching throughout western Atlantic coral reefs).

The FGB coral reefs remain in good condition and are healthy in comparison to reefs throughout the region. This may be due in part to their remote location. Continued monitoring of these reefs will document their long-term condition and be useful for studies focused on the dynamics of the robust benthic communities and the fish populations they support.

6.5.2 2005 Hurricane Impacts

Hurricane Rita, a Category 3 storm (Saffir–Simpson Index), passed within 50 miles of the Flower Garden Banks on September 23, 2005. Soon after the hurricane passed, FGBNMS staff rapidly assessed damage at the Banks and found large dislodged coral heads, gauged and damaged corals from waterborne projectiles, displacement of sand and

sediment, scouring of coral heads and an active and ongoing coral bleaching event. On November 13, 2005 a team was assembled to assess hurricane damage to the 100 m × 100 m monitoring site at the East Flower Garden Bank. Prior to the passing of Hurricane Rita, during the summer of 2005, seawater temperatures were elevated and coral bleaching was evident at the FGB and throughout much of the Caribbean. Both results of the hurricane damage and the thermal anomaly are reported from repetitive quadrat, video perimeter, and water quality analyses.

Eighteen percent of the East Bank repetitive quadrats photographed lost coral colonies due to the hurricane. These quadrats also showed ~7% bleaching of the coral cover. This is the highest level of bleaching reported since the bleaching event of 1990, when 5% of corals at the East Bank were bleached (Hagman and Gittings 1992). Bleaching was evident on most coral species, but most prevalent on the *Montastraea annularis* complex, *M. cavernosa*, and *Millepora alcicornis*. No corals at the deep stations were dislodged or missing. Deep site repetitive quadrats also exhibited bleaching, although to a lesser extent than the shallower sites (~2%). Water quality results showed that the reef cap experienced elevated water temperatures for 50 days and that the passage of Hurricane Rita brought relief in terms of lower water temperatures to the reef.

6.5.3 Human Impacts and Conservation Issues

The coral reefs of the Flower Garden Banks are fortunate in that their offshore location and depth have isolated them somewhat from many human-related impacts that affect so many other coral reefs in the world (Deslarzes and Lugo-Fernandez 2007). The reefs are generally outside of the range of direct coastal influences, such as nearshore runoff and sedimentation. However, they are still within the distal influence of run-off from the Atchafalaya–Mississippi Rivers that can cause variability in salinity but no variability in nutrient levels, since by the time the water reaches the Flower Garden Banks, its nutrients are already depleted (Deslarzes and Lugo-Fernandez 2007). Deslarzes and Lugo-Fernandez (2007) interpret changes in fluorescence banding in corals as a signal for input of riverine water, and Dodge and Lang (1983) found a negative correlation between river discharge and coral accretion at the Flower Garden Banks.

Water depth protects the Flower Gardens Banks from significant short-term temperature fluctuations and vessel groundings that affect shallow water coral reefs. Since monitoring programs have been maintained, the Flower Garden Banks have exhibited less impact from common coral maladies such as coral bleaching and disease than many other coral reefs (Aronson et al. 2005). However, there have been recent incidences of both bleaching and disease that cause significant concern, as discussed in the previous section. Historically, the primary human-related impacts of concern have been vessel anchoring and potential threats from offshore oil and gas activity. The National Oceanic and Atmospheric Administration (NOAA) and the Minerals Management Service (MMS) have largely addressed these issues through various management actions and through the designation of the Flower Garden Banks National Marine Sanctuary in 1992. In 2006 the FGBNMS initiated a process to review and revise its management plan. Through this process, managers and the public identified a number of priority issues that threaten the health and vitality of the Flower Garden Banks and vicinity (NOAA 2006b). Human activities that were identified as priority management concerns include fishing, boating and diving, and pollutant discharge. There is also a significant interest in protecting additional coral reef and coral community habitat in the northwestern Gulf of Mexico, especially those areas that harbor unique biological assemblages or are connected ecologically to the Flower Garden Banks. Enforcement, education and outreach were identified as overarching management concerns. In addition to those impacts that are directly related to human use, a number of other important concerns were raised that may be indirectly linked to human activity. These include water quality in the northern Gulf of Mexico, coral bleaching, coral disease and invasive species.

In the early 1970s, concern about potential negative impacts posed by offshore oil and gas exploration and development provided impetus to consider management protection of the reefs and banks of the northwestern Gulf of Mexico in general and the Flower Garden Banks in particular. Consequently, in 1974, the MMS designated a number of "No-Activity Zones" around most of the significant topographic features that prohibited direct oil and gas exploration activity (MMS 2007). In addition to the no-activity zones, a 4 mile buffer zone was created around the Flower Garden Banks within which oil and gas operations were required to comply with additional requirements to protect coral communities. One of the most important of these is the requirement to direct all drilling discharges through a "shunt" pipe to within 10 m (33′) of the seafloor instead of the standard practice of surface discharge. This significantly reduced the potential of sedimentation on nearby coral reefs. Mitigation of potential impacts related to oil and gas development is important due to the close proximity of significant levels of infrastructure and activities to the reefs of the northwestern Gulf of Mexico. For example, there is an operational natural gas production platform located less than 2 km from the reef cap at the East Flower Garden Bank (Fig. 6.28).

Furthermore, NOAA, through the Gulf of Mexico Fishery Management Council and the National Marine Fisheries Service, implemented a coral reef fishery management plan in 1982 that prohibited the direct take of stony corals and designated the Flower Garden Banks a "Habitat Area of Particular Concern" (HAPC), prohibiting anchoring of fishing vessels over 100 ft in length and other fishing activities. In spite of these management actions, injury from vessel anchoring and other impacts continued to occur, due to the fact that the regulations only applied to vessels engaged in oil and gas exploration or fishing activities. Others, such as freighters or recreational vessels, were not addressed. Recreational divers joined together in 1988 to install a number of mooring buoys at the reef to eliminate the need for dive boats to anchor at the Flower Garden Banks, but occasional incidents of freighter anchoring continued. The designation of the FGBNMS in 1992 carried with it a general prohibition on anchoring for all vessels greater than 100 ft in length. In 2001 these provisions were strengthened by the FGBNMS through the prohibition of anchoring by any vessel regardless of length. This prohibition mirrored a similar action implemented by the International Maritime Organization (IMO) in 2000, the first such anchor prohibition by the IMO for the purpose of protection of coral reefs (Johnson 2002).

FIG. 6.28. A natural gas production platform, High Island A389A, located within the boundaries of the East Flower Garden Bank, less than 2 km from the coral reef crest (Photo credit: FGBNMS/G.P. Schmahl)

The primary human related activities that may pose detrimental impacts to the coral reefs of the northwestern Gulf include fishing, diving and pollutant discharge. The potential impacts of fishing are addressed in the following section. Visitation by SCUBA divers is estimated to be between 2,500 and 3,000 divers/year (Gulf Diving LLC, personal communication 2007). Although relatively low, this number is likely to increase as the FGBNMS is becoming internationally known as a prime dive destination. Most divers access the reefs through commercial dive charter operations. Spearfishing and all other types of direct taking of marine organisms by divers are prohibited by sanctuary regulations. Nondirect impacts from divers can include accidental contact with living coral, deliberate contact with coral features (for stability or orientation), physical contact with marine animals (sea turtles, manta rays, whale sharks), harassment of marine animals (pursuit, underwater lights, flash photography) and the transmission of disease-causing pathogens from contaminated dive gear. Discharge of pollutants from sources inside and outside the sanctuary may have potential detrimental impacts on the reefs of the Flower Garden Banks. Within the FGBNMS, sewage discharge that passes through a US Coast Guard approved marine sanitation device (MSD) is allowed. Most MSD's do not remove nutrients or other possible pollutants that may be harmful

to coral reefs. Further, impacts from an oil spill or other hydrocarbon release is an ongoing concern. Major oil spills in the Gulf of Mexico have been very rare in recent years, but could cause significant injury if one did occur in the vicinity of sensitive coral communities. Ongoing discharges from oil and gas facilities include drilling lubricants and cuttings, produced water (water separated from oil and gas during production), and operational effluents (treated and untreated sewage, gray water and deck wash).

The decline of water quality in the northern Gulf of Mexico is a cause of significant concern, and is epitomized by the development of hypoxic areas of varying size each year, known as the "Dead Zone" (Rabalais et al. 2002). Scientific investigations indicate that oxygen stress in the northern Gulf of Mexico is caused primarily by excess nutrients delivered to Gulf waters from the Mississippi–Atchafalaya River drainage basin, in combination with localized nearshore stratification (Dodds 2006). Over 40% of the landmass of the contiguous United States drains into the Mississippi River watershed. Predominant current patterns direct much of this water away from the FGBNMS, but minor alterations in circulation patterns could result in contaminated water in the vicinity of the coral reefs. For example, in 2005, the combination of hurricanes Katrina and Rita resulted in a large plume of highly turbid water passing far offshore in the Gulf of Mexico, and over the Flower Garden Banks.

In 2002, the invasive coral species *Tubastraea coccinea* (orange cup coral – Fig. 6.29) was documented for the first time on natural substrates at the East Flower Garden Bank. This species is native to the Indo-Pacific and entered the western Atlantic and Caribbean region several decades ago and has expanded throughout the region (Fenner and Banks 2004). It may have arrived by attaching to a ship's hull or by its larvae being discharged in ships ballast water. The orange cup coral is common on oil and gas platforms in the northern Gulf of Mexico, and it is suspected that such artificial structures have played a major role in the spread of this species. In 2004, over 50 colonies of this invasive species were removed from natural substrates at Geyer Bank by FGBNMS staff. In May 2007, up to 100 colonies were observed on the same peak at Geyer Bank, and two colonies were documented at Sonnier Bank. In 2006, a species of nudibranch also native to the Indo-Pacific was photographed mating at Stetson Bank. It is clear that invasive species may become an important management concern in the future.

6.5.4   Fisheries Issues

There is a general lack of data to assess the status of local fisheries in the specific area of the Flower Garden Banks. Fishery statistics are maintained for the Gulf of Mexico, but not on a fine enough scale to analyze fish population status for individual reefs and banks. The fisheries of the Flower Garden Banks must therefore be viewed against the backdrop of the entire northern Gulf of Mexico. Primary species of importance to the fishery that utilize the Flower Garden Banks include reef fish within the snapper /grouper complex, including red snapper, vermilion snapper, deepwater groupers (yellowedge, snowy, speckled hind, and Warsaw) and shallow water groupers (gag, scamp, yellowfin, yellowmouth, black, rock hind, and red hind). A number of sharks and other pelagic fish, such as wahoo, mackerel and greater amberjack; and other reef fish, such as gray triggerfish, are also sought after species. In the Gulf of Mexico, many of the fish stocks have yet to be adequately assessed. For those stocks that have been assessed, it has been determined that goliath grouper has been subject to historical overfishing, gag grouper are undergoing overfishing, and the status of yellowedge grouper

FIG. 6.29. A newly recruited colony of an invasive species of orange cup coral (*Tubastraea coccinea*) at the East Flower Garden Bank (2006) (Photo credit: Joyce and Frank Burek)

is undetermined (NOAA 2006a). Red snapper has been overfished since the 1980's. Both red grouper and vermilion snapper have been categorized as subject to overfishing in the past, but have recently been removed from that designation. The greater amberjack and gray triggerfish are both overfished. The status of the large coastal shark complex is unknown, but one species, the dusky shark, has been determined to be subject to overfishing. NOAA Fisheries Office of Protected Species has designated the Nassau grouper, Warsaw grouper, speckled hind, dusky shark and sand tiger shark "species of special concern". Nassau and goliath grouper are very infrequent inhabitants of the Flower Garden Banks vicinity.

The status of "natural" populations of reef fish prior to the impacts of fishing is difficult to assess, because long-term data for most species do not exist. However, when historical data relating to the shark fishery in the Gulf of Mexico were analyzed, it was found that oceanic whitetip and silky sharks had declined by 99% and 90% respectively since the 1950s (Baum and Myers 2004). Neither of these species comprise a significant component of the shark fishery in recent years, and it is surprising to most that they were once commonly caught. The lack of recognition that the oceanic whitetip shark was the most prevalent species in the Gulf of Mexico at one time is clear evidence for the concept of "shifting baselines", and renders the true assessment of the impact of fishing on local fish populations very difficult. It is possible that species that are extremely rare at the Flower Garden Banks in recent years, such as the Nassau and goliath groupers, may have once been more abundant prior to fishing pressure.

Fishing is a very valuable component of the economy in the Gulf of Mexico. In 2005, the value of all species of fish landed in the Gulf exceeded $621 million, including $172.3 million in the state of Texas and $253.7 million in Louisiana (NOAA 2006a). The value of fish species landed in 2005 that are typically caught in the vicinity of the reefs and banks of the northwestern Gulf of Mexico (greater amberjack, Warsaw grouper, yellowedge grouper, king mackerel, scamp, blacktip shark, sharks (general), red snapper, vermilion snapper, tilefish, bluefin tuna and yellowfin tuna) totaled more than $7.5 million for the state of Texas and $14.5 million for Louisiana. Of these species, red

snapper, gag grouper and yellowfin tuna were the highest valued fisheries. Five of the top 20 US ports in terms of commercial fishery landings are located in the Gulf of Mexico. These include Chauvin, Venice and Intracoastal City in Louisiana, and Brownsville and Port Arthur in Texas.

NOAA is responsible for managing fisheries in US waters, and implements this through the Fisheries Service and the Gulf of Mexico Fishery Management Council (GMFMC). Fishery Management Plans (FMP's) have been promulgated for Reef Fish (including the snapper /grouper complex), Migratory Pelagics (mackerels and jacks) and Coral. The Coral Fishery Management Plan, developed in 1982, prohibits the take of stony corals or sea fans and sets strict limits on the harvest of octocorals. As part of the Coral FMP the Flower Garden Banks were designated as a "Habitat of Particular Concern" (HAPC), and within this area the use of bottom longlines, traps, pots, and bottom trawls, and the anchoring of vessels greater than 100 ft in length were prohibited. In 2006, the GMFMC implemented a generic amendment to all relevant fishery management plans to address "Essential Fish Habitat" in the Gulf of Mexico. This amendment designated four areas in the northwestern Gulf of Mexico that contained significant coral reefs and coral communities as HAPC's. These include the East and West Flower Garden Banks (expansion of the previously designated areas), Stetson Bank and McGrail Bank. Within these HAPC's regulations were established that (1) prohibit anchoring to protect coral reefs, and (2) prohibit the use of trawling gear, bottom longlines, buoy gear and all traps and pots within these areas.

Due to the distance offshore, the reefs and banks of the northwestern Gulf do not routinely receive intense levels of fishing effort. However, the areas are relatively small and prime habitats are concentrated in defined areas so that targeted fishing pressure could have a significant impact. Specific fishing-related impacts have been observed or reported within the FGBNMS and nearby reefs and banks. Some of these issues include: high levels of fishing effort within limited areas, damage to the coral reef and entanglement of marine organisms related to discarded fishing gear, targeted removal of rare species, fishing possible spawning aggregations, dumping of bycatch, mortality of non-target species, anchor damage, and poaching. There is evidence that areas of the FGBNMS and other reefs and banks are being utilized as aggregation sites and possibly spawning areas for deeper water species such as scamp and marbled grouper. Sanctuary staff have also witnessed entanglement of a loggerhead sea turtle in fishing line, accumulations of shrimp bycatch at Stetson Bank, dead sharks in reef areas, and evidence of taking of spiny lobster and queen conch.

6.6 Case Studies

6.6.1 Coral Spawning

The coral reefs of the Flower Garden Banks National Marine Sanctuary produce one of the most visually impressive mass coral spawning events in the Caribbean. This is due to the high density and cover of the broadcast spawning corals *Montastraea, Diploria* and *Colpophyllia*. Multiple species of coral and other organisms release gametes into the water over a period of several days, at specific times, separated by species, and sometimes by sex. This event typically occurs 7–10 days after the August full moon. However, in the event a full moon occurs in early August or late July, or if two full moons occur, mass spawning can happen in September instead of, or in addition to August. In some years, both months have culminated in a prolific spawning event. Since 1990, Flower Garden Banks researchers have been witnessing and documenting the mass coral spawning event, accumulating more precise data on timing and participant species. A summary of these observations is presented in Table 6.4.

Coral spawning at the FGB was first reported by a recreational diver in 1990. The coral spawning event has since been observed and documented by the Flower Garden Banks Research Team, and, since 1997, by a team of scientists led by Dr. Peter Vize (University of Calgary). The first published account of coral spawning was written by Dr. Tom Bright in 1991, and was published in a popular magazine, *Texas Shores*. Since then, numerous scientific and popular publications have been produced reporting the spawning event (Bright 1991; Bright et al. 1992; Gittings et al. 1992, 1994;

Hagman 1996; Boland 1998; Hagman et al. 1998a, b; Lugo-Fernandez et al. 2001; Vize 2006; Vize et al. 2005).

The primary mass spawning event typically begins on the 7th night after the full moon, in August (and/or September), with the release of gametes by the giant star coral *Montastraea cavernosa*, beginning shortly after sunset, at around 8:30 p.m. CST (although release has been observed as early as 6:30 p.m. CST). This species has separate sexes with male and female colonies releasing gametes within a similar time frame (Fig. 6.30). That same evening *Montastraea franksi* and *Diploria strigosa* release gametes beginning around 9:00 p.m. Both of these species are hermaphroditic, with eggs and sperm packaged together in bundles (Fig. 6.31). *M. franksi* can display a particularly spectacular reproductive behavior, where gamete release occurs in a "wave-like" progression over a colony (Fig. 6.32). After about 10:00 p.m., male and female colonies of the blushing star coral (*Stephanocoenia intersepta*) begin releasing their sperm and eggs. The females release eggs that look like champagne bubbles to a diver, and as the eggs are released, the chocolate brown tentacles of the coral are withdrawn, and the colony appears to "blush", revealing the white color of the underlying skeleton.

On the 8th night after the full moon, the sequence of spawning of the previous evening repeats itself, but with significantly more colonies releasing gametes. This is the peak night of the coral spawning event, and has often been described as witnessing an "underwater snowstorm". Also on the 8th night, the smooth star coral, *Montastraea faveolata*, releases its gamete bundles beginning soon after 10:00 p.m. The separation in the timing between *M. franksi* and *M. faveolata* demonstrates differing reproductive strategies of these two species.

Corals are not the only group to spawn at this time of year. Also on the 8th night after the full moon, beginning at around 8:30 p.m., ruby brittle stars (*Ophioderma rubicundum*) release their gametes. The males aggregate in piles of up to a dozen or more on top of coral heads (Fig. 6.33) for a synchronized release of red smoky sperm. Near the same time, individual female brittle stars climb to the tops of coral heads and stand on the very tips of their arms, releasing bright red eggs (Hagman and Vize 2003). Not yet well understood or well documented is what appears to be a mass spawning

TABLE 6.4 Spawning predictions based on historical records of intensity and timing of mass coral spawning event at the Flower Garden Banks. Columns refer to the number of nights after the full moon in August or September. Numbers refer to the time of day (local time) on a 24 hour clock, and the shading reflects the relative intensity of spawning activity

Species	6th night	7th night	8th night	9th night	10th night
Montastraea cavernosa female	1950–2115	2059–2138	2040–2200	2104–2230	
Montastraea cavernosa male	2100>	1830–2127	<2030–2145	2140–2200	
Montastraea franksi	2145–2210	2100–2245	2030–2245	2100–2250	
Diploria strigosa		2100–2215	1935–2230	1920–2230>	
Stephanocoenia intersepta male		2250	2208–2300	2200–2240	2215–2245
Stephanocoenia intersepta female		2200–2220	2218–2300	2200–2300	2215–2245
Montastraea faveolata		2221–2320	2113–2330	2300–5660	
Colpophyllia natans				2020–2132	2030–2110
Montastraea annularis		2223–2256	2246		2230
Xestospongia muta female	0850–0950	0840		0830–0950	
Xestospongia muta male	0915	0815–0820	0840	0830–0900	
Encrusting sponges				0915	
Ophioderma rubicundum male			2030–2150	×	
Ophioderma rubicundum female			2030–2150	2100–2220	
Ophioderma squamississimum female		×	2045	2030–2130	
Ophioderma squamississimum male			2045	2030–2130	

FIG. 6.30. *Montastraea cavernosa* male releases sperm into the water column during the mass spawning event at the Flower Garden Banks. (Photo credit: FGBNMS/G.P. Schmahl)

event of certain sponges. On several occasions, on the morning of the 9th day after the full moon in August, at around 9:30 a.m., mass spawning of sponges has been documented. In 2006, in both August and September, the barrel sponge, *Xestospongia muta*, has been observed spawning in large numbers. Females forcibly release eggs into the water column through their excurrent canals (Fig. 6.34a), which, because they are negatively buoyant, end up in "drifts" around the sponge colony (Fig. 6.34b).

FIG 6.31. Egg and sperm bundles line the mouths of brain coral polyps (*Diploria strigosa*) in readiness for release during the mass spawning event at the Flower Garden Banks. (Photo credit: FGBNMS/G.P. Schmahl)

FIG 6.32. *Montastraea franksi* releases egg and sperm bundles into the water column, sometimes in a wave-like motion across the coral head, during the mass spawning event at the Flower Garden Banks. (Photo credit: FGBNMS/G.P. Schmahl)

Males can release so much gametic material that the local visibility in the water column is reduced from well over 30 m (100 ft) to 3 m (10 ft) in a short period of time.

On the 9th night after the full moon, gamete release by all of the previous species continues, but at greatly reduced levels, with the exception of the blushing star coral, *Stephanocoenia intersepta*, which appears to peak on this night. A second, much larger species of brittle star, *Ophioderma squamosissimum*, spawns on this particular evening, between 8:30 and 9:30 p.m.

FIG 6.33. An aggregation of male ruby brittle stars (*Ophioderma rubicundum*) cluster together on top of a brain coral for a synchronized release sperm during the mass spawning event at the Flower Garden Banks. (Photo credit: FGBNMS/G.P. Schmahl)

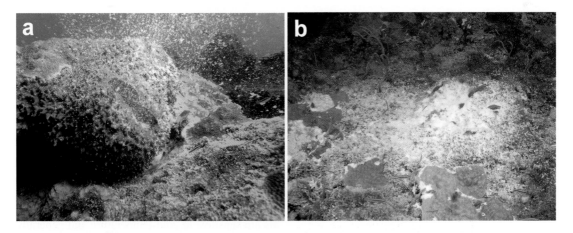

FIG 6.34. (**a**) A female barrel sponge, *Xestospongia muta*, releases eggs into the water column during a mass spawning event at the Flower Garden Banks. (Photo credit: FGBNMS/E.L. Hickerson) (**b**) A "drift" of eggs surrounds a female barrel sponge, *Xestospongia muta*, after a mass spawning event at the Flower Garden Banks (Photo credit: FGBNMS/G.P. Schmahl)

The final major episode of the annual mass coral spawning event occurs on the 10th night after the full moon. On this night, the brain coral *Colpophyllia natans* releases egg and sperm bundles beginning around 8:30 p.m. This appears to be the primary night for this species, and most of the other coral species, with the exception of *Stephanocoenia intersepta*, which

participates to a lesser extent than the previous night, appear to have completed their spawning effort.

Montastraea annularis also is a participant in the mass spawning event, however, it releases its gamete bundles intermittently throughout the spawning window, and the specific timing of this species is less obvious. The mass spawning event

FIG 6.35. FGBNMS Research Coordinator, Emma Hickerson, surfaces from a dive in a slick of orange coral gametes during a major mass coral spawning event at the Flower Garden Banks. (Photo credit: FGBNMS/G.P. Schmahl)

produces an enormous amount of the gametic material (Fig. 6.35). So far, no estimate has been attempted to quantify the biomass generated by this event. As odd as it seems, very little predation of the massive amount of gametic material has been observed. Besides the occasional observations of brittle stars (*O. rubicundum*) collecting gamete bundles with their arms as they retreat underneath a coral head, no other predation has been observed.

6.6.2 *Acropora palmata* at the Flower Garden Banks

The FGB are the northernmost reefs found in waters of the contiguous United States and are comprised primarily of a deeper-water (>20 m), *Montastraea*-dominated community. Living colonies of *A. palmata* have for the first time been reported by Zimmer et al. (2006) (Fig. 6.36). These discoveries are also the deepest reported records of *A. palmata,* occurring on the FGB to depths in excess of 23 m. It has been speculated that the absence of *Acropora* at the FGB is controlled primarily by: (1) the water column being periodically too cold during the winter months to allow colonies to establish and persist; (2) the tops of the banks being too deep for these shallow-

dwelling species to compete with better adapted, deeper-dwelling species; and (3) the remote and isolated nature of these banks requiring long-distance migration of viable larvae thus limiting the sexual recruitment potential of species such as the *Acropora* that broadcast their gametes into the water column.

Throughout the Late Quaternary, the two coral species of the genus *Acropora* have been the most important reef-builders in the Caribbean (Aronson and Precht 2001). Specifically, the branching elkhorn coral *Acropora palmata* was dominant in high-energy, shallow reef environments. These corals are sensitive to cold-temperature and generally do not occur in areas where the wintertime sea surface temperatures (SSTs) drop below 18°C. Along the Florida reef tract this element of the tropical reef-building community was historically absent north of Cape Florida (Miami). There is also no previous record of acroporid corals anywhere in the northern Gulf of Mexico.

Recent occurrences of *A. palmata* at the FGB could be in response to local increase in SST. In Florida, the fossil coral record reveals a precedent to a potential range expansion (Precht and Aronson 2004). During the Holocene Thermal Optimum (10,000–6,000 years ago), when SSTs were warmer than today, *Acropora*-dominated

Fig 6.36. This elkhorn coral (*Acropora palmata*), one of two known colonies at the Flower Garden Banks, was discovered by Beth Zimmer in 2005 (Photo credit: FGBNMS/G.P. Schmahl)

reefs were common along the east coast of Florida as far north as Palm Beach County (Lighty 1978; Lighty et al. 1978; Banks et al. 2007). The northern limits of the *Acropora* species subsequently contracted some 150 km south to Miami, in apparent response to climatic cooling in the late Holocene. Thus, the spatial response of the *Acropora* spp. to climate provides a context for interpreting their geographic distribution, past and present. While reefs at their latitudinal extremes have apparently responded rapidly to climate flickers, results from coring studies of Holocene reefs in the insular tropical Caribbean support the notion that tropical oceanic climates have been buffered from climatic variability (Gill et al. 1999).

In an open cave at a depth of approximately 21m, divers discovered a few meters of an exposed section of the ancient reef understory. Within this section, branches and trunks of fossil *A. palmata* were found in growth position. In addition, broken and transported *A. palmata* blades were found admixed in the adjacent rubble. This constitutes the first-ever fossil of *Acropora* reported from the FGB. Radiocarbon dating of one of the branches revealed a calibrated corrected age of 6,930–6,650 BP, which is within the window of dates recovered from the northernmost fossil *Acropora*-reefs

in Florida (ranging from 6,003 to 8,770 cal BP; Banks et al. 2007) and also corresponds with the later portion of the Holocene Thermal Optimum. Comparing the age of this sample with the corrected sea level curve of Toscano and Macintyre (2003) for the western Atlantic places this coral in approximately 10 m of water depth at the time of formation.

Recently (June 2007), evidence of an *A. cervicornis* fossil reef was discovered in growth position, at a depth close to 30 m. This may suggest that the development of the FGB reefs shows a deepening-upwards succession from a shallow reef community that evidentially lagged behind the rapidly rising sea level of the early-middle Holocene, drowned and was subsequently capped by the present reef community. While this discovery raises more questions than it answers, especially with regard to the local turn-on and turn-off mechanisms of the *Acropora* reef facies, it shows that *Acropora* species have a history of responding to local changes in environmental conditions throughout the Holocene. This discovery also confirms that reefs living under non-optimal conditions in more thermally reactive areas, including those in Florida and the northern Gulf of Mexico are more likely to show changes (range expansions and contractions) with climatic warming.

6.7 Conclusions

Hardgrounds in the Gulf of Mexico are home to rich and varied coral communities, such as those exemplified by the Flower Garden Banks, Stetson, Bright, Geyer, McGrail and Sonnier Banks. Only the Flower Garden Banks and McGrail Bank are "true" hermatypic coral reefs and as such are among the deepest z reefs in the USA dominated by zooxanthellate corals. The reef-related habitat can be differentiated into a "coral reef zone" that includes the actively accreting hermatypic coral assemblages, a "coral community zone" that contains hermatypic coral species at low densities, a "coralline algae zone" characterized by crustose coralline algae and rhodoliths, and a "deep coral zone" characterized by antipatharian and alcyonarian corals, crinoids, bryozoans, sponges, azooxanthellate branching corals and small, solitary hard corals. Although far from the mainland, the banks are still subjected to temperature and salinity variability controlled by mainland run-off and oceanographic conditions in the Gulf of Mexico. The banks provide essential habitat for a variety of fish species of commercial and recreational importance, and are increasingly being utilized by the recreational diving community, demonstrating the economic significance of the area.

References

Aronson RB, Precht WF (2001) White-band disease and the changing face of Caribbean coral reefs. Hydrobiologia 460:25–38

Aronson RB, Precht WF, Murdoch TJT, Robbart ML (2005) Long-term persistence of coral assemblages on the Flower Garden Banks, northwestern Gulf of Mexico : implications for science and management. Gulf Mex Sci 23:84–94

Badan A, Candela J, Sheinbaum J, Ochoa J (2005) Upper layer circulation in the approaches to Yucatan channel. In: Sturges W, Lugo-Fernandez A (eds) Circulation in the Gulf of Mexico : observations and models. Am Geophys Union, Washington DC, pp 57–69

Baum JK, Myers RA (2004) Shifting baselines and the decline of pelagic sharks in the Gulf of Mexico. Ecol Letters 7:135–145

Banks KW, Riegl B, Shinn EA, Piller WE, Dodge RE (2007) Geomorphology of the southeast Florida reef tract (Miami-Dade, Broward, and Palm Beach Counties, USA) Coral Reefs 26(3), DOI 10.1007/s00338–007–0231–0

Bernhardt SP (2000) Photographic monitoring of benthic biota at Stetson Bank, Gulf of Mexico. Thesis, Texas A&M University, College Station, TX, 73 pp

Biggs DC (1992) Nutrients, plankton and productivity in a warm-core ring in the western Gulf of Mexico. J Geophys Res 97:2143–2154

Bloom AL (1994) The coral record of late glacial sea level rise. In: Ginsburg RN (ed) Proc Colloquium on Global Aspects of Coral Reefs: Health, Hazards, and History. Rosenstiel School of Marine and Atmospheric Science, University of Miami, Miami, FL, pp 1–6

Boland GS (1998) Spawning observations of the scleractinian coral Colpophylia natans in the northwest Gulf of Mexico. Gulf Mex Sci XVI No. 2226

Borneman EH, Wellington GM (2005) Pathologies affecting reef corals at the Flower Garden Banks, northwestern Gulf of Mexico. Dedicated Issue, Flower Garden Banks National Marine Sanctuary, Gulf Mex Sci 25(1):95–106

Bright TJ (1977) Coral reefs, nepheloid layers, gas seeps and brine flows on hard-banks in the northwestern Gulf of Mexico. In: Taylor DL (ed) Proc 3rd Int Coral Reef Symp, Rosenstiel School Mar Atmos Sci, Miami, FL, 1:39–46

Bright TJ (1991) First direct sighting of star coral spawning. Texas Shores (Texas A&M University Sea Grant Program) 24:2

Bright TJ, Pequegnat LH (1974) Biota of the West Flower Garden Bank. Gulf Publishing Company, Houston, TX, 435 pp

Bright TJ, Pequegnat W, Dubois R, Gettleson D (1974) Baseline survey Stetson Bank Gulf of Mexico. Texas A&M University, College of Geosciences, College Station TX38 pp

Bright TJ, McGrail DW, Rezak R, Boland GS, Trippett AR (1985) The Flower Gardens: a compendium of information. OCS Studies/MMS 85–0024, Minerals Management Service, New Orleans, LA, 103 pp

Bright TJ, Gittings SR, Boland GR, Deslarzes KJP, C.L. Combs, Holland BS (1992) Mass spawning of reef corals at the Flower Garden Banks, NW Gulf of Mexico. Proc. 7th Int. Coral Reef Symp 1:500 (abstract)

Continental Shelf Associates (1985) Environmental monitoring program for Platform "A", Lease OCS-G-2759, High Island Area, South Extension, East Addition, Block A-389, near the East Flower Garden Bank. Rept. to Mobil Producing Texas and New Mexico,, The Woodlands, TX, 353 pp + appendices.

Dennis GD, Bright TJ (1988) Reef fish assemblages on hard banks in the northwestern Gulf of Mexico. Bull Mar Sci 43:280–307

Deslarzes KJP, Lugo-Fernandez A (2007) Influence of terrigenous runoff on offshore coral reefs : an example from the Flower Garden Banks, Gulf of Mexico.

In: Aronson RB (ed) Geological approaches to coral reef ecology. Springer, New York, pp 126–160

Dodds WK (2006)_ Nutrients and the "dead zone": the link between nutrient ratios and dissolved oxygen in the northern Gulf of Mexico. Frontiers in Ecology Environment 4(4):211–217

Dodge RE, Lang JC (1983) Environmental correlates of hermatypic coral (Montastrea annularis) growth on the East Flower Garden Bank, northwest Gulf of Mexico. Limnol Oceanogr 28:228–240

Dokken QR, MacDonald IR, Tunnell JW, Wade T, Withers K, Dilworth SJ, Bates TW, Beaver CR, Rigaud CM (2003). Long-term monitoring at the East and West Flower Garden Banks, 1998–2001: final report. US Dept. of the Interior, Minerals Management Service, Gulf of Mexico OCS Region, New Orleans, LA. OCS Study MMS 2003-031, 90 pp

Fairbanks RG (1989) A 17,000-year glacio-eustatic sea level record: influence of glacial melting rates on the Younger Dryas event and deep-ocean circulation. Nature 342:637–642

Fenner D, Banks K (2004) Orange Cup Coral Tubastraea coccinea invades Florida and the Flower Garden Banks, northwestern Gulf of Mexico. Coral Reefs 23:505

Gardner JV, Beaudoin J (2005) High-resolution multibeam bathymetry and acoustic backscatter of selected northwestern Gulf of Mexico outer shelf banks. Gulf Mex Sci 23:5–29

Gardner TA, Cote IM, Gill JA, Grant A,Watkinson AR (2003) Long-term region-wide declines in Caribbean corals. Science 301:958–960

Gavio B, Hickerson EL, Fredericq S (2005). *Platoma chrysymenioides* sp. nov. (Schizymeniaceae), and *Sebdenia integra* sp. nov. (Sebdeniaceae), two new red algal species from the northwestern Gulf of Mexico, with a phylogenetic assessment of the Cryptonemiales-complex (Rhodophyta). Dedicated Issue, Flower Garden Banks National Marine Sanctuary, Gulf Mex Sci 25(1):38–57

Geister J (1983) Holocene West Indian coral reefs : geomorphology, ecology and facies. Facies 9:173–284

Gill I, Hubbard D, Dickson JAD (1999) Corals, reefs and six millenia of Holocene climate. 11th Bathurst Meeting. J Conf Abst 4:921

Gittings SR (1998) Reef community stability on the Flower Garden Banks, northwest Gulf of Mexico. Gulf Mex Sci 16:161–169

Gittings SR, Boland GS, Deslarzes KJP, Combs CL, Holland BS, Bright TJ (1992) Mass spawning and reproductive viability of reef corals at the East Flower Garden Bank, northwest Gulf of Mexico. Bull Mar Sci 51:420–428

Gittings SR, Boland GS, Merritt CRB, Kendall JJ, Deslarzes KJP, Hart J (1994) Mass spawning by reef corals in the Gulf of Mexico and Caribbean Sea. A Report on Project Reef Spawn '94. Flower Gardens Fund Technical Series Report Number 94-03, 24 pp

Grimm DE, Hopkins TS (1977) Preliminary characterization of the octocorallian and scleractinian diversity at the Florida Middle Ground. In: Taylor DL (ed) Proc 3rd Int Coral Reef Symp, Rosenstiel School Mar Atmos Sci, Miami, FL, 1:31–37

Hagman DK (1996) Mass spawning, in vitro fertilization, and culturing of corals at the Flower Garden Banks. Proc. 15th Information Transfer Meeting, December 1995. US Dept. of Interior, Minerals Management Service, New Orleans, LA. OCS Study 96-0056, pp 62–67

Hagman DK, Gittings SR (1992) Coral bleaching on high latitude reefs at the Flower Garden Banks, NW Gulf of Mexico. Proc 7th Int Coral Reef Symp. 1:38–43

Hagman DK, Vize P (2003) Mass spawning by two brittle star species, *Ophioderma rubicundum* and *O. squamosissimum* (Echinodermata: Ophiuroidea), at the Flower Garden Banks, Gulf of Mexico. Bull Mar Sci 72:871–876

Hagman DK, Gittings SR, Deslarzes KJP (1998a) Timing, species participation, and environmental factors influencing annual mass spawning at the Flower Garden Banks (northwest Gulf of Mexico). Gulf Mex Sci 16:70

Hagman DK, Gittings SR, Vize PD (1998b) Fertilization in broadcast spawning corals of the Flower Garden Banks National Marine Sanctuary. Gulf Mex Sci 16:180

Hickerson EL, Schmahl GP (2005) The state of coral reef ecosystems of the Flower Garden Banks, Stetson Bank and other banks in the northwestern Gulf of Mexico. In: Waddell J (ed.) The state of coral reef ecosystems of the United States and Pacific Freely Associated States: 2005. NOAA Technical Memorandum NOS NCCOS 11, Silver Spring, MD, pp 201–221

Johnson LS (2002) New international measure to prohibit anchoring on coral reefs by large ships. In: Best BA, Pomeroy RS, Balboa CM (eds) Implications for coral reef management and policy: relevant findings from the 9th International Coral Reef Symposium. USAID, Washington, DC

Kramer PA (2003) Synthesis of coral reef health indicators for the western Atlantic: results of the AGRRA program (1997–2000). In: Lang JC (ed) Status of coral reefs in the western Atlantic: results of initial surveys, Atlantic and Gulf Rapid Reef Assessment (AGRRA) program. Atoll Res Bull 496, Smithsonian Institution, Washington DC, pp 1–75

Lang JC, Deslarzes KJP, Schmahl GP (2001) The Flower Garden Banks: remarkable reefs in the NW Gulf of Mexico. Coral Reefs 20:126

Lighty RG (1978) Relict shelf-edge Holocene coral reef: southeast coast of Florida. Proc 3rd Int Coral Reef Symp 2:215–221

Lighty RG, Macintyre IG, STuckenrath R (1978) Submerged early Holocene barrier reef south-east Florida shelf. Nature 275:59–60

Lugo-Fernandez A, Deslarzes KJP, Price JM, Boland GS, Morin MV (2001) Inferring probable dispersal of Flower Garden Banks Coral Larvae (Gulf of Mexico) using observed and simulated drifter trajectories. Cont Shelf Res 21:47–67

Lugo-Fernandez A (2006) Travel times of passive drifters from the western Caribbean to the Gulf of Mexico and Florida-Bahamas. Gulf Mex Sci:61–67

Matthews RK (1986) Quaternary sea-level change. In sea-level change – studies in geophysics, Nat Res Council, Washington DC, pp 88–103

Minerals Management Service (2007) Gulf of Mexico OCS oil and gas lease sales: 2007–2012: final environmental impact statement. MMS 2007–018, US Dept Interior, Gulf of Mexico OCS Region, New Orleans, LA, 922 pp

Minnery GA (1984) Distribution, growth rates and diagenesis of coralline algal structures on the Flower Garden Banks, northwestern Gulf of Mexico. Dissertation, Texas A&M University, College Station, TX, 177 pp

Moore HB (1969) Ecological guide to Bermuda inshore water. Bermuda Biological Station for Research, Special Publication 5, 25 pp

National Oceanic and Atmospheric Administration (2006a) NOAA Fisheries Service, commercial fishery landings data, website: http://www.st.nmfs.gov/st1/commercial/, Silver Spring, MD

National Oceanic and Atmospheric Administration (2006b) Flower Garden Banks: state of the Sanctuary Report 2006/2007. National Ocean Service, Silver Spring, MD, 22 pp

Pattengill CV (1998) The structure and persistence of reef fish assemblages of the Flower Garden Banks National Marine Sanctuary. Ph. D. dissertation. Texas A&M University, College Station, TX, 164 pp

Pattengill-Semmens CV (1999) Occurrence of a unique color morph in the smooth trunkfish (*Lactophrys triqueter* L.) at the Flower Garden Banks and Stetson Bank, northwest Gulf of Mexico. Bull Mar Sci, 65(2):587–591.

Precht WF, Aronson RB (2004) Climate flickers and range shifts of reef corals. Front Ecol Environ 2:307–314

Precht, WF, Robbart ML, Boland GS, Schmahl GP (2005) Establishment and initial analysis of deep reef stations (32–40 m) at the East Flower Garden Bank. Dedicated Issue, Flower Garden Banks National Marine Sanctuary, Gulf Mex Sci 25(1):124–127.

Precht WF, Aronson RB, Deslarzes KJP, Robbart ML, Murdoch TJT, Gelber A, Evans D, Gearheart B, Zimmer B (2006) Long-term monitoring at the East and West Flower Garden Banks, 2004–2005; Final report. US Dept. of the Interior, Minerals Management

Service, Gulf of Mexico OCS Region, New Orleans LA. OCS Study, 182 pp

Pulley TE (1963) Texas to the tropics. Houston Geol. Soc. Bull. 6:13–19.

Rabalais NN, Turner RE, Wiseman WJ (2002) Gulf of Mexico hypoxia, a.k.a. "The Dead Zone". Ann Rev Ecol Sys 33:235–263

Rezak R, McGrail DW, Bright TJ ((1982) Environmental studies at the Flower Gardens and selected banks: northwestern Gulf of Mexico, 1979–1981. Final Report, northern Gulf of Mexico Topographic Features Study, Technical Report No. 82–7-T, Texas A&M University, College Geosciences, College Station TX, 315 pp

Rezak R, Bright TJ, McGrail DW (1985) Reefs and banks of the northwestern Gulf of Mexico: their geological, biological and physical dynamics. Wiley, New York, 323 pp

Rezak R, Gittings SR, Bright TJ (1990) Biotic assemblages and ecological controls on reefs and banks of the northwest Gulf of Mexico. Am Zool 30:23–35

Schmahl GP, Hickerson EL, Weaver DC (2003) Biodiversity associated with topographic features in the northwestern Gulf of Mexico. Proc Twenty-second annual Gulf of Mexico Information Transfer Meeting, OCS Study MMS 2003–073, US Dept. of Interior, Minerals Management Service, Gulf of Mexico OCS Region, New Orleans LA, pp 84–88.

Schmahl GP Hickerson EL (2005) Planning for a network of marine protected areas in the northwestern Gulf of Mexico. Coastal Zone 05. Proc 14th Biennial Coastal Zone Conference, New Orleans, LA. NOAA, Coastal Services Center, Charleston, SC. NOAA/CSC/20518-CD

Schmahl GP, Hickerson EL (2006) McGrail Bank, a deep tropical coral reef community in the northwestern Gulf of Mexico. Proc 10th Int Coral Reef Symp:1124–1130

Schmitz WJ, Biggs DC, Lugo-Fernandez A, Oey L-Y, Sturges W (2005) A synopsis of the circulation in the Gulf of Mexico and on its continental margins. In: Sturges W, Lugo-Fernandez A (eds) Circulation in the Gulf of Mexico: observations and models. American Geophysical Union, Washington DC, pp 11–29

Stetson HC (1953) The sediments of the western Gulf of Mexico, Part 1 – The continental terrace of the western Gulf of Mexico: its surface sediments, origin, and development. Papers in Phys Oceanogr Meteorol, M.I.T./W.H.O.I. 12(4):1–45.

Stoddart DR (1969) Ecology and morphology of recent coral reefs. Biol Rev 44:433–493

Sturges W, Lugo-Fernandez A, Shargel MD (2005) Introduction to circulation in the gulf of Mexico. In: Sturges W, Lugo-Fernandez A (eds) Circulation in the Gulf of Mexico: observations and models.

Geophysical Monograph Series 161, Am Geophys Union, Washington DC, pp 1–10

Toscano MA, Macintyre IG (2003) Corrected western Atlantic sealevel curve for the last 11,000 years based on calibrated ^{14}C dates from *Acropora palmata* framework and intertidal mangrove peat. Coral Reefs 22:257–270

Viada ST (1980) Species composition and population levels of scleractinean corals within the *Diploria-Montastrea-Porites* zone of the East Flower Garden Bank., northwest Gulf of Mexico. M.S. thesis Dept of Oceanography, Texas A&M University, College Station, TX, 96 pp

Vize PD (2006) Deepwater broadcast spawning by *Montastraea cavernosa*, *Montastraea franksi*, and *Diploria strigosa* at the Flower Garden Banks, Gulf of Mexico. Coral Reefs 25:169

Vize PD, Embesi JA, Nickell M, Brown DP, Hagman DK (2005) Tight temporal consistency of coral mass spawning at the Flower Garden Banks, Gulf of Mexico, from 1997–2003. Dedicated Issue, Flower Garden Banks National Marine Sanctuary. Gulf Mex Sci 25:107–114

Wainwright SA (1965) reef communities visited by the Israel South Red Sea Expedition, 1962. Bull Sea Fish Res Sta Isr 38:40–53

Weaver DC, Rocha LA (in press) A new species of *Halichoeres* (Teleostei:Labridae) from the Western Gulf of Mexico. Copeia.

Weaver DC, Hickerson E, Schmahl GP (2006) Deep reef fish surveys by submersible on Alderdice, McGrail, and Sonnier Banks in the northwestern Gulf of Mexico. In Taylor JC (ed) Emerging technologies for reef fisheries research and management. NOAA Prof Pap NMFS 5, pp 69–83

Wicksten MK, McClure M (2003) A new species of *Alpheus* (Decapoda: Caridea: Alpheidae) from the Gulf of Mexico. Crustacean Research 32:26–31

Zimmer B, Precht W, Hickerson E, Sinclair J (2006) Discovery of *Acropora palmata* at the Flower Garden Banks National Marine Sanctuary, northwestern Gulf of Mexico. Coral Reefs 25:192

7

Coral-reef Geology: Puerto Rico and the US Virgin Islands

Dennis K. Hubbard, Randolph B. Burke, Ivan P. Gill,
Wilson R. Ramirez, and Clark Sherman

7.1 Introduction

7.1.1 Regional Geology

The distribution of Holocene coral reefs around Puerto Rico and the US Virgin Islands reflects the tectonic history of the region. By best estimates, the present-day Caribbean formed between 200 and 130 million years (MY) ago, when North and South America pushed apart and Pacific crust moved northeastward (Pindell 1994). By late Cretaceous time (80 MY: Fig. 7.1), Caribbean plate motion was starting to turn eastward, the Aves Ridge (presently southeast of St. Croix) was forming, and rudistid molluscs were the dominant reef fauna throughout the region (Fig. 7.2). By Oligocene time, the geography of the Caribbean looked much like it is today, and the Greater Antilles lay along a major fault separating the Caribbean and Atlantic plates. Caribbean coral reefs were populated by a much more cosmopolitan fauna than what exists today (Frost 1977; Frost and Weiss 1979). Throughout the Miocene, the Greater Antilles were being torn apart and twisted by both tensional and compressive forces between the opposing Caribbean and Atlantic Plates (Pindell and Barrett 1990; Masson and Scanlon 1991). This has resulted in a left-lateral offset of similar formations on different islands (e.g., the Ponce/Aymamon Formations on Puerto Rico and the Kingshill on St. Croix), as well as counterclockwise rotation of many individual islands (Reid et al. 1991; Gill et al. 2002). On both St. Croix and Puerto Rico, transtensional basins received shallow-water carbonates derived from shallow-water reefs of unknown origins. Movement along the more southerly Muertos Trough (Fig. 7.3) began some time around 10 MY years ago, widening the zone of interplate faulting to some 300 km. The resulting boundary between the Caribbean and North American plates is complex and may include a microplate sitting between the Puerto Rico Trench to the north and the Muertos Trough to the south (Fig. 7.3).

The result is a tectonically active region with generally narrow insular shelves and steep margins that plunge to oceanic depths (Fig. 7.4). The northern, fault-bound margin of St. Croix is one of the steepest submarine slopes in the world. A series of closely spaced faults results in a deep basin (5,500 m) that separates St. Croix from Puerto Rico and the northern Lesser Antilles. The Puerto Rico/Virgin Islands (PR/VI) Platform connects Puerto Rico and the northern Virgins, and was exposed during Quaternary glacial periods, allowing active mixing of terrestrial faunas. In contrast, St. Croix has been separate since its emergence some time after the Miocene and is geologically more related to the Aves Ridge (Fig. 7.3) than to the northern Virgin Islands. The latter are dominated by igneous rocks related to earlier collisional tectonics, whereas the basement rocks of St. Croix are sedimentary in origin (volcaniclastics), and are capped by Cretaceous and Neogene sediment -gravity flows containing numerous reef fossils.

Adey and Burke (1977) characterized St. Croix and the northern Virgin Islands as uplifted and subsided systems, respectively. While this resulted in differing platform histories over a longer timeframe,

B.M. Riegl and R.E. Dodge (eds.), *Coral Reefs of the USA*,
© Springer Science+Business Media B.V. 2008

both systems have been "inactive" more recently. Thus, the older underpinnings likely experienced different tectonic histories throughout Neogene time (23–1.8 million years), but Holocene reefs throughout Puerto Rico and the US Virgin Islands have probably developed under similar, and largely stable, tectonic conditions, with perhaps minor uplift (ca. 10–30 cm).

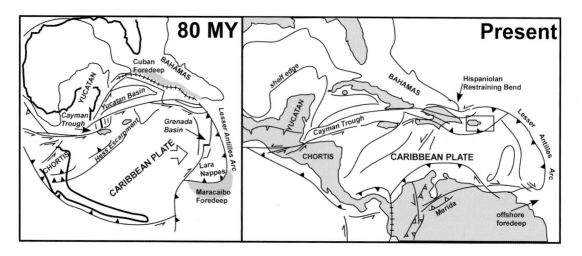

FIG. 7.1. The Caribbean Sea 80 million years ago and today. Before 80 million years ago, the Caribbean plate moved largely northeastward, but was eventually deflected to the east by the Atlantic Plate. The islands of the Greater Antilles lay along the northern boundary between the Caribbean and Atlantic Plates. Today, the Virgin Islands and Puerto Rico (box) lay between the Puerto Rico Trench to the north and the Muertos Trough to the south (Modified after Pindell et al. 1988 by permission of University of the West Indies Publishers)

FIG. 7.2. Photographs of Cretaceous rudists in central Jamaica. While the gregarious molluscs could reach 2 m across, most were more similar in size to those depicted here (coin ~2 cm)

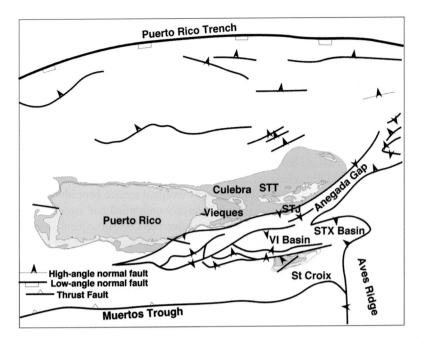

FIG. 7.3. Map of Puerto Rico and the Virgin Islands showing the major faults running through the region. Note the platform that connects Puerto Rico and the northern Virgin Islands (darker blue). This was exposed during glacial intervals, connecting Puerto Rico and the northern Virgin Islands. The V.I. Basin separates these islands from S. Croix and reaches depths of 5,500 m (Modified after Speed 1989 by permission of West Indies Laboratory)

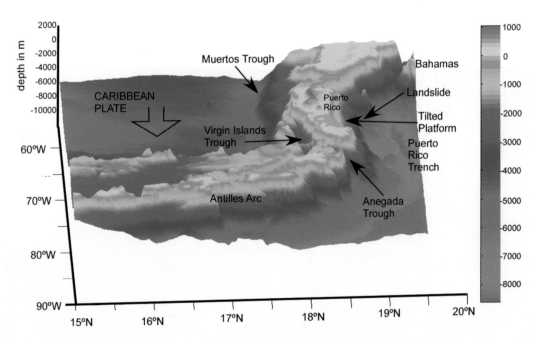

FIG. 7.4. Three-dimensional bathymetric view looking to the west across the eastern Greater Antilles and the northern Windward Islands. St. Croix sits on the ridge south of the Virgin Islands Trough and is separated by it from St. Thomas and St. John to the north. The Puerto Rico Trench (north: right) and the Muertos Trough (south: left) mark the extremes of the transition zone between the North American and Caribbean Plates. The Anegada Trough which cuts diagonally south of the PR/VI platform is the result of one of the many major faults that cut across the region. Diagram modified after Brink et al. (1999) using ETOPO2 data

7.1.2   Reef Types in the Area

Classification is the bane of every scientist, and agreeing on a common scheme for Caribbean reefs is no exception. The system of fringing reefs, barrier reefs and atolls (e.g., Darwin 1842) is unsatisfactory for a variety of reasons. The linear features on eastern St. Croix that are separated from the shoreline by a few kilometers fit the functional definition for barrier reefs. However, many might argue that their scale would relegate them to the fringing category when compared to reefs off Belize or eastern Australia. Horseshoe Reef off Anegada is of sufficient magnitude to qualify as a barrier reef, except much of the "lagoon" that is protected from northeasterly storms is in fact an open bank that is exposed to southeasterly swell and hurricanes passing along the main southerly track through the Caribbean. Circular reefs near South America owe their origins to decidedly different processes than those envisioned by Darwin. And finally, many of the reef buildups occurring near the shelf edge

have lagged behind rising sea level, and are not emergent "reef" features that easily fit into this ternary scheme.

Adey and Burke (1976) proposed a genetic classification based on their cores from reefs throughout the Caribbean. However, this approach suffers in many instances from a priori assumptions of internal reef character and underlying topography (e.g., a cemented terrace beneath "bench reefs", versus a largely unlithified bar associated with early "bank-barrier reefs"). Subsequent cores through at least some of the reefs considered below are not consistent with these earlier assumptions upon which this alternate classification was based. Also, we know less about others, and hesitate to speculate on their origins and histories without core data.

We have, therefore, chosen a morphologic approach that is based on large-scale reef geometry and location. This is used only for consistency with the discussion below, and is not proposed as an alernative classification that is any better than those just discussed; in fact, it is not without its own inherent flaws. Based on this system, the

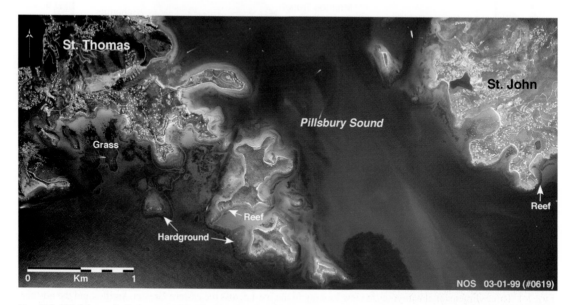

Fig. 7.5. NOAA vertical photograph of St. Thomas and St. John. The steep nearshore slope extends from the shore to depths greater than 10 m, resulting in fringing reefs close to land. The light-colored areas fringing the reefs are sand produced by bioerosion. Darker areas are either seagrasses or hardgrounds at ca. 15–20 m, with widely scattered hard and soft corals, and sponges. Pillsbury Sound is deeper and is dominated by sand that moves generally from north (top) to south

reefs of Puerto Rico and the US Virgin Islands generally fall into three categories. Where the shelf slopes steeply to depths exceeding 10–15 m, narrow *fringing reefs* hugging the steep shore are the norm (Fig. 7.5). In many cases, the reef crest is difficult to identify, and the reef forms a continuous veneer that slopes into the immediate nearshore zone (Fig. 7.6a). While more robust communities can occur along steep rocky headlands away from watershed axes, reef cover is generally low and of limited diversity (Fig. 7.6b). The proximity of these features to shore and the steep slopes above have most likely discouraged significant accretion, and they are generally considered to be relatively thin and concordant veneers over the underlying bedrock.

Where nearshore slopes are gentler, *barrier reefs* are separated from the coast by shallow lagoons (<5–10 m) that typically reach maximum widths of 2 km (Fig. 7.7). While the scale is much different from similar features in Belize or the Indo-Pacific region, the term "barrier" is applied, as the lagoon provides a significant buffer from terrestrial sedimentation, and the reef provides a barrier to open-ocean conditions.

Where the shelf gives way to a steep insular slope, well-developed *shelf-edge reefs* often occur. It has only recently been realized that accretion on these deeper features has been significant, and often exceeds that of emergent reefs (Hubbard et al. 1997, 2001). Holocene reef building has resulted in a present-day reef morphology that can be significantly different from that of the underlying Pleistocene substrate. These reefs can sit atop elevate Pleistocene rims (e.g., Lang Bank, SW Puerto Rico) or along simpler slope breaks that take advantage of proximity to deep water (e.g., Cane Bay, Salt River submarine canyon). Reefs at the shelf break are characterized by alternating reef buttresses separated by sand channels (Fig. 7.8) that show a clear relationship to Holocene wave processes (Roberts et al. 1977). The sand channels serve as short-term repositories for bioeroded (and mechanically derived) sediment (Hubbard et al. 1990), and are the main conduits through which sediment is exported during storms (Hubbard 1992b). Shelf-edge reefs can have some of the highest coral cover found on PR/VI reefs (Fig. 7.9), owing to active currents near the shelf break (Roberts et al. 1977) and the separation from terrestrial sediment stress.

FIG. 7.6. (**a**) Oblique aerial photograph looking northeast along inner Reef Bay, St. John. The steep, continuous slope encourages reefs close to shore with sand and seagrass to seaward. (**b**) Underwater photograph of the forereef in Reef Bay (December 1985). Note the lower coral cover and diversity

Fig. 7.7. Oblique aerial photograph looking west over eastern St. Croix. Note the general shift from reefs closer to shore near the steeper East Point to barriers separated from the island to the west. Buck Island National Park is at the far right of the photo. Salt River and Cane Bay are located along the highland that is visible in the distant background

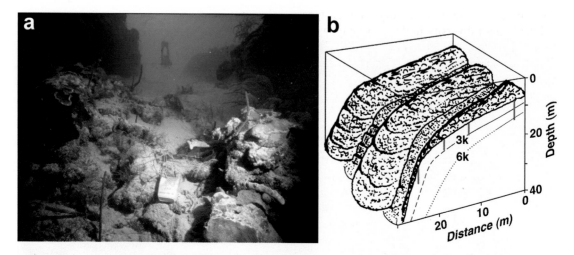

Fig. 7.8. Shelf-edge reef at Cane Bay on northwest St. Croix. (**a**) View looking up a sand channel that separates two reef buttresses (diver in background for scale). The sand channels are temporary storage areas for bioeroded sediment and are the main pathways for export during storms. The sediment trap in the foreground was part of a larger experiment to measure off-shelf sediment transport (Hubbard et al. 1981, 1990). (**b**) Diagram illustrating the character of the shelf edge at Cane Bay. The reef profile 6,000 and 3,000 years ago is shown (based on 4 cores). The present reef morphology reflects a progressive steepening of the forereef slope over time, but the main morphological control is still antecedent (After Hubbard 1989c by permission of West Indies Laboratory)

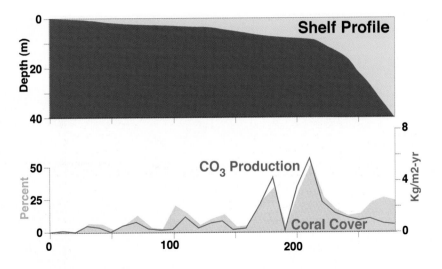

FIG. 7.9. Coral cover (yellow) and carbonate production (kg/m²-year – red line) across the shelf at Cane Bay (top) on northwest St. Croix. Higher coral cover and carbonate production are a result of lower sediment stress and greater current action near the break in slope (After Hubbard et al. 1981 by permission of West Indies Laboratory)

7.1.3 Relationship to Other Caribbean Reefs

Connell (1978) proposed that reefs are not as temporally stable (or spatially partitioned on a large scale) as has been generally accepted. As a result, the greatest variability in reef types and the highest diversity in those systems would occur where reefs are perturbed at intermediate levels. Adey and Burke (1977) showed a systematic increase in wind strength from Florida and the Bahamas to the Windward Islands. They tied the regional variability in reef type that they observed to changes in wind strength (and, by extension, wave energy) along the same geographic gradient (see their Fig. 7.4). Geister (1977) proposed a similar relationship. Hubbard (1989a) added storm intensity and return frequency to the list of factors that might tie into Connel's classic treatment. The strongest relationship exists along the reef crest, where wave energy is more pronounced, but effects related to swell can extend well down the forereef.

Figure 7.10 combines data for fair-weather wave energy with information on the dominant storm paths, and compares these to changes in reefs type. In the eastern Caribbean, well-developed algal ridges dominate the reef crest (Adey and Burke

1976, 1977). Frequent and strong hurricanes preclude *A. palmata* on the reef crest, where thick algal ridges can form due to the exclusion of grazers by high fair-weather waves. The importance of algal ridges throughout the eastern Caribbean contradicts earlier claims that they do not occur in the region (Stoddart 1969; Milliman 1973).

At the other end of the spectrum, Florida and the Bahamas experience fewer and less extreme hurricanes (Hubbard 1989a). Also, wave energy is generally much lower in the intervals between storms (Adey and Burke 1977; Geister 1977; Hubbard 1989a), allowing for an active grazing population on the reef crest. While lower wave energy results in less breakage of *A. palmata,* this branching coral cannot efficiently clear sediment on its own. As a result, it flourishes on only those more-exposed reef crests where turbulence is sufficient to remove sediment. A dominance of massive species on the reef crest is often the norm.

Puerto Rico and the US Virgin Islands generally experience intermediate fair-weather wind/wave intensity, and storm frequency falls between the extremes seen at either end of the Caribbean (Hubbard 1989a). This results in both high coral diversity and varied reef types, from well zoned systems with *A. palmata* near the crest (e.g., Buck Island, Tague Reef) to algal-ridge dominated reefs

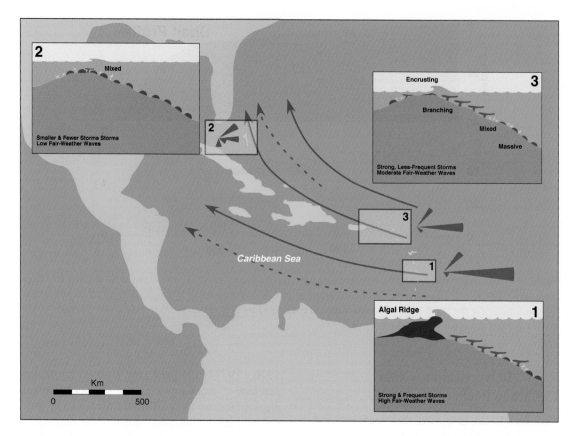

FIG. 7.10. Caribbean reef types relative to wave energy and hurricane intensity. Hurricane (solid) and tropical-storm (dashed) paths are shown by the red arrows. Note the main southerly path across the Windward islands and the secondary path just to the north. The arm lengths on the wave diagrams show the relative wave energy (fair-weather conditions) coming from that direction. The strong easterly trade-wind influence and northeasterly storm flow dominates. Wave energy generally decreases away from the Equator and from the Windward Islands toward Florida and the Bahamas. Algal ridges dominate the Windward Islands (Box 1), especially near Guadalupe and Marie Galante (Adey and Burke 1977) where frequent and strong storms break down *A. palmata,* while high intervening waves discourage grazing and allow thick coralline accumulations. In contrast, both storm and fair-weather conditions are more benign in the western Caribbean (Box 2). Without physical sediment removal by wave action, *A. palmata* has difficulty on all but the most energetic reef crests, and massive species often dominate. Intermediate disturbance levels and ambient wave energy in the northeastern Caribbean (Box 3) provide for high diversity of both organisms and reef types. The most complete Caribbean zonation is generally seen along these reefs (After Hubbard 1989a by permission of West Indies Laboratory)

facing southeasterly swell (e.g., Fancy Bay on St. Croix) or northeasterly storm waves (e.g., Boiler Bay; Adey 1975).

7.2 Pleistocene Reefs

The last time sea level was near its present elevation was ca. 125,000 years ago (125ka) (Fig. 7.11). Somewhat lower (10–20m) high-stands were centered on ca. 105 and 80 ka, with lower-yet oscillations in between (e.g., 55 and ca 40 ka). Both emergent and submerged Quaternary reef deposits have been described from Puerto Rico and St. Croix. The emergent reefs appear to be associated with the 125,000-year event (Hubbard et al. 1989; Taggart 1992; Muhs et al. 2005; Prentice et al. 2005) On western Puerto Rico, uplifted reefs at +2–3 m and + 10 m yielded ages of 114–134 ka (ka = thousand years)

(Taggart 1992). The lower terrace also contained Holocene corals, which Taggart interpreted as evidence for a regional sea-level highstand of 2–3 m above present between 1.4 and 3.3 ka. In this scenario, Pleistocene corals were eroded from the 10-m terrace just above. In contrast, Prentice et al. (2005) explained the elevated Holocene corals as a response to uplift. Based on the age and elevation of the emergent Pleistocene units, Taggart computed an average Quaternary uplift rate for Puerto Rico of 3.3 to 5.5 cm/ka. In-place corals from other elevated reef deposits along the northern and western coasts of Puerto Rico have been found between 0.5 and 1.5 m above sea level, and yielded dates ranging from ~127 to 114 ka. A *Strombus gigas* shell from the elevated (1–2 m) terrace on western St. Croix yielded a U/Th age of 125,000 years (Hubbard et al. 1989b). All these reefs fall within the timeframe of marine isotopic stage 5e and are consistent with a higher sea level at that time (Fig. 7.11).

Submerged Pleistocene strata underlying the shelf-edge reefs off Puerto Rico yielded a U/Th age of 102 ka (Hubbard et al. 1997), which is more consistent with substage 5c, as are similarly placed corals from other localities (Cutler et al. 2003;

Gallup et al. 1994; Toscano and Lundberg 1999). The present-day depth range of the Pleistocene sections in the Puerto Rico and St. Croix cores is likewise concordant with estimates of substage 5c sea-level elevations (Cutler et al. 2003; Toscano and Lundberg 1999). The character of these deposits varies from site to site. In some instances (e.g., the southern Puerto Rico shelf edge), the Pleistocene *Acropora palmata* appear identical, both macroscopically and microscopically, to those in the overlying Holocene section. In contrast, the Pleistocene deposits beneath many of the St. Croix reefs are often coral/mollusc grainstones and rudstones with a well-developed caliche cap, reflective of different environmental conditions at those sites.

7.3   Holocene Reefs

No reef cores have been recovered from the northern Virgin Islands. However, Adey and Burke (1976) provide observations on possible antecedent features that permit conjecture on the timing and extent of Holocene reef development in these areas. On St. Thomas and St. John, the proximity of reefs

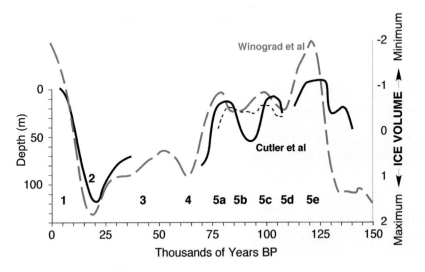

FIG. 7.11. Pleistocene sea-level record based on coral samples from the uplifted reefs of Barbados and New Guinea (Cutler et al. 2003: bold black) and submerged reefs in the Florida Keys (Toscano and Lundberg 1999: dashed, thin line). Also shown is the isotopic record from the SPECMAP foraminiferal ^{18}O time series of Imbrie et al. (1984: red dashed). Bold numbers above the abscissa are Shackleton's (1987) marine ^{18}O isotope stages (1–5) and substages (5a–5e). The known Pleistocene reefs on Puerto Rico and St. Croix are associated with substages 5c and 5e. (After Winograd et al. 1997 by permission of Elsevier)

to shore and the steep watersheds immediately behind argue against vertically extensive reefs. Off Puerto Rico, most available cores come from the southwest corner of the island. In part, this is related to the convenience of a permanent marine lab on nearby Isla Magueyes. However, general reconnaissance studies confirm higher coral cover and diversity in this area. Reefs along the western (e.g., Mayaguez) and southern coastlines (e.g., Ponce and Guayanilla) are generally subjected to higher sedimentation levels than near La Parguera, where our cores were recovered (Acevedo et al. 1989). The development of such diverse reefs in the southwest is largely a result of the broad shelf that provides a buffer from intense coastal sedimentation.

A coring investigation off Vieques, just east of mainland Puerto Rico, concluded that there was "relatively little present day reef growth" (Macintyre et al. 1983), similar to what we propose for the northern Virgin Islands and northern Puerto Rico. Long, linear features that are obvious from the air near San Juan on the north coast (Fig. 7.12) are not biologically built reefs. Rather, they are drowned dunes left by the last sea-level highstand. Shelf-edge features analogous to those described below probably occur, but the steep inshore slope and high wave energy would make significant accretion more difficult.

7.3.1 Reef Building and Holocene Sea-level Rise

The locations of reef cores upon which this chapter is based are shown in Fig. 7.13. The patterns of reef building at these sites are dependent on shelf geometry, the nature and proximity of adjacent landmasses, and the relationship between antecedent topography and rising sea level. As progressively shallower sites were flooded, the success of subsequent generations of reefs was dependent on both the rate of sea-level rise at the time and environmental conditions as those reefs tried to keep up. It, therefore, seemed most logical to describe reefs within a temporal context, rather than tie the discussion to reef type or location. We start with shelf flooding, and work our way through progressively shallower and more landward reef development.

Because we tie our story to rising sea level, we start our discussion there. While there is some disagreement over the details, it is generally accepted that Holocene sea level rose rapidly (>5–6 m/1,000 years – m/ka) until ca. 8,000–6,000 years ago, when it slowed to ca. 1 m/ka (Fig. 7.14). As will be discussed later, this decrease in the rate of rise triggered major changes in Caribbean reef development. The most widely accepted curve for Caribbean sea level was compiled by Lighty et al. (1982), using all available depth/age data for

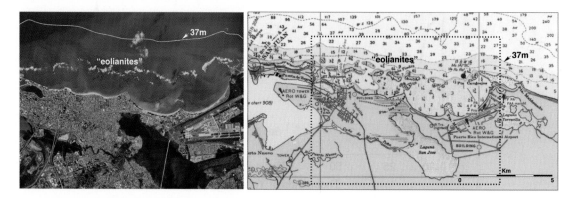

Fig. 7.12. NOAA Vertical air photo (left) and coastal chart (right) near San Juan, Puerto Rico. The breaking waves show the locations of remnant dunes ("eolianites") from the last sea-level highstand. The location of the air photo is shown by the dashed box on the chart. The steep nearshore slope (the 37 m contour is only 1–2 km offshore) makes it difficult to form extensive reefs with significant accretionary potential, a situation that is similar to the northern Virgin Islands. This is in contrast to the broad shelf along the southwestern insular margin of Puerto Rico, where some of the reefs described in this chapter were cored

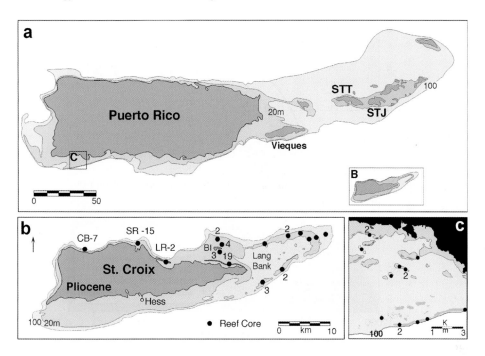

FIG. 7.13. (**a**) Map showing the locations of sites in Puerto Rico and the Virgin Islands discussed in this chapter. The boxes show the locations of B and C. (**b**) Core locations around St. Croix. Numbers next to black circles indicate multiple cores. The three cores along SW Lang Bank are those of Adey et al. (1977). BI = Buck Island; CB = Cane Bay; SR = Salt River. C) Cores off La Parguera on the southwest corner of Puerto Rico

A. palmata. They assumed that the upper boundary of the plotted samples represented sea level (i.e., reefs do not normally build above sea level), and the vertical spread in samples reflected either the depth range (ca. 5–10 m) of this fast-growing branching coral or down-slope transport. The curve shown in Fig. 7.14 is based on isotopic corrections of the Lighty et al. curve using either a proprietary algorithm developed by Beta Analytic, Inc. (Hubbard et al. 2000, 2005) or a similarly structured freeware program, *Calib* (Toscano and Macintyre 2003). The standard for referring to these corrected ages is "calendar years before present" (Cal. BP). All ages mentioned below refer to Cal. BP, even though the shorthand ka is used throughout the text for "thousand years".

The Lighty et al. curve is similar to a peat-based curve for Bermuda (Neumann 1977),[1] corrected

for differential biological fractionation (corals vs. mangroves) and ocean reservoir effects using the methods described above. However, recent challenges have come from Blanchon (2005), Gischler and Hudson (2004), and Gischler (2006). Blanchon (2005) argues for three steps in the curve, similar to the two "meltwater pulses" described for Barbados by Fairbanks (1989). However, Blanchon's curve (red in Fig. 7.14) in inconsistent with the number of radiocarbon dates that plot above it, as well as the lack of any interruption in reef accretion that such jumps should have presumably caused. Also, isotopic data do not support a sudden pulse of meltwater into the world ocean as was the case for the Barbados cores (Fairbanks 1989). Gischler (2006) reports a number of *A. palmata* samples from Belize that plot above the curve used here. These were accompanied by peat samples from Belize, Florida and Jamaica. Toscano and Macintyre (2003) interpreted these as intertidal to supratidal peat, and drew the curve beneath them, while Gischler (2006) argued for a subtidal origin, and a curve that sits above. Fortunately for the following discussion,

---

[1] This curve originally appeared in Adey (1975: Fig. 13), with the sole reference of "Neumann (in ms)". It subsequently occured in Adey and Burke (1976; Fig. 3) without citation. Bloom (1977) published it with the permission of the author, again citing it as "unpublished data".

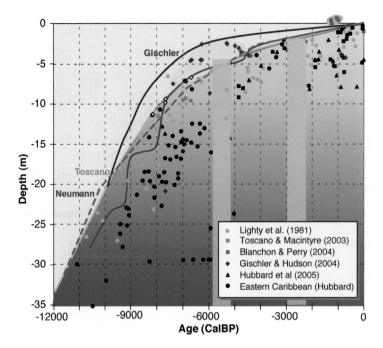

Fig. 7.14. Holocene sea-level curves for the Caribbean. Reef development described in this chapter is referred to the sea-level curve of Lighty et al. (1982), corrected for metabolic and sea-water effects (Hubbard et al. 2000, 2005: light blue; Toscano and Macintyre 2003: orange). Each data point reflects the corrected age of an *A. palmata* sample and its depth below present sea level (see legend for data source). Curves suggested by Neumann (1977: dashed green), Blanchon (2005: red) and Gischler (2006: dark blue) are also shown. The *A. palmata* samples used to derive the curves are provided (see key for sources). The yellow bars represent two intervals in which *A. palmata* is either rare (three samples between 5.9–5.3 ka) or absent (3–2.2 ka) from the record of dated samples for the entire region. After Hubbard et al. (2005) by permission of SEPM

whichever curve one chooses in Fig. 7.14, the general pattern remains the same: rapid sea-level rise before 8–6 ka and a much slower rise thereafter. We will take no side here, inasmuch as the change in the rate of sea-level rise between 8 and 6 ka is the only point that is critical to our arguments that follow.

7.3.2 Shelf Flooding

Prior to 12,000 years ago, most of the shelf margins around Puerto Rico and the Virgin Islands were exposed. The deeper southern margin near St. Thomas and St. John (at ca. 35 m) would have flooded earlier, when sea level was rising very quickly. Because Atlantic sea level is slightly higher than its Caribbean counterpart, tidal flow through Pillsbury sound is dominantly to the south, a pattern that probably also existed 12,000 years ago. The southerly off-shelf transport, in

combination with rapidly rising sea level, probably discouraged significant reef development along the southern edge of the PR/VI Platform, especially opposite breaks between islands. Today, these areas support scattered corals and large demosponges. A similar fauna is common along the southern and western margins of St. Croix at similar depths.

Where the shelf-edge reefs reach present depths less that 15 m, depths along the underlying Pleistocene surface generally fall between 20 and 35 m. Cores along deeper antecedent shelves yield older ages near their base, owing to earlier flooding. Overwhelmingly, these early reefs were dominated by branching *A. palmata* (Fig. 7.15). Based on the sea-level curve in Fig. 7.16, the underlying Pleistocene surface would have flooded between 12,000 and 10,000 years ago. The oldest radiometric ages for corals in the lower sections of the cores post-date flooding by ca. 1,500 years. Timelines

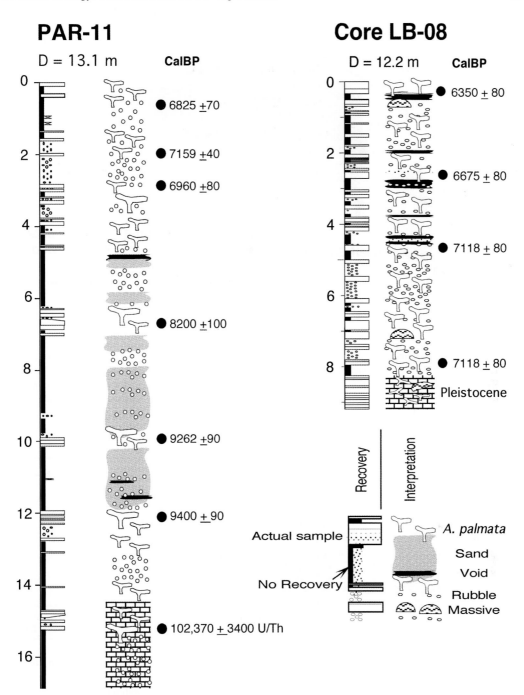

FIG. 7.15. Core logs from typical shelf-edge reefs off southwestern Puerto Rico near La Parguera (PAR-11) and eastern St. Croix (LB-08). Actual core recovery is summarized in the left column of each log; the interpreted section (including coral species and the character of intervening sections, i.e., void, sand, rubble) make up the right half. See key at lower right for details. The dominant corals at both sites were branching *A. palmata*. In general the antecedent shelf was shallower on St. Croix and, therefore, flooded later.

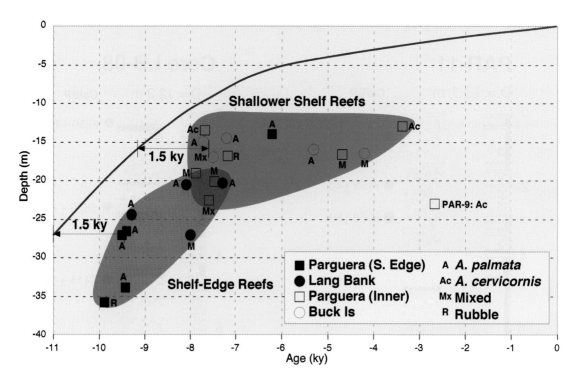

FIG. 7.16. "Start-up" ages and depths for reefs where cores penetrated through the entire Holocene section. Based on the corrected Lighty et al. curve (black line), the oldest preserved samples lag behind flooding by ca. 1,500–2,000 years. The lag associated with branching *A. palmata* (A) was roughly half that associated with more sediment-tolerant massive species (M). If this lag were a response to sediment stress, more tolerant massive corals should have been able to adapt more quickly, contrary to the pattern shown by the cores. Ac = *A. cervicornis*; Mx = mixed species.

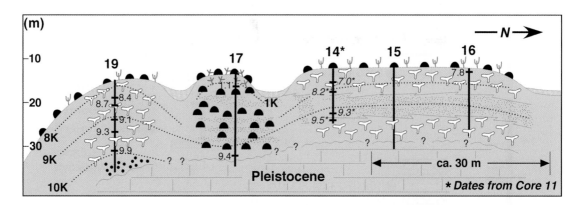

FIG. 7.17. Profile across the shelf-edge reef near La Parguera, SW Puerto Rico (Caribbean Sea is to the left). Reef composition and time lines across the feature are based on multiple cores (see Fig. 7.15 for coral key). The morphology at this site does not allow for significant downslope coral transport that might bias age/depth relationships discussed here. The innermost ridge (Cores 14–16) was cored at two other places along the shelf edge, and demonstrates along-shelf continuity of the lithology and timing shown here. The outer feature (C19) formed first, and was dominated by *A. palmata*. A reef inhabited by massive species formed ca. 20 m landward (C17). A broad platform (ca. 30 m across) covered by *A. palmata* formed the shallowest part of the reef complex. Water depth atop the reef (12–13 m) is identical to the rim around Lang Bank (St. Croix), where a nearly identical history was revealed. The *A. palmata* reefs quit around 6,800 – 6,500 years ago, but the intermediate ridge dominated by massive species continued to build more slowly after the *A. palmata* reefs had succumbed (note the 1.1 ka date 3 m below the present surface in core PAR-17). The present-day cover by massive and soft corals contrasts with the species found in the immediate subjacent strata except on the middle reef

drawn along core transects show that the reefs were relatively narrow and flat-topped features with deeper water both in front and behind (Fig. 7.17). This precludes samples having rolled down from upslope and younger reefs.

The lag between shelf flooding and the oldest ages in the cores has traditionally been related to the suppression of coral recruitment by turbid water, high sedimentation, and elevated nutrient levels derived from Pleistocene soil horizons that were eroded by wave action as the platform was overtopped (Adey et al. 1977; Schlager 1981; Neumann and Macintyre 1985; Macintyre 1988). While re-disturbed sediments undoubtedly played some role in discouraging early coral communities, it is difficult to envision this inhibition lasting on such a large scale for 1,500 years or longer. Also, in many cases massive-coral reefs lagged behind similar structures dominated by more susceptible acroporids by as much as 3,000 years. If this lag

were related to turbid water flowing off the flooded platform, then one would expect more tolerant massive species to be associated with shorter lag times, opposite to the pattern shown here.

An alternative scenario recognizes that "the oldest reef dates are simply the first *preserved* corals" (R. Buddemeier). Early shelf-edge reefs had to deal with not only turbid water, but also the possibility of an unstable substrate (Adey and Burke 1976). The slowly developing reefs were undoubtedly subjected to bioerosion that may have been more severe, given the elevated nutrient levels that probably existed during earliest flooding. Under this scenario, the lag reflects not so much poor water quality as it does a near balance between construction and destruction along fledgling shelf-edge reefs. Seven of the eight shelf-edge cores from St. Croix and Puerto Rico that penetrated into the underlying Pleistocene strata contained basal Holocene intervals dominated largely by sediment and reef rubble,

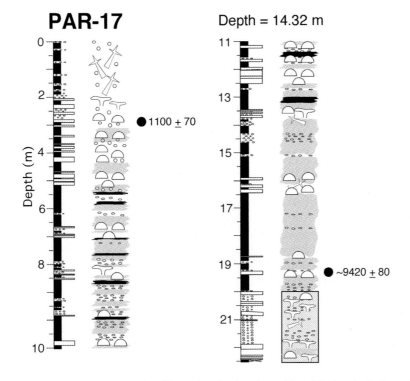

FIG. 7.18. Shelf-edge core from SW Puerto Rico illustrating the abundance of reef detritus in the lower 7 m of the reef that "started up" ca. 9,420 years ago (the black bar in the left column denotes no recovery; stippled pattern in right column is an interpretation of sandy sediment in the non-recovered interval). Note the general increase in recovered coral framework up-core. This pattern is typical of nearly all of the shelf-edge cores that penetrated through into the underlying Pleistocene strata. See Fig. 7.15 for key

in some instance several meters thick (Fig. 7.18). The abundance of bioclastic debris logically reflects a productive system being largely reduced to rubble. Whether this material was part of an allochthonous spit/bar that served as a focal point for reef development (e.g., Adey and Burke 1976; Burke et al. 1989) or a largely autochthonous reef fabric reflecting increasing coral preservation over time is not clear. Regardless of the origins of this bioclastic debris, something had to be there to bioerode, and the most likely sources are the antecedent Pleistocene surface and Holocene corals. The preserved caliche surface at the top of many Pleistocene sections precludes the antecedent surfaces as a bioclastic sediment source at these sites. Thus the "lag" after shelf flooding appears to have been related more to poor initial preservation than to delayed coral recruitment.

7.3.3 Shelf-edge Reef Building

Reef building throughout the early Holocene implies conditions that were ideal for *A. palmata*. Even though sea level was rising at its maximum rate, extensive branching-coral reefs had been able to form on deep topographic benches off Barbados (70–120 m below present sea level), and flourished intermittently between 20,000 and 11,000 years ago (Fairbanks 1989). It is likely that these features occur widely throughout the region. As upslope Caribbean shelf margins were flooded, this dominance of *A. palmata* continued.

The oldest radiometric age reported from Caribbean shelf margins (11.1 ka) comes from a reef in Salt River submarine canyon (Fig. 7.13) on the north shore of St. Croix at a depth slightly greater than 30 m below present sea level (Hubbard et al. 1986). The reef was dominated by branching corals and sat within 2–3 m of sea level at the time. Reefs also formed along an antecedent rim around Lang Bank (the 10.3 ka reef of Adey et al. 1977; see their Fig. 4) and an elevated shelf margin off southwestern Puerto Rico (Hubbard et al. 1997; Fig. 7.19, Table 7.1). These are coincident with *A. palmata* reefs forming in Florida (10,700 Cal. BP: corrected from Lighty et al. 1982) at similar depths along the edge of the Gulf Stream.

As sea level rose further, reefs formed along the tops of antecedent Pleistocene ridges (e.g. cores 14–16: Fig. 7.17). While massive corals did occur in more protected areas (Core 17: Fig. 7.17), the reef crest was generally dominated by *A. palmata* (only 4 of 18 cores through shelf-edge reefs in Puerto Rico and St. Croix yielded massive species, and all of these were landward of other cores dominated by *A. palmata*). By 9 ka, widespread branching-coral reefs flourished at both shelf-edge sites, and along other shelf margins at similar depths throughout the Caribbean. Water depth over the St. Croix and Puerto Rico reefs was generally less than 10 m, but did reach as much as 16 m (Fig. 7.16). Stands of *A. palmata* can be found at similar depths on Lang Bank today, but these are isolated colonies that are probably not contributing to in-place reef framework. However, modern reef pinnacles at similar depths off SW Puerto Rico supported well-developed stands of *A. palmata* that appear to have been contributing to a largely in-place reef structure until the colonies were killed by bleaching and/or disease, probably within the past 2–3 decades.

Despite the considerable depth over most of the early Holocene shelf-edge reefs, accretion rates averaged between 4 and 5 m/ka. These reefs were, therefore, capable of keeping pace with rapidly rising sea level, which was advancing at a similar rate. Thus, while the reefs were not shoaling over time, they were not falling further behind either.

7.3.4 Backstepping and the Formation of Shallower Reefs

7.3.4.1 Puerto Rico

By 8 ka, flooding of the inner shelf triggered the development of shallower reefs closer to shore (Fig. 7.19). These started from elevated, antecedent surfaces between 13 and 23 m below present sea level (most commonly between 14 and 17 m: Fig. 7.16: "shallower shelf reefs"; Table 7.1; Hubbard et al. 1997). There was a roughly 2,000-year overlap between the early development of inshore reefs 8,000 years ago and the subsequent demise of their deeper counterparts. Thus, the demise of the shelf-edge reefs logically had no causative effect on the shifting of reefs toward shore. Unlike the shelf-edge reefs that were largely dominated by *A. palmata*, the newly formed inshore structures were built by massive species, and only shifted to branching

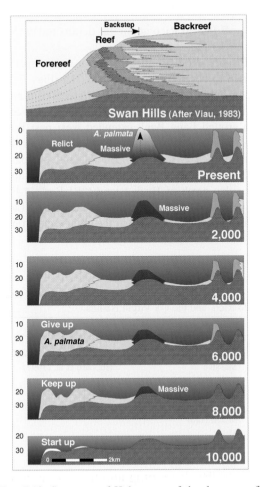

FIG. 7.19. Summary of Holocene reef development off La Parguera in SW Puerto Rico. The earliest reefs date to slightly before 10 ka, and were dominated by *A. palmata*. The next 4,000 years saw expansion of the shelf-edge reefs and the subsequent formation of new reefs closer to shore. The latter were dominated by massive corals, probably a response to higher levels of sedimentation closer to the island. Development along the shelf edge is shown in greater detail in Fig. 7.17. After 6 ka, the inshore reefs continued to build, but accretion near the shelf edge had ceased. The abandonment of shelf-edge reefs in favor of shallower and more landward sites was similar in both character and scale to reef "backstepping" that has occurred at many sites in the geologic past. The top panel shows initial outbuilding of Devonian reefs in Canada followed by a sudden "backstep", interpreted as a response to a sudden and rapid rise in sea level. The cross sections have all been drawn at the same vertical and horizontal scales. Similar phenomena occurred in Australia at about the same time (Playford 1980) (After Hubbard 1992a; Hubbard et al. 1997 by permission AAPG and ISRS)

acroporids much later. This difference was probably a response to higher terrestrial sediment influx closer to shore.

By 6,000 years ago, extensive inner-shelf reefs were well-established along two primary trends (Fig. 7.20), but the mid-shelf reefs had not yet caught up to rising sea level (Fig. 7.19). Antecedent topography played a significant role in where reefs developed. The larger mid-shelf reefs and reef islands (e.g., Turromote, Corral, Enrique: Fig. 7.20.) all sit atop elevated Pleistocene features that occupied the same locations the last time sea level was near its present elevation (Fig. 7.21). Similarly the series of inner shoals and mangrove islands appear to have nucleated on elevated features that were present 125,000 years ago. Our longest core through one prominent mud shoal ("Jack's Mound": Fig. 7.22) passed through the entire Pleistocene interval and penetrated older, well-indurated silts and clays. These lowermost sediments in the core are terrestrial in origin, and may have been part of an alluvial-fan complex built at the mouths of dissected watersheds that remain today (see the rugged upland areas in Fig. 7.20). The arcuate shape of the innermost reef-island trend is suggestive of such a feature beneath.

"Backstepping" is a common phenomenon throughout the geologic record of reefs. Between 370 and 350 million years ago, reefs generally built out under a regime of slow sea-level rise. It has been proposed that near the end of that interval, sea level suddenly rose quickly, causing the position of the dominant reefs to shift dramatically landward (Playford 1980; Viau 1983). This phenomenon has been described from other reefs at different times, and has likewise been attributed to sudden and dramatic increases in the rate of sea-level rise. Schlager's (1981) "Drowning Paradox" is based on similar assumptions. The top panel in Fig. 7.19 characterizes the reefs of northern Canada that backstepped in the late Devonian (after Viau 1983). The cross section is plotted at the same vertical and horizontal scale as the Holocene cross sections across the Puerto Rico shelf. Clearly, the spatial scales of the two scenarios are similar. However, the sudden rise in sea level used to explain backstepping in the fossil Canadian reefs was clearly not involved in the Holocene reefs off Puerto Rico.

7.3.4.2 Virgin Islands

The shelf-edge reefs around St. Croix have a history similar to that just described for Puerto Rico (Hubbard et al. 2000, 2001). On Lang Bank (Fig. 13b), the outermost ridge cored by Adey et al. (1977) yielded a maximum age of 10.3 ka, coincident with the start-up date for a similar feature off Puerto Rico (Core 19: Fig. 7.17). The Pleistocene surface beneath the shallow rim of Lang Bank sits at 25–30 m, slightly shallower than its counterpart off SW Puerto Rico (up to 34 m). As a result, the shelf-edge reefs off eastern St. Croix started slightly later than those sit-

ting off Puerto Rico at similar depths today (8–9 vs. 10 ka Fig. 7.23).

Inshore reefs dominated by massive species started to develop between 8 and 7 ka (Figs. 7.16, 7.24a, Table 7.1), while A. palmata reefs continued to flourish along the shelf edge. The Pleistocene rim around Lang Bank extended west past Buck Island (Fig. 7.13b), generally shallowing in that direction (ca. 15 m north of Buck Island, compared to 20–30 m to the east). As a result, accretion along the ridge north of Buck Island ("BI Bar" in Fig. 7.24) started much later than was the case on Lang Bank, and is more related to the inner reefs around

TABLE 7.1. Information on start-up conditions for reefs of Puerto Rico and the US Virgin Islands.

SHALLOW REEFS

Core	Location	D (m)	Age (Cal. BP)	Substrate	Source
PAR-01	SW PR	20.2	7,355	Massive	Hubbard et al. (1997)
PAR-04	SW PR	13.3	6,180	Sediment	
PAR-05	SW PR	19.5	7,910	Massive, A. cervicornis	
PAR-06	SW PR	13.5	7,725	A. cervicornis	
PAR-07	SW PR	13.3	3,350	Mixed coral	
PAR-08	SW PR	16.1	4,720	Mud	
PAR-09	SW PR	23.0	2700	Sand; massive	
PAR-10	SW PR	22.5	7,625	Massive; sand	
PAR-12	SW PR	16.7	7,150	Rubble	
BI-01	Buck Is (N)	15.9	5,270	Mixed	Hubbard et al. (2005)
BI-02	Buck Is (N)	16.0	5,270	A. palmata	
BI-04	Buck Is (S)	13.6	>7,175	Rubble; mixed	
BI-05	Buck Is (N)	16.5	4,210	Massive	
BI-07	Buck Is (S)	16.01	2,840	Massive; rubble	
BB-01	Buck Is Bar	15.3	7,770	A. palmata	
TB-17	Tague Reef	13.5	6,975	Massive; sand	Burke et al. (1989)

SHELF-EDGE REEFS

Core	Location	D (m)	Age (Cal. BP)	Substrate	Source
PAR-11	SW PR	28.0	9,453	A. palmata	Hubbard et al. (1997)
PAR-14	SW PR	29.6	no date	A. palmata	
PAR-17	SW PR	34.2	9,420	Massive; rubble	
PAR-19	SW PR	~36.8	9,900	A. palmata	
LB-02	Lang Bnk (N)	20.9	8,075	A. palmata	Hubbard (unpublished)
LB-03	Lang Bnk (S)	24.2	9,254	A. palmata	
LB-06	Lang Bnk (E)	32.0	>8,048	Mixed	
LB-08	Lang Bnk (N)	20.5	7,310	A. palmata	
USGS-2	Sand Key	12.8	8,900	Massive	Toscano and Lundberg (1999)
CDR-3	Carysfort	20.0	6,600	Massive	
CSFT-4	Carysfort	15.5	6,300	Massive	
CDR-1	Carysfort	15.9	7,100	Massive	
GR-4	Grecian Rks	7.5	6,750	Massive	Shinn et al. (1989)
N/A	Looe Key	10.0	7,440	A. palmata	

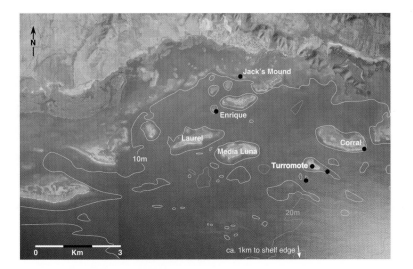

FIG. 7.20. Vertical NOAA aerial photograph of the mid-shelf reefs (Laural, Media Luna, Enrique and Corral) and inner reefs (just inside the 10-m contour) off SW Puerto Rico. Generalized bathymetry (10- and 20-m contours, based on soundings by Jack Morelock, in Hubbard et al. 1997) is provided for reference. The shelf-edge reefs are just off the bottom of the photo. Core sites are shown by the black dots. Antecedent topography is an important control of reef location. The mid-shelf reefs and the submerged shoals associated with them sit atop remnant Pleistocene highs (e.g., Fig. 7.21 on Turromote). The inner belt of muddier shoals and rubble islands that parallel the coast appear to sit on muddy, alluvial deltas which pre-date the Pleistocene (e.g., "Jack's Mound" near the northernmost black dot: Fig. 7.22)

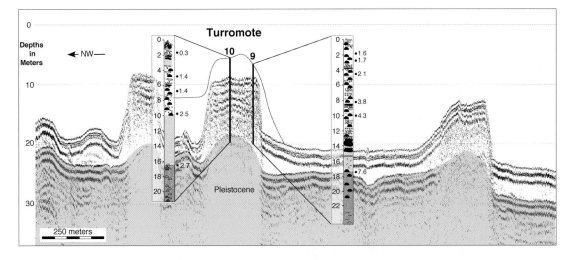

FIG. 7.21. Seismic line across the reef at Turromote (Fig. 7.20). Note the strong correspondence between the present-day reefs and highs in the antecedent Pleistocene surface beneath (gray shading). The seismic line was run over a low spot in Turromote Ridge; the generalized shape of the shallower reef where the cores were actually recovered is shown by the solid line above. Generalized core logs are also shown (after Hubbard et al. 1997). Massive corals dominated throughout, and ages ranged from ca. 7.6 ka to the present. See Fig. 7.15 for key to cores

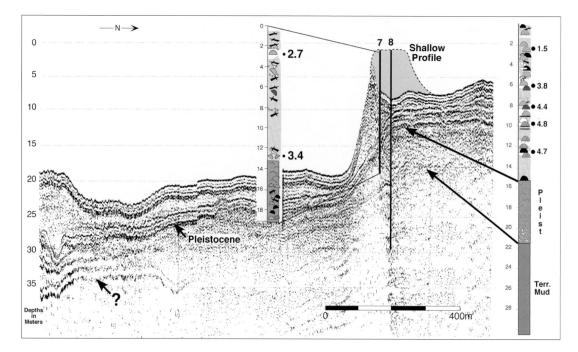

FIG. 7.22. Seismic line across the inner mud shoals near "Jack's Mound" (innermost core in Fig. 7.20). Massive corals and branching *A. cervicornis* dominated the Holocene section of the shoals, which range in age from ca. 4.7 ka to the present (cores after Hubbard et al. 1997 by permission of ISRS; see Fig. 7.15 for key). Note that the shoal sits atop an antecedent Pleistocene ridge. Beneath that, an older (Tertiary?) feature occurred at the same site (lower arrow), and was dominated by terrestrially derived mud. This is interpreted as an alluvial fan built by heavy runoff from the steep and heavily dissected watershed to the north. See Fig. 7.15 for key to logs

Buck Island (Hubbard et al. 2005), and Tague Reef on St. Croix (Burke et al. 1989) than to those along the Lang Bank shelf edge to the east. As the shallower reefs formed along the northern St. Croix shelf, the greater wave exposure and distance from shore allowed branching *A. palmata* to form along Buck Island Bar, in contrast to the massive-coral dominated reefs closer to shore. Reefs built steadily all across the northern shelf through to the present. As the reefs built and slopes steepened over time, zonation formed and was characterized by *A. palmata* near the reef crest, giving way to massive species below (Fig. 7.24b). This trend continued, except in the interval between 3 ka (Fig. 7.24c) and 2 ka, when *A. palmata* was absent for reasons that will be discussed below.

The general timing of the shallower St. Croix reefs (Burke et al. 1989; Hubbard et al. 2005) is strikingly similar to what has been documented

for similar reefs off Puerto Rico (Hubbard et al. 1997), Florida (Lighty et al. 1982; Shinn et al. 1989), the Bahamas, Antigua (Macintyre et al. 1985), Belize (Gischler and Hudson 2004) and Panama (Macintyre and Glynn 1976). The fringing reefs in the northern Virgins probably share a history starting after 7 ka and following the scenario just described. All of the shallower and now-emergent reefs throughout the Caribbean appear to have started at about the same time as sea level was slowing, and underwent the same general patterns of development.

7.3.4.3   St. Croix Algal Ridges

As described earlier, the character of reefs in the northeastern Caribbean is highly varied, owing to intermediate levels of fair-weather wave action and storm disturbance. Algal ridges

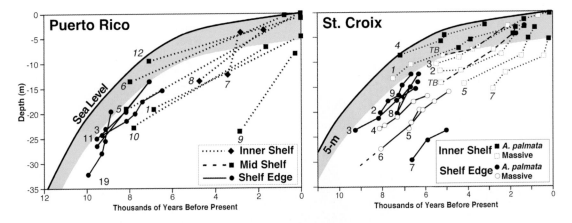

Fig. 7.23. Accretion by shelf-edge and inner-shelf reefs on Puerto Rico and St. Croix. Off Parguera (left), the oldest shelf-edge reefs, dominated by *A. palmata* (closed circles; solid lines) started 10,000 years ago and thrived until ca. 6,500 Cal. BP. Reefs closer to shore had already been actively building for ca. 1,500 years (closed squares and diamonds; dotted line). Similarly, the shelf-edge reefs around Lang Bank on eastern St. Croix (right – circles and solid lines) date back to ca. 9,500 Cal. BP. The reefs made up of *A. palmata* (closed circles) stopped accreting slightly later than those off Puerto Rico, but massive-coral reefs (open circles) continued to build for another 1,000 years. Inshore reefs around Buck Island (dotted lines; core numbers in *italics*) and Tague Bay (dashed lines: *TB*) started as early as 7,500 Cal. BP, but the oldest ages from most cores were on the order of 6,000–5,500 Cal. BP. Note that the accretion rates for *A. palmata* reefs are not significantly different than those for massive species, regardless of water depth. Also, when the shelf edge reefs quit, they were either no deeper (or, in some cases, shallower) than when they initiated. Clearly, they were not building slower that sea level was rising, despite being below the 5-m-depth envelope (gray shading) that is generally considered to be a threshold for *A. palmata*-driven accretion. Thus, being outpaced by sea level is clearly not an explanation for the demise of reefs at the shelf edge

can be found at many sites along the eastern end of St. Croix. On the north shore, at Boiler Bay (BB: Fig. 7.25), an inner series of degraded ridges sits near the beach in a protected lagoon behind eastern Tague Reef. In contrast, active algal ridges can be found at several sites along the south-facing shore where there is no seaward reef to block incoming wave energy (Robin Bay (RB): Fig. 7.25).

Adey and Burke (1976) proposed that the Boiler Bay algal ridges formed on alluvial fans in shallow water, and were still exposed to high waves 2,000–3,000 years ago (Fig. 7.25). As Tague Reef built up in front of Boiler Bay, wave energy over the algae ridges dropped, and they fell victim to increased grazing pressure and macroalgal buildup. They proposed a slow initial buildup along Tague Reef associated with slower-growing massive corals at greater water depths (see the tan "reef envelope" in Fig. 7.25).

In this scenario, Tague Reef built into shallower water, and *A. palmata* gradually took over the reef crest, triggering accelerated reef accretion. This concave-upward accretion curve reflects reef-accretion rates tied to water depth and coral type (i.e., branching *A. palmata* vs. massive corals). Once Tague Reef built to sea level, wave energy to the Boiler Bay algal ridges was cut off, and they began a cycle of degradation, characterized by increased bioerosion and overgrowth by fleshy algae.

Core data for Tague Reef (Burke et al. 1989) corroborate the general story of early algal-ridge development close to shore, followed by wave attenuation as Tague reef built to seaward. However, the accretionary patterns within Tague reef are substantially different than those summarized in Adey and Burke (1976; Fig. 7.25). Unlike the convex-upward accretion curve envisioned by Adey and Burke (1976), the reef

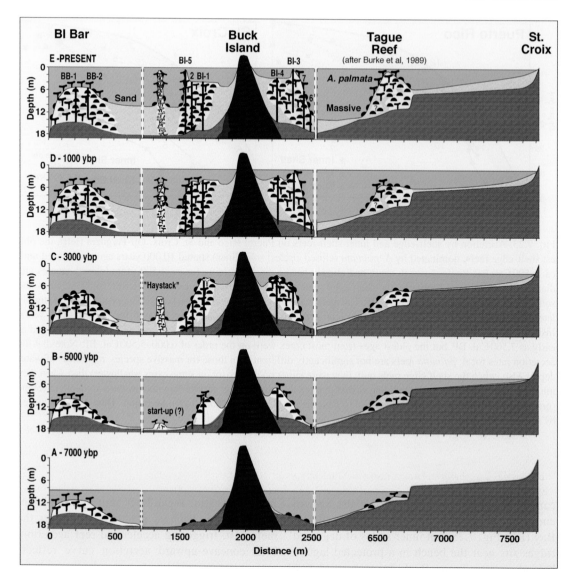

Fig. 7.24. Reef history on NE St. Croix. Reefs closer to the shelf edge generally started up first and were dominated by branching *A. palmata* (A). Reefs closer to Buck Island and St. Croix were subject to greater sediment stress, and were dominated by more resistant massive-coral species. Reefs along the south side of Buck Island (Hubbard et al. 2005) and northern St. Croix (Burke et al. 1989) reflect greater quantities of sediment than their higher-energy counterparts on northern Buck Island and Buck Island Bar (Macintyre and Adey 1990). As reef topography developed, zonation from *A. palmata* near the surface to massive species at depth became the norm, with the exception of the period starting 3,000 years ago (C). Today, the reefs have reached sea level with the exception of Buck Island Bar, which appears to have been depressed by high wave action along this more-exposed margin (Macintyre and Adey 1990) (Figure after Hubbard et al. 2005 by permission of SEPM)

appears to have built to sea level quickly and at a generally uniform rate, regardless of water depth or coral type (red curves in Fig. 7.25; based on

data from Burke et al. 1989). This pattern is counter to widely held presumptions of species- and depth-dependant reef accretion.

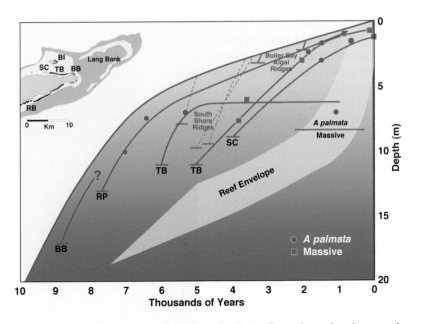

FIG. 7.25. Shallow-water reef development on northeastern St. Croix. General core locations are shown on the map inset (BB = Boiler Bay; TB = Tague Bay; RP = Romney Pt (between BB and TB); SC = Sandy Cay; BI = Buck Island; RB = Robin Bay). Sea level over the past 10,000 years is summarized by the blue curve. Reef accretion along northeastern St. Croix is shown in red (the horizontal bars are the measured or inferred Pleistocene basement; data from Burke et al. 1989). Dated core samples are differentiated between *A. palmata* (red circles) and massive corals (green squares). The brown dashed lines are inferred accretion patterns for algal ridges on the north (Boiler Bay) and south shore (Robin Bay). The concave-up "reef envelope" (tan-gray) summarizes the general pattern of accretion envisioned by Adey and Burke (1976) for Tague Reef. The envelope reflects different histories starting on a deeper substrate to the east (base of the "envelope") and a shallower antecedent surface to the west. They proposed that Tague Reef initially built slowly because it sat in deeper water, and was dominated by slower-growing massive corals. As the reef built into shallower water, faster-growing acroporids presumably took over, and the reef built quickly to sea level, cutting off wave energy to the algal ridges in Boiler Bay. Core data from Burke et al. (1989: red lines) show that Tague Reef actually built rapidly to sea level much earlier, and at rates that did not vary in response to either water depth or coral type

7.3.4.4 Cane Bay and Salt River

Two noteworthy and morphologically distinct reef systems occur at Salt River and Cane Bay along the narrow northwestern margin of St. Croix (SR and CB: Fig. 7.13b). Cane Bay sits along a narrow shelf that reaches depths of 20–30 m within 250 m from shore. From the shelf break, the bottom drops precipitously to abyssal depths (5,500 m), at an average slope of 45° (the upper 2,500 m are vertical). To the west, the slope break deepens to as much as 80 m.

The reefs along this insular margin share attributes of both fringing and shelf-edge reefs. Cores have not reached the underlying Pleistocene

strata, but it appears that the shape of the present reef surface reflects both Holocene reef-building and antecedent topography (Fig. 7.8b; Hubbard et al. 1984, 1986). Steep hills overlook the bay, and sediment stress has been naturally high. As a result, massive corals have dominated the 6,400 years for which our core record exists. Because of constant terrestrial input, the high coral cover and carbonate production rates occur near the shelf edge (Fig. 7.9).

The reef at Cane Bay is organized into alternating reef buttresses and intervening sand channels (Figs. 7.8, 7.26). The channels occur at regular intervals, suggesting an extrinsic control on their periodicity.

A similar pattern has been identified off Jamaica (Goreau and Land 1974), in the Bahamas (Hubbard et al. 1976), and on Grand Cayman Island (Roberts et al. 1977). The latter authors proposed that channel patterns were responding to Holocene wave climate. More recently, Hubbard et al. (1981, 1982, 1990) and Sadd (1984) identified the role of the channels as short-term repositories for bioclastic sediment that might otherwise overwhelm the reef. Storms periodically flush these stored sediments from the reef (Fig. 7.26; Hubbard 1992b).

Salt River (SR in Fig. 7.13b), where Christopher Columbus made first landfall on his second voyage, sits at the head of a reef-lined submarine canyon (Figs. 7.27, 7.28). For a decade, this was home to *Hydrolab*, the world's longest-lived, open-water saturation-diving facility. It served as a base of operations to well over a hundred reef scientists from myriad research institutions. This program provided a wealth of baseline information on the biology, geology and oceanography of this unusual marine system.

Both the larger canyon morphology and the distribution of coral species along its walls are a response to the pathways of sediment movement under the influence of trade-winds circulation (Hubbard 1986). Sediment generally moves to the west and into the canyon along its eastern side (larger gray arrows in Figs. 7.27 and 7.28). As a result, coral recruitment is discouraged, and a rubble - and oncoid-covered slope occurs along the inner portion of the eastern margin (Fig. 7.28). In contrast, bioclastic sediment is moved either down-canyon (smaller arrows) or away from the west side of the canyon along the adjacent shelf, and a steep reef wall has formed in response to lower stress. Storm processes also play a profound role in the sediment budget of the canyon. Detailed transport measurements under a variety of conditions have demonstrated that nearly half of the sediment removed from the canyon in the past century was the result of a single storm, Hurricane Hugo in 1989 (Hubbard 1992b). Like the smaller reef channels at Cane Bay and similar shelf-edge sites throughout the Caribbean, the canyon acts as a temporary holding area for bioclastic and terrigenous sediment that could otherwise overwhelm reef development.

Vertical cores along the western shelf (depth ~10 m), combined with horizontal cores into the

Fig. 7.26. Shelf-edge channel at Cane Bay 1 month after the direct it by Hurricane Hugo, a Category 4–5 storm. Strong return flows scoured most of the sand from this and other channels, exposing anchors that were probably left by eighteenth-century schooners picking up cane, molasses and rum from plantations along the shore. The arrows show the elevation of the sand before the storm. A small coral from a similar anchor in Salt River submarine canyon yielded an uncorrected age of 300 years before present (After Hubbard 1992b by permission of SEPM)

canyon wall (circles in Fig. 7.28) have been used to reconstruct reef development along the west side of Salt River submarine canyon over the past 11,000 years (Fig. 7.29). The oldest coral date from anywhere in the Caribbean (11,100 Cal. BP) comes from the deepest core on the west wall (present depth ~ 30 m). The first 7 m of this core reflects a reef dominated by branching *A. palmata* over the following 4,000 years (Hubbard et al. 1986, 1989b; Hubbard 1992a). After that, massive and platy corals took over. Shortly after 7 ka the shallow reef that separated Salt River Bay from the canyon was

FIG. 7.27. Air photo looking east across Salt River Bay (right) and the submarine canyon (dark reentrant) fronting it. The larger blue arrows show general sediment -transport in from the east and away from the canyon to the west. During storms, strong down-canyon flows (smaller arrows) export sediment along the base of the west wall and into the deep basin north of St. Croix. The small white "dot" near the head of the canyon is the support boat above *Hydrolab*

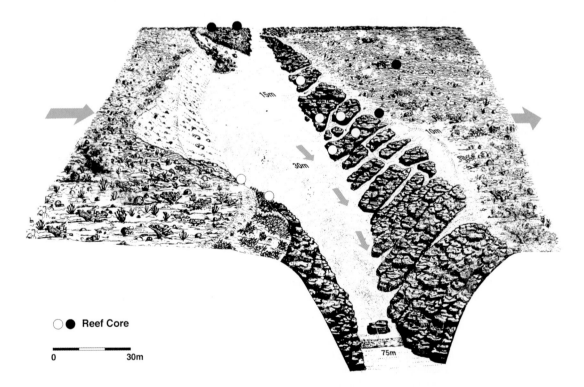

○● **Reef Core**

0   30m

FIG. 7.28. Stylized block diagram looking up-canyon (south) at Salt River. This diagram is based on bathymetry and over 300 dives by the first author and colleagues. Sediment-transport pathways (gray arrows) and cores (black and white circles) are also shown. Figure 7.29 is based on the three seawardmost cores along the west (right) wall (After Hubbard 1989b by permission of West Indies Laboratory)

forming. This gradually created a sheltered lagoon environment in which both carbonate sediment and terrestrial silts and clays coming down Salt River were deposited. Tidal current increasingly moved this material into the canyon during the ebbing tide. As is the case today, easterly waves probably pushed this dirty water against the western canyon margin, signaling the end of *A. palmata* as the dominant reef builder in the canyon. At about the same time, this species was beginning to struggle throughout the region, so it is hard to separate the relative impact of local sedimentation and regional decline. However, when branching corals recovered at other sites around the Caribbean ca. 5,000 years ago, the reef on the west wall did not follow suit, implying a primacy of local sediment stress.

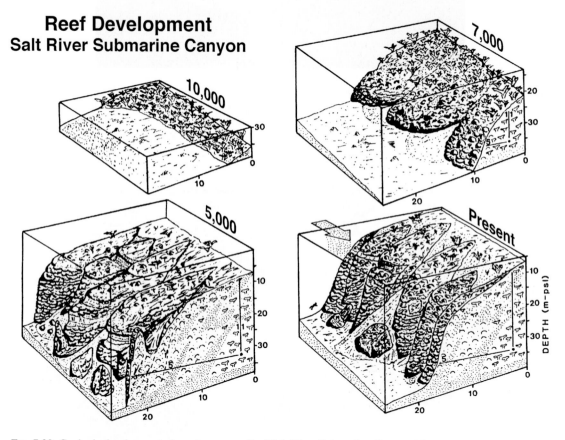

FIG. 7.29. Geologic development along the west wall of Salt River Submarine Canyon. The canyon axis is a remnant channel that probably developed during the last lowstand of sea level, when Salt River crossed the shelf before emptying into the deep basin to the north. As the river valley flooded, *A. palmata* reefs developed along the western side, where sediment pressure from the adjacent shelf was lower (Figs. 7.27, 7.28). Over the next 3,000 years, an increasingly complex and steeper reef flourished, and was dominated by branching corals. By 6 ka reefs had formed near the head of the canyon, creating a sheltered bay, behind which muddy sediments accumulated. Tidal flows increasingly carried fine-grained sediment into the canyon, and wave action pushed that toward the west wall, creating an environment that was increasingly inhospitable to branching *A. palmata*. At about the same time, *A. palmata* appears to have experienced difficulty throughout the region, contributing to the dramatic turnover in the reef community. When branching corals recovered throughout the Caribbean 5,000 years ago, the dominance of massive corals on the west wall persisted, due to their higher tolerance to elevated sedimentation that remains today (Hubbard et al. 1986) (From Hubbard 1989 by permission of West Indies Laboratory)

7.4 Revisiting Long-held Presumptions

7.4.1 The Role of Reef "Framework"

Considering that it has only been a little more than three decades since Ian Macintyre (1975) introduced his revolutionary portable drilling system, the advances in our understanding of coral-reef geology are remarkable. Virtually every coring system that has followed mimics his original design. We have all tinkered with power levels and changed core barrels or tripods to fit the particulars of our objectives and our field areas, but the fundamentals have remained largely unchanged. The development of this system remains as perhaps the single-most significant event in the evolution of our approach to studying Holocene reef geology.

In 1975, our perception of geologic reefs was still one of in-place and interlocking corals creating rigidity and structural framework (e.g., Lowenstam 1950; Newell et al. 1953). Thus, biological production enjoyed a position of primacy, with biological and chemical alterations relegated to subordinate roles. Discussions often focused on why modern reefs were so fundamentally different than their fossil forebears, and whether we should even use Holocene reefs as models for the past – hence, the comment (ca. 1960) "The Present is the Key to the Late Pleistocene … perhaps" (M. Lloyd, personal communication 2003).

These opinions were largely based on extrapolations from the surface of modern reefs to their interiors. The advent of affordable reef drilling and the myriad coring investigations that followed have provided a view of Holocene coral-reef interiors over a wide variety of oceanographic settings. What has emerged is a realization that 60–70% of the carbonate ending up in the reef is typically bioclastic debris (Hubbard et al. 1981, 1990, 1998; Burke et al. 1989; Conand et al. 1997). It is interesting that the notion of largely in-place reefs (Fagerstrom 1987) was not shaken until recently by the long list of cored reefs that inferred otherwise.

The occurrence of bioerosion has been recognized at least since the early nineteenth century (Grant 1826), but the magnitude of its importance has only recently been appreciated (Neumann 1966). As early as 1888, Johannes Walther (translated in Ginsburg et al. 1994) estimated coral abundance in raised reefs of the Sinai at only 40%, with the remainder being bioclastic sediment. Recovery from cores in Panama (Macintyre and Glynn 1976) averaged roughly 70% sand, rubble, and void (Ian Macintyre, personal communication 1998), as was the case in the reefs off Cane Bay, Salt River (Hubbard et al. 1984, 1986) and eastern St. Croix (Burke et al. 1989). Nevertheless, the popular "framework" models of Lowenstam (1950) and Newell et al. (1953) prevailed, despite an apparent later recantation by Newell in 1971. The idea of reefs built less by organisms in growth position than by the encrusted and cemented remains of broken and bioeroded corals was not widely embraced until the 1990s (Hubbard et al. 1990, 1998).

R.N. Ginsburg has provided us with an elegant analogy for reefs through time by way of a long-running Shakespearian play He pointed out that, despite the vagaries of changing taxonomy over evolutionary time, as well as the pervasive impacts of taphonomy, we can make useful comparisons between modern reefs and their fossil counterparts through an understanding of the "plot" and the "roles" of individual players – not in the form of specific actors, but rather the characters they played. This idea was echoed by the "reef guild" concept of Fagerstrom (1987). Looking at moderns reefs, calcification by corals is no longer the "star", but must share the billing with bioerosion (and physical breakage), transport, encrustation and cementation. Looking back through time, we can not only recognize these roles, but can use the varied reef fabric across deep time to consider the ever-changing roles of diversity and competition, ocean chemistry, and local paleoenvironment as they related to the "arms race" between organisms that built reefs and those trying to tear them down (Hubbard et al. 1990).

7.4.2 Coral Growth and Reef Accretion

Schlager (1981) proposed that reef accretion would mirror the fundamentally depth-related pattern of calcification. This was based on an assumption that, while bioerosion and sediment redistribution would occur in any reef, calcification would still dominate. In this scenario, "reef growth" would be roughly an order of magnitude slower than "coral growth", but would follow a pattern that was fundamentally linked to coral type (branching-coral reefs "grow" faster than massive-coral reefs) and water depth (lower light = less calcification = slower accretion). Adey (1977) had asked the question, "If *A. palmata* is capable of rates of reef building of 8 m/1,000 years, then why is it that most shelf margins in the Caribbean are not built to sea level by Holocene reef frameworks?" Schlager's (1981) "Drowning Paradox" is based on these same rates of accretion, and asks the same question. He proposed that Quaternary reefs would drown ("give-up" reefs of Neumann and Macintyre 1985) only if the rate of sea level increased suddenly and dramatically (the "melt-water pulses" of Fairbanks 1989), or water quality suddenly degraded to the point where the potential for "reef growth" was significantly reduced. "Inimical bank waters" associated with sea level overtopping the platform provided the link between Schlager's "Drowning Paradox" and the models of Caribbean reef development based on the landmark study on Lang Bank by Adey et al. (1977). The fundamental principles shared by all these studies are: (1) reef accretion is irrevocably tied to water depth and coral species and (2) many (most?) Holocene reefs are capable of accretion rates in excess of sea-level rise, and must be otherwise compromised in order for them to be left behind.

A synthesis of available reef-accretion data from all across the Caribbean paints a fundamentally different picture (Hubbard 2005). Calibrated radio-carbon ages and paleo-depth reconstructions based on the corrected Lighty et al. (1982) sea-level curve (Fig. 7.14) were used to compute reef accretion in 151 core intervals from reefs off St. Croix, Puerto Rico and the larger Caribbean. The results of this synopsis clearly show a striking lack of dependence on either water depth or coral type within the upper 25 m of the water column (Fig. 7.30). It appears that depth-related patterns of bioerosion and sediment transport may effectively offset the well-documented decrease in carbonate production with depth (Hubbard 2005). Therefore, while the bulk of the raw materials is provided by corals, reef building is as much a physical process as it is a biological one. The term "reef growth" unrealistically implies a dominance of biological "growth" in controlling patterns of reef building. Recent findings argue for reconsidering the term "reef growth": in short, corals grow and reefs accrete.

7.4.3 The Present Caribbean Coral-reef Model

While this chapter focuses on local reefs, the studies detailed above have provided new data that inform our understanding of reef development

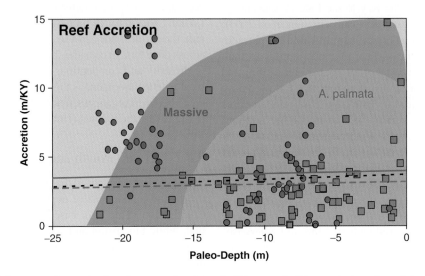

FIG. 7.30. Holocene reef accretion based on core data from Caribbean reefs. The greatest number of *A. palmata* reefs occurs in shallow water (red circles), but the depth range is broad and extends into paleo-water depths similar to those for massive species (green squares). The average accretion rates for *A. palmata* reefs (solid red line) and those dominated by massive corals (green dashed line) are not significantly different. Also, depth-related changes for all reefs (black dashed line) are not statistically significant ($r^2 \sim 005$) (After Hubbard 2005 by permission of GSA)

for the wider Caribbean. Our prevailing model began with three cores from southwestern Lang Bank (Fig. 7.31). According to Adey et al. (1977), reefs dominated by *A. palmata* first formed along the steep platform margins 11,000 years ago (Fig. 7.32a), when rising sea level was rapidly approaching the bank top. As the platform margin flooded, waves and currents mobilized soils that developed during the previous lowstand. Sediment-laden waters flowed off the platform, shutting down reef development (Fig. 7.32b). As sea level continued to rise quickly, once-thriving shelf-edge reefs were left behind. By the time the waters cleared (ca. 8 ka: Fig. 7.32c), water was too deep for branching acroporids, and slower-growing massive species dominated the recovery. The result was a slowly accreting reef that has yet to reach sea level (Fig. 7.32d). Reef "backstepping" was thus caused by compromised calcification in the face of rapidly rising sea level.

This study formed the fundamental underpinning of Caribbean reef models that were to follow. Adey (1978) proposed that for Cenozoic reefs "at

depths greater than 10–20 m, a major shelf rim is likely to experience a significant period of minimal reef growth (1,000–2,000 years)". Following suit, Macintyre (1988) proposed a model for high-energy Caribbean reefs based on "relict give-up reefs along the upper slopes and shelf edges (due to 'inimical bank waters'), and relatively young late Holocene reefs fringing most coastlines". Rapid coral growth logically translated into rapid reef accretion (Schlager 1981), so much so that the failure of reefs to keep up with rising sea level could be explained only by short-lived, upward "sprints" of sea level or episodes of "inimical bank waters" that slowed down calcification and the reef accretion that depended on it. Our present Caribbean reef model relies on the latter scenario.

7.5   Lessons Learned

At the 1977 Miami meeting of the International Society for Reef Studies, Adey (1977) summarized our understanding of Caribbean coral-reef geology,

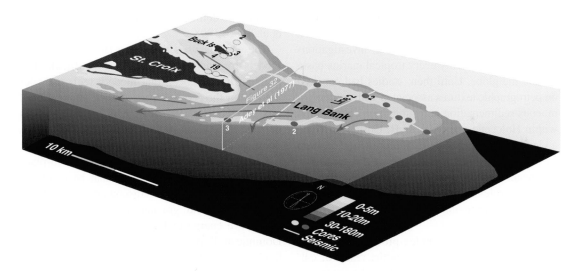

FIG. 7.31. Three dimensional view looking northwest across Lang Bank. Cores shown in red are from early (ca. 10 ka), shelf-edge reefs. Numbers next to core sites indicate multiple cores. The westernmost site on southern Lang Bank (marked "3") was described by Adey et al. (1977). Yellow circles show cores from inshore reefs that formed after 7 ka. Seismic lines are shown in yellow. The location of the cross section in Fig. 7.32 is shown by the yellow box. The brown arrows show likely pathways for turbid bank waters and bioclastic reef debris off the bank. Note how the elevated reef rim all around the bank (light blue) would have protected newly formed shelf-edge reefs from "inimical bank waters" derived from the deeper bank center, except where breaks occurred in the southern margin and along the large channels passing westward. It is likely that the hiatus described by Adey et al. (1977) was as much the result of terrestrial sediments from the bank interior as it was bioclastic debris from flourishing updrift reefs

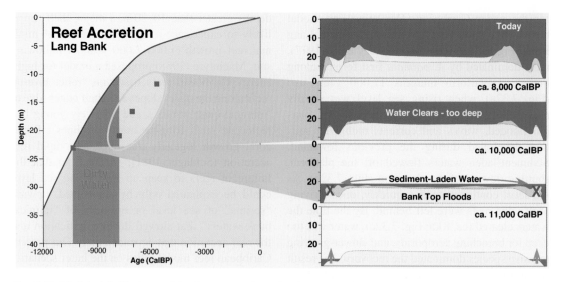

FIG. 7.32. Model of shelf-edge reef development on Lang Bank proposed by Adey et al. (1977). The profile described in panels A-D is located in Fig. 7.31 (yellow box). In this scenario, early *A. palmata* reefs were killed by turbid waters soon after bank flooding ca. 10,000 years ago. These "inimical bank waters" inhibited reef development until 8,000 years ago, by which time water depth was too deep for *A. palmata*. By that time, reef development had shifted to shallower areas closer to shore, as deeper reefs now dominated by massive corals reestablished on Lang Bank

and set the stage for integrating a growing body of information into an overarching reef model. Since then, Holocene reef-coring investigations have spread to virtually every tropical marine basin around the globe. Until then, studies had centered on petrographic textures associated with contemporaneous sedimentation and submarine cementation within Holocene and older reefs. Coring provided a view of the larger-scale reef interior, and allowed us to consider the processes that were responsible. The result has been a greatly improved understanding of the roles of biology, geology and chemistry under changing regimes of physical oceanography and rising sea level.

Reef-building is a complex process that can no longer be understood simply as the result of calcium carbonate set in place solely by the organisms that secrete it. Our prevailing models have focused on the role of in-place and interlocking organisms in providing the rigidity that is the central tenet of reef characterizations (e.g., Fagerstrom 1987). Our new-found ability to sample modern reef interiors has resulted in a realization that much more of the reef edifice is made up of detritus than has been widely held. And, not all of the recognizable coral

remains are demonstrably in life position (Hubbard et al. 1990, 1998).

The good news is that using Holocene reefs as models for ancient ones is far less difficult. The problem has not been that modern reefs are poor models for the geologic past. Rather, our *perceived* modern models have been poor approximations of what they actually look like inside. Structures dominated by in-place corals certainly occur. Algal ridges in the eastern Caribbean are largely in-place constructions of coralline algae that have formed in an environment that discourages branching-coral growth and grazing, in this case, one of high wave energy and frequent hurricanes. In the Dominican Republic, an exposed Holocene reef has a significant component of in-place massive corals (Hubbard et al. 2004).

Grazing has been a problem for reef builders (early stromatolites) since the advent of the articulated and skeletonized jaw. Likewise, infaunal bioerosion has reduced substrate to sediment, leaving a weakened skeleton that is more susceptible to breakage. The general absence of significant framework build-ups in the rock record no longer poses a problem. Modern reefs are no different;

they often comprise significant volumes of bound bioclastic debris – just like their ancient forebears.

What then does this say about our present model of Caribbean coral-reef development? As the bank margins of Puerto Rico and the US Virgin islands flooded between 12,000 and 10,000 years ago, wave action undoubtedly encountered sediments formed during the long interval of exposure associated with lower sea level. Based on the initial study of Lang Bank, it was assumed that once additional cores were recovered from other Caribbean sites, a picture akin to Fig. 7.33a would emerge: an earlier reef sequence dominated by *A. palmata* and separated by a 2,000-year hiatus from overlying and disconformable massive species. However, more recent core data paint a decidedly different picture (Fig. 7.33b). Neither the reef gap between 10 and 8 ka related to "inimical bank waters" nor the subsequent shift to massive corals occurred at other sites in the Caribbean. In fact, reefs all around Lang Bank thrived throughout the period when *A. palmata* was absent from the reefs at the Adey et al. (1977) site. New seismic data from eastern St. Croix (Fig. 7.34) show an elevated Pleistocene rim that would have directed turbid bank-top sediments from the deeper center of platform to the west and away from most of the reefs atop the elevated

margin. The depression of central Lang Bank connected to Buck Island Channel north of St. Croix and a similar feature along the southern shelf (Fig. 7.31). The southern branch would have been an effective conduit to move sediment through breaks in the rim and toward the core site where our Caribbean shelf-edge model began (brown arrows in Fig. 7.31). As is the case with modern reefs, storm waves would have moved large volumes of water into what was then a deep lagoon behind the shelf-edge reefs. The transport of this water to the west would have carried terrestrial sediment languishing in the usually quiet and protected environment. Ironically, the reefs described by Adey et al. (1977) were probably impacted by not only this terrestrial sediment from the bank, but by bioclastic debris from thriving updrift reefs on Lang Bank as well.

Four important and new points emerge from Fig. 7.33. First, while the "inimical bank water" story of Adey et al. (1977) is a valid explanation for the reefs near the southwestern corner of Lang Bank, extrapolation to a larger scale seems unwarranted. The hiatus they described occurred neither elsewhere on Lang Bank nor at many other Caribbean and western Atlantic sites. This runs counter to the core principles upon which our pre-

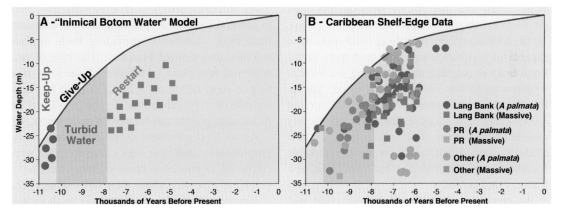

FIG. 7.33. Comparison between the "Inimical Bank Waters" model for Caribbean reef development extrapolated from Lang Bank (A: Adey et al. 1977) and more recent data from shelf-edge reefs throughout the Caribbean (B: Lighty et al. 1982; Hubbard et al. 1997, 2005; Toscano and Macintyre 2003; Hubbard, unpublished data). In contrast to the hiatus envisioned by Adey (1978) and Macintyre (1988) *A. palmata* flourished throughout the proposed "Lang Bank gap". Also, corals after the proposed hiatus (10–8 ka) were not confined to massive species, as was proposed on the southern bank. Most reefs around Puerto Rico and St. Croix were catching up throughout the proposed "turbid water" interval and, by 6,000 years ago, were facing a slowing sea level (After Hubbard 2005)

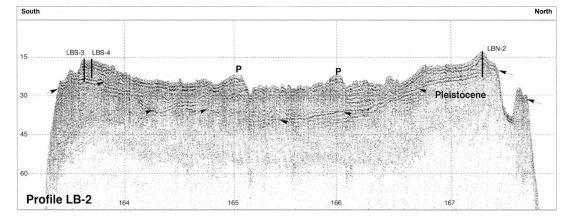

FIG. 7.34. North–south seismic line across Lang Bank showing the elevated Pleistocene rim that isolated the deeper, central lagoon from the outer reefs that were forming 10,000 years ago. Arrows show the Pleistocene surface, which has been confirmed by cores LBN-2 and LBS-3. The reefs, which have built to present depths of ca. 13 m, started along the top of this feature, and thrived until ca. 6,000 years ago, well after the proposed hiatus related to "inimical bank waters". Deeper reefs on the northern ridge appear thinner than those on the shallow rim. The location of this profile (yellow line labeled "Line 2") is shown in Fig. 7.31 (Data were collected with Chuck Holmes of the US Geological Survey)

vailing Caribbean reef models are based. Second, *A. palmata* built extensive reefs at water depths (up to 25–27 m) well beyond the 10-m figure we are used to seeing as the lower limit for extesive branching-coral framework. The deepest reported observation of live *A. palmata* to date is from 23 m of water in the Gulf of Mexico (Zimmer et al. 2006) and that colony subsequently died (William Precht, personal communication, 2006). Third, even the deeper branching coral reefs were capable of significant accretion and were either keeping up or catching up with rising sea level at a time when it was rising at rates of ca. 5 m/ka. Finally, these *A. palmata* reefs "quit" around 6,000 years ago, despite their having largely caught up to a then slowing sea level. A new model is needed that builds on the considerable foundations of earlier studies but takes these new findings into account.

We propose a model of Caribbean coral reef development that is far simpler than the current one. Shelf margins were flooded between 12 and 10 ka. Coral recruitment began soon after, with "inimical bank waters" having less impact than has been widely held. Carbonate production was high, but this was effectively offset by storm damage and bioerosion, as reflected in the abundant bioclastic

debris within the undated lower sections of most shelf-edge cores. Proximity to the shelf edge would have facilitated off-bank transport. Thus, initial carbonate production and subsequent degradation may have been even higher than what is reflected in the abundant sediment accumulation at the base of each core.

As bioclastic sediment progressively gave way to preserved corals, *A. palmata* was the dominant framework contributor, providing both in-place colonies and toppled branches. The rate of reef accretion was close to the rate of sea-level rise, despite most of the reefs being at what has been considered to be suboptimal depths (>5–10 m) for *A. palmata* growth and framework production.

As sea level continued to rise ever more slowly, progressively shallower sites closer to shore were colonized, in particular those sites that were topographically elevated above their surroundings. The dominance of massive species closer to shore was probably a function of sediment input from nearby highlands. Reefs at the shelf edge continued to be dominated by fast-growing acroporids, as were shallower reefs further removed from terrestrial effects (e.g., Buck Island Bar: Macintyre and Adey 1990; Hubbard et al. 2005). As waters cleared closer to shore, *A. palmata* increasingly took over the role as the primary frame-builder. A major tipping point in this scenario was the slowing of sea-level rise ca. 7 ka. Most of the reefs closer to shore started around this time and built not only upward, but also outward (Fig. 7.35) as carbonate production exceeded accommodation space being created by an ever slower sea-level rise. This pattern can be seen in many other Caribbean reefs forming at the same time.

The one wrinkle in this otherwise simpler story is the appearance of two gaps in the recorded history of *A. palmata* during the Holocene development of Caribbean coral reefs (yellow bars in Fig. 7.14). Over 200 samples of *A. palmata* from virtually every region of the Caribbean reflect abundant branching corals over the past 10,000 years. Yet, only three samples have been dated between 5.8 and 5.2 ka. No dated samples have been reported between 3 and 2.2 ka (Hubbard et al. 2005).

The first of these two gaps coincides with the abandonment of shelf-edge reefs throughout the region. Based on available core data, reefs off

Florida appear to have started their decline earliest (8–6.8 ka), followed by those off Puerto Rico (7.9–6.9 ka), and eventually St. Croix (6.6–5.9 ka) (Table 7.2). Whether this temporal progression has any significance or simply reflects the vagaries of preservation or sampling, it appears that branching-coral reefs throughout the region had by-and-large given up by ca. 5.8 ka (at both shelf edge and inshore sites). A few areas already inhabited by massive species continued to accrete along the shelf edge until 5,000–3,000 years ago, but where *A. palmata* had dominated, a return to massive species was conspicuously absent. Inshore reefs switched to massive corals and never skipped a beat. Despite present-day coral cover up to 60% at some sites, little or no reef accretion has subsequently occurred at any of the shelf-edge sites where cores have been taken.

The widely accepted models of Adey et al. (1977) and Macintyre (1988) linked "inimical bank waters" at the time of platform flooding to the demise of the shelf-edge reefs and eventual backstepping. However, shelf-edge reefs cored on

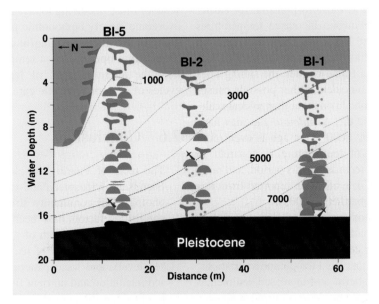

FIG. 7.35. Core transect across the northern Buck Island reef. Note that the reef built both upward and outward continually throughout its history. The slower sea-level rise after 7 ka allowed branching- and massive-coral reefs alike to produce carbonate at rates faster than accommodation space was being made available. Also note that the patterns of accretion appear to be independent of the dominant coral type. Similar patterns can be seen in many other Caribbean reefs formed in shallow water after 8 ka (After Hubbard et al. 2005 by permission of SEPM)

St. Croix and Puerto Rico (and elsewhere in the Caribbean) not only survived the period of bank flooding, but appear to have flourished throughout its presumed duration (Hubbard et al. 1997, 2001). Based on the sea-level curve and samples plotted in Fig. 7.14, water depth over the abandoned reefs varied from as much as 20–25 m off Puerto Rico and St. Croix to as little as 0.4–8.0 m off Florida. All had accreted at rates equal or close to sea-level rise. By the time the shelf-edge reefs quit, they had survived the "inimical bank waters" that were supposed to have killed them, had tracked rapidly rising sea level, and were entering a period of slowed sea-level rise that should have facilitated their catching up. Nevertheless, the apparently vibrant *A. palmata* community at each site stopped building framework, not only along the shelf edge but along shallower reefs closer to shore. Despite the absence of rapidly growing branching corals, shallow-water reefs not only continued to accrete at the same rate, but began to build seaward as accommodation space was decreasingly available.

By ca. 5.2 ka, *A. palmata* appears to have returned as a major frame-builder, but only along inshore reefs. The reasons for this recovery is as poorly understood as those responsible for earlier difficulties. If we invoke increased temperatures associated with the Holocene Thermal Maximum (Kaufman et al. 2004) to explain the sudden loss of *A. palmata* as a frame-builder, the timing does not appear to be coincident. Other possible causes are difficult to reconcile on other than a local scale. However enigmatic this *A. palmata* gap may have been, its repetition 3,000 years ago is even more so. No elevation of temperature is apparent in published climate records for this period, and not a single *A. palmata* date has been reported from this time interval (Hubbard et al. 2005).

The likelihood of these gaps being a sampling artifact seems small, considering the number of sites and investigators involved. Hubbard et al. (2005) concluded that local temperature variations, changing storm patterns and increased sedimentation or nutrients could not have occurred at all sites with the synchronicity required to explain these two gaps on a Caribbean-wide scale. Blanchon and Shaw (1995) and Blanchon (2005) have argued for a sudden jump in the rate of sea-level rise, but this is not supported by isotopic data, and is difficult to

envision given the position of sea level at the time (i.e., too close to present).

Regardless of the cause, major community shifts appear to have occurred at least twice in the late Holocene. Moreover, they involved the same species that signaled the start of the most recent reef decline – *A. palmata*. While we might be tempted to argue that the recent situation is simply the latest in a possibly cyclic phenomenon, this is most likely not the case. Even if the root causes were similar, the additional stresses today related to overfishing, increased sediment and nutrients, and probably rapidly rising global temperature were not present either 6,000 or 3,000 years ago. Thus, even if there are similarities in some of the triggering mechanisms of past and present events, factors that might limit or encourage subsequent recovery are assuredly greater today. Nevertheless, identifying the root causes of these past gaps is of considerable importance. How this might inform not only our understanding of why reef communities change, but also why and how they recover is critical and demands attention. The signal preserved in the recent geologic record is clearly sensitive enough to allow us to identify past turnover events on the scale of those occurring today. The challenge is to overcome largely taphonomic problems if we are going to provide a meaningful context for natural change that operates on a temporal scale longer than the attention span of legislators, funding cycles of NSF or scientific careers.

7.6 Conclusions

1. The reefs of Puerto Rico and the US Virgin Islands are widespread and varied in their morphology and community structure. The major controlling factors have been location, tectonic regime, sea level, depth of the antecedent substrate (and the timing of its flooding by rising sea level), and local patterns of current flow, sedimentation and nutrient flux.
2. Reefs along submerged shelf margins in Puerto Rico and the Virgin Islands started 10,000–12,000 years ago on antecedent rims or slope breaks 20–35 m below present sea level, and in paleo-water depths up to 25 m. These early reefs were dominated by branching *A. palmata*.

3. As sea level rose, reefs formed closer to shore 7,000–8,000 years ago. These reefs were largely dominated by massive corals, in response to higher levels of terrestrial sediment /nutrient input and possibly less stable substrates. Over time, their forereef slopes steepened, and *A. palmata* increasingly dominated the reef crest.

4. Reefs starting on shallower antecedent surfaces since 8 ka by and large accreted in a regime of slowing sea-level rise. They built continuously, and presently thrive at or near sea level. Exceptions include algal ridges in Boiler Bay (St. Croix) that were cut off by the subsequent development of Tague Reef (Adey and Burke 1976), and Buck Island Bar where unusually high wave energy has prevented the reef from building to sea level (Macintyre and Adey 1990).

5. Schlager's "Drowning Paradox" was built on the fundamental assumption that reef accretion follows a depth- and species-related pattern that mirrors coral growth (i.e., shallow *Acropora* reefs build faster than deeper reefs dominated by massive species). Also, most reefs are capable of accreting at rates faster than Holocene sea-level rise. Following these assertions, reefs should not "give up" unless sea level suddenly accelerates above ca. 10 m/ka or environmental conditions suddenly deteriorate to the point where calcification is compromised or terminated. Our prevailing Caribbean model invokes "inimical bank waters" as the solution to this "paradox", and presumes that abandonment of Caribbean shelf-edge reefs and backstepping to shallower sites was linked to this phenomenon.

6. Accretion data from all Caribbean cores reveal a pattern that contradicts Schlager's presumption that most reefs have been capable of outpacing even early Holocene sea-level rise. Regardless of water depth or dominant coral type, most reefs build at rates generally falling between 3 and 5 m/ka. Thus, reefs that were building prior to 7,000 years ago would have had difficulty keeping up with rising sea level under all but the most ideal conditions. The shelf edge apparently provided such an environment, and reefs on St. Croix and Puerto Rico kept pace with early an early Holocene sea-level rise on the order of 5 m/ka.

7. In contrast to earlier models, reef accretion continued throughout the flooding interval in Puerto Rico and the Virgin Islands. Core data from cored sites in Puerto Rico and the Virgin Islands, appear to invalidate the "inimical bank waters" scenario for all but a few specific sites. It seems more likely that, in the larger Caribbean, most shelf-edge reefs not only survived flooding, but thrived through it.

8. Commonplace and thick (up to several meters) detrital sequences at the base of most shelf edge cores imply that the lags of up to 3,000 years (more typically 1,500 years), previously attributed a lack of corals due to "inimical bank waters" (IBW) as the bank flooded, more likely represent poor framework preservation. High bioerosion related to elevated nutrient levels, combined with lower recruitment rates due to sedimentation and turbidity, may have been the cause. In this scenario, IBW had a bearing on reef accretion, but less so on coral recruitment. A flourishing community capable of reef building probably existed and gradually increased in its ability to build preservable framework. Most important, once reef building started, it was largely uninterrupted, and backstepping was unrelated to IBW developing along the platform margin.

9. Shallow reefs (barrier and fringing) started after 8 ka off Puerto Rico and the Virgin Islands, and throughout most of the Caribbean. They have developed under a regime of slowing sea level, and have had no difficulty in keeping up. As a result, they have built both vertically and laterally as carbonate production exceeded accommodation space.

10. Enigmatically, shelf-edge reefs eventually quit ca. 6,000 years ago when *A. palmata* ceased building framework. The loss of *A. palmata* framework rom the record also occurred on shallower reefs closer to shore, and apparently throughout the Caribbean. Many of the affected reefs continued to build, but the frame-building community was limited to massive species. Along the shelf edge, *A. palmata* reefs never recovered. The hiatus in *A. palmata* accretion lasted some 800 years and was repeated 3,000 years ago. The reasons for these changes in Caribbean community structure are unclear, but their synchronicity requires a regional explanation. Both sea level and "inimical bank waters" offer poor explanations. Whether or not these intervals of

TABLE 7.2. Data for "give-up" reefs along Caribbean shelf margins.

Core/site	Location	Age (ka)	D (m)[a]	SL (m)	PaleoD (m)	Source
LBN-1	Lang Bank	6.2	13.0	−6.0	7.0	Hubbard (unpublished)
LBN-2	Lang Bank	6.6	12.5	−6.5	6.0	Hubbard (unpublished)
LBS-3	Lang Bank	6.4	15.0	−6.3	8.8	Hubbard (unpublished)
LB-7	Lang Bank	5.9	25.5	−5.5	20.0	Hubbard (unpublished)
LB-8	Lang Bank	6.4	12.5	−6.3	6.3	Hubbard (unpublished)
LB-9	Lang Bank	6.4	13.7	−6.3	7.5	Hubbard (unpublished)
PAR-2	SW PR	6.9	14.0	−7.1	6.9	Hubbard et al. (1997)
PAR-3	SW PR	7.0	13.4	−7.5	5.9	Hubbard et al. (1997)
PAR-11	SW PR	6.9	13.5	−7.1	6.4	Hubbard et al. (1997)
PAR-16A	SW PR	7.8	12.5	−10.0	2.5	Hubbard et al. (1997)
PAR-A	SW PR	7.9	17.6	−10.3	7.4	Hubbard et al. (1997)
SR-5	Salt R	6.2	31.0	−6.0	25.0	Hubbard et al. (1985)
	Fla	7.0	7.9	−7.5	0.4	Precht (in Toscano and Macintyre 2003)
	Fla	8.0	17.5	−10.9	6.6	Lighty et al. (1982)
	Sand Key	6.8	9.5	−6.8	2.7	Toscano and Macintyre (2003)
	Carysfort	7.0	9.7	−7.5	2.2	Toscano and Macintyre (2003)

[a] Depth in meters below present sea level

poor *A. palmata* preservation share anything with the recent epidemic of branching-coral loss, understanding the causes of such large-scale community shifts provides both opportunities and challenges with respect to unraveling both natural and anthropogenic change.

Acknowledgements Trying to adequately acknowledge all the appropriate people in a summary paper like this is one of the few things more daunting then reef classification. Many of the five authors have worked together for 3 decades, and the number of individuals who have either impacted on our careers, helped us come together, or supported us substantially in our research is truly staggering. Countless undergraduates and colleagues at West Indies Lab probably come to the fore, and we count among these all the support staff who were no less important to our coring efforts than anyone else – especially in the wake of Hurricane Hugo, when it would have been so much easier to take care of matters at home than to help us core. The efforts to rebuild the facility were ultimately thwarted, and it is, thus, safe to say that these studies were successfully completed despite the level of support by Fairleigh Dickinson University. We are also obliged to staff and colleagues at the UPR Marine Lab at Isla Maguayes. The boat crews, the dive-support staff, and the students were unfailing in their efforts.

While a list of specific supporters is too long to include, there are a few individuals who have been so influential in our collective careers and in creating an environment in which these studies have been made possible, that we list them individually at the risk of insulting others who have been incredibly important: Walter Adey, Bob Dill, Ian Macintyre and Jack Morelock. This work has enjoyed extensive support of the National Science Foundation, NOAA 's National Undersea Research Program and Sea Grant Program, the National Park Service, the National Institute for Global Environmental Change, the Petroleum Research Fund of the American Chemical Society, and internal funds from the home institutions of all the authors.

In a larger sense, this paper owes everything to those earlier reef workers who challenged us to change the scale at which we view modern reefs. Walter Adey's broad view of Caribbean -wide patterns showed us the tremendous value of regional context. Our favorite reef no longer existed in a vacuum. Reefs varied along gradients of energy such that no single model was adequate. Not only was it OK that our reef didn't fit the Jamaica model – it shouldn't. Ian Macintyre's submersible coring system allowed, for the first time, access to modern reef interiors at a cost that was affordable and did not require a massive expeditionary force and a large, expensive drilling ship. This heralded a revolution in the way we have come to view modern reefs and their relationship to their ancient forebears. Peter Davies and David Hopley carried the technique to Australia, and the focus gradually shifted away from who's reef was bigger or better – they were either different or similar, and what was important was understanding why. Names have been added to this list: Cabioch, Camoin, Dullo, Fairbanks, Fletcher, Geister, Grossman, Montaggioni, Shinn, my co-authors, and many more. But, whatever any of us might add to the revolutionary ideas laid out at the ISRS Meeting in Miami in 1977, we all draw from them.

References

Acevedo R, Morelock J, Olivieri RA (1989) Modification of coral reef zonation by terrigenous sediment stress. Palaios 4:92–100

Adey WH (1975) The algal ridges and coral reefs of St. Croix : their structure and Holocene development. Atoll Res Bull 187:1–67

Adey WH (1977) Shallow water Holocene bioherms of the Caribbean Sea and West Indies. Proc 3rd Int Coral Reef Symp 2:xx1–xxiv

Adey WH (1978) Coral reef morphogenesis: a multidimensional model: Science 202:831–837

Adey WH, Burke RB (1976) Holocene bioherms (algal ridges and bank barrier reefs) of the eastern Caribbean. Geol Soc Am Bull 87:95–109.

Adey WH, Burke RB (1977) Holocene bioherms of Lesser Antilles – geologic control of development, in: Frost SH, Weiss MP, Saunders JB (eds), Reefs and Related Carbonates – Ecology and Sedimentology, AAPG Stud Geol 4:67–81

Adey WH, Macintyre IG, Stuckenrath R, Dill RF (1977) Relict barrier reef system off St. Croix : its implications with respect to late Cenozoic coral reef development in the western Atlantic. Proc 3rd Int Coral Reef Symp 2:15–21

Blanchon P (2005) Comments on "Corrected western Atlantic sea-level curve for the last 11,000 years based on calibrated ^{14}C dates from *Acropora palmata* framework and intertidal mangrove peat". Coral Reefs 24:183–186

Blanchon P, Shaw J (1995) Reef drowning during the last deglaciation; evidence for catastrophic sea-level rise and ice-sheet collapse. Geology 23:4–8

Bloom AL (1976) Atlas of sea level curves. International Geological Correlation Programme, Project 61:C18

Burke RB, Adey WH, Macintyre IG (1989) Overview of the Holocene history, architecture and structural components of Tague reef and lagoon, in: Hubbard DK (ed), Terrestrial and Marine Geology of St. Croix,

U.S. Virgin Islands. Spec Publ No. 8, West Indies Laboratory:105–109

Conand C, Chabanet P, Cuet P, Letourneur Y (1997) The carbonate budget of a fringing reef in La Reunion Island (Indian Ocean): sea-urchin and fish bioerosion and net calcification. Proc 8th Int Coral Reef Symp 1:953–958

Connell JH (1978) Diversity in tropical rain forests and coral reefs. Science 199:1302–1309

Cutler KB, Edwards RL, Taylor FW, Cheng H, Adkins J, Gallup CD, Cutler PM, Burr GS, Bloom AL (2003) Rapid sea-level fall and deep-ocean temperature change since the last interglacial period. Earth Planet Sci Lett 206:253–271

Darwin C (1842) The Structure and Distribution of Coral Reefs. Reprinted by the University of Arizon Press, 1984, 214 pp

Fagerstrom A (1987) The Evolution of Reef Communities, Wiley 600 pp

Fairbanks RG (1989) A 17,000 year glacio-eustatic sea level record: influence of glacial melting rates on the Younger Dryas event and deep-ocean circulation. Nature 342:637–642

Frost SH (1977) Cenozoic reef systems of Caribbean: prospects for paleoecologic synthesis, in: Frost, S, Weiss M, Saunders J (eds), Reefs and Related Carbonates – Ecology and Sedimentology. AAPG Stud Geol 4:93–110.

Frost SH, Weiss MP (1979) Patch-reef communities and succession in the Oligocene of Antigua, West Indies. Geol Soc Am Bull 90:I 612–I 616

Gallup CD, Edwards RL, Johnson RG (1994) The timing of high sea levels over the past 200,000 years. Science 263:796–800

Geister J (1977) The influence of wave exposure on the ecology and zonation of Caribbean coral reefs. Proc 3rd Int Coral Reef Symp 1:23–29

Gill IP, Hubbard DK, McGlaughlin, PP, Moore CH (2002) Distribution and types of porosity within the Kingshill aquifer, in: Renkin RA et al. (eds), Geology and Hydrogeology of the Caribbean Islands Aquifer System of the Commonwealth of Puerto Rico and the U.S. Virgin Islands. U.S. Geol Surv Prof Pap 1419:123–131

Ginsburg RN, Gischler E, Schlager W (1994) Johannes Walther on reefs: Comparative Sedimentology Laboratory, University of Miami, Geological Milestones II, 141 pp

Gischler E (2006) Comments on "Corrected western Atlantic sea-level curve for the last 11,000 years based on calibrated ^{14}C dates from *Acropora palmata* framework and intertidal mangrove peat". Coral Reefs 25:273–279

Gischler E, Hudson JH (2004) Holocene development of the Belize barrier reef. Sedim Geol 164:223–236

Goreau TF, Land LS (1974) Fore-reef morphology and depositional processes, north Jamaica, in: Laporte LF (ed), Reefs in Time and Space. Soc Econ Petrol Mineral Spec Publ No. 18:77–98

Grant RE (1826) Notice of a new zoophyte (*Cliona celata* Gr.) from the Firth of Forth. Edinburg. New Philos J, Apr–Oct:78–81

Hubbard DK (1986) Sedimentation as a control of reef development : St. Croix, U.S.V.I. Coral Reefs 5:117–125

Hubbard DK (1989a) Modern carbonate environments of St. Croix and the Caribbean : a general overview, in: Hubbard DK (ed), Terrestrial and Marine Geology of St. Croix, U.S. Virgin Islands. Spec Publ No. 8, West Indies Laboratory:85–94

Hubbard DK (1989b) Depositional environments of Salt River estuary and submarine canyon, St. Croix, U.S.V.I., in: Hubbard DK (ed.), Terrestrial and Marine Geology of St. Croix, U.S. Virgin Islands. Spec Publ No. 8, West Indies Laboratory:181–196

Hubbard DK (1989c) The shelf-edge reefs of Davis and Cane Bay, northwestern St. Croix, in: Hubbard DK (ed), Terrestrial and Marine Geology of St. Croix, U.S. Virgin Islands. Spec Publ No. 8, West Indies Laboratory:167–180

Hubbard DK (1992a) A modern example of reef backstepping from the eastern St. Croix shelf, U.S. Virgin Islands. Proc Amer Assoc Petrol Geol, Ann Meet:57

Hubbard DK (1992b) Hurricane-induced sediment transport in open-shelf tropical systems – an example from St. Croix, U.S. Virgin Islands. J Sedim Pet 62:946–960

Hubbard DK (2005) Holocene reef development in the Caribbean, geologic perspectives revisited: abstracts with programs. Geol Soc Am:400

Hubbard DK, Ward LG, FitzGerald, DM (1976) Reef morphology and sediment transport, Lucaya, Grand Bahama Island. Proc Am Assoc Petr Geol Ann Meet, New Orleans, LA

Hubbard DK, Sadd JL, Miller AI, Gill IP, Dill RF (1981) The Production, Transportation and Deposition of Carbonate Sediments on the Insular Shelf of St. Croix, U.S. Virgin Islands : Technical Report No. MG-1, West Indies Laboratory, St. Croix, USVI, 145 pp

Hubbard DK, Sadd JL, Roberts HH (1982) The role of physical processes in controlling sediment -transport patterns on the insular shelf of St. Croix, U.S.V.I. Proc 4th Int Coral Reef Symp 1:399–404

Hubbard DK, Burke RB, Gill IP (1984) Accretionary styles in shelf-edge reefs, St. Croix, U.S.V.I. Proc Am Assoc Petr Geol Ann Meet, San Antonio, TX

Hubbard DK, Burke RB, Gill IP (1985) Accretion in shelf-edge reefs, St. Croix, U.S.V.I. in: Crevello PD, Harris PM (eds), Deep-water carbonates. SEPM Core Workshop No. 6:491–527

Hubbard DK, Burke RB, Gill IP (1986) Styles of reef accretion along a steep, shelf-edge reef, St. Croix, U.S. Virgin Islands. J Sedim Petrol 56:848–861

Hubbard DK, Venger L, Parsons K, Stanley D (1989) Geologic development of the west-end terrace system on St. Croix, U.S. Virgin Islands, in: Hubbard DK (ed), Terrestrial and Marine Geology of St. Croix. Spec Pub No. 9, West Indies Laboratory:73–84

Hubbard DK, Miller AI, Scaturo D (1990) Production and cycling of calcium carbonate in a shelf-edge reef system (St. Croix, U.S. Virgin Islands): applications to the nature of reef systems in the fossil record. J Sedim Petrol 60:335–360

Hubbard DK, Gill IP, Burke RB, Morelock J (1997) Holocene reef backstepping – southwestern Puerto Rico shelf. Proc 8th Int Coral Reef Symp 2:1779–1784

Hubbard DK, Burke RB, Gill IP (1998) Where's the reef: the role of framework in the Holocene. Carbonate Evaporite 13:3–9

Hubbard DK, Gill IP, Burke RB (2000) Caribbean -wide loss of A. *palmata* 7,000 yr ago: sea-level change, stress, or business as usual? Abstracts and Programs. 9th Int Coral Reef Symp:57

Hubbard DK, Gill IP, and Burke RB, 2001, The Coral Reefs revisited: inimical bottom waters turned friendly. Geol Soc Am Abstracts with Programs:A-408

Hubbard DK, Ramirez W, Davis A, Lawson GR, Oram J, Parsons-Hubbard K, Greer L, Cuevas D, and del Coro M (2004) A preliminary model of Holocene coral-reef development in the Enriquillo Valley, SW Dominican Republic. Geol Soc Am Abstracts with Programs:313

Hubbard DK, Zankl H, Van Heerden I, Gill IP (2005) Holocene Reef Development Along the Northeastern St. Croix Shelf, Buck Island, U.S. Virgin Islands. J Sedim Res 75:97–113

Imbrie J, Hays JD, McIntyre A, Mix AC, Morley JJ, Pisias NG, Prell W, Shackleton NG (1984) The orbital theory of Pleistocene climate : support from a revised chronology of the marine $\delta^{18}O$ record, in: Berger A, Imbrie J, Hays J, Kukla G, Saltzman B (eds), Milankovitch and Climate, Part 1. Reide, Boston, MA, pp 269–305

Kaufman DS, Ager TA, Anderson NJ, Anderson PM, Andrews JT, Bartlein PJ, Brubaker LB, Coats LL, Cwynar LC, Duvall ML, Dyke AS, Edwards ME, Eisner WR, Gajewski K, Geirsdottir A, Hu FS, Jennings AE, Kaplan MR, Kerwin MW, Lozhkin AV, MacDonald GM, Miller GH, Mock CJ, Oswald WW, Otto-Bliesner BL, Porinchu DF, Ruhland K, Smol JP, Steig EJ, Wolfe BB (2004). Holocene thermal maximum in the western Arctic (0–180 W). Quat Sci Rev 23:529–560

Lighty RG, Macintyr, IG, Stuckenrath R (1982) *Acropora palmata* reef framework : a reliable indicator of sea level in the western Atlantic for the past 10,000 years. Coral Reefs 1:125–130

Lowenstam HA (1950) Niagaran reefs of the Great Lakes area geological society of America. Memoir 67:215–248

Macintyre IG (1975) A diver operated hydraulic drill for coring submerged substrates. Atoll Res Bull 185:21–25

Macintyre IG (1988) Modern coral reefs of western Atlantic : new geologic perspective. Amer Assoc Petrol Geol Bull 72:1360–1369

Macintyre IG, Adey WH (1990) Buck Island Bar, St. Croix, USVI: a reef that cannot catch up with sea level. Atoll Res Bull 336:1–7

Macintyre IG, Glynn PW (1976) Evolution of modern Caribbean fringing reef, Galeta Point, Panama. Amer Assoc Petrol Geol Bull 60:1054–1072

Macintyre, IG, Raymond, B, Stuckenrath R (1983) Recent history of a fringing reef, Bahia Salina del Sur, Vieques Island, Puerto Rico. Atoll Res Bull 268:1–9

Macintyre IG, Multer HG, Zankl H, Hubbard DK (1985) Growth and depositional facies of a windward reef complex (Nonsuch Bay, Antigua, W.I.). Proc 5th Int Coral Reef Symp 6:605–610

Masson DG, Scanlon KM (1991) The neotectonic setting of Puerto Rico. Geol Soc Am Bull 103:144–154.

Milliman JD (1973) Caribbean coral reefs, in: Jones OA, Endean R (eds), Biology and Geology of Coral Reefs, Academic

Muhs DR, Simmons KR, Taggart BE, Prentice, CS, Joyce J, Troester JW (2005) Timing and duration of the last interglacial period from U-series ages of unaltered reef corals, Northern and Western Puerto Rico. Abstracts of 17th Caribbean Geological Conference, San Juan, PR

Neumann AC (1966) Observations on coastal erosion in Bermuda and measurements of the boring rate of the sponge Cliona lampa. Limnol Oceanogr 11:92–108

Neumann AC (1977) A sea-level curve for Bermuda, in: Bloom AL (ed), Atlas of Sea Level Curves. International Geological Correlation Programme, Project 61, p C18

Neumann AC, Macintyre I (1985) Reef response to sea-level rise: keep-up, catch-up, or give-up. Proc 5th Int Coral Reef Symp 3:105–110

Newell ND (1971) An outline history of tropical organic reefs. Am Mus Novit 2465:1–37

Newell ND, Rigby JK, Fischer AG, Whiteman AJ, Hickox JE, Bradley JS (1953) The Permian Reef Complex of the Guadalupe Mountains Region, Texas and New Mexico: A Study in Paleoecology. W.H. Freeman, San Francisco, CA, 236 p

Playford PW (1980) Devonian "Great Barrier Reef" of Caning Basin, Western Australia. Amer Assoc Petrol Geol Bull 64:814–840

Pindell JL (1994) Evolution of the Gulf of Mexico and the Caribbean, in: Donovan SK, Jackson T (eds),

Caribbean Geology: And Introduction, University of W Indies Publishers Association/University of the West Indies Press, Kingston, Jamaica, pp 13–39

Pindell JL, Barrett SF (1990) Geologic evolution of the Caribbean region: a plate-tectonic perspective, in: Dengo G, Case JE (eds), The Caribbean Region. Geol Soc Am, Geol North Am Ser H:405–432

Pindell J, Cande S, Pitman W, Rowley D, Dewey J, Labrecque J, Haxby W (1988) A plate-kinematic framework for models of Caribbean evolution. Tectonophysics 115:121–138

Prentice CS, McGeehin J, Simmons KR, Muhs, DR, Roig C, Joyce J, Taggart B (2005) Holocene marine terraces in Puerto Rico: evidence for tectonic uplift?. Abstracts of 17th Caribbean Geological Conference, San Juan, PR

Reid JA, Plumley, PW Schellekens, JH (1991) Paleomagnetic evidence for Late Miocene counter-clockwise rotation of north coast carbonate sequence, Puerto Rico. Geophys Res Lett 18:565–568

Roberts HH, Murray SP, Suhayda JN (1977) Physical processes in a fore-reef environment. Proc 3rd Int Coral Reef Symp 2:507–515

Sadd JL (1984) Sediment transport and $CaCO_3$ ≃ budget on a fringing reef, Cane Bay, St. Croix, U.S. Virgin Islands. Bull Mar Sci 35:221–238

Schlager W (1981) The paradox of drowned reefs and carbonate platforms. Geol Soc Am Bull 92:197–211

Shackleton NJ (1987) Oxygen isotopes, ice volume and sea level. Quat Sci Rev 6:183–190

Shinn EA, Lidz BH, Kindinger JL, Hudson JH, Halley RB (1989) Reefs of Florida and the Dry Tortugas. US Geological Survey, St. Petersburg, FL, 53 p

Speed R (1989) Tectonic evolution of St. Croix, Implications for tectonics of the northeastern Caribbean, in: Hubbard DK (ed.), Terrestrial and marine Geology of St. Croix, U.S. Virgin Islands, Spec Publ No. 8, West Indies Laboratory:9–36

Stoddart D (1969) Ecology and morphology of recent coral reefs, Biol Rev 44:433–498

Taggart BE (1992) Tectonic and eustatic correlations of radiometrically dated marine terraces in northwest Puerto Rico and Isla de Mona: Puerto Rico. Ph.D. dissertation, University of Puerto Rico, Mayagüez, 252 p

ten Brink, US, Dillon WP, Frankel A, Rodriguez R, Mueller C (1999) Seismic and tsunami hazards in Puerto Rico and the Virgin Islands. US Geological Survey Open-file Report 99–353

Toscano MA, Lundberg J (1999) Submerged late Pleistocene reefs on the tectonically stable S.E. Florida margin: high-precision geochronology, stratigraphy, resolution of substage 5a sea-level elevation, and orbital forcing. Quat Sci Rev 18:753–767

Toscano MA, Macintyre IG (2003) Corrected western Atlantic sea-level curve for the last 11,000 years based on calibrated ^{14}C dates from *Acropora palmata* framework and intertidal mangrove peat. Coral Reefs 22:257–270

Viau C (1983) Depositional sequences, facies and evolution of the upper Devonian Swan Hills buildup, central Alberta, Canada, in: Harris PD (ed), Carbonate Buildups – A Core Workshop.

Soc Econ Petrol Mineral Core Workshop No. 4:113–143

Winograd IJ, Landwehr JM, Ludwig KR, Coplen TB, Riggs AC (1997) Duration and structure of the past four interglaciations. Quat Res 48:141–154

Zimmer B, Precht W, Hickerson E, Sinclair J (2006) Discovery of *Acropora palmata* at the Flower Garden Banks Marine Sanctuary, northwestern Gulf of Mexico. Coral Reefs 25:192

8
Ecology of Coral Reefs in the US Virgin Islands

Caroline S. Rogers, Jeff Miller, Erinn M. Muller, Peter Edmunds, Richard S. Nemeth, James P. Beets, Alan M. Friedlander, Tyler B. Smith, Rafe Boulon, Christopher F.G. Jeffrey, Charles Menza, Chris Caldow, Nasseer Idrisi, Barbara Kojis, Mark E. Monaco, Anthony Spitzack, Elizabeth H. Gladfelter, John C. Ogden, Zandy Hillis-Starr, Ian Lundgren, William Bane Schill, Ilsa B. Kuffner, Laurie L. Richardson, Barry E. Devine, and Joshua D. Voss

This chapter is dedicated to Judith and Ed Towle

8.1 Introduction

The US Virgin Islands (USVI) in the northeastern Caribbean, consist of St. Croix (207 km²), St. Thomas (83 km²), St. John (52 km²) and numerous smaller islands (Dammann and Nellis 1992). They are part of the Lesser Antilles and Leeward Islands on the eastern boundary of the Caribbean plate (Fig. 8.1). An extensive platform underlies St. Thomas and St. John and connects these islands to Puerto Rico and the British Virgin Islands. This platform extends about 32 km north of the islands and then slopes gradually to depths of over 300 m and eventually descends into the 8,000 m deep Puerto Rican Trench. South of the islands, the platform extends about 13 km and then abruptly drops off to over 4,000 m. St. Croix, about 60 km to the south, is on a separate platform which is much shallower than the northern Virgin Islands' platform and extends less than 5 km from shore except on the east end of the island. The deepest part of the Virgin Islands Trough that separates St. Thomas and St. John from St. Croix is 4,200 m.

Fringing, bank-barrier, patch, spur and groove reefs, algal ridges, and a submarine canyon are all present in the US Virgin Islands (Ogden 1980). Corals are found from the shoreline to depths of about 50 m (Figs. 8.2–8.4). Coral communities, as opposed to true coral reefs, are found growing on boulders and mangrove prop roots in shallow water around most of the island shorelines. St.

Croix, St. Thomas, and St. John have 113, 85 and 80 km of shoreline, respectively (Dammann and Nellis 1992). Some reefs have grown off of rocky points and across the mouths of bays, creating salt ponds, for example, in Newfound Bay, St. John (Robinson and Feazel 1974). Reefs are absent directly offshore of the mouths of intermittent streams (Hubbard 1987). The most developed reefs in general are found off the eastern, windward ends of the islands. Algal ridges occur off the eastern end of St. Croix (Adey 1975). The steep, lower forereefs of the fringing reefs around the islands tend to have higher coral cover than other habitats at depths less than 20 m around the islands, although high coral cover is found on deeper offshore reefs such as those that are part of the Mid-shelf Reef complex and the Red Hind Bank which lie south of St. Thomas and St. John. Well-developed reefs dominated by *Montastraea annularis* complex (*M. annularis, M. franksi,* and *M. faveolata*) occur at depths of 33–47 m south of St. Thomas (Armstrong et al. 2006; Herzlieb et al. 2006). Sand halos from the grazing of herbivorous fishes and sea urchins, notably *Diadema antillarum* (zones that separate the reef from nearby seagrass beds), are often seen at the base of the lower forereefs and around patch reefs (Randall 1965; Ogden et al. 1973).

Some reefs are close to seagrass beds and mangroves, e.g., Salt River Submarine Canyon and Tague Bay Reef (St. Croix), reefs in Benner Bay and around Cas Cay (St. Thomas), and reefs in

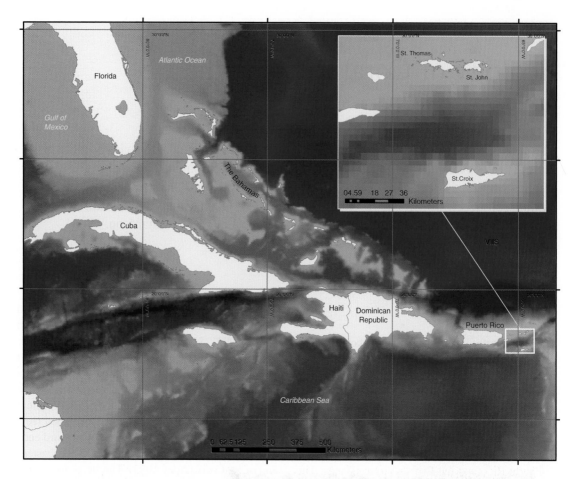

FIG. 8.1. Index map shows location of St. Thomas, St. John, and St. Croix relative to other islands in the wider Caribbean region

Great Lameshur Bay (St. John), although mangroves are not extensive in the Virgin Islands.

Many shallow reefs have extensive stands of dead *Acropora palmata* (elkhorn coral), although high density stands of living elkhorn occur in some areas. In general, *Montastraea annularis* complex is dominant on many mid-depth and deeper reefs (Armstrong et al. 2006, Herzlieb et al. 2006, Rogers and Miller 2006), although *Agaricia* species become relatively more abundant with depth. The deepest reefs often have higher coral cover and more plate-like coral growth.

Estimates of total reef area in the USVI (and wider Caribbean) vary widely because of differences in definitions of reefs, depth zones, and mapping approaches (Burke and Maidens 2004). The World Atlas of Coral Reefs (Spalding et al. 2001) estimates a total area for USVI reefs as 200 km^2, 1%

of the wider Caribbean. They define reefs as "shallow structures built by corals and other hermatypic organisms". Rohmann et al. (2005) included mangroves, seagrass beds, and other habitats in their estimates of potential coral reef ecosystem area (or extent) within the 10-fathom (~18 m) depth curve and arrived at a total area of 344 km^2. Based on detailed mapping from aerial photographs and, like Rohmann et al. (2005), including other associated habitats, Kendall et al. (2001) estimated a total reef area of 485 km^2 to a depth of 30 m. US Virgin Islands coral reefs also include large areas of mostly unexplored reefs at depths below 30 m (Armstrong et al. 2006). In 2005, multibeam and ROV video surveys conducted off the north side of Buck Island, St. Croix, revealed colonies of the deep water coral *Lophelia* at depths greater than 1,000 m.

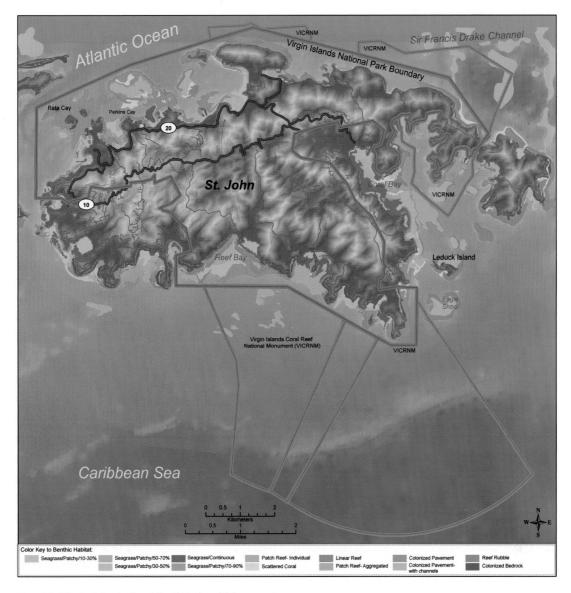

FIG. 8.2. Virgin Islands Coral Reef National Monument

8.2 History of Research

Early studies of coral reefs and reef organisms in the USVI, as elsewhere before the development of snorkeling and scuba gear, were restricted to what someone pulled up in a dredge or found in very shallow water near shore. Bayer (1969) reviewed findings from numerous research cruises beginning in the late 1800s, although many focused more on Puerto Rico than the USVI. HMS *Challenger* crossed the Atlantic and reached St. Thomas in 1873. The ship's scientists collected 40 species of fishes, corals, and other organisms, including seven new species. Among these were *Diadema setosum* (presumably *D. antillarum*), *Madrepora palmata* and *Madrepora cervicornis* (*Acropora palmata* and *Acropora cervicornis*). A few references to corals, brittle stars, and other organisms collected in

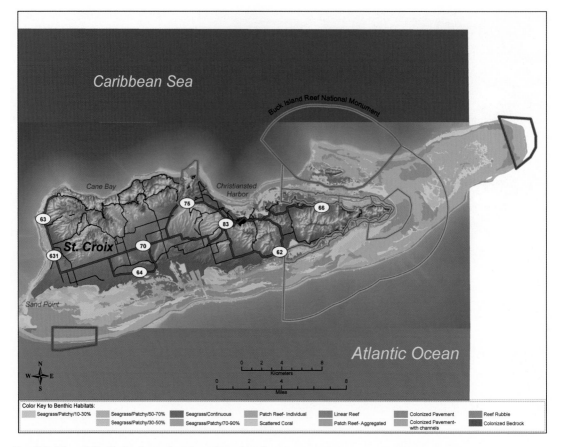

FIG. 8.3. Map of St. Croix shows location of coral reefs

the 1840s from the USVI appear in Wolff (1967). Early descriptions of fishes from Puerto Rico and the Virgin Islands appear in Evermann and Marsh (1902) and Nichols (1929, 1930). Fiedler and Jarvis (1932) described the VI fishery in 1930, but few other substantive reports on marine resources were written until those by Jack Randall and his associates beginning in the late 1950s and 1960s.

Although some scientists had hoped for the establishment of a permanent biological research station in the Danish West Indies in the early 1900s (Wolff 1967), the first station was established at the Virgin Islands Environmental Resource Station (VIERS) in Lameshur Bay, St. John, in 1966. From 1958 to 1961, Randall and others were based there. A comprehensive review by Dammann (1969) includes studies of St. John reefs in Chocolate Hole and Mary Creek (Stoeckle et al. 1968 cited in Dammann 1969). Some of the early studies done in

the USVI included some of the first experimental manipulations such as tagging of fishes (Randall 1962), use of artificial reefs (Randall 1963), and fish exclusion cages (Earle 1972, Mathieson et al. 1975).

The Caribbean Research Institute of the College of the Virgin Islands (now University of the Virgin Islands, UVI) produced many reports in the 1970s and 1980s, primarily focusing on water quality. Island Resources Foundation prepared an inventory of the marine environments in the USVI (IRF 1977) with an emphasis on oceanography, climatology, and marine ecology.

Research increased following the establishment of Fairleigh Dickinson University's West Indies Laboratory (WIL) on St. Croix, in 1971. The faculty and students produced numerous papers (>300 papers plus numerous technical reports) including some which provided a baseline for

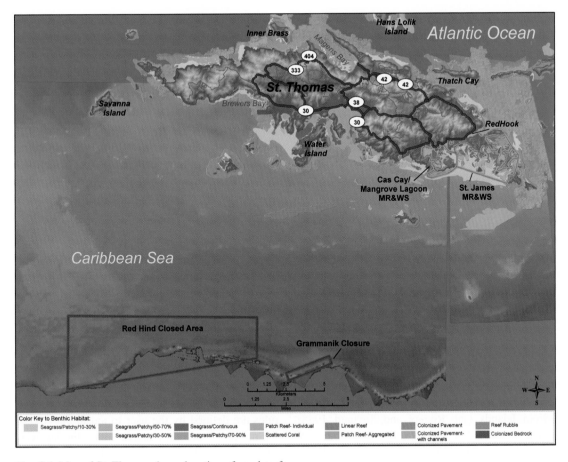

Fig. 8.4. Map of St. Thomas shows location of coral reefs

long-term monitoring (see annotated bibliography of papers on research at Buck Island Reef National Monument by Gladfelter 1992). Research conducted at WIL includes a variety of early studies, many of which stimulated subsequent research on key issues in tropical marine ecology including herbivore–plant interactions, coral morphology/physiology/ecology, chemical and mechanical defenses, seagrass ecology, invertebrate ecology, fish community structure, recruitment and dispersal in reef fishes, resource partitioning in reef fishes, behavioral ecology of fishes, microbiology, nutrient dynamics and productivity.

A series of projects (Tektite I and II) was conducted by scientists operating out of an underwater laboratory/"habitat" in Great Lameshur Bay, St. John, in 1969 and 1970 (Collette and Earle, 1972; Earle and Lavenberg 1975; see textbox). Later,

from 1977 to 1989, the West Indies Laboratory and the National Oceanic and Atmospheric Administration (NOAA) ran a saturation diving program to support research out of two other underwater laboratories (first Hydrolab and then Aquarius) in Salt River Canyon, St. Croix (see Appendix A in Kendall et al. 2005). There were studies of fishes, fish behavior, corals, squids, octopuses, crinoids and ophiuroids, black corals, gorgonians, sponges, seagrasses, queen conch, and coral metabolism.

In 1981, three dives were made in the *DSRV ALVIN* to 2455–3950 m depth along the northern St. Croix slope, and a strong connection between the shallow insular shelf and deep basin was discovered (Hubbard et al. 1981). Shallow-water sediments and detrital seagrasses were collected, along with three species of holothurians and two species of urchins. The analysis of stable carbon C^{13}:C^{12} ratios of seagrass detritus and animal tissue revealed that a significant proportion of the nutrition of both groups is derived from detrital seagrasses either by direct consumption or by feeding on sediments enriched by decomposed seagrasses. One urchin species fed almost exclusively on *Syringodium* (Suchanek et al. 1985).

In 1985, the submersible Johnson Sea-Link II was used to make observations and record video of areas 36–758 m deep off of St. Thomas, St. Croix, and St. John, partly to determine the potential for a deep water commercial fishery and to learn more about the deepwater corals and macroalgae (Nelson and Appeldoorn 1985). In one dive off the 100 fathom contour south of St. John, scientists described "a spectacular wall with ambient light penetrating to ca. 272 m (800')" and provided a list of fishes that were observed.

The Virgin Islands Resource Management Cooperative (VIRMC), established in 1983 with support primarily from the National Park Service and under the direction of Island Resources Foundation, produced a series of reports that established baseline data for the USVI and British Virgin Islands (see Rogers and Teytaud 1988). These included descriptions and maps of bays and fish habitats within Virgin Islands National Park (VINP) and Buck Island Reef National Monument (BIRNM) (Anderson et al. 1986, Beets et al. 1986, Boulon 1986a, b), assessment of fish and shellfish stocks in the VINP and BIRNM (Dammann 1986, Tobias et al. 1988), early fish and reef monitoring efforts (Boulon et al. 1986, Rogers and Zullo 1987), and observations of white band disease on elkhorn coral (Davis et al. 1986).

8.2.1 Mapping of USVI Coral Reef Ecosystems

Mapping of the coral reef ecosystems found around the US Virgin Islands has a long history ranging from simple visual assessments by divers on tow boards to analysis of multispectral imagery. Over time the maps of corals and other benthic habitats, such as seagrass and algae, have increased in spatial resolution and thematic content. The first map of reefs, seagrass beds, and sand bottom to a depth of about 20 m around St. John was produced in 1958 from observations made by scientists

towed on a sled behind a boat (Kumpf and Randall 1961). Maps of St. Croix reefs are presented in Ogden et al. (1972). Rosemary Monahan and Betsy Gladfelter made a detailed map of Buck Island in 1977 (Gladfelter et al. 1977). As part of the VIRMC studies, maps of St. John and Buck Island were produced from visual interpretation of aerial photos by Beets et al. (1986), Boulon et al. (1986), and Anderson et al. (1986). In 2000, an extensive seafloor mapping project around all of the USVI was completed by NOAA (Kendall et al. 2001). This project mapped 485 km² of benthic habitats in the USVI to a nominal depth of 30 m based on visual interpretation of color aerial photography (Jeffrey et al. 2005). Analyses of these maps revealed that coral reef and hard-bottom habitats make up 61%, submerged aquatic vegetation 33%, and unconsolidated sediments 4% of the shallow water areas (Kendall et al. 2001). Most recently, Harborne et al. (2006) collected 410 km² of multispectral imagery around St. John and St. Thomas and generated benthic habitat maps using unsupervised classification in conjunction with contextual decision rules (Mumby et al. 1998) to classify the digital imagery. Contextual decision rules are defined by the user to aid in classifying the imagery based on features within the imagery, such as dramatic changes in bathymetry and the presence of bright reflectance areas indicating sand habitat.

All of the above mapping studies addressed the shallow water (<30 m) coral reef ecosystems surrounding the USVI, but several sonar-based efforts have recently characterized deeper reefs in the USVI (20–1,000 m). Sonar is useful for mapping benthic habitats that are too deep or turbid to map using either aerial or satellite - based sensors. Side scan sonar has been used by the Caribbean Fishery Management Council to map the Marine Conservation District south of St. Thomas. Recently, the NOAA Biogeography Team (http://ccma.nos.noaa.gov/ecosystems/coral-reef/usvi_nps.html) has collected multibeam sonar data and underwater video to define the bathymetry and associated habitats for selected areas of the USVI with emphasis on Virgin Islands Coral Reef National Monument (VICRNM) and Buck Island Reef National Monument. These data are currently being processed into digital bathymetry maps, and analysis of the "backscatter" return signal from the sonar is being classified into benthic habitats and confirmed by the visual inspection of underwater video. Extensive sidescan sonar data have been collected south of St. John to the shelf edge and in select areas south of St. Thomas. Detailed geo-referenced bathymetric maps exist for most of the marine protected areas and seasonally closed areas along the shelf margin such as the Marine Conservation District (41 km^2 area), the Grammanik Bank (1.5 km^2), and Lang Bank (12 km^2). Between 2004 and 2006, NOAA's Research Vessel Nancy Foster conducted extensive sea floor mapping around Buck Island, Salt River Bay, and Lang Bank (St. Croix). In 2006 NOAA collected 143.3 km^2 of multibeam coverage from 14.7 m water depth to 1,000 m. Total ROV data for the mission included 22 linear km of video transects from 20 to 830 m water depth. For St. Croix, approximately 81 km^2 of multibeam bathymetry and backscatter data were collected from Salt River to the northeast end of Lang Bank north of the island to greater than 1,000 m (Battista 2006).

8.2.2 Long-term Monitoring

Over the last 20 years, there has been an increasing investment in long-term monitoring in the USVI by many different agencies. The National Park Service (NPS), University of the Virgin Islands (UVI), US Geological Survey (USGS), the Division of Fish and Wildlife (DFW), the West Indies Laboratory (until it was closed after Hurricane Hugo in 1989), the National Oceanic and Atmospheric Administration (NOAA), The Nature Conservancy (TNC), The Ocean Conservancy (TOC), the National Science Foundation, the Sea Grant Program of the University of Puerto Rico, and others have conducted or supported research on coral reefs in the USVI. As a result, some of the oldest continuous records of anywhere in the Caribbean on water quality, coral reefs, and reef fishes are available for the USVI. Over the last few decades there has been increasing use of technology, including GPS, digital video cameras, *in situ* recording stations such as those in the Integrated Coral Observatory Network, Acoustic Doppler Current Profilers, CTD (Conductivity, Temperature, and Depth) sensors, ROVs, and multibeam sonar. Extensive long-term monitoring of coral reefs has been augmented by experimental research

including studies of ecological processes such as coral and fish recruitment (Edmunds 2000, 2004, 2006; Nemeth 1998, Tolimieri et al. 1998, Rogers et al. 1984, Rogers and Garrison 2001), herbivory (Carpenter 1990a, b; Steneck 1993), calcification and coral growth rates (Gladfelter 1982, 1984, Gladfelter et al. 1978), coral and reef metabolism (Adey et al. 1981, Rogers and Salesky 1981).

The establishment of relatively large marine reserves off St. Thomas in 1999 and St. John and St. Croix in 2001 (see below) has led to efforts to evaluate the effectiveness of these areas in reversing degradation of marine resources in the USVI.

Island Resources Foundation has assembled and maintained a comprehensive library with references pertaining to research and resource management in the USVI and wider Caribbean (IRF 1989, www.irf.org).

8.3 Physical Oceanography/climate

Easterly trade winds predominate in the USVI. The wind varies in direction and intensity, with maximum winds usually occurring in winter and minimum in the fall. Hurricanes usually occur between the months of June and November, with a peak in August and September (US Naval Oceanographic Office 1963, Hubbard 1989, Kendall et al. 2005). Rainfall is variable with no well-defined wet and dry seasons, although most rainfall occurs between August and December (see NOAA Climatological Bulletin). Rainfall can be very intense, with significant amounts of rain falling during very short time periods. For example, on the north side of St. John, in April 1983, 19 in. (483 mm) of rain fell in 21 h; in November 2003, 11.2 in. (284 mm) fell in 144 h with a total of 23.79 in. (604 mm) for the month, one of the wettest months recorded in the past several decades. Annual totals from 1984 to 2005 for this site ranged from 27.2 (691) to 69.6 in (1768 mm). (R. Boulon, unpublished data 2006). Rainfall can be localized and is often associated with tropical storms and hurricanes. Although there are no permanent streams or rivers, even brief (but intense) rains result in runoff from the intermittent streambeds on the islands.

Coastal currents within the USVI are usually less than 10 cm/s (range: 0–40 cm/s) and are primarily

Tektite Program

As early as 1972, Great Lameshur Bay was referred to as "one of the best known marine communities in the world" (Collette 1972) because it was the focus of a great deal of research by Jack Randall (e.g., Randall 1963, 1965) and the site of the Tektite underwater habitat program (Fig. 8.5).

The Tektite Program took place in Great Lameshur Bay in 1969 and 1970 (Collette and Earle 1972; Earle and Lavenberg 1975). Divers lived in an underwater habitat at a depth of 15 m for up to 60 days. Although the objective of the program was to learn how divers could function safely and effectively under saturation conditions and not to conduct an integrated study of a single coral reef or provide a baseline for future research (Collette 1972), some of the papers provide intriguing insights into changes that have taken place at Tektite Reef in the last 4 decades. One scientist noted "maximum visible range exceeded 30 m on exceptionally clear days, and rarely fell short of 10 m under the worst conditions" (Clifton and Phillips 1975). Now underwater horizontal visibility rarely exceeds 10–15 m.

Some observations made during Tektite missions show how fish assemblages have changed. For example, Collette and Talbot (1972) noted seven species of groupers, some quite numerous, all of which are now rarely seen. (It is currently rare to see even a single Nassau grouper during an hour dive in most locations). They noted that most of the fishes they saw on their study reefs were benthic carnivores, particularly the larger groupers, whereas now most of the fishes at the Tektite Reef are herbivores. They reported that yellowtail snappers and bar jacks (schools of several of these up to 20 cm long) were the most common predators near the Habitat. Schools of up to 200 lane snappers were seen. Threespot damselfish (*Stegastes planifrons*) which are now very abundant (J. Beets and A. Friedlander, personal communication 2007) were only "moderately common" when the Tektite missions were conducted.

Some of the projects involved experimental manipulation, such as the use of cages to exclude herbivorous fish from small areas on the reef (Earle 1972; Mathieson et al. 1975). Earle (1972) found 35 species of herbivorous fishes in 14 families, and recorded feeding behavior. She also found 154 species of marine plants, including 26 never reported before from the USVI. Other projects included several on fish behavior, cleaner shrimps, and lobsters (Mahnken 1972; Smith and Tyler 1972; Herrnkind et al. 1975).

Tektite Reef is now the subject of long-term monitoring by NPS and other researchers and has been the focus of intensive research on changes in coral cover resulting from bleaching and disease (see below).

FIG. 8.5. The Tektite underwater habitat on a barge in Great Lameshur Bay (Photo: G. Davis)

driven by wind and tides. Within the semi-enclosed bays of the islands, the currents are dominated by the tides and secondarily by entrainment into the bays from the eastern ends and detrainment at the western ends of the bays. Entrainment and detrainment into and out of the bays are influenced by the easterly trade winds that force surface currents to advect from east to west (Halliwell and Mayer 1996). The complex components of current regimes near the coasts and especially in embayments lead to complex retention/dispersal dynamics of larval transport (Cowen 2002). In offshore waters surface currents (5 m depth) average 23 cm/s but can range from 12 to 65 cm/s whereas near the bottom (30 m depth) current speeds average 16 cm/s and range from 10 to 27 cm/s [from University of the Virgin Islands (UVI) Acoustic Doppler Current Profiler (ADCP) data from November 2005 to March 2006]. During the passage of Hurricane Hugo (1989), the most severe hurricane to ever hit the USVI, currents and water levels were measured in Salt River Submarine Canyon, on the north shore of St. Croix. Currents reached a maximum velocity of 5 m/s (9.7 knots) (Hubbard et al. 1991).

Tides within the region are mixed semi-diurnal, and ranges typically are <20 cm (Hubbard 1989), but can reach 40 cm during spring tides. Circular tidal currents dominate the unidirectional wind-driven currents during the full and new moons. At other times, wind-driven currents dominate the tidal influences (http://tidesandcurrents.noaa.gov, N. Idrisi, personal communication 2007).

The University of Miami is developing the HYCOM-ROMS general circulation ocean model for the Interamerican Seas region that includes the Caribbean (see Cowen et al. 2000, 2006; Kourafalou et al. 2006). To ground-truth the model for the Virgin Islands and increase the resolution, the University of the Virgin Islands is using data from ADCPs deployed in 2005–2006 and bathymetry data from the Environmental Protection Agency (EPA) research cruise in 2006 (N. Idrisi, personal communication).

The reefs around the Virgin Islands are influenced by freshwater lenses migrating from the Amazon and Orinoco Rivers as anticyclonic rings. These signals peak in June to August (Hu et al. 2004). UVI conductivity-temperature-depth (CTD) data indicate the salinity signal from the south of St. Thomas and St. John (Caribbean side) is lower (34–35 ppt) than to the north of the islands (Atlantic side: 36 ppt). Along with lower salinity waters, these anticyclonic rings are nutrient-rich with greater ocean color signals as seen from SeaWiFS data (Hu et al. 2004).

In 2002, NOAA installed a Coral Reef Early Warning System Station (CREWS, now referred to as ICON for Integrated Coral Observing Network) near the west wall of Salt River Canyon, St. Croix. This station provides hourly data on wind direction, wind speed, air and water temperature, salinity, photosynthetically active radiation (PAR), and ultraviolet radiation at the water surface and 1 m depth (see www.coral.noaa.gov/prototype). Recently, NOAA installed a tsunami warning system on St. John.

The USVI Government Department of Planning and Natural Resources collects water quality data quarterly from 135 stations around St. Thomas, St. Croix, and St. John. The National Park Service collects data quarterly from 16 stations around St. John. All of these data are entered into the Environmental Protection Agency's database, STORET.

One of the longest *in situ* water temperature records is from Great Lameshur Bay, a semi-enclosed bay on the south side of St. John, where data have been collected from 9 and 14 m for almost 20 years. Here mean monthly seawater temperature has increased gradually by 0.6°C/decade since 1989, and the number of days exceeding 29.3°C, the bleaching threshold defined by NOAA for the Virgin Islands, has increased as well (Edmunds 2004). In 2005, sea water temperatures exceeding 30°C were associated with the most severe bleaching event on record in the USVI. Data on subsurface sea water temperature from Saba Island south of St. Thomas have been collected since 1990 and compared to AVHRR satellite records (Quinn and Kojis 1994a, b). Water temperature data from the USVI are available from near the surface to a depth of 40 m. Along the shelf edge south of St. Thomas at 40 m depth sea water temperature averaged 27.1°C and ranged from 26°C to 29°C from February 2003 to May 2005 (Nemeth et al. 2007).

Dust from the Sahel/Saharan region in Africa affects the USVI frequently, primarily in the summer months (Prospero and Lamb 2003; Griffin et al. 2003). The possible role of African dust in causing reef degradation is the subject of ongoing research (Shinn et al. 2000; Garrison et al. 2003;

Griffin et al. 2003). Large dust clouds can dampen hurricane activity (Dunion and Velden 2004).

Volcanic ash from the active Soufriere Hills volcano on Montserrat periodically reaches the USVI. Effects on marine ecosystems are not known.

8.4 Biodiversity of USVI Coral Reefs

Similar to other Caribbean reefs, reefs in the USVI have over 40 species of scleractinian corals and three species of *Millepora* (Appendix 8.2). A comprehensive inventory of octocorals, sponges, and other invertebrates has not been prepared, but lists from particular locations are found in several papers, including Gladfelter (1993a, b), Kendall et al. (2005), and Idjadi and Edmunds (2006).

Randall's "Caribbean Reef Fishes" (1968) describes 300 fish species, over half of which were collected and photographed from the Virgin Islands. Clavijo et al. (1980) listed 400 species of fishes in 93 families from around St. Croix. NOAA fish surveys from 2001 to 2006 list a total of 215 fish species from St. John and 202 fish species from St. Croix (combined 236 species) from visual censuses (http://www.ccma.nos.noaa.gov/ecosystems/coral-reef/reef_fish.html). A recent study of sharks and shark nursery habitats in the USVI lists nine shark species: great hammerhead, scalloped hammerhead, Caribbean reef, tiger, blacktip, lemon, blacknose, Caribbean sharpnose and nurse (DeAngelis 2006). Great white, thresher, white-tip, and mako sharks have also been reported from USVI waters. Several whale sharks were seen south of St. Thomas and St. John in 2006.

Buck Island Reef Sea Turtle Research: Program Summary

There are very few places in the Caribbean where any large numbers of hawksbill turtles (*Eretmochelys imbricata*) remain today (NMFS/USFWS 1993) (Fig. 8.7). Today throughout their range, hawksbill turtles nest in low density; nesting aggregations consist of a few dozen to at most a few hundred individuals (NMFS/USFWS 1993). Buck Island Reef National Monument (BIRNM) is one of the most significant areas under US jurisdiction where hawksbill sea turtles are still nesting in any numbers (50–75/season) (data as of 2006). The Hawksbill Recovery Plan (NMFS/USFWS 1993) identified BIRNM as an index beach for hawksbill turtle recovery in the Eastern Caribbean. The Monument also provides critical habitat for post-pelagic to subadult sea turtles that shelter in the reefs and feed on zoanthids, sponges, and seagrasses.

Endangered hawksbill sea turtles, threatened green sea turtles, occasional leatherback turtles and most recently loggerheads nest on Buck Island. Throughout the peak summer months a saturation tagging program records nesting behavior and fidelity, remigration period, individual fecundity, size, and hatching and emergence success. Tissue samples are taken for genetic analysis. Threats to hatching success such as predation, poaching, inundation by seawater, and desiccation are monitored and mitigated (Phillips and Hillis-Starr 2002).

In the course of 19 years of conducting basic research on hawksbill sea turtle nesting behavior, several other projects have been initiated including radio-, sonar-, and satellite telemetry to determine the movements of nesting hawksbill turtles during their inter-nesting period, and after nesting. Buck Island was the site of a study to develop a non-lethal method of determining the sex of sea turtle hatchlings. Incubation temperature, not X or Y chromosomes, determines the sex of sea turtles, and the results of this study along with records of nesting beach temperatures, enables determination of the sex ratio of hatchlings without sacrificing them.

Buck Island nesting hawksbill turtles are not part of a larger population, but genetically distinct and isolated from hawksbill turtles nesting in Puerto Rico, Antigua, and Barbados. However, they show strong genetic identity with hawksbill turtles sampled in Belize and Nicaragua; additionally, three tag recoveries for Buck Island nesting hawksbill turtles are from Central America and Cuba.

Hawksbill turtles may take 30 years to reach sexual maturity. In light of the increasing number of new recruits encountered from 1996 to 2006, Buck Island may be starting to see the results of 30 years of nesting beach protection and conservation. The genetic analyses indicate that the island's nesting hawksbill population may be distinct in the Caribbean and therefore should be afforded as much protection as possible.

FIG. 8.6. Brokenbar blenny, *Starksia smithvanizi*. This small blenny (15–25 mm total length) was first recognized as a new species as a result of studies conducted at Buck Island Reef National Monument

FIG. 8.7. A hawksbill sea turtle (Photo: C. Rogers)

In a study of cryptic fishes around Buck Island, St. Croix, using controlled rotenone treatments, Smith-Vaniz et al. (2006) found 228 species (in 55 families), with 60 of these documented for the first time from St. Croix. These included 13 additional species in the family Gobiidae, 12 in the Labrisomidae, five each in the Chaenopsidae and Bythitidae, and four each in the Gobiesocidae and Ophidiidae (Fig. 8.6).

Earle (1972) recorded 154 species of algae from Great Lameshur Bay, St. John, including 26 never reported before for the US VI. Gladfelter and Gladfelter (2004) documented 164 species of molluscs (seashells) from Southgate Beach, St. Croix.

Four species of sea turtles are found in the USVI, with hawksbills and greens the most abundant (see Side bar).

In 2005, a wide diversity of habitat types was characterized from multibeam and ROV data off of Buck Island. Seafloor features included rock precipices, ledges, limestone caves, boulders, rock outcroppings, and flat expanses of mud. The biota below 200 m, never visually characterized before, included *Lophelia* coral, black coral, sea whips, feather stars, sea pens, sea anemones, sea stars, brittlestars, urchins, sponges, isopods, sea cucumbers, albino lobsters, shrimps, crabs, conch, Orange Roughys, roundnose grenadiers, tripod fish, and several types of snappers.

Especially useful guides to the identification of marine organisms in the USVI include Gladfelter (1988), Suchanek (1989), and Beets and Lewand (1986), and for the Caribbean in general Warmke and Abbott (1962), Voss (1976), Kaplan (1982), Colin (1978), Humann and DeLoach (2002a, b, c), and Littler and Littler (2000).

8.5   Marine Protected Areas in the Virgin Islands

Many Marine Protected Areas (MPAs), including some marine ("no-take") reserves, are found in the USVI. Buck Island Reef National Monument (BIRNM) was established in 1961 and consisted

Federal Marine Reserves/National Monuments

In 1999, former Secretary of the Department of the Interior Bruce Babbitt conceived of a marine protected area for the Virgin Islands that would encompass all representative ecosystems and provide further protection for the area's marine resources. The original proposal recommended a marine protected area of about one-million acres (404,700 ha). When the Virgin Islands Coral Reef National Monument (VICRNM) was ultimately established in 2001, it included 12,708 acres (1,096 ha) of submerged land. At the same time Buck Island Reef National Monument (BIRNM) was expanded from 880 (356 ha) to 19,015 acres (7,695 ha). The process which resulted in the VICRNM and the expanded BIRNM is complex.

In 1974 the Submerged Lands Act transferred all submerged lands out to three nautical miles (5.6 km) from the US Government to the USVI. However, within the Act, there was an exception for "submerged lands adjacent to US owned above-tideland uplands" out to the extent of the 5.6 km (3 nm) Territorial Sea. As the National Park Service owned coastlines around St. John and Buck Island at that time, this exception applied. The Minerals Management Service was asked in 1999 to determine what this included, and they mapped approximately 37,000 acres (14,974 ha) of submerged lands around St. John, Buck Island and Water Island that met this exception. Only the submerged lands contiguous with NPS lands were considered for monument status.

In 2001 President William Clinton used the Antiquities Act of 1906 which allows for Presidential Proclamation of National Monuments to establish VICRNM and expand BIRNM. These proclamations were challenged twice by the USVI Governor and Delegate to Congress but were upheld both times by the US General Accounting Office in late 2002. Rules and regulations for both monuments were enacted in May 2003. Both BIRNM and VICRNM are no-take/no-anchoring areas with the exception of regulated harvest of Blue Runner and baitfish (two migratory species) in VICRNM.

One result of using the Submerged Lands Act exception to map the monument areas is that the boundaries of these areas are defined politically rather than ecologically. Therefore, some of the essential marine habitats necessary for ecosystem balance are not included in the monuments. This also produced non-contiguous sections of VICRNM on the south side of St. John due to private coastal lands. This issue is being resolved by exchanging an equivalent amount of submerged land within the eastern boundary of VICRNM for the USVI owned strip of submerged lands in the middle of the monument. This will eliminate confusion for users, improve enforcement, and include a significant reef structure within the VICRNM.

The Marine Conservation District (MCD), also referred to as the Red Hind Bank, 10 km south west of St. Thomas encompasses deep water (35–50 m) shelf-edge habitats and contains extensive well-developed *Montastraea* spp. dominated coral reefs, patch reefs, colonized hard-bottom, algal plains and sand flats. Two deep-water coral ridges 50–100 m wide and over 15 km long run parallel to the southern edge of the insular platform. The outer ridge is immediately adjacent to the drop-off and varies between coral reef and colonized hard bottom habitats. The inner ridge about 300 m from the drop-off is wider and deeper than the outer ridge and is primarily coral reef habitat. The two coral ridges are separated by a deeper 5–100 m wide channel (50 m) composed of sand, patch reef, and rubble (Nemeth et al. 2007). Patch reef and soft bottom habitats extend for several km north of the inner ridge before giving way to extensive *Montastraea* reefs especially near the northwestern corner of the MCD. The Red Hind Bank was closed seasonally in December 1990 and established as the permanently closed Red Hind Bank Marine Conservation District (MCD) in December 1999 (Federal Registers 55(213), November 2, 1990 and 64(213), November 4, 1999, respectively).

FIG. 8.8. Buck Island Reef National Monument

of 356 ha. Virgin Islands National Park (VINP) was established in 1956, with the marine portions (2,286 ha) added in 1962. In 1999 the Marine Conservation District (MCD, also known as the Red Hind Bank) was established to protect 41 km² of deep reef habitats south of St. Thomas. The MCD includes an important spawning aggregation site of the red hind grouper. Local and federal fisheries regulations and small marine reserves such as the Marine Garden at BIRNM (188 ha) and Trunk Bay (21 ha) were not effective in protecting marine resources in the Virgin Islands (Rogers and Beets 2001; Rogers et al. 2007), and in 2001 President Clinton established the Virgin Islands Coral Reef National Monument (VICRNM) and expanded BIRNM by 7,339 ha (total now is 7,695 ha) (See textbox; Figs. 8.2 and 8.8). Although the boundaries were based on ownership of federal property

and not on ecological considerations, the monuments are relatively large and could play a vital role in reversing marine resource degradation in the USVI (Rogers et al. 2007).

In addition to these federal MPAs, there are several territorial MPAs, most notably the recently established East End Marine Park (Fig. 8.9) and several seasonally closed areas (see textbox).

8.6   Changes in USVI Coral Reefs

Coral reefs in the Virgin Islands have changed dramatically in the last three decades. Insights into these changes come from long-term monitoring of sites ranging in depth from sea level to 40 m. Live coral cover has declined; coral diseases have become more numerous and prevalent; macroalgal cover

East End Marine Park, Marine Reserves and Wildlife Sanctuaries, and Salt River Bay National Historic Park and Ecological Preserve

The East End Marine Park (EEMP) was established by the 24th Legislature of the USVI in 2003 through Act No. 6572 of the VI Code Title 12, Chapter 1. This act not only established the EEMP but also gave the Virgin Islands Department of Planning and Natural Resources (DPNR) the authority to establish other Territorial Marine Parks. The DPNR Coastal Zone Management Division has management responsibility for the EEMP. The legislative authority establishing the park states that its goal is "to protect territorially significant marine resources, promote sustainability of marine ecosystems, including coral reefs, seagrass beds, wildlife habitats and other resources, and to conserve and preserve significant natural areas for the use and benefit of future generations…." The website for the EEMP is www.stxeastendmarinepark.org.

A comprehensive management plan for the park was developed and formally adopted in 2002. The plan was formulated by the Virgin Islands chapter of The Nature Conservancy (TNC) based on a participatory process involving many different stakeholders on St. Croix.

EEMP is comprised of four different types of managed areas or zones. These are: Recreational Management Areas, a Turtle Wildlife Preserve Area, No-take Areas, and Open Areas. Allowable activities in the Recreational Management Areas include snorkeling, diving, catch and release fishing, cast net bait fishing and boating. The primary intention of the Turtle Wildlife Preserve Area is to protect index turtle nesting beaches (as defined in species-specific Recovery Plans) for green, hawksbill, and leatherback turtles. A prohibition on the use of gill and trammel nets in this area also offers protection for turtles in the park waters. Approximately 8.6% of the EEMP is made up of No-take Areas established to protect critical habitats for important reef species. All commercial and recreational fishing is prohibited within these areas. Over 80% of the EEMP has been designated as Open Area where existing USVI fishing and other marine activity regulations apply and the removal of coral or live rock is prohibited.

There are five Marine Reserve and Wildlife Sanctuaries (MRWS) in the USVI. Three of these sites are located on the east end of St. Thomas, and St. Croix and St. John each have one. All five sites were authorized by both the Wildlife and Marine Sanctuaries Act of 1980 (Act No. 5229) and the Virgin Islands Code Title 12, Chapter 1, and were officially designated between 1992 and 2000. The Department of Planning and Natural Resources is responsible for the management of these protected areas which have one primary goal in common: the protection of fish and wildlife resources and the habitats on which they depend.

The Compass Point Pond MRWS on St. Thomas was established in 1992 to protect this important wetland area on St. Thomas and prevent any further degradation of its natural resources. The pond is connected to the sea and fringed with mangroves that filter sediment from a large watershed. All plants and animals are protected, and alterations to habitat are prohibited.

The Cas Cay/Mangrove Lagoon MRWS was established in 1994 in Benner Bay, St. Thomas, to protect essential habitat for juvenile reef fish, lobsters, birds and wetland plants and animals and to support the restoration of these populations within the protected area. It is illegal to take any living organism from this protected area with the exception of baitfish within 50 ft of the shoreline of Cas Cay by permit only.

The St. James MRWS, established in 1994 on the southeast coast of St. Thomas, includes all the waters from Cas Cay around Great St. James Island to Cabrita Point. It is closed to all harvest of marine species except for baitfish and fish caught by hook and line.

The Frank Bay MRWS (2000) essentially protects the salt pond at Frank Bay on St. John. It is illegal to harvest or disturb any wildlife or plant species around or in the pond, similar to the Compass Point MRWS.

The Salt River MRWS on St. Croix was established in 1995. Proposed Rules and Regulations were signed in 2002. These regulations make it unlawful to remove any marine or other wildlife from the Salt River MRWS or to anchor outside of designated areas.

Although the regulations, or proposed regulations, that exist for each of these sites are comprehensive and seek to effectively protect marine and wildlife resources, none of these sites has a complete management plan, and no staff is dedicated to education and management of this MRWS system. There is little enforcement of the regulations described above.

The Salt River Bay National Historic Park and Ecological Preserve, established in 1992, is co-managed by the federal and territorial government. It is comprised of 224 acres of land. The NPS area is part of the larger Salt River Marine Reserve and Wildlife Sanctuary established in 1995 and is jointly managed by the NPS and the VI Government. Whereas the NPS has jurisdiction over the land under its ownership, DPNR has jurisdiction over adjacent wetlands and the marine portion of the protected area.

Seasonally Closed Areas

The Grammanik Bank is located east of the MCD and is a relatively small and recent (2005) seasonal closure designed to protect a multi-species spawning aggregation site used by a variety of groupers (yellowfin, Nassau, tiger, yellowmouth) and snappers (cubera, dog and schoolmaster). A similar reef structure to the MCD (i.e., two parallel *Montastraea* coral ridges) exists at the Grammanik Bank in 35–50 m depth. Below 55 m depth large patch reefs of *Agaricia* spp. extend beyond scuba diving limits (R. Nemeth, personal observation 2006). East and west of the Gammanik Bank, the ridges transition into a shallower (30 m) hard bottom habitat. The Grammanik Bank was closed to all fishing, except for migratory pelagics, from February 1 to April 30 each year, and all bottom fishing gear, including trap fishing, is prohibited year round (Federal Register 70(208), October 28, 2005).

The Lang Bank seasonal closure is composed largely of colonized hard-bottom and patch reef habitats. Near the eastern margin of the shelf and about 16 km east of St. Croix, *Montastraea* -dominated coral reef ridges exist on old spur and groove formations in 30–40 m depth. A deep-water basin (50–60 m deep) separates the inner and the outer coral ridges (Nemeth et al. 2007). The eastern end of Lang Bank is closed seasonally from December 1 to February 28 to protect a red hind spawning aggregation. This closure was implemented December 1993 (Federal Register 58(197), October 14, 1993). On October 28, 2005 (Federal Register 70(208)), the Lang Bank seasonally closed area was closed to all bottom fishing gear year-round.

The Mutton Snapper seasonal closure, which was established in 1993 to protect a mutton snapper spawning aggregation, is located off the southwest corner of St. Croix and encompasses 2.5 km² of both territorial and federal waters. The habitats within the closure are composed of linear reefs, patch reefs, hardbottom and sand habitats from 20 to 30 m deep. The closed season extends from March 1 to June 30 each year and was designed to protect a mutton snapper spawning aggregation, although enforcement is lacking and poaching prevalent (Federal Register 58(197) October 14, 1993).

To further protect several species during their seasonal spawning aggregations, harvest of yellowfin, tiger and yellowmouth grouper and mutton snapper are prohibited in federal and territorial waters during most or all of the period of the seasonal closures (Federal Register 58(197); VI Rules and Regulations T.12, Chapter 9A). Harvest of Nassau grouper is prohibited year round in federal and territorial waters (VIRR T.12, Chapter 9A).

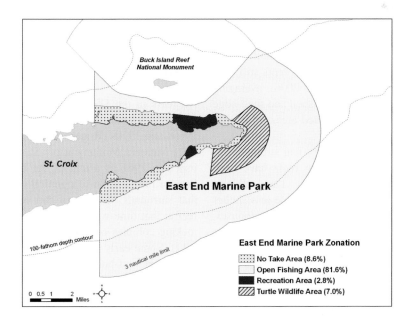

FIG. 8.9. St. Croix East End Marine Park Zonation (DPNR)

has increased; fish of some species are smaller, less numerous or only rarely seen; and the long-spined black sea urchins *Diadema antillarum* are less abundant.

Coral cover has declined on most if not all reefs in the USVI for which there are quantitative data. In the 1970s and 1980s coral cover on some reefs was over 40% and even higher in some shallow elkhorn coral zones (Gladfelter et al. 1977, Gladfelter 1982, Rogers et al. 1983, Edmunds 2002). At that time, algal turf typically made up a large component of the benthic cover, while macroalgae were absent or scarce. Hurricanes caused significant losses in coral cover and changes in the physical structure of many reefs (Hubbard et al. 1991). For example, Hurricane David (1979) caused a reduction in mean coral cover along transects at Flat Cay Reef (St. Thomas) from 65% to 44% (Rogers et al. 1983). In addition, Hurricane Hugo (1989) caused a 30–40% decline in coral cover along transects and within quadrats in Great Lameshur Bay, St. John (Edmunds and Witman 1991, Rogers et al. 1991).

By the 1990s, many long-term monitoring sites had coral cover of about 25% or less, and mac-roalgal cover, although variable, often reached much higher values than in the past. Coral cover continued to decline or remain stable until the major 2005 bleaching /disease event (described more fully below). Now coral cover is less than 12% on many reefs, including five long term study sites in St. John and St. Croix covering over 10 ha of reefs that formerly had high coral cover and diversity. Even deep reefs have been affected by bleaching and disease (Herzlieb et al. 2006). Some deeper reefs still have high coral cover (Herzlieb et al. 2006), averaging over 30% for deep (>30 m) mid-shelf and shelf-edge reefs inside and outside protected areas, even after the extensive bleaching in 2005.

In shallow zones (<6 m), physical structure has changed remarkably as elkhorn reefs have been decimated by storms and disease. Dead elkhorn branches litter the bottom and provide less shelter than intact colonies for parrotfishes, octopuses, hawksbill turtles and other organisms. *Porites* patch reefs that have little live coral can be found in many bays around St. John (Rogers 1999) as well as St. Thomas (Magens Bay) and St. Croix (fore reef of Tague Bay).

Overall, the most significant cause of coral mortality on Virgin Islands reefs has been disease

following the bleaching event of fall 2005 (Miller et al., 2006; T. Smith, personal communication 2006). Hurricanes have been very destructive in localized areas, especially in shallow water. White band disease in the 1970s and 1980s affected just two species, *Acropora palmata* and *Acropora cervicornis*, but the effects were devastating and widespread, and the losses of these reef-building species have had lasting effects on the USVI coral reefs. In contrast to the effects of storms and white band disease, the bleaching/disease episode in 2005 and 2006 affected most coral species to depths of over 30 m. The coral losses from the 2005 bleaching event and subsequent disease outbreak were especially well documented at long-term monitoring sites maintained by NPS and USGS around St. John and St. Croix.

The following discussion provides greater detail on both the shallow (*Acropora palmata*-dominated; mostly <6 m deep) and deeper reefs around the USVI. The term "deeper reefs" refers here to reefs that are not characterized by living or dead *Acropora palmata* (elkhorn coral) and which occur mostly at depths >6 m, although they range all the way to the shoreline in some locations. Many of these reefs are dominated by *Montastraea annularis* complex (Fig. 8.10). These different reef types have somewhat different ecological histories and have been studied using different methods.

8.7 *Acropora palmata* Reefs in the Virgin Islands

No reefs in the USVI currently have densities of *Acropora palmata* as high as those recorded in the 1960s and 1970s (Rogers et al. 2002). Buck Island Reef National Monument off St. Croix was established in 1961 as a unit of the US National Park Service primarily in recognition of the barrier reef that surrounds the eastern end of the island. The reef at that time was characterized by dense, interlocking colonies of living *Acropora palmata* (Fig. 8.11). Some early studies of this species, the most significant reef-building species in the Caribbean and western Atlantic, took place at Buck Island and at Tague Bay Reef, 1.6 km to the south off the north shore of St. Croix. These included studies of growth rates (Gladfelter et al. 1978), metabolism (Rogers

FIG. 8.10. *Montastraea annularis* is the most abundant coral species on many USVI reefs (Photo: J. Miller)

FIG. 8.11. *Acropora palmata,* Buck Island, 1970 (Photo: W. Gladfelter)

and Salesky 1981), effects of hurricanes (Rogers et al. 1982), and white band disease (Gladfelter 1982, 1991; Davis et al. 1986). Gladfelter et al. (1977) first described white band disease in 1977 and followed its progression through a reef area. This disease had devastating effects at Tague Bay and Buck Island, and it is thought to be the cause of extensive mortality of *A. palmata* throughout the Caribbean (Aronson and Precht 2001; Bruckner 2002). In 2006,

Acropora palmata (and *Acropora cervicornis)* were listed as threatened under the Endangered Species Act (*Acropora* Biological Review Team 2005).

Hurricanes also have killed elkhorn corals in the USVI (e.g., Rogers et al. 1982). In surveys of reefs around St. John in 1984, Beets et al. (1986) noted active white band at several sites and large areas with dead *A. palmata* from disease and storm damage. Gladfelter documented a decrease from white band disease of 85% to 5% elkhorn cover in a 200 m² study plot at Buck Island, and then a further decrease to less than 1% after Hurricane Hugo in 1989 (Gladfelter 1991). In the fall of 2005, elkhorn coral bleached for the first time on record in the USVI, causing some mortality.

Informal observations around St. John (and videotape documentation around Buck Island) showed some increase in number of elkhorn colonies in the 1990s although densities were very low.

8.7.1 *Acropora palmata* Reefs: St. John

Scientists with USGS, UVI, and NPS began intensive monitoring of elkhorn colonies in 2003 on the fringing reef in western Haulover Bay, St. John, in a zone that once had one of the most impressive elkhorn stands in the USVI (Beets et al. 1986). Initial surveys located 67 colonies in the area (17,627 m²), dispersed widely over a distance of about 500 m parallel to the shoreline. These colonies were monitored and photographed every month from February 2003 to December 2006 for signs of disease (primarily white pox and white band), bleaching, physical breakage, and predation (Fig. 8.12). Identification of diseases in the field is problematic, although white pox and white band are relatively well-defined (Fig. 8.13). There is no evidence so far that the bacterium

Serratia marcescens found in human sewage and elsewhere is associated with white pox around St. John, although it has been reported as the cause of white pox on Florida reefs (Patterson et al. 2002). Other, undescribed "white diseases" which do not resemble either white pox or white band have also been observed.

Randomly selected elkhorn colonies at Hawksnest Bay, St. John (n = 60), were monitored almost monthly from May 2004 to December 2006. During each survey, complete and partial mortality were estimated as well as the cause of any recent mortality, defined as areas of recently exposed skeleton absent of filamentous algae, sediments, and sessile invertebrates. The causes of recent mortality included physical damage, predation, sedimentation, abrasion, bleaching, as well as disease. This location is less than 3 m deep, and 78% of the colonies experienced physical damage, most likely from snorkelers and high wave action. Although fewer colonies suffered from disease (73%), it was the most frequent cause of tissue loss. There were a total of 180 separate disease incidences with over 500 disease-induced lesions, causing much more damage than the number of broken branches (72). The prevalence of disease also showed an increasing trend during times of higher sea surface temperatures (Fig. 8.14). Higher prevalence during

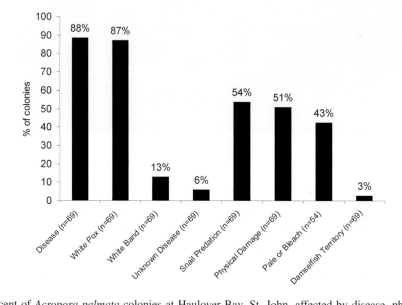

Fig. 8.12. Percent of *Acropora palmata* colonies at Haulover Bay, St. John, affected by disease, physical damage, bleaching, and damselfish territories

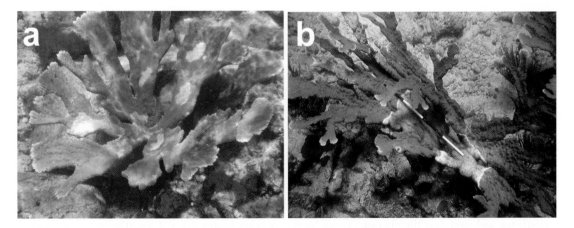

FIG. 8.13. White pox (**a**) and white band (**b**) disease affecting *Acropora palmata* colonies [Photo (**a**): C. Rogers; (**b**): P. Mayor]

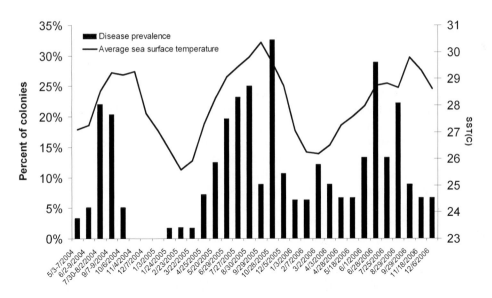

FIG. 8.14. The relationship of diseases and bleaching of *Acropora palmata* and sea-surface temperature (SST) at Hawksnest Bay

warm water conditions may occur from either a more compromised host or an increase in virulent pathogens within the reef area. Regardless of the cause, as global temperatures rise and the oceans continue to warm, an increase in mortality from disease is likely.

Complete colony mortality was highest during the summer/fall 2005 Caribbean bleaching event. Bleaching of colonies, from the loss of zooxanthellae or their pigments, began in late July 2005 and peaked in late September 2005 when the monthly temperature averaged 30.4°C. The next survey in late October revealed the highest amount of disease prevalence (33%) recorded during the 32 month study (Fig. 8.14). Approximately half of the colonies showed some sign of thermal stress through paling or bleaching between July and December 2005. A combination of bleaching and/or disease caused 16% (9) of the colonies to completely die and 23% (13) to die partially. All of the colonies that survived had regained full coloration by January 2006.

To complement the monthly surveys at Haulover and Hawksnest, from August 2004 to May 2005, 13 elkhorn zones (11 within Virgin Islands National Park and two outside) with a total of 3,628 colonies were surveyed (Rogers et al. 2005). Densities at these 13 sites ranged from 0.05 to 9.4 elkhorn colonies/10 m². The focus of this work was disease occurrence and size class distribution. White band disease was noted on only one coral. White pox prevalence (number of colonies with this disease divided by the total number of colonies surveyed within the reef area) ranged from less than 1% to 34.7% and was more often found on colonies greater than 50 cm in maximum dimension. Saltpond Bay had the highest disease prevalence of any elkhorn site around St. John, with a prevalence of 34.7% during the first survey in September 2004.

Surveys were done almost every month at 2 of the original 13 sites (Saltpond on the south side of St. John and Trunk on the north side) from July 2005 to July 2006 (Rogers et al. 2006). All *A. palmata* colonies were surveyed 17 times over a 22-month period, from September 2004 (Saltpond only) to July 2006. Disease, bleaching, mortality, and predation were documented and photographed for all colonies (including fragments) encountered. A fragment was defined as any coral not attached by tissue to the substrate. Trunk Bay is the site within VINP that receives the greatest number of snorkelers, while few snorkelers go to reefs within Saltpond Bay. No apparent correlation was found between disease prevalence and visitation.

Overall, bleaching associated with high water temperatures in the fall of 2005 (over 30°C) caused more complete mortality at Trunk and Saltpond than disease, predation, physical breakage, and competition. Twenty-one colonies completely died at Saltpond Bay with the majority (11) dying directly from bleaching and only two from disease. Thirteen (13) colonies died at Trunk Bay, with four dying from bleaching and four from disease.

At Saltpond the prevalence of disease increased from approximately 3% before bleaching began, to 4.9% during the month following the height of the bleaching event. This slightly higher level of disease was sustained for 5 months, until March 31, 2006. Although there was a small increase in the amount of disease, the overall prevalence of disease during the 12 months of study was much lower (1.4–4.9%) than the initial survey at Saltpond in September 2004 (34.7%). Disease prevalence at Trunk Bay did not increase during the months following the bleaching event. White pox or recent mortality caused by disease that could not be categorized as either white pox or white band (referred to as "unknown disease") was present during every survey at Saltpond, with prevalence ranging from 1.4% to 4.9%. The highest amount of white pox and unknown disease (4.9%) was found in late October 2005 when water temperatures were approximately 29.6°C. The prevalence of disease at Trunk Bay ranged from 0% to 10.7% with the highest number of *A. palmata* colonies with disease occurring in February 2006 when water temperatures were relatively low (26.06°C). White band disease affected only 15 colonies at Saltpond and two at Trunk Bay.

Physical damage to elkhorn at Trunk and Saltpond was more from heavy seas than from careless snorkelers. Although broken branches of *A. palmata* can re-attach and grow as separate colonies, research at these sites revealed that about 40–50% of the observed fragments died. The reef at Trunk Bay experiences the heaviest visitation in VINP. Here, 47.4% of the fragments (total n = 19) were alive at the end of the study in July 2006, and no fragments had attached to the substrate.

In comparison, at Saltpond, which has far fewer snorkelers, 59.5% of the fragments (total n = 205) were still alive when the study ended in July 2006, although 33.3% of them had lost over half of their tissue. The causes of mortality included disease and bleaching.

Overall, bleaching caused more mortality than disease, predation, and physical breakage at Saltpond and Trunk. In general, unlike on deeper reefs dominated by *Montastraea* spp. (see below), bleached elkhorn corals regained normal coloration by January 2006, and then only minor outbreaks of disease were observed. Out of a subset of 467 elkhorn colonies being monitored monthly from late 2005 to July 2006 at Saltpond, Trunk, Hawksnest, and Haulover, 48% bleached, 13% died partially, and only 8% died completely.

The ability to determine the genotypes of elkhorn coral colonies (Baums et al. 2005a) creates opportunities to explore some interesting research questions. For example, do corals with different genotypes have different susceptibility to bleaching and disease? At Haulover, one elkhorn coral bleached and died while the immediately adjacent colony did not (Fig. 8.15). These colonies had different genotypes although they had the same zooxanthellae clade (data from I. Baums and B. Schill). At Haulover, 43 of 48 colonies had different genotypes.

Although *Acropora cervicornis* is an important reef-building coral throughout the Caribbean and was listed as a threatened species along with *A. palmata* in 2006, it has received much less attention than *A. palmata*. *A. cervicornis* grows over a much larger depth range and often exists as isolated and widely dispersed colonies in the USVI, unlike *A. palmata* that is often in depths less than 6 m and in nearly monospecific stands. In 2005 and 2006, *A. cervicornis* populations were surveyed in a 28,824 m² area at Haulover Bay. Quantifying colony size is difficult in this species. Three-dimensional size measurements were made by measuring the height, length and width to determine a volume for each colony. The total volume decreased by 19.3% from 2005 to 2006. The total number of colonies increased from 358 to 655, for respective densities of 0.012 and 0.023 colonies/m². However, the average volume per colony decreased by 55.9%. The increase in number of colonies coupled with the decrease in colony size, suggests that remnant patches of tissue from the original colonies were isolated from each other by mortality, resulting in several smaller colonies where there was originally one. Alternatively, some colonies may have experienced physical damage that broke an individual colony into several smaller ones. However, no evidence of physical damage was seen on the majority of colonies, and the decrease in total volume suggests considerable mortality occurred between years. Incidence of white band disease did not change substantially, with 27 colonies in 2005 and 30 colonies in 2006 affected. However, the proportion of colonies with white band disease dropped from 7.5% to 4.6%. The number of coral-eating snails found on the colonies more than doubled from 40 in 2005 to 82 in 2006.

FIG. 8.15. Adjacent elkhorn colonies with bleaching of the right colony no bleaching of the left by (Photo: E. Muller)

8.7.2 *Acropora palmata* Reefs: St. Croix

In 2002, nine reef sites around the eastern tip of St. Croix were surveyed that had formerly been dominated by *Acropora palmata* were surveyed, (Rogers et al. 2002). These included six north shore reefs, of which two had measured planar cover of 62% (Buck Island barrier eastern fore reef; Gladfelter et al. 1977) and 47% (Tague Bay forereef; Gladfelter 1982) and the others an estimated 25–35% during the mid 1970s; the three south shore reefs had measured planar cover of between 7% and 33% (Adey et al. 1981). In 2002, the north shore reefs had between 0.1% and 3.6% cover, while the south shore reefs were between <0.1% and 1% cover. Several sites (with an areal extent of hundreds of square meters), had numerous young, healthy *A. palmata* colonies, many of which were the result of more than one successful episode of sexual recruitment. These populations were capable of recovery, barring other sources of mortality like storms, bleaching, and disease (E. Gladfelter, personal communication, 2007).

In 2004, 2,492 large elkhorn colonies (greater than 1 m maximum dimension) were recorded along randomly selected transects at depths of 10 m or less within Buck Island Reef National Monument (Mayor et al. 2006). Density ranged from 0.004 to 0.160/m². The overall prevalence of white band disease was 3%, but along transects with white band disease an average of 15% of the colonies were affected. Gladfelter et al. (1977) found 3% prevalence within the initial boundary of BIRNM but 42% at Tague Bay. White pox disease was not quantified in the 2004 study but appeared more common than white band.

At BIRNM, *Acropora palmata* experienced extensive bleaching in 2005 (Fig. 8.16). National Park Service staff quantified the extent of the bleaching and the subsequent mortality of *Acropora palmata*. Shallow *Acropora palmata* habitat is present on reef crest formations and "haystack" features in addition to the barrier reef surrounding Buck Island. However, the majority of *Acropora palmata* habitat at BIRNM is found on the Buck Island bar to the north of Buck Island, at a depth of 5–10 m. In general, *Acropora palmata* colonies located on the shallow barrier reef bleached earlier and suffered greater tissue loss than those in deeper water outside the barrier reef.

The extent of bleaching among *A. palmata* colonies at BIRNM was measured in two ways: (1) by continuing to monitor 44 colonies at three sites ("Selected Sites"); and (2) by a rapid assessment of survey plots (250 m²) at 62 random sites in suitable habitat throughout the Monument ("Monument-wide Colonies"). Two of the three Selected Sites were located on the barrier reef (referred to as the backreef and south forereef sites), and the other was located on the north bar. The colonies at these sites were

FIG. 8.16. Extensive bleaching of *Acropora palmata* at Buck Island Reef National Monument, November 2005 (Photo: E. Muller)

monitored monthly before, during, and after the bleaching event (beginning in March 2005). The Monument-wide survey based primarily on planar photographs of colonies was initiated in November 2005 and repeated once in February 2006 to augment observations from the Selected Sites. Photographs of a subset (65) of the 277 colonies originally surveyed were analyzed for percent tissue bleached and unbleached, and for percent algal-covered skeleton to assess bleaching and mortality. Since shaded portions (undersides) of *Acropora palmata* colonies are less likely to bleach, and would not be recorded in planar photographs, these results may overestimate bleaching and may not be comparable to results from studies where bleaching was quantified *in situ*.

Among the 321 colonies (277 + 44) examined for bleaching, 113 colonies (35%) showed no bleaching. Maximum bleaching on the barrier reef (66.5% of all live tissue for the backreef and forereef sites) occurred in November 2005. At the north bar, outside the barrier reef, 65% of the live tissue was bleached in November 2005. However, colonies in the Monument-wide survey, which also showed a peak in bleaching in November 2005, showed a much lower level of bleaching (an average of 41%).

Interestingly, colonies located on the backreef were impacted before colonies located on both the forereef and on the north bar. Already by August 2005 the backreef site was experiencing bleaching levels of 25%, whereas the forereef site was experiencing only 11% bleaching. It is possible that decreased current and wave action caused colonies located on the backreef to be exposed to higher levels of thermal stress than sites outside the barrier reef. Most of the sea water temperature measurements that exceeded 30°C were recorded in September, with the highest (30.6°C) on September 29, 2005 on the backreef.

Mortality, like bleaching, was higher on the barrier reef than throughout the rest of the Monument, and the backreef experienced more mortality and experienced it sooner than the forereef. The backreef site experienced the highest average mortality (66.4%) during the event, followed by the south forereef (58.1%), and the north bar site (36.4%). The Monument-wide sites experienced 21% average tissue mortality, however mortality was only recorded from November 2005 to February 2006.

8.7.3 *Acropora palmata* Reefs: St. Thomas

In general, *A. palmata* and *A. cervicornis* reefs have been much less studied on the island of St. Thomas. Six reefs dominated by (living or dead) *Acropora* were surveyed in 2003. Impressive mixed stands of these species (and the hybrid *A. prolifera*) occur around St. Thomas with percent cover of living *Acropora* spp. varying from 11% to 13% at Hans Lollik, Flat Cay, and Coculus Point and 6% to 8% at Botany Bay, Inner Brass Island, and Caret Bay (at Vluck Point). White band disease affected an average of 7% of colonies, but prevalence was highest (28%) for *A. palmata* colonies at Caret Bay, suggesting white band as the cause of substantial recent mortality at this reef (Nemeth et al. 2004).

8.8 Deeper Reefs

8.8.1 St. John

Since 1987 Edmunds (2002, 2006) has monitored photoquadrats along haphazardly selected transects at two sites in Great Lameshur Bay (GLB) (three 10 m transects/site), one located just north of the NPS site at Yawzi Point (9 m depth), and the other at Tektite Reef (14 m depth) near the mouth of GLB (Fig. 8.17). At the Yawzi site, declines in coral cover, also of *Montastraea annularis*, were recorded that began with Hurricane Hugo in 1989 and continued throughout the 1990s (Edmunds 2002); these trends are similar to those documented at the nearby site monitored by NPS researchers (Rogers et al. 1991; Rogers and Miller 2006). In contrast, transects on the deeper Tektite reef escaped damage by Hurricane Hugo, and staged a 34% increase in coral cover from 1987 (37% cover) to 1998 (43% cover) that continued at a modest rate until about 2004. The 2005 bleaching /disease event killed >20% of the coral at this deeper site, but the losses at the shallower site at Yawzi Point were barely detectable, largely because the coral cover already had declined to ca. 9.0% by 2005 and fell to 8.6% in 2006 (P.J. Edmunds, unpublished data 2007).

Yawzi Point

Tektite

Quadrat 1   Quadrat 2

Quadrat 1   Quadrat 2

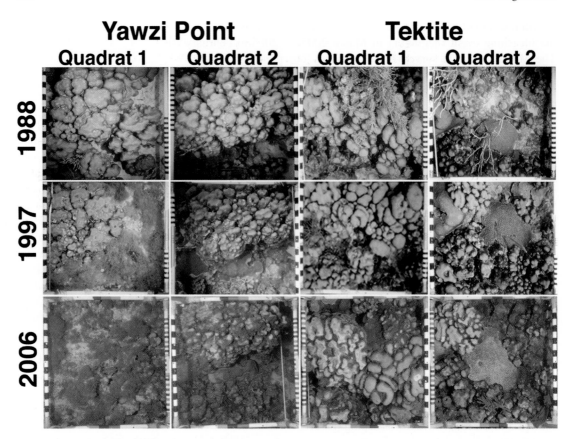

Fig. 8.17. Representative quadrats from the Yawzi Point (9 m depth) and Tektite (14 m depth) study sites that have been monitored with an annual frequency since 1987 (Edmunds 2002; Edmunds and Elahi 2007)

NPS scientists began long-term monitoring of coral, algae, and other benthic substrate on haphazardly selected transects on reefs of high coral cover, diversity and complexity in Great Lameshur Bay (Lameshur Reef) and Newfound Bay in 1989 and 1990, respectively (Rogers et al. 1991; Rogers and Miller 2006). Lameshur Reef is a fringing reef off Yawzi Point, which separates Great and Little Lameshur Bays on the south side of the island and falls within the boundary of Virgin Islands National Park. The reef extends seaward from a nearshore, shallow *Acropora palmata* zone in to deeper water. The base of the reef occurs at about 15 m where there is a sand halo adjacent to an algal plain. Newfound Reef is on the northeastern side of St. John, outside the boundary of Virgin Islands National Park. Although outside the national park, the watershed associated with this

reef has no development. The reef crest is wider and better defined than at Lameshur Reef. It parallels the east and west shores of Newfound Bay and extends partway across the mouth of the bay from either direction, creating a shallow lagoon with a channel to the outer reef. The reef drops to about 14 m where it ends abruptly in a sand halo near an algal plain. Haphazardly selected "chain" transects (average depth ca. 12 m at Lameshur, 7.6 m at Newfound) were supplemented with randomly selected video transects (average depth 13.7–15.0 m at Lameshur, 6.5–9.5 m at Newfound) from larger areas of the reefs at both sites in 1999 (see below).

Montastraea annularis is the most abundant coral at each reef. Hurricane Hugo in 1989 caused a 40% decline in coral cover (from about 20% to 12%) along the five transects off Yawzi Point, with a loss of *M. annularis* (from 7.5% to 5.2% cover). No increase in coral cover was noted up through 2002. There was a significant though small increase in coral cover along the single 100 m chain transect at Newfound from 1990 to 2002.

Coral cover (~8%) did not change significantly along the randomly selected video transects at Lameshur from 1999 to 2003. However, at Newfound, coral cover declined significantly from 18% to 14% from 1999 to 2000, with declines seen along each video transect. The suspected cause is disease (Rogers and Miller 2006).

The randomly selected video transects at Newfound and Lameshur were established during pilot studies in protocol development under the NPS/USGS Inventory and Monitoring (I&M) program. Digital video monitoring was used, along with a newly developed Random Sample Selection Protocol (Fig. 8.18, Miller and Rogers 2002; Rogers et al. 2002). This represented a large change in sampling strategy for monitoring within the Virgin Islands, and coral reef monitoring in general (Lewis 2004). Traditionally (including most studies presented within this chapter), coral reef and other marine habitat sampling is conducted with haphazardly selected study plots or sampling units (quadrats or transects) which

provide excellent data on the selected units, but those data may not have inference over any area other than the quadrats or transects and can not be said to be representative of the entire reef or other habitat that contains them. The Random Sample Selection Protocol used a sonar mapping system to:

1. Accurately define or map the study area
2. Identify the entire population (given defined spacing between sample points)
3. Randomly select the sample points from within the sample population (origins of transects in the case of using the Video Monitoring Protocol)

This allowed every point within the defined sample area to have an equal chance to be chosen for sampling, thus allowing the results obtained to be inferred over the entire defined area (given a large enough sample size). Results obtained using these methods are identified with the reef "name" and size of the study area from which the samples were chosen (domain) so the area to which the data may be inferred is identified.

The I&M program operating at VINP was absorbed into the South Florida /Caribbean Network (SFCN) in 2002 and additional long-term

Fig. 8.18. Videotaping along randomly selected transects, Newfound Reef (Photo: C. Rogers)

328   C.S. Rogers et al.

monitoring sites were established at the South Fore Reef (Buck Island 2002), Haulover (2003) and Tektite (2005) (Table 8.1). Note that trends will be provided for data prior to and through the bleaching /disease event in 2005/06. (The data for the South Fore Reef and Western Spur and Groove sites off Buck Island are included here for comparison.)

8.8.2 Effects of Bleaching and Disease

Some of the warmest sea temperatures on record for the Caribbean, with temperatures reaching over 31°C, occurred in 2005, and USVI and Puerto Rico coral reefs were particularly affected by bleaching (Fig. 8.19). More than 90% of the coral cover bleached at five long-term monitoring sites (Miller et al. 2006). In early October 2005, 279 mm of rain fell in St. John. The rainfall and overcast conditions lowered the seawater temperatures, and many corals began to regain their normal coloration. However, a severe outbreak of white plague disease led to significant coral mortality (Fig. 8.20).

Intensive monitoring throughout the bleaching /disease outbreak revealed that at the peak intensity, the number of disease lesions increased an average of 40 fold (range: 16.4 to 72.9) and total

TABLE 8.1. NPS long-term monitoring sites in St. John and Buck Island, St. Croix, and trends in coral cover prior to the bleaching /disease event (NPS, unpublished data).

Site	Location	Study area (m²)	Annual monitoring began	Pre-bleaching /disease Coral cover trend
Newfound	St. John	13,786	1999	Decrease ($p = 0.0002$)
Yawzi	VINP	7,125	1999	increase ($p = 0.05$)
Mennebeck	VINP	12,495	2000	increase ($p = 0.0432$)
Haulover	VINP	13,568	2003	No change
Tektite	VINP	18,711	2005	Not applicable
S. Fore Reef	BIRNM	40,753	2002	Increase ($p = 0.0006$)
W. Spur and Groove	BIRNM	26,365	2000	No change

FIG. 8.19. Bleached corals off Scott Bay, St. John (Photo: C. Rogers)

FIG. 8.20. A severe outbreak of white plague disease (Photo: E. Muller)

coral tissue area killed increased an average of 25.4 times (range: 2.1 to 80.0) across all sites. Tektite Reef had the highest levels (area killed and number of lesions) of disease, and high levels of disease while corals were bleached so an outbreak may have been underway at Tektite Reef prior to other sites. For more discussion of coral disease findings during this outbreak, see section on coral diseases.

The combination of extremely severe bleaching followed by unprecedented levels of mortality from coral disease caused catastrophic losses in coral cover at all sites averaging 51.3% decline (range 34.1–61.8; data through SFCN annual monitoring for 2006, see Table 8.2).

8.8.3 Coral Species Effects: Changes in Relative Abundance

Montastraea annularis complex was and remains the dominant coral within these reefs but its abundance relative to other coral species dropped during the bleaching/disease event from an initial average cover of 79.2% (SD=7.1) to 71.8% (SD=9.4) (Table 8.3). A reef building species, *Colpophyllia natans*, although a smaller component of the reef community, also decreased relative to other corals. *Agaricia agaricites* declined dramatically in cover and relative abundance, due to mortality from

bleaching, as 93% of *A. agaricites* bleached (base on cover), the corals were rarely affected by disease. *Montastraea cavernosa*, *Siderastrea siderea* and poritids which bleached less than *M. annularis* (complex), *C. natans*, and agariciids, had "relatively" moderate disease levels and have increased in abundance compared to other corals.

8.9   The Deepest Reefs

Most information on coral reefs around St. John comes from relatively shallow (0–20 m deep) study sites. Surveys by NOAA and the NPS in 2005 using a remotely operated vehicle expanded the sampling range to deeper waters (200 m). Their data revealed that in general deep zooxanthellate coral reefs are less deteriorated than their shallower counterparts. Coral cover in deep reefs often exceeded 40% and estimates of algae cover were relatively low. However, deep reefs are not categorically invulnerable. NOAA/NPS surveys in early 2005 found a massive coral mortality event on a 30–40 m deep reef. The mortality event was distinguished by a high amount of dead coral covered by turf algae. As much as 50% of the reef within the transect (500 m^2) was affected and estimates of coral loss exceeded 30%.

TABLE 8.2. Bleaching at long-term monitoring sites in St. John and at Buck Island, and losses in coral cover following severe disease outbreak (NPS and USGS, unpublished data 2007).

Island or Park	Site	Pre-bleaching/disease			Peak of disease outbreak			Mean percent coral cover		
		Mean % coral cover bleached	Total # lesions	Total area (sp. cm.)	Total # lesions	Total area (sp. cm.)	Month of peak disease	Pre-bleaching/ disease	Latest 2006 data	Relative % decrease in coral cover
STJ	Newfound	92	17	1162	569	5318.0	Mar. 2006	13.3	6.2	−53.4
VINP	Mennebeck	94	No sampling prior to event		641[a]	5718.5[b]	Apr. 2006 and Dec. 2005	26.7	10.2	−61.8
VINP	Haulover	96	8	149.5	583	3968.4	Apr. 2006	22.5	12.4	−44.9
VINP	Tektite	97	No sampling prior to event		1213	26025.5	Nov. 2005	24.7	11.1	−55.1
BIRNM	South Fore Reef	96	10	108.3	627	8635.0	Mar. 2006	19.8	11.4	−42.4
VINP	Yawzi	71	Not sampled episodically during bleaching/disease outbreak					8.5	5.6	−34.1
BIRNM	W. Spur and Groove	4[c]						5.1	3.1	−39.2
	Average values:	95.0	11.7	473.3	726.6	9933.1		21.4	10.3	−51.5
	Range values:	71 – 97	8 – 17	108.3 – 1162.0	569 – 1213	3968.4 – 26025.5		5.1 – 26.7	3.1 – 12.4	−34.1 to −61.8

[a] Peak in number of lesions observed in April 2006

[b] Peak in total area disease observed in December 2005

[c] Sampling occurred in May of 2005 and 2006; bleaching data not comparable with other sites or used in ranges/averages.

TABLE 8.3. Changes in relative abundance of coral species. MACX = *Montastraea annularis* complex (calculated as MA = *Montastraea annularis*, MFAV = *Montastraea faveolata*, MFRA = *Montastraea franksi*), SD = standard deviation.

Species/family/group	Initial			Latest 2006 sample			Relative
	Avg.	SD	Rank	Avg.	SD	rank	Loss/gain
MACX = MA + MFAV + MFRA	79.2	7.1	1	71.8	9.4	1	−9.3
Colpophyllia natans	1.3	0.8	7	0.9	1.4	5	−30.0
Diploria spp.	1.8	1.5	5	2.3	2.1	4	29.3
Montastraea cavernosa	1.5	1.8	6	2.4	2.2	3	55.0
Siderastrea siderea	1.9	0.7	4	4.7	1.3	6	143.4
Agaricia spp.	2.0	0.6	3	0.3	0.5	7	−82.4
Porites spp.	9.5	6.1	2	12.4	3.9	2	30.2

Monitoring surveys in the summer of 2006 revealed significant degradation at depths of 30 m on reefs that are part of the Mid-shelf Reef complex, 2–8 km south of St. Thomas and St. John. Coral cover varies greatly on these reefs. UVI scientists surveyed transects at one *Montastraea*-dominated site and found that coral cover dropped from 56% during the 2005 bleaching event to 41% following the bleaching and subsequent white plague outbreak (T. Smith et al., personal communication 2006). This was a 26% drop in coral cover on one of the highest coral cover sites known in the USVI. This reef area is less than a kilometer away from the site of the mortality event noted in 2005 and described above. Additionally, random surveys conducted by NOAA and NPS within this reef system, inside the Virgin Islands Coral Reef National Monument (VICRNM), suggested a significant decrease in coral cover between 2005 and 2006. Coral cover decreased from 4.0 ± 0.4% (range 1–9) in 2005 to 2.3 ± 0.3% (range 1–5) in 2006 ($\chi^2 = 9.21$, P = 0.0024, n = 20). Interestingly, the decline in coral cover between 2005 (15.0%, range 3–45) and 2006 (12.3%, range 0–25) on the same reef system but outside VICRNM boundaries was not significant ($\chi^2 = 0.19$, P = 0.6609, n = 15).

8.9.1 St. Croix

At Buck Island Reef National Monument, permanent, haphazardly selected transects and individual colonies of *Montastraea annularis, Diploria strigosa*, and *Porites astreoides* have been monitored since 1988 (Bythell et al. 1993, 2000a,

b). Hurricane Hugo (1989) was responsible for the most significant changes up through 2000. The three sites with permanent transects differed greatly from each other in terms of coral cover and coral species composition. Coral cover increased from 32% to 40% at one site, but decreased from 25% to 15% at another site. Although Hurricane Hugo caused more loss of *Porites porites* than of the more abundant *M. annularis*, bleaching in 1998 caused more loss of *M. annularis* than of other species along the transects. Within 5 years of Hurricane Hugo an increase in coral cover was noted, but bleaching in 1998 and additional hurricanes caused more declines. Bythell et al. (2000a) suggested that in spite of the variations along the transects the reef assemblages over a scale of tens to hundreds of meters were relatively stable, with changes over time being less than differences among sites.

Observations of 303 individual coral colonies from 1988 to 2000 showed that mortality (or removal) of entire colonies was 6% for *Montastraea annularis*, 20% for *Diploria strigosa*, and 49% for *Porites astreoides*. The major reef-building species *M. annularis* sustained greater losses from bleaching in 1998 than from hurricanes while *Porites astreoides* and *D. strigosa* had more severe losses from the storms (Bythell et al. 2000b). Disease incidence was low throughout the entire study period and was associated with bleaching.

Twelve permanent coral reef sites ranging in depth from 4 to 23 m have been monitored on St. Croix by researchers from the University of the Virgin Islands. Of these sites, ten were established in 2001, with the remaining shelf-edge

Mutton Snapper site and the nearshore Great Pond site established in 2002 and 2003, respectively. Overall coral cover averaged 15%, with 13% at nearshore sites, similar to the 16% coral cover at nearshore St. Thomas sites. In common with St. Thomas coral cover, there were no apparent trends in coral cover over time prior to the 2005 bleaching (Nemeth et al. 2005; Smith et al., personal communication). With few exceptions, *Montastraea* spp. (MACX = *Montastraea annularis* complex) were most abundant (average = 52%) at all sites combined (62% with *M. cavernosa*). Dead coral with algae (dca) consistently contributed the highest cover ranging from 20% to 80%, while macroalgae ranged from 2% to 40%. Less than 10% of the cover was sponges and gorgonians.

The prevalence of disease on St. Croix reefs prior to the 2005 bleaching event was generally low. However, there were low level (less than 1% prevalence) infections of white plague at the high *Montastrea* spp. cover sites, Sprat Hall and Mutton Snapper. Low-level severity, chronic bleaching was also low, from only a few colonies affected at Mutton Snapper to 6% at Sprat Hall.

8.9.2   St. Thomas

In 2001, University of the Virgin Islands scientists began the monitoring of nearshore (6–20 m depth) reefs and expanded the program in 2003 and again in 2005 to include mid-shelf (5–30 m depth – including reefs which fringe offshore cays and non-emergent reefs on the island platform), and shelf-edge reefs (>30 m depth) (Fig. 8.21). Of the 15 sites monitored before the 2005 bleaching event, coral cover averaged 20% and ranged from 4% to 53%. The lowest coral cover values were typically found on nearshore reefs (average = 14%) with the lowest cover (4%) occurring in a relict coral reef just outside the highly impacted Charlotte Amalie Harbor. Coral cover generally increased with depth and distance from shore and averaged 26% on mid-shelf reefs and 36% on shelf edge reefs (Herzlieb et al. 2006). The highest coral cover was typically found on the deep shelf-edge reefs, attaining a maximum value of 53% at College Shoal. Prior to the 2005 bleaching event, coral cover tended to remain constant at all the monitoring sites sampled over

multiple years. Coral species composition varied by reef location with the percent of *Montastraea* spp. increasing with depth and distance from shore. Nearshore reefs contained an average of 34% *M. annularis* complex [(MACX) – or 44% with *M. cavernosa*], 61% MACX (or 79% with *M. cavernosa*) at mid-shelf sites and 85% MACX (or 88% with *M. cavernosa*) at shelf-edge sites (Nemeth et al. 2004). Macroalgae cover averaged 33% and ranged from 4% to 60%, with the lowest values corresponding with high coral cover sites and highest values corresponding with exposed areas on the mid-shelf. Macroalgae cover was the most abundant component at all sites except at two of three shelf-edge sites, and trended upward at three nearshore (Benner, Botany, and Magen's Bays) and one 23 m deep mid-shelf site (S. Capella). Dead coral covered with turf algae averaged 30% and ranged from 20% to 50%. Combined sponge and gorgonian cover averaged 8% and ranged from 3% to 12%, with no apparent cross-shelf trends.

One would predict that reefs closer to shore and therefore more likely to be affected by humans would be more degraded than those in deeper water and/or farther from shore. For St. Thomas, evidence of a nearshore to offshore trend in increasing coral cover supports this. In 2003, Herzlieb et al. (2006) assessed a total of eleven reefs within three categories based upon reef position along the insular platform south of St. Thomas: near-shore reefs (5–30 m depth, <1 km from the shoreline of St. Thomas); mid-shelf reefs (5–30 m depth, 1–10 km from shore); and shelf-edge reefs (≥30 m depth, 10–15 km from shore). Percent cover of biotic and abiotic substrata, coral species composition, and levels of bleaching and disease were compared among the near-shore, mid-shelf, and shelf-edge reef systems. Nearshore reefs had significantly lower live coral cover and higher cover of dead coral with turf algae than the other two reef systems. In addition, nearshore reefs had a significantly lower relative abundance of a coral species sensitive to terrigenous stress (*Montastraea annularis*) and significantly higher percent composition of a coral species resistant to terrigenous stress (*Siderastrea siderea*) than the other reef systems.

Smith et al. (personal communication 2007) found that coral diseases on these St. Thomas reefs

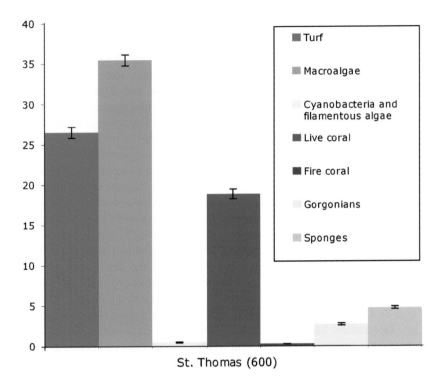

FIG. 8.21. The mean percent cover of benthic categories from 600 transects at 16 coral reefs surveyed off St. Thomas over the years 2001–2006. Bars represent the standard error of the mean. Locations included nearshore reefs (6–13 m), mid-shelf reefs associated with islands (9–15 m), mid-shelf reefs not associated with islands (19–30 m), and shelf-edge reefs (30–40 m)

from 2003 to 2005 (before the bleaching began) were more prevalent at nearshore monitoring sites (8% of colonies affected) as opposed to mid-shelf and shelf edge sites (2–4%). Two coral diseases (dark spots syndrome and yellow blotch disease) were more prevalent at nearshore coral reef sites, independent of coral species competition, and tended to drive the onshore to offshore differences, with a combined average of 7% of colonies affected nearshore and less than 1% at offshore sites. Furthermore, bleaching, predominantly low-level partial bleaching potentially associated with chronic stress, was highest nearshore (30% of colonies affected) and trended downward offshore to the shelf-edge sites (12%). The proportion of coral colonies with old partial mortality (i.e., degraded skeleton and algal cover) was highest at nearshore sites (36%), double the average at offshore sites (17%). Distressingly, recent partial mortality was

highest at the shelf edge sites around St. Thomas, and, Lang Bank, the deepest and most remote St. Croix site, had the highest incidence of disease at about 17% and the highest bleaching (ca. 22%). This suggests that degradation may have begun in some of the most remote coral reefs in the USVI. Herzlieb et al. (2006) speculated that increased disease of deep coral reefs on the Grammanik Bank might be due to fish traps which act as vectors for pathogens when they are moved from shallow to deep water sites on a seasonal basis.

8.9.3   Benthic Composition of Reefs in St. John and St. Croix: NOAA

Since 2001, NOAA has been using a random sampling design to characterize benthic composition on reef and hardbottom areas around St. John and Buck Island, and along the northeastern shore of St. Croix including the East End Marine Park. Data from 788 sites indicate that reef and hardbottom areas in both St. Croix and St. John generally are dominated by algae (Fig. 8.22). The most abundant of three algal categories observed was turf and crustose algae with a mean cover of $36 \pm 1.2\%$ in St. Croix and $30.4\% \pm 1.7\%$ in St. John. Macroalgae cover averaged $12.1 \pm 0.5\%$ and $13.9 \pm 0.9\%$ in St. Croix and St. John, respectively.

In St. Croix, the macroalgae with the highest observed cover were *Dictyota* spp., *Halimeda* spp., *and Sargassum* spp. In St. John, the most common macroalgae observed were *Dictyota* spp., *Lobophora variegata, and Halimeda* spp. Filamentous algae and Cyanobacteria accounted for $5.2 \pm 0.6\%$ cover in St. Croix and $1.7 \pm 0.4\%$ cover in St. John (Fig. 8.22).

Live scleractinian coral cover was low and averaged $6.4 \pm 0.5\%$ in St. Croix and $5.8 \pm 0.5\%$ in St. John (Fig. 8.22). Gorgonians had higher crown cover in St. John when compared with reef and hardbottom areas in St. Croix (Wilcoxon/Kruskall-Wallis One-way χ^2 test, $p < 0.0001$). Milleporid

corals and sponges also had higher cover in St. John than in St. Croix ($p < 0.0002$).

Patterns in the relative cover of benthic organisms were consistent across reef types (Kendall et al. 2001), with two algal categories (turf/crustose algae and macroalgae) dominating all six reef types (Fig. 8.23). Cyanobacteria and filamentous algae had the highest cover and were most variable on reef rubble and scattered coral and rock sites. The mean percent cover of live scleractinian coral was significantly highest on patch reefs ($11.8 \pm 1.2\%$, $p < 0.05$, Dunn's multiple comparison test) and lowest on reef rubble and scattered coral and rock sites ($1.59 \pm 0.6\%$, Fig. 8.23). Gorgonians had the lowest cover on reef rubble sites. The percent cover of sponges and fire corals were similar among benthic habitats.

Live scleractinian coral cover in St. Croix and St. John comprised 18 coral genera (Fig. 8.24). The three most abundant genera were *Montastraea* spp., *Porites* spp., and *Diploria* spp. Some significant differences in coral composition on reefs and hardbottom areas were observed between St. Croix

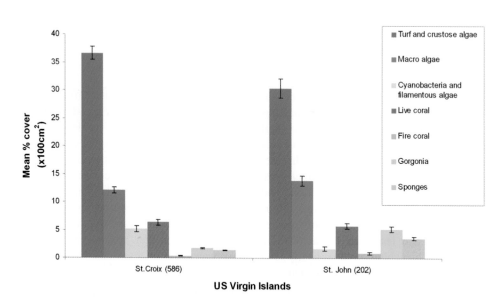

FIG. 8.22. Mean percent cover of benthic organisms on reefs in St. John and St. Croix. Bars represent the standard error of the mean. The number of sites (n) surveyed on each island is shown in parentheses. Sites ranged in depth from 0 to 28 m. In St. John, sites were located around the entire island, whereas in St. Croix, only sites in the north shore of East End Marine Park and the Buck Island Reef National Monument were surveyed. Benthic composition was estimated visually from five replicate 1 m² quadrats within a randomly chosen 100 m² belt transect at each site (NOAA Biogeography Program, http://www8.nos.noaa.gov/biogeo_public/query_main.aspx)

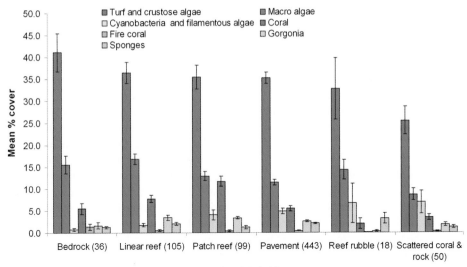

FIG. 8.23. Mean percent cover of benthic organisms found in different reef habitats off St. John and St. Croix. Other sessile invertebrates include anemones, tunicates, zooanthids, and tubeworms. Habitat types were classified based on digital benthic maps (Kendall et al. 2001). Bars represent the standard error of the mean. The number of sites (n) surveyed for each habitat is shown in parentheses. Benthic composition was estimated visually from five replicate 1 m² quadrats within a randomly chosen 100 m² belt transect at each site (NOAA Biogeography Program, http://www8. nos.noaa.gov/biogeo_public/query_main.aspx)

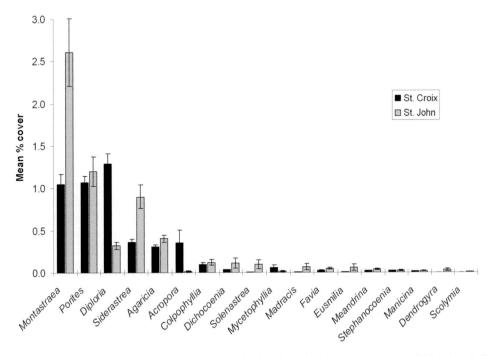

FIG. 8.24. Mean percent live cover of coral genera on randomly selected reef sites between 0 and 28 m deep in St. John and St. Croix. Bars represent the standard error of the mean. The percent cover of live coral was determined visually from five replicate 1 m² quadrats at each site (n = 768)

and St. John. *Montastraea* spp., *Siderastrea* spp., and *Agaricia* spp. had higher average cover in St.

John compared with St. Croix (*p*<0.0001, Wilcoxon/Kruskall Wallis One-way χ^2 test). However, *Diploria* spp. and *Acropora* spp. had higher cover in St. Croix than in St. John, (*p*<0.04). The cover of other coral genera was similar between St. Croix and St. John.

8.10   Coral Diseases

All of the coral diseases reported for the Caribbean (Weil et al. 2006, Sutherland et al. 2004) have been seen on reefs in the USVI. White plague and white band disease have been the most severe. In the late 1990s white plague appeared on reefs in St. John. From December 1997 through November 2005, monthly surveys of disease, coral cover, and macroalgal cover were conducted using 1 m^2 quadrats along eight 10 m transects in a portion of Tektite Reef which initially had very high coral cover (ca. 66%) (Miller et al. 2003; NPS, unpublished data 2007). Disease incidence was estimated by percent planar cover within quadrats and by size of disease patches (lesions). The disease had the gross appearance of white plague, and samples confirmed the presence of *Aurantimonas coralicida,* the reported pathogen (Denner et al. 2003). Disease was present each month and was not correlated with seawater temperature. The mean live coral cover declined significantly over the study period, with significant losses in seven of the eight transects. Especially severe outbreaks were seen in August 2000 and August 2005. Most of the coral in the transects was *Montastraea annularis*, although other species such as *Colpophyllia natans* and *Diploria labyrinthiformis* were present. The dead coral was primarily covered with algae, although some recruits of *Agaricia* and *Porites* were observed.

In addition to analysis of digital videotapes for changes in percent coral cover over time (described above), two other approaches have been used to examine the responses of corals at the NPS long-term study sites to the 2005 bleaching /disease event. First, the amount of disease affecting the coral reefs was estimated on each sampling date by measurement of lesions (areas that have recently been killed by disease) on coral colonies one meter

on either side of the permanent transects. At all locations, disease was more extensive following the bleaching than before bleaching began (based on videotapes and quantitative data). Second, videotapes from successive time periods at each long-term site have been compared side by side to follow the condition and fate of individual coral colonies. This analysis showed that some coral species, including the larger, major framework - building species (*Montastraea annularis* complex, *Colpophyllia natans*, and others) exhibited the most severe bleaching. Recovery from bleaching varied by species, with some corals such as those in the *M. annularis* complex showing significant recovery followed by severe disease and colonies of other species dying directly from bleaching.

A total of 6,061 disease lesions were recorded on 23 coral species from September 2005 to July 2006 at all sites. Five diseases/syndromes were observed including black band disease, dark spots syndrome, white band disease, and yellow blotch, but 99% of the lesions and of the total area killed was due to white plague. Ninety-three percent of the disease occurred on colonies within the genus *Montastraea* (*M. annularis* complex primarily, with *M. cavernosa* to a much lesser extent). Other affected genera included *Colpophyllia*, *Siderastrea*, *Diploria*, and *Porites*.

Samples of healthy and diseased corals are being analyzed to determine if there are shifts in the associated microbial communities that occur when corals become diseased. In August 2005, just before the severe bleaching event, samples of diseased and apparently healthy corals (mostly *Montastraea annularis*) were taken along transects at Tektite Reef, within Virgin Islands National Park, using a non-destructive swabbing method. Sterile foam swabs were used to sample corals, and material was transferred to Whatman FTA cards for storage and transport to the laboratory. Bacterial 16S ribosomal genes and zooxanthellae ITS-1 genes were readily amplified by polymerase chain reaction (PCR) from card samples. PCR products were further analyzed to examine the diversity of bacteria and zooxanthellae colonizing the corals sampled. No obvious associations between disease status and zooxanthellae clades were noted, however investigations of bacterial associations were informative. Both healthy and diseased corals had *Aurantimonas coralicida,* an alphaproteobacterium. Further analysis revealed different communities of alphaproteobacteria in diseased vs healthy *M. annularis* (Pantos and Bythell 2006; USGS, unpublished data 2007).

Members of the alphaproteobacteria are extremely diverse in form, function, and ecological role. The subdivision includes symbionts as well as serious plant and animal pathogens. Several studies from multiple researchers have detected shifts in coral-associated alphaproteobacteria that seem to be associated with heath/disease status. To further study this possible relationship, a set of PCR primers was developed that direct the amplification of a highly variable sequence stretch in the alphaproteobacterial gene that codes for 16S ribosomal RNA. Sixty-four samples taken from the apparently healthy Tektite Reef *Montastraea annularis* colonies and 31 samples taken from diseased colonies were amplified and examined by melting curve analysis. Of the 64 apparently healthy colonies, 56 harbored a common, single type of alphaproteobacterium, while the remaining eight harbored either distinctly different single types or multiple types of these microbes. This contrasts with the 31 diseased samples that all harbored alphaproteobacterial types that appeared to be different from any of those found in the healthy colony samples. The species identifications of characteristic alphaproteobacteria associated with healthy and diseased coral colonies are currently being determined by genetic sequence analysis. The data suggest that there may be an alphaproteobacterium that forms either a commensal or a symbiotic relationship with *Montastraea annularis* and that this relationship is disturbed in the disease process. The unusual alphaproteobacterial signatures found in 8 of 64 of the apparently healthy colony samples may represent either rarely occurring normal flora, or alternatively, may be the first representation of the onset of disease in which case these sorts of analyses could be predictive in assessment of reef vulnerability.

Black band disease (BBD) has been seen at low levels on reefs in the USVI for at least 2 decades (Edmunds 1991) and affects fewer species than the more recently appearing white plague (Fig. 8.25). Edmunds (1991) documented the proportion of corals infected on shallow reefs (<10 m deep) in Great Lameshur Bay, St. John, between August 1988 and September 1989. BBD

infections were most common on *Diploria strigosa*, *D. labryinthiformis*, *Montastraea annularis*, *Siderastrea siderea*, and *Colpophyllia natans*, but only 0.2% of 6908 colonies were infected in the autumn of 1988. Infection rates were lower in February, when the seawater temperature was the coolest, compared to September and November. Edmunds (1991) estimated that the disease could remove 3.9% of the living tissue of *Diploria strigosa* colonies each year.

Recently a study was carried out to compare the microbial communities in BBD on corals from three regions of the wider Caribbean, including the USVI (Voss et al. 2007). BBD consists of a migrating, cyanobacterial-dominated microbial mat that moves across corals at rates up to 1 cm/day, completely degrading coral tissue and exposing pure coral skeleton. It can kill an individual coral colony in a matter of months. No primary pathogen has been identified, and the disease may by a polymicrobial infection that requires a specific microbial community. It has been shown that four major physiological groups are always present in BBD – phototrophs, heterotrophs, sulfate reducers, and sulfide oxidizers.

In this study 97 samples were analyzed from 19 reef sites within the three regions. These consisted of three sites at St. John (Haulover Bay, Hawknest, and Watermelon Cay), seven sites at Lee Stocking Island, Bahamas, and nine sites at the northern Florida Keys. Depths ranged from 2 to 6 m at St. John, 3 to 20 m at Lee Stocking Island, and 2 to 6 m on the northern Florida Keys. Five of the 97 samples were from St. John, with individual samples from BBD on *D. stokesi*, *D. labyrinthiformis*, *M. annularis*, *M. cavernosa*, and *S. siderea*.

Data analysis was carried out by profiling the BBD microbial community using molecular techniques that targeted the 16S rRNA gene. The BBD microbial communities were statistically discriminate ($p < 0.05$) among the three regions and between host species. The variability was driven by differences in cyanobacteria within the community as well as alphaproteobacteria, a heterotrophic group. These results suggest that, if BBD is a true polymicrobial infection, different members of the major physiological groups may be represented by different species that perform the same physiological role within the BBD consortium.

Fɪɢ. 8.25. Progression of black band disease on *Diploria strigosa*, St. John, from July 2004 to July 2005 (Photos: C. Rogers)

8.11 Sedimentation

Runoff is recognized as one of the most serious stressors affecting coral reefs in the USVI (Fig. 8.26; Hubbard 1987). Steep slopes, with more than 80% of them on St. Thomas and St. John over 35% grade, drenching rainfall, shallow easily eroded soils, and numerous drainage guts combine to increase the amounts of erosion, sedimentation and non-point source pollution that

Fig. 8.26. Development is leading to increased rates of runoff in the USVI

reach downstream marine communities. Hubbard (1987) reviewed effects of sedimentation on coral reefs and highlighted the potential for reef damage from many development projects in the USVI (see also Hubbard et al. 1987). However, almost 20 years later we still lack quantitative studies that show sedimentation rates before, during, and after upland and coastal construction that can be conclusively linked to reef degradation. Some studies show lower coral cover in areas that have higher sedimentation rates, but these correlations cannot pinpoint sedimentation as a cause of the existing benthic composition or relative abundance of corals. While there is no doubt that increasing amounts of sediment are entering nearshore waters, the effects of chronic sedimentation are harder to document than the more conspicuous results of hurricanes or coral diseases.

Development of steep hillsides in the USVI and cutting of new, unpaved roads has caused severe runoff of silt-laden water into the bays. No permanent streams or rivers occur in the islands, but runoff is a significant problem because the islands slope steeply to the coast and many reefs are so close to shore. [The highest points of land on St. Croix, St. Thomas, and St. John are 355, 472, and 389 m, respectively (Dammann and Nellis 1992)].

Plumes of silt are typically seen after short but intense rains. Runoff of sediments is an increasing concern because of the accelerating pace of development in the USVI and the very steep hillsides. New roads and driveways are being cut in areas with heavy vegetation. Measurement of erosion rates on St. John (in a study from 1998 to 2001) indicated that unpaved roads contribute up to four orders of magnitude more sedimentation than undisturbed hillsides (Ramos-Scharron and MacDonald 2005).

Sediment core testing for terrestrial based sediments deposited in nearshore wetland and coastal embayments from around St. Thomas and St. John, show that over the past 15–25 years, sedimentation rates have increased from 1 to 2 orders of magnitude (Brooks et al. 2004). Unpaved roads and altered drainage contribute the most sediment, but excavation for home and driveway constructions also contribute significantly to the sediment load. While the pace of construction development on heavily populated St. Thomas and St. Croix is moderate, St. John has seen a significant increase in the cutting of new roads and the excavation of home sites on steep inclines. As a result of physical topography and human activities on these islands, downstream marine communities, reefs especially, are subjected to multiple stressors with unknown long term synergistic impacts.

Sedimentation rates in the Virgin Islands vary considerably among sites and seasons. Nearshore waters adjacent to highly developed watersheds (Fish Bay, Magens Bay) typically average over 10 mg/cm^2/d. In contrast reefs adjacent to less developed watersheds (i.e., Lameshur Bay, Sprat Bay) or offshore cays (i.e., Flat Cay, Buck Island) receive less than 4 mg/cm^2/d. Offshore reefs not associated with a land mass typically receive less than 0.5 mg/cm^2/d (R. Nemeth, unpublished data 2007). Seasonal variation in sedimentation rates are usually highest during the rainy season when sediment load can increase from less than 2 mg/cm^2/d during the dry season to greater than 30 mg/cm^2/d during a severe rain event (Nemeth and Sladek Nowlis 2001). Sometimes, however when terrigenous sediments are deposited in channels between reefs and become re-suspended during large swells not associated with storm events. For example, sedimentation rates can increase from less than 2 mg/cm^2/d to over 15 mg/cm^2/d when a north swell hits the north coast of St. Thomas (Nemeth and Sladek Nowlis 2001). During one study coral cover was monitored before, during, and after development in Caret Bay, St. Thomas, from July 1997 to March 1999. A weak correlation between bleaching and sedimentation and decline in coral was found, but the study took place at the time of the October 1998 bleaching event in the USVI and coral losses cannot be conclusively attributed to sedimentation. Coral cover was less than 5% on the study reef and declined along five transects by an average of 14%. This study also found a significant correlation between sedimentation rate and bleaching (Nemeth and Sladek Nowlis 2001).

8.12 Fisheries and Fish Assemblages

Here we present a brief overview of the fisheries in the Virgin Islands to provide a context for subsequent discussion of changes in reef fish assemblages over the last several decades (Fig. 8.27). The area available for fishing around the USVI is relatively small, an estimated 5,180 km^2 (Dammann 1969). In 1930, the population of the USVI was 22,012 and approximately 405 fishers used about 1,600 traps (Fiedler and Jarvis 1932). In the late 1950s, Idyll and Randall (1959) reported over 500 traps were in use around St. John alone. In 1961, there were 400 fishers using 838 traps (Anonymous 1961 cited in Dammann 1969). In 1968, the estimated number of fishers remained the same, but the population of the USVI had more than doubled to 55,000. In 2003 there were 383 licensed commercial fishers in the USVI, 160 in the St. Thomas /St. John District and 223 in the St. Croix District, and the USVI population had increased to almost 110,000 (Kojis 2004).

In the 1980s and 1990s, the USVI fishery greatly capitalized, and the effort increased offshore to the shelf edge. By 2003, fishers used a wide variety of gear including pots, handlines, a variety of nets, vertical set lines and scuba. Based on a census conducted by the DPNR/Division of Fish and Wildlife in 2003 (Kojis 2004), commercial fishers owned approximately 1,234 fish traps in the St. Croix District and 7,407 fish and lobster traps in the St. Thomas /St. John District. Traps were still an important fishing gear in the St. Thomas/St. John District, but had largely been replaced by other types of fishing equipment on St. Croix where fishers had experienced severe trap loss from hurricanes. Traps are more vulnerable to storm damage on St. Croix because of the narrow, shallower shelf.

Boat size changed little between 1930 and 2003 for the majority of fishers. Most commercial boats in 2003 were between 5 and 8 m long (Kojis 2004) compared to 4.6 to 6 m in 1930 (Fiedler and Jarvis 1932). However, boat ownership increased from 50% of commercial fishers in 1930 (Fiedler and Jarvis 1932) to 99% in 2003 (Kojis 2004). In 1930 very few boats had engines (Fiedler and Jarvis 1932) while by 1968 100% of boats were powered by an engine (Swingle et al. 1970). In 1967, there was a fleet of large vessels with inboard engines in the St. Croix District that ventured up to 100 miles (160 km) to catch and sell seafood (Swingle et al. 1970). This fleet declined as Caribbean countries claimed jurisdiction of their 200 nm (370 km) Exclusive Economic Zone. In 2003, only 11 boats (4.4% of the USVI fishing fleet) were >30 ft (9 m) in length (Kojis 2004). However, fishers often used new technology such as GPS, echo sounders, winches and electric or

Fig. 8.27. Fishing with traps off St. John, 1909 (NPS files)

hydraulic reels to increase their fishing efficiency (Kojis 2004).

8.12.1 Changes in Fish Assemblages

Fish assemblages have been characterized and monitored intensively in the USVI. All studies have shown low abundance of fishes that are targeted by the trap fishery in the USVI. There have been losses of shelf-edge spawning aggregations, declines in fish species sizes, and changes in fish assemblage structure (Olsen and LaPlace 1978; Appeldoorn et al. 1992; Beets 1997; Beets and Friedlander 1999). Strong evidence suggests that fishing pressure had already changed the fish assemblage decades before sustained monitoring began in the late 1980s. Reef fish assemblages have changed since the 1950s and 1960s as a result of deterioration or loss of reef, seagrass, and mangrove habitats and intense fishing pressure. Even before the loss of habitats from coastal development, coral diseases, hurricanes, and other stresses, some signs of overfishing were evident (J. Randall's field notes 1958–1961; Olsen et al. 1975). Jack Randall's observations from the late 1950s and early 1960s indicate that the fishes targeted by the fishery were already in decline. He noted: "The trapping of reef fishes in pots is the major commercial fishery of the Virgin Islands.

Most fishing takes place over the narrow fringing reef that surrounds much of the islands. The limited fringing reef area receives nearly all of the fishing effort, and as a consequence the effect of overfishing is evident." (Randall 1963). In reference to Lameshur Bay, St. John, he wrote: "Impressed by the lack of food fishes such as groupers and snappers. *Cephalopholis fulvus* are occasional, but I saw only two small Nassau groupers, one tiger rockfish, and no other groupers, a couple of gray snappers and schoolmasters. It would seem that there has been considerable fishing effort".

However, many commercially important fishes, including Nassau groupers, were undoubtedly more abundant in the 1960s than at present. For example, Randall speared over 100 Nassau groupers around St. John over 2.5 years (1958–1961), and this species was the most abundant grouper in his samples. In addition, he tagged 124 adult Nassau groupers in Lameshur Bay during a study between February 1959 and June 1961 (Randall 1962). A major Nasssau grouper spawning aggregation site was fished out in the 1970s (Olsen and LaPlace 1978). In 1994–1999 surveys of groupers in 32 sample plots (each 5,000 m²) on four reefs around St. John, only 37 Nassau groupers were observed (Beets and Friedlander, unpublished data 2007).

Randall also mentioned midnight parrotfish as "moderately common" and spadefish as "ubiquitous". Both of these species are very rare around St. John now. No midnight parrotfishes and only a few spadefish have been observed in annual visual point count samples taken from 1988 to 2006 (Beets and Friedlander, unpubublished data). It is also unusual to see rainbow parrotfishes and hogfish. All of these fish are readily caught in fish traps and are attractive to spearfishers.

An experimental trapping study at Yawzi Point Reef over 6 months in 1993–1994 clearly showed that even a small number of traps fished over a relatively short time period caused statistically significant declines in several trophic groups (Beets 1996). Results from this investigation were also compared to records from 6 to 8 traps of similar design set by a fisher in 1982–1983 on the same reef. A comparison of the data from the two time periods 11 years apart suggests alarming changes (Beets 1997). The species composition had changed with large increases in the proportion of herbivorous fishes and decreases in the proportion of groupers and snappers. The average size of fishes in all trophic groups captured was smaller in the 1993–1994 samples. Four species of groupers caught in traps hauled during 1982–1983 were not trapped in 1993–1994. These findings are in stark contrast to the results Randall obtained from poison stations within the same bay, which showed that groupers and other related species (Serranidae) were the second most abundant group of fishes (Randall 1963).

Two studies conducted in the 1990s on St. John offer further evidence of the present scarcity of preferred predatory fish species and the increase in relative abundance of herbivorous fishes. Garrison et al. (1998) recorded the number and sizes of individuals of each species observed in traps set by fishers in 1992, 1993, and 1994 inside and outside VINP waters. Only 6 out of a total of 1,340 fish observed in traps in their study were Nassau groupers. The most abundant family of fishes observed in traps was the Acanthuridae.

In 1994, Wolff (1996) used visual censuses (Bohnsack and Bannerot 1986) and experimental trapping in four habitats (patch reef, rocky reef, gorgonian hardbottom, and seagrass) to compare species composition and vulnerability of fishes to trapping. No Nassau groupers were seen in any of the 159 visual censuses in Wolff's study, and this species comprised less than 1% of the catch

when present in trap hauls, with most caught in gorgonian not stony coral habitat. Three herbivorous species, the redband parrotfish (*Sparisoma aurofrenatum*), blue tang (*Acanthurus coeruleus*), and ocean surgeon (*Acanthurus bahianus*) were the most abundant species observed in visual censuses and in traps, accounting for over 50% of the individuals recorded. The scarcity of Nassau groupers, large snappers, and queen triggerfish and the dominance of herbivorous species in these two studies are striking and indicative of overfishing.

Dominance of herbivorous fishes was also found in surveys conducted from 1998 to 2001 on St. John, St. Thomas and St. Croix where herbivore densities represented 70% of the fish on a typical reef in the Virgin Islands (Table 8.4). However, significantly higher densities of carnivores (primarily grunts and snappers) occurred on St. John reefs ($p < 0.03$) relative to St. Thomas and St. Croix (Table 8.4).

Recruitment of large predatory fishes, such as groupers, is presently very low on St. John reefs. Very few juvenile groupers (<10 cm) were observed in monthly samples of juvenile reef fishes on St. John from July 1997 to July 2000 (Miller et al. 2001). However a survey of shallow shoreline habitats in 2006 found Nassau groupers (n = 46) to be the most numerically abundant grouper species around St. John (search time = 1,388 min) followed by red hind (*Epinephelus gutattus, n* = 36) and rock hind (*E. adscensionis, n* = 25) (R. Nemeth, unpublished data). Of 11 sites on St. Thomas (search time = 958 min), the three most numerically dominant grouper species were red hind (n = 36), Nassau (n = 31) and graysby (*Cephalapholis cruentatus,* n = 11). On both islands Nassau were most common on rubble covered with macroalgae or rocky reef habitats adjacent to seagrass beds whereas red hind were found on *Porites porites* coral or coral rubble and patch reef habitats.

Reef fish assemblages within VINP, which is not a marine reserve, do not differ substantially from those outside the park (Rogers and Beets 2001). For example, Garrison et al. (1998) found no significant differences in the species or number of fishes observed in traps inside vs outside the park. Visual point count samples from reefs around St. John from 1989 to 1994 demonstrated no significant differences in the number of fishes, number of species or biomass of fishes per sample and for

TABLE 8.4. Density (#/100 m^2) and percent composition of herbivores (Acanthuridae and Scaridae) and carnivores (Serranidae, Lutjanidae, and Haemulidae) on St. John (STJ), St. Thomas (STT) and St. Croix (STX) between 1998 and 2000 (Modified from Nemeth et al. 2003a).

Island	Herbivore density (%)	Carnivore density (%)
STJ	38.9±23.02 (60.9%)	61.1±12.91 (39.1%)
STT	27.3±11.43 (81.7%)	32.8±3.97 (18.3%)
STX	9.8±6.87 (72.6%)	35.8±5.13 (27.4%)
Virgin	14.6±18.10 (69.6%)	47.5±12.20 (30.4%) Islands

mean size of fishes observed inside vs outside park boundaries.

Standardized fish trap samples conducted inside and outside park boundaries during 1993 also documented no significant difference in number of fishes caught per trap haul (n = 145 trap hauls, t-stat: 1.24, P = 0.22; Beets 1996).

All of these studies documented the failure of federal and territorial regulations to protect reef fishes or reverse the declines in abundance of preferred species such as the large groupers and snappers. Lack of enforcement played a role; over 50% of the traps set by fishers observed during 1993–1995 had no functioning biodegradable panels (required by territorial legislation) to allow fish to escape if traps were lost or abandoned (Garrison et al. 1998). Enforcement is difficult since many trap lines are set without buoys, or with buoys located across park boundaries. It is also confounded because park legislation allows traditional fishing with traps, and distinguishing between commercial and traditional fishing is problematic. However, it is unlikely that even full compliance with existing regulations would be adequate to reverse the alarming trends.

8.13 Reef Fish Monitoring Trends at Long-time Sites in St. John: 1989–2006

Trends in reef fish assemblage characteristics in VINP over the past 17 years have been dominated primarily by storm effects. Monitoring of four reference sites (Yawzi Point, Haulover Bay, Tektite Reef, and Newfound Bay, outside the park) began

following Hurricane Hugo in 1989, the largest storm to pass the Virgin Islands in decades, which had a large impact on reef substrate, encrusting organisms (especially corals), as well as reef fishes. Similar impacts were documented following the second largest storm that passed the Virgin Islands during the past 20+ years, Hurricane Marilyn (1995). Although these large storms damaged reef structure and decreased coral cover in shallow water, reef fish abundance and species richness recovered within 3–5 years following these impacts (Fig. 8.28).

During the past several years, the most profound changes in the reef fish assemblage have been shifts in trophic structure. From 2000 to 2005, the abundance of planktivorous fishes has increased, along with their proportion of total abundance that has surpassed the previously dominant guild of herbivorous fishes (Fig. 8.29). The plantivorous damselfishes (*Chromis* spp.) are the dominant species responsible for the increase. Numerous factors may contribute to this shift, but the changes in benthic cover with the large decrease in coral cover and subsequent increase in algal cover are probably large contributors. Additionally, reduction in habitat complexity associated with these biotic changes has likely affected the distribution and abundance of many reef fish taxa.

The massive coral bleaching /disease event in 2005 apparently had an effect on trophic structure, with a decline in planktivorous fishes and increases in herbivorous fishes on all four reference reefs. Abundance increases were noted for small benthic herbivores (benthic damselfishes) and large mobile herbivores (parrotfishes and surgeonfishes). These increases in herbivore abundance are likely correlated with the increase in macroalgae cover as a result of the bleaching and disease mortality.

Predatory fishes provide strong regulatory effects in reef systems (Hixon 1991; Bascompte et al. 2005) and have experienced large changes in abundance over decades throughout the Caribbean (Jackson et al. 2001; Pandolfi et al. 2005). Large fishes, particularly the intensively harvested grouper and snappers, declined in the USVI prior to the establishment of NPS monitoring programs (Beets and Rogers 2002; Beets and Friedlander unpubl. data 2007). During the 17-year monitoring period, the frequency of occurrence of large groupers in

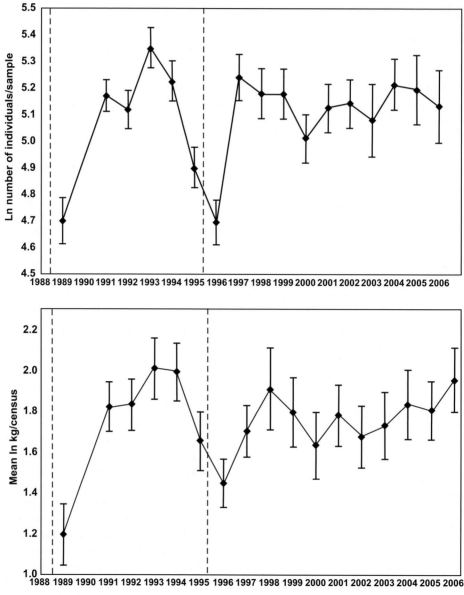

FIG. 8.28. Trends in average reef fish abundance and biomass (+SD) over 17 years of monitoring on four reference reefs, St. John, US Virgin Islands, 1989–2006 (Beets and Friedlander, personal communication)

samples has declined and remained very low since 2000 (Fig. 8.30). A mid-sized grouper, red hind (*Epinephelus guttatus*), has shown an increase during recent years, likely in response to the spawning aggregation closure enacted in 1990. Small groupers have increased during recent years, probably due to ecological release in response to sustained low numbers of larger groupers.

The reef fish assemblage in the USVI has suffered the loss of large predators and declines in abundance across all trophic levels resulting from decades of overfishing, prior to the 2005/06 bleaching and disease mortality. This release from top-down control has likely increased the importance of bottom-up processes such as disturbance events and habitat loss.

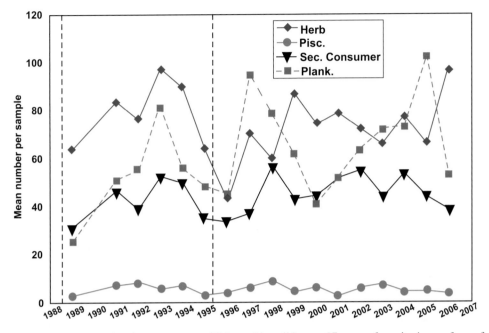

FIG. 8.29. Trends in average abundance among reef fish trophic guilds over 17 years of monitoring on four reference reefs, St. John, US Virgin Islands, 1989–2006. Herb = herbivores, Pisc = piscivores, Sec. Consumer = secondary consumers, and Plank = planktivores

8.14 Monitoring Trends for Commercially Important Reef Fish Species: St. Croix and St. Thomas

Nemeth et al. (2004) found at St. Croix that commercially important species (e.g., groupers, snapper, angelfishes, triggerfishes) are rarely seen but are more frequently observed at mid-shelf and shelf-edge sites where average fish size also tends to be larger. Species richness and diversity of fishes did not appear to be correlated with the amount of living coral or algal cover. Wrasses, damselfishes, parrotfishes and surgeonfishes were the most numerically abundant fishes on both St. Croix and St. Thomas with all other families representing less than 2% each. A comparison of relative abundance of eight commercially and ecologically important fish families in St. Croix showed little change (mean = 0.01%, range = −2.75 to 3.05%) between 2001 and 2004 surveys. Average relative abundance for these families between St. Croix and St. Thomas in 2004 were: Scaridae (40% vs 43%), Acanthuridae (25% vs 15%), Haemulidae (10% vs 8%), Serranidae (10% vs 4%), Chaetodontidae

(7% vs 14%), Balistidae (5% vs 1%), Pomacanthidae (2% vs 2%), and Lutjanidae (2% vs 13%).

Between 2001 and 2004, most commercially important species on St. Croix increased an average of 2 cm in length with the exception of parrotfishes and grunts that were smaller by 3.0 and 2.5 cm, respectively. In 2004, commercial species on St. Thomas were, on average, 5 cm larger (range = 0–9 cm) than on St. Croix. This trend in fish size was also found by Nemeth et al. (2006a) for spawning populations of red hind on St. Thomas and St. Croix.

Known grouper spawning sites in the USVI include Red Hind Bank, Grammanik Bank and Lang Bank, and snapper spawning sites include Seahorse Cottage Shoal, Red Hind Bank, Mutton Snapper and Grammanik Bank. The Red Hind Bank, also known as the MCD, prohibits all fishing year round. The Grammanik Bank, which has been seasonally protected since 2005, is a unique multi-species spawning aggregation site that supports at least four species of groupers and three species of snappers. Heavy fishing on the Grammanik Bank spawning aggregation removed about 10,000 pounds of yellowfin grouper in March 2000 and 2001 (USVI DFW, unpublished data 2007) and

346   C.S. Rogers et al.

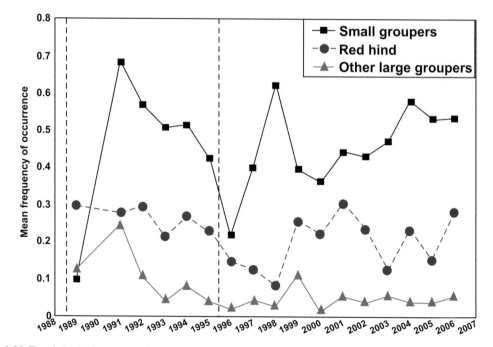

FIG. 8.30. Trends in the frequency of occurrence of groupers in samples over 17 years of monitoring on four reference reefs, St. John, US Virgin Islands, 1989–2006

is suspected to have caused the aggregation not to form in 2002 (R. Nemeth, personal observation 2006). Continued fishing on the Grammanik Bank is thought to have caused decreases in yellowfin and Nassau groupers between 2003 and 2004 as well. However, since the seasonal closure was implemented in 2005, spawning population estimates of yellowfin grouper have increased from ca. 600 to over 1,000 fish, and Nassau grouper have increased from ca. 100 to nearly 200 fish in 2006, the first potential recovering spawning aggregation in the Caribbean. Spawning aggregations of tiger grouper, cubera snapper and dog snapper contain up to 100, 800 and 1,000 fish, respectively. All aggregating species use similar sections of the reef and frequently overlap in time (Nemeth et al. 2006b).

8.15   Reef Fish Monitoring at Randomly Selected Sites Among Different Benthic Habitats: St. John and Buck Island

Reef fish data collected by the NOAA Biogeography Program between 2001 and 2005 around St. John and Buck Island showed that community structure

and fish assemblages varied considerably among different benthic habitats (Menza et al. 2006). The fish community was defined as the compilation of all observed fish species, community structure as indices of diversity or density for the community, and fish assemblages as components of the fish community categorized by trophic group or taxonomic family. Fish-habitat relationships were identified by grouping spatially explicit fish data according to the benthic habitat type in which the data were collected and examining the mean and variance of samples. Benthic habitat types were differentiated using regional benthic habitat maps (Kendall et al. 2001).

Nonparametric analysis of variance indicated that habitat types significantly explained some of the variance in species richness, species diversity (Shannon-Weaver), grouper density, snapper density, herbivore density and piscivore density (Table 8.5). At Buck Island the highest species richness and assemblage densities were typically found in linear reef, aggregated patch reef, and individual patch reef habitats and were lowest in sand, seagrass, and scattered coral/rock in sand habitats (Fig. 8.31). At St. John, the highest species richness and assemblage densities were not found to be as consistently associated with habitat types as around Buck Island. Species richness, community density and grouper density were highest at mid-shelf reef sites, but densities of snappers and piscivores (all species combined) were conspicuously low (Fig. 8.32). The distinction among relative densities of piscivores and groupers at Mid-Shelf Reef sites suggests that groupers were not a large component of the piscivore assemblage. Aggregated patch reefs, individual patch reefs, and colonized bedrock habitats also possessed high densities for some of the tested assemblages (e.g., snapper, herbivores, piscivores), but this pattern was not consistent across assemblages (Fig. 8.32). As in the Buck Island study area, sand and seagrass habitats were associated with low assemblage densities and species richness.

Most reef fish community measures, except density, showed little annual change between 2001 and 2005 (Fig. 8.33) (Menza et al. 2006). Multiple comparisons using 95% confidence intervals (with sequential Bonferroni correction) indicated that significant changes occurred in community density at BIRNM (2002>2003; 2002>2004; 2002 >2005) and species richness in VINP (2003> 2005). In 2002, the community density estimate at BIRNM and VINP had an abnormally large confidence interval. The large interval in both parks is an indication that the increase may have been a regional phenomenon (Menza et al. 2006).

Metrics for trophic or taxonomic components of the fish community were more variable than for the whole community, yet changes in grouper density (2002>2005), snapper density (2002> 2004), piscivore density (2002>2005) in BIRNM and grouper density and frequency of occurrence (2005>2004) in VINP were found (C.I.=0.95, Menza et al. 2006). Density estimates for grouper, snapper, and piscivore assemblages were all larger in 2002 than in other years, partly explaining high community density in 2002. Grouper, snapper, and piscivore density decreased monotonically from 2002 to 2005 in BIRNM and snapper density in VINP decreased from 2001 to 2005.

Temporal changes were also observed in total number of red hind and Nassau groupers (*Epinephelus guttatus* and *E. striatus*) between 2001 and 2006 (Table 8.6; NOAA Biogeography Program, unpublished data 2007). The observed increase was greater in St. John, where the total number of red hinds (<35 cm) increased steadily from 21 individuals in 2001 to 90 in 2006. There was greater variability in the number of red hinds at Buck Island, with total observed ranging between 42 and 52 individuals during the same period. Very few larger red hind and Nassau groupers (>35 cm) were observed during the 5-year study, but more of them were seen at St. John than at Buck Island (Table 8.6).

TABLE 8.5. The results from a nonparametric analysis of variance (Kruskal-Wallis test) for species richness, community density, and assemblage densities among 12 habitat types in the (A) Buck Island and (B) St. John study areas.

(A) Buck Island.

Community or Assemblage (Metric)	Kruskal-Wallis H	P [H] $< \chi^2_{0.05,10}$
Species richness	494.89	<0.0001
All species (density)	394.21	<0.0001
Groupers (density)	393.24	<0.0001
Snappers (density)	81.12	<0.0001
Herbivores (density)	452.96	<0.0001
Piscivores (density)	24.36	0.0113

(B) St. John

Community or assemblage (Metric)	Kruskal-Wallis H	P [H] $< \chi^2_{0.05,10}$
Species richness	318.14	<0.0001
All species (density)	256.01	<0.0001
Groupers (density)	168.73	<0.0001
Snappers (density)	31.65	0.0016
Herbivores (density)	299.47	<0.0001
Piscivores (density)	22.37	0.0335

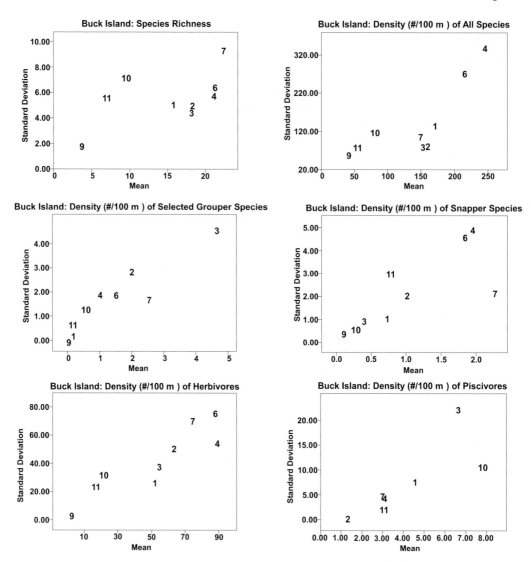

FIG. 8.31. Plots of density (fish per unit area) or species richness against variance among distinct habitat types in the Buck Island study area. Benthic habitat types are defined as: 1– colonized bedrock, 2 – colonized pavement, 3 – colonized pavement with sand channels, 4 – linear reef, 5 – macroalgae, 6 – aggregated patch reefs, 7 – individual patch reefs, 9 – sand, 10 – scattered coral/rock in unconsolidated sediment, and 11 – seagrass

8.16   Spawning Aggregations

Research on reproduction in reef fishes has a long history in the Virgin Islands starting with Randall's observations of parrotfish spawning aggregations off St. John in the late 1960s (Randall and Randall 1963; see also Colin 1996). Early work on Nassau grouper spawning aggregations off St. Thomas in the 1970s documented their vulnerability to overfishing and extirpation (Olsen and LaPlace 1978).

Extensive studies of bluehead wrasse (*Thalassoma bifasciatum*) by Warner and others on St. Croix in the 1980s and 1990s laid the foundation for theoretical and empirical studies of reproductive strategies in reef fishes (Warner 1988, 1990; Warner and Swearer 1991). Most recently a renewed interest in the importance of spawning aggregations (Fig. 8.34) to sustaining local fisheries has resulted in the use of new techniques (Whiteman et al. 2005) and provided new information on the reproductive

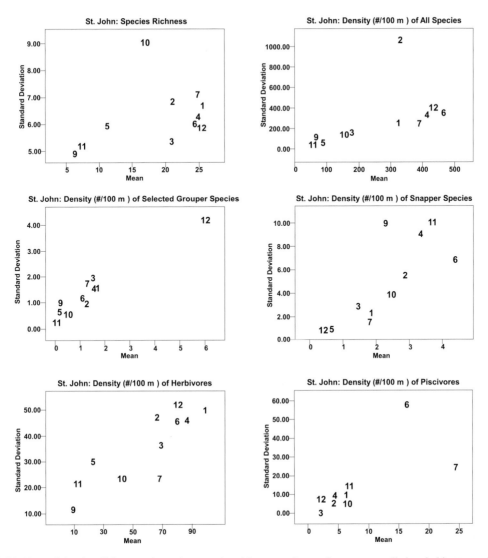

FIG. 8.32. Plots of density (fish per unit area) or species richness against variance among distinct habitat types in the St. John study area. Benthic habitat types are defined as: 1 – colonized bedrock, 2 – colonized pavement, 3 – colonized pavement with sand channels, 4 – linear reef, 5 – macroalgae, 6 – aggregated patch reefs, 7 – individual patch reefs, 9 – sand, 10 – scattered coral/rock in unconsolidated sediment, 11 – seagrass, and 12 – MSR (Mid-Shelf Reef)

characteristics and movement patterns of commercially important grouper and snapper species (Beets and Friedlander 1999; Nemeth 2005; Kadison et al. 2006; Nemeth et al. 2006a, b, 2007).

Since 1999, a long term study of a previously fished red hind spawning aggregation in St. Thomas documented that protection during the spawning season can result in population recovery. Nemeth (2005) found that the average size of red hind increased 10 cm during 12 years of seasonal closure. From 2000 to 2003 average density and biomass of spawning red hind increased over 60% and maximum spawning density more than doubled following permanent closure. Nemeth (2005) estimated that total population size increased dramatically from ca. 11,000 red hind in 1997, to 26,000 in 2000, 38,000 in 2001 to over 84,000 red hind in 2003. Strong recruitment into the spawning population and protection from fishing mortality of resident fish within the Marine Conservation

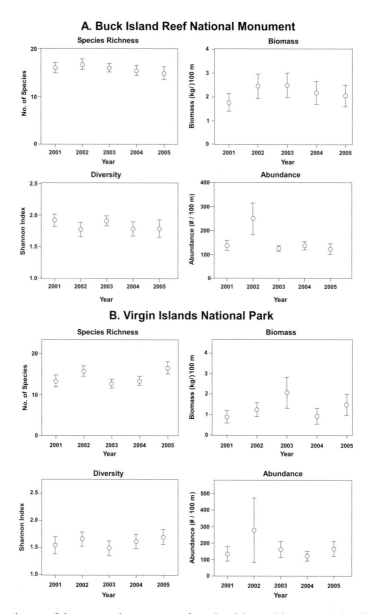

FIG. 8.33. Annual estimates of the community measures of species richness, biomass, species diversity, and density within (**A**) BIRNM and (**B**) VINP during 2001–2005. Error bars are 95% confidence intervals

District most likely contributed to these dramatic increases. Data from St. Thomas port landings show the average length of red hind from the commercial catch has also steadily increased since the season closure was established (Fig. 8.35). Moreover, interviews with commercial and recreational fishermen around St. Thomas during the past several years highlighted a general perception that the red hind being caught now are larger and more abundant than before the MCD was established (Pickert et al. 2006).

However, a seasonally protected area may not have the same effect on every spawning population. For example, a comparative study of two spawning aggregations in the USVI found that 10 years of seasonal protection resulted in significant increases in length and biomass of *E. guttatus* on St. Thomas but little change on St. Croix (Nemeth et al. 2006a).

TABLE 8.6. Number of observed *Epinephelus guttatus* and *E. striatus* individuals for two different size groups observed at Buck Island, St. Croix, and St. John, USVI, between 2001 and 2006 (Data: NOAA Biogeo, http://www8. nos.noaa.gov/biogeo_public/query_fish.aspx).

Region	Size class (cm)	Species	2001	2002	2003	2004	2005	2006
Buck Island, St. Croix	1–35 cm	*Epinephelus guttatus*	42	56	60	34	87	52
		E. striatus	0	0	0	0	1	2
St. John		*E. guttatus*	21	57	60	78	68	90
		E. striatus	0	0	0	2	2	2
Buck Island, St. Croix	36–45 cm	*E. guttatus*	0	0	0	1	3	1
		E. striatus	0	0	0	0	1	0
St. John		*E. guttatus*	0	1	1	6	0	4
		E. striatus	0	0	0	0	1	0

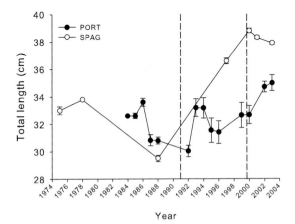

FIG. 8.34. Red hind length data (±SD) from port sampling (PORT) and spawning aggregation catches (SPAG) from 1975 to 2003. Seasonal closure of aggregation site was in 1990 and permanent closure was in 1999 (From Nemeth 2005)

Although a variety of factors may have influenced these differences, it was found that the Lang Bank, St. Croix, spawning site was only 600 m from the closure boundary while on St. Thomas the closure boundary was over 3 km distant (Nemeth et al. 2006a, 2007). The close proximity of the spawning aggregation to the closure boundary may not have been sufficient to protect the aggregation from poaching or fishing activity on the boundary edge during daily or monthly movements of spawning *E. guttatus*. Tagging studies have shown red hind can migrate 2–30 km to spawning aggregation sites and remain on the spawning site from 1 to 8 weeks. In addition, their spawning is highly synchronized with the lunar cycle (Nemeth 2005; Nemeth et al. 2007). All these factors make them very vulnerable to fishing mortality.

In addition to the two red hind spawning aggregation sites, larger groupers (Nassau, yellowfin, and tiger), snappers (cubera, dog, mutton and schoolmaster) and possibly jacks (black, permit, horse-eye) form spawning aggregations on the Grammanik Bank and within the MCD south of St. Thomas (Nemeth et al. 2006b; Kadison et al. 2006). Mutton snapper also spawn off southwest St. Croix. While these sites typically occur close to the shelf edge other sites on the mid-shelf (i.e., Seahorse Cottage Shoal) host spawning aggregations of lane and gray snapper. Year-round protection exists for the Red Hind Bank while seasonal closures include Lang Bank (December 1–February 28), Grammanik Bank (February 1–April 30) and Mutton Snapper (March 1–June 30).

FIG. 8.35. Aggregation of spawning red hind (*Epinephelus guttatus*) located on Lang Bank, St. Croix, at 30 m depth on old spur-and-groove reef (Photo: R. Nemeth)

8.17   Acoustic Tracking of Reef Fishes

A current project to examine movement of fishes in space and time among management units around St. John uses an array of *in situ* acoustic receivers to track fishes implanted with "pinging" tags. Results to date show consistent diel movement of grunts off-reef to adjacent seagrass beds just after sunset, returning to the reef just before dawn. Other species such as groupers have shown strong site fidelity and little within reef movement. Such studies provide a better understanding of the linkages between ecosystem components and potential benefits of the new monument (VICRNM) to adjacent areas from enhanced reproductive output and adult spillover into VINP and adjacent harvested areas (Friedlander and Monaco, personal communication 2007).

8.18   Evaluating the Effectiveness of Marine Reserves

Recognition of the changes in the fish assemblages in the USVI was a primary basis for the establishment of the National Park Service national monuments. Evaluation of the effects of these monuments is a primary consideration in recent research. For example, Monaco et al. (2007) sampled fishes and habitats along the mid-shelf reef (17–35 m) within the southern portion of the VICRNM using belt transects in July 2002, 2003 and 2004. They compared benthic habitat and fish assemblage characteristics (species richness, numerical density, biomass density) inside and outside the monument. Rugosity and live coral cover were greater outside the VICRNM than inside. Fish biomass, species richness and fish density were all significantly greater outside the national monument. Of the few economically important groupers that were observed, more were seen outside the monument.

8.19   Conchs and Lobsters

Queen conchs (*Strombus gigas*) support valuable fisheries in the Caribbean (Brownell and Stevely 1981). Historically, queen conchs were extremely abundant around St. John. Randall (1964) and colleagues collected and observed hundreds of queen conchs during investigations of their biology. Schroeder (1965, p. 8) mentioned "conchs by the thousands in Salt Pond Bay", St. John during his work with John Randall. Conch are not abundant today. Concerns over overharvesting of this species led to a moratorium in St. Thomas and St. John from 1988 to 1992. In spite of this moratorium, additional regulations in 1994, and a limit of two conchs per person per day for VINP waters, as of 1996, conch populations in general appeared to be decreasing and density of conchs inside park waters was not significantly higher than outside the park (Friedlander 1997). Conchs were surveyed along transects around St. John and St. Thomas in 1981, 1985, 1990, 1996 and 2001 (Wood and Olsen 1983; Boulon 1987; Friedlander et al. 1994; Friedlander 1997; Quinn and Hanrahan 1996; Gordon 2002). Surveys in 1996 showed that conchs were usually found in seagrass beds. This habitat has been reduced greatly as a result of hurricane and anchor damage.

Transects surveyed in 1996 were re-evaluated in 2001 to compare densities of juvenile and adult conchs around St. Thomas, St. Croix, and St. John at several different depths and in different habitats. A total of only 244 conchs were found in 22 transects (Gordon 2002). Conch abundance and density were examined in 1998/99 in six shallow backreef bays around St. Croix. Most of the conchs were juveniles and were found in seagrass habitats, suggesting these bays are important nursery areas (Tobias 2005).

Because conchs have patchy distributions and move among several habitats and over a gradient of depths, it is difficult to document and interpret changes in their abundance.

Wolff (1998) noted that lobster densities at Fish Bay and Reef Bay in 1996 were similar to those in 1985 (Boulon 1987), and densities in Lameshur Bay and Tektite Reef in 1996 were similar to estimates made in 1970 (Cooper et al. 1975). However, overall data suggest that there has been a large decline in the average size of the lobsters within the park since 1970 (Olsen et al. 1975). Virgin Islands National Park regulations allow only two lobsters to be taken per person per day; however, it is widely known that harvest exceeds this amount.

Tobias (2000) noted that lobsters accounted for 6% of total reported fishery landings for the USVI in 1998–1999. He reported a 10% decrease in mean size between 1997 and 2000, which suggests that overfishing is occurring (Tobias 2000; Mateo and Tobias 2002; see also Bohnsack et al. 1991). The USVI Division of Fish and Wildlife routinely monitors commercial lobster landings (weight and carapace length) and has periodically monitored lobster recruitment around St. Thomas where recruitment appears to be highly variable but generally low (Gordon and Vasques, personal communication).

Florida Fish and Wildlife Conservation Commission (Florida Marine Research Institute) scientists conducted spiny lobster surveys annually from 2004 to 2006 at BIRNM (unpublished data 2006). Timed lobster surveys were conducted on scuba inside and outside the park boundaries in hardbottom habitat (depth range 3–40 m). On average 150–190 lobsters were found on 35–45 surveys. Sizes ranged from an individual with 10 mm carapace length (CL) found at the base of a linear reef to one with 130 mm CL found on a deep reef west of the park boundary. Most spiny lobsters were found in backreef habitats. The mean size of adult *Panulirus argus* found inside the park was 99 mm CL (N = 64). Juvenile habitat was not found in the park but located inside Tague Bay among near shore patch reefs covered with rubble and algae. Although spiny lobster habitat at BIRNM has been severely degraded by coral disease and hurricanes, the Monument remains an important refuge for this commercially important species. The designation of the East End Marine Park also provides critical habitat protection for juvenile spiny lobsters not found in the Monument (Florida Marine Research Institute, unpubl. data 2006).

8.20 Ecological Relationships and Processes

Coral reefs, mangroves, and seagrass beds in the Caribbean are linked through the movement of larval and adult organisms and the transport of nutrients (Ogden and Gladfelter 1983; Meyer et al. 1983; Ogden 1997). Numerous research projects in the USVI have explored relationships among habitats, among habitats and organisms, and ecological processes such as recruitment, herbivory, and calcification. The West Indies Laboratory on St. Croix was a key study site for the Seagrass Ecosystem Study (SES), one of the programs of the International Decade of Ocean Exploration (IDOE) funded by the US National Science Foundation from 1974 to 1979. The SES put seagrasses on the global map, emphasized the interconnection with coral reefs, and produced a few hundred publications and several books.

At the start of the SES in the early 1970s the prevailing view was that detritus was the primary pathway for seagrass production into higher trophic levels. By the end of the program, however, a large number of grazers had been identified and studied but the percentage of seagrass production going to direct grazers was still estimated to be relatively small. Research interest in seagrass herbivores and their importance has grown steadily (Larkum et al. 2006). The global estimated percentage of production grazed is still <30%. Recent historical ecology studies have concluded that prior to extensive harvesting of Caribbean marine resources by humans the seagrass fauna was dominated by large herbivores, especially green turtles and fishes (Jackson et al. 2001).

Seagrass beds in the USVI are not as extensive as they once were because of severe anchor damage and hurricane effects (Rogers and Beets 2001). Data on seagrass densities and changes over time appear in Williams (1988) and Muehlstein and Beets (1999). The installation of mooring buoys in VINP has resulted in some increase in density of seagrasses (NPS, unpublished data).

Scientists with the West Indies Laboratory produced some key papers on urchin and fish herbivory and the relationship of different levels of herbivory to productivity (Ogden et al 1973; Carpenter 1990a, b). Carpenter (1990a) showed a dramatic increase in macroalgae after the sea urchin (*Diadema antillarum*) die-off in 1983/84 (Lessios et al. 1984) and a decrease in rates of primary productivity. He also showed an increase in herbivorous fish population densities and an increase in their grazing intensity after the die-off (Carpenter 1990b). Levitan (1988) showed a 30-fold increase in algal biomass in St. John after the mass mortality of the sea urchins. In an early study, Rogers et al. (1984) showed a decrease in abundance with depth from the surface to 37 m on both walls of Salt River Submarine

Canyon, St. Croix (density 2.6 ind/m^2 at 9 m and 0 at 37 m).

Although there are signs of a population recovery of *Diadema antillarum* in St. Croix (Fig. 8.36) and elsewhere throughout the Caribbean (Miller et al. 2003; Carpenter and Edmunds 2006), population densities generally remain low on shallow reefs, and even lower at greater depths. Between 1998 and 2000, Nemeth et al. (2003b) reported a range of 0.3–11.5 *Diadema*/100 m^2 for shallow reefs around St. John, St. Croix, and St. Thomas, while Kuffner (unpublished data 2007) noted an average of 0.86/m^2 from several shallow sites surveyed around St. John in 2004. In a more recent survey (July–September 2006), densities of 1–3/m^2 were found from sites about 2 m deep around St. Thomas. Although these urchins are becoming more abundant around St. Thomas, their densities are still lower than those reported by Hay and Taylor (1985) for Brewers Bay, St. Thomas before the die-off (Walters et al., unpublished data 2006).

UVI also monitored populations of the important herbivorous urchin *Diadema antillarum* at coral reef monitoring sites and found an average of 2.3 urchins/100 m^2, with a maximum at Great Pond, St. Croix, of 28.4 urchins/100 m^2. Most urchins were found in shallower sites and were patchy in distribution as 70% of urchins encountered at 34 sites were at shallow (<12 m) nearshore reefs, and 60% of all urchins were concentrated at only four of 34 sites.

8.20.1 Calcification and Herbivory

By quantifying rates of photosynthesis, respiration, calcification, and dissolution of reef communities under two different levels of *Diadema antillarum* herbivory (no urchins and high density) Kuffner et al. (personal communication) explored effects of grazing on reef metabolic processes. They used the Submersible Habitat for Analyzing Reef Quality (SHARQ). Biogeochemical measurements were made before and after urchins were transplanted into reef areas (plots) enclosed within the SHARQ (two pairs of plots, one control and one with urchins added). The reef plots where urchins were introduced (at densities similar to those before the *Diadema* die-off-4.3 urchins/m^2) showed statistically significant reductions in the standing crop of algae and reef productivity in one set of plots,

FIG. 8.36. The sea urchin, *Diadema antillarum,* is becoming more abundant in shallow water in the USVI (Photo: P. Mayor)

and reduction in respiration rates in the other set. Though replication was limited due to the large scale of the experiment, this work supports the hypothesis that algal-dominated reefs may experience higher area-specific rates of productivity and/or respiration compared to reefs that are regularly grazed by herbivores.

8.20.2 Connectivity

The degree of connectivity among the marine reserves and other MPAs in the USVI has significant management implications not only for the USVI but for the entire Caribbean region. Recent papers by Cowen et al. (2000, 2006) present a model of current patterns and potential dispersal of reef fish larvae which demonstrate that St. Croix could be more isolated than many other land masses, indicating that local management could be more significant for this island than "upstream" areas. A finer resolution model for the USVI, with more shallow water bathymetry and current data, is needed.

Significantly, although most marine organisms have planktonic larvae that can be dispersed over large distances, evidence to date suggests that most larvae of reef-associated animals come from within tens of kilometers rather than hundreds of kilometers away. This finding suggests the need for more closely linked reserves throughout the region. Cowen's et al. (2000, 2006) models focused on fish larvae. A similar model of the transport of elkhorn coral larvae throughout the Caribbean has been presented by Baums et al. (2005b). It suggests some separation of elkhorn populations in eastern and western regions of the Caribbean, with mixing near Puerto Rico.

If the reserves are largely dependent on local habitats and fish assemblages for larvae for replenishment, local management becomes even more critical. For example, bluehead wrasse are locally retained around St. Croix (Warner et al. 2000; Hamilton et al. 2006). Also further research on movement of adult fishes and patterns of habitats use is needed for a better understanding of connectivity among reefs, reserves, and adjacent exploited areas. NOAA has just begun a tagging and telemetry study of several species of fishes within VINP and VICRNM (see above).

The connections between seagrass ecosystems and reefs have been examined by many scientists working in the USVI, including Randall (1963, 1967), Meyer et al. (1983), Robblee and Zieman (1984), Beets et al. (2003), Kendall et al. (2003), and Grober-Dunsmore et al. (2006). In general, proximity of seagrass beds appears to increase diversity and abundance of reef fishes.

Mangroves in the USVI are little-researched. Mangroves filter sediments from runoff, provide nursery habitat, and have trophic/nutrient links to seagrass beds and coral reefs (Ogden 1988). Mangrove forests are not extensive in the islands, and many have been removed by filling and dredging, or killed by droughts and hurricanes. Most are primarily narrow fringes around sheltered bays and salt ponds, although large mangrove areas are found near Salt River, St. Croix, and Benner Bay/Cas Cay, St. Thomas. They are important nurseries for grunts, snappers, and other reef fishes (Boulon 1992; Adams and Tobias 1994).

8.20.3   Fish/reef Interactions

In response to the devastating coral losses in the USVI, the National Park Service and USGS have begun censusing fish along randomly selected transects at long-term study sites in St. John. Other, ongoing studies are exploring the correlation between fish assemblage characteristics and the amount of living coral and reef topographical complexity ("rugosity"). NOAA's fish monitoring program suggested that fish community structure on reefs in St. John were influenced by the amount of live coral cover and structural complexity. Total fish abundance and fish species richness increased, whereas fish diversity decreased significantly with an increase in live coral cover and reef rugosity (Figs. 8.37 and 8.38). The decrease in fish diversity as percent live coral cover increased may have resulted from an overwhelming abundance of a few species (e.g., *Chromis* spp.) at sites with high coral cover, which could have reduced the overall fish diversity of those sites compared with sites having lower coral cover. The percent live coral cover was significantly correlated with reef rugosity ($r^2 = 0.37$,

$p = 0.00$). Additionally, the overall average percent live coral cover measured at 144 reef sites in St. John was 7.5% and has not changed significantly since 2001, although data on coral cover since the 2005/06 bleaching/disease event have not yet been analyzed ($r^2 = 0.02$, $P = 0.09$). Thus, fish community structure and reef condition (coral cover) in St. John were spatially rather than temporally correlated, such that reefs with more live coral supported greater numbers of fish individuals and species than did less healthy reefs.

In another study, a positive relationship ($r^2 = 0.85$, $P < 0.005$) between coral cover and fish density was found (Fig. 8.39) at 16 sites around St. Thomas and St. John, but no relationship between coral cover and fish diversity (H') (R. Nemeth, unpublished data). Also AGRRA (Atlantic and Gulf Rapid Reef Assessment Program) surveys between 1998 and 2000 in the USVI and the British Virgin Islands found a weak relationship between fish species richness and percent live coral cover ($r^2 = 0.30$, $p < 0.01$) (Nemeth et al. 2003a).

8.20.4   Coral and Fish Recruitment

Research on coral recruitment has taken place on St. John, St. Croix, and St. Thomas. Rogers et al. (1984) examined different recruitment rates at depths from 30' to 120' (9–37 m) at Salt River Submarine Canyon, St. Croix. Rogers and Garrison (2001) showed relatively high recruitment in the scar created on a reef by a cruise ship within VINP but no increase in coral cover over at least 10 years. Many of the studies show similar densities in coral recruits, 15–25 juveniles/m² (Rogers and Garrison 2001; Edmunds 2000, 2004). Juvenile corals in Great Lameshur Bay (GLB), St. John, have been studied extensively by Edmunds (2000, 2004, 2006). Starting in 1994, the density of juvenile corals – defined as colonies <40 mm diameter – has been documented at six sites within and near GLB, and starting in 1996, individual juvenile colonies have been tagged to track their fates (e.g., growth, mortality). These annual surveys are in shallow water (5–9 m deep) and are biased towards the subset of species that are encountered in large numbers as small colonies: *Porites*, *Agaricia*, *Siderastrea* and *Favia*. Over 14 years, the density of juvenile corals generally has remained high

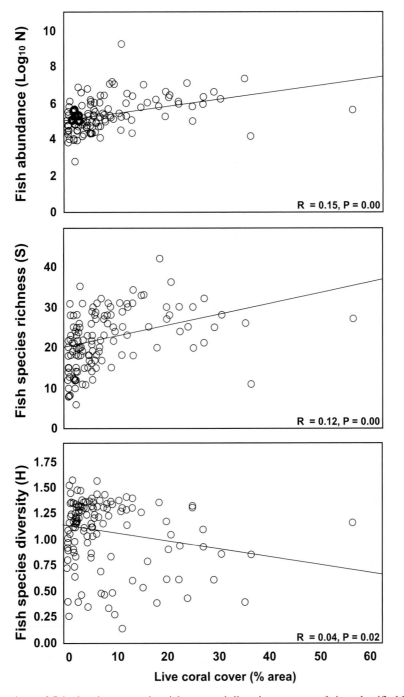

FIG. 8.37. Comparison of fish abundance, species richness, and diversity among reef sites classified by coral cover in St. John, USVI

(e.g., up to a mean of 22 juveniles/m^2) relative to other Caribbean locations (Edmunds 2000, 2004), although there has been a high degree of spatio-temporal variability (Edmunds 2000).

Seawater temperature can account for much of the variation in density of juvenile corals among years with, somewhat surprisingly, the density increasing in years characterized by warmer

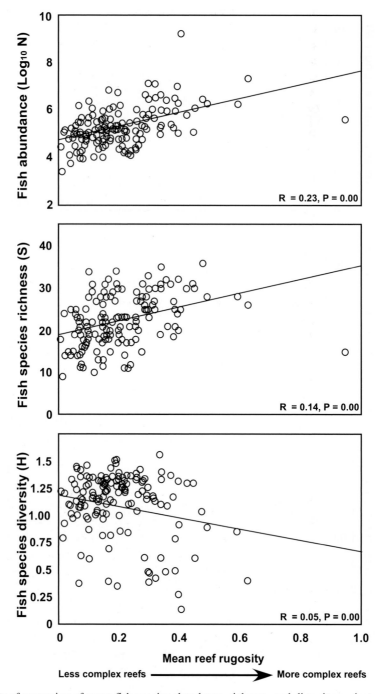

Fig. 8.38. Results of regression of mean fish species abundance, richness, and diversity against mean reef rugosity for reefs in St. John, USVI

average seawater temperatures (Edmunds 2004). It is unclear, however, to what extent this effect represents the consequence of high temperature on larval development, or post-settlement events that influence the growth and survivorship of juvenile colonies.

While elevated temperatures of extreme magnitude clearly result in coral death, sublethal

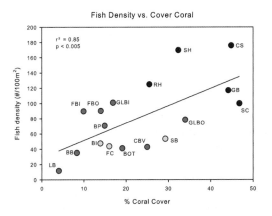

Fig. 8.39. The relationship between coral cover and fish density at 16 sites around St. Thomas and St. John. Blue = mid-shelf and shelf-edge reefs, yellow = mid-shelf cays, red = nearshore. BP = Black Point, LB = Long Bay, BB = Benner Bay, FBI = Fish Bay inner, FBO = Fish Bay outer, GLBI = Great Lameshur Bay inner, GLBO = Great Lameshur Bay outer, BOT = Botany Bay, CBV = Caret Bay Vluck, BI = Buck Island, SB = Sprat Bay, FC = Flat Cay, SC = South Capella, SH = Sea Horse, CS = Collage Shoal, RH = Red Hind Bank, GB = Grammanik Bank

increases in temperature would be expected to accelerate the development of coral larvae (Edmunds et al. 2001; O'Conner et al. 2007), and probably also would affect larval settlement, metamorphosis, and post-settlement success. Small increases in temperature could enhance coral recruitment by accelerating larval development, restricting larval dispersal (O'Conner et al. 2007), and promoting larval settlement and metamorphosis, or they could modify post-settlement success by reducing the growth of new recruits and juvenile colonies, and perhaps increasing their mortality rates (Edmunds 2004). Teasing apart the effects of temperature on pre- and post-settlement events will be difficult, but one approach with considerable promise is the use of settlement tiles to sample competent coral larvae in local habitats differing in thermal regimes. One potential mechanism underlying the reduced growth of juvenile corals at higher temperatures in St. John recently was identified by a more detailed analysis, which revealed that warm water favored isometric growth (i.e., growth that is independent of size), while cool water favored positive allometry for growth (i.e., growth rates were accelerated at greater size) (Edmunds 2006).

Kojis (1997) examined annual scleractinian and milleporan recruitment to terracotta tiles at three sites around St. Thomas : Saba Island, Fortuna Bay, and Hans Lollik Island. Twenty terracotta tiles were deployed on PVC arrays at each site and retrieved annually over a 2-year period (mid-1992–mid-1994). The deployment depth ranged from 9 to 12 m at Saba Island and Fortuna Bay and 1–5 m at Hans Lollik. Scleractinia comprised 62.3% of the recruits at the two southern sites and 91.8% of the recruits at the northern site, while *Millepora* comprised 37.9% of the recruits at the southern sites and 8.2% at Hans Lollik. The dominant Scleractinia recruiting to the tiles were the Poritidae, Agariciidae, and Faviidae. Most of the poritid recruits were *Porites astreoides* and most of the faviid recruits were *Favia fragum*. The agariciid recruits could not be identified to species. Only one *Acropora* spat recruited to the tiles, indicating a low recruitment rate for this genus. Recruits from the family Agariciidae were more common at the northern and shallower Hans Lollik site than at the two sites south of St. Thomas, comprising 44–48% of the recruits at Hans Lollik compared to 8.6–13.2% at the Fortuna and Saba sites. Poritids dominated recruitment at the two southern sites, comprising 65–82% of all scleractinian recruits. In contrast, poritids comprised only 38–44% of the recruits at Hans Lollik. Faviid recruitment varied from 8.1% to 22.6% over all sites and showed no difference with locale.

Nemeth et al. (2003b) reported coral recruit densities ranging from 4.4 to $10.3/0.25\,m^2$ quadrat for shallow reefs in St. John, St. Thomas, and St. Croix. The most abundant species were *Siderastrea* spp, *Agaricia* spp., and *Porites* spp.

Steneck (unpublished data 2005) examined coral recruitment to terracotta tiles in St. John, St. Croix, Belize, Mexico, Bonaire and the Bahamas. The average number of coral recruits per plate was lowest in St. Croix (<1), within Buck Island Reef National Monument, which at the time of the study was not a "no take" zone. Recruitment for the Newfound and Haulover sites in St. John (ca. 2.5) was lower than in Bonaire but higher than all other locations.

Recruits of *Acropora palmata* can be impossible to distinguish from remnants of colonies that have suffered partial mortality from bleaching, disease, and other causes. In general, few recruits of this species or the other major reef-building coral in the USVI, *Montastraea annularis* complex, are found in settling plate studies. Nemeth et al. (2004) found that density of *Acropora* recruits within sampling quadrats ranged from $0.1/m^2$ at Caret Bay and Flat Cay to $0.9/m^2$ at Coculus Point. Spawning of *A. palmata* was observed in 2004 and 2005 in St. John, but not in 2006.

Most of the coral recruits observed in the USVI studies have been from the Poritidae and Agariciidae, as has been shown for other Caribbean locations. *Agaricia* was particularly hard hit by the 2005 bleaching event (NPS, unpublished data), and recruits from these species could possibly decrease in abundance.

Several studies of fish recruitment, in addition to those referred to in this chapter, have been conducted in the USVI (Booth and Beretta 1994; Nemeth 1998; Tolimieri 1995, 1998a, b; Tolimieri et al. 1998; Risk 1997, 1998). Damselfishes, stoplight parrotfishes, and ocean surgeonfishes have been the focus of this research.

8.21   Conclusions

In the last four decades, coral reefs in the USVI have been affected by acute and chronic stressors including hurricanes, high seawater temperatures, coral diseases, elevated sedimentation rates, and fishing pressure. The most conspicuous changes to the physical structure of the reefs have been from: 1) white band disease (affecting *A. palmata* and *A. cervicornis*) and hurricanes in the late 1970s through the mid-1990s; and 2) the extreme bleaching/disease episode that began in September 2005. Listed as threatened under the Endangered Species Act in May 2006, *A. palmata* showed signs of increasing in abundance within the last 10 years but then bleached for the first time on record in 2005 in the USVI with total mortality of some colonies. White band disease is now rare, but white pox and other undescribed diseases are very common, and along with physical damage from storms and boat groundings are hindering the regrowth of this species. Unlike white band disease which affected only two (albeit key reef-building) coral species (*A. palmata* and *A. cervicornis*) with most pronounced effects in depths less than 10 m, the 2005 disease outbreak (most likely white plague) affected virtually all coral species to depths over 30 m.

At deeper long-term monitoring sites, most of the (non-acroporid) coral colonies surviving the bleaching and disease that began in September 2005 had begun to regain some of their normal coloration by February/March 2006. However, many coral colonies remained pale as late as December 2006. The severity of disease peaked on the deeper reefs (dominated by *M. annularis* complex) within 2 to 7 months after the maximum seawater temperatures and associated bleaching. Disease following the severe 2005 bleaching event caused drastic losses (an average of 50% within one year) in living coral, the most substantial declines on these reefs in the last 40 years. Prevalence of active disease varies, but disease now seems to be present on the reefs year-round. Bleaching events (and major storms) are expected to become more frequent in the future. More research is urgently needed on these diseases and their interaction with bleaching, both of which could undermine the benefits of the marine reserves and other protected areas in the USVI; and on the effects of the reef degradation on fishes and other organisms.

Reef fish assemblages in the USVI have changed because of habitat degradation (not only of coral reefs but also of seagrass beds, mangroves, and deeper algal plains) and fishing pressure. Intensive and extensive monitoring of reef fishes, which began after fishing pressure had already caused changes, shows low abundance and reduced size of fish targeted by the fisheries (including Nassau and other groupers), and the absence or scarcity of some fish species.

The coral reefs of the USVI have been the subject of research for 40 years. Intensive and extensive monitoring of reefs, fishes and other reef organisms has been augmented by studies of ecological processes. Future research should focus on the role of the USVI reserves and the recently established Research Natural Area in Dry Tortugas National Park in the overall region (Western Atlantic and Caribbean), as well as the possible links among the three major US Virgin Islands, between VINP and VICRNM, and between BIRNM and the East End Marine Park. Connectivity among mangroves, seagrass beds, and coral reefs inside and outside the marine protected areas is an important area needing further study, along with an evaluation of the effectiveness of the marine reserves. Marine reserves are more likely to support the recovery of reef fish assemblages than the benthic resources. Recovery to previous levels of coral cover and to former relative abundance and diversity of coral and fish species in the USVI seems very unlikely.

Acknowledgements We wish to express our appreciation to S. Wright, NPS, for producing the maps. I. Baums provided data on elkhorn genotypes. S. Rohmann assisted with interpretations of estimates of total reef area. Thanks also to W. Toller for his help.

APPENDIX 8.1. References pertaining to research at West Indies Laboratory, St. Croix.

Subject	References
Herbivore–plant interactions	Ogden et al. 1973; Carpenter 1979, 1984a, 1985, 1986, 1988; Robblee 1983; Steneck 1983, 1989; Adey and Steneck 1985
Coral morphology/physiology/ecology	Gladfelter W. 1982; Gladfelter E. 1982; 1983a, b, 1984, 1985; Bythell 1988; Sebens and Miles 1988; Porter and Targett 1988; Patterson et al. 1991; Gleason 1993
Chemical and mechanical defenses	Targett et al. 1986; Hay 1985; Hay et al. 1988; Harvell and Fenical 1989; Harvell et al. 1993, 1996; West et al. 1993
Seagrass ecology	Ogden and Zieman 1977; Thayer et al. 1984
Invertebrate ecology	Gladfelter W 1975, 1978; Scheibling 1979; Suchanek 1983, 1985; Nowlis 1993
Fish community structure	Shulman 1983; Gladfelter and Gladfelter 1978; Gladfelter et al. 1980, Ogden and Ebersole 1981; Clarke 1988, 1996
Recruitment and dispersal in reef fishes	Shulman et al. 1983, 1984; Shulman 1984, 1985; McFarland and Ogden 1985, McFarland et al. 1985; Shulman and Ogden 1987; Caselle and Warner 1996; Warner 1997; Swearer et al. 1999; Warner et al. 2000; Swearer et al. 2002; Shulman and Bermingham 1995; Shulman 1998
Resource partitioning in reef fishes	Moese 1980; Grippo 1981; Robblee 1983; Gladfelter and Johnson 1983; Clarke 1989, 1992, 1994
Behavioral ecology of fishes	Ogden and Erlich 1977; Quinn and Ogden 1984; Wolf et al. 1983; McFarland et al. 1979; Gladfelter 1979; Fallows 1985; Helfman et al. 1982; Helfman 1983, 1989; Henson and Warner 1997; Petersen and Warner 1998; Warner and Dill 2000
Microbiology	King et al. 1990; Fenchel et al. 1979
Nutrient dynamics and productivity	Adey et al. 1981; Rogers and Salesky 1981; Carpenter 1990a, b; Gladfelter 1977; Bythell 1988; Szmant-Froelich 1983; Meyer et al. 1983; Williams et al. 1985; Williams 1984; Williams and Fisher 1985

APPENDIX 8.2. Stony coral species occurring in the US Virgin Islands.

Stephanocoenia michelinii	Diploria clivosa
Madracis decactis	Diploria labyrinthiformis
Madracis mirabilis	Diploria strigosa
Acropora palmata	Manicina areolata
Acropora cervicornis	Colpophyllia natans
Acropora prolifera (a hybrid)	Colpophyllia breviserialis
Agaricia agaricites	Cladocora arbuscula
(several different forms)	
	Montastraea annularis
Agaricia tenuifolia	Montastraea franksi
Agaricia undata	Montastraea faveolata
Agaricia lamarcki	Montastraea cavernosa
Agaricia grahamae	Solenastrea bournoni
Agaricia fragilis	Oculina diffusa
Helioseris cucullata	Meandrina meandrites
(= Leptoseris cucullata)	
	Dichocoenia stokesi
Siderastrea siderea	Dichocoenia stellaris
Siderastrea radians	Dendrogyra cylindrus
Porites astreoides	Mussa angulosa
Porites branneri	Scolymia lacera
Porites porites	Scolymia cubensis
Porites divaricata	
Porites furcata	
Isophyllia sinuosa	
Isophyllastrea rigida	
Mycetophyllia lamarckiana	
Mycetophyllia ferox	
Mycetophyllia aliciae	
Eusmilia fastigiata	
Millepora alcicornis	
Millepora complanata	
Millepora squarrosa	
Tubastraea aurea	
Favia fragum	

References

Acropora Biological Review Team (2005) Atlantic Acropora Status Review Document. Report to National Marine Fisheries Service. Southeast Regional Office, 152 pp

Adams AJ, Tobias WJ (1994) Red mangrove prop-root habitat as a finfish nursery area: a case study of Salt River Bay, St. Croix, USVI Proc 46th Gulf and Caribbean Fisheries Institute, pp 22–45

Adey WH (1975) The algal ridges and coral reefs of St. Croix : their structure and Holocene development. Atoll Res Bull 187:1–67

Adey WH, Steneck RS (1985) Highly productive eastern Caribbean reefs: synergistic effects of biological, chemical, physical and geological factors. In: Reaka ML (ed) The Ecology of Coral Reefs, Symposia Series Undersea Research, NOAA Undersea Research Program. Rockville, MD 3:169–187

Adey WH, Rogers CS, Steneck RS, Salesky N (1981) The south shore St. Croix reef. Report to Department of Conservation and Cultural Affairs, Virgin Islands Government. West Indies Laboratory, Fairleigh Dickinson University, St. Croix, VI, 64 pp

Anderson M, Lund H, Gladfelter EH, Davis M (1986) Ecological community type maps and biological community descriptions for Buck Island Reef National Monument and proposed marine park sites in the British Virgin Islands. Biosphere Reserve research report no. 4 VIRMC/NPS, 236 pp

Anonymous (1961) Report of meeting of Caribbean Fishery officers in Puerto Rico

Appeldoorn RS, Beets J, Bohnsack JA, Bolden S, Matos D, Meyers S, Rosario A, Sadovy Y, Tobias T (1992) Shallow water reef fish stock assessment for the U.S. Caribbean. National Oceanic and Atmospheric Administration. Technical Memorandum NMFS-SEFSC-304, 70 pp

Armstrong RA, Singh H, Torres J, Nemeth RS, Can A, Roman C, Eustice R, Riggs L, Garcia-Moliner G (2006) Characterizing the deep insular shelf coral reef habitat of the Hind Bank marine conservation district (US Virgin Islands) using the Seabed autonomous underwater vehicle. Continental Shelf Res 26:194–205

Aronson RB, Precht WF (2001) White-band disease and the changing face of Caribbean coral reefs. Hydrobiologia 460:25–38

Bascompte J, Melian CJ, Sala E (2005) Interaction strength combinations and the overfishing of a marine food web. Proc Natl Acad Sci 1023:5443–5447

Battista T (2006) Data Acquisition and Processing report: Benthic Habitat and Hydrographic Survey of Buck Island, St Croix, U.S. Virgin Islands and La Parguera, Puerto Rico. NOAA's Center for Coastal Monitoring and Assessment Project no. NF-06–03, S-I911-NF-06. Silver Spring, MD, 104 pp

Baums IB, Hughes CR, Hellberg M (2005a) Mendelian microsatellite loci for the Caribbean hard coral *Acropora palmata*. Mar Ecol Prog Ser 288:115–127

Baums IB, Miller MW, Hellberg ME (2005b) Regionally isolated populations of an imperiled Caribbean coral, *Acropora palmata*. Mol Ecol 14:1377–1390

Bayer FM (1969) A review of research and exploration in the Caribbean Sea and adjacent waters. Symposium on investigations and resources of the Caribbean sea and adjacent regions. Contribution no. 1127 from the School of Marine and Atmospheric Sciences, 91 pp

Beets J (1996) The effects of fishing and fish traps on fish assemblages within Virgin Islands National Park and Buck Island Reef National Monument. US National Park Service. Technical Report, 44 pp

Beets JP (1997) Can coral reef fish assemblages be sustained as fishing intensity increases? Proc 8th Int Coral Reef Symp 2:2009–2014

Beets J, Friedlander A (1999) Evaluation of a conservation strategy: a spawning aggregation closure for red hind, *Epinephulus guttatus*, in the U.S. Virgin Islands. Environ Biol Fishes 55:91–98

Beets J, Lewand L (1986) Collection of common organisms within the Virgin Islands National Park/Biosphere Reserve. Virgin Islands Resource Management Cooperative. Biosphere Reserve Research Report no. 2, 45 pp

Beets J, Rogers CS (2002) Changes in fishery resources and reef fish assemblages in a Marine Protected Area in the US Virgin Islands : the need for a no take marine reserve. Proc 9th Intl Coral Reef Symp 1:449–454

Beets J, Lewand L, Zullo E (1986) Marine community descriptions and maps of bays within the Virgin Islands National Park/Biosphere Reserve. Biosphere Reserve Research Report no. 2 VIRMC/NPS, 118 pp

Beets J, Muehlstein L, Haught K, Schmitges H (2003) Habitat connectivity in coastal environments: patterns and movements of Caribbean coral reef fishes with emphasis on bluestriped grunt, *Haemulon sciurus*. Gulf and Caribbean Res 14:29–42

Bohnsack JA, Bannerot SP (1986) A stationary visual census technique for quantitatively assessing community sturcture of coral reef fishes. NOAA Technical Report NMFS 41, 15 pp

Bohnsack JA, Meyers S, Appeldoorn RS, Beets J, Matos D, Sadovy Y (1991) Stock assessment of spiny lobster, *Panulirus argus*, in the U.S. Caribbean. Report to the Caribbean Fishery Management Council. Miami Laboratory Contribution no. MIA-90/91-49

Booth DJ, Beretta GA (1994) Seasonal recruitment, habitat associations, and survival of pomacentrid reef fish in the US Virgin Islands. Coral Reefs 13:81–89

Boulon RH (1986a) Map of fishery habitats within the Virgin Islands Biosphere Reserve. Biosphere Reserve Research Report no. 8 VIRMC/NPS, 70 pp

Boulon RH (1986b) Fisheries habitat of the Virgin Islands region of ecological importance to the fishery resources of the Virgin Islands Biosphere Reserve. Biosphere Reserve research report no. 9 VIRMC/NPS, 22 pp

Boulon RH (1987) Basis for long-term monitoring of fish and shellfish species in the Virgin Islands National Park. Biosphere Reserve Research Report no. 22 VIRMC/NPS, 66 pp

Boulon RH (1992) Use of mangrove prop root habitats by fish in the northern US Virgin Islands. Proc Gulf and Caribbean Fishery Institute 41:189–204

Boulon RH, Beets J, Zullo E (1986) Long-term monitoring of fisheries in the Virgin Islands Biosphere Reserve. Biosphere Reserve Research Report no. 13 VIRMC/NPS, 32 pp

Brooks G, Larson R, Devine B (2004) A depositional framework for St. Thomas and St. John, USVI. Virgin Islands Wetlands and Riparian Areas Inventory. Island Resources Foundation, Washington, DC, 81 pp

Brownell WN, Stevely JM (1981) The biology, fisheries, and management of the queen conch, *Strombus gigas*. Mar Fish Rev 43:1–12

Bruckner AW (ed) (2002) Proc Caribbean *Acropora* Workshop: Potential Application of the US Endangered Species Act as a Conservation Strategy. NOAA Tech Memo NMFS-OPR-24, Silver Spring, MD, 199 pp

Burke L, Maidens J (2004) Reefs at Risk in the Caribbean. World Resources Institute, Washington, DC, 8 pp

Bythell JC (1988) A total nitrogen and carbon budget for the elkhorn coral *Acropora palmata* (Lamarck). Proc 6th Intl Coral Reef Symp 2:535–540

Bythell JC, Gladfelter EH, Bythell M (1993) Chronic and catastrophic natural mortality of three common Caribbean reef corals. Coral Reefs 12:143–152

Bythell JC, Hillis-Starr Z, Phillips B, Burnett WJ, Larcombe J, Bythell M (2000a) Buck Island Reef

National Monument, St. Croix, US Virgin Islands: Assessment of the impacts of Hurricane Lenny (1999) and status of the reef 2000. Final Report, US Department of Interior, National Park Service. University of Newcastle, 38 pp

Bythell JC, Hillis-Starr ZM, Rogers CS (2000b) Local variability but landscape stability in coral reef communities following repeated hurricane impacts. Mar Ecol Prog Ser 204:93–100

Carpenter RC (1979) The foraging strategy of *Diadema antillarum* (Phillipi). University of the Pacific, 37 pp

Carpenter RC (1984a) Herbivores and Herbivory on Coral Reefs: Effects on Algal Community Biomass, Structure, and Primary Production (PhD Thesis). University of Georgia, Georgia, 175 pp

Carpenter RC (1984b) Predator and population density control of homing behavior in the Caribbean echinoid *Diadema antillarum*. Mar Biol 82:101–108

Carpenter RC (1985) Sea urchin mass mortality: effects on reef algal abundance, species composition and metabolism and other coral reef herbivores. Proc 5th Int Coral Reef Cong 4:53–60

Carpenter RC (1986) Partitioning herbivory and its effects on coral reef algal communities. Ecol Monogr 56:345–363

Carpenter RC (1988) Mass-mortality of a Caribbean sea urchin: immediate effects on community metabolism and other herbivores. Proc Natl Acad Sci 85:511–514

Carpenter RC (1990a) Mass mortality of *Diadema antillarum* I. Long-term effects on sea urchin population-dynamics and coral reef algal communities. Mar Biol 104:67–77

Carpenter RC (1990b) Mass mortality of *Diadema antillarum* II. Effects on population densities and grazing intensity of parrotfishes and surgeonfishes. Mar Biol 104:79–86

Carpenter RC, Edmunds PJ (2006) Local and regional scale recovery of *Diadema* promotes recruitment of scleractinian corals. Ecol Lett 9:271–280

Caselle JE, Warner RR (1996) Variability in recruitment in coral reef fishes: importance of habitat at large and small spatial scales. Ecology 77:2488–2504

Clarke RD (1988) Chance and order in determining fish-species composition on small coral patches. J Exp Mar Biol Ecol 115:197–212

Clarke RD (1989) Population fluctuation, competition and microhabitat distribution of two species of tube blennies, *Acanthemblemaria* (Teleostei: Chaenopsidae). Bull Mar Sci 44:1174–1185

Clarke RD (1992) Effects of microhabitat and metabolic rate on food intake, growth and fecundity of two competing coral reef fishes. Coral Reefs 11:199–205

Clarke RD (1994) Habitat partitioning by chaenopsid blennies in Belize and the Virgin Islands. Copeia: 398–405

Clarke RD (1996) Population shifts in two competing fish species on a degrading coral reef. Mar Ecol Prog Ser 137:51–58

Clavijo I, Yntema J, Ogden JC (1980) An annotated list of the fishes of St. Croix, US Virgin Islands. West Indies Laboratory, St. Croix, 57 pp

Clifton EH, Phillips RL (1975) Physical setting of the Tektite Experiments. In: Earle S, Lavenberg RJ (eds) Results of the Tektite Program: Coral Reef Invertebrates and Plants. Natural History Museum of Los Angeles County Science Bulletin 20, pp 7–10

Colin PI (1978) Caribbean Reef Invertebrates and Plants: A Field Guide to the Invertebrates and Plants Occurring on Coral Reefs of the Caribbean, the Bahamas, and Florida. TFH, Neptune City, NJ, 512pp

Colin PI (1996) Longevity of some coral reef fish spawning aggregations. Copeia 1996:189–192

Collette BB (1972) Conclusions. In: Collete BB, Earle S (eds) Results of the Tektite Program: Ecology of Coral Reef Fishes, pp 171–174

Collette BB, Earle SA (1972) Results of the Tektite Program: Ecology of Coral Reef Fishes. In: Collette BB, Earle SA (eds) Natural History Museum, Los Angeles County. Sci Bull 14: 18

Collette BB, Talbot FH (1972) Activity patterns of coral reef fishes with emphasis on nocturnal-diurnal changeover. In: Collette BB, Earle SA (eds) Results of the Tektite Program: Ecology of Coral Reef Fishes. Natural History Museum, Los Angeles County. Sci Bull 14: 98–124

Cooper RA, Ellis R, Serfling S (1975) Population dynamics, ecology and behavior of spiny lobsters, *Panulirus argus*, of St. John, USVI (III) Population estimation and turnover. In: Earle SA, Lavenberg RJ (eds) Results of the Tektite Program: Coral Reef Invertebrates and Plants. Natural History Museum of Los Angeles County. Sci Bull 20:23–30

Cowen RK (2002) Chapter 7: Oceanographic influences on larval dispersal and retention and their consequences for population connectivity. In: Sale PF (ed) The Ecology of Fishes on Coral Reefs. Academic, San Diego, CA, pp 475–508

Cowen RK, Lwiza KMM, Sponaugle S, Paris CB, Olson DB (2000) Connectivity of marine populations: Open or closed? Science 287:857–859

Cowen RK, Paris CB, Srinivasan A (2006) Scaling of connectivity in marine populations. Science 311:522–527

Dammann AE (1969) Special Report: study of the fisheries potential of the Virgin Islands. Contribution Number 1. Virgin Islands Ecological Research Station, 197 pp

Dammann AE (1986) Assessment of fish and shellfish stocks produced in the Biosphere Reserve. US National Park Service. Biosphere Reserve Research Report no. 10 VIRMC/NPS, 30 pp

Dammann AE, Nellis DW (1992) A Natural History Atlas to the Cays of the US Virgin Islands. Pineapple, Sarasota, Florida, 160 pp

Davis M, Gladfelter EH, Lund H, Anderson M (1986) Geographic range and research plan for monitoring white band disease. Biosphere Reserve Research Report no. 6 VIRMC/NPS, 28 pp

DeAngelis BM (2006) The distribution of elasmobranchs in St. Thomas and St. John, United States Virgin Islands with an emphasis on shark nursery areas. M.S. thesis, University of Rhode Island, 110 pp

Denner EBM, Smith G, Busse HJ, Schumann P, Narzt T, Polson SW, Lubitz W, Richardson LL (2003) *Aurantimonas coralicida* gen.nov., sp. nov., the causative agent of white plague type II on Caribbean scleractinian corals. Int J Syst Evol Microbiol 53:1253–1260

Dunion JP, Velden CS (2004) The impact of the Saharan air layer on Atlantic tropical cyclone activity. Amer Metereol Soc, pp 353–365

Earle S (1972) The influence of herbivores on the marine plants of Great Lameshur Bay, with an annotated list of plants. In: Collette BB, Earle SA (eds) Results of the Tektite Progam: Ecology of Coral Reef Fishes, Natural History Museum of Los Angeles County, Sci Bull 14:17–44

Earle S, Lavenberg RJ (eds) (1975) Results of the Tektite Program: Coral Reef Invertebrates and Plants. Natural History Museum of Los Angeles County. Sci Bull 20: 1–103

Edmunds PJ (1991) Extent and effect of black band disease on a Caribbean reef. Coral Reefs 10:161–165

Edmunds PJ (2000) Patterns in the distribution of juvenile corals and coral reef community structure in St. John, US Virgin Islands. Mar Ecol Prog Ser 202:113–124

Edmunds PJ (2002) Long-term dynamics of coral reefs in St. John, US Virgin Islands. Coral Reefs 21:357–367

Edmunds PJ (2004) Juvenile coral population dynamics track rising seawater temperature on a Caribbean reef. Mar Ecol Prog Ser 269:111–119

Edmunds PJ (2006) Temperature-mediated transitions between isometry and allometry in a colonial modular invertebrate. Proc. R Soc London 273:2275–2281

Edmunds PJ, Elahi R (2007) The demographics of a 15-year decline in cover of the Caribbean reef coral *Montatraea annularis*. Ecol Mono 77:3–18

Edmunds PJ, Witman J (1991) Effect of Hurricane Hugo on the primary framework of reefs along the south shore of St. John, US Virgin Islands. Mar Ecol Prog Ser 1991:201–204

Edmunds PJ, Gates, RD, Gleason DF (2001) The biology of larvae from the reef coral *Porites astreoides*, and their response to temperature disturbances. Mar Biol 139:981–989

Evermann BW, Marsh MC (1902) The fishes of Puerto Rico. Bull US Fish Comm for 1900 XX, Part 1:57–350, pls. i–xlix

Fallows J (1985) Behavioral ecology of the yellowtail snapper *Ocyurus chrysurus*. Ph.D. thesis, University of Newcastle

Fenchel T, McRoy CP, Ogden JC, Parker PL, Rainey VE (1979) Symbiotic cellulose degradation in the green turtle *Chelonia mydas*. J Appl Environ Microbiol 37:348–350

Fiedler RH, Jarvis ND (1932) Fisheries of the Virgin Islands of the United States. US Department of Commerce, Bureau of Fisheries. Investigational Report 14, 32 pp

Friedlander A (1997) Status of the queen conch populations around the northern USVI with management recommendations for the Virgin Islands National Park. Report prepared for US Geological Survey, St. John, USVI, 40 pp

Friedlander A, Appeldoorn RS, Beets J (1994) Spatial and temporal variations in stock abundance of queen conch, *Strombus gigas,* in the US Virgin Islands. In: Appeldoorn RS, Rodriguez B (eds) Queen Conch Biology, Fisheries and Mariculture. Fundacion Cientifica Los Roques, Caracas, Venezuela, pp 51–60

Garrison VH, Rogers C, Beets J (1998) Of reef fishes, overfishing and in situ observations of fish traps in St. John, USVI. Rev Biol Trop 46:41–59

Garrison VH, Shinn EA, Foreman WT, Griffin DW, Holmes CW, Kellogg CA, Christina A, Majewski MS, Richardson LL, Ritchie KB, Smith GW (2003) African and Asian Dust: from desert soils to coral reefs. Bioscience 53:469–480

Gladfelter EH (1977) Primary productivity and calcification in the reef coral *Acropora palmata*. M.S. thesis, University of Pacific, 51 pp

Gladfelter EH (1982) Skeletal development in *Acropora cervicornis*: I Patterns of calcium carbonate accretion in the axial corallite. Coral Reefs 1:45–52

Gladfelter EH (1983a) Skeletal development in *Acropora cervicornis*: II Diel patterns of calcium carbonate accretion in the axial corallite. Coral Reefs 2:91–100

Gladfelter EH (1983b) Spatial and temporal patterns of mitosis in the cells of the axile polyp of the reef coral *Acropora cervicornis*. Biol Bull 165:811–815

Gladfelter EH (1984) Skeletal development in *Acropora cervicornis*. III. A comparison of monthly rates of linear extension and calcium carbonate accretion measured over a year. Coral Reefs 15:51–57

Gladfelter EH (1985) Metabolism, calcification and carbon production. II. Organism-level studies. Proc 5th Intl Coral Reef Cong 4:527–539

Gladfelter EH (1992) Annotated bibliography of marine research at Buck Island Reef National Monument, St. Croix, US Virgin Islands. Report for the National Park Service, US Department of Interior, 34 pp

Gladfelter EH, Monahan RK, Gladfelter WB (1978) Growth rates of five reef-building corals in the northeastern Caribbean. Bull Mar Sci 28:728–734

Gladfelter WB (1975) Sea anemone with zooxanthellae: simultaneous contraction and expansion in response to changing light intensity. Science 189:570–571

Gladfelter WB (1978) General ecology of the cassiduloid urchin *Cassidulus caribbearum*. Mar Biol 47:149–160

Gladfelter WB (1979) Twilight migrations and foraging activities of the copper sweeper, *Pempheris schomburgki*. Mar Biol 50:109–119

Gladfelter WB (1982) White-band disease in *Acropora palmata*: implications for the structure and growth of shallow reefs. Bull Mar Sci 32:639–643

Gladfelter WB (1988) Tropical Marine Organisms. St. Croix, VI, 149 pp

Gladfelter WB (1991) Chapter 5: Population structure of *Acropora palmata* on the windward forereef, Buck Island Reef National Monument: seasonal and catastrophic changes 1988–1989. Ecological studies of Buck Island Reef National Monument, St. Croix, US Virgin Islands: a quantitative assessment of selected components of the coral reef ecosystem and establishment of long term monitoring sites, Part 1. National Park Service Coral Reef Assessment Program, St. John, 22 pp

Gladfelter WB (1993a) Report on annual change in sponge and gorgonian communities at Newfound Bay: comparison with Yawzi Point. Report to the National Park Service, St. John, USVI, 20 pp

Gladfelter WB (1993b) Annual changes in sponge and gorgonian communities at Yawzi Point, St. John, USVI 1991 to 1993. Report to the National Park Service, St. John, USVI, 28 pp

Gladfelter WB, Gladfelter EH (1978) Fish community structure as a function of habitat structure on West Indian patch reefs. Rev Biol Trop 26:65–84

Gladfelter WB, Gladfelter EH (2004) Molluscs of Southgate and Green Cay Sound: Seashells found on the Barrier Beach. Coast and Harbor SCR Technical Report no. 8, 8 pp

Gladfelter WB, Gladfelter EH, Monahan RK, Ogden JC, Dill RF (1977) Environmental studies of Buck Island Reef National Monument, St. Croix, USVI, US Department of Interior, National Park Service Report, 140 pp

Gladfelter WB, Ogden JC, Gladfelter EH (1980) Similarity and diversity among coral reef fish communities: a comparison between tropical western Atlantic (Virgin Islands) and tropical central Pacific (Marshall Islands) patch reefs. Ecology 61:1156–1168

Gladfelter WB, Johnson WB (1983) Feeding niche separation in a guild of tropical reef fishes (Holocentridae). Ecology 64:552–563.

Gleason DF (1993) Differential effects of ultraviolet radiation on green and brown morphs of the Caribbean coral *Porites astreoides*. Limnol Oceanogr 38:1452–1463

Gordon S (2002) USVI queen conch assessment. Final report to the Southeast Area Monitoring and Assessment Program-Caribbean. Division of Fish and Wildlife, Department of Planning and Natural Resources, USVI, 65 pp

Griffin DW, Kellogg CA, Garrison VH, Lisle JT, Borden TC, Shinn EA (2003) Atmospheric microbiology in the northern Caribbean during African dust events. Aerobiologia 19:143–157

Grippo RS (1981) Resource partitioning among nocturnal planktivorous fishes in a tropical reef ecosystem. M.S. thesis, Fairleigh Dickinson University, St. Croix, VI, 44 pp

Grober-Dunsmore R, Frazer TK, Lindberg WJ, Beets J (2006) Reef fish and habitat relationships in a Caribbean seascape: the importance of reef context. Coral Reefs 26:201–216

Halliwell G, Mayer DA (1996) Frequency response properties of forced climatic SST anomaly variability. J Climate 9:3575–3587

Hamilton SL, White JW, Caselle JE, Swearer SE, Warner RR (2006) Consistent long-term spatial gradients of population replenishment for an island population of a coral reef fish. Mar Ecol Prog Ser 306:247–256

Harborne AR, Mumby PJ, Zychaluk K, Hedley JD, Blackwell PG (2006) Modeling the beta diversity of coral reefs. Ecology 87:2871–2881

Harvell CD, Fenical W (1989) Chemical and structural defense of Caribbean gorgonians (*Pseudopterogorgia* spp.): intracolony localization of defense. Limnol Oceanogr 34:382–389

Harvell CD, Fenical W, Roussis V, Ruesink JL, Griggs CC, Greene CH (1993) Local and geographic variation in the defensive chemistry of a West Indian gorgonian coral (*Briareum asbestinum*). Mar Ecol Prog Ser 93:165–173

Harvell CD, West J, Griggs CC (1996) Chemical defense of embryos and larvae of a West Indian gorgonian coral, *Briareum asbestinum*. Invertebrate Reproduction and Development 30:239–246

Hay ME (1985) Spatial patterns of herbivore impact and their importance in maintaining algal species richness. Proc 5th Intl Coral Reef Cong 4:29–34

Hay ME, Taylor PR (1985) Competition between herbivorous fishes and urchins on Caribbean reefs. Oecologia 65:591–598

Hay ME, Paul VJ, Lewis SM, Gustafson K, Tucker J, Trindell RN (1988) Can tropical seaweeds reduce herbivory by growing at night? Diel patterns of growth, nitrogen content, herbivory, and chemical versus morphological defenses. Oecologia 75:233–245

Helfman GS (1983) Resin coated fishes: a simple model technique for in situ studies of fish behavior. Copeia 1983:547–549.

Helfman GS (1989) Threat-sensitive predator avoidance in damselfish-trumpetfish interactions. Behav Ecol Sociobiol 24:47–58

Helfman GS, Meyer JL, McFarland WN (1982) The ontogeny of twilight migration patterns in grunts (Pisces: Haemulidae). Anim Behav 30:317–326

Henson S, Warner RR (1997) Male and female alternative reproductive behaviors in fishes: a new approach using intersexual dynamics. Ann Rev Ecol Syst 28:571–592

Herrnkind WF, VanDerwalker JA, Barr L (1975) Population dynamics, ecology and behavior of spiny lobsters, Panulirus argus, of St. John, USVI (IV) Habitation, patterns of movement and general behavior. In: Earle S, Lavenberg RJ (eds) Results of the Tektite Program: Coral Reef Invertebrates and Plants. Natural History Museum of Los Angeles County. Sci Bull 20:31–45

Herzlieb S, Kadison E, Blondeau J, Nemeth R (2006) Comparative assessment of coral reef systems located along the insular platform of St. Thomas, US Virgin Islands, and the relative effects of natural and human impacts. Proc 10th Intl Coral Reef Symp 4:1144–1151

Hixon MA (1991) Predation as a process structuring coral reef fish communities. In: Sale PF (ed) The Ecology of Fishes on Coral Reefs. Academic, San Diego, CA, pp 475–508

Hu C, Montgomery ET, Schmitt FE, Muller-Kargar FE (2004) The dispersal of the Amazon and Orinoco River water in the tropical Atlantic and Caribbean Sea: observations from space and S-PALACE floats. Deep Sea Resh II, pp 1151–1171

Hubbard DK (1987) A general review of sedimentation as it relates to environmental stress in the Virgin Islands Biosphere Reserve and the Eastern Caribbean in general. Biosphere Reserve Research Report no. 20 VIRMC/NPS, 41pp

Hubbard DK (1989) Modern carbonate environments of St. Croix and the Caribbean : a general overview. In: Hubbard DK (ed) Terrestrial and Marine Geology of St. Croix, USVI, pp 85–94

Hubbard DK, Suchanek TH, Gill IP, Cowper SW, Ogden JC, Westerfield JR, Bayes JS (1981) Preliminary studies of the fate of shallow-water detritus in the basin north of St. Croix, US Virgin Islands. Proc 4th Int Coral Reef Symp 1:383–387

Hubbard DK, Stump J, Carter B (1987) Sedimentation and reef development in Hawksnest, Fish and Reef Bays, St. John, US Virgin Islands. Biosphere Reserve research report no. 21 VIRMC/NPS, 99 pp

Hubbard DK, Parsons KM, Bythell JC, Walker ND (1991) The effects of Hurricane Hugo on the reefs and associated environments of St. Croix, US Virgin Islands – A preliminary assessment. J Coast Res 8:33–48

Humann P, DeLoach N (2002a) Reef Coral Identification: Florida, Caribbean, Bahamas. New World, Jacksonville, FL, 278 pp

Humann P, DeLoach N (2002b) Reef Creature Identification: Florida, Caribbean, Bahamas. New World, Jacksonville, FL, 420 pp

Humann P, DeLoach N (2002c) Reef Fish Identification: Florida, Caribbean, Bahamas. New World, Jacksonville, FL, 481 pp

Idjadi JA, Edmunds PJ (2006) Scleractinian corals as facilitators: evidence for positive interactions between scleractinian corals and other reef invertebrates. Mar Ecol Prog Ser 319:117–127

Idyll CP, Randall JE (1959) Sport and commercial fisheries potential of St. John, Virgin Islands. Fourth International Gamefish Conference, 10 pp

Island Resources Foundation (1977) Marine environments of the Virgin Islands. Technical Supplement no. 1. Prepared for the Virgin Islands Government Coastal Zone Management Progam, 188 pp

Island Resources Foundation (1989) Eastern Caribbean Parks and Protected Areas Bibliography. Biosphere Reserve Research Report no. 31. Report to National Park Service, 43 pp

Jackson JBC, Kirby MX, Berer WH, Bjorndal KA, Botsford LW, Bourque BJ, Bradbury RH, Cooke R, Erlandson J, Estes JA, Hughes TP, Kidwell S, Lange CB, Lenihan HS, Pandolfi JM, Peterson CH, Steneck RS, Tegner MJ, Warner RR (2001) Historical over-fishing and the recent collapse of coastal ecosystems. Science 293:629–637

Jeffrey C, Anlauf U, Beets J, Caseau S, Coles W, Friedlander A, Herzlieb S, Hillis-Starr Z, Kendall M, Mayor V, Miller J, Nemeth R, Rogers C, Toller W (2005) Chapter 4: The state of coral reef ecosystems of the US Virgin Islands. In: Waddell JE (ed) The State of Coral Reef Ecosystems of the United States and Pacific Freely Associated States. NOAA Technical Memorandum NOS NCCOS 11. NOAA/NCCOS Center for Coastal Monitoring and Assessment's Biogeography Team, Silver Spring, MD, pp 45–83

Kadison E, Nemeth R, Herzlieb S, Blondeau J (2006) Temporal and spatial dynamics of Lutjanus cyanopterus (Pisces: Lutjanidae) and L. jocu spawning aggregations in the USVI. Revista de Biologia Tropical 54:69–78

Kaplan EH (1982) A Field Guide to Coral Reefs: Caribbean and Florida. Houghton-Mifflin, Boston, MA, 289 pp

Kendall MS, Kruer CR, Buja KR, Christensen JD, Finkbeiner M, Warner RA, Monaco ME (2001) Methods used to map the benthic habitats of Puerto Rico and the US Virgin Islands. NOAA, NOS, NCCOS. Silver Spring, MD, 45 pp

Kendall MS, Christensen JD, Hillis-Starr Z (2003) Multi-scale data used to analyze the spatial distribution of French grunts, Haemulon flavolineatum, relative to hard and soft bottom in a benthic landscape. Environ Biol Fishes 66:19–26

Kendall MS, Takata LT, Jensen O, Hillis-Starr Z, Monaco ME (2005) An ecological characterization of the Salt River Bay National Historical Park and Ecological Preserve, US Virgin Islands. NOAA Technical Memorandum NOS NCCOS 14, 115 pp

King GM, Carlton RG, Sawyer TE (1990) Anaerobic metabolism and oxygen distribution in the carbonate sediments of a submarine canyon. Mar Ecol Prog Ser 58:275–285

Kojis BL (1997) Baseline data on coral recruitment in the Northern US Virgin Islands. Caribbean Fishery Management Council. San Juan, PR, 17 pp

Kojis BL (2004) Census of the marine commercial fishers of the US Virgin Islands. Report to the Caribbean Fishery Management Council. Division of Fish and Wildlife, Department of Planning and Natural Resources, USVI, 87 pp

Kourafalou VH, Balotro RS, Peng G, Lee TN, Johns E, Ortner PB, Wallcraft A, Townsend T (2006) Seasonal variability of circulation and salinity around Florida Bay and the Florida keys: SoFLA-HYCOM results and comparison to in-situ data. /UM/RSMAS Tech Rep, 102 pp

Kumpf HE, Randall HA (1961) Charting the marine environments of St. John, USVI. Bull Mar Sci 11:543–551

Larkum AWD, Orth RJ, Duarte CM (eds) (2006) Seagrasses: Biology, Ecology and Conservation. Springer, The Netherlands, 691 pp

Lessios HA, Robertson DR, Cubit JD (1984) Spread of Diadema mass mortality through the Caribbean. Science 226:335–337

Levitan DR (1988) Algal-urchin biomass responses following the mass mortality of the sea urchin Diadema antillarum Philippi at St. John, US Virgin Islands. J Exp Mar Biol Ecol 119:167–178

Lewis JB (2004) Has random sampling been neglected in coral reef faunal surveys? Coral Reefs 23:192–194

Littler DS, Littler MM (2000) Caribbean Reef Plants. An Identification Guide to the Reef Plants of the Caribbean, Bahamas, Florida, and the Gulf of Mexico. Offshore Graphics, Washington, DC, 542 pp

Mahnken C (1972) Observations on cleaner shrimps of the genus Periclimenes Results of the Tektite Program: Ecology of Coral Reef Fishes. Natural History Museum, Los Angeles County. Sci Bull 14:71–83

Mateo I, Tobias WJ (2002) Preliminary estimations of growth, mortality, and yield per recruit for the spiny lobster Panulirus argus. Proc Gulf and Caribbean Fisheries Institute 53:58–75

Mathieson AC, Fralick RA, Burns R, Flahive W (1975) Phycological studies during Tektite II, at St. John, USVI. In: Earle SA, Lavenberg RJ (eds) Results of the Tektite Program: Coral Reef Invertebrates and Plants. Natural History Museum, Los Angeles County. Sci Bull 20:77–103

Mayor P, Rogers C, Hillis-Starr Z (2006) Distribution and abundance of elkhorn coral, Acropora palmata, and preva-lence of white-band disease at Buck Island Reef National Monument, St. Croix, USVI. Coral Reefs 25:239–242

McFarland WN, Ogden JC (1985) Recruitment of young coral reef fishes from the plankton. In: Reaka ML (ed) The Ecology of Coral Reefs. NOAA Symp Ser Undersea Res, Washington, DC, 3:37–51

McFarland WN, Ogden JC, Lythgoe JN (1979) The influence of light on the twilight migrations of grunts. Env Biol Fishes 4:9–22

McFarland WN, Brothers EB, Ogden JC, Shulman MJ, Bermingham EL, Kotchian-Prentiss NM (1985) Recruitment patterns in young French grunts, Haemulon flavolineatum (Family Haemulidae), at St. Croix, Virgin Islands. Fish Bull 83:413–426

Menza C, Ault J, Beets J, Bohnsack J, Caldow C, Christensen J, Friedlander A, Jeffrey C, Kendall M, Luo J, Monaco M, Smith S, Woody K (2006) A Guide to Monitoring Reef Fish in the National Park Service's South Florida /Caribbean Network. NOAA Technical Memorandum NOS NCCOS 39, 166 pp

Meyer JL, Schultz ET, Helfman GS (1983) Fish schools: an asset to corals. Science 220:1047–1049

Miller J, Rogers CS (2002) A new approach to tracking change on coral reefs: using videotape to monitor coral reefs, and using Aqua Map^{tm} at a study site. US Geological Survey, Inventory and Monitoring protocol, 71 pp

Miller J, Beets J, Rogers C (2001) Temporal patterns in fish recruitment on a fringing reef in Virgin Islands National Park, St. John, USVI. Bull Mar Sci 69:567–577

Miller J, Rogers C, Waara R (2003) Monitoring the coral disease, plague type II, on coral reefs in St. John, US Virgin Islands. Rev Biol Trop 51:47–55

Miller J, Waara R, Muller E, Rogers C (2006) Coral bleaching and disease combine to cause extensive mortality on reefs in the US Virgin Islands. Coral Reefs 25:418

Miller RJ, Adams AJ, Ogden NB, Ogden JC, Ebersole JP (2003) Diadema antillarum 17 years after the mass mortality; is recovery beginning on St. Croix ? Coral Reefs 22:181–187

Moese MD (1980) Distribution, abundance, and resource partitioning of Diodon holocanthus L., the balloonfish, in Tague Bay, St. Croix, USVI, M.S. thesis, Fairleigh Dickinson University, St. Croix, VI, 43 pp

Monaco M, Friedlander AM, Caldow C, Christensen JD, Rogers C, Beets J, Miller J, Boulon RH (2007) Characterising reef fish populations and habitats within and outside the US Virgin Islands Coral Reef National Monument: a lesson in marine protected area design. Fish Mgmt Ecol 13:1–8

Muehlstein L, Beets J (1999) Disturbance weakens the positive enhancement of seagrass-coral interactions. Abstract. Fifteenth Biennial International Conference of the Estuarine Research Federation, September 1999, New Orleans

Mumby PJ, Green EP, Clark CD, Edwards AJ (1998) Digital analysis of multispectral airborne imagery of coral reefs. Coral Reefs 17:59–69

National Marine Fisheries Service and US Fish and Wildlife Service. (1993) Recovery Plan for Hawksbill Turtles in the US Caribbean Sea, Atlantic Ocean, and Gulf of Mexico. National Marine Fisheries Service, St. Petersburg, FL, 52 pp

Nelson WR, Appeldoorn RS (1985) Cruise Report. R/V Seward Johnson: a submersible survey of the continental slope of Puerto Rico and the US Virgin Islands October 1–23, 1985. NOAA, University of Puerto Rico, CODREMAR, Government of the USVI, Caribbean Fishery Management Council, 76 pp

Nemeth R (1998) The effect of natural variation in substrate architecture on the survival of juvenile bicolor damselfish. Environ Biol Fishes 53:139–141

Nemeth R (2005) Population characteristics of a recovering US Virgin Islands red hind spawning aggregation following protection. Mar Ecol Prog Ser 286:81–97

Nemeth RS, Sladek Nowlis J (2001) Monitoring the effects of land development on the near-shore reef environment of St. Thomas, USVI. Bull Mar Sci 69:759–775

Nemeth RS, Whaylen LD, Pattengill-Semmens C (2003a) A rapid assessment of coral reefs in the Virgin Islands (Part 2: fishes). Atoll Res Bull 496:566–589

Nemeth RS, Quandt A, Requa L, Rothenberger P, Taylor M (2003b) A rapid assessment of coral reefs in the Virgin Islands (Part 1: stony corals and algae). Atoll Res Bull 496:544–565

Nemeth RS, Herzlieb S, Kadison E, Taylor M, Rothenberger P, Harold S, Toller W (2004) Coral reef monitoring in St. Croix and St. Thomas, United States Virgin Islands. Year Three Final Report, Submitted to Department of Planning and Natural Resources USVI, 79 pp

Nemeth RS, Smith T, Taylor M, Herzlieb S, Kadison ES, Blondeau J, Carr L, Allen-Requa L and Toller W (2005) Coral reef monitoring in St. Croix and St. Thomas, United States Virgin Islands. Year Five Final Report, Submitted to Department of Planning and Natural Resources USVI, 65 pp

Nemeth RS, Herzlieb S, Blondeau J (2006a) Comparison of two seasonal closures for protecting red hind spawning aggregations in the US Virgin Islands. Proc 10th Intl Coral Reef Symp 4:1306–1313

Nemeth RS, Kadison E, Herzlieb S, Blondeau J, Whiteman E (2006b) Status of a yellowfin grouper (*Mycteroperca venenosa*) spawning aggregation in the US Virgin Islands with notes on other species. Proc 57th Gulf and Caribbean Fisheries Institute 57:543–558

Nemeth RS, Blondeau J, Herzlieb S, Kadison E (2007) Spatial and temporal patterns of movement and migration at spawning aggregations of red hind, *Epinephelus guttatus*, in the US Virgin Islands. Environ Biol Fishes 78:365–381

Nichols JT (1929) The fishes of Puerto Rico and the Virgin Islands: Branchiostomidae to Sciaenidae Scientific Survey of Puerto Rico and the Virgin Islands. New York Academy of Sciences, New York 10:161–295

Nichols JT (1930) The fishes of Puerto Rico and the Virgin Islands: Pomacentridae to Ogcocephalidae Scientific Survey of Puerto Rico and the Virgin Islands. New York Academy of Sciences, New York 10:299–399

Nowlis J (1993) Mate- and oviposition-influenced host preferences in the coral-feeding snail *Cyphoma gibbosum*. Ecology 74:1959–1969

O'Conner MI, Bruno JF, Gaines SD, Halpern BS, Lester SE, Kinlan BP, Weiss JM (2007) Temperature control of larval dispersal and the implications for marine ecology, evolution, and conservation. Proc Natl Acad Sci 104:1266–1271

Ogden JC (1980) The major marine environments of St. Croix. In: Multer EG, Gerhard LC (eds) Guidebook to the Geology and Ecology of some Marine and Terrestrial Environments, St. Croix, US Virgin Islands. Special publication no. 5 West Indies Laboratory, Fairleigh Dickinson University, St. Croix, VI, pp 5–19

Ogden JC (1988) The influence of adjacent systems on the structure and function of coral reefs. Proc 6th Int Coral Reef Symp, pp123–129

Ogden JC (1997) Ecosystem interactions in the tropical coastal seascape. In: Birkeland C (ed) Life and Death of Coral Reefs. Chapman & Hall, New York, 536 pp

Ogden JC, Ebersole JP (1981) Scale and community structure of coral reef fishes: a long-term study of a large artificial reef. Mar Ecol Prog Ser 4:97–103

Ogden JC, Ehrlich PR (1977) The behavior of heterotypic resting schools of juvenile grunts (Pomadasyidae). Mar Biol 42:273–280

Ogden JC, Gladfelter EH (1983) Coral reefs, seagrass beds, and mangroves: their interaction in the coastal zones of the Caribbean. UNESCO Reports in Marine Science 23, 133 pp

Ogden JC, Zieman JC (1977) Ecological aspects of coral reef-seagrass bed contacts in the Caribbean. Proc 3rd Int Coral Reef Symp, pp 377–382

Ogden JC, Helm D, Peterson J, Smith A, Weisman S (eds) (1972) An ecological study of Tague Bay Reef, St. Croix, USVI, West Indies Laboratory Special Publication 1, 51 pp

Ogden JC, Brown RA, Salesky N (1973) Grazing of the echinoid *Diadema antillarum* Philippi: formation of halos around West Indian patch reefs. Science 182:715–717

Olsen DA, Herrnkind WF, Cooper RA (1975) Population dynamics, ecology and behavior of spiny lobsters, *Panulirus argus*, of St. John, USVI (I) Introduction and general population characteristics. In: Earle SA, Lavenberg RJ (eds) Results of the Tektite Program: Coral Reef Invertebrates and Plants. Natural History Museum of Los Angeles County. Sci Bull 20:11–16

Olsen DA, LaPlace JA (1978) A study of a Virgin Islands grouper fishery based on a breeding aggregation. Proc 31st Gulf and Caribbean Fisheries Institute 31:130–144

Pandolfi JM, Jackson JBC, Baron N, Bradbury RH, Guzman HM, Hughes TP, Kappel CV, Micheli F, Ogden JC, Possingham HP, Sala E (2005) Are US coral reefs on the slippery slope to slime? Science 307:1725–1726

Pantos O, Bythell JC (2006) Bacterial community structure associated with white band disease in the elkhorn coral *Acropora palmata* determined using culture-independent 16srRNA techniques. Dis Aquat Org 69:79–88

Patterson KL, Porter JW, Ritchie KB, Polson SW, Mueller E, Peters EC, Santavy DL, Smith GW (2002) The etiology of white pox, a lethal disease of the Caribbean elkhorn coral, *Acropora palmata*. Proc Natl Acad Sci 99:8725–8730

Patterson MR, Sebens KP, Olson RR (1991) *In situ* measurements of flow effects on primary production and dark respiration in reef corals. Limnol Oceanogr 36:936–948

Petersen CW, Warner RR (1998) Sperm competition and sexual selection in fishes. In: Birkhead TR, Møller AP (eds) Sperm competition and sexual selection. Academic, London, pp 435–463

Phillips B, Hillis-Starr Z (2002) Sea turtle nesting research and monitoring protocol. US Department of the Interior, US Geological Survey, 129 pp

Pickert P, Kelly T, Nemeth RS, Kadison E (2006) Seas of change: spawning aggregations of the Virgin Islands. In: Pickert P, Kelly T (eds) DVD by Friday's films, San Francisco, CA

Porter J, Targett MM (1988) Allelochemical interactions between sponges and corals. Biol Bull 175:230–239

Prospero JM, Lamb PJ (2003) African droughts and dust transport to the Caribbean: climate change implications. Science 302:1024–1027

Quinn NJ, Hanrahan M (1996) Status of queen conch resources in St. Thomas and St. John, US Virgin Islands: Is there hope for recovery? Proc 44th Session of the Gulf and Caribbean Fisheries Institute: 439–458

Quinn NJ, Kojis BL (1994a) Evaluation of the use of AVHRR satellite imagery and *in situ* obtained subsurface sea water temperatures for monitoring coastal marine communities in the Caribbean Sea. Proc 2nd Thematic Conf Remote Sensin Coastal Marine Environment, New Orleans 1:653–664

Quinn NJ, Kojis BL (1994b) Monitoring sea water temperatures adjacent to shallow benthic communities in the Carribean Sea: A comparison of AVHRR satellite records and *in situ* subsurface observations. Mar Technol Soc J 28:22–27

Quinn TP, Ogden JC (1984) Field evidence of compass orientation in migrating juvenile grunts (Haemulidae). J Exp Mar Biol Ecol 81:181–192

Ramos-Scharron CE, MacDonald LH (2005) Measurement and prediction of sediment production from unpaved roads, St. John, US Virgin Islands. Earth Surface Processes and Landforms 30:1283–1304

Randall JE (1962) Tagging reef fishes in the Virgin Islands. Proc Gulf and Caribbean Fisheries Institute 14:201–241

Randall JE (1963) An analysis of the fish populations of artificial and natural reefs in the West Indies. Carib J Sci 3:31–47

Randall JE (1964) Contributions to the biology of the queen conch, *Strombus gigas*. Bull Mar Sci 14:246–295

Randall JE (1965) Grazing on sea grasses by herbivorous reef fishes in the West Indies. Ecology 46:255–260

Randall JE (1967) Food habits of reef fishes of the West Indies. Stud Trop Oceanog 5:665–847

Randall JE (1968) Caribbean Reef Fishes. TFH, Neptune City, NJ, 318 pp

Randall JE, Randall HA (1963) The spawning and early development of the Atlantic parrotfish, *Sparisoma rubripinne*, with notes on other scarid and labrid fishses. Zoologica 48:49–60

Risk A (1997) Effects of habitat on the settlement and post-settlement success of the ocean surgeonfish, *Acanthurus bahianus*. Mar Ecol Prog Ser 161:51–59

Risk A (1998) The effects of interactions with reef residents on the settlement and subsequent persistence of ocean surgonfish, *Acanthurus bahianus*. Env Biol Fish 51:377–389

Robblee MB (1983) Resource partitioning by coral reef fishes while active nocturnally over a tropical *Thalassia* feeding ground. Ph.D. thesis, University of Virginia, Virginia

Robblee MB, Zieman JC (1984) Diel variation in the fish fauna of a tropical seagrass feeding ground. Bull Mar Sci 34:335–345

Robinson AH, Feazel CT (1974) Fringing reef, enclosing bays, and salt ponds of St. John, Virgin Islands. Oceans: 40–43

Rogers C (1999) Dead *Porites* patch reefs, St. John, US Virgin Islands. Coral Reefs 18:254

Rogers CS, Beets JP (2001) Degradation of marine ecosystems and decline of fishery resources in marine protected areas in the US Virgin Islands. Environ Conserv 28:312–322

Rogers CS, Garrison VH (2001) Ten years after the crime: lasting effects of damage from a cruise ship anchor on a coral reef in St. John. Bull Mar Sci 69:793–803

Rogers C, Miller J (2006) Permanent "phase shifts" or reversible declines in coral cover? Lack of recovery of two coral reefs in St. John, USVI. Mar Ecol Prog Ser 306:103–114

Rogers CS, Salesky N (1981) Productivity of *Acropora palmata* (Lamarck), macroscopic algae, and algal turf from Tague Bay reef, St. Croix, USVI. J Exp Mar Biol Ecol 49:179–187

Rogers CS, Teytaud R (1988) Marine and Terrestrial Ecosystems of the Virgin Islands National Park and Biosphere Reserve. Biosphere Reserve Research Report no. 29 VIRMC/NPS, 112 pp

Rogers CS, Zullo E (1987) Initiation of a long-term monitoring program for coral reefs in the Virgin Islands National Park. Biosphere Reserve research report no. 17 VIRMC/NPS, 33 pp

Rogers CS, Davis GE, McCreedy C (2007) National Parks and Caribbean Marine Reserves Research and Monitoring Workshop. NPS Water Resources Division Technical Report. NPS/NRWRD/NRTR-2007/362

Rogers CS, Gilnack M, Fitz I, HC (1983) Monitoring of coral reefs with linear transects: a study of storm damage. J Exp Mar Biol Ecol 66:285–300

Rogers CS, McLain L, Tobias C (1991) Effects of Hurricane Hugo (1989) on a coral reef in St. John, USVI. Mar Ecol Prog Ser 78:189–199

Rogers CS, Miller J, Waara R (2002a) Tracking changes on a reef in the US Virgin Islands with videography and SONAR: a new approach. Proc 9th Int Coral Reef Symp 2:1065–1071

Rogers CS, Miller J, Waara R (2002b) Tracking changes on a reef in the US Virgin Islands with videography and sonar: a new approach. Proc 9th Int Coral Reef Symp 2:1065–1071

Rogers CS, Suchanek T, Pecora F (1982) Effects of Hurricanes David and Frederic (1979) on shallow *Acropora palmata* reef communities: St. Croix, USVI. Bull Mar Sci 32:532–548

Rogers CS, Fitz HC III, Gilnack M, Beets J, Hardin J (1984) Scleractinian coral recruitment patterns at Salt River submarine canyon, St. Croix, USVI. Coral Reefs 3:69–76

Rogers CS, Gladfelter W, Hubbard D, Gladfelter E, Bythell J, Dunsmore R, Loomis C, Devine B, Hillis-Starr Z, Phillips B (2002) *Acropora* in the US Virgin Islands: a wake or an awakening? In: Bruckner AW (ed) Proceedings of the Caribbean *Acropora* workshop: Potential application of the US Endangered Species Act as a Conservation Strategy NOAA Tech Memorandum, NMFS-OPR 24:99–122

Rogers C, Muller E, Devine B, Nieves P (2005) Mapping the spatial distribution of diseases affecting elkhorn coral within Virgin Islands National Park: a closer look at an endangered reef-building species. Report to Disney Wildlife Conservation Fund, 6 pp

Rogers C, Muller E, Spitzack T, Devine B, Nieves P, Gladfelter E (2006) A closer look at elkhorn coral (*Acropora palmata*) on two reefs within Virgin Islands National Park: the role of disease, physical breakage, predation, and competition. Report to Disney Wildlife Conservation Fund, 8 pp

Rohmann SO, Hayes JJ, Newhall RC, Monaco ME, RW G (2005) The area of potential shallow-water tropical and subtropical coral ecosystems in the United States. Coral Reefs 24:370–383

Scheibling RE (1979) The ecology of *Oreaster reticulatus* (Echinodermata: Asteroidea) in the Caribbean. Ph.D. thesis, McGill University, Canada, 361 pp

Schroeder R (1965) Something Rich and Strange. Harper & Row, New York 184 pp

Sebens KP, Miles JS (1988) Sweeper tentacles in a gorgonian octocoral: morphological modifications for interference competition. Biol Bull 175:378–387

Shinn EA, Smith GW, Prospero JM, Betzer P, Hayes ML, Garrison VH, Barber RT (2000) African dust and the demise of Caribbean coral reefs. Geo Res Lett 27:3029–3032

Shulman MJ (1983) Species richness and community predictability in coral reef fish faunas. Ecology 65:1308–1311

Shulman MJ (1984) Resource limitation and recruitment patterns in coral reef fish assemblages. J Exp Mar Biol Ecol 74:85–109

Shulman MJ (1985) Recruitment of coral reef fishes: effects of distribution of predators and shelter. Ecology 66:1056–1066

Shulman MJ (1998) What can population genetics tell us about dispersal and biogeographic history of coral reef fishes? Aust J Ecol 23:216–225

Shulman MJ, Bermingham EL (1995) Early life histories, oceanographic currents, and the population genetics of Caribbean reef fishes. Evolution 49:897–910

Shulman MJ, Ogden JC (1987) What controls reef fish populations: recruitment or benthic mortality ? An example in the Caribbean reef fish, *Haemulon flavolineatum*. Mar Ecol Prog Ser 39:233–242

Shulman MJ, Ogden JC, Ebersole JP, McFarland W, Miller SL, Wolf NG (1983) Priority effects in the recruitment of juvenile coral reef fishes. Ecology 64:1508–1513

Shulman MJ, Ogden JC, Ebersole JP, McFarland W, Miller S, Wolf NG (1984) Timing of recruitment and species composition in coral reef fishes. BioScience 34:44–45

—Smith CL, Tyler JC (1972) Space resource sharing in a coral reef fish community. In: Collette BB, Earle SA (eds) Results of the Tektite Program: Ecology of Coral Reef Fishes. Natural History Museum, Los Angeles County. Sci Bull 14:125–170

Smith-Vaniz WF, Jelks HL, Rocha LA (2006) Relevance of cryptic fishes in biodiversity assessments: a case study at Buck Island Reef National Monument, St. Croix. Bull Mar Sci 79:17–48

Spalding MD, Ravilious C, Green EP (2001) World Atlas of Coral Reefs. Prepared at the UNEP World Conservation Monitoring Centre. University of California Press, Berkeley, CA

Steneck RS (1983) Quantifying herbivory on coral reefs: just scratching the surface and still biting off more than we can chew. In: Reaka ML (ed) The ecology of deep and shallow coral reefs. NOAA, pp 103–111

Steneck RS (1989) Herbivory on coral reefs: a synthesis. Proc 6th Int Coral Reef Symp 1:37–49

Steneck RS (1993) Is herbivore loss more damaging than hurricanes? Case studies from two Caribbean reef systems (1978–1988). In: Ginsburg RN (ed) Global aspects of coral reefs - health, hazards, and history. University of Miami, pp 220–226

Stoeckle D, Rytuba J, Foose T (1968) The Mary Creek Reef Complex – St. John, US Virgin Islands, M.S. thesis, Amherst Geological Expedition to St. John

Suchanek TH (1983) Control of seagrass communities and sediment distribution by *Callianassa* (Crustacea, Thalassinidae) bioturbation. J Mar Res 41:218–298

Suchanek TH (1985) Callianassid shrimp burrows: ecological significance as species-specific architecture. Proc 5th Int Coral Reef Cong 5:205–210

Suchanek TH (1989) A Guide to the Identification of the Common Corals of St. Croix. In: Hubbard DK (ed) Terrestrial and Marine Geology of St. Croix, USVI. Special Publication no. 8, West Indies Laboratory, St. Croix, VI, pp 197–213

Suchanek TH, Williams SL, Ogden JC, Hubbard DK, Gill IP (1985) Utilization of shallow-water seagrass detritus by Caribbean deep-sea macrofauna. Deep Sea Res 32:201–214

Sutherland KP, Porter JW, Torres C (2004) Disease and immunity in Caribbean and Indo-Pacific zooxanthellate corals. Mar Ecol Prog Ser 266:273–302

Swearer SE, Caselle J, Lea DW, Warner RR (1999) Larval retention and recruitment in an island population of a coral-reef fish. Nature 402:799–802

Swearer SE, Shima JS, M. E., Hellberg ME, Thorrold SR, Jones GP, Robertson DR, Morgan SG, Selkoe KA, Ruiz GM, Warner RR (2002) Evidence of self-recruitment in demersal marine populations. Bull Mar Sci 70:251–271

Swingle WE, Dammann AE, Yntema J (1970) Survey of the commercial fishery of the Virgin Islands of the United States. Proc 22nd Gulf and Caribbean Fisheries Institute, pp 110–121

Szmant-Froelich A (1983) Functional aspects of nutrient cycling on coral reefs. In: Reaka ML (ed) The Ecology of Deep and Shallow Coral Reefs, Symposia Series Undersea Research, no. I. NOAA Undersea Research Program, Rockville 1:133–139

Targett NM, Targett TE, Vrolijk NH, Ogden JC (1986) Effect of macrophyte secondary metabolites on feeding preferences of the herbivorous parrotfish, *Sparisoma radians*. Mar Biol 92:141–148

Thayer GW, Bjorndal KA, Ogden JC, Williams SL, Zieman JC (1984) Role of larger herbivores in seagrass communities. Part A. Faunal relationships in seagrass and marsh ecosystems. Estuaries 7:351–376

Tobias W (2000) US Virgin Islands/National Marine Fisheries Service inter-jurisdictional fisheries program final progress report, 25 pp

Tobias W (2005) Assessment of conch densities in backreef embayments on the northeast and southeast coast of St. Croix, US Virgin Islands. Report to the Southeast Area Monitoring and Assessment Program-Caribbean. Division of Fish and Wildlife, Department of Planning and Natural Resources, USVI, 31 pp

Tobias WJ, Telemaque E, Davis M (1988) Buck Island fish and shellfish populations. Biosphere Reserve research report no. 26 VIRMC/NPS, 27 pp

Tolimieri N (1995) Effects of microhabitat characteristics on the settlement and recruitment of a coral reef fish at two spatial scales. Oecologia 102:52–63

Tolimieri N (1998a) The relationship among microhabitat characteristics, recruitment, and adult abundance in the stoplight parrotfish, *Sparisoma viride*, at three spatial scales. Bull Mar Sci 62:253–268

Tolimieri N (1998b) Effects of substrata, resident conspecifics and damselfish on the settlement and recruitment of the stoplight parrotfish, *Sparisoma viride*. Env Biol Fish 53:393–404

Tolimieri N, Sale PF, Nemeth RS, Gestring KB (1998) Replenishment of populations of Caribbean reef fishes: are spatial patterns of recruitment consistent through time? J Exp Mar Biol Ecol 230:55–71

US Naval Oceanographic Office (1963) Oceanographic Atlas of the North Atlantic Ocean.

Voss GL (1976) Seashore Life of Florida and the Caribbean. A Guide to the Common Marine Invertebrates of the Atlantic from Bermuda to the West Indies and of the Gulf of Mexico. EA Seemann, Miami, FL, 168 pp

Voss JD, Myers JL, Mills DK, Remily ER, Richardson LL (2007) Black band disease microbial community variation on corals in three regions of the wider Caribbean. Microb Ecol 54:730–739

Warmke G, Abbott RT (1962) Caribbean Seashells. Livingston, Narbeth, PA, 344 pls 346 pp

Warner RR (1988) Traditionality of mating-site preferences in a coral reef fish. Nature 335:719–721

Warner RR (1990) Resource assessment versus tradition in mating-site determination. Am Nat 135:205–217

Warner RR (1997) Evolutionary ecology: how to reconcile pelagic dispersal with local adaptation. Coral Reefs 16S:115–128

Warner RR, Dill LM (2000) Courtship displays and coloration as indicators of safety rather than of male quality: the safety assurance hypothesis. Behav Ecol 11:444–451

Warner RR, Swearer SE (1991) Social control of sex change in the bluehead wrasse, *Thalassoma bifasciatum* (Pisces: Labridae). Biol Bull 181:199–204

Warner RR, Swearer SE, Caselle J (2000) Larval accumulation and retention: implications for the design of marine reserves and essential fish habitat. Bull Mar Sci 66:821–830

Weil E, Smith G, Gil-Agudelo DL (2006) Status and progress in coral reef disease research. Dis Aquat Org 69:1–7

West J, Harvell CD, Walls AM (1993) Morphological plasticity and variation in reproductive traits of a gorgonian coral over a depth cline. Mar Ecol Prog Ser 94:61–69

Whiteman EA, Jennings CA, Nemeth RS (2005) Sex structure and potential female fecundity in a red hind (*Epinephelus guttatus*) spawning aggregation: applying ultrasonic imaging. J Fish Biol 66:983–995

Williams SL (1984) Uptake of sediment ammonium by rhizoids and translocation in a marine green macroalga *Caulerpa cupressoides*. Limnol Oceanogr 29:374–379

Williams SL (1988) *Thalassia testudinum* productivity and grazing by green turtles in a highly disturbed seagrass bed. Mar Biol 98:447–455

Williams SL, Fisher TP (1985) Kinetics of nitrogen-15 labeled ammonium uptake by *Caulerpa cupressoides* (Chlorophyta). J Phycol 21:487–296

Williams SL, Gill IP, Yarish SM (1985) Nitrogen cycling in backreef sediments, St. Croix, USVI. Proc 5th Int Coral Reef Cong 3:389–394

Wolff N (1996) The fish assemblages within four habitats found in the nearshore waters of St. John, USVI : with some insights into the nature of trap fishing. MSc thesis, University of Rhode Island, 207 pp

Wolff N (1998) Spiny lobster evaluation within Virgin Islands National Park (summer of 1996). Report to US Geological Survey St. John, USVI, 16 pp

Wolff T (1967) Danish Expeditions of the Seven Seas. Rhodos International Science and Art, Copenhagen, Denmark, 336 pp

Wolf NG, Bermingham EB, Reaka ML (1983) Relationships between fishes and mobile benthic invertebrates on coral reefs. In: Reaka ML (ed) The Ecology of Deep and Shallow Coral Reefs. Symposia Series Undersea Research, NOAA Undersea Research Program, Rockville, MD 1:69–78

Wood RW, Olsen DA (1983) Application of biological knowledge to the management of the Virgin Islands conch fishery. Proc Gulf and Caribbean Fish Institute 35:112–131

9
Biology and Ecology of Puerto Rican Coral Reefs

David L. Ballantine, Richard S. Appeldoorn, Paul Yoshioka,
Ernesto Weil, Roy Armstrong, Jorge R. Garcia, Ernesto Otero,
Francisco Pagan, Clark Sherman, Edwin A. Hernandez-Delgado,
Andrew Bruckner, and Craig Lilyestrom

9.1 Introduction

Puerto Rico, the easternmost island (18°15′ N and 66°30′ W) of the Greater Antilles, is about 50 km wide and 180 km long on its east/west axis, and has a coastline of 1,384 km including the adjacent islands of Vieques, Culebra, Desecheo, and Mona (Fig. 9.1). Puerto Rico is a "high" island with a central mountain range running east/west with peak elevations greater than 1,000 m. As with most Caribbean islands, the coastal climate of Puerto Rico is usually dictated by oceanic influences. Seasonal rainfall and wind patterns largely follow north/south movements of the Intertropical Tropical Convergence Zone (ITCZ) and the atmospheric high pressure area in the North Central Atlantic (the "Bermuda High"). Tradewinds are strongest in winter/spring and weakest in summer/fall because of the south/north positions of the Bermuda High, respectively. Conversely, rainfall is greatest in summer/fall and least in winter/spring due to the north/south positions of the ITCZ, respectively. Due to the relative location of mountainous terrain, annual rainfall is high (1,550 mm) on the east, north and west coasts. In contrast, the south coast (especially the southwest coast in the vicinity of La Parguera) receives annual rainfalls of less than 800 mm. As a result of these geographic and rainfall patterns, major rivers are found on the west (e.g., Añasco River) and north (e.g., Manati, Plata, and Loiza Rivers) coasts. Rivers on the south coast are few with intermittent and low flows except for heavy rainfall events (López Marrero et al. 2006).

Compared to the other coasts, the north coast of Puerto Rico has a narrow shelf (<2 km wide), no offshore islands, is subject to sustained wave action due to northeast trade winds, and often experiences large swells (>6 m tall) generated by North Atlantic storms. To a lesser degree the south coast is also subject to trade wind induced waves because winds are deflected landward (northward) during the day due to solar heating of the insular air mass. Thus, except for hurricanes and tropical storms, wave action is highest on the north coast, intermediate on the south coast and least on the west and east coasts. Tides in Puerto Rico are diurnal on the south coast and semi-diurnal on the north and west coasts with a mean tidal range less than 30 cm (Kjerfve 1981).

Water currents along the open coasts are driven by winds, tides, and shelf topography. Except for oscillatory tidal effects, coastal currents are generally westward along the north and south coasts and variable on the west and east coasts (Capella and Laboy 2003; Appeldoorn et al. 1994; Hensley et al. 1994; Ojeda 2002).

Puerto Rico was first settled by the Ortoiroid culture at 3000–2000 BC and subsequently populated by Igneri, Arawak, and Taino Indians. Christopher Columbus was the first European to arrive on the island in 1493 and Spanish colonization began in 1508. Puerto Rico remained under Spanish control until nearly the end of the nineteenth century. The island became a territory of the United States at the end of the Spanish–American War in 1898 and assumed local jurisdiction after drafting its own constitution in 1952. Local waters extend 9 nmi

B.M. Riegl and R.E. Dodge (eds.), *Coral Reefs of the USA*,
© Springer Science+Business Media B.V. 2008

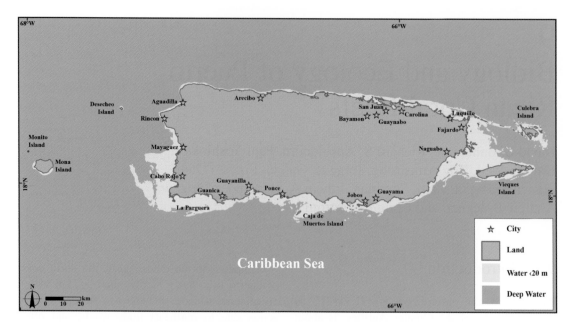

Fig. 9.1. Map of Puerto Rico (Waddell 2005 reproduced by permission of NOAA)

from the coast and thus encompass the whole of the insular platform except for the outer margin of the western shelf, which extends into the US Exclusive Economic Zone.

9.2 Biota of Puerto Rican Reefs

The earliest studies of reefs in Puerto Rico were primarily taxonomic in nature and date to the collections of Gundlach in the 1870s on mollusks (Gundlach 1883), crustaceans (Gundlach 1887), and fishes (Poey 1881) and those of Stahl (1883) on polyps, worms, fishes, and crustaceans. Studies of Puerto Rican algae date to the latter part of the last century. Hauck's (1888) report was the first published account of Puerto Rican algae, reporting 90 species from the island. It was not until 1898, with the Fish Hawk expedition that the first organized scientific study targeting coral reefs occurred, including the first in situ descriptions. For example, the reefs off Mayagüez were reported to consist chiefly of the scleractinian corals *Acropora palmata* and *A. cervicornis* mixed with brain corals *Platygyra viridis* (=*Diploria strigosa*) plus patches of alcyonarian corals *Pterogorgia acerosa* (=*Pseudopterogorgia acerosa*) and *Rhipidogorgia*

flabellum (=*Gorgonia ventalina*). Elkhorn coral was reported to grow from a few centimeters to 1–2 m, but in some areas masses of corals were exposed at low tide. The hydrocoral, *Millepora alcicornis* was also present, and in the interstices of the reef were starfishes, crustaceans, and urchins, the latter dominated by *Diadema antillarum* (Evermann 1900).

At the present time 69 shallow-water (<40 m) scleractinian species have been identified in Puerto Rican waters (Table 9.1). The marine fish fauna includes 260 species (Dennis et al. 2004). Forty-six shallow-water alcyonarian species are additionally known from Puerto Rico (Table 9.2). The benthic marine algal flora of Puerto Rico is the best known among Caribbean islands and consists of approximately 500 species, excluding cyanobacteria (Ballantine and Aponte 2002; Ballantine et al. 2002, 2004; Ballantine and Ruiz 2005; Ballantine and Abbott 2006).

Largely as a consequence of the climatic and geomorphological features described above, coral reefs and colonized hard bottoms account for 15.1% (757 km²) of shallow-water (<18 m) benthic habitats (Kendall et al. 2001). Other prominent habitats include seagrass meadows (12.8%), macroalgal beds (1.9%), and mangrove fringes (1.5%). In general,

TABLE 9.1. List of shallow water (<35 m depth) scleractinian corals known from Puerto Rico (Prepared by E. Weil).

Acroporidae	M. danae
Acropora cervicornis	Dendrogyra cylindrus
A. palmata	Dichocoenia stokesi
A. prolifera (hybrid)	D. stellaris
Agariciidae	Mussidae
Agaricia agaricites	Mycetophyllia ferox
A. purpurea	M. aliciae
A. humilis	M lamarckiana
A. carinata	M. danaana
A. danae	M. reesi
A. tenuifolia	Scolymia cubensis
A. grahamae	S. lacera
A. lamarcki	S. wellsi
A. fragilis	Mussa angulosa
Helioseris cucullata	Isophyllia sinuosa
Leptoseris cailleti	I. multiflora
Astrocoeniidae	Isophyllastrea rigida
Stephanocoenia intersepta	Oculinidae
Caryophylliidae	Oculina diffusa
Eusmilia fastigiata	O. valenciennesi
Caryophyllia smithii*	O. varicosa
Rhizosmilia maculata*	Pocilloporidae
Tubastrea aurea*	Madracis decactis
Faviidae	M. formosa
	M. auretenra
Cladocora arbuscula	M. pharensis luciphilla
Colpophyllia natans	M. pharensis luciphogous
C. amaranthus	M. senaria
Diploria clivosa	Poritidae
D. labyrinthiformis	
D. strigosa	Porites astreoides
Favia fragum	P. divaricata
Manicina areolata	P. furcata
M. mayori	P. porites
Montastraea annularis	Rhizangiidae
M. faveolata	
M. franksi	Astrangia solitaria*
M. cavernosa	Colangia immersa*
Solenastrea bournoni	Phyllangia americana*
Gardineriidae	Siderastreidae
Gardineria minor	Siderastrea radians
Meandrinidae	S. siderea
Meandrina meandrites	

* Azooxanthellate

TABLE 9.2. List of alcyonarian species known from Puerto Rico (Prepared by P. Yoshioka).

Holaxonia	E. succinea
	E. tourneforti
Ellisella barbadensis	Gorgonia mariae
Eunicea asperula	G. ventalina
E. calyculata	Muricea elongata
E. clavigera	M. muricata
E. fusca	M. pinnata
E. knighti	Muriceopsis. flavida
E. laciniata	M. sulphurea
E. laxispica	
E. mammosa	

TABLE 9.2. (continued).

Plexaura flexuosa
P. homomalla
P. kuna
P. nina
Plexaurella dichotoma
P. fusifera
P. grandiflora
P. grisea
P. nutans
Pseudoplexaura crucis
P. porosa
P. flagellosa
P. wagenaari
P. crucis
Pseudopterogorgia acerosa

	P. albatrossae
	P. americana
	P. bipinnata
	P. rigida
	Pterogorgia anceps
	P. citrina
Scleraxonia	
	Briareum asbestinum
	Diodogorgia nodulifera
	Erythropodium caribaeorum
	Icilogorgia schrammi
Telestacea	
	Carijoa reesei
	Iciligorgia schrammi
Telestacea	
	Carijoa riisei

the marine biota of Puerto Rico is similar to other areas in the Caribbean. For instance, numerically dominant hard bottom alcyonarians in Puerto Rico such as *Pseudopterogorgia americana*, *Gorgonia ventalina*, *Plexaura homomalla*, and *Plexaurella dichotoma* occur throughout the Caribbean including Jamaica (Kinzie 1973), Florida (Opresko 1973; Goldberg 1973), Belize (Muzik 1982), and Mexico (Jordan 1989; Jordan-Dahlgren 2002). A similar biogeographic pattern holds for scleractinians (Alcolado et al. 2003; Geraldez 2003; Weil and Locke 2005; Weil 1999, 2006), except for higher biodiversity in southern regions (e.g., Venezuela) (Weil 2003; Guzmán 2003). On smaller spatial scales, distributional patterns of the coral reef biota depend on the complex interplay among a variety of habitat-specific features such as wave action, light levels, reef rugosity, pollution, etc. It is largely this general aspect of coral reef biology that current ecological research in Puerto Rico is focused.

9.3 North Coast Reefs

Descriptions of hard bottom communities on the north coast include McKenzie and Benton (1972), Goenaga and Cintrón (1979), Goenaga et al. (1985), Goenaga (1988, 1989), Goenaga and Boulon (1992), Hernández-Delgado (1992, 2000, 2005), García et al. (2003), and Weil et al. (2003). Compared to other areas in Puerto Rico, north coast reefs are dominated by macroalgae, with low abundances of scleractinian corals (1–5% cover) such as

Porites astreoides, *Dichocoenia stokesii*, *Diploria strigosa*, *D. clivosa*, *Meandrina meandrites*, *Oculina diffusa*, and *Cladocora arbuscula*. Large oceanic swells are undoubtedly a major causal factor. Alcyonarians favoring hardbottom high energy areas, such as *Muriceopsis sulphurea*, *Gorgonia mariae*, *Pterogorgia* spp., and *Erythropodium caribaeorum*, as well as the hydrocoral *Millepora squarrosa* are common in this region but essentially absent on other coasts. In addition, Kaye (1959) attributes the absence of coral reefs to the lack of platforms raised above the bottom indicating that smothering of sessile fauna by wave-related sediment transport (bedload) is a contributing factor.

The most notable exceptions to poor coral reef development on the north coast are reefs off Luquillo and other areas on the northeast coast (Hernández-Delgado 2005). A wider shelf (~5km) and the consequent reduction of ocean swells probably play a role in this pattern. Reefs off the northeast coast may extend from 0.3 to 1.3km offshore, with well-defined back reef and reef front zones. Many of these reefs show evidence of substantial scleractinian populations of *Montastraea* spp., *Diploria* spp., *Siderastrea siderea*, *Acropora palmata*, and *Porites porites* in the recent past (Hernández-Delgado 2000; Weil et al. 2003). Hernández-Delgado (2005) attributes the recent decline in some instances to increased water turbidity and frequent sediment pulses from river runoff. At the present time these reefs are dominated by macroalgae and other non-reef building taxa (Hernández-Delgado 2000).

9.4 East Coast Reefs

Coral reefs are often well-developed in this region (Goenaga and Boulon 1992; Hernández-Delgado 2000; Hernández-Delgado et al. 2000; Hernández-Delgado and Rosado-Matías 2003). Other descriptions of coral reefs on the east coast include Almy and Carrión Torres (1963), McKenzie and Benton (1972), Goenaga and Vicente (1990) and García et al. (2003).

Fringing reefs along the mainland of Puerto Rico are the most common reef type in this region. Many are discontinuous and have grown on rocky outcrops from soft bottom (Goenaga and Cintrón 1979; Hernández-Delgado 2005). In shallow areas, zonation patterns are often poorly defined but dominated by zoanthids: *Zoanthus pulchellus*, *Z. sociatus*, and *Palythoa caribbaeorum*. Scleractinian coral taxa that are relatively rapid colonizers (e.g., *Porites astreoides*, *Siderastrea radians*) are also abundant, as well as alcyonarians such as *Erythropodium caribaeorum*, *Pseudopterogorgia* spp., and *Gorgonia* spp. on the reef front. Percent cover of scleractinian corals rarely exceeds 5%, although extensive stands of *Acropora palmata* occur occasionally (Hernández-Delgado et al. 1996).

Fringing reefs on offshore islands have variable (often high) scleractinian coral cover (Hernández-Delgado 2000). The reef flat and crest in the windward zones of these reefs are typically 0.3–1.0 m deep with low scleractinian and hydrocoral cover (<10%), principally *Porites porites*, *P. astreoides*, *Siderastrea radians*, *Agaricia humilis*, and *Millepora* spp., as well as sporadic aggregations of small *Acropora palmata*. The deeper (3–5 m) windward zone has structures resembling spur and groove formations with live cover between 5% and 20% dominated by *Diploria clivosa* and *D. strigosa* with abundant dead stands of *A. palmata*. At greater depths, live scleractinian cover varies between 30% and 80% dominated by *Montastraea annularis*, *M. faveolata*, *Colpophylla natans*, *S. siderea*, *D. clivosa*, *D. strigosa*, and *Agaricia* spp. Leeward areas are well-developed with high scleractinian diversity and cover, the latter reaching 64% at some sites (Hernández-Delgado 2003, 2005) dominated by *M. annularis*, *M. faveolata*, *Diploria* spp., and *S. siderea* (Hernández-Delgado et al. 2000). Leeward reefs in some areas drop abruptly to 3–6 m and often form overhangs and crevices that support abundant reef fish communities.

Coral assemblages also occur on shallow (<3 m deep) basalt outcrops exposed to wave action. These outcrops often have steep slopes with abundant cracks, caverns, and pinnacles. Scleractinian cover is patchy (1–35%) dominated by *Diploria* spp. and *Dendrogyra cylindrus*. High abundances of non-reef building taxa such as alcyonarians and zoanthids are also present. Coral reef communities on basaltic outcrops are common in several locales. Invertebrate cover in shallow areas (<6 m) of these formations is often low (1–20%) and dominated by hydrocorals, *Millepora* spp., zoanthids, *Palythoa caribaeorum*, and alcyonarians. Scleractinian cover is low but basaltic outcrops in Culebra often have high abundances of coral recruits, mostly *Porites astreoides*, *P. porites*, *Diploria* spp., *Siderastraea radians*, and *Favia fragum*. Deeper areas (10–20 m) of basaltic outcrops with steep slopes subject to strong water circulation have low coral cover (<5%), but colonies of the deep water sea fan, *Icilogorgia schrammi*, and antipatharians, *Antipathes* spp., occur occasionally. At 25–40 m, antipatharians, *Antipathes* spp., and *Stichopathes* spp., as well as the alcyonarian, *Ellisella barbadensis*, become more abundant. Shelf edge reefs, occurring at depths of 20–40 m, are also extensive on the east coast. These reefs often possess spur and groove formations and exhibit high scleractinian species richness and cover, the latter approaching 85–100% (Torres 1975).

Coral reefs on the east coast have declined significantly in recent years (Fig. 9.2; Hernández-Delgado 1992, 2005; Hernández-Delgado and Alicea Rodríguez 1993; Hernández-Delgado and Sabat 2000). While military activities have been reported to have degraded coral reefs at Vieques and Culebra (Carrera-Rodríguez 1978; Porter 2000), this issue remains controversial (see also Hernandez-Cruz et al. 2006; Riegl et al. 2008). Although some activities only have short-term effects (Dodge 1981; Antonius and Weiner 1982; Deslarzes et al. 2006; Evans et al. 2006), bomb craters, reef framework damage, etc. have long-term impacts (Rogers et al. 1978; Macintyre et al. 1983). For instance, there has been only limited recovery of localized bomb-cratered reefs after more than three decades (Hernández-Delgado 2005). Hurricane David in 1979 also caused severe damage to windward reefs especially to extensive stands of *Acropora palmata* that have failed to recover after nearly 3 decades. Other threats to the coral reefs are discussed in greater detail below.

9.5   South and West Coast Reefs

Compared to the north coast, the south and west coasts of Puerto Rico have a relatively broad insular platform (up to 16 km wide on the south coast and 26 km on the west coast) that supports extensive sea grass, coral reef, sand, and algal communities. Scattered mangrove forests and coralline keys line the coast in protected areas and the platform is often fringed by well-developed bank reefs. Detailed descriptions of the reefs here are given below in the section on La Parguera.

9.6   Mona Island

Mona Island is an uninhabited carbonate island (Gonzalez et al. 1997) surrounded by deep water (370–1,160 m depth). It assumes particular

importance due to its isolated location. Oceanic influences are prominent because Mona is small (5,486 ha) and located relatively far from both Puerto Rico (73 km) and Hispaniola (65 km). Due to the absence of rivers and anthropogenic impacts, water clarity is high and land-based sources of pollution are virtually non-existent. Because of circulation patterns and distance from the main islands, the Mona Channel acts as a partial zoogeographic barrier to coastal marine species (Rojas-Ortega and García Sais 2002; Taylor and Hellberg 2003), particularly for those organisms that spawn during warmer months (Baums et al. 2006), while for other species Mona may act as an important biogeographic "stepping stone" for connectivity across the channel (Ojeda 2002).

The north shore of Mona descends to depths of 27–30 m as a vertical wall that is mostly colonized by encrusting sponges, alcyonarians, hydrozoans, and isolated scleractinians. The west, south, and east shores are surrounded by a narrow insular platform that support coral reef and limited seagrass communities (Cintron et al. 1975; Canals et al. 1983).

The west shore has emergent reef crests with narrow (<50 m wide) and shallow (<3 m deep)

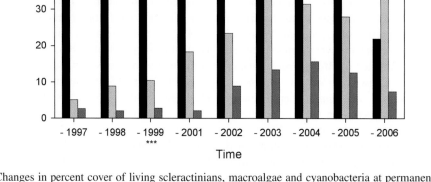

FIG. 9.2. Changes in percent cover of living scleractinians, macroalgae and cyanobacteria at permanent monitoring station CR1 within the Luis Peña Channel, a no-take Natural Reserve, Culebra Island. Asterisks indicate initiation of no-take policy (E.A. Hernández-Delgado and A.M. Sabat, unpublished data 2007)

back reef zones. The back reef is primarily sand and coral rubble with low coral cover (<5%), except for large thickets (>10 m diameter) of *Acropora palmata*. Large colonies of *Diploria strigosa* and *D. clivosa* (1–2 m diameter), small stands of *Montastraea annularis* and mounds of *Porites porites* also occur here. The fore reef includes a shallow (<8 m deep), high relief spur and groove system with numerous caves mostly colonized by clionid sponges and *Palythoa caribbaeorum* with low (<5%) live coral cover; principally *Porites astreoides*, *M. annularis*, and *A. palmata*. The deeper (5–15 m) fore reef slopes gently and is dominated by large colonies (2–3 m diameter) of *M. faveolata* and *M. annularis*, and other massive and plating corals, with live cover of 30–50% during the 1990s. At 15–25 m depth is a primarily hardground system (<5% coral cover) colonized mostly by large sponges (e.g., *Xestospongia*) and macroalgae. However, several extensive stands of *A. cervicornis* (20–100 m in length) occur at depths of 20–30 m.

The southern platform slopes gently to depths of 20–25 m before dropping precipitously into deeper water. The nearshore zone is primarily hardground with gorgonians and small (<50 cm diameter) isolated coral colonies. At depths of 12–25 m, corals are more abundant in the western sector while alcyonarians and sponges are abundant in the east. The deep terrace (>15 m) is characterized by numerous mounds, ridges, and pinnacles with high scleractinian cover (particularly large *M. faveolata*).

The eastern shore has an emergent reef crest located several hundred meters offshore. The shallow (<3 m) back reef lagoon has extensive sandy areas with small seagrass, *Thalassia testudinum*, patches. Isolated corals (primarily *Diploria strigosa*, *Acropora palmata*, and *Porites porites*) are found immediately behind the reef crest. The shallow fore reef is dominated by *A. palmata* and large *D. strigosa* colonies and low abundances of other corals. A series of low relief, hardground spurs with narrow sand channels extend seaward to depths ~15 m. Shelf edge reefs at depths of 15–30 m are the best-developed coral reefs at Mona with the typical high relief spur and groove formation. Scleractinian species richness and cover (30–80%) are high in these reefs.

Significant changes have occurred to Monas reefs since 1995, including decreased coral cover from disease, bleaching, hurricanes, and other factors, in addition to declines in reef fish abundances (Bruckner and Bruckner 2001, 2006a). These include extensive mortalities to large *Acropora palmata* colonies (3–6 m diameter) since 2001, primarily due to white band disease, and a widespread disease-related decline to the *Montastraea* complex, post-bleaching mortality, and overgrowth by clionid sponges, macroalgae, and cyanobacteria (Bruckner and Bruckner 2006a, b).

9.7 Deep Reefs

Deep reefs occur at depths greater than 30 m along the insular slope of Puerto Rico and associated islands. Initial ecological observations of deep-water hermatypic corals and other taxa in Puerto Rico were based on submersible observations (Nelson and Appeldoorn 1985), primarily off La Parguera and Guánica on the south coast. Thirty-one fish species were recorded at Guánica (Nelson and Appeldoorn 1985) and a total of 70 fish species elsewhere in Puerto Rico and the Virgin Islands on the bases of these dives. Of the major predators observed, only one (*Lutjanus buccanella*) was not a species also characteristic of shallow reefs.

Subsequent observations at various locales are reported by Armstrong et al. (2002, 2006), Singh et al. (2004), and Nemeth et al. (2004). In general, scleractinian cover decreases with depth in deep reefs. For instance, the maximum scleractinian cover (10.7%) occurs at depths of 25–35 m at the shelf edge in La Parguera and consists of a scleractinian–sponge–alcyonarian hardground. The maximum depth of hermatypic scleractinians is between 70 and 100 m (C. Arneson, personal communication 1996). Similarly, abundances of alcyonarians peak at depths of 25–35 m and noticeably decrease after 65 m. However, peak abundances of antipatharians (*Antipathes* spp.) occurred at 54 m (Nelson and Appeldoorn 1985; Singh et al. 2004). Based on the Guánica observations, sponges are the major structural component of the deep reef fauna to depths of 100 m. However, at both Guánica and La Parguera hard ground areas, algae accounted for >40% of the benthic cover at depths less than 100 m. Based on collections from 50 to 60 m in southwest Puerto Rico, the algal community is particularly diverse (Ballantine and Wynne 1986, 1987; Wynne and Ballantine 1986; D.L. Ballantine, personal communication 2007).

García-Sais et al. (2004, 2005, 2006) described deep reefs off the south coast of Vieques and Desecheo Islands (Figs. 9.3 and 9.4) and the Bajo de Sico seamount in the Mona Passage. The reef off Vieques occurs in waters of exceptional transparency and is dominated by turf algae (57% cover). However, scleractinian cover (29%) and diversity (25 species) are high, consisting mainly of *Porites astreoides, Agaricia grahamae, Montastraea cavernosa* and particularly *M. franksi* (22% cover). Two antipatharians including *Antipathes caribbeana* and one hydrocoral were also noted, as were the fleshy brown alga *Lobophora variegata* and calcareous algae (including *Halimeda* spp.). A total of 54 reef fish species dominated by *Coryphopterus personatus* (71%), *Clepticus parrae* (17%) and *Chromis cyanea* (7%) were identified with a mean fish abundance of 18.3 ind/m^2.

Drop-off walls are found in the deep reefs off Desecheo Island and the Bajo de Sico seamount. The wall at Desecheo is between 30 and 40 m deep and dominated by macroalgae (mostly *Lobophora*

variegata), sand, sponges, and massive corals. Twenty-five scleractinian species are present, dominated by *Montastraea cavernosa* and the *M. annularis* complex, as well as three hydrocoral and two antipatharian species. Erect and branching sponges are prominent (17.3% cover) and provide substantial structural habitat for fishes and invertebrates. In many instances, sponges grow attached to stony corals, forming sponge–coral bioherms of considerable size. One of the most common associations consists of brown tube (*Agelas conifera* and *A. sceptrum*) and row pore (*Aplysina* spp.) sponges with *M. cavernosa* and *M. annularis*. Seventy fish species were identified, numerically dominated by planktivorous taxa, suggesting that planktonic food webs are critical to deep reefs. The observation that drop-off walls are also important habitats for many predatory fishes, such as groupers (*Epinephelus striatus, E. guttatus, Mycteroperca venenosa*) and snappers (*Lutjanus cyanopterus*), as well as smaller reef fish species including *Chromis cyanea, Gramma loreto, Centropyge argi, Chaetodon* spp. and *Opistognathus* spp. further supports the idea that productivity in these areas is supported by nutrient inputs from other areas.

Fɪɢ. 9.3. Coral dominated deep reef (Black Jack Reef), Vieques Island, 34 m (Photo reprinted from García-Sais et al. 2004)

FIG. 9.4. Coral dominated reef, Desecheo Island, 50 m (Photo reprinted from García-Sais et al. 2005)

At still greater depths (>40 m) rhodolith reefs occur on gently sloping terraces with very low topographic relief on Desecheo and Bajo de Sico. The rhodolith reef at Desecheo appears to have developed on rhodoliths that have rolled down the terrace slope from shallower areas and became stabilized in the absence of water motion. The crustose algal formation is colonized by brown algae (*Lobophora variegata*), erect and branching sponges (*Agelas* spp.), and scleractinian lettuce corals (*Agaricia* spp.). The biota cover over the rhodoliths is high (>95%) and consists of 18 scleractinian, two hydrozoan (*Millepora alcicornis* and *Stylaster roseus*), and an antipatharian (*Stichopathes lutkeni*) species. Mean scleractinian coral cover was 13%, dominated by *A. lamarcki* and *A. grahamae* (9% cover) at depths of 45–53 m.

9.8 Coral Reef Fisheries

The exploitation of Puerto Rico's reefs dates to pre-Columbian times. While commercial harvesting occurred during the Spanish colonial period, overall production was low due to a combination of limited technology for preservation of fish and low demand due to large imports of salted cod. By 1900, fishing

was well developed using hook and line, traps and a variety of nets. West coast boats had live-wells and fished as far as Mona Island. Fishing not only targeted shallow waters (snappers, grunts, Nassau grouper) but also the insular slope, with silk snapper (*Lutjanus vivanus*) being one of the most abundant species (Wilcox 1900). However, it was not until the 1960s, with the increasing development of Puerto Rico's tourist industry, that commercial fishing grew to significant proportions. Sustained increases in landings occurred though the 1970s, with the latter portion driven by the large-scale targeting of the deepwater snapper–grouper complex. Landings of demersal reef fishes peaked in 1979 at 2,400 mt then rapidly declined to 394 mt in 1988 (Appeldoorn and Meyers 1993). During the same period, landings of spiny lobster (*Panulirus argus*) showed a concomitant decline (Bohnsack et al. 1991). All assessments of the Puerto Rican fishery over the past 30 years have reported overfishing to various degrees (Stevenson 1978; Colin 1982; Appeldoorn and Lindeman 1985; Dennis 1988; Acosta and Appeldoorn 1992; Appeldoorn et al. 1992; Bohnsack et al. 1991; Appeldoorn and Posada 1992). The fishery has not changed substantially over the past century, being still artisanal in nature, using small boats, traps, nets, and hook and line. Major innovations have included replacement of sails with outboard motors, use of rebar and vinyl-coated wire mesh in traps, monofilament for lines and nets, and the introduction of power reels and SCUBA.

Extensive fishing has lead to substantial changes in species composition, both in landings and in community structure on reefs. All large predatory fishes have been greatly reduced in abundance. For example, the Nassau grouper, once among the most important species (Wilcox 1900) even up to 1970 (Suárez-Caabro 1970), was reduced to effective commercial extinction by the mid 1980s (Sadovy 1999). No studies exist, however, on the impact of species removals on coral communities in Puerto Rico. Nevertheless, the vast loss of predatory and herbivorous fishes, along with key macro-invertebrates (including the die-off of the long-spined sea urchin, *Diadema antillarum* (Lessios et al. 1984), can be expected to have significant impact, as observed elsewhere in the Caribbean (McClanahan and Muthiga 1998; Hughes 1994; Jackson et al. 2001). For example, spiny lobsters (*Panularis argus*) are key predators that probably play an important role in controlling gastropods that feed on corals (e.g., *Coralliophila abbreviata*). Once so abundant on reefs that they were fished by hand while walking on reef flats at night, lobsters are now uncommon in shallow reef areas.

9.9   Threats to the Coral Reefs of Puerto Rico

Puerto Rico has a population of nearly four million people and one of the highest population densities (475 ind/km^2) in the Caribbean (López Marrero et al. 2006). Of the neighboring islands, only Vieques and Culebra are permanently inhabited but with significantly lower population densities. Also, over four million tourists visit the island annually increasing the pressure on coastal ecosystems. Human activities undoubtedly have adverse effects on coral reefs in many ways. For instance, the loss of populations of keystone species (Gladfelter 1982; Lessios et al. 1984) has caused a cascade of changes to coral reef communities (Hughes 1994; Lessios 1998; Aronson and Precht 2001; Weil et al. 2002).

9.9.1   Disease and Bleaching

Although comprising only approximately 8% of the coral reef area of the world (Spalding and Greenfeld 1997), over 65% of all diseases/syndromes are reported from the Caribbean (Green and Bruckner 2000). Bleaching (Fig. 9.5g) and disease (Figs. 9.5 and 9.6) have been implicated in the deterioration of Caribbean coral reefs over the past 2 decades (Hughes 1994; Smith et al. 1996; Peters 1997; Hoegh-Guldberg 1999; Glynn et al. 2001; Weil et al. 2002; Richardson and Aronson 2000; Miller et al. 2003; McClanahan 2004; Rosenberg 2004; Weil 2004; Ward et al. 2006). The Caribbean is small, enclosed, highly interconnected, and surrounded by dense human populations. Interactions with anthropogenic impacts, global warming and fast moving water currents may have facilitated the emergence, spread, virulence, frequency, and host range of pathogens in the Caribbean (Epstein et al. 1998; Goreau et al. 1998; Richardson 1998; Hayes and Goreau 1998; Harvell et al. 1999, 2002; Richardson and Aronson 2000; Smith and Weil 2004; Sutherland et al. 2004; Weil et al. 2006). Thus, the Caribbean has been dubbed a "disease hot spot".

White plague-II (WP-II) and Caribbean yellow band disease (YBD) are the most widespread and damaging diseases/syndromes affecting reef-building scleractinian coral in Puerto Rico. Other diseases affecting corals include white band (WBD), black band (BBD), white pox (WPX), red band (RBD), Caribbean ciliate infection (CCI), dark spots disease (DSD), and tumors (TUM). The fungal infection, aspergillosis (ASP), still has a high prevalence on populations of the alcyonarian

FIG. 9.5. Common scleractinian/alcyonarian diseases in Puerto Rico and the Caribbean for which the causative agent is known (**a**) black band disease in *Diploria strigosa*, (**b**) white plague-II in *D. labyrinthiformis* (**c**) white band disease in *Acropora palmata,* (**d**) hyperplasia in *D. strigosa*, (**e**) white Pox in *A. palmata*, (**f**) aspergillosis in *Gorgonia ventalina* and (**g**) a totally bleached colony of *Montastraea faveolata* (From E. Weil, unpublished data 2007)

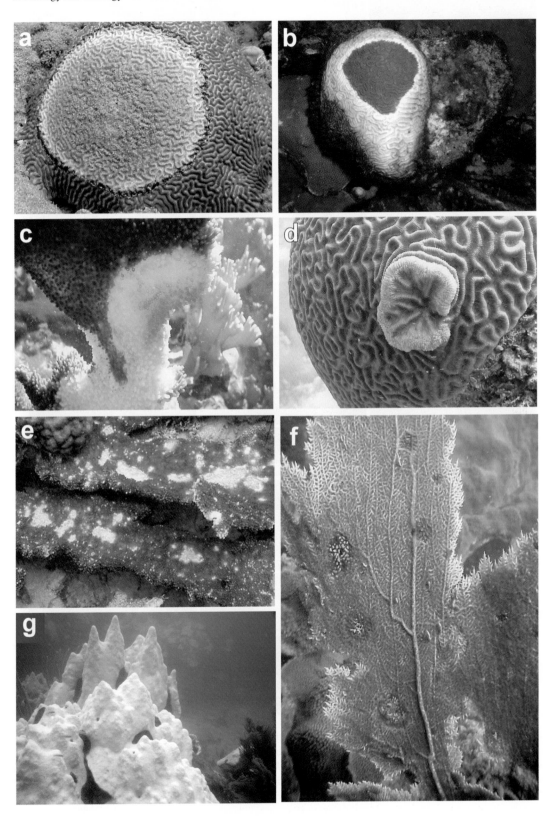

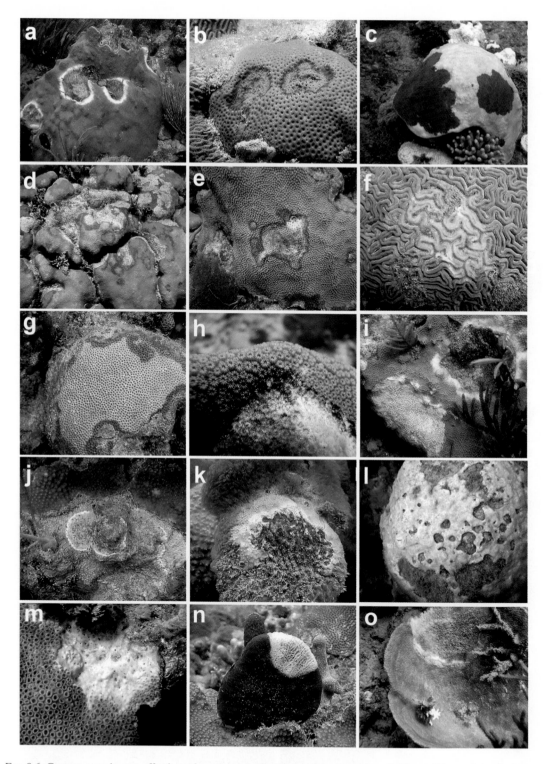

Fɪɢ. 9.6. Common syndromes affecting scleractinians and other coral reef organisms in Puerto Rico and the Caribbean. (**a**) yellow band disease, (**b** and **c**) dark spots disease, (**d** and **e**) dark bands, (**f**) tissue "necrosis", (**g**) pigmentation response, (**h**) dark line of ciliates (*Halofoliculina* sp.) on *Montastraea,* (**i**) Coral colony with multiple diseases, (**j**) Coralline white band in crustose red algae (*Neogoniolithon accretum*) (**k** and **l**) decomposing-necrotic tissue in crustose octocorals and (**m**) zoanthids, and (**n** and **o**) diseased sponges overgrowth (From E. Weil, unpublished data 2007)

Gorgonia ventalina (Weil et al. 2002, 2006). Populations of *Acropora palmata, A. cervicornis* and the hybrid *A. prolifera* have declined significantly probably due to WBD (Gladfelter 1982) and other diseases (Figs. 9.5 and 9.6; see also Weil and Ruiz 2003), as well as hurricanes, bleaching, and the deleterious effects of anthropogenic activities (Weil et al. 2003).

Other syndromes, including "tissue necrosis" (TNS), dark bands (DBS) and white blotches (WB) have become more prevalent and widespread, while additional syndromes have been found that produce tissue and/or colony mortalities not only in corals and alcyonarians, but also in other coral reef taxa including hydrocorals, sponges, zoanthids, and calcareous crustose algae (Fig. 9.6) (Weil 2004; Ballantine et al. 2005; Weil et al. 2006). In addition, global warming may play a contributing role because disease outbreaks, which generally occur during warmer periods, have been increasing in incidence. Pathogens have been identified for only seven Caribbean coral diseases, WBD, WP-II, WPX, BBD, TUM, RBD, and CCI, and one alcyonariandisease, ASP. Koch's postulates (a set of four criteria established by Koch to determine the causal agent of a disease) have only

been fulfilled for four of these (WP-II, WBD, WPX, and ASP).

Beginning in 1987, widespread bleaching (a non-infectious disease) has occurred in the Caribbean at various intervals. In general, bleaching has resulted in low colony mortalities among scleractinians and variable effects among other taxa. For instance, bleaching in 2003 affected up to 25% of corals in inshore reefs but caused low (less than 1.0%) scleractinian colony mortalities. At this time many alcyonarians, zoanthids, hydrocorals, anemones, and sponges completely bleached for the first time and caused up to 90% colony mortalities in the hydrocoral *Millepora*. In contrast, bleaching in 2005 was the most intense ever recorded in Puerto Rico, with significant colony mortalities of several scleractinian genera (including *Agaricia, Montastraea, Colpophyllia, Acropora, Mycetophyllia*, etc.) and for the first time, the alcyonarian *Briareum asbestinum* (E. Weil, personal communication 2007). This bleaching event and post-bleaching mass mortality followed record-breaking sea surface temperatures up to 31.8°C at 30m depths and 33.1°C at reef crests (Fig. 9.7; E. Weil, personal communication 2007; E. Hernández-Delgado, personal communication 2007). Strong water motion

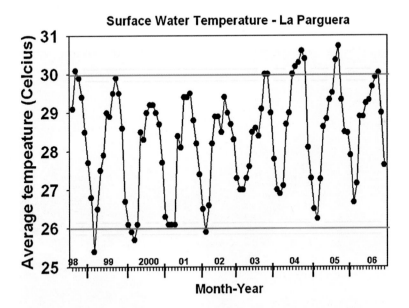

FIG. 9.7. Average surface water temperature at La Parguera. Puerto Rico from July 1998 to November 2006. Red lines mark the 26°C "normal" minimum average temperature and the 30°C upper tolerance level for most zooxanthellate corals before bleaching. Extensive bleaching occurred in 1998, 2003 and 2005 (From E. Weil, unpublished data 2007)

may ameliorate the impact of temperature (Nakamura and Van Woesik 2001; Nakamura et al. 2003, 2005) as bleaching (and mortality) was observed to be relatively low (20–60%) in areas experiencing moderate/strong water circulation and high in leeward (80–97%) and protected (60–80%) areas (Hernández-Delgado et al. 2006).

The 2005 bleaching event was compounded by intensive outbreaks of WP-II and YBS mainly affecting *Montastraea, Diploria,* and *Colpophyllia* species. The combination of bleaching and disease produced significant losses of live scleractinian cover. For example, *M. faveolata* colonies lost 50–60% of live tissue at intermediate and deep habitats at Turrumote Reef, La Parguera. Also, the prevalence of YBS in *Montastraea* species has since increased, and 40% of colonies are now infected by this syndrome. Furthermore, this syndrome has become permanent in colonies, and no longer disappears during cooler months. This change in temporal pattern could be related to the warming trend registered in sea surface temperatures in the area (Fig. 9.7) during the last 10 years, supporting the hypothesis that the emergence of new infectious diseases and epizootic events could be related to global warming (Harvell et al. 1999, 2002;

Hayes et al. 2001; Weil et al. 2002; Rosenberg and Ben Haim 2002; Willis et al. 2004).

The 2005 bleaching event had similar effects in Culebra. A total of 82 cnidarian species, 52 of them scleractinian, were bleached (E. Hernández-Delgado et al., personal communication 2007). The extent of bleaching varied among taxa. For instance, bleaching affected all colonies in 19 scleractinian species, and <50% of the colonies for ten species. Bleaching was followed by a white plague like outbreak and the latter or the synergistic effects of both resulted in a 20–60% decline in living scleractinian cover. Partial mortality affected nearly all colonies of *Montastraea annularis, M. faveolata, M. franksi,* and *Acropora cervicornis* with many *Montastraea* colonies losing between 50% and 95% of live tissue (E. Hernández-Delgado et al., personal communication 2007). The overall result of the 2005 bleaching event and disease was a 67% decline in live coral cover in the Natural Reserve at Culebra. In contrast, cyanobacterial and macroalgal cover increased two- to fivefold during this time indicating possible onset of a phase shift in the reef community (See Fig. 9.8).

The combined effect of the 2005 bleaching event and subsequent epizootics, led to a failure of sexual reproduction in *Acropora* and *Montastraea* species in 2006. Very few colonies

Fɪɢ. 9.8. Cyanobacteria, *Schizothrix* sp., dominating substratum between scleractinians, shelf edge off La Parguera, 18 m (Photo by H. Ruiz)

spawned, producing minimal numbers of egg bundles, which were not viable (E. Weil, personal communication 2007; E. Hernández-Delgado, personal communication 2007).

9.9.2   Water Quality

Tropical coastal marine ecosystems are increasingly facing water quality deterioration due to rapid urban and industrial development and poor land-use practices (NOAA 2000; GESAMP 2001; Burke and Maidens 2004; García-Sais et al. 2005). Suspended sediments and eutrophication are main factors responsible for the high turbidity levels observed in Puerto Rico 's coastal waters (Acevedo and Morelock 1988; Burke and Maidens 2004; García-Sais et al. 2005). Increases in turbidity reduce light penetration and limit its availability for photosynthesis by zooxanthellate corals and algae (Souter and Linden 2000) and consequently, for their normal growth and survival (Kinzie et al. 1984).

The important role that turbidity has in affecting reef communities was demonstrated by Bejarano Rodríguez (2006), who studied the effects of turbidity on coral and fish communities along the southwest coast of Puerto Rico and Mona Island, using the extinction coefficient of light ($K_{d\,PAR}$) as a measure of turbidity. This was a strong predictor of percent live coral cover, but the nature of this relationship depended on additional factors, such that patterns for hard bottom areas and reefs at Mona

Island did not fit the relationship defined by other sites along the southwest shelf of Puerto Rico (Fig. 9.9). She also found that fish biomass, abundance, and species richness were all reduced as turbidity increased. The value of K_d was closely correlated to distance from shore

9.9.3   Algal Epizoism

Algal blooms on coral reefs are probably attributable to a combination of decreased herbivore pressure and nutrient enrichment (Valiela et al. 1997). The associated overgrowth of sessile invertebrates by macroscopic algae has been recognized as a threat to coral reefs for several years (Morand and Briand 1996; Lapointe 1997). The extreme manifestation of this, a phase shift from coral to macroalgal assemblages (Done 1992; Hughes 1994; McCook 1999), has become increasingly common in coral reefs (Smith et al. 2001; McCook 1999; Ferrier-Pages et al. 2000; Lirman 2001). In recent years blooms of filamentous cyanobacteria have been observed on coral reefs (Golubic et al. 1999). On the southwest coast of Puerto Rico, cyanobacterial blooms, principally *Schizothrix mexicana* (Fig. 9.8) have been noted since 2004 at shelf edge sites that were previously regarded as relatively undisturbed. These cyanobacterial blooms were often extensive (>40% cover) and resulted in overgrowth on several corals and other substrata.

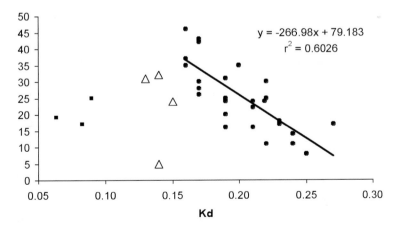

FIG. 9.9. Relationship between the extinction coefficient of light and percent live coral cover for reefs located at 10 m depth. Solid squares are reefs at Mona Island, open triangles are hardbottom sites, solid circles are for reefs along the southwest coast of Puerto Rico, for which the linear regression is shown (Adapted from Bejarano Rodríguez 2006)

Another example of algal epizoism involves the red alga, *Metapeyssonnelia corallepida*. This species, which has recently been reported in Florida and the broader Caribbean including Puerto Rico, overgrows and kills hydrocorals, *Millepora* spp., as well as the scleractinian genus *Porites* (Antonius 1999; Antonius and Ballesteros 1998; Ballantine et al. 2004). *Metapeyssonnelia* can form extensive carpets and has been measured to overgrow *Millepora* tissue at a rate of 0.012 mm/day in shallow reef front environments at La Parguera (Ballantine et al. 2004; K. Raimundi, personal communication 2006). Moreover, because *Millepora* has replaced *Acropora palmata* as the dominant reef crest coral species in many reefs in southwest Puerto Rico, the high abundances of *Metapeyssonnelia* may have significant ecological effects on the reef community related to loss of three-dimensional habitat.

9.9.4 Ship Groundings

At least ten groundings of large ships (>100 ft) have occurred in Puerto Rico and associated islands since 1985 (See Table 9.3). Damage from groundings can be devastating to coral reefs by (1) fracturing the reef framework, (2) crushing, pulverizing, or fragmenting coral colonies, (3) displacing coral colonies, (4) forming berms through vessel movement, and (5) creating blowholes, berms and exposure of the reef framework by propeller wash. Recovery efforts by tugboats can also damage reefs by (6) prop wash and (7) abrasion of the seafloor by towlines. Finally, transfer of toxic antifouling paint from the ship bottom to the sea floor can hinder recolonization of impacted areas.

The *Morris Berman* grounding off San Juan probably caused the greatest damage to a Puerto Rican reef, reducing corals to rubble and discharging over 800,000 gal of fuel oil into nearshore waters (Marine Resources Inc. 2006). The *Fortuna Reefer* that grounded on an *Acropora palmata* reef at Mona Island is notable because of damage caused by improper removal operations. Tug boat tow lines used to remove this ship scraped across the bottom, causing far greater damage to coral colonies than the grounding itself. Similarly, removal of the *Margara* off Guayanilla caused significant damage to the coral reef from the drifting and pivoting of the vessel for a considerable distance along the reef.

Relatively few groundings have involved significant restoration efforts. The largest restoration effort undertaken followed the *Magara* grounding, where over 9,000 coral colonies including ~1,000 *Acropora cervicornis* colonies were reattached. Smaller scale restoration efforts have included the *Fortuna Reefer* (>1,857 *A. palmata* fragments reattached) and the *Isla Grande Ferry* (>100 coral fragments reattached). Results of restoration efforts have been mixed. Unfortunately, only ~190 (10%) of the originally restored *A. palmata* fragments at the *Fortuna Reefer* grounding site survived for 9 years. Of these only 121 colonies had live tissue covering most of their original upper branch surfaces (Bruckner and Bruckner 2006b, c). However, greater than 75% of the latter colonies had successfully reattached to the substratum and had multiple new branches 40–80 cm in length. Major sources of fragment mortality included loss due to disease, breakage of wire used in reattachment, overgrowth by the sponge *Cliona*, and predation by the gastropod *Coralliophila abbreviata*. Although fragment mortality was high, *A. palmata* populations outside of the grounding site experienced catastrophic

TABLE 9.3. Major vessel groundings in Puerto Rico since 1985.

Vessel	Location	Date
M/V Regina	Mona Island	Feb. 15, 1985
Barge Morris J. Berman	Escambrón, San Juan	Jan. 7, 1994
M/V Fortuna Reefer	Mona Island	July 23, 1997
M/V Kapitan Egorov	Guayanilla Bay	Jun. 20, 1998
M/V Author	Ponce	Sept. 8, 1998
M/V Sergo Zakariadze	El Morro, San Juan	Nov. 18, 1999
Ferry Isla Grande	Cayo Caballo Blanco, Vieques	Nov. 2002
M/V Kent Reliant	Isla de Cabras, Cataño	Sept. 18, 2003
M/T Sperchios	Tallaboa Bay	Nov. 28, 2005
M/T Margara	Tallaboa Bay	Apr. 27, 2006

levels of mortality (>90%) from an outbreak of white band disease. As a result, the living cover of *Acropora* within the grounding site now exceeds surrounding areas.

In addition to the groundings of large ships, recreational vessels frequently run aground on reefs. Unfortunately, many are removed before they can be properly documented. In most incidents, damage from recreational vessels is relatively limited. However, damage can be great because of improper removal techniques and cumulative impacts, especially in areas of high vessel traffic.

9.10   Case Study: La Parguera

The best-developed coral reefs of Puerto Rico occur in the La Parguera area of southwest Puerto Rico (Fig. 9.10) because of (1) lower wave action and a broad shelf (8–10 km wide) compared to the north coast, (2) lower rainfall and terrestrial runoff compared to the other coasts, and the (4) lowest human population density among coastal regions. Additionally the La Parguera region, compared to other south coast sites, possesses coastal hills close to the shoreline that limit terrestrial runoff to a narrow coastal zone.

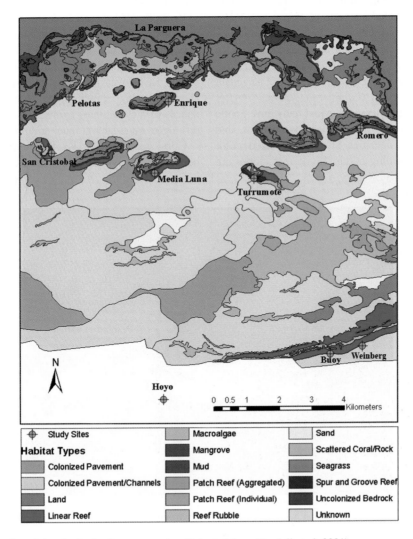

FIG. 9.10. Coral reef sites in the La Parguera region (Adapted from Kendall et al. 2001)

Two lines of emergent reefs (henceforth called the inner and mid-shelf reefs) roughly parallel the coastline, and offer some protection from wind/wave action. The best-developed coral reefs are located at the shelf edge, under the influence of oceanic conditions. Inner and mid-shelf reefs are visually dominated by algal, alcyonarian, or scleractinian–sponge communities depending upon specific environmental factors. Excluding areas of urban development, mangroves dominate shorelines of the coast and keys. Seagrass beds often dominate the coastal fringe at depths of less than 4 m.

Fringing reefs bordering the shoreline are the dominant inshore coral reef habitat in La Parguera. These reefs extend from the surface to 20 m depth, and often have a wide platform with water depths of 1–4 m usually dominated by alcyonarian, but also with scleractinian colonies of *Acropora palmata, Dendrogyra cylindrus, Montastraea faveolata, M. cavernosa, Diploria clivosa, D. strigosa, D. labyrinthiformis, Siderastrea siderea, Porites astreoides,* and milleporids. Most reefs have gentle slopes with dense alcyonarian populations and relatively high densities of medium sized colonies of reef-building corals. Some reef slopes are

vertical and short with walls mostly covered with crustose sponges, corals, and turf algae.

Fringing reefs in the mid-shelf zone also display variable profiles depending on specific conditions. In general, shallow platforms are dominated by alcyonarian, encrusting scleractinians, and stands of dead hydrocoral, *Millepora complanata,* killed by recent bleaching events. Stands of *Acropora palmata* still dominate exposed shallow areas of several reefs. Intermediate depths are dominated by large scleractinian colonies of *Montastraea faveolata, M. annularis, M. cavernosa, M. franksi, Diploria strigosa, D. labyrinthiformis, Siderastraea siderea,* and *Colpophyllia natans.* However, many of these populations are now highly reduced due to disease (WP-II and YBS) and the bleaching event in 2005.

Reefs in offshore areas occur at depths of 18–20 m and are mostly low relief bank reefs or well-developed spur and groove formations (Fig. 9.11) with relatively high scleractinian cover and biodiversity. Live cover of sponges and crustose coralline algae is also higher in this zone. Alternatively, abundances of alcyonarians, hydrocorals, zoanthids and turf algae are greater at inner and mid-shelf reefs.

Coral reefs in the La Parguera region have been studied by individual scientists for nearly

FIG. 9.11. Spur and groove formation at edge of insular shelf, Weinberg site, La Parguera, 20 m (Photo: Emmanuel Irizarry)

50 years, principally from the marine lab of the Department of Marine Sciences, University of Puerto Rico. Beginning in 2002 these reefs have been the subject of an intensive integrated, multi-disciplinary research program. This reef study is focused on eight sites (Fig. 9.10) involving inshore reefs (Romero, Enrique, Pelotas), mid-shelf reefs (Turrumote, Media Luna, San Cristobal), and off-shore reefs (Weinberg, El Hoyo, with some additional observations at the Buoy). Quantitative, non-destructive observations are being made at various time intervals for scleractinian corals, alcyonarians, and algae in permanent quadrats, as well as fish assemblages along fixed transects.

While aiming to provide a high-resolution account of the region's coral reefs, present research is centered on several hypotheses, including:

- The integrity and structure of coral reef ecosystems depend upon low transports of watershed-based materials to the marine environment.
- Internal (in situ) processes play a major role in the structure and function of coral reef communities.

As implied by the preceding descriptions of the coral reefs of Puerto Rico, environmental features involving geological, physical, and chemical factors as well as anthropogenic impacts play a critical implicit (but essential) role in the above hypotheses. Thus, knowledge of the spatial/temporal interplay among various biotic and abiotic components of coastal ecosystem is critical for the resolution of these hypotheses. For example, spatial/temporal variations in temperature, turbidity, and chlorophyll a concentrations in the water column can provide insights into circulation patterns and their consequent effects on the transport and fate of watershed-based materials as well as the retention/dispersal of planktonic larvae. For this reason, considerable effort is being directed to study of geological, physical, chemical factors and human-related influences.

9.10.1 Circulation Patterns

Similar to other south coast areas, currents in the La Parguera region are generally westward (e.g., Sanderson et al. 1995; Ojeda 2002) with velocities ~10–15 cm/s (Hensley et al. 1994). Within this overall pattern, focus is placed on spatial/tempo-ral variations in circulation patterns arising from interactions among shelf topography, tidal excursions, wave action, wind conditions, and other factors. For instance, Tyler (1992) and Tyler and Sanderson (1996) found that offshore winds cause a subsurface eastward current on the La Parguera shelf. Similarly, field observations (Appeldoorn et al. 1994, R. Appeldoorn, personal communication 2007; Sanderson et al. 1995) and computer models (Pagán 2003) indicate that surface flow is eastward/southeastwardly when trade winds abate. Giese et al. (1990) and Chapman and Giese (1990) reported seiching of shelf waters induced by internal waves impacting and breaking up and over the insular slope, while observations by Yoshioka et al. (1985) and Ojeda (2002) indicate that current directions may vary at the shelf edge due to oceanic eddies.

9.10.2 Watershed and Sedimentation

Sediments from terrestrial runoff represent a major threat to Caribbean reefs (Rogers 1990). In La Parguera, there has been an increase in terrestrial runoff over the last 100 years (J.P. Walsh, personal communication 2006), with urban development in the last 20 years greatly accelerating this process (Ramos, personal communication). Average sedimentation rates are higher at the inner (cores ~4.8 mm/year; traps ~8 mg/cm^2/d) and mid-shelf reefs (cores ~2 mm/year; traps ~9 mg/cm^2/d) compared to the shelf edge reefs (traps ~1 mg/cm^2/d). Long-term accumulation at nearshore sites is variable being lower at sites immediately off large mangrove stands, indicating their potential role in sediment retention. Preliminary x-ray diffraction and bulk carbon composition analyses also indicate that percentages of terrigenous material decrease offshore, averaging ~21%, ~17%, and ~5% at the inshore, mid-shelf and shelf edge sites, respectively, strongly suggesting that little runoff from La Parguera reaches the outer shelf. In addition, short-term sedimentation rates vary considerably through time, and temporal peaks result from sediment resuspension by wave action rather than the influx of terrigenous material, based on higher proportions of coarser (>63 μm) sediments during periods of high sedimentation.

9.10.3   Water Quality

Nutrient concentrations are generally lower during the dry season. Inorganic nitrogen (IN, nitrates plus nitrites), total dissolved phosphorous (TDP), and inorganic phosphorous (IP) are often undetectable ($<0.25\,\mu M$) in January. At other times, higher nutrient concentrations usually occur inshore. Inorganic nitrogen decreases from 2.6 to $15.1\,\mu M$ at sites adjacent to shore to $1.5\,\mu M$ at a near-shore reef (Enrique) while total dissolved nitrogen (TDN) decreases from $6.5–9.7\,\mu M$ to $4.5–5.8\,\mu M$, respectively. Similarly, inorganic phosphorous decreases from 1.9 to $3.9\,\mu M$ at inshore areas to $0.3\,\mu M$ at Enrique.

As is the case with nutrients, chlorophyll a (chl a) concentrations and turbidity decrease from inshore to offshore with the exception of events such as major storms. However, considerable variability is evident within these general patterns. For instance, at inshore areas during June 2003, turbidity and chl a averaged <10 NTU and ~0.5 µg/l respectively at all stations except at one where NTU ranged between 10 and >200 and chl a averaged about 4.9 µg/l (Otero and Carberry 2005).

The preceding suggests that nutrient pulses, Chl a, and turbidity are closely associated with precipitation/runoff. However, their ultimate effect on coastal ecosystems may be highly dependent upon interactions with other factors. For instance, phytoplankton growth from the assimilation of inorganic nutrients can reduce water transparency, hence photosynthesis in zooxanthellate corals. The association between high transparency and high coral cover (Fig. 9.9) and species richness (D. Armstrong and M. Cardona, personal communication 2007) may be related to this relationship. More interestingly, the effects of the microbial mineralization of organic nutrients on reef systems are unknown. In contrast to the conclusions of Corredor et al. (1985), these effects may not be restricted to the immediate coastal area but may extend to mid shelf reefs.

9.10.4   Scleractinian Corals

Scleractinian coral cover and colony abundances differ significantly among inshore, mid-shelf and offshore reefs, as well as among depth zones within reefs. Offshore reefs (Fig. 9.10) as Weinberg and the Buoy generally have higher average coral cover (39%) compared to inshore reefs as Romero (13%). However, a major exception to this pattern was at El Hoyo, which had the lowest coral cover and colony abundances of all sites as it is a hard ground area with scattered scleractinian colonies and not a coral reef formation. In terms of depth, average cover is significantly higher at the intermediate (10 m) depth zones compared to the shallower and deeper areas. The overall average cover for La Parguera is 18%, which is low but similar to reefs in the northern Caribbean. (It is important to note that this average is based on all depth zones thus representing the entire reef instead of specific reef zones with high coral populations.)

Montastraea was the most abundant scleractinian taxa, averaging 5% live cover in the inner reefs and 28% at the offshore Weinberg and Buoy reef sites. Other scleractinians showing an offshore increase in abundance include *Porites* and species of the *Agaricia* complex, while *Diploria, Colpophyllia,* and *Siderastrea* showed no inshore-offshore patterns. Acroporids (mostly *Acropora palmata*) were most abundant in shallow habitats of Enrique (5.2% cover) and San Cristobal (3.9% cover) where large stands of live colonies were present. No significant stands of *A. cervicornis* are found in the areas sampled, but scattered colonies, up to 1 m in diameter, are common at all reefs. However, most of these are short-lived due to damselfish, snail predation, and disease (WBD). The only extensive (70×25 m) live stand of *A. cervicornis* (with stands of *A. prolifera*), which is located on the back, shallow (1 m) platform of San Cristobal Reef, suffered partial mortality after the bleaching event of 1998 and the impacts of hurricanes Hortense and Georges in the 1990s. This stand recovered and even though there was partial mortality produced by damselfish and snail predation, boat and fishermen activities and WBD, the population was presumably healthy until February 2006, when close to 100% mortality was observed following the 2005 bleaching event.

Because the biological fate of coral reefs is highly dependent upon successful coral reproduction, considerable attention has been paid to juvenile/recruit corals (Fig. 9.12). A total of 882 juvenile colonies from 26 scleractinian species averaging 1.5 colonies/m^2 have been found at the La Parguera sites, which is comparable to densities (1.2–3.4 col/m^2) at other Caribbean reefs (Chiappone and Sullivan

1996; Miller et al. 2000; Ruiz-Zarate and Gonzalez 2004). Important reef-building genera *Siderastrea, Agaricia, Diploria, Porites, Montastraea,* and *Madracis* represented the majority of juveniles.

Juvenile densities differ significantly among inshore–offshore reefs and among depths within reefs. In general, significantly higher juvenile densities are found at the shelf-edge reefs, with the highest densities at Media Luna and the Buoy. Higher densities are also more common in deeper habitats (15–20 m) including Pelotas, an area presumably unfavorable for corals because of high levels of turbidity and sedimentation. With respect to small scale features, no significant correlation has been found between juvenile densities and substratum rugosity, coral cover, abundance of parental colonies, species composition, and survivorship of juvenile corals. Thus, juvenile densities may be influenced by other factors such as predation, overgrowth, sediment load, and substratum orientation. This is supported by recent experimental work (A. Szmant and E. Weil, personal communication 2007) showing coral settlement of *Acropora palmata* and *Montastraea flaveolata* to be dependent on the reef specific microflora, with greater settlement occurring at inshore reefs, but greater short-term survival occurring at offshore reefs.

Juvenile survivorship, when evaluated from August 2003 to August 2005, was variable with no consistent inshore–offshore pattern. For instance, survivorship at inshore reefs was low at Pelotas (48%) but high at Enrique (79%). Similarly, survivorship at offshore reefs was low at El Hoyo (49%) and high at Weinberg (63%). Survivorship was also variable with depth but notably higher at intermediate depths at Enrique and Weinberg. With respect to taxa, survivorship was relatively high for both brooding (*Porites, Agaricia*) and broadcasting (*Montastraea, Siderastrea, Diploria*) groups.

9.10.5   Alcyonarians

In general, the alcyonarian species found at the La Parguera reef sites are present elsewhere in Puerto Rico as well as the entire Caribbean region. The major exceptions include species found in other habitat types. For instance, *Carijoa* (*Telesto*) *reesei* is commonly observed on mangrove roots rather than the rock substrata of the reef sites. More interestingly, the hard bottom species *Muriceopsis sulphurea* and *Gorgonia mariae* are essentially absent from these sites (and the south coast in general) because they are restricted to areas in the north coast subject to high wave action.

FIG. 9.12. In situ photo showing 19 juvenile colonies belonging to 8 scleractinian species in an area of 0.25 m² at Media Luna Reef, off La Parguera (Photo: E. Weil)

No significant inshore–offshore patterns in alcyonarians, based on Reciprocal Averaging Analyses, were revealed. In addition, most species occur on the terrace, break and slope reef zones (where applicable) at each sampling site. As an example, *Pseudopterogorgia americana* occurs in more than 93% of the transects, regardless of depth. In general this pattern supports the contention that hard bottom, shallow-water alcyonarians are ecologically similar with similar habitat requirements (Yoshioka and Yoshioka 1989a). However, some exceptions to overall pattern of similarities in distribution patterns of alcyonarian species are evident. *Pterogorgia anceps* is restricted to the reef terrace with relatively low topographic relief and relatively high wave action, while *Muricea pinnata* and *Pseudopterogorgia bipinnata* are restricted to the reef slope with high relief and low wave action. These patterns are consistent with the inference that differences in distribution patterns are subtle and are related to an environmental gradient involving the combination of low relief and high wave action versus low relief and high wave action (Yoshioka and Yoshioka 1989b).

9.10.6 Algae

On healthy coral reefs, macroscopic algae are relatively inconspicuous (Wanders 1977; Borowitzka 1981). Nevertheless on closer inspection, algae are relatively abundant and, in fact, have higher species richness and percent cover than might be expected (Van Den Hoek et al. 1975; Littler and Littler 1994). In the La Parguera region, coral reef-associated algal species diversity generally increases in an inshore/offshore gradient. However, there is considerable variability within this overall pattern. For instance, among offshore sites algal diversity is higher at the coral dominated Weinberg site as opposed to the hard ground, El Hoyo site. Species diversity also varies with depth at each site. Many algae associated with coral reefs occur in "turf habitats" which generally comprise 50–66% of the cover at all depths. Encrusting coralline red algae (including *Porolithon* sp., *Lithophyllum* sp. and *Neogoniolithon* sp.) plus encrusting Peyssonneliaceae, *Dictyota* spp., and *Lobophra variegata* comprise the majority of the remaining algal cover.

While corals and sponges grow relatively slowly, the benthic alga community is highly dynamic, as exemplified by substantial short-term changes observed to occur in permanent quadrats. As an example, Fig.9.13 shows changes in average algal cover at the Weinberg site at the edge of the insular shelf over 2.5 years. Virtually all of the major recognizable constituents showed dramatic cover changes frequently over short intervals. *Dictyota* peaked at the II.2004 sampling and then declined while *Lobophora variegata* increased in cover during the entire period. Average cover from 2003 to 2006 at El Hoyo and Weinberg sites (Fig. 9.14)

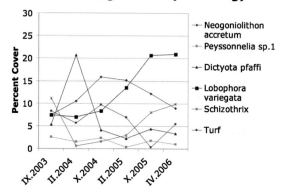

FIG. 9.13. Temporal percent cover variation of macro algal species at the edge of the insular shelf (Weinberg site) off La Parguera (20 m)

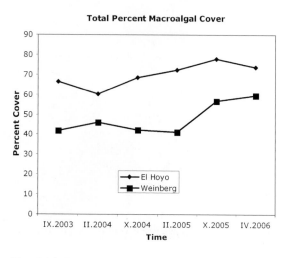

FIG. 9.14. Increasing trend of macroalgal cover at two shelf edge stations off La Parguera (20 m)

reveal average algal cover increased from 66.6% to 77.9%, and 42.0% to 59.6% respectively. The increase of algal cover at El Hoyo is mostly due to a steady increase of the phaeophyte *Lobophora variegata*. With the exception of cyanobacteria, other algal taxa experienced a steady decline in cover during this period. Nevertheless, the steady increase in algal cover at the offshore sites (in addition to presence of cyanobacterial blooms) may well indicate the development of an alternative state in these reefs.

9.10.7 Fishes

Reef fish communities of La Parguera are fairly typical of the wider Caribbean (Sullivan Sealy and Bustamante 1999). Within this overall framework, specific characteristics of reef fish communities depend on a variety of parameters operating on differing spatial scales including water quality, reef rugosity, coral cover, reef area and proximity to juvenile habitats. Moreover, effects of these factors are not independent and interact in complex ways (Ault and Johnson 1998; Prada Triana 2002; Appeldoorn et al. 2003; Aguilar Perera 2004; Bejarano Rodríguez 2006; Cerveny 2006; Foley and Appeldoorn 2007).

Fore reef fish communities off La Parguera show distinct differences associated with depth and distance from shore. Most apparent are several species unique to the deeper offshore communities including the triggerfishes *Balistes vetula* and *Melichthys niger*, the princess parrotfish *Scarus taeniopterus*, and the longspine squirrelfish *Holocentrus rufus*. With respect to the inshore and mid-shelf reefs, a Reciprocal Averaging Analysis indicates that the distribution of fish is related primarily to depth (Axis 1 in Fig. 9.15) and rugosity (Axis 2). For example, with respect to depth, three damselfishes, while occurring at all depths, show distinct depth preferences, with *Stegastes adustus*, *S. planifrons*, and *S. leucostictus* preferring shallow (3 m), intermediate (10 m) and deep (15 m) habitats, respectively. Also, although herbivorous fishes such as parrotfishes and surgeonfishes dominate fish communities across the shelf, the three dominant parrotfishes (*Sparisoma aurofrenatum*, *S. iseri*, *S. viride*) and all surgeonfishes tend to be more abundant at

shallower depths. Finally, there is a difference between mid-shelf and nearshore reefs in that the decline in abundance with depth is greater at the inshore reefs.

The effect of rugosity is suggested by the distribution of the threespot damselfish (*Stegastes planifrons*), which is typically found at intermediate depths. However, *S. planifrons* is relatively abundant at shallow (3 m) depths at Enrique, San Cristobal and Pelotas reefs (see dashed circle in Fig. 9.15) where rugosity is relatively high due to the presence of the scleractinian genera *Acropora* or *Montastraea* as well as coralline algae and rubble. In contrast, shallow depths at Romero, Media Luna and Turrumote reefs are characterized by low relief pavement with sparse coral cover.

9.11 Conclusions – Patterns, Ecological Processes and the Future of Coral Reefs in Puerto Rico

The patterns observed off La Parguera represent the ecological 'stage' on which the interplay of causal processes occurs and demonstrate that abiotic and biotic features vary on a variety of spatial and temporal scales. Coral reefs off La Parguera historically developed under a regime of low transport of watershed-based materials to the marine environment. Sediments, organic matter and nutrients derived from the La Parguera watershed are restricted to the immediate nearshore waters behind the inner reef line, and subsequent dispersal is westward, along shore. Even waters from upstream typically wrap along the eastern, nearshore waters, leaving the outer platform and reefs unexposed to upstream sources of runoff. Coral communities showed distinct inshore–offshore differences for both adults and juveniles, with greater abundances found in the clearer offshore waters, except for the low values at El Hoyo, which are probably due to the topography of the antecedent rock platform. The biotic and abiotic processes (e.g., keystone predation and wave action/water motion, respectively) responsible for the observed inshore–offshore patterns of species richness of fish and corals are probably limited in

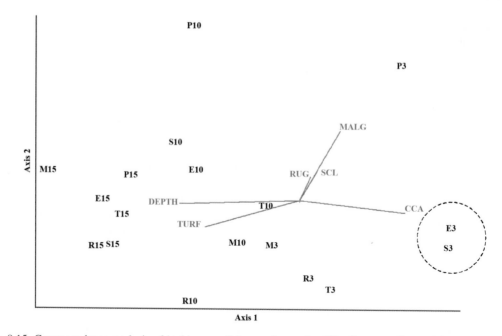

Fig. 9.15. Correspondence analysis of herbivorous fishes on forereefs off La Parguera, Puerto Rico, excluding the shelfedge sites, overlaid with second matrix of environmental data. Numbers refer to depth (m). RUG = rugosity, SCL = Scleractinian corals, MALG = Magroalgae, CCA = Crustose Coralline Algae, TURF = Turf Algae

number. However, their interactions are likely to be numerous, complex and reef-specific.

The complex of interactions among causal processes is illustrated by considering the functional role of *Diadema antillarum* as a keystone bioeroder and its consequent effects on the sediment transport (bedload) regime. In addition to its direct grazing effects, *D. antillarum* may indirectly increase mortalities of newly settled sessile organisms through sediment smothering. Spatial patterns of bedload transport are highly dependent upon water motion, reef slope and substratum rugosity. Bedload transport is also affected by sediment consolidation by microfloral populations that, in turn, are influenced by nutrient concentrations and light levels. Reef topography also plays an important role because field observations indicate that *D. antillarum* primarily grazes within a few meters of crevices that probably provide hiding places from predators during the day (Hay 1984). Finally, abundances of *D. antillarum* differ among reefs because of varying patterns of recovery from the mass mortality in early 1984 (Weil et al. 2005).

Long-term changes are evident in the coral reefs of Puerto Rico as a result of natural and anthropogenic causes. The former include the Caribbean -wide die off *D. antillarum* (Lessios et al. 1984) and the decline of *A. palmata* due to disease (Gladfelter 1982; Weil and Ruiz 2003). Local anthropogenic activities have clearly resulted in increased levels of sedimentation and nutrient runoff over time. While some of this no doubt disperses to the outer shelf, this area is more threatened with sources further upstream (e.g., Guanica, Guayanilla, Ponce), especially after major storm events. Recent increases in cyanobacterial overgrowth at the shelf edge may reflect higher levels of nutrients than in the historical, or even immediate past, suggesting that the cumulative effects of greater coastal development and eutrophication occurring along the south coast may overcome the natural circulation patterns that historically have protected sections of the coast and allowed coral reef development. The changes, and the rate at which they are occurring, bode ill for the future of Puerto Rico's coral reefs in the face of global scale stresses such as global warming (and resultant bleaching and disease) (McWilliams et al. 2005) and ocean acidification (Kleypas et al. 1999). Aggressive management will be required to significantly reduce local anthropogenic stresses through better land-use practices and watershed scale activities to control sediment and nutrient loading, coupled with reductions in fishing effort, including the establishment of no-take reserves over a significant portion of the shelf, to restore top down control of reef communities (Littler et al. 2006). Continued research on coral reef processes will be critical to determining the most efficient mechanisms for management intervention and their success in conserving Puerto Rico's coral reefs.

Acknowledgements Much of the work included in this chapter is the result of research funded by the National Oceanographic and Atmospheric Administration Center for Sponsored Coastal Ocean Research under award #NA17OP2919 to the University of Puerto Rico – Mayagüez. Dr. Carlos E. Ramos-Scharron, Island Resources Foundation, provided information on terrestrial sources of sedimentation, and Dr. John P. Walsh, University of East Carolina, provided data on long-term sedimentation rates. A number of graduate students of the Department of Marine Sciences participated in the research discussed here including: Kelly K. Carbery, Belyneth Deliz López, Raquel Hernández, Carolina Hincapié, Emmanuel Irizarry, Katie Flynn, Nilda Jiménez, Elizabeth McLean, Alex Mercado, Michael Nemeth, Carlos Prada, Luisa Fernanda Ramírez, Esther Rodríguez, Hector Ruíz, and Stacey Williams. Financial support of bioptical studies by NOAA and the Department of Natural and Environmental Resources (DRNA) of the Commonwealth of Puerto Rico provided graduate student assistantships to Ivonne Bejarano, Maria Cardona, Rene Esteves, Sara Rivero, Janeth Rojas, and Juan Torres. This publication is partly a result of funding from the National Oceanic and Atmospheric Administration, Center for Sponsored Coastal Ocean Research, under award NA04NOS4260206 to the University of Puerto Rico (UPR) for the Caribbean Coral Reef Institute. Also, partial funding to E.A. Hernández-Delgado was provided by the Sea Grant College Program (UPR), Caribbean Marine Research Center, Institutional Funds for Research (UPR), and Environmental Defense.

References

Acevedo R, Morelock J (1988) Effects of terrigenous sediment influx on coral reef zonation in southwestern Puerto Rico. Proc 6th Int Coral Reef Symp 2:189–194

Acosta A, Appeldoorn RS (1992) Estimation of growth, mortality and yield per recruit for *Lutjanus synagris* (Linnaeus) in Puerto Rico. Bull Mar Sci 50: 282–291

Aguilar Perera JA (2004) Coastal habitat connectivity of reef fishes from southwestern Puerto Rico. Ph.D. thesis, University of Puerto Rico, Puerto Rico, p 159

Alcolado PM, Claro-Madruga R, Mendez-Macias G, Garcia-Parrabo P, Martinez-Daranas B, Sosa M (2003) The Cuban Coral Reefs. In: Cortez J (ed) Latin American Coral Reefs. Elsevier, Amsterdam, The Netherlands, pp 53–76

Almy Jr CC, Carrión Torres C (1963) Shallow-water stony corals of Puerto Rico. Carib J Sci 3:133–162

Antonius A (1999) *Metapeyssonnelia corallepida*, a new coral-killing red alga on Caribbean reefs. Coral Reefs 18:301

Antonius A, Ballesteros E (1998) Epizoism: a new threat to coral health in Caribbean reefs. Rev Biol Trop 46 Supl 5: 145–156

Antonius A, Weiner A (1982) Coral reefs under fire. Mar Ecol 3:255–277

Appeldoorn R, Beets J, Bohnsack J, Bolden S, Matos D, Meyers S, Rosario A, Sadovy Y, Tobias W (1992) Shallow water reef fish stock assessment for the U.S. Caribbean. NOAA Tech. Mem NMFS/SEFSC/304, p 70

Appeldoorn RS, Lindeman KC (1985) Multispecies assessment in coral reef fisheries using higher taxonomic categories as unit stocks, with an analysis of an artisanal haemulid fishery. Proc 5th Int Coral Reef Cong 5:507–514

Appeldoorn RS, Meyers S (1993) Puerto Rico and Hispaniola. In: Marine fishery resources of the Antilles. FAO Fish Tech. Paper 326:99–158

Appeldoorn RS, Posada JM (1992) The effects of mesh size in Antillean fish traps on the catch of coral reef fish. Report to the Caribbean Fishery Management Council, Hato Rey, PR

Appeldoorn RS, Hensley DA, Shapiro DY, Kioroglou S, Sanderson BG (1994) Egg dispersal in a Caribbean coral reef fish, *Thalassoma bifasciatum*. II. Dispersal off the reef platform. Bull. Mar Sci 54: 271–280

Appeldoorn RS, Friedlander A, Sladek Nowlis J, Ussegilo P, Mitchell-Chui A (2003) Habitat connectivity on the insular platform of Old Providence-Santa Catalina, Colombia: mechanisms, limits and ecological consequences relevant to marine reserve design. Gulf Carib Res 14:61–77

Armstrong RA, Singh H, Torres J (2002) Benthic survey of insular slope coral reefs using the SeaBed AUV. Backscatter fall/winter, pp 22–25

Armstrong RA, Singh H, Torres J, Nemeth RS, Can A, Roman C, Eustice R, Riggs L, Garcia-Moliner G (2006) Characterizing the deep insular shelf coral reef habitat of the Hind Bank marine conservation district (US Virgin Islands) using the Seabed autonomous underwater vehicle. Cont Shelf Res 26:194–205

Aronson RB, Precht WF (2001) White-band disease and the changing face of Caribbean coral reefs. In: Porter JW (ed) The Ecology and Etiology of Newly Emerging marine Diseases. Developments in Hydrobiology. Kluwer, Dordrecht, The Netherlands, pp 23–35

Ault T, Johnson C (1998) Spatial variation in fish species richness on coral reefs: habitat fragmentation and stochastic structuring processes. Oikos 82:354–364

Ballantine DL, Abbott IA (2006) *Ganonema vermiculare* (Liagoraceae, Rhodophyta), a new species from Puerto Rico, Carib Sea Bot Mar 49:122–128

Ballantine DL, Aponte NE (2002) A checklist of the benthic marine algae known to Puerto Rico, second revision. Constancea 83, online continuation of California Publications in Botany (1902–2002) http://ucjeps.berkeley.edu/constancea/83/ballantine_aponte/checklist.html

Ballantine DL, Ruiz H (2005) Two *Peyssonnelia* species (Peyssonneliaceae, Rhodophyta) from Puerto Rico including *P. flavescens* nov. sp. Phycologia 44:328–334

Ballantine DL, Wynne MJ (1986) Notes on the marine algal flora of Puerto Rico I. Additions to the flora. Bot Mar 29:131–135

Ballantine DL, Wynne MJ (1987) Notes on the marine algae of Puerto Rico III. *Branchioglossum pseudoprostratum* sp. nov. and *B. prostr*atum Schneider (Rhodophyta: Delesseriaceae). B Mar Sci 40:240–245

Ballantine DL, Ruiz H, Wynne MJ (2002) Notes on the Benthic Marine Algae of Puerto Rico VII. Further Additions of Rhodophyta Species. Carib J Sci 38:252–256

Ballantine DL, Ruiz H, Aponte NE (2004) Notes on the benthic marine algae of Puerto Rico VIII. Additions to the flora. Bot Mar 47:335–340

Ballantine DL, Weil E, Ruiz H (2005) Coralline white band syndrome, a coralline algal affliction in the tropical Atlantic. Coral Reefs 24:117

Baums IB, Paris CB, Chérubin LM (2006) A bio-oceanographic filter to larval dispersal in a reef-building coral. Limnol Oceanogr 51:1969–1981

Bejarano Rodríguez I (2006) Relationships between reef fish communities, water and habitat quality on coral reefs. M.S. thesis, University of Puerto Rico, Puerto Rico

Bohnsack J, Meyers S, Appeldoorn R, Beets J, Matos D, Sadovy Y (1991) Stock assessment of spiny lobster in the U.S. Caribbean. Caribbean Fishery Management Council, Hato Rey, PR

Borowitzka MA (1981) Algae and grazing in coral reef ecosystems. Endeavor 5:99–106

Bruckner AW, Bruckner RJ (2001) Condition of restored *Acropora palmata* fragments off Mona Island, Puerto Rico, 2 years after the Fortuna Reefer ship grounding. Coral Reefs 20:235–243

Bruckner AW, Bruckner RJ (2006a) Consequences of yellow-band disease (YBD) on *Montastraea annularis* (species complex) populations on remote reefs off Mona Island, Puerto Rico. Dis Aquat Org 69:67–73

Bruckner AW, Bruckner RJ (2006b) Chapter 19: Restoration outcomes of the Fortuna Reefer Grounding at Mona Island, Puerto Rico. In: Precht WF (ed) Coral Reef Restoration Handbook – The Rehabilitation of an Ecosystem Under Siege. CRC, Boca Raton, FL, pp 257–270

Bruckner AW, Bruckner RJ (2006c) Survivorship of restored *Acropora palmata* fragments over six years at the M/V *Fortuna Reefer* ship grounding site, Mona Island, Puerto Rico. Proc 10th Int Coral Reef Symp: 1645–1650

Burke L, Maidens J (2004) Reefs at Risk in the Caribbean. Worlds Resources Institute, Washington, DC, p 81

Canals M, Ferrer H, Merced H (1983) Los arrecifes de coral de Isla Mona. Octavo Simposio de Recursos Naturales (12/7/81). Departamento de Recursos Naturales, ELAPR, pp 1–26

Capella J, Laboy E (2003) Chapter 2: Environmental Settings. In: Jobos Bay Estuarine Profile, pp 14–39

Carrera-Rodríguez CJ (1978) Evaluación sobre las consecuencias de las actividades militares llevadas a cabo en Vieques, PR por las Fuerzas Armadas de los EE. UU. de América en los sistemas marinos del este de Vieques, con atención principal a los arrecifes de coral. September, 1978. Departamento de Recursos Naturales, San Juan, PR, p 8

Cerveny K (2006) Distribution patterns of reef fishes in southwest Puerto Rico, relative to structural habitat, cross-shelf location, and ontogenetic stage. M.S. thesis, University of Puerto Rico, Puerto Rico

Chapman DC, Giese GS (1990) A model for the generation of coastal seiches by deep-sea internal waves. J Phys Oceanogr 20:1459–1467

Chiappone M, Sullivan KM (1996) Distribution, abundance and species composition of juvenile scleractinian corals in the Florida reef tract. Bull Mar Sci 58:555–559

Cintron G, Thurston J, Williams J (1975) Results of exploratory dives in the Carabinero shelf area, Mona Island. Depart of Natural Resources, Commonwealth of Puerto Rico, San Juan, PR

Colin PL (1982) Aspects of the spawning of Western Atlantic reef fishes. In: Huntsman GR, Nicholson WR, Fox Jr WW (eds) The Biological Bases for Reef Fishery Management. NOAA Tech Mem NMFS-SEFC-80, pp 69–78

Corredor JE, Morell J, Nieves F, Otero E (1985) Studies on eutrophication of the marine ecosystem of La Parguera, Puerto Rico. In: Mem Undecimo Simp Rec Nat Departamento de Recursos Naturales, San Juan, PR

Dennis GD (1988) Commercial catch length-frequency data as a tool for fisheries management with an application to the Puerto Rico trap fishery. Mem Soc Cien Nat La Salle 48 supl 3:289–310

Dennis GD, Hensley D, Colin PL, Kimmel JJ (2004) New records of marine fishes from the Puerto Rican plateau. Carib J Sci 40:70–87

Deslarzes KJP, Nawojchik R, Evans DJ, McGarrity CJ, Gehring P (2006) The condition of fringing reefs of former military training areas at Isla de Culebra and Isla de Vieques, Puerto Rico : preliminary results. Proc 10th Int Coral Reef Symp: 1152–1159

Dodge RE (1981) Growth characteristics of reef-building corals within and external to a naval ordnance range: Vieques, Puerto Rico. Proc 4th Int Coral Reef Symp 2:241–248

Done TJ (1992) Phase shifts in coral reef communities and their ecological significance. Hydrobiologia 247:121–151

Epstein PR, Sherman K, Spanger-Siegfried W, Langston A, Prasad S, McKay B (1998) Marine Ecosystems: Emerging Diseases as Indicators of Change. Health Ecological and Economic Dimensions (HEED), NOAA Global Change Program, p 85

Evans DJ, Nawojchik R, Deslarzes KJP (2006) The status off coral reef fish populations of former military ranges at the Islands of Culebra and Vieques, Puerto Rico : preliminary data. Proc 10th Int Coral Reef Symp: 1105–1109

Evermann BW (1900) General report on the investigations in Puerto Rico of the United States Fish Commission Steamer Fish Hawk in 1899. US Fish Comm Bull: 3–26

Ferrier-Pages C, Gattuso JP, Dallot S, Jaubert J (2000) Effect of nutrient enrichment on growth and photosynthesis of the zooxanthellate coral *Stylophora pistillata*. Coral Reefs 19:103–113

Foley KA, Appeldoorn RS (2007) Cross-shelf habitat-fish associations in La Parguera, Puerto Rico : factors affecting essential fish habitat and management applications. Proc Gulf Carib Fish Inst 58:21–28

García JR, Morelock J, Castro R, Goenaga C, Hernández E (2003) Puerto Rican reefs: research synthesis, present threats and management perspectives. In: Cortés J (ed) Latin American Coral Reefs. Elsevier, Amsterdam, The Netherlands, p 497

García-Sais J, Appeldoorn R, Bruckner A, Caldow C, Christensen J, Lilyestrom C, Monaco ME, Sabater J, Williams E, Diaz E (2005) The state of coral reef ecosystems of the Commonwealth of Puerto Rico. In: Jeffrey C et al. (eds) The State of Coral Reef Ecosystems of the United States and Pacific Freely Associated States : 2005. NOAA Technical Memorandum NOS NCCOS 11, pp 91–134

García-Sais J, Castro R, Sabater J, Carlo M (2004) Monitoring of coral reef communities from Isla de Vieques, Puerto Rico. Final Report submitted to the Department of Natural and Environmental Resources of Puerto Rico, San Juan, PR, p 118

García-Sais J, Castro R, Sabater J, Carlo M (2005) Inventory and atlas of corals and coral reefs, with emphasis on deep-water coral reefs from the U. S. Caribbean EEZ (Puerto Rico and the United States Virgin Islands). Final Report submitted to the Caribbean Fishery Management Council. Coral Grant 2003 NAO3NMF4410352, p 215

García-Sais J, Castro R, Sabater J, Esteves R, Carlo M (2006) Monitoring of coral reef communities from natural reserves in Puerto Rico : Isla Desecheo, Rincon, Mayaguez Bay, Guanica, Ponce and Caja de Muerto. Final Report submitted to DNER-NOAA, San Juan, PR, p 145

Geraldez JX (2003) The coral reefs of the Dominican Republic. In: Cortes J (ed) Latin American Coral Reefs. Elsevier, Amsterdam, The Netherlands, pp 77–110

GESAMP (2001) Protecting the oceans from land-based activities. Land-based sources and activities affecting the quality and uses of the marine, coastal and associated freshwater environment. United Nations Environment Program, Nairobi, p 168

Giese GS, Chapman DC, Black PG, Fornshell JA (1990) Causation of large-amplitude coastal seiches on the Caribbean coast of Puerto Rico. J Phys Oceanogr 20:1449–1458

Gladfelter WB (1982) White-band disease in *Acropora palmata*: implications for the structure and growth of shallow reefs. B Mar Sci 32:639–643

Glynn PW, Maté JL, Baker AC, Calderón MO (2001) Coral bleaching and mortality in Panamá and Ecuador during the after de 1997–98 El Niño Southern-Oscillation event: Spatial/temporal patterns and comparison with the 1982–83 event. B Mar Sci 69:79–109

Goenaga C (1988) The distribution and growth of *Montastrea annularis* (Ellis and Solander) in Puerto Rican inshore platform reefs. Ph.D. thesis, University of Puerto Rico Department of Marine Sciences, Puerto Rico, p 186

Goenaga C (1989) Informe sobre las comunidades marinas de Punta Picúa. Sometido a Servicios Legales de Puerto Rico, Canóvanas, PR, p 10

Goenaga C, Boulon Jr RH (1992) The State of Puerto Rican and U.S. Virgin Islands Corals: An Aid to Managers. Report submitted to the Caribbean Fishery Management Council, Hato Rey, PR, p 66

Goenaga C, Cintrón G (1979) Inventory of the Puerto Rican Coral Reefs. Report submitted to the Coastal Zone Management of the Department of Natural Resources, San Juan, PR, p 90

Goenaga C, Vicente V (1990) Apéndice 4. Informe de observaciones de campo sobre corales y organismos asociados. In: Suplemento técnico para el Plan de Manejo de la Reserva Natural La Cordillera, Fajardo. Departamento de Recursos Naturales, San Juan, PR, p 13

Goenaga C, García JR, Vicente VP (1985) Appendix 6. Survey of seagrass and coral reefs near the Carolina discharge site. Center for Energy and Environmental Research, University of Puerto Rico, Puerto Rico, p 16

Goldberg WM (1973) The ecology of the coral-octocoral communities off the southeast Florida coast: geomorphology, species composition, and zonation. Bull. Marine Sci 23:465–488

Gonzalez LA, Ruiz HM, Taggart BE, Budd AF, Monell V (1997) Geology of Isla de Mona, Puerto Rico. In: Vacher HL, Quinn TM (eds) Geology and Hydrogeology of Carbonate Islands. Developments in Sedimentology 54, Elsevier, Amsterdam, The Netherlands, pp 327–358

Golubic S, Le Campion-Alsumard T, Cambell SE (1999) Diversity of marine cyanobacteria. Bull Inst Oceanogr Monaco, N⁰ special 19:53–76

Goreau TJ, Cervino J, Goreau M, Hayes R, Hayes M, Richardson LL, Smith GW, DeMeyer G, Nagelkerken I, Garzón-Ferreira J, Gill D, Peters EC, Garrison G, Williams EH, Bunkley-Williams L, Quirolo C, Patterson KL (1998) Rapid spread of diseases in Caribbean coral reefs. Rev Biol Trop 46:157–171

Green EP, Bruckner AW (2000) The significance of coral diseas epizootiology for coral reef conservation. Biol Cons 96–461

Gundlach J (1883) Apuntes para la fauna Puerto Riqueña. Anales de la Sociedad Española de Historia Natural 12:5–58, 441–484

Gundlach J (1887) Apuntes para la fauna Puerto Riqueña. Anales de la Sociedad Española de Historia Natural 16:115–133

Guzmán HM (2003) Caribbean coral reefs of Panamá: present status and future perspectives. In: Cortes J (ed) Latin American Coral Reefs. Elsevier, Amsterdam, The Netherlands, pp 241–274

Harvell CD, Kim K, Burkholder JM, Colwell RR, Epstein PR, Grimes DJ, Hofmann EE, Lipp EK, Osterhaus ADME, Overstreet AM, Porter JW, Smith GW, Vasta GR (1999) Emerging marine diseases̆climate links and anthropogenic factors. Science 285:1505–1510

Harvell CD, Mitchell CE, Ward JR, Altizer S, Dobson AP, Ostfeld RS, Samuel MD (2002) Climate warming and disease risk for terrestrial and marine biota. Science 296:2158–2162

Hauck F (1888) Meeresalgen von Puerto-Rico. Botanische Jahrbücher für Systematik. Pflanzengeschichte und Pflanzengeographie 9:457–470

Hay ME (1984) Patterns of fish and urchin grazing on Caribbean coral reefs: are previous results typical? Ecology 65:446–454

Hayes ML, Bonaventura J, Mitchel TP, Prospero M, Shinn EA,Van Dolah F, Barber R (2001) How are climate and marine biological outbreaks linked? In: Porter JW (ed) The Ecology and Etiology of Newly Emerging Marine Diseases Developments in Hydrobiology. Kluwer, Dordrecht, The Netherlands, pp 213–220

Hayes RL, Goreau NI (1998) The significance of emerging diseases in the tropical coral reef ecosystem. Rev Biol Trop 46:172–185

Hensley DA, Appeldoorn RS, Shapiro DY, Ray M, Turingan RG (1994) Egg dispersal in a Caribbean coral reef fish, *Thalassoma bifasciatum*, I- Dispersal over the reef platform. Bull Mar Sci 54:256–270

Hernandez-Cruz LR, Purkis S, Riegl B (2006) Documenting decadal spatial changes in seagrass and *Acropora palmata* cover by aerial photography analysis in Vieques, Puerto Rico : 1937–2000. Bull Mar Sci 79(2):401–414

Hernández-Delgado EA (1992) Coral reef status of northeastern and eastern Puerto Rican waters: Recommendations for long-term monitoring, restoration and a coral reef management plan. Submitted to the Caribbean Fishery Management Council, Hato Rey, PR, p 87

Hernández-Delgado EA (2000) Effects of anthropogenic stress gradients in the structure of coral reef fish and epibenthic communities. Ph.D. thesis, University of Puerto Rico Department of Biology, San Juan, PR, p 330

Hernández-Delgado EA (2003) Suplemento técnico al Plan de Manejo para la Reserva Natural del Canal Luis Peña, Culebra, Puerto Rico. I. Caracterización de habitáculos. Informe Técnico sometido al Programa de Manejo de la Zona Costanera, DRNA. San Juan, PR, p 109

Hernández-Delgado EA (2005) Historia natural, caracterización, distribución y estado actual de los arrecifes de coral Puerto Rico. In: Joglar RL (ed) Biodiversidad de Puerto Rico: Vertebrados Terrestres y Ecosistemas. Serie Historia Natural. Editorial Instituto de Cultura Puertorriqueña, San Juan, PR, pp 281–356

Hernández-Delgado EA, Rosado-Matías BJ (2003) Suplemento técnico al Plan de Manejo para la Reserva Natural del Canal Luis Peña, Culebra, Puerto Rico. II. Inventario biológico. Informe Técnico sometido al Programa de Manejo de la Zona Costanera, DRNA. San Juan, PR, p 60

Hernández-Delgado EA, Sabat AM (2000) Ecological status of essential fish habitats through an anthropogenic environmental stress gradient in Puerto Rican coral reefs. Proc Gulf Carib Fish Inst 51:457–470

Hernández-Delgado EA, Alicea Rodríguez L (1993) Estado ecológico de los arrecifes de coral en la costa este de Puerto Rico : I. Bahía Demajagua, Fajardo, y Playa Candelero, Humacao. Memorias del XII Simposio de la Fauna y Flora del Caribe, 30 de abril de 1993, Depto. Biología, Recinto Universitario de Humacao, Humacao, PR, pp 2–23

Hernández-Delgado EA, Rodríguez Class EO, Martínez Suárez JE (1996) Evaluación biológica del arrecife Cayo Ahogado, Bahía Algodones, Naguabo, Puerto Rico. Informe sometido a la Junta de Calidad Ambiental y la Junta de Planificación, San Juan, PR, p 41

Hernández-Delgado EA, Alicea Rodríguez L, Toledo Hernández CG, Sabat AM (2000) Baseline characterization of coral reef epibenthic and fish communities within the proposed Culebra Island Marine Fishery Reserve, Puerto Rico. Proc Gulf Carib Fish Inst 51:537–556

Hernández-Delgado EA, Toledo CG, Claudio H, Lassus J, Lucking MA, Fonseca J, Hall K, Rafols J, Horta H, Sabat AM (2006) Spatial and taxonomic patterns of coral bleaching and mortality in Puerto Rico during year 2005. Coral Bleaching Response Workshop, NOAA, St. Croix, USVI, p 16

Hoegh-Guldberg O (1999) Climate change, coral bleaching and the future of World's coral reefs. Mar Freshwater Res 50:839–866

Hughes TP (1994) Catastrophes, phase shifts and large scale degradation of a Caribbean coral reef. Science 265:1547–1549

Jackson JBC, Kirby MX, Berger WH, Bjorndal KA, Botsford LW, Bourque BJ, Bradbury R, Cooke R, Erlandson J, Estes JA, Hughes TP, Kidwell S, Lange CB, Lenihan HS, Pandolfi JM, Peterson CH, Steneck RS, Tegner MJ, Warner R (2001) Historical over-fishing and the recent collapse of coastal ecosystems. Science 293: 629–638

Jordan-Dahlgren E (2002) Gorgonian distribution patterns in coral reef environments of the Gulf of Mexico: evidence of sporadic ecological connectivity? Coral Reefs 21:205–215

Jordan E (1989) Gorgonian community structure and reef zonation patterns on Yucatan coral reefs. Bull Mar Sci 45:678–696

Kaye C (1959) Shoreline Features and Quaternary Shoreline Changes, Puerto Rico. US Geol Survey Prof Paper 317-B, p 140

Kendall MS, Monaco ME, Buja KR, Christensen JD, Kruer CR, Finkbeiner M, Warner RA (2001) Methods used to map the benthic habitats of Puerto Rico and the U.S. Virgin Islands. NOAA, Silver Spring, MD, p 45

Kleypas JA, Buddemeier RW, Archer D, Gattuso JP, Langdon C, Opdyke BN (1999) Geochemical consequences of increased atmospheric CO_2 on coral reefs. Science 284:118–120

Kinzie R (1973) The zonation of West Indian gorgonians. Bull Mar Sci 23:93–155

Kinzie RA, Jokiel PL, York R (1984) Effects of light of altered spectral composition on coral zooxanthellae associations and on zooxanthellae in vitro. Mar Biol 78:239–248

Kjerfve B (1981) Tides of the Caribbean Sea. J Geophy Res 86:4243–4247

Lapointe BE (1997) Nutrient thresholds for bottom up control of macroalgal blooms in coral reefs in Jamaica and southeast Florida. Limnol Oceanogr 42:1119–1131

Lessios HA (1998) Diadema antillarum 10 years after mass mortlity: still rare, despite help from a competitor. Proc R Soc London S B 259:331–337

Lessios HA, Robertson DR, Cubit JD (1984) Spread of Diadema mass mortality throughout the Caribbean. Science 226:335–337

Lirman D (2001) Competition between macroalgae and corals: effects of herbivore exclusion and increased algal biomass in coral survivorship. Coral Reefs 19:392–399

Littler MM, Littler DS (1994) Essay: tropical reefs as complex habitats for diverse macroalgae. In: Lobban CS, Harrison PJ (eds) Seaweed Ecology and Physiology. Cambridge University Press, New York, pp 72–75

Littler MM, Littler DS, Brooks BL (2006) Harmful algae on tropical coral reefs: bottom-up eutrophication and top-down herbivory. Harmful Algae 5:565–585

Macintyre IG, Raymond B, Stuckenrath R (1983) Recent history of a fringing reef, Bahía Salinas del Sur, Vieques Island, Puerto Rico. Atoll Res Bull 268:1–9

Marine Resources Inc. (2005) Habitat Suitability Analysis: Compensation for Injured Reef in Support of Restoration Planning for the Berman Oil Spill, San Juan, PR, p 29

López Marrero, del Mar T, Villanueva Colón N (2006) Atlas Ambiental de Puerto Rico. La Editoral Universidad de Puerto Rico. San Juan, PR, p 160

McClanahan T (2004) Coral bleaching, diseases and mortality in the western Indian Ocean. In: Rosemberg E, Loya Y (eds) Coral Reef Health and Diseases. Springer, Berlin, pp 243–258

McClanahan TR, Muthiga N (1998) An ecological shift in a remote coral atoll of Belize over 25 years. Environ Conserv 25:122–130

McCook LJ (1999) Macrolagae, nutrients and phase shifts on coral reefs: scientific issues and management consequences for the Great Barrier Reef. Coral Reefs 18:357–367

McKenzie F, Benton M (1972) Biological inventory of the waters and keys of north-east Puerto Rico. Final report submitted to the Division of Natural Resources, Department of Public Works, San Juan, PR, p 90

McWilliams JP, Côté IM, Gill JA, Sutherland WJ, Watkinson AR (2005) Accelerating impacts of temperature-induced coral bleaching in the Caribbean. Ecology 86:2055–2060

Miller J, Rogers CS, Waara R (2003) Monitoring the coral disease plague type II on coral reefs in St. John, US Virgin Islands. Proc of the 30th Scientific Meeting of the Association of Marine Laboratories of the Caribbean, PR. Rev Biol Trop 51:47–55

Miller MW, Weil E, Szmant AM (2000) Coral recruitment and juvenile mortality as structuring factors for reef benthic communities in Biscayne National Park. Coral Reefs 19:115–123

Morand P, Briand X (1996) Excessive growth of macroalgae: a symptom of environmental disturbance. Bot Mar 39:491–516

Muzik K (1982) Octocorallia (Cnidaria) from Carrie Bow Cay, Belize. In: Rutzler K, Macintyre I (eds) The Atlantic Barrier Reef Ecosystem at Carrie Bow Cay, Belize. Smithsonian Institute Press, Washington, DC, pp 309–316

Nakamura T, Van Woesik R (2001) Water-flow rates and passive diffusion partially explain differential survival of corals during the 1998 bleaching event. Mar Ecol Progr Ser 212:301–304

Nakamura T, Yamasaki H, Van Woesik R (2003) Water flow facilitates recovery from bleaching in the coral Stylophora pistillata. Mar Ecol Progr Ser 256:287–291

Nakamura T, Van Woesik R, Yamasaki H (2005) Photoinhibition of photosynthesis is reduced by water flow in the reef-building coral Acropora digitifera. Mar Ecol Progr Ser 301:109–118

Nelson WR, Appeldoorn RS (1985) Cruise Report R/V Seward Johnson. A submersible survey of the continental slope of Puerto Rico and the U. S. Virgin Islands. Report submitted to NOAA, NMFS, SEFC, Mississippi Laboratories. University of Puerto Rico, Department of Marine Sciences, Puerto Rico, p 76

Nemeth RS, Herzlieb S, Kadison ES, Taylor M, Rothenberger P, Harold S, Toller W (2004) Coral reef monitoring in St. Croix and St. Thomas, United States Virgin Islands. Final report submitted to the Department of Planning and Natural Resources, US Virgin Islands, 79 p

NOAA Coastal Ocean Program Decision Analysis Series No.21 (2000) COASTAL The Potential Consequences of Climate Variability and Change. A report of the National Coastal Assessment Group for the U.S. Global Change Research Program

Ojeda E (2002) Ontogenic development of the red hind Epinephelus guttatus, a spatial and temporal ichthyoplankton distribution study during a spawning aggregation event in "El Hoyo" La Parguera Puerto Rico. Ph.D. thesis, University of Puerto Rico, Puerto Rico

Opresko DM (1973) Abundance and distribution of shallow water gorgonians in the area of Miami, Florida. Bull Mar Sci 23:535–558

Otero E, Carbery KK (2005) Chlorophyll a and turbidity patterns over coral reef systems of La Parguera, reserve. Int J Trop Biol 53(Suppl. 1):25–32

Pagán F (2003) Modeling Egg and Larval Dispersal in the Nearshore Waters of Southwestern Puerto Rico. Ph.D. thesis, University of Puerto Rico, Puerto Rico

Peters EC (1997) Diseases of coral-reef organisms. In: Birkeland C (ed) Life and Death of Coral Reefs. Chapman & Hall, London, pp 114–136

Poey F (1881) Apuntes para la fauna Puerto Riqueña. Anales de la Sociedad Española de Historia Natural 10:317–350

Porter JW (2000) The effects of naval bombardment on the coral reefs of Isla Vieques, Puerto Rico. Submitted to King and Spalding, Atlanta, GA, p 38

Prada Triana MC (2002) Mapping benthic habitats on the south west of Puerto Rico as determined by side scan sonar. Ph.D. thesis, University of Puerto Rico, Puerto Rico

Richardson LL (1998) Coral diseases: what is really known? Trends Ecol Evol 13:438–443

Richardson LL, Aronson RR (2000) Infectious diseases of reef corals. Proc 9th Int Coral Reef Symp Bali, Indonesia 2:1225–1230

Riegl B, Moyer RP, Walker B, Kohler K, Dodge RE, Gilliam D (2008) A tale of germs, storms, and bombs: geomorphology and coral assemblage structure at Vieques (Puerto Rico) compared to St. Croix (U.S. Virgin Islands). J Coastal Res 24(4)

Rogers CS, Cintron G, Goenaga C (1978) *The impact of military operations on the coral reefs of Vieques and Culebra*. Report to Department of Natural Resources, San Juan, PR.

Rogers CS (1990) Responses of coral reefs and reef organisms to sedimentation. Mar Ecol Prog Ser 62:185–202

Rojas-Ortega J, García Sais J (2002) Characterization of the ichthyoplankton within the Mona Channel with emphasis on coral reef fishes. Proc Gulf Carib Fish Inst (in press)

Rosenberg E (2004) The bacterial diseases hypothesis of coral bleaching. In: Rosenberg E, Loya E (eds) Coral Reef Health and Diseases. Springer, Berlin, pp 445–462

Rosenberg E, Ben Haim Y (2002) Microbial diseases of corals and global warming. Environ Microbiol 4:318–326

Ruiz-Zarate MA, Gonzalez JE (2004) Spatial study of juvenile corals in the Northern region of the Mesoamerican Barrier Reef System. Coral Reefs 23:584–594

Sadovy Y (1999) The case of the disappearing grouper: *Epinephelus striatus*, the Nassau grouper, in the Caribbean and Western Atlantic. Proc Gulf Carib Fish Inst 45:5–22

Sanderson BG, Okubo A, Webster IT, Kioroglou S, Appeldoorn RS (1995) Observations and idealized models of dispersion on the south western Puerto Rican insular shelf. Mathl Comput Model 21:39–63

Singh H, Armstrong R, Gilbes F, Eustice1 R, Roman C, Pizarro O, Torres J (2004) Subsurface Sensing Technologies and Applications 5:25–42

Smith GW, Weil E (2004) Aspergillosis of gorgonians. In: Rosemberg E, Loya Y (eds) Coral Reef Health and Diseases. Springer, Berlin, pp 279–288

Smith GW, Ives ID, Nagelkerken IA, Ritchie KB (1996) Aspergillosis associated with Caribbean sea fan mortalities. Nature 382:487

Smith JE, Smith CM, Hunter CL (2001) An experimental analysis of the effects of herbivory and nutrient enrichment on benthic community dynamics on a Hawaiian reef. Coral Reefs 19:332–342

Souter D, Linden O (2000) The health and future of coral reef systems. Ocean Coast Manage 43:657–688

Spalding DL, Greenfeld A (1997) New estimates of global and regional coral reef areas. Coral Reefs 16:225–230

Stahl A (1883) Fauna de Puerto Rico. Clasificación sistemática de lo animales que corresponden á esta fauna, y catálogo del gabinete zoológica del Dr. A. Stahl en Bayamon. San Juan, PR, p 249

Stevenson DK (1978) Management of a tropical fish pot fishery for maximum sustainable yield. Proc Gulf Carib Fish Inst 30:95–114

Suárez-Caabro (1970) Puerto Rico 's fishery statistics 1968–1969. Contribuciones Agropecuarias y Pesqueras 2:1–38

Sullivan Sealey K, Bustamante G (1999) Setting geographic priorities for marine conservation in Latin American and the Caribbean. The Nature Conservancy, Arlington, VA

Sutherland KP, Porter JW, Torres T (2004) Diseases and immunity in Caribbean and Indo-Pacific zooxanthellate corals. Mar Ecol Prog Ser 266:273–302

Taylor M, Hellberg M (2003) Genetic evidence for local retention of pelagic larvae in a Caribbean reef fish. Science 299:107–109

Torres F (1975) Notas de campo y manuscrito no publicado sobre el estado de los arrecifes de coral en Puerto Rico, Unpublished manuscript, p 14

Tyler R (1992) The wind driven pressure and flow fields around a cylindrical island in a 1.5-layer ocean: Comparison with observations from the island of Puerto Rico. M.S. thesis, University of Puerto Rico, Puerto Rico, p 146

Tyler R, Sanderson BG (1996) Wind-driven pressure and flow around an island. Cont Shelf Res 16:469–488

Valiela I, McClelland J, Hauxwell J, Behr PJ, Hersh D, Foreman K (1997) Macroalgal blooms in shallow estuaries: controls and ecophysiological and ecosystem consequences. Limnol Oceanogr 45:1105–1118

Van Den Hoek C, Cortel-Breeman AM, Wanders JBW (1975) Algal zonation in the fringing coral reef of Curaçao, Netherlands Antilles, in relation to zonation of corals and gorgonians. Aquat Bot 1:269–308

Waddell JE (ed.) (2005) The State of Coral Reef Ecosystems of the United States and Pacific Freely Associated States: 2005. NOAA Technical Memorandum NOS NCCOS 11. NOAA/NCCOS Center for Coastal Monitoring and Assessment's Biogeography Team. Silver Spring, MD, p 522

Wanders JBW (1977) The role of benthic algae in the shallow reef of Curaçao (Netherlands Antilles) III: significance of grazing. Aquat Bot 3:357–390

Ward JR, Rypien KL, Bruno JF, Harvell CD, Jordan-Dahlgren E, Mullen KM, Rodriguez-Martinez RE, Sanchez J, Smith G (2006) Coral diversity and disease in Mexico. Dis Aquat Org 69:23–31

Weil E (1999) Diversidad y abundancia de corales, octocorales y esponjas en el Parque Nacional Jaragua, República Dominicana. In: Ottenwalder JA (ed) Biodiversidad Marino-Costera de la República Dominicana, p 31

Weil E (2003) The corals and coral reefs of Venezuela. In: Cortéz J (ed) Latin American Caribbean Coral Reefs. Elsevier, Amsterdam, The Netherlands, pp 303–330

Weil E (2004) Coral reef diseases in the wider Caribbean. In: Rosenberg E, Loya Y(eds) Coral Reef Health and Diseases. Springer, Berlin, pp 35–68

Weil E (2006) Coral, Ocotocoral and sponge diversity in the reefs of the Jaragua National Park, Dominican Republic. Rev Biol Trop 54:423–443

Weil E, Locke JM (2005) Scleractinian coral biodiversity in the Caribbean Revisited: are there more species? 32nd Scientific Meeting of the Association of Marine Laboratories of the Caribbean, Curacao, The Netherlands Antilles

Weil E, Ruiz HM (2003) Tissue mortality and recovery in *Acropora palmata* (scleractinian, Acroporidae) after a patchy necrosis outbreak in southwest Puerto Rico. 32nd Scientific Meeting of the Association of Marine Laboratories of the Caribbean, Curacao, The Netherlands Antilles

Weil E, Urrieztieta I, Garzon-Ferreira (2002) Local and geographic variability in the incidence of disease in western Atlantic coral reefs. proc 9th Int Coral Reef Sym 2:1231–1238

Weil E, Hernández-Delgado EA, Bruckner AW, Ortiz AL, Nemeth M, Ruiz H (2003) Status of Acroporid populations in Puerto Rico. Proc Caribbean Acropora Workshop: Potential Application of the US Endagered Species Act as a Conservation Strategy. NOAA technical Memorando NMFS-OPR-24, pp 71–92

Weil E, Torres JL, Ashton M (2005) Population characteristics of the black sea urchin *Diadema antil-larum* (Philippi) in La Parguera, Puerto Rico, 17 years after the mass mortality event. Special Issue: Equinodermos de Latino América. Rev Biol Trop 53:219–231

Weil E, Smith GW, Agudelo DG (2006) Status and progress in coral disease research. Dis Aquat Org Special Issue I 69:1–7

Wilcox WA (1900) The fisheries and fish trade of Puerto Rico. US Fish Comm Bull: 29–48

Willis BL, Page CA, Dinsdale DA (2004) Coral disease on the Great Barrier Reef. In: Rosenberg E, Loya Y (eds) Coral Reef Health and Diseases. Springer, Berlin, pp 69–104

Wynne MJ, Ballantine DL (1986) The genus *Hypoglossum* Kützing (Delesseriaceae, Rhodophyta) in the tropical western Atlantic, including *H. anomalum* sp. nov. J Phycol 22:185–193

Yoshioka PM, Owen GP, Pesante D (1985) Spatial and temporal variations in Caribbean zooplankton near Puerto Rico. J Plank Res 7:733–751

Yoshioka PM, Yoshioka BB (1989a) Effects of water motion, topographic relief and sediment transport on the distribution of shallow-water gorgonians of Puerto Rico. Coral Reefs 8:145–152

Yoshioka PM, Yoshioka BB (1989b) A multispecies, multiscale analysis of spatial pattern and its application to a shallow-water gorgonian community. Mar Ecol Prog Ser 54:257–264

10
Reef Geology and Biology of Navassa Island

Margaret W. Miller, Robert B. Halley, and Arthur C. R. Gleason

10.1 Introduction

Navassa is a small oceanic island (5.2 km² in size) located ~30 km west of the southwest tip of Haiti, 160 km south of the US Naval Base at Guantanamo Bay, Cuba, and in the heart of the Windward Passage. Navassa was claimed in 1856 by the United States. Navassa has also been claimed by Haiti since its independence in 1825 and, prior to that, was considered part of colonial Haitian territory. The current Haitian constitution (1987) claims Navassa by name as Haitian territory (Wiener 2005). This disputed sovereignty is a basis of much resource management challenge.

From the earliest geological investigations of Navassa (Leibig 1864; Gaussoin 1865) Navassa was recognized to be a carbonate island with abundant fossil corals and a thin cap of oolitic mineral phosphate. Although not guano, the phosphate-rich surficial deposits were mined and shipped to Baltimore to be converted to phosphoric acid and used to make fertilizer. Skaggs (1994) provides both a fascinating and disturbing historical account of 30 years of harsh phosphate mining on Navassa after the US Civil War.

Mining at Navassa proved to be a logistically challenging endeavor and ceased in 1898 after a labor uprising. The US Coast Guard built and manned a lighthouse at Navassa from 1917 to 1929 when it was replaced by an automatic beacon. The US Coast Guard ceased operation of the lighthouse in 1996 and jurisdiction passed to the US Department of Interior, which named Navassa, along with a 12-mile radius of marine territory, as a US National Wildlife Refuge in 1999. The Refuge is administered from an office ~800 km away in Puerto Rico and is logistically difficult to visit.

Although Navassa has been technically uninhabited since 1929, it is frequented by Haitian fishers and has been for generations (Wiener 2005; Miller 1977). Since the US Fish and Wildlife Service has had jurisdiction, observations have been made of small scale burns and cultivation on the island, presumably tended by the transient fishers. Navassa is an extremely inhospitable place (hot, rocky, no natural surface water sources, no food, and difficult access) and this fact alone has prevented full scale colonization and occupation by Haitians (Wiener 2005).

Remoteness and inhospitable conditions have also contributed to the general scientific neglect of Navassa. Scientific assessments of marine resources are summarized in Table 10.1. The first known scientific observations of Navassa marine resources were made in 1977 by the NOAA-National Marine Fisheries Service at the request of the Caribbean Fishery Management Council.[1] This survey was curtailed from a planned 10 days to 3 due to ship mechanical issues. A flurry of scientific attention was directed to Navassa following the Coast Guard pullout in 1996. The Center for Marine Conservation (now called the Ocean Conservancy) sponsored several expeditions, with various partners for both terrestrial and marine ecological inventories. A total of three expeditions were conducted between 1998 and 2000 which focused on island geology and taxonomy for a wide range

---

[1] Miller G.C. (1977) Navassa Island resource assessment survey. Cruise results for Oregon II 77–08 (80).

B.M. Riegl and R.E. Dodge (eds.), *Coral Reefs of the USA*,
© Springer Science + Business Media B.V. 2008

TABLE 10.1. Timeline of important historical and marine scientific events at Navassa Island.

Date	Event	Notes
1856	US Guano Act	Claim to Navassa formalized in 1857
1857–1898	Active mining by Navassa Phosphate Co.	~one million pounds of phosphate deposits removed
1917	US Coast Guard asserts claim and builds Navassa lighthouse	Residence by one keeper and two assistants
1929	USCG retrofits lighthouse with automatic beacon	Periodic visits only for maintenance
1977	NMFS/Oregon II cruise	Three working days at Navassa; tumbler dredge, line and trap fishing, snorkel observations, plankton tows
1996	USCG closes lighthouse and pulls out	Jurisdiction passes to US Dept of the Interior, Office of Insular Affairs
1998 (July–Aug)	CMC cruise (R/V Mago de Mar)	Focus on algae, ROV and SCUBA surveys (Littler et al. 1999)
1998 (Sept)	NMFS/Oregon II visit	One day, long line and drop camera video (Grace et al. 2000)
1999	CMC cruise/island survey (R/V Mago de Mar; Quest)	Comprehensive fish inventory including cryptic fishes (Collette et al. 2003); terrestrial inventory including geological assessments and sampling reported here
2000	CMC cruise (R/V Coral Reef II)	Focus on echinoderms, molluscs, crustatceans, and general reef status (Miller and Gerstner 2002)
2002	NMFS cruise (R/V Coral Reef II)	Comprehensive reef assessment including stationary fish counts, benthic community description and coral colony condition; fishery description (Miller 2003; Miller et al. 2005). First observation of net fishing
2004	NMFS cruise (R/V Coral Reef II)	Similar with addition of single-beam acoustical mapping; coral disease outbreak observed and characterized (Miller and Williams 2007); sponges characterized; intense fishing activity observed
2006 (Apr)	NOS cruise (NOAA Ship Nancy Foster)	Multibeam mapping completed (Piniak et al. 2006), temperature loggers installed, minimal fishing activity observed
2006 (Nov)	NMFS cruise (R/V Coral Reef II)	Severe coral bleaching event underway; moderate fishing activity with no net fishing observed

of terrestrial (plants, lizards, spiders, and insects) and marine groups including echinoderms, molluscs, algae, and fishes. Notable publications from this effort include Littler et al. (1999), Miller and Gerstner (2002), and Collette et al. (2003), but many results remain unpublished. Starting in 2002, NOAA/NMFS, in cooperation with the US Fish and Wildlife Service, has conducted more comprehensive reef assessment cruises on a biennial basis. A sociocultural study of the Haitain fishing communities affecting Navassa was conducted in 2004–2005 (Wiener 2005). Most recently, a distinct NOAA effort completed multibeam mapping of the Navassa shelf in 2006 (Fig. 10.1).

10.2   Regional Setting

Navassa is located near the geographic center of the Caribbean and in the middle of the Windward Passage. Both of these aspects contribute to a highly energetic physical-oceanographic regime, including periodic subjection to hurricane disturbance. In recent years, at least five hurricanes have passed within influential distance of Navassa (Ernesto 2006; Dennis 2005; Charlie 2004; Ivan 2004; Lili 2002). Also, the windward east coast of the island is subject to chronic large swells and this high energy regime is reflected in the submerged habitats and assemblages (Littler et al. 1999). This windward coast hosts massive boulders along the coastal cliffs, extensive rubble fields a bit further offshore, with minimal reef coral development.

Although the near-shelf oceanography has not been characterized, it is clear from observation that intermittent, presumably tidal, currents are fastmoving along the southwest coast. Regional oceanography, too, is not well characterized in this area, though general inflow from the Atlantic into the Caribbean Sea throughout the Lesser and Greater Antilles is known (Johns et al. 2002). The Windward Passage is one of the deepest of

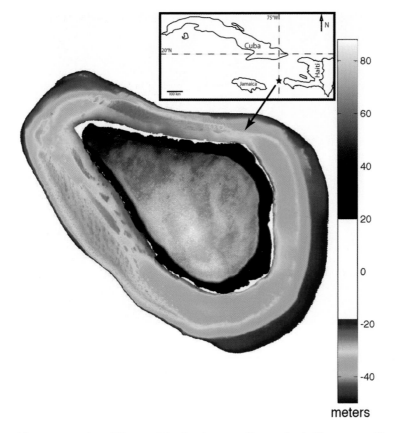

FIG. 10.1. Topographic representation of Navassa Island and surrounding marine habitats, out to 50 m depth. (LIDAR and multibeam bathymetry data from NASA and Mike Stecher (Solmar Hydro), respectively). Note that the two terraces above sea level tilt down toward the northwest corner (darker shades of gray to NW), but the submerged shelf is relatively level

these passages at 1700 m. Though the transport has not been well quantified, it is believed to be in the range of 3–6 Sv (1 Sv = 10^6 m³/s (Johns et al. 2002)). For comparison, total transport of the Florida Current at 27° N is ~32 Sv (Baringer and Larsen 2001).

The other important physiographic factor about Navassa Island is its topography. Figures 10.1 and 10.2 illustrate the 'wedding cake' morphology of progressive shelves, cliffs, and plateaus that constitute Navassa Island and its nearshore coastal region. Cliffs, often severely undercut by erosion and surrounding the island, reach down to a first shelf at 20–30 m depth. Navassa was believed to be a small uplifted atoll, complete with a fringing reef and volcanic core, an idea that persisted contrary to existing evidence until the 1980s (Fairbridge

1983). Some saw the first terrace (Fig. 10.2), with its slightly elevated rim, as atoll-like (Gaussoin 1866). The second terrace (Fig. 10.2) has the geomorphology of a fringing reef. But D'Invilliers (1891) recognized that this was an erosional terrace that he attributed to uplift (being unaware of sea-level changes of this magnitude). Sheer coastal cliffs and sharp bathymetric drop off precludes the possibility of beaches, lagoons, seagrass beds, and mangroves that typify the coasts of most islands in the Caribbean.

This habitat mosaic clearly has implication for ecosystem structure and function. We would hypothesize that species that use these connected systems for nursery or foraging habitat might be reduced in abundance, that potential nutrient runoff from the island (natural or anthropogenic) might

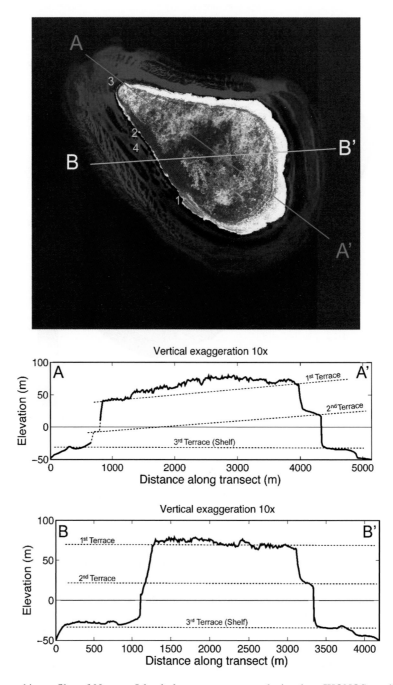

FIG. 10.2. Topographic profiles of Navassa Island along two transects depicted on IKONOS satellite image. Note different distance scales. Numbers on map indicate sites with time series of reef condition data (see Fig. 10.16); (1) Lulu Bay, (2) West Pinnacles, (3) Northwest Point, (4) Deep Patch. Dashed lines on profiles A and B indicate the three terraces mentioned in the text (Data are from the same sources as Fig. 10.1)

reach reef organisms more directly without the buffering of uptake by mangroves and seagrasses (Littler et al. 1999), and that overall reef trophic structure might lack allochtonous inputs from these adjacent sources of primary productivity. While the virtual lack of grunts (family Haemulidae) on Navassa reefs bears out this first hypothesis (see discussion below), the others have not been tested.

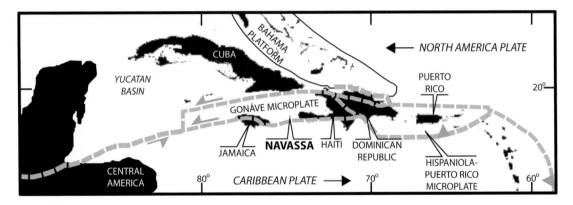

FIG. 10.3. Location of Navassa Island in relation to Haiti, Jamaica, Cuba, and the fault boundary between the Caribbean and North American Plates. The Gonave Microplate is considered the latest addition to the North America Plate (After Mann et al. (1995) by permission of Elsevier)

10.3 Geologic Setting

Navassa is located on the Gonave microplate (De Mets and Wiggins-Grandison 2007), a piece of the earth's crust caught in the great strike-slip faults between the Caribbean and North American plates (Fig. 10.3). The Gonave microplate is the latest in a series of three crustal wedges that originated on the Caribbean plate and have been transferred to the North American plate as the former moves eastward. The Gonave plate developed in the Miocene, causing uplift and folding in the region. Some areas became mountain ranges in southwestern Hispaniola. The southwestern peninsula of Haiti extends below sea level westward as the Navassa Ridge, Navassa being the only portion of the ridge above sea level. However, there are many other bathymetrically shoal regions in the area, the result of complicated folding and faulting of the crust. One of these, Navassa Knoll, lies only 6.4 km to the east of the island and rises to within 400 m of sea level. The region is seismically active today. Late Pleistocene and Holocene uplifted fossil reefs occur in southwestern Haiti, although they are absent at Navassa. Mann et al. (1995) found a decrease in uplift rates from north to south in western Haiti, with uplift rates of 0.37 mm/year at the Northwest Peninsula to 0 mm/year in the south-central part of western Haiti. It seems possible that Navassa may have been stable or even subsiding during this time period.

Proctor (1959) reported a range in age from Miocene to recent based primarily on foraminifera from surface samples of the shallow-water carbonate deposits on the island. They found that Navassa was not an atoll and that the island was capped with red–brown lateritic loam, poor in phosphate (2–8%) and rich in calcium and aluminum. They suggested that phosphate resided in a mineral similar to sokolovite, known to be associated with bauxite, a similarly reddish deposit commonly found in laterites. Burke et al. (1974), using biostratigraphic methods, reported fossil coral from Navassa as Eocene in age. They concluded that the ooids and pisolitic surface material contained volcanic ash, accounting for the unusual chemistry reported by Proctor (1959). Interestingly, Burke et al. (1974) found no evidence of Pleistocene marine deposits or reefs on the second terrace (Fig. 10.2), a feature common on many other Caribbean islands.

10.3.1 Geomorphology

The most prominent geomorphic features of Navassa are two terraces above sea level and a third that forms a small shelf below sea level (Figs. 10.1 and 10.2). The surrounding narrow shelf is 1–2 km wide and currently supports reef growth. It is likely that the shelf is underlain by an erosional terrace, similar to the lower terrace of the island, overlain by Pleistocene and Holocene deposits. At a depth of 30–40 m, the shelf is covered by an unknown thickness of Holocene reef sediments and may have been a site of active carbonate production during previous Pleistocene sea-level highstands. The shelf was most recently submerged about 11,000 years ago, 10,000 years after the glacial maximum (Chappell and Shackleton 1986; Toscano and Macintyre 2003). The spur and groove structure seen at the southwest and north margins of the shelf (Fig. 10.1) is evidence of past reef-building.

Above sea level, Navassa presents a fascinating view, with cliffed shores around most of the island and two prominent terraces (Figs. 10.1, 10.2 and 10.4). The lower (second) terrace is tilted to the northwest, forming 20 m cliffs on the east side of the island and a submerged bench off the northwestern tip to a depth of about 10 m (Figs.

10.1, 10.2, and 10.4). The higher island plateau is similarly tilted about 30 m down to the northwest across the length of the island. The bathymetry of the narrow shelf does not exhibit any tilt, being uniformly 30–40 m deep at both ends of the island (Figs. 10.1 and 10.2). These relationships suggest that the tilting predates Holocene shelf sedimentation, that shelf sedimentation has compensated for island tilting, and that there are at least 30 m of Pleistocene and Holocene deposits on the northwestern shelf.

10.3.2   Lithology

The island has a simple stratigraphy consisting of approximately 60 m of gray dolostone capped

Fig. 10.4. (**a**) View along the eastern edge of Navassa. The lower terrace extends about 300 m to cliffs along the island edge. The upper terrace is 60 m above sea level. (**b**) View of northwest Point of Navassa showing a submerged, eroded remnant of the lower terrace extending seaward from the island. Inset: Highly eroded dolomite karst on the submerged portion of the lower terrace (see also Fig. 10.11c)

by thin veneer of red–brown phosphorite. There are some lateritic soil deposits on the island, now mostly preserved in caves, but the bulk of the soil was stripped away during phosphate mining. There are also numerous karst features and cave deposits that will be referred to here only as they relate to the long history of island emergence. The cliffs around the island exhibit some horizons of greater cave frequency suggesting locations of past water tables. Cursory exploration of the current cave system indicates continued karst development with some wet caves being observed on the island. A vertical exploratory shaft dug during the mining era proved the continuity of the dolostone from the surface of the upper terrace to sea level. Bedding seems to be absent and the dolostone is highly jointed and fractured. Although an exhaustive search was not possible, evidence of unconformities, disconformities, or soil horizons were not observed within the dolostone. However, these features can be subtle and may be found during more extensive surveys.

10.3.2.1 Dolostone

The dolostone is highly fossiliferous, but weathered surfaces obscure the internal structures of the rock. This prevents facies mapping in the field. Where depositional fabrics can be observed, reef and reef-associated sediments dominate. Macrofossils include a variety of corals, mollusks, foraminifera, green and red (coralline) calcareous algae (Fig. 10.5). Reef fabrics, including voids, internal sediment and marine cements are evident on some surfaces. Some areas are very course-grained others muddy suggesting the presence of a variety of microfacies that might be mappable with extensive petrographic effort. Thin sections reveal a typical suite of reef-associated organisms and sediments (Fig. 10.6). These sediments were originally deposited as calcium carbonate (calcite and aragonite) skeletons of the organisms. All of the original sediment has been replaced by dolomite (calcium – magnesium carbonate). Fossils are replaced by fine-grained dolomite that preserves many details of skeletal architecture and allows identification of classes of organisms but generally not enough detail to identify species. The magnesium needed for dolomitization was derived from seawater moving through the sediments and altering mineralogy as well as the minor and trace elements of the

rocks. Some rock samples contain a last generation of calcite cement, perhaps related to the long and continuing exposure and karstification of the island.

Navassa dolomite has much in common with other Tertiary dolomites of the northern Caribbean. Similar dolomites are found on Jamaica, the Cayman Islands, and Haiti. Budd (1997) reviewed the origin of dolomite on carbonate islands, including those near Navassa. Of six dolomitization models reviewed by Budd (1997), five require some degree of subaerial exposure to develop brines and/or subsurface seawater flow. Only thermal driven dolomitization does not require exposure, a mechanism that probably could not function at Navassa because of its small size. After the deposition of Navassa carbonate sediments, subaerial exposure resulted in one or more periods of dolomitization, leading to a massive, fine-grained, replacement dolomite in the terminology of Budd (1997).

Strontium isotope ratios were obtained from dolomite of the lower terrace, slope, and upper terrace respectively (Fig. 10.7). These values progressively increase upslope and can be related to seawater composition of the Late Miocene, roughly 5–10 million years ago (McArthur et al. 2001) As discussed above, this period relates to the time of dolomitization, not the deposition of the original reefal sediments.

The contact of the dolomite with the overlying phosphorite is disconformable and complicated by the development of karst prior to phosphorite deposition. Evidence of this exposure includes: (1) phosphorite fills caves and solution enlarged fractures and joints in the dolostone ; (2) phosphorite overlies travertine in the dolostone; (3) pieces of stalactites are encased in phosphorite; and (4) weathered clasts of dolostone occur in the base of the phosphorite (Fig. 10.8). The extensive karst on the dolostone surface was exhumed over broad areas during phosphate mining. The karst surface is similar on both the upper and lower terraces, although the lower terrace appears to have been more completely covered by phosphorite than the upper terrace. The lower terrace has been very thoroughly mined exposing previously filled solution holes generally 0.3–1 m wide, 1–2 m deep resulting in a pitted surface (Fig. 10.9a). The upper terrace still retains undisturbed areas where phosphorite covers the surface (Fig. 10.9b) and weathered

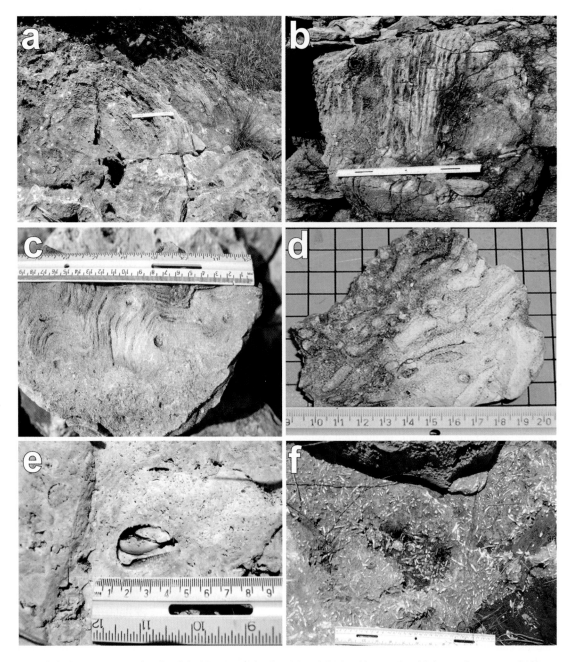

FIG. 10.5. Common macrofossils of the Navassa dolomite. (**a**) and (**b**) boulder corals with internal structure similar to *Montastraea* and *Diploria*; (**c**) and (**d**) corals with internal structure similar to *Porites* and *Stylophora*; (**e**) gastropod (cowrie) mold; (**f**) fragments of coralline algae

dolomite ridges and knolls (Fig. 10.9c) that appear to have been exposed by natural processes. A prominent ridge (Fig. 10.9d) surrounds the edge of the upper terrace and was cited by geologists in the nineteenth century as evidence that Navassa

was an atoll. However, Purdy (1974) demonstrated that a raised rim was the common product of karst processes operating on carbonate islands. In the absence of facies patterns that support the atoll hypothesis, it seems unlikely that Navassa carbonate

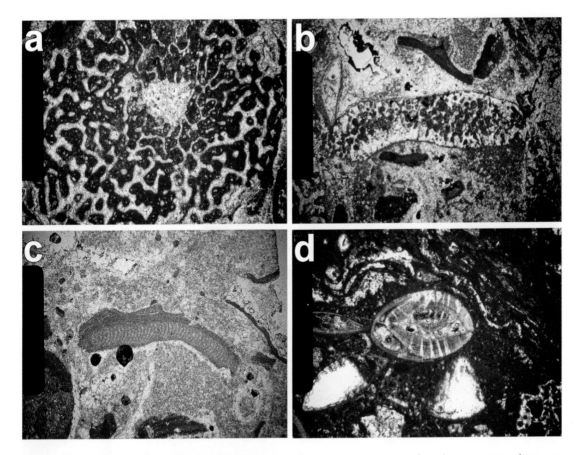

FIG. 10.6. Photomicrographs of Navassa dolomite textures. (**a**) coral, e.g. *Porities*; (**b**) calcareous green algae, e.g. *Halimeda*; (**c**) coralline algae, e.g. *Goniolithon*; (**d**) foraminifera, e.g. *Amphistegina*. Field of view is 3 mm in all images

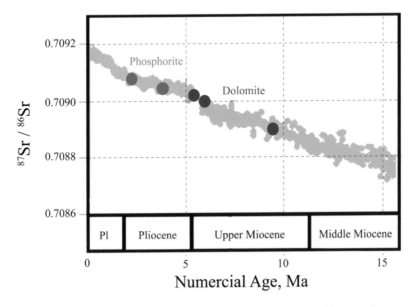

FIG. 10.7. Strontium isotope ratios for five samples (solid circles) from Navassa. The three lowest values are from dolomite (blue), indicating Late Miocene dolomitization; two phosphorite samples (red) indicate Pliocene ages. Background (gray) is the seawater strontium isotope curve after McArthur et al. (2001) (Figure adapted from DePaolo (1986), by permission via the fair use policy of the Geological Society of America)

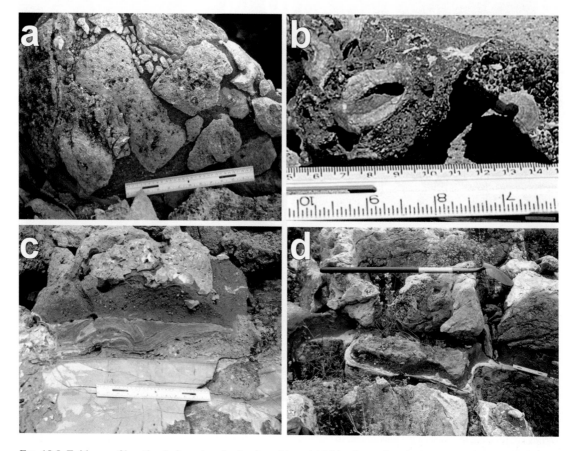

FIG. 10.8. Evidence of karsting before phosphorite deposition. (**a**) Lithoclasts of eroded dolomite are incased in phosphorite at the dolomite/phosphorite contact; (**b**) pieces of tubular, fractured stalactites are incased in oolitic pebble phosphate; (**c**) phosphorite overlying truncated travertine; (**d**) phosphorite infilling solution-enlarged joint

sediments originated as an atoll. The island is more likely the eroded remnant of a larger carbonate bank shaped by chemical and physical erosion to its present form.

Navassa dolostone probably has counterparts among the Tertiary carbonates of western Hispaniola but these units are described in insufficient detail to know if they are correlative. There are detailed similarities between the dolostone of the well-described Bluff Group in the Cayman Islands (Jones et al. 1994) that provide some confidence in suggesting they are correlative in age with Navassa dolostone. First, both units are late Tertiary and uncertainties in the ranges of the fossils identified from Navassa do not preclude these rocks from being the same age as those

identified on Grand Cayman by Jones and Hunter (1989). Second, both the dolostone of Navassa and Cayman are finely crystalline dolomite preserving much of the skeletal architecture of enclosed fossil corals, bivalves, coralline algae, gastropods, forams, and the green calcareous alga *Halimeda*. Third, both units yield strontium isotope ratios indicative of major episodes of dolomitization that took place in the late Miocene (Jones and Luth 2003). Fourth, both units were exposed above sea level and developed paloekarst during the Late Miocene (Jones and Smith 1988). Fifth, assuming that the shelf is eroded into dolomite similar to the terraces, the unit is at least 90–100 m thick, comparable with the thickness of dolomite encountered in wells on Grand Cayman (Jones and Luth 2003).

Fɪɢ. 10.9. (**a**) Severe karst surface of the lower terrace. The solution holes of this surface were originally filled by phosphorite that was arduously stripped out by miners during the nineteenth century. (**b**) Unconsolidated oolitic phosphorite filling in karsted dolomite on an undisturbed portion of the upper terrace (sample bags and geologic hammer for scale). (**c**) Dolomite knoll on the upper terrace. Some areas of exposed dolomite on the upper terrace are not as severely karsted as the lower terrace. (**d**) Raised rim along the margin of the upper terrace (**e**) Cemented oolitic phosphorite. This phosphorite layer is cemented with calcite cement and is reverse graded, that is, the grain size increases upward. Reverse grading is a characteristic of some cave deposits. Scale on right in millimeters. (**f**) Mound of phosphate sand ore left from mining activities in the nineteenth century. Note the lack of trees or bushes on the mound

The Bluff Group contains areas of preserved lime-stone and several disconformities not identified at Navassa, but these differences may be the result of a lack of thorough examination of all the outcrops on Navassa.

10.3.2.2 Phosphorite

Phosphorite on Navassa occurs in a variety of forms including fine-grained fissure deposits, sand, pebbles, nodules, and irregular clastic conglomerates, but the most common form is as an oolitic sand deposit. It ranges in color from orange to red/brown. The phosphorite is lithified near its contact with the underlying dolomite, thoroughly cemented with

either phosphate or calcite cement (Fig. 10.9e). Where the phosphorite appears to be undisturbed, it occurs as oolitic sand, an unconsolidated deposit that was the ore of choice during the 1800s. Piles of this mined material remain along railway beds, still waiting to be shipped from the island (Fig. 10.9f).

Extensive searching revealed no marine fossils associated with the phosphate deposit. A few small bone fragments were found, but nothing sufficient to identify a genus. However, a reptile tooth, turtle fragments, land snails, and possible bird bones were found preserved within the phosphate rock.

The oolitic phosphate is typically poorly sorted with phosphate grains ranging from about 100 µm to 2 mm in diameter (Fig. 10.10a). In thin section, two

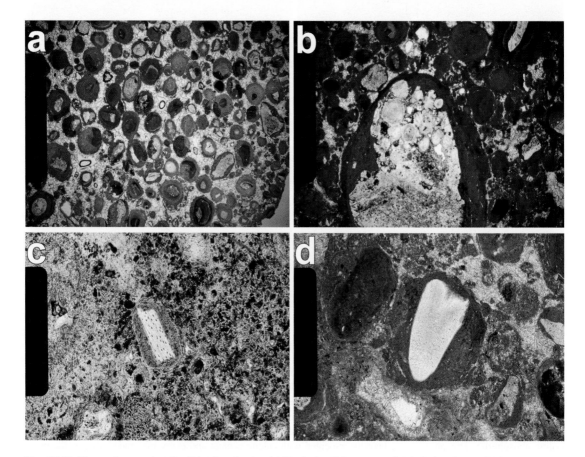

FIG. 10.10. Photomicrographs of oolitic phosphorite. (**a**) Typical oolitic texture, field of view 3 mm; (**b**) composite phosphate grain in poorly sorted oolitic matrix, field of view 3 mm; (**c**) phosphate ooid grain with bone fragment nucleus in fine-grained phosphate matrix, field of view 1.5 mm; (**d**) phosphate ooid with bone-fragment nucleus in poorly sorted phosphate matrix, field of view 1.5 mm

to eight layers surround nuclei that may be other fragments of phosphate or occasionally bone fragments (Fig. 10.10b–d). Pebbles and nodules are typically microcrystalline in thin section. X-ray diffraction analyses indicate the primary mineral is calcium fluorapatite with trace amounts of quartz, calcite, hematite and bauxite. The samples we obtained did not contain volcanic glass or feldspar, as reported by (Burke et al. 1974). This is not surprising as volcanic glass weathers quite quickly to clay.

Chemical analyses of fourteen phosphorite samples are shown in Table 10.2, four from the lower terrace and ten from the upper terrace. These analyses agree with early mining assays that reported 20–40% phosphate. Burns, Versey and Williams reported in Proctor (1959) found only 1–7% phosphate in Navassa samples and suggested that a mineral similar to sokolvite was the major constituent, but our analyses confirm the phosphate content reported during the 1800s and are consistent with the mineral calcium fuorapatite.

Two strontium isotope ratio values from the oolitic phosphorite are shown on Fig. 10.7. If it is assumed that the strontium was ultimately marine, these values indicate a Pliocene age for the phosphorite. These values suggest that the exposure and karstification of the underlying dolomite occurred during the latest Miocene, perhaps during the Messinian sea-level low stand, the same event that resulted in exposure recorded in the Cayman Islands (Jones and Hunter 1994).

The abundant phosphorite, the absence of marine fossils, and the karst surface beneath the phosphorite suggest that Navassa was enriched with guano during the Pliocene and that subsequent weathering and diagenetic alteration has produced the present deposit. Phosphorite deposits are not known on nearby islands, and perhaps the presence of rodents or other predators prevented birds from occupying those islands in sufficient numbers to produce guano. Similar deposits on Pacific islands are generally accepted to be the result of chemical interactions between solutions leaching seabird guano and the underlying reef carbonate (Braithwaite 1980; Aharon and Veeh 1984). During weathering, nitrogen and organic matter are removed from the guano and phosphatic ooids are formed diagenetically by soil and vadose (cave) processes.

The mineral phosphate deposits are highly insoluble, much less soluble than dolomite or calcite. It may be that the oolitic phosphate was originally cemented with calcite and that the calcite has dissolved away with continued weathering of the island surface. In this scenario, the oolitic sand is an insoluble residue left on the surface of the island after millions of years of karstification. The phosphate mineral is even less soluble in seawater. The phosphorous is so tightly bound by the mineral that it is not bioavaiable. Thus the deposit contributes little to either the productivity of the island or the surrounding marine shelf.

TABLE 10.2. Chemical composition of 14 phosphorite samples from Navassa.

Sample	SiO_2(%)	AlO_2(%)	Fe_2O_3(%)	MgO(%)	CaO(%)	P_2O_5(%)	Sr (ppm)
Lter1	1.71	7.38	4.24	0.23	40.99	35.29	4,657
Lter2	1.53	6.62	3.78	0.23	42.31	36.01	4,269
Lter3	2.81	6.45	2.94	1.06	42.08	11.51	4,516
Lter4	1.59	3.56	1.69	0.48	47.39	16.39	2,414
Uter1	1.91	7.52	3.52	2.77	39.64	11.52	3,099
Uter2	3.15	11.88	5.44	1.12	34.43	13.47	4,804
Uter3	2.24	8.75	4.05	1.72	39.24	12.82	3,581
Uter4	3.43	13.32	6.11	1.69	31.71	14.41	5,394
Uter5	2.27	10.51	4.75	0.99	37.05	16.24	4,241
Uter6	2.15	9.53	4.31	0.74	38.40	21.74	4,046
Uter7	3,61	15.14	7.22	0.22	28.04	28.72	7,208
Uter8	3.75	15.18	7.27	0.22	27.07	28.44	7,473
Uter9	3.88	15.54	7.41	0.22	28.29	28.92	7,388
Uter10	4.18	15.9	7.42	0.22	26.90	28.21	7,528

10.3.3   Brief Geological History

Navassa carbonates may have been part of a much larger carbonate bank dominated by shallow water deposits from the Eocene – Oligocene to the Miocene (30–6 Ma). During this time reef growth was widespread, but probably did not construct an atoll similar to the present geomorphology of the island. During the late Miocene, sea-level regression and tectonic uplift initiated exposure and dolomitization followed by erosion and karstification that began to shape the island. The absence of marine deposits younger than Miocene suggests Navassa has been an island for perhaps the past 5–7 million years. This provided a long period of exposure, sufficient time for the island to be extensively karsted, to become a guano island, for the guano to be weathered and altered to phosphorite, and for the island to be tilted and perhaps subside. The apparent absence of emergent late Pleistocene marine deposits, particularly those from isotope stage 5e (125,000 years BP), is problematic. At that time sea level was 2–6 m higher than today (Fruijtier et al. 2000; Schellmann et al. 2004) and reefs could have established on the northwestern rim of the island where the lower terrace is near sea level. Perhaps reefs of this age did exist and were eroded away, or perhaps the island is subsiding and those deposits are hidden beneath Holocene sediments on the shelf. During the last glacial maximum, about 20,000 years BP, sea level was more than 130 m below present, exposing the entire shelf and enlarging the island to more than twice its present size. At about 11,000 years BP sea level had risen sufficiently to flood Navassa's narrow shelf, and coral reef growth could resume.

10.4   Biogeography and Biodiversity

In the classic understanding of Island Biogeographic Theory (MacArthur and Wilson 1967) we would predict Navassa, as a very small, relatively isolated island, to host a relatively depauperate community, based on the difficulty of colonization from distant source populations and the likelihood that small islands offer more restricted niche or habitat options. The latter (restricted niche) characteristic is likely borne out in that Navassa is surrounded by cliffs and rocky shores, with no beach, mangrove or seagrass habitats. Clearly, species dependent upon these interconnecting habitats would be less likely to persist at Navassa.

This aspect of niche limitation is evident in the Navassa fish fauna. Several authors have noted the rarity of groups that utilize seagrass beds for nursery or forage grounds such as grunts and hogfish (Miller and Gerstner 2002; McClellan and Miller 2003) and the absence of forage fishes such as herring and anchovies (Miller 1977; Collette et al. 2003). Similarly, apparent absence or rarity of certain groups of echinoderms such as sea cucumbers and burrowing urchins has been suggested to result from limitation of fine sediment, sheltered habitat and appropriate nursery substrates (G. Hendler, personal communication, 2000)

The best biogeographic analysis for Navassa reef organisms relates to fishes. Collette et al. (2003) used rotenone sampling in addition to more traditional methods such as visual surveys, light/dip netting, hand lining, and trolling to perform a quite comprehensive inventory of the Navassa fish fauna on the M/V Quest expedition in 1999 (Table 10.1). These authors recorded 224 fish species, from 66 families. Their collections added 160 species that had not previously been recorded from Navassa, and brought the total species list for Navassa fishes to 237. Collette et al. (2003) analyze the Navassa fish fauna in comparison to Bermuda and several oceanic atolls off of Colombia and describe the Navassa assemblage as "relatively depauperate", noting the lack of continentally associated families such as toadfishes, searobins and snooks as well as seagrass associated foraged fishes such as herring and anchovies. Collette et al. (2003) cite for comparison the large island of Cuba, the nearest island for which recent fish faunal records are available, with 806 'shore fishes'. Interestingly, 17 fish species were found in Navassa which have not been reported for Cuba, but Collette et al. (2003) caution that these are cryptic fishes collected via rotenone which would likely be encountered in Cuba if similar sampling methods were applied. Subsequent visual census on following expeditions have added at least 40 additional species (McClellan and Miller 2003) to the total list bringing the total to over 275.

The only direct analysis of connectivity of Navassa reef organisms utilized population genetic markers of the imperiled Elkhorn coral, *Acropora palmata*. Although shallow habitats occupied by *A. palmata* are scarce at Navassa, this species appears to be on an increasing trend at Navassa (from a reported dozen colonies limited to Lulu Bay in 1997 (Littler et al. 1999) to over 1,800 colonies mapped in 2006 distributed along the entire southwest and north coasts (D.E. Williams, personal communication 2006). The Navassa population of this imperiled species belongs to a 'western Caribbean' population, genetically distinct from areas east of the Mona Passage (Baums et al. 2005). Baums et al. (2006) showed the highest levels of *A.palmata* clonal diversity (hence, inferred rates of sexual recruitment) of any region sampled across the Caribbean occurred at Navassa Island. Along with highly successful sexual recruitment, this extreme level of genotypic diversity also indicates a lack of successful asexual recruitment in the Navassa population, the reproductive mode which is presumed dominant for this species overall (Highsmith 1982; Bruckner 2002). This may be related to the predominantly vertical bathymetry which precludes the retention of fragments that do occur. An alternative explanation is that the lack of genotypes with lots of ramets is an artifact of a rapidly recovering population, i.e., asexual recruits have simply not had time to accumulate in a relatively 'new' population.

10.5   Benthic Habitats

Systematic efforts to map the benthic habitats of Navassa began in 2004. The objective was to produce classified maps of the seabed that could be used (a) to focus limited field time by stratifying sampling effort, and (b) to support management plans for the refuge. Two independent methods were used: manual image interpretation and automated classification of acoustic echoes. The resulting set of maps provides a basis for improved fish and benthic invertebrate population surveys and suggests that physical forcing mechanisms are the dominant control on Navassa's benthic communities.

10.5.1   Manual Image Interpretation

The objective of manual image interpretation was to produce a product consistent with other NOAA maps of coral reef habitats (see http://ccma.nos. noaa.gov/about/biogeography/ for specific initiatives). Inputs to the classification were an IKONOS satellite image from 3 March 2001, diver photoquadrats (on the shelf) from 2004, diver observations (near shore) from 2004, and drop camera images from April 2006. The interpretation used a classification scheme very similar to NOAA (2005) to produce maps of cross shelf geomorphologic zone, substrate, and benthic cover.

Two data processing steps were required. First, the IKONOS satellite image was corrected for sun glint artifacts following Hochberg et al. (2003). The corrected image was then used as a base on top of which polygons demarking geomorphologic zone and substrate were hand digitized with ArcGIS9 using a minimum mapping unit (MMU) of 1 acre. Second, benthic cover data were clustered using haphazardly placed 1 m^2 photoquadrats from 28 sites sampled in 2004. These photoquadrats (n = 6–10/site) were analyzed via standard point count methods (CPCe, NCRI, Nova Southeastern University). Non-parametric similarity and clustering were then calculated on percent cover of major benthic groups (e.g., hard corals, macroalgae, sponges, etc.) using PRIMER (v.5).

Navassa lacks the typical patterns of reef zonation, which were classically described for Jamaican reefs in the 1960s–1970s (Goreau 1959), since no reef crest, shallow lagoon, or back reef habitats are present. Nevertheless, four cross shelf geomorphologic zones were identified: shallow shelf, wall, wall base, and shelf. The shallow shelf zone is mostly limited to a small area on the NW point where the second terrace dips below sea level to a depth of about 10 m. Tiny "shoulders" nestled along the western wall also fall in this zone, but were smaller than the MMU. Vertical cliff walls completely circle the island (Fig. 10.11a). The wall base forms a transition zone between the wall and the shelf and can be distinguished by intermediate slopes and the presence of boulders fallen from the cliffs. The shelf is the largest geomorphological zone at Navassa, here defined as the area of the third terrace (Fig. 10.2) extending from the edge of the wall base zone out to the 50 m isobath.

Six major substrate classes were identified from IKONOS image interpretation (Fig. 10.12): boulders, calves (house-sized boulders; Fig. 10.11b), hardbottom, reef, rubble, and sand. Hardbottom sites were divided into two subclasses based on the

FIG. 10.11. Reef habitat types. (**a**) Vertical cliff surrounding the island, including encrusting *Acropora palmata* colonies; (**b**) House-sized boulders at the wall base, note three small divers for scale (Photo credit: K. Pamper); (**c**) Eroded rib structure in the shallow reef shelf at the northwest point; (**d**) Patch reef along southwest coast; (**e**) Hardbottom with sand veneer and substantial sponge structure; (**f**) Rubble fields along the east and southeast coasts (Photo credit: B. Yoshioka), inset shows conch with substantial seaweed growth

presence or absence of a veneer of sediment and a third subclass for the shallow area of the NW point covered by sharp, small scale ridges (Fig. 10.11c) composed of eroded dolomite as discussed above. Reef sites were divided into four subclasses: aggregate (continuous, larger than the MMU, lacking sand channels); large patches between sand channels (resembling extremely large spur and groove formations); small patches containing sand channels (resembling spur and groove formations); and small isolated patches in sand (collections of high relief patch reefs that are smaller than the MMU). The class names "patches containing sand channels" were chosen rather than "spur and grove" because most of these formations are not oriented perpendicular to the shoreline or shelf break.

Sites were grouped into five classes based on benthic cover from the photoquadrat clustering analysis (Fig. 10.13). The windward east coast contains an extensive gravel/rubble bed (Fig. 10.11d). Conch appear to serve as the primary substrate for

seaweeds and sessile benthic invertebrates in this habitat (Fig. 10.11d, inset). High abundances of large sponges and gorgonians are common in the "hardbottom with sand veneer" substrate along the northwest, north, and east portions of the shelf (Fig. 10.11e). Macroalgae dominates the inner portion of the reef on the north shelf. The greatest coral cover and reef development is found on reefs along the more protected southwest coast (Fig. 10.11f). Sites with "intermediate" cover, a mix of coral and macroalgae, are found on the margins of the areas that were classified as macroalgal dominated or high coral cover (Fig. 10.13).

10.5.2 Automated Acoustic Classification

The goal of the acoustic survey was to investigate the potential for automated seabed classification of a coral reef environment using a commercial single beam mapping system, the Quester Tangent Series

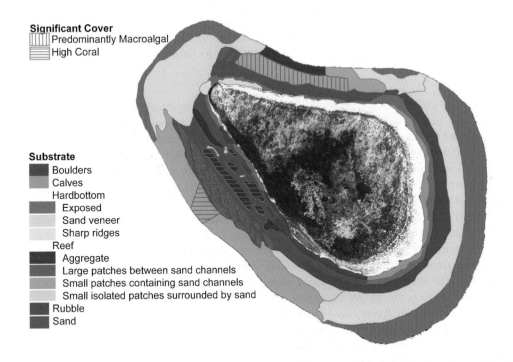

FIG. 10.12. Benthic habitats derived from manual image interpretation. Solid colors reflect the substrate identified for each polygon (reef, rubble, sand, etc.). The horizontal and vertical grey stripes superimposed on some polygons represent the benthic cover class identified from photoquadrats or drop camera observations. For clarity, benthic cover data are shown only for the areas dominated by macroalgae (vertical stripes) and with high coral cover (horizontal stripes). The main point is that the distribution of benthic cover is spatially coherent, with continuous portions of reefs on the northern side dominated by macroalgae and on the southwest side by high coral cover

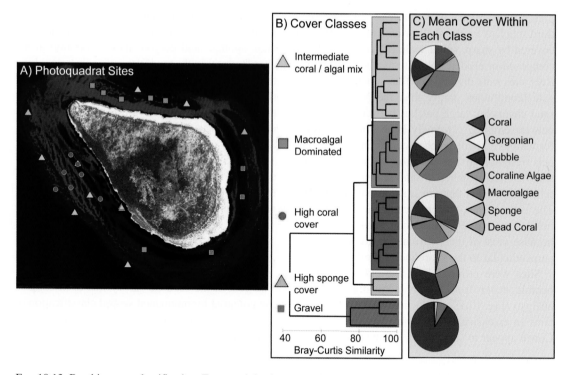

FIG. 10.13. Benthic cover classification. Twenty-eight sites around the Navassa shelf were sampled with photoquadrats in 2004 (**A**) depicted on IKONOS satellite image (**B**) and classified according to similarity in percent cover of major benthic groups (**C**) Pie graphs depict the mean percent cover for the sites sampled within each class

V (QTCV). The classification scheme chosen for this work was based on Franklin et al. (2003), which was used by NOAA and the University of Miami for a fisheries stock assessment of the Dry Tortugas National Park (Ault et al. 2002). The classification scheme is based on patchiness, defined as the fraction of the seabed covered in sediment, and relief, defined as the local variability in depth.

Approximately 100 km of ship tracks were run in 2004 to acquire the raw single beam data using a QTCV interfaced with a Suzuki 2025 50 kHz echo sounder. The standard QTC processing steps (Gleason et al. 2006) produced five clusters of echoes, three of which were sediment classes and two of which were rocky (reef) classes, based on inspection with the IKONOS image. The classified data and depth measurements were then used to compute continuous values of patchiness and relief along the track lines. Franklin et al (2003) used multiple thresholds to define ten classes (nine with rocky bottom plus bare sediment). In the case of Navassa, however, we used only five classes (four with rocky bottom plus bare sediment) due to limited field time. A majority filter

was used to aggregate the individual classified points along the track lines up to the desired 100 m grid size for stratifying random fish and coral surveys.

The classified acoustic map (Fig. 10.14) shows that the major areas of rocky seabed occur along the northern and southwestern portions of the Navassa shelf. The majority of the northwest and southeast portions of the shelf are classified as sediment, which may be due to a thin veneer reducing the roughness, and therefore backscatter strength, of the hard bottom in these areas. Confirmation of this speculation, however, requires additional experiments.

10.5.3 Habitat Classification Comparison

There is a striking similarity between the image-derived and acoustic-based depictions of the northern and southwestern reefs (Figs. 10.12 and 10.14). In both cases, the northern reef is mapped as two solid bands parallel to the shoreline. The southwest reef, on the other hand, is depicted as a complex mosaic of patches of differing size and relief separated by

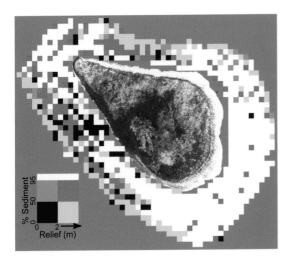

% Sediment
95
50
0
Relief (m)
0   2

FIG. 10.14. Benthic habitats derived from automated acoustic classification. Two continuous variables are derived from the acoustics: "patchiness", the percent of the seabed covered with sediment, and "relief", the local variability in depth. Arbitrary thresholds are used to bin these variables to discrete classes for visualization and sample stratification purposes. Data were acquired as points along track lines, classified, then gridded to 100 m cells using a majority filter. No data is shown as gray; note the area of no data close to the island is larger on the windward north and east shores

sediment channels of varying widths. The coherent picture from these two classification methods, which are based on completely different physical processes, supports the theory that Navassa's benthic communities are strongly controlled by physical forcing.

The comparison of optical and single beam acoustic methods of seabed classification reveals strengths and limitations of each approach. Practical limitations on how closely ship tracks can be spaced imply that traditional image interpretation will generally resolve smaller features than single beam acoustics. In this study, the acoustic method did not differentiate sediment from gravel or the hardbottom communities on the east/west parts of the shelf, but these were identified from the imagery. The acoustic method, in contrast with image interpretation, is fully automated and objective. Furthermore, the ability to reliably map reef from non-reef areas using acoustics, while useful at Navassa, is potentially more important for other areas where water depth or clarity limits the ability to use overhead imagery to map seabed habitats.

The difficulty of proportionally representing vertical habitats is a challenge which is common

to any mapping method and is acute in places like Navassa. Vertical (cliff) habitats likely represent a large component of habitat for both benthic and fish assemblages and these assemblages may be qualitatively distinct from their horizontal cousins a few meters away. The inability to proportionally represent cliff area on a map implies the risk that the importance of vertical habitats is under represented in map-based interpretation of habitat function and/or management planning.

10.6 Environmental Factors Influencing Reef Biology

The great impact of physical energy in determining Navassa's reef communities and patterns has been described above. Extreme events such as hurricanes are clearly also influential. Obvious physical disturbance (upended soft corals and sand movement scouring) were observed (MWM, personal observation) in November 2004 following hurricanes Ivan and Charlie. However, the long gaps between observations of reef condition at Navassa (Table 10.1) make it difficult to attribute relative impacts of storms versus other intermittently observed disturbances (see below).

One striking characteristic of Navassa reef environments is the consistently high water clarity (Fig. 10.11c). Extended depth distributions, with occurrences much deeper than reported in other locations, have been noted for diverse groups at Navassa, including certain species of benthic symbiont-bearing foraminifera (Williams 2003), crustose coralline algae (Begin and Steneck 2003), and Elkhorn coral (Miller 2003, Appendix 2) and this pattern may be attributable to enhanced light availability at depth. On the other hand, certain Ophioroid species have been observed at Navassa in waters much shallower than expected, perhaps because of the proximity of abyssal depths off the shelf break (G. Hendler, personal communication, 2000).

Many environmental factors that are expected to influence reef organisms and communities (including temperature, nutrients, other aspects of water quality) are not well characterized. Littler et al. (1999) hypothesized a gradient in phosphate exposure with higher levels along the northeast and lower levels along the southwest leading to inhibition of coral development along the northeast and east coast by phosphate (inhibiting coral calcification). This hypothesized nutrient gradient also coincides with gradients of physical energy (which may provide a better explanation of reef community distribution) as already discussed. The authors have observed extensive nesting of red-footed boobies and considerable avian droppings on the island, but most extensively along the southwest coast and not enough to currently accumulate as guano.

An early author described Navassa as resembling a 'petrified sponge' (Putnam 1918). Given this porous karst topography, the abundance of birds, and the lack of buffering habitats, it is possible that Navassa reefs are influenced by a naturally high nutrient regime. The only indications of nutrient levels on Navassa reefs are a few water samples taken in and around a cave along the southwest island cliff (~10 m depth) in 2000. High nutrient discharge was detected inside the cave (Fig. 10.15) whereas ambient nutrient concentrations just

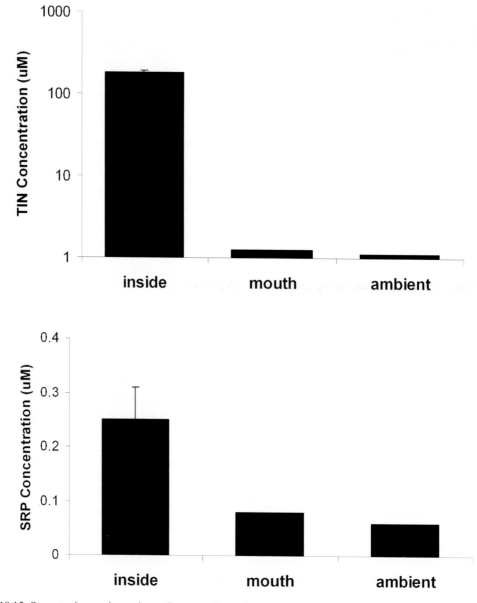

FIG. 10.15. Seawater inorganic nutrients. Concentrations of total inorganic nitrogen (TIN; note log scale) and soluble reactive phosphate (SRP) measured in and adjacent to a cave (~10 m depth) along the southwest coast in April 2000

outside the cave were just over 1 μM Total Inorganic Nitrogen and around 0.06 μM for Soluable Reactive Phosphate. On the other hand, the consistent and radically high water clarity observed at Navassa belies the expected impacts of nutrient enrichment on water quality. Sandin (2003) even suggested that a relatively low biomass in the Navassa planktivorous fish guild may be indicative of food limitation (i.e., lack of plankton) which is counter to expectation for an unbuffered, nutrient enriched, nearshore reef system. This counterintuitive pattern may simply result from high rates of oceanographic advection of introduced nutrients away from this island 'point source'. The nutrient and trophic dynamics of Navassa reefs remain to be satisfactorily characterized.

10.7 Present Status of Reef Health

Reef monitoring efforts at Navassa have been hindered by inaccessibility. Preliminary reef assessment data was collected during the 2000 Center for Marine Conservation (CMC) sponsored cruise (Table 10.1; Miller and Gerstner 2002), but data collection was severely hindered by air diving limitations.

Subsequent cruises have utilized Enriched Air NITROX diving and this has allowed much improved scope of sampling. A biennial schedule of reef assessment cruises has been maintained since then, allowing only episodic observations. Nonetheless, several obvious disturbance events have been observed 'in the act'.

Mean percent cover over time for several sites (mostly the shallow nearshore ones) is given in Fig. 10.16. An apparent overall increase in macroalgal cover has taken place at these sites, particularly evident in 2006. Although change in coral cover is not as obvious in the nearshore/shallow sites (Northwest Point, Lulu Bay, and West Pinnacles), loss of live coral is more dramatic at the deep patch reef site. This loss is likely attributable to a coral disease outbreak which was observed at Navassa during the 2004 cruise (Fig. 10.17b, d) when up to ~35% of coral colonies were affected by coral disease at some sites (Miller and Williams 2007). Many recently dead or mostly dead (estimated 6–18 months based on colonization by macroalgae but intact corallites allowing clear identification at least to genus, Fig. 10.17f) coral colonies were observed across a wide range of sites in 2006. Also in 2006, an extensive coral bleaching event (Fig. 10.17a and c) was observed with mean prevalence of 60–80% of all colonies (>4 cm in diameter) displaying some degree of bleaching across all shelf sites (20–37 m

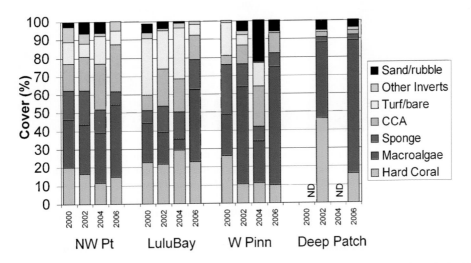

Fig. 10.16. Reef cover over time. Four sites along the southwest coast that were sampled over time via *in situ* point-intercept transects. Approximate locations are given in Fig. 10.2

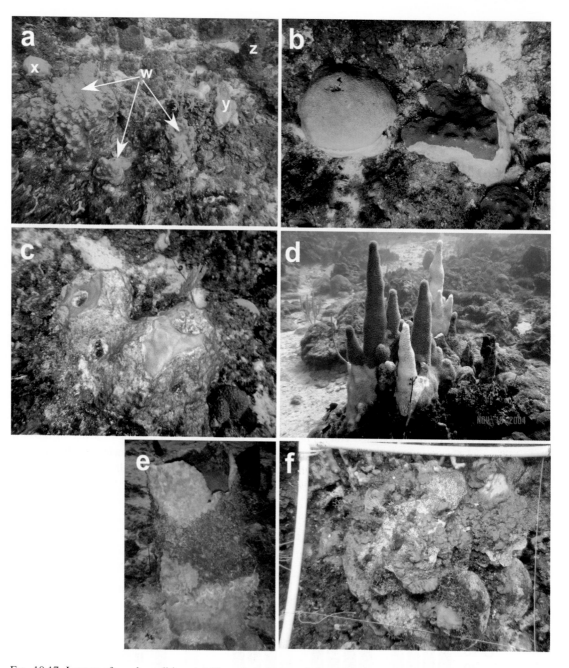

Fig. 10.17. Images of coral condition. (a) General reef scene during bleaching event observed in November 2006 showing slightly bleached (w), fully bleached (x and y also partially dead) and unaffected colonies (z); (b) colonies affected by rapid tissue loss in Nov. 2004; (c) large colony (*Montastraea faveolata*, ~1 m in total size) with remnant tissue patches yet alive, but bleached in Nov. 2006; (d) Rapidly dying *Dendrogyra cylindrica* in 2004; (e) large barrel sponge (*Xestospongia muta*) with substantial bleached/dead tissue; (f) Colony of *M. faveolata* observed in 2006 that has died within the last 6–18 months

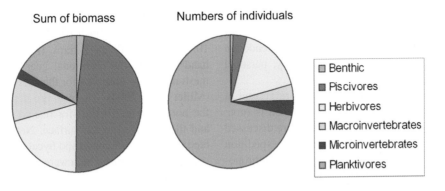

FIG. 10.18. Reef fish assemblage. Trophic composition of fish assemblage by density and biomass as sampled in November 2002 (n = 110 stationary point counts; McClellan and Miller 2003)

depth) and lesser extent of bleaching (mean 15–55% prevalence) in the fewer shallow (7–10 m) sites (M.W. Miller and D.E. Williams, unpublished data, 2006). Apparent disease conditions have also been noted in the sponges of Navassa (Fig. 10.17e) but not quantified.

Navassa reef fish assemblages have been assessed on a similar temporal scheme as benthic communities; preliminary assessments at a few sites in 2000 (Miller and Gerstner 2002) with expanding scope of sampling in subsequent biennial assessments from 2002 to 2006. Overall, Miller and Gerstner (2002) described a fish assemblage in 2000 which compared favorably (in terms of density and sizes) with others around the Caribbean. However their sampling was restricted to five high relief (shallow shelf and cliff) sites, which may confound temporal comparison with later, more spatially complete sampling events. Miller and Gerstner (2002) report ~200 g/m² for total reef fish biomass sampled in 2000 and four of the five sites surveyed had density of large (<30 cm) grouper greater than one-half fish per 100 m².

Subsequent sampling around these same sites in 2002 and 2004 showed only one of the five sites to have such high density of large grouper (D.B. McClellan et al., personal communication, 2005). Sampling conducted in 2002 across a range of reef habitats (n = 110 point samples) resulted in a relatively low overall biomass of reef fish of 58 g/m² using stationary point counts (McClellan and Miller 2003) compared to a long term average of 72 g/m² reported for the Florida Keys (Bohnsack et al. 1999). The same approach conducted over a

similar range of habitats (n = 123 point samples) yielded an overall 40.4% decline in fish biomass per point sample between 2002 and 2004 (D.B. McClellan et al., personal communication, 2005). Though several hurricanes and other natural disturbances (e.g., coral disease event) occurred during this time interval, this decline also coincided with a dramatic increase in fishing activity (see below).

At Navassa, the overall reef fish assemblage is dominated by planktivores, which comprise ~70% of individuals in visual surveys and ~17% of total fish biomass (Fig. 10.18; McClellan and Miller 2003; Sandin 2003). For comparison, planktivores comprise only 44.2% of numbers and 5.1% of biomass in the Florida Keys (Bohnsack et al. 1999). Piscivores are the most massive trophic group (~48% of fish biomass, Fig. 10.18) though comprising only ~5% of individuals. Invertivore guilds are particularly under-represented (Fig. 10.18; McClellan and Miller 2003), reflecting the rarity of grunts and snappers.

10.8 Human Impacts and Conservation Issues

Local human impacts at Navassa have varied greatly over the historical period. Vestiges of the former 'guano' mining and lighthouse maintenance operations are clear in the reef environment in the form of debris in the shallow reef flats at Lulu Bay. Surface mining devegetated much of the island and human occupation brought feral goats, rats, dogs, and cats, likely resulting in the apparent extinction of four of the reported eight endemic lizard species (Powell 1999) and possibly one of the endemic palm species on Navassa (R.B. Halley, personal observation, 1999).

Currently, only transient humans are present at Navassa: Haitian fishers on a regular basis (note fishing activities, trends and impacts are discussed in the next section) and the occasional expedition of scientists. No local development or anthropogenic land-based sources of impact are occurring. Since coral declines cannot be attributed to local land-based anthropogenic impacts in this case, either fishing (which is ongoing, and likely impacting Navassa's reef ecosystem) or large scale (regional or global) changes must be considered as driving patterns of reef change.

10.8.1 Fisheries Issues

Though the island is uninhabited, transient fishers from Haiti are active at Navassa and fishing is the primary anthropogenic threat to reef condition. Individuals currently involved in the Navassa fishery state that they have been fishing at Navassa for at least 40 years (Wiener 2005) so the impact is not new. However, on-island observations since 2000 indicate that acceleration in Navassa fishing activities is clearly underway (Miller and Gerstner 2002; Miller et al. 2004, 2007; Wiener 2005). This acceleration is evident in the introduction of novel, more destructive gear types (i.e., nets first observed in 2002) and the overall abundance of boats and gear buoys deployed on the Navassa shelf. The first socio-cultural assessments of the Haitian fishing communities was recently completed in 2005 and presents the first estimates of effort and trends in the Navassa fishery (Wiener 2005). This report emphasizes that economic resource limitation (e.g., boats, fuel) is the only thing preventing even further acceleration in fishery exploitation and that the general hostility of the Navassa landscape is the only thing preventing these transient fishers from establishing permanent residence on the island.

The Navassa fishery involves small wooden vessels (~6 m) with four to six fishers on board, mostly with small outboard motors (Fig. 10.19a, c). However, while at Navassa, vessels are often propelled via oars or sail to save fuel. Boats travel over from Haiti for 4–8-day trips. Most catch is salted/dried,

though a few boats have ice boxes and act as 'runners' transiting fresh fish back to Haiti (Fig. 10.19a, b). Antillean Z-traps (Fig. 10.19c) and hand lines are the traditional gear types. Triple-mesh nets were observed for the first time in 2002 (Miller et al. 2004). The nets are deployed along the bottom (Fig. 10.19d) for one to several days and then dragged to the surface. Net fishing enabled the capture of conch and bycatch of hawksbill sea turtles neither of which were observed in 2000. High densities of gear buoys were also observed during 2004, hindering navigation of the ~25 m research vessel.

It appears likely that fishing activity has been altering reef fish assemblages at Navassa on a very short time scale (e.g., 2002–2004; Miller et al. 2007). However, preliminary observations in 2006 suggest that fishing effort may have abated somewhat. No net fishing and fewer boats and gear buoys were observed by both Navassa expeditions in 2006 (Table 10.1; Wiener 2006; Piniak et al. 2006). A total of only 34 traps were being fished during the first 2 weeks of November 2006 (Wiener 2006), while over 175 gear buoys (traps and nets) were observed in November 2004 (Miller et al. 2007).

Due to the fact that fishing is *the* local anthropogenic impact on Navassa coral reef ecosystems, it has been suggested that it represents a unique opportunity to examine fishing and potential trophic cascade impacts in isolation from land-based water quality impacts (Miller and Gerstner 2002). However, the recent impact of other, regional-scale disturbances such as hurricane impacts, coral disease outbreaks, and bleaching events complicate this straightforward interpretation. That is, high levels of coral mortality have been observed coincident with escalating and now, possibly stabilized fishing pressure. Thus, even in this isolated, uninhabited island, there remains some ambiguity in attributing cause and effect for various aspects of reef decline.

Acknowledgments Interest in and work at Navassa for MWM and RBH was instigated by the Center for Marine Conservation (now The Ocean Conservancy). This organization deserves great credit for initiative in sponsoring research at Navassa in its transitional period of the mid to late 1990s. Most of the recent

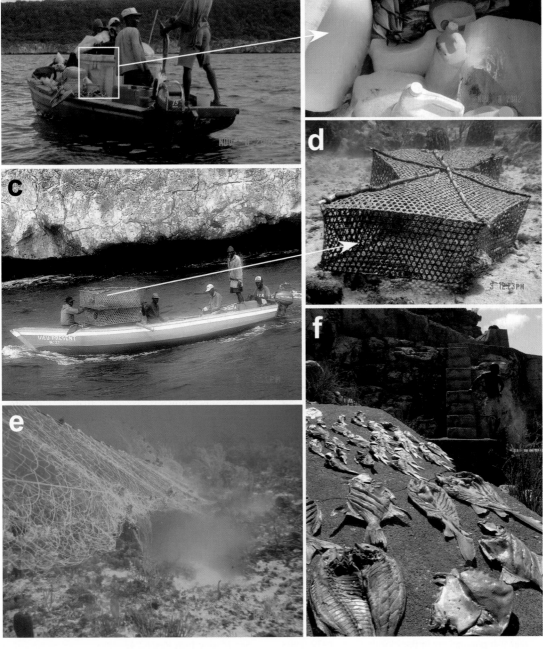

FIG. 10.19. Fishing activities as Navassa. (**a**) Haitian fishing vessel with ice chest. Barracuda caught via hand line. (**b**) Catch contained in the ice chest. (**c**) Typical vessel hauling traps. (**d**) Trap (~2m in size) as deployed. (**e**) Barrier net being hauled to the surface from hardbottom (~25m depth). (**f**) Most catch is salted/dried for preservation and a crew member is sometimes left on the island during daily fishing activities to tend to this task (Photo credit: K. Pamper)

reef and fishery assessment work has been conducted by the Southeast Fisheries Science Center with funding from the NOAA Coral Reef Conservation Program and partnership and permitting from the US Fish and Wildlife Service, Caribbean Islands National Wildlife Refuge. The John G. Shedd Aquarium has provided partnership and logistic support on four expeditions via its Research Vessel Coral Reef II and her crew. LIDAR data was acquired by NASA. Multibeam data was made available by Mike Stecher, Solmar Hydro. Drop camera video was made available by Greg Piniak, NCCOS. B. Mason, C. Fasano, and L. Johnston assisted in analyses of data presented in this chapter. J.Boyer (Florida International University) provided the nutrient analyses.

References

Aharon P, Veeh HH (1984) Isotope studies of insular phosphates explain atoll phosphatization. Nature 309:614–617

Ault JS, Smith SG, Luo J, Meester GA, Bohnsack JA, Miller SL (2002) Baseline Multispecies Coral Reef Fish Stock Assessments for the Dry Tortugas. Report No. NMFS-SEFSC-487

Baringer MO, Larsen JC (2001) Sixteen years of Florida Current transport at 27°N. Geophys Res Lett 28:3179–3182

Baums IB, Miller MW, Hellberg ME (2005) Regionally isolated populations of an imperiled Caribbean coral, Acropora palmata. Mol Ecol 14:1377–1390

Baums IB, Miller MW, Hellberg ME (2006) Geographic variation in clonal structure of a reef-building Caribbean coral, Acropora palmata. Ecol Monogr 76:503–519

Begin C, Steneck R (2003) Crustose coralline algae and juvenile scleractinian corals of Navassa Island. In: Miller MW (ed) Status of Reef Resources of Navassa Island: Nov 2002 NOAA Technical Memorandum NMFS-SEFSC-501, pp 57–65

Bohnsack JA, McClellan DB, Harper DE, Davenport GS, Konoval GJ, Eklund AM, Contillo JP, Bolden SK, Fischel PC, Sandorf GS, Javech JC, White MW, Pickett MH, Hulsbeck MW, Tobias JL, Ault JS, Meester GA, Smith SG, Luo J (1999) Baseline data for evaluating reef fish populations in the Florida Keys, 1979–1998. NOAA Techn Mem NMFS-SEFSC-427, p 61

Braithwaite CJR (1980) The petrology of oolitic phosphorites from Esprit (Aldabra), western Indian Ocean. Phil Trans R Soc Lond B 288:511–540

Bruckner AW (2002) Proceedings of the Caribbean Acropora Workshop: potential application of the U.S. Endangered Species Act as a conservation strategy: April 16–18, 2002, Miami, FL, Silver Spring, MD

Budd DA (1997) Cenozoic dolomites of carbonate islands: their attributes and origin. Earth Sci Rev 42:1–47

Burke RV, Horsfield WT, Robinson K (1974) The geology of Navassa Island. J Carib Sci 14:109–114

Chappell J, Shackleton NJ (1986) Oxygen isotopes and sea level. Nature 324 137–140

Collette BB, Williams JT, Thacker CE, Smith ML (2003) Shore fishes of Navassa Island, West Indies: a case study on the need for rotenone sampling in reef fish biodiversity studies. Aqua 6:89–131

De Mets C, Wiggins-Grandison M (2007) Deformation of Jamaica and motion of the Gonave microplate from GPS and seismic data. Geophysics J Int 169:362–378

DePaolo DJ (1986) Detailed record of the Neogene Sr isotopic evolution of seawater from DSDP site 590B. Geology 14:103–106

D'Invilliers EV (1891) The phosphate deposits of the island of Navassa. Geol Soc Am Bull 2:75–84

Fairbridge RW (1983) Syndiagenesis-anadiagenesis-epidagenesis: phases in lithogenesis. In: Larsen G, Chilingar GV (eds) Diagenesis in Sediments and Sedimentary Rocks, 2. Elsevier, Amsterdam, The Netherlands, p 572

Franklin EC, Ault JS, Smith SG, Luo J, Meester GA, Diaz GA, Chiappone M, Swanson DW, Miller SL, Bohnsack JA (2003) Benthic Habitat Mapping in the Tortugas Region, Florida. Mar Geod 26:19–34

Fruijtier C, Elliot T, Schlager W (2000) Mass-spectrometric 234U-230Th ages from the Key Largo Formatoin, Floirda Keys, United States: constraints on diagenetic age disturbance. Geol Soc Am Bull 112:267–277

Gaussoin E (1865) Memoir of the island of Navassa. JB Rose, Baltimore, MD, p 32

Gaussoin E (1866) Memoir of the island of Navassa. JB Rose, Baltimore, MD, p 32

Gleason, ACR, A-M. Eklund, RP Reid and V Koch (2006). Acoustic Seabed Classification, Acoustic Variability, and Grouper Abundance in a Forereef Environment. NOAA Professional Papers NMFS 5: 38–47.

Goreau TF (1959) The ecology of Jamaican reef corals: I. Species composition and zonation. Ecology 40:67–90

Grace M, Bahnick M, Jones L (2000) A preliminary study of the marine biota at Navassa Island, Caribbean Sea Marine Fish Rev 62:43–48

Highsmith RC (1982) Reproduction by fragmentation in corals. Mar Ecol Progr Ser 7:207–226

Hochberg EJ, Andrefouet S, Tyler MR (2003) Sea surface correction of high spatial resolution IKONOS images to improve bottom mapping in near-shore environments. IEEE Trans Geosci Remote Sens 41:1724–1729

Johns WE, Townsend TL, Fratantoni DM, Wilson WD (2002) On the Atlantic inflow to the Caribbean Sea. Deep-Sea Res I 49:211–243

Jones B, Hunter IG (1989) The Oligocene-Miocene bluff formation of the Cayman Islands. J Carib Sci 25:71–85

Jones B, Hunter IG (1994) Messinian (Late Miocene) karst on Grand Cayman, British West Indies: an example of an erosional sequence boundary. J Sedim Res A64:530–541

Jones B, Luth RW (2003) Temporal evolution of the Tertiary dolostones on Grand Cayman as determined by 87Sr/86Sr. J Sedim Res 73:187–205

Jones B, Smith DS (1988) Open and filled karst features on the Cayman Islands: implication for the recognition of paleokarst. Can J Earth Sci 25:1277–1291

Jones B, Hunter IG, Kyser K (1994) Revised stratigraphic nomenclature for Tertiary strata of the Cayman Islands, British West Indies. J Carib Sci 30:53–68

Leibig GA (1864) Report to the corporations of the Navassa Phosphate Company of New York, Baltimore, MD, p 15

Littler MM, Littler DS, Brooks BL (1999) The first oceanographic expedition to Navassa Island, USA: Status of marine plant and animal communities. Reef Encounter 25:26–30

MacArthur RH, Wilson EO (1967) The theory of island biogeography. Princeton University Press, Princeton, NJ

Mann P, Taylor FW, Edwards RL, Ku TL (1995) Actively evolving microplate formation by oblique collision and sideways motion along strike-slip faults: An example from the northeastern Caribbean plate margin. Tectonophysics 246:1–69

McArthur JM, Howarth RJ, Bailey TR (2001) Strontium isotope stratigraphy: LOWESS Version 3: best fit to the marine Sr-isotope curve for 0–509 MA and accompanying look-up table for deriving numerical age. J Geol 109:155–170

McClellan DB, Miller GM (2003) Reef fish abundance, biomass, species composition and habitat characterization of Navassa Island. In: Miller MW (ed) NOAA Techn Mem NMFS-SEFSC-501, pp 24–41

Miller GC (1977) Navassa Island resource assessment survey. Cruise results for Oregon II 77–08 (80), p 12

Miller MW (2003) Status of Reef Resources of Navassa Island: Nov 2002. NOAA Technical Memorandum NMFS-SEFSC-501, p 119

Miller MW, Gerstner CL (2002) Reefs of an uninhabited Caribbean island: fishes, benthic habitat, and opportunities to discern reef fishery impact. Biol Conserv 106:37–44

Miller MW, Williams DE (2007) Coral disease outbreak Navassa, a remote Caribbean island. Coral Reefs 26:97–101

Miller MW, McClellan DB, Begin C (2004) Observations on fisheries activities at Navassa Island. Mar Fish Rev 65:43–49

Miller MW, McClellan DB, Wiener JW, Stoffle B (2007) Apparent rapid fisheries escalation at a remote Caribbean island. Environ Conserv 34:92–94

National Oceanic and Atmospheric Administration (NOAA) (2005) Shallow-water Benthic Habitats of American Samoa, Guam, and the Commonwealth of the Northern Mariana Islands (CD-ROM), NOAA Nat Cent Coast Ocean Sci, Silver Spring, MD

Piniak GA, Addison CM, Degan BP, Uhrin AV, Viehman TS (2006) Characterization of Navassa National Wildlife Refuge: a preliminary report for NF-06–05 (NOAA ship Nancy Foster, April 18–30 2006). NOAA Techn Mem NOS NCCOS #38, p 48

Powell R (1999) Herpetology of Navassa Island, West Indies. Carib J Sci 35:1–13

Proctor GR (1959) Observations on Navassa Island. J Geol Soc Jamaica (Geonotes) 2:49–54

Purdy EG (1974) Karst-determined facies patterns in British Honduras: Holocene carbonate sedimentation model. Am Assoc Petrol Geol 58:825–855

Putnam GR (1918) An important new guide for shipping: Navassa light. Nat Geogr 34:401–405

Sandin SA (2003) Reef fish trophic analysis from Navassa Island: exploring biotic and anthropogenic factors. In: Miller MW (ed) Status of Reef Resources of Navassa Island: November 2002, Vol NOAA Techn Mem NMFS-SEFSC-501, pp 43–56

Schellmann G, Radtke U, Potter E-K, Esat TM, McCulloch MT (2004) Comparison of ESR and TIMS U/Th dating of marine isotope stage (MIS) 5e, 5c, and 5a coral from Barbados-implications for palaeo sea-level changes in the Caribbean. Quat Internat 1120:41–50

Skaggs JM (1994) The Great Guano Rush, St. Martin's Griffin, New York

Toscano MA, Macintyre IG (2003) Corrected western Atlantic sea-level curve for the last 11,000 years based on calibrated 14C dates from Acropora palmata framework and intertidal mangrove peat. Coral Reefs 22:257–270

Wiener JW (2005) Oral history and contemporary assessment of Navassa Island fishermen. Report for The US Department of Commerce NOAA/NMFS. Available at http://www.sefsc.noaa.gov/PDFdocs/Navassa%5FFishers%5FReport%5FFinal%5FFoProBIM.pdf

Wiener JW (2006) Navassa November 2006 Cruise: Daily Log Report to NOAA-Fisheries SEFSC, p 11

Williams DE (2003) Symbiont-bearing foraminifera of Navassa Island. In: Miller MW (ed) Status of Reef Resources of Navassa Island: Nov 2002 NOAA Techn Mem NMFS-SEFSC-501, pp 84–89

11
Geology of Hawaii Reefs

Charles H. Fletcher, Chris Bochicchio, Chris L. Conger, Mary S. Engels,
Eden J. Feirstein, Neil Frazer, Craig R. Glenn, Richard W. Grigg, Eric E. Grossman,
Jodi N. Harney, Ebitari Isoun, Colin V. Murray-Wallace, John J. Rooney,
Ken H. Rubin, Clark E. Sherman, and Sean Vitousek

11.1 Geologic Framework of the Hawaii Islands

11.1.1 Introduction

The Hawaii hot spot lies in the mantle under, or just to the south of, the Big Island of Hawaii. Two active subaerial volcanoes and one active submarine volcano reveal its productivity. Centrally located on the Pacific Plate, the hot spot is the source of the Hawaii Island Archipelago and its northern arm, the Emperor Seamount Chain (Fig. 11.1).

This system of high volcanic islands and associated reefs, banks, atolls, sandy shoals, and seamounts spans over 30° of latitude across the Central and North Pacific Ocean to the Aleutian Trench, and contains at least 107 separate shield volcanoes (Clague and Dalrymple 1987). The trail of islands increases in age with distance from the hot spot (Fig. 11.2) and reflects the dynamic nature of the Pacific Plate, serving as a record of its speed and direction over the Hawaii hot spot for the last 75–80 MY (Clague and Dalrymple 1987). A major change in plate direction is marked by a northward kink in the chain at the end of the Hawaii Ridge approximately 3,500 km from the site of active volcanism (Moore 1987). On the basis of dredged basalts, Sharp and Clague (2006) assign an age of 50 Ma to this shift from northern to northwestern plate motion, thought to be a result of changes in the movement of neighboring plates to the west. Today the Pacific Plate migrates northwest at a rate of about 10 cm/year (Moore 1987).

The eight main islands in the state: Hawaii, Maui, Kahoolawe, Lanai, Molokai, Oahu, Kauai, and Niihau, make up 99% of the land area of the Hawaii Archipelago. The remainder comprises 124 small volcanic and carbonate islets offshore of the main islands, and to the northwest. Each main island is the top of one or more massive shield volcanoes (named after their long low profile like a warriors shield) extending thousands of meters to the seafloor below. Mauna Kea, on the island of Hawaii, stands 4,200 m above sea level and 9,450 m from seafloor to summit, taller than any other mountain on Earth from base to peak. Mauna Loa, the "long" mountain, is the most massive single topographic feature on the planet.

11.1.2 Island Evolution

In traditional lore, the demigod Maui navigated to the Hawaii Islands using an alignment of stano shaped like a fishook. He discovered each island in sequence from oldest to youngest, demonstrating the Hawaii people's remarkable recognition of the islands' age gradient.

When an island is born, molten rock pours onto the seafloor, piling upon itself and solidifying under the cooling influence of the surrounding ocean water. Slowly, over hundreds of thousands of years, volcanic rock accumulates and eventually breaches the sea surface to become a high volcanic island. At first, the volcano does not grow as a neat layer-cake of lava beds. Rather, submarine eruptions break into boulders of glass and pyroclastics that accumulate as

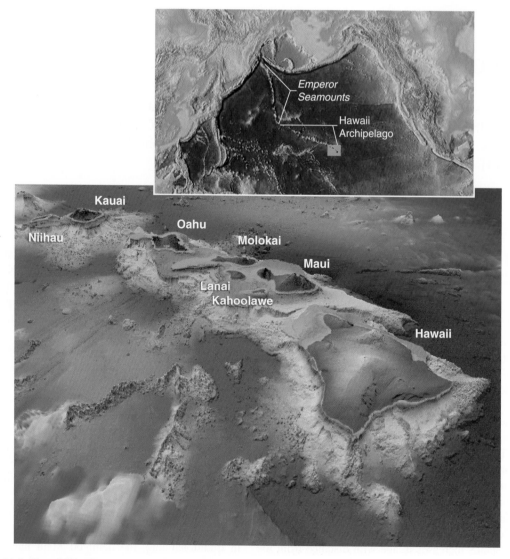

FIG. 11.1. Hawaii Islands

a great pile of broken volcanic talus on the seafloor. As the volcano erupts, it accretes steep aprons of broken glass, ash, and pillow basalts forming the core of the edifice (Grigg 1997).

The main bodies of the Hawaii Islands are built by successive flows of fluid, ropy pahoehoe and viscous, jagged aa basalt lavas that evolve through a recognized sequence of morphological and chemical stages (Clague and Dalrymple 1987). Alkalic basalts dominate the earliest submarine pre-shield building stage; with a transition to tholeiitic basalt that occurs as the volcano grows and approaches sea level. Loihi Seamount, on the southeast flank of Mauna Loa Volcano, is an example of this stage.

Loihi currently rises over 3,000 m from the sea floor, erupting hydrothermal fluids at its summit and south rift zone. Loihi's current growth rate may allow it to emerge above sea level in several thousands of years.

The subaerial main shield-building stage produces 98–99% of the lava in Hawaii. Kilauea Volcano, on the southeast flank of Mauna Loa on the Big Island, is actively undergoing shield building. Kilauea Volcano alone has produced ~2 km^3 of lava since 1983 (Garcia et al. 2000). The main shield-building stage produces tholeiitic lavas of basalt and volcanic glass with olivine, clinopyroxene, and plagioclase as the primary crystalline components.

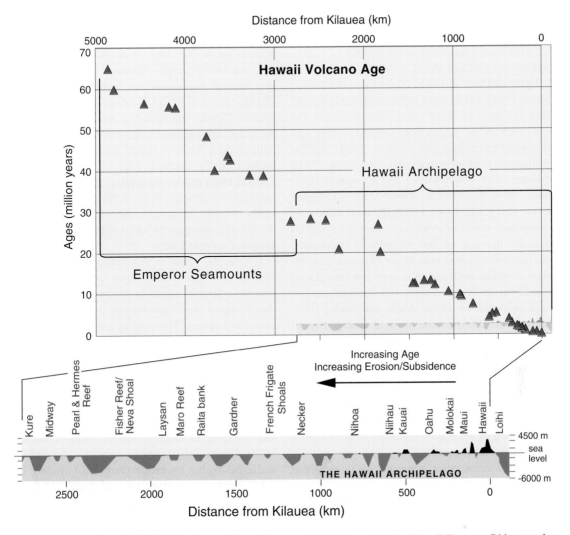

FIG. 11.2. The series of shield volcanoes, atolls, and seamounts that comprise the Hawaii-Emperor Ridge are the product of Pacific Plate movement across the Hawaii hotspot over 75–80 million years

Declining eruption rates accompany the onset of a post shield-building stage. This marks the end of active volcanism and is dominated by massive flows of fluid aa, and marked by a shift back to alkalic lavas that typically cap the top of the shield.

From the moment of their birth, Hawaii volcanoes grow into ponderous giants, too heavy for the underlying lithosphere to support without bending under the weight. As the Pacific Plate flexes down the volcanic pile subsides, resulting in an island-wide rise in sea level upon its shores. Local relative

sea-level rise on the subsiding Big Island is presently measured at ~4 mm/year.

Isostatic subsidence over the hot spot leads to plate flexure, a process responsible for an arch of uplifting lithosphere in an aureole surrounding the depression (Watts and ten Brink 1989). As an island moves off the hot spot on the ever-shifting Pacific Plate, it evolves from a regime of subsidence (Big Island and Maui) to one of uplift (Oahu) as it passes over this flexing arch. The question of which islands are experiencing uplift by this mechanism is somewhat controversial. The island of Lanai is thought to be experiencing uplift (Rubin et al. 2000) but the evidence for this (gravel corals) has also been interpreted as the deposit of a large tsunami (Moore and Moore 1984).

Uplift of Oahu is widely conjectured (Muhs and Szabo 1994) and thought to be responsible for the high elevation of fossil coral framestone (*in situ* reef) such as the Kaena Reef (approximately +24 to +30 m) ca. 532 ka (Szabo et al. 1994) and the Waimanalo Reef (approximately +8.5 m) ca. 125 ka (Moore 1970, 1987; Moore and Campbell 1987; Jones 1994; Szabo et al. 1994; Muhs and Szabo 1994; Grigg and Jones 1997). In their review of Quaternary sea-level evidence, Muhs et al. (2003) describe the age of last interglacial framestone and associated coral gravels on Oahu as consistently showing evidence of a long last interglacial epoch. An uplift rate ranging 0.3–0.6 mm/year has been proposed for Oahu (Muhs and Szabo 1994; Muhs et al. 2003) on the basis of emerged fossil coral framestone and gravels. Alternatively, Hearty (2002) contends that the high position of the Kaena limestone is due to a combination of much slower uplift and a higher sea level (~ +20 m) ca. 400 ka resulting from the "complete disintegration of the Greenland and West Antarctic ice sheets and partial melting of the East Antarctic ice sheet during the middle Pleistocene".

In some cases, after lying quietly in erosional conditions for thousands to millions of years, Hawaii volcanoes experience a rejuvenation stage of eruption, from a distinctly different magma source. These silica-poor alkalic lavas typically contain combinations of nepheline, olivine, and clinopyroxene with either melilite or plagioclase. Post-erosional rejuvenated eruptions are more explosive than the less viscous eruptions of the shield building stage, commonly producing cinder and tuff cones around active rift zones and vents atop eroding older shields. Three cones on the island of Oahu: Diamond Head, Koko Head, and Koko Crater are the result of this process. Although we have described the classic sequence for Hawaii shields, an individual volcano may become extinct during any of the phases, ending the evolution of that particular volcano (Moore 1987).

Each island represents a cycle of growth and destruction; passing through life's stages from the youth of volcanic eruption as on the expanding Big Island, to the mature dormancy of an eroding Maui, to the old age and extinction of Nihoa and Necker islets (north of Kauai) dating back 7–10 million years. Yet even in its burial beneath the sea, volcanic pedestals sustain a thriving population of reef organisms that build atolls upon its slopes. Atolls (Fig. 11.3), rings of reef with an interior lagoon, maintain the march to the northwest and subside finally beneath the waves at the aptly named "Darwin Point" (Grigg 1982). Today located on the northern edge of Kure Atoll, the Darwin Point marks the limit of reef tolerance for northerly conditions. Beneath the waves, drowned atolls become seamounts, the oldest of which is the 80 million-year-old Meiji at the far end of the Hawaii-Emperor Chain.

It is in the midst of this evolutionary progression of the high islands, after volcanic outpourings have largely waned yet before the atoll stage, that fringing reefs develop on the slopes of the shield volcanoes. These reefs constitute the shallow modern shelf of the Hawaii Islands, yet their origin belies this simple description. Pleistocene oscillations in eustatic sea level, superimposed on the flexing plate complicate the internal stratigraphy and history of fringing reefs. High sea levels provide optimal conditions for new reef accretion and flooding of older limestone units. Low sea levels expose limestones to subaerial dissolution and erosion that lower the platform. Flooding by the subsequent highstand leads to recrystallization and renewed accretion. Unraveling this complex mosaic of interwoven skeletal components, and the environmental history they record, is the subject of the remaining pages of this chapter.

11.2   Winds, Waves, and Reef Builders

Two seasons dominate the Hawaii climate: summer and winter. These were first defined by ancient Hawaiians as "kau wela" (hot season) and "ho'oilo" (to cause growth, referring to the nourishing rains of winter). While the warmest daytime summer temperatures in Hawaii infrequently exceed 35°C, the chilliest nighttime temperatures in winter rarely fall below 13°C. The difference in average daytime temperature at sea level throughout the year is less than 10°C, making the Hawaii Islands home to Earth's most temperate climate.

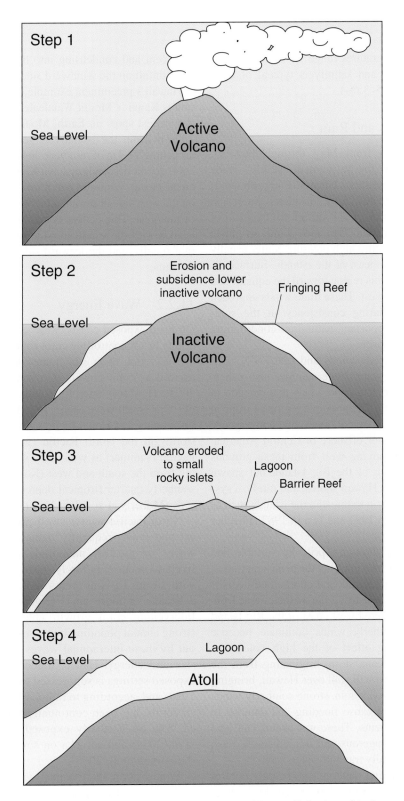

FIG. 11.3. Atolls are large, ring-shaped reefs with a central lagoon. Most atolls begin as fringing reefs on a subsiding shield volcano. The reef continues to grow vertically as the volcano subsides and erodes. Eventually, all surface evidence of the volcano disappears and a central lagoon marks the former position of the caldera

Sea surface temperatures range 24°C in winter to 27°C in summer and salinity is typical of open ocean values (34.5–35‰).

11.2.1 Winds and Rain

Summer extends from May through October, accompanied by the misnamed "gentle" trade winds. On any other shore, consistent 40 km/h winds gusting to 56 km/h are hardly considered gentle and many days, the seas of Hawaii are frothy foam of whitecaps as a result. The trade winds originate with the North Pacific high pressure center usually located to the northeast of the islands. Surface air flows away from this region toward the equator and turns to the southwest. These trade winds approach Hawaii with enduring consistency in the summer and provide constant, natural air-conditioning throughout the season.

Beyond the steady winds and seasonally tranquil north shore waters, late summer also ushers in threatening tropical depressions that develop in the eastern Pacific. The principle region of tropical storm genesis is an area of trade-wind convergence north of the equator offshore of Central America. Storm systems moving west from these grounds usually pass south of the Big Island, but spawn waves that impact Hawaii shores, and rarely, a system may curve north to threaten the islands such as hurricanes Iniki in 1992 and Iwa in 1982 (Fletcher et al. 1995).

From October through April, the variable and southerly Kona winds interrupt the persistent trades, bringing the frequent rain and cool, cloudy conditions of winter. Kona storms derive their name from the typically sheltered west coast of the Big Island where local southerly winds dominate because of the shadowing effect of the high volcanoes. Occasionally, a Kona storm originating from the south or southwest will stall over Hawaii, bringing persistent, island-wide rain, strong southerly winds and high surf that cause flooding and damage to nearshore ecosystems. These wet events can lead to remarkably dangerous and unpredictable flash flooding in normally tranquil watersheds.

Not only do wind patterns significantly control island weather and microclimates, so do the high shield volcanoes. In areas where onshore trade winds are obstructed by tall mountains, the moist tropical air rises up the mountain slope, cooling on its ascent and condensing into heavy cloud cover and rainfall on the windward side (Fig. 11.4).

Hawaii's preeminent example of this orographic rain is Kauai 's Mount Waialeale. Famed as one of the wettest spots on Earth, Mount Waialeale rises 1,587 m from the island's center, and receives a drenching 11.7 m of rain/year. In sharp contrast, the cool, dry air descending down Kauai's leeward side creates local, semi-arid conditions. Polihale Beach, on the west side of the island receives a mere 20 cm of rain/year. This pattern is similar for most of the islands with the windward side receiving frequent rain squalls, and the leeward side boasting eternal sunshine.

11.2.2 Wave Energy

On open exposed coasts, wave energy is the governing factor controlling coral community structure and accretion of *in situ* framework (Grigg 1998) as well as coral gravels (Grossman and Fletcher 2004). Hawaii is microtidal with an annual range of 0.8 m. Swell wave energy impinges on all shorelines under a seasonal regime governed by distant storms in the North Pacific (winter) and South Pacific (summer) as well as by systems approaching from the south and west (Kona storms) or any southerly quarter (tropical depressions) in all seasons (Moberly et al. 1965; Fig. 11.5).

North Pacific winter swell produces the largest and most frequently damaging energy. Yet waves of greatest magnitude and impact are likely to occur only rarely, associated most often with strong El Niño years (e.g., 1998) perhaps a decade or more apart (Rooney et al. 2004). Intervening coral growth able to survive the strong annual pounding by waves are often wiped out by these interannual waves of extraordinary size and energy. Hence, modern framework in exposed settings is suppressed to a veneer (Grigg 1998) and, according to radiocarbon dates of fossil corals, has been continually suppressed since ca. 5 ka on northerly exposed coasts (Rooney et al. 2004) and ca. 3 ka on southern shorelines (Grossman et al. 2006).

Typical north shore annual waves have periods of 14–20 s and breaking face heights of 2–15 m. Waves of this magnitude are able to generate and transport a coarse bedload of carbonate gravel that scours and abrades the reef surface toppling living

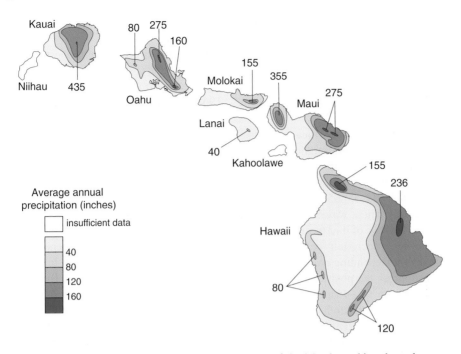

FIG. 11.4. Trade winds deliver moisture to the northeastern sides of the islands, making these the wettest regions. Large-scale storm systems passing near the islands, usually from the south and southwest, provide rainfall to southern shores that are otherwise dry (Oki et al. 1999). Modified from Oki et al. (1999) and Giambelluca et al. (1986)

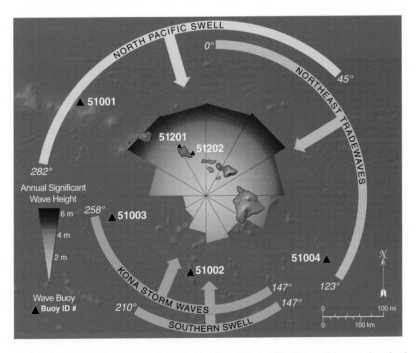

FIG. 11.5. Long period swell impacts coral growth on all sides of Oahu. North swell is prevalent in the winter, south swell in the summer. Swell waves from both directions refract around the island to adjacent shorelines. Trade wind swell and local seas occur over 75% of the year, and 90% of summer months

colonies and interrupting the recruitment process. Northerly swell can refract around islands and impact eastern and western shores. Consequently, nearly half the Oahu shelf is periodically swept clear of significant modern coral growth by large northern waves during years characterized by strong El Niños.

Trade winds are incident to northeastern, eastern, and southeastern shores throughout most of the year. Trade wind waves are the most common wave type in Hawaii and persist for 90% of the summer months and 55–65% of the winter (Fletcher et al. 2002). These waves typically have periods of 5–9 s, heights of 1–3 m and generally do not impede reef accretion below approximately −10 m depth. Depending on exposure, they can, however, suppress the position of maximum accretion to depths up to −15 m. Extreme, high trade wind events (>50–65 km/h) occur for 1–2 weeks in many years, generating seas with high waves (3–5 m) that break across the shallow shelf on windward shores.

Relative to northern swell-dominated Hawaii coasts, most windward margins experience moderate wave energy, but compared with many Caribbean and Indo-west Pacific regions, these rank as high-energy shorelines (Grossman and Fletcher 2004). Where north swell wraps around the island east side, trade wind waves and north swell combine to deliver a double punch of converging wave energy on the shelf.

Long south swell occurs in summer with periods 14–22 s and heights 1–5 m and greater. Generally detrimental to reef accretion in shallow water, these limit the hydrodynamic profile of reef structures and the accretion of coral framestone in late Holocene time (Grossman et al. 2006). Southerly shores are also exposed to infrequent but highly damaging hurricane-generated waves that periodically cause high mortality to exposed reef-building corals (Grigg 1995). Also characteristic of southerly shores are Kona-storm waves generated locally by winter low-pressure systems that pass over the islands from the south and west. Kona waves have periods of 6–12 s and heights up to 4 m that are occasionally destructive to reefs. Kona weather systems also bring intense rainfall that is channeled into coastal waters sometimes impacting coral communities closest to shore.

11.2.3 Reef Builders

Like all Hawaii life forms, ancestors of the organisms that build reef communities traveled across the vast Pacific to arrive on island shores (Scheltema 1968, 1986). Roughly half of native Hawaii marine species are indigenous also to the waters of Indonesia, the Philippines, and other islands of the Indo-West Pacific region. Another 10–15% are shared with the west coast of the Americas; about 13% are ubiquitous tropical marine species found across the oceans, and 15–20% are endemic to Hawaii alone (Kay and Palumbi 1987; Hourigan and Reese 1987).

These organisms were dispersed to Hawaii as floaters, swimmers, and hitch-hikers on currents that circulate the North Pacific. Biologists consider that all Hawaii corals originated in the Indo-west Pacific region, hence easterly currents are required to populate Hawaiian reefs. The Kuroshio Current from the Philippines and southern Japanese islands carries into the east-flowing Sub-tropical Counter Current (SCC) and the North Pacific Current that spans the mid-latitude waters to the north of Hawaii. While the North Pacific Current is thought too cold to successfully carry coral larvae, the SCC consists of a train of warm anticyclonic eddies 300–600 km across that break off from the Kuroshio and head east as far as the Hawaiian Islands. Occasionally accelerating to over 50 cm/s this system is probably responsible for the delivery of reef biota to the island chain (Grigg 1981). Some attached to debris and grew to maturity on the journey. Others traveled in larval stage gambling to hit shore before reaching adulthood.

Although perhaps some species endured sustained journeys, most probably took advantage of short-cuts using fortuitous "stepping stone" islands harboring abundant reef communities at intermediate positions across the ocean. Studies show, for instance, that currents may span the distance from Johnston Atoll in 30–50 days and from Wake Atoll in 187 days – both these journeys from the west are well within the survival spans of larvae that characterize many marine species (e.g., Scheltema 1986).

The trip is arduous, and only the hardiest individuals of the most appropriately adapted species survived. Eastern Pacific reef communities (including Hawaii) are notable for their low species count in comparison to those of the west Pacific. The numbers of marine types steadily decrease along an eastwardly extending line. This biotic attenuation marks a natural filter ensuring that those arriving in Hawaii waters, and able to secure a safe position within the ecology, are winners of a survival lottery in which a great many organisms begin the journey but only those with great luck, fortitude, and the appropriate set of physical attributes complete it.

As a result of Hawaii's isolated position in the Central Pacific, reef communities experience relatively low coral species diversity in comparison to other Indo-Pacific sites. Of the approximately 57 coral species documented in the main Hawaii Islands (Maragos 1977, 1995), fewer than 25% typically constitute dominant components of the ecology. Of these, a few highly plastic species (e.g., *Porites lobata*, *Montipora patula* and *M. capitata*; Fig. 11.6) employ multiple growth forms to exploit and dominate wave-controlled niches in the reef system. In protected and deep environments (below wave base) the dominant coral species is *Porites compressa*. In more exposed settings, *P. lobata* is the dominant coral. These species may construct reef framework rapidly if not limited by wave impact, bottom scour, or turbidity. Coralline (e.g., *Porolithon gardineri*, *Hydrolithon onkodes*) and "calcareous" algae (e.g., *Halimeda discoidea*, *H. incrassata*) constitute important members of all Hawaii reefal ecosystems.

On exposed shores, wave energy tends to suppress reef development in shallow depths. Windward Kailua Bay (Fig. 11.7) on Oahu is an example of such a reef system. Moderate wave energy back-reef areas are dominated by an abraded fossil reef surface with thin veneers of encrusting and massive *P. lobata*, occasional stout *Pocillopora meandrina* and the coralline alga *Porolithon gardineri*. In higher energy back-reefs, coral growth is limited and the skeletal limestone surface is dominated by crustose coralline algae (*H. onkodes*) with occasional stout branching *P. gardineri*.

Encrusting *P. lobata*, *M. patula*, *M. capitata*, and *H. onkodes* populate persistently wave-swept regions near the reef crest and mixed colonies of massive *P. lobata* and encrusting *P. lobata*, *M. patula*, and *M. capitata* are found in regions of the reef flat where there is reduced wave stress. A vibrant community of delicate branching *Porites compressa*, platy *M. patula*, *M. capitata* and *P. lobata* can be found in deeper central fore-reef regions where there is reduced wave stress. Platy forms of coral exploit the vertical walls of drowned channels and karst holes throughout many reefs.

11.2.3.1 Reef Communities and Shear Stress

Now classic studies by Dollar (1982) and Dollar and Tribble (1993) identified physical disturbance from waves as the most significant factor determining the structure of Hawaiian coral reef communities. Expanding on this, Grigg (1983) based on Connell (1978) articulated how the intermediate disturbance hypothesis fit on community succession in Hawaii and presented two models of coral community succession: (1) an undisturbed (lack of wave impact) community that reaches peak diversity due to recruitment followed by a reduction due to competition and (2) a disturbed community where diversity is set back to zero in the case of a large disturbance, or diversity is ultimately increased in the case of intermediate disturbance (substrate is opened for new recruitment). In the case of geological studies, interpreting drill cores of paleo-communities must be grounded in an understanding of the roles of succession and disturbance. Hence, it is common to develop community assemblage models during studies of Hawaiian reef stratigraphy.

To improve understanding of reef community assemblage, Harney et al. (2000), Harney and Fletcher (2003), Grossman and Fletcher (2004), Engels et al. (2004), and Grossman et al. (2006) employed surveys of benthic communities to develop coral assemblage models marking distinct environments. Surveys were conducted using a modified line-intercept technique in water depths from 2 to 25 m after Montebon (1992). Researchers laid a 10 m line along the benthic surface and recorded the position of every change in substrate type, surface morphology, coral species, coral morphology, coral associations, algal species, algal morphology, algal associations, and presence of bioeroders and zooanthids. Species diversity and morphologic diversity were calculated after Harney and Fletcher (2003) using the Shannon-Weaver diversity index with the equation $Hc = -\Sigma\pi (\log \pi)$ where Hc is diversity, π is the present cover on the ith species or morphologic form.

In their work along the south shore of the island of Molokai, Engels et al. (2004) employ this

444

C.H. Fletcher et al.

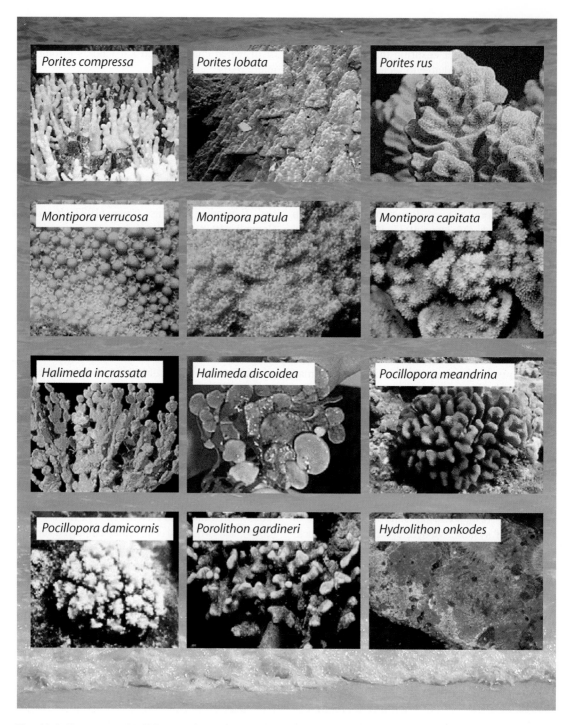

Fig. 11.6. Common reef-building corals and algae in Hawaii

FIG. 11.7. Kailua Bay, Oahu has been the site of several studies of reef structure, community assemblage, and sediment production (Isoun et al. 2003, (reproduced by permission of Springer)

method to develop a community zonation model related to wave-generated bed shear stress as modeled by Storlazzi et al. (2002). Engels et al. define three assemblages (Fig. 11.8); (1) a low-energy assemblage, (2) a mid-energy assemblage, and (3) a high-energy assemblage. The zonation model relates bed shear stress in Newtons per square meter with percent living coral cover, relative percent coralline algae cover, dominant coral species, dominant coral morphologies, and water depth. Each assemblage is divided into three depth zones, <5, 5–10, and >10 m. All observed coral types that account for at least 10% of living coral cover are represented in the model.

Engels et al. (2004) find that percent coralline algae cover is inversely related to water depth and

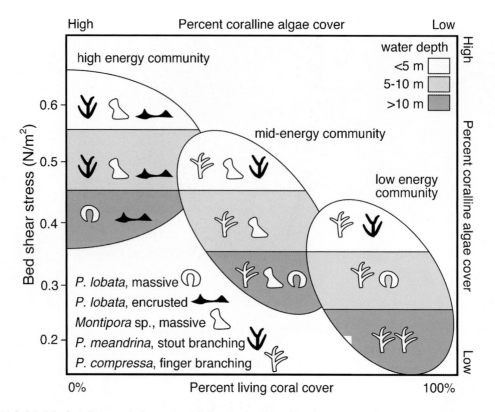

FIG. 11.8. Model of modern coral-algae assemblage zonation from Engels et al. (2004) relating wave-generated bed shear stress (Storlazzi et al. 2002) with percent living coral cover, dominant coral species, coral morphologies, and relative percent coralline algae cover. The low energy community is dominated by *Porites compressa*, especially at depths >5 m. As bed shear stress increases with increasing exposure to wave energy, the ecosystem shifts to a mid-energy community dominated by *P. compressa*, and *Montipora* sp. At higher shear stresses, especially with exposure to north swell, a high energy community emerges dominated by *Porites lobata*, *Pocillopora meandrina*, and *Montipora* sp.

directly related to bed shear stress. In all three assemblages coralline algae had their highest representation in depths <5 m where bed stress is highest. Also, corallines have greater representation in the high-energy assemblage than in the low-energy assemblage. This suggests that coralline algae are out-competed by coral under conditions well-suited to coral growth (lower wave energy) and flourish in conditions adverse to coral growth.

Low-energy assemblages with bed shear stresses ranging from <0.2 to ~0.4 N/m², are dominated by the columnar coral *P. compressa*. In Hawaii this species is generally regarded as a climax species in areas protected from heavy wave action as it tends to out-compete other species with its rapid growth and thin columnar morphology. Coral cover in this assemblage is high and increases with depth often

approaching 100% at depths >15 m. The other two coral species that appear in this assemblage are robust stout branching *Pocillopora meandrina* in depths <5 m, and massive *Porites lobata* in depths 5–10 m.

Mid-energy assemblages experience ~0.3–0.5 N/m². Here, encrusting *Montipora* species compete with *P. compressa* for space. *Montipora* are generalists (Maragos 1977) inhabiting depths from 0 to 50 m, occurring in a range of energetic conditions, and a multitude of growth forms. Dominant coral types change in waters >5 m where *Pocillopcia meandrina* becomes a significant assemblage component, though in depths >10 m *P. lobata* in massive form takes over. As with the low-energy assemblage, percent living coral cover in the mid-energy assemblage increases with depth.

The high-energy assemblage experiences bed shear stresses from ~0.4 to over 0.6 N/m². Fragile branching *P. compressa* disappears completely from the dominant corals and is replaced by *Montipora* sp., *Poc. meandrina*, and encrusting forms of *P. lobata*. At depths >10 m, *P. lobata* in encrusting and massive forms take over as the dominant coral as *Montipora* sp., and *Poc. meandrina* diminish. This assemblage is characterized by low living coral cover and, unlike low-energy and mid-energy assemblages, coral growth is optimized between 5 and 10 m water depth where a balance between encrusting/stout morphology, shear stress, and ambient light are achieved.

It is worth noting that the high-energy assemblage modeled by Engels et al. (2004) lives on skeletal limestone (fossil reef) radiocarbon dated ca. 5 ka. This suggests an absence of net accretion since mid-Holocene time where reef is exposed to north swell. We infer that survey data supporting the model represent a community that is periodically interrupted by high energy events that clear the substrate of living community. Hence, we conceive of an "extreme-energy assemblage" consisting of temporally restricted living coral and algae on an antecedent seafloor, as described by Rooney et al. (2004) and Grossman et al. (2006). We return to this issue later in the chapter.

The reef community at Kailua Bay is exposed to three general types of waves: (1) direct trade wind swell from the east-northeast (Hs = 2 m, T = 8 s, Dir = 60°) during much of the year; (2) refracted north swell (Hs = 3 m, T = 16 s, Dir = 10°) at times in the winter; and (3) unusually large swell events (Hs = 6 m, T = 19 s, Dir = 45°) approaching from the northeast interannually (i.e., "decadal"). The back reef at Kailua slopes gently from the shoreline to approximately −15 to −20 m where it meets the fore reef and drops steeply to approximately −30 to −35 m. Using the commercial software Delft3D it is possible to model shear stresses impacting the Kailua reef community under these three types of waves (Fig. 11.9).

Modeled results indicate that reef area shallower than −5 m is subject to high shear stress (0.4 to >0.6 N/m²) under annual trade wind swell. This correlates to the high energy community of Engels et al. (2004).

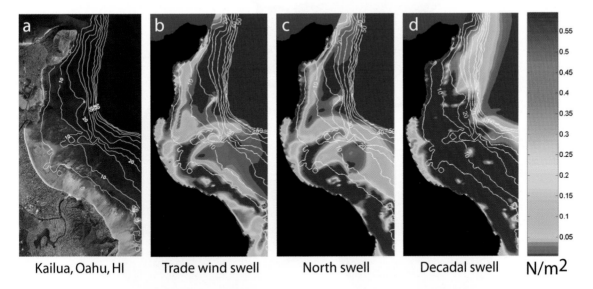

Kailua, Oahu, HI   Trade wind swell   North swell   Decadal swell   N/m²

Fig. 11.9. Wave generated shear stress, Kailua Bay, windward Oahu (contour interval 5 m). (**a**) Kailua Bay has a broad fringing reef with a gently sloping back reef (0 to −20 m), and steeply sloping reef front (−20 to −30 m). (**b**) The reef is subject to annual trade wind swell (Hs = 2 m, T = 8 s, Dir = 60°) that generate high shear stresses in the shallowest portions. (**c**) Many years swell generated by storms in the North Pacific will refract into Kailua (Hs = 3 m, T = 16 s, Dir = 10°) and generate shear stress exceeding 0.6 N/m² across the back reef. (**d**) More rarely, unusually large events (Hs = 6 m, T = 19 s, Dir = 45°) approach from the northeast and generate shear stresses >> 0.6 N/m² shallower than −20 m

However, deeper than −5m, bed stress ranges from 0.15 to <0.05N/m² suggesting that a low energy community should be stable across most of the reef surface under typical trade swell. However, in most years north swell refracts into Kailua, generating shear stress exceeding 0.6N/m² across broad areas of the back reef. This suppresses the depth of framework accretion to −10m or more, leaving shallower areas devoid of long-term coral cover. Extreme energy also occurs on a low frequency, interannual basis (i.e., decadal) generating damaging stresses across the entire back reef and into depths of −20 to −25m. Presumably, it is these events that have suppressed widespread Holocene accretion since ca. 5ka (Grossman and Fletcher 2004; Rooney et al. 2004). Localized sheltering and deeper topography experience stress levels below 0.5N/m² and framework accretion continues despite the wave environment.

Careful collection of reef community survey data recording percent coral and algal coverage, diversity, and substrate type are needed to verify community response to these model results. Until then, the relationship between modeled shear stress and reef community organization constitute a testable hypothesis.

11.3 Physical Characteristics of the Oahu Shelf

The geology of Hawaii reefs has been most intensively studied on the island of Oahu. Offshore of island beaches, the Oahu shelf dips gently seaward to near the −20m contour (Fig. 11.10). There, a limestone drop-off usually marks the end of the

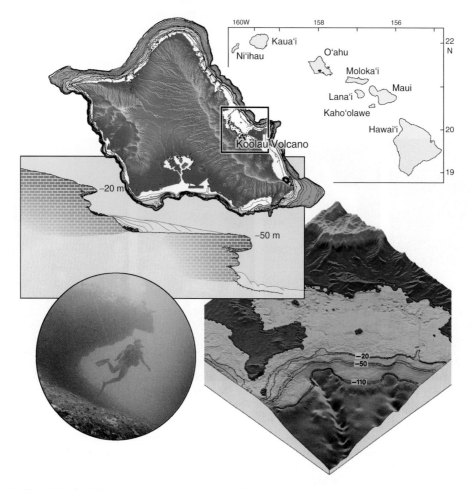

FIG. 11.10. The Oahu shelf is a series of terraces gently dipping seaward with sharp vertical faces often undercut with paleoshorelines

shelf. The base of this wall typically occurs near −30 m depth where a deeper, usually sand-covered terrace extends seaward to approximately −50 m. Below −50 m a second wall and third terrace are found (Fletcher and Sherman 1995).

11.3.1 General Bathymetry

The Oahu shelf is swept by large swells that generate high shear stress limiting modern reef development. But in the past, reef accretion was apparently more widespread because this shelf is almost entirely built of Pleistocene skeletal limestone sequences where it has not been interrupted by volcanic activity. A major problem driving scientific research among paleoreef workers in Hawaii has been the question how and when and under what conditions such massive fossil reefs could develop despite the fact that little modern reef accretion was occurring.

A mosaic of paleoreefs that comprise the island shelf has been worked out by researchers and is

presented in a later section. Because shelf construction spans three interglacial periods and two glacial periods, karstification, recrystallization, cementation, and accretion have all left their mark and added geologic components to the limestone mosaic. As discussed below, these geologic features are evident in the bathymetry, rugosity, nature of sand fields, and pattern of reef development.

The majority of the shallow seafloor above −30 m consists of well-lithified limestone with a veneer (occasionally thick) of loose carbonate sand and patchy occurrences of coral and coralline algae growth. On much of the shallow shelf a combination of subaerial karst processes, paleo-stream channel incision, and occasional living coral veneer produces a highly rugose and complex seafloor. For instance, at Waimanalo Bay on windward Oahu, the shallow seafloor displays a sand-filled bathymetric depression in the morphology of a barrier-lagoon (Fig. 11.11). Whether this depression has been incised during past interglacials

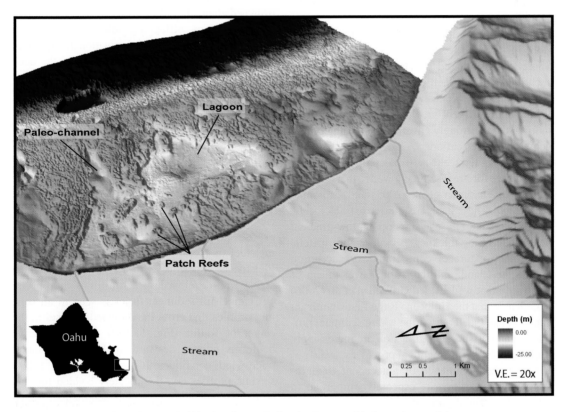

FIG. 11.11. Waimanalo carbonate shelf in the shape of a barrier-lagoon of indeterminate origin

wherein the shelf was exposed to meteoric waters leaving paleo-channel, karst, and doline type depressions (Purdy 1974) or it is the product of Holocene framework growth, or a combination, can only be determined by dating and interpreting surrounding components. Depressions such as these are a common feature on Hawaii carbonate shelves (Stearns 1974); in Waimanalo Bay they range in size from <5 to ~2,700 m^2 and are visible to depths of −40 m, and are likely found deeper. Sediment fills a majority of these features burying a 2–3 m wall as well as circular to oval shaped paleo-patch reefs. In Fig. 11.11 the sediment fill of the depression has been removed in order to display the antecedent topography.

The seaward edge of the Oahu shelf is marked by a wall that drops to depths of −30 m or more. In many areas a prominent erosional notch marks the face of the wall at approximately −18 to −25 m. Because it is clearly a former intertidal feature, Stearns (1974) named the notch the "Kaneohe Shoreline" and proposed that it formed in late Pleistocene time ca. 80 ka. Figure 11.10 shows a researcher SCUBA diving within the Kaneohe Shoreline.

Where living corals aggregate on the seaward slope of the shelf (e.g., Kailua Bay), the Kaneohe Shoreline is obscured by Holocene accretion. At such locations the shelf edge marks the top of the fore-reef and the Kaneohe Shoreline likely underlies modern fore-reef growth. Hence, modern reef morphology is strongly governed by the antecedent geometry of the seafloor.

The second terrace (> −30 m depth) is covered in an extensive fore-reef sediment wedge that has been studied by Hampton et al. (2003) and Grossman et al. (2006). These sands are inferred to be Holocene in age, and found in seaward sloping, low-gradient deposits reaching thicknesses over tens of meters. The patchy nature of coral framework accretion around the island, and the strong development of these sediment wedges, led Grossman et al. (2006) to conclude that the Holocene is the first significant epoch in late Quaternary time where sediment production and offshore progradation are increasingly important shelf constructional processes.

Given the dominance of antecedent topography, seismic surveys are especially important in revealing surficial and subsurface aspects of the shallow shelf.

11.3.2 Acoustic Facies

Grossman et al. (2006) define four seismic facies (Fig. 11.12) on the basis of acoustic reflection profiles from the shelf of Oahu (principally Kailua Bay and the south shore of Oahu): (1) coral reef/mound, (2) pavement, (3) channel-fill sediment, and (4) fore-reef wedge.

The coral reef/mound facies forms positive topographic features on the otherwise low-gradient shallow terrace. Internal reflectors tend to be concentric and continuous. These are parallel to subparallel features that are moderate in acoustic strength despite the potential energy attenuation of the highly rugose and porous limestone.

The pavement facies is marked by continuous to discontinuous high amplitude reflectors at or within a few meters of the seafloor. This facies is marked by significant attenuation of acoustic energy due to the high density, low magnesium calcite mineralogy of the limestone resulting from meteoric recrystallization of aragonitic coral.

Channel-fill facies display variable internal structure characteristic of sediment infilling of drowned channels and valleys. The fore-reef wedge facies displays offlapping reflectors representing bedding surfaces and depositional sequences within the thick sediment apron burying the deep terrace offshore of the −20 m contour.

Seismic reflection profiles of Kailua Bay (Fig. 11.13; Grossman et al. 2006) reveal that framework accretion is characterized by all four seismic facies. Reef/mound facies are found only in the central portion of the bay in depths −8 to −15 m (Fig. 11.13a, b). Core samples from the same site (Grossman and Fletcher 2004) show the Holocene reef in this locality is ~11 m thick, ranging in age ca. 7.9–5.3 ka. The majority of the Kailua shelf exposes pavement facies indicating that the fringing reef is composed largely of recrystallized Pleistocene limestone with a Holocene veneer, in places well-developed.

Channel-fill sediment facies is seen in all seismic profiles that cross the drowned channel in the center of Kailua (Fig. 11.13d). Although active bedforms migrate seaward in the channel under trade-wind conditions (Cacchione et al. 1999), seismic profiles reveal intricately bedded internal structures indicating a variable sedimentation history. Seaward of the shelf profiles reveal a large

Seismic character

Sedimentary facies

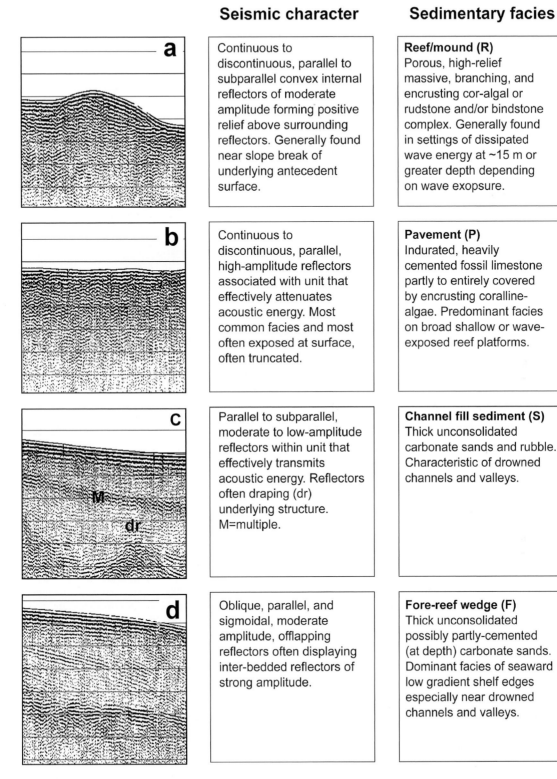

a
Continuous to discontinuous, parallel to subparallel convex internal reflectors of moderate amplitude forming positive relief above surrounding reflectors. Generally found near slope break of underlying antecedent surface.

Reef/mound (R)
Porous, high-relief massive, branching, and encrusting cor-algal or rudstone and/or bindstone complex. Generally found in settings of dissipated wave energy at ~15 m or greater depth depending on wave exopsure.

b
Continuous to discontinuous, parallel, high-amplitude reflectors associated with unit that effectively attenuates acoustic energy. Most common facies and most often exposed at surface, often truncated.

Pavement (P)
Indurated, heavily cemented fossil limestone partly to entirely covered by encrusting coralline-algae. Predominant facies on broad shallow or wave-exposed reef platforms.

c
Parallel to subparallel, moderate to low-amplitude reflectors within unit that effectively transmits acoustic energy. Reflectors often draping (dr) underlying structure. M=multiple.

Channel fill sediment (S)
Thick unconsolidated carbonate sands and rubble. Characteristic of drowned channels and valleys.

d
Oblique, parallel, and sigmoidal, moderate amplitude, offlapping reflectors often displaying inter-bedded reflectors of strong amplitude.

Fore-reef wedge (F)
Thick unconsolidated possibly partly-cemented (at depth) carbonate sands. Dominant facies of seaward low gradient shelf edges especially near drowned channels and valleys.

Fig. 11.12. Major acoustic facies of the windward and southern Oahu shelf including seismic and sedimentary characteristics (Grossman et al. 2006; reproduced by permission of Elsevier Ltd.)

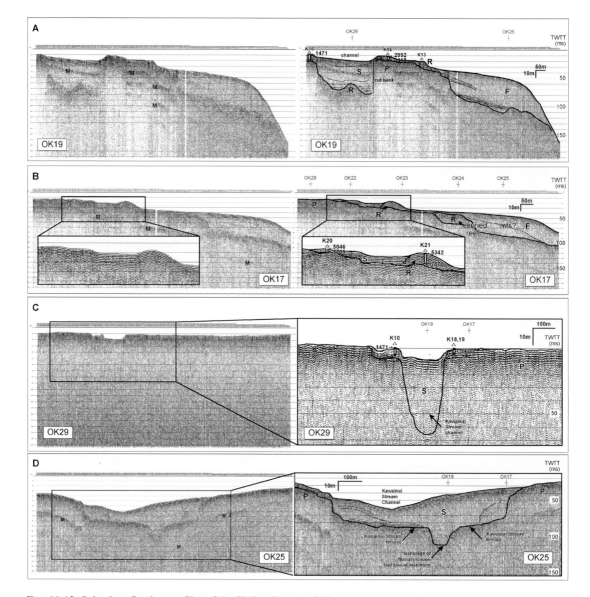

FIG. 11.13. Seismic reflection profiles of the Kailua Bay shelf (Grossman et al. 2006; reproduced by permission of Elsevier Ltd.) showing both uninterpreted (left) and interpreted (right) sections. Drill core sites (triangles) are shown with radiometric age results (Grossman and Fletcher 2004), acoustic multiples (M), line crossings (crosses), reef facies (R), pavement facies (P), channel fill sediment facies (S), and fore-reef wedge facies (F)

sediment wedge with internal structures characteristic of interbedded sediment/rubble units. The deposit has a thickness reaching ~40 m and its age is unclear. Lack of major internal reflectors suggest that subaerial exposure, as would have occurred at the last glacial maximum, did not produce a distinct horizon, or the unit has largely accumulated during Holocene time.

11.3.3   Shelf Substrate

Grossman et al. (2006) describe shelf substrate on the basis of camera tows (Fig. 11.14) and seismic profiles across the south and east margin of Oahu. They define three general types: (1) coral reef, (2) pavement (limestone and volcanic), and (3) unconsolidated sediment.

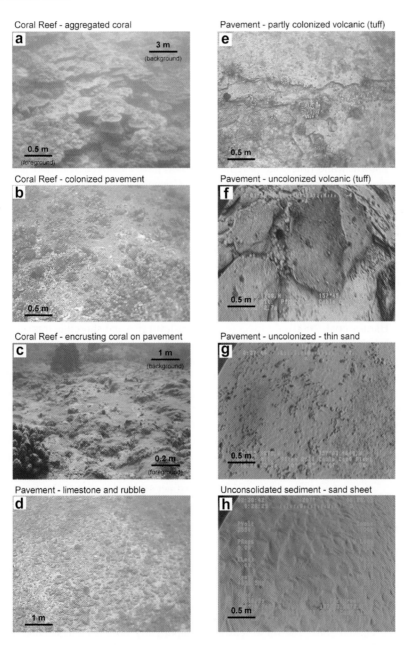

FIG. 11.14. Modern shelf substrates (Grossman et al. 2006; reproduced by permission of Elsevier Ltd.).
(**a**) Aggregated coral reef; (**b**) Colonized pavement; (**c**) Encrusting coral reef; (**d**) Limestone pavement with rubble;
(**e**) Partially colonized volcanic pavement; (**f**) Uncolonized volcanic pavement; (**g**) Uncolonized pavement;
(**h**) Unconsolidated sediment

Reef substrates occur primarily in shelf settings below wave base and are generally thin veneers except in wave-protected environments. They have high rugosity except where the community is dominated by encrusting forms and display spur and groove morphology. Modern coral reef substrates are ephemeral features on Oahu due to the interannual occurrence of large swell events, tsunami, and tropical depressions. Periodically, extensive tracts of surface coral are entirely removed by passing hurricanes and high waves (Dollar and Tribble 1993; Grigg 1995, 1998).

Pavement substrates occur between 0 and −120 m depth and are low-gradient surfaces comprised of fossil reef limestone or volcanic basalts. Volcanic pavements commonly display locally high rugosity in the form of ledges, pedestals, and meter-size plates or boulders, whereas limestone pavements are often of low relief. The age of volcanic pavements is poorly known, while limestone pavement ages range ca. 5–210 ka (Sherman et al. 1999; Grossman and Fletcher 2004; Rooney et al. 2004).

Unconsolidated sediments are primarily marine carbonate sands found in channels and fields across the inner and middle shelf (Harney et al. 2000; Harney and Fletcher 2003; Conger 2005). Sands are also common along the outer shelf in the form of thick sediment deposits (Hampton et al. 2003) often supporting stands of the green calcareous alga *Halimeda* (Harney et al. 2000).

Variations to this simple classification result from the temporal colonization of these substrates by coral and algae (coralline and fleshy green and brown algae). Other substrates including volcanic boulder fields are known along portions of the south Oahu shelf (Makapuu Pt.) that have not been studied.

Sandy nearshore substrate is important as a sand and gravel resource, habitat, and dynamic region of the bathymetry and hence worthy of further discussion.

11.3.3.1 Sand Fields

Conger (2005) investigated shallow, reef flat sand fields on the Oahu shelf and analyzed their geometry and relationship to depth. Sand deposits, and their distribution on fringing reefs have a significant effect on shoreline stability and the geologic framework of the coastal zone. For these reasons it is important to improve understanding of sand storage in shallow water.

Most shelf sands are carbonate with only a small percentage of terrigenous content (Moberly et al. 1965; Harney et al. 2000; Harney and Fletcher 2003). These biogenic sands accumulate in relatively thin patches, fields, and linear deposits perched on the shallow shelf. Their presence results from a state of semi-equilibrium among various processes controlling the sand budget including biologic production, temporary and permanent

storage, and loss (including abrasion, dissolution, bioerosion, and offshore transport). A combination of wave energy, water quality, biologic productivity, and storage space all control creation, destruction, and storage of carbonate sands. Changes in sea level mean geomorphologies from subaerial exposure and modern reef accretion play key roles in available sand storage space on the reef. Only two studies, (Moberly et al. 1975; Sea Engineering 1993) have cataloged nearshore sands in Hawaii.

Conger (2005) focused on sandy fields extending from 0 to −20 m depth. Both coral and algal growth rates are highest in these depths (Stoddart 1969) because of water circulation, nutrient availability, and available light (Grigg and Epp 1989; Grigg et al. 2002),. Most sediment on the reef is produced by reef builders, reef dwellers, and reef bioeroders, making this zone the primary source of nearshore sands. Only in the last 8,500 years has sea-level rise and shoreline transgression led to the inundation of this portion of the shelf (Grigg 1998) and allowed for modern carbonate production.

Most waves reach wave base within the zone 0 to −20 m and convert their wave energy into shear stress across the sea floor, providing a means for mechanical abrasion of both carbonate framework and direct sediment producers (Storlazzi et al. 2002). On Oahu, −20 m marks the approximate edge of the nearshore shelf that terminates in a distinct seaward facing wall (Stearns 1974; Fletcher and Sherman 1995). By extension of the Hawaii eolianite model proposed by Stearns (1970) and modified by Fletcher et al. (2005) where this is a barrier to upslope sand transport by winds during times of lowered sea level, it may also act as a barrier for shoreward submarine transport except where channelized, similar to the fossil barrier reef off southeast Florida (the "leaky valve" of Finkl 2004). Importantly, airborne and satellite sensors are capable of accurately imaging the sea floor within this depth range allowing researchers to prospect for sands via satellite imagery (Isoun et al. 2003, Conger et al. 2006; Fig. 11.15).

Conger (2005) analyzed 125 km² of fringing reef and identified sand deposits totaling about 25 km² or ~20% of the studied area. He used a supervised classification algorithm on multispectral QuickBird Satellite scenes (2.4 m pixel resolution) to identify five discrete classes of sand deposits: (1) Channels and Connected Fields, (2) Complex

Fig. 11.15. Variable nature of sand fields (red) on the fringing reef at Kailua Bay. The 10 m contour is outlined in green

Fields and Very Large Depressions, (3) Large Depressions and Fields, (4) Linear Deposits, and (5) Small Depressions and Simple Fields. This yielded a total of 14,037 sand deposits of five connected pixels or larger. He split these five classes into three depth groups, providing insight into sand storage variability on the reef flat: (1) 0 to −10 m, (2) −10 to −20 m, and (3) deposits that straddle the −10 m contour. The −10 m contour approximates the boundary of two reef sub-environments: (1) shallow reef limited by wave-generated shear stress where bathymetry largely reflects antecedent karst morphology, and (2) deeper reef where wave forces are less significant and the bathymetry is more likely to reflect modern coral framework accretion related to reef growth.

Channels and Connected Fields account for the majority (64%) of all sand deposit surface area, and Complex Fields and Very Large Depressions account for 18%. Just over 72% of all sand deposit surface area straddles the −10 m contour line, and 24% is shallower than −10 m. Combined sands crossing or shallower than −10 m represent more than 96% of all sand deposit surface area. When deposit classes are distinguished by depth range, Channels and Connected Fields that straddle the −10 m contour account for 63%, Complex Fields and Very Large Depressions shallower than −10 m account for 10%, and Complex Fields and Very Large Depressions that cross the −10 m contour account for 7%. Together, these three subgroups total 80% of all surface area for sand deposits.

Conger (2005) summarized his findings in terms of the physical morphology of the Oahu shelf. He concludes that the first order control in sand storage is general shape of the reef (Fig. 11.16): wide reef vs. deep reef. Wide and shallow back reefs with well-defined reef crests have more surface area covered by sand, while deeper fringing reefs and barrier reef fronts have less percent coverage. The second order control on sand fields is the energy environment on similar reef

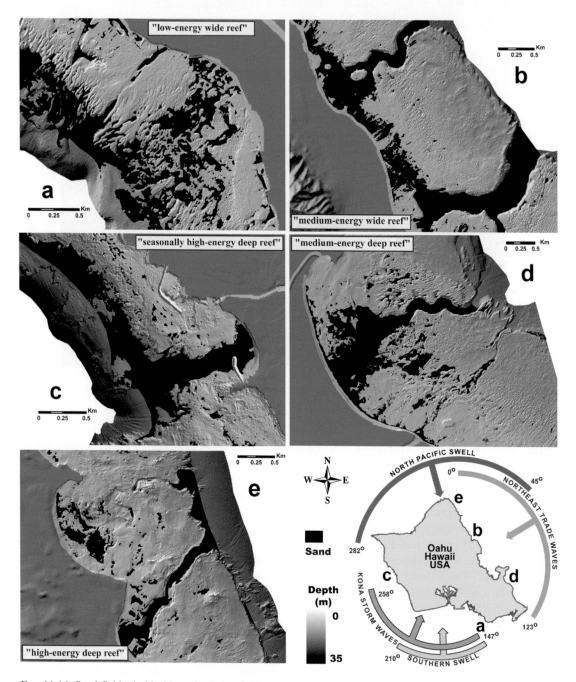

FIG. 11.16. Sand fields (in black) on the Oahu shelf shallower than −20 m (Conger 2005)

geomorphologies. Sand storage is highest in the general reef morphology Conger calls "low-energy wide reef" located on the south shore of Oahu. His reef types "medium-energy wide," "seasonal high-energy deep," "medium-energy deep reef," and "high-energy deep reef" have decreasing sand storage respectively. Hence, shallow, low energy shelf environments promote sand storage.

11.3.3.2 Sediment Production

Harney and Fletcher (2003) quantify a shelf sediment budget at Kailua Bay, Oahu. Although Kailua Bay has a well-developed modern coral community compared to the majority of the Oahu shelf, aspects of the sedimentology provide insight to island-wide patterns. Reefal sediments are primarily composed of carbonate skeletal fragments (>90%) derived from two sources: (1) biological and mechanical erosion of the coral–algal reef framework and (2) direct sedimentation upon the death of organisms such as *Halimeda*, articulated coralline algae, molluscs, and benthic foraminifera. Harney et al. (2000) report that shelf sands in Kailua range in age from modern to 4.5 ka, averaging ca. 1.5 ka, and that sand age varies with proximity to source, size fraction, and composition.

Harney and Fletcher (2003) estimate total annual calcareous sediment production over the 12 km^2 reef system between 0 and −20 m water depth is 4,048 (±635) m^3/year. Of this, bioerosion of coral and coralline algae species annually releases approximately 1,911 (±436) m^3/year of unconsolidated sediment, and mechanical erosion (coral breakage) releases another 315 m^3/year. Direct production of sediment by *Halimeda*, branching coralline algae, molluscs, and benthic foraminifera contribute a combined 1,822 (±200) m^3/year. Total production of sediment in Kailua Bay corresponds to an average rate of 0.53 (±0.19) kg m^{-2}/year. Of this, erosion of coralgal framework is responsible for approximately 0.33 (±0.13) kg m^{-2}/year and direct sediment production for 0.20 (±0.06) kg m^{-2}/year.

Sediment composition reflects the relative importance of calcareous sediment producers (Harney et al. 2000). Most beach and submarine sediment assemblages are dominated by coralline algae (e.g., *Porolithon,* up to 50%) and *Halimeda* (up to 32%). Coral is generally a minor constituent (1–24%),

as are molluscs (6–21%), benthic foraminifera (1–10%), and echinoderms (<5%).

11.4 Lithostratigraphy of Reefs

In the presence of suitable environmental conditions including sufficient sunlight and nutrients, low hydrodynamic energy, water quality, and temperature, low latitude shelves accrete reefal limestones under high sea levels on amenable substrate (Purdy 1974; Stoddart 1969, Scholle et al. 1983).

Past studies of reefal limestones in Hawaii have focused on either: (1) subaerial mapping of carbonate exposures such as coral-algal framestone, coral-algal gravels, or calcareous grainstone (i.e., eolianite) on the islands of Oahu, Molokai, Maui, and Lanai (Stearns 1978; Moore and Moore 1984; Muhs and Szabo 1994; Grigg and Jones 1997; Rubin et al. 2000) or (2) drilling submerged limestones constituting the insular shelf of Oahu and Molokai (Easton and Olson 1976; Grigg 1998; Sherman et al. 1999; Engels et al. 2004; Rooney et al. 2004). Nearly all limestones in Hawaii date from sea-level high stands associated with interglacial periods of the late Quaternary. To define the chronostratigraphy of these deposits, researchers utilize the oxygen isotopic stage (MIS) system of nomenclature first proposed by Emiliani (1955) and further defined by Shackleton and Opdyke (1973) and Shackleton (1987) (Fig. 11.17).

Among the Hawaii Islands, reefs on the island of Oahu have the most thoroughly studied geologic history. Past workers document the pattern and timing of sea-level highstands (Ku et al. 1974; Stearns 1974, 1978; Jones 1993; Sherman et al. 1993; Szabo et al. 1994; Fletcher and Jones 1996; Grossman and Fletcher 1998; Grossman et al. 1998; Hearty 2002) and island tectonics (Muhs and Szabo 1994). Other studies integrate submarine with subaerial carbonate deposits (Moberly and Chamberlain 1964; Lum and Stearns 1970; Coulbourn et al. 1974; Harney et al. 2000; Harney and Fletcher 2003; Fletcher et al. 2005). Grigg (1998), Sherman et al. (1999), Grossman and Fletcher (2004), Engels et al. (2004), Rooney et al. (2004), and Grossman et al. (2006) describe the submerged complex of carbonates characterizing the insular shelf and relate accretion style and history to environmental conditions.

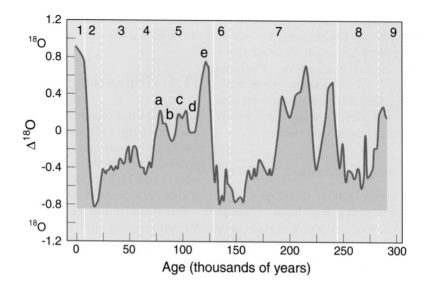

FIG. 11.17. Marine isotopic stages (MIS) of the late Quaternary (After Shackleton and Opdyke 1973)

Over 50 drill cores ranging in length from 1 to 20 m have been acquired from the submerged Oahu and Molokai shelf by workers at the University of Hawaii. Cores were collected from both windward and leeward sides of the islands at elevations ranging from +2 to −35 m. Workers employed open-bit and wireline drilling techniques, both powered by surface-supplied hydraulic pressure (Fig. 11.18). Cored facies are associated after Embry and Klovan (1971).

11.4.1 Pleistocene Reefs

Oahu 's coastal plain is underlain by reefal limestones, examples are the Waimanalo (Stearns 1974, 1978; Sherman et al. 1993) and Waianae (Sherman et al. 1999) formations. The Waimanalo reef dates to the last interglacial (MIS 5e) sea-level highstand ca. 134–113 ka (Muhs et al. 2002). These consist of *in situ* framestones rising to +8.5 m elevation and rudstone deposits reaching as high as +12.5 m elevation (Ku et al. 1974; Muhs and Szabo 1994; Szabo et al. 1994).

Resting unconformably on lower-lying portions of the coastal plain are moderately to poorly lithified carbonate sands. These include eolianites dating from the late last interglacial (MIS 5a–d; Fletcher et al. 2005) as well as unlithified dune and beach deposits dating from withdrawal of the Kapapa sea-level highstand of late Holocene time

(discussed in more detail in a later section) ca. 1.5–4 ka, +2 m in elevation (Fletcher and Jones 1996; Grossman et al. 1998). Hearty et al. (2000) assign last interglacial dunes to MIS 5e, thus differing with the interpretation of Fletcher et al. (2005) on the timing of their deposition.

Offshore of island beaches, the Oahu shelf is characterized by a distinct stair-step bathymetry created by reefal limestone units organized in association with past sea-level stillstands. On the basis of cored facies, Sherman et al. (1999) found the main stratigraphic component of the shallowest terrace is a massive limestone they name the Waianae Reef, dating from the penultimate interglacial, MIS 7. They identify a unit of MIS 5a–d reefal limestone accreted on the seaward-facing front wall of the Waianae Reef that is consistent with formation during periods when sea level was below present. The age of this reef correlates with carbonate units described by Stearns (1978) that he named Leahi and so this name is applied to the MIS 5a–d reef.

11.4.2 Mineralogy and Cementation

Limestone mineralogy on the Oahu shelf reveals that it has been exposed to subaerial conditions followed by marine inundation on more than one occasion. Carbonate-secreting organisms and cements are composed principally of aragonite and magnesium calcite (Morse and Mackenzie 1990).

Fig. 11.18. Cored samples are obtained using both wireline and open-bit coring. The University of Hawaii jack-up drill barge provides access to shallow (<2 m depth) reef sites

These phases are metastable and in most instances will convert to calcite when exposed to fresh-water environments (Tucker and Wright 1990). In Hawaii, the high volcanic islands produce a powerful orographic effect yielding abundant fresh ground water that converts fossil corals to calcite.

Grossman (2001) found that mineralogical data from the mixed Holocene /Pleistocene reef system found in Kailua Bay strongly segregate into two groups. The dominant Pleistocene mineralogy is recrystallized, low-magnesium calcite (<5 mole% $MgCO_3$). Bioskeletal coral, coralline-algae and bulk grainstones of the MIS 5e Waimanalo Reef typically range 1–4 mol% $MgCO_3$. The Waianae Reef (MIS 7) skeletal components typically contain <1 mol% $MgCO_3$, except rare fragments of *Porolithon gardineri*.

Elsewhere on Oahu, Sherman et al. (1999) discovered a generally seaward trend of increasing aragonite content and greater abundances of high magnesium calcite. That is, shoreward samples are predominantly calcite stabilized by exposure to meteoric conditions. Seaward of these, samples have higher percentages of metastable aragonite and magnesium calcite (Fig. 11.19). This is consistent with a seaward decrease in the age of shelf limestones associated with the shift from mid-shelf MIS 7 limestones to MIS 5a–d limestones at the shelf edge.

Interstitial cements document a history of alternating submarine and subaerial exposure through sea-level stands of the late Quaternary. Evidence of early shallow-marine diagenesis is found in first-generation interstitial aragonite and magnesium calcite cementation. Aragonite is found exclusively as acicular aggregates, whereas magnesium calcite is found in a variety of forms, including microcrystalline, peloidal, and bladed spar. All are common shallow-marine reef cements (Macintyre 1977).

During periods of subaerial exposure, limestones of the Oahu shelf underwent cementation by calcite, neomorphism, and dissolution (Sherman et al. 1999). Meteoric alteration is patchy on all scales and preservation ranges from pristine to massive alteration. Meteoric calcite forms needle fibers, anastomosing micritic networks (alveolar texture), and equant calcite. In addition to carbonate products, red iron-rich, noncarbonate clays are incorporated into the limestones and form void and grain coatings perhaps reflecting soil development. Evidence of marine diagenesis occurring after subaerial exposure is found in last-generation highly unstable magnesium calcite cements and internal sediments that have otherwise been almost wholly stabilized to calcite.

Mineralogic and cementation histories of the Oahu shelf provide evidence of alternating periods of marine and meteoric diagenesis attributed to glacioeustatic fluctuations (Sherman et al. 1999). Extensive early marine cementation in the MIS 7 Waianae Reef is consistent with a high wave energy environment characterized by strong circulation. Marine cementation is favored and most pervasive in the active marine phreatic zone near the sediment -water interface in high-energy environments where water can be flushed through the porous structure of the reef (Tucker and Wright 1990). Other components of the Waianae Reef display micritization and absence of early marine cementation. These are consistent with accretion in a back-reef, low energy (lagoonal) environment, a stagnant marine phreatic zone.

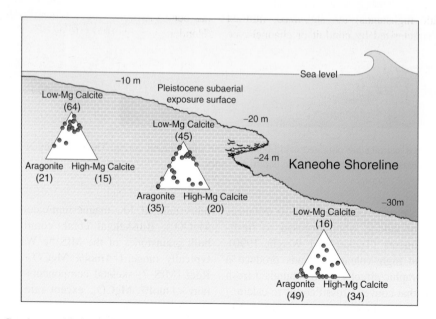

FIG. 11.19. Consistent with decreasing limestone age, the abundance of metastable aragonite and magnesium calcite phases increases in the seaward direction across the Oahu shelf (Modified from Sherman et al. 1999 by permission of SEPM)

Notably, the diagenetic record in Waianae limestones appears to reflect local meteorological conditions. The Oahu shelf was subaerially exposed during sea-level lowerings of MIS 6 and again following MIS 5a. However, there is no direct evidence of meteoric vadose zone alteration of western Oahu shelf limestones. Sherman et al. (1999) speculate that this was probably a result of the rain shadow effect that impacts the leeward side of Oahu where many of their deeper cores were obtained. This effect would be magnified during low sea-level conditions related to increased rain shadowing (aridity) in that region. However, in windward Kailua Bay, Grossman (2001) records a distinct pattern of lowered magnesium content in older (Waianae Reef – MIS 7) samples reflecting the influence of MIS 6 exposure in a wetter environment where orographic rainfall would be enhanced by sea-level lowering.

Post-meteoric marine cementation is extensive in Waianae Reef samples that experienced stagnant marine phreatic zone conditions (lagoonal deposition) during their early accretion. During periods of subaerial emergence (during glacioeustatic lowstands) these limestones were stabilized and lithified to calcite in a meteoric environment and underwent partial solution and creation of vug and channel porosity. When subsequently drowned again (during glacioeustatic highstands) the stabilized, lithified substrate characterized by conduit or channel-type porosity, hosted high marine flow rates induced by channelized seawater. This resulted in precipitation of thick isopachous rims of bladed magnesium calcite spar lining the walls of large voids.

The general trend of decreasing mineralogic stabilization progressing seaward across the terrace is consistent with Th-U ages of fossil corals from these deposits, i.e., less mineralogic stabilization in the younger limestones. The age of carbonate systems is discussed in the next section.

11.4.3   Age of Carbonate Systems

Th-U and radiocarbon ages of carbonate systems document their age and history on both the coastal plain and shallow shelf. Fossil corals fall into four age groupings. (1) Where the shelf is a fossil limestone surface in the absence of modern coral growth ("pavement facies" of Grossman et al. 2006), samples from the outer edge of the shelf

range 82.8–110.1 ka (MIS 5a–d); (2) those from the middle and inner portion of the shelf range 206.4–247.2 ka (MIS 7). A third group (3), found composing most rocky limestone coasts (and the offshore islet of Popoia in Kailua Bay), document last interglacial MIS 5e ages when sea level was above present. The fourth group (4), Holocene in age, is found on the middle to outer shelf (occasionally accompanied by modern coral growth), and on some very shallow fringing reefs of Oahu and Molokai (Rooney et al. 2004). These typically document early to middle Holocene ages (MIS 1) where exposed to north swell and middle to late Holocene ages in protected settings and where exposed to south swell (Grossman et al. 2006). Calibrated radiocarbon ages of *in situ* Holocene framework corals range modern to >7 ka. This range is confirmed by two Th-U ages ca. 5.3–7.8 ka (Grossman and Fletcher 2004).

Age-corrected $^{234}U/^{238}U$ in coral should compare to modern seawater if the coral has remained chemically pristine since fossilization and if $^{234}U/^{238}U$ in seawater has remained constant through the Quaternary. $\delta^{234}U$ in modern corals and $\delta^{234}U_i$ in Holocene corals are typically indistinguishable from modern seawater (Chen et al. 1986; Edwards et al. 1986; Rubin et al. 2000). Further, various workers have suggested that $^{234}U/^{238}U$ in seawater has been essentially constant over the past 250 ka (Henderson et al. 1993). Elevated $\delta^{234}U_i$ observed in some older corals likely indicates open-system behavior (Bard et al. 1996a, b; Sherman et al. 1999; Gallup et al. 1994; Hamelin et al. 1991), although some have suggested that $^{234}U/^{238}U$ in seawater may have differed in the past (Hamelin et al. 1991). Most workers apply a "working definition" of $^{230}Th-^{234}U-^{238}U$ age quality based on $\delta^{234}U_i$ relative to modern: 145–153‰ is considered "highly reliable", 139–159‰ or 165‰ is "moderately reliable", and $\delta^{234}U_i$ >165‰ is "less reliable" (Bard et al. 1996a; Szabo et al. 1994; Stirling et al. 1998). These ranges are somewhat arbitrary because "acceptable" $\delta^{234}U_i$ varies with how and when open-system behavior occurred and the absolute age of the sample, and because not all open-system events that modify $\delta^{234}U_i$ also affect sample age (Chen et al. 1991; Hamelin et al. 1991).

Sherman et al. (1999) document the age-corrected $^{234}U/^{238}U$ values of their dated coral samples. Older corals from the Oahu shelf have $\delta^{234}U_i$ that range from 149‰ to 254‰ indicating diagenetic alteration in some samples and probable age biasing. Sample MAI5–1S1 ($^{234}U/^{238}U$ = 149.0 (±1.0) ‰; Sherman et al. 1999) provides the most reliable age for the Waianae Reef, 223.3 (±1.5) ka. With the exception of two samples, the younger corals collected at the seaward margin of the shelf have $\delta^{234}U_i$ values that cluster within the range 149 (±10) ‰ and thus provide reliable ages (ca. 82–110 ka).

Fletcher et al. (2005) interpret amino acid racemisation data from both Molokai and Oahu coastal plain deposits and correlate calcarenite formation to late MIS 5. The current mean annual temperature in Hawaii is about 25°C hence diagenetic temperatures are likely to have been relatively stable over the late Quaternary. A geochronological framework for assessing AAR data is provided by independently dated *Periglypta reticulata* (mollusc) remains from Barbers Point and Kapapa Island, Oahu (Sherman et al. 1993; Grossman and Fletcher 1998). Also available are two submerged coral samples from Oahu independently dated by the TIMS U-series method, and an electron spin resonance (ESR) age of 562 (±96) ka on Middle Pleistocene coral from Barbers Point (Sherman et al. 1993).

The low extent of racemisation in *P. reticulata* samples from Kapapa Island is consistent with their middle Holocene age determined by radiocarbon (Fletcher and Jones 1996; Grossman and Fletcher 1998), and by analogy with extensive comparisons of Holocene AAR and radiocarbon ages (Murray-Wallace 1993, 1995). In contrast, a significantly higher extent of racemisation is evident for *P. reticulata* from a last interglacial coral rudstone unit at Barbers Point (Unit III of Sherman et al. 1993). These correlate well with last interglacial molluscs in Australia that have equivalent current mean annual temperatures (Murray-Wallace 1995).

11.4.4 Lithofacies

Skeletal components of cored limestones on the Oahu shelf (Fig. 11.20) are typical of reefal environments and include coralline algae, coral, molluscs, echinoderms, and benthic foraminifers. Terrigenous input is limited to rare volcanic clasts, grain and void coatings of iron-rich clay.

In past work on the Oahu shelf (Sherman et al. 1999), Hawaii reefal limestones were classified into biolithofacies on the basis of their dominant skeletal component and fabric. Two facies were described: (1) massive coral, and (2) branching coral. Each facies included both an autochthonous (*in situ*) and allochthonous component. The massive coral facies incorporated coarse skeletal grainstones and rudstones, as well as encrusting bindstones predominately of crustose coralline algae. The dominance of massive corals (i.e., *Porites lobata*) and encrusting algae indicates a shallow marine high-energy environment of deposition. The branching coral facies included delicate branching corals (i.e., *Pocillopora damicornis*), coralline algae, and other biota set in a lime-mud matrix forming *in situ* bafflestones, floatstones, or wackestones. The presence of delicate branching corals and lime-mud matrix indicate a low energy environment of deposition such as a lagoonal or embayed setting, or the inner parts of large reef flats away from breaking waves.

Grossman and Fletcher (2004) adopted a classification with five facies (Fig. 11.21) encountered in their work in Kailua Bay. In order of decreasing depositional energy, these are: (1) encrusting coral-algal bindstone facies, (2) coral rudstone facies, (3) grainstone facies, (4) massive coral framestone facies, and (5) branching coral framestone facies. Engels et al. (2004), working on the nearby island of Molokai with cores from the south shore fringing reef, encountered the same bindstone, rudstone, massive framestone, and branching framestone facies described by Grossman and Fletcher (2004). However, they failed to find a grainstone facies, and they describe an "unconsolidated floatstone" facies in their work.

Because the distribution of many coral and coralline algae species is governed by temperature, light levels, nutrient levels, and wave energy, facies can serve as paleoecologic indicators of the Quaternary (Adey 1986; Cabioch et al. 1999).

As described by Grossman and Fletcher (2004), the encrusting coral-algal bindstone facies consists of *in situ* encrusting forms of coral and coralline algae with occasional grainstones, rudstones and algal rhodoliths. Typical corals include *Montipora patula*, *Cyphastrea ocellina*, *P. lobata*, and *M. capitata*. These are found either as *in situ* bindstones with a semifriable coarse grainstone to rudstone matrix, or as unconsolidated subrounded to angular, oblate to bladed, pebble-size clasts. Although *M. patula*

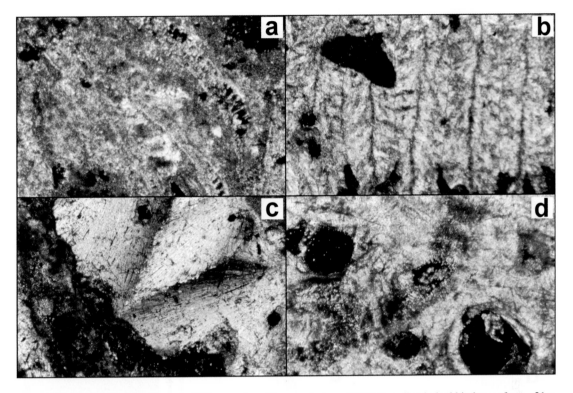

Fig. 11.20. Lithofacies (Engels et al. 2004; reproduced by permission of SEPM). **a** Coral-algal bindstone from −21 m depth with encrusting coral and heavy laminar micritic cement displaying knobby texture. **b** Massive coral framestone from −10.7 m depth. **c** Mixed skeletal rudstone from −8.5 m depth showing brachiopod fragment and knobby micritic cement (lower left). **d** Branching coral framestone-bafflestone from −17.7 m depth showing bored, branching *Porites compressa* with aragonitic cement

can be found from the intertidal zone to depths of −15 m on modern Hawaii reefs, it is most frequently found high on the reef slope or in shallow bays with moderate wave action (Gulko 1998). *C. ocellina* is usually found near shore in shallow water, frequently in areas that have moderate wave action. The co-occurrence of these two species and their encrusting morphologies suggest that they grew in a shallow, moderate energy environment. *Hydrolithon onkodes* is the dominant encrusting coralline alga. The crustose coralline algae *Tenarea tessellatum* is also common. This assemblage is abundant on shallow, wave-agitated reef platforms along windward Hawaii shores. Also present is the encrusting foraminifer *Homotrema* as well as vermetid gastropods, boring molluscs (*Lithophaga*) and (rarely) serpulids.

The coral-algal bindstone facies is indicative of high-energy wave conditions characteriz-

ing shallow reef platforms that are persistently scoured by shoaling waves. Studies by Littler and Doty (1975) on Hawaii algal ridges showed that *H. onkodes* and *P. gardineri* dominate the seaward margin of the reef. *P. gardineri* dominates subtidal portions of the crest. *H. onkodes* dominates intertidal portions of the ridge crest, inshore flat, and seaward front. The ecological specificity of coralline algae, especially when core samples are collected in thick monospecific communities provide for tracking the position, (± 1–2 m), of slowly rising sea level over the century to millennia scale.

The coral rudstone facies is characterized by unsorted, angular to round clasts of *Porites compressa*, *Pocillopora meandrina*, *Poc. eydouxi*, and *P. lobata*. *P. compressa*, the most abundant component, is typically found growing in moderately high-energy reef environments characterized by active circulation. This facies is found in upper

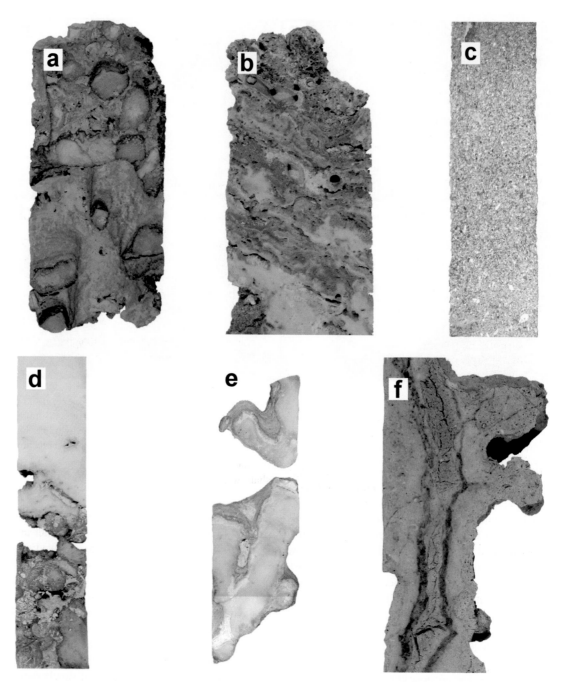

Fig. 11.21. Limestone lithofacies in Hawaii studies (Sherman et al. 1999; Grossman and Fletcher 2004; Engels et al. 2004; Rooney et al. 2004; reproduced by permission of Elsevier and SEPM). **a** branching coral rudstone dominated by clasts of *Porites compressa* and massive peloidal micrite crusts (ca. 3 ka); **b** encrusting coral-algal bindstone formed of alternating layers of *Montipora patula* and *Hydrolithon onkodes*; **c** grainstone from stranded mid-reef beachrock outcrop; **d** massive coral framestone of *Porites lobata* with borings by *Lithofaga* overlying branching coral rudstone with coarse shallow platform skeletal debris of *Halimeda* and molluscs (ca. 6.5 ka); **e** branching coral framestone of delicate-branching *Porites compressa* with fine laminar micrite (ca. 4 ka); **f** mudstone/wackestone with desiccation cracks lined by coralline algae and infilled with skeletal debris and peloidal micrite converted to calcite

sections of a central reef platform. Thin crusts of *H. onkodes* frequently envelope single and multiple coral clasts indicating encrustation preceded final deposition. Additional skeletal components include fragments of *Halimeda*, molluscs, branching coralline algae, echinoderma and foraminifera. Cored samples display burrows, borings and secondary encrustation by foraminifera and bryozoans. The rudstone facies represents a high-energy depositional environment generally consisting of coral fragments originally derived from protected (deeper fore reef) settings.

Relatively rare, grainstone facies is composed of medium to coarse, rounded skeletal fragments of coralline algae, coral and molluscs. This facies is formed by cementation of former moderate to high-energy beach ridges stranded by shoreline retreat in reef flat settings. *Halimeda* grains are present and typically fine to medium size. Isopachous magnesium calcite rim cements coat grains and partially fill interstitial void space.

Typical of the massive coral framestone facies, *P. lobata* is the most common and widespread of Hawaii corals and can occur anywhere from the intertidal zone to depths of −40 m. However, *P. lobata* is most common high on wave-exposed reef slopes just below highest wave action between depths of −3 to −15 m (Maragos 1977; Gulko 1998). Grigg (1998) showed where exposed to high wave energy *P. lobata* is the dominant reef builder. Crustose coralline algae (*H. onkodes*) are important members of the massive-coral facies. It occurs as sheet-like encrustations on the upper surfaces of corals, lining voids in coral framework, over previously lithified rudstone, and coating coral clasts in rudstones. The dominance of massive-corals along with encrusting algae indicates a shallow, high-energy environment of deposition (Tucker and Wright 1990; James and Bourque 1992). The combination of a grainstone and rudstone matrix with *in situ* framework is also common in high-energy, shallow water settings. This distribution is consistent with the expected zonation of lithofacies in a marginal reef complex, where rudstones and framestones are most common in reef flat, reef crest, and reef front environments (James and Bourque 1992).

The branching coral framestone facies is composed of delicate, *in situ* branching corals (*Poc. damicornis* and *P. compressa*), coralline algae, and associated biota set in a lime-mud matrix forming *in situ* bafflestones, floatstones, or wackestones. In Hawaii, *Pocillopora damicornis* is usually found in protected bays or upon the inner parts of large reef flats away from breaking waves (Maragos 1977). Likewise *P. compressa*, the most competitively superior coral species in low wave energy environments, is often found on the fore reef where it can monopolize substrate until disturbance intervenes. The presence of *Poc. damicornis* and dominance of a lime-mud matrix indicate a low-energy environment of deposition typically along inner parts of a reef flat landward of the massive-coral facies. In the case of *P. compressa*, the assemblage may indicate reef flat or fore reef settings near or below wave base.

Mudstone/wackestone units are relatively rare in Hawaii fringing reefs. They are characterized by infilled cavities with a fine, white semifriable powder or as a brown, clotted, indurated lime mud. This lithology is restricted to the lower sections of shallow core sites on the fringing reef flat but has also been found in at least one deeper core site reoccurring throughout the sequence.

11.4.5 Assembly of the Oahu Shelf

A complex history of reef, dune, and coastal plain accretion during the late Quaternary has produced a mosaic of stratigraphic components comprising the shallow coastal plain and shelf of Oahu (Fig. 11.22).

11.4.5.1 MIS 7 – Waianae Reef

The earliest reef accretion on the Oahu shelf for which there is widespread evidence dates to middle MIS 7 (Fig. 11.23a). The duration of MIS 7 is 182–242 ka (Bassinot et al. 1994). Four well-preserved coral samples from both windward and leeward sides of Oahu provide absolute ^{230}Th-^{234}U-^{238}U ages dating 206–247 ka within acceptable limits of $\delta^{234}U_i$ ($\delta^{234}U_i < 165‰$).

On leeward Oahu, cored facies reveal MIS 7 fossil reef crest communities of *in situ* stout branching and massive *P. lobata* comprising the reef framework. Windward Oahu cores of the Waianae Reef display a high-energy community of encrusting crustose algae (e.g., *Hydrolithon onkodes*) characteristic of shallow subtidal to intertidal algal ridges at the seaward margin of fringing reefs throughout the central and eastern Pacific. Post-glacial flooding by Holocene sea level is responsible for Holocene

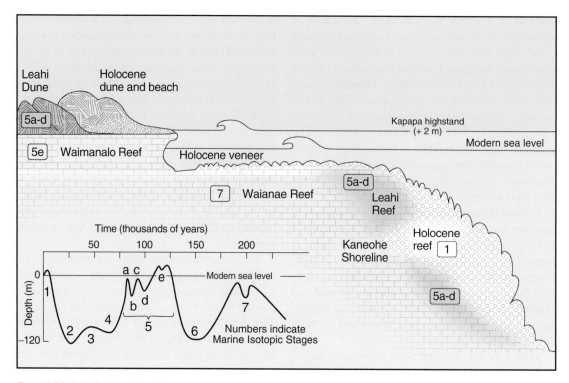

Fig. 11.22. Principal stratigraphic members of the Oahu carbonate shelf

age marine cements, isopachous rims of bladed Mg calcite spar (ca. 2.8–5.6 ka) found within the framework matrix of the Waianae Reef (Sherman 2000).

Cores shoreward of the fossil reef crest contain branching corals of delicate *Poc. damicornis* in a lime mud matrix, characteristic of a lagoonal or back-reef community. Sherman (2000) analyzes the position of paleo-sea level during accretion of the Waianae Reef. Corrected for island uplift (0.03–0.06 mm/year; Muhs and Szabo 1994) he concludes that sea level in Hawaii was −9 to −20 m below present when the Waianae Reef formed. This is consistent with workers who have placed MIS 7 at 0 to −20 m below present in other locations (Chappell and Shackleton 1986; Harmon et al. 1983; Gallup et al. 1994).

Cored facies of the Waianae reef indicate an ecologic response to hydrodynamic forcing similar to modern conditions. Massive *P. lobata* comprising the reef framework is consistent with modern high wave stresses generated by winter north and west swell originating from North Pacific low-pressure systems. Delicate *Poc. damicornis* in a

lime mud matrix found behind the protection of the fossil reef crest of the time are consistent with lagoonal conditions of reduced wave stress. On windward shores the predominance of encrusting crustose algae, *H. onkodes*, is consistent with modern persistent trade winds. Although formed at somewhat lower sea level, overall the marine floral and faunal record of MIS 7 in Hawaii reflects a marine climatology not different from modern conditions.

Bassinot et al. (1994) indicate a chronology of MIS 7 extending over the period 182–242 ka. The best date for the Waianae Reef is 223 ka with a $\delta^{234}U_i$ value of approximately 149±3. The next best acceptable $\delta^{234}U_i$ values, 164 (±3) and 166 (±2), correspond to dates of 211 ka, and 206 ka respectively. Taken together these dates indicate accretion of the main body of the Waianae Reef at the end of isotopic event 7.2 and during events 7.3–7.4 of Bassinot et al.'s chronology.

The wide geographic range of MIS 7 samples from the Oahu shelf (leeward and windward shores), the range of their depths (−5 to

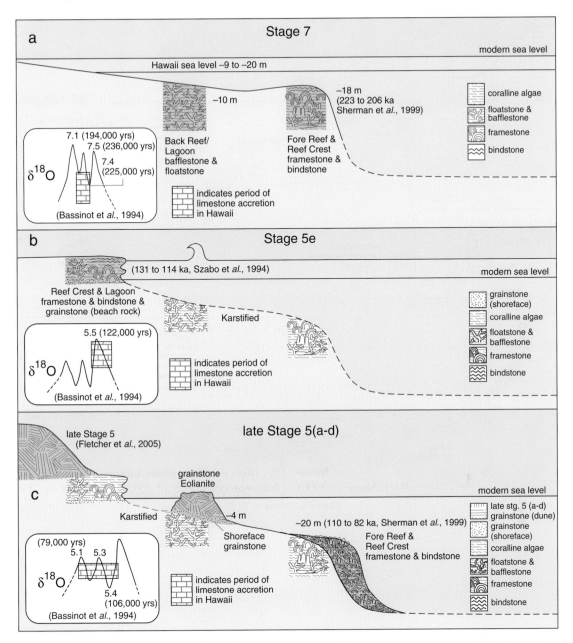

FIG. 11.23. Assembly of the Oahu shelf: (**a**) The Waianae Reef, MIS 7. Corrected for island uplift, facies indicate sea-level −9 to −20 m below present (Sherman 2000). (**b**) Last interglacial Waimanalo Reef is found on many rocky Oahu shores. An unconformable contact at −5 m defines the contact between Waimanalo and underlying Waianae reefal units (Grossman and Fletcher 2004). (**c**) Workers find late interglacial framestone accretion (Leahi Reef) ca. 110–82 ka on the seaward front of the Oahu shelf. Large-scale dune deposition ensued during the period of general sea-level fall (as first hypothesized by Stearns 1974) at the end of stage 5

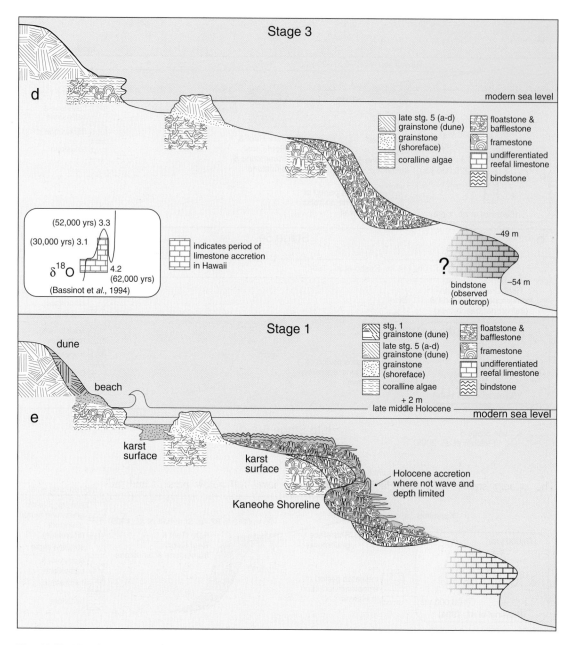

FIG. 11.23. (Continued) (**d**) A limestone unit of unknown age is exposed on the Oahu slope in the range −49 to −54 m. (**e**) Holocene accretion may bury exposures of earlier limestone but only where wave energy is limited. Highest rates of accretion correlate with *in situ* framestone during early Holocene time when sea level was rising >2–3 mm/year. Rooney et al. (2004) document the end of fringing reef accretion at 5 ka on shores exposed to north swell

−20 m), and observations of the continuity of the shelf extending nearly unbroken around the island have led Sherman et al. (1999) to infer that the Waianae Reef is the largest and most significant stratigraphic component of the Oahu shelf. They conclude that other limestone units are of secondary importance by volume, superimposed as they are, upon the main Waianae Reef body. It is likely that the Waianae Reef was able to accrete largely unimpeded by certain modern limitations on reef accretion (i.e., shallow substrate above wave base). That is, the Waianae Reef was evidently the first major carbonate accretion episode in the post-shield building era.

Grossman (2001) infers from the shallow depth of the Waianae Reef in Kailua Bay (Popoia Island, −5 m; see next section) and the routine acquisition of long limestone cores (e.g., >30 m) around the Oahu coastal plain for commercial purposes (i.e., water wells, geoengineering studies) that the Waianae Reef is an important, previously unrecognized stratigraphic unit in the Oahu coastal plain underlying the Waimanalo Reef. Final confirmation of this hypothesis will come when future researchers obtain: (1) sample ages from longer cores that fully penetrate the submerged Oahu shelf and coastal plain, as well as (2) sample ages from cores of unstudied portions of the Oahu shelf (i.e., the north and south shores).

But an important question remains unanswered: Why and how did the Waianae Reef form under wave conditions similar to present? Alternatively, is the Waianae Reef evidence that wave climate during MIS 7 differed from today?

11.4.5.2 MIS 5e – Waimanalo Reef

The widely studied last interglacial *in situ* reef (Waimanalo Reef) is exposed along rocky carbonate shores on Oahu (Fig. 11.23b). Waimanalo exposures are widespread due to long-term island uplift and deposition under a higher than present paleo-sea level (Stearns 1978; Muhs and Szabo 1994; Muhs et al. 2002). Last interglacial limestones on Oahu have been the subject of several studies that focus exclusively on surficial subaerial exposures (Ku et al. 1974; Sherman et al. 1993; Jones 1994; Grigg and Jones 1997; Muhs and Szabo 1994; Szabo et al. 1994; Muhs et al. 2002). The Waimanalo Reef is primarily composed of *in situ* coral-algal framestone with locally important bindstone and bioclastic grainstone facies variants. It is unconformably overlain by calcarenite (eolianite and rudstone deposits) of the Leahi Formation (Stearns 1974).

The only identification of the base of the Waimanalo Reef is described in Grossman (2001). He cored through an offshore windward islet in Kailua Bay, Popoia Island, an outcrop of the Waimanalo Reef, to determine the stratigraphic relationship between the superposed Waimanalo and Waianae reef units. An unconformable contact, encountered at approximately −5 m below sea level, is characterized by subaerial diagenetic alteration to low-magnesium bladed sparry calcite formed as isopachous rims. Both units are characterized by varying massive and stout branching framework coral growth with abundant crustose coralline algae that fills voids and binds carbonate sediments.

The age of coral samples from the Waimanalo reef (ca. 131–114 ka) reported by Szabo et al. (1994) indicate the duration of the last interglacial sea-level highstand was approximately 17,000 years. This contrasts to a duration of about 8,000 years inferred from the orbitally "tuned" marine oxygen isotope record (Martinson et al. 1987). Muhs et al. (2002) report new ages of the Waimanalo reef (ca. 134–113 ka) that essentially confirm the early start of MIS 5e high sea levels and their long duration. They conclude that orbital forcing may not have been the only controlling factor on global ice sheet growth and decay during the last interglacial.

11.4.5.3 MIS 5a–d – Leahi Reef and Eolianite

At the close of the 5e highstand, ca. 113 ka, sea level fell below present and the Waimanalo Reef was stranded. The fossil substrate of the Waianae Reef proved to be suitable for continued framestone accretion along the seaward margin (Fig. 11.23c). Late interglacial framestone accretion (ca. MIS 5a–d) is dated with four samples of pristine *in situ* coral ca. 82–110 ka collected −25 to −30 m depth from the leeward side of Oahu. The present depth of this paleo-reef crest is approximately −20 m marking the seaward edge of the insular shelf. Sample ages of the 5a–d reef (ca. 110 ka) overlap with published dates (Szabo et al. 1994) of the Waimanalo Formation sampled *in situ* at Kaena Point above present sea level, indicating simultaneous coral growth at offshore locations contemporaneous with the last vestiges of Waimanalo Reef accretion.

Samples from the shelf front indicate that accretion ensued with falling sea level across the substage 5e/5d boundary. There is no evidence of subaerial exposure, suggesting that sea level stayed above −25 m (approximately −27 to −31 m corrected for island uplift) between ca. 110 and 104 ka. Cored facies consist of massive head corals grading upwards to coral bindstone and encrusting coral rudstone. Workers infer from this, and the general trend of decreasing age with distance offshore, that accretion was occurring during general sea-level fall in the latter part of MIS 5. Falling sea level caused a shift in the reef community toward a relatively shallow moderate to high-energy environment. Following Stearns (1974) this late interglacial reef unit is referred to as the Leahi Reef.

On the adjacent coastal plain, vast deposition of calcareous eolianite characterized the margin of Oahu, as well as the islands of Maui, Kauai and Molokai. AAR values of both whole rock and mollusc samples indicate these units formed during the MIS 5a–d interval (Fletcher et al. 2005). Analysis of the racemisation history of coastal eolianites composed of reefal carbonate skeletal fragments (analyzed with both whole rock and mollusc samples) on windward and leeward Oahu indicates that large-scale dune deposition ensued during the period of general sea-level fall (as hypothesized by Stearns 1974) at the end of stage 5. Although sediments on the Waianae Reef exposed by falling sea level following the 5e highstand would surely compose the extensive eolianite units of the Oahu coast, the volume and extent of the dunes requires continuous sediment production during the 5a–d interval, as would be provided by the Leahi Reef that was accreting immediately offshore. Lithified dune units not only comprise significant stratigraphic members of the subaerial coastal plain, they are important components of the shallow submerged terrace, in places forming a substrate for Holocene coral growth in the form of a major barrier reef (Kaneohe Bay) and multiple offshore islets.

Many locations on Oahu display a deep intertidal notch at approximately −18 to −25 m depth (named the Kaneohe Shoreline; Stearns 1974) carved into the front of the island shelf into both Waianae-age and Leahi -age reefal limestones. Whether this feature records the former position of sea level at the end of the last interglacial as the sea dropped into glacial MIS 4, or is a product of sea-level movements associated with MIS 3 or MIS 1 is difficult to determine as a notch is an erosional feature and therefore its exact age can only be correlated, not measured. However, Fletcher and Sherman (1995)

note the correspondence of the Kaneohe Shoreline to the sea-level rise event ca. 8 ka proposed by Blanchon and Shaw (1995). Additionally, a reef lithofacies record from Molokai showing rapidly rising sea level ca. 8 ka in the depth range of the Kaneohe Shoreline (approximately −20 to −25 m) suggests the notch is early Holocene in age (Engels et al. 2004). Notably, the presence of a pronounced and submerged notch with an intact visor is suggestive of rapid drowning, not abandonment; hence post-glacial timing is most likely. Holocene patterns are discussed in the next section.

A broad and well-developed reefal limestone unit is exposed on the Oahu slope in the range of −49 to −54 m depth (Fig. 11.23d). We infer from its elevation that this unit correlates with MIS 3; confirmation of this hypothesis waits further testing.

11.5 Holocene Accretion and Sea-level Change

Framework accretion during Holocene time (Fig. 11.23e) is recorded on the Oahu shelf largely in response to wave energy, sea-level position, and proximity to acceptable water quality. Rooney et al. (2004) document a statewide pattern of framework accretion prior to ca. 5 ka on exposed coastlines that has since ended. In the time since ca. 5 ka, locations of low antecedent topography below wave base (nominally −10 to −30 m) may host Holocene framework, and modern growth is still found in many of these locations. But such sites tend to be geographically restricted and dependent on reduced incident wave energy. Examples of such locales include: lagoon settings (i.e., Kaneohe Bay); on vertical walls carved into Pleistocene carbonates (i.e., drowned stream channels); low energy oceanic settings (i.e., Hanauma Bay, Kahana Bay, Lanikai); and (most importantly) the seaward front of larger fringing reefs (i.e., Waimanalo, Kailua). The region of relatively sheltered waters among the islands of Molokai, Lanai, West Maui, and Kahoolawe are also likely to have hosted late Holocene accretion, though research is lacking on this point.

Generally speaking, southern and windward portions of the Oahu shelf below wave base (i.e., reef front, paleochannels), both settings that are protected from direct northerly winter swell, preserve the most complete Holocene sequences. As broadly described by Grigg (1998), breakage, scour, and abrasion of living corals during high wave events appears to be the major source of coral mortality and ultimately limits accretion to restricted settings. Much Holocene reef development in waters shallower than (nominally) −10 m is a veneer on the Pleistocene foundation and is limited by lack of accommodation space.

11.5.1 Overall Holocene Pattern

Coastal systems in Hawaii are subject to high energy levels related to large swell, tsunami, tropical cyclones, extreme tides, and periods of intense rainfall. These are modulated by climatic factors such as El Niño, the Pacific Decadal Oscillation, and stochastic shifts in North Pacific currents and SST's such that from 1 year to the next wave energy fluctuates and storm incidence varies. As a consequence, the living reef community consists of hardy species able to adapt to rapidly changing conditions. For example, typically only three to five species dominate up to 75% of living coral cover on typical wave exposed Oahu reefs and these vary in proportion and depth to reflect a continuum of seasonal wave stress, decadal disturbance, and ecological succession (Grigg 1998). From one reef to another the same species (e.g., *Porites lobata*) exhibit high plasticity and may change growth morphology in response to bed shear stress, thrive at a range of depths, and compete successfully among community members. A single species may be found in massive, encrusting, and knobby growth forms at various depths on the same reef. In this setting carbonate accretion can record past sea-level positions, but in general these are not sensitive records because they are contaminated by environmental and ecological uncertainty resulting from these processes. Interpreting cored sequences of species assemblages can be challenging because they do not necessarily indicate single environmental conditions (Grossman 2001).

Nonetheless, reef communities tend to form assemblages as shown by Engels et al. (2004) that when encountered in drill cores of skeletal limestone, are indicative of certain conditions. From such records, and others, researchers have demonstrated that the end Pleistocene to Holocene pattern of reef accretion and sea-level change is characterized by

six significant periods (Fletcher et al. 2006): (1) ca. 21–8.1 ka when the deep reef formed between −120 and −30–40 m (Grigg et al. 2002; Grigg 2006); (2) ca. 8.1–7.9 ka an acceleration in the rate of sea-level rise is apparent in the coral framework and geomorphological record; (3) a widespread end of coral framework accretion occurring ca. 5 ka possibly related to strengthened El Niño and onset of extreme interannual swell as well as stabilized sea level producing prolonged wave disturbance; (4) ca. 3 ka the Kapapa sea-level highstand culminating at approximately 1–2 m above present mean sea level; (5) late Holocene coastal plain widening related to shoreline regression; (6) modern sea-level rise. Below we discuss the evidence and geologic framework for these events (Fig. 11.24).

Radiocarbon dates of Holocene corals in Engels et al. (2004), Grossman and Fletcher (2004), Rooney et al. (2004), and Grossman et al. (2006) indicate the earliest reefal Holocene limestones date ca. 8–8.3 ka at depths of approximately −18 to −25 m. Older dates (ca. 8.9–9.6 ka) have been published on samples from −52 to −58 m depth offshore of Maui, but these were exposed on the seafloor and not associated with net reef accretion (Grigg 2006). Samples contributing to reef development include mostly delicate branching framework corals (*P. compressa*) grading upward to encrusting algae and stout coral assemblages reflecting shallowing conditions (*Poc. meandrina*). Textures and cement composition are well-preserved and micritization of skeletal grains is limited. Where studied, Holocene accretion is restricted to subaerially eroded portions of Pleistocene platforms (e.g., paleostream channels) reaching more than 11 m in thickness. Elsewhere Holocene accretion is limited to thin veneers of encrusting coral-algal bindstone <1 m.

11.5.2 Holocene Coral Accretion on a Drowned Maui Reef

Twenty-one thousand years ago, at the sea-level nadir of the last ice age, Maui, Lanai and Molokai were interconnected by two subaerial limestone fossil bridges creating a super-island known as Maui-Nui (Grigg et al. 2002). Sea level then was approximately −120 m lower than it is today. Since that time, melt-water and expansion due to seawater warming near the surface, have driven a rise in sea-level that has drowned both bridges producing

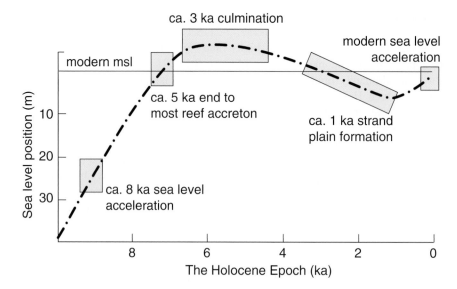

FIG. 11.24. The Holocene pattern of reef accretion and sea-level change is characterized by five significant periods: (1) rapid acceleration in sea level and reef drowning ca. 8 ka; (2) end to coral framestone accretion on exposed coasts ca. 5 ka; (3) culmination of sea level ca. 3 ka forming the Kapapa sea-level highstand approximately 1–2 m above present mean sea level; (4) sea-level fall and coastal plain widening; (5) modern sea-level rise

three separate islands and two drowned fossil reefs on which Holocene and present day reef accretion have taken place. Multi-beam high-resolution bathymetric surveys of these fossil and modern reefs have revealed the existence of numerous drowned features including solution basins, solution ridges (rims), sand, sediment plains and conical shaped pinnacles, some capped with modern coral growth (Fig. 11.25). The concentric basins are floored by flat lagoon-like sand bottoms and are rimmed by steep-sided limestone walls. Most of the walls contain under-cut notches marking paleo-lake levels or sea-level stillstands. Conical pinnacles are all found in wave-sheltered locations on the fossil limestone bridges. Many pinnacles peak at −50 to −70 m suggesting they drowned between 14,000 and 10,000 years ago when sea-level rise averaged 15 mm/year (Grigg et al. 2002). Virtually all of the hard bottom topography below −50 m is fossilized limestone, some karstified and some with thin but temporary patches of modern reef growth. These thin patches eventually spall off due to bio-erosion.

At depths above −50 m in the Au'au Chanel, accretion gradually increases in cover and thickness demarking a depth limit for the permanent accre-

tion of reef-building corals at this site in the present day. Recent measurement of coral growth using *P. lobata* as a proxy for the reef community as a whole, show that individual corals commonly grow at depths of −100 m or even slightly deeper but that reef accretion and permanent attachment are found only at depths shallower than −50 m (Fig. 11.26; Grigg 2006). In deeper water, rates of bio-erosion on colony holdfasts equal or exceed the growth of basal attachments, causing colonies to detach from the bottom. Continued bio-erosion further erodes and dislodges these colonies leading to their breakup and ultimately to the formation of coralline rubble and sand. Conceptually viewed, the −50 m threshold for reef accretion is analogous to a site-specific vertical Darwin Point. More importantly, it explains the history of reef building in the Au'au Channel. The hiatus at −50 m separates shallow water modern coral growth from the deeper reef which is a limestone foundation that formed during the early Holocene and late Pleistocene epochs.

At depths shallower than −50 m, the growth of *P. lobata* increases gradually in correlation with the exponential increase in down-welling photosynthetically active radiation reaching a maximum rate of linear extension of 13.5 mm/year at 6 m depth.

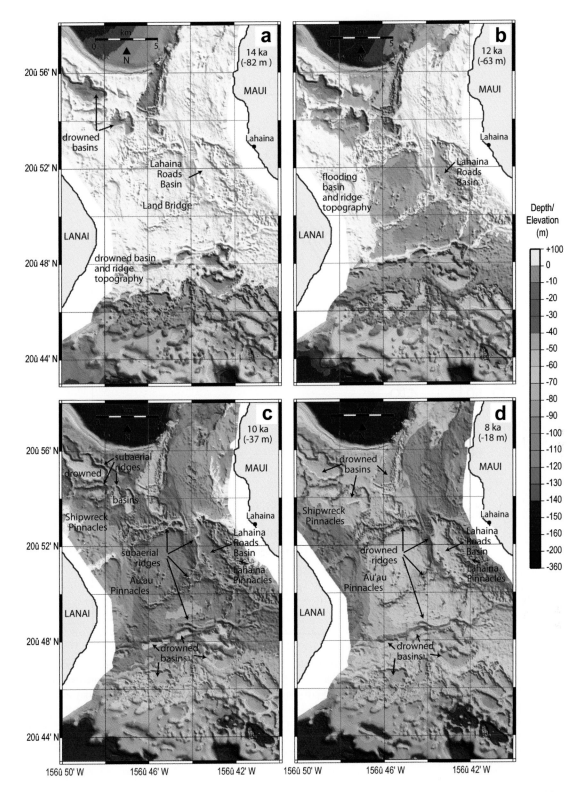

FIG. 11.25. Post-glacial sea-level reconstructions in the Au'au Channel of Maui (from Grigg et al. 2002; reproduced by permission of Springer) using the sea-level history of Bard et al. (1996): **a** Land bridge between Maui and Lanai 14,000 years ago when sea level was −82 m lower than present. **b** Land bridge between Maui and Lanai 12,000 years ago when sea level was −63 m below present. By this time, most of the channel was flooded. Notice wave protected lagoon -position of reef pinnacles inside ridges that extend across the channel in three places and served as barrier reefs. **c** Land bridge between Maui and Lanai 10,000 years ago when sea level was −37 m below present. Almost the entire channel was flooded by this time. **d** Land bridge between Maui and Lanai 8,000 years ago when sea level was −18 m

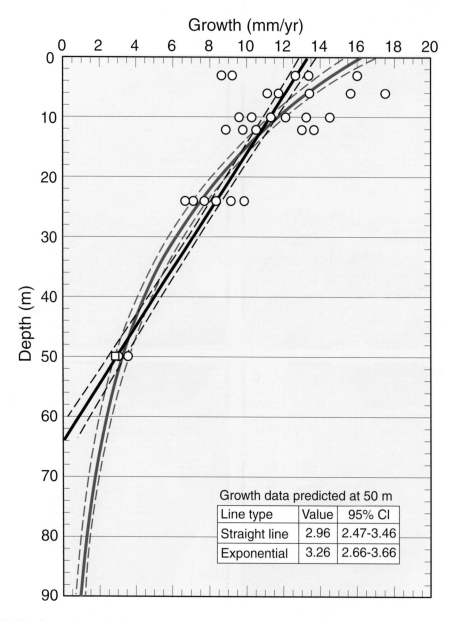

FIG. 11.26. This figure shows the growth rate (linear extension) of *Porites lobata* colonies vs. depth at stations off Lahaina, Maui in the Au'au Channel. Because colony age (*P. lobata* skeleton samples) jumps from modern to early Holocene at −50 m, this suggests that the rate of bio-erosion below −50 m exceeds the rate of present day accretion, and in this sense, the −50 m hiatus represents a vertical Darwin Point. These are best-fit regressions with 95% confidence limits (dotted lines) for *P. lobata* growth. The linear plot (black) includes all data points (n = 384); r² = 0.64. The exponential plot (red line) excludes the data from 3 m (n = 345); r² = 0.68. The open circles are the mean growth data for colonies. Values for growth predicted from both regressions at −50 m are shown in the box

At shallower depths, growth is slightly inhibited (14% at −3 m), possibly due to high levels of UV radiation, increased turbidity, or episodic sedimen-

tation events. The growth curve for *P. lobata* versus depth in the Au'au Channel can be considered optimal for the southeastern Hawaiian Islands. This is because light transmission is optimal for a coastal environment, run-off from the land is minimal and the channel is totally protected from all sources (and directions) of long period waves except in very shallow water (< −5 m).

11.5.3   Ca. 8 ka Sea-Level Acceleration

Holocene climate has been remarkably stable relative to major reorganizations dominating the Pleistocene (Alley et al. 1997). Climate during the Holocene has not, however, remained static. Bond et al. (1997) discovered millennial-scale climatic swings in the North Atlantic, and a notable climatic reversal ca. 8.2 ka has been the subject of investigation (Barber et al. 1999; Tinner and Lotter 2001). Sea-level change may be associated with climatic shifts and has been documented in coral reefs (Fairbanks 1989; Blanchon and Shaw 1995; Bard et al. 1996a). However, it remains unclear if and how sea level responded to the 8.2 ka climatic reversal. Blanchon and Shaw (1995) propose a rapid rise ca. 7.6–7.2 ka ("CRE3") they attribute to West Antarctic ice sheet instability. Using assumptions about erosional lowering, Blanchon et al. (2002) support CRE3 with a relic reef off Grand Cayman. However, Toscano and Lundberg (1998) document a decrease in the rate of rise over the same period. Tornqvist et al. (2004) interpret the stratigraphy of Mississippi Delta sediments as recording the 8.2 ka jump in sea level at less than 2 m. A final understanding of this event awaits additional research.

Hawaii reefs hold a record of rapid drowning ca. 8 ka that approximately correlates to this event. Coral framework stratigraphy documenting rapid sea-level rise ca. 8.1–7.9 ka has been discovered in drill cores from Hale O Lono, Molokai (Engels et al. 2004; Fig. 11.27). This record is consistent with Kayanne et al. (2002) from Palau, and with the measured ages of the termination of reef accretion found by Blanchon et al. (2002) absent their erosional lowering model.

In Fig. 11.27, cores 8–10 (−21 to −25 m depth), composed of high energy bindstone and rudstone lithofacies, are age-dated ca. 8.1 ka and have accretion rates of ~5 mm/year. Cores 5–7, obtained 125 m landward of cores 8–10 at a depth of −17 m, consist of low energy framestone accreting at ~23 mm/year. Landward of these, in −14 m depth, cores 3–4 record a return to high-energy condi-

tions (< −5 m depth) with bindstone and rudstone lithofacies and decreased accretion (~9 mm/year). Of particular interest are the laminated layers of encrusting coralline algae that dominate the bottom ~10 cm of cores 1–3, suggesting deposition within a meter or so of sea level in the form of an encrusting algal ridge. Cores 1 (−5 m depth) and 2 (−8 m depth) are composed of high-energy lithofacies accreting at ~2 mm/year.

Of special significance is the lithofacies change ca. 7.9 ka approximately −21 m below modern sea level. Calm water conditions are necessary to grow columnar *P. compressa*. This delicate species lacks the skeletal strength to withstand repeated wave impacts. *P. compressa* in Hawaii and on Molokai are found in deep fore-reef slopes or in protected back reef setting (Maragos 1977; Engels et al. 2004).

Was core site 5–7 a protected lagoonal setting? The site is 3 m shallower than site 8–10 indicating a lack of appropriate relief to provide protection for growth of *P. compressa* in a back-reef setting unless subsequent differential erosion has ensued, for which there is no evidence. Additionally, accretion terminated at cores 8–10 before the start of accretion at cores 5–7. If site 8–10 had sheltered a deep lagoon one would expect some simultaneous accretion at the two locations. Water depths at site 5–7 must have been sufficient (> −10 m) for *P. compressa* growth in a reef-front setting. Change in sea level is the most likely mechanism for increasing water depth.

Above, differential erosion is mentioned as an alternative explanation for the observed lithofacies architecture. Blanchon et al. (2002) employ an erosional model to reconstruct the timing and elevation of a relict reef at Grand Cayman. The gentle seaward gradient at Hale O Lono is consistent with a history of smooth undifferentiated erosion or simple non-deposition caused by increased wave energy. The upland is arid and lacks pronounced channelization, reducing the possibility of differential channeling in the bathymetry due to watershed retreat during the Holocene transgression. Straight and parallel depth contours and general hydrodynamic conditions do not suggest differential scouring. Importantly, the uppermost units in cores 5–10 consist of *in situ* accretion, and therefore provide no direct evidence of erosion. Differential erosion and therefore differential exposure of lithofacies is unnecessary and not supported by any direct evidence.

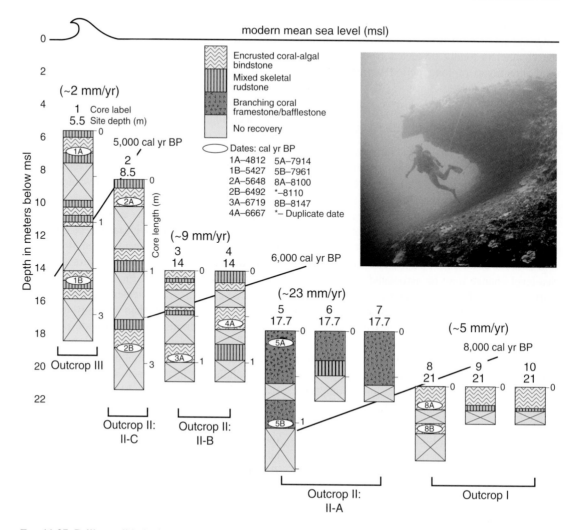

FIG. 11.27. Drill core lithologies, accretion rates, and radiometric dates from Hale O Lono, Molokai where a record of sea level acceleration has been recovered. The age of exposed sea floor increases with distance offshore showing that no net accretion has taken place since ca. 4.8 ka. Lithofacies associated with cores 5–7 are low-energy (> −10 m depth) lithofacies, distinctly different from high-energy lithofacies (< −5 m depth) that comprise the rest of the cores and implying a rapid sea-level rise prior to ca. 7.9 ka. A drowned intertidal notch, the Kaneohe shoreline, is found at the same depth as the stratigraphic record. Together, lithofacies changes and the drowned notch document 2–5 m of upward sea-level movement between 8.1 and 7.9 ka

Coseismic vertical movement could cause rapid local sea-level rise. However, several workers have interpreted the tectonic history of Molokai as related solely to long-term lithospheric flexure lacking any evidence for coseismic movement (Fletcher et al. 2002; Clague 1998; Grigg and Jones 1997).

The lithofacies sequence at Hale O Lono records early to middle Holocene transgression characterized by rapid drowning immediately prior to 7.9 ka. The magnitude of this drowning must be sufficient to place core site 5–7 in a depth >−10 m, the upper range for growth of monospecific stands of *P. compressa* in West Molokai (Engels et al. 2004). If cores 8–10 were originally < −5 m depth, this suggests a minimum sea level movement of 5 m between 8.1 and 7.9 ka and explains the high accretion rates at cores 5–7 (~23 mm/year) as the result of a "catch-up" mode of reef growth following sudden drowning.

This is an important section of the Hawaiian sea-level record because the reef platform at Kailua Bay on Oahu was likely flooded by early Holocene sea-level rise, as was Hanauma Bay (Easton and Olson 1976; Grossman and Fletcher 2004) and by implication other shallow platforms in Hawaii. Notably, this period of time, and the position of sea level, is consistent with the preservation of a distinct drowned notch ringing Oahu at −18 to −25 m depth, the "Kaneohe Shoreline", mentioned earlier and interpreted by Fletcher and Sherman (1995) and first named by Stearns (1978). Good preservation of the notch visor suggests that sea level rose upward out of the notch at a high rate (Pirazzoli 1986) – correlating with the core record from Molokai. As pointed out in Fletcher and Sherman (1995), this period is consistent with the timing identified for the breakout of Laurentide glacial lakes (Dyke and Prest 1987; Barber et al. 1999). The Hale O Lono, Molokai record, and a similar record in Palau (Kayanne et al. 2002), suggests that forcing of the cold event ca. 8.2 ka by catastrophic drainage of Laurentide lakes was accompanied by a sea-level response with global impact.

11.5.4   Ca. 5 ka End of Widespread Framework Accretion

The most detailed core sampling of Holocene reef accretion on Oahu comes from two studies: (1) Easton and Olson (1976) who cored protected Hanauma Bay fringing reef, reviewed and integrated with data from elsewhere on the island by Grigg (1998) and (2) Grossman (2001) and Grossman and Fletcher (2004) who cored Kailua Bay, open to trade wind seas and refracting north swell. The Hanauma Bay research was seminal and has become a classic work among the literature on Holocene reefs. Kailua Bay results benefit from improvements in dating technology and more attention paid to analyzing cored lithofacies in light of paleo-sea level position.

Grossman and Fletcher (2004) report on cored lithofacies in Kailua Bay. Corals are entirely aragonite and coralline algae exhibit the normal range of 15–19 mole% $MgCO_3$. Occasionally, coralline algae encrust interskeletal coral cavities

that may also be partly infilled with magnesium-calcite microcrystalline cement (rarely exceeding 2% by weight of total $CaCO_3$). Massive peloidal micrites, grain coatings, and void lining cements of magnesium-calcite characterize most cements and aragonite cement is rare. It is restricted to interskeletal coral cavities where it occurs as thin acicular fibrous needles. The most abundant cement is massive peloidal micrite characterized by knobby club-shaped columns ranging 0.1–1 cm in height. These often occur immediately above laminar crusts, creating thick (2–20 cm) sequences of massive lithified peloidal micrite. Comprising a major portion of branching framestones, micrite characterizes internal sediment trapped within inter-and intraskeletal cavities and significantly reduces porosity.

At Kailua, early Holocene accretion ca. 8–6 ka, approximately 14–24 m below sea level, is typically restricted to the reef front or paleochannels below wave base. Mixtures of encrusting and massive forms of P. lobata colonized sandy and rudstone substrate or the antecedent Waianae Reef surface. One long core from the outer reef at Kailua records 3–4 m of massive growth until ca. 6.5 ka succeeded by branching colonies of P. compressa that accreted another 7.5 m to ca. 5.3 ka.

Middle Holocene accretion is more complex and reflects the role of several processes. Because of Kailua's partially exposed/partially protected orientation to damaging north swell, it is difficult to definitively isolate controlling influences on reef development. Accretion in this period is less common in Kailua and characterized by a shift from widespread framestone development in topographically low areas to localized algal ridges, rudstone pavements, and spotty framestone accretion. These localized patterns typically developed over the period ca. 4.7–3.2 ka, with thicknesses of only 1–2 m. Grossman and Fletcher (2004) conclude that the highest rates of accretion correlate with in situ framestone accretion during the early middle Holocene when sea level was rising more than 2–3 mm/year (or faster). As sea-level rise slowed, accretion also slowed, but persisted (in the form of rudstone accumulation) at 1–2 mm/year even as sea-level fell at 1.5–2.0 mm/year following the Kapapa highstand.

Alternatively, Rooney et al. (2004) examined a data set of reef growth in more exposed settings (outer Kailua Bay, Molokai, Kauai, windward Oahu) and found a remarkably consistent end to reef accretion ca. 5 ka. They found that framestone accretion during early and middle Holocene time occurred in areas where today it is precluded by the wave regime, suggesting an increase in wave energy. They conclude the restricted nature of Middle Holocene reef development is a reaction to heightened north swell activity associated with stronger El Niño episodes beginning ca. 5 ka.

Analysis of reef cores by Rooney et al. (2004) reveal patterns of rapid early Holocene accretion in several locations terminating by middle Holocene time, ca. 5 ka (Fig. 11.28). Previous analyses have suggested that changes in Holocene accretion were a result of reef growth "catching up" to sea level whereas their new data and interpretations indicate that the end of reef accretion in the middle Holocene may be influenced by factors in addition to sea level. Rooney et al. reason that the decrease in reef accretion occurred prior to best estimates

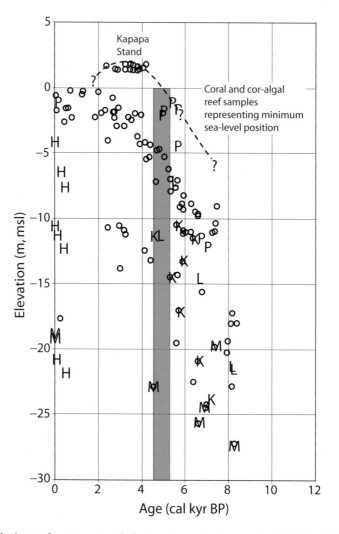

FIG. 11.28. Corals and other reef components (circles), and a sea-level curve for Oahu (dashed line) from Grossman and Fletcher (1998). The red band indicates the approximate cut-off time of framework accretion at reef sites exposed to north swell. K-Kailua, Oahu; L-Hale O Lono, Molokai; H-Hikauhi, Molokai; P-Punaluu, Oahu; M-Mana, Kauai. Punaluu samples near the sea-level curve are from non-*in situ* corals; and samples from Hikauhi, Molokai are sheltered from North swell. Circles marked "Kapapa Stand" from Grossman and Fletcher (1998), other circles from sheltered Hanauma Bay, Oahu (Easton and Olsen 1974)

of the decrease in relative sea-level rise during the mid-Holocene high stand of sea level in the main Hawaii Islands. If the end of accretion is a result of slowing sea-level rise, then it should decrease at or following the highstand (ca. 3 ka), especially if reef growth were "catching up" to sea level. Instead, core evidence from exposed reefs indicates widespread reef demise 2,000 years earlier. This pattern persists today despite the availability of hard substrate suitable for colonization at a wide range of depths between −30 m and the intertidal zone.

Rooney et al. (2004) analyzed the record of large, interannual waves and found that numbers of both extraordinarily large North Pacific swell events and hurricanes in Hawaii are greater during El Niño years. They infer that if these major reef-limiting forces were suppressed, net accretion would occur in some areas in Hawaii that are now wave-limited. Studies have shown that El Niño /Southern Oscillation (ENSO) was significantly weakened during early-mid Holocene time, only attaining an intensity similar to present ca. 5 ka (McGlone et al. 1992; Rodbell et al. 1999; Moy et al. 2002). They speculate that this shift in ENSO may assist in explaining patterns of Holocene Hawaiian reef accretion that are different from those of the present and apparently not solely related to sea-level position. Whereas a "sea level only" model suggests that early reef accretion reached a maximum in middle Holocene time as a result of reef growth "catching up" to sea level, Rooney et al. (2004) propose that the modern period of wave energy -limited accretion began ca. 5 ka, possibly related to Pacific-wide enhancement of the ENSO phenomenon.

Most likely, the apparent conflict between "sea-level restricted accretion" and "wave restricted accretion" models is more interpretive than real. In settings fully restricted from north swell (i.e., Hanauma Bay) Holocene accretion proceeds through middle Holocene time without regard to wave energy changes and is largely controlled by an available water column determined by sea-level position. In exposed settings (i.e., Punaluu, Oahu and other northerly exposures), a wave energy limitation beginning ca. 5 ka is consistent with observations of coral framework accretion. Settings that fall between these two end members (i.e., portions of Kailua Bay) are likely to experience limitations originating from both processes and the data can be interpreted as such.

Late Holocene reef development is characterized by rudstone accumulation and encrusting coral-algal growth with isolated head corals. At Kailua, a 2–3 m topographic ridge of branching coral *P. compressa* rudstones accreted ca. 3.3–1.8 ka under the Kapapa highstand. Although modern coral and coralline algae growth is prolific in Kailua Bay, the only significant reef accretion in the late Holocene is these cemented "pile-up" reefs of wave-broken debris dating from the Kapapa highstand.

11.5.5 Ca. 3 ka to Present, Sea-Level Highstand and Coastal Plain Development

Kapapa Islet lies on the windward edge of Kaneohe Bay, Oahu (Fig. 11.29). The island is formed by an outcropping of eolianite, part of a broad submerged ridge of cemented carbonate sand deposited during MIS 5a–d (Fletcher et al. 2005) and forming the core of a "barrier reef" defining the lagoon of Kaneohe Bay. The rock surface of the island is planed off, apparently by wave abrasion and inter-tidal bioerosion, at an elevation of 1–2 m above present mean sea level. An unconsolidated, 2–3 m thick deposit of beach sand blankets the middle of the island and has been interpreted as a paleo-beach that is evidence of a higher than present sea level ca. 3 ka (Fletcher and Jones 1996). Grossman and Fletcher (1998) excavated the sand deposit and found it contains stratigraphically distinct units recording littoral deposition ca. 3.4–1.5 ka. Stearns (1935), originator of the highstand hypothesis, named this sea level event the "Kapapa Stand of the Sea" on the basis of geomorphological evidence.

The Kapapa highstand is a Pacific-wide phenomenon resulting from realignment of the geoid due to changes in Earth's gravity field between glacial and interglacial states (Mitrovica and Milne 2002). In the original description, Stearns (1935) cited an emerged bench approximately 2 m above mean sea level on Kapapa Island and Hanauma Bay (Oahu) as evidence of wave abrasion under a sea-level highstand. Fletcher and Jones (1996) investigated the chronostratigraphy of deposits on Kapapa Island and elsewhere on windward Oahu and find agreement with the concept of a late Holocene highstand. Grossman et al. (1998) reviewed published evidence for the highstand on other islands of Polynesia and map paleotopog-raphy of the central Pacific sea surface showing

Fɪɢ. 11.29. Kapapa Island, Oahu, type locality of the middle Holocene "Kapapa Stand of the Sea"

trends consistent with geoid shifts modeled by Mitrovica and Peltier (1991) as a process termed "equatorial oceanic siphoning" also called "geoid subsidence". Grossman and Fletcher (1998) modeled the post-glacial rise of sea level on Oahu using dated coral samples and paleo-shoreline data.

The Kapapa highstand provided for increased wave energy across fringing reefs moving sand shoreward for the construction of beaches and eolianites of middle to late Holocene age. As sea-level dropped at the end of the highstand, shoreline regression built wide sandy strand plains burying the shoreward margins of fringing reefs. Sediment production by reefs fueled this process as wave-generated currents delivered carbonate sands to the coast. Sands consisting of fragments of lithic limestone, coral and coralline algae, molluscs and echinoderms, *Halimeda*, and foraminifera built beaches and dune fields on the island coastal plain (Harney and Fletcher 2003).

If modern sediment production in Kailua Bay (approximately $4,048 +/- 635 \, m^3$/year; Harney and Fletcher 2003) is applied over the middle to late Holocene period that the embayment has been completely inundated by post-glacial sea level, an estimated $20,239(+/-3,177) \times 10^3 \, m^3$ of calcareous sediment (at 40% porosity) could have been produced in the system. Some of this sediment has been lost offshore and to dissolution and abrasion (~25%), some is stored in paleochannels,

sand fields, and other submarine storage sites (~19% +/-5%), and reconstructions by Harney and Fletcher (2003) indicate that approximately 5% or $1,000 +/-1,000 \times 10^3 \, m^3$ is stored in the beach. The Kailua coastal plain, a broad sandy accretion strand plain originally marked by dune crests and beach ridges is calculated to store $10,049 +/-1,809 \times 10^3$ m^3 (or 51% +/-17%).

Holocene coastal dune and beach accretion were enhanced under the Kapapa highstand as characterized by radiocarbon dates of shallow water/beach carbonate sand grains that tend to cluster ca. 0.5–2 ka (Harney et al. 2000). Sea level subsequently fell prior to the tide gauge era where today a consistent century-long rise of 1.5–2.0 mm/year is recorded at the Honolulu station. Radiocarbon ages of sand grains (ca. 0.5–5 ka; Harney et al. 2000) from broad tracts of living reef display a strongly dominant antecedent component reflecting an era of enhanced carbonate production under the Kapapa highstand. Notably, the oldest dates (ca. 4–5 ka) were acquired from the modern dynamic beach face indicating the active role that fossil grains play in modern beach processes. General lack of modern sand grains in an otherwise healthy coral-algal reef complex also reflects a seaward shift in modern carbonate grain production to the reef front and subsequent offshore loss of sediment.

The Kapapa highstand, as high as 2 m above modern sea level, flooded most low-lying coastal plains around Oahu. Where Waimanalo Reef is prevalent, flooding at the time was limited by the > +3 m elevation of the old limestone surface. However, at other locations lacking last interglacial deposits, low-lying coastal lands were flooded by Kapapa seas and blanketed with a layer of late Holocene carbonate sands. These locations developed into accretion strand plains as sea-level fell over the period ca. 1.5–0.5 ka (e.g., Hanalei and Kailua coastal plains; Calhoun and Fletcher 1996). The recent period has been characterized by modern dune development over former shorelines on the strand plains and adjustment of littoral sand budgets to rising sea level, largely through the process of shoreline recession.

Today, Hawaii reefs continue to experience significant natural limitations in exposed settings. In their study of shelf stratigraphy and the influence of the antecedent substrate, Grossman et al. (2006) conclude that whereas Holocene coral framestone accretion terminated on the windward Kailua shelf ca. 5 ka, it was maintained until 3–2.4 ka offshore of Waikiki and elsewhere on the southern shelf of Oahu, but has since ended. Grigg (1995) documents the destruction of coral beds at Waikiki (Oahu south shore) during Hurricane Iwa in 1982, exhuming fossil mid-Holocene pavement dating 2.5–6 ka. Little coral growth has occurred since. The lack of framestone accretion despite coral colony growth rates >1 cm/year on Oahu (Grigg 1983) suggests that regular and periodic wave scouring associated with wave base has been a primary control on reef accretion since the middle Holocene.

Modern coral growth in Hawaii exploits the narrow window between wave base and the antecedent surface by utilizing a range of ecotypic growth forms (species "plasticity") maximizing exposure to irradiance and minimizing exposure to hydraulic forces. Despite this, the vast majority of the modern Oahu shelf is largely devoid of Holocene accretion as observed by Sherman et al. (1999) who found no evidence of Holocene reef accretion in 30 separate cores from both windward and leeward settings in water depths between 5.5 and 35 m.

11.6   Conclusions

A complex history of reef, dune, and coastal plain accretion on Oahu during the late Quaternary has produced a mosaic of stratigraphic components comprising the shallow coastal plain and shelf of Oahu. Where Holocene accretion is prolific, Fig. 11.30 depicts our model of the stratigraphic relations among reefal limestones. However, as stated throughout this chapter, modern growth is limited by wave-induced shear stresses, and reefal facies above shallow Pleistocene units is largely absent.

By volume and geographic extent, the most significant stratigraphic component of the Oahu shelf is the Waianae Reef dating from MIS 7. Four well-preserved coral samples from both windward and leeward sides of Oahu provide absolute ^{230}Th-^{234}U-^{238}U ages dating 206–247 ka within acceptable limits of $\delta^{234}U_i$ ($\delta^{234}U_i$ <165‰). The unit displays limestone facies documenting paleo-reef crest and lagoonal environments as well as distinct leeward and windward accretion patterns. These indicate that marine paleoclimatologic conditions similar to today controlled the Oahu shore during MIS 7. Holocene-age marine cements, isopachous rims of bladed magnesium calcite spar (ca. 2.8–5.6 ka) from within the framework matrix of the Waianae Reef, reflect post-glacial flooding by sea level during the Kapapa highstand. Analysis of contemporaneous sea level, corrected for island uplift (0.03–0.06 mm/year) indicates a position of −9 to −20 m below present during accretion of the Waianae Reef.

The Waimanalo Reef represents peak last interglacial time on Oahu. Szabo et al. (1994) and Muhs et al. (2002) identify a discrepancy between the start and duration of the last interglacial and the timing of peak insolation as represented by orbital tuning of the marine isotope record. A long core through this unit reveals the contact of Waianae and Waimanalo limestones at approximately −5 m below modern sea level. The shallow depth of this contact and the routine acquisition of long limestone borings (e.g., >30 m) around the Oahu coastal plain for commercial purposes (i.e., water wells, foundation studies) indicates the Waianae Reef is an important stratigraphic unit in the Oahu coastal plain underlying Waimanalo Reef.

The fossil substrate of the Waianae Reef accreted MIS 5a–d framestones along its seaward margin following the peak of the last interglacial. This growth is documented with four samples of pristine *in situ* coral ca. 82–110 ka collected −25 to −30 m depth from the leeward side of Oahu. The present depth of the paleo-reef crest from this time is approximately −20 m. Following Stearns

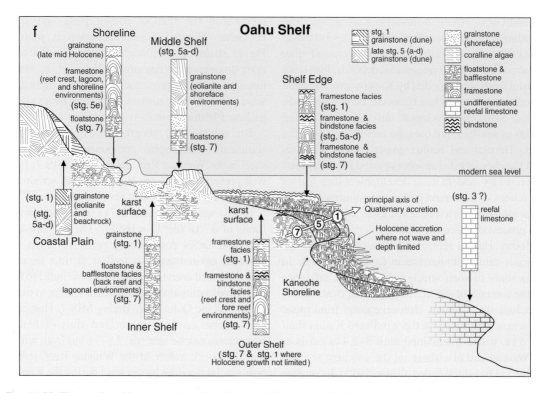

FIG. 11.30. The stratigraphic complexity of the Oahu shelf is modeled in this illustration. Coastal plain and shoreline stratigraphy consists of Holocene and late last interglacial (Leahi age, MIS 5a–d) eolianite grainstone units. These grade seaward to late Holocene shoreface grainstones (beachrock) resting on MIS 5e Waimanalo Reef framestones and bindstones. Waimanalo units in turn rest unconformably at a depth of −5 m on MIS 7 Waianae Reef limestone. On the inner shelf, late Holocene unconsolidated (and consolidated beachrock) grainstones rest unconformably on Waianae reefal limestone consisting of lagoonal floatstones and bafflestones, the major stratigraphic component of the Oahu shelf. Mid-shelf settings are characterized by Leahi-age (MIS 5a–d) eolianites lying unconformably on Waianae Reef, or karstified exposures of Waianae Reef. Where not limited by wave stress, outer shelf units consist of early to middle Holocene framestones over Waianae-age (MIS 7) framestones and bindstones. On most of the Oahu shelf, modern framestone accretion is severely restricted by wave stress and the shelf consists of karstified Waianae Reef framestones and bindstones representing former reef crest and fore reef environments. The outer shelf may consist of Holocene framestones where not wave limited, or, more commonly on Oahu, a sequence of Leahi-age (MIS 5 a–d) reefal limestones (framestones and bindstones) resting unconformably upon Waianae Reef framestones and bindstones of the MIS 7 reef crest and fore reef. The front of the shelf displays an intertidal notch (approximately −18 to −25 m depth) correlating to rapid sea-level rise in the early Holocene ca. 7.9–8.1 ka

(1974) this unit is named the Leahi Reef. Samples (Sherman 2000; Szabo et al. 1994) indicate that early Leahi Reef accretion and late Waimanalo Reef accretion were contemporaneous at the end of MIS 5e. Facies changes, and the general trend of decreasing age with distance offshore, indicate that Leahi accretion continued over a period of general sea-level fall during the latter part of MIS 5 causing a shift in the reef community toward a relatively shallow moderate to high-energy environment.

Exposures of calcareous eolianite characterize the margin of Oahu during Leahi time (MIS 5a–d) as well as the islands of Maui, Kauai, and Molokai. Racemisation history indicates that large-scale dune deposition ensued during the period of general sea-level fall (as hypothesized by Stearns 1974) at the end of stage 5. The volume and extent of Leahi Dunes suggests continuous sediment production during the 5a–d interval, as would be provided by the Leahi Reef. Lithified dunes comprise significant stratigraphic members of the subaerial coastal plain and they are important components of the shallow submerged shelf. In places, they form a substrate for Holocene coral growth in the form of a major "barrier reef" (Kaneohe Bay) and multiple offshore islets (e.g., Kapapa Island).

Long-term flexural uplift of Oahu coupled with heavy wave stress and reduced accommodation space above the Pleistocene surface placed severe natural limitations on Holocene and modern reef accretion. Most modern accretion occurs on the front, deep (−10 to −30 m) slope of the Pleistocene shelf where ambient light and nutrient levels permit coral growth in areas protected from wave stress. Other accretion centers are found infilling paleo-channels and other types of protected environments such as the Au'au Channel. Holocene framework is largely a veneer on the wave-scoured shallow surface of the shelf.

Reefal limestones sampled from earliest Holocene time date ca. 7.9–8.1 ka at depths of approximately −18 to −25 m. These include mostly delicate branching framework corals that grade upward to encrusting algae and stout coral assemblages reflecting shallowing conditions. A major "jump" in sea level at this time produced a drowned shoreline in the islands (a submerged intertidal notch) and a stratigraphic record of sudden drowning of a fringing reef crest facies.

The majority of Holocene coral framework accretion terminated ca. 5 ka on Oahu. Middle Holocene reef accretion is uncommon in exposed regions and characterized by a shift from widespread framestone development in topographically low areas to localized algal ridges, rudstone pavements, and spotty framestone accretion. These localized accretion patterns typically developed ca. 4.7–3.2 ka, with thicknesses of only 1–2 m. Late Holocene reef development is characterized by rudstone accumulations and encrusting coral-algal growth with isolated head corals. The late Holocene Kapapa highstand was a time of Holocene rudstone accumulation on reef flats termed "pile-up" reefs. The post highstand fall of sea level produced regression strand plains composed of middle to late Holocene-age sands. These sandy coastal plains are fronted by beaches that rely heavily on fossil sand stores from late Holocene highstand production rather than modern sand production.

In conclusion, the stratigraphic and environmental complexity of the Oahu shelf has produced severe limitations on accommodation space for continued coral framework accretion. Flexural uplift of Oahu, the shallow antecedent surface, and widespread high wave stress presently limit modern reef accretion which in turn restricts carbonate sand production. This geologic history therefore inhibits the ability of the existing Oahu coral communities to withstand future negative environmental factors such as sedimentation, localized eutrophication, and other types of human impacts. Much of the modern coral growth observed on the shelf (i.e., coral heads, stout branching forms, etc.) is attached to an ancient limestone pavement ranging in age ca. 200–5 ka. This implies that most modern growth is temporary, constantly inhibited or removed by the large energy events that regularly sweep the shelf, and the time for widespread reef accretion has largely passed other than in isolated and protected settings.

Acknowledgements The authors extend sincere appreciation for research funding to the National Geographic Society, the Office of Naval Research, the US Geological Survey Coastal and Marine Geology Program, the National Science Foundation Earth Systems History (ESH) Program, the Sea Grant College of Hawaii, the NOAA Coastal Services Center, the Hawaii Department of Land and Natural Resources, the NOAA Hawaii Coral Reef Initiative Research Program, The US Army Corps of Engineers, and the Khaled bin Sultan Living Oceans Foundation. We especially acknowledge the hard work of the following individuals for assistance in field, laboratory work, and stimulating discussion: Dolan Eversole, Matt Barbee, Chyn Lim, Captains Alan Weaver and Joe Reich, Jane Schoonmaker, Scott Calhoun, Bruce Richmond, Abby Sallenger, Gordon Tribble, and Mike Field.

References

Adey WH (1986) Coralline algae as indicators of sea level. In: Van de Plassche O (ed) Sea Level Research. Geo Books, Norwich, pp 229–280

Alley RB, Mayewski PA, Sowers T, Stuiver M, Taylor KC, Clark PU (1997) Holocene climatic instability: a prominent, widespread event 8200 yr ago. Geology 25:483–486

Barber DC, Dyke A, Hillaire-Marcel C, Jennings AE, Andrews JT, Kerwin MW, Bilodeau G, McNeely

R, Southon J, Morehead MD, Gagnon J-M (1999) Forcing of the cold event of 8,200 years ago by catastrophic drainage of Laurentide lakes. Nature 400:344–348

Bard E, Hamelin B, Arnold M, Montaggioni L, Cabioch G, Faure G, Rougerie F (1996a) Deglacial sea-level record from Tahiti corals and the timing of global meltwater discharge. Nature :382:241–244

Bard E, Jouannic C, Hamelin B, Pirazzoli P, Arnold M, Faure G, Sumususastro P, Syaefudin C (1996b) Pleistocene sea levels and tectonic uplift based on dating of corals from Sumba Island, Indonesia. Geophys Res Lett 23:1473–1476

Bassinot FC, Labeyrie LD, Vincent E, Quidelleur X, Shackleton NJ, Lancelot Y (1994) The astronomical theory of climate and the age of the Brunhes-Matuyama magnetic reversal. Earth Planet Sci Lett 126:91–108

Blanchon P, Shaw J (1995) Reef drowning during the last deglaciation : evidence for catastrophic sea-level rise and ice-sheet collapse. Geology :4–8

Blanchon P, Jones B, Ford DC (2002) Discovery of a submerged relic reef and shoreline off Grand Cayman : further support for an early Holocene jump in sea level. Sediment Geol 147:253–270

Bond G, Showers W, Cheseby M, Lotti R, Almasi P, deMenocal P, Priore P, Cullen H, Hajdas I, Bonani G (1997) A pervasive millennial-scale cycle in North-Atlantic Holocene and glacial climates. Science 278:1257–1266

Cabioch G, Montaggioni LF, Faure G, Ribaud-Laurenti A (1999) Reef coralgal assemblages as recorders of paleobathymetry and sea level changes in the Indo-Pacific province. Quat Sci Rev 18:1681–1695

Cacchione D, Richmond B, Fletcher C, Tate G, Ferreira J (1999) Sand transport in a reef channel off Kailua, Oahu, Hawaii. In: Fletcher C, Matthews J (eds) The Non-Steady State of the Inner Shelf and Shoreline: Coastal Change on the Time Scale of Decades to Millennia in the Late Quaternary, Abstracts with Programs, IGCP Project #437. University of Hawaii, Honolulu, HI, ,pp 42

Calhoun RS, Fletcher CH (1996) Late Holocene coastal plain stratigraphy and sea level history at Hanalei, Kauai, Hawaii Islands. Quat Res 45:47–58

Chappell J, Shackleton NJ (1986) Oxygen isotopes and sea level. Nature 324:137–140

Chen JH, Edwards RL, Wasserburg GL (1986) ^{238}U, ^{234}U and ^{232}Th in seawater. Earth Planet Sci Lett :241–251

Chen JH, Curran HA, White B, Wasserburg GJ (1991) Precise chronology of the last interglacial period: ^{234}U-^{230}Th data from fossil coral reefs in the Bahamas. Bull Geol Soc Am 103 :82–97

Clague DA (1998) Moloka'i and Lanai,. In: Juvik SP, Juvik JO (eds) Atlas of Hawaii, third edition. ,University of Hawaii Press, Honolulu, HI, pp 11–13

Clague DA, Dalrymple GB (1987) The Hawaii-emperor volcanic chain, part i, geologic evolution. In: Decker RW, Wright TL, Stauffer PH (eds) Volcanism in Hawai'i. US Geol Surv Prof Pap 1350:5–54.

Conger C (2005) Identification and characterization of sand deposit distribution on the fringing reefs of Oahu, Hawaii, M.S. thesis, University of Hawaii, Geology and Geophysics, 150 p.

Conger CL, Hochberg E, Fletcher C, Atkinson M (2006) Decorrelating remote sensing color bands from bathymetry in optically shallow waters. IEEE Trans Geosci Remote Sens 44(6):1655–1660

Connell J (1978) Diversity in tropical rain forests and coral reefs. Science 199:1302–1310

Coulbourn WT, Campbell JF, Moberly R (1974) Hawaii submarine terraces, canyons and Quaternary history evaluated by seismic -reflection profiling. Mar Geol 17:215–234

Dollar SJ (1982) Wave stress and coral community structure in Hawaii. Coral Reefs 1:71–81

Dollar SJ, Tribble GW (1993) Recurrent storm disturbance and recovery: a long-term study of coral communities in Hawaii. Coral Reefs 12:223–233.

Dyke AS, Prest VK (1987) Late Wisconsinan and Holocene history of the Laurentide ice sheet. Geographie Physique et Quaternaire 41:237–264

Easton WH, Olson EA (1976) Radiocarbon profile of Hanauma Bay, Oahu, Hawaii. Geol Soc Am Bull 87:711–719

Edwards RL, Chen JH, Wasserburg GJ (1986) ^{238}U-^{234}U-^{230}Th-^{232}Th systematics and the precise measurement of time over the past 500000 years. Earth Planet Sc Lett 81:175–192

Emiliani C (1955) Pleistocene temperatures. J Geol 63:539–578

Embry AF, Klovan JE (1971) A late Devonian reef tract on northeastern Banks Island NWT. Bull Canad Petrol Geol 19:730–781

Engels MS, Fletcher CH, Field ME, Storlazzi CD, Grossman EE, Rooney JJB, Conger CL, Glenn C (2004) Holocene reef accretion : southwest Molokai Hawaii USA. J Sediment Res 742:255–269

Fairbanks RG (1989) A 17000-year glacio-eustatic sea level record: Influence of glacial melting rates on the Younger Dryas event and deep-ocean circulation. Nature 342:637–642

Finkl CW (2004) Leaky valves in littoral sediment budgets: loss of nearshore sand to deep offshore zones via chutes in barrier reef systems southeast coast of Florida USA. J Coastal Res 20(2):605–611

Fletcher CH, Richmond BM, Barnes GM, Schroeder TA (1995) Marine flooding on the coast of Kaua'i during

Hurricane Iniki: hindcasting inundation components and delineating washover. J Coastal Res 11:188–204

Fletcher CH, Sherman C (1995) Submerged shorelines on Oahu Hawaii: archive of episodic transgression during the deglaciation ? J Coastal Res Spec Issue 17:141–152

Fletcher CH, Jones AT (1996) Sea-level highstand recorded in Holocene shoreline deposits on Oahu Hawaii. J Sediment Res 663:632–641

Fletcher CH, Grossman EE, Richmond BM, Gibbs AE (2002) Atlas of Natural Hazards in the Hawaii Coastal Zone. US Geological Survey, Geologic Investigations Series I-2761, 182 pp

Fletcher CH, Murray-Wallace C, Glenn C, Sherman C, Popp B, Hessler A (2005) Age and origin of late Quaternary eolianite Kaiehu Point (Moomomi) Molokai Hawaii. J Coastal Res SI42:97–112

Fletcher CH, Engels MS, Grossman EE, Rooney JJ, Sherman CE (2006) Decoding the geologic record of Holocene sea-level change in the Hawaii Islands. Abstracts with Programs, Geological Society of America, Philadelphia Annual Meeting Pardee Keynote Symposia P4 – Holocene Sea-Level Change in North America, A Post-Katrina Assessment

Gallup CD, Edwards RL, Johnson RG (1994) The timing of high sea levels over the past 200000 years. Science 263:796–800

Garcia MO, Pietruszka AJ, Rhodes JM, Swanson K (2000) Magmatic processes during the prolonged Pu'u 'O'o eruption of Kīlauea Volcano Hawai'i. J Petrol 41:967–990

Giambelluca TW, Nullet MA, Schroeder TA (1986) Rainfall atlas of Hawaii: State of Hawaii, Department of Land and Natural Resources Report R76, 267 pp

Grigg RW (1981) *Acropora* in Hawaii. Part 2. Zoogeography. Pac Sci 35:15–24

Grigg RW (1982) Darwin Point: a threshold for atoll formation. Coral Reefs 1:29–34

Grigg RW (1983) Community structure, succession and development of coral reefs in Hawaii. Mar Ecol Progr Ser 11:1–14

Grigg RW (1995) Coral reefs in an urban embayment in Hawaii: a complex case of history controlled by natural and anthropogenic stress. Coral Reefs 14:253–266

Grigg RW (1997) Benthic communities on Loihi submarine volcano reflect high-disturbance environment. Pac Sci 51:209–220

Grigg RW (1998) Holocene coral reef accretion in Hawaii: a function of wave exposure and sea level history. Coral Reefs 17:263–272

Grigg RW (2006) Depth limit for reef building corals in Au'au channel, S.E. Hawaii. Coral Reefs 25:77–84

Grigg RW, Epp D (1989) Critical depth for the survival of coral islands: effects on the Hawaii Archipelago. Science 243(4891):638–641

Grigg RW, Jones AT (1997) Uplift caused by lithospheric flexure in the Hawaii Archipelago as revealed by elevated coral deposits. Mar Geol 141:11–25

Grigg RW, Grossman EE, Earle SA, Gittings SR, Lott D, McDonough J (2002) Drowned reefs and antecedent karst topography Au'au Channel SE Hawaii Islands. Coral Reefs 21:73–82

Grossman EE (2001) Holocene sea level history and reef development in Hawaii and the central Pacific Ocean. Unpublished Ph.D. thesis, University of Hawaii Geology and Geophysics, 257 p

Grossman EE, Fletcher CH (1998) Sea level 3500 years ago on the Northern Main Hawaii Islands. Geology 26:363–366

Grossman EE, Fletcher CH, Richmond BM (1998) The Holocene sea-level highstand in the Equatorial Pacific: analysis of the insular paleosea-level database. Coral Reefs 17:309–327

Grossman EE, Fletcher CH (2004) Holocene reef development where wave energy reduces accommodation space Kailua Bay windward Oahu Hawaii USA. J Sediment Res 741:49–63

Grossman EE, Barnhardt WA, Hart P, Richmond BM, Field ME (2006) Shelf stratigraphy and influence of antecedent substrate on Holocene reef development south Oahu Hawaii. Mar Geol 226:97–114

Gulko D (1998) Hawaii Coral Reef Ecology Honolulu. Mutual Publishing, 245 pp

Hamelin B, Bard E, Zindler A, Fairbanks RG (1991) $^{234}U/^{238}U$ mass spectrometry of corals: how accurate is the U-Th age of the last interglacial period Earth and Planetary. Sci Lett 106:169–180

Hampton MA, Blay CT, Murray C, Torresan LZ, Frazee CZ, Richmond BM, Fletcher CH (2003) Data report geology of reef-front carbonate sediment deposits around Oahu Hawaii. US Geological Survey Open-file Report 03–441 http://geopubswrusgsgov/open-file/of03–441/

Harmon RS, Mitterer RM, Kriausakul N, Land LS, Schwarcz HP, Garrett P, Larson GJ, Vacher HL, Rowe M (1983) U-series and amino acid racemization geochronology of Bermuda: implications for eustatic sea-level fluctuation over the past 250000 years. Paleogeogr Paleoclim Paleoecol 44:41–70

Harney JN, Grossman EE, Richmond BM, Fletcher CH (2000) Age and composition of carbonate shoreface sediments Kailua Bay Oahu Hawaii. Coral Reefs 19:141–154

Harney JN, Fletcher CH (2003) A budget of carbonate framework and sediment production Kailua Bay Oahu Hawaii. J Sediment Res 736:856–868

Hearty PJ, Kaufman DS, Olson SL, James HF (2000) Stratigraphy and whole-rock amino acid geochronology of key Holocene and Last Interglacial carbonate deposits in the Hawaii Islands. Pac Sci 544:423–442

Hearty PJ (2002) The Kaena highstand on Oahu Hawaii: further evidence of Antarctic Ice collapse during the middle Pleistocene. Pac Sci 56(1) 65–82

Henderson GM, Cohen AS, O'Nions RK (1993) ^{234}U/^{238}U ratios and ^{230}Th ages for Hateruma atoll corals: implications for coral diagenesis and seawater ^{234}U/^{238}U ratios earth and planetary. Sci Lett 115:65–73

Hourigan TF, Reese ES (1987) Mid-ocean isolation and the evolution of Hawaii reef fishes. Trends Ecol Evol 2:187–191

Isoun E, Fletcher CH, Frazer N, Gradie J (2003) Multispectral mapping of reef bathymetry and coral cover. Coral Reefs 22:68–82

James NP, Bourque P-A (1992) Reefs and Mounds. In: Walker RG, James NP (eds) Facies Models: Response to Sea Level Change. St John's Newfoundland, Geological Association of Canada, pp 323–347

Jones AT (1993) Elevated fossil coral deposits in the Hawaii Islands: a measure of island uplift in the Quaternary. Unpublished Ph.D. thesis, University of Hawaii, Honolulu, HI, 274 pp

Jones AT (1994) Review of the chronology of marine terraces in the Hawaii Archipelago. Quat Sci Rev 12:811–823

Kay EA, Palumbi (1987) Endemism and evolution in Hawaii marine invertebrates. Trends Ecol Evol 2:183–186

Kayanne H, Yamano H, Randall RH (2002) Holocene sea-level changes and barrier reef formation on an oceanic island Palau Islands western Pacific. Sediment Geol 150:47–60

Ku T-L, Kimmel MA, Easton WH, O'Neil TJ (1974) Eustatic sea level 120000 years ago in Oahu Hawaii. Science 183:959–962

Littler MM, Doty MS (1975) Ecological components structuring the seaward edges of tropical Pacific reefs: the distribution communities and productivity-ecology of Porolithon. J Ecol 63:117–129

Lum D, Stearns HT (1970) Pleistocene stratigraphy and eustatic history based on cores at Waimanalo Oahu Hawaii. Geol Soc Am Bull 81:1–16

Macintyre IG (1977) distribution of submarine cements in a modern Caribbean fringing reef Galeta Point, Panama. J Sediment Petrol 47:503–516

Maragos JE (1977) Order Scleractinia. Stony corals. In: Devaney DM, Eldredge LG (eds) Reef and Shore Fauna of Hawaii Section 1: Protozoa Through Ctenophora. Honolulu, HI, Bishop Museum Press, 278 pp

Maragos JE (1995) Revised checklist of extant shallow-water stony coral species from Hawaii (Cnidaria: Anthozoa: Scleractinia) Bishop Mus Occ Pap 2:54–55

Martinson DG, Pisias NG, Hays JD, Imbrie J, Moore Jr TC, Shackleton NJ (1987) Age dating and the orbital theory of the ice ages: development of a high resolution 0 to 300,000-year chronostratigraphy. Quatern Res 27:1–29

McGlone M, Kershaw AP, Markgraf V (1992) El Niño/Southern Oscillation climatic variability in Australia and South American paleoenvironmental records. In: Diaz H, Markgraf V (eds) El Niño: Historical and Paleoclimatic Aspects of the Southern Oscillation. New York, Cambridge University Press, pp 434–462

Mitrovica JX, PeltierWR (1991) On postglacial geoid subsidence over the equatorial oceans. J Geophys Res 96:20053–20071

Mitrovica JX, Milne GA (2002) On the origin of late Holocene sea-level highstands within equatorial ocean basins. Quaternary Sci Rev 21(20–22):2179–2190

Moberly R, Chamberlain T (1964) Hawaiian Beach Systems, Honolulu, Hawaii Institute of Geophysics, Honolulu HIG 64–233 pp

Moberly R, Baver LD, Morrison A (1965) Source and variation of Hawaii littoral sand. J Sediment Petrol 35:589–598

Moberly R, Campbell JF, Coulbourn WT (1975) Offshore and other sand resources for Oahu, Hawaii. UNIHI-SEAGRANT-TR-75–03 Sea Grant and the Hawaii Institute of Geophysics, Honolulu, HI

Montebon ARF (1992) Use of the line intercept technique to determine trends in benthic cover. Proc 7th Int Coral Reef Sym, Guam 1:151–155

Moore JG (1970) Relationship between subsidence and volcanic load Hawaii. Bull Volcanol 34:562–576

Moore JG (1987) Subsidence of the Hawaii Ridge. In: Decker RW, Wright TL, Stauffer PH (eds) Volcanism in Hawaii. Washington, US Government Printing, Office US Geological Survey Professional Paper 1350, pp 85–100

Moore JG, Moore GW (1984) Deposit from a giant wave on the island of Lana'i Hawai'i. Science 226:1312–1315

Moore JG, Campbell JF (1987) Age of tilted reefs Hawaii. J Geophys Res 92:2641–2646

Morse JW, Mackenzie FT (1990) Geochemistry of Sedimentary Carbonates. New York, Elsevier, 707 pp

Moy CM, Seltzer GO, Rodbell DT, Anderson DM (2002) Variability of El Niño/Southern Oscillation activity at millennial timescales during the Holocene epoch. Nature 420:162–165

Muhs DR, Szabo BJ (1994) New uranium-series ages of the Waimanalo limestone Oahu Hawaii: implications for sea level during the last interglacial period. Mar Geol 118:315–326

Muhs DR, Simmons KR, Steinke B (2002) Timing and warmth of the Last Interglacial period: new U-series evidence from Hawaii and Bermuda and a new fossil compilation for North America. Quat Sci Rev 21:1355–1383

Muhs DR, Wehmiller JF, Simmons KR, York LL (2003) Quaternary sea-level history of the United States. In: Gillespie AR, Atwater BF, Porter SC (eds) Quaternary Period in the United States: Developments in Quaternary Science. Elsevier, pp 147–153

Murray-Wallace CV (1995) Aminostratigraphy of Quaternary coastal sequences in southern Australia – an overview. Quat Int 26:69–86

Murray-Wallace CV (1993) A review of the application of the amino acid racemisation reaction to archaeological dating. The Artefact 16:19–26

Oki DS, Gingerich SB, Whitehead RL (1999) Ground water atlas of the United States: Hawaii HA 730-N: http://cappwaterusgsgov/gwa/ch_n/N-HIsummary2html

Pirazzoli PA (1986) Marine Notches. In: Plassche de VO (ed) Sea-Level Research: A Manual for the Collection and Evaluation of Data. Norwich Geo Books, pp 361–400

Purdy EG (1974) Karst determined facies patterns in British Honduras: Holocene carbonate sedimentation model. Am Assoc Petr Geol B 58:825–855

Rodbell DT, Seltzer GO, Anderson DM, Abbott MB, Enfield DB, Newman JH (1999) Science 283: 516–520

Rooney JJR, Fletcher CH, Grossman EE, Engels M, Field ME (2004) El Niño influence on Holocene reef accretion in Hawaii. Pac Sci 58:305–324

Rubin KH, Fletcher CH, Sherman CE (2000) Fossiliferous Lana'i deposits formed by multiple events rather than a single giant tsunami. Nature 408:675–681

Scholle PA, Bebout DG, Moore CH (1983) Carbonate depositional Environments. American Association of Petroloeum Geologists Memoir 33, Tulsa, 708 pp

Sea Engineering Inc (1993) Beach Nourishment Viability Study. Sea Engineering Inc, Waimanalo

Shackleton NJ, Opdyke ND (1973) Oxygen isotope and paleomagnetic stratigraphy of equatorial Pacific core V28-238: oxygen isotope temperatures and ice volumes on a 10^5 year and 10^6 year scale: Quat Res 3:39–55

Shackleton NJ (1987) Oxygen isotopes ice volume and sea level. Quat Sci Rev 6:183–190

Sharp WD, Clague DA (2006) 50-Ma Initiation of Hawaii-Emperor bend records major change in Pacific plate motion. Science 313:1281–1284

Scheltema RS (1968) Dispersal of larvae by equatorial ocean currents and its importance to the zoogeography of shoal water tropical species. Nature 217:1159–1162

Scheltema RS (1986) Long-distance dispersal by planktonic larvae of shoal-water benthic invertebrates among central Pacific islands. Bull Mar Sci 39:241–256

Sherman CE (2000) Accretion and diagenesis of a submerged Pleistocene reef Oahu Hawaii. Ph.D. thesis, University of Hawaii Geology and Geophysics, 96 pp

Sherman CE, Glenn CR, Jones A, Burnett WC, Schwarcz HP (1993) New evidence for two highstands of the sea during the last interglacial oxygen isotope substage 5e. Geology 21:1079–1082

Sherman CE, Fletcher CH, Rubin KH (1999) Marine and meteoric diagenesis of Pleistocene carbonates from a nearshore submarine terrace Oahu, Hawaii. J Sed Res 69:1083–1097

Stearns HT (1935) Shore benches on the island of Oahu, Hawaii. Bull Geol Soc Am 46:1467–1482

Stearns HT (1970) Ages of dunes on Oahu, Hawaii. Occas Pap Bernice P. Bishop Mus 24:50–72.

Stearns HT (1974) Submerged shoreline and shelves in the Hawaii Islands and a revision of some of the eustatic emerged shoreline. Geol Soc Am Bull 85:795–804

Stearns HT (1978) Quaternary shorelines in the Hawaii Islands. Bernice P Bishop Museum Bulletin, vol. 237, Honolulu, HI, 57 pp

Stirling CH, Esat TM, Lambeck K, McCulloch MT (1998) Timing and duration of the last interglacial: evidence for a restricted interval of widespread coral reef growth. Earth Planet Sci Lett 160:745–762

Stoddart DR (1969) Ecology and morphology of recent coral reefs. Biol Rev 44:433–498

Storlazzi CD, Field ME, Dykes JD, Jokiel PL, Brown E (2002) Wave control on reef morphology and coral distribution: Molokai Hawaii, WAVES 2001 Conference Proceedings, American Society of Civil Engineers, San Francisco, CA, vol. 1, pp 784–793

Szabo BJ, Ludwig KR, Muhs DR, Simmons KR (1994) Thorium-230 ages of corals and duration of the last interglacial sea-level high stand on Oahu Hawaii. Science 266:93–96

Tinner W, Lotter AF (2001) Central European vegetation response to abrupt climate change at 82 ka. Geology :551–554

Tornqvist TE, Bick SJ, González JL, Van der Borg K, De Jong AFM (2004) Tracking the sea-level signature of the 82 ka cooling event: new constraints from the Mississippi Delta. Geophys Res Lett 31:L23309, doi:10.1029/2004GL021429

Toscano MA, Lundberg J (1998) Early Holocene sea-level record from submerged fossil reefs on the southeast Florida margin. Geology 26:255–258

Tucker ME, Wright PV (1990) Carbonate Sedimentology. Blackwell, Oxford, 482 p

Watts AB, ten Brink US (1989) Crustal structure, flexure, and subsidence history of the Hawaii Islands. J Geophys Res 94:10473–10500.

12
Biology and Ecological Functioning of Coral Reefs in the Main Hawaiian Islands

Paul L. Jokiel

12.1 Introduction

12.1.1 Geographic Location

The eight main Hawaiian Islands (MHI) are the emergent volcanic islands of the Hawaiian Archipelago (Fig. 12.1). The Hawaiian Archipelago is located in the center of the north Pacific Ocean and consists of islands, atolls, submerged banks and shoals, trending northwest by southeast in the between latitudes 19° N and 29° N. The island chain extends approximately 2,400 km (1,500 miles) from the Island of Hawaii in the southeast to Kure Atoll in the northwest. Hawaii is among the most isolated island groups on the planet being located approximately 3,000 km (1,860 miles) from the nearest continent. These islands are the emergent portion of the undersea Hawaiian-Emperor seamount chain that was formed continuously over the past 70–75 million years as the Pacific tectonic plate moved north and northwest over a stationary magma "hot spot" (Clague and Dalrymple 1994) at a rate of from 5 to 10 cm/year. Molten lava breaking through the thin rigid crust slowly creates volcanic mountains that eventually reach the surface of the ocean and emerge as islands (Macdonald et al. 1983). As the islands move off the hot spot they undergo erosion and subsidence. Eventually they are worn down and gradually sink, forming low islands and atolls and ultimately submerged seamounts. The eight MHI at the southeastern end of the archipelago represent approximately 5 million years of that cycle. The island of Hawaii is the youngest island with the oldest rocks dating to only about 430,000 years ago. Its youngest volcano, Kilauea, is currently active, along with a newly forming submerged volcano named Loihi located to the southeast of the Island of Hawaii.

The Hawaiian Archipelago has traditionally been split into two artificially defined groups of islands: the Northwestern Hawaiian Islands (NWHI) and the Main Hawaiian Islands (MHI) for various political, administrative and biogeographic purposes. The MHI consist of populated, high volcanic islands with non-structural reef communities and fringing reefs, while the (NWHI) consist of uninhabited atolls and banks as discussed by Grigg et al. in Chapter 14. However, Jokiel and Rodgers (2007) observed that the ten NWHI and the eight MHI are not parts of two disjointed systems, but rather are inseparable components of a single highly isolated ecosystem. Most fish, corals and other marine species found in Hawaii occur on reefs throughout the archipelago. Green Sea Turtles are a single genetic population with individuals that migrate from forage areas in the MHI to nest in the NWHI (Balazs and Chaloupka 2004a, b). Sharks and other large fish are known to move freely throughout the archipelago and do not observe the artificial boundary created by humans (K. Holland and C. Meyer 2006, personal communication). During past decades, the endangered Hawaiian Monk Seal was largely restricted to the NWHI, but has not respected this arbitrary division and recently has begun to re-colonize the MHI (Baker and Johanos 2005).

B.M. Riegl and R.E. Dodge (eds.), *Coral Reefs of the USA*,
© Springer Science+Business Media B.V. 2008

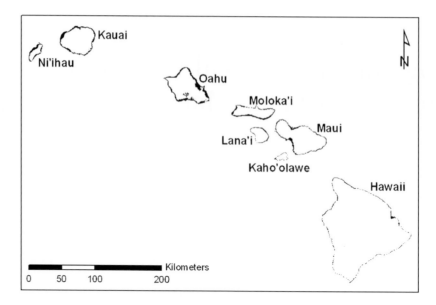

Fig. 12.1. Major features of the main Hawaiian Islands relevant to coral reef development include steep volcanic slopes running into very deep oceanic waters. The dark line in the figure represents the area between the high tide mark and the 30 m depth contour which generally exceeds the lower limit of reef coral development in Hawaii. Note more extensive fringing reefs on older islands or south facing shores, especially S. Molokai

12.1.2 Political and Social History

12.1.2.1 Background

Detailed accounts regarding the history of Hawaii are presented elsewhere (Daws 1968; Kuykendall and Day 1976), but are summarized here in relation to human impact on coral reefs. The Hawaiian Islands were initially settled by Polynesians. The earliest immigration appears to have been from the Marquesas as early as AD 500, followed by continued immigration from Tahiti after AD 1300. During 1778 Captain James Cook made the first western contact. Kamehameha I unified the Hawaiian Islands in battle and formally established the Kingdom of Hawaii in 1810. In 1819, Kamehameha II ascended to the throne and abolished the traditional kapu system of laws and regulations. In 1820, missionaries from New England arrived and converted some of the highest-ranking chiefs. Commoners followed the example of their leaders and converted to Protestant Christianity. Under the traditional system all land was held by the chiefs. Westerners pushed for private ownership of land. The ruling chiefs eventually allowed the land to be divided between the king, the chiefs, and the commoners. The Great

mahele (land division) was signed into law in 1848 by King Kamehameha III. In 1874 Hawaii signed a treaty with the United States granting Americans exclusive trading rights. The 1876 Reciprocity Treaty between the Kingdom of Hawaii and the United States allowed for duty-free importation of Hawaiian grown cane sugar. Sugar cane and plantation agriculture expanded. Asian immigrants were brought in to work the plantations. The traditional Hawaiian staple crop (taro) was replaced by rice growing to satisfy an expanding local market for the latter. In 1887, a group of cabinet officials and advisors to King David Kalukaua used armed militia to force the king to accept what is now known as the Bayonet Constitution, which stripped the monarchy of much of its authority. When Kalakaua died in 1891, his sister Liliuokalani took the throne. The queen proposed a new constitution that would restore the monarchy's authority. A group of European and American Hawaiian citizens and residents responded in 1893 by forming a committee of safety to prevent the queen from abrogating the 1887 constitution. United States Minister Stevens, concerned about possible threats to American lives and property, summoned a company of US Marines and US sailors who came ashore on January 16, 1893. A provisional government was established under threat of force by a local militia group. Liliuokalani gave up her throne. The Republic of Hawaii was established July 4, 1894 under the presidency of Sanford Dole with the intent of annexation to the United States. A subsequent investigation established by President Cleveland concluded that the United States diplomatic and military representatives had abused their authority. The request for annexation of Hawaii was initially denied, but changes in the political situation eventually led to Hawaii becoming a United States territory in 1900. The attack on Pearl Harbor on 7 December 1941 by the Empire of Japan was a trigger for the United States' entry into World War II. Hawaii became a major staging area for the conquest of the Pacific during the war with a large buildup of troops. The Island of Kahoolawe was declared a target island for ships and aircraft to practice bombing before deploying into the Pacific, with massive impact on the island and its coral reefs (Jokiel et al. 1993b). The strategic role of Hawaii in the areas of military activity and transportation combined with its appeal as a tourist destination and an attractive place to live led to rapid increases in human population and rapid development following the war. Hawaii formally became the 50th state of the Union on August 21, 1959.

12.1.2.2 Historical Changes in the Condition and Management of Reefs

Political and social changes in Hawaii over the past centuries as noted above had a direct impact on natural resource management and the condition of Hawaii's reef resources (Jokiel PL, Rodgers KS, Walsh WJ, Polhemus D, personal communication 2007). After the initial colonization of Hawaii by Polynesians, the increase in human population and depletion of resources eventually led to the development of a carefully regulated and sustainable "ahupua'a" system. This system was based on an integration of watershed, streams and near shore resources and involved local control and use of adaptive management practices that were triggered by subtle changes in fisheries resources. Sophisticated social-spiritual controls on resource utilization were an important component of the system. After western contact, many rapid changes occurred. Introduction

of livestock and plantation agriculture resulted in overgrazing and rapid erosion of watersheds (Roberts 2001). Breakdown of the "ahupua'a" system due to the initiation of private land ownership resulted in open access to reef resources with little or no control. New and more efficient fishing gear and methods were rapidly introduced. Advances in technology led to dredging and filling of reef areas. With increasing urbanization sewage outfalls were constructed. The present "western management system" gradually evolved and replaced much of the traditional Hawaiian system. Both systems target sustainable productivity of the reefs. However, major differences exist in the areas of management practices, management focus, knowledge base, dissemination of information, resource monitoring, legal authority, access rights, stewardship and enforcement (Jokiel PL, Rodgers KS, Walsh WJ, Polhemus D, personal communication 2007). Failures of the western system of marine resource management in maintaining productivity of inshore fisheries are leading to a re-evaluation of current practices in reference to the traditional system (Jokiel PL, Rodgers KS, Walsh WJ, Polhemus D, personal communication 2007). There has been a recent trend toward incorporation of major features of the traditional scheme using methods and terminology acceptable and appropriate to present day realities (e.g. community based management, marine protected areas, environmental education and outreach). The major strength of the present western system appears to be an ability to adapt to changing social, political, and economic conditions, and to the impacts of invasive species. The continued integration of several basic principles found in the traditional system may further strengthen the ability of contemporary mangers to insure sustainability in a manner that complements and supports the growing interest in ancient Hawaiian culture.

12.1.3 History of Biological Research

Prior to western contact, the traditional fisheries knowledge of the native Hawaiians may have surpassed that of modern marine biologists in some areas (Gosline and Brock 1960; Lowe 2004). Hawaiians in pre-historic times possessed an understanding of the life histories of fishes. Careful observations led to an intimate knowledge of the physical, biological, and ecological factors that influence fisheries. This knowledge was transmitted orally through the generations but was largely lost as western practices replaced the traditional Hawaiian system. The first contemporary scientific information on Hawaiian inshore marine biology in Hawaii began with the slow gathering of information and specimens obtained during infrequent expeditions into the region. Such accumulated data were periodically synthesized and published. For example, Vaughan (1907) compiled taxonomic and distributional information on Hawaiian scleractinian corals. In 1907 the College of Hawaii, which later became the University of Hawaii, was formed. From the beginning there was a strong interest in marine science and the history of coral reef studies in Hawaii. The first president, John Gilmore, recommended construction of a marine biology laboratory as part of the Department of Zoology as early as 1910. A temporary facility was in operation at Pier 6 in Honolulu Harbor by 1917 which moved to the Waikiki Aquarium in 1919. Charles Howard Edmondson came to Hawaii in 1920 under a joint appointment between the Bishop Museum and the University of Hawaii and served as Director of the marine laboratory at Waikiki which had been named the Cook Memorial Laboratory. Edmondson is regarded as the father of Hawaii marine biology. Over the years he published numerous scientific papers and reports and synthesized much information in the classic book "Reef and Shore Fauna of Hawaii" (Edmondson 1933). The major figure in the continued development of marine science was Robert W. Hiatt who expanded the UH Department of Zoology curriculum to include marine coursework in 1951. Notable faculty involved in the growth of the marine biology program at that time included Albert Testor, Vernon Brock and Albert (Hank) Banner. That same year Hiatt moved the marine laboratory at he Waikiki Aquarium to Coconut Island in Kaneohe Bay which was renamed the Hawaii Marine Laboratory (Hiatt 1955). The laboratory grew and later became the Hawaii Institute of Marine Biology (HIMB) in 1965 with construction of a major new building. In addition to HIMB, Hiatt founded the University of Hawaii Department of Oceanography in 1964, along with the Hawaii Institute of Geophysics and the Pacific Biomedical Research Center among other programs. These units have continued a substantial involvement in coral reef studies. The expansion in marine sciences and coral reef studies initiated by Hiatt has continued at an accelerating pace over the past 40 years (Karl 2004).

12.2   Regional Setting

12.2.1   Geography

Hawaii is truly an ocean state, being located in the middle of the North Pacific Ocean. Hawaii's coral reef communities provide food and recreation to the people of Hawaii and are critically important to the State's approximately $800 million/year marine tourism industry (Cesar and van Beukering 2004). Over 70% of the state's 1.2 million people live on Oahu, and are mostly concentrated in Honolulu. In addition to the resident population, nearly seven million tourists visit Hawaii each year. Increasing population has put anthropogenic pressure on Hawaii's coral reefs through various direct and indirect means.

12.2.2   Climate

The Hawaiian Islands are located at the northern edge of the tropics and experience a mild subtropical climate due to the persistent northeasterly trade winds. The average wind velocity varies between 10 and 20 knots with extended velocities of over 20 knots for periods of a week or more (Patzert et al. 1970). Length of day and temperature are relatively uniform throughout the year. Hawaii's longest days are about 13.5 h and the shortest days about 11 h, which translates into small seasonal variations in incoming solar radiation and air temperature. The major features of Hawaii's climate are: year round mild and equitable temperature, persistence of northeasterly trade winds, significant differences in rainfall within short distances due to orographic effects on the steep slopes of the high islands, and infrequent hurricanes or severe storms (Bach and Daniels 1973; Price 1983). Evaporation exceeds precipitation between 15° N and 36° N, but extremely high rainfall occurs over some areas of the high islands. Hawaii's steep volcanic topography influences local weather and climate. The volcanic mountains block, deflect, and accelerate the flow of air through channels and passes and alter local conditions on the reefs.

When warm, moist air is forced over windward coasts and slopes, clouds are formed with high rainfall. After being stripped of its moisture the air descends into leeward areas that tend to be sunny and dry. Due to the seasonal lag in ocean temperature, Hawaii's warmest months are not June and July, but rather August and September. Likewise, the coolest months, are February and March rather than December and January. Seasonal variation in climate is related to storm tracks and the Pacific High which follow the seasonal shift of the sun, moving north in summer and south in winter. The Pacific High tends to be stronger and more persistent in summer than in winter. Therefore, in winter, the trade winds may be blocked for days or weeks by fronts or migratory cyclones from the northern latitudes and by "Kona" storms forming south of the islands. Therefore, the winter season in Hawaii has more frequent clouds and rainstorms, as well as southerly and westerly winds. Hawaii's heaviest rains occur during winter storms between October and April.

Flash floods are an important feature of the climate and are often highly localized, intense and of short duration. According to Ramage (1971), factors leading to heavy rains include a large-scale disturbance (e.g. weather front), plentiful moisture supply (e.g. warm humid warm air as found around Hawaii) and a surface discontinuity (e.g. steep slopes of a high island). The coastline in the main Hawaii Islands (MHI) fits all of these features with steep topographic relief, humid air and frequent exposures to frontal movements. Therefore the MHI are and their offshore reefs are extremely vulnerable. Flooding often occurs when convective cells are formed or enhanced by orographic effects, and become anchored against the high-vertical relief features of the MHI. Such floods can trigger mudflows and landslides which transport mud onto coral reefs directly or via stream flow. Jokiel (2006) estimated from 5 to 10 flash floods per year from available data. Such flood waters may contain up to 90% sediment by some estimates (Jones et al. 1971).

12.2.3 Biogeography

The Hawaiian marine fauna is depauperate with a large percentage of endemics. In general, those species that have reached Hawaii are derived from the Indo-west Pacific region and have a broad geographic distribution. Approximately 30% of invertebrates, corals and fish are endemic (Kay and Palumbi 1987; Jokiel 1987; Hourigan and Reese 1987). A striking feature of the Hawaiian reef-fish fauna is that the majority of the abundant and ubiquitous species are endemics (Gosline and Brock 1960). Genera containing multiple endemic species generally are derived from separate Indo-west Pacific species rather than radiating from a common ancestor. Thus, while the Hawaiian Archipelago is severely isolated from other reef areas, the geographic barriers between the different islands of the Hawaiian archipelago are insufficient to isolate marine populations long enough to allow speciation within most of the taxonomic groups. However, sub-populations can be detected at the genetic level in some cases. Thus the NWHI and the MHI comprise a single biogeographic region. Perhaps this generalization should include nearby Johnston Atoll which has a coral fauna very similar to Hawaii (Maragos and Jokiel 1986).

Factors contributing to the low number of species in Hawaii include a northerly subtropical location with lower temperatures and irradiance, remoteness, lack of favorable currents to transport larvae from the southwest Pacific, lack of reef stepping stones in the region since the Cretaceous, and possible defaunation during eustatic sea-level rise and fall. For example, there are approximately 40 species of stony corals in Hawaii (Maragos 1995), whereas the Philippines, Palau and Japan each have over 400 species of reef corals (Veron 1993). Over 7,000 marine species have been recorded from the Hawaiian Islands (Paulay 1997). There are approximately 450 species of inshore fishes on Hawaiian reefs (Gosline and Brock 1960; Randall 1996).

Currents as well as distance control dispersal. Jokiel and Cox (2003) used pumice as a geological tracer of oceanic dispersal patterns by comparing the elemental signatures of drift pumice collected from Christmas Island and Hawaii in relation to prevailing oceanic currents and pumice source areas. Both sites lie isolated in the middle of the Pacific Ocean, far from sources of volcanic pumice or potential colonizers of the reefs. Hawaii lies in the persistent westward-flowing North Equatorial Current. Christmas Island, in contrast, is influenced by the highly variable westward-flowing South Equatorial Current and the eastward-flowing

Equatorial Counter Current. Analyses of pumice collected from the two locations revealed that Christmas Island pumice and Hawaii pumice distributions are dissimilar. Pumice is very abundant at Christmas Island in the beach drift. Pumice from Christmas Island is derived from the western Pacific Ocean (Krakatau), southwestern Pacific Ocean (Tonga Trench), east Pacific Ocean (Mexico), South Atlantic Ridge and an unknown source. In contrast, pumice is rare in Hawaii. Pumice from Hawaii originates primarily from the South Sandwich Islands, Mexico (Isla San Benedicto) and Krakatau. The currents that control dispersal of pumice also control dispersal of larvae and rafted organisms. Christmas Island has a higher coral diversity (31 genera, 81 species) than Hawaii (17 genera, 50 species). Hawaii receives only small amounts of pumice drift from a limited area to the east and has a more restricted coral diversity, while Christmas receives massive amounts of pumice (and presumably larvae and rafted organisms) from the area of high coral diversity to the west.

Johnston Atoll lies 800 km southwest of the nearest reefs of Hawaii and appears to fall into the same faunal region as Hawaii (see also Chapters 17 by Lobel and Lobel, 15 and 16 by Maragos et al.). Only 33 species and 16 genera and subgenera of shallow water stony corals have been reported from the atoll, with most of them being the same species found in Hawaii (Maragos and Jokiel 1986). Despite low species diversity, coral coverage is high in most environments. The coral reefs of Johnston Atoll are dominated by several species of *Acropora* that occur rarely in Hawaii. Several reefs in the center of the Hawaiian archipelago appear to have been colonized by *Acropora valida*, *Acropora cytherea* and *Acropora humilis* larvae from Johnston Atoll (Grigg 1981; Grigg et al. 1981). It has been proposed that larvae are transported from Johnston Atoll to Hawaii by the Subtropical Countercurrent. Available evidence suggests that the larvae could reach Hawaii in 50 days under optimum current conditions. Barkley (1972) computed currents from long line and ship drift data in the region. These data suggested the existence of 60 cm s^{-1} northerly current that produced an eddy wake moving downstream of Johnston to the northwest (320° at 45 cm s^{-1}). This wake probably extended at least 600 km downstream of the atoll (Barkley 1972). Journey

time for such eddies moving north into Hawaiian waters could be as little as 21 days if maximum speeds were sustained. The possibility of reciprocal exchange of larvae between Hawaii and Johnston is thus established. The similarity of the species lists of the two locations certainly supports this interpretation. Lack of common Hawaiian coral species such as *Montipora flabellata*, *Porites compressa* and *Tubastraea coccinea* at Johnston and the absence of common Johnston Atoll species, including all three hydrozoan corals (*Millepora*, *Distichopora* and *Stylaster*) in the Hawaiian Archipelago suggest that larvae of some species might not be able to bridge this gap. Kobayashi (2006) used computer simulation and high-resolution ocean current data to identify two potential larval transport corridors between Johnston Atoll and the Hawaiian Archipelago. One corridor connects Johnston Atoll with the middle portion of the Hawaiian Archipelago in the vicinity of French Frigate Shoals in the Northwestern Hawaiian Islands with the second connection to Kauai in the MHI.

A notable inhabitant of Hawaiian coral reefs is the Hawaiian monk seal (*Monachus schauinslandi*) which is an endemic endangered species. The major part of the population resides in the NWHI, but increasing numbers of seals are observed in the MHI. In other parts of the world similar species of tropical seals have been extirpated or greatly reduced in numbers. This last major population of monk seals is a valuable resource from the standpoint of biodiversity, conservation and the ecology in the Hawaiian Archipelago.

12.3   Natural Forces Influencing Reef Biology

The major natural factors influencing reef coral community structure and reef fish community structure in the Hawaiian Islands include currents, waves, substrate type, depth, island age, and rugosity (Friedlander et al. 2003; Jokiel et al. 2004). Some of these important parameters show correlations with each other. For example, water motion and light penetration are both negatively correlated with depth and positively correlated with each other.

12.3.1 Ocean Currents

The north Pacific gyre, centered at about 28° N latitude controls the general oceanographic conditions in the Hawaiian region. The dominant waves in Hawaii are generated by the prevailing Northeast trade winds which also drive surface currents to the west at speeds of from 15 to 30 cm s^{-1} (Flament et al. 1998). The resulting North Equatorial Current (NEC) flows past Hawaii and reaches an average westward speed of 17 cm s^{-1} to the south of with current speed gradually decreasing to the north along the island chain where the currents are strongly influenced by the islands. The NEC forks at the island of Hawaii with the northern branch becoming the North Hawaiian Ridge Current that intensifies as it moves along the eastern edge of the archipelago. West of the islands a clockwise circulation is evident with a centered at 19° N, merging to the south with the southern branch of the NEC.

Patterns of wind, currents and waves produce cooler surface temperatures in the channels and warmer surface temperatures in the lee of the larger islands. These variations in wind speed induce divergent and convergent surface currents, which in turn lift or depress the thermocline and form clockwise (anticyclonic) and counterclockwise (cyclonic) eddies in the lee of the major islands. The large counterclockwise average circulation is believed to result from the repeated occurrence of eddies spun off by the shear lines of the islands of Maui and Hawaii. Eddies can also be generated by intense currents such as the NEC impinging on the islands, much like swirls formed in a swift river downstream of a bridge abutment. The large clockwise circulation southwest of the Island of Hawaii appears to be caused by many such clockwise eddies that repeatedly form near South Point on the island of Hawaii. Geostrophic currents result from these variations of thermocline depth, in the form of intense counterclockwise eddies under northern shear lines, and (somewhat less intense) clockwise eddies under southern shear lines. The depth of the mixed layer in the lee of the island of Hawaii can vary from less than 20 m in the counterclockwise eddy, to more than 120 m in the clockwise eddy.

Localized currents resulting from tides and other oscillations are produced by the combination of diurnal, semidiurnal and fortnightly tidal components control sea level in the Hawaiian Islands (Flament et al. 1998). Mean tidal range for various stations in the MHI vary from 0.3 to 0.5 m with maximum diurnal change varying between 0.3 and 0.8 m. The small tidal range and steep nature of Hawaiian shorelines results in a very narrow intertidal zone and poorly developed intertidal fauna. Local bathymetry affects the ranges and phases of tides along the shore as the tidal waves wrap around the islands. Nearshore tidal currents are often stronger than caused by the large scale circulation and have an effect on local coral and fish communities. Semi-diurnal and diurnal tidal currents tend to move parallel to the shoreline. Localized accelerated currents on reefs often result from tidal currents flowing around points and headlands.

12.3.2 Waves

Wave damage is often cited as the single most important factor in determining the community structure and composition of exposed MHI reef communities (Dollar 1982; Dollar and Tribble 1993; Dollar and Grigg 2004; Jokiel et al. 2004). Response of Hawaiian coral community structure to periodic wave damage has been shown to fit Connell's (1978) 'intermediate disturbance hypothesis' (Grigg, 1983). Moderate coral cover and high diversity results from a continual cycle of intermediate intensity disturbances. High coral cover with low species diversity occurs in sheltered embayments and reefs in the wave shadow of other islands. Studies of coral communities on dated lava flows on the island of Hawaii suggest that it takes about 50 years for Hawaiian reefs to reach peak diversity following a catastrophic event (Grigg and Maragos 1974). A study of the impacts of storm waves of varying intensity on the west coast of the island of Hawaii has been conducted over a period of 30 years (Dollar 1982; Dollar and Tribble 1993; Dollar and Grigg 2004). Results indicate that shallow areas populated primarily by a wave-resistant pioneering species of cauliflower coral (*Pocillopora meandrina*) can recover completely within 20 years. However, deep reef slope zones populated by more delicate branching and plating species showed only the initial stages of recovery during the same period. Recovery does not always result in immediate replacement of the same dominant species in a particular zone. This cycle of repetitive impact and recovery is believed to be the major factor responsible for the present-day lack of reef accretion in exposed areas throughout the Hawaiian Islands. In wave sheltered areas Holocene reef accretion is on the order of 10–15 m thick. At wave exposed stations, Holocene accretion is represented by only a thin veneer of living corals resting on antecedent Pleistocene limestone foundations (Grigg 1998). The lack of coral reef accretion along open ocean coastlines may explain the absence of mature barrier reefs in the high Hawaiian Islands. However, extensive accretionary pre-Holocene reefs did form throughout the Hawaiian chain approximately 11,000 years ago. Rooney et al. (2004) proposed that storm wave intensity now is much greater compared to this earlier time and that the present wave regime is preventing formation of massive carbonate reef structures.

Reef corals can modify growth form within limits in order to adjust to a particular wave regime. For example, the Hawaiian coral *Montipora capitata* can assume encrusting, massive, plate-like or branching growth forms depending on the light and wave environment. In high wave energy environments robust species such as *Pocillopora meandrina* and *Porites lobata* dominate (Fig. 12.2). Delicate fast growing forms of species such as *Montipora capitata* or *Porites compressa* dominate in low wave environments. The normal biological processes of recruitment, growth, mortality and competition lead to the orderly development of a reef community adapted to the prevailing wave energy regime. However, an unusual storm wave event that occurs rarely (on the order of 10–50 years) can fragment the corals and totally alter the community composition within a matter of hours. Coral reef fish communities are influenced by reef coral development, so a relationship between wave exposure and fish community structure has been documented (Friedlander et al. 2003).

A hydrodynamic force-balance model was developed to calculate wave-induced forces on stony corals and predict the hydraulic conditions under which the skeletons of four dominant species of Hawaiian reef forming corals would fail and break (Storlazzi et al. 2005). The robust high-energy corals *Porites lobata* and *Pocillopora meandrina* are found in high wave energy environments. The more delicate branching corals such as *Porites*

compressa and *Montipora capitata* are found in areas that are deeper or more sheltered from storm waves (Fig. 12.2). The model was tested against observed species distribution along the south shore of Molokai, Hawaii. Results suggest that wave-induced forces are the primary control on coral species distribution and that the transition from one species to another is very likely due to the corals' strength in relation to prevailing wave conditions. Overall, the model appears to accurately define coral species distribution in the Hawaiian Islands based solely on the region's general wave climate and the corals' strength and morphologies; these results further support the long-standing ideas that waves are the dominant control on coral species zonation.

Jokiel et al. (2004) surveyed sites throughout the MHI and showed that mean wave height had a positive relationship with species richness, an observation consistent with the intermediate disturbance hypothesis (Connell 1978). Mean wave direction (compass bearing) shows a negative relationship with coral cover and diversity (Jokiel et al. 2004) because major storm surf in Hawaii arrives along a gradient that roughly diminishes in a counter clockwise direction from the North (Table 12.1). The largest and most frequent storm surf arrives during the winter North Pacific Swell (bearing 315°) with the less frequent and less damaging storm waves during the summer from the South Swell (bearing 190°) to the less severe Trade Wind Swell (bearing 45°). Maximum wave height is the most prominent factor with a negative relationship with coral cover, diversity and species richness (Jokiel et al. 2004). Maximum wave height is a good index of destructive wave events that damage Hawaiian reefs. In general, a reef's optimum growth and coral cover exists between 10 and 20 m, reflecting the trade-off between reduced wave induced stress at depth with decreased light available for photosynthesis (Storlazzi et al. 2002).

Events classified as storm waves strike Hawaiian shores from various directions with a frequency of from 2 to 7 times each month (Jokiel 2006). These waves come from different directions and are generated by various sources (Table 12.1).

The NE Trade Winds predominate throughout the year in Hawaii, but reach maximum intensity between spring and fall. These winds can build substantial waves as they move across the

FIG. 12.2. (**a**) *Pocillopora meandrina* community, Kauai. (**b**) *Porites lobata* community, Hawaii. (**c**) Diverse community in shallow water with high complexity, Molokai. (**d**) *Porites compressa* community, Maui. (**e**) Algal community on deep shelf, South Molokai. (**f**) *Montipora capitata* community, Kaneohe Bay, Oahu

Pacific toward Hawaii. Trade Winds diminish during the night and gradually increase throughout the morning to maximum wind speeds in the afternoon. Increased wind speed results in an increase in the size of wind-driven waves. Offshore wind-generated wave heights of 0.4–1.1 m are typical with periods of only 5–8 s (Table 12.1). These offshore waves break and dissipate along the north and east shores of the MHI. The high islands are a barrier to the surface winds, which

TABLE 12.1. Waves influencing the main Hawaiian Islands (Summarized from Moberly 1974 and Pat Caldwell National Oceanographic Data Center, Honolulu, personal communication 2006).

Wave type	Typical			Extreme			Direction	
	Height		Period	Height		Period		
	(m)	(ft)	(s)	(m)	(ft)	(s)	Mean	Range
NE Trade wind waves	1.2–3.7	4–12	5–8	4.0–5.5[a]	13–18[a]	9–12	NE 45°	0–90°
North Pacific swell	2.4–4.6	8–15	10–17	4.9–7.6	16–25	18–25	NW 315°	282–45°
Southern swell	0.3–1.2	1–4	12–17	1.5–3.1	5–10	14–25	SSW 190°	236–147°
Kona storm waves	0.9–1.5	3–5	8–10	1.8–3.1	6–10	11–14	SW 210°	258–247°

[a]Fully developed seas

increase in velocity as they form a jet of high velocity wind and funnel through gaps between the islands. The NE Trades are forced between the MHI and produce considerable wave chop in the channels with sharp boundaries known as shear lines. During late November 2003 an extremely destructive NE wave event struck Hawaii, which produced wave heights well above the NE Trade Wind wave height that is normally encountered. The November 2003 storm is believed to be the cause of damage observed on a section of reef flat at Pilaa, Kauai (Jokiel and Brown 2004b). Coral cover in this area declined from 14% to 6% with extensive breakage of coral colonies into rubble. A similar decline as the result of this storm was observed on the NE facing shore at Sandy Beach, Oahu (S.J. Dollar, personal communication 2005).

North Pacific Swell is generated by Northern Hemisphere winter storms in the North Pacific (Table 12.1). Breaking waves inshore with faces over 15 m have occasionally been observed. Wave energy of this magnitude prevents coral reef development along the north shores of the islands.

Southern Swell is generated by Antarctic Southern Hemisphere winter storms and is generally encountered in summer and early autumn (Table 12.1). Waves generated in the southern Pacific take 6–8 days to reach Hawaii and lose much of their energy due to spreading before they arrive in the islands. Southern Swell rarely approaches the heights of North Pacific Swell seen on the northwest shores in winter. The largest southern waves on record (June 1955) had faces over 6 m breaking in shallow water.

Kona Storm Waves can occur throughout the year, but are most common from October through April. During this time, waves may be generated by southerly or southwesterly winds that precede the northerly winds of cold fronts. Typical wave heights are from 0.9 to 1.5 m with periods of 8–10 s. Under extreme conditions these waves can exceed 3 m in height. A 1980 Kona storm generated inshore plunging breakers of up to 6 m, reduced living coral cover at a site off the west coast of the Island of Hawaii from 46% to 10% (Dollar and Tribble 1993).

Hurricane Waves are infrequent and unpredictable events that can have profound effects on reefs. Limited historical information exists on the location, size of waves and amount of reef damage on Hawaiian reefs caused by Hurricane waves. Recorded hurricanes in Hawaii have followed trajectories that led to direct impact on the islands of Kauai and Oahu, with less impact on the other islands (Schroeder 1998). Most central Pacific hurricanes originate near Central America or southern Mexico. Many of these storms die out if they move northwestward over cooler water or encounter unfavorable atmospheric conditions. Of those that survive, most pass far to the south of Hawaii. Hurricane season begins in June and lasts through November in the Hawaiian Islands. Hurricane Iniki in 1992 produced waves powerful enough to break and abrade corals over much of south Kauai. Homes, appliances, furnishings, trees and other objects were carried into the surf and added to the mechanical damage to the reefs. Re-colonization and recovery of the reef corals over the past decade has been substantial, and most of the reefs have returned to their pre-hurricane condition. Waves generated by Hurricane Iniki had a small but measurable impact on reefs hundreds of kilometers to the east at Kona, Hawaii where coral cover decreased from 15% to 11% (Dollar and Tribble 1993). Hurricane Iniki also impacted coral reef communities in Mamala Bay, Oahu (Brock 1996). The greatest change attributable to the hurricane was the loss of topographical relief and shelter habitat for many fish and invertebrate species. This loss was caused by the movement of loose materials (rubble and sand) across the bottom with infilling of holes and depressions resulting in a less heterogeneous habitat. However, in other areas storm waves removed accumulated sediment and rejuvenated certain reef environments.

Tsunamis or seismic sea waves consist of a series of immense waves caused by violent movement of the sea floor during an earthquake, underwater landslide, or volcanic eruption. These waves are characterized by great speed (up to 950 km h^{-1}), substantial wave length (up to 190 km), long period between successive crests (varying from 5 min to a few hours, generally 10–60 min), and low height in the open sea. A tsunami event can last several hours and destroy everything in its path. The first tsunami recorded in Hawaii occurred in 1819, and since then 85 others have been observed, 15 of which resulted in significant damage along the coastline (Curtis 1998). Tsunamis have accounted for more lost lives than the total of all other local disasters. Damage to reefs resulting from past tsunamis has not been documented on Hawaiian coral reefs, but may have been considerable as evidenced by the level of destruction to property and human life.

The impact of waves on a given reef involves a complex interaction between wave direction, island topography, and bathymetry (Dollar and Tribble 1993; Storlazzi et al. 2002, 2005). The islands block waves and create a "wave shadow" that moderates the impact of waves on reefs in the lee of islands (Storlazzi et al. 2005). Islands such as Molokai with an elongate E–W morphology create a large wave shadow along the south coast that blocks wave energy from the North Pacific Swell although wave refraction does impact the reefs on the east and west ends of the south coast. Other islands to the south and east further protect the south coast of Molokai. In contrast, circular islands such as Kauai that do not fall into the wave shadow of other major islands are vulnerable to waves and wave refraction from all compass directions.

12.3.3 Solar Radiation

The growth of coral reefs is limited by light due to the dependence of reef corals and algae on photosynthesis (Wells 1957). Corals are plant–animal symbioses that require sunlight as a primary source of energy. Therefore, their lower depth distribution is set by light penetration. Light is attenuated by sea water, so eventually a depth is reached where photosynthesis cannot support reef building. The exceptional clarity of offshore oligotrophic waters off Hawaii allow sufficient light penetration for the development of coral communities of *Leptoseris* spp. to develop to depths as great as 150 m (Kahng and Maragos 2006). However, these corals exist as thin plates and are not major reef builders. In clear offshore Hawaiian waters not influenced by severe swell, impressive reef communities develop in deeper water. For example, the offshore ocean reefs off south Moloka'i have rich *Porites compressa* communities to depths of 30 m (100 ft). At greater depths, the substrate is unsuitable for corals as the reef gives way to a sand terrace dominated by the alga *Halimeda* (Fig. 12.2e). High turbidity inshore due to fine sediment dramatically reduces light penetration and consequently reduces coral reef development in many areas.

Sunlight is necessary for photosynthesis, but the short ultraviolet wavelengths of solar radiation are potentially lethal. Hawaiian reef corals thrive under levels of solar ultraviolet radiation that would kill or severely damage many forms of marine life (Jokiel 1980). Early work suggested that an undefined substance termed "S-320" had such characteristics and might act as a protective screen against UV radiation (Shibata 1969). Concentration of this material was shown to decrease in Hawaiian corals with increasing depth, presumably as a result of attenuating UV radiation (Maragos 1972). Growth of zooxanthellae in vitro is inhibited by solar UV (Lesser and Shick 1989), but zooxanthellae in vivo apparently are not affected (Jokiel and York 1982), unless a non-acclimatized coral is subjected to a sudden increase in UV irradiance (Scelfo 1984). The S-320 material in corals was subsequently identified by Dunlap and Chalker (1986) as a group of compounds known as mycosporine-like amino acids (MAAs). Effects of UV radiation can be subtle. For example, Kuffner (2001) found that larvae of the Hawaiian reef coral *Pocillopora damicornis* delay settling to the substrate when UV radiation levels are high.

12.3.4   Water Temperature

Temperature is a primary physical factor governing reef coral distribution (Wells 1957). The optimum growth temperature for Hawaiian reef coral is 27°C, although they can tolerate prolonged temperature of from 20°C to 29°C, and short exposure to temperatures of from 18°C to 31°C (Jokiel and Coles 1990). Surface water temperature shows a strong north-to-south gradient along the Hawaiian Archipelago, with lowest sea surface temperature in March, and highest in September. The seasonal range of sea surface temperatures near Hawaii is only about 3° with the average surface water temperature around O'ahu ranging from 24°C in winter to 27°C in summer. In the open ocean, the surface waters are mixed by the wind and have uniform properties. The depth of this warm mixed layer can vary from nearly 120 m in winter to less than 30 m in summer (Flament et al. 1998).

Field investigations on a reef impacted by thermal discharge from a power generation station (Jokiel and Coles 1974) and controlled experiments with Hawaiian corals (Jokiel and Coles 1977) demonstrated that loss of symbiotic zooxanthellae, or "bleaching" is one of the first visible signs of thermal stress. Bleaching in Hawaiian corals can be induced by short-term exposure (i.e. 1–2 days) at temperature elevations of from 3° to 4° above normal summer ambient or by long-term exposure (i.e. several weeks) at elevations of 1–2° above normal long term summer maxima (Jokiel and Coles 1990). Temperature and light interact synergistically; high light accelerates bleaching caused by elevated temperature (Jokiel and Coles 1990). Bleaching threshold is determined by level of solar irradiance, degree of heating over summer maximum and duration of exposure (Jokiel 2004). Critical threshold temperatures for coral bleaching vary geographically, but can be expressed universally as fixed increments of 1–2°C relative to the historical mean local summer maximum (Coles et al. 1976; Jokiel 2004). Temperature elevations above summer ambient, but still below the bleaching threshold, can impair growth and reproduction in the Hawaiian coral *Pocillopora damicornis* (Jokiel and Guinther 1978). Bleaching susceptibility is correlated with respiration rate. Any factor that increases respiration (such as high incident solar radiation) accelerates bleaching at higher temperatures.

Hawaiian surface waters have shown a trend of increasing temperature over the past several decades (Jokiel and Brown 2004a) that is consistent with observations in other coral reef areas of the world (Coles and Brown 2003). Jokiel and Coles (1990) observed a warming trend in Hawaiian waters and warned that Hawaiian corals were perilously close to their bleaching threshold during the summer months. The first documented regional coral bleaching in the main Hawaiian Islands occurred in Hawaii in late summer of 1996 (Jokiel and Brown 2004a). Bleaching was recorded at a number of locations, with the most severe impact observed on Oahu (Kaneohe Bay, Kailua Bay) and lesser bleaching reported on Maui and Hawaii. On Maui, weekly temperatures on reefs along the southwest coastline experienced temperature of 28.0–28.5°C in late August and early September, with peak temperatures approaching 29°C. Corals began to bleach at Olowalu, Maui in late August, but the extent and severity of bleaching was minor, with less than 10% of the corals being affected. Recovery occurred after several months. Documented bleaching events in Hawaii were all triggered by prolonged regional oceanic positive oceanic sea surface temperature anomalies greater than 1°C that developed offshore during the time of the annual summer temperature maximum. High summer solar energy input and low winds further elevated inshore water temperatures by 1–2°C in reef areas with restricted water circulation and in areas in the lee of the larger islands where mesoscale eddies retain water masses close to shore for prolonged periods of time. Major bleaching events were observed in the NWHI during the summer of 2002 (Aeby et al. 2003; Jokiel and Brown 2004a) and again in the summer of 2004 (Kenyon et al. 2004; Kenyon and Brainard 2006).

12.3.5   Depth

As noted above, vertical distribution of corals is controlled largely by light and water motion which both diminish with depth. Grigg (2006) used growth rate of the abundant Hawaiian reef building coral *Porites lobata* as a proxy to describe the relationship between reef development and depth. The lower limit for this species is 80–100 m, yet reef accretion ceases at approximately 50 m depth. Below this depth the rate of bio-erosion exceeds coral growth, causing coral colonies to fragment and break down into rubble.

12.3.6   Substrate Type

Substrate type is a major factor that defines Hawaiian habitats (Coyne et al. 2003). The reef building crustose coralline algae and reef corals generally require a hard substratum. Sand and mud substratum is unsuitable for coral settlement and growth. Corals can, however, develop in these areas through settlement on isolated outcrops of hard rock. Accretion of skeletal material over many generations of coral will produce large coral "mounds" or "ridges" surrounded by mud and/or sand. Sand and mud are transported onto and off the reef in response to currents and wave regime. Passage of sediment through coral communities can smother new coral settlements and encrusting corals, and can even bury large coral colonies, which are killed and sometimes uncovered later by subsequent erosion.

12.3.7   Land Impact

Terrigenous runoff carries fresh water, fine sediment and nutrients onto the reef. Fine sediments can reduce light penetration to the corals and can smother and kill colonies. Fine sediment prevents settlement of coral larvae. High nutrient levels carried off the watershed or seeping in with ground water stimulate growth of macroalgae that can choke out corals.

12.3.8   Salinity

Reef corals have been described as having very narrow salinity tolerance (Wells 1957), but corals and coral reefs are known to occur under natural conditions at salinities ranging from 25‰ to 42‰ (Coles and Jokiel 1992). Hawaiian corals can tolerate salinity of 15‰ for several days before they die (Coles and Jokiel 1992; Jokiel et al. 1993a). This represents a 50% dilution of seawater with fresh water. Steep bathymetry and exposure to high wave energy characterize most of the coastline of the Hawaiian Archipelago and the resulting flushing by waves and currents maintains salinity on reefs within the range of tolerance for reef corals. Salinity on ocean reefs in Hawaii is generally close to that of the open ocean which is on the order of 34–35‰. Low salinity can occur near stream mouths on reef flats, but the fresh water is rapidly mixed with seawater in most open coastal situations. Nevertheless, discharge of fresh water from rivers and streams limits the development of reefs at river mouths, forming breaks in otherwise continuous fringing reefs (Stoddart 1969). Suppression of reefs in these areas is due to input of nutrients and sediment as well as fresh water. Fresh water is much more buoyant than seawater, so it often forms a surface layer that does not impact corals in deeper water.

"Reef kills" caused by low salinity associated with flood events have occurred in Hawaii (Jokiel et al. 1993a). Such floods represent a major but episodic stress for corals in highly enclosed embayments such as Kaneohe Bay (Oahu), Hilo Bay (Island of Hawaii), Kahalui Bay (Maui), Pearl Harbor (Oahu) and Nawiliwili Bay (Kauai), where circulation is restricted and salinity can be reduced to levels lethal to corals. The best documented events occurred in Kaneohe Bay, Oahu which is a large (4 × 10 km) estuarine system with relatively unrestricted exchange of water with the open ocean. Consequently a rich coral reef complex has developed in the bay. Salinity in the bay is normally close to oceanic conditions of 34–35‰, but commonly surface waters drop to 29‰ during flood events. However, during extreme flood conditions, salinity can be reduced to the point of killing corals on the reefs. For example, rainfall exceeding 60 cm in 24 h on the Kaneohe Bay watershed during May 1965 produced a surface layer of low salinity water that killed corals and invertebrates to a depth of 1–2 m on reef flats throughout the inner portion of the bay (Banner 1968). Storm floods again occurred in late December 1987 and early January 1988 that reduced salinity in the surface waters of Kaneohe Bay to 15‰ and produced massive mortality of coral reef organisms in shallow water. Virtually all coral was killed to a depth of 1–2 m in the inshore regions of the Bay (Jokiel et al. 1993a).

12.3.9   Sedimentation

Hawaiian corals and coral reefs are sensitive to sediment loading (Maragos 1972; Banner 1974).

The detrimental effects of sediment on corals, their larvae and other reef organisms have been reviewed by Rogers (1990). Sedimentation buries or smothers coral, blocks light needed for photosynthesis and inhibits settlement of corals. Nutrients associated with the sediment can stimulate algal blooms and toxic materials contained in the sediments can be harmful to delicate marine life. Maragos (1972) reported a significant negative relationship between light extinction coefficient and growth for the Hawaiian coral *Montipora verrucosa* during the time when Kaneohe Bay was undergoing severe impact from sewage outfalls.

In Hawaii, factors such as wave exposure, water motion and localized topography influence sediment stress on coral reefs. Sediment deposits on the coral reefs of Hawaii are dynamic features that are continually being replenished by terrestrial runoff as well as by biogenic carbonate material from the reef. Roy (1970) estimated that more than 70% of the sediment in Kaneohe Bay, Oahu is internally derived from the breakdown of calcium carbonate materials. These deposits are, in turn, being altered and depleted by resuspension and removal of fine fractions during periods of strong surf and currents. On reef slopes there is a continual resuspension and movement of sediments into deeper water that is enhanced by waves and currents. For example, Bothner et al. (2006) deployed sediment traps on the south Molokai reefs and found that storms with high rainfall, floods, and exceptionally high waves resulted in sediment trap collection rates greater than 1,000 times higher than during non-storm periods, primarily because of sediment resuspension by waves. Floods recharged the reef flat with land-derived sediment, but had a low potential for burying coral on the fore reef when accompanied by high waves. The high trapping rate and low sediment cover indicate that coral surfaces on the fore reef are exposed primarily to transient resuspended sediment. Studies were undertaken on a shallow fringing reef flat on Molokai, Hawaii to determine the temporal and spatial dispersal patterns of terrigenous suspended sediment (Presto et al. 2006). Trade-wind conditions produced strong currents and resuspended moderate amounts of sediment on the reef flat on a daily basis, resulting in an overwhelming contribution to the total sediment flux. Sediment resuspension and transport was found to be controlled by state of tide and magnitude and direction of the trade winds relative to the orientation of the coastline. Locations where sediment moves offshore appear to be correlated with areas of low coral coverage on the fore reef.

The high wave energy regime of Hawaiian coastal reefs coupled with the existence of very deep ocean waters offshore of the reefs allows rapid removal of sediments from the reefs. The dynamic relationship between sediment input, deposition, resuspension and removal favors rapid regeneration of costal reefs once the terrigenous input has been reduced or eliminated. For example, a major program resulted in the removal of 20,000 goats from the island of Kahoolawe so that re-vegetation could occur and stabilize the soils that were rapidly being eroded into the sea. Following corrective action, Jokiel et al. (1993b) noted uncovering of old buried reefs with intact dead skeletons of delicate corals at many locations previously subjected to high sediment loading. Rapid recruitment of new coral colonies onto the recently uncovered reef surfaces was noted at all sites. The reefs appeared to be undergoing recovery as sediment input diminished and as waves and currents removed the existing deposits. Similar observations were made by Grigg (1985), who reported a recovery period of 5–10 years for sediment damaged reefs after the discharge from sugar plantations was halted on the island of Hawaii. Extensive mud flows from an illegal grading project covered the reef at Pilaa, Kauai. Measures were taken to stabilize the graded area and stop the chronic flow of mud. Subsequently, winter storms over a period of years mobilized and removed much of the mud from the damaged reef (Jokiel and Brown 2004b).

12.4 Zonation and Community Patterns

Gosline and Brock (1960) presented a zonation scheme in relation to distributions of Hawaiian fishes including offshore pelagic, littoral, reef sub-surge zone, reef surge zone and supra-surge zones. They concluded that zonation of reef fishes is controlled largely by wave regime and substrate type. The habitat and zone descriptions presented by Maragos (1998) are more detailed in three dimensional space and time. This approach is based on a number of interacting ecological factors and is consistent with the nomenclature of contemporary field ecologists working on Hawaiian coral reefs. The habitat and zone classification scheme of Coyne et al. (2003) for Hawaiian reefs is a simplified two-dimensional mapping approach necessitated by the limitations of producing maps from remote sensing images. The various classification approaches are of value in describing ecological features of Hawaii coral reefs and can be readily modified and adapted to meet the requirements of specific investigations. A very wide range of reef types occurs throughout the MHI (Fig. 12.3). A synthesis of information relevant to descriptions of habitats and zonation on Hawaiian coral reefs is as follows:

12.4.1 Physical Factors Controlling Community Composition

Major factors that control zonation and community patterns (Fig. 12.2) include reef type (morphology), wave exposure (Fig. 12.3f), depth, substrate type, sedimentation, island age, latitude and nutrient availability. Freshwater intrusion either through surface runoff or groundwater seepage can also influence community composition. Light penetration,

FIG. 12.3. (a) Fringing reef, South Molokai. (b) Highly modified reef, Honolulu, Oahu. (c) Vertical walls, Napali Coast, Kauai. (d) Molokini Island, Maui. (e) Sediment-impacted fringing reef, Lanai. (f) Wave-exposed coastline, North Molokai

wave energy and temperature diminish with depth and create environmental gradients leading to vertical zonation over a depth range (Friedlander et al. 2003).

12.4.2 Major Reef Types

As volcanic islands in the Hawaiian Archipelago form and move to the northeast with the Pacific Plate they gradually erode to sea level and eventually subside to form atolls, banks and eventually deep water seamounts. The formation of reefs around the perimeter of the main Hawaiian Islands follows the early phases of atoll formation in the general scheme first proposed by Charles Darwin in the nineteenth century (Darwin 1976). During the first stages of development, Hawaiian reefs are nothing more than a thin veneer of corals and crustose coralline algae overlaying the volcanic rock of a high island. As carbonate is accumulated the reefs grow outward to form wide fringing reefs. The next stage is the gradual submergence of the reef with rapid growth on the seaward margin. Rapid outward growth forms a barrier reef. As the island subsides, upward growth of the outer barrier reefs keeps pace with subsidence to form an atoll as the volcanic core sinks below the surface. Such atolls occur in the NWHI.

12.4.2.1 Apron Reefs

Apron reefs are the first phase of reef development and consist of a thin veneer of calcifying organisms overlaying the basalt substratum. A newly emergent Hawaiian island starts as a series of volcanic eruptions on the deep ocean floor. As the island builds into the photic zone, calcifying organisms such as corals and coralline algae can colonize the basalt and begin the process of reef building. Marine habitats on the young volcanic islands of Hawaii and Maui are characterized by steep unstable volcanic slopes, rocky shorelines and beaches consisting of basalt cobble. Reef coral communities form on these basalt surfaces as discontinuous "apron reefs" that consist of a thin veneer of corals and calcifying organisms. Sandy carbonate beaches are uncommon and where beaches occur, they consist largely of black volcanic sand. Lava flows reaching the sea create new basalt surfaces that are devoid of life. As the lava cools it is colonized by bacteria, algae, various invertebrates and

eventually by corals. Development rate of the coral community on a fresh lava flow depends largely on exposure to sea and swells (Grigg and Maragos 1974). In wave-exposed areas, recovery time (in terms of number of species, percent cover and diversity) of coral communities is approximately 20 years. In sheltered embayments more than 50 years is required for complete recovery. Succession in exposed areas is continually interrupted by large wave events which keeps these communities in perpetual pioneer stages. In wave-sheltered areas the reefs become more fully developed and therefore require more time for recovery to their original condition after being covered by a lava flow. Coral coverage and diversity increase over time, but diversity shows a decline as climax is approached. The downturn is due to interspecific competition for space, with dominance by fast-growing species that can crowd out competitors.

12.4.2.2 Fringing Reefs

If conditions are favorable to growth of carbonate-secreting organisms the pioneer apron reefs will continue to accrete carbonate materials and eventually grow seaward and form true biogenic fringing reefs, provided that the shoreline is not undergoing submergence, slumping or is being impacted by extremely heavy wave action (Fig. 12.3f). The older islands of Maui, Molokai, Oahu and Kauai have areas of well developed fringing reefs that are lacking on the younger islands of Maui and Hawaii. The structure of a fringing reef is developed by accretion of the skeletons of calcifying organisms rather than taking the form of the underlying basalt. The outer slope of fringing reefs typically extends to a depth of 30–50 m.

12.4.2.3 Barrier Reefs

An ancient drowned barrier reef exists at Mana off northwest Kauai at a depth of about 30 m with a lagoon that is presently at a depth of 50 m. Kaneohe Bay, Oahu is a lagoon that is formed by a structural barrier reef on the seaward margin and is the largest sheltered body of water in the main Hawaiian Islands (Holthus 1986). The bay is about 13 km long and 4 km wide. This barrier reef is not a true biogenic barrier reef, but rather is a ridgeline formed along a drowned river valley that has developed a veneer of carbonates. Two major channels connect the lagoon to the open ocean. Patch reefs have developed within the lagoon and fringing reefs have formed along the landward side of the lagoon. The area is unique and atypical of Hawaii's coastline but is noteworthy because of the complex structural reef development, diversity of habitats and organisms, occurrence of patch reefs, presence of a well developed fringing reef along the landward margin, blue holes, and presence of a well-flushed lagoon. Further, this system has been studied extensively for the past 50 years by scientists working at the Hawaii Institute of Marine Biology at Coconut Island.

12.4.3 Zones

12.4.3.1 Supra-tidal

Anchialine ponds are salt water ponds located in the supra-tidal with no surface connection to the ocean. However, they have a subsurface connection to the sea through the porous volcanic rock as evidenced by the rising and falling of water level with changes in tide. There are more than 700 of these ponds in Hawaii with most of them occurring on young lava flows on the Island of Hawaii and Maui. These are unique Hawaiian ecosystems are inhabited by a number of exotic organisms including endemic species of small red shrimps (Brock and Bailey-Brock 1998; Santos 2006). Some of the shrimp species feed on algae and bacteria, while others are predators on the herbivorous shrimp.

Tide pools are depressions along rocky coastlines that flood at high tide and are intermittently continuous with the open ocean. They experience extreme fluctuations in temperature and salinity and typically are inhabited by hardy species capable of withstanding high solar radiation, temperature extremes, desiccation, periods of low salinity and episodes of high wave energy. A wide variety of algae, echinoderms, mollusks, barnacles, crustaceans and worms occur in this habitat. Fish fauna include blennies, gobies and juveniles of certain species (Gosline and Brock 1960) that use the area as a nursery grounds to avoid predators.

12.4.3.2 Intertidal

The area between the mean high water line and lowest spring tide level is defined as the intertidal.

Typically, this zone is narrow due to the small tidal range in the main Hawaiian Islands. The intertidal along sheltered shorelines typically consists of sand and/or mud substratum. In open coastal situations wave action scours rocky coastlines and creates beaches. Sandy beaches are most common on the older islands where carbonate reefs have developed (Kauai, Niihau, and Oahu,) and less common on the younger islands (Hawaii and Maui). Surf run up caused by waves that may reach a height of several meters overshadows the importance of tidal range which is only the order of one meter.

Mangroves were introduced on the island of Molokai in 1902, primarily for the purpose of stabilizing coastal mud flats. Throughout the main Hawaiian Islands mangroves have become a conspicuous component of the intertidal and shallow subtidal in sheltered embayments or along shorelines with well developed fringing reefs or barrier reefs. Mangroves are considered to be valuable components of coral reef ecosystems in the tropics, but in Hawaii they are invasive and have negative ecological impacts (Allen 1998) such as reduction in habitat quality for endangered water birds and colonization of habitats to the detriment of native species (e.g. in anchialine pools).

12.4.3.3 Subtidal

Reef Flats

Reef flats are shallow (semi-exposed) areas between the shoreline intertidal zone and the reef crest of a fringing reef. This zone is protected from the high-energy waves commonly experienced on the fore reef and reef crest. Typical reef flat habitats have a substrate ranging from mud to sandy reef rubble with coral mounds or ridges depending on water motion, circulation and rate of terrigenous input of mud. Typical assemblages include coral communities on mounds or ridges of hard substratum, algal communities and occasionally sea grass in mud/sand areas. Blue holes are interesting features found on wide reef flats off south Molokai (Fig. 12.3a) and Kaneohe Bay Oahu.

Lagoons

Lagoons are shallow areas inshore of a fringing or barrier reef. Kaneohe Bay, Oahu is the best developed example in Hawaii. The lagoon lies between the shoreline intertidal zone and the back reef of a barrier reef. This zone is protected from the high-energy waves by the barrier reef. Typical habitats include mud and sand substrate due to lack of wave action. Seagrass may be present, along with algal communities, patch reefs, and rich coral communities.

Lagoon Patch Reefs

Small isolated reefs that rise from lagoon bottom and break the surface of the sea are often called patch reefs. These are unique features in the MHI that are restricted largely to Kaneohe Bay, Oahu.

Back Reef

This is the area between the seaward edge of a lagoon floor and the landward edge of a reef crest. This zone is present only when a reef crest and lagoon exist. Typical habitats include sand, reef rubble, seagrass beds, and algal communities.

12.4.4　Seaward Ocean Reef Zones

12.4.4.1　Reef Crest

The term refers to the flattened, emergent (especially during low tides) or nearly emergent segment of a reef. This zone lies between the back reef and fore reef zones. Breaking waves will often be visible in aerial images at the seaward edge of this zone. The major feature is an algal ridge that breaks the surface and is built by crustose coralline algae and other organisms. Small channels, depressions, areas of sand and cobble, areas with wave-resistant corals and flat areas with attached macro algae are typically present.

12.4.4.2　Fore Reef

This is the area from the seaward edge of the reef crest that slopes into deeper water down to the bank/shelf platform in typical situations. The term applies to reefs without an emergent reef crest but still having a seaward-facing slope that is significantly greater than the slope of the bank/shelf.

12.4.4.3　Bank or Shelf

This refers to a deep water area extending offshore from the seaward edge of the fore reef to the beginning of an escarpment where the insular shelf drops off into deep, oceanic water. Such flattened platforms between the fore reef and deep open ocean waters are common in Hawaii. Typical Habitats in these areas (Fig. 12.2e) include sand deposits, algal beds (*Halimeda* spp., *Padina* spp., sea grass, etc.), colonized and uncolonized pavement, and sand channels.

12.4.4.4　Shelf Escarpment

In this zone the depth increases rapidly into deep, oceanic water. The zone begins at approximately 20–30 m depth and extends into much deeper water.

12.4.5　Other Zones

12.4.5.1　Vertical Walls

In many locations throughout Hawaii vertical walls of basalt remain (Fig. 12.3c) where sections of islands have fallen away in slumps and prodigious submarine avalanches (Moore et al. 1989). These coastlines are characterized by near-vertical slope from shore to shelf or shelf escarpment. Such areas are typically narrow and may not be distinguishable in remote sensing imagery, but are important habitats along the coastlines of Hawaii. The resulting steep cliffs rise above sea level and can extend vertically to great depths. These stretches of coastline typically contain sea caves where waves rapidly erode areas of softer lava rock or where lava tubes occur. Such shorelines often have wave cut benches with tide pools. A typical intertidal develops on the upper rock face grading into a heavily wave impacted subtidal that typically has low coral coverage of wave resistant *Pocillopora meandrina* (Fig. 12.2a). Deeper on the face encrusting *Porites* spp. (Fig. 12.2b) become more abundant with *Montipora* spp. (Fig. 12.2f) increasing in abundance in deeper water as wave action is less and light becomes limiting.

12.4.5.2　Naturally Formed Channels

These are a common feature of coral reefs. Such channels often are drowned river valleys. Channels can cut across several other zones. On fringing reefs breaking waves on the reef crest transport water onto the reef flat. This water eventually drains off the reef through channels (Fig. 12.3c).

Channels are characterized by strong out-flowing currents, especially during periods of high waves and at falling tide. The typical habitat in the bottoms of the channels is unconsolidated cobble, sand or mud depending on current speed and wave action. Channel sides typically are colonized by corals and crustose coralline algae. Fish often are abundant due to the vertical relief of the walls and the strong currents which carry algae, small organisms and detritus that serves as food for the fishes.

12.4.5.3 Pinnacles, Stacks and Offshore Islets

Such features occur throughout the MHI and possess a surprisingly rich and diverse marine fauna. For example, the islet of Moku Manu off Mokapu Peninsula (east Oahu) has an extensive system of undersea caves with extremely high abundance and diversity of sponges and associated organisms. Numerous habitats such as spur and groove, vertical wall, boulder and coral communities are represented in a small area. The sea stack called Mokapu which is located off Kalupapa, Molokai has a very rich benthic fouling community along with diverse associated organisms such as nudibranchs and starfish that are seldom seen in other locations. The sheltered side of this sea stack has been colonized by the soft coral *Sinularia,* which is uncommon in Hawaii. Antipatharians normally found in deeper water occur at 10 m depth. Another well known example is Molokini, Maui (Fig. 12.3d). The crescent-shaped islet represents the emergent one third of the rim of a small tuff cone about 400 m in diameter. The submerged crater area of the cinder cone forms a "lagoon" and has extremely high coral coverage and a rich reef fish assemblage. Maximum lagoon depth is 30 m with the surrounding waters is being approximately 150 m deep. Rich coral reefs in the crater support a diverse and abundant fish fauna. Coral coverage in this area can reach over 70% of available hard substratum. Boulder, sand and coral habitats are present. The south face of the crater is a nearly vertical wall that extends down to over 100 m (300 ft) depth and is subjected to seasonal south swells. The area is a marine life conservation district and attracts a large number of visitors.

12.4.5.4 Modified Shorelines

This zone is the result of human impact in which natural geomorphology is disrupted or altered by excavation, dredging, filling or construction of sea walls (Fig. 12.3c). Such zones are commonly encountered throughout Hawaii due to land fills, construction of harbors, sea walls and modification of beached for recreational. Sand and mud habitats are generally present along the bottom of dredged areas. Vertical dredged faces or boulder rip-rap are common habitats along the margins. These often become coral habitats such as has occurred at Honokahau Harbor on the Island of Hawaii (Maragos 1991) and Maalaea Harbor on Maui (Jokiel and Brown 1998).

12.4.6 Habitat Mapping

Rohmann et al. (2005) estimated a coral reef habitat area of 1,231 km^2 within 10 fathom (18 m) depth curve for the MHI. Within this area a hierarchical classification scheme was created by Coyne et al. (2003) to define and delineate reef habitats in the MHI. The classification scheme was designed to be as consistent as possible with NOS's coral reef mapping in the Florida Keys and Caribbean as well as with existing classification schemes for the Pacific. The minimum mapping unit (MMU) was set at 1 acre for visual photo-interpretation based on time and resource constraints. The hierarchical scheme allows users to expand or collapse the thematic detail to suit their needs. Habitat polygons smaller than the MMU can be delineated or habitat polygons already delineated using this scheme can be further attributed with other information such as dominant species of coral. The classification scheme defines the benthic communities on the basis of two attributes: large geographic "zones" which are composed of smaller "habitats". Zone refers only to benthic community location and habitat refers only to substrate and/or cover type. Eleven mutually exclusive zones were identified from land to open water corresponding to typical insular shelf and coral reef geomorphology. These zones include: land, vertical wall, shoreline intertidal, lagoon, reef flat, back reef, reef crest, fore reef, bank/shelf, bank/shelf escarpment, channel, and dredged. Zone refers only to each benthic community's location and does not address substrate or cover types. For example, the lagoon zone may include patch reefs, sand, and seagrass beds. Twenty-seven distinct and non-overlapping habitat types were identified that could be mapped by visual interpretation of remotely collected imagery. Habitat refers only to substrate and/or cover type and does not address location or depth. Habitats are defined in a collapsible hierarchy ranging from four broad classes (unconsolidated sediment, submerged vegetation, coral reef and hardbottom, and other), to more detailed categories (e.g. emergent vegetation, seagrass, algae, individual patch reefs, uncolonized volcanic rock), to patchiness of some specific features (e.g. 50–90% cover of macroalgae).

Approximately 60–70% of the total reef area of the MHI was mapped (total = 813 km^2), being limited by factors such as high turbidity, cloud cover or lack of imagery. Approximately 27% of the mapped area was classified as "soft" bottom (21% sand, 6% mud). Macraoalgae communities account for 19% of the total area mapped with 15% falling into the 10–50% macroalgae category and 4% in the 50–90% macroalgae category with only a fraction of a percent in the greater than 90% coverage category. The remaining hard bottom in various categories accounts for most of the remaining area (50% total). Hard bottom areas with less than 10% coral cover classified as "uncolonized" and comprise a total of 29% of the mapped area with 13% as uncolonized pavement, 12% as volcanic rock/boulder and 4% as uncolonized pavement with channels. Areas with greater than 10% coral cover were classified as "colonized" and account for 21% of the total distributed over 5 subcategories. On a broad scale for the entire MHI this translates roughly into 27% soft bottom, 19% algal communities, 29% uncolonized (less than 10% coral), 21% colonized and the remaining 14% in various other categories such as emergent vegetation, fishponds, man-made structures, etc. Average reef coral coverage for 152 reef stations measured on hardbottom substrate throughout the MHI was 21%, with six species accounting for most of the coverage (Jokiel et al. 2004a). The six dominant species were: *Porites lobata* (6%), *Porites compressa* (5%), *Montipora capitata* (4%), *Montipora patula* (3%), *Pocillopora meandrina* (2%) and *Montipora flabellata*.

12.5   Human Impact on Local Coral Reefs and the Present Status of Reef Health

In general, Hawaii's coral reefs are in better condition than many other reefs around the world. Coral ecosystems in the MHI are in fair to excellent condition, but are threatened by continued population growth, over fishing, runoff, and development (Friedlander et al. 2004, 2005). There is clear evidence of overexploitation of many food fishes and invertebrates. Introduced aquatic alien species have an impact on the structure and function of Hawaii's reefs and may out-compete endemic species. Human activity has already taken a toll on the reefs of the populated high islands. Jokiel et al. (2004) showed that human population within 5 km of a reef had a negative relationship in Hawaii with coral cover, diversity and species richness, suggesting that anthropomorphic stressors are important contemporary forces shaping Hawaii coral reef community structure along with natural factors. Also there was a significant relationship between reef condition and degree of management protection (Friedlander et al. 2003; Jokiel et al. 2004).

The unpopulated islands and reefs of the NWHI have fared better than the MHI. For example, results of a recent analysis of "reef health and reef value" indicate that the "worst island" of the NWHI ranks with the "best island" of the MHI (Jokiel and Rodgers 2007). The quantitative numerical ranking devised in this study is based on extensive data on fish, corals and endangered species. The result mirrors the personal experience of marine biologists and others who visit the area and note how different the NWHI islands appear to be compared to the MHI in terms of biological abundance and diversity on the shallow coral reefs

12.5.1   Land Use and Pollution

Areas of reef decline appear to be concentrated on islands with high human population or in areas suffering from extensive land runoff and sedimentation (Jokiel et al. 2004). Reefs receiving high terrigenous runoff contain sediment deposits with high organic content. Spatial analysis shows an inverse relationship between percent organics and coral species richness and diversity (Jokiel et al. 2004), suggesting that organic addition is a negative factor on Hawaiian coral reefs. Improper land use increases sedimentation and freshwater runoff. Major anthropogenic impacts such as increased sedimentation and nutrification dominate Hawaiian reef environments where waves are not a major controlling factor (Dollar and Grigg 2004). These environments are typically bays and lagoons that do not receive sufficient wave energy to flush fine sediments from the system. Thus we observe a paradox that areas vulnerable to storm waves are less vulnerable to storm floods and areas impacted by storm floods are less vulnerable to storm waves. Large-scale sugar cane and pineapple agriculture, which periodically exposes land to erosion, is being phased out throughout Hawaii and may result in a decrease in sediment delivery to the ocean. However, many low-lying coastal areas that were once wetlands and flood plains have been altered and continued development of these areas is underway. These low-lying areas once served as settling basins and filters, removing sediments and nutrients from runoff before it entered the ocean. Development increases the amount of impervious surface and causes increased runoff which is generally diverted to storm drain systems. The drains transport sediment, trash, and chemical pollutants directly into coastal waters. As coastal areas are developed, the floodplains are filled, storm drains are constructed, and streams are "channelized", resulting in more sediment being delivered to the reefs. More coastal construction is planned in order to accommodate new large cruise ships, container ships, and an inter-island ferry. Harbor improvements involve dredging which can influence adjacent reefs.

12.5.1.1 Anthropogenic Increases in Sedimentation

Historically the major cause of erosion, runoff and accelerated sedimentation on Hawaiian coral reefs has been overgrazing on watersheds. Roberts (2001) has reviewed the importance of this process on the reefs of south Molokai. Serious overgrazing by feral ungulates (pigs, goats, deer) continues to cause severe damage to watersheds on Molokai, Lanai, west Maui and the north coast of Kauai. A serious overgrazing problem over the past two

centuries led to massive erosion on the island of Kahoolawe. The Kahoolawe situation was corrected with the complete eradication of over 20,000 goats in 1990 (Jokiel et al. 1993b). Elimination of the goats and efforts to reestablish vegetation on the island appear to be having a positive effect on the reefs. Currently sediment deposits are being winnowed off the reefs by wave action faster than new sediments are being deposited.

Muddy runoff pollution of coral reefs from development sites is a frequent occurrence in Hawaii. In 1996, rivers of mud filled Maalaea harbor on Maui, causing hundreds of thousands of dollars in damage. In 2000, a torrent of mud flowed off acres of land graded for a golf course just north of Kealakekua Bay on the island of Hawaii. Similar incidents took place off Lanai in 2002.

12.5.1.2 Sewage

Starting in the early 1960s, raw sewage discharged into the south basin of Kaneohe Bay, Oahu had a dramatic effect on the reefs (Maragos 1972; Banner 1974; Smith et al. 1981; Hunter and Evans 1993). High nutrient levels led to blooms of phytoplankton which reduced water transparency and blocked light to the photosynthetic benthos. Massive mats of the "green bubble algae" overgrew and choked out living corals. The benthic community became dominated by macro-algae and filter feeding invertebrates. Sediments became anoxic and seaweed washed ashore to form large rotting berms of organic matter. Removal of sewage outfalls in Kaneohe Bay in 1979 led to dramatic decrease in nutrient levels, turbidity and phytoplankton abundance (Smith et al. 1981) and a rapid recovery of reef coral populations (Maragos et al. 1985). By 1983 coral coverage had more than doubled from 12% to 26% (Hunter and Evans 1993). However, initial planning in the early 1960s should have placed outfalls in deep water outside the bay, avoiding the impact and cost to relocate them in the late 1970s.

A major reef kill occurred in Kaneohe Bay in May 1965 due to heavy rains (Banner 1968). However, conditions of heavy sewage pollution prevented recovery of the reefs until after sewage abatement in 1979. The same coral reefs were subjected to a similar reef kill in late 1987, but showed substantial recovery within 5 years (Jokiel et al. 1993a). It appears that coral reefs can recover quickly from major natural disturbances, but not under polluted conditions.

A recurring problem in Hawaii occurs when storm floods overwhelm wastewater treatment facilities. For example, a recent storm flood event in early January 2004 caused sewage spills at 14 locations on Oahu and forced the closing of beaches off Honolulu, Kailua and Waimanalo (Hoover 2004). Another storm flood in early February 2004 again resulted in wastewater spills and beach closures (Honolulu Advertiser 2004).

12.5.1.3   Dredging, Filling and Other Shoreline Construction.

Much of Hawaii's shoreline has been modified by human activity. Notable examples include Kaneohe Bay, Oahu, Hawaii. Major dredging activities in the late 1930s and early 1940s removed reefs and patch reefs to create sea plane runways and cut a deep ship channel (Devaney et al. 1982). The 1972 Reef Runway Project at Honolulu International Airport (Fig. 12.3b) involved the dredging and fill of some 11 million cubic meters of coral reef material, impacting over 480 ha of coral reef (Chapman 1979).

12.5.2   Fisheries

Management of the biological resources of Hawaiian coral reefs is primarily the responsibility of the State of Hawaii Division of Aquatic Resources (DAR), but with many overlapping areas of responsibilities with other agencies. DAR utilizes several management tools including full or partial closure of a reef area as a marine protected area (MPA), rotational and seasonal closures, restrictions on fishing gear or methods, size and bag limits, and rules preventing the take of certain species.

Even though it is likely that a much smaller proportion of the population presently fishes relative to ancient times, marine resources in Hawaii have steadily declined over at least the last century (Maly and Maly 2004; Shomura 1987). Comparison of fish abundances in the Main Hawaiian Islands with those of the relatively unexploited Northwestern Hawaiian Islands also points to major fishery declines in the populated islands (Friedlander and DeMartini 2002). Early in the century Jordan

and Evermann (1902) noted that the fisheries of Honolulu were declining rapidly due to localized overfishing. In 1927 it was reported that the fish fauna of Hawaiian reefs was much less abundant than several decades earlier and many common species had become rare (Jordan et al. 1927). Declining marine resources were acknowledged again by resource managers in the 1950s when they reported that desirable food and game fishes were on a "declining trend and have deteriorated to such an extent that the need for sound conservation measures is urgent" (Division of Fish and Game 1956). Fishermen and other ocean users are well aware of declining reef resources. Surveys of both commercial and noncommercial fishers (Harman and Katekaru 1988; Division of Aquatic Resources 1998) have documented this perception. In the 1998 survey 57% of respondents felt inshore fishing was now poor to terrible. Over fishing is most often cited as the prime cause of resource depletion (Division of Aquatic Resources 1998). Increased fishing pressure is due to increased human population, introduction of new fishing technology and loss of traditional conservation practices (Lowe 2004; Birkeland and Friedlander 2002).

Perhaps the major factor behind the historical decline has been improved fishing techniques (gill nets, skin diving and scuba equipment, geographic positioning systems, power boats, sonar fish finders). The introduction of inexpensive monofilament gill nets enables further exploitation of fish stocks that are already depleted and stocks in deeper water that had previously escaped overharvesting (Clark and Gulko 1999).

Recreational and subsistence fishers in Hawaii are not licensed and have no reporting requirements, so data on their catch are lacking. The possibility of under-reporting by the commercial fishermen further increases the uncertainty of catch statistics for the state. The recreational and subsistence catch use a wider range of fishing techniques and targets a wider range of species than the commercial fishermen and is probably equal to the commercial fisheries catch (Everson and Friedlander 2004). The commercial catch underwent a 70% decline between 1950 and 2002 (Zeller et al. 2005).

It is against the law in Hawaii to take or have stony coral and "live rock" (marine substrate with live attached organisms). This law is being enforced as evidenced by numerous convictions for violations. However, the collecting of aquarium fish and certain invertebrates is allowed. A relatively small number of species dominates the aquarium trade catch with only ten species constituting over 70% of the total (Walsh et al. 2004). The most commonly caught fish species include surgeonfishes, butterflyfishes, and wrasses. The yellow tang (*Zebrasoma flavescens*) accounts for nearly 40% of the total catch. Feather duster worms, hermit crabs, and shrimp are the main targets among the invertebrates. The commercial aquarium fishery in Hawaii has grown into one of the state's major inshore fisheries, with landings of over 708,000 specimens with a reported value of over $1 million (Walsh et al. 2004). Cesar et al. (2002) estimated industry gross sales at over $3 million. There is a general belief that aquarium fishery catch is underreported throughout the state. Such collecting activities in Hawaii can deplete targeted species (Tissot and Hallacher 2003). In response to the growing problem, a network of fish replenishment areas was established on the Kona coast on the island of Hawaii in 2000. Within 4 years there were increases in the abundance of several targeted species, and the overall value of the fishery reached an all-time high (Walsh et al. 2004).

12.5.3 Alien Species

Increasing human population and resulting human activity have led to both accidental and intentional species introductions. Although most of these don't survive, a few persistent species have become a source of serious ecological and economic impacts to the state. In the MHI, 343 alien marine species have been documented and inventoried (Eldredge and Carlton 2002). Invertebrate species dominate with 287 species, followed by algae (24), fishes (20), and flowering plants (12). The mechanisms of introduction and the number of estimated introductions into Hawaii are as follows: Hull fouling (212), solid ballast (21), ballast water (18), intentional release for fisheries (18), parasites associated with alien introduced species (8), organisms associated with commercial oyster shipments (7) and aquarium release (3). A more comprehensive summary of alien species in Hawaii can be found in Godwin et al. (2006). None of these species are viewed as desirable by the public or by marine

managers. For example, the blue-lined snapper or Ta'ape, (*Lutjanus kasmira*) was introduced from the Marquesas in 1958 and although only 3,200 Ta'ape were released on the island of Oahu, they have increased their range to include the entire Hawaiian archipelago. The fish has become locally abundant but is not seen as a desirable food fish and might be displacing locally more desirable fish. Another example is the soft coral *Carijoa riisei*. This species was originally observed in Pearl Harbor and is thought to have arrived through ship hull fouling or in ballast water. In its native Caribbean, it is found in shallow waters as part of the fouling community on pier pilings. Initially, it was not considered a threat since it was thought to be restricted to such underutilized low-light habitats. *C. riisei* has since expanded its range to all islands in the MHI chain. Results from a 2001 found *C. riisei* had spread into waters up to 110 m deep and is competing with the native black corals. They both feed and compete for the same zooplankton and are competing for space. *C. riisei* can grow up to 12 times faster than the valuable black coral. This species blankets everything in its path and drastically reduces the biodiversity of the area. Grigg (2003) has reported that this species has completely decimated black coral beds in areas in the deep trench (75–110 m) between West Maui and Lanai.

12.5.4 Disease

All organisms are subject to disease, and coral reef organisms are no exception. Interest in the diseases of corals has grown tremendously due to serious outbreaks of coral disease in many parts of the world (Richardson 1998). Environmental stress renders corals more prone to disease, as is the case with other organisms. Reef corals in the main Hawaiian Islands have not experienced the massive epidemic disease outbreaks reported from other areas. Baseline surveys show low prevalence of disease (Hunter 1999; Work and Rameyer 2001).

Disease is a major cause of concern for population of the endangered Hawaiian green sea turtle. Turtles are being affected by fibropapillomatosis, which causes large external and internal tumors and occurs in 40–60% of observed Hawaiian turtles (Balazs and Pooley 1991). This may pose a significant threat to the long-term survival of the species.

An alpha-herpes virus appears to be the cause (Lu et al. 2004).

12.5.5 Crown-of-thorns Starfish

There has been a long standing suspicion that outbreaks of crown-of-thorns starfish, *Acanthaster planci,* are related to human activity (Birkeland and Lucas 1990). Such outbreaks have not been a major problem in Hawaii. The only documented outbreak occurred during 1969–1970, a large aggregation estimated to consist of 20,000 *A. planci* was observed off south Moloka'i (Branham et al. 1971). The State of Hawaii Department of Fish and Game undertook extensive surveys and eradication efforts after discovery of the infestation (Onizuka 1979). Divers killed a total of approximately 26,000 starfish between 1970 and 1975 by injecting them with ammonium hydroxide. Additional surveys were conducted throughout the State of Hawaii, but no other infestations were detected at that time, nor have any been found since.

12.5.6 Landmark Case Study – Illegal Grading and Coral Reef Damage

A landmark case in Hawaiian coral reef protection occurred as the result of damage to a pristine coral reef at Pilaa in a secluded area on the northeast coast of Kauai. Jimmy Pflueger, a prominent figure in Hawaii, conducted grading without permits on his 378 acre property which included extensive massive grading of a coastal plateau, building a road just above the beach and creating a 40-ft cliffside cut. The grading work did not include measures to control storm water in case of heavy rain. During a rainstorm on November 26, 2001, tons of graded mud slid from the hillside onto the beach, engulfing a waterfront private home and spread across the white sand beach and onto the reefs. In continuing rains, mud ran repeatedly into the sea. Studies showed significant damage to the reef which included extensive coral mortality, high turbidity, high percentages of mud in reef sands, anoxic substratum and formation of algal mats over the reef (Jokiel et al. 2002; Jokiel and Brown 2004b).

Legal action against Pflueger was initiated by US Department of Justice; the Environmental Protection Agency; Hawaii Department of Land and Natural Resources, State Department of Health; Kauai County; and Earthjustice, representing the Limu Coalition and Kilauea Neighborhood Association. Pflueger was sentenced to a $500,000 fine after pleading guilty to ten felony water pollution counts and a county fine of $310,000 for coastal zone violations. He also agreed to pay nearly $8 million to settle claims under the federal Clean Water Act. The settlement is the largest storm-water settlement in the country for violations at a single site by a single landowner according to Wayne Nastri, regional administrator for the U.S. Environmental Protection Agency, which enforces the Clean Water Act (Leone 2006). A lawsuit against Pflueger by neighbors whose property was damaged by the landslide is still pending.

The response at all levels of government as well as by community groups in defense of the reef was noteworthy. However, a major precedent in the protection of Hawaiian reefs was set the State of Hawaii DLNR when they determined that the reef had been damaged in the amount of $4 million and ordered payment together with administrative costs to the State of Hawaii special land and development fund. The ruling was subsequently appealed and upheld in the Hawaii State Circuit Court. This action marks the first time that the State of Hawaii has taken action against a developer responsible for damaging reef resources. Further, the ruling established the dollar value of coral reefs in Hawaii and a mechanism to recover the dollar value of the lost resources.

12.5.7 Reef Restoration and Mitigation

Numerous mitigation and restoration projects have been conducted in Hawaii as discussed in detail by Jokiel et al. (2006). Reef restoration efforts using transplanted corals have failed along exposed coastlines due to destructive storm waves and other factors. However, there have been transplant successes in sheltered embayments. One of the major conclusions of the Jokiel et al. (2006) review is that the cost of reef repair and coral transplantation is generally high but effectiveness is usually very low. Protection and conservation, rather than restoration of damaged reefs, is the preferred priority.

Nevertheless, there have been a number of successful mitigation efforts in Hawaii. Mitigation of sewage pollution in Kaneohe Bay as discussed in

Section 12.5.1.2 is one example. The efforts taken at Kahoolawe discussed in Section 12.5.1.1 is another. An extensive area of reef off Kahe Point, Oahu, was damaged by thermal effluent from a power generation station (Jokiel and Coles 1974). When generating capacity of the plant was increased from 270 to 360 MW, the area of dead and damaged corals increased from 0.38 to 0.71 ha. The requirement for plant expansion and further increases in discharge led to installation of a new outfall pipe in 1976 in deeper offshore waters. This pipe is over 100 m in length, is protected from wave action by heavy rock riprap, and now carries heated effluent offshore and away from the reef. Colonization of the damaged area and the riprap was dramatic, with coral colonization rates among the highest reported in the literature (Coles 1984). Discharge of silt-laden water and crushed cane (bagasse) from sugar mills along Hawaii's Hamakua coastline for many decades caused extensive damage to coral reefs (US Environmental Protection Agency 1971). Termination of discharges led to a rapid clearing of the sediment and bagasse waste by wave action and subsequent regeneration of coral reefs in the former discharge zones (Grigg 1985).

Acknowledgement This work partially supported by USGS cooperative agreement 04WRAG001 and EPA Star Grant R832224-010.

References

Aeby GS, Kenyon JC, Maragos JE, Potts DC (2003) First record of mass coral bleaching in the Northwestern Hawaiian Islands. Coral Reefs 22:256

Allen JA (1998) Mangroves as alien species: the case of Hawaii. Glob Ecol Biogeogr Lett 7:61–71

Bach W, Daniels AP (1973) Climate, pp 53–62. In: Armstrong RW (ed) Atlas of Hawaii. University of Hawaii Press, Honolulu., HI, 222 pp

Banner AH (1968) A fresh water "kill" on the coral reefs of Hawaii. Hawaii Inst Mar Biol Tech Rep 15:1–29

Banner AH (1974) Kaneohe Bay, Hawaii: urban pollution and a coral reef ecosystem. Proc 2nd Int Coral Reef Symp 2:685–702

Baker JD, Johanos TC (2005) Distribution and abundance of Hawaiian Monk Seals in the main Hawaiian Islands. Values are from 2001 survey reported in (http://www.mmc.gov/reports/workshop/pdf/baker.pdf)

Balazs GH, Chaloupka M (2004a) Thirty-year recovery trend in the once depleted Hawaiian green sea turtle stock. Biol Cons 117:491–498

Balazs GH, Chaloupka M (2004b) Spatial and temporal variability in somatic growth of green sea turtles (*Chelonia mydas*) resident in the Hawaiian Archipelago. Mar Biol 145:1043–1059

Balazs GH, Pooley SG (eds) (1991) Research plan for marine turtle fibropapilloma. NOAA Technical Memorandum NMFSSWFSC-156,113 pp

Barkley RA (1972) Johnston Atoll 's wake. J Mar Res 30:201–216

Birkeland C, Friedlander AM (2002) The importance of refuges for reef fish replenishment in Hawaii. The Hawaii Audubon Society, 19 pp

Birkeland C, Lucas J (1990) *Acanthaster planci*: major management problem of coral reefs. CRC, Boca Raton, FL, 272 pp

Bothner MH, Reynolds RL, Casso MA, Storlazzi CD, Field ME (2006) Quantity, composition, and source of sediment collected in sediment traps along the fringing coral reef off Molokai, Hawaii. Mar Pollut Bull 52:1034–1047

Branham JM, Reed SA, Bailey JH, Caperon J (1971) Coral-eating sea stars *Acanthaster planci* in Hawai'i. Science 172:1155–1157

Brock RE (1996) A study of the impact of hurricane Iniki on coral communities at selected sites in Mamala Bay, Oahu, Hawaii. Project report PR-96–09. University of Hawaii Water Resources Research Center, Honolulu, HI

Brock RE, Bailey-Brock JH (1998) An unique anchialine pool in the Hawaiian Islands. Int Rev Hydrobiol 83:65–75

Cesar H, van Beukering P (2004) Economic valuation of the coral reefs of Hawaii. Pac Sci 58:231–242

Cesar H, van Beukering P, Pintz S, Dierking S (2002) Economic valuation of the coral reefs of Hawaii. Final report. Hawaii Coral Reef Initiative Research Program, 117 pp

Chapman GA (1979) Honolulu International Airport reef runway post-construction environmental impact report, Vol. 2. Technical report to the Department of Transportation, Air Transportation Facilities Division, State of Hawaii by Parsons Hawaii, 137 pp

Clague DA, Dalrymple GB (1994) Tectonics, geochronology and origin of the Hawaiian Emperor Volcanic Chain, pp 5–40. In: Kay EA (ed) A Natural History of the Hawaiian Islands. University of Hawai'i Press, 274 pp

Clark AM, Gulko DA (1999) Hawaii's state of the reef report, 1998. State of Hawaii Department of Land and Natural Resources, Honolulu, HI, 41 pp

Coles SL (1984) Colonization of Hawaiian reef corals on new and denuded substrata in the vicinity of a Hawaiian power station. Coral Reefs 3:123–130

Coles SL, Brown BE (2003) Coral bleaching -capacity for acclimatization and adaptation. Adv Mar Biol 46:183–223

Coles SL, Jokiel PL (1992) Effects of salinity on coral reefs. In: Connell DW, Hawker DW (eds) Pollution in tropical aquatic systems. CRC, London pp 170–191

Coles SL, Jokiel PL, Lewis CR (1976) Thermal tolerance in tropical versus subtropical Pacific reef corals. Pac Sci 30:156–166

Connell JH (1978) Diversity in tropical rainforests and coral reefs. Science 199:1302–1310

Coyne MS, Battista TA, Anderson M, Waddell J, Smith W, Jokiel P, Kendell MS, Monaco ME (2003) NOAA Technical Memorandum NOS NCCOS CCMA 152 (On-line). Benthic habitats of the Main Hawaiian Islands. National Oceanic and Atmospheric Administration, Silver Spring, MD, 48 pp and CD-ROM

Curtis G (1998) Tsunamis. In: Juvik S, Juvick JO (eds) Atlas of Hawai'i, 3rd edition. University of Hawaii Press, Honolulu, HI, pp 76–78

Darwin C (1976) The structure and distribution of coral reefs. (Reprinted from Geological observations on coral reefs, volcanic islands, and on South America published by Smith, Elder and Co., London 1851.) University of California Press, Berkeley, CA, 214 pp

Daws G (1968) Shoal of time; history of the Hawaiian Islands. University of Hawaii Press, Honolulu, HI, 494 pp

Devaney D, Kelly MM, Lee PJ, Motteler LS (1982) Kane'ohe a history of change. The Bell Press, Honolulu, HI

Division of Aquatic Resources (1998) Fishing survey summary report. Division of Aquatic Resources, State of Hawai'i, 9 pp

Division of Fish and Game (1956) Annual report of the Board of Agriculture and Forestry, Territory of Hawaii. FY July 1, 1955; June 30, 1956

Dollar S (1982) Wave stress and coral reef community structure in Hawaii. Coral Reefs 1:71–81

Dollar SJ, Grigg RW (2004) Anthropogenic and natural stresses on selected coral reefs in Hawaii: a multi-decade synthesis of impact and recovery. Pac Sci 58:281–304

Dollar SJ, Tribble GW (1993) Recurrent storm disturbance and recovery: a long-term study of coral communities in Hawaii. Coral Reefs 12:223–233

Dunlap WC, Chalker BE (1986) Identification and quantification of near-UV absorbing compounds (S-320) in a hermatypic scleractinian. Coral Reefs 5:155–159

Edmondson CH (1933) Reef and shore fauna of Hawaii. Bernice P. Bishop Museum Special Publication 22. Honolulu, HI, 295 pp

Eldredge LG, Carlton JT (2002) Hawai'i marine bioinvasions: a preliminary assessment. Pac Sci 56:211–212

Everson A, Friedlander AM (2004) Catch, effort, and yields for coral reef fisheries in Kaneohe Bay, Oahu and Hanalei Bay, Kauai : comparisons between a large urban and a small rural embayment. In: Friedlander AM (ed) Status of Hawaii's coastal fisheries in the new millennium. Proceedings of a symposium sponsored by the American Fisheries Society, Hawaii Chapter, Honolulu, HI, pp 110–131

Flament P, Kennan S, Lumpkin R, Sawyer M, Stroup E (1998) The Ocean. In: Juvik S, Juvick JO (eds) Atlas of Hawai'i, 3rd edition. University of Hawaii Press, Honolulu, HI, pp 82–86

Friedlander A, Aeby G, Brainard R, Brown E, Clark A, Coles S, Demartini E, Dollar S, Godwin S, Hunter C, Jokiel P, Kenyon J, Kosaki R, Maragos J, Vroom P, Walsh W, Williams I, Wiltse W (2004) Status of coral reefs in the Hawaiian Archipelago. In: Wilkinson C (ed) Coral reefs of the world, Vol. 2, Australian Institute of Marine Science, Townsville, pp 411–430

Friedlander A, Aeby G, Brown E, Clark A, Coles S, Dollar S, Gulko D, Hunter C, Jokiel P, Smith J, Walsh W, Williams I, Wiltse W (2005) The status of the coral reef ecosystems of the Main Hawaiian Islands. In: Waddell JE (ed) The status of the coral reef ecosystems of the United States and Pacific Freely Associated States. National Oceanic and Atmospheric Agency, Silver Spring, MD, pp 222–269

Friedlander AM, DeMartini EE (2002) Contrasts in density, size, and biomass of reef fishes between the northwestern and the main Hawaiian Islands : the effects of fishing down apex predators. Mar Ecol Prog Ser 230:253–264

Friedlander AM, Brown EK, Jokiel PL, Smith WR, Rodgers KS (2003) Effects of habitat, wave exposure, and marine protected area status on coral reef fish assemblages in the Hawaiian archipelago. Coral Reefs 22:291–305

Godwin S, Rodgers KS, Jokiel PL (2006) Reducing potential impact of invasive marine species in the northwestern Hawaiian Islands marine national monument. Report submitted to northwest Hawaiian Islands Marine National Monument Administration, Honolulu, HI, August 2006, 66 pp http://cramp.wcc.hawaii.edu/

Gosline WA, Brock VE (1960) Handbook of Hawaiian fishes Honolulu,. University of Hawaii Press, Honolulu, HI, 372 pp

Grigg RW (1981) *Acropora* in Hawaii Part 2. Zoogeography. Pac Sci 35:15–24

Grigg RW (1983) Community structure, succession and development of coral reefs in Hawai'i. Mar Ecol Progr Ser 11:1–14

Grigg RW (1985) Hamakua coast sugar mills revisited: an environmental impact analysis in 1983. University of Hawaii, Sea Grant Pub. No. UNIHI-SEAGRANT-TR-85–02, Honolulu, HI, 81 pp

Grigg RW (1998) Holocene coral reef accretion in Hawaii: a function of wave exposure and sea level history. Coral Reefs 17:263–272

Grigg RW (2003) Invasion of a deep black coral bed by *Carijoa riisei* off Maui, Hawai'i. Coral Reefs 22:121–122

Grigg RW (2006) Depth limit for reef building corals in the Au'au Channel, S.E. Hawaii. Coral Reefs 25:77–84

Grigg RW, Maragos JE (1974) Recolonization of hermatypic corals on submerged lava flows in Hawaii. Ecology 55:387–395

Grigg RW, Wells J, Wallace C (1981) *Acropora* in Hawaii. Part 1. History of the scientific record, systematics and ecology. Pac Sci 35:1–13

Harman RF, Katekaru AZ (1988) 1987 Hawai'i commercial fishing survey. Division of Aquatic Resources, State of Hawai'i, 71 pp

Hiatt RW (1955) The Hawaii marine laboratory. AIBS Bull 5(1):26–27

Holthus P (1986) Coral reef communities of Kaneohe Bay, Hawaii: an overview. In: Jokiel PL, Richmond RH, Rogers RA (eds) Coral reef population biology. Hawaii Inst Mar Biol Tech Rep 37:19–34

Honolulu Advertiser (2004) City dealing with wastewater, sewage spills. Honolulu Advertiser 9 Feb 2004 edition

Hoover W (2004) Oahu beaches, streams spoiled by sewage spills. Honolulu Advertiser 4 Jan 2004 edition

Hourigan TF, Reese RS (1987) Mid-ocean isolation and the evolution of Hawaiian reef fishes. Trends Ecol Evol 2:187–191

Hunter CL (1999) First record of coral disease and tumors on Hawaiian reefs. In: Maragos JE, Gober-Dunsmore R (eds) Proceedings of the Hawaii coral reef monitoring workshop, June 7–9, 1998, pp 73–98

Hunter CL, Evans CW (1993) Reefs of Kaneohe Bay, Hawai'i: two centuries of western influence and two decades of data. pp 339–345. In: Ginsburg RN (ed) Global aspects of coral reefs. University of Miami, Rosenstiel School of Marine and Atmospheric Science, pp 339–345

Jokiel PL (1980) Solar ultraviolet radiation and coral reef epifauna. Science 207:1069–1071

Jokiel PL (1987) Ecology, biogeography and evolution of corals in Hawaii. Trends Ecol Evol 2:179–182

Jokiel PL (2004) Temperature stress and coral bleaching. In: Rosenberg E, Loya Y (eds) Coral health and disease. Springer, Heidelberg, pp 401–425

Jokiel PL (2006) Impact of storm waves and storm floods on Hawaiian reefs. Proc 10th Int Coral Reef Sym:390–398

Jokiel PL, Brown EK (1998) Coral baseline survey of Ma'laea Harbor for light draft vessels, Island of Maui. Final report for DACW83–96-P-0216. US Army Engineer District, Environmental Resources Branch, Fort Shafter, HI, 44 pp

Jokiel PL, Brown EK (2004a) Global warming, regional trends and inshore environmental conditions influence coral bleaching in Hawaii. Global Change Biol 10:1627–1641

Jokiel PL, Brown EK (2004b) Reef coral communities at Pila'a reef. Results of the 2004 survey Hawaii coral reef assessment and monitoring program report, May 20. Kaneohe, HI, 23 pp

Jokiel PL, Coles SL (1974) Effects of heated effluent on hermatypic corals at Kahe Point, Oahu. Pac Sci 28:1–18

Jokiel PL, Coles SL (1977) Effects of temperature on the mortality and growth of Hawaiian reef corals. Mar Biol 43:201–208

Jokiel PL, Coles SL (1990) Response of Hawaiian and other Indo-Pacific reef corals to elevated temperatures associated with global warming. Coral Reefs 8:155–162

Jokiel PL, Cox EF (2003) Drift pumice at Christmas Island and Hawaii: evidence of oceanic dispersal patterns. Mar Geol 202:121–133

Jokiel PL, Guinther EB (1978) Effects of temperature on reproduction in the hermatypic coral *Pocillopora damicornis*. Bull Mar Sci 28:786–789

Jokiel PL, Rodgers KS (2007) Ranking coral ecosystem "health" and "value" for the islands of the Hawaiian Archipelago. Pac Conserv Biol 13:60–68

Jokiel PL, York RH (1982) Solar ultraviolet photobiology of the reef coral *Pocillopora damicornis* and symbiotic zooxanthellae. Bull Mar Sci 32:301–315

Jokiel PL, Hunter CL, Taguchi S, Watarai L (1993a) Ecological impact of a fresh water "reef kill" in Kaneohe Bay, Oahu, Hawaii. Coral Reefs 12:177–184

Jokiel PL, Cox EF, Crosby MP (1993b) An evaluation of the nearshore coral reef resources of Kahoolawe, Hawaii. Final report for co-operative agreement NA27OM0327. Hawaii Institute of Marine Biology, Kaneohe, HI 185 pp

Jokiel PL, Hill E, Farrell F, Eric K, Rodgers K (2002) Reef coral communities at Pila'a reef in relation to environmental factors. Hawaii coral reef assessment and monitoring program report, December 12. Kaneohe, HI, 100 pp

Jokiel PL, Brown EK, Friedlander A, Rodgers SK, Smith WR (2004) Hawaii coral reef assessment and monitoring program: spatial patterns and temporal dynamics in reef coral communities. Pac Sci 58:145–158

Jokiel PL, Kolinski, SP,Naughton J, Maragos JE (2006) Review of coral reef restoration and mitigation in Hawaii and the US- affiliated Pacific Islands. In: Precht WF (ed) Coral reef restoration handbook – the rehabilitation of an ecosystem under siege. CRC, Boca Raton, FL, pp 271–290

Jones BL, Nakahara RH, Chinn SSW (1971) Reconnaissance study of sediment transported by streams, Island of Oahu. Circular C33, US Geological Survey, Honolulu, HI 45 pp

Jordan DS, Evermann BW (1902) Preliminary report on an investigation of the fishes and fisheries of the Hawaiian Islands. In: Report of the Commissioner for the year ending June 30, 1901. U.S. Commission of Fish and Fisheries, Government Printing Office, Washington, DC, pp 353–380

Jordan DS, Evermann BW, Tanaka S (1927) Notes on new or rare fishes from Hawai'i. Calif Acad Sci Proc 4:649–680

Kahng SE, Maragos JE (2006) The deepest zooxanthellate scleractinian corals in the world? Coral Reefs 25:254

Karl DM (2004) UH and the sea. SOEST report 04–01. University of Hawaii School of Ocean and Earth Science and Technology, Honolulu, HI, 104 pp

Kay EA, Palumbi SR (1987) Endemism and evolution in Hawaiian marine invertebrates. Trends Ecol Evol 2:183–186

Kenyon JC, Aeby GS, Brainard RE, Chojnacki JD, Dunlap MD, Wilkinson CB (2004) Mass coral bleaching on high-latitude reefs in the Hawaiian Archipelago. Proc 10th Int Coral Reef Sym, Okinawa:631–643

Kenyon JC, RE Brainard (2006) Second recorded episode of mass coral bleaching in the Northwestern Hawaiian Islands. Atoll Res Bull 543:505–523

Kobayashi DR (2006) Colonization of the Hawaiian Archipelago via Johnston Atoll : a characterization of oceanographic transport corridors. Coral Reefs 25:407–417

Kuffner I (2001) Effects of ultraviolet (UV) radiation on larval settlement of the reef coral *Pocillopora damicornis*. Mar Ecol Progr Ser 217:251–261

Kuykendall RS, Day AG (1976) Hawaii: a history, from Polynesian kingdom to American state. Prentice-Hall, Englewood Cliffs, NJ, 331 pp

Leone D (2006) Pflueger is fined more than $7.8M. Honolulu Star Bull 11(79) March 20, 2006

Lesser MP, Shick JM (1989) Effects of irradiance and ultraviolet radiation on photoadaptation in the zooxanthellae of *Aiptasia pallida*: primary production, photoinhibition, and enzymatic defenses against oxygen toxicity. Mar Biol 102:243–255

Lowe MK (2004) The status of inshore fisheries ecosystems in the Main Hawaiian Islands at the dawn of the millennium: cultural impacts, fisheries trends and management challenges. In: Friedlander AM (ed) Status of Hawai'i's coastal fisheries in the new millennium. Hawai'i Audubon Society, Second Printing, Honolulu, HI, pp 12–107

Lu Y, Wang Y, Yu Q, Aguirre AA, Balazs GH, Nerurkar VR, Yanagihara R (2004) Detection of herpes viral sequences in tissues of green turtles with fibropapilloma by polymerase chain reaction. Arch Virol 45:1885–1893

Macdonald GA, Abbott AT, Peterson FL (1983) Volcanoes in the sea. University of Hawaii Press, Honolulu, HI, 517 pp

Maragos J (1991) Two decades of coral monitoring surveys following construction of Honokohau Harbor, Hawaii. In: Studies of water quality, ecology, and mixing processes at Honokohau and Kawaihae Harbors on the Island of Hawaii. Oceanic Institute, Makapu Point, Waimanalo, HI, 28 pp

Maragos JE (1972) A study of the ecology of Hawaiian reef corals. Ph.D. dissertation, University of Hawaii, 290 pp

Maragos JE (1995) Revised checklist of extant shallow-water stony coral species from Hawaii (Cnideria: Anthozoa: Scleractinia). Occ Pap Bishop Mus 42:54

Maragos JE (1998) Marine Ecosystems. In: Juvik S, Juvick JO (eds) Atlas of Hawai'i, 3rd edition. University of Hawaii Press, Honolulu, HI, pp 111–120

Maragos JE, Jokiel PL (1986) Reef corals of Johnston Atoll : one of the world's most isolated reefs. Coral Reefs 4:141–150

Maragos JE, Evans C, Holthus P (1985) Reef corals in Kaneohe Bay six years before and after termination of sewage discharges. Proc 5th Int Coral Reef Symp 4:189–194

Maly K, Maly O (2004) Ka Hana Lawai'a a me n? Ko'a o n? kai 'Ewalu. A history of fishing practices and marine fisheries of the Hawaiian Islands. Prepared for the nature conservancy by Kumu Pono Associates LLC, Hilo, HI, 506 pp

Moberly RM (1974) Types and sources of Hawaiian waves, p 48. In: Atlas of Hawaii. University of Hawaii Press, Honolulu, HI, 222 pp

Moore JG, Clague DA, Holcomb RT, Lipman PW, Normark WR, Torresan ME (1989) Prodigious submarine landslides on the Hawaiian ridge. J Geophys Res 94:17465–17484

Onizuka E (1979) Studies on the effects of crown-of-thorns starfish on marine game fish habitat. Final report of project F-17-R-2. State of Hawai'i Department of Fish and Game, Honolulu, HI, 25 pp

Patzert WC, Wyrtki K, Santamore HJ (1970) Current measurements in the central North Pacific Ocean. Hawaii Inst Geophys HIG-70- 31:1–26

Paulay G (1997) Diversity and distribution of reef organisms. In: Birkeland C (ed) Life and death of coral reefs. Chapman & Hall, New York pp 298–353

Presto MK, Ogston AS, Storlazzi CD, Field ME (2006) Temporal and spatial variability in the flow and dispersal of suspended-sediment on a fringing reef flat, Molokai, Hawaii. Est Coast Shelf Sci 67:67–81

Price S (1983) Climate. In Armstrong RW (ed) Atlas of Hawaii, 2nd edition. University of Hawaii Press, Honolulu, HI, 238 pp

Ramage CS (1971) Monsoon meteorology. Academic, pp 112–116

Randall JE (1996) Shore fishes of Hawaii. Natural World Press, Vida, OR, 216 pp

Richardson LL (1998) Coral diseases: what is really known? Trends Ecol Evol 13:438–443

Roberts L (2001) Historical land use, coastal change and sedimentation on South Molokai reefs. In: Saxena N (ed) Recent advances in marine science and technology, 2000. PACON International, Honolulu, HI, pp 167–176

Rogers CS (1990) Responses of coral reefs and reef organisms to sedimentation. Mar Ecol Prog Ser 62:185–202

Rohmann SO, Hayes JJ, Newhall RC, Monaco ME, Grigg RW (2005) The area of potential shallow-water tropical and subtropical coral ecosystems in the United States. Coral Reefs 24(3):370–383

Rooney J, Fletcher C, Grossman E, Engels, Field M (2004) El Nino influence on Holocene Reef Accretion in Hawaii. Pac Sci 58:305–324

Roy KJ (1970) Changes in bathymetric configuration, Kaneohe Bay, Oahu. 1882–1969. Hawaii Institute of Geophysics report 13 70–15. University of Hawaii, 226 pp

Santos SR (2006) Patterns of genetic connectivity among anchialine habitats: a case study of the endemic Hawaiian shrimp Halocaridina rubra on the island of Hawaii. Molec Ecol 15:2699–2718

Scelfo G (1984) Relationship between solar radiation and pigmentation of the coral Montiproa verrucosa and its zooxanthellae. In: Jokiel PL et al. (eds) Coral reef population biology. Hawaii Institute of Marine Biology Technical Report 37, pp 440–451

Schroeder TA (1998) Hurricanes. In: Juvik S Juvick JO (eds) Atlas of Hawai'i, 3rd edition, University of Hawaii Press, Honolulu, HI, pp 74–75

Shomura RS (1987) Hawai'i's marine fisheries resources: yesterday (1900) and today (1986). Southwest fisheries administrative report H-87–21, 15 pp

Shibata K (1969) Pigments and a UV-absorbing substance in corals and a blue-green alga living in the Great Barrier Reef. Plant Cell Physiol 10:325–335

Smith SV, Kimmerer WJ, Laws EA, Brock RE, Walsh TW (1981) Kaneohe Bay Sewage Diversion Experiment: perspectives on ecosystem response to nutritional perturbation. Pac Sci 34:279–402

Stoddart DF (1969) Ecology and morphology of recent coral reefs. Biol Rev 44:433–498

Storlazzi CD, Field ME, Dykes JD, Jokiel PL, Brown E (2002) Wave control on reef morphology and coral distribution: Molokai, Hawaii. Proc 4th Int Symp Waves, pp.784–793

Storlazzi CD, Field ME, Jokiel PL, Rodgers SK, Brown E, Dykes JD (2005) A model for wave control on coral breakage and species distribution: southern Molokai, Hawaii. Coral Reefs 24:43–55

Tissot BN, Hallacher LE (2003) Effects of aquarium collectors on coral reef fishes in Kona, Hawaii. Conserv Biol 17:1759–1768

US Environmental Protection Agency (1971) The Hawaii sugar industry waste study. US Environmental Protection Agency, Region IX, San Francisco, CA, US Government Printing Office Publication, Washington, DC, pp 81–150

Vaughan TW (1907) Recent Madreporaria of the Hawaiian Islands and Laysan. Bull US Natl Mus 59:1–427

Veron JEN (1993) A biogeographic database of hermatypic corals. Species of the central Indo-Pacific, genera of the world. Aust Inst Mar Sci Monogr Ser 10:1–433

Walsh WJ, Cotton SP, Dierking J, Williams ID (2004) The commercial marine aquarium fishery in Hawaii 1976–2003. In: Friedlander AM (ed) Status of Hawaii's coastal fisheries in the new millennium. Proceedings of a symposium sponsored by the American Fisheries Society, Hawaii Chapter, Honolulu, HI, pp 132–159

Wells J (1957) Corals. Geol Soc Am Mem 67:1087–1104

Work T, Rameyer R (2001) Evaluating coral health in Hawaii. US Geological Survey, National Wildlife Health Center, Hawaii Field Station, 42 pp

Zeller D, Booth S, Pauly D (2005) Reconstruction of coral reef- and bottom-fisheries catches for US flag island areas in the Western Pacific, 1950 to 2002. Report to the Western Pacific Regional Fisheries Management Council, August 2005

13
Geology and Geomorphology of Coral Reefs in the Northwestern Hawaiian Islands

John J. Rooney, Pal Wessel, Ronald Hoeke, Jonathan Weiss,
Jason Baker, Frank Parrish, Charles H. Fletcher, Joseph Chojnacki,
Michael Garcia, Russell Brainard, and Peter Vroom

13.1 Introduction

The Northwestern Hawaiian Islands (NWHI) comprise a portion of the middle of the 6,126 km long Hawaiian–Emperor seamount chain, considered to be the longest mountain chain in the world Grigg (1983) (Fig. 13.1). Located in the middle of the North Pacific Ocean, the Hawaiian Islands have been referred to as the most geographically isolated archipelago in the world. The islands are ~3,800 km from the nearest continental landmass, the west coast of North America. The nearest other island, Johnston Atoll, is located almost 900 km southwest of the NWHI. A distance of 1,500 km separates the island of Hawai'i at the southern end of the archipelago from the next nearest island, Kingman Reef in the Line Islands. Kure Atoll at the northwestern end of the Hawaiian Islands is the closest point in the archipelago to the northernmost of the Marshall Islands, 2,000 km to the southwest.

Because of their physical isolation and low levels of marine biodiversity, Hawaiian reefs feature high levels of endemism. Approximately a quarter of the species present are found nowhere else, and the Hawaiian Archipelago has approximately twice as many endemic coral species as any other area its size in the world (Fenner 2005). This unique area is home to over 7,000 marine species (Hawai'i DLNR 2000), the critically endangered Hawaiian monk seal (*Monachus schauinslandi*, Fig. 13.2), and provides nesting grounds for 14 million seabirds and 90% of Hawai'i's threatened green sea turtles (*Chelonia mydas*) (NWHI MNM 2006). Reefs in the NWHI are

dominated by top predators such as sharks and jacks, which make up more than half of the overall fish biomass (Fig. 13.3). In contrast, top predators in the heavily fished main Hawaiian Islands (MHI) make up approximately 3% of the overall fish biomass, which is more typical of coral reefs worldwide (Friedlander and Demartini 2002; Maragos and Gulko 2002).

Although resources in the Northwestern Hawaiian Islands have been exploited by humans since their discovery by ancient Polynesians, their isolation has afforded some protection and helped preserve their coral reef ecosystems. Additionally, most of the atolls and reefs of the NWHI have been afforded some level of environmental protection for almost a century. As a result, the NWHI are the only large-scale coral reef ecosystem on the planet that is mostly intact, a marine wilderness that provides us with insights on what other coral reefs may have been like prior to human exploitation (Fenner 2005).

13.1.1 History and Resource Management

13.1.1.1 The Political History of the NWHI

The first humans to discover and settle the Hawaiian Islands were Polynesian voyagers, possibly from the Marquesas Islands. Evidence from multiple sites indicates that permanent settlements were established in the MHI by at least AD 600, and perhaps several centuries earlier (Kirch 1998). However, it was not until approximately AD 1000 that early Hawaiians first arrived in the NWHI. A

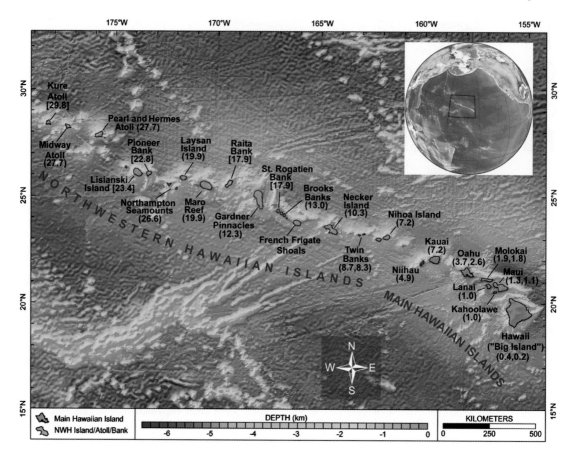

Fig. 13.1. Islands of the Hawaiian Archipelago. The subaerial extents of the main eight islands are shown in green. The prominent islands, atolls, and banks in the NWHI, and the shelves around them are shown in pink, overlain on color-coded bathymetry data from Smith and Sandwell (1997). Numbers next to the islands are their ages in millions of years from Clague (1996) that have been measured (in parentheses), or calculated ages (in brackets). The extents of the map are indicated by the box on the globe in the upper right corner

permanent community was established on Nihoa Island that survived for an estimated 700 years, but disappeared prior to Western European contact, the arrival of Captain Cook, in 1778. Mokumanamana, or Necker Island (Fig. 13.4), was never permanently settled, but remains of temporary habitation and numerous cultural sites are found on the island (Cleghorn 1988). Mokumanamana is located approximately 9 km from the present position of the Tropic of Cancer at the northern limit of the sun's path throughout the year. A thousand years ago however, it was directly in line with the rising and setting of the equinoctial sun and a place of spiritual and ceremonial significance to the ancient Hawaiian culture (NWHI MNM 2006).

Multiple lines of evidence strongly suggest that the settlement of the Hawaiian Islands by Polynesian navigators was not limited to the participants of a single voyage. Voyages between the Hawaiian Islands and the Cook, Society, and Marquesas Islands apparently continued for centuries but stopped after about AD 1200 (Kirch 1998). It seems unlikely that a society with such a strong tradition of voyaging thousands of kilometers across the open ocean would not have found other islands and atolls of the NWHI, which lie in a fairly straight line and are a few hundred kilometers apart or less. In particular, the resource-rich French Frigate Shoals is less than 150 km from Necker Island. However, the ancient Hawaiians apparently

FIG. 13.2. Endangered Hawaiian Monk Seal (*Monachus schaunslandi*) swims around spur and groove structures at Pearl and Hermes Atoll (Photograph by Molly Timmers)

FIG. 13.3. Scientist studies a coral reef at Lisianski Island under the watchful eyes of a shark and an ulua, two of the top predators of the NWHI. Although this is a typical scene in the NWHI, years of fishing pressure make a situation like this quite unusual for most of the world's coral reefs (Photograph by Jean Kenyon)

FIG. 13.4. Mokumanamana, also called Necker Island, was a place of spiritual and cultural significance to the native Hawaiian culture a thousand years ago, and is again today (Photograph by Jean Kenyon)

did not create permanent settlements at any of the other islands, and signs of their visits have not been reported.

Nihoa Island, the southernmost Northwestern Hawaiian Island, was seemingly forgotten prior to its rediscovery in 1788 by Captain Colnett of the British vessel *Prince of Wales* (Rauzon 2001). In 1822, Queen Ka'ahumanu traveled to Nihoa Island and claimed it for the Kamehameha Monarchy, which ruled the Kingdom of Hawai'i from 1795 to 1872. One of her successors, King Kamhameha III, claimed Pearl and Hermes Atoll in 1854. In 1857, King Kamehameha IV sent Captain William Paty to explore the NWHI and later that year claimed the recently discovered islands of Laysan and Lisianski. Finally, in 1886, King David Kalakaua claimed Kure Atoll. A group of local businessmen, predominantly of American and European descent and backed by the US military, overthrew the Kingdom of Hawai'i in 1893. In 1898, the entire Hawaiian Archipelago, including the NWHI, was ceded to the United States (NWHI MNM 2006; Rauzon 2001).

13.1.1.2 Midway – the Most Famous Atoll in the NWHI

Thanks to the pivotal World War II battle fought nearby, Midway Atoll is the best known of the Northwestern Hawaiian Islands. Discovered in 1859

by Captain Brooks of the *Gambia*, it was claimed under the US Guano Act of 1856, making Midway the only atoll or island in the NWHI that does not today belong to the State of Hawai'i (Rauzon 2001). The US Navy began construction of a channel into the lagoon in 1870 and then abandoned the effort until 1940. In the meantime, The Pacific Commercial Cable Company developed Sand Island at Midway as a station for a cable between Guam and San Francisco that was completed in 1903. Pan American Airways built an airport on the island in 1935, and it became a refueling stop for their Clipper floatplanes on the Manila to San Francisco route. These commercial ventures, especially the cable company, imported over 8,000 t of topsoil to Sand Island and planted extensively. They transformed the once mostly desolate and unvegetated sand island into the "garden" that it is today, but introduced hundreds of alien species to the NWHI in the process. In 1939, the Navy decided that an air base at Midway was of national strategic importance and commissioned the Midway Naval Air Station on August 1, 1941. A few months later, directly after the attack on Pearl Harbor in Honolulu on December 7, the station was bombed. The naval battle 6 months later near Midway was the turning point of World War II in the Pacific theater. In 1957 Midway became a key base in the Distant Early Warning line, a radar screen between Alaska and Hawai'i that was an important component of US strategy in the Cold War. Midway's last military function was as an important refueling station during the Vietnam War, before control of the island was turned over to the US Fish and Wildlife Service in 1996 (Rauzon 2001).

13.1.1.3 Exploitation and Management of Natural Resources

Several research expeditions passed through the Archipelago during the period of rediscovery of the NWHI, contributing to base of knowledge about the area. The naturalist James Dana, on US Exploring Expedition, visited the islands in the late 1840s, followed by the British Challenger Expedition between 1872 and 1876. The Albatross Expedition of 1902 mostly dredged deep waters around the Archeiplago, and the Tanager Expedition of 1923–1924 primarily collected specimens and data (Grigg 2006). These expeditions were primarily driven by scientific inquiry, but most other efforts in the area were focused on resource extraction. The previously near-pristine NWHI were heavily exploited in the nineteenth and early twentieth centuries for seabirds, albatross eggs, monk seals, turtles, guano and other resources resulting in widespread ecological damage and causing populations to plummet (NWHI MNM 2006; Rauzon 2001). The damage was compounded by the introduction of a number of plants, insects, and other animals.

Recognizing the importance of the islands to seabirds, President Theodore Roosevelt put Midway Island under control of the Navy Department in 1903 to prevent poaching by Japanese feather hunters. In 1909, in response to public uproar over the killing of millions of seabirds throughout the NWHI for the feather trade, he established the Hawaiian Island Bird Reservation through Executive Order 1019. This was the first of several steps taken over the following century to protect the living and other resources of the NWHI (Fig. 13.5). In 1940, President Franklin Delano Roosevelt changed the name of the Bird Reservation to the Hawaiian Islands National Wildlife Refuge, increasing the

FIG. 13.5. Although marine life in the NWHI has enjoyed nearly a century of legal protection, a turtle trapped in a derelict fishing net and the remains of a dead bird with its digestive tract full of plastic are grim reminders that anthropogenic threats still exist. Marine debris including nets and other fishing gear, and a wide range of discarded plastic items from across the entire North Pacific Ocean accumulate in the waters of the subtropical front, continuing to entangle marine life and damage benthic habitats across the NWHI (Turtle photograph by Jake Asher; photograph of bird by Jean Kenyon)

level of protection afforded to wildlife and enabling many populations to return to levels similar to those when the islands were first discovered.

Commercial fishing in the NWHI by vessels from the Main Hawaiian Islands resumed after World War II. This practice continued until the Honolulu market declined in the late 1950s. In an effort to relieve pressure on marine resources in the Main Hawaiian Islands, in 1969 the Governor's Task Force on Oceanography recommended developing fisheries in the NWHI. The Governor's Advisory Committee on Science and Technology urged State, Federal, and academic research agencies to conduct collaborative research in the NWHI to further that goal. In response the National Marine Fisheries Service conducted research cruises to the area in 1973 and 1975 and that agency, the Fish and Wildlife Service, and the Hawai'i Department of Land and Natural Resources entered in to the Tripartite Cooperative Agreement to survey and assess the living resources of the NWHI (Hawai'i DLNR 2000). Aided by tens of millions of dollars of "extended jurisdiction" funding, the Tripartite partners, joined by the University of Hawai'i Sea Grant program, conducted a number of studies between 1976 and 1981, eventually resulting in research symposia in 1980 and 1983, and the publicating of 115 papers or abstracts in three volumes of proceedings (Grigg 2006).

In 1983 the US Exclusive Economic Zone (EEZ) was established by President Ronald Reagan. This gave the US jurisdiction over all resources between 4.8 and 322 km (3 and 200 nm) from any of their shorelines and also provided Extended Jurisdiction funds for research within the newly designated EEZ. Midway Atoll National Wildlife Refuge was established in 1988, followed by the Kure Atoll State Wildlife Sanctuary in 1993. In June 1998 President William J. Clinton issued Executive Order 13089 establishing the US Coral reef Task Force (CRTF), whose mission is to lead, coordinate, and strengthen US government actions to better preserve and protect coral reef ecosystems (CRCP 2007). The creation of the CRTF and passage of the Coral Reef Conservation Act of 2000 has since 2001 made millions of dollars available annually for research and resource management activities in the NWHI, ushering in a new era of discovery. President Clinton also issued Executive Orders 13178 and 13196 in 2000 and 2001, creating

the Northwestern Hawaiian Islands Coral Reef Ecosystem Reserve and initiating a process that was expected to redesignate the Reserve as a National Marine Sanctuary. Hawai'i Governor Linda Lingle established a marine refuge in 2005 and signed a law to prohibit all extractive uses in NWHI nearshore waters out to 4.8 km (3 miles) offshore, except at the federally managed Midway Atoll. Finally, on June 15, 2006, President George Bush issued Presidential Proclamation 8031, creating the Northwestern Hawaiian Islands Marine National Monument. This proclamation forms the largest fully protected marine conservation area in the world. The National Oceanic and Atmospheric Administration, US Fish and Wildlife Service, and State of Hawai'i Department of Land and Natural Resources serve as co-trustee management agencies of this 360,000 km^2 tract of islands and surrounding ocean (NWHI MNM 2006). On March 2, 2007 the monument was renamed the Papahanaumokuakea Marine National Monument. The name "Papahanaumokuakea" is composed of four Hawaiian words: Papa, Hanau, Moku, and Akea. Papa can be considered the Hawaiian deity who is the equivalent of Mother Earth, "hanau" means to give birth, "moku" means island, and "akea" means a broad expanse (Nakaso 2007).

13.1.2 Island and Atoll Descriptions

The following section provides information about all of the atolls and islands in the NWHI, and some of the submerged banks, running in order from the northwest to the southeast down the chain. Table 13.1 contains basic information for each of the major islands, atolls, and shallow banks in the NWHI, including the name of an island or other feature, its location, land area and lagoon volume (if any), etc. Additional facts or descriptions of some of them are also included below.

Kure Atoll (Fig. 13.6) is the farthest north of all the atolls and islands in the NWHI and presumably the oldest, and has the distinction of being the northernmost atoll in the world. The circular-shaped atoll is about 10 km in diameter and includes two small islets, Sand and Green Islands. The latter was home to a Coast Guard-operated long range navigation (LORAN) station from the early 1960s until 1992, and was supported by the 1,200 m long runway that is still occasionally

TABLE 13.1. Characteristics of the islands, atolls, and some submerged banks in the NWHI, listed in order from the northwest down to the southeast. Note that most of the subaerially exposed islands, sea stacks, and atolls are surrounded by extensive shallow banks. Island ages are from Clague (1996), with values in brackets from K-Ar dated basalt samples, and other ages estimated from geophysical calculations. Lagoon water volumes are from Hoeke et al. (2006).

Island, atoll, or bank	Type of feature	Longitude	Latitude	Age (Ma)	Reef Emergent land (km^2)	Lagoon habitat <100 m (km^2)	Bank volume (10^4 m^3)	summit depth (m)
Kure Atoll	Cosed atoll	178° 19.55'	28° 25.28'	29.8	0.86	167	141,000	–
Nero Seamount	Bank	177° 57.07'	27° 58.88'	29.1	0.00	17	–	68
Midway Atoll	Closed atoll	177° 22.01'	28° 14.28'	[27.7], 28.7	1.42	223	213,000	–
Pearl & Hermes Atoll	Closed atoll	175° 51.09'	27° 51.37'	[20.6], 26.8	0.36	1,166	2,930,000	–
Lisianski Island/ Neva Shoal	Open atoll	173° 58.12'	26° 4.23'	23.4	1.46	979	242,000	–
Pioneer Bank	Bank	173° 25.58'	26° 0.71'	22.8	0.00	390	–	26
North Hampton Seamounts	Bank	172° 14.08'	25° 26.84'	[26.6], 21.4	0.00	430	–	5
Laysan Island	Carbonate island	171° 44.14'	25° 46.13'	[19.9], 20.7	4.11	57	3,600	–
Maro Reef	Open atoll	170° 38.34'	25° 30.2'	19.7	0.00	1,508	611,000	–
Raita Bank	Bank	169° 30.04'	25° 31.72'	17.9	0.00	650	–	16
Gardner Pinnacles	Basalt sea stacks	167° 59.82'	25° 0.04'	[12.3], 15.8	0.02	1,904	–	–
St. Rogatien Banks	Banks	164° 7.26'	24° 20.0'	14.7	0.00	500	–	22
Brooks Banks	Banks	166° 49.31'	24° 7.03'	[13.0], 13.6	0.00	320	–	20
French Frigate Shoals	Open atoll*	166° 10.75'	23° 45.99'	12.3	0.23	733	1,910,000	–
Bank 66	Bank	165° 49.37'	23° 51.86'	11.9	0.00	0	–	120
Necker Island	Basalt island	164° 41.90'	23° 34.64'	[10.3], 10.6	0.21	1,538	64.2	–
Twin Banks	Bank	163° 3.78'	23° 13.08'	8.7, 8.3	0.00	95	–	53
Nihoa Island	Basalt island	161° 55.25'	23° 3.73'	[7.2], 7.3	0.82	246	–	–

used today. Claimed by the Hawaiian Kingdom in 1886, Kure was placed under Naval jurisdiction by Executive Order in 1936 and then after World War II was inadvertently returned to the Territory of Hawai'i instead of the US Department of the Interior as were most of the NWHI (Rauzon 2001). The atoll was designated as a state wildlife refuge under the control of the Hawai'i Department of Land and Natural Resources and is

also now part of the Papahanaumokuakea Marine National Monument.

Midway Atoll includes three islands: Sand Island, Eastern Island, and the tiny and often changing Spit Islet, encircled by a roughly round atoll rim approximately 10 km in diameter. There is a dredged pass into the lagoon on its southern side that is navigable by ships and another pass on the western side.

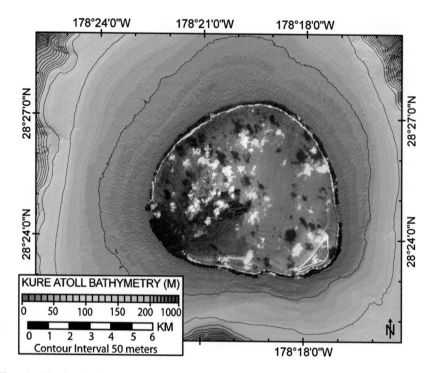

FIG. 13.6. Kure Atoll's classic circular rim structure and Green Island in the southeastern corner of the lagoon are visible in this satellite image. Distinctive spur and groove morphology outside the atoll rim is evident in recently collected multibeam bathymetry from around many of the islands and atolls of the NWHI (http://www.soest.hawaii.edu/pibhmc/)

Sand and Eastern Islands have been extensively modified for commercial ventures and military use, as discussed earlier. Despite the long and often ecologically damaging modifications to the island, Midway has the world's largest nesting colony of Laysan Albatrosses (*Phoebastria immutabilis*), and almost a million birds of a range of species visit the atoll annually (Rauzon 2001). With a century of heavy traffic of humans and cargo, Midway is also home to over 200 invasive species of plants and numerous insects. Black rats had been introduced to Midway and had a significant and negative effect on the flora and fauna there, until they were finally removed in an eradication program in the 1990s. Nineteen species of cetaceans have been found in the waters around the NWHI, with Bottlenose and Spinner Dolphins (*Tursiops truncates* and *Stenella longirostris*, respectively) the most common and often found in Midway's lagoon (Rauzon 2001). With the departure of the military, monk seals have become a common sight on Midway beaches.

Pearl and Hermes Atoll is a roughly oval-shaped atoll approximately 34 km long and half that distance across, with seven small islets, including several with low vegetation. Much of the atoll's lagoon features a well developed and shallow reticulate reef structure nicknamed "the maze," which provides a challenge to small boat navigation (Fig. 13.7). The atoll is named for the British whaling vessels *Pearl*, which ran aground there in 1822, and her sister ship *Hermes* that foundered when trying to aid the *Pearl* (Rauzon 2001). In 1928, a population of the black-lipped pearl oyster (*Pinctada margaritifera*) was discovered at Pearl and Hermes Atoll. For the next 2 years the shells were heavily harvested, mostly to be made into "mother of pearl" buttons, which devastated the oyster population and resulted in a ban on further harvesting (Keenan et al. 2006). A survey in 2003 reported that the current population apparently now has a sustained level of reproduction, but that the species has failed to recover to pre-exploitation

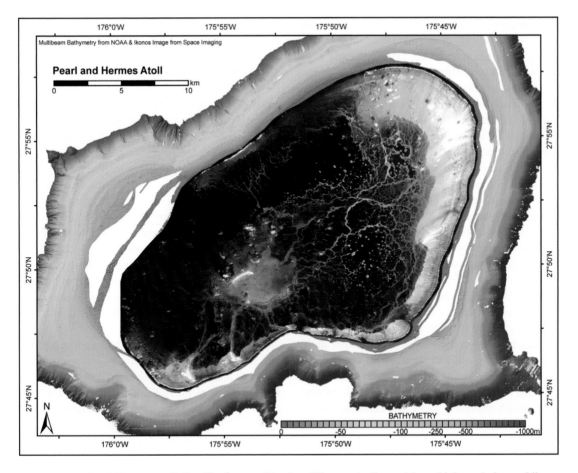

Fig. 13.7. Pearl and Hermes Atoll. Satellite image of Pearl and Hermes Atoll, overlain on high resolution multibeam bathymetry (http://www.soest.hawaii.edu/pibhmc/). The extensive reticulate reef structure evident within the lagoon at Pearl and Hermes is the most well-developed and clearly visible example of this morphology within the NWHI, but it is unclear why this morphology develops

levels despite more than 70 years of protection (Keenan et al. 2006). On the brighter side however, the atoll has the highest biomass of reef fish in the NWHI today (Friedlander and DeMartini 2002).

Lisianski, one of the larger islands in the NWHI, features a 12 m high sand dune on its northern side, an expanse of dry grass, and a few low shrubs. Numerous seabirds are now found on the island, although more than a million birds were harvested from Lisianski for the feather trade, the introduction of mice, and the release of rabbits destroyed the island's ecology for a time (Rauzon 2001). The island is surrounded by an open atoll called Neva Shoals, which lacks a distinct atoll rim. It does,

however, contain a network of reticulate and linear reefs which offer protection to the waters within them and are home to what has been described as one of the more scenic "coral gardens" in the entire NWHI (Maragos and Gulko 2002).

Laysan Island is the only low coral island in the NWHI that is not associated with what can be classified as an atoll today, although it may have been during earlier periods. A shallow saline lake covers approximately 0.4 km^2 of the island's interior and serves as a food, water, and rest stop for a wide variety of birds and is a critical habitat for one of the two remaining populations of the endangered Laysan Duck (*Anas laysanensis*) (Maragos and Gulko 2002;

Rauzon 2001). Laysan Island is described by many as the "crown jewel" of the NWHI, but suffered from severe exploitation in the late 1800s and early in the twentieth century. Hundreds of thousands of tons of guano were mined from the island, and activities of the miners living and working on the island had serious impacts on the island's ecosystem. Many thousands of albatross eggs were harvested from the island, and feather poachers killed hundreds of thousands of birds as well. Rabbits introduced to Laysan Island as potential livestock for a meat canning business rapidly destroyed practically all of the island's vegetation, forever altering its ecology and facilitating widespread erosion (Rauzon 2001). The combined impact of these stresses led to the extirpation of 26 plant species and several bird species. The bird population fell from 10 million in 1903 to about 30,000 by 1923 (Maragos and Gulko 2002; Olson 1996). However, almost a century of wildlife protection and years of effort by the US Fish and Wildlife Service to remove alien species and restore native ones are paying off, and there are many signs that the island is recovering (Maragos and Gulko 2002).

Maro Reef is, similar to Neva Shoals, an open atoll which presently lacks the classical circular atoll rim structure and is composed instead of linear and reticulate reef structure. This type of atoll structure is not reported from any other location in the world. The only emergent land at Maro Reef consists of a few large blocks of reef rock, but at 1,508 km^2, its potential coral reef habitat shallower than 100 m is the second largest in the NWHI. Many of the reef structures at Maro are narrow and unconsolidated in places, and gaps in the reef structure enable wave energy to penetrate into the lagoonal waters, keeping fine sediments suspended in the water column most of the time. Despite the turbidity, Maro Reef is one of the most fertile marine areas in the Hawaiian Archipelago and was formerly one of the areas targeted by a commercial lobster fishery. It has a high diversity of both reef fish and corals, very high coral cover in some areas, and other areas dominated by crustose coralline algae (Maragos and Gulko 2002; Maragos et al. 2004; Rauzon 2001).

Gardner Pinnacles consists of two steep basalt sea stacks, the largest of which is approximately 50 m tall and 180 m long, the last vestiges of basalt above sea level moving to the northwest up the chain (Fig. 13.8). Marine habitats shallower than a

Fig. 13.8. Gardner Pinnacles, the last (most northwesterly) vestige of basalt above sea level in the Hawaiian Archipelago (Photograph by Jean Kenyon)

depth of 20 m are restricted to the immediate vicinity of the stacks themselves, but Gardner Pinnacles has 1,904 km² of habitat shallower than 100 m, the most of any island atoll or bank in the NWHI. The shallow shelves around the sea stacks are exposed to wave energy from all directions, so there is fairly low coral cover on them, but markedly more on the southwestern (leeward side) of the stack, and more than at Necker and Nihoa Islands (Maragos et al. 2004). Reef fish diversity is high, presumably due to the large shelf area, and the area formerly provided a significant portion of the commercial lobster catch in the NWHI. Gardner Pinnacles and Necker Island host relatively large concentrations of the Giant Opihi (*Cellana talcosa*), the largest of the Hawaiian limpets, which requires a basalt substrate. Prized as an abalone-like seafood delicacy, the species has been severely overfished in the MHI (Maragos and Gulko 2002).

French Frigate Shoals is the most southerly atoll in the Hawaiian Islands. It is a large open atoll and features a well-developed rim only along the northeastern half, with reticulate reef prevalent inside. The atoll features a basalt sea stack, La Perouse Pinnacle, Tern and East Islands, and eight islets, some of which have mostly eroded away.

The continued loss of beach area at French Frigate Shoals has potentially serious consequences for the monk seals and green sea turtles that breed and nest on them and will be discussed in detail later. Tern Island was enlarged by dredging and filling by the US Navy just prior to World War II and features an airstrip that is still in use. The US Coast Guard operated a LORAN station on the island that was abandoned in 1979, but the USFWS operates a year-round field station there now. Magnificent stands of *Acropora* table corals are found in the lagoon (Fig. 13.9) and are more common at French Frigate than at any other location in the archipelago (Maragos et al. 2004). They are believed to have originated from Johnston Atoll, 865 km to the southwest, along with a number of other species, along an occasionally active oceanographic transport corridor from Johnston to French Frigate Shoals. This and another corridor connecting Johnston with islands in the vicinity of Kauai in the MHI are believed to be related to the subtropical countercurrent and the Hawaiian Lee countercurrent (Kobayashi 2006; Maragos et al. 2004).

Necker Island is a small, steep, and hooked-shaped ridge of basalt with a summit 84 m above sea level, surrounded by the second largest marine

FIG. 13.9. One of many luxuriant stands of *Acropora* table corals found at French Frigate Shoals (Photograph by Jean Kenyon)

habitat shallower than 100 m in the NWHI. The island supports only five species of flowering plants, but 60,000 seabirds of 16 species roost there, and endangered Hawaiian monk seals are known to haul out on the island and forage on the marine terraces there (Rauzon 2001). The broad shelves off Necker have also been commercial fishing grounds for lobster and bottomfish.

Nihoa, a small basalt island of 0.6 km^2 with a summit 275 m above sea level, is the only area in the NWHI known to have been permanently settled by native Hawaiians prior to western contact. The lack of protection from large waves makes it difficult for corals and other species to survive. In shallow waters, the substrate around Nihoa is composed mostly of heavily eroded and wave scoured basalt, with live coral reef limited to depths greater than about 12 m. The surrounding shelf is one of the smaller ones in the NWHI. Heavily exposed to wave energy, stony corals are less abundant and diverse at Nihoa than at atolls and islands in the middle and northern end of the NWHI (Maragos et al. 2004). However, the island does host one of the highest reef fish biomass densities of any island in the NWHI (Friedlander and DeMartini 2002). The area was sufficiently productive to support the ancient Hawaiian community that lived there and, presumably, relied on seafood for protein to supplement their diet of sweet potato. Spared the ravages of guano miners, Nihoa has the most intact coastal ecosystem left in Hawai'i. For example, at least 40 of the terrestrial insect, spider, and crustacean species on Nihoa are endemic to the island, and six species of land snails, extinct in the MHI, are still found on Nihoa. The island hosts half a million seabirds of a number of species, including 17 that breed on Nihoa alone (Rauzon 2001). The rats, and many of the invasive insects and plants that have displaced native species on other islands with greater natural resources, have not become established on Nihoa, leaving us with a glimpse of how different the ecology of the NWHI may have been prior to extensive human exploitation. It also highlights the threat of how quickly the remaining ecosystems, marine and terrestrial, can be disrupted in the face of invasive species.

13.1.3   Climate and Oceanography

The NWHI experience high surface gravity wave events which are arguably among the highest of any tropical or subtropical island archipelago. The vigorous Aleutian Low atmospheric pressure system, large scale of the North Pacific Ocean, and the NWHI's central location all combine to create exceptionally high waves, often with long periods, in the region. Extreme wave events (deep water wave heights in excess of 7 m) occur several times in an average year; 4–6 m heights occur on the order of ten or more times a year (Fig. 13.10). Associated wave periods may be as long as 25 s, but are more typically 8–18 s (Fig. 13.11). These episodes are generated from two primary sources, including the aforementioned Aleutian Low, which are mid-latitude cyclones spawned as waves on the polar front (Graham and Diaz 2001; Bromirski et al. 2005). Extreme waves are also occasionally generated from subtropical cyclones known as Kona Lows, which generally form in the vicinity of the NWHI themselves (Caruso and Businger 2006). Ocean waves associated with the Aleutian Low tend to be long period swell from the northwest quadrant. Kona Lows generate extreme waves much less frequently, and these waves tend to be of shorter period and from a more westerly or southwesterly direction. Wave events from both of these mesoscale weather systems are seasonal, with almost all extreme episodes occurring between October and April. These weather systems sometimes bring strong winds (15–30 + m/s (30–60 + knots)) and most of what little rain occurs in the NWHI (Rauzon 2001). In between episodes, easterly trade winds associated with the North Pacific Subtropical High tend to dominate, particularly during the boreal summer. These are the modal conditions, and typically bring waves with 1–3 m wave heights and 7–11 s periods from the east (Figs. 13.10 and 13.11). Atmospheric pressure gradients in the NWHI tend to be less than in the MHI during summer, so the resulting trade winds are often somewhat weaker, particularly towards the northern end of the chain. Long period southern-hemisphere swell, which can be significant in the MHI, generally decreases moving north and west. Kure, Midway, and Pearl and Hermes (the northern atolls) experience very little of this swell; it is mostly absent in the wave climatology shown in Fig. 13.11 for Midway Atoll.

In addition to the large seasonal change between cyclone frequency (high in winter) and trade wind conditions (predominantly in summer), the mesoscale

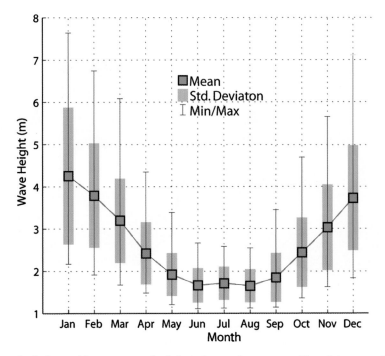

FIG. 13.10. Climatological monthly mean, standard deviation, and mean monthly minimum and maximum wave heights derived from NOAA Wave Watch III computations at 6 h time steps at Midway Atoll from January 1997 through December 2006

and synoptic weather features discussed above exhibit great variability in intensity, frequency, mean path, and location over interannual to decadal time scales. At interannual time scales, the Aleutian Low tends to be more intense and track farther to the south (closer to the NWHI) during positive ENSO phase (El Niño) periods (Bromirski et al. 2005). This brings higher winds and larger waves to the NWHI region than are typical in negative ENSO-phase (La Niña) years (Rooney et al. 2004; Wang and Swail 2001). There is also evidence that the mean intensity of the Aleutian Low has increased in the last several decades. Given the currently short span of human observations in the region it is difficult to determine if anthropogenic climate change is a factor, or if it reflects decadal time scale oscillations (Graham and Diaz 2001).

The Pacific Decadal Oscillation (PDO) can be described as an ENSO-like pattern of Pacific climate variability with each phase lasting perhaps 20–30 years. A key indicator of PDO warm (positive) phases is decreased pressure in the central North Pacific, which is generally reflected

by increased number of cyclones on 20–50 year timescales (Bond and Harrison 2000). Hurricanes occasionally strike the Hawaiian Archipelago, usually during summer months, and are significantly more frequent during El Niño periods and warm phases of the PDO. However, even during these periods, hurricane tracks rarely reach as far west and north as the NWHI (Chu 2002). There are some indications that both positive PDO phases and positive ENSO phases tend to weaken the North Pacific Subtropical High atmospheric pressure system and perhaps shift it southward, bringing lighter trades to the NWHI, but there has been little research on the subject (Bond and Harrison 2000; Hoeke et al. 2006).

The location and variability of these mesoscale and synoptic weather patterns are important in defining oceanographic structure and its variability. The ocean, in turn, influences the atmosphere through multiple feedback mechanisms, creating a coupled system. The boundary between the oligotrophic (low in nutrients and high in dissolved oxygen) surface waters of the North Pacific

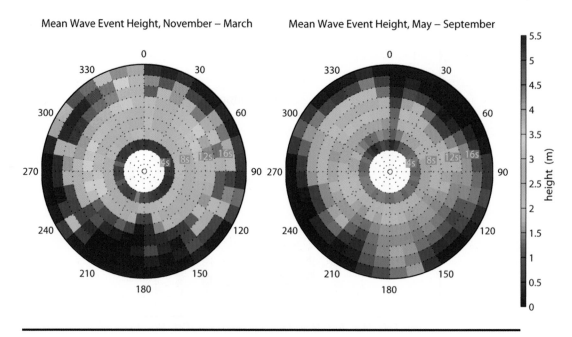

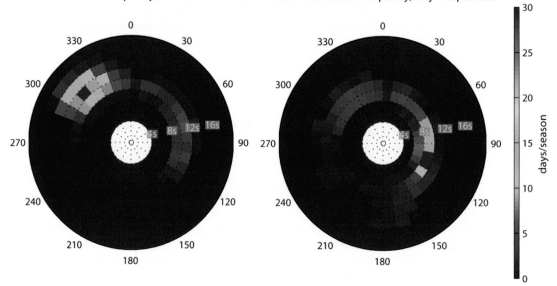

FIG. 13.11. Seasonal wave height, direction, and period event climatology derived from NOAA Wave Watch III computations at 6 h time steps at Midway Atoll from January 1997 to December 2006. Upper panel (a) represents mean seasonal wave event height in given directional and period bin (e.g., between November and March an average wave event with 16–18 s period from 300° has a height of 5.5 m). Lower panel (b) represents mean frequency (likelihood) of occurrence in days per season (e.g., between May and September, 1.5 m wave with 8–10 s periods from 100° occur, on average, 30 or more days in a season)

Subtropical Gyre associated with the North Pacific Subtropical High, and the nutrient-rich surface waters of the North Pacific Subpolar Gyre, is often termed the "subtropical front." It is frequently defined as the surface expression (outcropping) of the 17°C isotherm and/or 0.2 mg/m³ surface chlorophyll concentration (Leonard et al. 2001; Polovina et al. 2001). During the winter, this front is typically located at about 30–35° N latitude and in the summer at about 40–45° N (Polovina et al. 2001). However, subsurface expressions of the front as shallow as 30 m have been recorded as far south as 28° N (Leonard et al. 2001). The location of this front is likely to exert a significant influence on the ecology of the NWHI. Evidence suggests that during southward extensions of the front, waters around the northern atolls at 28° N are far less oligotrophic than they, or waters further south in the archipelago, typically are. This front migrates significant distances on interannual and decadal time scales, in concert with atmospheric fluctuations; positive ENSO appears to correlate to southern extensions of the front (Leonard et al. 2001). There is evidence that climatic conditions favoring southern extensions of the subtropical front have large decadal time-scale impacts on the ecosystem. Polovina et al. (1994) noted biomass changes on the order of 30–50% associated with decadal climate indices; it appears that this is at

least partially due to fluctuations in nutrient enrichment associated with the migration of the subtropical front. Similarly, Antonelis et al. (2003) suggest that the body condition of weaned Hawaiian monk seal pups may be improved during El Niño events due to the enhanced availability of prey species. Besides impacting nutrient availability, the location of this front is associated with the concentration of marine debris (Kubota 1994), which has been shown to most severely impact the northern atolls (Boland et al. 2006; Dameron et al. 2007; Donohue et al. 2000).

During the winter, the NWHI experience much cooler sea surface temperatures (SSTs) relative to the MHI, due to the proximity of the subtropical front and enhanced vertical mixing of surface waters by increased winds and waves. During the summer, however, NWHI surface waters tend to become highly stratified, frequently causing higher SSTs in the northern and central portions of the NWHI than are found in the MHI (Fig. 13.12).

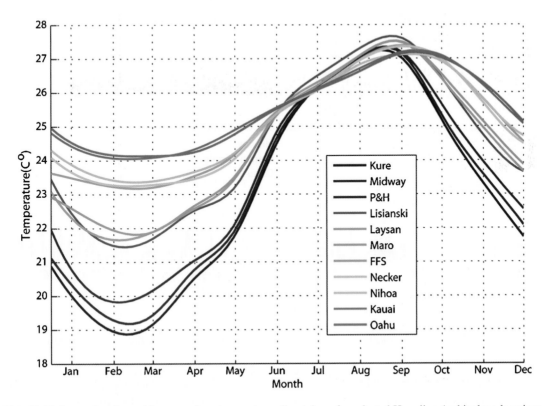

FIG. 13.12. Interpolated monthly sea surface temperature climatology for selected Hawaiian Archipelago locations, derived from Pathfinder SST (Vazquez et al. 2002)

This is attributed partially to the summertime position of the North Pacific Subtropical High. Large atmospheric pressure gradients south of the high generate strong trade winds and tend to keep SSTs cooler through wind mixing in the MHI. Further north towards the mean center of the high, weaker pressure gradients lead to lighter trade winds which tend to increase stratification and elevate SSTs. Climatological observations indicate that these processes are common.

These processes contributed to mass coral bleaching events which occurred in the NWHI during the summers of 2002 and 2004 (Hoeke et al. 2006). The northern atolls were particularly affected (up to 90% bleached coral in some areas), and severity decreased moving southeast down the chain (Kenyon and Brainard 2006; Kenyon et al. 2006). Figure 13.13 shows that SSTs were anomalously

warm during these two events, which were among the highest in an extended record at Midway. French Frigate Shoals and Oahu remained much cooler. Maximum summertime SSTs at Oahu, and the rest of the MHI, are generally cooler than the higher latitude northern atolls of the NWHI. In addition to large-scale oceanographic regimes favoring coral bleaching in the northern NWHI, the large relatively sheltered lagoons and backreef areas of the northern atolls enhance further stratification and elevated localized SSTs. This contributed to the significantly greater coral bleaching at the northern atolls documented in 2002 and 2004 (Hoeke et al. 2006; Kenyon and Brainard 2006).

Jokiel and Brown (2004) noted increasing trends in SST over the 1981–2004 period in the Hawaiian Archipelago, particularly at Midway Atoll. This trend is not as apparent in longer time series (Fig. 13.13).

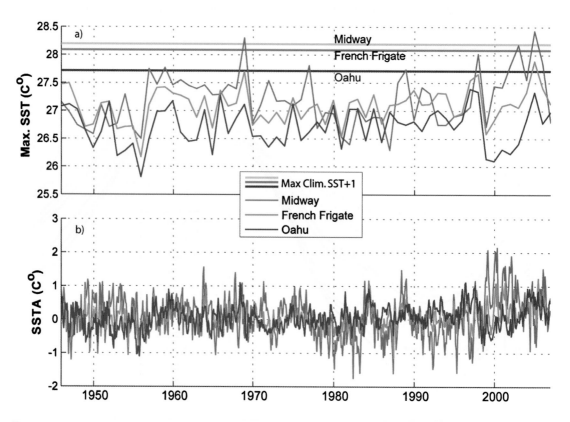

FIG. 13.13. Extended–reconstructed sea surface temperature (SST) and sea surface temperature anomaly (SSTA) at Midway, French Frigate Shoals, and Oahu. Upper panel (a) represents maximum monthly SST at each location; grey bar are maximum monthly climatological SST + 1°C, often used as an indication of bleaching conditions (Wellington et al. 2001). Lower panel (b) represents SSTA; greater inter-annual to decadal variability in SSTA at Midway compared to Oahu is evident

An extreme SST event at Midway, similar to 2002 and 2004, occurred in 1969 and was followed by several decades of cooler temperatures. These episodes and trends appear to be related to the PDO signal, but large uncertainties exist in the SST records before the advent of satellite SST measurements in 1981. Links between summertime SST in the NWHI, long-term SST trends, the mean position and intensity of the North Pacific Subtropical High, and major climate indices (e.g., PDO and ENSO) have not been well studied.

The information above, drawn from many sources, suggests that the northern atolls of Kure, Midway, and Pearl and Hermes are oceanographically distinct from the rest of the archipelago. The reefs and banks south of Pearl and Hermes in the NWHI generally experience conditions that are increasingly similar to those of Kauai and Oahu the farther to the southeast they are located. The northern atolls however, experience much colder water temperatures, more vigorous waves and winds, and sometimes greatly enhanced nutrient enrichment in the winter. In the summer, they tend to experience lighter winds and higher SSTs relative to the rest of the Hawaiian Archipelago. Two documented manifestations of

these differences are the high levels of marine debris recruitment (Boland et al. 2006; Dameron et al. 2007) and the severity of coral bleaching in the northern atolls (Kenyon and Brainard 2006; Kenyon et al. 2006). The reefs and banks in the vicinity of Lisianski and Laysan appear to be a transition area between the oceanographic regimes of the northern atolls and the central-south parts of the chain. The observations discussed above represent broad generalizations: seasonal and interannual variations are large and often deviate from these climatological means. Changes imposed by climate variability and change remain largely unknown and are worthy areas of study.

13.2 Pre-Holocene Reef History

13.2.1 Age and Evolution of Hawaiian Volcanoes

The Hawaiian–Emperor chain is anchored in the central Pacific basin at 19° N, the locus of current volcanism (Fig. 13.14). It includes at least 129 massive shield volcanoes that formed over the past

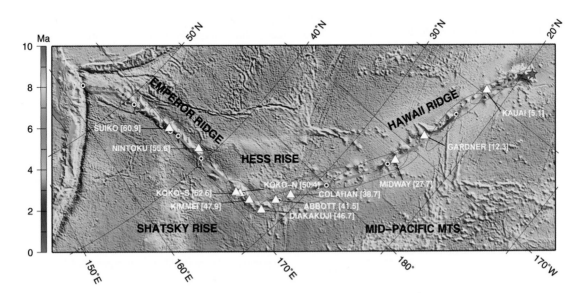

FIG. 13.14. Geometry of the Hawai'i–Emperor seamount chain. The major bend in the chain has now been dated to 47–50 Ma (Chron 21–22) based on ages (in white) from Sharp and Clague (2006) and Clague (1996); Nintoku age taken from Tarduno et al. (2003). Predicted hotspot chain track (rainbow line) from the absolute Pacific plate motion model of Wessel et al. (2006) shows ages modulo 10; dotted circles indicate start of each 10 Myr section. Hotspot location is located near Kilauea (star). Red ellipses indicate the uncertainty of the reconstruction at selected times (red crosses)

85 Myr, with volcano ages generally decreasing in age towards the southeast (Jackson et al. 1975; Clague 1996). The early Hawaiians were the first to recognize this age trend, which was recorded in their oral tradition of the fire goddess Pele. She was reported to have migrated with her fire southward along the island chain (Westerveldt 1916) causing successively younger eruptions to the south. Early western explorers to Hawai'i also noted the apparent decreasing age of the islands to the south (Dana 1891). The overall age progression of the islands (Figs. 13.1 and 13.14) has been confirmed in several studies using radiometric isotopes (Clague and Dalrymple 1987; Garcia et al. 1987), although major gaps remain in our knowledge of the formation time of NWHI volcanoes because suitable samples for radiometric dating are difficult to obtain. It has been found, however, that the frequency of volcano formation has increased over time while the spacing between them has decreased, as shown in Table 13.2. The islands at the younger end of the chain are also significantly higher than those that formed earlier (Clague 1996).

Various evolutionary sequences have been proposed for the growth of Hawaiian volcanoes starting with Stearns (1946). A current popular scheme divides the growth period into four major stages: preshield, shield, postshield, and rejuvenation (Fig. 13.15). The first three stages are responsible for building a massive shield volcano that may reach a maximum thickness of 13 km and a volume of up to 80,000 km³ (Mauna Loa, the largest volcano on Earth; Lipman 1995). The preshield stage lasts ~250,000–300,000 years (Guillou et al. 1997) and is distinct from the shield in producing alkalic magmas (i.e., magmas containing a relatively high percentage of sodium and potassium alkali) (Moore et al. 1982; Garcia et al. 1995). Although

observed only at Lō'ihi and possibly Kīlauea volcanoes (Lipman et al. 2003), this stage is thought to be at the core of all Hawaiian volcanoes (Clague and Dalrymple 1987). Loihi Seamount, the youngest volcano in the Hawaiian chain, rises more than 3,000 m above the floor of the Pacific Ocean and is located approximately 40 km south of Kīlauea. Volumetrically, the preshield stage is minor, forming about 1–2 vol% of a typical overall Hawaiian volcano (Fig. 13.16), although it may create an edifice that is 4–5 km tall (Garcia et al. 2006).

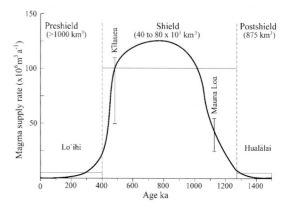

FIG. 13.15. Growth history model for a typical Hawaiian shield volcano. This composite model is based on volume estimates for each stage (large boxes). Magma supply estimates (vertical bars) are given for Kīlauea (early shield stage; Pietruszka and Garcia 1999) and Mauna Loa (late shield stage; Wanless et al. 2006). The preshield stage is represented by Loihi (Garcia et al. 2006; modified by permission of Elsevier) and the postshield by Hualalai (Moore et al. 1987) volcanoes

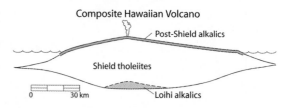

Fig. 13.16. Cartoon cross section of a composite Hawaiian volcano at the end of the postshield stage showing rock type proportions. The preshield stage is represented by Lō'ihi alkalic lavas, comprising 1–2 vol% of the shield. The shield stage is based on Mauna Loa and is composed of tholeiites that form the bulk of the volcano (~98%). The postshield stage forms a cap on the shield, comprising up to 1% of the volcano, based on Mauna Kea volcano (Frey et al. 1990). Note the section is two times vertically exaggerated

TABLE 13.2. Frequency and spacing of volcanoes in the Hawaiian–Emperor chain, from Clague (1996). The Hawaiian Ridge includes the NWHI and the seamounts between Kure and the Emperor Chain.

	Number of volcanoes per million years	Spacing between volcanoes (km)
Emperor Chain	1.1	57
Hawaiian Ridge	1.7	45
Main Hawaiian Islands	4.0	30

As the volcano moves closer to the center of the hotspot and its source experiences higher temperatures and degrees of partially melting, the magma composition switches to tholeiitic (containing less sodium and potassium at similar concentrations of silica compared to alkali basalt) and volcanism becomes more voluminous (Garcia et al. 1995; Guillou et al. 1997). Depending on whether the volcano forms on the flanks of a pre-existing volcano (like most Hawaiian volcanoes, e.g., Necker), or was isolated (e.g., Kure), the volcano emerges above sea level perhaps 50,000–300,000 years after the preshield, forming a subaerial shield volcano. Coral reefs and other aquatic life are destroyed or their growth slowed during this emergence stage as molten rock reacts with seawater in a shallow pressure environment creating fragmental debris (e.g., Moore and Chadwick 1995). However, eruption rates are high during this stage (Fig. 13.15) and a stable island is quickly formed as growth rates exceed rates of marine erosion and subsidence (Garcia et al. 2007). Subaerial flows on the new island are a mixture of pahoehoe and a'a, with a'a becoming more abundant as the island grows in size (Garcia et al. 2007). After another ~100,000 years, the growing volcano may reach the size of Kīlauea volcano (Quane et al. 2000). Vigorous activity persists for another ~600,000 years creating giant shield volcanoes. Where several shield volcanoes are clustered together (e.g., the French Frigate Shoals area), large islands may form. These islands are subject to giant landslides that may remove large sections of the volcano (e.g., up to 40% for Ko'olau volcano on the island on Oahu; Garcia et al. 2006). Landslides may occur at any stage during the volcano's growth or afterwards, although the largest slides tend to occur during the shield stage.

By the end of the shield-building stage, a Hawaiian volcano is about 1.25 Myr old and has drifted off the center of the hotspot. As it enters the post-shield stage, mantle melting temperatures progressively decrease causing magma compositions to gradually switch back to alkalic (Feigenson et al. 1983; Frey et al. 1990). The volcano is now subsiding at a faster rate (~2.5 mm/year; Moore and Chadwick 1995) than it is growing and the surface area of the volcano shrinks. A record of the volcano's maximum shoreline extent is recorded as an inflection in its slope (Moore and Campbell 1987). The submarine slope of these volcanoes is steeper, generally >15°, than the subaerial slopes, typically 3–7°, with steeper slopes on volcanoes that have undergone postshield volcanism (Mark and Moore 1987). The submerged subaerial lavas form a shelf around the volcano. A rapid decline in eruption rate occurs over the next 250,000 years, which is accompanied by an abrupt shift to more fractionated, more viscous and less dense lava compositions (Hawaiites to trachytes; Macdonald et al. 1983), as magmas pond at greater depths (~30 km) before eruption (Frey et al. 1990). These late stage lavas are responsible for the somewhat steeper slopes than on shield volcanoes, although these lavas form only a thin veneer (Fig. 13.16). The volcano dies after about 1.5 million years of growth (Fig. 13.15), following which it continues to subside and marine erosion shrinks the size of the island.

Many, but not all, of Hawaiian volcanoes experience a period of renewed volcanism 0.6–2.0 Myr after the end of post-shield (Tagami et al. 2003). The lavas produced during this 'rejuvenated' stage are generally strongly alkalic and tend to be explosive where they intersect groundwater or coral reefs (Winchell 1947; Walker 1990). Rejuvenated volcanism may last briefly (over several thousand years) producing a few vents and flows. Various models have been proposed to explain the origin of rejuvenated volcanism. One of the two currently favored models suggests that a zone of secondary melting 100–300 km northwest of the hotspot forms as the plume of magma is deflected laterally and rises (Ribe and Christensen 1999). The other model invokes flexural uplift associated with the rapid subsidence in the area of rapid volcano growth (e.g., Bianco et al. 2005), but neither model satisfactorily explains all aspects of rejuvenation volcanism.

It should be noted that not all Hawaiian volcanoes follow this sequence, as some lack post-shield and or rejuvenated stages. Following rejuvenated volcanism, if it occurred, the volcano continues to be eroded by the ocean and to subside under the weight of their massive shields. Eventually, all Hawaiian volcanoes will disappear beneath the waves forming seamounts. The tops of the once high-standing Emperor volcanoes are now more than a kilometer below sea level. Such will be the fate of the NWHI.

13.2.2 Dynamics of the Pacific Plate and Hawaiian Hotspot

13.2.2.1 Horizontal Displacement Due to Plate Motion

Tuzo Wilson (1963) hypothesized that the distinctive linear shape of the Hawaiian–Emperor seamount (Fig. 13.14) chain resulted from the Pacific plate moving over a deep, stationary hotspot in the mantle, located beneath the present-day position of the Island of Hawai'i (Fig. 13.17). He further suggested that continuing plate movement eventually carries the island beyond the hotspot, cutting it off from the magma source, and volcanism ceases. As one island volcano becomes extinct, another develops over the hotspot, and the cycle is repeated. This process of volcano growth and death, over many millions of years, has left a long trail of volcanic islands and seamounts across the Pacific Ocean floor.

One of the most characteristic features of the Hawaiian chain is the prominent bend near longitude 172° E, named the Hawai'i–Emperor Bend

(HEB). Since the suggestion by Wilson (1963) this feature has been attributed to a 60° change in direction of Pacific plate motion over a fixed hotspot in the mantle beneath Hawai'i (Morgan 1971). A lingering problem, however, was posed by the age of the bend, initially dated to 43 Ma (Dalrymple et al. 1987). Intuitively, such a large change in the motion of the dominant Pacific plate should have left clear and unequivocal evidence of contemporaneous tectonic and magmatic events at the plate boundary; however, careful examination of the record failed to find the expected correlation (Atwater 1989; Norton 1995). Furthermore, rock samples recovered by deep-sea drilling from several Emperor seamounts revealed a frozen-in paleomagnetic field that suggested they were formed at a latitude significantly further north (5°–10°) than that of present-day location of Hawai'i (Kono 1980; Tarduno and Cottrell 1997). Finally, efforts to project the absolute plate motion of Africa, via the global plate circuit, into the Pacific failed to reproduce the characteristic bend at 43 Ma (Cande et al. 1995; Raymond et al. 2000). These

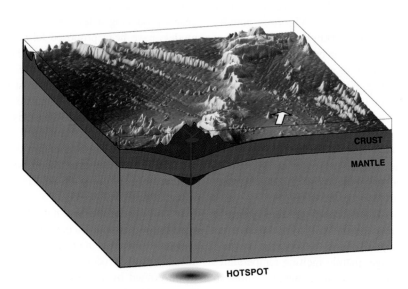

Fig. 13.17. Formation of the Hawai'i–Emperor seamount chain over the Hawai'i hotspot. Given the thick and strong lithosphere near Hawai'i (~90 Myr old), individual seamounts can grow very tall and breach sea level to form oceanic islands. The large volcanic piles deform the lithosphere, which responds by flexure. The plume beneath the plate feeds the active volcanoes by a network of feeder dikes; some magma may pond beneath the crust as well (Watts and ten Brink 1989). As the plate motion carries the volcanoes away from the hotspot (arrow indicates current direction of motion) they cease to be active and form a linear seamount/island chain

concerns gave impetus to alternative models in which the plume responsible for the volcanism was drifting in the mantle "wind" (Steinberger 2000; Steinberger and O'Connell 1998). Such models could simultaneously fit the changing latitude of the hotspot (constrained by paleomagnetics) and the geometry and age-progression of the seamount chain. Some researchers went as far as to conclude that no change in Pacific plate motion was needed at all to generate the HEB. Since the trail reflects the vector sum of plate and plume motion, it could be completely explained by a rapid slowdown in the southward motion of the plume while the plate motion remained unchanged in direction and magnitude (Tarduno et al. 2003).

The reconciliation of absolute plate motion models inferred from the Indian, Atlantic, and the Pacific Oceans requires their propagation via the global plate circuit (Acton and Gordon 1994). Due to incomplete knowledge of the history of all relative plate motion changes between conjugate plate pairs, the projection of the African absolute plate motion into the Pacific is subject to uncertainties that are difficult to quantify. By choosing a slightly different plate circuit for connecting the Pacific to Australia via the Lord Howe Rise, Steinberger et al. (2004) were able to show that the HEB did appear to have a plate motion component and could not simply be due to plume drift. Most recently, an effort to redo the dating of the rock samples taken from around the Hawai'i–Emperor Bend region resulted in the startling discovery that the HEB is considerably older than previously thought. Sharp and Clague (2006) reported that the rocks around the bend were formed during the 47–49 Ma interval (Chron 21–22) and, by allowing for 1–2 Myr of construction time, that the bend itself might have formed closer to 50 Ma – fully 7 Myr earlier than the conventional wisdom. These new dates undermine many of the previous conclusions about the lack of correlation between the bend and plate boundary processes since Chron 21–22 is known to have been an exceedingly active tectonic period in the Pacific and elsewhere (Cande and Kent 1995; Rona and Richardson 1978). Sharp and Clague (2006) also argued that the details of the age progression as we approach the bend from the north make it very unlikely that the plume was moving during the formation of the bend, thus strengthening the original explanation (purely a plate motion

change over a stationary hotspot) for the origin of the HEB.

Absolute plate motion studies have revealed much of interest in the last decade and new discoveries are happening at a fast pace. Recently, Whittaker and Müller (2006) made a preliminary announcement that they had identified tectonic evidence for a similar bend in plate motion between Australia and Antarctica that appears to be contemporaneous with the Hawai'i–Emperor bend. This change is expected to greatly improve the quality of the plate circuit and strengthen the argument for a plate kinematics origin even further. If confirmed, this observation would suggest that the paleomagnetic evidence of more northerly latitudes might have to be explained by true polar wander.

13.2.2.2 Vertical Displacement Due to Plate Flexure

As seamounts and islands are constructed, complex interactions between the loads and the responses of the supporting lithosphere and mantle take place. These responses are first and foremost vertical isostatic adjustments and reflect both the rheology and dynamics of the lithosphere–asthenosphere system (Zhong and Watts 2002). To first order, the lithosphere will respond to surface loads as an elastic plate, flexing downward beneath the loads and bulging upward further away. Walcott (1970), who used a broken plate analogy with the break oriented along the ridge, pioneered early modeling of such flexural deformation beneath the Hawaiian Ridge, suggesting an elastic thickness of almost 60 km. Later, Watts and coworkers used Hawai'i as a case study for numerous flexure studies, including a two-ship, multichannel seismic experiment collecting images of crustal structure across the island chain (ten Brink and Brocher 1987; Watts 1978; Watts et al. 1985; Watts and Cochran 1974; Wessel 1993). These modeling efforts using continuous plates revealed a smaller plate thickness in the 25–35 km range.

The seismic experiment demonstrated beyond any doubt that the lithosphere indeed was flexed downward beneath Hawai'i by several km, confirming directly what had been inferred indirectly from the free-air gravity anomalies across the chain. Furthermore, the detailed study of the seismic images allowed for a better understanding of the evolution of the island chain as evidenced by the stratigraphy of the sediments in the flexural moat (Rees et al. 1993; ten Brink and Watts 1985). Four main lithostratigraphic units were observed within the moat. At the base lies a unit of approximately constant thickness of pelagic sediments, presumably predating the islands. Above it lies a thick wedge of lense-shaped units that are onlapping the flexural arch; these are internally chaotic and thought to represent buried landslide deposits. Next comes a sequence of continuous horizons that offlap the arch and are tilted towards the islands, finally topped by a ponded unit in the deepest part of the moat which contain the most recent sediment deposits. This stratigraphy clearly reflects the competing effects of mass-wasting on a large scale (Moore et al. 1989; Smith and Wessel 2000) and the ongoing flexural subsidence driven by the island construction (Rees et al. 1993; Watts and ten Brink 1989). The bulk of the deformation takes place during the extended shield-building phase after which the lack of additional surface loads causes the flexural subsidence to cease (Moore and Clague 1992).

Once shield-building and rejuvenated volcanism ceases, the islands are passively carried on the back of the Pacific plate away from the center of volcanic activity. Over the last 30 Myr, the plate has on average moved 95 km/Myr over the Hawaiian hotspot (Wessel et al. 2006). However, vertical displacements will continue to moderate the evolution of the islands and seamounts. It has been known since the time of Darwin that volcanic islands are eroded down to sea level and eventually drown altogether (Darwin 1842). The reason for this slow subsidence is the continued cooling of the oceanic lithosphere (Turcotte and Schubert 1982). Created hot and thin at the mid-ocean spreading center, the lithospheric plate loses heat through the seafloor and the depth-averaged reduction in plate temperature promotes a corresponding increase in density due to thermal contraction. Consequently, the plate must sink deeper to maintain its isostatic balance. Unlike the flexural vertical motions driven by shield-building (Moore 1970) or mass-wasting (Smith and Wessel 2000), which may reach several mm/year, the slow thermal subsidence is orders of magnitude smaller, typically in the range or 0.02–0.03 mm/year for the Northwestern Hawaiian Islands, as predicted by plate-cooling models (DeLaughter et al. 1999;

Stein and Stein 1992). It was long assumed that the passing of the plate directly over the hotspot would impart significant heat into the lithosphere, which would "reset" its thermal age and thus explain the rapid shoaling of the seafloor near Hawai'i (Detrick and Crough 1978). However, no heat flow anomaly has yet to be conclusively demonstrated (McNutt 2002), and numerical modeling has shown that the swell uplift is more likely a combined effect of chemical buoyancy and dynamic uplift (Phipps Morgan et al. 1995; Ribe and Christensen 1999). Nevertheless, the subsidence is steady and cumulative and will submerge any island given enough time to act. In equatorial regions, coral growth can generally keep up with this rate of subsidence, but natural variabilities in the climate and the eustatic sea level may combine to overwhelm this ability, as evidenced by drowned reefs along the Hawaiian Ridge (Grigg and Epp 1989).

13.2.3   Sea Level and Reef Development

13.2.3.1   Sea Level

For the purposes of this chapter, the term "coral reefs" refers to areas of biogenic carbonate accretion, some portion of which is composed of scleractinian or stony coral components. Most reef-building scleractinian corals contain and derive a significant portion of their energy budget from zooxanthellae, single-celled algal symbionts residing within coral tissue that utilize photosynthesis to produce carbohydrates and also generate oxygen as a by-product. As a result, scleractinian corals are most commonly found within the top few tens of meters of the photic zone. Over time scales of decades or longer, the growth and accretion of coral reefs in the NWHI and elsewhere are closely tied to the position of sea level.

Over the entire history of the NWHI they have been impacted by rapid fluctuations in sea level of up to perhaps as much as 150 m, which have been largely driven by glacio-eustatic processes (Cronin 1999). Sea-level changes resulting from oscillations of the Antarctic ice sheet have occurred since at least the beginning of the Oligocene Epoch ca. 34 million years ago (Barron et al. 1991). More recently, Moran et al. (2006) document a cooling of waters in the Arctic Ocean and the formation of sea ice approximately

45 Ma that is consistent with a global shift in from "greenhouse" to "icehouse" conditions, with the latter characterized by ice sheets periodically growing and receding at both poles.

Sea-level oscillations from the latter half of the Tertiary Period (ca. 1.8–45 Ma) are believed to have been both less frequent and of lower amplitudes than those of the last 800 kyr or so. Based on analyses of cores collected by the Ocean Drilling Program (ODP) off the New Jersey coast on the eastern seaboard of the US, sea-level changes of 20–34 m have been reported for the Oligocene (24–34 Ma) (Pekar and Miller 1996). Additional data suggest that fluctuations have occurred over million-year time scales throughout much of the period between 36 and 10 Ma (Miller et al. 1996). Although there is considerable debate about the timing and magnitude of ice sheet fluctuations over the last 10 Ma, several lines of evidence suggest that multiple sea-level oscillations of several tens of meters occurred on time scales of 100 ka–1 Ma (Cronin et al. 1994; Jansen and Sjøholm 1991; Naish 1997; Wilson 1995).

During the late Pliocene and early Pleistocene, approximately 2.7–0.8 Ma, the magnitude of oscillations in proxy records of ice volume grew larger than they were during the preceding period. The records show a principal periodicity of ~40 kyr

and suggest that fluctuations in sea level were on the order of 40–60 m (Cronin et al. 1994; Lambeck et al. 2002; Naish 1997; Ruddiman et al. 1989). Starting about 800 ka, a strong 100 Kyr cycle of oscillations in paleoclimate and ice volumes, and fluctuations in sea level with maximum amplitudes of 120–140 m, were established.

Ice volume changes were first derived more than half a century ago from the ratio of $^{18}O/^{16}O$ recorded in the calcium carbonate shells of foraminifera recovered from deep-sea sediment cores (Emiliani 1955). The total change in the ratio of $^{18}O/^{16}O$ between interglacial and glacial periods is approximately 1.5‰ (Mix 1992). The heavier ratios correspond to periods of high ice volume and reflect the preferential evaporation of the lighter isotope from the oceans and its subsequent deposition in ice sheets, leaving ocean waters with a higher concentration of heavier isotopes. By convention, interglacial periods or stages in the marine isotope record are assigned odd numbers, and even numbered stages represent glacial periods. The period of lighter isotopic ratios characterizing the current interglacial period is referred to as Marine Isotope Stage 1 (MIS 1). The last glacial period is referred to as MIS 2, the interglacial period before that is called MIS 3, and so on, as seen in Fig. 13.18.

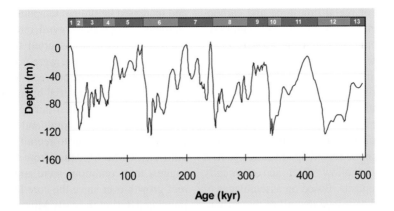

FIG. 13.18. Late Quaternary Sea Level. Estimated sea level in the NWHI for the period from 0–500 Kya, adapted from Webster et al. (2004). Modern sea level is shown at 0 m, and the position of sea level through time is adjusted for uplift of Oahu and subsidence of the NWHI. Data were also adjusted upward to match the position of MIS 5 deposits in Hawai'i. Boundaries of the numbered Marine Isotope stages in the bar at the top of the graph are from Imbrie et al. (1984)

The height of MIS 2, the last glacial period, is often referred to as the Last Glacial Maximum or LGM. During the LGM, from ~30 ka to ~19 ka, the ice-volume-equivalent sea level is inferred to have been −130 to −135 m below present sea level (Lambeck et al. 2002; Yokoyama et al. 2000). Starting at ~19 ka, sudden and strong deglaciation resulted in a massive marine transgression in which sea level rose rapidly. The transgression continued on into the Holocene, reaching present levels ca. 6 ka, and will be discussed in greater detail later.

13.2.3.2   Pre-Holocene Reef Development

Geologists long suspected that the NWHI were still found near sea level due to the upward growth of coral reef, and that the islands themselves were the tops of carbonate caps built on ancient volcanoes that had subsided long ago. Their suspicions were confirmed in the summer of 1965 when two holes were drilled from the surface of Midway Atoll down through an expanse of reef limestone and into basalt (Ladd et al. 1967, 1970). Their results provide the most comprehensive overview of pre-Holocene reef development in the NWHI currently available.

The first hole drilled at Midway was started on Sand Island inside the lagoon and went through 135 m of carbonate material overlying 17 m of basaltic sedimentary layers before hitting basalt rock. The second hole was drilled just inside the atoll rim on the eastern side of the atoll and penetrated 300 m of reefal limestone and carbonate sediments over 81 m of mostly volcanic clays before entering basalt. Ladd et al. (1967, 1970) analyzed core material and found, based on textural and geochemical information, that the basalts they recovered were erupted above sea level. They also concluded that the volcanic mound was partially truncated by wave action in pre-Miocene time (i.e., prior to 24 Ma), prior to subsiding below sea level. Weathered clays on the subsiding volcanic pedestal were reworked in shallow water and eventually covered by reef deposits. Based on assemblages of mollusks, it was determined that all of the carbonate material was deposited in shallow, warm, reef and lagoonal environments. Coral fragments were found throughout the carbonate sections of the core and even in the volcanic clay layer lying directly above the basalt at the base of the reef hole. These observations led Ladd et al. (1967, 1970) to conclude that corals were at least as important earlier in Midway's history as they are today. Miocene coral fauna was found to be more diversified than later assemblages suggesting a more favorable climate regime during that epoch. Coralline algae were abundant and widespread throughout both cores, suggesting that they have functioned as important contributors to reef building throughout the history of the NWHI.

Reflecting the numerous fluctuations in sea level reported in earlier sections, unconformities in the Midway cores show evidence of at least two periods of emergence during the Miocene and at least one during the Pleistocene (Ladd et al. 1967, 1970). Furthermore, the lithologies they describe suggest the occurrence of a number of others. Endodontid land snails, which are common on the larger volcanic islands in the Hawaiian Archipelago but are not presently found on the low carbonate islands of the NWHI, were also found in the upper 50 m of both cores, providing further evidence of their temporary emergence.

Given sufficient time and under the appropriate conditions of water temperature and clarity, irradiance, wave energy, and availability of substrate, coral reefs will tend to accumulate off low latitude shorelines. Grigg (1988, 1997) presents evidence that coral reefs have been continuously present in the Hawai'i–Emperor chain since the early Oligocene. He argues that closure of the Tethys Seaway, isolation of Antarctica, global cooling, and increased latitudinal temperature gradients collectively strengthened gyral circulation, enabling coral larvae to be transported to Hawaii. Clague (1996) reports that there were no islands with land areas above sea level between 34 and 30 Ma. All species requiring terrestrial habitat for part of their lifecycle were extirpated during this period. Few, if any, scleractinian coral and reef fish species would have been able to recruit to the H–E chain until ca. 30 Ma. Evidence from the Midway cores suggest that conditions were appropriate for coral reef growth ever since the core locations were first inundated during the early Miocene, at least during interglacial periods of relatively high sea level.

Holster and Clark (2000) found that sea surface temperatures (SSTs) in Hawai'i were 3.5–6.6°C cooler during the LGM, while Lea et al. (2000) report glacial-interglacial temperature differences as great as 5°C for the tropical Pacific over the last 450,000 years. SSTs are already so low at Kure that they are on the threshold below which coral reef accretion is precluded (Grigg 1982). It is likely, therefore, that coral reef growth did not occur at Kure during glacial periods, at least during the late Quaternary. As seen in Fig. 13.12, winter season SSTs are about 4°C cooler at Kure today than they are at Nihoa Island, Necker Island, and French Frigate Shoals, so it is conceivable that coral reefs may have been able to accrete at the southern end of the NWHI during late Quaternary glacial periods. Also, assuming a constant velocity of the Pacific plate of 7.6 cm/year since the beginning of the Miocene (Clague 1996), the NWHI would have ranged between approximately 400 and 1,800 km southeast of their present locations over the duration of the Miocene. Their more southerly locations during this period enhance the likelihood that at least the southernmost islands were accreting reef material during glacial low stands of sea level.

It is apparent from the Midway cores that environmental conditions were favorable for coral growth in the NWHI, at least during interglacial periods of high sea level. There is little if any other evidence currently available on reef development in the NWHI in the Pliocene and early Quaternary, but it is likely that reef accretion did occur, especially during warmer interglacial periods and on the southern islands and atolls. A number of studies have investigated late Quaternary fossil reefs in the Hawaiian Archipelago, but have usually focused on emerged deposits in the MHI. The relatively few studies that also investigate submerged reef deposits are discussed below and provide a model of Main Hawaiian Islands reef development that can be adapted to the NWHI.

On Oahu, interglacial periods of coral reef accretion were followed by subaerial erosion and diagenesis during the ensuing glacial low sea level stand. Previous reef deposits were, at times, eroded by the next sea level high stand, which also enabled new rounds of coral reef accretion in locations where conditions of sea level, environmental conditions, and accommodation space allowed (Fletcher and Sherman 1995; Grossman and Fletcher 2004; Sherman et al. 1999). As a result, late Quaternary reef deposits on Oahu are a complex and interwoven mix of multiple episodes of accretion and erosion.

The primary structural unit of the Oahu insular shelf and an important underlying unit of the coastal plain is a massive limestone shelf, the Waianae Reef (Stearns 1974) that accreted during MIS-7 and is found there today at depths from −5 to −20 m. Analysis of cores recovered from the Waianae Reef reveal differences in accretion on different sides of the island, reflecting variations in hydrodynamic forcing that are similar to current conditions. Predominantly coralline algal bindstone and some massive coral facies are observed on the windward (northeasterly) side of the island. Cores from the leeward side contain massive coral facies at a more seaward location and branching coral floatstone, interpreted to be from reef flat and back reef environments, respectively. The suite of sample ages taken from all cores ranges between 206 and 247 ka (Sherman et al. 1999). Fletcher et al. (2008, Chapter 11) contend that other limestone deposits on Oahu are of secondary importance to the Waianae Reef, and that its formation was the result of the first major episode of reef accretion following the volcanic shield-building phase of the island's history. The Waianae Reef was subaerially exposed and heavily eroded and karstified during MIS 6, and in some locations, during later glacial periods as well.

Lying unconformably on the Waianae Reef is interglacial reef accretion from MIS 5e, identified by Stearns (1974) as the Waimanalo Reef. Reef rock from this formation is subaerially exposed in a number of locations around Oahu and other MHI and has been well studied as a result. It is primarily composed of *in situ* coral-algal framestone with some bindstone and grainstone facies present as well. Sample dates from Oahu indicate that the reef was accreting as early as 143 ka and until at least 126 ka. On the leeward side of the island, massive coral framestone of the Waimanalo Reef accreted along the face and base of the seaward wall of the MIS 7 Waianae Reef. A contact between the Waimanalo Reef and underlying MIS 7 Waianae Reef was reported from one core recovered from windward Oahu at −5 m (Chapter 11 Fletcher et al., Sherman et al. 1999). The Waimanalo Reef has been reported from a number of locations under the Oahu coastal plain (Stearns 1974, 1978), but the shallow depth of the contact between the Waianae and Waimanalo Reefs and evidence from long limestone cores recovered from the coastal plain for commercial purposes indicate that the Waimanalo Reef is underlain by the more massive Waianae Reef complex.

As sea level dropped at the end of MIS 5e, framestone accretion from MIS 5a–d occurred on the seaward margin of the MIS 5e reef over the period between ca. 79–106 ka and is referred to as the Leahi Reef (Stearns 1974). Continuous accretion of massive coral between MIS 5e and 5d from a core from leeward Oahu suggests that sea level never fell below −25 m relative to modern sea level (Chapter 11, Fletcher et al.). The MIS 5a–d period also featured massive deposits of calcareous eolianite, which are found today on the coastal plains of at least the MHI of Maui, Molokai, Oahu, and Kauai. They are interpreted as resulting from large-scale dune deposition occurring during the drops in sea level between MIS 5e and MIS 5a–d, as well as at the end of MIS 5, and from extensive calcareous sediment production during MIS 5a–d (Chapter 11, Fletcher et al.; Stearns 1974). The eolianite outcrops on the coastal plains of these islands today have a significant impact on modern sediment dynamics. They also provide substrate for Holocene reef accretion in some locations and form small islets offshore of modern coastlines. Subaerial outcrops of both MIS 5e reef rock and MIS 5a–d eolianite have not been reported in geologic investigations in the northern atolls of the NWHI (Gross et al. 1969) or in general descriptions of all of the NWHI (Maragos and Gulko 2002; Rauzon 2001). Presumably, this is a result of subsidence of the NWHI since MIS 5.

A broad limestone shelf off Oahu has been observed at depths of −49 to −54 m. Based on its depth, it is hypothesized to have been deposited during MIS 3, ca. 30–60 ka (Chapter 11, Fletcher et al.), but no samples have been recovered from the structure to confirm this hypothesis. At the end of MIS 3 (ca. 30 ka), sea level dropped 50 m or more as the last glacial period, MIS 2, started. During which time, as discussed earlier, sea surface temperatures may have been warm enough to enable coral reef accretion, at least on the more southerly of the NWHI, although evidence of this has yet to be published.

When trying to compare late Quaternary reef development in the Main and Northwestern Hawaiian Islands, it is important to consider the effects of their different rates of vertical motion. The NWHI are well away from the mantle hotspot

and from the effects of lithospheric loading from the islands near it. As a result, the vertical motion of the NWHI follows a trajectory that is generally consistent with that predicted by the Parson-Sclater subsidence curve, at an estimated rate of ca. −0.024 mm/year at French Frigate Shoals (Grigg 1997), or 0.02–0.03 mm/year for the NWHI in general (DeLaughter et al. 1999; Stein and Stein 1992). Oahu on the other hand, where most of the MHI reef development data are from, is estimated to have uplifted at 0.02 mm/year from 0 to 125 kya, and at a rate of 0.06 mm/year between 125 and 500 kya (Grigg and Jones 1997). Using these rates and depth or elevation data from coral reef deposits on Oahu (Fletcher and Jones 1996; Grigg and Jones 1997; Jones 1993), crude estimates of the present depth/elevation of paleo-shorelines can be made. However, in the absence of data on magnitudes of net reef accretion during different interglacial periods, e.g., from cores or possibly from seismic studies, it is difficult to calculate past elevations or depths of the different islands in the NWHI. A general observation that can be made, however, is that relative to Oahu, the stratigraphy of fossil reefs of the NWHI is likely to be less complex and have more of a "layer cake" morphology. Younger deposits will have a greater tendency to overlie older limestone, thanks to the subsidence of the NWHI throughout the Quaternary, and undoubtedly throughout most of the Miocene for the older islands and atolls. Of course, sequences of carbonate reef are also much thicker and older in the NWHI than they are in the MHI. A program of fossil reef coring in the NWHI would help to address the above and other questions regarding patterns of late Quaternary reef accretion and relative sea level change.

13.3   Holocene Reef History

13.3.1   Holocene Sea Level

13.3.1.1   Post-glacial Sea-Level Rise and the Holocene Transgression

At the end of MIS 2, about 19 ka, eustatic sea level began to rapidly rise from its inferred position of −130 to −135 m to near modern sea level by about 6 ka, at a mean rate of 10 mm/year or more. As the sea level curves from Fig. 13.19 indicate however, that mean rate is composed of a series of irregularly alternating higher and lower rates. Considerable effort has been made to correlate the accelerations and decelerations of the overall sea level rise to specific events associated with deglaciation. During the first stage of deglaciation, until 14.5 ka, sea level rise was relatively slow at approximately 4 mm/year. Evidence suggests that it was caused by the decay of the Barents Ice Sheet on the continental shelf of the eastern Arctic Ocean (Elverhoi et al. 1993; Jones and Keigwin 1988; Peltier 1988; Vogt et al. 1994).

Results from ^{14}C dating of *Acropora palmata* corals from Barbados, and a model of their ecology and growth relative to sea level, led Fairbanks (1989) to propose a sudden 24 m jump in sea level. Called Meltwater Pulse 1A (MWP 1A), it is estimated to have started ca. 14.5 ka and lasted less than 1,000 years. The source of MWP 1A has generally been attributed to meltwater released by the decay of the Laurentide Ice Sheet which covered the northern portion of North America (Bond et al. 1992; Keigwin et al. 1991; Teller and Kehew 1994), but has also been attributed to decay of the Antarctic Ice Sheet (Clark et al. 1996).

MWP 1A was followed by a period of reduced sea level rise during a climatic cooling event known as the Younger Dryas (YD), which occurred between approximately 12.5–11.5 ka (Alley et al. 1993; Kennett 1990). There is ongoing debate about the causes, timing, effects on sea-level rise, and other details of this event. Blanchon and Shaw (1995) estimate sea-level rise at 13 mm/year during the YD, and most researchers agree that there was a several fold decrease in sea level rise rates around this time.

Another meltwater pulse, MWP 1B, raised sea level about 28 m beginning about 11 ka and at rates perhaps as high as 45 mm/year (Blanchon and Shaw 1995; Chappell and Polach 1991; Fairbanks 1989). Sources of this meltwater are believed to include both the Laurentide and Fennoscandinavian Ice Sheets. Following MWP 1B, sea level rise declined to a rate of ca. 10 mm/year, reaching or slightly exceeding modern sea levels by 6 ka.

13.3.1.2 Mid-Holocene Highstand

In the equatorial Pacific, Holocene sea level rise continued past 6 ka, reaching levels of 1–3 m above modern sea levels between 1.5 and 5 ka (Fletcher and Jones 1996; Grossman and Fletcher 1998;

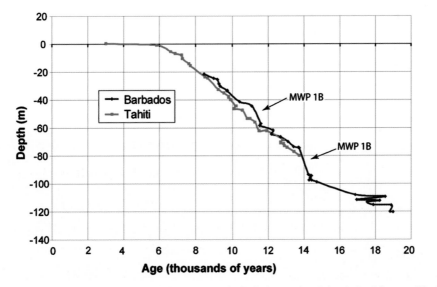

FIG. 13.19. Sea level rise during the last deglaciation. A record of relative sea level rise derived from uplifted coral reef deposits on the islands of Barbados and Tahiti (Bard et al. 1990, 1998) that we expect is similar to the sea level rise pattern from the NWHI. Note the two major increases in the rate of sea level rise during Melt Water Pulses 1A and 1B

Grossman et al. 1998). Geophysical models suggest that the high stand resulted from an ongoing process of glacial isostatic adjustment (GIA) in areas distant from Late Pleistocene ice cover (Clark et al. 1978). Mitrovica and Peltier (1991) developed the term "equatorial ocean siphoning" to describe a sequence of events consistent with geologic observations of the high stand of sea level. They propose that during a glacial period, the weight of continental ice sheets causes downward deformation of the crust, essentially squeezing the earth at the poles and forcing mantle material to flow towards central equatorial ocean basins. This process results in a gravitational anomaly in these areas and sea levels that are higher than they otherwise would be. Deglaciation resulted in a shifting of mass from the high-latitude continental ice sheets to ocean basins, enabling continents to viscoelastically rebound and material sequestered in the mantle under equatorial ocean basins to flow back to higher latitudes. The timing of the sea level high stand and its decay are attributed to the viscosity and flow dynamics of mantle material.

Sea level notches and beach and reef deposits elevated above the present position of mean sea level and dating from the mid-Holocene are found at a number of locations throughout the equatorial

Pacific and the MHI (Calhoun and Fletcher 1996; Fletcher and Jones 1996; Grossman and Fletcher 1998; Grossman et al. 1998). They have also been documented at Kure and Midway Atolls (Stearns 1941). Gross et al. (1969) dated several samples of reef rock from the emergent atoll rims at Midway and Kure Atolls (Fig. 13.20), finding a range of dates between 2.4 and 1.2 ka. They report that the emergent fossil reef must have been deposited at, or slightly below, low tide level and interpret their results and observations as indicating that sea level was 1–2 m above present levels during at least the period from which they have dated samples.

Fossil reef samples recovered from the MHI provide data with which a Holocene sea-level curve can be estimated (Fig. 13.21). Results are consistent with the patterns of eustatic and equatorial sea level change discussed above, but more data from the late Pleistocene and early Holocene would be useful. Additional data between the present and about 3 ka would help to better constrain sea level changes following the mid-Holocene high stand. Summing rates of NWHI subsidence and uplift on Oahu reported earlier, the total vertical offset of the NWHI relative to Oahu at 8 ka is estimated to be 0.35 m. Keeping that offset in mind, it is reasonable to use the Oahu sea level curve as a proxy for

FIG. 13.20. A segment of the atoll rim from Midway Atoll. Composed of in situ carbonate reef material, it was obviously deposited during a period of higher than present sea level (Photograph by Jean Kenyon)

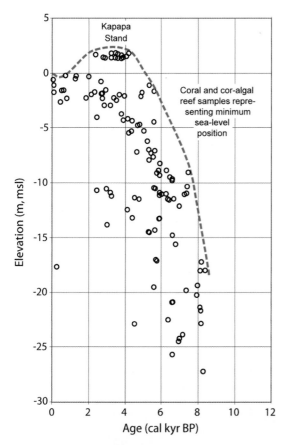

Fig. 13.21. Estimated Holocene sea level curve for the northern MHI and NWHI. Each circle represents the age and elevation of an *in situ* coral or other reef deposit (Easton and Olson 1976; Engels et al. 2004; Gross et al. 1969; Grossman and Fletcher 2004; Rooney et al. 2004). The red dashed line fit to the dated samples shows the minimum position of sea level

relative sea level (RSL) in the NWHI over the same period. These sea level data provide constraints on Holocene reef development for the islands and atolls between Oahu and Kure.

13.3.1.3 Modern Sea-Level Rise and Loss of Terrestrial Habitat

As with most regions where coral reefs are located, RSL has been rising over recent decades in the NWHI. There are two tide gauges in the NWHI, located at Midway Atoll and French Frigate Shoals which are part of the Permanent Service for Mean Sea Level (PSMSL 2006). Table 13.3 shows rates

of change in RSL, with positive values indicating a sea level rise for both islands and for Honolulu in the MHI calculated from monthly sea level observations.

As expected, the records show a general trend of RSL rise, but with a wide range of rates. It is very unlikely that the changes in rates are a result of variations in subsidence, especially considering the island's ages. The longest Midway record shows a slower rate of RSL rise relative to the same period at Honolulu, but over more recent periods shows increasingly faster rates. These results are consistent with those reported by Caccamise et al. (2005), who find that steric sea level trends (those attributed to thermal expansion of the water column) show falling sea levels to the north of the MHI and rising levels to the southeast between 1945 and 1995. However, between 1993 and 2002 they note an opposite pattern in satellite-derived sea surface heights and attribute the reversal to multi-decadal fluctuations in the spatial structure and magnitude of upper ocean temperature associated with the Pacific Decadal Oscillation.

Whatever the cause, RSL in the NWHI has been rising at moderate rates over the last few decades and several times faster more recently, with potentially dramatic impacts to the ecosystems there. Terrestrial habitats in the Northwestern Hawaiian Islands (NWHI) consist largely of low-lying oceanic sand islands (cays) and atolls, which are home to 4 land bird species, 3 terrestrial snail species, 12 plant species and over 60 species of terrestrial arthropods found nowhere else in the world (Conant et al. 1984). As all of these spend their entire lives on land, the NHWI terrestrial habitat represents their only tether to existence.

The NWHI are also important for large marine vertebrates including sea birds, green sea turtles,

TABLE 13.3. Rates of relative sea level rise for different periods, at different tide gauge stations in the Hawaiian Islands (PSMSL 2006).

Period	French Frigate Shoals	Midway Atoll	Honolulu
1905–2003	NA	NA	1.47 mm/year
1947–2003	NA	0.58 mm/year	1.37 mm/year
1974–2001	1.35 mm/year	2.95 mm/year	0.65 mm/year
1992–2001	7.42 mm/year	7.71 mm/year	−3.97 mm/year

and Hawaiian monk seals, all of which feed at sea but require terrestrial habitat with few or no predators to either nest (turtles and seabirds) or raise offspring (seabirds and seals). The endangered Hawaiian monk seal is one of the rarest marine mammals in the world, with a declining population of only approximately 1,200–1,300 individuals, primarily found in the NWHI (Antonelis et al. 2006). Over 90% of Hawaiian green sea turtles breeding females nest on beaches at French Frigate Shoals (Balazs and Chaloupka 2004). Terrestrial areas in the NWHI are also habitat for some 14 million seabirds of 18 species (Harrison 1990). Nesting of Laysan albatross (*Phoebastria immutabilis*) and black-footed albatross (*Phoebastria nigripes*) occurs almost entirely in the NWHI (Harrison 1990). The sooty, or Tristram's, storm petrel (*Oceanodroma tristrami*) has its most populous remaining breeding

sites in the NWHI (Harrison 1990). A significant proportion of the world population of Bonin petrels (*Pterodroma hypoleuca*) also breeds in the NWHI (Fefer et al. 1984; Harrison 1990).

The low-lying land areas of the NWHI are highly vulnerable to sand erosion due to storms and sea-level rise. Sea-level rise reduces cays by passive flooding and active coastal erosion, particularly during periods of seasonal high swell. As a result, the subaerial land area supporting these important littoral and coastal ecologies is at risk. Demonstrating this, cays such as Whaleskate Island, which had been important habitat for green turtles and monk seals at French Frigate Shoals, have been greatly reduced in size during roughly the past 40 years (Fig. 13.22) (Antonelis et al. 2006).

Using handheld Global Positioning System receivers and a theodolite, Baker et al. (2006)

FIG. 13.22. Whaleskate Island at French Frigate Shoals, NWHI. Once an important nesting island for Hawaiian green sea turtles and a primary pupping site for endangered Hawaiian monk seals, pictured from the air in 1963 (above) and from a small boat in 2002 (below)

produced the first topographic maps in three NWHI locations (Lisianski Island, Pearl and Hermes Reef, and French Frigate Shoals). They then used passive flooding scenarios to estimate the area that would be lost if islands maintained their current topography and the sea were to rise by various amounts. The Intergovernmental Panel on Climate Change (IPCC) evaluation of a number of model scenarios predicted low (9 cm), median (48 cm), and high (88 cm) estimates of sea level rise by 2100 (Church et al. 2001). Thus, passive flooding scenarios were run for each of these levels at mean low water (MLW) and at spring tide.

The uncertainty of predictions increases over time, but the expectation is that sea level will continue to rise beyond 2100 (Church et al. 2001).

Moreover, recent evidence suggests that sea level may rise more rapidly than previous models have predicted, due in part to an accelerated rate of ice loss from the Greenland (Ragnat and Kanagaratnam 2006) and West Antarctic (Shepherd and Wingham 2007) ice sheets.

The projected effects of sea level rise on island surface area varied considerably among the islands examined and depending upon the sea level rise scenario. For example, Lisianski Island is projected to be the least affected of the islands surveyed, losing only 5% of its area even under the maximum rise scenario. In contrast, the islets at French Frigate Shoals and Pearl and Hermes Reef are projected to lose between 15% and 65% of their area under the median sea level rise scenario, as shown in Fig. 13.23.

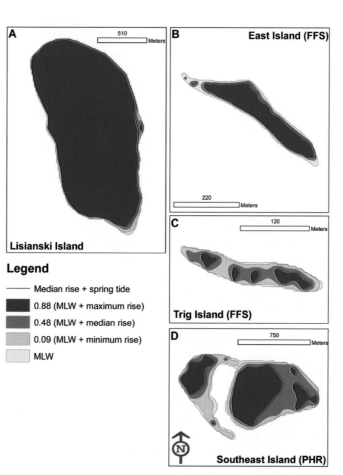

FIG. 13.23. Current and projected maps of four Northwestern Hawaiian Islands at mean low water (MLW) with minimum (9 cm), median (48 cm) and maximum (88 cm) predicted sea level rise. The median scenario at spring tide is also shown. (A) Lisianski Island; (B) East Island, French Frigate Shoals; (C) Trig Island, French Frigate Shoals; (D) Southeast Island, Pearl and Hermes Reef (Baker et al. 2006; reproduced by permission of Inter-Research)

Loss of terrestrial habitat to sea level rise will likely have considerable impacts on some species and no effect on others. Monk seals require islands for resting, molting, and, most importantly, parturition and nursing. Seals may experience more crowding and competition for suitable landing sites when islands shrink. French Frigate Shoals also hosts nests for ninety percent of Hawaiian green sea turtles, primarily at East Island (Balazs 1976). This may be fortunate since the island is projected to lose a smaller percentage of its area than the others analyzed. A small population of the Laysan finch *Telespiza cantans*, an endangered Hawaiian honeycreeper, occurs on Southeast Island at Pearl and Hermes Reef (the primary population of these birds is on Laysan Island). Considerable habitat would be lost from Southeast Island under the Baker et al. (2006) median scenario (Fig. 13.23), which could greatly increase this population's extinction risk (McClung 2005).

Baker et al.'s (2006) projected scenarios treat the islands' current configurations as static, though some, especially the smaller islets, are more likely to be dynamic. Therefore, their projections should be viewed as the currently best available demonstration of the potential effects of sea-level rise. Furthermore, passive flooding scenarios do not take into account ancillary factors that could substantially influence the future of the NWHI. These include erosive recession of the shoreline causing land loss, redistribution of sediments by long-shore drift (resulting in both gains and losses of land area), net permanent loss of sand volume offshore, and onshore sand deposition by overwash during high wave activity. A rise in the groundwater table during sea level rise could also displace seabird burrows. If reef growth does match sea level rise, this could result in increased sand accretion, thereby mitigating losses. For all these reasons, the impact of factors other than simply passive flooding as a result of increased sea level could lead to greater or lesser loss of habitat than presented here.

There are many uncertainties associated with the projected effects of sea level rise in the NWHI. A key issue is whether coral reef accretion will be able to occur at both the appropriate locations and at sufficient rates to enable them to offset future rises in sea level and continue to buffer lagoonal areas. If accretion is sufficient, land areas in the NWHI may continue to enjoy whatever current protection

they have from ocean swells. We recommend that measurement of site-specific rates of reef accretion be made a management priority.

13.3.2   Holocene Reef Development

13.3.2.1   Coral Growth Rates, Cover, and Reef Accretion

There are a handful of studies that discuss aspects of Holocene reef development in the NWHI and others from the MHI whose results are relevant to the NWHI. Grigg (1982) reported mean linear growth rates and densities of ten colonies of *Porites lobata* coral collected from a depth of −10 m on the fore reef of the southwest side of a number of islands and atolls across the Hawaiian Archipelago. He also reported percentages of coral cover from each sampling location and derived an estimate of reef accretion. Using his values of island-specific coral community measures, vertical rates of reef accretion can also be estimated and are shown for four islands in Table 13.4. As the table suggests, Grigg (1982) found that growth rates show a strong latitudinal dependence, with corals in cooler more northerly islands growing more slowly. He also noted that, based on the work of Gross et al. (1969), at least at the more northerly atolls the carbonate contribution from coralline algae is likely to be significantly more than that from corals.

Leveraging results collected by teams of researchers during visits between 2000 and 2003, Siciliano (2005) reports that the percentage of coral cover at Kure Atoll varies between different geomorphic habitats, with a mean value for the atoll of 20%, as seen in Table 13.5. In contrast and rather surprisingly, another study found no significant difference between benthic communities at most windward fore reef and back reef sites, although they did confirm differences between these areas and lagoonal communities at French Frigate Shoals (Vroom et al. 2005). Siciliano (2005) also collected coral colonies or short cores through coral colonies from all ten islands in the NWHI to measure annual growth rates. In that study corals in wave-protected back reef and lagoonal environments are found to grow at similar rates of ca. 5–9 mm/year regardless of latitude. However, in wave-exposed locations results are similar to Grigg's (1982), with

TABLE 13.4. Latitudinal variations in coral colony growth rates and reef accretion across the Hawaiian Archipelago, from Grigg (1982). Note that coral cover percentages are from high coral cover areas and have been shown by more recent and extensive data collection to not be representative of most areas around these islands.

	Hawai'i	Oahu	FFS	Kure
Coral Growth Rate (mm/year)	13	12	8	3
Density (mg/mm^3)	1.42	1.47	1.4	1.6
Coral Cover (%)	93	61	66	10
Reef Accretion (kg CaCO^3/m^2/year)	15	10	8.9	0.3
Reef Accretion by Corals (mm/year)	10.6	6.8	6.4	0.2

TABLE 13.5. Coral cover and predicted net reef accretion at Kure Atoll, from Siciliano (2005). The range of percentages of coral measured at French Frigate Shoals (FFS) and Pearl and Hermes Atoll (P & H) from visits between 2001–2005 are from Kenyon et al. (in press).

Geomorphic habitat	Kure % Coral cover	Predicted net accretion (mm/year)	FFS % Coral cover	P & H % Coral cover
Fore Reef/Pass	16	1.7	7–9	6–9
Back Reef	27	1.5	10–19	10–15
Lagoonal Patch and Reticulate Reef	27	3.9	8–28	14–20
Mean (weighted) for Kure	20	2.1	–	–

rates in the NWHI ranging from ca. 2–8 mm/year and showing a statistically significant decline with increasing latitude.

Both of these studies attempted to extrapolate coral growth rates and other data to estimate rates of reef accretion. Grigg's (1982) estimates are for accretion due to corals only. They reflect the linear dependence of coral growth rates on latitude and range from 6.4 mm/year at French Frigate Shoals to 0.2 mm/year at Kure, although he suggests that the latter value might be about 0.6 mm/year when accretion from coralline algae is included. Siciliano's estimate of 1.7 mm/year for net reef accretion in fore reef environments (Table 13.5) at Kure is a more sophisticated approach which attempts to quantify all sources of carbonate accretion and erosion. Although there is a several fold difference in their rates, it is important to note that both studies conclude that net reef accretion is presently occurring today at Kure Atoll, even in fore reef environments. By extension, their results can be considered to apply to fore reef areas across the entire NWHI, or at least to those on the southwestern sides of islands and atolls.

This conclusion is contrary to results from the MHI. On Oahu, Holocene reef accretion is well documented to have primarily been limited to areas which are sheltered to some degree from high wave energy, particularly from long period North Pacific swell that strikes the Hawaiian Archipelago in the winter (Dollar 1982; Dollar and Tribble 1993; Grigg 1983, 1998; Sherman et al. 1999; Storlazzi et al. 2002). At locations exposed to wave energy, Holocene accretion is limited to a thin and patchy veneer of living coral and coralline algae resting on an eroding Pleistocene-age fossil reef surface. Even this limited growth is episodically removed by large swell and storms. In more sheltered areas however, such as Hanauma Bay and Kailua Bay, 10–15 m thick sequences of Holocene coral reef have accreted (Easton and Olson 1976; Grossman and Fletcher 2004). We suspect that accretion of corals and coralline algae on most fore reef environments in the NWHI is ephemeral and periodically removed by high wave events. The small size of coral communities (5–20 cm) on NHWI fore reefs supports this hypothesis (Kenyon et al. in press). However, accretion is likely to be occurring at back reef and lagoonal patch and reticulate reef areas. In the absence of data from whole-reef cores Siciliano's (2005) accretion rates for these areas are the best estimates available at this time.

13.3.2.2 Potential Future Impacts of Ocean Acidification on Reef Accretion

Though studies to date have not yet adequately examined the influences of changing seawater carbonate chemistry on past reef accretion processes in the Northwestern Hawaiian Islands, research findings of the past decade have led to mounting concern that rising atmospheric carbon dioxide (CO_2) concentrations will cause changes in the ocean's carbonate chemistry system, and that those changes will affect some of the most fundamental biological and geochemical processes in the sea (Boyd and Doney 2003; Kleypas et al. 2006). Anthropogenic activities have driven atmospheric carbon dioxide (CO_2) concentrations to levels greater than, and increasing at a rate faster than, experienced for at least the last 650,000 years (IPCC 2001; UK Royal Society 2005). Global oceans are the largest natural reservoir for this excess CO_2, absorbing ~1/3 of that emitted each year (Kleypas et al. 2006). According to recent predictions, dissolved CO_2 in the surface ocean is expected to double over its preindustrial value by the middle of this century (UK Royal Society 2005; Kleypas et al. 2006). Oceanic uptake of CO_2 drives the carbonate system to lower pH (acidification) and lower saturation states of the carbonate minerals calcite, aragonite, and high-magnesium calcite. These are the materials used to form supporting skeletal structures in many major groups of marine organisms (Kleypas et al. 2006). Ocean acidification has already reduced surface ocean pH by 0.1 units (equivalent to a 30% increase in the concentration of hydrogen ions) and is expected to reduce surface ocean pH by 0.3–0.5 units over the next century (UK Royal Society 2005).

A growing number of laboratory controlled experiments now demonstrate that ocean acidification adversely affects many marine organisms. Organisms that construct their shell material from calcium carbonate (marine calcifiers) are expected to be especially vulnerable. In particular, ocean acidification has been shown to hamper the ability of reef-building corals and reef-cementing crustose coralline algae to calcify, thereby affecting their growth

and accretion and making them more vulnerable to erosion (Fabry 1990; Langdon et al. 2000, 2003). By mid-century, corals may erode faster than they can be rebuilt potentially making them less resilient to other environmental stressors (e.g., coral bleaching and disease) (Langdon 2007, personal communication). This could compromise the long-term viability of these ecosystems, perhaps impacting the thousands of species inhabiting reef habitats. In addition, ocean acidification may elicit broad physiological responses from non-calcifying organisms through less obvious and unknown pathways.

For the NWHI, which (as described above) have generally low net accretion rates, the predicted decreases in calcification rates (or even reversal to dissolution and increased erosion) in response to ocean acidification could seriously impact the ability of many islands, reefs, and banks to accrete fast enough to keep up with predicted sea level rise. Furthermore, the combined influences of sea level rise and acidification could fundamentally alter the biological and geochemical processes of the NWHI ecosystems. With the potential magnitude and consequences of the predicted changes, we strongly recommend significant new research to observe the spatial and temporal patterns of changing carbonate chemistry, calcification rates, and ecological and geochemical responses in the Northwestern Hawaiian Islands.

13.3.2.3 Patterns of Holocene Reef Development

As discussed above, long-term reef accretion is unlikely at locations in the NWHI exposed to high wave energy from seasonal long period swell. At sites that are sheltered to some degree however, reef growth and accretion is occurring. Detailed studies of reef accretion at several sites on Oahu provide insights into patterns of Holocene coral reef development that are likely to be occurring at sheltered locations in the NWHI. Easton and Olson (1976) analyzed ten cores across an actively growing fringing reef at the moderately low-energy environment of Hanauma Bay on Oahu's southeastern shore. Grossman and Fletcher (2004) report on their findings based on 32 cores and benthic transects from the moderately high wave energy setting of Kailua Bay on the island's northeastern shore. Both studies found that modern reef growth was initiated

7–8 kya after these areas were inundated during the Holocene transgression. The initial stage of "catch-up" (to sea level, Neumann and Macintyre 1985) accretion was rapid, up to 6 mm/year, and created a reef composed predominantly of coral framestone. Around 5–6 kya, both studies found that accretion rates stopped altogether or only resulted from rubble accumulation (Kailua Bay), or slowed down and began to include more calcareous algae (Hanauma Bay). A final stage of slow "keep-up" style bindstone accretion is reported at both sites, along with rapid horizontal seaward progradation of the reef at Hanauma Bay. As a whole, Holocene accretion patterns have been controlled by the availability of accommodation space for reef growth, beneath the wave base (Grossman and Fletcher 2004).

The above two studies suggest likely scenarios to describe coral reef accretion in low to moderate wave energy environments in the NWHI. Rates of coral colony growth reported by Siciliano (2005) indicate that the more northerly latitudes would not preclude similar results in lagoonal environments. However, wave-sheltered locations from more northerly atolls in the NWHI which are exposed to cooler non-lagoonal waters are likely to feature slower rates of accretion, lower percentages of coral, and higher percentages of calcareous algae.

Another study from Oahu was conducted at the Punaluu Reef, which is exposed to high wave energy from North Pacific swell (Rooney at al. 2004). Due to its morphology and wave exposure, the Punaluu Reef is more likely than the other two studies to have accretion patterns that are representative of fore reef and reef crest environments in the NWHI. Portions of the Punaluu reef crest are emergent during most of the tidal cycle and were obviously deposited during periods of higher than modern sea level, as was reported above for locations in the NWHI (Gross et al. 1969). These findings indicate that reef crests were active areas of accretion during the mid-Holocene, and suggests that they may be active sites of long-term accretion today.

Results of benthic ecological surveys from Punaluu are consistent with geologic evidence of reef crest accretion. They show that portions of the reef crest below the low tide level have up to 95% live cover of encrusting coral and coralline algae. This is a markedly higher percentage of seafloor inhabited by reef-accreting organisms than was found anywhere else across the reef, and indicates that the reef crest is an area of particularly active growth, and potentially, of accretion. Although quantitative surveys have not been reported for very shallow, high wave energy NWHI reef crests, surveys do report high coverages (37%) of crustose coralline red algae at fore reef areas on the northwestern side of Pearl and Hermes Atoll (Page et al., in review). Growth on the reef crest is possible despite high wave energy from winter swell, apparently because the sloping fore reef breaks the worst of the incident wave energy, and background trade wind swell provides a continuous flow of ocean water from offshore over the reef crest.

Geological and ecological data collectively suggest that shallow reef crest areas, including those in the NWHI, may be able to keep pace with at least moderate levels of relative sea level rise. In the face of predictions of accelerating sea level rise, determining accretion rates in the vicinity of the reef crest has significant management implications for the fate of terrestrial and shallow marine ecosystems in the NWHI.

13.3.2.4 El Niño Influence of Holocene Reef Accretion

As mentioned earlier, wave energy has been shown to be the predominant control of reef accretion in the MHI, both today and throughout the Holocene. Although direct evidence is lacking, patterns of coral growth and wave measurements suggest that this is true for the NWHI as well. However, data from several locations in the MHI show that rapid accretion during early to middle Holocene time occurred in areas where today it is precluded by the wave regime, suggesting an increase in wave energy (Rooney et al. 2004). Analysis indicates that accretion terminated at these locations ca. 5 kya, preceding the best available estimates for the decrease in relative sea-level rise during the mid-Holocene high stand of sea level at ca. 3 kya (Grossman et al. 1998). No permanent accretion is occurring at these sites today, despite the availability of hard substrate at depths between −30 m and the intertidal zone, suggesting that some factor other than relative sea-level rise has altered the ability of these sites to support accretion.

Numbers of both large North Pacific swell events and hurricanes in Hawai'i are statistically greater during El Niño years because of the increased intensity of the Aleutian Low and its more southward locus during these times. The study infers that if those major reef-limiting forces were suppressed, net accretion would occur in some areas in Hawai'i that are now wave-limited. Studies have shown that El Niño/Southern Oscillation (ENSO) was significantly weaker during early-mid-Holocene time, only gaining an intensity similar to today's ENSO signal ca. 5 kya (e.g., Clement et al. 2000; McGlone et al. 1992; Moy et al. 2002; Rodbell et al. 1998). It is hypothesized that this shift in ENSO may at least partially explain patterns of Holocene reef accretion in the Hawaiian Archipelago, and especially in the NWHI, which experience markedly higher wave energies from North Pacific swell than do the MHI.

As discussed earlier, southward extensions of the subtropical front appear to correlate with positive ENSO phases and expose the northernmost atolls to waters that are both cooler than normal and contain higher nutrient concentrations. What impact this ENSO-related phenomena may have on reef accretion at these atolls is unknown, but may be significant and should be investigated.

13.4 Geomorphology of Holocene Reefs

13.4.1 Atolls

13.4.1.1 Atoll Formation

In the NWHI, Laysan Island is the only low coral island in the NWHI that is not associated with an atoll. The term atoll can be defined as a "ring-shaped coral reef that surrounds a lagoon without projecting land area and that is surrounded by open sea" (Parker 1997). The "projecting land area" refers to original volcanic basalt structures, and not to low coral islands, which are common features of atolls. There are approximately 425 atolls in the world, with over 300 in the Indo-Pacific region (Guilcher 1988). Charles Darwin's (1842) famous theory of atoll formation postulates that fringing reef will grow around the periphery of a volcanic island in tropical latitudes. As the reef grows outward and the island subsides, the fringing reef is eventually transformed into a barrier reef. This progression results

in an atoll once "the last and highest" volcanic pinnacle disappears. That progression is clearly visible in the Hawaiian Archipelago moving from the island of Hawai'i, at the southeastern end of the chain, to the northwest. The other MHI have fringing reef in many areas, some of which is hundreds of meters or more offshore. Small remnants of volcanic peaks can be seen at Nihoa and Necker Islands, at French Frigate Shoals, and at Gardner Pinnacle. True atolls, as well as low carbonate islands, are found further north in the chain.

Darwin's theory was generally confirmed in 1951 when the first of several holes drilled through the former Engebi Island (later obliterated by nuclear weapons testing) at Eniwetok Atoll reached volcanic basement rock (Ladd and Schlanger 1960). However, glacio-eustatic sea level fluctuations were an unknown concept in Darwin's time, although subsequent research has shown their importance on the modern geomorphology of atolls. A general correlation between atoll area and maximum atoll lagoon depth has been recognized for decades (e.g., Emery et al. 1954; Purdy 1974; Purdy and Winterer 2001; Umbgrove 1947). A positive correlation has also been reported for annual rainfall and the maximum depth of atoll lagoons (Purdy 1974; Purdy and Bertram 1993; Purdy and Winterer 2001). Assuming that modern rainfall is a reasonable proxy for precipitation during glacial low stands, the correlations suggest that maximum lagoon depth is primarily a function of preferential subaerial dissolution of the surface of the central portion of an atoll. However, Purdy and Winterer (2001) point out that subsequent interglacial sea levels high enough to flood atoll surfaces but not so high as to drown them would provide accommodation space for new reef growth. Atoll rims, elevated above the sediment-filled basins would preferentially accrete new reef growth. This alternating pattern of atoll rim accretion during high stands and preferential lagoonal erosion during low stands is likely to have been occurring on NWHI atolls throughout the late Quaternary, and to have made a significant contribution to their present morphology.

13.4.1.2 Open Versus Classical Atolls in the NWHI

In general, atolls tend to be the best developed on their windward sides (Guilcher 1988), and this is true for those in the NWHI. The degree of variation between windward and leeward morphologies have led to atolls in the NWHI being classified as either classical or open atolls (Maragos and Gulko 2002). Classical atolls feature an emergent or very shallow atoll rim around a deeper lagoon. Although there may be passes or breaks in it, the rim encircles more of the atoll than it does not. Kure, Midway, and Pearl and Hermes Atolls are all considered classical atolls.

The perimeter reef or carbonate rim structure is not present around much or all of an open atoll. Since the atoll rim may not be present at all in an open atoll, the presence of a lagoon is critical to defining whether a carbonate structure is an atoll, island, or reef. The term lagoon can be defined as, "a shallow sound, pond, or lake generally near but separated from or communicating with the open sea" (Parker 1997). The Northwestern Hawaiian Islands of French Frigate Shoals, Lisianski Island/Neva Shoal, and Maro Reef are classified as open atolls and are listed in order of least to most open. Presumably, open atolls are formed in the same manner as classical atolls, but with a definitive atoll rim initially forming only on the windward side, in response to dominant seas and currents from that direction.

The positive feedback described above for atoll formation during repeated subaerial exposure followed by submergence during glacial–interglacial cycles would tend to reinforce the original open atoll morphology.

13.4.1.3 Reticulate Reef

Lisianski Island/Neva Shoal and Maro Reef are classified as open atolls because they lack an atoll rim, but still have waters that are in communication with but separated from the open ocean. At these locales reticulate reef, rather than an atoll rim, isolates lagoonal waters. Reticulate reef, so named for its net-like structure, is also found inside the lagoon at Pearl and Hermes Atoll (see Fig. 13.7), and smaller extents of reticulate reef are also found at French Frigate Shoals. Reticulate reef in the NWHI appears to often have high coral cover and species diversity (Fig. 13.24). This is the case along the sides of the reef structures, and on their tops as well, if the water is deep enough and the reef is protected to some degree from wave energy. At shallower and more exposed locations, encrusting coralline red algae are common (Maragos and Gulko 2002; Page et al., in review).

Fig. 13.24. High coral cover is evident on the side of this reticulate reef structure at Maro Reef, despite the often relatively turbid water conditions (Photograph by Jean Kenyon)

Reticulate structure is found in many coral reef regions, including for example French Polynesia, Kiribati, and the Great Barrier Reef (e.g., Guilcher 1988; Woodroffe et al. 2004). On the Great Barrier Reef, reticulate reef has been attributed purely to patterns of Holocene reef development on a flat and homogeneous Pleistocene basement (Collins et al. 1996). On the other hand, reticulate reef at Matiaiva Atoll in the Tuamotu Archipelago has been attributed to karstification of fossil reef during a lower stand of sea level, followed by Holocene accretion (Guilcher 1988). The origin of reticulate reef in the NWHI is unknown, although it does appear that the structures are active areas of coral reef accretion, at least in some areas. Modern rates of accretion across reticulate reef networks in the NWHI have yet to be investigated, but this is a research topic of interest from both academic and resource management perspectives.

13.4.2 Spur and Groove Morphology

13.4.2.1 Spur and Groove Formation

One structural pattern, which emerges in many areas across the globe where scleractinian corals are found, is spur and groove reef. Named for the corrugated pattern of alternating coral ridges and troughs that make up the reef's surface, these features are oriented in the direction of wave travel during typical wave conditions. Spur and groove can be seen in aerial and satellite photographs to curve in accordance with the refraction of waves as they approach shore.

Found in fore reef environments which experience moderate to high wave intensity, spur and groove morphology is the result of complementary processes: elevated rates of carbonate accretion on the coral spurs, and accelerated erosion of reef framework in the grooves. The relative importance of these processes to spur and groove formation varies between locations and through time depending on environmental conditions. Regardless of the balance at any given reef, as suggested by Fig. 13.25, elevated wave energy accelerates spur and groove formation; both accretion and erosion have been shown to independently accelerate spur and groove formation with increased wave energy (Cloud 1959; Goreau 1959; Kan et al. 1997; Shinn et al. 1981; Sneh and Friedman 1980; Wood and Oppenheimer 2000).

The spacing of spurs is controlled by wave power, a function of wave height and period (Munk and Sargent 1948; Roberts et al. 1975; Storlazzi

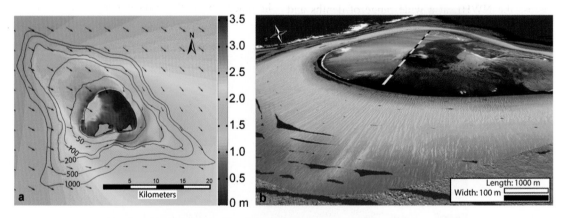

Fig. 13.25. (**a**) A plot of significant wave height generated by a typical northwest storm swell at Midway Atoll, as calculated by the Simulating Waves Nearshore (SWAN, http://vlm089.citg.tudelft.nl/swan/index.htm) model, version 40.51. The northwest swell is the most powerful wave field in the north central Pacific, and consistent with that, model results show high wave energy concentrated on the northwest side of the atoll. Note the wave shadow (in blue) to the southeast of the atoll. (**b**) A perspective view of Midway from the northwest illustrates preferential spur and groove development in the area most exposed to the northwest swell. Color indicates depth. Spurs extend to a depth of 60 m (blue). For scale, the 1 km long black and white bars shown in the foreground are overlain across the top of the atoll and indicate a lagoon diameter of approximately 10 km

et al. 2003) and may be a harmonic response to the dominant frequency of breaking waves (Shinn et al. 1981). The spur and groove zones found in shallow and more energetically exposed reef tracts tend to have a low amplitude, high frequency morphology. In contrast, spur and groove structures which are more protected by depth and or wave exposure angle develop a high amplitude, low frequency profile (Roberts et al. 1975; Storlazzi et al. 2003). Storlazzi et al. (2003) also found that spur and groove dimensions are relatively stable across depths on high energy reef tracts, but exhibit great variability across depths in more sheltered environments. The tallest, most widely spaced spurs were found in deep, relatively quiescent conditions where only the occasional large swell influences the bottom.

13.4.2.2   Spur and Groove Morphology in the NWHI

Spur and groove geomorphology has been identified at seven locations in the Northwestern Hawaiian Islands, although further mapping may identify more locations. From south to north they are: West Nihoa Bank, French Frigate Shoals, Middle Brooks Bank, Lisianski Island, Pearl and Hermes Atoll, Midway Atoll, and Kure Atoll. Spur and groove features are found on islands and atolls distributed across the NWHI, at a wide range of depths and wave intensity regimes. They can be divided into deep, relict features and shallow, actively developing features. Spur and groove reefs restricted to depths greater than 20–30 m and unconnected to shallow spur and groove on the fore reef are likely relict features from lower sea level stands. Strong wave action may be forcing sediment scour sufficient to perpetuate pre-existing spur and groove structures, but it is unlikely that their development was initiated at such depths. Deep spur and groove has been observed on West Nihoa and Middle Brooks Banks as well as offshore of some of the atolls, at depth ranges of approximately 40–60 m.

Shallow spur and groove features become better defined and exhibit greater vertical relief to the northwest in the NWHI. French Frigate Shoals has one small area of well-developed spur and groove and a perimeter of moderate-amplitude spur and groove restricted to shallow depths. Lisianski exhibits only small, isolated areas of low amplitude

spur and groove limited to shallow depths. The fore reef perimeters at Pearl and Hermes, Midway, and Kure Atolls on the other hand are almost entirely encircled by well-defined spur and groove features which extend down to 60 m at Midway and Kure (Fig. 13.6). Furthermore, the large expanse of high-amplitude spur and groove across the bank tops of Kure and Midway are unmatched anywhere else in the Northwest Hawaiian Islands.

13.4.3   Sea Level Notches and Terraces

13.4.3.1   The Distribution of Sea-Level Notches and Terraces in the NWHI

Morphological features related to former sea-level stands including terraces and notches have long been recognized in the Hawaiian Archipelago on land and beneath the sea. Terraces are characterized by gently sloping shelves created when sea level remained constant long enough for wave action to carve them into the host substrate. In some cases, carbonate-secreting marine organisms may have accreted on top or along the seaward or outer edge of the eroded surface, widening or otherwise modifying its shape. Sea-level notches may be found along the landward or inner edges of terraces, with some of the original host material remaining in place above the eroded notch. The presence of an overhanging visor above a notch indicates a period of relatively constant sea level during which the notch was formed by a combination of biological and mechanical erosion. The still stand was then followed by a rapid rise in sea level that was large enough to flood the overhanging visor to a depth sufficient to remove it from the zone of active erosion, thereby preserving it (Fletcher and Sherman 1995). The identification and cataloging of sea-level notches and terraces are important for developing a record of paleo-sea level in the Hawaiian Islands, which may in turn contribute to our understanding of global climate change.

Attempts have been made to correlate sea-level terraces and notches across the MHI to investigate a wide range of topics including Quaternary sea-level change and tectonic subsidence related to volcanic loading. However, little attention has been paid to similar features in the NWHI because of difficulties associated with accessing the remote banks and atolls. Recent efforts have been made to collect high resolution multibeam bathymetry and backscatter imagery from the atolls and banks in the NWHI (Miller et al. 2004, http://www.soest. hawaii.edu/pibhmc/) and have surveyed approximately 30% of the seafloor between depths of 20 and 500 m. These data have enabled the following identification of sea-level terraces and associated features.

The low-elevation carbonate islands, atolls, and shallow flat-topped banks in the NWHI all have a major break in slope that ranges between depths of −100 to −200 m at different islands. The break is characterized by a significant change from gently sloping shallow seafloor to the steep and heavily eroded submarine flanks of the islands. The islands, atolls, and flat top banks shallower than −50 m in the NWHI also have terraces at a variety of depths shallower than −100 m eroded into the side of the islands by prolonged periods of relatively stable sea level.

At both Kure and Midway Atolls a series of discontinuous terraces at depths of −30 to −60 m with steep slopes along their outer edges can be traced only part way around the atolls. The terraces are best preserved on the south and east sides where exposure to large northwest swell energy is reduced. On both atolls spur and groove features originate at the fringing reef edge and often correlate further down slope with gullies that cross the shallow terrace ledges.

At Pearl and Hermes Atoll, a terrace at ca. −90 m encircles almost the entire atoll, but lacks a prominent seaward ledge which makes the feature hard to distinguish. A series of terraces at −30 to −70 m on the west side of the bank become narrower and eventually disappear on the northwest side of Pearl and Hermes where the atoll flank is locally very steep.

Multibeam data are currently available only from the southwestern side of Maro Reef, but indistinct terraces are present just above the major slope break at ca. −100 m and at −45 to −55 m. A prominent 10–15 m high ledge on its seaward side and steep slope with a notch cut at its base helps to distinguish a terrace at −28 to −32 m. Linear gullies and channels cross the terrace edge and the terrace top exhibit varied seafloor morphologies, including expansive sand fields and channels dotted with pinnacles and hard bottom areas with rough topography.

On the southwestern bank at French Frigate Shoals, two terraces exist above the major slope break. The deeper terrace at ca. −50 m is discontinuous and does not have a pronounced ledge, whereas the more prominent terrace at −25 to −35 m terminates seaward at a 5–15 m high ledge. Gullies cross the ledge and isolated blocks of material and notches are present at the base of the scarp on its landward side. The morphology of the shallow terrace suggests that coral reef has accreted on top of it. The seaward edges of the prominent shallow terraces at Maro Reef and French Frigate Shoals are distinguished in the backscatter data by a distinct change in acoustic intensity. Intensity ranges from high on the outer edge of the terrace and on the steeper slopes above and below it to low along the inner margin of the terrace. Backscatter results suggest that the shallow terrace serves as a sink for sediment that is washed out of the atoll.

Flat submarine banks in the NWHI were planed off by erosion when their tops were subaerial or just below sea level. The Brooks Banks are four neighboring flat-topped banks northwest of French Frigate Shoals that include, from southeast to northwest, Southeast Brooks, East Brooks (EB), Middle Brooks (MB), and West Brooks (WB) Banks (Fig. 13.26). The major break in slope on all three is at −100 to −120 m except on the southwestern side of EB where a portion of the bank edge is missing. We interpret this as a probable landslide head wall because comparable smaller features occur on the northwest flank of EB, and blocky material lies down slope at −350 m. Equivalent features in the MHI were similarly interpreted by Mark and Moore (1987). EB is the smallest of the three banks, with a prominent terrace at −58 to −65 m, marked by a 10–15 m high steep slope on its seaward margin. Both MB and WB are shallower and larger than EB. Outer terrace edges are well defined on both banks at −60 to −65 m and −35 to −40 m. The tops of all three banks gradually deepen to the north, possibly reflecting more erosion due to the impact of north and northwest swell energy, or karst processes during periods of subaerial exposure. The origin of curious linear and arcuate ridges on the tops of MB and WB is uncertain, but we hypothesize that they may be reef or beachrock deposits.

In summary, on the more northerly NWHI such as Kure and Midway Atolls sea-level terraces and notches are difficult to distinguish but generally

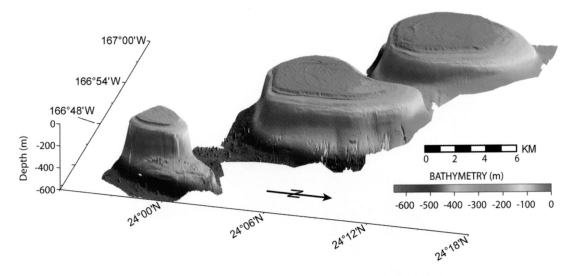

FIG. 13.26. Perspective view of Brooks Banks as seen in Multibeam bathymetry (http://www.soest.hawaii.edu/pibhmc/pibhmc_nwhi_utm3n_Brooks.htm) looking from the northeast. The flat-topped and terraced nature of submarine banks in the NWHI is exemplified by East, Middle, and West Brooks Banks. The major break in slope on each is below 100 m. A prominent terrace at a depth of 58–65 m on East Brooks can be correlated with similar features on Middle and West Brooks banks. A terrace at 35–40 m is present only on Middle and West Brooks banks. We hypothesize that arcuate and linear ridges present on top of the shallow terrace are drowned reef or beach rock deposits. Blocky material interpreted as probable mass wasting and landslide debris is present at 350 m on the northeast flank of East Brooks

group in the −30 to −60 m depth range. On the more southerly banks and atolls, wave-cut terraces and notches are more easily distinguished and group into depth ranges similar to those identified in the MHI. One group of features lies in the −45 to −65 m range and may correlate with the Penguin Bank shoreline complex (pbsc), while another group at −25 to −45 m appear in a similar depth range to the Kaneohe shoreline complex (ksc) (Colbourn et al. 1974; Fletcher and Sherman 1995). Fletcher and Sherman (1995) suggest that the pbsc and ksc may have been formed during MIS 3 (ca. 50–64 ka) and MIS 5 (ca. 79–110 ka), respectively.

As with similar features in the MHI, sea-level terraces and notches in the NWHI are discontinuous and difficult to recognize and correlate between different islands. The problems with identifying a single feature on all sides of all islands are probably due to a combination of erosion, burial, and because terraces may have never formed in certain locations (Colbourn et al. 1974). Additionally, a single terrace may have been occupied by sea level on more than

one occasion, and our understanding of paleosealevel is rudimentary, making it difficult to correlate the formation of any given terrace or notch to a specific period of time. The large distances between the islands and banks of the NWHI and the drastic differences in island and bank size further complicate the situation. However, sea-level features in the NWHI group into general depth ranges that, especially in the younger NWHI, may correlate with features at similar depths in the MHI. Further analyses and additional data are needed to better understand the history of sea-level terraces and notches across the Hawaiian Archipelago.

13.4.3.2 Drowned Sea-Level Notches and the Endangered Hawaiian Monk Seal

Monk seals were first documented using shallow submerged caves in the early 1980s when divers encountered them during surveys of the shallow reefs of the Northwestern Hawaiian Islands region as part of the multi-agency NWHI tripartite

Fɪɢ. 13.27. Monk seal and French Frigate Shoals outfitted with a CRITTERCAM (Photograph by Frank Parrish)

investigation. The seals were observed wedged into tidal and sub-tidal cracks and sea caves of the basalt islands of Gardner Pinnacles, La Perouse Pinnacle, and the islands of Necker, Nihoa, Kaula, and Lehua. Similar sheltering by seals has been seen in shallow reef caves just offshore from the sand islets of the atolls. Seals were thought to use these caves as a refuge from predators and to avoid social competition with other seals for the limited beach area on the few sand islets of the atoll.

Between 1995 and 2001, computer controlled video cameras (National Geographic Television's CRITTERCAMs (Fig. 13.27)) were attached to monk seals at French Frigate Shoals to identify key foraging habitats. A total of 42 cameras were deployed on adult males and 9 juveniles (male and female) collecting 69 h of underwater surveillance. Twenty-eight of the seals were seen to take up stationary positions on the bottom. Half would make a series of prolonged broadcast vocalizations, and the rest appeared to be sleeping. Vocalizations were limited to the adults (only males were instrumented) and occurred on open bottom. The resting seals used overhangs and ledges of the bottom as dens. The resting episodes were as long as 15 h during which the seals were motionless except for periodic subdued trips to surface to take a breath.

Documentation of this underwater resting redefines surface-based perspectives of the monk seal foraging. Early foraging estimates used the seals time away from the beach as a measure of foraging effort when in fact considerable time spent away from the beach is spent resting in underwater caves. More interesting is the fact that underwater caves are clearly a resource the seals exploit and are an important component in the seal's submarine landscape. The underwater caves include reef caves and sea level notches (Fig. 13.28). Reef caves are generally ledges and voids formed as part of the configured structure of the reef matrix. Monk seals were seen resting in these structures at depths as shallow as the intertidal and as deep as 20 m. Sea level notches are submerged ledges and sea caves that occur considerably deeper and have been carved out of the base carbonate by ocean waves and bioerosion during previous low stands of prehistoric sea level.

Monk seals are generally solitary animals and space is limited on the sand islets of the atoll. Use of the underwater caves provides a means to rest at sea without drifting off site. It also affords concealment from predators and other seals. Most of the resting was seen inside of the atoll close to the seals haulout beaches, but some were at the neighboring banks including the sea level notch of SE Brooks

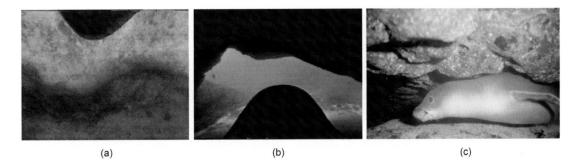

(a)   (b)   (c)

FIG. 13.28. Monk seal resting habitat on Brooks Banks. (a) Overhead view of a sea level notch or overhang imaged by CritterCam. (b) CritterCam view of the entrance of a cavern used as a resting spot by Monk seals. (c) Monk seal at rest in a cave

and the summit reef caves of Middle Brooks. In general, the banks of the NWHI are low relief tracts of seafloor (Parrish and Boland 2004; Kelley et al. 2006). Thus, areas with caves represent a limited resource that seals are likely to identify and revisit to rest and extend their foraging range to more remote grounds. Future studies should assess the degree to which such undersea features influence the foraging behavior and at sea movements of the endangered Hawaiian monk seal.

13.4.4 Drowned Islands and Banks

13.4.4.1 Darwin Point

Kure Atoll has the farthest north emergent land in the Hawai'i–Emperor Chain, with all of the 63 seamounts and guyots further north than Kure (Clague 1996) now permanently below the depth at which coral reef can accrete. Rotondo (1980) determined that over at least the past 20 Ma the latitude at which guyots north of Kure drowned, or subsided below the depth at which coral reefs can accrete, ranged no further north than between 27° N and 31° N. Building on Rotondo's results, Grigg (1982) coined the term Darwin Point to refer to a threshold for atoll formation and stated that a Darwin Point exists at the northern end of the Hawaiian Archipelago at 29° N latitude, where Kure Atoll is located. Grigg (1982) states that the reef at Kure has kept pace with sea level, but that the mean growth rate for corals there (0.2 mm/year) represents 22–60% of the carbonate necessary to keep pace with sea level rise. Grigg (1982) further suggested that the additional 40–78% of the required carbonate

is produced by other calcareous organisms such as coralline algae, molluscs, bryozoa, etc.

Results from studies in the MHI mentioned earlier suggest that the fore reef at Kure and other islands and atolls of the NWHI are generally not accreting. However, rapid growth rates of corals from inside the reef crest (Siciliano 2005), and the presence of mid-Holocene age emergent fossil reef (Gross et al. 1969) suggest that accretion in these areas may be able to keep up with moderate rates of sea level rise. A program of reef coring at Kure would help to elucidate Holocene patterns of reef accretion and evaluate the atoll's potential to adapt to rates of accelerated sea level rise projected for the future by some studies (Ragnat and Kanagaratnam 2006).

13.4.4.2 Drowned Banks

In addition to the islands and atolls in the NWHI that have emergent land, a number of drowned banks are also found there at a range of depths. It has been suggested that the reason that some islands manage to keep up with sea level rise while others do not may be related to erosion by high wave energy. Grigg (1982) points out that there is a negative correlation between the surface area of bank tops and drowning of banks. Exploring that idea further, Grigg and Epp (1989) identified a total of 37 islands, atolls, and banks in the NWHI with summits shallower than −200 m. They calculated the summit area of these banks as the area enclosed within the −180 m (100 fathom) contour and their summit depth as the shallowest flat surface area exceeding 1.0 km^2. Plotting results for all 37 islands and banks, they found that the logarithm of bank summit area is inversely related to depth. Noting that the flat tops of the banks suggests that they were all truncated by wave action, Grigg and Epp (1989) hypothesize that smaller banks are truncated faster and to deeper depths than larger banks. Once eroded to deeper depths, banks are more likely to be found below the critical depth of −30 to −40 m. Below these depths, it is unlikely that they will be able to accrete significant volumes of coral reef and maintain the bank's position near the surface under even very moderate rates of relative sea level rise.

We note from Grigg and Epp's (1989) data that there are no bank tops between depths of ca. −70 and −110 m. A cluster of bank summits are found at depths of ca. −50 to −70 m, and additional bank tops are found below −110 m and above −50 m. We hypothesize that the banks in the shallower cluster (−50 to −70 m) were above the −30 to −40 m critical depth of drowning for coral reefs (Grigg and Epp 1989) during MIS 3 and so were able to actively accrete coral reef. This hypothesized accretion may have enabled the banks to both increase their summit area and decrease their depths during the 30 kyr duration of MIS 3. Banks below the gap (i.e., currently deeper than 110 m) were likely below the critical depth during MIS 3 and unable to accrete coral reef. Vertical reef accretion at a modest 1.3 mm/year is sufficient to generate a 40 m depth difference between the tops of bank summits above and below the −70 to −110 m gap. This scenario is reasonably consistent with estimated depths of sea level during MIS 3 of approximately −50 to −80 m below present sea level (Lea et al. 2002). This issue could be addressed by sampling and dating carbonate from the cluster of bank tops in the −50 to −70 m depth range, and from those below −110 m.

13.5   Future Research

The paucity of work on geologic aspects of reef development in the NWHI is not surprising considering the logistical difficulties associated with work in this remote area. However, as a result, a number of significant questions remain to be answered regarding late Quaternary reef accretion and relative sea level change in the NWHI. With unprecedented rates of climate change predicted to

fundamentally alter many of the biogeochemical processes both globally and in the NWHI, the urgency of addressing these questions and improving our understanding of these processes is greater than it has ever been. Although the NWHI are well protected from direct human impacts, they remain highly vulnerable to the significant climate change-related threats of coral bleaching and disease, ocean acidification and decreased calcification and accretion rates, sea level rise, and others. It is important to begin improving our understanding of past and present biogeochemical processes in the NWHI as soon as possible. Through improved knowledge and understanding of these processes, past, present, and future, our ability to identify and assess options to increase resilience and long-term adaptability and sustainability of these ecosystems will be enhanced. A few of the most important research needs have been pointed out in the preceding pages and are reviewed below.

The collection and processing of high-resolution multibeam bathymetry for shallow to moderate (−20 to −500 m) depths in the NWHI is only about 30% completed, but should clearly be continued. Most of the seafloor shallower than −20 m is characterized by occasional soundings. Depths have been estimated using satellite imagery, but those data have numerous gaps due to cloud cover and are found to be unreliable below about −12 m. The development of complete and seamless digital terrain and bathymetry models for all the NWHI to a depth of at least −500 m would be a significant asset for unraveling patterns of reef development and for a host of other research and management needs. There is an active program of multibeam mapping in the NWHI (http://www.soest.hawaii.edu/pibhmc/pibhmc_nwhi.htm). A complementary effort to map emergent land and shallow water areas using bathymetric Light Detection and Ranging (LIDAR) is recommended.

Results of the Oahu relative sea level curve are reasonably consistent with the eustatic and equatorial sea level curves discussed above and are used as a proxy for relative sea level in the NWHI. However, the limited temporal extents of the Oahu curve and lack of data from the NWHI make it difficult to interpret patterns of coral reef and coral island development in the NWHI and to predict future morphologies. The development of a sea level curve for the NWHI would provide an important tool for these efforts, as well as a representative dataset for the central North Pacific gyre, which would be of interest to paleoclimatologists and others. A program of fossil reef coring such as has been conducted on Oahu and other MHI in recent years (http://www.soest.hawaii.edu/coasts/research/index.html) would be very helpful for developing a sea level curve.

Further mapping and analysis of sea level terraces and notches, and constraining their ages of formation through core analysis, would also contribute to our understanding of past sea levels in the NWHI. Mapping of these features would also be a critical step towards better understanding their influence on the foraging patterns of the endangered Hawaiian monk seal.

In the face of predictions of accelerating sea level rise and ocean acidification, determining "whole reef" accretion rates in different geomorphic zones at a range of islands and atolls across the NWHI would be useful. In particular, accretion rates from reef crest areas have significant implications for the fate and management of terrestrial and lagoonal habitats in the NWHI which, in turn, are important to the many marine species with links to these habitats. Reef crests may prove to be "keystone areas" for the protection of environments inside atoll rims from some of the effects of sea level rise. Concurrent with examining whole reef accretion rates, it is necessary to understand the spatial and temporal structure of seawater carbon chemistry cycles and their impact on calcification and accretion processes. If accretion is sufficient to keep pace with future rates of relative sea level rise, terrestrial and lagoonal environments in the NWHI may continue to enjoy whatever current protection they have from ocean swells. Determining long-term rates of reef crest accretion through core analysis, and monitoring of reef crest benthic communities, are required to effectively address this important issue and should be a management priority.

Rates of terrestrial habitat loss and the factors that cause it are critical for several endangered or threatened species in the NWHI. A study of historical shoreline change using existing survey charts and aerial photographs may be able to provide a longer-term perspective of the evolution of islands and islets of the NWHI than currently exists. Such a perspective may be of particular importance to shoreline dynamics, given the multi-decadal nature of climatic fluctuations such as the Pacific Decadal Oscillation (PDO), which have been shown to influence shoreline processes in the MHI (Rooney and Fletcher 2005). Also, an understanding of the patterns and causes of long-term change are critical to the design of effective measures to mitigate terrestrial habitat loss.

Continuous time series of surface and subsurface temperatures, salinity, and wave energy from existing instrument networks in the NWHI (http://www.pifsc.noaa.gov/cred/oceanography.php) are of only a few years duration. It is hard to interpret the geological significance of measurements, such as temperature spikes associated with recent bleaching events (Hoeke et al. 2006; Kenyon and Brainard 2006), without a longer-term context in which to evaluate them. Records of SST and salinity covering periods of decades to as much as a century or longer can be developed using stable isotope analyses of cores extracted from large living coral colonies. These data would be of use to both local resource managers as well as to the climate change research community.

It was noted earlier that there are no bank tops between depths of ca. −110 and −70 m in the NWHI, but that a cluster of banks do exist at depths of ca. −70 to −50 m. We hypothesize that the banks in that cluster were above the −30 to −40 m critical depth of drowning for coral reefs (Grigg and Epp 1989) and so were able to actively accrete coral reef during MIS 3. This issue could be addressed by sampling and dating carbonate from this cluster of bank tops in the −50 to −70 m depth range and from bank tops below −110 m.

The studies suggested above are but a small sample of some of the more obvious and pressing of a wide range of other geologically related research topics waiting to be addressed in the NWHI. This chain of islands, banks, and atolls is unique in terms of the high latitude of its northwesterly atolls, the wide geographic range it covers, the high percentages of endemic species that inhabit it, and the high degree of protection from anthropogenic disturbance it has received over the last century. We anticipate that work in these extraordinary islands over the next few decades will make significant contributions to our understanding and management of coral reefs around the world.

Acknowledgements Salary support from the NOAA Coral Reef Conservation Program for J. Rooney, R. Hoeke, J. Weiss, and J. Chojnacki while writing this manuscript is gratefully acknowledged. We thank Chris Kelley, Rick Grigg, Scott Ferguson, Joyce Miller, Emily Lundblad, Francis Lichoswki, and Jamie Smith for their extensive efforts in the NWHI that contributed to this work and for many helpful discussions. Jean Kenyon graciously gave us access to her extensive library of photographs from the NWHI, and others were provided by Molly Timmers and Jake Asher. Edouard Bard and Jody Webster generously sent us copies of their sea level data. We thank the captains, officers, and crew members past and present of the NOAA ships *Oscar Elton Sette*, *Hi'ialakai*, and *Townsend Cromwell* for outstanding field support during cruises to the NWHI. The contents of this chapter were improved by the careful reviews of Jochen Halfar, Francine Fiust, Joyce Miller, and an anonymous reviewer.

References

Acton GD, Gordon RG (1994) Paleomagnetic tests of Pacific plate reconstructions and implications for motion between hotspots. Science 263:1246–1254

Alley RB, Meese DA, Shuman CA, Gow AJ, Taylor KC, Grootes PM, White JWC, Ram M, Waddington EB, Mayewski PA, Zielinski GA (1993) An abrupt increase in Greenland snow accumulation at the end of the Younger Dryas event. Nature 362:527–529

Antonelis GA, Baker JD, Polovina JJ (2003) Improved body condition of weaned Hawaiian monk seal pups associated with El Nino events: potential benefits to an endangered species. Mar Mammal Sci 19:590–598

Antonelis GA, Baker JD, Johanos TC, Braun RC, Harting AL (2006) Hawaiian monk seal (*Monachus schauinslandi*): staus and conservation issues. Atoll Res Bull 543:75–101

Atwater T (1989) Plate tectonic history of the Northeast Pacific and western North America. In: Winterer EL, Hussong DM, Decker RW (eds) The Eastern Pacific Ocean and Hawaii. The Geology of North America. Geological Society of America, Boulder, CO, pp 21–72

Baker JD, Littnan CL, Johnston DW (2006) Potential effects of sea level rise on the terrestrial habitats of endangered and endemic megafauna in the Northwestern Hawaiian Islands. Endangered Spec Res 4:1–10

Balazs GH (1976) Green turtle migrations in the Hawaiian Archipelago. Biol Conserv 9:125–140

Balazs GH, Chaloupka M (2004) Thirty-year recovery trend in the once depleted Hawaiian green sea turtle stock. Biol Conserv 117:491–498

Bard E, Hamelin B, Fairbanks RG, Zindler A (1990) Calibration of the 14C timescale over the past 30,000 years using mass spectrometric U-Th ages from Barbados corals. Nature 345:405–410

Bard E, Arnold M, Hamelin B, Tisnerat-Laborde N, Cabioch G (1998) Radiocarbon calibration by means of mass spectrometric 230Th/234U and 14C ages of corals. An updated data base including samples from Barbados, Mururoa and Tahiti. Radiocarbon 40:1085–1092

Barron JA, Larsen B, Baldauf JG (1991) Evidence of late Eocene to early Oligocene Antarctic glaciation and observations on late Neogene glacial history of Antarctica: results from Leg 119. Proc ODP Sci Res 119:869–891

Bianco T, Ito G, Becker J, Garcia MO (2005) Secondary Hawaiian volcanism formed by flexural arch decompression. Geochem, Geophys, Geosys 6:Q08009, doi:10.1029/2005GC000945

Blanchon P, Shaw J (1995) Reef drowning during the last deglaciation: evidence for catastrophic sea level rise and ice-sheet collapse. Geology 23:4–8

Bond G, Heinrich H, Broeker W (1992) Evidence of massive discharges of icebergs into the North Atlantic ocean during the last glacial period. Nature 360:245–249

Bond NA, Harrison, DE (2000) The Pacific Decadal Oscillation, air-sea interaction and central north Pacific winter atmospheric regimes. Geophys Res Lett 27(5):731–734

Boland R, Zgliczynski B, Asher J, Hall A, Hogrefe K, Timmers M (2006) Dynamics of debris densities and removal at Northwestern Hawaiian Islands coral reefs. Atoll Res Bull 543:461–470

Boyd P, Doney SC (2003) The impact of climate change and feedback process on ocean carbon cycle. In: Fasham MJR (ed) Ocean Biogeochemistry: The Role of the Ocean Carbon Cycle in Global Carbon Cycle in Global Change. Springer, Germany, pp 157–193

Bromirski PD, Cayan DR, Flick RE (2005) Wave spectral energy variability in the northeast Pacific. J Geophys Res 110:C03005, doi:10.1029/2004JC002398

Caccamise DJ, Merrifield MA, Bevis M, Foster J, Firing YL, Schenewerk MS, Taylor FT, Thomas DA (2005) Sea level rise at Honolulu and Hilo, Hawaii: GPS estimates of differential land motion. Geophys Res Lett 32:L03607, doi:10.1029/2004GL021380

Calhoun RS, Fletcher CH (1996) Late Holocene coastal plain stratigraphy and sea-level history at Hanalei, Kauai, Hawaiian Islands. Quat Res 45:47–58

Cande SC, Kent DV (1995) Revised calibration of the geomagnetic polarity timescale for the Late Creatceous and Cenozoic. J Geophys Res 100:6093–6095

Caruso SJ, Businger S (2006) Subtropical cyclogenesis over the Central North Pacific. Weather Forecast 21(2):193–205

Chappell J, Polach H (1991) Post-glacial sea-level rise from a coral record at Huon Peninsula, Papua New Guinea. Nature 349:147–149

Chu P (2002) Large-scale circulation features associated with decadal variations of tropical cyclone activity over the central North Pacific. J Climate 15:2678–2689

Church JA, Gregory JM, Huybrechts P, Kuhn M, Lambeck K, Nhuan MT, Qin D, Woodworth PL (2001) Changes in sea level in climate change 2001: the scientific basis. In: Houghton JT, Ding Y, Griggs DJ, Noguer M, van der Linden P, Dai X, Maskell K, Johnson CI (eds) Contribution of Working Group I to the Third Assessment Report of the Intergovernmental Panel on Climate Change. Cambridge University Press, Cambridge, pp 641–693

Clague D, Dalrymple GB (1987) The Hawaiian-Emperor Volcanic Chain: Part I. Geologic Evolution. US Geological Survey Professional Paper 1350, pp 5–54

Clague DA (1996) The growth and subsidence of the Hawaiian-Emperor volcanic chain. In: Keast A, Miller SD (eds) The Origin and Evolution of the Pacific Island biotas, New Guinea to Eastern Polynesia: Patterns and Processes. SPB Academic Publishing, Amsterdam, The Netherlands, pp 35–50

Clark JA, Farrell, WE, Peltier WR (1978) Global changes in postglacial sea level: a numerical calculation. Quat Res 9:265–287

Clark PU, Alley RB, Keigwin LD, Licciardi JM, Johnsen SJ, Wang H (1996) Origin of the first global meltwater pulse following the last glacial maximum. Paleoceanography 11:536–577

Cleghorn PL (1988) the settlement and abandonment of two Hawaiian outposts: Nihoa and Necker Islands. Bishop Mus Occas Pap 28:35–49

Clement AC, Seager R, Cane MA (2000) Suppression of El Niño during the mid-Holocene by changes in the Earth's orbit. Paleoceanography 15(6):731–737

Cloud, PE Jr (1959) Geology of Saipan Mariana Islands, Part 4 – submarine topography and shoal-water ecology. US Geol Surv Prof Pap 280-K:361–445

Colbourn WT, Campbell JF, Moberly R (1974) Hawaiian submarine terraces, canyons, and Quaternary history evaluated by seismic-reflection profiling. Mar Geol 17:215–234

Collins LB, Zhu ZR, Wyrwoll KH (1996) The structure of the Easter Platform, Houtman Abrolhos Reefs: Pleistocene foundations and Holocene reef growth. Mar Geol 135:1–13

Conant S, Christensen CC, Conant P, Gange WC, Goff ML (1984) The unique terrestrial biota of the northwestern Hawaiian Islands. In: Grigg, RW and Tanoue, KY (eds) Proceedings of the Second Symposium on Resource Investigations in the Northwestern Hawaiian Islands. Volume 1 Sea Grant Miscellaneous Report UNIHI-SEAGRANT-MR-84-01, Honolulu, HI, pp 77–94

CRCP (2007) U.S. Department of Commerce, NOAA Coral Reef Conservation Program. http://www.coral-reef.noaa.gov/about/welcome.html

Cronin TM (1999) Principles of Paleoclimatology. Columbia University Press, New York

Cronin TM, Kitamura A, Ikeya N, Kamiya T (1994) Mid-Pliocene paleoceanography of the Sea of Japan. Palaeogeogr, Palaeoclimatol, Palaeoecol 108:437–455

Dalrymple GB, Clague DA, Vallier TL, Menard HW (1987), 40Ar/39Ar age, petrology, and tectonic significance of some seamounts in the Gulf of Alaska. In: Keating BH, Fryer P, Batiza R, Boehlert (eds) Seamounts, Islands, and Atolls. American Geophysical Union Monograph 43, Washington, DC, pp 297–315

Dameron OJ, Parke M, Albins M, Brainard R (2007) Marine debris accumulation in the Northwestern Hawaiian Islands: an examination of rates and processes. Mar Pollut Bull 54(4):423–433

Dana JD (1891) Characteristics of Volcanoes from Hawaiian Islands. Dodd, Mead and Company, New York, 399 p

Darwin C (1842) The Structure and Origin of Coral Reefs. Smith Elder & Co., London, pp 214

DeLaughter J, Stein S, Stein CA (1999) Extraction of a lithospheric cooling signal from oceanwide geoid data. Earth Planet Sci Lett 174:173–181

Detrick RS, Crough ST (1978) Island subsidence, hot spots, and lithospheric thinning. J Geophys Res 83:1236–1244

Dollar SJ (1982) Wave stress and coral community structure in Hawaii. Coral Reefs 1:71–81

Dollar SJ, Tribble GW (1993) Recurrent storm disturbance and recovery: a long-term study of coral communities in Hawaii. Coral Reefs 12:223–233

Donohue M, Brainard R, Parke M, Foley D (2000) Mitigation of environmental impacts of derelict fishing gear through debris removal and environmental monitoring. Proceedings of the 4th International Marine Debris Conference on Derelict Fishing Gear and the Marine Environment. NOAA, Honolulu, HI, pp 58–78

Easton WH, Olson EA (1976) Radiocarbon profile of Hanauma Reef, Oahu, Hawaii. Geol Soc Am Bull 87:711–719

Elverhoi AW, Fjeldskaar W, Solheim A, Nyland-Berg M, Russwurm L (1993) The Barents Sea Ice Sheet – a model of its growth and decay during the last ice maximum. Quat Sci Rev 12:863–873

Emery KO, Tracey JI, Ladd HS (1954) Geology of Bikini and Nearby Atolls. US Geological Survey Professional Paper 260-A, Washington, DC, p 265

Emiliani C (1955) Pleistocene temperatures. J Geol 63:538–568

Engels M, Fletcher C, Field M, Storlazzi C, Grossman E, Rooney J, Conger C (2004) Holocene reef accretion: Southwest Molokai, Hawaii. J Sedim Res 74(2):255–269

Fabry VJ (1990) Shell growth rates of pteropod and heteropod molluscs and aragonite production in the open ocean: implications for the marine carbonate cycle. J Mar Res 48:209–222

Fairbanks RG (1989) A 17000 year glacio-eustatic sea level record: influence of glacial melting rates on the Younger Dryas event and deep-ocean circulation. Nature 260:962–968

Fefer SI, Harrison CS, Naughton MB, Shallenberger RJ (1984) Synopsis of results of recent seabird research conducted in the Northwestern Hawaiian Islands. In: Grigg, RW and Tanoue KY (eds) Proceedings of the second symposium on resource investigations in the northwestern Hawaiian Islands, Volume 1. Sea Grant Miscellaneous Report UNIHI-SEAGRANT-MR-84-01, Honolulu p 9–76

Feigenson MD, Hofmann AW, Spera FJ (1983) Case studies on the origin of basalt. II. The transition from tholeiite to alkalic volcanism on Kohala Volcano, Hawaii. Contrib Mineral Petr 84:90–405

Fenner D (2005) Corals of Hawaii. Mutual Publishing, Honolulu, HI, 143 pp

Fletcher CH, Jones AT (1996) Sea-level highstand recorded in Holocene shoreline deposits on Oahu, Hawaii. J Sedim Res 66:632–641

Fletcher CH, Sherman C (1995) Submerged shorelines on Oahu, Hawaii: archive of episodic transgression during the deglaciation? J Coastal Res, Special Issue 17:141–152

Fletcher CH, Bochicchio C, Conger CL, Engels M, Feierstein EJ, Frazer LN, Glenn CR, Grigg RW, Grossman EE, Harney JN, Isoun E, Murray-Wallace CV, Rooney JJ, Rubin K, Sherman CE, Vitousek S (2008) Geology of Hawaii Reefs. in: Riegl B, Dodge RE (eds) Coral Reefs of the USA, Springer, pp 431–483

Friedlander AM, Demartini EE (2002) Contrasts in density, size, and biomass of reef fishes between the northwestern and main Hawaiian Islands: the effects of fishing down apex predators. Mar Ecol Progr Ser 230:253–264

Frey FA, Wise WS, Garcia MO, West H, Kwon ST, Kennedy A (1990) Evolution of Mauna Kea volcano, Hawaii: petrologic and geochemical constraints on postshield volcanism. J Geophys Res 95:1271–1300

Garcia M, Grooms D, Naughton J (1987) Petrology and geochronology of volcanic rocks from seamounts along and near the Hawaiian Ridge. Lithos 20: 323–336

Garcia M, Foss D, West H, Mahoney J (1995) Geochemical and isotopic evolution of Loihi Volcano, Hawaii. J Petrol 36:1647–1674

Garcia MO, Caplan-Auerbach J, De Carlo EH, Kurz MD, Becker N (2006) Geology, geochemistry and earthquake history of L'ihi Seamount, Hawai'i's youngest volcano. Chemie der Erde 66:81–108

Garcia MO, Haskins EH, Stolper E (2007) Stratigraphy of the Hawaiian scientific drilling project. Anatomy of a Hawaiian volcano. Geochem, Geophys, Geosys (G3) 8:Q02G20 DOI 10.1029/2006GC001379

Goreau TF (1959) The ecology of Jamaican coral reefs. I. Species composition and zonation. Ecology 40:67–90

Graham NE, Diaz HF (2001) Evidence for intensification of North Pacific winter cyclones since 1948. Bull Am Meteorol Soc 82(9):1869–1893

Grigg RW (1982) Darwin point: a threshold for atoll formation. Coral Reefs 1:29–34

Grigg RW (1983) Community structure, succession and development of coral reefs in Hawaii. Mar Ecol Progr Ser 11:1–14

Grigg RW (1988) Paleoceanography of coral reefs in the Hawaiian-Emperor Chain. Science 240:1737–1743

Grigg RW (1997) Paleoceanography of coral reefs in the Hawaiian-Emperor Chain – revisited. Coral Reefs 16(Suppl):S33–S38

Grigg RW (1998) Holocene coral reef accretion in Hawaii: a function of wave exposure and sea level history. Coral Reefs 17:263–272

Grigg RW (2006) The history of marine research in the Northwestern Hawaiian Islands: lessons from the past and hopes for the future. Atoll Res Bull 593:13–22

Grigg RW, Epp D (1989) Critical depth for the survival of coral islands: effects on the Hawaiian Archipelago. Science 243:638–641

Grigg RW, Jones AT (1997) Uplift caused by lithospheric flexure in the Hawaiian Archipelago as revealed by elevated coral deposits. Mar Geol 141:11–25

Gross MG, Milliman JD, Tracey JI, Ladd HS (1969) Marine geology of Kure and Midway Atolls, Hawaii: a preliminary report. Pac Sci 23:17–25

Grossman EE, Fletcher CH (1998) Sea level 3500 years ago on the northern main Hawaiian Islands. Geology 26:363–366

Grossman EE, Fletcher CH (2004) Holocene reef development where wave energy reduces accommodation space, Kailua Bay, Windward Oahu, Hawaii, USA. J Sedim Res 74:49–63

Grossman EE, Fletcher CH, Richmond B (1998) The Holocene sea-level highstand in the Equatorial Pacific: analysis of the insular paleosea-level database. Coral Reefs 17:309–327

Guilcher A (1988) Coral Reef Geomorphology. Wiley, Chichester, UK, pp 139–165

Guillou H, Garcia M, Turpin L (1997) Unspiked K-Ar dating of young volcanic rocks from the Loihi and Pitcairn seamounts. J Volc Geotherm Res 78:239–250

Harrison CS (1990) Seabirds of Hawaii: Natural History and Conservation. Cornell University Press, Ithaca, NY, 249 pp

HI DLNR (2000) The Northwestern Hawaiian Islands: What Makes These Islands Special? http://www.hawaii.gov/dlnr/exhibits/nwhi/NWHI_3.htm. Hawai'i Department of Land and Natural Resources

Hoeke R, Brainard R, Moffitt R, Merrifield M (2006) The role of oceanographic conditions and reef morphology in the 2002 coral bleaching event in the northwestern Hawaiian Islands. Atoll Res Bull 543:489–503

Hostetler SW, Clark PU (2000) Tropical climate at the last glacial maximum inferred from glacier mass-balance modeling. Science 290:1747–1750

Imbrie J, Hays JD, Martinson DG, Macintyre A, Mix AC, Morley JJ, Pisias NG, Prell WL, Shackleton NJ (1984) The orbital theory of Pleistocene climate: Support from a revised chronology of the marine d10O record. in: Berger AL (ed) Milankovich and Climate. D. Reidel, Norwell, Mass., pp 269–305

IPCC (2001) The Third Assessment Report of the Intergovernmental Panel on Climate Change (IPCC). Cambridge University Press, Cambridge and New York, 880 pp

Jackson ED, Shaw HR, Bargar KE (1975) Stress fields in central portions of the Pacific plate delineated in time by lenear volcanic chains. J Geophys Res 80:1861–1874

Jansen E, Sjøholm J (1991) Reconstruction of glaciation over the last 6 Myr from ice-borne deposits in the Norwegian Sea. Nature 349:600–603

Jokiel PL, Brown EK (2004) Global warming, regional trends and inshore environmental conditions influence coral bleaching in Hawaii. Global Change Biol 10:1627–1641

Jones AT (1993) Review of the chronology of marine terraces in the Hawaiian Archipelago. Quat Sci Rev 12:811–823

Jones GA, Keigwin LD (1988) Evidence from Fram Strait (78°N) for early deglaciation. Nature 336:56–59

Kan H, Hori N, Ichikawa K (1997) Formation of a coral reef-front spur. Coral Reefs 16:3–4

Keenan EE, Brainard RE, Basch LV (2006) Historical and present status of the pearl oyster, *Pinctada margaritifera*, at Pearl and Hermes Atoll, northwestern Hawaiian Islands. Atoll Res Bull 543:333–344

Keigwin LD, Jones GA, Lehman SJ (1991) Deglacial meltwater discharge, North Atlantic deep circulation, and abrupt climate change. J Geophys Res 96(C9):16811–16826

Kelley C, Moffitt R, Smith J (2006) Mega- to microscale classification and description of bottom fish essential fish habitat on four banks in the northwestern Hawaiian Islands. Atoll Res Bull 543:319–332

Kennett JP (1990) The Younger Dryas cooling event: an introduction. Paleoceanography 5:891–895

Kenyon JC, Brainard RE (2006) Second recorded episode of mass coral bleaching in the northwestern Hawaiian Islands. Atoll Res Bull 543:505–523

Kenyon JC, Brainard RE, Hoeke RK, Parrish FA, Wilkinson CB (2006) Towed diver surveys, a method for mesoscale spatial assessment of benthic reef habitat: a case study at Midway Atoll in the Hawaiian Archipelago. Coast Manag 34:339–349

Kenyon JC, Dunlap MJ, Wilkinson CB, Page KN, Vroom PS, Aeby GS (2007) Community structure of hermatypic corals at Pearl and Hermes Atoll, northwestern Hawaiian Islands: unique conservation challenges within the Hawaiian archipelago. Atoll Research Bulletin (in press)

Kirch PV (1998) Archaeology. In: Juvik SP, Juvik JO (eds) Atlas of Hawaii. Third edition. University of Hawai'i Press, Honolulu, HI, pp 161–168

Kleypas JA, Feely RA, Fabry VJ, Langdon C, Sabine CI, Robbins LI (2006) Impacts of Increasing Ocean Acidification on Coral Reefs and Other Marine Calcifiers: A Guide for Future Research. report of a workshop held 18–20 April 2005, St. Petersburg, FL, sponsored by NSF, NOAA, and the U.S. Geological Survey, 88 pp

Kobayashi DR (2006) Colonization of the Hawaiian Archepalago via Johnston Atoll: a characterization of oceanographic transport corridors for pelagic larvae using computer simulation. Coral Reefs 25:407–417

Kono M (1980) Paleomagnetism of DSDP Leg 55 basalts and implications for the tectonics of the Pacific plate. In Jackson ED, Koisumi I (eds) Initial Reports of the Deep Sea Drilling Project 55. U.S. Govt. Printing Office, Washington, DC, pp 737–752

Kubota M (1994) A mechanism for the accumulation of floating marine debris north of Hawaii. J Phys Oceanogr 24:1059–1064

Ladd HS, Schlanger SO (1960) Drilling Operations on Eniwetok Atoll. U.S. Geological Survey Professional Paper 260-Y, Washington, DC, p 265

Ladd HS, Tracey JI, Gross MG (1967) Drilling on Midway Atoll, Hawaii. Science 156:1088–1094

Ladd HS, Tracey JI, Gross MG (1970) Drilling on Midway Atoll. U.S. Geol Surv Prof Pap 680-A, Washington, DC, 22 pp

Lambeck K, Tezer ME, Potter EK (2002) Links between climate and sea levels for the past three million years. Nature 419:199–206

Langdon C, Takahashi T, Marubini F, Atkinson M, Sweeney C, Aceves H, Barnett H, Chipman D, Goddard I (2000) Effect of calcium carbonate saturation state on the calcification rate of an experimental coral reef. Global Biogeochem Cyc 14:639–654

Langdon C, Broeker WS, Hammond DE, Glenn E, Fitzsimmons K, Nelson SG, Peng TH, Hajdas I, Bonani G (2003) Effect of elevated CO_2 on the community metabolism of an experimental coral reef. Global Biogeochem Cyc 17(1):1011, doi:10.1029/2002GB001941

Lea DW, Pak DK, Spero HJ (2000) Climate impact of late Qaternary equatorial Pacific sea surface temperature variation. Science 289:1719–1724

Lea DW, Martin PA, Pak DK, Spero HJ (2002) Reconstruction of a 350 ky history of sea-level using planktonic Mg/Ca and oxygen isotope records from a Cocos Ridge core. Quat Sci Rev 21:283–293

Leonard CL, Bidigare RR, Seki MP, Polovina JJ (2001) Interannual mesoscale physical and biological variability in the North Pacific Central Gyre. Progr Oceanogr 49(1–4):227–244

Lipman PW (1995) Declining growth rate of Mauna Loa during the last 100,000 years: rates of lava accumulation versus gravitational subsidence. In: Rhodes JM, Lockwood JP (eds) Mauna Loa Revealed: Structure, Composition, History and Hazards. Geophysical Monograph 92. American Geophysical Union, Washington, DC, pp 45–80

Lipman PW, Eakins BW, Yokose H (2003) Ups and downs on spreading flanks of ocean-island volcanoes: evidence from Mauna Loa and Kilauea. Geology 31:841–844

Macdonald GA, Abbot AT, Peterson FL (1983) Volcanoes in the Sea, 2nd Ed. University of Hawai'i Press, Honolulu, HI, 528 pp

Maragos J, Gulko D (eds) (2002) Coral Reef Ecosystems of the Northwestern Hawaiian Islands: Interim Results Emphasizing the 2000 Surveys. U.S. Fish and Wildlife Service and the Hawai'i Department of Land and Natural Resources, Honolulu, HI, 49 pp

Maragos JE, Potts EC, Aeby G, Gulko D, Kenyon J, Siciliano D, VanRavenswaay D (2004) 2000–2002 Rapid ecological assessment of corals (Anthozoa) on shallow reefs of the northwestern Hawaiian Islands. Part 1: species and distribution. Pac Sci 58:211–230

Mark RK, Moore JG (1987) Slopes of the Hawaiian ridge. In: Volcanism in Hawaii. U.S. Geological Survey Professional Paper 1350, pp 101–107

McClung A (2005) A count-based population viability analysis of the Laysan finch (Telespiza cantans). Ph.D. thesis, University of Hawaii, Honolulu, 105 pp

McGlone M, Kershaw AP, Markgraf V (1992) El Niño/Southern Oscillation climatic variability in Australia and South American paleoenvironmental records. In:

Diaz H, Markgraf V (eds) El Niño: Historical and Paleoclimatic Aspects of the Southern Oscillation. Cambridge University Press, New York, pp 434–462

McNutt MK (2002) Heat flow variations over Hawaiian swell controlled by near-surface processes, not plume properties. In Takahashi E (ed) Hawaiian Volcanoes: Deep Underwater Perspectives. Geophysical Monograph Series 128. American Geophysical Union, Washington, DC, pp 365–372

Miller JE, Hoeke RK, Appelgate TB, Johnson PJ, Smith JR, Bevacqua S (2004) Bathymetric Atlas of the Northwestern Hawaiian Islands, Draft. NOAA, Honolulu, http://www.soest.hawaii.edu/pibhmc/

Miller KG, Mountain GS, the Leg 150 Shipboard Party, Members of the New Jersey Coastal Plain Drilling Project (1996) Drilling and dating New Jersey Oligocene-Miocene sequences: ice volume, global sea level, and Exxon records. Science 271:1092–1094

Mitrovica JX, Peltier WR (1991) On postglacial geoid subsidence over the equatorial oceans. J Geophys Res 96:20053–20071

Mix AC (1992) The marine oxygen isotope record: Constraints on timing and extent of ice-growth events (120–65 ka). In: Clark PU and Lea PD (eds) The late Interglacial-glacial Transition in North America, Geol Soc Am Spec Pap 270. GSA, Boulder, pp 19–30

Moore JG (1970) Relationship between subsidence and volcanic load, Hawaii. Bull Volcanol 34:562–576

Moore JG, Campbell JF (1987) Age of tilted reefs, Hawaii. J Geophys Res 92:2641–2646

Moore JG, Chadwick WW (1995) Offshore geology of Mauna Loa and adjacent area, Hawaii. In: Rhodes JM, Lockwood JP (eds) Mauna Loa Revealed: Structure, Composition, History, and Hazards. Am Geophys Union Geophys Monogr 92:21–44

Moore JG, Clague DA (1992) Volcano growth and evolution of the island of Hawaii. Geol Soc Am Bull 104:1471–1484

Moore JG, Clague DA, Normark WR (1982) Diverse basalt types from Loihi seamount, Hawaii. Geology 10:88–92

Moore JG, Clague DA, Holcomb RT, Lipman PW, Normark WR, Torresan ME (1989) Prodigious submarine landslides on the Hawaiian ridge. J Geophys Res 94:17465–17484

Moore RB, Clague DA, Rubin M, Bohrson WA (1987) Volcanism in Hawaii. Hualalai Volcano: a preliminary summary of geologic, petrologic, and geophysical data. U.S. Geol Surv Prof Pap 1350:571–585

Moran K, Backman J, Brinkhuis H, Clemens SC, Cronin T, Dickens GR, Eynaud F, Gattacceca J, Jakobsson M, Jordan RW, Kaminski M, King J, Koc N, Krylov A, Martinez M, Matthiessen J, McInroy D, Moore TC, Onodera J, O'Regan M, Pälike H, Rea B, Rio D, Sakamoto T, Smith DC, Stein R, St John K, Suto I, Suzuki N, Takahashi K, Watanabe M, Yamamoto M,

Farrell J, Frank M, Kubik P, Jokat W, Kristoffersen Y (2006) The Cenozoic palaeoenvironment of the Arctic Ocean. Nature 441:601–605

Morgan WJ (1971) Convection plumes in the lower mantle. Nature 230:43–44

Moy CM, Seltzer GO, Rodbell DT, Anderson DM (2002) Variability of El Niño/Southern Oscillation activity at millennial timescales during the Holocene epoch. Nature 420:162–165

Munk WH, Sargent MC (1948) Adjustment of Bikini Atoll to ocean waves. Trans Amer Geophys Union 29:855–860

Naish T (1997) Constraints on the amplitude of late Pliocene eustatic sea-level fluctuations: new evidence from the New Zealand shallow-marine sediment record. Geology 25:1139–1142

Nakaso D (2007) Papahanaumokuakea: A New Monument Name. Honolulu Advertiser, March 2, 2007, http://www.honoluluadvertiser.com/apps/pbcs.dll/article?AID = 2007703030341

Neumann AC, Macintyre I (1985) Reef response to sea level rise: keep-up, catch-up or give-up. Proc 5th Int Coral Reef Congr, Tahiti 3:105–110

Norton IO (1995) Plate motions in the North Pacific: the 43 Ma nonevent. Tectonics 14:1080–1094

NWHI MNM (2006) A Citizen's Guide to the Northwestern Hawaiian Islands Marine National Monument. NOAA/NOS Northwestern Hawaiian Islands Marine National Monument, 27 pp

Olson SL (1996) History and ornithological journals of the *Tanager* expedition of 1923 to the northwestern Hawaiian Island, Johnston and Wake Islands. Atoll Res Bull 433:1–210

Parker SP (ed) (1997) Dictionary of Earth Science. McGraw-Hill, New York, pp 29 and 197

Parrish FA, Boland RC (2004) Habitat and reef-fish assemblages of banks in the Northwestern Hawaiian Islands. Mar Biol 144:1065–1073

Pekar S, Miller KG (1996) New Jersey Oligocene "Icehouse" sequences (ODP Leg 150) correlated with global del ^{18}O and Exxon eustatic records. Geology 24:567–570

Peltier WR (1988) Global sea level and Earth rotation. Science 240:895–901

Phipps Morgan J, Morgan WJ, Price E (1995) Hotspot melting generates both hotspot volcanism and a hotspot swell? J Geophys Res 100:8045–8062

Pietruszka AP, Garcia MO (1999) The size and shape of Kilauea Volcano's summit magma storage reservoir: a geochemical probe. Earth Planet Sci Lett 167:311–320

Polovina JJ, Mitchum GT, Graham NE, Craig MP, DeMartini EE, Flint EN (1994) Physical and biological consequences of a climate event in the central North Pacific. Fish Oceanogr 3:15–21

Polovina JJ, Howell E, Kobayashi DR, Seki MP (2001) The transition zone chlorophyll front, a dynamic global feature defining migration and forage habitat for marine resources. Progr Oceanogr 49(1–4):469–483

Purdy EG (1974) Reef configurations: Cause and effect. In: Laporte LF (ed) Reefs in Time and Space. Society of Economic Paleontologists and Mineralogists Special Publication 18: 9–76

Purdy EG, Bertram GT (1993) Carbonate concepts from the Maldives, Indian Ocean. Amer Assoc Petrol Geol Studies in Geology 34:1–56

Purdy EG, Winterer EL (2001) Origin of atoll lagoons. Geol Soc Am Bull 113:837–854

PSMSL (2006) Permanent Service for Mean Sea Level, http://www.pol.ac.uk/psmsl/. Proudman Oceanographic Laboratory

Quane S, Garcia MO, Guillou H, Hulsebosch T (2000) Magmatic evolution of the east rift zone of Kilauea volcano based on drill core from SOH. 1. J Volcanol Geotherm Res 102:319–338

Ragnat E, Kanagaratnam P (2006) Changes in the velocity structure of the Greenland Ice Sheet. Science 311:986–990

Rauzon MJ (2001) Isles of Refuge: Wildlife and History of the Northwestern Hawaiian Islands. University of Hawai'i Press. Honolulu, HI, 207 pp

Raymond CA, Stock JM, Cande SE (2000) Fast Paleogene motion of the Pacific hotspots from revised global plate curcuit constraints. In: Richards MA, Gordon RG, Van der Hilst RD (eds) The History and Dynamics of Global Plate Motions. Geophysical Monograph 121, American Geophysical Union, Washington, DC, pp 359–375

Rees BA, Detrick RS, Coakley BJ (1993) Seismic stratigraphy of the Hawaiian flexural moat. Geol Soc Am Bull 105:189–205

Ribe NM, Christensen UR (1999) The dynamical origin of Hawaiian volcanism. Earth Planet Sci Lett 171:517–531

Roberts HH, Murray SP, Suhayda JN (1975) Physical processes in a fringing reef system. J Mar Res 33:233–260

Rodbell DT, Seltzer GO, Anderson DM, Abbott MB, Enfield DB, Newman JH (1998) An ~15,000-year record of El-Niño alluviation in southwestern Ecuador. Science 283:516–520

Rona PA, Richardson ES (1978) Early Cenozoic global plate reorganization. Earth Planet Sci Lett 40:1–11

Rooney J, Fletcher C, Engels M, Grossman E, Field M (2004) El Niño Control of Holocene Reef Accretion in Hawaii. Pac Sci 58(2):305–324

Rooney JJ, Fletcher CH (2005) Shoreline change and Pacific climatic oscillations in Kihei, Maui, Hawaii. J Coastal Res 21(3):535–547

Rotondo G (1980) A reconstruction of linear island chain positions in the Pacific: a case study using the Hawaiian Emperor chain. M.S. thesis, University of Hawaii, Honolulu, 58 pp

Ruddiman WF, Raymo ME, Martinson DG, Clement BM, Backman J (1989) Pleistocene evolution: northern hemisphere ice sheets and North Atlantic Ocean. Paleoceanography 4:353–412

Sharp WD, Clague DA (2006) 50-Ma initiation of Hawaii-Emperor bend records major change in Pacific plate motion. Science 313:1281–1284

Shepherd A, Windham D (2007) Recent sea-level contributions of the Antarctic and Greenland ice sheets. Science 315:1529–1532

Sherman C, Fletcher CH, Rubin K (1999) Marine and meteoric diagenesis of Pleistocene carbonates from a nearshore submarine terrace, Oahu, Hawaii. J Sedim Res 69:1083–1097

Shinn EA, Hudson JH, Robbin DM, Lidz B (1981) Spurs and grooves revisited: construction versus erosion Looe Key Reef, Florida. Proc 4th Int Coral Reef Sym, Manila 1:475–483

Siciliano D (2005) Latitudinal Limits to coral reef Accretion: testing the Darwin Point Hypothesis at Kure Atoll, northwestern Hawaiian Islands, using new evidence from high resolution remote sensing and *in situ* data. Ph.D. dissertation, University of California, Santa Cruz, pp 30–154

Smith JR, Wessel P (2000) Isostatic consequences of giant landslides on the Hawaiian Ridge. Pure Appl Geophys 157:1097–1114

Smith WFH, Sandwell DT (1997) Global sea floor topography from satellite altimetry and ship depth soundings. Science 277:1956–1962

Sneh A, Friedman GM (1980) Spur and groove patterns on the reefs of the northern gulf of the Red Sea. J Sedim Petrol 50:981–986

Stearns HT (1941) Shore benches on North Pacific islands. Geol Soc Am Bull 52:773–780

Stearns HT (1946) Geology of the Hawaiian Islands. Hawai'i Div Hydrography Bull 8: 1–105

Stearns HT (1974) Submerged shoreline and shelves in the Hawaiian Islands and a revision of some of the eustatic emerged shoreline. Bull Geol Soc Am 85:795–804

Stearns HT (1978) Quaternary shorelines in the Hawaiian Islands. Bernice P. Bishop Museum Bull 237, Honolulu, 57 pp

Stein CA, Stein S (1992) A model for the global variation in oceanic depth and heat flow with lithospheric age. Nature 359:123–129

Steinberger B (2000) Plumes in a convecting mantle: models and observations for individual hotspots. J Geophys Res 05:11127–111152

Steinberger B, O'Connell RJ (1998) Advection of plumes in mantle flow: implications for hot spot motion, mantle viscosity and plume distributions. Geophys J Int 132:412–434

Steinberger B, Sutherland R, O'Connell RJ (2004) Prediction of Emperor-Hawai'i seamount locations from a revised model of global plate motion and mantle flow. Nature 430:167–173

Storlazzi CD, Field ME, Dykes JD, Jokiel L, Brown E (2002) Wave control on reef morphology and coral distribution: Molokai, Hawaii. In: Waves 2001 Conference Proceedings Volume 1. American Society of Civil Engineers, San Francisco, CA, pp 784–793

Storlazzi CD, Logan JB, Field ME (2003) Quantitative morphology of a fringing reef tract from high-resolution laser bathymetry: southern Molokai, Hawaii. Geol Soc Am Bull 115:1344–1355

Tagami T, Nishimitsu Y, Sherrod DR (2003) Rejuvenated stage volcanism after 0.6-m.y. quiescence at West Maui volcano, Hawaii: new evidence from K–Ar ages and chemistry of Lahaina Volcanics. J Volcanol Geotherm Res 120:207–214

Tarduno JA, Cottrell RD (1997) Paleomagnetic evidence for motion of the Hawaiian hotspot during formation of the Emperor seamounts. Earth Planet Sci Lett 153:171–180

Tarduno JA, Duncan RA, Scholl DW, Cottrell RD, Steinberger B, Thordarson T, Kerr BC, Neal CR, Frey FA, Torii M, Carvallo C (2003) The Emperor seamounts: southward motion of the Hawaiian hotspot plume in earth's mantle. Science 301:1064–1069

Teller JT, Kehew AE (1994) Introduction to the last glacial history of large proglacial lakes and meltwater runoff along the Laurentide ice sheet. Quat Sci Rev 13:795–799

ten Brink US, Watts AB (1985) Seismic stratigraphy of the flexural moat flanking the Hawaiian islands. Nature 317:421–424

ten Brink US, Brocher TM (1987) Multichannel seismic evidence for a subcrustal intrusive complex under Oahu and a model for Hawaiian volcanism. J Geophys Res 92:13687–13707

The Royal Society (2005) Ocean Acidification Due to Increasing Atmospheric Carbon dioxide. The Royal Society. London

Turcotte DL, Schubert G (1982) Geodynamics. Wiley, New York, 456 pp

Umbgrove JHF (1947) Coral reefs of the East Indies. Geol Soc Am Bull 58:729–778

Vazquez J, Perry K, Kilpatrick K (2002) NOAA/NASA AVHRR Oceans Pathfinder Sea Surface Temperature Data Set User's Reference Manual Version 4.0. JPL Publication D-14070, http://podaac.jpl.nasa.gov/podaac_web/sst/sst_doc.html

Vogt PR, Crane K, Sundvor E (1994) Deep Pleistocene iceberg plowmarks on the Yermak Plateau: sidescan and 3.5 kHz evidence for thick calving ice fronts and a possible marine ice sheet in the Arctic Ocean. Geology 22:403–406

Vroom PS, Page KN, Peyton KA, Kukea-Shultz JK (2005) Spatial heterogeneity of benthic community assemblages with an emphasis on reef algae at French Frigate Shoals, northwestern Hawai'ian Islands. Coral Reefs 24:574–581

Walcott RI (1970) Lithospheric flexure, analysis of gravity anomalies, and the propagation of seamount chains. In Sutton GH, Manghnani MH, Moberly R (eds) The Geophysics of the Pacific Ocean Basin and Its Margin. Geophysical Monograph 19. American Geophysical Union, Washington, DC, pp 431–438

Walker GPL (1990) Geology and volcanology of the Hawaiian Islands. Pac Sci 44:315–347

Wang SL, Swail VR (2001) Changes of extreme wave heights in Northern Hemisphere oceans and related atmospheric circulation regimes. J Climate 14:204–2221

Wanless VD, Garcia MO, Trusdell FA, Rhodes JM, Norman MD, Weis D, Fornari DJ, Kurz MD, Guillou H (2006) Submarine radial vents on Mauna LoaVolcano, Hawai'i. Geochem, Geophys, Geosys 7:Q05001, doi:10.1029/2005GC001086.

Watts AB (1978) An analysis of isostasy in the world's oceans. 1. Hawaiian-Emperor seamount chain. J Geophys Res 83:5989–6004

Watts AB, Cochran JR (1974) Gravity anomalies and flexure of the lithosphere along the Hawaiian-Emperor seamount chain. Geophys J R Astronom Soc 38:119–141

Watts AB, ten Brink US (1989) Crustal structure, flexure, and subsidence history of the Hawaiian Islands. J Geophys Res 94:10473–410500

Watts AB, ten Brink US, Buhl P, Brocher TM (1985) A multichannel seismic study of lithospheric flexure across the Hawaiian-Emperor seamount chain. Nature 315:105–111

Webster JM, Wallace L, Silver E, Appelgate B, Potts D, Braga JC, Riker-Coleman K, Gallup C (2004) Drowned carbonate platforms in the Huon Gulf, Papua New Guinea. Geochem, Geophys, Geosys 5:Q11008, doi:10.1029/2004GC000726

Wellington GM, Glynn PW, Strong AE, Navarrete SA, Wieters, Hubbard D (2001) Crisis on coral reefs linked to climate change. EOS 82(1):1

Wessel P (1993) A reexamination of the flexural deformation beneath the Hawaiian islands. J Geophys Res 98:12177–112190

Wessel P, Harada Y, Kroenke LW (2006) Towards a self-consistent, high-resolution absolute plate motion model for the Pacific. Geochem, Geophys, Geosys 7:Q03L12, doi:10.1029/2005GC001000

Westervelt WD (1916) Hawaiian Legends of Volcanoes (mythology) Collected and Translated from the Hawaiian. Ellis, Boston, MA, 205 pp

Whittaker JM, Müller RD (2006) New 1-minute Satellite Altimetry Reveals Major Australian–Antarctic Plate Reorganisation at Hawaiian-Emperor Bend Time. EOS, Transactions, American Geophysical Union, Fall Meeting Supplement 87: Abstract T51F-08

Wilson GS (1995) The Neogene East Antarctic ice sheet: a dynamic or stable feature? Quat Sci Rev 14:101–123

Wilson JT (1963) A possible origin of the Hawaiian islands. Can J Physics 41:863–870

Winchell H (1947) Honolulu series, Oahu, Hawaii. Bull Geol Soc Am 58:1–48

Wood R, Oppenheimer C (2000) Spur and groove morphology from a Late Devonian reef. Sedim Geol 133:185–193

Woodroffe CD, Kennedy DM, Jones BG, Phipps CVG (2004) Geomorphology and Late Quaternary development of Middleton and Elizabeth reefs. Coral Reefs 23:249–262

Yokoyama Y, Lambeck K, De Deckker P, Johnston J, Fifield LK (2000) Timing of the last glacial maximum from observed sea-level minima. Nature 406:713–716

Zhong S, Watts AB (2002) Constraints on the dynamics of mantle plumes from uplift of the Hawaiian Islands. Earth Planet Sci Lett 203:105–116

14

Biology of Coral Reefs in the Northwestern Hawaiian Islands

Richard W. Grigg, Jeffrey Polovina, Alan M. Friedlander, and Steven O. Rohmann

14.1 Geography and History

The Northwestern Hawaiian Islands (NWHI) represent the northern three-quarters of the Hawaiian Archipelago. This part of the Hawaiian chain stretches across 2,000 km of the North Pacific between 23 and 29 degrees north latitude and consists of nine major islets, coral islands and/or atolls. Numerous reefs, submerged banks and seamounts also exist between and around the main islands. Together with the Main Hawaiian Islands (MHI), the Hawaiian Archipelago is the longest, oldest and best studied archipelago on earth. The NWHI are also considered the most isolated and pristine group of coral reef ecosystems in the world (Figs. 14.1, 14.2).

The NWHI were first discovered and explored by the ancient Hawaiians during pre-historic time before European contact. Numerous archaeological remains on Nihoa and Necker islands demonstrate that both islands were inhabited by early Hawaiians but exactly when, and for how long, is unknown. References to the islands further west in the archipelago also exist in early Hawaiian chants and legends. When the Kingdom of Hawaii was overthrown by the US Government in 1893, the NWHI (except Midway) were included within the new provisional government that immediately was established and renamed the Republic of Hawaii. In 1898, the US Congress passed a resolution to annex the Republic of Hawaii as a Territory. A final step in the political history of the NWHI was inclusion into the Union as a State bestowed by the US Congress in 1959 and approved by a Plebiscite held in Hawaii that same year (Atlas of Hawaii 1998). Midway Atoll was actually claimed separately and much earlier by the US Government in 1859 by a US Captain, N.C. Brooks, under the authority granted to him by the Guano Act of 1856. Midway is therefore the only island in the NWHI that does not actually belong to the State of Hawaii *per se* (Rauzon 2001).

The history of biological research in the NWHI has recently been reviewed and summarized by Grigg (2006). In this paper, the contributions in natural history by the early Hawaiians are recognized and many of these go back over 1,000 years. The first organized science in the NWHI did not begin until the nineteenth century and consisted of numerous expeditions by the Explorer-Naturalists. The first of these was the US Exploring Expedition in 1840, followed by the British Challenger Expedition in 1872–1876, The Albatross Expedition in 1902, and the Tanager Expedition in 1923. The Tanager Expedition was more intensive than the previous campaigns that focused primarily on biological collections. Tanager scientists occupied the islands and documented numerous ecological impacts associated with guano mining, harvesting seals, birds, bird eggs, pearl oysters and the logging of sandalwood. By 1923, several birds had already been driven to extinction and the pearl oyster population at Pearl and Hermes Atoll had been devastated. Except for these extinction events and the lack of full recovery of the pearl oyster at Pearl and Hermes Atoll, hardly any of the human impacts associated with this early period are visible today.

B.M. Riegl and R.E. Dodge (eds.), *Coral Reefs of the USA*,
© Springer Science+Business Media B.V. 2008

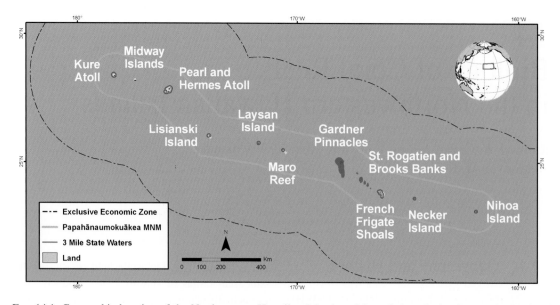

F<small>IG</small>. 14.1. Geographic location of the Northwestern Hawaiian Islands and boundaries of what is now the National Monument (Map courtesy of National Wildlife Refuge Association)

Over the past 30 years, biological research in the NWHI has greatly increased. A major study program in the 1970s and 1980s (Grigg and Pfund 1980; Grigg and Tanoue 1984) produced over 100 published scientific papers, as well as two management plans for endangered or threatened species, the Hawaiian Monk Seal and the Hawaiian Green Turtle. Several Fishery Management Plans (FMPs) have also been developed for commercial species by the Western Pacific Fishery Management Council (WPFMC). Building on this foundation, a multi-year NOAA study (NOWRAMP) focusing primarily on coral reefs began in 2002 and is ongoing at the present time.

14.2   Paleoceanography of Coral Reefs in the NWHI

The Hawaiian Archipelago originated over a relatively stationary melting anomaly or hotspot in earth's mantle located below the floor of the Pacific Plate. In this region of the Pacific, the plate is drifting over the hotspot to the northwest at a rate of about 8 cm/year. Each island or volcanic edifice in the NWHI is therefore moving to

the northwest at this rate and because the plate is slowly cooling as it moves away from the hotspot, the overriding islands and other volcanic features are slowly subsiding (Fig. 14.3). Once the islands reach sea-level, the upward growth of coral reefs is all that maintains them near the ocean surface as either coral islands (Laysan and Lisianski) or atolls (Pearl and Hermes, Midway and Kure). At Kure Atoll, the upward growth or accretion of the surrounding reefs due to coral growth is almost exactly balanced by negative losses due to bio-erosion, mechanical erosion and subsidence (Grigg 1982, 1997). Thus, as islands drift northwest of this point, losses exceed gains and the islands drown. This threshold for atoll formation has been dubbed the Darwin Point (Grigg 1982, 1997) (Fig. 14.4).

One consequence of this history is that every island in the NWHI is different; different in age, different in elevation, in land area, in underwater topography, in shelf area, in shelf depth, different in oceanography and to some degree, different in biological composition in both the marine and terrestrial realms. The range in age of the NWHI is 7.2 ma for Nihoa, the youngest island to about 28 ma for Kure Atoll, the oldest (Figs. 14.1, 14.2). The youngest islands are actually islets (small volcanic outcrops with no

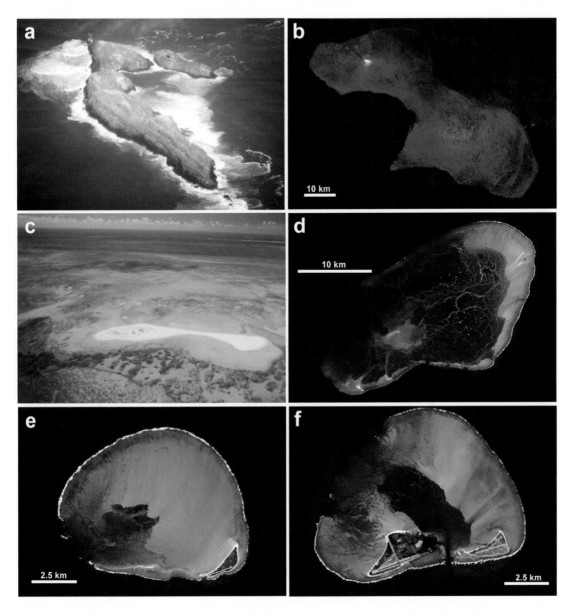

FIG. 14.2. The Northwestern Hawaiian Islands. (**a**) Aerial photograph of Necker Island oriented from southeast to northwest. The total surface area of Necker Island is 0.2 km^2 (**b**) Landsat 7 photograph of Necker Island showing the island and its surrounding bank visible to a depth of about 40 m. (**c**) Aerial photograph of French Frigate Shoals showing Sand Island in the foreground. Twelve islets are present at FFS but collectively they only represent 0.2 km^2 of surface area. (**d**) Landsat 7 photograph of Pearl and Hermes Atoll. (**e**) Landsat 7 photograph of Kure Atoll (0.3 km^2). (**f**) Landsat 7 photograph of Midway Islands (Atoll). Midway Atoll does not officially belong to the State of Hawaii, having been officially claimed by the US Government in 1859 under the authority of the Guano Act

barrier reefs), followed by near-atolls, reefs and coral islands of intermediate age and finally the oldest islands which are all atolls. Most of these differences reflect differences in the chronological position of each island in the chain notwithstanding unique attributes associated with island shape and edifice mass originally produced over the hotspot.

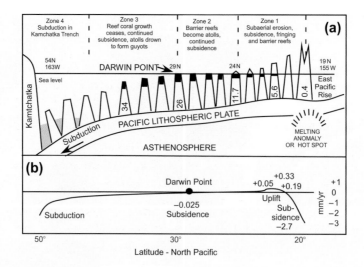

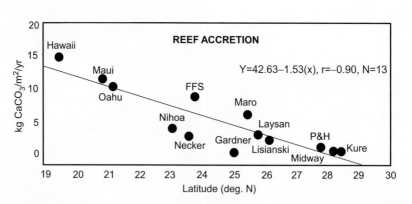

FIG. 14.3. Schematic representation of the origin, subsidence, drowning and subduction history of the Hawaiian Islands and the location of the Darwin Point (Grigg 1997)

FIG. 14.4. Reef accretion across the Hawaiian Archipelago. Note that the net accretion rate diminishes to zero just beyond Kure Atoll, thereby defining a threshold for atoll formation known as the Darwin Point beyond which atolls drown (Grigg 1997)

In contrast to the differences, there are also many similarities which depend on geographic and geomorphic similarity and the degree to which the various islands are interconnected by oceanic circulation patterns (Section 14.4). In brief, the biological composition of coral reefs on each island or reef in the NWHI represents a complex interplay of both unique differences and physical and biological similarities. A more detailed discussion of the biological composition and community structure of coral reefs in the NWHI is presented in the next two sections.

14.3 Modern Ecology of Coral Reefs in the NWHI

In the most recent surveys of the NWHI, 57 species of zooxanthellate stony corals were recorded, although at least 11 of these need to be verified by detailed taxonomic analysis (Maragos et al. 2004). Even if all new records are confirmed, the coral reefs of the NWHI must be considered depauperate relative to the Indo-West-Pacific where up to 700

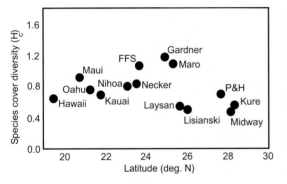

Fig. 14.5. Coral species cover diversity plotted against latitude within the Hawaiian Archipelago. Note peak values occur at FFS, Gardner Island and Maro Reef (Grigg 1983)

species have been reported (Veron 1995). The most plausible cause is geographic isolation (Grigg 1983). High endemism (30%) of the coral fauna is also very likely due to isolation in which rare colonization events are followed by evolutionary radiation, the so-called founder effect.

Within the archipelago, coral abundance and species diversity both peak at French Frigate Shoals (FFS) and Maro Reef, reflecting optimal conditions in terms of both habitat (a large open atoll) and environmental conditions (wave shelter, temperature and low disturbance (Fig. 14.5). Southeast of French Frigate Shoals, the small islets of Necker and Nihoa islands are openly exposed to severe wave events particularly during winter months. Compared to French Frigate Shoals and Maro Reef, one third fewer species are found at Necker and Nihoa islands. The same degree of reduction of coral species diversity is found at Midway and Kure Atolls, although the causal factors are not the same. In the case of Midway and Kure Atolls, the decrease in diversity is linked to lower winter water temperatures and lower average annual solar radiation (Grigg 1982).

Rohmann et al. (2005) assessed the area potentially occupied by coral ecosystems in US waters, based on substrata suitable for coral growth situated within 10 and 100 fathom depth curves (isobaths). Calculations show that for all of the US, the potential area covered by coral ecosystems is 36,813 km^2 inside 10 fathoms and 143,059 km^2 inside 100 fathoms. The percent of the total area in US waters occupied by potential coral ecosystems in the NWHI is about 4.8% inside 10 fathoms and

9.6% inside 100 fathoms. These values depart dramatically from an earlier estimate produced by Miller and Crosby (1998) who concluded that the coral ecosystems in the NWHI represented about 70% of coral reefs in US waters. The major source of confusion in the Miller and Crosby report was the almost complete omission of the wide shelf areas off Southern Florida that either are the locale of, or have the potential for coral and reef growth.

Regarding the overall diversity supported by coral reef ecosystems in the NWHI, NOWRAMP surveys have recently produced species lists for benthic inhabitants of the NWHI (other than corals) that include 355 species of algae (Vroom and Page 2006) and 838 species of other invertebrates (Friedlander et al. 2005). These numbers will likely increase as a result of future sampling efforts, however, relative to the Indo-West Pacific flora and fauna, these counts are still very low, again consistent with the geographic isolation of the NWHI. The evaluated non-benthic marine fauna of the NWHI also includes the fish fauna, Hawaiian Green Turtles (*Chelonia mydas*) and Hawaiian Monk seals (*Monachus schauinslandi*) (Fig. 14.6).

14.4 Oceanography of the NWHI Relative to Coral Reefs

The most important factors controlling the distribution and abundance of coral reefs in the NWHI are depth and shelter from large open ocean winter swell (Grigg 1983; Friedlander et al. 2005). The combined effect of these factors translates ecologically into differences in habitat that range from rocky wave swept islets (Nihoa and Necker Islands) to islands that afford varying degrees of wave shelter by way of coastline shape or the presence of barrier reefs and lagoons.

Every winter, the NWHI are exposed to extreme wave events produced by deep low-pressure storms that generally originate at mid-latitudes in the far western Pacific and typically move to the northeast along various paths depending on the configuration of the jet stream. As these storms migrate eastwards, they generally pass between 1,000 and 3,000 km to the north of the NWHI. In the process, they generate large, long-period swell that sweeps down the island chain. Approximately 12–18 of these large

Fig. 14.6. The Hawaiian Monk Seal is endemic to the Hawaiian Archipelago and it is on the endangered species list (Photo by R. Grigg)

swell events are generated every winter, commonly producing waves up to 10–12 m in vertical height. Even larger swells between 15 and 20 m occur about once every decade. Since waves of this magnitude break in a depth of water about equal to their height (vertical face between crest and trough), the maximum breaking depth of these waves is about 20 m (Caldwell 2005). This depth represents the wave base where concussion forces from breaking waves and bottom wave surge re-suspends sand and other particulate material that in turn produces abrasion and scour on bottom communities. Both of these factors have limiting effects on the growth and abundance of coral communities living between depths of 20 m and the surface. This zone, particularly on the north and western sides of all the islands (outside the barrier reefs in the case of the atolls), is characterized by coral communities with very low coral cover ranging between 5% and 10%. In addition, barrier reef morphology is barely present on the northwest sectors of Pearl and Hermes, Midway and Kure Atolls (Fig. 14.1 and Fig. 14.2). Coral accretion below the 20 m isobath is either lacking altogether or minimal at best. The best developed reefs on all the islands exist either in the lagoons or in southwestern exposures (Grigg 1982).

Much of the shelf area that surrounds the islets and tops of many banks in the NWHI range in depth from 30 to 60 m. The vast majority of shelf habitat at these depths consists of carbonate rubble, sand and isolated barren outcrops that are below critical depth (~20 m) for the accretion of reef building corals. This depth range also corresponds to the depth of sea-level for about 30,000 years during the last ice age or glacio-eustatic cycle (Grigg and Epp 1989), suggesting that shelf architecture in the NWHI may be a function of wave-base erosion from long period swell during this early time period. Hurricanes are extremely infrequent in the NWHI (Friedlander et al. 2005) and do not produce long period swell, and therefore do not significantly impact coral reefs in the NWHI.

Water temperature is another important oceanographic factor that controls the distribution and abundance of coral reefs in the NWHI. Seasonal differences in temperature and the depth to the thermocline are both important and affect the reefs in different ways. Summer temperatures at the surface (SSTs) along the island chain are generally similar, peaking at about 28°C. Short excursions slightly above 28°C, however, do occur in very shallow lagoon habitats mainly in the northern atolls where they are associated with recurrent bleaching of corals, e.g., *Pocillopora* spp. and *Montipora* spp. In 2002 and again in 2004, two bleaching events in these depth zones were documented by Aeby et al. (2003) and Hoeke et al. (2006).

In contrast to summer temperatures, winter sea surface temperatures (SSTs) are much cooler at the northern end of the chain where peak lows dip to 17°C or 18°C. While these extremes may explain the drop in coral species diversity north of Pearl and Hermes Atoll (Fig. 14.5), the response of most corals that exist there is simply slower growth as mentioned in Section 14.2. This partially explains the Darwin Point phenomenon. The difference between summer and winter SSTs at the northern end of the chain is about 10°C (18–28°C). In comparison, the summer/winter difference in SST at the southern end of the NWHI is only half as great: 5°C (22–27°C).

The depth to the top of the thermocline can and does set lower depth limits for coral growth. At the southern end of the NWHI, near Nihoa Island, data archived by the Hawaii Undersea Research Laboratory (based on 5 years of submersible dives in September and October months) shows a drop in temperature from 26°C to 20°C at about 80m,

whereas at Kure Atoll, over the same time period, an almost identical drop occurs at 40m (Fig. 14.7). Observations and *in situ* measures of temperature at Midway and Kure Atolls in 1977–1982 documented periods when the top of the thermocline was as shallow as 20m, corresponding to the depth limit of reef accretion at Midway and Kure Atolls (Grigg 1983).

Solar radiation, as a limiting factor for coral growth, exhibits a declining gradient moving northwestward within the NWHI. From Nihoa Island to the north, annual radiation steadily declines with latitude. Sadler et al. (1976) calculated the average daily solar radiation (corrected for cloud cover) and found a 10% difference between Nihoa Island and Kure Atoll (381 mean langleys/day versus 343 mean langleys/day).

In addition to wave forcing, temperature and light, current also plays an important role in shaping the

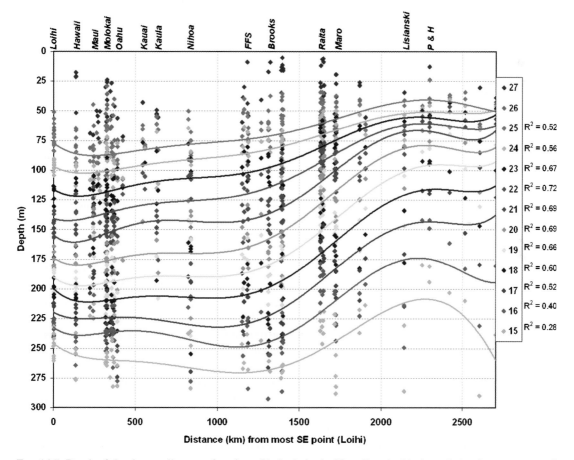

FIG. 14.7. Depth of the thermocline as a function of latitude in the Hawaiian Archipelago during late summer and early fall months (HURL data achive). Axis on the left is depth in meters, axis on the right shows water temperature in °C. The position of the thermocline is interpolated. Clear shoaling of the thermocline can be seen towards the right of the graph (near Lisianski and Pearl and Hermes)

community structure of coral reefs in the NWHI. Colonization from larval settlement between the islands and from outside the archipelago is largely dependent on surrounding currents. Based on 10 years of data collected by NOAA ships using Acoustic Doppler Current Profilers (ADCPs), from about 20 cruises along the chain, Firing et al. (2004) showed that the average mean flow of surface waters is predominately from east to west in response to the prevailing northeasterly tradewinds. The data also demonstrated a high degree of variability both in speed and direction due to island scale eddies. Kobayashi (2006) recently modelled larval transport by currents into the Hawaiian Islands and demonstrated a likely pathway via Johnston Atoll, which has strong faunal affinity to the NWHI with regards to corals (Grigg 1981; Grigg et al. 1981; Maragos and Jokiel 1986; Maragos et al. 2004) and also fishes (Rivera et al. 2004).

Coral species diversity peaks midway in the NWHI at French Frigate Shoals, Gardner Pinnacles and Maro Reef (Fig. 14.5) and correlates with episodic flow of the subtropical counter-current between Johnson Atoll and the archipelago (Seckel 1962; Grigg 1981). A sharp drop in abundance of three species of *Acropora* southeast of French Frigate Shoals is indicative of lack of southeastward larval transport within the NWHI (Fig. 14.8). Grigg (1981) found only one colony of *Acropora cytherea* at both Nihoa and Kauai and recently Kenyon et al. (2007) reported another two colonies of the same species from Kauai. No live colonies of any species of *Acropora* have ever been observed southeast of Kauai. Only one large, but dead colony of nearly 1 m diameter was observed off Sandy Beach, Oahu in 1990 (S.J. Dollar, personal communication 1990) providing evidence that the absence of *Acropora* in the Main Hawaiian Islands in not due to its inability to survive there, but rather due to its lack of recruitment suggesting a lack of larval transport between the NWHI and the MHI.

Water quality as measured by other physical, chemical or biological factors including salinity, oxygen concentration, turbidity, nitrate, particulate carbon, chlorophyll a, phaeopigments and zooplankton are relatively uniform at least in surface waters along the NWHI (Hirota et al. 1980). The only exception to this pattern appears to be higher oceanic productivity in waters surrounding the northernmost islands (Fig. 14.15). The observed (but not quantified) high rates of bio-erosion seen within the lagoons and nearshore reefs at Midway and Kure Atolls may be related to enhanced production of algal slimes and phytoplankton that favor bio-eroding organisms such as urchins and boring bivalves (Grigg 1997) particularly within lagoonal environments.

14.5 Fish Community Structure and Fisheries in the NWHI

A total of 258 species of reef and shore fishes have been reported from Midway Atoll, NWHI compared with 557 in the MHI (Randall 1995). The few species found in the NWHI but not the MHI include the Japanese angelfish (*Centropyge interruptus*) known only from the NWHI and Japan while the blotcheye soldierfish (*Myripristis murdjan*) is an Indo-Pacific species but restricted to the NWHI in the Hawaiian archipelago (Mundy 2005). Two species, the slingjaw wrasse (*Epibulus insidiator*) and the chevron butterflyfish (*Chaetodon trifascialis*), are associated with *Acropora* corals that are only found in the central portion of the NWHI (Grigg 1981), although these fish are occasionally observed in the MHI (Mundy 2005). Despite the taxonomic similarly with the MHI fauna, the NWHI fish assemblage differs somewhat from that of the MHI at various ecological and demographic levels owing to oceanography (water temperature) and habitat (coral species and lagoons).

Many warmer water shallow-water fish species are restricted from occurring in the NWHI by cooler winter temperatures (Mundy 2005). Other reasons for the lower number of species in the NWHI include: a much lower overall sampling effort and the lack of many high island habitats such as estuaries and rocky shorelines. Some shallow-water species are adapted to cooler water and can be found in deeper waters at the southern end of the archipelago. This phenomenon known as tropical submergence is exemplified by species such as the yellowfin soldierfish (*Myripristis chrysonemus*), the endemic Hawaiian grouper (*Epinephelus quernus*), and the masked angelfish (*Genicanthus personatus*) which are found in shallower water at Midway but are restricted to

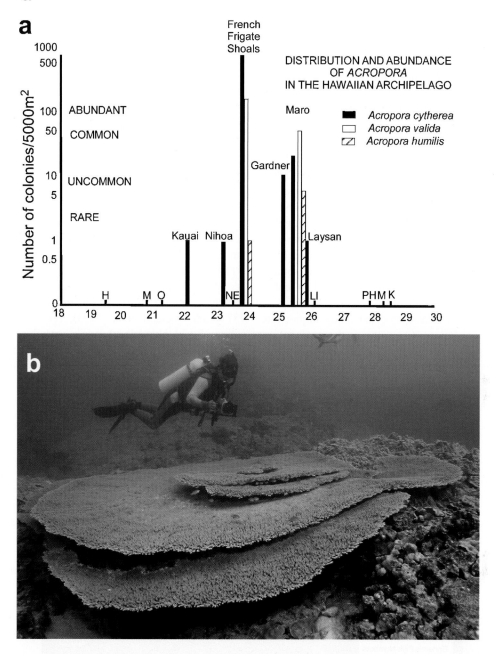

FIG. 14.8. (**a**) Species of *Acropora* are limited to the central portion of the Archipelago, suggesting a route of colonization from Johnston Island via the sub-tropical counter-current (see text for more detail). H = Hawaii, M = Maui, O = Oahu, NE = Necker, LI = Lisianski, PH = Pearl and Hermes, M = Midway, K = Kure. (**b**) A large colony of *Acropora cytherea* at FFS. Depth approximately 12 m (Photo by James Watt)

deeper water in the MHI (Randall et al. 1993; Mundy 2005). Some of the non-endemic species abundant at higher latitude reefs in the NWHI have anti-tropical distributions, such as the Hawaiian morwong (*Cheilodactylus vittatus*), and may have established themselves in the archipelago when surface waters were previously cooler (Randall 1981).

14.6   Endemism

The Hawaiian Island chain is among the most isolated on earth and exhibits the highest level of marine fish endemism of any archipelago in the Pacific (Randall 1987, 1995, 1998; Randall and Earle 2000; Allen 2002). Overall fish endemism based on species number is higher in the NWHI (30%) than the MHI (25%) (Friedlander et al. 2003; DeMartini and Friedlander 2004). Endemism based on numerical densities averages more than 52% in the NWHI, and increases with latitude throughout the Hawaiian Archipelago (Fig. 14.9). Examination of endemism based on numerical density gives greater insight in ecological processes since endemics tend to be some of the most common individuals observed on reefs in Hawaii. Greater endemism towards Midway and Kure appears related to consistently higher rates of replenishment by young-of-the-year following dispersal as pelagic larvae and/or juveniles (DeMartini and Friedlander 2004). Similar trends

were observed in the 1990s at French Frigate Shoals (FFS) and Midway, where consistently higher recruitment of young-of-the-year at Midway Atoll occurred despite the generally greater densities of older-stage fishes at FFS (DeMartini 2004). Because of a decline in global marine biodiversity, endemic "hot spots" like Hawaii are important areas for global biodiversity conservation.

14.7   Introduced Species

A number of nearshore fishes were intentional introductions to the MHI (Main Hawaiian Islands) in the late 1950s and 1960s. The blueline snapper (*Lutjanus kasmira*) has been by far the most successful fish introduction (Friedlander et al. 2002) reaching Midway Atoll, 1,180 nmi from Oahu in 37 years (Randall et al. 1993). From a total of about 3,170 bluestripe snapper introduced from the Marquesas Islands to Hawaii beginning in 1955 (Oda and Parrish 1981), this species spread to

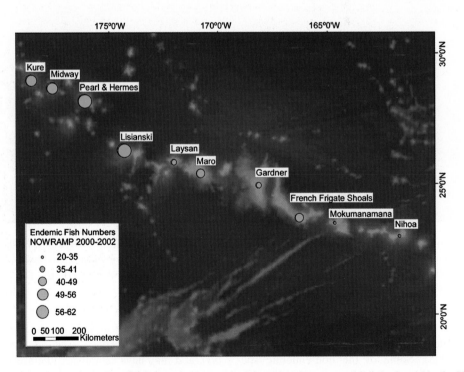

FIG. 14.9. Percent endemism of reef fish (based on numerical densities) increases with latitude within the Hawaiian Archipelago (DeMartini and Friedlander 2004)

French Frigate Shoals in the NWHI, 490 nmi from Oahu, sometime between 1977 and 1982, and was sighted in June 1979 another 330 nmi farther up-chain at Laysan Island (820 nmi from Oahu) (Oda and Parrish 1981). It was first observed at Midway Atoll in 1992 (Randall et al. 1993), suggesting dispersal rates of about 18–70 nmi/year over this time period. Over this same period, another well established introduction, the peacock grouper (*Cephalopholis argus*), has spread only as far as French Frigate Shoals, 490 nmi from Oahu (NWHI-RAMP, unpublished data 2004) with dispersal rates of 5–17 nmi/year. Compared to *C. argus*, *L. kasmira* is much more abundant and is more aggressive and may have displaced, to some extent, the Hawaiian Kumu (*Parupeneus porphyreus*).

14.8 Comparisons with the Main Hawaiian Islands

Unlike the reefs of the NWHI, the MHI reefs experience substantial fishing pressure from commercial and ornamental trade fisheries, as well as from large subsistence and recreational fishing activities. Comparison between the large, remote, and lightly fished NWHI, and the urbanized, heavily fished MHI, reveal dramatic differences in numerical density, size, and biomass of the shallow reef fish assemblages (Friedlander and DeMartini 2002) (Fig. 14.10a, b). Overall fish standing stock in the NWHI is more than 360% greater than in the MHI with large apex predators (primarily sharks and jacks) comprising more than 54% of the total fish biomass in the NWHI, but less than 3% of the fish biomass in the MHI (Fig. 14.10a, b). The strongly inverted biomass pyramid found in the NWHI results from short-lived, lower trophic level fishes which serve as energetic shunts, converting food (e.g., plankton, algae) into fish biomass, which is quickly transferred up the food chain and stored in long-lived, higher trophic level fish through predation. Most of the dominant species by weight in the NWHI are either rare or absent in the MHI and the species that are present, regardless of trophic level, are nearly always larger in the NWHI (Friedlander and DeMartini 2002). These differences represent both near-extirpation of apex predators and heavy exploitation of lower trophic levels in the MHI compared to the largely unfished NWHI. These differences in fish assemblage structure are evidence of the high level of exploitation in the MHI and the pressing need for ecosystem-level management and enforcement of existing regulations of reef systems in the MHI.

The relative lack of fishing in the NWHI can serve as a reference to assess individual fish stocks in the MHI. Estimates of fished and unfished abundances

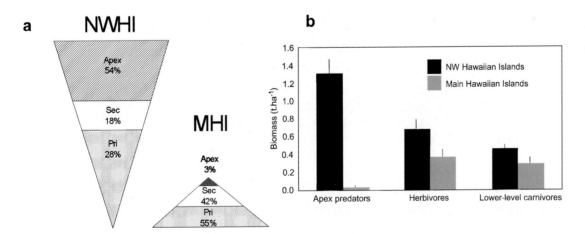

FIG. 14.10. (**a**) Fish biomass pyramids for the northwestern (NWHI) and main (MHI) Hawaiian Islands based on visual surveys. Total mean biomass in the NWHI was 2.44 t ha^{-1} compared to 0.68 t ha^{-1} for the MHI (data from Friedlander and DeMartini 2002). Note the extreme abundance of apex predators in the NWHI. Apex = top predators, sec = secondary, pri = primary. (**b**) Estimates of total biomass of apex predators, herbivores and lower-level carnivorous fishes in the NWHI compared with the MHI (Friedlander and DeMartini 2002)

provide a basis for a preliminarily assessment of the status of 49 previously un-assessed food fish and aquarium fish stocks (Sladek Nowlis et al., unpublished data 2002). Of the 49 species examined in the MHI, 17 (30%) were depleted below overfished thresholds (<25% of unfished biomass), and nearly half had biomass levels that were less than 10% of the unfished NWHI. The species showing the greatest differences in biomass density between the NWHI and MHI were ulua (carangids) sharks, snappers (lutjanids), soldier-fishes (*Myripristis* spp), groupers (serranids) and other large bodied food fishes.

14.9 Importance of the NWHI as a Refuge for the MHI

Because of poor condition of reef fish resources in the MHI, the NWHI has been often considered a refuge to replenish depleted fish stocks in the MHI. Genetic evidence shows connectivity at the evolutionary-scale for a number of species throughout the Hawaiian Islands and Johnston atoll, however, current patterns show mean westward flow from Oahu to Necker (Firing and Brainard 2006) and larval distribution models indicate high retention around individual islands (Kobayashi 2006) in the upper 100–200 m (Boehlert et al. 1992; Boehlert and Mundy 1992). Even in the Caribbean, ecologically significant levels of dispersal have been documented to be on the scale of 50–100 km for most species with a relatively high rate of local retention or recruitment from adjacent locations. Hence, based on the genetic evidence, current patterns, and scales of ecological connectivity, the NWHI are not likely a sufficient or consistent source to replenish stocks in the MHI although sporadic contributions are possible and need to be investigated more thoroughly.

14.10 The Importance of Predation in the NWHI Ecosystem

Several studies in the 1970s and 1980s documented the extent and magnitude of piscivory on shallow NWHI reefs (Hobson 1984; Norris and Parrish 1988; Parrish et al. 1985). The dominance of apex predators in the nearshore NWHI ecosystem exerts a strong "top-down" control on the size, composition, and spatial distribution of prey species (DeMartini and Friedlander 2006). The influence of apex predators on the distribution and abundance on prey species has been documented for the shallow water fish assemblages (reviewed in DeMartini and Friedlander 2006) as well as the deeper banks (30–40 m) in the NWHI (Parrish and Boland 2004).

Large-bodied parrotfishes are among the most abundant coral reef fishes in the NWHI (Friedlander and DeMartini 2002) and are a preferred prey item of Giant Trevally (*Caranx ignobilis*), the dominant predator in the NWHI. The size at sex change of parrotfishes, as noted by the changes in color patterns between females and terminal phase males, was observed to be inversely related to the density of Giant Trevally at individual atolls and islands in the NWHI and is likely related to predation (DeMartini et al. 2005). The size distributions of select labroids (parrotfishes and wrasses) (Fig. 14.11) and all other prey reef fishes were also strongly and negatively correlated with densities of Giant Trevally among all atolls and partial atolls (Lisianski Island/Neva Shoals, Maro Reef, French Frigate Shoals) (DeMartini et al. 2005). Juveniles and other small-bodied fishes particularly susceptible to predation have been documented to occupy lagoonal patch reef, and other sheltered (wave-protected) habitats as nursery areas that reduce the risk of predation (DeMartini 2004).

14.11 Fishing Impacts

Records of Polynesians fishing in the NWHI date back to the eleventh century. More recently, the harvest of guano, seabirds, monk seals, turtles, and various other natural resources occurred from the mid-1800s to the earliest twentieth century (Shallenberger 2006). Limited harvesting by fishing boats from the MHI took place until after WWII when a commercial fishing base with an airstrip was established at French Frigate Shoals to serve the Honolulu market. This fishery failed after several years due to financial constraints and the remote location of French Frigate Shoals. Large-scale commercial harvest of groundfish by Japan and Soviet Union occurred in the 1960s and 1970s at Hancock Seamount for Alfonsin (*Beryx*

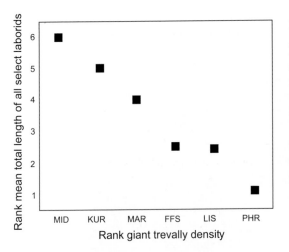

FIG. 14.11. Inverse relationship between rank giant trevally (*Caranx ignobilis*) density and rank median total length (TL) of select labroids (parrotfishes and wrasses) among atolls and partial atolls in the NWHI (Data from DeMartini et al. 2005)

decadactylus). Unfortunately, this fishery soon collapsed due to overfishing (Rutka 1984).

In the 1970s, due to the decline in fish stocks around the MHI, a state of Hawaii task force recommended the development of fisheries in the NWHI. Lobster and bottomfish fisheries expanded in the NWHI and catches for both peaked by the mid-1980s. However, the NWHI Lobster Trap Fishery was closed in 2000 because of large decreases in catch per unit effort along with a growing uncertainty in the population models used to assess stock status (DeMartini et al. 2003). Bottomfish stocks are primarily deep-slope (75–100 fanthoms) snappers (Lutjanidae) and one endemic species of grouper (Serranidae). Unfortunately, these stocks also declined since the 1980s and the fishery is scheduled to close in 2011 as mandated in the 2006 Presidential Proclamation establishing a new NWHI Marine National Monument. These declines are likely the result of fishing combined with a possible natural decline in ecosystem productivity associated with the Pacific-wide inter-decadal oscillation (see next section).

The lower abundances of Giant Trevally at Midway and Kure atolls likely reflect the result of many decades of recreational fishing by military personnel (DeMartini et al. 2002, 2005). In 1996, a recreational catch-and-release fishery was begun at Midway after the Midway Naval

Air Station was closed and the Atoll became a US Fish and Wildlife Service National Wildlife Refuge. Further decreases in Giant Trevally abundance occurred after this time at Midway, including changes in Ulua (carangids) behavior such as conditioned aversion to boats and divers (DeMartini et al. 2002).

In spite of sporadic but limited human disturbance events over the past 100 or more years, today the reefs in the NWHI are nearly pristine and represent one of the last remaining intact predator-dominated, large-scale coral reef fish assemblages on earth and remind us of what the reefs in the main Hawaiian Islands (MHI) may have looked like before human contact (Friedlander and DeMartini 2002). The coral reef ecosystem in the NWHI being close to a natural baseline, may also serve as a natural laboratory for comparison with conditions in the MHI and possibly elsewhere.

14.12 NWHI Coral Reef Ecosystem and Its Dynamics

The coral reefs of the NWHI provide habitat for a diverse marine ecosystem (Fig. 14.12; see also Sections 14.3 and 14.4). During the early 1980s, this ecosystem was described in considerable detail at French Frigate Shoals based on a focused multi-year field program by dozens of scientists (Grigg and Tanoue 1984) and a model (Ecopath) was developed to describe the flow of energy through the ecosystem (Polovina 1984) (Fig. 14.13). The Ecopath Model creates a static mass-balanced snapshot of the ecosystem and its trophic web. The trophically linked species groups consist of a single species, or a group of species representing ecological guilds. For each species group, biomass, production/biomass ratio (or total mortality), consumption/biomass ratio, and ecotrophic efficiency are measured. The orginal Ecopath Model has now been updated and extended to include a new dynamic version known as Ecosim (http://www.ecopath.org).

An application of Ecopath to French Frigate Shoals, based on data from the 1970s, partitioned the ecosystem into 12 species groups (Fig. 14.13). Sharks, jacks, monk seals, sea birds, and tuna at the top trophic level, reef fishes at the center, and 90%

FIG. 14.12. A mixed school of reef fishes among *Acropora cytherea* and *Porites lobata* corals at FFS. The blue line snapper, *Lujanus kasmira*, present in the middle of the school, is an introduced species (see text for more detail) (Photo by James Watt)

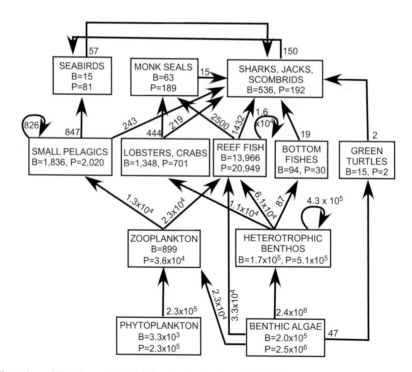

FIG. 14.13. Illustation of the Ecopath Model for the food web at FFS. The trophic pathway, annual production (P), and mean annual biomass (B) (kg/km^2) is given for 12 species groups based on an area of 1,200 km^2. Figure from Polovina (1984). The FFS ecosystem, as modelled by the Ecopath Model, is the only island in the world where the whole reef ecosystem has been analyzed

of the productivity from benthic algae. Modification of the species groups divided the large reef fishes group in four feeding guilds showed an ecosystem spanning almost five trophic levels with sharks, jacks, and piscivorous reef fish representing the top predators (Fig. 14.14). Benthic carnivorous reef fishes compose the mid-trophic level, with small benthic organisms at the lower trophic level and productivity largely from benthic algae. Except for some handline fishing for deepwater snappers and trapping for slipper and spiny lobster, the NWHI have been little fished. This model provides a picture of a coral reef ecosystem dominated by an abundance of apex predators (Fig. 14.14).

The Ecosim Model was used to simulate temporal ecosystem dynamics in response to top-down or bottom-up forcing (Christensen and Walters 2004) which was modelled by assuming 30 years of high benthic primary productivity, followed by 30 years of low benthic primary productivity (Fig. 14.16).

Considerable temporal lags, differing by as much as a decade, were observed in the responses of the various trophic levels both under an increase and a decrease in benthic productivity. Planctivorous reef fishes decreased during periods of high benthic productivity in response to increases in their predators (other reef fishes and jacks) even though prey plankton was unchanged. Also sharp increases were followed by decline in benthic carnivorous reef fishes immediately after the change from low to high benthic productivity (Fig. 14.16). The reef fishes quickly increased in abundance in response to higher prey availability, but 5 years later as their predators increased, their abundance declined. Potentially ecosystem dynamics can be more complicated than this and it is not always coherent with physical forcing.

These productivity-forced models are of relevance because the intensification of the Aleutian Low Pressure System during 1977–1988, characteristic of the cold phase of the Pacific Decadal

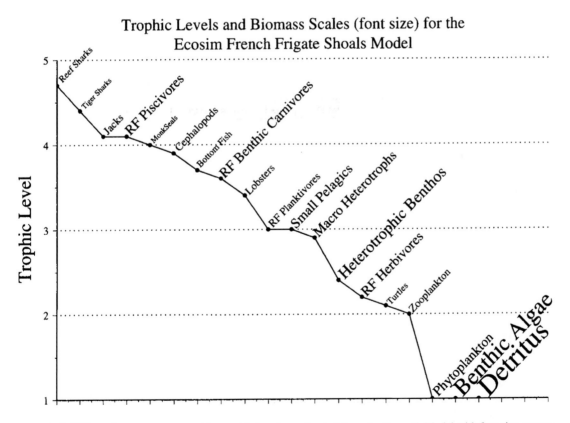

FIG. 14.14. FFS species groups arranged by trophic level as estimated from the Ecopath Model with font size proportional to biomass. See text for further explanation

Oscillation (PDO), appears to have resulted in increased winter deep mixing leading to more nutrients in the mixed layer and enhanced ecosystem productivity (Polovina et al. 1994). After this phase ended, productivity of the NWHI ecosystem returned to lower levels; and major declines in sea bird reproductive success (30–50%), monk seal pup survival, and spiny lobster recruitment were documented in the late 1980s (Polovina et al. 1994). More recently the biological impacts of changes in vertical stratification and mixing in the NWHI have been studied with an ocean model coupled with a plankton model. The results support earlier work that showed changes in vertical stratification producing changes in nutrients in the euphotic zone with impacts on plankton production. During El Niño periods vertical mixing is enhanced, resulting in higher plankton productivity while during La Niña events, vertical stratification increases, resulting in lower plankton productivity (Polovina et al., 2001). The Transition Zone Chlorophyll Front (TZCF) is a basin-wide feature separating the vertically mixed high surface chlorophyll water Transition Zone water and the vertically stratified subtropical water (Fig. 14.15). It is used as a trans-Pacific migratory pathway for some pelagic animals (Polovina et al. 2001). During the winter months, this boundary lies close to the northern atolls of the NWHI (in particular Kure, Midway, and Laysan). In some winters, often during El Niño periods, the TZCF shifts sufficiently south to include the northern atolls within the more productive northern waters. In other years, during La Niña periods, this boundary remains north of the atolls and the islands experience oligotrophic subtropical water (Fig. 14.16). The impact of the inter-annual variation of the TZCF on the endangered Hawaiian monk seals residing on the northern atolls is significant. For example, a positive relationship between the winter position of the TZCF and monk seal pup survival 2 years later, has been documented (Baker et al., 2002). A southward shift in the TZCF resulted in more productive Transition Zone waters impinging on the atolls and increased monk seal pup survival 2 years later.

An important future direction for biological research in the Hawaiian Archipelago will be advancing our understanding of metapopulation dynamics and connectivity, especially for coral reef species. The archipelago is composed of islands, reefs, and banks with intervening depths exceeding 2,000 m. The adults of most coral reef species are associated with shallow habitat and do not move between isolated shallow-water topographic features, however, it is very likely that mixing of pelagic larvae of many of these species does occur between banks. Some preliminary work to describe metapopulation dynamics is already underway. A transport model, driven by estimates of ocean currents derived from satellite altimetry, has been used to simulate the advection and retention of spiny lobster, *Panulirus marginatus*, larvae in the NWHI (Polovina et al. 1999). The results suggest there are differences between banks in the proportion of larvae retained from resident spawners as well as the proportion of larvae received from other banks (Polovina et al. 1999). These differences are due to the location of banks relative to broad current patterns and dynamics of local eddies. Results are consistent with the variation in mean densities of spiny lobster fishery catches between banks. Attention is drawn again to the issue of extent of larval transport between the Hawaiian Archipelago and other islands or banks outside the archipelago. The patchy distribution of corals of the genus *Acropora* in the NWHI with a concentration center at French Frigate Shoals (Fig. 14.8), led to a hypothesis that Johnston Atoll, 720 km southwest of the NWHI, was a source of colonization (Grigg 1981; Rivera et al. 2004). Simulation of larval transport from Johnston Atoll using an ocean circulation model provided support for this hypothesis, e.g., for a corridor of general larval transport between Johnston Atoll and the NWHI in the vicinity of FFS (Kobayashi 2006).

14.13   Environmental Impacts and Threatened and Endangered Species

Ecosystems are not static and respond as much to fisheries as to the climatic impacts described in the previous section. Changes in the system can be driven by commercial fisheries like the ones that have harvested lobsters and deepwater snappers in the NWHI. The trap fishery for spiny and slipper lobsters began in the late 1970s and continued

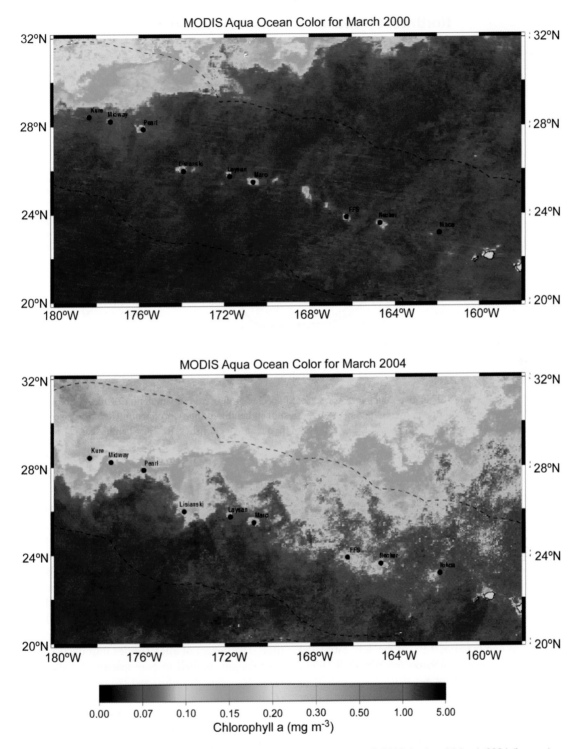

Fɪɢ. 14.15. Surface chlorophyll estimated from SeaWiFS ocean color March 2000 (top) and March 2004 (bottom)

Northwest Hawaiian Islands Ecosystem
30 Year Regime Shift in Benthic Productivity

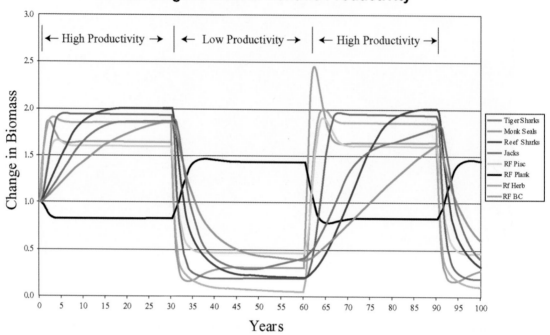

FIG. 14.16. Time series of relative abundance for selected species groups from a simulation of the FFS coral reef using the Ecosim Model driven by a hypothetical regime shift in benthic primary production (RF Pic = piscivorous reef fishes, RF Plank = planktivorous reef fishes, RF Herb = herbivorous reef fishes, RF BC = benthic carnivorous reef fishes (From Polovina 2005)

until it was closed in 2000. Landings reached their peak of about two million lobsters (slipper and spiny lobsters combined) from about 1 million trap hauls in 1985 (Polovina 2000). Management regulations included a limited-entry system with a maximum of 15 vessels, an annual fleet harvest quota, a closed season January to June to protect spawning biomass prior to summer spawning, and closed areas (Polovina 2000). The annual harvest quota was computed as 13% of the July estimated exploitable biomass (DiNardo and Wetherall 1999). However, the NWHI lobster fishery was closed in 2000 as a precautionary action due to concerns over low catch rates and a scientific need to improve the stock assessment model to address metapopulation dynamics.

A handline fishery for deepwater snappers and grouper in the NWHI has existed since about 1950. The maximum sustainable yield for the fishery has been estimated at about 450,000 lbs and typically

the fishery landings were less than 400,000 lbs (Moffitt et al. 2006).

One of the most blatant examples of past fishing pressure is the over-harvesting of pearl oysters at Pearl and Hermes Atoll in the early 1900s and the lack of full recovery even to the present day (Rauzon 2001). Fortunately, only lobsters and bottomfish are the two other species groups exhibiting signs of adverse fishing impact at the present time. The impact of alien species introductions has been largely limited to two species of fishes, the taape (*Lutjanus kasmira*) and the peacock grouper (*Cephalopholis argus*), and a small number of encrusting invertebrates likely introduced via ship hulls that are restricted to Midway harbor.

Other anthropogenic stressors in the NWHI include marine debris, ship groundings, disease, sporadic aquarium fish collecting, pollution and, of course, global warming. Fortunately, at the present time, the cumulative effects of these sources of stress on the reef ecosystem in the NWHI are apparently not yet significant. Marine debris has been described as one of the greatest anthropogenic impacts to the reefs of the NWHI (Friedlander et al. 2005) and much money has been spent for debris clean-up and removal (mostly lost fishing nets). Fish collecting and pollution could be problematic at a larger scale but both are presently highly localized and sporadic and unlikely to be of regional significance.

Of 52 known historic shipwrecks in the NWHI none has posed a critical threat to the reefs (Friedlander et al. 2005), despite the potential destruction such events can cause. Perhaps the greater concern should be about accidental introduction of alien species, such as rats, to terrestrial ecosystems on the islands. Coral disease is another perceived threat to coral reef ecosystems in the NWHI. A NOAA survey in 2003 estimated disease prevalence on the reefs to be only ~0.5% on average (Friedlander et al. 2005). Two mass coral bleaching events were reported in back-reef lagoon habitats at Pearl and Hermes, Midway and Kure Atolls in 2002 and again in 2004 (Aeby et al. 2003; Hoeke et al. 2006). In the impacted areas, coral coverage ranged between 1% and 11% which was presumably a result of recurrent disturbance events, potentially bleaching among them, giving the hope that only a few areas are susceptible to large-scale bleaching.

On the positive side, Hawaiian Green Turtles have increased significantly in the past 30 years and the population overall appears to be approaching carrying capacity (Balaz and Chaloupka 2006). Monk seals on the other hand remain endangered with a total population at about 1,300. The largest sub-population located at French Frigate Shoals has declined since the late-1970s, but in recent years a small number of monk seals has been observed in the main Hawaiian islands (Antonelis et al. 2006), giving hopeful signs that they may be increasing there.

The assessment of anthropogenic impacts to coral reef ecosystems in the NWHI as presented here is a reaffirmation of the overall pristine condition of the marine ecosystems in the NWHI. However, the examples of the pearl oyster and the lobster fishery collapse and the successful colonization of the alien Taape (*Lutjanus kasmira*) serve as reminders of the inherent ecological vulnerability of the NWHI ecosystems. Isolation, uniqueness, remoteness and small size combine to add importance to this concern. In spite of the overall pristine condition of these islands, there is no question about the need to continue to protect and manage the NWHI.

14.14 Past, Present and Future Management

The first formal step to protect wildlife in the NWHI (except Midway) was an executive order signed by president Roosevelt in 1909 that created a bird reservation entitled "The Hawaiian Islands Reservation". Midway, being under military control and not part of the Territory of Hawaii, was afforded similar protection by president Roosevelt under an earlier executive order in 1903. These orders had been prompted near the turn of the century by uncontrolled and excessive exploitation of seabirds taken for sale of their feathers and eggs (Shallenberger 2006). Hundreds of thousands of albatross and other Hawaiian seabirds were being sold at weekly auctions in America and Europe. A presidential proclamation in 1940 renamed the reservation "The Hawaiian Islands National Wildlife Refuge" (HINWR) and placed management control under the US Fish and Wildlife Service (USFWS) which did not establish a permanent presence in Hawaii until 1964. Exploitation of fish and Green Turtles continued in the NWHI into the early 1960s. Intermittent surveys of the islands were carried out but were limited in scope due to their remoteness. The USFWS faced problems with introduced species during the early century and particularly during World War II, such as rabbits on Laysan Island, rats, ants and mosquitoes on Midway and Kure Atolls and grasshoppers on Nihoa Island (Shallenberger 2006). Little active marine management occurred in the NWHI due to wars and concentration of terrestrial invasives but luckily the marine ecosystems of the NWHI (except for marine birds) were less impacted than the land.

During the latter half of the twentieth century, management of the resources in the NWHI has been split between a number of agencies including the State of Hawaii, the Western Pacific Fishery Management Council (WPRMC), NOAA, the USFWS, the US military and the US Coast Guard. Fishery regulations of the State of Hawaii apply to the HINWR and Kure Atoll 3 miles from the shore-line. From its inception in 1976, the WPRMC has been active in developing management plans for fishery resources within the entire US controlled Western Pacific from 3–200 miles. In the NWHI, the WPRMC plans are based on five management units: bottomfish, lobsters, pelagic fish, seamount ground fish and precious corals (Rutka 1984). Of these, only the first three have been applied to active fisheries and in the year 2000, the lobster fishery was closed. NOAA has been active in resource management in the NWHI through its National Marine Fishery Service (NMFS) whose primary role has been research and over-sight over the FMPs. The role of the US Coast Guard has traditionally been enforcement.

While all of these agencies continue to collaborate in management and protection, a new set of regulations emerged in 2000 by Executive Order 13196 when President W. Clinton created a Coral Reef Ecosystem Reserve (CRER) for the NWHI. This act created an overlap of new Federal restrictions in waters surrounding the islands (except Midway) from 3 to 50 nmi offshore. This legislation established inter-alia, upper limits on some commercial fisheries and created fifteen Reserve Preservation Areas. And while the ink was barely dry on this executive order, Presidential Proclamation 8031 by President G.W. Bush created the NWHI National Marine Monument on June 26, 2006, under the authority of the Antiquities Act. The Proclamation applies to all lands and submerged lands and waters to 50 nmi and includes the NWHI CRER, the Midway Atoll National Wildlife Reserve and the HINWR. Primary management authority now resides under the Departments of Commerce and Interior through NOAA and the USFWS, while the State of Hawaii is given management responsibility over state waters within the Monument. Together, the three agencies are considered co-trustees of the NWHI and are mandated to work in cooperation.

It is clear that the future management of the NWHI will be an overlapping and multi-agency responsibility requiring extensive collaboration and cooperation. The unique attributes of the NWHI must be re-emphasized. The NWHI ecosystem not only represents one of the few remaining large-scale "intact" marine tropical ecosystems in the world, but it's remoteness, isolation and – paradoxically – relatively small area of coral habitat combine to add to its vulnerability to anthropogenic disturbance. Maintained in its natural state, it is not only a unique natural laboratory for ecological research but it is also relevant to the understanding and management of many other comparable tropical island ecosystems. It is the crown jewel of US coral reefs.

References

Aeby GS, Kenyon JC, Maragos JE, Potts DC (2003) First record of mass coral bleaching in the NWHI. Coral Reefs 22:256

Allen GR (2002) Indo-Pacific coral-reef fishes as indicators of conservation hotspots. Proc 9th Int Coral Reef Symp 2:921–926

Antonelis GA, Baker JD, Johanos TC, Braun RC, Harting AL (2006) Hawaiian Monk seal (*Monachus schauinslandi*): status and conservation issues. Atoll Res Bull 543:75–101

Atlas of Hawaii (1998) 3rd edn, University of Hawaii Press, Honolulu, HI

Balaz GH, Chaloupka M (2006) Recovery trend over 32 years at the Hawaiian green turtle rookery of French Frigate Shoals. Atoll Res Bull 543:147–158

Boehlert GW, Mundy BC (1992) Distribution of ichthyoplankton around Southeast Hancock Seamount, central North Pacific in summer 1984 and winter 1985: data report. US Department of Commerce, NOAA Technical Memorandum NMFS-SWFSC-176, 109 p

Boehlert GW, Watson W, Sun LC (1992) Horizontal and vertical distribution of larval fishes around an isolated oceanic island in the tropical Pacific. Deep Sea Res 39:439–466

Caldwell PC (2005) Validity of North Shore, Oahu, Hawaiian Islands surf observations. J Coastal Res 21:1127–1138

Christensen V, Walters CJ (2004) Ecopath with Ecosim: methods, capabilities and limitations. Ecol Model 172:109–139

DiNardo GT, Wetherall JA (1999) Accounting for uncertainty in the development of harvest strategies for the Northwestern Hawaiian Islands lobster trap fishery. ICES J Mar Sci 56:943–951

DeMartini EE (2004) Habitat and endemism of recruits to shallow reef fish populations: selection criteria for no-take MPAs in the NWHI Coral Reef Ecosystem Reserve. Bull Mar Sci 74:185–200

DeMartini EE, Friedlander AM (2004) Spatial patterns of endemism in shallow reef fish populations of the Northwestern Hawaiian Islands. Mar Ecol Progr Ser 271:1–296

DeMartini EE, Friedlander AM (2006) Predation, endemism, and related processes structuring shallow-water reef fish assemblages of the Northwestern Hawaiian Islands. Atoll Res Bull 543:237–256

DeMartini EE, Parrish FA, Boland RC (2002) Comprehensive evaluation of shallow reef fish populations at French Frigate Shoals and Midway Atoll, Northwestern Hawaiian Islands (1992/93, 1995–2000). NOAA Technical Memorandum NMFS SWFSC 347:1–54

DeMartini EE, DiNardo GT, Williams HA (2003) Temporal changes in population density, fecundity, and egg size in the Hawaiian spiny lobster (*Panulirus marginatus*) at Necker Bank, Northwestern Hawaiian Islands. Fish Bull 101:22–31

DeMartini EE, Friedlander AM, Holzwarth S (2005) Size at sex change in protogynous labroids, prey body size distributions, and apex predator densities at NW Hawaiian atolls. Mar Ecol Progr Ser 297:259

Firing JB, Brainard RE (2006) Ten years of shipboard ADCP measurements along the Northwestern Hawaiian Islands. Atoll Res Bull 543:347–363

Firing JB, Hoeke R, Brainard R, Firing E (2004) Connectivity in the Hawaiian Archipelago and beyond: potential larval pathways. Proceedings of the10th International Coral Reef Symposium, Okinawa, 310 pp

Friedlander AM, DeMartini EE (2002) Contrasts in density, size, and biomass of reef fishes between the Northwestern and the Main Hawaiian Islands: the effects of fishing down apex predators. Mar Ecol Prog Ser 230:253–264

Friedlander AM, Parrish JD, DeFelice RC (2002) Ecology of the introduced snapper Lutjanus kasmira in the reef fish assemblage of a Hawaiian bay. J Fish Biol 60:28–48

Friedlander AM, Brown EK, Jokiel PL, Smith WR, Rodgers KS (2003) Effects of habitat, wave exposure, and marine protected area status on coral reef fish assemblages in the Hawaiian archipelago. Coral Reefs 22:291–305

Friedlander AM, Aeby G, Brainard R, Clark A, Godwin E, Godwin S, Kenyon J, Kosaki R, Maragos J, Vroom P (2005) The State of Coral Reef Ecosystems of the NWHI. In: The State of Coral Reef Ecosystems of the United States and Pacific Freely Associated States: 2005. National Oceanic and Atmospheric Administration, Center for Coastal Monitoring and Assessment, Silver Spring, MD, p 270

Grigg RW (1981) Acropora in Hawaii. Part II. Zoogeography. Pac Sci 35:15–24

Grigg RW (1982) Darwin Point: a threshold for atoll formation. Coral Reefs 1:29–34

Grigg RW (1983) Community structure, succession and development of coral reefs in Hawaii. Mar Ecol Prog Ser 11:1–14

Grigg RW (1997) Paleoceanography of coral reefs in the Hawaiian-Emperor Chain – Revisited. Coral Reefs 16:S33–S38

Grigg RW (2006) The history of marine research in the NWHI: lessons from the past and hopes for the future. Atoll Res Bull 543:13–22

Grigg RW, Epp D (1989) Critical depth for the survival of coral islands: effects on the Hawaiian Archipelago. Science 243:638–640

Grigg RW, Pfund RT (1980) Proceedings of the Symposium on status of Resource Investigations in the NWHI. UNIHI-SEAGRANT-MR-80–4, 333p

Grigg RW, Tanoue KY (1984) Proceedings of the Second Symposium on Resource Investigations in the Northwestern Hawaiian Islands. UNIHI-SeaGrant-MR84_01, 191p

Grigg RW, Wells J, Wallace C (1981) Acropora in Hawaii. Part 1. History of the scientific record, systematics and ecology. Pac Sci 35:1–13

Hirota J, Taguchi S, Shuman RF, Jahn AE (1980) Distribution of plankton stocks, productivity, and potential fishery yield in Hawaiian water. In: Proceedings of the Symposium on Status of Resource Investigations in the NWHI. UNIHI-SEAGRANT-MR-80–04, pp 191–203

Hobson ES (1984) The structure of reef fish communities in the Hawaiian Archipelago. In: Grigg R, Tanoue K (eds) Proceedings of the 2nd Symposium on Resource Investigations in the Northwestern Hawaiian Islands, Vol. 1. UNIHI-SEAGRANT-MR-84–01. University of Hawaii Sea Grant College Program, Honolulu, HI, pp 101–122

Hoeke RK, Brainard R, Moffitt R, Merrifield M, Skirving W (2006) The role of oceanographic conditions and reef morphology in the 2002 coral bleaching event in the NWHI. Atoll Res Bull 543:489–504

Kenyon J, Godwin S, Montgomery A, Brainard R (2007) Rare sighting of *Acropora cytherea* in the main Hawaiian islands. Coral Reefs 26:309

Kobayashi DR (2006) Colonization of the Hawaiian Archipelago via Johnston Atoll: a characterization of oceanographic transport corridors for pelagic larvae using computer simulation. Coral Reefs 25:407–417

Maragos JE, Jokiel PL (1986) Reef corals of Johnston Atoll: one of the world's most isolated reefs. Coral Reefs 4:141–150

Maragos JE, Potts DC, Aeby G, Gulko D, Kenyon J, Sicilano D, Van Ravensway D (2004) 2000–02 rapid ecological assessment of corals (Anthozoa) on shallow reefs of the NWHI. Part I. Species and distribution. Pac Sci 58:211–230

Miller SL, Crosby MP (1998) The extent and condition of US coral reefs. In: NOAA's State of the Coast Report. Silver Spring, MD, pp 1–34

Moffitt RB, Kobayashi DR, DiNardo GT (2006) Status of bottomfish stocks, 2004. Pacific Islands Fisheries Science Center Administrative Report H-06–01 NOAA Honolulu, HI

Mundy BC (2005) Checklist of the fishes of the Hawaiian Archipelago. Bishop Museum Bulletin in Zoology. BM Press, Honolulu, Hi

Norris JE, Parrish JD (1988) Predator-prey relationships among fishes in pristine coral reef communities. Proc 6th Int Coral Reef Symp, Townsville 2:107–113

Oda DK, Parrish JD (1981) Ecology of commercial snappers and groupers introduced to Hawaiian Reefs. Proc 4th Int Coral Reef Symp 1:59–67

Parrish FA, Boland RC (2004) Habitat and reef-fish assemblages of banks in the Northwestern Hawaiian Islands. Mar Biol 144:1065–1073

Parrish JD, Callahan MW, Norris JE (1985) Fish trophic relationships that structure reef communities. Proc 5th Int Coral Reef Symp, Tahiti 4:73–78

Polovina JJ (1984) Model of a coral reef ecosystem. Part I. ECOPATH and its application to French Frigate Shoals. Coral Reefs 3:1–11

Polovina JJ (2000) The lobster fishery in the Northwestern Hawaiian Islands. In: Phillips BF, Kittaka J (eds) Spiny Lobsters. Fisheries and Culture, 2nd edn, Blackwell, Oxford, pp 679

Polovina JJ (2005) Climate variation, regime shifts, and implications for sustainable fisheries. Bull Mar Sci 76:233–244

Polovina JJ, Mitchum GT, Graham NE, Craig MP, Demartini EE, Flint EN (1994) Physical and biological consequences of a climate event in the central North Pacific. Fish Oceanogr 3:5–21

Polovina JJ, Kleiber P, Kobayashi D (1999) Application of TOPEX/POSEIDON satellite altimetry to simulate transport dynamics of larvae of the spiny lobster (Panulirus marginatus), in the Northwestern Hawaiian Islands, 1993–96. Fish Bull 97:132–143

Polovina JJ, Howell E, Kobayashi DR, Seki MP (2001) The transition zone chlorophyll front, a dynamic global feature defining migration and forage habitat for marine resources. Progr Oceanogr 49(1–4):469–483

Randall JE (1981) Examples of antitropical and antiequatorial distribution of Indo-West-Pacific fishes. Pac Sci 35:197–209

Randall JE (1987) Introductions of marine fishes to the Hawaiian Islands. Bull Mar Sci 41:490–502

Randall JE (1995) Zoogeographic analysis of the inshore Hawaiian fish fauna. In: Maragos JE, Peterson MNA, Eldredge LG, Bardach JE, Takeuchi HF (eds) Marine and Coastal Biodiversity in the Tropical Island Pacific Region, Vol. 1. Species Systematics and Information Management Priorities. East-West Center, Honolulu, HI, pp 193–203

Randall JE (1998) Zoogeography of shore fishes of the Indo-Pacific region. Zool Stud (Taiwan) 37:227–268

Randall JE, Earle JL (2000) Annotated checklist of the shore fishes of the Marquesas Islands. Occas Pap Bishop Museum 66:1–39

Randall JE, Earle JL, Pyle RL, Parrish JD, Hayes T (1993) Annotated checklist of the fishes of Midway Atoll, Northwestern Hawaiian islands. Pac Sci 47:356–400

Rivera MAJ, Kelly CD, Roderick G K (2004) Subtle population genetic structure in the Hawaiian grouper, Epinephelus quernus, as revealed by mitochondrial DNA analyses. Biol J Linn Soc 81:449–468

Rutka J (1984) Management plans for fishery resources in the NWHI. In: Proceedings of 2nd symposium on resource investigations in the NWHI. UNIHI-SEAGRANT-MR-84–01, pp 463–476

Rauzon MJ (2001) Isles of Refuge. University of Hawaii Press, 205 pp

Rohmann SO, Hayes JJ, Newhall RC, Monaco ME, Grigg RW (2005) The area of potential shallow-water tropical and subtropical coral ecosystems in the United States. Coral Reefs 24:370–383

Sadler J, Oda L, Kilonsky B (1976) Pacific Ocean Cloudiness from Satellite Observations. University of Hawaii, Department of Meteorology, pp 76–101

Seckel GR (1962) Atlas of the oceanographic climate of the Hawaiian Islands region. Fish Bull US Fish and Wildlife Serv 61:371–427

Shallenberger RJ (2006) History of management in the NWHI. Atoll Res Bull 543:23–31

Veron JEN (1995) Corals in Space and Time. Cornell University Press, Ithaca, NY, 321 pp

Vroom PS, Page JC (2006) Relative abundance of macroalgae on NWHI reefs. Atoll Res Bull 543:533–548

15

US Coral Reefs in the Line and Phoenix Islands, Central Pacific Ocean[1]: History, Geology, Oceanography, and Biology

James Maragos, Joyce Miller, Jamison Gove, Edward DeMartini,
Alan M. Friedlander, Scott Godwin, Craig Musburger, Molly Timmers,
Roy Tsuda, Peter Vroom, Elizabeth Flint, Emily Lundblad, Jonathan Weiss,
Paula Ayotte, Enric Sala, Stuart Sandin, Sarah McTee, Todd Wass,
Daria Siciliano, Russel Brainard, David Obura, Scott Ferguson,
and Bruce Mundy

15.1 Introduction (by J. Maragos and J. Miller)

Pacific remote island areas (PRIA) are sovereign United States unincorporated and unorganized territories not falling within the jurisdiction of any other US territory or State (GAO 1997; US DOI 2003). There are eight PRIA and all are under the jurisdiction of the US Department of Interior (DOI). All are low reef islets or atolls in the central Pacific Ocean. Table 15.1 and Fig. 15.1 present the size, location, and regional geography of the five PRIA that are the primary focus in this chapter: Baker and Howland Islands in the Phoenix Islands; and Jarvis Island, Kingman Reef, and Palmyra Atoll in the Line Islands. All five are located between Hawai'i and Samoa and are National Wildlife Refuges (NWR) administered by the US Fish and Wildlife Service (USFWS). The remaining three PRIA: Johnston Atoll NWR, Midway Atoll NWR, and Wake Atoll, are the subject of other chapters in this volume (Chapter 13, Rooney et al.; Chapter 17, Lobel and Lobel). Johnston Atoll is mentioned in this chapter because it is geologically part of the Line Islands archipelago. Wake Atoll, administered by the US.

Air Force, is north of the Republic of the Marshall Islands. Midway Atoll is located at the northwest end of the Hawaiian Islands and is a PRIA because it was excluded from state jurisdiction in the Hawaii Statehood Act of 1959. This chapter covers the cultural, geological, and biophysical characteristics and history of the five PRIA. The following chapter (Chapter 16, Maragos et al.) covers the status, threats and significance of the five PRIA.

Howland and Baker Islands are geologically part of the northwest–southeast trending Tokelau submarine ridge and located just 68 and 53 km north of the Equator, respectively, with Howland about 66 km northwest of Baker. They are normally referred to as outliers of their nearest island neighbors: the eight Phoenix Islands to the southeast that are within the jurisdiction of the Republic of Kiribati. The Tokelau submarine ridge extends further southeast through the western Phoenix Islands to the Tokelau Islands (a New Zealand Protectorate of three atolls) and Swains Island (within the US Territory of American Samoa). The nearest population center from Howland and Baker is Kanton Atoll (the only inhabited island of the Phoenix group) about 350 km to the southeast. The next closest inhabited islands are Tarawa Atoll, capital of Kiribati in the Gilbert Islands about 900 km to the west, and Funafuti Atoll, capital of Tuvalu (Ellice Islands), 900 km to the southwest of Howland and Baker.

---

[1] The opinions of the authors included in this report do not necessarily represent those of the agencies or institutions at which they are employed.

B.M. Riegl and R.E. Dodge (eds.), *Coral Reefs of the USA*,
© Springer Science+Business Media B.V. 2008

TABLE 15.1. Emergent and submerged areas within the NWR boundaries of the five PRIA (GAO 1997; 2003–2004 Quickbird Imagery for Baker, Howland and Jarvis; US DOI 2003).

Name	Emergent reef[a]/land	Submerged lands (to 3[b] or 12 nmi[c])
Baker Island NWR[b]	172.5 ha (426.2 ac)	12,680 ha (31,332 ac)
Howland Island NWR[b]	183.3 ha (453 ac)	12,989 ha (32,096 ac)
Jarvis Island NWR[b]	447.5 ha (1107 ac)	14,744 ha (36,432 ac)
Kingman Reef NWR[c]	1.2 ha (3 ac)[a]	195,752 ha (483,703 ac)
Palmyra Atoll NWR[c]	275.2 ha (680 ac)	208,512 ha (515,233 ac)

nmi = nautical mile = 1.828 km; ha = hectare; ac = acre
[a] Emergent reef area only for Kingman which lacks land
[b] Areas based on 3 nm offshore boundaries for Baker, Howland and Jarvis
[c] Areas based on 12 nm offshore boundaries for Palmyra and Kingman

Palmyra Atoll, Kingman Reef, and Jarvis Island are part of the 12 Line Islands and associated with the northwest–southeast trending Line Islands submarine ridge. Kingman and Palmyra are near the north and Jarvis at the center of the ridge. Despite Johnston's locale at the far north end of the submerged Line Islands ridge, it is closest to the Hawaiian Islands in terms of distance (700 km vs 1,600 km from Kingman Reef) and biogeography (Maragos and Jokiel 1986). The remaining eight emergent Line Islands belong to the Republic of Kiribati with three inhabited (Teraina, Tabuaeran, and Kiritimati) between Palmyra and Jarvis and the remaining five uninhabited (Malden, Starbuck, Vostok, Millennium, and Flint) south of Jarvis and accounting for the southern portion of the archipelago. Jarvis is located near the Equator approximately 1,600 km due east of Howland and Baker. Kingman and Palmyra are at latitude ~6° N, with Kingman about 61 km northwest of Palmyra. Their nearest population centers are Honolulu 1,880 km to the north, and Teraina, Tabuaeran, and Kiritimati about 220–600 km to the south. Jarvis is about 365 km southwest of inhabited Kiritimati and about 730 km south of Palmyra and Kingman.

Before Kiribati assumed jurisdiction of 16 Line and Phoenix Islands in 1979, Teraina, Rawaki, Tabuaeran,

Kiritimati, Millennium, and Kanton were previously named Washington Island; Phoenix Islands; and Fanning, Christmas, Caroline, and Canton Atolls, respectively, and are referred to the latter in earlier scientific literature cited in this chapter.

15.2 Political and Cultural History (by J. Maragos, M. Timmers, and J. Miller)

Archaeological evidence at neighboring islands and atolls in the Line and Phoenix Islands indicate that early Polynesians and Micronesians were likely the first visitors to the five PRIA during the previous millennium, although brief archaeological surveys to date at Palmyra (C. Streck 1987) and Baker and Howland (Shun 1987) have not revealed supporting evidence. The five PRIA were all uninhabited at their time of rediscovery by European and American explorers, whalers, and miners during the past two centuries. All five are presently uninhabited except for caretakers and a dozen or more researchers at Palmyra during the field season. The lack of land at Kingman; arid conditions at Baker, Howland, and Jarvis; and remote locations for all five are factors that probably discouraged or prevented their long-term occupation in the past.

The five PRIA were claimed for the United States under the Guano Islands Act of 1856. Guano consists of seabird droppings rich in phosphates that were commercially valuable as fertilizer for agriculture and for military production of explosives. The Act provided the discoverer of uninhabited islands, unclaimed by other nations, the exclusive right to mine guano for US citizens, and these rights extended to their heirs (GAO 1997). Baker, Howland, and Jarvis Islands were also claimed by the United Kingdom as British Overseas Territories from 1886 to 1934, and guano mining was conducted by both British and American companies through the end of the nineteenth century, after which guano deposits were largely depleted. Palmyra Atoll and Kingman Reef were not actively mined because Palmyra was too wet for guano accumulation while Kingman was mostly submerged and lacked guano.

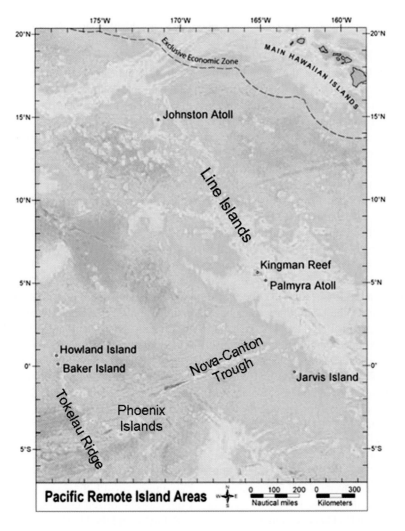

FIG. 15.1. Location of the five US Line and Phoenix Islands PRIA covered in this chapter (Baker, Howland, Jarvis, Kingman and Palmyra) plus Johnston Atoll. Map from Coral Reef Ecosystem Division (CRED) of NOAA's Pacific Islands Fisheries Science Center (PIFSC), with seafloor topography derived from satellite altimetry and ship-based soundings (Smith and Sandwell 1997)

15.2.1 Baker, Howland and Jarvis Islands

Baker and Jarvis were first sighted in 1832 by Michael Baker, the captain of an American whaling ship. After discovering guano deposits at Jarvis, he claimed Jarvis for the United States. Captain Baker returned to Baker Island in 1838 to claim it for the United States as well. In 1855 Captain Baker sold his interest in Baker and Jarvis Islands to the American Guano Company (AGC), who in turn claimed both islands under the Guano Islands Act of 1856. The company mined guano at both islands until the 1880s, and then abandoned them. Howland was discovered in 1842 by a New England whaler, George Netcher who reported it overrun by rats from an earlier shipwreck. Later the AGC and a competing firm owned by Arthur Benson claimed Howland. Both companies mined guano there between 1857 and 1878 and then abandoned the island (GAO 1997).

In the interest of developing the islands for commercial aviation stopovers and firmly establishing American jurisdiction, the US organized military personnel and Hawaiian high school graduates of Kamehameha Schools from Honolulu, later to be known as the *Hui Panala'au* (Bryan 1974; Kikiloi and Tengan 2002), to occupy the three islands and others beginning in 1935. In 1936 President Roosevelt moved to place all three islands under the jurisdiction of the Secretary of Interior. Airfields were constructed at Howland in 1937 prior to Amelia Earhart's and Fred Noonan's failed attempt to land there. After a 1942 Japanese military attack that killed two of the Hawaiians at Howland, the US evacuated all civilians from the islands and destroyed all facilities at Jarvis. In 1943, American troops continued to occupy Howland and Baker and built an airstrip at Baker in preparation for the American assault on Japanese forces at Tarawa Atoll. The island housed up to 120 officers and 2,000 men. By March 1944, after the successful outcome of war efforts, the US military abandoned Howland and Baker. In 1974, the US Fish and Wildlife Service assumed administrative responsibility for all three islands and established them as National Wildlife Refuges. Later, the USFWS surveyed the islands and sponsored the eradication of cats from Howland and Jarvis, promoting the return of ground-nesting seabirds (see Rauzon 1985).

15.2.2 Kingman Reef

Kingman was first discovered by Captain Fanning, an American whaler, in 1798 and again by Captain Kingman in 1853. The US Guano Company claimed the atoll reef in 1860. Kingman was claimed again in 1922 by Lorrin Thurston on behalf of the Palmyra Copra Company for use as a fishing base. Later the Fullard-Leo family of Hawai'i claimed that the Palmyra Copra Company had ceded its interest in Kingman to the family. Despite the lack of guano at Kingman Reef, it was the fact that an American had first discovered the reef and no other nation had claimed it. Consequently, in 1934 President Roosevelt placed Kingman under the control of the US Navy, formally asserting American rights to it. In 1941, Kingman was among several of the PRIA around which Naval Defensive Seas and over which National Airspace Reservations were declared by President Roosevelt to exclude their use or occupation by foreign military powers. In 1937, Kingman was used as a temporary landing area for Pan American Airways seaplanes traveling between Hawai'i and Samoa. In January 2001 the Navy agreed to convey Kingman Reef to the US Department of Interior, and Kingman was established as a NWR under administration of the USFWS. In contrast to the earlier designated PRIA refuges, the Kingman and Palmyra NWR boundaries extended to 12 nm due to President Reagan's 1988 decision to extend US territorial waters to 12 nm offshore. In retrospect, a 3 nm boundary around Kingman would have been insufficient to protect all shallow reefs. At present (August 2007), a 2006 lawsuit filed by the Fullard-Leo heirs and associates in federal court contesting federal ownership of Kingman Reef was decided in favor of the federal owners.

15.2.3 Palmyra Atoll

Palmyra was discovered by the captain of the American ship *Palmyra* in 1802, but was not claimed until 1862 when ownership was asserted by Captain Zenas Bent and J.B. Wilkinson for the Kingdom of Hawai'i (GAO 1997). The British also claimed Palmyra in 1889. The Pacific Navigation Company bought Palmyra in 1885 and the company's interests were conveyed in 1911 to Judge Henry Cooper via petition to the Land Court of the Territory of Hawai'i. Judge Cooper sold all of Palmyra except two islets to the Fullard-Leo family in 1922. In preparation for possible war, the US Navy attempted to lease Palmyra from the Fullard-Leo family in 1938. However, in 1939 the US Congress authorized construction of a naval base at Palmyra, and the US filed suit to annex the atoll. Up to 6,000 servicemen occupied Palmyra Atoll Naval Air Station during the World War II (WWII) era. In 1947 the US Supreme Court, returned ownership of the atoll to the Fullard-Leo family. The 1959 Hawai'i Statehood Act specifically excluded Palmyra, and by that time US Navy occupation had ceased and all other federal presence at the atoll ended. Subsequently, the atoll remained abandoned except for resident caretakers supported by the Fullard-Leo family (GAO 1997). In 2000, The Nature Conservancy (TNC) purchased Palmyra, and later the US Department of Interior bought all reefs and islands from TNC except for the main island (Cooper), its airfield, and dock. In January 2001, the atoll was established as a NWR under administration of the USFWS. In 2005, TNC established a field research station capable of housing up to 20 scientists and staff on Cooper Island which is now maintained by several caretakers.

15.3   History of Biophysical Research (by J. Maragos and R. Tsuda)

Early scientific research focused on botanical and seabird surveys at all islands except Kingman which lacked land. The *Whippoorwill* expedition of 1924 and others sponsored terrestrial surveys and marine collections at many of the PRIA (G.C. Munro 1924, unpublished data; Christopherson 1927; Fowler 1927). The Pacific Ocean Biological Survey Program of the Smithsonian Institution sponsored several bird and vegetation surveys at the PRIA islands during the 1960s (Sibley et al. 1965). Based on specimens collected during the Smithsonian Institution surveys in July and October 1964, Tsuda and Trono (1968) reported 11 and 16 marine algae species, respectively, from Howland Island and Baker Island. In a monographic treatment of the Pacific red algae *Polysiphonia* and *Herposiphonia*, Hollenberg (1968a, b) included two species of *Polysiphonia* and three species of *Herposiphonia* from Jarvis Island; and one species of *Herposiphonia* from Howland Island. The only previous records of marine algae from Palmyra were 12 species reported by Howe and Lyon (1916).

The outbreak of ciguatera fish poisoning during the WWII Navy occupation of Palmyra stimulated extensive investigations beginning in the 1950s on marine algae and fish carrying the toxicity (Dawson et al. 1955; Goe and Halstead 1955; Halstead and Schall 1958; Helfrich et al. 1968). During this period marine phycologist E.Y. Dawson (1959) published an early assessment of environmental impacts on lagoon and vegetation caused by construction of the US Naval base at Palmyra, the first scientific documentation of anthropogenic impacts on an atoll reef. Other marine research included oceanography (Barkley 1962; Wyrtki 1967 and others later reviewed in this report), geology (later reviewed in this report), meteorology (Vitousek et al. 1980), and commercial fisheries, mostly by the National Oceanic and Atmospheric Administration and its predecessors. In 2000 research on bonefish populations was initiated at Palmyra because the atoll offered the opportunity to evaluate the effects of its low-intensity catch-and-release sport fishery on bonefish population structure and function in a protected and largely unaltered environment (Friedlander et al., in press).

The first coral surveys at any of the five PRIA began at Palmyra (Maragos 1979, 1987), and collection and identification of a few dozen corals were accomplished for Baker and Howland (J. Schmerfeld and J.E. Maragos 1998, unpublished data). Comprehensive research on coral reefs began in 2000, under the mandate of the US Coral Reef Task Force. NOAA has sponsored joint Pacific Reef Assessment and Monitoring Program (RAMP) investigations carried out by scientists from NOAA, USFWS, and several other institutions, to establish baseline parameters, collect oceanographic data, accomplish towed diver surveys around the perimeters of all reefs, and conduct biennial monitoring of the coral reef ecosystems at all eight PRIA, other US Pacific territories, and Hawai'i. The regular use of NOAA research vessels has facilitated these expeditions, and all five PRIA covered in this chapter have been investigated during expeditions in 2000, 2001, 2002, 2004, and 2006, with the next expedition scheduled in early 2008. The ecosystem research investigations have adopted uniform scientific methods to assess, map and monitor all other US-affiliated islands in the Pacific, including the Hawaiian, Samoan, and Mariana archipelagos and remaining PRIA, thus allowing comparisons of the status and health of all US coral reef ecosystems in the Pacific. These methods are detailed in earlier reports by Turgeon et al. (2002) and Brainard et al. (2005) and in the companion state of the reefs report (SORR) to be concurrently released at the 2008 International Coral Reef Symposium in Florida (Anon., in preparation), with Miller et al. (in preparation) authoring the PRIA chapter.

The Palmyra Atoll Research Consortium (PARC), established in 2005, is operated by TNC and primarily funded by the Gordon and Betty Moore Foundation and nine member research institutions. At present the American Museum of Natural History, California Academy of Science, Scripps Institution of Oceanography of the University of California (UC) at San Diego, Stanford University, UC at Irvine, UC at Santa Barbara, UC at Santa Cruz, University of Hawai'i, and Victoria University of Wellington are all presently members or affiliates of PARC. In addition, Stanford, Scripps, UC Santa Barbara and other PARC institutions have sponsored recent expeditions to Kingman Reef and neighboring Kiribati islands and atolls in the northern Line Islands (Teraina, Tabuaeran, and Kiritimati).

15.4   Regional Setting

15.4.1   Geology and History of Reef Building in the Central Pacific (by J. Miller)

The Central Pacific seafloor in the vicinity of the Line and Tokelau ridges is a complex area, not well understood from a geological perspective. The Line Island ridge (Fig. 15.2) has numerous parallel and NW–SE trending features and is intersected by three fracture zones: Molokai, Clarion, and Clipperton. The volcanic structures of the Line Islands are diverse in shape, size, composition, and morphology (Davis et al. 2002). The Tokelau ridge lies near the termination of the Clipperton Fracture

Zone and is intersected by the Nova Canton trough (Fig. 15.1). The Tokelau ridge has two bends in its lineation and may be a possible hotspot trace or the result of plate extension, as discussed by Koppers and Staudigel (2005). Limited geological drilling and dredging has been done in both the Line (Schlanger et al. 1976, 1984; Garcia et al. 1983; Haggerty et al. 1982; Davis et al. 2002) and Tokelau (Koppers and Staudigel 2005) submarine ridges. Recent multibeam mapping around all of the US Pacific remote island areas (Vroom et al. 2006a; Ferguson et al. 2006) provides a detailed look at the morphology of the Line and Phoenix Islands covered here and provide the potential for a better understanding of their individual geological histories.

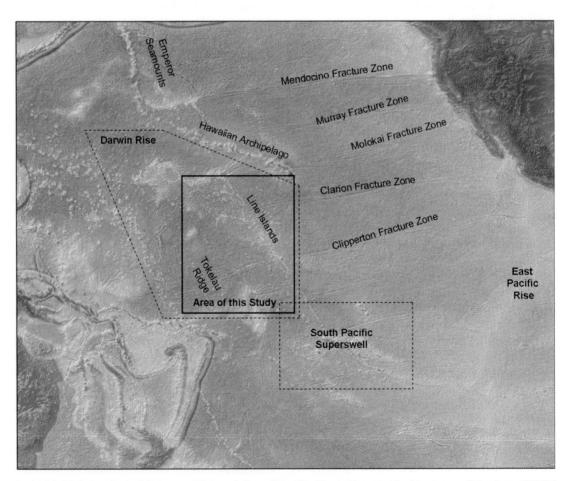

FIG. 15.2. Major geological features of East and Central Pacific Ocean discussed in this chapter (Map from CRED/ PIFSC with seafloor topography from Smith and Sandwell (1997))

15.4.1.1 Tectonic Origins and Seafloor Ages of the Central Pacific

McNutt and Fisher (1987) identified the South Pacific Superswell (Fig. 15.2) and argued that the Darwin Rise area (Menard 1964) in the NW and Central Pacific, was located over the same Superswell area during the Cretaceous (144–65 Ma) (Jordahl et al. 2004). The northwestern part of the Darwin Rise is also called the West Pacific Seamount Province by Koppers et al. (2003). The Darwin Rise and the Superswell are both areas that are 200–700 m shallower than would be predicted by lithospheric subsidence models proposed by Sclater and Francheteau (1970), and Parsons and Sclater (1997). Various mechanisms were originally proposed to explain these depth anomalies, including an anomalously thin 75 km thick thermal plate (McNutt and Fisher 1987), uplift of the Darwin Rise during the Cretaceous (McNutt et al. 1990), and/or a Cretaceous super-plume (Larson, 1991). Koppers et al. (2003) point out that concentrated intra-plate volcanism in several island chains, including the Line Islands, was causally related to the thermal rejuvenation that formed the Darwin Rise and that this ancient Superswell resulted in lithospheric uplift of the seafloor. Lithospheric thinning and superplume models would predict that heat flow in the Darwin Rise and Superplume areas would be higher than normal, but heat flow has been shown to be no higher in these areas than for comparable age lithosphere in other areas (Stein and Abbott 1991; Stein and Stein 1993). Thus, alternate mechanisms that are currently proposed for the anomalously shallow areas include the presence of a buoyant volcanic layer beneath the crust-mantle boundary (Moho) (McNutt and Bonneville 2000), reheating events (Smith and Sandwell 1997), the dynamic effects of mantle plumes (Sleep 1992), and areas of the asthenosphere that are either more fertile or warmer than surrounding mantle (Natland and Winterer 2005).

The Line and Tokelau ridges lie within the Darwin Rise, on magnetically "quiet" seafloor formed during the Cretaceous Normal Superchron (120–83 Ma) (Atwater et al. 1993). The underlying seafloor along the Line Islands chain ranges from mid-Cretaceous (119 Ma) at the northern end to late Cretaceous at the southern end (83 Ma) (Davis et al. 2002; Fig. 15.3). According to Natland and Winterer's (2005) model, the Line Islands were formed near a spreading center on the eastern edge of the Pacific Plate between 70 and 122 Ma. The seafloor underlying the northern part of the Tokelau ridge is dated between 120.4 and 131.9 Ma (Early Cretaceous) (Muller et al. 1997; Clouard and Bonneville 2005).

In addition to the underlying seafloor ages, another critical element in understanding the evolution of island chains is accurate dating of dredge, drilling, and paleontological samples from the individual seamounts and islands. Davis et al. (2002) presented revised age estimates in the northern Line Islands (6–20° N) based upon previous studies (Schlanger et al. 1984; Garcia et al. 1983) and new $^{40}Ar/^{39}Ar$ geochronological evidence from nine volcanic edifices. These authors concluded that there were two major episodes of volcanism in the northern Line Islands, which lasted approximately 5 Myr and were separated by approximately 8 Myr. The older episode (81–86 Ma) occurred along the eastern part of the northern islands over a distance of 1,200 km, while volcanism during the younger episode (68–73 Ma) was concentrated along the western edge of the chain and may have extended over 4,000 km. Haggerty et al. (1982), using paleontological techniques, reported Cretaceous (70–75 Ma) ages at the southern end of the Line Island chain (9° S) near Caroline Atoll (Millennium) as well as Eocene sediments (33.7–54.8 Ma) engulfed by a volcanic eruption in the same area, indicating the volcanism in both the Cretaceous and the Eocene.

Because of these studies showing multiple, synchronous episodes of volcanism over large segments of the Line Island chain in multiple time periods, Morgan's (1972) hypothesis, that the Line Islands are a simple hotspot trace similar to the Hawaiian–Emperor chain that increases steadily in age from east to west, has been repeatedly challenged. A number of alternate models for formation of the Line Island chain and for central Pacific Cretaceous volcanoes in general have been proposed. Jackson and Schlanger (1976) suggested a mid-ocean ridge crest origin for the Line Islands, but petrologic results from DSDP (Deep Sea Drilling Project) Leg 33 showed that the Line Island basalts were similar to Hawaiian basalts, which argues against a ridge crest origin.

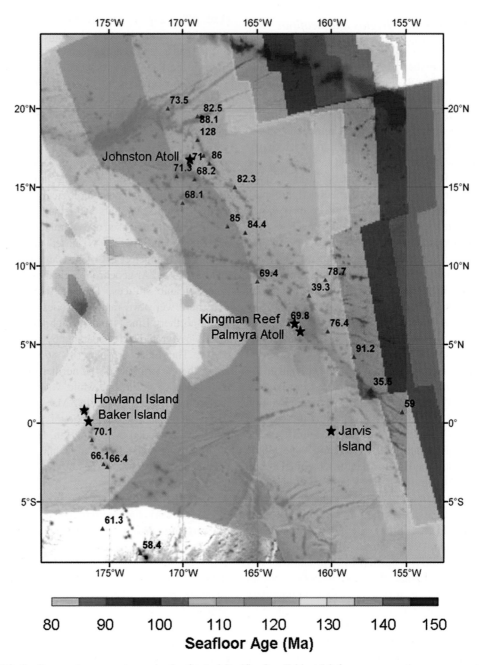

FIG. 15.3. Seafloor and seamount ages in the Central Pacific. See Table 15.2 for sources and error parameters for seamount ages. Triangles indicate locations of dredge or drill age data, and stars indicate location of multibeam data at six PRIA locations (Baker, Howland, Jarvis, Johnston, Kingman and Palmyra; Figs. 15.4–15.10) (Seafloor ages from Muller et al. 1997; Map by CRED/PIFSC)

Winterer (1976) and Orwig and Kroenke (1980) first proposed fractures in the Line Islands as mechanisms for volcanic activity. Schlanger et al. (1984) and Garcia et al. (1983) suggested two or possibly more hotspots as the origin of the Line Islands. Epp (1984) suggested a variety of mechanisms to explain the age distributions in the Line Islands including lithospheric melting and interaction with transform faults. Davis et al. (2002) discuss lithospheric extension as the mechanism for formation of these islands, while Koppers et al. (2003) propose that plate tension, hotlines, faulting, wetspots, and self-propagating volcanoes may work in conjunction with hot spot volcanism over larger areas of the Pacific. Natland and Winterer (2005) discuss five sets of stresses in the upper mantle that combine to produce the patterns of great fissures and systems of fissures through the lithosphere along which seamount provinces and linear chains form.

Recent research by Koppers and Staudigel (2005) along the Tokelau ridge and Gilbert Island chain establishes seamount ages from dredge samples in these areas, particularly focusing on dating the bends that occur in the two chains. These seamount ages are combined with predicted hotspot traces from extinct hotspots (Wessel et al. 2003; Kroenke et al. 2004) to compare the age of Tokelau (57 Ma) and Gilbert (67 Ma) bends with the 47 Ma age of the Hawaiian–Emperor Bend (Dalrymple and Clague 1976), which has recently been revised to 50 Ma (Sharp and Clague 2006). Koppers and Staudigel (2005) argue that because the ages of the three bends are asynchronous, these data do not support a stationary hot spot paradigm, but may point to either hot spot motion or magmatism caused by short-term local lithospheric extension as the origin of the Tokelau ridge.

15.4.1.2 Reef Growth and History

The history of reef growth and subsidence in the Line Islands was researched by Schlanger et al. (1984) based upon stratigraphy from three DSDP sites (165, 315, 316) in the north and central sections of the Line Islands and four dredge hauls in the southern Line Islands (Fig. 15.3, Table 15.2). DSDP site 315 was drilled near Fanning Atoll (Tabuaeran) which lies on the eastern (older)[2] side of the Line Islands, approximately 330 km south of Kingman Reef. The oldest mid-Cretaceous reefs in the Line Islands are from this site and were dated from Cenomanian-Albian age fossils (93.5–112 Ma) in basalts found in turbidites that were determined to be 93.3 Ma. Similar mid-Cretaceous reefs are also found at DSDP site 217 near Horizon

Guyot, which lies approximately 550 km to the north of Johnston Island (Winterer et al. 1973) at the far northern end of the Line Island chain.

Schlanger et al. (1984) identified Late Cretaceous (70–75 Ma) reefs along 2,500 km of the Line Island chain – at Kingman Reef (DSDP Site 165), Fanning Atoll (DSDP Site 315), south of Christmas Atoll (Kiritimati) (DSDP Site 316), and on seamounts just north of Caroline Atoll (Millennium) (Dredges 44 and 45). The only evidence for post-Cretaceous reef formation was found at Kingman Reef and north of Caroline Atoll (Millennium) at the opposite ends of the Line Island chain. Late Eocene (33.9–40.4 Ma) fossils were found in Oligocene (23.0–33.9 Ma) turbidites at Kingman Reef. Dredge samples from north of Caroline Atoll (Millennium) contained reef limestones of Eocene through Plio–Pleistocene (0.1–55.8 Ma) age. Schlanger et al. (1984) also note that no reef limestone samples were found in eight dredge hauls taken between Johnston Atoll and Kingman Reef.

Reef growth, emergence, and subsidence on the six PRIA including Johnston Atoll since the Plio–Pleistocene have not been extensively documented, and only a few studies exist on any of the Central Pacific reefs, reef islands, and atolls. All six of the PRIA for which multibeam data are presented in this paper are very low lying with little remaining subaerial evidence of previous sea stands. In addition, the surface and/or lagoons of the islands have been extensively modified by either guano mining (Baker, Howland, and Jarvis) or by military construction activities (Johnston and Palmyra).

On Christmas Atoll (Kiritimati), which lies ~600 km southeast of Kingman Reef, a previous +0.5–1.0 m sea-level stand was documented with evidence of lagoon reefs that flourished 1.5–4.5 ka (1,000s of years before present) (Woodroffe and McLean 1998). An outcrop of limestone was dated at 130 ka and interpreted as being from the Last Interglacial period. Twenty-four cores were taken for a water resources survey. These cores consisted of highly fractured Pleistocene limestones to depths of 20–30 m, overlain by 1–4 m of sand and/or gravel. Radiocarbon dates from these cores ranged in age from 1.9 to ~40 ka. Woodroffe and McLean (1998) concluded that there has been little Holocene reef accumulation

---

[2] However, according to the isochron map in Fig. 15.3, the eastern side of the Line Island chain is younger.

TABLE 15.2. Date estimates in Ma (millions of years before present) from seafloor samples, error estimates and sources of information (From Clouard and Bonneville 2005).

Long.	Lat.	Date	Error	Name	Source
−150.4	−9	70.5	1.1	RD45-26D	Schlanger et al. 1984
−151.5	−7.5	71.9	1.4	RD44-3	Schlanger et al. 1984
−155.28	0.7	59.0	0.8	RD43-1	Schlanger et al. 1984
−157.35	2.1	35.5	0.9	RD41-1	Schlanger et al. 1984
−158.5	4.2	91.2	2.7	DSDP-315	Lanphere and Dalrymple 1976
−160.25	5.83	76.4	0.5	123D-15	Schlanger et al. 1984
−162.9	6.3	69.8	0.3	SO33 D29	Davis et al., 2002
−161.5	8.1	39.3	1.5	RD33	Schlanger et al., 1984
−165	9	69.4	0.3	SO33 D49	Davis et al. 2002
−160.4	9.1	78.7	1.3	128D-11	Schlanger et al. 1984
−165.8	12.1	84.4	0.9	133D	Saito and Ozima 1976
−167	12.5	85.0	1.1	RD59-12	Schlanger et al. 1984
−170	14	68.1	0.2	SO37 D10	Davis et al. 2002
−166.5	15	82.3	0.6	RD61-1	Schlanger et al. 1984
−169.2	15.5	68.2	0.2	SO33 D43	Davis et al. 2002
−170.4	15.7	71.3	0.2	SO33 D28	Davis et al. 2002
−169.5	16.4	71.0	0.2	Johnston	Davis et al. 2002
−168.22	16.5	86.0	0.9	RD63-7	Schlanger et al. 1984
−168.6	17	84.8	0.2	Karin Ridge	Davis et al. 2002
−169	18	128.0	5	142D	Saito and Ozima 1976
−169	19.5	88.1	0.4	143D-102	Schlanger et al. 1984
−168.75	19.5	82.5	0.4	Horizon	Davis et al. 2002
−171	20	73.5	0.1	S033 D72	Davis et al. 2002
−176.19	−1.07	70.1	0.5	Lelei	Koppers and Staudigel 2005
−175.4	−2.6	66.1	0.6	Siapo	Koppers and Staudigel 2005
−175.15	−2.78	66.4	0.6	Polo	Koppers and Staudigel 2005
−175.47	−6.7	61.3	0.6	Matai	Koppers and Staudigel 2005
−172.88	−8.27	58.4	0.3	Ufiata	Koppers and Staudigel 2005

and that Christmas Atoll (Kiritimati) reached its present form in the Middle Pleistocene or earlier. The Last Interglacial and the Holocene Interglacial seem to have deposited only a minor veneer of coral over these Pleistocene surfaces. Woodroffe and McLean (1998) also concluded that Christmas Island (Kiritimati) has not undergone significant subsidence through the Late Quaternary. Local residents have claimed the contrary, rapid emergence over the past 30 years based upon lagoon shrinking, land expansion and shoreline shifts in the inner lagoon areas (U. Bukareiti, personal communication 2004).

Emery (1956), Ashmore (1973), and Keating (1987, 1992) all concluded that Johnston does not have a true atoll structure with emergent reef surrounding a central lagoon and described the anomalous structure with the only emergent reef being on the leeward (NW) side of the edifice, rather than on the windward side as more commonly found on oceanic islands and atolls.

Ashmore (1973) suggested that Johnston Atoll is tilted to the south-east, and Keating's research supports this hypothesis. Carbonate samples, but no volcanic rocks, were collected during four submersible dives on Johnston Island (Keating 1987) down to depths of 400 m, and all ages were determined to be less than 11,000 ybp. Keating concluded that all of the samples collected during the dives were formed on the shallow carbonate bank, dislodged and deposited at these deeper sites.

15.4.2 Geomorphology of the PRIA (by J. Miller, E. Lundblad, J. Weiss, and S. Ferguson)

Recently collected, high-resolution multibeam bathymetric data provide a detailed look at the submarine geomorphology of six islands, atolls and reefs in the US Line and Phoenix Islands (NOAA 2008 State of the Reefs Report) and can provide insight into the origin of the volcanic edifices and the history of the reefs. Bathymetric and backscatter data were collected in early 2006 using EM3002D and EM300 multibeam sonars aboard the NOAA Ship *Hi'ialakai* and a Reson 8101ER sonar aboard the survey launch R/V *AHI* (Vroom et al. 2006a; Ferguson et al. 2006).

Statistics for each of the islands, reefs or atolls to highlight possible similarities and differences among the six volcanic edifices are presented in Table 15.3. While no ages are available for Baker Island or Palmyra Atoll, these features lie very close to Howland Island (66 km) and Kingman Reef (61 km), respectively, and support the hypothesis that Baker is similar in age to Howland (70–72 Ma) and that Palmyra is similar in age to Kingman Reef. If this assumption can be made, five of the six PRIA were formed between 69 and 72 Ma during the second major pulse of volcanism in the area, although they lie on seafloor with a much wider range of ages (111–126.7 Ma).

As shown in Table 15.3, the summit areas of these islands range in size from 2–57 km², yet all still remain near or above sea level approximately 70 Ma after their formation. These islands are considerably older than Kure Atoll (29.8 Ma), which is the oldest emergent edifice in the Hawaiian/Emperor chain. According to seafloor subsidence models (Sclater and Francheteau 1970; and Parsons and Sclater 1977), none of these old, very small islands should remain above sea level. Their continued emergence also contradicts work in the Hawaiian Archipelago by Grigg and Epp (1989) that showed that the depth of drowned banks is inversely related to summit area, and that the summits of smaller banks were progressively deeper.

All of these island summit areas are considerably smaller than any emergent summit in the Hawaiian Archipelago, except for Nihoa Island, which is the youngest of the NWHI submerged banks with an age of approximately 7.2 Ma (Dalrymple et al. 1974). This might be explained by faster reef growth in areas further south than the Hawaiian Archipelago; however, Grigg and Epp (1989) indicate that Hawai'i and Johnston Island have similar coral growth rates of up to 9 mm/year, with the fastest known growth rate in the world being only 15 mm/year at St. Croix, USVI. In addition, there is little evidence of significant Holocene reef accretion at Christmas Atoll (Kiritimati), a process that is not possible if the atoll is emerging rather than subsiding.

The summits of Baker, Howland (Fig. 15.4), and Jarvis Islands are the smallest of the PRIA, and the highest point on any of these lies just 8 m above sea level. Their tops are all isolated peaks rising from the deep ocean (3,000 m or greater) with no surrounding ridges or other connected features (refer to Fig. 15.1). Although the flanks of these islands are very steep with slopes between 30° and 40° down to a depth of about 1,000 m, all three also have small shallow terraces that probably reflect previous sea level stands. The terraces (Fig. 15.5), marked by the areas of low slope, are present between ~7 and 17 m and between ~90 and 130 m around all three islands. Additional terraces are present at multiple depths off Howland Island.

Whereas Howland, Baker, and Jarvis Islands are isolated features with no underlying ridge structures, Kingman Reef and Palmyra Atoll are located on the same, very large shallow ridge structure that stretches

TABLE 15.3. Characteristics and age estimates of Baker, Howland, and Jarvis Islands, Kingman Reef,; and Palmyra and Johnston Atolls. Emergent area, total area, and maximum elevation from the CIA World Fact Book 2007. Summit area is calculated as area above the 100 fathoms/182 m depth curve from recently collected multibeam data. Island ages from Clouard and Bonneville 2005. Volume, ocean bottom and seafloor ages from the Seamount Catalogue (Koppers 2007).

Island	Land area (km²)	Summit area (km²)	Total area (km²)	Volume (km³)	Ocean bottom (m)	Max. Elev. (m)	Seafloor age (Ma)	Island age (Ma)
Baker Island	2.1	4.21	129	2,302	5,015	8	123.8–124.3	n/a
Howland Island	2.6	2.37	139	2,889	4,850	3	124.6–125.2	70–72
Jarvis Island	5	5.07	152	4088	4580	7	111.6	n/a
Kingman Reef	0.01	32.52	1958	12,352	3,495	1	111.7–113.0	69.76
Palmyra Atoll	3.9	20.48	1,950	2,168	2,715	2	111.7–112.2	n/a
Johnston Atoll	2.6	56.61	276	n/a	n/a	10	120.4–126.7	71.01

FIG. 15.4. Perspective view of Baker and Howland Islands showing steep slopes on all sides and mass wasting primarily on the eastern side of both islands (Shallow (0–1,000 m) bathymetric data from CRED/PIFSC and deeper data from the Seamount Catalog (Koppers 2007))

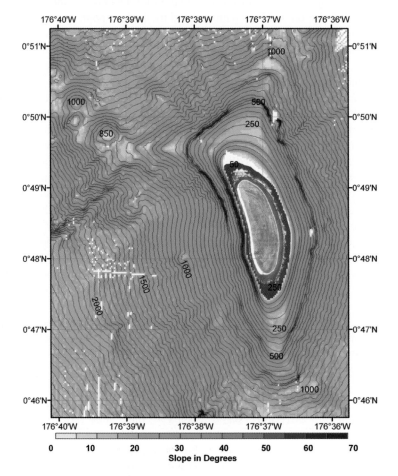

FIG. 15.5. Slope map of Howland Island with 50 m bathymetric contours. Multiple terraces are seen on the NE, NW, and S rift zones. Ikonos satellite data are displayed inside of multibeam coverage (Bathymetric data collected by CRED/PIFSC; Ikonos image courtesy of Space Imaging)

from Kingman Reef at 6° N to past Christmas Atoll (Kiritimati) at 1° N (Fig. 15.1). This ridge structure is intersected by the Clipperton Fracture Zone, which stretches on towards the Tokelau ridge as the Nova Canton trough. Ages from dredge and drilling samples along this ridge range from 35 to 91 Ma and indicate three periods of volcanism. Both Kingman Reef and Palmyra Atoll connect to nearby secondary peaks on platforms in the 1,000–2,500 m depth range. Kingman is anomalous in that it has numerous conical features (Fig. 15.6), not seen on any of the other five formations that follow the rift zone structures. Because there have been three different periods of volcanism in the Kingman area, we hypothesize that these cone structures might represent resurgent volcanism in the area. Further dredging or submersible work on Kingman Reef would be necessary to test this hypothesis and to date the features.

Unexpectedly, Kingman Reef (Fig. 15.7) lacks shallow terrace structures outside of the central lagoons, such as those on Howland, Baker, and Jarvis Islands, although Palmyra Atoll (Fig. 15.8) has broad terraces off the east and west ends of the atoll. On both Kingman Reef and Palmyra Atoll, only limited multibeam data were collected in the central lagoon area due to operational and time constraints. Further mapping in the lagoons or on the islands and sample collection for age dating would be necessary to reconstruct the sea level records of the areas. On the western edge of Palmyra there is a shallow shelf or terrace where depths between 13 and 20 m were mapped. In the large area inside Kingman Reef (Fig. 15.8), the few multibeam lines that were run showed rapidly varying depths down to over 100 m. The central lagoon area of Palmyra also has rapidly varying depths,

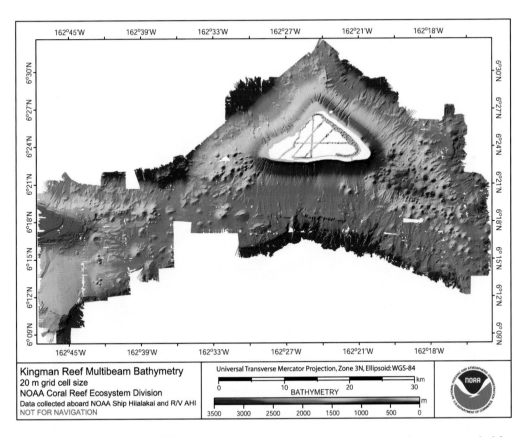

Fig. 15.6. Multibeam bathymetry of Kingman Reef showing a shallow area to the east and numerous conical features on the rift zones surrounding the atoll reef (Bathymetric map from CRED/PIFSC)

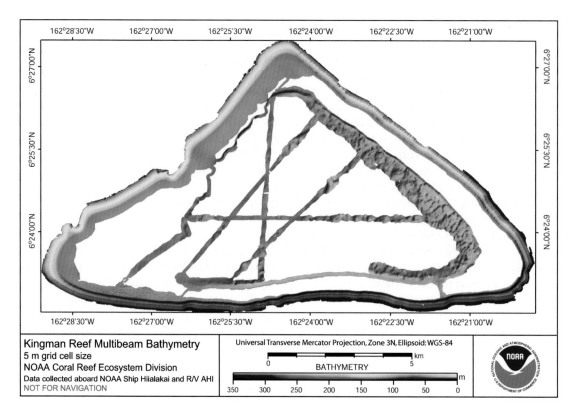

Kingman Reef Multibeam Bathymetry
5 m grid cell size
NOAA Coral Reef Ecosystem Division
Data collected aboard NOAA Ship Hiialakai and R/V AHI
NOT FOR NAVIGATION

Universal Transverse Mercator Projection, Zone 3N, Ellipsoid: WGS-84

BATHYMETRY

FIG. 15.7. Shallow multibeam data collected by CRED/PIFSC around and inside of Kingman Reef

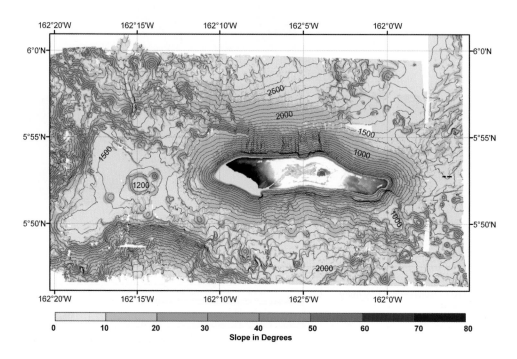

Slope in Degrees

FIG. 15.8. Slope map from CRED/PIFSC with 50 m bathymetric contours for Palmyra Atoll. Ikonos image (courtesy of Space Imaging) is shown in the center of the multibeam data

but significant dredging in this lagoon during the WWII era has modified its bathymetry.

Multibeam bathymetric data were also collected around Johnston Atoll in 2006 but no data were collected inside the lagoon, which has also been extensively dredged. Johnston is located at the northern end of the Line Island chain and is in an elevated area much less extensive than the Kingman Reef to Christmas Island ridge structure. Johnston is underlain by a smaller ridge structure that was discussed by Keating (1987). The multibeam bathymetry data (Fig. 15.9) show extensive evidence of mass wasting on the S and NE side of the atoll and a broader shelf on the N/NW. The major slope break around the entire island is at a depth of 1,000 m. Overall, the geomorphology of Johnston Atoll is much more similar to Kingman Reef and Palmyra Atoll than it is to Jarvis, Howland, or Baker Islands. However, Johnston has much more

evidence of previous sea-level stands on very narrow, but almost continuous terraces around the atoll at 20–33 m, 50–90 m, and 115–120 m (Fig. 15.10).

15.4.3 Local Shallow-Water Geomorphology of the PRIA Coral Reefs (by R. Brainard, J. Maragos and D. Siciliano)

15.4.3.1 Reef Classification

Baker, Howland, and Jarvis are classified as low reef islands, Palmyra as an atoll, and Kingman as an atoll reef. The geological processes that formed all five are essentially the same: reef development began when the volcanic foundations were still emergent islands in the Cretaceous to Eocene periods, followed by subsidence and upward reef

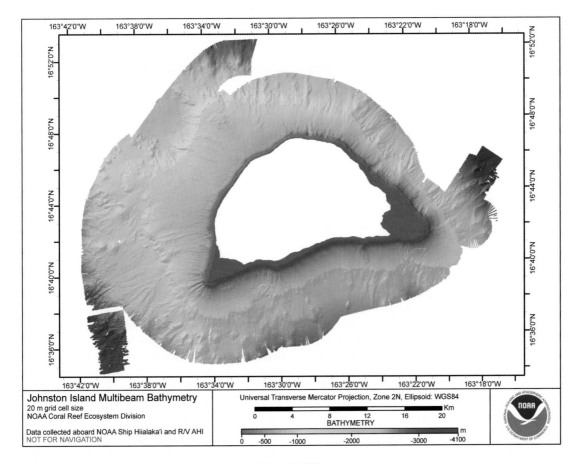

FIG. 15.9. Multibeam bathymetry of Johnston Atoll from CRED

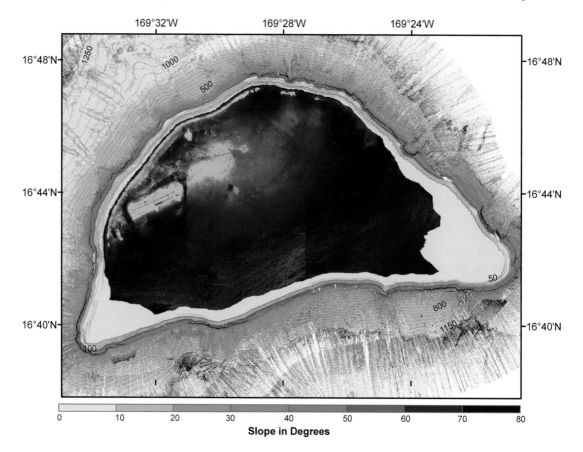

FIG. 15.10. Johnston Atoll slope map with 50 m bathymetric contours. Bathymetric data from CRED. Superimposed multi-spectral "color" Ikonos image shows emergent reef only on NW side of island (Image courtesy of Space Imaging)

growth maintaining proximity to the sea surface during prolonged time periods. The differences among the five are the contemporary nature of their islands and reefs. Kingman is an atoll reef because it lacks islands, but its shallow reefs encircle a large deep lagoon. Palmyra is a true atoll because it has reefs encircling three sub-lagoons and supporting many islets. Baker, Howland, and Jarvis are low reef islands resting atop and encircled by shallow reef summits too narrow in breadth to create contemporary lagoons at present sea level stands.

Towed divers have been able to survey the perimeters of all PRIA reefs, plot the positions of all surveys, and quantitatively characterize local oceanographic processes and bottom habitats during NOAA RAMP expeditions. These data have been supplemented by the observations of RAMP divers at fixed sampling sites, multibeam surveys, towed camera surveys, and interpretation of Ikonos satel-

lite imagery acquired in 2000 and Quickbird satellite imagery acquired in 2003–2004.

15.4.3.2 Baker, Howland and Jarvis Islands

For illustrations see Figs. 15.4, 15.5, 15.11–15.13 and 15.17. Baker and Jarvis are elongated in an east–west direction, have similar geological features, and experience similar oceanographic processes. Narrow fringing reef crests emerge at extreme low tides and encircle all three islands. Reef slopes are steep and descend to great depths off the west, north and south sides of Baker (Fig. 15.11) and Jarvis (Fig. 15.13). Broad submerged fore reef terraces are found off the eastern sides of both islands that gradually descend from 3 to 20 m depth, beyond which they plummet to great depths. Howland (Fig. 15.12) is elongated along a north–south axis, and only towed divers have been

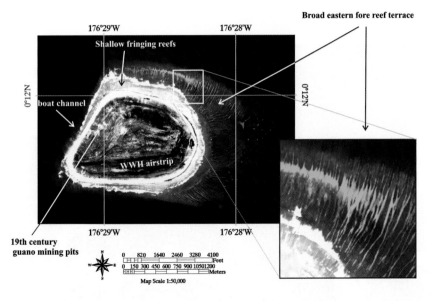

FIG. 15.11. Satellite image of Baker Island acquired on January 9, 2004 by Quickbird imager. Image processed to highlight the submerged reef terraces (Image courtesy of DigitalGlobe)

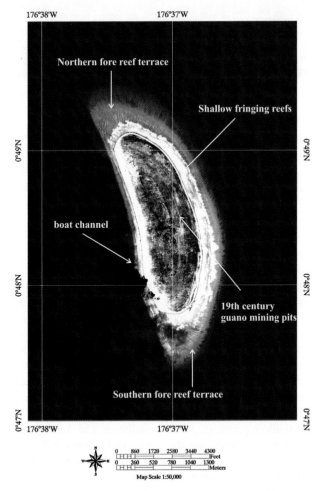

FIG. 15.12. Satellite image of Howland Island acquired on October 16, 2003 by Quickbird sensor (courtesy of DigitalGlobe). Image processed to highlight the submerged reef terraces. The blue spots over the southern fore reef terrace are clouds

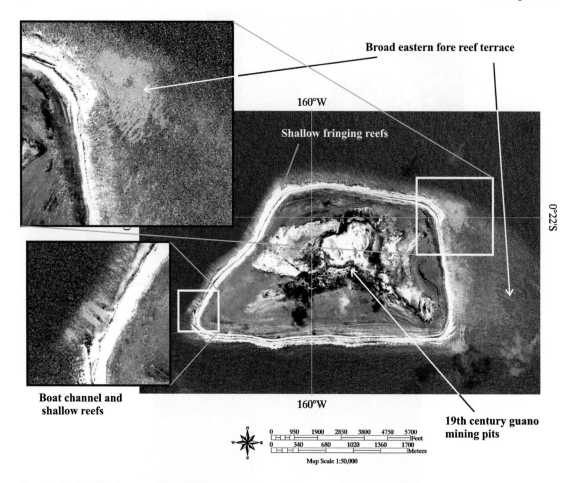

FIG. 15.13. Satellite image of Jarvis Island acquired on August 24, 2004 by Quickbird imager. Image processed to highlight the submerged shallow fringing reefs and reef terrace. The blue spots at the south edge of the fore reef are clouds (Image courtesy of DigitalGlobe)

able to characterize the reefs along its east coast, north tip and south tip due to persistent onshore winds, associated swells, and strong currents. Satellite imagery and multibeam bathymetric data indicate Howland lacks the broad fore reef terraces characterizing the eastern sides of Baker or Jarvis, although narrow pointed fore reef terraces are present off the north and south ends of Howland. Fringing reef crests on the western side of Howland are narrow, and the reef slope beyond plummets to great depths. Small channels have been blasted through the shallow perimeter reefs off the western sides of all three islands and the south side of Jarvis. These allowed small boat transport of supplies and personnel during the guano and WWII eras.

15.4.3.3 Kingman Reef

Kingman (Figs. 15.6–15.7) is a roughly triangular reef 20 km long in an east–west direction, with its acute apex facing due east into windward seas (Fig. 15.6). The perimeter reefs are shallowest and broadest at the eastern apex, and their crests gradually deepen to the west to depths of 10–15 m. The elongated south perimeter reef is bisected by one named natural pass (La Paloma Channel) with a depth of 5–7 m near the eastern apex, and another shallower and narrower pass located about 2 km west of La Paloma. Towards the southwest corner, a much broader pass bisects the south perimeter reef and has sill depths of about 10–15 m. Most of the shorter western perimeter reef (10 km in a north–south direction) itself is submerged and mostly a broad pass varying in depth from 12 to 35 m. There is one pass cutting through the elongated northern perimeter reef and a few narrow rills and irregular depressions. The lagoon is divided into a smaller reef pool, 5–7 m in depth, inside the eastern apex of the perimeter reef, with a shallow (0–2 m) north–south linear reef separating the pool from the rest of the lagoon that gradually widens and deepens to the west. The deep lagoon floor, 50 to >100 m deep, is unexplored and punctuated with about two dozen known pinnacles and patch reefs whose crests reach near sea level in its shallower east and central sections. All accessible fore reefs appear to have steep slopes although those off the northeast perimeter and eastern apex reefs have terraces at ~10 m depths that gradually descend to seaward.

15.4.3.4   Palmyra Atoll

Palmyra (Fig. 15.8) is an elliptical reef 20 km in length with unusual elongated fore reef terraces extending 3–5 km in breadth off the east and west ends of the atoll from depths of 7–25 m. The central part of the atoll is encircled by shallower perimeter reefs that supported up to 50 islets and three sub lagoons separated by shallow north–south linear reefs before 1940. Beyond the eastern end of the lagoon and ring of islets are three large reef pool complexes (3–15 m deep), one deeper and circular in shape, just east of East Lagoon and the other two elongated depressions farther east in the otherwise expansive shallow eastern reef flat (1–5 m depth). No natural deep passes originally connected the lagoons to ocean areas.

Early visitors characterized the atoll as one of the "Pearls" of the Pacific and Polynesia and a "necklace of emerald islets" (Boddam-Whetham 1876; Bryan 1940; Wright 1948). Beginning in 1939, the US Navy dredged a 9 m deep and 60 m wide ship channel through the shallow perimeter reef between the southwest ocean reef and West Lagoon, dredged the reef separating the West and Center Lagoons to a depth of 3 m for a seaplane runway, and connected all but a few islets by constructing elevated road causeways on the shallow perimeter reefs around most of West and Center Lagoons. East Lagoon, including a north–south road causeway on the elongated reef wall separating West and East Lagoons was completely encircled and cut off from surface water circulation with adjacent reefs (Dawson 1959; Maragos 1993). Two small runways were also constructed in Center Lagoon. The main island (Cooper-Menge) was also greatly enlarged to support a larger, 1,800 m long runway, a deep draft dock and other facilities. The lingering effects of these modifications and associated oceanographic changes are covered in Chapter 16 and are being investigated by the Victoria University of Wellington (J. Gardner, personal communication 2007) and other PARC institutions.

15.5   Climate, Oceanography, and Ocean Currents in the Central Pacific (J. Gove)

The equatorial region is dominated by easterly Trade Winds spanning the Pacific basin. These winds force a complex system of westward-flowing surface currents and eastward-flowing surface and subsurface counter-currents (Fig. 15.14). Baker Island (0°12′ N, 176°29′ W), Howland Island (0°48′ N, 176°37′ W), and Jarvis Island (0°22′ S, 160°03′ W) all lie within one degree latitude of the equator in the central Pacific and are in the mean flow path of two major ocean currents: the Equatorial Undercurrent (EUC) and South Equatorial Current (SEC). The EUC, a cold and nutrient rich (Chavez et al. 1999) eastward-flowing subsurface counter-current, lies along the equator and within the thermocline, with Undercurrent core depths ranging from 200 m in the western Pacific to 20 m in the east, and with an extraordinarily wide range of velocities between that average about $1.0\,\mathrm{ms^{-1}}$ (Yu and McPhaden 1999). The South Equatorial Current (SEC), is a westward-flowing current, lying directly above the EUC. It is generally weaker (fluctuating around a mean of $\sim 0.3\,\mathrm{m\ s^{-1}}$) and warmer than the EUC (Yu and McPhaden 1999; Keenlyside and Kleeman 2002). Kingman Reef (6°24′ N, 162°24′ W) and Palmyra Atoll (5°52′ N, 162°6′) are located farther north and lie predominantly in the North Equatorial Countercurrent (NECC); a warm, eastward flowing surface counter-current (Fig. 15.14).

Oceanographic conditions in the equatorial Pacific display a prominent seasonal cycle. The EUC and upper thermocline undergo large seasonal

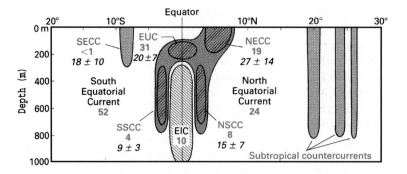

FIG. 15.14. Cross-sectional sketch of the equatorial current system in the central Pacific Ocean (170° W). Shown in crosshatch are the North and South Equatorial Countercurrent (NECC and SECC), subsurface Equatorial Intermediate Current (EIC), North and South Subsurface Countercurrents (NSCC and SSCC), and Equatorial Undercurrent (EUC). Eastward flow is colored green or brown, and all westward flow is white, including the North Equatorial Current (NEC) north of 5° N and the South Equatorial current (SEC) south of 5° N and outside the EIC. Black numbers in italics were observations from January 1984 to June 1986 (latitude 165° E), and bold red numbers were observations from April 1979 to March 1980 (latitude 155° W), with both representing transports in Sverdrups (Sv = 10^6 m^3/s) (Modified from Tomczak and Godfrey (2003))

vertical migrations in the central Pacific beginning in January, with maximum shoaling occurring during boreal springtime (Yu and McPhaden 1999; Keenlyside and Kleeman 2002). Sea surface heights seasonally vary 180° out of phase with thermocline depth, with minima (maxima) surface heights corresponding roughly to periods of shallowest (deepest) thermocline depth (Wyrtki and Eldin 1982; Yu and McPhaden 1999). Eastward zonal subsurface currents are strongest in boreal spring as a result of Trade Wind relaxations in the eastern Pacific, which cause a basin-wide adjustment of the pressure gradient (McPhaden et al. 1998). From the surface to the EUC core, a pronounced basin wide eastward surge occurs around April–July causing the EUC to strengthen and shoal, and the SEC to weaken (McPhaden et al. 1998; Yu and McPhaden 1999; Keenlyside and Kleeman 2002). The NECC, just north of the SEC, also varies seasonally in strength and position. During February–April, monsoonal variability in the western Pacific prevents the South Equatorial Current from feeding into the NECC at the western boundary; the countercurrent is fed only from the north as is then restricted to 4–6° N with maximum flow speeds of 0.2 m s^{-1}. During May–January, NECC surface flow increase to 0.4–0.6 m s^{-1} and lies between 5° N and 10° N as the current is fed from both hemispheres (Tomczak and Godfrey 2003)

Mean sea surface temperature (SST) across the equatorial Pacific has two prominent features: the western warm pool, characterized by temperatures greater than 28°C located east of 170° W, and the equatorial cold-tongue, a narrow band of cooler waters in the eastern Pacific derived from Trade-wind-forced equatorial upwelling (Yu and McPhaden 1999) (Fig. 15.15). Located further west than Jarvis, Howland and Baker Islands are consistently bathed in the hot temperatures (28°C) of the western warm pool, with little seasonal variation. Jarvis resides in the transition zone between the warm pool and the equatorial cold-tongue, where mean SST is slightly cooler (~27°C). Kingman Reef and Palmyra Atoll, too far north to be influenced by equatorial upwelling, experience similar annual SSTs as Howland and Baker Islands.

Sea surface temperatures, ocean currents, precipitation, winds, and biological production, are highly variable in the vicinity of the PRIA on inter-annual time scales due to the El Niño Southern Oscillation (ENSO) (Philander 1990; McPhaden et al. 1998). ENSO has two distinct signatures: El Niño and La Niña, which are defined by sustained SST anomalies of magnitude greater than 0.5°C across the equatorial Pacific (Trenberth 1997). During El Niño conditions, trade winds weaken and occasionally reverse, resulting in anomalously warm SSTs in the central and eastern Pacific, eastward surface transport, a deepening of the thermocline (Yu and

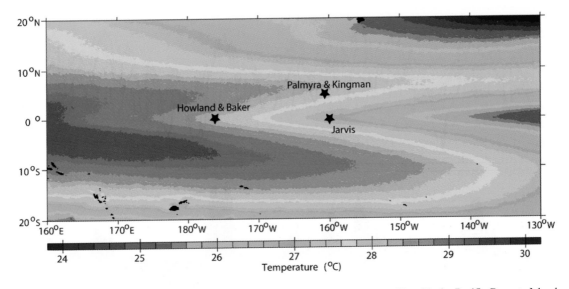

FIG. 15.15. Pathfinder sea surface temperature climatology of the equatorial Pacific with the Pacific Remote Island Areas; Howland, Baker, and Jarvis Islands, Kingman Reef, and Palmyra Atoll, shown in black stars (After NOAA/NASA Pathfinder radiometric satellite data)

McPhaden 1999), and a considerable weakening of the EUC (Firing et al. 1983; Roemmich 1984). Conversely, La Niña conditions are characterized by anomalously strong trade winds, cool SSTs, enhanced westerly surface transport, a shallow thermocline, and a strong EUC. Due to the location of the PRIA, surface temperatures there are highly variable on interannual time scales, particularly at Jarvis, where SSTs can range over 7°C from year to year due to ENSO forcing (Fig. 15.16).

Embedded in the large scale forcing and regional oceanographic variability of the PRIA are local hydrographic affects which can have substantial influence on near-shore dynamics. Jarvis Island, and to some extent Howland and Baker Islands, experience variable intrusions of cool, nutrient rich waters to the near surface due to the physical interaction with EUC (Fig. 15.17). Jarvis has been the focus of two historical surveys which showed that the blocking of the EUC by the island results in current flow stagnation and positive vertical isotherm displacement to the upstream or western side of the island (Hendry and Wunsch 1973; Roemmich 1984). More recently, a study by Gove et al. (2006) focused on the time dependency of near shore temperature fluctuations and showed that upwelling at Jarvis can be highly variable on

seasonal to interannual timescales due to fluctuations in the depth and strength of the EUC. On seasonal timescales, upwelling is observed strongest at Jarvis (and by proxy, Howland and Baker Islands) during boreal spring (April–June) when a weakening of the Trade Winds in the eastern and central Pacific coincides with a locally shallow thermocline and a shallow and strong EUC. Year-to-year differences in upwelling are associated with Trade Wind strength variations in the western Pacific caused by ENSO; a strengthening (weakening) of the Trades associated with La Niña (El Niño) conditions favor the strengthening (weakening) and shoaling (deepening) of the EUC at Jarvis. Additionally, the study found that superposition of internal tides on EUC driven upwelling can produce rather remarkable temperature changes near the surface several times a day, some as great as 4°C.

Variable intrusions of upwelled water to the surface layer serve as an important source of nutrients and suspended particles to the local ecosystem (Leichter and Miller 1999), thereby enhancing overall productivity. The EUC is known to be laden with nutrients when compared to surface waters (Chavez et al. 1999), but until a recent survey of Jarvis, no historical nutrient or chlorophyll data

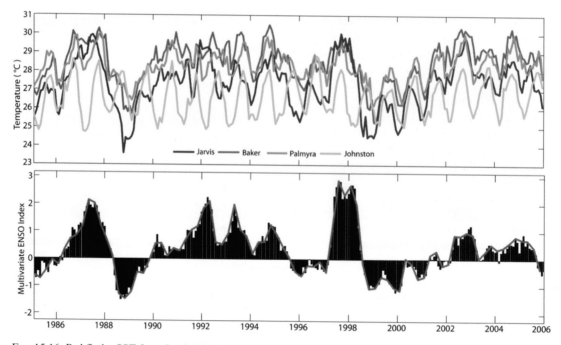

FIG. 15.16. Pathfinder SST from Jarvis Island, Baker Island, Palmyra Atoll and Johnston Atoll (top) with the ENSO Multivariate Index (MEI) (bottom) from 1985 to 2006. NOAA Pacific Islands Fisheries Science Center and Coral Reef Ecosystem Division unpublished data. The blue bars in the lower graph show the temporal variability, and the red line reflects the central tendency of the data shifts for the multivariate ENSO index (After NOAA/NASA Pathfinder radiometric satellite data)

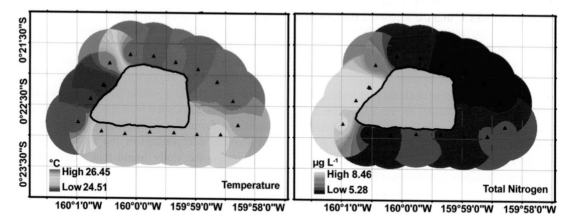

FIG. 15.17. Temperature (left panel) and total Nitrogen (right panel) at 30 m depth at Jarvis derived from nearest neighbor interpolation of near shore CTD casts (locations indicated by triangles) (PIFSC-CRED, 2006, unpublished data)

has been collected at the equatorial PRIA. Based on the new data, the spatial variation of nutrient levels at Jarvis Island corresponds markedly with the spatial upwelling pattern; with the greatest nutrient concentrations occurring on the western side (Fig. 15.17).

In terms of importance to higher trophic levels, ENSO has a rather profound affect on EUC-driven upwelling, and therefore nutrient availability and productivity at Jarvis, Howland, and Baker Islands. The Equatorial Undercurrent can be influenced greatly in both depth and strength during each phase of ENSO. The 1982–1983 El Niño, for example, was the strongest recorded El Niño on record and resulted in a significant weakening and eventual shutting down of the EUC from July 1982 to January 1983 (Firing et al. 1983). A considerable weakening in the EUC causes a cessation in local-island upwelling, resulting in a significant increase in sea surface temperatures around the entire island (Roemmich 1984). No biological observations were made during this time, but with upwelling no longer providing copious amounts of nutrients and cooler waters to the surface, biological production was likely reduced. No physical or biological data have ever been recorded during a strong La Niña phase at any of the equatorial PRIA; the greatest upwelling and subsequent enhancement of productivity has yet to be observed.

Kingman Reef and Palmyra Atoll, located farther north, are influenced by disparate oceanic conditions and ocean–atmospheric interactions when compared to the other PRIA. As mentioned above, the NECC seasonally varies in both position and strength; so too does the boundary between the NECC and the SEC. At times, the horizontal shear between the NECC and the westward flowing SEC is so large that wave-shaped instabilities develop between the two currents, causing enhanced mixing and an uplift of isotherms along the boundary (Tomczak and Godfrey 2003). Kingman and Palmyra are located near this latitudinal boundary and are likely affected by temperature and nutrient fluctuations stemming from this interaction. Additionally, the NECC is normally positioned below the Inter-tropical Convergence Zone, the climatic and oceanographic equator where the southeast trade winds of the southern hemisphere collide with the northeast trade winds of the northern hemisphere. The collision forces upward movement of air masses and heavy precipitation at the ITCZ as the masses cool. Thus, Kingman and Palmyra often experience light, variable winds and a humid tropical climate. Palmyra averages approximately 175 in. of rain per year. Precipitation is likely the same at Kingman; however, the heavy rainfall affects the two reefs differently. Kingman lacks permanent land and vegetation while Palmyra contains several large, vegetated islands supporting significant bird populations. Terrigenous runoff and associated nutrient input is likely substantial at Palmyra when compared to Kingman, potentially enhancing productivity and affecting local coral reef ecosystem dynamics.

15.5.1 Local Oceanographic Processes at PRIA Coral Reefs (by R. Brainard, J. Gove, and J. Maragos)

15.5.1.1 Baker, Howland and Jarvis

The largest swells to periodically strike reefs off all three islands originate from the north and northwest, usually during winter months. The most persistent influences are the trade winds and associated swells approaching all islands from the east, with the western sides of all islands protected from onshore winds and seas. When the wind shifts to the northeast, the south coast of Baker and Jarvis Islands form lee shores. Howland is elongated in a north–south direction and thus lacks any lee when winds shift to the north or south. Normally, easterly trade winds and associated swells strike Howland along the broad eastern shoreline generating strong wave setup, eddies and rip currents at both the north and south ends of the island. Rip currents often occur in the small blasted channels of all three islands when swell breaks on adjacent reefs. Alongshore currents also run to the west off the north and south sides of Baker and Jarvis during seas from the east, but are not as strong as those off Howland.

As noted earlier, upwelling from the EUC occurs at all three islands along their western sides and appears to be strongest at Jarvis. The upwelling subsidizes higher levels of marine productivity that likely benefits fisheries and reef biota. This phenomenon may be rare since there are few other Pacific islands subjected to subsurface currents as strong as the EUC. The Galapagos Islands straddle the Equator and the only other Pacific islands within one degree latitude of the Equator are Banaba Island, Nonouti Atoll, Abemama Atoll, and Aranuka Island in the Gilbert Islands of Kiribati and Nauru Island. In addition to upwelling by the EUC which is limited to the western sides of the three islands, equatorial upwelling characteristic of the entire Pacific also increases productivity in the waters around Baker, Howland and Jarvis.

15.5.1.2 Kingman Reef

Wave action and associated wave driven currents control the shallow biotic composition of Kingman reef. Large swell from the NW Pacific periodically strike the northern shallow reefs, especially during the winter months. Because the apex reef points due east, it receives chronic wave action during prevailing trade wind conditions, with wave set-up driving oceanic waters over the reef and into the eastern pool, and alongshore currents accelerating as they move westward along the eastern ocean-facing sides of both the northern and southern perimeter reefs. As these reefs deepen to the west, the alongshore currents continue to gain speed and then begin moving over the perimeter reef crests into the lagoon.

During NOAA RAMP cruises towed divers have explored the western half of Kingman. Ordinarily, during trade wind driven wave action, this is too dangerous for stationary dives because of strong currents everywhere. In the central and western lagoon, during trade wind conditions and the NEC approaching from the east, a continuous mass of lagoon waters sets up to the west and exits the lagoon over broad passes and deeper perimeter reefs at the western end of the atoll reef. Kingman may also experience flow from at least one other separate ocean boundary current: the eastward flowing NECC when the Inter-tropical Convergence Zone (ITCZ) is overhead. Kingman may also be exposed to the westward flowing SEC when the ITCZ shifts well north of the atoll reef.

Changes in Palmyra's ocean circulation over the past 60 years are covered in the next chapter. During trade wind conditions and the presence of the NEC, Palmyra's eastern reefs receive ocean swells from the east that refract around both the elongated north and south sides of the atoll, generating alongshore currents that move to westward. As with Kingman, Palmyra also is affected by the eastward flowing NECC and possibly the westward flowing SEC, depending on the position of the ITCZ. Periods of heavy rainfall and corresponding runoff/discharges from the islands have been observed by J. Maragos to temporarily modify lagoon and near-shore circulation close to the islands.

15.5.2 Global Climate and Ocean Change in the PRIA (by R. Brainard and J. Maragos

Field observation of corals at Palmyra in 1987 and 1998, and at all five PRIA during five separate expeditions from 2000 to 2006 indicate that corals appear to be recovering from a massive die-off of unknown origin (possibly a bleaching event) that likely took place around 1997. However, corals have not recovered off the southwest reefs near the dredged channel and West Lagoon where ambient water temperatures are higher (Friedlander 2004). Since 2000, *Acropora* (known to be sensitive to bleaching) has increased in size and abundance at Howland and Baker, although it is still uncommon at Jarvis. Although coral bleaching was predicted to occur along the Equator in 2003 based upon NOAA satellite-based temperature and wind data, no evidence of bleaching was reported during the early 2004 visits to Baker, Howland or Jarvis.

Sea level rise is well documented throughout the world's oceans (Vitousek 1994), but local data at the PRIA are lacking. Thus, the magnitude of changes in sea level and their impacts on PRIA ecosystems are currently speculative. The impacts of changes in atmospheric and oceanic chemical concentrations are also mostly unknown. Many of the impacts of global climate change are not currently being documented at the equatorial Pacific island refuges (especially Howland, Jarvis and Baker). However, the refuges could provide unique baselines for monitoring future changes at the Equator. In addition, analysis of coral cores from reefs never affected by prolonged human use or habitation may reveal historical patterns of variation in environmental factors.

15.5.2.1 Ocean Acidification

Ocean acidification, in response to increasing levels of atmospheric carbon dioxide, has been projected to have widespread and chronic negative impacts on marine calcifiers (Kleypas et al. 2006). The calcifiers on PRIA reefs, including corals, crustose coralline algae, sand-producing *Halimeda* algae, and many shell producing invertebrates are likely to be affected by these global carbon chemistry changes.

15.6   Reef Zonation and Community Patterns

15.6.1   Benthic Community Patterns for Algae, Coral and Other Invertebrates

Earlier companion state of the reefs reports (SORR) (Turgeon et al. 2002; Brainard et al. 2005) and the upcoming SORR to be released by NOAA in 2008 provide comprehensive quantitative information on the abundance, distribution and temporal changes in reef biota at all US reefs to 2006. Hence, this report provides only a brief, descriptive and contemporary overview of the distribution of reef biota at the five PRIA.

In general, stony corals dominate the upper ocean-facing reef slopes of all five PRIA from depths of 1 to 15 m. Crustose coralline algae dominate all shallow reef crests and the upper windward reef slopes from depths of 0 to 5 m and remain abundant to depths of 10 m. Fleshy and turf algae become progressively more abundant and corals less abundant at depths greater than 10–15 m. Little is known of the benthic biota of the deep lagoons (>30 m) of Kingman and Palmyra.

15.6.1.1   Baker, Howland and Jarvis

Corals on the upper reef slopes of all three islands are recovering from a likely 1997 coral bleaching event based upon permanent transect data collected over a 6-year period by J. Maragos (Anon., in preparation). Together with coralline algae, corals are the dominant benthic reef life at shallow depths (<5 m). Other algae are more abundant and diverse at greater depths. Table corals (*Acropora*) are increasing in abundance and size on the western reef slopes of all three islands, while staghorn corals of *Acropora* cover much of the eastern reef terrace at Baker to depths of 15 m. Towed diver surveys revealed that similar *Acropora* formations are present on the southeastern terrace of Howland. Plate corals (*Montipora*) flourish off most reef slopes at Jarvis to similar depths. The soft coral *Sinularia* dominates the southwestern facing reef slope and may be stimulated by elevated nutrient levels in upwelled waters.

Although the fringing reef habitat of these three islands is very similar, there are differences in the associated invertebrate species. Echinoids were the most common echinoderm class present in surveys at all islands. *Diadema sauvignyi* is the most common echinoid at Howland and Baker, while *Echinothrix calamaris* is most common at Jarvis. Holothurians are uncommon at all islands and only three species are recorded as conspicuous: *Holothuria atra*, *Actinopyga obesa*, and *Actinopyga mauritiana*. The asteroid *Linckia multifora* is most common and *Linckia guildingi* also common at Jarvis. Noticeable differences in the composition of species of molluscs were observed between all three islands. The gastropod *Turbo argyrostomus* is common at Baker but extremely rare at Howland and Jarvis. A similar disparity exists for the trochid gastropod *Astralium rhodostomum*, which is common at Baker and Jarvis but is rare at Howland. The giant clam *Tridacna maxima* is found at all three islands but is extremely rare at Jarvis and most abundant at Howland. The only mollusk commonly found at all islands is the cephalopod *Octopus cyanea*. The hermit crab *Dardanus longior* are seen at all islands, as is *Calcinus haigae* and *Calcinus isabellae*, while *Calcinus lineapropodus* is only recorded from Howland and Baker.

15.6.1.2   Kingman Reef

Fleshy algae and corals co-dominate the shallow lagoon reef slopes at Kingman to depths of 15 m below which algae and sand become progressively more dominant. The highest coral cover and diversity at Kingman occurred off the southeastern and eastern apex fore reefs where table corals *Acropora* and a large variety of other corals and other cnidarians predominated in 2005. However, by the time of the next surveys in early 2006, the *Acropora* colonies were nearly all preyed upon by the crown-of-thorns sea star, *Acanthaster planci*. The shallow eastern reef pool is unique in being dominated by crustose coralline algal substrates and multitudes of mushroom corals (*Fungia*), giant clams (*Tridacna*), anemones (*Heteractis*), fire corals (*Millepora*), and table corals (*Acropora*). The presence of *Tridacna squamosa* at Kingman Reef represents its northernmost record in the Pacific. Corals are more abundant and varied on seaward slopes vis-à-vis lagoon slopes, but attenuate to the west with increasing exposure to currents and predation by *Acanthaster*. The common genera *Acropora*, *Pocillopora*, and *Montipora* diminish in succession to the west eventually leaving *Porites* as the only consistently common shallow water coral along the western shallow reefs of the atoll. Interestingly, a single stationary dive to 35 m in the deep wide pass off the western rim of Kingman in 2006 revealed a reef community entirely covered by foliose forms of stony corals (especially *Pachyseris*) and sea fans (*Subergorgia*) not seen elsewhere at the atoll reef. More extensive towed diver surveys in 2006 revealed that soft corals predominated along most of the western fore reef terrace and northeastern facing fore reefs.

Other common mollusks at Kingman included the bivalve *Spondylus*, coralliophilid snails associated with *Porites* coral, and the large predatory species *Lambis lambis*, *Cassis cornuta* and *Charonia tritonis*. The echinoderms are less abundant than mollusks, but more diverse, with holothurians the most abundant and varied, followed by echinoids and asteroids. Holothuroids are the dominant echinoderms on patch reefs and noteworthy in fore reef habitats and eastern back reef habitats. *Acanthaster planci* is patchy across all habitats, but appears to be the main driver of disturbance within the Kingman Reef atoll system. This is especially true in back reef and lagoon patch reef habitats. There is obvious evidence of incremental predation by *A. planci* beginning with favored prey (*Acropora*, *Pocillopora*, and *Montipora*) and followed by prey-switching to less favored species of stony corals (*Fungia*, *Porites*, *Turbinaria*, *Pavona* and Faviidae) and soft corals (*Sinularia*). Hermit crabs were commonly seen in the fore reef/terrace habitats with *Calcinus haigae*, *Calcinus lineapropodus* the most common.

15.6.1.3 Palmyra

Numerous observations by J. Maragos reveal that corals, crustose coralline algae, and other typical lagoon reef species are almost entirely absent on lagoon reefs, a lasting legacy of earlier military dredging and filling operations. The most unique and healthy habitats are the two shallowest and easternmost of the reef pools off the southeast and northeast corners of the shallow eastern reef flat. These are dominated by crustose coralline algae and many species of *Acropora*, *Montipora* and *Astreopora* and *Pocillopora*, and are commonly

referred to as the "coral gardens". Other healthy reef substrates occur along the entire north ocean-facing reef slopes of the atoll where many species of corals, algae, reef fish and invertebrates appear to thrive and were apparently unaffected by the recent coral die-off event. *Halimeda* sands are prevalent in the fore reef habitats of Palmyra although nearly absent from comparable habitats at Kingman.

Stony and soft corals on the southern half of the western reef terrace and southern ocean-facing reef slopes are recovering rapidly from a massive die-off (possibly caused by bleaching, *Acanthaster* predation, or storm damage) that occurred between surveys by Maragos in 1987 and 1998. Continuous platforms of staghorn *Acropora* covering more than 90% of the bottom stretched along the entire southern half of the western reef terrace in September 1987. These were reduced to rubble fields before the next survey in November 1998. Other corals, mostly massive *Porites* and *Pocillopora* (J.E Maragos 1998, unpublished data) remain and appear to be increasing. Although there have been some recolonization of *Acropora* thickets on the terrace since 2000 reported by towed divers (R. Brainard personal communication 2007), they are nowhere near their dominance reported by Maragos (1987). Corals are recovering much more slowly on the "Penguin Spit" and neighboring reefs near the southwest dredged channel and western perimeter reef crests. Crustose coralline algae are abundant and cover many large dead standing corals that offer lingering evidence of the magnificent coral communities reported on these reefs by Maragos (1987).

The low abundance of macro-invertebrates fauna on the reef habitats at Palmyra Atoll is presently inexplicable, although under investigation by the Victoria University at Wellington (J. Gardner, personal communication 2007). Lingering effects from WWII era construction may be one explanation, especially from elevated temperatures and water residence times in the lagoon. The abundance of macro-invertebrate on the fore reef is low throughout the atoll, while it is higher in restricted littoral and shallow subtidal locations in the lagoon. Echinoderms, usually the dominant macro-invertebrate component of coral reefs, are rare in Palmyra Atoll's fore reef habitats. Rare occurrences of the sea cucumber *Holothuria atra*, the echinoid *Echinothrix calamaris*, and the asteroid *Echinaster luzonicus* have been recorded on the fore reef. The holothurians *Holothuria atra*, *Holothuria edulis*, and *Stichopus chloronotus* can be found commonly in restricted shallow habitats in the lagoon. The giant clam *Tridacna maxima* is present in low numbers on the southern fore reef and at the southeastern back reef and shallow "coral gardens" pool habitat. The most common mobile macro invertebrates are trapezid crabs and diogenid hermit crabs, which are mostly associated with *Pocillopora* corals. The most commonly seen hermit crab species is *Calcinus haigae*. More speciose communities made up of sponges, crustaceans, echinoderms, and gastropod mollusks are associated with supralittoral and littoral zones in the lagoon. The gastropods *Cypraea moneta* and *Strombus maculatus* are extremely abundant in shallow intertidal zones associated with lagoon islands, as were a variety of brachyuran crustaceans (Xanthidae).

15.7 Biogeography of the Central Pacific Ocean

Within the region stretching from the Red Sea east to the central Pacific Ocean, Ekman (1953) proposed abroad biogeographic region for warm water fauna, termed the Indo-west Pacific marine faunal region. Within the Pacific Ocean, Ekman identified four sub-regions based upon endemism and distribution patterns of key fauna: (1) the islands of the Central Pacific excluding Hawai'i, (2) Hawai'i, (3) subtropical Japan, and (4) Australia. The dramatic west-to-east attenuation in the diversity of shallow water stony corals across the Pacific Ocean was further quantified by Stehli and Wells (1971) and Veron (1986, 1993, 2000). Stoddart (1992) compiled similar diversity attenuation patterns for several other groups of reef biota and wetland vegetation, and proposed several geological provinces to further characterize and explain the regional biogeography of the tropical Pacific. The largest of these provinces is the Pacific geological province of low reef islands and atolls that stretches from Micronesia in the northwestern tropical Pacific to French Polynesia in the southeastern tropical Pacific. The Line

and Phoenix Islands lie near the middle of this geological province at the geographic center of the Pacific Ocean, straddling the Equator and just east of the International Dateline. The two chains account for 23 of the approximately 300 atolls and low reef islands within the Pacific Ocean (Bryan 1953; Dahl 1991; Maragos and Holthus 1995).

Based upon the above considerations and the distribution of atolls, reef islands and coral taxa, the tropical Pacific Ocean can be subdivided into five biogeographic provinces, excluding adjacent continents (Fig. 15.18).

The provinces consist of:

- *The southeast tropical Pacific*: including scattered volcanic islands and 83 atolls and reef islets in French Polynesia and a few United Kingdom atolls
- *The central tropical Pacific*: including collectively about 47 mostly isolated atolls and reef islets in eastern American Samoa, Phoenix and Line Island (including the 5 PRIA), northern Cook Islands, Tokelau, and Ellice Islands (Tuvalu)
- *The northwest tropical Pacific*: including a few volcanic islands, many raised reef islands, and approximately 90 atolls and reef islets in the Marshall Islands (including Wake Atoll), Gilbert Islands of Kiribati, Caroline Islands (Palau and Federated States of Micronesia), Mariana Islands, and Nauru
- *The southwest tropical Pacific*: dominated by many large volcanic and continental islands, numerous raised reef islands, and approximately 69 atolls and reef islets in Papua New Guinea, Solomon Islands, Vanuatu, Fiji, Tonga, Samoan Islands, and New Caledonia
- *Hawaiian Islands*: including ~20 large and small volcanic islands: seven atolls and reef islets; and Johnston Atoll in the isolated tropical/subtropical north Pacific

Tropical marine biogeographic provinces are arbitrary and largely based upon the chosen reef taxa and other criteria. Provinces based upon fish, non-coral invertebrates, or benthic algae criteria would likely have different bio-geographical patterns and boundaries based on differential capacities and strategies for larval and spore dispersal, settlement and survival.

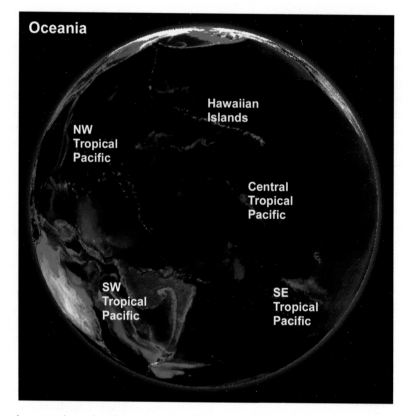

FIG. 15.18. Ellipses the approximate locations of coral reef bio-geographic provinces in the tropical Pacific Ocean basin (Base map courtesy of NASA Earth Wind)

15.7.1 Biodiversity and Distribution of the Flora and Fauna at the PRIA

The major biota of the US Line and Phoenix Islands from emergent land to the depth limits of the NWR includes the following groupings:

- Terrestrial vegetation
- Terrestrial invertebrates and insects
- Resident seabirds and migratory shorebirds
- Shallow-water corals and other macro-cnidarians
- Shallow benthic invertebrates other than cnidarians
- Marine algae
- Deep reef corals and other benthos
- Inshore fishes
- Marine reptiles and mammals

All occur at the five PRIA except vegetation is absent at Kingman. Terrestrial mammals are absent at the five PRIA except introduced rats

(*Rattus rattus*) at Palmyra, and the house mouse (*Mus musculus*) at Jarvis and Baker Islands. Most marine biological surveys in the PRIA have not extended to depths below 30 m. Marine biological data at greater depths have been mostly based on towed camera and video surveys at all PRIA during NOAA Pacific RAMP surveys and 16 submersible dives to depths of 200–1,000 m sponsored by Hawai'i Undersea Research Laboratory (HURL) off Kingman, Jarvis and Palmyra in 2005. It is beyond the scope of this paper to address the deep reef biota (due to lack of adequate information) and terrestrial biota except for seabirds that spend much of their time near the reefs and seasonally nest on most PRIA.

15.7.1.1 Seabirds (by E. Flint)

Very few islands in the entire insular Pacific have been spared from the introduction of rats, a group of species that have dramatically modified the distribution and diversity of breeding seabirds (Flint 1999). All of the islands of the Line and Phoenix archipelagos except for Phoenix (Rawaki) and McKean islands, and some islets of Kiritimati have been infested with *Rattus* species since human contact first occurred. Rats and cats are now eradicated from Jarvis, Howland, and Baker, leading to the return of ground-nesting terns, shearwaters and petrels vulnerable to mammalian predators. Kingman Reef has limited emergent reef not suitable for permanent vegetation or seabird nesting, although birds rest on the three small emergent features. Table 15.4 is a historical listing of all breeding seabirds reported at the five PRIA though it is possible that there were a few more central tropical Pacific species that occurred but were extirpated prior to the first ornithological documentation.

Baker

No records of birds at Baker existed for the time before rats were introduced, and only a few birds were left at the time of the first ornithological surveys. Today 11 seabird species nest at Baker including almost one million pairs of sooty terns. Baker (along with Howland and Jarvis) is also one of the few known areas where lesser Frigate birds nest on the ground, rather than in trees. Several migratory shorebirds use Baker including Wandering Tattlers (*Tringa incana*), Sharp-tailed Sandpipers (*Calidris acuminata*), Bristle-thighed Curlews (*Numenius tahitiensis*), Ruddy Turnstones (*Arenaria interpres*), and Pacific Golden Plovers (*Pluvialis fulva*), the last three being species of High Concern in the US Shorebird plan (Engilis and Naughton 2004).

Howland

Nesting seabird species have increased from 4 to 12 species since cats were eradicated from Howland in the early 1990s, and restoration potential exists for seven more species, including the Blue Noddy (the world's smallest tern and limited to the central Pacific), the Phoenix Petrel, and White-throated Storm-petrel. These last two have been extirpated from most of their range in the Pacific due to vulnerability to rats and other mammalian predators. Howland is also a regular wintering station for eight species of arctic migrant shorebirds. Three of these, the Pacific Golden Plover, Ruddy Turnstone and Bristle-thighed Curlew are species of High Concern in the US conservation priority plan for shorebirds (USFWS 2002, 2005; Engilis and Naughton 2004).

TABLE 15.4. Seabirds historically reported on land from the US Line and Phoenix Islands (Data compiled by E. Flint).

Seabirds species	Scientific name	HOW	BAK	KIN	PAL	JAR
Wedge-tailed shearwater	*Puffinus pacificus*	X			X	X
Christmas shearwater	*Puffinus nativitatis*					X
Audubon's shearwater	*Puffinus lherminieri*					X
White-throated Storm-petrel	*Nesofregetta fuliginosa*					X
White-tailed Tropicbird	*Phaethon lepturus*				X	
Red-tailed Tropicbird	*Phaethon rubricauda*	X	X		X	X
Masked Booby	*Sula dactylatra*	X	X	X	X	X
Brown Booby	*Sula leucogaster*	X	X	X	X	X
Red-Footed Booby	*Sula sula*	X	X		X	X
Great Frigate bird	*Fregata minor*	X	X		X	X
Lesser Frigate bird	*Fregata ariel*	X	X			X
Gray-backed Tern	*Onychoprion lunatus*	X	X			X
Sooty Tern	*Onychoprion fuscatus*	X	X		X	X
Black Noddy	*Anous minutus*				X	
Brown Noddy	*Anous stolidus*	X	X	X	X	X
Blue Noddy	*Procelsterna cerulea*	X	X			X
White Tern	*Gygis alba*	X	X		X	X
	Totals	12	11	3	11	15

HOW = Howland Island, BAK = Baker Island, KIN = Kingman Reef, PAL = Palmyra Atoll, and JAR = Jarvis Island

Jarvis

Cats, rats, goats and mice brought by human settlers modified the landscapes and limited seabird use to three species at the time of the ornithological surveys in 1966. Elimination of these, including eradication of the last cats by the FWS, was completed by 1990, and since then the numbers of birds and species have climbed dramatically. With 15 species of breeding seabirds, Jarvis is second only to Kiritimati Atoll in the Line Islands. Its Sooty Tern colony is one of the largest left in the world estimated at three million pairs. The importance of Jarvis continues to increase as human pressures at nearby inhabited Kiribati reef islets and atolls continues to diminish the numbers and species of seabirds (Pierce et al. 2007). White-throated Storm-petrels (*Nesofregetta fuliginosa),* a species that has been extirpated from much of its range by rodent introductions, have recently been observed on the ground at Jarvis Island. The strong upwelling off Jarvis may be supporting hundreds of pairs of breeding Blue Noddies that forage just offshore (M. Rauzon and J. Gove 2007). Several species of shorebirds over-winter at Jarvis, including the Bristle-thighed Curlew.

Palmyra

Palmyra supports the third largest Red-footed Booby colony in the world. The atoll also supports healthy populations of Brown Noddy, Black Noddy, and Brown Booby. A total of 227 Bristle-thighed Curlews were counted at Palmyra in 1992, a large number considering that the entire world population totals only 7,000 birds. Other numerous seabird nesters include the Great Frigate bird, Sooty Tern, White Tern, Red-tailed Tropicbird, and Masked Booby. Seabird species that prefer trees for nesting are dominant at this atoll. The atoll's proximity to the higher primary productivity of the convergence zone may enhance foraging opportunities for breeding seabirds there.

15.7.1.2   Marine Mammals

Marine mammals have been commonly observed near the PRIA during NOAA Pacific RAMP and Scripps Institution sponsored ship expeditions since 2000. Species observed include Bottle-nose dolphins *Tursiops truncatus* and spinner dolphins *Stenella longirostris.* Pods of melon-headed whales *Peponocephala electra* appear to be residents off Palmyra, were also observed at Kingman in 2005, and have been regularly reported off Kiritimati (K. Andersen, personal communication 2004). The endangered Hawaiian monk seal, *Monachus shauinslandi*, is occasionally seen off Palmyra (E. Flint, personal communication 1999). According to Rice (1977) approximately 30 species of marine mammals are residents in the insular Pacific, and Eldredge (1999) has compiled a list of species of all marine mammals known from the tropical Pacific including associated references. Since 2005 three small beaked whales have stranded and been collected at Palmya Atoll. Originally identified as the Ginkgo-toothed Beaked Whale *Mesoplodon ginkgodens* these specimens are now being re-evaluated by marine mammalogists and may represent a new species (Dalebout et al., in press)

15.7.1.3   Sea Turtles

There have been few recent reports of turtle nesting at Howland, Jarvis and Baker (NMFS and USFWS 1998; Gyuris 1999). At Jarvis, low density of nesting of the green turtle *Chelonia mydas* was reported in the 1930s by Bryan (1974) and Balazs (1982) at Jarvis, but information from the other islands is largely absent. There are unpublished records of a few green turtles nesting along the north coast of Palmyra in 1987 (S. Fefer, 1998, in NMFS and USFWS, personal communication). Earlier military construction may have caused extensive beach erosion now evident along the north coast, in turn leaving very little sandy beach habitat to support nesting (see Palmyra case study in the next chapter). The small emergent reefs at Kingman are not considered suitable for nesting sea turtles, although sea turtles may occasionally rest on them.

Earlier inhabitants and visitors to the Line and Phoenix Islands may have harvested nesting turtle populations at the three islands and Palmyra Atoll, as documented at Flint Island (Fig. 15.19), a neighboring reef island in the southern Line Islands. Prolonged harvesting of large sea turtles for consumption by local residents (miners, soldiers) or trade with visiting ships during the

FIG. 15.19. Hand-hauled rail trucks carry guano, copra, timber and large green turtles in this anonymous 1908 photo taken at Flint Is. (Photo courtesy of the Mary Lea Shane Archives, Lick Observatory, University of California, Santa Cruz)

guano mining and WWII eras may have depleted or extirpated local nesting populations. However, swimming green turtles *Chelonia mydas* (a threatened species) and hawksbill turtles *Eretmochelys imbricata* (an endangered species) have been reported during recent dive surveys at all five PRIA, and swimming green turtles at Palmyra are especially numerous.

15.7.1.4 Corals and Other Macro Cnidarians (by J. Maragos, S. Godwin, and D. Obura)

Approximately 200 species of stony corals and two dozen other conspicuous cnidarians (soft corals, anemones, black corals, corallimorphs, and hydrozoans) have collectively been reported at the five PRIA. However, the taxonomy of stony corals (Order Scleractinia) to the species level is still largely unresolved which in turn compromises accurate estimates of the distribution and diversity of stony coral species among the five PRIA. For this reason, the description and distribution of the corals and other cnidarians will be focused at the generic level, including estimates of the number of observed species per genus (Table 15.5). Although all five PRIA have been visited five to six times

by coral biologists since 2000, mostly as a part of NOAA's Pacific RAMP expeditions, for safety reasons most diving has been limited to depths of 30 m or less. Moreover, surveys along windward reefs have sometimes been thwarted by dangerous waves, swells, currents, and onshore winds, especially at Howland and Kingman.

Additional coral data collected at neighboring Kiribati atolls and reef islets in the Line and Phoenix Islands are similarly compiled for comparison in Table 15.5. Data on non-stony cnidarians were not collected during some earlier surveys at Kanton and McKean in the Phoenix Islands (Dana 1975; Maragos and Jokiel 1978), and data for non-stony coral taxa were not collected at Tabuaeran until the latest surveys in 2005.

Stony Corals

A total of 51 genera of stony corals have been reported collectively at six of the Line Island (from north to south: Kingman, Palmyra, Teraina, Tabuaeran, Kiritimati, and Jarvis) and four of the Phoenix Islands (Howland, Baker McKean and Kanton), with 48 of these genera reported in the five PRIA of this chapter. Table 15.5A reveals that stony corals range from 34 to 38 genera for each atoll or atoll reef, regardless of atoll size, location,

TABLE 15.5. Line and Phoenix Islands corals and benthic cnidarian genera and number of species reported at depths <30 m.

15.5A Stony corals

Coral genera	Phoenix Islands				Line Islands					
Stony corals	**BAK**	**HOW**	KAN	MCK	**KIN**	**PAL**	TER	TAB	KIR	**JAR**
Acanthastrea	1				1			1		
Acropora	24	22	13	5	38	45	1	31	19	7
Alveopora					1	1		1		
Astreopora	1		1		3	5	3	6	6	
Balanophyllia						1				
Barabattoia									1	
Cladopsammia	1									1
Coscinaraea	2	1	1		1		1	2	1	1
Cycloseris	1	1			2	2	1	2	6	
Ctenactis						1				
Culicia			1					1	1	
Cyphastrea	1	1	1	1	1	1			1	
Diploastrea	1									
Distichopora[a]			1		1	1		1	1	1
Echinophyllia			1		1	2		1	1	1
Echinopora	1	1	1	1						
Favia	3	5	4	2	9	8	2	6	6	2
Favites	2	5	2	2	6	8	4	6	4	2
Fungia	4	3	5	2	10	9	7	7	10	2
Gardineroseris	1	1	1		1	1	1	1	1	
Goniastrea		2	1		2	3				1
Halomitra	1	1	1	1	1	1	1	1	1	
Herpolitha	1	1	1		2	1	1	1	2	
Hydnophora	1	3	2	2	3	3	1	2	3	1
Leptastrea	2	4	2	2	4	8	5	5	5	
Leptoria			1					1		
Leptoseris	1	1	2	1	2	1	1	2	3	1
Lobophyllia	1		1	1	3	3	1	3	3	
Merulina					1	1	1	1		
Millepora[a]	1	1	1	2	1	1	1	1	1	1
Montastrea	2	2			2	3	1	3	2	
Montipora	10	13	4	6	21	21	7	16	16	7
Oxypora					1				1	
Pachyseris	1	1	1		1	1		1	1	1
Pavona	4	4	6	5	7	8	11	8	6	5
Platygyra	2		2	1	6		2	6	7	
Plerogyra			1		1				1	
Plesiastrea				1		1		1		
Pocillopora	3	3	6	5	6	8	3	8	6	5
Podabacia		1	1							
Porites	5	10	7	8	12	12	5	7	6	4
Psammocora	4	4	4	1	4	6	3	5	2	3
Rhizopsammia	1									
Sandalolitha	1		1	1	1	1	1	1	1	
Scapophyllia					1					
Stylaster[a]			1		1	1		1	1	1
Stylocoeniella					1					
Stylophora					1	1	1	1		
Symphyllia	1	1			1					
Tubastraea	1	1	2			1	1	1	1	1
Turbinaria	1		1	1	2	4	1	1	2	

TABLE 15.5. (continued)

15.5A Stony corals

Coral genera	Phoenix Islands					Line Islands				
Stony corals	**BAK**	**HOW**	KAN	MCK	**KIN**	**PAL**	TER	TAB	KIR	**JAR**
Total surveys	**35**	**31**	100	30	**59**	**90**	2	61	32	**30**
Reef areas	**11.6**	**8.4**	9.1	1.5	**105**	**65**	7.4	34	328	**7.4**
Total genera	**33**	**26**	34	21	**38**	**36**	27	36	34	**20**
Total species	**87**	**93**	81	51	**162**	**176**	68	143	136	**48**

[a]Order Scleractinia, Class Hydrozoa **15.5B** Other benthic Cnidaria

Other cnidarians		Phoenix Islands			Line Islands			
	BAK	**HOW**	**KIN**	**PAL**	TER	TAB	KIR	**JAR**
Aiptasia[a]	**1**							
Antipathes[b]								
Cirrhipathes[b]		**1**					1	
Cladiella		**1**		**1**		1	1	
Cryptodendrum[a]		**1**		**1**		1	1	
Dendronephthya		**1**					1	
Discosoma[d]		**1**		**1**				
Entacmaea[a]	**1**							
Gorgonians		**1**		**1**	2	2		
Gymnagium[e]				**1**	1	1		
Heteractis[a]	**1**		**4**	**2**				
hydrozoan #1[e]							1	
hydrozoan #2[e]		**1**		**1**	1			
Lobophytum		**1**	**2**		1	1	1	**1**
Pachyclavularia		**1**		**1**	1			
Palythoa[c]	**1**	**1**	**2**	**2**	1	1	1	
Phymanthus[a]		**1**						
Rhodactis[d]	**1**	**1**	**1**	**1**		1		
Sarcophyton		**1**		**1**		1	1	
Sinularia		**1**		**1**		1	1	**1**
Sterionephthya				**1**		1		
Stichodactyla[a]		**1**		**1**	1			
Subergorgia		**1**		**1**				
Zooanthus[c]							1	
Total surveys	**35**	**31**	**59**	**69**	2	12	31	**30**
Total genera	**5**	**4**	**16**	**15**	6	11	10	**2**
Total species	**5**	**4**	**22**	**17**	7	12	10	**2**

Orders Alcyonacea, [a]Actiniaria, [b]Antipatharia, [c]Zoanthidea, [d]Corallimorpharia, and [e]Class Hydrozoa
bold = US PRIA. Non-bold = Kiribati islands. **BAK** = Baker I., **HOW** = Howland I., KAN = Kanton A. (Maragos and Jokiel 1978), MCK = McKean I. (Dana 1975), **KIN** = Kingman R., **PAL** = Palmyra A., TER = Teraina I. (TAB = Tabuaeran A. (Maragos 1974), KIR = Kiritimati A. (Maragos 1974, 1997), and **JAR** = Jarvis I. One unpublished record by Kenyon, *Acanthastrea*, is included in the totals (Shallow reef area estimates (in km²) from Dahl 1991, Maragos and Holthus 1995, Hunter 1995 and USFWS 2007)

or sampling intensity in the Line or Phoenix Islands. These levels are higher than would be expected for the SE Pacific (French Polynesia) but lower than would be expected at atolls and islands in the northwest (Micronesia) and southwest Pacific (Melanesia and western Polynesia). The few genera that were not reported at the five PRIA are cryptic or rare in the central Pacific (*Barabattoia, Culicia, Leptoria*). The numbers of genera are consistent with what has been reported by Stehli and Wells (1971) and Veron (1986, 2000) for the Central Pacific region, and reflect the consistency and success of stony corals colonizing many remote atolls.

The numbers of stony coral genera observed at individual reef islets were generally lower, and more variable (Table 15.5A). For example, Teraina I. in the Line Islands surveyed at only two sites to date, revealed substantially more genera (27) than Jarvis I. (20) surveyed at 30 sites since 2000. In the Phoenix Islands, Baker had the highest total for stony coral genera (33) compared to the other reef islands, including neighboring Howland (26) and McKean (21). However, McKean, about 480 km to the southeast of Baker, is much smaller than the other two, and the constantly inaccessible windward (eastern) half of Howland has not yet been adequately surveyed, a habitat expected to yield additional genera. Eight genera out of a combined total of 35 for both islands were reported only at Baker, with a ninth reported only at Howland. The number of genera at Baker (33) was comparable to that at Kanton (34), the only atoll in the Phoenix Islands with published coral data. The highest numbers of species and genera of stony corals have been recorded at Kingman reef and Palmyra Atoll. The lowest numbers of corals and genera were reported at Jarvis Island. Overall, there was higher atoll generic diversity in the Line Islands, but higher reef island generic diversity in the Phoenix Islands, although further surveys at Teraina will likely disprove the latter generalization.

Other Cnidarians

A total of 24 genera of non-stony coral cnidarians were reported collectively at two Phoenix Islands (Howland and Baker) and at seven Line Islands (Table 15.5B). As noted above, surveys of non-stony cnidarians were limited, and these genera may be underrepresented at some reefs. However, four Line Islands atolls reported 10–16 genera for these cnidarians. The low reef islands ranged from two to six genera of non-stony coral cnidarians, much lower than the levels reported for the atolls. As with the stony corals, the highest numbers of species and genera of non-stony coral cnidarians were recorded at Palmyra and Kingman, and the lowest numbers recorded at Jarvis.

15.7.1.5 Marine Algae (by P. Vroom and R. Tsuda)

Approximately 200 species of marine turf and macroalgae have been documented from the PRIA.

Comprehensive species lists did not exist for these locations before the efforts of NOAA's Pacific RAMP surveys that began in 2000, and previous collections of algae were limited to relatively few shallow water, incidental samples collected by researchers sent to various islands principally to conduct terrestrial and avian surveys. No published records of algae from Kingman Reef exist in the literature, and all collections for this reef represent new records. Floras discussing the marine plants for Howland, Baker, and Jarvis Islands and Kingman Reef are currently being written. NOAA algal collections from Palmyra Atoll are being processed and have not yet been taxonomically identified.

A total of 120 genera of cyanophytes (blue-green algae), red algae, brown algae, and green algae have been reported cumulatively from Howland Island, Baker Island, Jarvis Island, Tabuaeran Atoll (Tsuda et al. 1973), and Kingman Reef, with each island containing from 34 to 71 genera (48–147 species) that typically cover 50–80% of suitable substrate (Vroom et al. 2006b). As is expected for tropical marine floras, diversity among red algae is high while brown algal diversity is low. The Line Islands appear to support higher algal diversity than the Phoenix Islands. Despite being of similar size and geomorphology, Jarvis Island (Line Islands) supports 1.4–2.3 times the number of species found at Howland and Baker Islands (Phoenix Islands). Of biogeographic interest is that geographically close islands (e.g. Howland and Baker, 66 km apart) contain substantially different species. For instance, in Table 15.6, it might be assumed that because both Baker and Howland Islands contain species in genera such as *Lyngbya*, *Jania*, *Peyssonnelia*, *Gelidiopsis* that the same species occur at both islands. However, species that occur at one island often do not occur at neighboring islands.

Algal colonization of remote islands across the tropical Pacific is not well understood. As discussed by Silva (1992), Pacific islands vary in structure, age, stability, and degree of isolation. Large volcanic islands support more diverse algal populations than small sand islets, but the specific mechanisms by which algal propagules reached the remote PRIA remain unknown. Algal spores have not been found to survive for exceptionally long periods of time in the water column. It is hypothesized that ancestral algal populations reached

TABLE 15.6. Number of species recorded for each algal genus for Howland and Baker Is. (US Phoenix Islands), Tabuaeran Atoll (Kiribati Line Islands), and Kingman Reef and Jarvis Island (US Line Islands).

	Phoenix Islands			Line Islands		
Algal genera	**BAK**	**HOW**	Other 8 Is.	**KIN**	TAB	**JAR**
Blue-green algae						
Merismopedia					1	
Chroococcus					1	
Entophysalis					2	
Lyngbya	**2**	**2**	4	5		**3**
Blennothrix	**1**		1	**1**	1	
Calothrix				1	1	
Scytonematopsis					1	
Hormothamnion					1	
Hydrocoleum	1					
Spirocoleus	**1**	**1**				**1**
Schizothrix	**1**		1	2	2	**1**
Oscillatoria			1	1		
Phormidium				1		
Rivularia				1		
Red algae						
Bangia					1	
Chroodactylon			1			
Stylonema	**1**	**1**		2	1	**2**
Sahlingia				1		
Acrochaetium	**1**			1	1	**2**
Galaxaura	**1**		1			
Liagora				1		
Gelidiella			1	4	2	**2**
Gelidium				5	2	**2**
Parviphycus					1	
Wurdemannia					1	
Pterocladiella	**1**	**1**	1	1		**4**
Asparagopsis			1			
Gracilaria	**1**				1	**1**
Hydropuntia					1	
Hydrolithon			1		3	
Amphiroa						**1**
Jania	**2**	**2**	1	4	2	**2**
Lithothamnion			1			
Cryptonemia	**1**			2		
Grateloupia						**1**
Halymenia	**1**			1		**1**
Hypnea	**3**	**1**	1	2	5	**2**
Predaea				3		
Peyssonnelia	**1**	**1**	1	1	1	**1**
Asteromenia				1		
Champia				2		
Coelarthrum				1		
Chrysymenia						**1**
Rhodymenia					1	**2**
Gelidiopsis	**2**	**1**	1	1	1	**1**
Lomentaria	**2**	**2**		1		**1**
Aglaothamnion	**2**		1	1		**1**
Anotrichium		**1**	1	2	1	**1**
Antithamnionella	**1**		1	2		**1**
Antithamnion	**1**			3		**3**

(continued)

TABLE 15.6. (continued).

Algal genera	Phoenix Islands			Line Islands		
	BAK	HOW	Other 8 Is.	KIN	TAB	JAR
Callithamnion					2	
Centroceras					1	
Ceramium	**6**	**2**	3	**5**	2	**9**
Corallophila	**3**	**2**	1	**3**	2	**3**
Crouania	**1**		1	**2**		
Diplothamnion				**1**		
Griffithsia			1	**2**	1	**2**
Lejolisea	**1**	**1**		**1**		**1**
Monosporus				**1**		
Ptilothamnion	**1**	**1**	1	**2**		**1**
Spyridea				**1**		
Tiffaniella				**1**		1
Wrangelia	**1**	**1**		**1**	1	
Dasya	**1**	**1**	1	**4**		**3**
Heterosiphonia	**1**	**1**	1			**1**
Branchioglossum						**1**
Cottoniella	**1**			**1**		**1**
Dotyella				**3**		
Hypoglossum			2	**4**		**5**
Taenioma				**1**		
Chondria	**2**		2	**3**	2	**5**
Chondrophycus					1	
Digenia	**1**					
Herposiphonia	**2**	**2**	1	**3**	1	**7**
Laurencia	**1**	**2**		**2**		**1**
Lophocladia			1	**2**		
Neosiphonia	**1**				2	
Polysiphonia	**5**		1	**9**	2	**6**
Spirocladia						**1**
Brown algae						
Asteronema	**1**				1	
Cutleria				**1**		**1**
Dictyopteris	**1**	**1**	1	**1**		**1**
Dictyota	**2**	**2**	2	**1**	1	**2**
Feldmannia					1	
Hincksia				**1**		
Lobophora	**1**	**1**	1	**1**	1	**1**
Padina	**1**					
Sphacelaria					2	
Turbinaria	**1**		1		1	
Green algae						
Palmophyllum		**1**				
Ulvella		**1**				
Ulva	**3**	**2**		**2**	2	**1**
Microdictyon	**1**			**2**		**2**
Chaetomorpha					1	
Cladophora	**3**	**2**	1	**1**	6	**1**
Rhizoclonium	**1**					
Boodlea	**1**			**1**	2	
Cladophoropsis	**1**	**2**	1	**4**	2	**2**
Dictyosphaeria	**2**	**1**	2 **3**	**2**	3	
Rhipdophyllon				**1**		
Valonia	**1**	**1**	1 **2**	**1**	2	
Valoniopsis			**1**			

(continued)

TABLE 15.6. (continued).

Algal genera	Phoenix Islands				Line Islands		
	BAK	**HOW**		Other 8 Is.	**KIN**	TAB	**JAR**
Ventricaria	1			1			
Bryopsis	1	1	2	2	2	3	
Caulerpa	2	2	4	3	4	1	
Caulerpella			1		1		
Codium	1					1	1
Derbesia	1				1	1	
Halimeda	3	1		7	7	7	3
Ostreobium						1	
Avrainvillea		2			2	1	
Chlorodesmis				1			
Pseudochlorodesmis				1			1
Rhipidosiphon				1			
Rhipilia				3			
Siphonogramen		1					
Udotea	1						
Bornetella					1		
Neomeris				1	2	1	
Parvocaulis				1		2	
Total genera	**56**	**34**		48	71	59	**54**
Total species	**86**	**48**		69	147	97	**115**

Bold = US islands. Non-bold = Kiribati islands. **BAK** = Baker, **HOW** = Howland, "Other 8 Islands" = combined totals for eight Kiribati Phoenix Islands (after South et al. 2001), TAB = Tabuaeran (after Tsuda et al. 1973), **KIN** = Kingman, and **JAR** = Jarvis

remote islands through drifting fragments of adult individuals, but this remains speculative (van den Hoek 1987). For the PRIA, the age of each island and the existing ocean currents that existed at the time of their formation would have greatly influenced the type of ancestral algal populations present, and help explain the differences in algal populations that exist on these islands today. These colonizing populations would have, in turn, influenced subsequent species introductions through competition.

15.7.1.6 Macro-invertebrates Other than Cnidarians (by S. Godwin and M. Timmers)

If only conspicuous non-coral macro-invertebrates are considered (Actiniaria, Mollusca and Echinoderms), there have been roughly 370 species identified by various survey efforts in the Phoenix and Line Islands (Kay 1971; Townsley and Townsley 1973; Godwin 2002; Brainard et al. 2005; Ferguson et al. 2006). This number would reach into the thousands if other representative groups and cryptic mobile and sessile invertebrates were included. Only a brief overview of the conspicuous mollusks and echinoderms at the five PRIA is presented in Table 15.7 below. Table 15.5 above covers corals and other cnidarians.

The majority of the data are from fore reef and adjacent terrace habitats that dominate the ocean side all six reefs. Only Kingman Reef, Palmyra Atoll, and Tabuaeran Atolls have lagoon habitats, and quantitative survey data are provided only for Kingman. Palmyra's lagoon remains in a degraded state since the WWII era that has discouraged invertebrate surveys therein.

The giant clam, *Tridacna maxima*, is found at all six of the reefs, and is most abundant in Kingman's lagoon (especially in the shallow eastern reef pool) and also more abundant on Kingman's fore reefs compared to the other PRIA. A second and larger giant clam, *T. squamosa*, has been reported only from Kingman. Records of conspicuous echinoderms include 10–12 species at five reefs, but at Kingman they total 25 species. Furthermore, the densities of the echinoderms are generally highest at Kingman compared to the other reefs. The most common echinoderms across all reefs are the sea

TABLE 15.7. Inventory and abundance of conspicuous giant clams and echinoderms for Howland and Baker Is. (US Phoenix Islands), Kingman Reef, Palmyra Atoll, and Jarvis I. (US Line Islands), and Tabuaeran Atoll (Kiribati Line Islands) (After Kay 1971; Townsley and Townsley 1973; Godwin 2002; Godwin 2004 in Brainard et al. 2005; Godwin in Ferguson et al. 2006).

PHYLA, Class and species	Phoenix Islands			Line Islands		
	BAK	HOW	KIN	PAL	TAB	JAR
Habitat	F	F	F & L	F & B	all	F
MOLLUSCA						
Bivalvia						
Tridacna maxima	0.0008	0.015	0.8–222	1.8	X	X
Tridacna squamosa			X			
ECHINODERMATA						
Holothuroidea						
Holothuria atra	0.0005	0.002	0.35–1.98	0.146	X	X
Holothuria edulis			X	X	X	
Holothuria whitmaei			0.250.22	X		
Actinopyga mauritiana	X	X	X	X	X	
Actinopyga obesa			0.3–2.64			0.0009
Stichopus chloronotus			X	X	X	
Stichopus cf. *horrens*			X			
Bodhadschia argus			X		X	
Bodhadschia graeffei			X			
Bodhadschia paradoxa			X			
Thelanota ananas			X			
Euapta sp.			X			
Polyplectana kerfersteini			X		X	
Other holothuroid			0.65–1.36			0.0004
Echinoidea						
Diadema sauvignyi	0.043	0.108	0.844			0.013
Echinothris calamaris	X	X	X	X		X
Echinothrix diadema	X	X	X		X	X
Echinothrix sp.	0.002	0.01	7.3–6.0			0.038
Echinostrephus sp.	X	X	X	X	X	X
Echinometra sp.	X	X	X	X	X	X
Heterocentrotus mammillatus	X		0.133		X	
Toxopnuestes pileolus	X	X	X		X	
Tripnuestes gratilla			0.125		X	
Asteroidea						
Acanthaster planci			1.311			
Fromia milleporella			X	X		
Mithrodia fisheri						
Echinaster luzonicus				X		
Linckia multifora	0.052	0.0003	0.044	X		0.009
Linckia guildingi	X	X	X			0.005
Other asteroid				0.036		
Total species	12	11	27	12	13	11

BAK = Baker, HOW = Howland, JAR = Jarvis, PAL = Palmyra, KIN = Kingman, and TAB = Tabuaeran (Fanning) Atoll. F = fore reef, B = back reef, and L = lagoon. X = species present at reef but abundance not estimated. Numbers are estimates of densities of species present at reefs (number per 100 m^2)

cucumber, *Holothuria atra*, and the sea urchin, *Diadema*. The predatory asteroid, *Acanthaster planci*, is common in lagoon habitats at Kingman. Towed diver surveys in 2002 and 2004 revealed that the sea star was abundant on southern fore reefs and northeastern back reefs and slopes. By 2006 it was still abundant on both northeastern and southeastern fore reefs.

*15.7.1.7 Inshore Fishes (by C. Musburger,
E. DeMartini, A. Friedlander, E. Sala,
S. Sandin, P. Ayotte, S. McTee, T. Wass,
and B. Mundy)*

Over 200 species of nearshore, reef-associated fishes have been documented at each of the US Phoenix and Line Islands through 2006. Palmyra Atoll, with records of 418 species to date from all sources, is currently known as the most diverse of these islands. As discussed above, there are habitats present at Palmyra that are not represented at the other islands that likely enable additional species to inhabit this atoll. However, fish collection activity has been considerably more intense at Palmyra than at any of the other locations (B. Mundy et al. 2002, unpublished manuscript). As such, many of the records of species recorded at Palmyra and not currently known from the other islands likely occur at other islands as well. This is especially likely for cryptic species that have been collected using chemical techniques at Palmyra but not elsewhere. Similarly, substantially less survey effort has been made at Kiritimati and Tabuaeran Atolls and this may explain why their family and species totals are slightly lower than those reported from the other islands. Table 15.8 includes records of all known species from each island to date with the number of species and families recorded from all reliable sources totaled at the bottom.

Despite differences in past survey effort, there are patterns in fish diversity which have emerged at these islands even when including all known records. Six families make up about half of the known species records. Surgeonfish (Acanthuridae), butterflyfish (Chaetodontidae), wrasses (Labridae), damselfishes (Pomacentridae), parrotfishes (Scaridae), and groupers (Serranidae) are extremely diverse at each island. The high diversity among these families is similar to patterns reported in coral reef fish communities at other locations. Some typically diverse fish families are composed of many cryptic species which have likely gone unobserved at most of these locations. For example, the recorded number of 29 goby species from Palmyra – much more than the number known from any of the other locations – is largely the result of extensive collection there. The 2006 survey team observed only three goby species there. This disparity is largely a result of the previ-

ous extractive surveys (chemical and otherwise) that have been conducted at Palmyra and highlights the limitations of accurately documenting cryptic species with the visual census method employed in 2005 and 2006. Other families that will undoubtedly yield higher species counts as survey effort increases include blennies (Blenniidae), moray eels (Muraenidae), and scorpionfishes (Scorpaenidae).

Detailed discussion of species-level patterns in fish diversity at these locations cannot be included here due to space limitations. For a more complete discussion, B. Mundy et al. (2002, unpublished manuscript) is a recommended reference.

15.7.2 Regional Diversity Patterns for Reef Biota (by J. Maragos, P. Vroom, and E. DeMartini)

Several factors could explain the geographic patterns for the reef biota described above:

- *Geographic isolation*: Five of the six Line Islands and atolls (except Kingman) rank between 3rd and 16th as the most isolated in the Pacific, based upon indices of isolation values between 121 and 129 reported by Dahl (1991). Kingman is a reef (and not an island or atoll), and it is not included in the index. Baker, Howland, Kanton and McKean in the Phoenix Islands also ranked high with indices of isolation values from 103 to 105. The long distances to neighboring reefs may prevent the larvae of many species from successfully establishing at smaller low reef islands, leading to lower levels of diversity, such as that documented for cnidarians at Jarvis. However, marine algae species diversity at Jarvis was highest among all reef islands, perhaps related to the enhanced upwelling off the island.

- *Habitat variety*: Atolls have a greater variety of reef habitats including sheltered lagoon pinnacles, patch reefs, sandy bottom, and passes that are lacking on low reef islands. As a consequence species adapted to sheltered habitats have a better chance of surviving at atolls. Atolls also have areas of semi-enclosed circulation that could retain larvae longer, increasing the likelihood of colonization. As a group, the atolls/atoll reef supported greater numbers of cnidarian genera compared to the reef islands.

TABLE 15.8. Number of species recorded for inshore fish families for Howland and Baker Island (US Phoenix Islands), Kingman Reef, Palmyra Atoll and Jarvis Island (US Line Islands) during 2006 surveys conducted by Musburger, Ayotte, McTee and Wass (2006) and from all surveys and number of species and families for Tabuaeran Atoll and Kiritimati Atoll (Kiribati Line Islands) from 2005 surveys conducted by DeMartini et al., submitted, and surveys in 1972 at Tabuaeran (Chave and Eckert 1974).

	Phoenix Islands				Line Islands		
Family	BAK	HOW	KIN	PAL	TAB	KIR	JAR
Acanthuridae	28	28	32	31	29	27	26
Albulidae					1		
Antennariidae						1	
Apogonidae	2	1	2	2	7	3	1
Aulostomidae		1	1	1	1	1	
Balistidae	8	10	9	10	9	9	11
Belonidae			1	1	1		
Blenniidae	4	7	2	1	9	8	5
Bothidae						1	
Caesionidae	3	2	2	2	2	2	1
Callionymidae		1			1	3	
Caracanthidae	1				1	1	1
Carangidae	6	6	6	6	8	6	6
Carcharhinidae	3	4	2	3	2		3
Chaetodontidae	11	12	19	17	16	18	12
Chanidae			1	1	1	1	
Cirrhitidae	5	5	4	4	5	5	7
Dasyatidae		1					1
Diodontidae	1	1			1	1	
Echeneidae	1						
Ephippidae	2						
Fistulariidae			1	1	1	1	1
Gobiidae	2	1	3		10	4	
Holocentridae	6	6	8	7	11	5	7
Kyphosidae	2	2	3	3	3	1	3
Labridae	29	38	42	43	42	37	33
Lethrinidae	4	4	1	4	5	5	1
Lutjanidae	8	8	4	6	7	7	4
Malacanthidae		1	2	1	1	1	1
Microdesmidae	1	2	5	2			
Mobulidae		1	1	1			1
Monacanthidae	4	3	5	4	3	4	3
Moringuidae				1			
Mugilidae	2			5	2		
Mullidae	5	4	5	6	7	8	4
Muraenidae	4	3	2	1	5	9	3
Myliobatidae		1	1	1	1	1	
Ostraciidae		1	1	1	1	1	1
Pempheridae	1	2	1	1	1	1	1
Pinguipedidae					1	2	
Pomacanthidae	8	8	5	4	4	3	5
Pomacentridae	13	19	15	15	18	15	13
Priacanthidae			1		1		
Pseudochromidae					1		
Ptereleotridae					4	5	
Scaridae	4	7	16	19	17	11	9
Scombridae	2	2			1	1	2
Scorpaenidae	1	2	1	1	2	5	2
Serranidae	22	22	16	14	19	19	19
Sphyraenidae	1		3	3	3	1	2

TABLE 15.8. (continued).

Family	Phoenix Islands			Line Islands			
	BAK	**HOW**	**KIN**	**PAL**	TAB	KIR	**JAR**
Sphyrnidae		1					
Syngnathidae					1		
Synodontidae			2		1	4	
Tetraodontidae	1	2	2	1	4	4	3
Tripterygiidae			1	1	2		
Zanclidae	1	1	1	1	1	1	1
Totals:							
2006 species	194	220	229	220			193
2006 families	34	38	40	37			34
All species	311	342	297	418	274	243	284
All families	46	52	47	63	46	41	46

Bold = US PRIA and non-bold = Kiribati areas. BAK = Baker, HOW = Howland, KIN = Kingman, PAL = Palmyra, TAB = Tabuaeran, KIR = Kiritimati, and JAR = Jarvis I

Algal species and generic diversity at Kingman was highest for all sites, with Tabuaeran Atoll ranking second in generic diversity and third in species diversity behind Kingman and Jarvis. However, Kingman was the only one of three atolls/atoll reef to support more species of conspicuous echinoderms and mollusks, and Palmyra was the only atoll to report high numbers of inshore fish families, although the latter is likely due to greater sampling effort.

- *Habitat quantity*: The total quantity of reef habitat on the atolls is greater than at any of the low reef islands. The larvae of reef biota have a better chance of reaching larger reefs compared to smaller reefs, and the leeward sides of larger reefs provide more shelter from wave exposure. Again as a group, the larger reefs (all atolls) supported more cnidarian genera than the smaller reefs. McKean, the smallest reef island, had fewer genera and species compared to Howland and Baker, its closest neighbors. Two atolls were highest in number of genera of algae and ranked first (Kingman) and third (Tabuaeran) in species totals.

- *Proximity to the biodiversity rich western Pacific*: The Phoenix Islands are nearly 1,300 km closer to the western Pacific than the Line Islands. The three reef islands in the Phoenix Islands as a group supported greater numbers of cnidarian genera than Jarvis Island in the Line Islands, but the Line Islands atolls supported higher diversity than the one atoll, Kanton in the Phoenix Islands, for which data are available.

Biodiversity of marine algae was also higher in the Line Islands and biodiversity of echinoderms higher at Kingman.

- *Proximity to the North Equatorial Countercurrent (NECC)*: The eastward-moving NECC, at roughly 5° N latitude, can carry the larvae of corals and other reef life from the biodiversity-rich western Pacific over distances of thousands of kilometers downstream to reefs close to the current in the Central Pacific. Seasonally both Kingman and Palmyra are in the path of the NECC, with Teraina only one half degree latitude farther south. Teraina registered 27 genera and 68 species of cnidarians after surveys at just two sites, in contrast to 20 genera and 50 species at Jarvis after surveys at 30 sites. Palmyra and Kingman recorded the highest number of genera and species of corals, and Kingman supported the most species and genera of marine algae and conspicuous echinoderm compared to any of the other atolls or reef islands. Similar patterns were not evident for inshore fishes. Marine algae data are not available for Palmyra.

- *Proximity to the Equatorial Undercurrent*: The EUC and associated upwelling may have contributed to the greater diversity of marine algae and higher abundance of soft corals at Jarvis.

- *El Niño, La Niña and tropical cyclones*: Previously documented and undocumented coral bleaching, cooling water, and tropical cyclone events have affected the composition and diversity of corals and other reef species. These effects may continue. Generally, tropical cyclones weaken towards the Equator as the Coriolis force approaches zero, but there have been some intense tropical cyclones at low latitudes in the central Pacific that greatly impacted coral reefs (Blumenstock 1958; Maragos et al. 1973; Holthus et al. 1993).

- *Variation in survey intensity*: Survey intensity varied considerably among the reefs and biotic groups and among representative habitats. Inadequate sampling of some reefs and habitats likely led to lower numbers of species, genera, and families. Clearly, more of all three would be expected at Teraina after additional sites are surveyed. The same may apply to Howland after windward sites can be surveyed. Species and genera totals might also be higher at Kiritimati, and possibly Tabuaeran after additional surveys. The lower totals of species and genera of marine algae at the eight Kiribati Phoenix Islands vis-à-vis other islands may also be related to less sampling intensity in Kiribati. Certainly the higher numbers of species and families of inshore fishes at Palmyra can be attributed to greater collection and survey effort there.

The preceding interpretation of species richness patterns is limited by complications of uneven survey effort among habitats and generally among the PRIA island reefs and atolls and should be considered preliminary. A more formal analysis is needed that controls for survey method, stratifies diversity by major habitat type within each island reef and atoll, and adjusts for sampling effort (both area and number of stations surveyed). S. Sandin et al. (personal communication 2007) describes richness patterns in the northern Line Islands based on data collected at Kiritimati, Tabuaeran, Palmyra Atoll, and Kingman Reef on the 2005 Scripps Line Islands Expedition. It provides an example of such an approach and insights into what might be expected at an expanded geographic scale including all of the PRIA.

15.8 Local Monitoring and Assessment Programs

The companion NOAA report: "State of the Coral Reefs 2008" includes a full chapter on the assessment and monitoring programs in all PRIA including the five covered here. The first report (Turgeon et al. 2002), emphasized assessment, abundance and distribution of reef biota at the PRIA and most other US reefs. The second report (Brainard et al. 2005) increased emphasis on monitoring. These reports cover corals, algae, other invertebrates, fish, geology and oceanography. It is beyond the space limitations here to address these further. In addition, the Pacific Islands Fisheries Science Center's Coral Reef Ecosystem Division is preparing a series of comprehensive Coral Reef Ecosystem Monitoring Reports for the PRIA to be published in 2009. These reports will be based on the findings of Pacific RAMP surveys between 2000 and 2008.

15.9 Conclusion

The Pacific Remote Islands Area (PRIA) encompasses in total eight islands and atolls, of which five are discussed in detail in this chapter (Palmyra, Kingman, Howland, Baker, and Jarvis). These reefs form part of Line and the Phoenix Islands, and are associated with the Line and Tokelau Ridges of Cretaceous age. They consist of a basaltic core with limestone caps dated to variable age, some to the Cretaceous but not all are sufficiently cored. The atolls harbor a typical Indo-Pacific and rich coral reef fauna. All atolls are protected as National Wildlife Refuges and include some of the most pristine coral reef ecosystems in the United States.

Acknowledgements We thank the crews of the NOAA research vessels, *Townsend Cromwell, Oscar Elton Sette*, and *Hi'ialakai* for fantastic support, safety, and teamwork during the five expeditions to the PRIA between 2000 and 2006 and thank the leadership of CRED, the Pacific Islands Fisheries Science Center of NOAA, and principal investigators of these cruises for sponsorship, including those not contributing directly to this chapter. We also thank the Center for its critical review of the draft report. We thank the organizers and crew of the 2005 Scripps Institution of Oceanography cruise to the northern Line Islands. We also thank the following for contributing information and helping us during preparation of this report: Greta Aeby, Jake Asher, Bill Austin, Steve Barclay, Suzanne Case, Joe Chojnacki, John Collen, Charles Cook, Bonnie

DeJoseph, Elizabeth Dinsdale, Ainsley Fullard-Leo, Jonathan Gardiner, Phyllis Ha, Amy Hall, Kyle Hogrefe, Ron Hoeke, Jeremy Jackson, David Johnson, Elizabeth Keenan, Jean Kenyon, Nancy Knowlton, Matt Lange, Roger Lextrait, Anders Lyons, Jo Ann MacIsaac, Nancy MacKinnon, Karla McDermid, Megan Moews, Michael Molina, Robin Newbold, Chris Olsen, Donald Palawski, Olga Pantos, John Randall, Melissa Roth, John Schmerfeld, Robert Schroeder, Jennifer Smith, Robert Smith, William Smith, Bernardo Vargas-Angel, Martin Vitousek, Richard Wass, Alex Wegmann, Shawn White, Lee Ann Woodward, and Brian Zgliczynski. The first author expresses his utmost appreciation for the outstanding contributions of the co-authors, all who responded on short notice and produced magnificently and collaboratively for this report.

References

Ashmore SA (1973) The geomorphology at Johnston Atoll, Technical Report TR-237. Naval Oceanographic Office, Washington, DC, 1–25

Atwater T, Sclater J, Sandwell D, Severinghaus J, Marlow MS (1993) Fracture zone traces across the North Pacific Cretaceous Quiet Zone and their tectonic implications. In Pringle MS, Sager WW, Sliter WV, Stein S (eds) The Mesozoic Pacific: Geology, Tectonics and Volcanism. Geophys Monogr Ser, 77, AGU, Washington, DC, pp 137–154

Balazs G (1982) Status of sea turtles in the central Pacific Ocean. In Bjorndal KA (ed) Biology and Conservation of Sea Turtles. Smithsonian Institution, Washington, DC, pp 243–252

Barkley RA (1962) A review of the oceanography of the Central Pacific Ocean in the vicinity of the Line Islands. US Bureau of Commercial Fisheries, Biological Laboratory, Honolulu, HI, 22 p

Blumenstock DI (1958) Typhoon effects at Jaluit Atoll in the Marshall Islands. Nature 182:1267–1269

Boddam-Whetham JW (1876) Pearls of the Pacific. Hurst & Blackett, London

Brainard R, Maragos J, Schroeder R, Kenyon J, Vroom P, Godwin S, Hoeke R, Aeby G, Moffitt R, Lammers M, Gove J, Timmers M, Holzwarth S, Kolinski S (2005) Pacific remote island areas. In Waddell JE (ed) The State of Coral Reef Ecosystems of the United States and Pacific Freely Associated States. NOAA Tech Memo. NOS NCCOS 11, pp 338–372

Bryan EH Jr (1940) Palmyra, necklace of emerald islets. Honolulu Advertiser Feb. 4, 1940, magazine section:116

Bryan EH Jr (1953) Check list of atolls. Atoll Res Bull 19:1–39

Bryan EH Jr (1974) Panala'au Memoirs. Pacific Science Information Center. Bernice P Bishop Museum, Honolulu, HI

CIA World Fact Book (2007) https://www.cia.gov/cia/publications/factbook

Chave EH, Eckert DB (1974) Ecological aspects of the distribution of fishes at Fanning Island. Pac Sci 28:297–317

Chavez F, Strutton P, Friederich G, Feely R, Feldman G, Foley D, McPhaden M (1999) Biological and chemical response of the equatorial Pacific Ocean in the 1997–1998 El Niño. Science 286:2126–2131

Christopherson E (1927) Vegetation of Pacific equatorial islands. Bernice P Bishop Museum Bulletin 44:1–79

Clouard V, Bonneville A (2005) Ages of seamounts, islands, and plateaus on the Pacific Plate. In Plate, Plumes and Paradigms, Geol Soc Am Bull, Special Paper 388:71–90

Dahl AL (1991) Island Directory. UNEP Regional Seas Directories and Bibliographies No 35. United Nations Environment Programme

Dalebout ML, Baker CS, Steel D, Robertson KM, Chivers SJ, Perrin WF, Mead JG, Grace RV, Schofield D (2008) A divergent mtDNA lineage among Mesoplodon beaked whales: molecular evidence for a new whale in the Trophical Pacific? Marine Mammal Science (in press)

Dalrymple GB, Lanphere MB, Jackson ED (1974) Contributions to the petrology and geochronology of volcanic rocks from the leeward Hawaiian Islands. Geol Soc Am Bull 85:727–738

Dalrymple GB, Clague DA (1976) Age of the Hawaiian Emperor bend. Earth Planet Sci Lett 31:313–329

Dana TF (1975) Ecological aspects of hermatypic coral distribution in three different environments. Ph.D. thesis. University of California, San Diego, CA

Davis AS, Gray LB, Clague DA, Hein JR (2002) The Line Islands revisited: new ^{40}Ar/^{39}Ar geochronologic evidence for episodes of volcanism due to lithospheric extension. Geochem Geophys Geosyst 3:1018

Dawson EY (1959) Changes in Palmyra Atoll and its vegetation through the activities of man 1913–1958. Pacific Naturalist 1:1–51

Dawson EY, Aleem AA, Halstead BW (1955) Marine algae from Palmyra Island with special reference to the feeding habits and toxicology of reef fishes. A. Hancock Found Pub Occ Paper 17:1–39

Ekman S (1953) Zoogeography of the Sea. Sidgwick & Jackson, London, xiv + 417 pp.

Eldredge LG (1999) Marine mammals in the tropical island Pacific, pp 211–212. In Eldredge LG, Maragos JE, Holthus PF, Takeuchi, HF (eds) Marine and Coastal Diversity in the Tropical Island Pacific. Program on

Environment, East-West Center and Pacific Science Association, Bishop Museum, Honolulu, HI, 456 pp

Emery KO (1956) Marine geology of Johnston Island and its surrounding shallows, Central Pacific Ocean. Geol Soc Am Bull 67:1505–1519

Engilis, A Jr, M Naughton (2004) U.S. Pacific Islands Regional Shorebird Conservation Plan. US Department of the Interior, Fish and Wildlife Service. Portland, OR

Epp D (1984) Possible perturbations to hotspot traces and implications for the origin and structure of the Line Islands. J Geophys Res 89:11273–11286

Ferguson S, Musburger C, Ayotte P, Wass T, Vargas-Angel B, Maragos J, Godwin S, Tribollet A, DeJoseph B, Hall A, Timmers M, Coccagna E, Dobbs E, Hogrefe K, Merritt D, Young C, Lino K, Lundblad E, Jones J, Weiss J, Charette S, Bostick J (2006) *Hi'ialakai* HI0604 Cruise Report. NOAA Publication. www.soest.hawaii.edu/pibhmc/

Firing E, Lukas R, Sadler J, Wyrtki K (1983) Equatorial Undercurrent disappears during 1982–1983 El Niño. Science 222:1121–1123

Flint E (1999) Status of seabird populations and conservation in the tropical island Pacific. In Eldredge LG, Maragos JE, Holthus PF, Takeuchi HF (eds) Marine and Coastal Diversity in the Tropical Island Pacific. Program on Environment, East-West Center and Pacific Science Association, c/o Bishop Museum, Honolulu, HI, pp. 189–212

Fowler HW (1927) Fishes of the tropical Central Pacific. Bernice P Bishop Museum Bulletin 38:1–32

Friedlander A (2004) Unpublished temperature data collected in the lagoons at Palmyra between 2003–2004.

Friedlander A, Caselle J, Beets J, Lowe C, Bowen B, Ogawa T, Kelly K, Calitri T, Lange M, Anderson B (2007) Aspects of the biology, ecology, and recreational fishery for bonefish at Palmyra Atoll National Wildlife Refuge, with comparisons to other Pacific Islands. In Ault J (ed) Biology and Management of the World's Tarpon and Bonefish Fisheries. CRC, Boca Raton, FL, pp 28–56

(United States General Accounting Office) (1997) U.S. Insular Areas, application of the U.S. Constitution. GAO/OGC-98-5. Washington, DC, 71 p

Garcia MO, Park KH, Davis GT, Staudigel H, Mattey DP (1983) Petrology and isotope geochemistry of lavas from the Line Islands Cahin, Central Pacific Basin. In Pringle MS, Sager WW, Sliter WV, Stein S (eds) The Mesozoic Pacific: Geology, Tectonics and Volcanism. Geophys Monogr Ser 77, Amer Geophys Union, Washington, DC, pp 217–231

Godwin S (2002) Rapid ecological assessment of the marine invertebrate fauna of American Samoa and the U.S. Phoenix and Line Islands. Report submitted to the NOAA-NMFS Coral Reef Ecosystem Investigation, 17 p

Goe DR, Halstead BW (1955) A case of fish poisoning from *Caranx ignobilis* Forskal from Palmyra Island, with comments on the sensitivity of the mouse-injection technique for the screening of toxic fishes. Copeia 1955(3):238–240

Gove JM, Merrifield MA, Brainard R (2006) Temporal variability of current driven upwelling at Jarvis Island. J Geophys Res 111(C12):C12011

Grigg RW, Epp D (1989) Critical depth for the survival of coral islands: effects on the Hawaiian Archipelago. Science 243:638–641

Gyuris E (1999) Marine turtles. In Eldredge LG, Maragos JE, Holthus PF, Takeuchi HF (eds) Marine and Coastal Diversity in the Tropical Island Pacific. Program on Environment, East-West Center and Pacific Science Association, Bishop Museum, Honolulu, HI, pp 177–188

Haggerty JA, Schlanger SO, Premoli-Silva I (1982) Late Cretaceous and Eocene volcanism in the southern Line Islands and implications for hotspot theory. Geology 10:433–437

Halstead BW, Schall DW (1958) A report on the poisonous fishes of the Line Islands. Acta Tropica 15(3):193–223

Helfrich P, Piyakarchana T, Miles P (1968) Ciguatera fish poisoning. Part I. The ecology of ciguatera reef fish in the Line Islands. Occ Pap Bernice P Bishop Museum 23(14):305–369

Hendry R, Wunsch C (1973) High Reynolds number flow past an equatorial island, J Flu Mech 58:97–114

Hollenberg GJ (1968a) An account of the species of *Polysiphonia* of the central and western tropical Pacific Ocean. I. Oligosiphonia. Pac Sci 22:56–98

Hollenberg GJ (1968b) An account of the species of the red alga *Herposiphonia* occurring in the central and western tropical Pacific Ocean. Pac Sci 22:536–559

Holthus PF, Maragos JE, Naughton J, Dahl C, David D, Edward A, Gawel M, Liphei S (1993) Oroluk Atoll and Minto Reef Resource Survey. East-West Center Program on Environment. Honolulu, HI, 94 p

Howe MA, Lyon HL (1916) Algae. In Rock JF (ed) Palmyra Island with a description of its flora. Honolulu Star Bulletin: 31–32.

Hunter CL (1995) Review of status of coral reefs around American flag Pacific Islands and assessment of need, value and feasibility of establishing a coral reef fishery management plan for the Western Pacific Region. Prepared for Western Pacific Regional Fishery Management Council, Honolulu, HI

Jackson ED, Schlanger SO (1976) Regional Synthesis, Line Islands Chain, Tuamotu Island Chain and Manihiki Plateau, Central Pacific Ocean, Initial Rep Deep Sea Drill Proj 33:915–927

Jordahl K, Caress D, McNutt M, Bonneville A (2004) Seafloor topography and morphology of the Superswell region. In Hekinian R, Stoffers P, ChemineeJ-L (eds) The Pacific Ocean Hot Spots. Springer, Berlin, pp 9–28

Kay EA (1971) The littoral marine mollusks of Fanning Island. Pac Sci 25(2):260–281

Keating B (1992) Geology and offshore mineral resources of the Central Pacific Basin. Springer, New York

Keating BH (1987) Structural Failure and Drowning of Johnston Atoll, Central Pacific Basin, in Seamounts, Islands, and Atolls. In Keating BH, Fryer P, Batiza R, Boehlert GW (eds) Amer Geophys Union Geophys Monogr 43:49–60

Keenlyside N, Kleeman R (2002) Annual cycle of equatorial zonal currents in the Pacific. J Geophys Res 107:1–13

Kikiloi SK, Tengan TK (2002) Hui Panal 'au. (pamphlet). Bernice P Bishop Museum, Honolulu, HI, 15 p

Kleypas JA, Feely RA, Fabry VJ, Langdon C, Sabine CL, Robbins LL (2006) Impacts of ocean acidification on coral reefs and other marine calcifiers: a guide for future research. Rep Workshop NSF, NOAA, USGS, 85 pp

Koppers AAP (2007) The Seamount Catalog. http://earthref.org/cgi-bin/er.cgi?s = sc-s0-main.cgi

Koppers AAP, Staudigel H (2005) Asynchronous Bends in Pacific Seamount Trails: a Case for Extensional Volcanism? Science 307:904–907

Koppers AAP, Staudigel H, Pringle MS, Wijbrans JR (2003) Short-lived and discontinuous intraplate volcanism in the South Pacific: hot spots or extensional volcanism? Geochem Geophys Geosys 4:1089

Kroenke LW, Wessel P, Sterling A (2004) Motion of the Ontong Java plateau in the hot-spot frame of reference: 122 Ma-present. In Fitton JG, Mahoney JJ, Wallace PJ, Saunders AD (eds) Origin and Evolution of the Ontong Java Plateau. Geol Soc Lond Spec Pub 229: 9–20

Lanphere M, GB Dalrymple (1976) Identification of excess 40Ar by the ^{40}Ar/^{39}Ar age spectrum technique. Earth Planet Sci Lett 32:141–148

Larson RL (1991) Latest pulse of earth: evidence for a mid-Cretaceous superplume. Geology 19:547–550

Leichter J, Miller S (1999) Predicting high-frequency upwelling: spatial and temporal patterns of temperature anomalies on a Florida coral reef. Cont Shelf Res 19:911–928

Maragos JE (1974) Reef corals of Fanning Island. Pac Sci 28:247–255

Maragos JE (1979) Palmyra. Appendix. Preliminary Environmental Survey and Assessment. U.S. Army Corps of Engineers, Pacific Ocean Division, Fort Shafter, HI, 31 p + App.

Maragos JE (1987) Notes on the Abundance and Distribution of Reef Corals and Reef Features at Palmyra Atoll, Line Islands. U.S. Army Corps of Engineers, Pacific Ocean Division, Defense Environmental Restoration Program. Fort Shafter, HI, 17 p

Maragos JE (1993) Impact of coastal construction on coral reefs in the U.S.-affiliated Pacific Islands. Coast Manage 21:235–69

Maragos JE (1997) Ecological surveys of corals and reef environments at Kiritimati Atoll for the proposed HOPE-X landing site project. Dames & Moore, Honolulu, HI, 36 p

Maragos JE, Holthus PF (1995) A status report on the coral reefs of the insular tropical Pacific. In Eldredge LG, Maragos JE, Holthus PF, Takeuchi HF (eds) Marine and Coastal Diversity in the Tropical Island Pacific. Program on Environment, East-West Center and Pacific Science Association, Bishop Museum, Honolulu, HI, pp 47–118

Maragos JE, Jokiel PL (1978) Reef corals of Canton Atoll: I. Zoogeography. Atoll Res Bull 221:57–70

Maragos JE, Jokiel PL (1986) Reef corals of Johnston Atoll: one of the world's most isolated reefs. Coral Reefs 4:141–150

Maragos JE, Baines GBK, Beveridge PJ (1973) Tropical cyclone Bebe creates a new land formation on Funafuti Atoll. Science 181:1161–1164

McNutt MK, Fisher KM (1987) The South Pacific Superswell. In Keating B, Fryer P, Batiza R, Boehlert G (eds) Seamounts, Islands and Atolls. Amer Geophys Union, Washington, DC, pp 25–34

McNutt MK, Winterer EL, Sager W, Natland JH, Ito G (1990) The Darwin Rise: a cretaceous superswell? Geophys Res Lett 17:1101–1104

McNutt MK, Bonneville A (2000) Chemical origin for the Marquesas swell. Geochem Geophys Geosys 1–6, doi:10.1029/1999GC000028

McPhaden MJ, Busalacchi AJ, Cheney R, Donguy J-R, Gage KS, Halpern D, Ji M, Julian P, Meyers G, Mitchum G, Niiler PP, Picaut J, Reynolds JR, Smith N, Takeuchi K (1998) The Tropical Ocean-Global Atmosphere (TOGA) observing system: a decade of progress. J Geophys Res 103:14169–14240

Menard HW (1964) Marine Geology of the Pacific. McGraw-Hill, New York, 271 pp

Morgan WJ (1972) Deep mantle convection plumes and plate motion. Am Assoc Petrol Geol Bull 56:203–213

Muller RD, Roest WR, Royer JY, Gahagan LM, Sclater JG (1997) Digital isochrons of the world's ocean floor. J Geophys Res 102:3211–3214

National Marine Fisheries Service and US Fish and Wildlife Service (1998) Recovery Plan for the U.S. Pacific Populations of the Green Turtle (*Chelonia mydas*). National Marine Fisheries Service. Silver Spring, MD, 84 pp

Natland JH, Winterer EL (2005) Fissure control on volcanic action in the Pacific. In Plates, Plumes and Paradigms. Geol Soc Am Special Pap 388: 687–710

Orwig TL, Kroenke LW (1980) Tectonics of the eastern central Pacific basin, Mar Geol 34:29–43

Parsons B, Sclater JG (1977) An analysis of the variation of ocean floor bathymetry and heat flow with age. J Geophys Res 82:803–827

Philander SG (1990) El Nino, La Nina, and the Southern Oscillation. Academic, San Diego, CA

Pierce RJ, Etei T, Kerr V, Saul E, Teatata A, Thorson M, Wragg G (2007) Phoenix Islands conservation survey and assessment of restoration feasibility: Kiribati. (draft of 12–9–06). Preparation for Conservation International Samoa and Pacific Islands Initiative New Zealand. Prepared by Oceania, Whangarei, New Zealand, 122 p + appendices 25 p.

Rauzon MJ (1985) Feral cats on Jarvis Island: their effects and their eradication. Atoll Res Bull 282–292:1–32

Rice DW (1977) A list of the marine mammals of the world. NOAA Technical Report NMFS SSRF-711, 15 pp

Roemmich D (1984) Indirect sensing of equatorial currents by means of island pressure measurements. J Phys Oceanogr 14:1458–1469

Saito K, M Ozima (1976) (super 40) Ar/ (super 39) Ar ages of submarine rocks from the Line Islands; implications on the origin of the Line Islands. In Sutton, GH, Manghnani MH, Moberly R (eds) The Geophysics of the Pacific Ocean Basin and its Margins, Geophysical Monograph Series 19. American Geophysical Union, Washington, DC, pp 369–374

Schlanger SO, Jackson ED, Boyce RE, Cook HE, Jenkyns HC, Johnson DA, Kaneps AG, Kelts KR, Martini E, McNulty CL, Winterer CL (1976) Initial Reports of the Deep Sea Drilling Project 33, US Government Printing Office, Washington, DC

Schlanger SO, Garcia MO, Keating BH, Naughton JJ, Sager WW, Haggerty JA, Philpotts JA, Duncan RA (1984) Geology and geochronology of the Line Islands., J Geophys Res 89(11):261–272

Sclater JG, Francheteau J (1970) The implications of terrestrial heat-flow observations on current tectonic and geochemical models of the crust and upper mantle of the earth. Geophys J R Astr Soc 20:509–542

Sharp WD, Clague RA (2006) 50-Ma Initiation of Hawaiian–Emperor Bend Records Major Change in Plate Motion. Science 313:1281–1284

Shun K (1987) Archaeological reconnaissance site survey and limited subsurface testing of Baker and Howland Islands. Final Report. Prepared for the US Army Engineer District under contract No. DACA83–86-M-0120, April 1987, 55 pp

Sibley FC, Clapp RB, Long CR (1965) Biological Survey of Howland Island, March 1963–May 1965. Unpublished Report of Pacific Ocean biological Survey Program. Division of Birds Smithsonian Institution, Washington, DC

Silva PC (1992) Geographic patterns of diversity in benthic marine algae. Pac Sci 46:429–437

Sleep NH (1992) Hotspots and mantle plumes. Ann Rev Earth Planet Sci 20:19–43

Smith WHF, Sandwell DT (1997) Global sea floor topography from satellite altimetry and ship depth soundings. Science 277:1956–1962

South GR, Skelton PA, Yoshinaga A (2001) Subtidal benthic marine algae of the Phoenix Islands, Republic of Kiribati, Central Pacific. Bot Mar 44:559–570

Stehli FG, Wells JW (1971) Diversity and age patterns in hermatypic corals. Syst. Zool. 20(2):115–126

Stein C, Abbott D (1991) Heat flow constraints on the South Pacific Superswell, J Geophys Res 96:16038–16100

Stein CA, Stein S (1993) Constraints on Pacific midplate swells from global depth-age and heat flow-age models. In Pringle M, Sager WW, Sliter W, Stein S (eds) The Mesozoic Pacific. Amer Geophys Union Geophys Monogr Ser 77, Washington, DC, pp 53–76

Stoddart DR (1992) Biogeography of the tropical Pacific. Pac Sci 46:276–293

Tomczak M, Godfrey JS (2003) Regional oceanography: an Introduction, 2nd edition, Daya, Delhi.

Townsley SJ, Townsley M (1973) A preliminary investigation of the biology and ecology of the holothurians at Fanning Island, pp 173–186. In Fanning Island Expedition, July and August, 1972. Hawaii Institute of Geophysics, University of Hawaii, Hawaii, HIG-73–13, 320 pp

Trenberth KE (1997) The Definition of El Niño. Bull Am Meteorol Soc 78 (12):2771–2777

Tsuda RT, Trono G Jr (1968) Marine benthic algae from Howland Island and Baker Island, Central Pacific. Pac Sci 22:194–197

Tsuda RT, Russell DJ, Doty MS (1973) Checklist of the marine benthic algae from Fanning Atoll, Line Islands, pp 61–67. In Fanning Island Expedition, July and August, 1972. Hawaii Institute of Geophysics, University of Hawaii, Hawaii, HIG-73–13, 320 pp

Turgeon DD et al. (37 additional co-authors) (2002) Condition of U.S. Pacific Remote Insular Reef Ecosystems. In The State of Coral Reef Ecosystems in the United States and Pacific Freely Associated States. NOAA, in cooperation with partners from federal, state, territorial and commonwealth agencies and the freely associated states. Silver Spring, MD, p 88

US Department of the Interior (2003) Protecting the Nation's Coral Reefs. Office of the Assistant Secretary for Fish and Wildlife and Parks. Washington, DC, 13 pp

US Fish and Wildlife Service (2002) Birds of Conservation Concern. Arlington, VA

US Fish and Wildlife Service (2005) Regional Seabird Conservation Plan, Pacific Region. US Fish and

Wildlife Service, Migratory Birds and Habitat Programs, Pacific Region, Portland, OR

Van den Hoek C (1987) The possible significance of long-range dispersal for the biogeography of seaweeds. Helgol Meeresunters 41:261–272

Veron JEN (1986) Corals of Australia and the Indo-Pacific. Australian Institute of Marine Science. University of Hawaii Press. Honolulu, HI, 644 pp

Veron JEN (1993) A biogeographic database of hermatypic corals. Australian Institute of Marine Science Monograph Series 10, Australia, 433 pp

Veron JEN (2000) Corals of the World. Australian Institute of Marine Science, 3 vols, Cape Fergeson, Australia

Vitousek MJ, Kilonsky B, Leslie WG (1980) Meteorological Observations in the Line Islands, 1972–1980. Honolulu, HI, 74 pp

Vitousek PM (1994) Beyond global warming: ecology and global change. Ecology 75:1861–1876

Vroom P, Dailer M, Timmers M, Maragos J, Vargas-Angel B, Musburger C, Ayotte P, McTee S, Richards B, Keenan E, Charette S, Hall A, Gove J, Hogrefe K, Hoeke R, Jones J, Miller J, Chojnacki J, Woodward L, Eggleston C, Cooper-Alletto S, Heikkenin R (2006a) Hi'ialakai HI0601 Cruise Report. NOAA Publication available at www.soest.hawaii.edu/pib-hmc/

Vroom P, Page KN, Kenyon JC, Brainard RE (2006b) Algae-Dominated Reefs. Am Sci 94:429–437

Wessel P, Harada Y, Kroenke LW, Sterling A (2003) The Hawaii-Emperor Bend: Clearly a Record of Pacific Plate Motion Change. EOS 84, V32A

Wessel P, Harada Y, Kroenke LW (2006) Towards a self-consistent, high-resolution absolute plate motion model for the Pacific. Geochem Geophys Geosys 7: Q03L12, doi:10.1029/2005GC001000

Winterer EL (1976) Bathymetry and regional tectonic setting of the Line Islands chain. Initial Reports of Deep Sea Drilling Project 33. US Government Printing Office, Washington, DC, pp 731–748

Winterer EL, Ewing JI et al. (1973) Initial Reports Of Deep Sea Drilling Project 17. US Government Printing Office, Washington, DC

Woodroffe CD, McLean RF (1998) Pleistocene morphology and Holocene emergence of Christmas (Kiritimati) Island, Pacific Ocean. Coral Reefs 17:235–248

Wright TT (1948) Palmyra, pearl of Polynesia. Honolulu Advertiser. Sunday Polynesian Section June 6, 1948: 1–2

Wyrtki K (1967) Oceanographic observations during the Line Islands Expedition, February-March 1967. University of Hawaii, Hawaii Institute Geophysics HIG-67–19, Hawaii

Wyrtki K, Eldin G (1982) Equatorial upwelling events in the central Pacific. J Phys Oceanogr 12:984–988

Yu X, McPhaden M (1999) Seasonal variability in the equatorial Pacific. J Phys Oceanogr 29:925–947

16

US Coral Reefs in the Line and Phoenix Islands, Central Pacific Ocean: Status, Threats and Significance[*]

James Maragos, Alan M. Friedlander, Scott Godwin, Craig Musburger,
Roy Tsuda, Elizabeth Flint, Olga Pantos, Paula Ayotte, Enric Sala,
Stuart Sandin, Sarah McTee, Daria Siciliano, and David Obura

16.1 Introduction

This is the second of two chapters on the coral reefs of the five US Line and Phoenix Islands, consisting of Baker, Howland and Jarvis Islands, Kingman Reef,; and Palmyra Atoll (Fig. 16.1). The previous chapter (Chapter 15, Maragos et al.) covers the history, geology, oceanography and biology, while this chapter covers the status, threats and significance of the five. All are low reef islets or atolls in the central Pacific Ocean administered by the US Fish and Wildlife Service as National Wildlife Refuges. These 5 Refuges are among 20 within the tropical Pacific and among 10 that protect coral reefs. Together they are geographically a part of the largest series of fully protected marine areas under unified management in the world.

16.2 Historical Anthropogenic Effects on PRIA Islands and Reefs

Early Micronesians and Polynesians left behind few traces at the five PRIA (Pacific Remote Island Areas), except possibly for the presence of Pacific rats on some islands and coconut palms (*Cocos nucifera*) at Palmyra. Most of the impacts of nineteenth and

early twentieth century visitors were terrestrial, leaving alien mammalian predators on the islands that destroyed most seabird populations, introducing alien weed and alien ornamental vegetation to some PRIA, and perhaps depleting sea turtle nesting populations at most. Small boat channels were blasted through the reefs at the three Equatorial islands; and iron debris from anchors, chains, and possibly sunken vessels, have caused localized deterioration of reefs at the landing sites. WWII era modification and occupation of Palmyra left lasting impacts that remain unresolved to this day (see case study below, Figs. 16.2–16.4). These include the introduction of numerous alien species, degradation of land and lagoon habitats, dredging and filling of reefs, and dumping of toxic, hazardous and unsightly military waste. To a much lesser extent, toxic waste was also left at Howland and Baker by the US military.

16.3 Present Threats to PRIA Coral Reefs

Present threats include the residual impacts of guano mining and WWII era military construction at Palmyra, Baker and Howland. Coral bleaching events at the PRIA to date have not had a severe or chronic affect on coral populations (except at Palmyra), perhaps because of the healthy state of the reef environments and associated resilience of the reef and coral communities. More recently, an invasive corallimorph,

[*] The options of the authors included in this report do not necessarily represent those of the agencies or institutions at which they are employed

B.M. Riegl and R.E. Dodge (eds.), *Coral Reefs of the USA*,
© Springer Science+Business Media B.V. 2008

FIG. 16.1. (**a**) Jarvis Island seen from the air (**b**) Kingman Reef seen from the air. There is only one small rubble ridges that rise about 2 m above sealevel (**c**) the N–S causeway on Palmyra Atoll. It is easy to see why this atoll earned itself the nickname "Pearl of the Pacific" (**d**) Forest of *Pisonia grandis* trees on Palmyra Atoll (Images a and b by the USFWS, and c and d by J.E. Maragos)

Rhodactis howseii, has established itself at all sites where shipwrecks, marker buoys and other iron debris and equipment release dissolved iron, causing the corallimorph to spread and cover corals, algae and other benthos.

16.3.1 Palmyra Case Study: Lingering WWII Construction Impacts on Atoll Circulation (A. Friedlander and J. Maragos)

Before military construction at Palmyra (Fig. 16.2: year 1939), wave set-up created currents over windward reefs which likely pushed cool oceanic waters between the islets from northeast to southwest into the lagoon. Woodbury (1946) notes "The lagoon,

with its encircling barrier, acted like a storage tank with a one-way valve. Great waves thundering in from the north, poured millions of tons of water over the reef into the enclosure, where it was impounded until it had risen high enough to flow out again over the reef to the south. This kept the normal [lagoon] level at least a foot higher than the ocean side". The cooler oceanic waters then would have likely descended to the floor of East Lagoon and displaced warmer waters upward that would move towards Center Lagoon. This process would then be repeated between Center and West Lagoons, with the warmest waters eventually spilling over the shallow southern and western perimeter reefs.

World War II era construction disrupted the atoll's water flow regime (Fig. 16.2: year 1946), completely isolating East Lagoon, periodically

1939

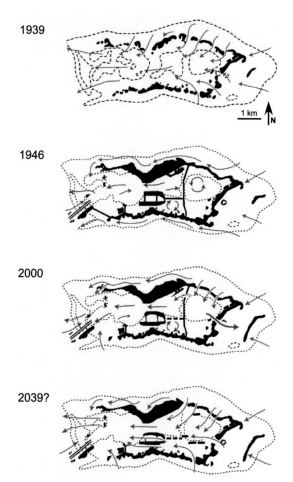

1 km ↑ N

1946

2000

2039?

FIG. 16.2. Time sequence diagrams showing hypothetical changes in water circulation patterns at Palmyra since 1939 and projected to 2039 with restoration measures. Solid black areas are islands and dashed lines are edges of shallow reefs. Blue arrows approximate net direction of cooler waters, and red arrows approximate net directions of warmer waters with suspended sediments. Constructed with the assistance of Phyllis Ha and John Collen from Woodbury (1946), Dawson (1959), Maragos (1993, unpublished observations), Ikonos 2000 image (courtesy of Space Imaging), and educated predictions for the 2036 scenario

had accumulated over thousands of years on the lagoon floor. Furthermore, Woodbury (1946) noted that when the channel was opened, the considerable head of water in the lagoon "released the water in a swift river that brought all but the most powerful boats to a standstill". In turn, this likely permanently lowered water levels in the lagoon. Wave set-up currents no longer entered the lagoon between windward islets after being blocked by the causeways. Instead ocean-side currents strengthened and moved westward and along the ocean shores of both the north and south sides of the atoll, resulting in beach erosion that is documented by loss of islets and several WWII concrete fortifications now well seaward of the shorelines of surviving islets. Possibly in 1997, catastrophic death of staghorn and table corals occurred on the southern half of the western terrace and shallower reefs near the southwest dredged channel. Warmer water discharges from West Lagoon may have exacerbated the effects of the bleaching event, further elevating sea surface temperatures to lethal levels.

Ikonos imagery of Palmyra (Fig. 16.3) and field observations add to our understanding of the residual effects of WWII construction. Several sections of the northern causeway have breached, with wave set up funneling water flow into East lagoon, which in turn elevates water levels in the lagoon that must have contributed to the large breach through the eastern causeway. Although there are small breaches that have developed in the north–south causeway separating East and Center Lagoons, the amount of net water discharge from them to Center Lagoon is small in comparison the net discharge of heated sediment-laden water from East Lagoon towards the outside eastern reefs and reef pools, degrading the turtle hole and inshore sections of the coral gardens.

Temperature data collected in the lagoons in late 2003 confirm the consistent increase in average lagoon water temperatures from east to west (Friedlander 2004; Fig. 16.4). This is consistent with observations of current and tidal patterns. Temperature variability was greatest in the main ship channel next to Big Eddies flat. The lowest temperature variability was observed in the eastern lagoon near the Downeast receiver (Fig. 16.3 bottom panel and Fig. 16.4). Higher water residence time most likely correlates with higher water temperature levels in the lagoons.

reversing water flow from West into Center Lagoon (Dawson 1959), and leading to increases in water residence times and temperatures in all lagoons. Moreover, the dredged channel connected the deeper ocean to West Lagoon, exposing the latter to stronger tidal and sub-surface currents, possibly leading to stratification and the stirring up of substantial amounts of fine sediments that

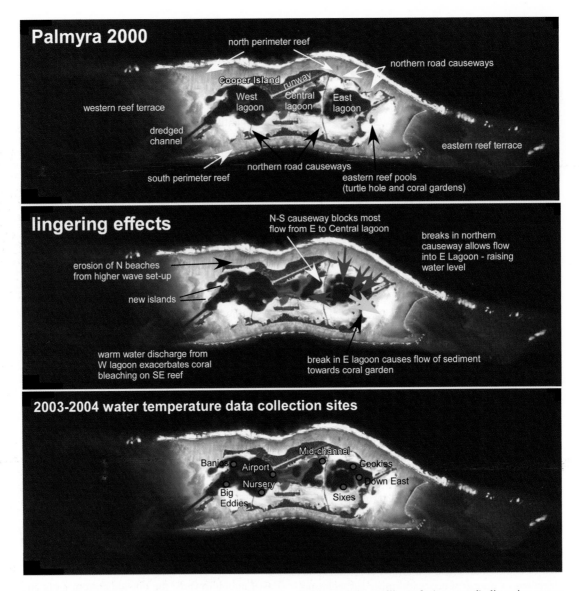

Fig. 16.3. Ikonos image of Palmyra Atoll showing major features of the atoll's reefs (top panel), lingering water circulation effects from WWII construction in 2000 (middle panel), and bone-fishing site names and locations of temperature data collection by Friedlander, 2004 (bottom panel) (Satellite image from Space Imaging)

A futuristic restoration scenario is shown in Fig. 16.2, bottom panel (marked 2039), that begins with removal of the north–south causeway to reduce water residence time and temperatures in the lagoon. The excess fill would be used to create small islets to replace seabird nesting habitat previously provided by the causeway for rare ground-nesting seabirds threatened by alien rats on the main islands. Other waterways would be created or widened to re-establish the predominant westward flow of water through the lagoon, reduce SSTs, and minimize discharge of warm turbid waters on the eastern reef pools. An additional breach would be cut through the NE causeway to reduce alongshore currents and shoreline erosion, and beach protection measures

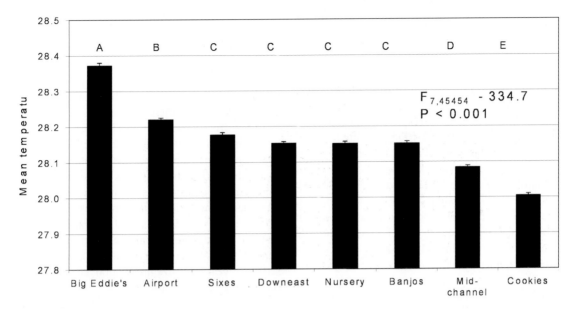

FIG. 16.4. Mean water temperatures at lagoon receiver sites between November 21 2003 and March 19 2004. Receiver sites are listed from west to east. Locations with the same letter are not significantly different at $\alpha = 0.05$ (After Friedlander 2004)

would be initiated on surviving NW islands. Other benefits of these measures would be to soften the impact of future bleaching events, restore sea turtle nesting habitat, reverse the degradation of the eastern reef pools, and promote more rapid recovery of corals near the dredged channel and western perimeter reefs.

This is one of many possible plans, and implementation of any plan would entail some risk. Simulation of the outcomes should be modeled before any physical change to the atoll is initiated, with and without restoration measures. For example, shorelines and islands are already eroding, and restoration measures may hasten the rates of erosion and mobilize chronic sedimentation over many years. Moreover, the recovery of lagoon habitats to pre-WWII conditions is uncertain and as yet undefined. Perhaps a less severe and complicated option would be to first model and, if justified, remove the north–south causeway, and evaluate the effects of that measure before pursuing additional restoration. Victoria University of Wellington is now embarking on a long term program to estimate previous conditions and to model changes to the lagoon without restoration

measures (J. Gardner, personal communication 2007). Hopefully these and other PARC initiatives will eventually lead to a consensus on restoring Palmyra to a stature once called one of the "Pearls of the Pacific". Ironically Dawson's (1959) assessment of the construction impacts to Palmyra was the first scientific documentation of anthropogenic impacts on an atoll reef. Let us hope that Palmyra is not among the last to be relieved of those impacts.

16.3.2 Fishery Issues at the PRIA Coral Reefs

An emerging threat to the five PRIA is the potential for increases in commercial fishing activities within the established boundaries of the five National Wildlife Refuges despite refuge laws and regulations that clearly prohibit such fishing. Illegal and unauthorized shark finning by mostly foreign fishers and bottom and lobster fishing by American fishers within the NWR are both serious contemporary threats that may occur unnoticed at uninhabited reefs or beyond sight of on-island caretakers.

16.3.2.1 Evidence of Reef Shark Finning at the PRIA (J. Maragos)

Palmyra

Randall and Helfman (1973) had reported two researchers wading on shallow reef flats being attacked by black tip reef sharks *Carcharhinus melanopterus* in 1959 and 1965 in Palmyra Atoll. During his initial visit to Palmyra, Maragos (1979) was a passenger on the seaplane that brought a scientific team to the atoll and noted about 20 grey reef sharks *Carcharhinus amblyrhynchos*, circling the floating seaplane. The species was common throughout West and Center Lagoons and the dredged entrance channel. Black tip reef sharks were so numerous along all shorelines (averaging one shark per linear distance of 15 m) that several bumped his feet during a trek around the atoll.

When Maragos (1987) returned to uninhabited Palmyra, two of three dogs abandoned at Palmyra in 1979 were still alive and each morning would hunt in tandem for black-tip reef sharks in the shallows of West Lagoon, often successful in trapping and dragging sharks out of the water and killing them. Although there were fewer black-tip reef sharks in 1987 compared to the densities observed in 1979, they were still numerous in all sub lagoons. Grey reef sharks were also very abundant in the lagoons, southern fore reefs, and western reef terraces at most of the 20 sites surveyed in 1987.

Upon returning to Palmyra in late 1998, Maragos and M. Molina noted the complete absence of sharks anywhere at the atoll except for two 1.5 m long grey reef sharks reported at distances no closer than 30 m from the divers off the southern fore reef. Both sharks quickly sped off upon confirming the presence of divers. The on-island resident for the landowners could offer no explanation for the absence of the sharks (R. Lextrait, personal communication 1998). It is common for sharks to aggregate during mating and perhaps one of the two species were aggregating at sites not visited by the investigators. However, for both species of sharks to be aggregating at the same time is highly unlikely (R. Wass, personal communication 2007). Moreover, there is little information on mass die-off of sharks in the literature. The most credible explanation is intense fishing, particularly prolonged shark fining which led to the depletion of both species of sharks at Palmyra before November 1998.

Howland and Baker

Observations at Howland and Baker in 2000 revealed that only small grey reef sharks, *Carcharhinus amblyrhynchos* (0.3–1 m long) were present. One possible explanation is that the sharks may be genetically smaller there than at other locales or suffering from lack of food (R. Wass, personal communication 2007). However, prey species of fish seemed abundant at the time. Another explanation is that earlier shark finning had removed most sharks, and younger sharks that recruited to the reef after the finning incident were the ones observed in 2000. Casual observations by J. Maragos suggest that the average length and numbers of sharks appears to have increased at both reefs in the subsequent 5 years, but size distribution data collection on fish was initiated after the 2000 visit. Perhaps subsequent surveys in 2001, 2002, 2004, 2006 and beyond at Howland and Baker may help explain past and possibly future changes in the size distribution of sharks at the two reefs.

Shark finning continues to be a pervasive impact on many reefs in the tropical Pacific, especially at remote reefs away from the watchful eyes of residents. In fact, NOAA formally banned shark finning in US waters by US flagged vessels (NOAA 2002, 2004a). Often locals are paid large sums of money by Asians desiring shark fins, as has occurred recently in many Pacific islands, including the neighboring Kiribati Phoenix Islands (D. Obura, personal communication 2005) and the Marshall Islands (R. Thomas, personal communication 2007). In 2002, the US Coast Guard boarded the US fishing vessel *King Diamond II*, operated by a firm with offices in Honolulu and Hong Kong, off the coast of Mexico. This one vessel was carrying "32 tons of shark fins" equivalent to "the destruction of 30,000 sharks" (Raloff 2002). With the United Nations General Assembly's passage of a resolution "urging" member nations to ban "directed shark fisheries conducted solely for the purpose of harvesting shark fins", the issue has now been formally elevated to the global level (UNGA 2004).

16.3.2.2 The Continuing Search for Dwindling Commercial Stocks of Lobsters and Bottom Fish

The Western Pacific Fisheries Management Council (WESPAC) of the U S claims that Kingman Reef and Palmyra Atoll National Wildlife Refuges were established long after fishery management regimes at the two reefs were already in effect. WESPAC further claims that the public comment period for the two proposed refuges was insufficient prior to their establishment in the waning days of the Clinton Administration. The establishment of the northwestern Hawaiian Islands Marine National Monument in 2006 by President Bush has permanently closed commercial lobster fishing in a large majority of US waters and will close all bottom-fishing in the same region by 2011. Bottom fish stocks are now severely depleted in the main Hawaiian Islands and lobster stocks are depleted throughout the Hawaiian Islands.

Understandably, the affected commercial fishers are possibly looking elsewhere, and the options are now few within US waters. Fishers have supporters in administrative and legislative branches of the US government that may be looking to the PRIA as future commercial fishing grounds. In May 2007, the National Marine Fisheries Service (NMFS) issued several commercial lobster fishing permits in the PRIA. Since all PRIA, with the exception of Wake Atoll, are National Wildlife Refuges (NWR), this implies that commercial fishing would take place within NWR boundaries – which is prohibited by law. At the time of writing, the Regional Directors of NMFS and USFWS were planning to meet to resolve this conflict.

The judicial branch is presently evaluating the status of Kingman Reef since the Fullard-Leo family and associates filed a federal lawsuit contesting federal ownership in 2006. The suit has since been decided in favor of the federal government.

The five PRIA Refuges are part of the largest series of nationally protected areas in the Pacific under unified management, and the last of the largely non-fished reefs within the jurisdiction of the United States. Before their rediscovery two centuries ago, all eight PRIA had been uninhabited for hundreds of years or more or had never been inhabited at all. A century ago they were likely considered remote and unprofitable compared to numerous other fishing grounds in the tropical Pacific. However, fishing technologies have improved, markets have increased, fishery yields have declined, and the remoteness of distant fishing grounds is no longer an obstacle. The lack of on-island guardians and the lack of

effective surveillance and enforcement at remote, uninhabited reefs render them all vulnerable and the preferred targets of many fishers, many unaware of, or undeterred by established protective mandates. The application of remote surveillance technologies, including radar, satellites, video, and hydrophones could discourage unauthorized fishing at all uninhabited PRIA. For example, the People of Rongelap Atoll in the northern Marshall Islands are actively pursuing such technology for the protection of their adjacent uninhabited atoll, Ailinginae that is being nominated by the Republic of the Marshall Islands for World Heritage status in 2008 (J. Matayoshi, Mayor of Rongelap and A. Anjain-Maddison, National Senator for Rongelap, personal communication 2006). Also neighboring nations in the Pacific could benefit from such technology and organize a federation to help one another stem the ever increasing illegal harvest of their coral reef fisheries. NOAA leadership has already facilitated similar decisions by a fisheries federation of 63 member nations in the Atlantic and Caribbean (International Commission for the Conservation of Atlantic Tunas) that unanimously agreed to ban shark fishing (NOAA 2004b).

16.4 Ecological Significance of the PRIA

16.4.1 Ecology of Fish Assemblages and Implications for Conservation of the PRIA (A. Friedlander, E. Sala and S. Sandin)

Recent studies (Stevenson et al. 2007; J. Jackson personal communication 2007; E. Sala et al., personal communication 2007) indicate that some of the PRIA are among the most predator-dominated and biomass-rich reefs and atolls in the central, if not the entire, tropical Pacific. Fish assemblages at Howland, Baker, and especially Jarvis Islands rank among the most biomass-dense (3.4–8.0 mt/ha) and most piscivore-dominated (54–74%) reefs yet described (J. Jackson, personal communication 2007; Fig. 16.5). Kingman Reef in particular is recognized as a near-pristine relict of natural coral reef ecosystems, and now represents the new baseline standard (total biomass density > 5 mt/ha) against which to compare other central Pacific reefs degraded by human impact (Jackson et al., personal communication; E. Sala et al., personal communication 2007). Total reef fish biomass at Palmyra atoll in 2005 was over twice as great as that found along the inhabited coastline of Kiritimati, where the human population is relatively dense; total fish biomass at Kingman in turn was nearly twice that at Palmyra (E. DeMartini et al., personal communication 2007). Atoll differences in top predator biomass and composition were even more extreme: in August 2005 reef sharks comprised 62% of total fish biomass at Kingman (Fig. 16.5) – in stark contrast to their complete absence on qualitative surveys as well as quantitative transects conducted during the same month and year at Kiritimati (E. DeMartini et al., personal

communication 2007). J. Jackson et al. (personal communication 2007) examined all available evidence (mostly anecdotal) for the apparently marked historical declines in shark abundance at some central equatorial Pacific islands, noted the strong inverse relation between human population density and the standing biomass of sharks and other large predatory fishes, and concluded that overfishing has been the most likely cause for the observed declines in fish biomass.

Most of the PRIA island reefs and atolls truly are invaluable remnants of a natural world that must be conserved and protected from extraction and other human influence to all extent possible. The continued existence of PRIA as USFWS Wildlife

FIG. 16.5. (**a**) *Paracanthurus hepatus* and *Acropora* cf. *globiceps* on Howland (**b**) the Napoleon Wrasse *Cheilinus undulatus* on Palmyra. Throughout the Pacific, this species is over-harvested and large individuals, such as illustrated here, are very rare. (**c**) *Amphiprion chrysopterus* inside an *Entacmaea quadricolor* on Baker. (**d**) *Carcharhinus amblyrhynchus*, an unfortunately rare species on many fished Pacific reefs, is plentiful on Kingman Reef (Images by J.E. Maragos)

Refuges, protected against localized and roving human impacts including fisheries extraction, is a mandate that not only conserves populations within boundaries of the PRIA but also maintains them as a large fraction of the few remaining undisturbed areas of reference for studying the natural structure and function of coral reef ecosystems (E. DeMartini et al., personal communication 2007).

Similarly, recent bonefish studies at Palmyra (Friedlander et al., in press) offer a unique opportunity to understand how unaltered ecosystems are structured in a protected setting, how they function, and how they can most efficiently be managed. Not only did management of the small catch-and-release sport fishery at Palmyra benefit from the studies, but so did bonefish fisheries management elsewhere in the tropics where interpretations of research findings have been compromised by subsistence and commercial fishing and depletion of stocks.

16.4.2   Significance of Grazers on Coral Reefs (O. Pantos)

While coral reef grazer populations have been affected drastically worldwide by overfishing, the PRIA provide an exception and represent a true baseline for the study of reef ecosystem function. The almost universal nature of reductions in grazer populations on reefs around the world may have led to a general underestimation of the importance of grazers within the system, since scientists have rarely observed this natural baseline. Macro-grazers such as fish and sea urchins have traditionally been attributed to the control of macroalgae, whereas removal of this 'top-down' control through overfishing has been linked to the deterioration of reefs worldwide and the phase-shift from coral-dominated to macroalgal-dominated states. A recent study carried out on Palmyra Atoll (O. Pantos, personal communication 2007) demonstrated with the use of exclusion cages that in the absence of grazers the physico-chemical processes of the benthos were altered significantly, resulting in a change of both 'bottom-up' and 'top-down' control of macroalgae. In the absence of grazers, the benthic communities change and 'new' nitrogen is introduced into the system in the form of inorganic nutrients (ammonia, nitrate and nitrite)

as a result of increasing rates of nitrogen fixation. This change occurs within just a few days of the cessation of grazing. These nutrients are then available for algal growth (bottom-up control), and in the continued absence of grazers a positive feedback mechanism may be initiated which would ultimately lead to a shift in the system from coral-dominated to algal-dominated. Palmyra and the other PRIA refuges (Fig. 16.6) represent near-pristine reef systems, and provide a baseline for understanding the types of impacts that may occur on coral reefs on local and global scales due to changes in grazer populations.

Areas subject to human impacts that experience different chemical environments may respond differently to changes, such as to a mass coral mortality event which would result in increased areas of bare substrate available for algal settlement. An assessment of the links between human population density and water chemistry carried out along the Northern Line Islands (J. Jackson et al., personal communication 2007; O. Pantos, personal communication 2007) demonstrated that even in such remote locations, human influence is evident despite the relatively low population densities present there. Inorganic nutrient levels in the waters surrounding each island corresponded to human population size. Differences in the chemical environment due to human impacts may affect ecological processes on the reef, therefore not allowing a true representation of how a healthy reef might respond to change. Using the reefs of the PRIA as a baseline, it will be possible to control different factors and test how a reef might respond under different conditions. The knowledge that can be gained from this is imperative for the development and implementation of management strategies for the protection of near-pristine reefs and the remediation of those already damaged as a result of human activity.

16.4.3   Central Pacific World Heritage Project

Responding to the existence of less than ten marine World Heritage sites, UNESCO, IUCN, NOAA and others convened an international conference in Hanoi to identify priority marine areas for World Heritage consideration (UNESCO 2003). The Line

Fig. 16.6. (**a**) *Acropora* and *Pocillopora* corals at the SE reef pools, Coral Gardens, Palmyra Atoll (**b**) *Montipora aequituberculata* and Robin Newbold at Baker Island (**c**) corals on dredged channel wall at Palmyra Atoll in 1987 before they were killed-likely by a bleaching event – a decade later (**d**) Blue Plate Coral, *Montipora aequituberculata*, at Jarvis (Images by J.E. Maragos)

Islands and Phoenix Islands were among 23 priority areas identified for the Pacific Ocean basin. Despite covering a third of the globe, the Pacific has no marine World Heritage sites. Additional workshops and meetings in Honolulu, World Parks Congress in Durban, South Africa, and at Kiritimati in the Republic of Kiribati between 2003 and 2005 led to the selection of approximately 28 reef islands and atolls within four governments for possible inclusion in a single trans-boundary project entitled "The Central Pacific World Heritage Project" that included the five PRIA covered in this chapter. Other US sites include Johnston Atoll NWR, Rose Atoll NWR, and Swains Island, the latter two within American Samoa. The other candidates included all eight of the Phoenix Islands and all eight Line Islands under the jurisdiction of Kiribati, one atoll in the northern Cook Islands and three atolls within the westernmost of the Tuamotu group of French Polynesia.

At a Washington, DC presentation in March 2005, the US Assistant Secretary for Fish and Wildlife and Parks agreed to add the five PRIA and Rose Atoll NWR to the US Tentative List for World Heritage and further directed that all sites be prepared for nomination. However, at that time, the US Tentative List had not been opened for 20 years. In early 2007 the Department of Interior established procedures and guidelines for establishing a new tentative list, including a deadline of April 1, 2007 for the submission of all applications. Although the application entitled "Ancient US Reef Islets and Atolls of the Central Pacific" covering the five PRIA and Rose was completed shortly before the deadline and supported by the Governor of American Samoa and The Nature Conservancy, lengthy coordination within DOI could not be completed before the deadline because of the complexity of the nomination and need for added approvals. However, there are efforts underway to evaluate the listing of the PRIA on the 1971 Ramsar Convention, List of Wetlands of International Importance.

The Republic of Kiribati did successfully nominate its eight Phoenix Islands for World Heritage in 2007. If this nomination is eventually approved at the end of a process involving many steps, it is possible for other elements of the trans-boundary project (including the US PRIA) to be added to the World Heritage site, possibly involving more streamlined procedures. An excerpt from the Statement of Outstanding Universal Value proposed for the five PRIA and Rose Atoll in 2007 is stated below and summarizes many of the documented values of the PRIA reviewed in this chapter:

The Ancient US Reef Islets and Atolls in the Central Pacific are among the World's oldest living biogenic formations, in the World's oldest and largest ocean, forming first as volcanoes and then subsiding and evolving into reefs while moving slowly northwest ... to their present central Pacific locale over the past 70 million years or more. They serve as the last of the reef frontiers in the central Pacific, never permanently inhabited throughout their entire history and within the remotest part of the tropical Pacific Ocean. They were the last to be visited and occupied over the past several centuries and are the first of their kind to be afforded full protection. Consequently, they are among the most pristine coral reef ecosystems of Pacific, serving as critical components of flyways for seabirds and migratory shorebirds, and marine highways and stepping stones for many coral reef species and marine mammal colonizers that are now established in the southeastern Pacific. Dozens of marine and terrestrial threatened, endangered, depleted and endemic species thrive on the islets and reefs that are missing or rapidly vanishing elsewhere in the World. All major oceanic boundary currents in the tropical Pacific drift by these islands and reefs, subsidizing unique upwelling zones at some of the reefs .... Their ancient reef rock formations of thousands of feet thickness and laid down over millions of years, cap the tops of drowned volcanoes and have preserved the ancient climatic and oceanic history of the Pacific Ocean and World to this day, including the evolution of the marine species that now construct these features.

Acknowledgements We thank the crews of the NOAA research vessels, *Townsend Cromwell*, *Oscar Elton Sette*, and *Hi'ialakai* for fantastic support, safety, and teamwork during the five expeditions to the PRIA between 2000 and 2006 and thank the leadership of CRED, the Pacific Islands Fisheries Science Center of NOAA, and principal investigators of these cruises for sponsorship, including others aside from the contributors. We thank the organizers and crews of the 2005 and 2007 Scripps Institution cruises to the northern Line Islands. We also thank the following for contributing information and helping us during preparation of this report: Greta Aeby, Jake Asher, Bill Austin, Steve Barclay, Suzanne Case, Joe Chojnacki, John Collen, Charles Cook, Bonnie DeJoseph, Edward DeMartini, Elizabeth Dinsdale, Ainsley Fullard-Leo, Jonathan Gardiner, Phyllis Ha, Amy Hall, Annie Hillary, Kyle Hogrefe, Ron Hoeke, Jeremy Jackson, David Johnson, Elizabeth Keenan, Jean Kenyon, Marjaana Kokkonen, Nancy Knowlton, Matt Lange, Roger Lextrait, Anders Lyons, Jo Ann MacIsaac, Nancy MacKinnon, Karla McDermid, Megan Moews, Michael Molina, Robin Newbold, Chris Olsen, Donald Palawski, Olga Pantos, John Randall, Melissa Roth, John Schmerfeld, Robert Schroeder, Jennifer Smith, Robert Smith, William Smith, Salamat Tabbasum, Hans Thulstrup, Sarah Titchen, Bernardo Vargas, Martin Vitousek, Richard Wass, John Waugh, Alex Wegmann, Shawn White, Elspeth Wingham, Lee Ann Woodward, and Brian Zgliczynski. The first author expresses his utmost appreciation for the outstanding contributions of the co-authors, all who responded on short notice and produced magnificently and collaboratively for this report.

References

Dawson EY (1959) Changes in Palmyra Atoll and its vegetation through the activities of man 1913–1958. Pacific Naturalist 1(2):1–51

Friedlander A (2004) Unpublished temperature data collected in the lagoons at Palmyra between 2003–2004

Friedlander A, Caselle J, Beets J, Lowe C, Bowen B, Ogawa T, Kelly K, Calitri T, Lange M, Anderson B (2007) Aspects of the biology, ecology, and recreational fishery for bonefish at Palmyra Atoll National Wildlife Refuge, with comparisons to other Pacific Islands. In: Ault J (ed) Biology and management of the world's tarpon and bonefish fisheries. CRC, Boca Raton, FL, pp 28–56

Maragos JE (1979) Palmyra. Appendix. Preliminary environmental survey and assessment. US Army Corps of Engineers, Pacific Ocean Division, Fort Shafter, HI 31p + App

Maragos JE (1987) Notes on the abundance and distribution of reef corals and reef features at Palmyra Atoll,

Line Islands. US Army Corps of Engineers, Pacific Ocean Division, Defense Environmental Restoration Program. Fort Shafter, HI, 17 pp

NOAA - National Oceanic and Atmospheric Administration (2002) NMFS announces final rule to implement the Shark Finning Prohibition Act. National Marine Fisheries Service Press Release (Feb. 11). http://www. nmfs.noaa.gov/sfa/hms/shark_finning/fax_fr_f.pdf

NOAA – National Oceanic and Atmospheric Administration (2004a) Fact Sheet. Shark Management (Dec. 3). http:// www.nmfs.noaa.gov/sharks/FS_management

NOAA – National Oceanic and Atmospheric Administration (2004b) International commission adopts U.S. proposal for shark finning ban. US Department of Commerce/ National Oceanic and Atmospheric Administration press release (Nov.23). http://www.nmfs.noaa.gov/ docs/ICCAT_Conclusion.pdf

Raloff J (2002) No way to make soup – thirty-two tons of contraband shark fins seized on the high seas. Science News Online (Sept. 7). http://www.sciencenews.org/ articles/20020907/food.asp

Randall JE, Helfman GS (1973) Attacks on humans by the blacktip reef shark (Carcharhinus melanopterus). Pac Sci 27(3):226–238

Stevenson C, Katz LS, Micheli F, Block B, Heiman KW, Perle C, Weng K, Dunbar R, Whitting J (2007) High apex predator biomass on remote Pacific islands. Coral Reefs 26:47–51

UNESCO – United Nations Educational, Scientific and Cultural Organization, World Heritage Centre (2003) Proc World Heritage Marine Biodiversity Workshop. In: Hillary A, Kokkonen M, Max L (eds) UNESCO World Heritage Papers 4:1–92, Paris

UNGA – United Nations General Assembly (2004) Sustainable Fisheries Resolution A/59/L.23:59/25 of the Oceans and the Law of the Sea (Nov.17). http:// www.un.org/Depts/los/general_assembly/general_ assembly-resolutions.htm

Woodbury DO (1946) Builders for battle, how the Pacific naval air bases were constructed. Dutton, New York, 415 pp

17

Aspects of the Biology and Geomorphology of Johnston and Wake Atolls, Pacific Ocean

Phillip S. Lobel and Lisa Kerr Lobel

17.1 Introduction

Johnston Atoll and Wake Atoll are isolated and independent atolls in the North Pacific Ocean with a long history of military use. These atolls are also frequently referred to by the name of their main islands: Johnston Island and Wake Island. They have in common that they are both coral atolls located in the North Pacific Ocean (Fig. 17.1), both are exclusive properties of the US Department of Defense and neither had an indigenous islander population prior to military occupation. The two atolls are separated by 2,500 km (~1,600 miles) of open ocean. Both enjoyed the protection of being restricted access military bases until recently. In June 2003, Johnston Atoll was abandoned. Wake Atoll is currently in minimal caretaker status due to extensive damage from the last typhoon (Ioke on August 31, 2006). As of May 2007, the future of these atolls is uncertain because of a long-standing political quagmire among US Government agencies over natural resource management responsibilities and financing of the properties.

Johnston Atoll (Fig. 17.2) is isolated in the Central Pacific Ocean (16°45′ N 169°31′ W). The nearest landfall is French Frigate Shoals, 804 km (500 miles) north in the northwest Hawaiian Islands. It is about 1,287 km (800 miles) southwest of Honolulu, Hawaii. The Line Islands of Kiribati are about 1,440 km (900 miles) south and the Marshall Islands are about 2,560 km (1,600 miles) to the southwest.

Wake Atoll (Fig. 17.3) is isolated in the west-central Pacific (19°17′ N 166°36′ E). Wake is separated by 546 km (340 miles) of open ocean from the nearest reef system on Taongi Atoll in the Marshall Islands to the south. It is 2,000 km (1,250 miles) southwest of Midway Atoll, 1,400 km (840 miles) southeast of Minami-tori-shima (Marcus) Island, 2,260 km (1,350 miles) east of the Marianas Islands, and 1,100 km (660 miles) north of the Kwajalein Atoll.

Both atolls are in unique locations with respect to the biogeography of reef biota. Although, Johnston and Wake Atolls are not part of the Hawaiian–Emperor seamount chain as defined by position on the Pacific plate relative to the hotspot (Rotondo et al. 1981; see also Chapter 13, Rooney et al.). They both have intriguing patterns of marine species distributions that overlap with the Hawaiian Islands. The Hawaiian faunal connection has significant implications for interpreting biogeographic patterns that may have resulted either from dispersal of pelagic larvae or by vicariant events (Springer 1982). The stepping stone role of Johnston Atoll for reef animals, especially fishes from the equatorial Line Islands to the Hawaiian archipelago was defined by Gosline (1955), and subsequently further supported with additional data (Randall et al. 1985; Kosaki et al. 1991; Kobayashi 2006).

Until recently, The reef biota of both atolls was infrequently surveyed because access was/is limited due to their military missions (Lobel 2003; Lobel and Lobel 2004). Consequently, the marine biogeography and ecology of these atolls have not been completely detailed, especially with regard to more cryptic biota. The reef ecology and biota of Johnston Atoll has been more extensively studied

B.M. Riegl and R.E. Dodge (eds.), *Coral Reefs of the USA*,
© Springer Science+Business Media B.V. 2008

655

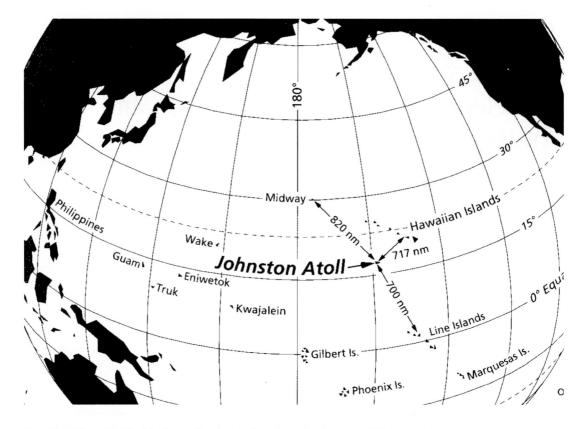

Fig. 17.1. Map of the Pacific Ocean showing the locations of Johnston and Wake Atolls

than has Wake. Even though these atolls were/are military bases and there were dramatic landscape alterations on land and in the water, today the coral reefs are in remarkably good condition. The positive side effect of the Department of Defense (DoD) mission was that fishing was very restricted at both atolls, at least until recently. The future of both atolls is somewhat uncertain and there are controversial political issues in play. The outcome of cooperative decisions to be made in 2007 and 2008 by DoD, DoI (Department of Interior) and NOAA (National Oceanographic and Atmospheric Administration) will ultimately define the future use of these atolls.

17.2   Johnston Atoll

Amerson and Shelton (1976) published a comprehensive scientific review of the natural history of Johnston Atoll. In some topics such as atoll geology, terrestrial

vegetation and ecology, there has been very little added since. We have tried to focus on additional information since Amerson and Shelton (1976) and refer readers to their excellent review. A recent photographic review of the marine life of Johnston Atoll shows many of the reef species and habitats (Lobel 2003).

17.2.1   History of Johnston Atoll as a US territory

Johnston Atoll was discovered on 2 September 1796 by a whaling ship, the brig *Sally* from Boston, MA. The Sally was accompanied by the British trading schooner, Prince William Henry captained by William Wake. The Prince William Henry also documented Wake Atoll during this voyage. Capt. Charles Johnston of the British ship HMS Cornwallis later documented Johnston Island on 14 December 1807 and it is for him that the atoll and main island are named.

FIG. 17.2. Johnston Atoll, view looking west. The largest island is Johnston Island. The nearest neighboring island is Sand Island. North Island is to right, north, in photo and East Island is to the left, east. The emergent reef borders the northern margin of the platform. The atoll is tilted to the south and submerged along the southern margin

FIG. 17.3. Wake Atoll, view looking southward. Peale Island is in the foreground to the right and is separated from Wake Island by a small channel. Wake Island is the large V shaped land on the left that wraps around to a causeway connecting to Wilkes Island. The shallow lagoon area that is dry at lowest tides is partially exposed in this photo, at left inside adjacent to Wake Island

Johnston Atoll has no natural source of freshwater and if islanders did visit, they could not reside for very long. Originally, there were two islands: the larger Johnston Island 0.2 km² (40 acres) and the smaller Sand Island 0.04 km² (10 acres). The islands began as mounds of guano on coral rubble and sand. No evidence has been found of Pacific Islanders visitations prior to the late 1800s. However, the large guano deposits attracted US attention. After the US passed the Guano Act that guaranteed US Military protection to citizens establishing guano-mining operations on uninhabited and unclaimed guano-laden islands, Americans collecting guano claimed Johnston Atoll on 19 March 1858. The Kingdom of Hawaii contested this claim on 27 July 1858 when King Kamehameha proclaimed ownership of Johnston Atoll (named Kalama in Hawaiian). The US annexed Hawaii in 1898 ending the dispute to the atoll. In addition to the visits by ships collecting guano, whaling ships hunted the region. One such whaler, the "Howland" out of New Bedford, Massachusetts wrecked onto the reef at Johnston Atoll in December, 1889 (Fig. 17.4).

The first official scientific survey was in 1923 by Dr. Alexander Wetmore from the Bishop Museum, Oahu, Hawaii. The expedition was financed by the Departments of Agriculture and Navy. The scientific report of the survey team described the incredible diversity of seabirds, fishes and other marine life. The political impact of this study resulted in Executive Order No. 4467, signed by President Calvin Coolidge on 29 July 1926. This placed Johnston Atoll under the control of the Department of Agriculture as a sanctuary for breeding seabirds.

President F.D. Roosevelt signed Executive Order No. 6935 on 29 December 1934, which then transferred Johnston Atoll to the Department of the Navy. At this time, the atoll still included two islands, Johnston Island and the smaller, Sand Island. Sand Island was kept as a bird sanctuary under the Department of Agriculture. However, increasing military activity between 1939 to 1941 resulted in a buildup of facilities on both islands with a small shallow channel dredged in-between. Sand Island became the main occupied island while dredging and filling the shallow lagoon expanded Johnston's size. Dredging also deepened the shipping channels into the lagoon allowing larger ships to enter. Between 1939 and 1942 that the island was leveled and enlarged from 0.2 km^2 (40 acres) to 0.9 km^2 (211 acres) (Fig. 17.5). At this point none of the original shoreline or vegetation on Johnston Island remained. Both Johnston and Sand Islands were virtually covered with buildings, roads, gun emplacements and the runway on Johnston Island. A Naval Air Station

was operational in August 1941. Executive order No. 8682 (February 1941) designated the area out to 3 miles from the atoll as a Naval Airspace Reservation and a Naval Defense Sea Area. During this time, Pan American Airways used the airfield between Hawaii and Asia.

The US forces on Johnston Atoll came under cannon fire by Japanese naval vessels on 15, 21 and 22 December 1941. Facilities were damaged but there were no casualties. During the war, the atoll served as a strategic site for refueling planes and submarines on patrol in defense of the Hawaiian Islands. During WWII, Johnston Island was further expanded in size to accommodate a longer runway and additional aircraft parking space.

After WWII, Johnston Atoll continued as an air station and was eventually transferred from the Navy to the US Air Force in July 1948. Lagoon dredging continued to create deeper and larger ship channels while slowly expanding the size of the islands by using the crushed coral as fill material (Fig. 17.5). The islands were in heavy use during the Korean War. The islands were built up in size and two new islands were created. Sand Island was expanded from about 0.04 to 0.09 km^2 (10–22 acres). Johnston Island was increased from 0.9 to 2.6 km^2 (211 to about 640 acres) (Fig. 17.6). By the mid-1960s, Johnston and Sand Islands were enlarged to their present sizes (Fig. 17.7). The two new artificial islands, North and East Islands, were built from dredge materials and are 0.1 and 0.07 km^2 (25 and 18 acres), respectively (Fig. 17.8). The overall lagoon is about 121.4 km^2 (30,000 acres).

A US Coast Guard Loran station was built on Sand Island in 1960 (Figs. 17.9 and 17.10). It was the master Loran station for the central Pacific until December 1992 when the Global Positioning Satellite navigational system became fully operational and rendered Loran obsolete. The former Loran station became a marine biological field station in 1993 and was used until base closure activities began in 2001.

During the nuclear test era of the 1960s, Johnston Atoll became a test site and staging area due to its isolation in all directions from other inhabited islands. In 1962, the US conducted a series of atomic tests. The USAF released operational control of the islands to the Atomic Energy Commission (AEC)

Fig. 17.4. The whaler ship, Howland, wrecked onto the reef at Johnston Atoll and sunk in 1889. Remains include iron ballast (photo), scattered anchors and various ship debris. The shipwreck was discovered by P. Lobel while conducting reef surveys

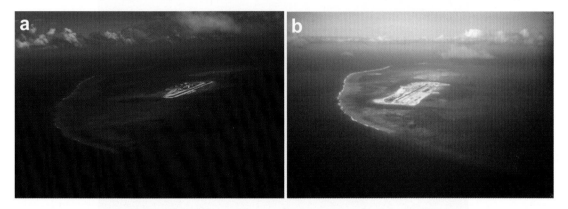

Fig. 17.5. (a) Johnston Atoll in 1952. Note that Johnston Island has been enlarged slightly. Sand Island had not yet been enlarged. North and East Islands did not exist yet (Photo by K.O. Emery and used with permission). (b) The atoll in 1993. When compared with the left photograph, notice how much closer the island is to the reef margin. The area of Sand Island has also been increased as well as the creation of two new islands, North and East. The semicircular shape of the atoll is evident. The emergent reef is to the north and the atoll platform tilts and is submerged to the south

and their military counterpart, Joint Task Force Eight (JTF-8). The atomic test series code-named DOMINIC included four high altitude and five low altitude successful tests, followed by a few THOR missile misfires. Radioactive contamination from fallout resulted. Today, the plutonium contaminated soils and other materials are buried in an EPA permitted landfill on Johnston Island. The Defense Threat Reduction Agency (formerly the Defense Nuclear Agency and then the Defense Special Weapons Agency) is responsible for any continuing radiological issues.

The comprehensive scientific study of Johnston Atoll's coral reefs was conducted by the University of Hawaii during 1963/64 (Brock et al. 1965, 1966). These biological surveys were conducted

FIG. 17.6. Johnston Island with the full development of the military–industrial complex in 1995

FIG. 17.7. Sand Island is in center, Johnston Island (partial) to right, North Island in foreground. The darker blue channel between Sand and Johnston was dredged from the reef to make a large ship channel and expand the islands. The coral carved from the rectangular basin adjacent North Island was crushed to create the artificial island

to characterize the reef ecosystem, measure the effects of dredging on the reefs and to estimate the prevalence of ciguatera and any changes in its occurrence associated with dredging. Ciguatera is a marine toxin from the dinoflagellate, *Gambierdiscus toxicus*. This dinoflagellate is epiphytic on other algae. Herbivorous fishes that consume algae with *G. toxicus* can then bioaccumulate the toxin. Depending on the level

of toxicity, effects on humans eating ciguateric fishes can range from mild illness to death. One hypothesis is that environmental disturbances such as dredging and severe storms foster the dinoflagellate blooms (Anderson and Lobel 1987). Johnston Atoll became the source for ciguatoxic fishes collected and used in testing by University of Hawaii scientists into the late 1970s. The diets and feeding ecology of herbivorous surgeonfishes

FIG. 17.8. East Island is also completely artificial. The buildings were removed with base closure

FIG. 17.9. This photo gives an interesting perspective of the dredged channel next to Sand Island. The dark blue channel is about 40 ft (14 m) deep and was carved from reefs that were growing to the surface that still exist adjacent to the channel. The bright sandy area is very shallow sand flats (10 ft, 3 m depth or less). The buildings were removed in 2001 and buried in place as a small hill (landfill). The north side of this west end of Sand Island is where most PCB contamination was found. Plutonium radiation was found on the south side. North Island is in background. The reticulated pattern of shallow lagoon reefs made boat travel a maze game

(Acanthuridae) at Johnston and in Hawaii were studied by Jones (1968).

A comprehensive study of the seabirds with extensive notes on other aspects of the atoll's natu-ral history was published by Amerson and Shelton (1976). This publication also included a detailed history and summary of research findings until that time.

Fig. 17.10. Sand Island. This photo from 1993 shows the LORAN tower (dismantled in 1995 after GPS become operational). The building on the causeway was the USCG station and later became the marine lab. The shore line around the east end of Sand Island (foreground with tower) is the only natural remaining shore geology remaining on the atoll with old emergent reef platform and tidepool habitat. Sand Island is the main breeding area for the migratory seabirds

In 1971, the US Army began using 0.2 km² (41 acres) of Johnston Island for chemical weapons storage. These weapons included nerve gas, mustard gas and other chemical agents contained in rockets, artillery shells, bombs and ton containers. In 1972, the USAF brought about 25,000 55 gal drums of the chemical Herbicide Orange (or Agent Orange) that came from Viet Nam and was being stored on Okinawa. This stock of Agent Orange was incinerated at sea in 1977 aboard the Dutch ship, *Vulcanus*. However, an unknown number of barrels leaked while stored on land and some barrels were dumped into the lagoon.

During the years 1970–1985, Johnston Atoll was maintained in caretaker status under the Defense Nuclear Agency, during which time the population averaged about 300 people including both military and civilian contractors. The atoll was in standby mode under "Safeguard C", which called for the resumption of nuclear weapons testing during the cold war era.

The building of the Johnston Atoll Chemical Agent Disposal System (JACADS) plant by the Army began in 1986 and with it, the island population grew to a level of about 1,200 personnel. The

US Fish and Wildlife Service stationed a full-time refuge manager on the island with Army funding in 1990 to help manage the natural resources. The increased number of people on the island also resulted in an increased exploitation of fish and marine invertebrates. Some species such as whitetip sharks and certain marine snails (with pretty shells) were severely over-collected. Shark fishing activity was eventually restricted in 1995. Whitetip sharks were the most impacted by this time. Now with the atoll abandoned since late 2003, these shark populations are again vulnerable.

The next series of reef ecological studies began in 1983 as part of the program to evaluate the potential environmental impact of the Army's JACADS chemical weapons incineration project. One of the first results was a checklist of fishes that increased the number of fishes known from the reefs from a previous count of 183 to a new total of 271 (Randall et al. 1985). Subsequent surveys added 30 new records for a new total of 301 fish species (Kosaki et al. 1991) and several more recently discovered new records of rare species remain to be described (P. S. Lobel and L. K. Lobel, personal communication 2003).

After the conclusion of the Army JACADS project it took about 1 year to demolish and remove all but one building (the JOC) from Johnston Atoll. The Johnston Atoll base once supported vast accommodations supporting a workforce of about 1,200 people. All freshwater was made by a desalination plant and power was generated from burning JP5 jet fuel. There were about 300 buildings and facilities in operation at the peak period from 1985 to 2002 on the nearly 1.4 million square feet of space which comprised Johnston Island. Today it is an abandoned atoll. Although the Air Force closed all atoll operations, Executive Order 8682, which established the Naval Sea Defense Area and Airspace Reservation, is still in effect. This executive order provides broad powers to DoD to control access to Johnston Atoll and to maintain it as a protected area

17.2.2 Geomorphology of Johnston Atoll

Johnston Atoll has an unusual geomorphology, which is unlike most atolls that have surrounding barrier reefs. Johnston Atoll has only a semicircular emergent barrier reef confined to the northern and western margins of the atoll platform (Fig. 17.5). Reefs to the east and around the south are submerged. The prevailing trade winds blow in from the east and, without an emergent reef barrier, drive current flow across the atoll.

Johnston Atoll is estimated to be about 85 MY in age (Keating 1985, 1992). Two distinct submarine platform surfaces are evident: one at about 10 m and the other at 2 m depth. The 10 m terrace has numerous ridges, knolls and coral growth. The 20 m terrace has many sink holes, often extending to more than 30 m depth. Numerous caverns have also been observed from submersible surveys at depths about 200–250 m. Large caves were also observed at 360 m which contained both stalagmites and stalactites (Keating 1985) and only limestone has been found at Johnston Atoll (Keating 1985, 1992). Within the 200 miles exclusive economic zone of Johnston Atoll, manganese crusts in seafloor plateau areas have been found to contain economically attractive concentrations of rare earth elements as well as cobalt, nickel and platinum (Wiltshire 1990).

Johnston Atoll rises from an abyssal plain at 4,950 m depth. On top, the atoll is a shallow platform of approximately 158 km^2 of reef habitat (Maragos and Jokiel 1986). The atoll is tilted to the southeast due to subsidence of the platform (Emery 1956; Amerson and Shelton 1976; Jokiel and Tyler 1992; Keating 1985). The shallow reef platform extends outside of the lagoon to about 19 km east–southeast and 8 km south of the main island, Johnston Island. The platform slopes gently to about 18 m depth then increases steeply to 180 m. Depths in the lagoon vary about from about 3 to 10 m. The lagoon habitat is composed of patches of sand, loose coral and large formations of live coral, especially *Acropora* species. The area of coral reef flats exposed during very low tide is approximately 5.2 km^2. The majority of this area is formed by the main northwestern reef, which forms an arc of about 4 km long. The outer reef has a gentle slope cut with many surge channels.

17.2.3 Climate of Johnston Atoll

The weather patterns at Johnston Atoll fall into two broad "seasons". The winter season extends from December through March and is characterized by slightly cooler temperatures, variable winds and heavier precipitation than during the summer months of April through November (Amerson and Shelton 1976). The mean average temperature is 26.3°C with daily ranges of about 13°C. The surface wind speed averages 15.1 mph with monthly means varying from 13.6 to 16.0 mph. These strong easterly trade winds (NE through ESE) are fairly constant throughout the summer months while during the winter months trade winds occur less than 80% of the time. Winter is characterized by light and variable winds, occasional westerlies associated with passing disturbances and weak cold fronts. Hurricanes do occasionally impact Johnston Atoll, for example, Celeste in 1972, John in 1994 and Ioke in 2006. Significant damage to island infrastructure occurred during hurricane John. During hurricane Ioke, a category 2 storm, portions of the Johnston Island seawall were breached increasing the erosion of the island. Mean annual precipitation is highly variable. The lowest annual rainfall recorded was 32.8 cm (12.9 in) in 1953 while 107.4 cm (42.3 in) fell during 1968. The mean is 66.3 cm (26.1 in).

17.2.4   Oceanography

The oceanographic environment around and within Johnston Atoll is one of well mixed oceanic water with surface salinities of 34.6–34.8‰ and annual temperatures ranging from 25°C to 27°C (Wennekens 1969). Our data show that sea surface temperatures (SST) generally vary less than 2°C during a daily cycle and seasonal SSTs range from lows of 23.2°C during February and March to highs of 29.2°C in late August to September (L. K. Lobel and P. S. Lobel 2000). SSTs within shallow portions of the lagoon exhibit higher variations due to solar warming (Fig. 17.11).

Ocean current flow patterns in and around Johnston Atoll were carefully studied by Kopenski and Wennekins (1966) using observational and empirical methods. The following oceanographic description is abstracted from their findings and supplemented with observations from another unpublished report included in the Army's JACADS EIA study (Lobel 1988).

The regional and local currents at Johnston Atoll are directly influenced by the flow of the North Equatorial Current. Grigg (1981) suggests that the atoll was also affected by the eastward flow of the Equatorial counter-current transporting water from tropical regions. The North Equatorial Current is driven primarily by the northeast trade wind system and is relatively strongest in January and weakest in July. The projection of the atoll into the flow of the North Equatorial Current produces two distinct but closely interrelated effects: (1) it deflects the current around and over the top of the atoll platform, and (2) it creates a wake on the downstream side of the platform (Fig. 17.12; Barkely 1972). The degree and intensity of these effects is driven by the mass flux of the North Equatorial Current.

The result of the North Equatorial Current flow interacting with the atoll geomorphology is a complex pattern of eddying and reversing currents in the shallow platform waters. These currents are also influenced by the tides. Although the prevailing current is transport to the west, the influence of the tide in shallow water makes the flow oscillate between the south and west, with short periods of flow to the south. The tidal effect is strongest in the summer, when the North Equatorial Current is weakest.

The circulation patterns outside the barrier reefs change with the variability in flow strength of the oceanic currents, which are strongest in winter. This water is forced through the reef and subsequently forms the almost continuous seaward flow out through the ship channels and reef cut. The oceanic currents are weakest in the summer. Because of lower current velocities at this time, strong convergence fronts behind the atoll were not evident. Although eddies also form during the summertime, it appears that their frequency and intensity is less than during the winter season.

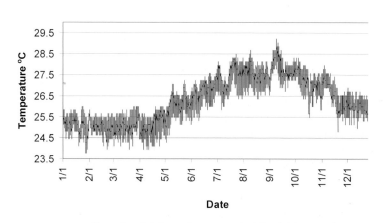

FIG. 17.11. Water temperature recordings (hourly) from the west end of Sand Island during 2000 (logger at 5 m)

FIG. 17.12. Schematic of current flow at Johnston Atoll. Flow is normally from the east to west and fluctuates with the strength of the trade winds and equatorial currents. The hatched area in the oval shows the area of highest species diversity and abundance located on the eastern side of the atoll which is flushed with incoming open ocean water. Ocean eddies form on the leeward side of the atoll and can retain drifting buoys etc., near the reefs (Lobel 1997). Intense water flow with changing tides occurs in the two main channels (shown by small double ended arrows) and through small reef channels (Atoll satellite imagery adapted from a 1985 NASA Space Shuttle photograph)

The current flowing along the southern margin of the platform appears to be the main contributor to eddy formation. This current flows past a sharp topographic projection on the southwest corner of the island platform. Eddy formation is probably associated with a tip vortex formed by fast water flowing past this projection. In addition to eddies, strong current convergence fronts have been observed along the seaward westerly margin of the island platform. Ocean eddies forming behind Johnston Atoll have been

observed by Barkley (1972), Boehlert et al. (1992) and Lobel (1997). These eddies can drift as far as 600 km downstream (Barkley 1972). A conceptual model of the current flow is shown in Fig. 17.12.

17.2.5 Seabirds

There are 13 seabird species that maintain annual breeding colonies on the islands and many more migratory birds that just stop over (Amerson and Shelton 1976). Over 20 years of careful census surveys by E.A. Schreiber and colleagues has shown that the populations have risen steadily (Schreiber et al. 2001; also Schreiber in Lobel et al. 2003). The population increase is due to the protection of the breeding colonies as well as increased space that was made available on the outer islands. Now that the main island is also vacant, seabirds have more room to expand. This raises the question of what will be the ecological consequences of an abundance of seabirds on the Johnston Atoll marine ecosystem. Before man's alterations of the atoll, only two islands of a total of approximately 0.2 km^2 (50 acres) existed, thereby limiting the space for the total number of seabirds. Today there is one 2.6 km^2 (640 acre) island and three more of 0.06–0.1 km^2 (15–25 acres) each. It is reasonable to predict that increased seabird abundances will result in greater feeding pressures on the atoll's fish and cephalopods and also increasing the nutrient loading in the lagoon from the increased guano production. The future ecological consequence of a rapid and unnaturally steep increase in abundance of seabirds on this small reef area has not been evaluated.

17.2.6 Coral Reef Biology

Johnston Atoll has had a long series of intermittent scientific surveys since the 1960s. Most of these surveys were focused efforts and to date there is still little detailed inventory of cryptic invertebrate species and the fauna in the offshore deeper habitats. Johnston Atoll and Hawaii have significant overlap in species distributions of shore fishes and reef corals. The majority of Johnston Atoll fish and invertebrate species are Hawaiian although some Line Islands species also occur at Johnston Atoll but not in Hawaii (Randall et al. 1985; Coles et al. 2001). The coral fauna at Johnston Atoll is essentially a subset of Hawaiian species with the majority of the Johnston species also occurring in Hawaii (Maragos and Jokiel 1986). For such close faunal affinities to occur between Johnston Atoll and Hawaii, ocean currents must exist, at least in some years, which transport pelagic larvae between locations. Johnston Atoll is isolated; however, it lies within the domain of an Archipelago-sized ocean current gyre feature during March–April (illustrated in Wyrtki 1975). Various evidence supports the hypothesis of an oceanographic link between Johnston Atoll and the Hawaiian Islands (Lobel 1989; Kobayashi 2006).

17.2.6.1 Algae

One hundred species of benthic algae have been identified from Johnston Atoll (Amerson and Shelton 1976; Coles et al. 2001). Buggeln and Tsuda (1966) found 12 algal species confined to the marginal reef while 33 species were only found within the lagoon. Of these 33 species, 11 exclusively occurred within the inshore areas of Johnston Island and two within the inshore area of Sand Island. Deep algae surveyed by submersible were reviewed by Agegian (1985). Lagoon dredging did impact algal distribution such that lagoon areas that were silted during the 1963/64 dredging were less diverse (Buggeln and Tsuda 1966; Brock et al. 1966). However, algal growth along the south side of Johnston Island is dense with green sea turtles feeding on abundant *Bryopsis pennata* and *Caulerpa racemosa*. During 2006, the NOAA Coral Reef Ecosystem Division survey teams (NMFS Pacific Islands Fisheries Science Center) reported low abundances of macroalgae with little change in algal communities since the previous survey in 2004 (PIFSC/CRED Cruise report NOAA 2006a). The most common macroalgae found included *Caulerpa*, *Dictyosphaeria*, and *Ventricaria* while turf algae was the most common component of the algal community (NOAA 2006a).

17.2.6.2 Corals

Coral coverage within the lagoon of Johnston Atoll is extensive, ranging from 80% to 100% of available habitat, although the diversity is low (Fig. 17.13). This low diversity results from low recruitment due to the natural geographic isolation of the atoll. Maragos and Jokiel (1986) reported 29 species of scleractinian and three species of hydrozoan corals in the first scuba based coral survey. Subsequent surveys have increased the number of coral species known to occur at Johnston to about 40 with the additional species mostly being uncommon (NOAA 2006a). The coral fauna is depauperate when compared to, for example, the Marshall Islands, with 366 species. Ecologically, Johnston's marine species are more closely related to those found in Hawaii, with 24 of the 29 Johnston species described in Maragos and Jokiel (1986) also occurring in Hawaii. However, some common Hawaiian corals are absent from Johnston and the most abundant coral at Johnston, *Acropora cytherea*, very rarely occurs in Hawaii (Kenyon et al. 2007). Even though strong zoogeographical affinities in coral faunas between Johnston and Hawaii exist, there are significant differences in the ecological relationships and dominant groups (Maragos and Jokiel 1986). At Johnston, the dominant corals in descending order of abundance are *Acropora*, *Montipora*, *Pocillopora* and *Porites*. In Hawaii the dominant corals are *Porites*, *Pocillopora* and *Montipora* (Maragos and Jokiel 1986). The 2006 NMFS Pacific Islands Fisheries Science Center (PIFSC) CRED Rapid Ecological Assessment (REA) surveys (by J. Maragos, B. Vargas Angel and collaborators) conducted at 18 locations in a variety of habitats found montiporid corals, specifically *Montipora capitata* and *M. patula* to be the most abundant (NOAA 2006a). Other important coral taxa from these surveys included *Acropora cytherea*, *A. valida*, *Pocillopora*, *Pavona*, and *Porites* (NOAA 2006a). Coral cover varied by exposure with cover ranging between 2% and 15% in exposed forereef areas compared to 65–80% coral cover found in protected back-reef areas (NOAA 2006a). Differences between the 2004 and 2006 rapid ecological assessment surveys in coral population parameters including declines in mean coral diameter, loss of large corals, and large declines in coral cover call for continued monitoring and study (NOAA 2006a).

The distribution of *Acropora* corals suggests that occasional larval transport occurs from Johnston Atoll to the northwest Hawaiian Islands and Kauai (Grigg 1981; Grigg et al. 1981; Kobayashi 2006; Kenyon et al. 2007). Initially, it was unclear if populations of *Acropora* coral living at French

Fig. 17.13. Typical lagoon reef at Johnston Atoll. Dominant coral is *Acropora cythera*

Frigate Shoals and nearby reefs were sexually reproductive and, had they all been asexual, would have all derived from larvae spawned at Johnston Atoll (Grigg 1981). However, later study found occasional sexual reproduction in these populations (Kenyon 1992). In contrast, recent discovery of *Acropora cytherea* on Kauai suggests Johnston Atoll as the larval source (Kenyon et al. 2007).

Many of the lagoon corals near Johnston Island were destroyed during the dredging between 1942 and 1964. Amerson and Shelton (1976) report that the dredging operations destroyed 4.5 km² (1,100 acres) of coral habitat. The dredging itself covered an area of 2.8 km² (700 acres) while newly deposited coral aggregate affected an additional 1.6 km² (400 acres). Additional impacts from the silt created during dredging reduced the percentage of living coral from 0% to 40% (10% average) in an area covering 28.3 km² (7,000 acres) (Amerson and Shelton 1976). Many areas have recovered to the extent that all available hard substrate in the formerly dredged channels now has extensive coral growth. However, the 2006 NMFS Pacific Islands Fisheries Science Center (PIFSC) CRED rapid ecological assessment (REA) surveys still found signs of sedimentation stress in corals at locations south of Johnston and East (Hikina) Islands (NOAA 2006a)

A major coral bleaching event was documented during the El Niño of 1996. Cohen et al. (1997) found that bleaching was confined to corals in the lagoon and that no bleaching occurred along the emergent reef (with the exception of one bleached colony *(Pocillopora meandrina)* noted on the inside of the eastern reef edge. They also found the bleaching to be species specific with all *Montipora* spp. and *Pocillopora* spp. affected, although to different degrees, while beaching was not observed in *Acropora cytherea*, the dominant coral species on Johnston Atoll.

Incidences of coral disease and altered coral morphology were documented during both CRED/ REA cruises (2004 and 2006). Morphological changes potentially indicative of poor coral health were observed at 11 of 12 sites surveyed during 2004 and at 14 of 18 sites surveyed during 2006 (NOAA 2006a). Data collected by G. Aeby in 2004 included 120 cases of coral disease ranging from bleaching, plague-like signs, tumors, patchy necrosis and ring syndrome (NOAA 2006a). Tissue loss and skeletal growth anomalies were the primary afflictions observed within 97 cases of "diseased" corals observed during 2006 (NOAA 2006a). Additionally, coral predation by *Acanthaster planci* and *Drupella* sp. were observed (NOAA 2006a).

Mass coral spawning usually occurs in late May–June with an occasional second spawning event in July. In some years, the coral spawn produced a very thick surface layer of planulae (Lobel 2003) with a permeating stringent odor that triggered the Army's chemical weapons sensors for mustard gas.

17.2.6.3   Macroinvertebrates

The first reports of invertebrate surveys from Johnston Atoll were from the Tanager expedition of 1923 (Edmondson et al. 1925; Clark 1949). Baseline invertebrate diversity data were collected by Brock et al. 1965 and 1966. A total of 182 species of invertebrates including the orders Annelida (12 spp), Echinodermata (37 spp), Crustacea (75 spp) and Mollusca (58 spp) were reported by Amerson and Shelton (1976). Cryptic invertebrate species have not yet been thoroughly documented. The Coles et al. (2001) survey and literature review increased the number of invertebrate species recorded from Johnston including one species of aschelminth, 20 polychaete species, 135 crustacean species, 221 mollusc species, 69 echinoderm species, 16 sponge species, 4 sipunculid species, 30 bryozoan species and 13 ascidian species.

17.2.6.4   Fishes

The fish fauna of Johnston Atoll has been the subject of surveys dating to the late 1880s (Smith and Swain 1882; Fowler and Ball 1925; Schultz et al. 1953; Halstead and Bunker 1954; Gosline 1955; Brock et al. 1965, 1966; Randall et al. 1985; Kosaki et al. 1991; Chave and Mundy 1994; Chave and Malahoff 1998; Lobel 2003). The known fish fauna of Johnston Atoll currently totals 301 species (Randall et al. 1985; Kosaki et al. 1991). In comparison, about 612 species are found in the Hawaiian Islands, about 844 species in the Mariana Islands and about 1,357 species in the Palau and Yap Island groups (Randall 1992, 2007). The lower number of species on Johnston Atoll is probably the simple result of there being fewer habitat types compared to Hawaii or other archipelagoes.

Johnston's fish fauna is dominated by Hawaiian species. Only 17 non-Hawaiian fish species compared to 53 endemic Hawaiian species occur at Johnston Atoll but not farther south in the Line Islands. Eleven other species are indigenous to Johnston Atoll and the Line Islands but are not found farther north in Hawaii (Randall et al. 1985; Kosaki et al. 1991; Lobel 2003).

Conspicuously absent from Johnston Atoll are any species of "algal farming" damselfish (Pomacentridae: *Stegastes* spp.), the blacktip shark (*Carcharhinus melanopterus*) and any of the goby–shrimp symbiotic species. There is one species (*S. fasciolatus*) of algal farming damselfish in Hawaii (Randall 1996) and two species: *Stegastes albofasciatus* and *S. nigricans,* in the Line Islands (Chave and Eckert 1974). The blacktip shark occurs in Hawaii but is rare (Gosline and Brock 1960). There is no evidence that either of these fishes were ever present at Johnston Atoll. This is an extremely interesting biogeographic anomaly. It is easy to consider that the ecological role of the blacktip shark may be compensated for by an abundance of other roving piscivores including large carangids and other sharks. However, algal farming damselfish influence fundamental reef processes including nitrogen cycling and micro-invertebrate densities (Lobel 1980). How the absence of any *Stegastes* spp. at Johnston Atoll may have (or not) altered reef processes there is still being pondered.

Twenty-five percent (of a total of 612 species) of the Hawaiian shore fishes (occurring to 200 m depth) are taxonomically recognized as endemic species (Randall 2007). Many of these species have planktonic larval phases lasting up to 3 months (Ralston 1981; Lobel 1997) and therefore have the potential for long range dispersal (Kobyashi 2006). One conceptual model proposes that the coincidence of fish spawning patterns in Hawaiian waters to ocean eddy current patterns entrains and retains pelagic larvae near home islands (Lobel 1978, 1989, 1997; Lobel and Robinson 1983, 1986). Limited evidence also suggests that Johnston Atoll may also have some degree of local retention of pelagic larvae and that fish populations on the atoll are mostly derived from the resident populations on the atoll rather than from larvae drifting in from other island groups (Lobel 1997).

Several Johnston Atoll fishes exhibit color patterns suggesting they are sub-species and therefore distinct from their sister species in Hawaii (Randall et al. 1985; Lobel 2003). These Johnston Atoll fish populations that appear distinctly colored from their nearest relatives in Hawaii include a surgeon fish (*Ctenochaetus strigosus*), a parrotfish (*Scarus perspicillatus*), a wrasse (*Labroides phthirophagus*) and two butterflyfishes (*Chaetodon multicinctus* and *C. tinkeri*) (Randall 1955; Randall et al. 1985; Kosaki et al. 1991). Only one fish, the pygmy angelfish, *Centropyge nahackyi*, is recognized as endemic to the atoll (Fig. 17.14; Kosaki

1989; Lobel 2003; Randall 2007). Additionally, another population unique to Johnston Atoll is a hybrid (Fig. 17.15) between two sister species; the Hawaiian endemic wrasse *Thalassoma dupperry* and its nearest relative in the Line Islands and West Pacific, *T. lutescens*.

Recent invasions to the Johnston fish fauna include the Hawaiian sergeant major damselfish (*Abudefduf abdominalis*) as well as the Indo-Pacific sergeant major (*A. vaigiensis*). Both species are recent immigrants as spawning populations were not historically present (Lobel 2003). *A. abdominalis* individuals were observed once at Johnston Atoll in 1986 but they were determined to be several waifs that eventually died off (Irons et al. 1990). A few *A. vaigiensis* were first observed in 1997 and in 1998, but in 1999 *A. vaigiensis* were observed spawning at several locations in the lagoon. By 2001, both species had well-established populations at multiple reef sites throughout the atoll.

17.2.7 Threatened and Endangered Species

The green sea turtle is the only threatened species commonly observed throughout the year. Other threatened or endangered species have been observed at Johnston Atoll, including the humpback whale, Cuvier's beaked whale, and the Hawaiian monk seal.

The green sea turtle, *Chelonia mydas*, has been frequently observed feeding on the south side of Johnston Island and swimming in the lagoon. The first known nest was observed by USFWS in 1996 on the south side of Johnston Atoll near the JACADS facility. The eggs in the nest did not hatch and reason why eggs were not viable is still to be determined (USFWS 1998, personal communication). An estimated 200 turtles use Johnston as a feeding area; this is one of the highest concentrations of green sea turtles in a non-nesting area in the Pacific (D. Forsell, 1990, USFWS, personal communication). Other studies have shown that the Hawaiian and Johnston populations are probably the same (Balazs 1986).

Reef sharks are at risk now that Johnston Atoll is without protection. The grey reef shark, *Carcharhinus amblyrhynchos*, and the reef whitetip shark, *Triaenodon obsesus*, are the most common sharks. An annual aggregation of female grey reef sharks (*Carcharhinus amblyrhynchos*) occurs between late February and May in the shallow waters off of Sand Island (Economakis and Lobel 1998).

17.2.8 Contamination Issues

Johnston Atoll was a very active military base beginning in the 1930s until June 2003. Lobel and

FIG. 17.14. This pygmy angelfish, *Centropyge nahackyi*, is endemic to Johnston Atoll. It lives in a rubble habitat along caves and ledges at depths below 60 ft (20 m)

FIG. 17.15. The wrasse, *Thalassoma dupperry × lutescens* hybrid, an incipient species? The Johnston Atoll population is dominantly composed of hydrids of the Hawaiian endemic, *T. dupperry*, and the widely distributed *T. lutescens* (also photos in Randall 1996; Lobel 2003)

Kerr (2002) reported the results of an atoll-wide survey of lagoon sediments that revealed the presence of a variety of contaminants including metals (antimony, arsenic, barium, chromium, copper, lead, mercury and zinc) and organics [polychlorinated biphenyls (PCBs), polycyclic aromatic hydrocarbons (PAHs), herbicides, dioxins and furans]. These contaminants were mainly concentrated near two particular sites: Sand Island (former US Coast Guard LORAN Station) and the northwest corner of Johnston Island (former "Agent Orange" storage site, burn pit, and fire training area, Fig. 17.16). The herbicides 2,4 D and 2,4,5 T as well as tetrachlorodibenzo-*p*-dioxin (TCDD) were of particular interest due to the storage and subsequent leakage of agent (herbicide) orange while it was being stored on Johnston Island. TCDD and other polychlorinated dibenzodioxins/furans (PCDDs/PCDFs) are extremely toxic contaminants found within the agent orange mixture. Sand Island was targeted for sampling due to the large amount of metal debris and electrical equipment, including PCB filled transformers, that were found in the lagoon. Large concentrations of disposed lead acid batteries used to power the aids to navigation were also found close to Sand Island.

The study was based upon 216 samples at 13 locations throughout the atoll (Lobel and Kerr 2002). In general, remote sites with no known adjacent sources of contamination had low levels of metals and organics. Metals at these remote sites usually included barium, chromium and zinc with an occasional detection of lead or arsenic. Organic contaminants at sites with no known contaminant history included low levels of PCBs, and PCDDs/PCDFs in some samples. Tetrachlorodibenzo-*p*-dioxin (TCDD) was not detected at any of the remote sites.

At sample locations offshore of Sand Island and Northwestern Johnston Island which were adjacent to potential contaminant sources, metal concentrations were low with antimony, arsenic, cadmium, chromium and mercury concentrations all falling below screening levels for marine sediments (Table 17.1). Lead, zinc, and copper rarely exceeded the screening guidelines. However, barium was found in 99% of sediment samples and exceeded the screening guideline in 53 samples (50%). PCBs were detected in most of the samples analyzed, but concentrations exceeded screening guidelines in only eight samples (14.5%) collected from Sand Island. PAHs exceeded the screening guideline for total PAHs in two (2%) samples collected offshore of the northwestern side of Johnston Atoll where the fire training and refuse burning pits were located. Herbicides were only detected in one sample. Dioxins and furans were detected in 80%

FIG. 17.16. The most contaminated reef area of the atoll was restricted close to shore of the northwest part (yellow arrow) of Johnston Island. Onshore was the Herbicide Orange storage, open burn pit for trash, fire training area, and other hazardous materials disposal operations. 1993 photo. Dredged channel is near island, thick reef area in foreground

of the samples although the majority of the detections were at fairly low concentrations. Dioxin/furan concentrations, expressed as toxic equivalents (TEQ), exceeded the screening guideline of 3.6 pg/g in nine (9%) samples collected directly adjacent to the shoreline of the former herbicide orange storage site. The most toxic dioxin isomer, 2,3,7,8 tetrachlorodibenzo-p-dioxin (TCDD) was detected in 28% of the samples.

PCB contamination was also found in sediments and fishes from localized areas around the northern side of Sand Island and within the small lagoon next to the former Navy pier on Johnston Island. In order to provide baseline monitoring criteria and evaluate sediment quality benchmarks used for ecological risk assessments in tropical regions, studies were conducted investigating PCB accumulation and adverse effects in an indicator species. The indicator species chosen was the damselfish, *Abudefduf sordidus*, due to its habit of consuming sediments along with its omnivorous diet, relatively small home range, long life, and demersal spawning (Kerr et al. 1997; Kerr Lobel 2005). Monitored adverse effects included the occurrence of embryonic abnormalities (offspring survival) within PCB contaminated and uncontaminated sites around

the atoll. Increased concentrations of PCBs were detected using immunohistochemical methods in embryos and larvae from contaminated areas (Fig. 17.17; Kerr Lobel and Davis 2002). Mean whole adult body concentrations of total PCBs ranged from 364.6 to 138,032.5 ng/g lipid and a significant residue-effect relationship was found between total PCB concentration and embryo abnormalities. The occurrence of embryo abnormalities was positively related to fish PCB concentration (Kerr 1997; Kerr Lobel 2005).

Overall, the northwest end of Johnston Island was the area of the atoll with the most variety of contaminants in fishes and sediments. This was the site of the island's open burn pit and trash dump, a fire training and explosives detonation area, and the former storage site of Agent (Herbicide) Orange (Fig. 17.16). After the Vietnam War, in 1972, 5.19 million liters (1.37 million gallons) of unused Agent Orange (AO) were transferred to Johnston Island for temporary storage. Due to corrosion of the metal drums, some AO product leaked onto the site. This required an active maintenance and re-drumming operation on site. It was estimated that approximately 49,000 lb (22.3 t) of AO escaped into the environment from 1972 to

TABLE 17.1. Summary of analyses completed on Johnston Atoll marine sediments showing average, minimum and maximum contaminant concentrations as well as the number of samples exceeding screening guidelines across the atoll. Samples were collected from 13 lagoon locations. Average, minimum and maximum concentrations were calculated using only those samples with detectable concentrations (Screening guidelines are from the NOAA SQuirTs website. Updated table from Lobel and Kerr 2002.

Analyte	Total # samples	% Detections	Average conc.	Minimum conc.	Maximum conc.	Screening guideline	# Samples >guidelines
Organics (ng/g)[a]							
PCBs	85	87	39.8	0.4	389.0	22.7	8
PAHs	105	30	589.6	9.2	7,243.0	4022.0	2
2,4-D	87	1	–	6.5	6.5	5	1
2,4,5-T	87	1	–	24.0	24.0	3	0
TEQs (pg/g)	105	80	11.849	0.001	901.286	3.6	9
TCDD (pg/g)	105	28	25.210	0.615	901.000	3.6	9
Metals (μg/g)							
Antimony	100	16	0.7	0.2	1.8	9.3	0
Arsenic	205	47	0.6	0.2	2.3	8.2	0
Barium	105	99	76.8	3.2	294.7	48.0	53
Cadmium	205	0	ND	–	–	1.2	0
Chromium	105	81	8.9	3.5	60.1	81.0	0
Copper	205	21	11.5	1.1	171	34.0	2
Lead	205	24	19.6	1.6	82.6	46.7	1
Mercury	205	8	0.02	0.005	0.078	0.15	0
Zinc	205	82	14.9	1.2	163.3	150.0	1

ND – None detected, NA – Not analyzed

[a]Note that units for TEQ and TCDD are pg/g

1977 at which time the AO was taken offshore and incinerated at sea. The former Agent Orange storage area encompassed approximately 0.02 km² (5.4 acres) of coastal property on the northwest corner of Johnston. The leaked AO product introduced dioxin into the adjacent reef area. It is assumed that the contamination of marine sediments by dioxin compounds was caused by soil transport (wind or rain erosion runoff) into the near-shore marine environment. A cement berm surrounding the former AO storage site was constructed in 1995 and was designed to prevent further contamination by erosion. The dioxin was removed from the entire terrestrial site surface material by soil remediation methods before the island base was closed.

The coral reef area immediately adjacent to the northwest end of Johnston Island (the AO-Burn Pit site) was calculated to be approximately 0.5 km² (116 acres). The total coral reef area of the atoll is approximately 157.8 km² (46.6 nm² or 39,000 acres; excluding landmass area), calculated using the 10 fathom isobath contour. Thus,

the affected area is about 0.3% of the total atoll reef area. Contamination in biota was found only at the sites within 20 m of the AO shore (Lobel and Kerr 1996, 1998). These reefs provide habitat to a diversity of fishes, invertebrates and algae. Upon visual reconnaissance, no readily apparent impacts from the presence of chemical contamination were obvious.

Less well studied is the plutonium contamination within the lagoon that occurred as a result of three failed atmospheric nuclear tests in the 1960s. These failed nuclear rockets tests resulted in radioactive fallout over Johnston Atoll. The Deptartment of Energy subsequently performed extensive cleanup operations with most terrestrial areas cleaned at the upper layer and the contaminated soils buried in the designated plutonium landfill on the north side of Johnston Island. In 1995–1996, key lagoon areas were surveyed for plutonium using an underwater radiation detector. Residual radiation was found mostly off of Sand Island (Johnson et al. 1997).

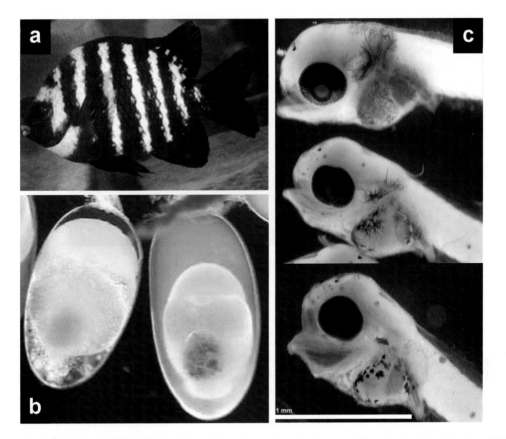

FIG. 17.17. (**a**) Adult male *Abudefduf sordidus* in breeding coloration. Immunohistochemical localization of PCBs in *A. sordidus* embryos and larvae. (**b**) PCB localization in embryos. PCBs are shown in pink found around the lipid rich oil globule of an early stage embryo (**c**) PCB localization in larvae. Top photo shows background staining as in gut; Middle photo shows little staining in larva from reference location with low PCB contamination; bottom photo shows increased staining in larva from PCB contaminated location (see Kerr Lobel and Davis 2002)

17.3   Wake Atoll

17.3.1   Introduction to Wake Atoll

Wake Island (Figs. 17.3, 17.8) is a possession of the United States under the ownership of the Department of the Interior by means of Executive Order No. 11048, Part I (5 September 1962). However, the property is managed by the US Air Force Pacific Command while atoll facilities are administered by the US Army Strategic and Missile Defense Command under a caretaker permit from the US Air Force. It is used as a support facility for testing intermediate range target missiles that are launched from Kwajalein Atoll in the Marshall Islands, located 1,100 km to the south. Current use of the atoll also includes housing a NOAA regional weather forecasting and data collection station and a University of Hawaii (Manoa) seismic monitoring station.

17.3.2   History of Wake Atoll as a US Territory

The first reported European sighting of Wake Atoll occurred on 2 October 1568 by Captain Alvaro de Mendaña. He named it San Francisco although his longitude coordinates were inaccurate. The atoll was renamed for Captain William Wake of the British trading schooner Prince William Henry who documented the atoll in 1796. Reports vary

Fig. 17.18. Wake Atoll viewed looking east. Wilkes Island is foreground to right, Peale Island in the back, Wake Island wraps around and connects to both

as to whether this was the first European landing on Wake Atoll or whether the islands were documented but not claimed. Additionally Wake was "discovered" and named in different positions by various explorers because of the difficulty of accurate navigation during the nineteenth century. Synonyms and misspellings found on charts included Halcyon, Helsion, Wilson, Waker's, Weeks and Wreck. The US Exploring Expedition led by Lt. Wilkes arrived on 19 December 1841 aboard the US warship Vincennes provided the first surveys, maps and detailed descriptions of the atoll. During the Spanish American War, Wake was formally claimed for the US by Major General Wesley Merritt and the US officially took possession of the atoll on 17 January 1899. A second scientific expedition followed in 1923 led by Alexander Wetmore of the Bishop Museum, Honolulu. During this expedition, the small western island was named for Lt. (later Commodore) Wilkes and the third Island for Titian Peale a naturalist and artist that accompanied Lt. Wilkes on the first expedition.

The US Navy was given responsibility for Wake in 1934. A proposed submarine base was under construction including a submarine channel through Wilkes Island that was never completed due to the outbreak of WWII. Wake was established as a transpacific refueling base by Pan American Airlines in 1935. The Japanese attacked Wake several hours after the attack on Pearl Harbor

and the atoll was captured on 23 December 1941 to remain in Japanese possession until 1945. Many of the captured civilians and military personnel were sent to prison camps, but the 98 civilians kept at Wake to construct Japanese fortifications were later executed by order of the Japanese commander. After WWII, the Federal Aviation Administration and later the US Air Force resumed use of the atoll as a transpacific refueling site.

Wake Atoll was designated a National Historic Landmark in 1985 (US Department of Interior 1984) to preserve the battlefield and Japanese and American structures where important WWII events occurred. Historic structures include several pillboxes, bunkers, aircraft revetments as well as the Pan American facilities and the US Naval submarine and aircraft base. The prisoner of war "rock" (actually a large coral head) inscribed with a US POW number remains a poignant reminder of WWII events (Fig. 17.19).

17.3.3   Geomorphology of Wake Atoll

The atoll is approximately 3 km wide by 6.5 km long and the three islands have a land area of approximately 6.5 km². The three individual islands, Wilkes, Wake and Peale have land areas of 0.8 km² (206 acres), 5.5 km² (1,350 acres) and 1.1 km² (270 acres) respectively (Fig. 17.19). The Islands are essentially flat with average elevations

FIG. 17.19. "POW rock". Coral on beach with the carved number of a US POW on Wake Island, WWII

of 3.7 m (12 ft). The maximum elevation of Wake and Peale islands is 6.4 m (21 ft) while Wilkes has a maximum elevation of 5.5 m (18 ft). The highest features are coral and shell ridges parallel to the shore found 3–15.2 m (10–50 ft) from the sand beaches. From these ridges, the terrain slopes gently to the lagoon. The area of the enclosed lagoon (water and sand flat) is approximately 9.7 km^2 (3.75 miles^2; Bryan 1959), while the lagoon itself is less than 7.8 km^2 (3 miles^2).

The atoll has an emergent reef on the western boundary (Fig. 17.3) over which most water exchange between the lagoon and open ocean takes place. Additional water flows during high tide through the channel between Peale and Wake Islands (Fig. 17.20) and the old submarine channel in the middle of Wilkes Island (Fig. 17.21). The small boat harbor was previously open to the lagoon but it was sealed off by construction of a seawall (Fig. 17.22). Water circulation in the lagoon was severely reduced when the causeway between Wake and Wilkes islands was built, resulting in the closure of the small boat harbor inlet into the lagoon. There have been reports of large-scale fish die-offs in the lagoon due to high temperatures when tidal flushing is low, resulting in low levels of dissolved oxygen.

Island soils consist of highly permeable sand and coral rubble covered in vegetated areas by a thin layer of organic material. The geological profile of the islands consists of alternating and often mixed layers of sand, shells, coral and soft limestone (US Army Space and Strategic Defense Command 1992). There is limited brackish groundwater due to the small land mass, flat topography and porous soils (US Army Space and Strategic Defense Command 1992).

17.3.4 Climate of Wake Atoll

Annual climate is uniform with average daily low and high temperatures of 24°C (75°F) and 30°C (85°F). Humidity averages 76% (NOAA 1992). The prevailing wind direction is east–northeast with a mean speed of 22.2 km/h (NOAA 1992, US Army Space and Strategic Defense Command 1992). Annual precipitation averages 89 cm (35 in.; NOAA 1992). Typhoons (tropical cyclones) threaten Wake from late summer to autumn. Severe flooding and wind damage occurred during the typhoons of 1957, 1967, 1978, 1986, 1992, and 2006. Typhoon Ioke of 2006 was one of the few storms both categorized as a hurricane in the eastern Pacific (also impacting Johnston Atoll) and a typhoon in the western pacific

FIG. 17.20. The channel between Wake Island and Peale Island (top). The main base complex is located in this area. Photo was taken at low tide and shows broad reef flats. At high tide water flows through the channel. The channel to lagoon is excellent bone fishing grounds

FIG. 17.21. The never-completed submarine channel started at the beginning of WWII, through the middle of Wilkes Island. Peale Island in background across the lagoon

("hurricane" and "typhoon" are regional names for tropical cyclones with winds of at least 33 m/s). It reached category 5 but passed close to Wake Atoll as a category 4 storm. Damage suffered by island infrastructure was so severe that the Department of Defense decided to place the atoll in caretaker status and reduced the workforce from 120 personnel to approximately a dozen as of February 2007.

Fig. 17.22. The small boat harbor was originally open to the lagoon but was sealed by construction of a seawall. The small boat channel that was once used is evident but was closed off when the causeway connecting Wake Island (left) to Wilkes Island (right, with the three fuel storage tanks) was constructed. The overall best diving was in the immediate area on the outside reefs by this harbor. The wreck of an oil tanker is in shallow water by the harbor entrance and is colonized by corals

17.3.5 Oceanography

There is very little data on the ocean current patterns around Wake Atoll. The first Reef Assessment cruise (PIFSC Cruise Report CR-06-024) to Wake deployed sea surface and subsurface temperature recorders and collected shallow and deep water oceanographic data around the atoll (NOAA 2006b). Initial data from conductivity-temperature-depth (CTD) profiles around the atoll show that surface waters (<50 m) in the forereef and offshore areas are well mixed with little stratification (NOAA 2006b). However, slightly lower temperatures (0.1–0.2 °C) and higher salinities (0.05 PSU) suggest very weak upwelling along the north and eastern margins of the atoll (NOAA 2006b). Interestingly, CTD data from within the lagoon (surface only due to shallow depths) varied by location within the lagoon and relative to forereef measurements indicating little exchange between oceanic and lagoon waters (NOAA 2006b). An intermittent ocean current flow connecting Hawaii and Wake Atoll seems to be caused by ocean eddies formed off of the island of Hawaii which flow westward eventually impinging on Wake (Mitchum 1995). Other evidence of current flow between these two locations includes reports of Hawaiian fish aggregation devices (FADS) recovered from Wake (Lobel and Lobel 2004).

17.3.6 Terrestrial Vegetation on Wake Atoll

Prior to human occupation, Wake Atoll was vegetated by only about 20 indigenous plant species and none were rare, unique or endemic to Wake (Bryan 1959). The relatively low diversity of terrestrial plant species on Wake has been correlated to the low overall annual rainfall in comparison to other Pacific Atolls (Wiens 1962).

17.3.7 Seabirds on Wake Atoll

There are 26 species of migratory seabirds listed for Wake Atoll (http://www.bsc-eoc.org/avibase/). Two of these birds are globally threatened species: the Laysan Albatross (*Phoebastria immutabilis*) and the black-footed Albatross (*Phoebastria nigripes*). Seabird nesting occurs mainly on Wilkes Island.

17.3.8 Coral Reef Biology on Wake Atoll

17.3.8.1 Marine Habitats

Four basic aquatic habitat types occur within Wake Atoll. A shallow and turbid lagoon with scattered patch reefs (Fig. 17.23) is land bound on one side and a large proportion of the lagoon habitat is occupied by sand flats that are fully exposed at lowest tides (Fig. 17.24). The lagoon is murky and shallow averaging 3 m (10 ft) depth with maximum depths reported of up to 4.5 m (15 ft) in 1923 (Bryan 1959). In all areas of the lagoon surveyed, the bottom was sand mixed with occasional corals and *Tridacna* clams. The lagoon was very turbid with water clarity scarcely a few feet. In sand areas that were regularly exposed at low tides, the bottom was sand with a limited variety of snails, crabs and other sand-burrowing invertebrates. The intertidal is a hard substrate, ocean reef flat, which is exposed at low tide (Figs. 17.25 and 17.26). The reef crest top is also exposed at very low tides (Figs. 17.27 and 17.28). The outer reef drops rapidly in depth on both the exposed windward and sheltered leeward sides (Figs. 17.27 and 17.29).

The ocean reefs had spectacular water clarity with visibility in excess of 30 m (100 ft). The bottom varied with mostly boulder type coral colonies growing over eroded older coral formations (Fig. 17.30). Reefs were in excellent ecological condition with no obvious signs of environmental problems when surveyed in 1997 and 1999 (Lobel and Lobel 2004).

17.3.8.2 Algae

A total of 40 species of benthic algae have been documented from Wake from a limited field collection (Tsuda et al. 2006). The algal species were common to both Hawaii and Micronesia, with a slightly closer affinity to Hawaii (Tsuda et al. 2006). Only 3 of the 40 species found at Wake have not been recorded from Hawaii. Sea grass beds, algal flats and mangroves habitats were not found at Wake Atoll during surveys in 1997–1999 (Lobel and Lobel 2004). The most common macroalgae observed in photoquadrats during the most recent reef surveys (PIFSC Cruise Report CR-06-024) were *Halimeda* sp., *Dictyota* sp., *Caulerpa* sp., and *Lobophora variegata* (NOAA 2006b). Common algae found within the lagoon included blue-green algal mats and *Caulerpa serrulata* (NOAA 2006b).

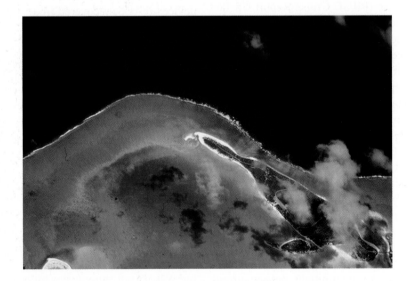

Fɪɢ. 17.23. The west margin of the atoll is a vast area of shallow reef that drops off rapidly. Inside the lagoon, scattered small patch reefs are spread over an even sand bottom. Peale Island is to the right

FIG. 17.24. This is the shallowest portion of the lagoon also shown in Fig. 17.3. This area is exposed during daily low tide. The dominant substrate type is sand mixed with many dead mollusks. It is unknown at this time, whether the mass amounts of dead mollusks is the result of long time natural accumulations or high mortality due to adverse ecological impacts. The lagoon is next to the airfield runway and air traffic and refueling operations may very possibly have added pollutants. The potential effects remain to be determined

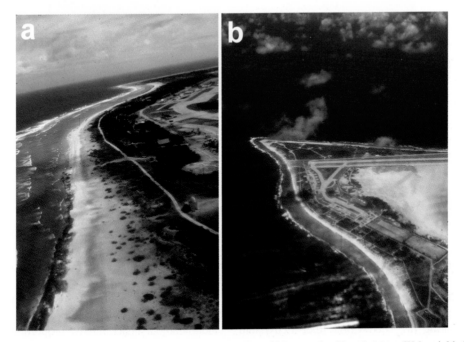

FIG. 17.25. Long sand beaches extend along most of the islands. This stretch of beach (a) on Wake yielded several washed up glass balls almost daily. A different view of the same coast (b) showing the broad reef flat, as well as the extent of development on the island

Fig. 17.26. The south seaward shoreline of Wake and Wilkes Islands are mostly exposed coralline platform with tidepools

Fig. 17.27. Wilkes Island in foreground was cleared of vegetation as part of a cleanup effort and to provide habitat for nesting seabirds. The shallow lagoon reef habitat on the western margin of the atoll is generally less turbid than the shallow lagoon nearest the airfield (back). The western reef area has abundant fishes and live coral with large groups of bumphead parrotfish and the Maori wrasse

17.3.8.3 Corals

Initial coral identifications from Wake Atoll generated a list of 50 scleractian corals (NOAA 2006b citing Maragos 1979, unpublished data; US Department of Interior, Fish and Wildlife Service 1999). During the 2005 Coral Reef Ecosystem Division cruise (OES-05-13; www.soest.Hawaii.edu/pibhmc/website/cruise.catalog.htm) the current number of scleractian corals reported from Wake Atoll increased to 102 (NOAA 2006b). Additionally two species of *Millepora*, a zooanthid and five genera of octocorals are reported from Wake (NOAA 2006b)

During the 2005 Coral Reef Ecosystem Division cruise, extensive data on percentage coral cover, cnidarian biodiversity, relative proportion live coral cover to macro/corraline algae cover, and size class distribution were collected. Within forereef sites, the dominant corals included *Favia, Montipora*

FIG. 17.28. The atoll is mostly landlocked on three sides with an emergent reef on the western boundary (Fig. 17.2). It is over this western reef that most water exchange between the lagoon and open ocean takes place although additional water flows during high tide through the channel between Peale and Wake Islands

FIG. 17.29. The leeward south side of Wake Island has a very narrow reef margin that drops rapidly

and *Pocillopora* (NOAA 2006b). *Montipora* dominated the single lagoon transect (NOAA 2006b). Percent live coral cover varied with exposure with the lowest coral cover found along southern and southwestern exposures (22.5–27.5%), higher cover along the western and eastern exposures (35.3–63.7%), and the highest coral cover found along northerly exposures (76.7–81.4%) (NOAA 2006b). Coral cover within the lagoon was 14.7% (NOAA 2006b). Biodiversity patterns reflected those of coral cover with the highest cnidarian biodiversity found along the northwest and northern reef margins (NOAA 2006b). Size class measurements found that 39.8% of lagoon coral colonies and 77.3% of forereef colonies were >20 cm maximum diameter (NOAA 2006b).

Fig. 17.30. Wake Atoll typical leeward reef

17.3.8.4 Macroinvertebrates

Preliminary results from the 2005 CRED cruise (PIFSC Cruise Report CR-06-024) found at least 92 species of macroinvertebrates at Wake Atoll (NOAA 2006b). These included 5 echinoids, 8 holothurian species, 4 ophiuroid species, 3 asteroid species, 32 crustacean species, and 36 mollusc species (NOAA 2006b). *Tridacna* were particularly abundant in shallow reef areas and within the lagoon. *Holothuria edulis* was also abundant in shallow soft-bottom reef areas but surprisingly not within the primarily shallow soft-bottom lagoon habitat (NOAA 2006b).

17.3.8.5 Fishes

A total of 323 fish species in 63 families were documented at Wake Atoll (Lobel and Lobel 2004). This species richness is similar to that found on other isolated Central Pacific atolls and islands including Johnston Atoll (N = 301; Randall et al. 1985; Kosaki et al. 1991), Rapa (N = 268; Randall et al. 1990) and Midway (N = 258; Randall et al. 1993). The low number of species at these locations is correlated to their small area, limited habitat diversity, geographic isolation and relatively large distance separating

them from other population sources (Randall 1992, 1998). Fish species that are found abundantly at Wake Atoll are typical of those on other Pacific Islands and include coastal shore fishes (299 spp.), elasmobranchs (6 spp.), and offshore pelagic fishes (18 spp.).

The central location of Wake in the Northern Pacific results in a fauna with a mixture of different zoogeographic affinities between east and west. Wake's fish fauna has the most commonality with the Marianas Islands and some similarity to Hawaii. Comparison with the four nearest neighboring archipelagos indicates that the greatest species overlap occurs with the Southern Marianas Islands (85%) and the Marshall Islands (81%). The Northern Marianas have 65% of species in common with Wake.

The wrasse *Ammolabrus dicrus* has only been found at Oahu, Hawaii and Wake (a similar or sister species was observed but not collected in the Ogasawara Islands, Japan (Randall and Carlson 1997). The spotted knifejaw, *Oplegnathus punctatus* is found in Hawaii, Japan, Johnston and Wake. The Hawaiian endemic *Sebastipistes ballieui* is also found at Wake. Two additional Hawaiian endemics (*Entomacrodus marmoratus* and *Eviota epiphanes*) were listed in the

Bernice P. Bishop Museum collection as being from Wake but the specimens could not be verified and none were found in later field collections (Lobel and Lobel 2004).

Absent from Wake are the trumpetfish *Aulostomus chinensis* (Aulostomidae) and the blacktip reef shark *Carcharhinus melanopterus* (Lobel and Lobel 2004). The blacktip reef shark was reported to have occurred at Wake in earlier literature. Our surveys, which were conducted mainly along the southern reefs found the grey reef shark, *Carcharhinus amblyrhynchos* to be present but rare. The most recent reef survey (PIFSC Cruise Report CR-06-024) found only 11 grey reef sharks at the 13 sites (belt transects) surveyed around the entire atoll (NOAA 2006b). However, the report states that during the towed diver surveys, grey reef sharks were observed both singly and in aggregations, and were in the top three most abundant species observed (NOAA 2006b). Reasons for the low abundance of grey reef sharks along the southern margin (and perhaps the absence of blacktip reef sharks) are not clearly understood. Lobel and Lobel (personal observation) noted that the island civilian work force frequently fished for sharks and dried their fins, a practice left uncontrolled by the DoD. Since the calmest sea conditions and small boat harbor are found along the southern reef margin, the greatest fishing pressure would occur in this area.

17.3.9 Contamination Issues on Wake Atoll

The shallowest portion of the lagoon is located next to the airfield and main military facilities on the eastern portion of the atoll. A large portion of this area is exposed daily during low tide. The dominant substrate type is sand mixed with dead seashells. It is unknown at this time, whether the mass amounts of dead mollusks is the result of long time, natural accumulations or the result of pollution in the lagoon. The lagoon is next to the airfield runway and long term air traffic and refueling operations may possibly have added pollutants (e.g. arsenic, PAHs and various solvents). Other sources of pollutants to the lagoon include the long term application of herbicides and pesticides to the runway area for vegetation and insect control. Pathways into the lagoon and nearshore habitats for these potential contaminants include runoff and percolation through the porous coral rubble of which much of the island is made up. The issue of contamination in the groundwater, lagoon and nearshore marine environment remains to be determined.

The oil tanker Stoner ran aground and broke up o the reef offshore from the small boat harbor in 1970. The oil spill resulted in significant reef damage along the ocean reef of Wake Island (Gooding 1971). No lingering effects of the spill are apparent (P.S. Lobel and L. K. Lobel 1997, 1999 pers. observ.).

Assessment of the contaminant situation is currently under study by the USAF. In February/March 2002, the Air Force conducted fish tissue sampling from Wake Island lagoon as part of a human health risk screening evaluation. Only lagoon areas were sampled and preliminary data from fish caught within Wake Island Lagoon indicated arsenic levels exceed screening values in fish tissue. The Wake Island Commander issued (5 February 2003) an advisory notice recommending that people not eat seafood caught in the lagoon.

17.4 Conservation Issues for Wake and Johnston Atolls

The protection of Wake Atoll's natural resources is under stewardship of the US Air Force. This is achieved primarily through base regulations that exclude commercial fishing ventures and ban reef fish spearfishing. Regulations can change and enforcement differs under each subsequent base commander. It is unclear what is currently being implemented for fish and wildlife management on the atoll. One reef fish of particular concern is the Maori wrasse (*Cheilinus undulatus*) which is listed on the IUCN redlist (www.iucnredlist.org) as endangered and in October of 2004 was listed under CITES Appendix II (http://www.cites.org). The Maori wrasse is a prized food fish and juveniles are targeted for the live reef fish food trade (LRFFT) (Donaldson and Sadovy 2001). This species is abundant at Wake and previous commanders had prohibited it being fished. However, fishing within the lagoon and on the reefs does occur by the residents and visitors of Wake. Specifically, the local workers (recruited from Thailand) reportedly fish for sharks as well as other reef species, presumably for their own consumption, but the extent to which this occurred and whether this has negatively impacted populations was not assessable.

In addition to the Maori wrasse, several teleost fishes and elasmobranchs on Johnston and Wake's reefs are threatened by over-exploitation elsewhere in the Pacific (Table 17.2). The most recent survey of Wake's reefs (PIFSC cruise 2005 and 2007; NOAA 2006b, 2007) found an abundance of Maori wrasse (*Cheilinus undulatus*) in addition to the large population of bumphead parrotfish (*Bolbometopon muricatum*) also observed during earlier surveys. Bumphead parrotfish sightings at other US Pacific reefs surveyed by CRED are rare (NOAA 2006b). Additionally, spotted eagle rays (*Aetobatis narinari*) were abundant at Wake (NOAA 2006b). Large biannual aggregations of female grey reef sharks (up to 160) were documented at Johnston Atoll (Economakis and Lobel 1998) and the Johnston Atoll pygmy angelfish, *Centropyge nahackyi*, warrants attention due to its being found exclusively at this atoll. The large numbers of these species that are often rare at other Pacific locations emphasizes the need for continued supervision and protection of these coral reef ecosystems.

Johnston Atoll is still a US military installation owned by the US Air Force Pacific Command (15th Air Wing, Hickam Air Force Base). The eventual disposition of the atoll is yet to be determined. One option was to transfer ownership to the US Fish and Wildlife Service. However, before USFWS was to accept the atoll, it required DoD to completely remove any traces of contamination from the islands and atoll lagoon including residual plutonium. Newspaper articles often questioned the future of Johnston Atoll as a USFWS Refuge because of the plutonium landfill on Johnston Atoll and plutonium contamination in the lagoon (*New York Times* 27 January 2003, *Honolulu Star Bulletin* 28 January 2003). A quote from a spokesperson for the Defense Threat Reduction Agency stated that they will continue to be responsible for the costs associated with the plutonium issue. One cost estimate places periodic seawall maintenance at approximately $1,000,000/100 ft to repair. With approximately 5 miles of seawall at risk of degradation at Johnston Atoll, the potential cost of maintenance and repair is $264 million. While all sources of contamination on the islands were cleaned up and the plutonium laced soil packed into a consolidated landfill, cleanup of the lagoon was not feasible and not warranted according to the ecological risk assessment. The cost for this cleanup would have ranged in the many millions of dollars and even then the remedial dredging of reef areas would likely have done more harm than good. However, if the island's seawalls were to deteriorate and allow the landfills to erode into the lagoon, sediment, turbidity and increased contaminant loading can be expected to have measurable impact on the adjacent coral reef communities.

A major issue in 2007 is that Johnston Atoll is abandoned (no people or facilities remaining) and without any effective measures in place to ensure that the atoll fishes and other natural resources are not illegally exploited. There is a concern that illegal shark fishing may have already occurred. Reef fish that are potentially tainted with PCBs or other contaminants could also be illegally fished and sold commercially.

17.5   Future of the Atolls

The future of Johnston and Wake Atolls is uncertain. At Johnston, a sea level increase of only a few inches will increase the frequency of the seawalls being breached by waves during storms and increase erosion. Land restrained by the seawall sequesters areas of landfill containing high concentrations of plutonium, lead paint, dioxins and other contaminants. It is unlikely that these landfills would represent serious ecological threats as long as seawall containment continues. But with the abandonment of

TABLE 17.2. Pacific fishes of special concern for conservation found at Wake and Johnston Atolls (Updated table from Lobel and Lobel 2004).

Species	Common name	IUCN category	Wake	Johnston
Manta birostris	Manta ray	NT	X	X
Aetobatus narinari	Spotted eagle ray	NT	X	X
Carcharhinus amblyrhynchos	Grey reef shark	LR/nt	X	X
Carcharhinus melanopterus	Blacktip reef shark	LR/nt	X	
Carcharhinus galapagensis	Galapagos shark	NT	X	
Galeocerdo cuvier	Tiger shark	LR/nt	X	
Triaenodon obesus	Whitetip reef shark	LR/nt	X	X
Cheilinus undulatus	Humphead/Maori Wrasse	EN CITES Appendix II	X	
Bolbometopon muricatum	Giant bumphead parrotfish		X	
Epinephalus lanceolatus	Giant grouper	VU	X	X
Epinephalus polyphekadion	Camouflage grouper	NT	X	
Epinephelus quernus	Hawaiian grouper	NT		X
Centropyge nahackyi	Rainbow pygmy angelfish			X

the atoll, a program for long-term maintenance no longer exists. If there were a catastrophic collapse of island seawalls and mass erosion of sediment into the lagoon, turbidity and silt on corals would be an adverse impact in itself, at least during initial stages.

The current state of affairs at Johnston is also a serious concern because it is without any on-site management and trespassing or illegal fishing thus cannot be prevented. The USAF has made several attempts to transfer the property to the US Fish and Wildlife Service or any other entity. The key issue for the Air Force is that they no longer have a military mission for the atoll property. However, no other federal agency is willing to inherit the financial burden from the legacy of contaminants and future seawall maintenance costs. Thus, what may eventually become of Johnston Atoll is uncertain. If no other federal agency is willing to step up and take a lead in its future, the Air Force will seek other uses.

Proposals for island uses have ranged from using the island as a rocket launch facility for shooting satellites into orbit, to ecotourist resort, to use as a hazardous waste storage site again. In this context it is worthwhile to review one of the repeated proposals for use of these remote pacific islands. There has been longtime historical consideration for the storage of spent nuclear fuel on these islands. Fosberg (1993) reviewed the advantages and planning considerations for placing such a facility at Wake and Johnston Atoll as well as several other atolls that are no longer possible considerations. In late 1978, the US Department of Energy circulated a draft Environmental Impact Statement entitled "Storage of Foreign Spent Power Reactor Fuel" which detailed the various options under which the United States would be prepared to accept a limited amount of spent fuel from foreign sources when such actions would contribute to non-proliferation (Department of Energy 1978. Environmental Impact Statement. Storage of Foreign Spent Power Reactor Fuel. DOE/EIS-0040-D). In 1979, the US Army Corps of Engineers, directed by the Department of Energy, conducted a preliminary environmental assessment of the construction and operation of a temporary storage facility for foreign nuclear spent fuel at Wake Island. The study was in response to the passage of the Non-Proliferation Act of 1978 which was developed to identify alternatives to large scale commercial reprocessing of spent power reactor fuel, particularly providing foreign nations with viable alternatives to commercial reprocessing. One specific action to implement the nonproliferation policy involved the offer by the US to establish an internationally managed temporary storage facility for nuclear spent fuel to service countries in the Western Pacific. The concept was for the USA to provide a temporary centralized storage facility for spent nuclear fuel at an isolated island in the Pacific. Given the world situation today, this may still be a political option. In the meanwhile, Midway was transferred from the Navy to DOI and became a National Wildlife Refuge and ecotourism destination with endangered monk seals. Palmyra was purchased by the US Fish and Wildlife Service in partnership with the Nature Conservancy and is also a wildlife refuge. Wake Atoll and Johnston Atoll remain as the last available remote atolls belonging to the USA. Let us hope that whichever agency or organization has stewardship responsibility for these atolls will accept coral reef and marine wildlife conservation as part of their agenda.

Acknowledgements Our studies of Johnston and Wake Atolls were sponsored by the Department of Defense Legacy Resource Management Program and the Departments of Army and Air Force as part of the program to implement coral reef conservation and protection. Funding was awarded by the Army Research Office (DAAG55-98-1-0304, DAAD19-02-1-0218), the Office of Naval Research (N00014-19-J1519, N00014-92-J-1969, N00014-95-1-1324), NOAA (DACA83-83-0051, NA90-AA-D-SG535) and we especially acknowledge the Department of Defense Legacy Resource Program for continued support of coral reef stewardship and research (DoD Legacy Resource Management Program DACA87-97-H-0006, DACA87-00-H-0021, DAMD17-93-J-3052). Photographs by Phillip and Lisa Lobel except Fig. 17.5 (a) which was taken by the late K.O. Emery in 1952 and given to the authors along with photos of the early days of Johnston Atoll. K.O. shared with us many fascinating discussions of the geology and history of the early days on Pacific atolls, for which we remember him fondly. We owe our many friends, supporters and colleagues at DoD as well as our scientific partners huge thanks for making our research in the Pacific Islands possible. While we cannot list everyone, the following provided us with essential help in our research at Johnston and Wake Atoll over the many years for which we are very grateful: Deetsie Chave, Mark Ingoglia, Jeff Klein, James Maragos, Gary McCloskey, John and Helen Randall, Lorri Schwatz, and David Shogren.

References

Agegian CR (1985) Deep water macroalgal communities: a comparison between Penguin Bank, Hawaii and Jonhston Atoll. Proc 5th Int Coral Reef Congr. Tahiti 5:47–50

Amerson AB, Shelton PC (1976) The natural history of Johnston Atoll, Central Pacific Ocean. Atoll Res Bull 192:1–479

Anderson DM, Lobel PS (1987) The continuing enigma of ciguatera. Biol Bull 172:89–107

Balazs GW (1986) Status and ecology of marine turtles at Johnston Atoll. Atoll Res Bull 285:1–46

Barkley RA (1972) Johnston Atoll's wake. J Mar Res 3:201–216

Boehlert GW, Watson W, Sun LC (1992) Horizontal and vertical distributions of larval fishes around an isolated oceanic island in the tropical Pacific. Deep Sea Res 39:439–466

Brock V, Jones RS, Helfrich P (1965) An ecological reconnaissance on Johnston Island and the effects of dredging. Tech Rpt 5, University of Hawaii, Honolulu, HI, 90 pp

Brock V, Van Heukelem W, Helfrich P (1966) An ecological reconnaissance on Johnston Island and the effects of dredging. Tech Rpt 11, University of Hawaii, Honolulu, HI, 56 pp

Bryan EH (1959) Notes on the geography and natural history of Wake Island. Atoll Res Bull 66:1–22

Buggeln RG, Tsuda RT (1966) A preliminary marine algal flora from selected habitats on Johnston Atoll. Hawaii Inst Marine Biol Tech Rept 9:1–29

Chave EH, Eckert DB (1974) Ecological aspects of the distribution of fishes at Fanning Island. Pac Sci 28:297–317

Chave EH, Malahoff A (1998) In deeper waters. Photographic studies of Hawaiian deepsea habitats and life-forms. University of Hawaii Press, Honolulu, HI, 125 pp

Chave EH, Mundy BC (1994) Deep-sea benthic fish of the Hawaiian Archipelago, Cross Seamount and Johnston Atoll. Pac Sci 48:36–404

Clark AH (1949) Ophiuroidea of the Hawaiian Islands. Bishop Mus Bull 195:1–133

Cohen A, Lobel PS, Tomasky GL (1997) An unusual event of coral bleaching on Johnston Atoll, Central Pacific Ocean. Biol Bull 193:276–279

Coles SL, DeFelice RC, Minton D (2001) Marine species survey of Johnston Atoll, Central Pacific Ocean, June 2000. Bishop Museum Tech Rept 19, 59 pp

Donaldson TJ, Sadovy Y (2001) Threatened fishes of the world: *Cheilinus undulatus* Ruppell 1835 (Labridae). Environ Biol Fishes 62:428

Economakis AE, Lobel PS (1998) Aggregations of grey reef sharks, *Carcharhinus amblyrhynchos*, at

Johnston Atoll,Central Pacific Ocean. Environ Biol Fish 51(2):129–139

Edmondson CH, Fisher WK, Clark HL, Treadwell AL, Cushman JA (1925) Marine zoology of the tropical central Pacific. Tanager Expedition. Bishop Mus Bull 27:1–148

Emery KO (1956) Marine geology of Johnston Island and its surrounding shallows, Central Pacific Ocean. Geol Soc Amer Bull 97:1505–1520

Fosberg CW (1993) An ocean island geological repository – a second generation option for disposal of spent and high-level waste. Nuclear Techn 101:40–53

Fowler HW SC Ball (1925) Fishes of Hawaii, Johnston Island, and Wake Island. Bull BP Bishop Mus 26:1–31

Gooding RM (1971) Oil pollution on Wake Island from the Tanker RC Stoner. Spec Scient Rep US fish Wildl Serv 636:1–10

Gosline WA (1955) The inshore fish fauna of Johnston Island, a central Pacific atoll. Pac Sci 9:442–480.

Gosline WA, Brock VE (1960) Handbook of Hawaiian fishes. University of Hawaii Press, Honolulu, HI, 372 pp

Grigg RW (1981) Acropora in Hawaii. Part 2. Zoogeography. Pac Sci 35:15–24

Grigg, RW, Wells JW, Wallis C (1981) Acropora in Hawaii. Part 1. History of the scientific record, systematics and ecology. Pac Sci 35(1):1–14

Halstead BW, Bunker NC (1954) A survey of the poisonous fishes of Johnston Island. Zoologica 39:67–81

Irons D, Kosaki R, Parrish J (1990) Johnston Atoll Resource Survey Final Report, phase six (21 Jul 1989–20 Jul 1990). Tech Rep Army Corps Engineers, Honolulu, HI, 147 pp

Johnson O, Osborne D, Wildgoose C, Worstell W, Lobel P (1997) Development and testing of a large-area underwater survey device. IEEE Trans Nuclear Sci 44(3):792–798

Jokiel PL, Tyler WA (1992) Distribution of stony corals in Johnston Atoll lagoon. Proc 7th Int Coral Reef Symp 2:683–692

Jones RS (1968) Ecological relationships in Hawaiian and Johnston Island Acanthuridae surgeonfishes). Micronesica 4:309–361

Keating B (1985) Submersible observations on the flanks of Johnston Island (Central Pacific Ocean): Proc 5th Int Coral Reef Symp 6:413–418

Keating B (1992) Insular geology of the Line Islands. In: Keating B, Bolton BR (eds) Geology and offshore mineral resources of the central Pacific basin, Circum-Pacific Council for Energy and Mineral Resources Earth Science Series 14, Springer, New York, 77–99

Kenyon, JC (1992) Sexual reproduction in Hawaiian Acropora. Coral Reefs 11:37–43

Kenyon JC, Godwin S, Montgomery A, Brainard R (2007) Rare sighting of Acopora cytherea in the main Hawaiian Islands. Coral Reefs 26:309

Kerr LM (1997) Embryonic abnormalities and reproductive output for the damselfish Abudefduf sordidus (Pomacentridae) relative to environmental contamination at Johnston Atoll, Central Pacific Ocean. Masters thesis, Boston University, Boston, MA, 144 pp

Kerr LM, Lang K, Lobel PS (1997) PCB contamination relative to age for a Pacific damselfish, Abudefduf sordidus (Pomacentridae). Biol Bull 193:279–281

Kerr Lobel L (2005) Evaluating developmental and reproductive effects of chemical exposure in the coral reef fish, Abudefduf sordidus (Pomacentridae). Ph. D. thesis. University of Massachusetts, Boston, MA

Kerr Lobel LM, Davis EA (2002) Immunohistochemical detection of polychlorinated biphenyls in field collected damselfish (Abudefduf sordidus; Pomacentridae) embryos and larvae. Environ Poll 120:529–532

Kobayashi DR (2006) Colonization of the Hawaiian Archipelago via Johnston Atoll: a characterization of oceanographic transport corridors for pelagic larvae using computer simulation. Coral Reefs 25:407–417

Kopenski RP, Wennekins MP (1966) Circulation patterns Johnston Island, US Naval Oceanographic Office (N00 SP-93) Washington, DC

Kosaki RK (1989) Centropyge nahackyi, a new species of angelfish from Johnston Atoll (Teleostei: Pomacanthidae). Copeia 4:880–886

Kosaki RK, Pyle RL, Randall JE, Irons DK (1991) New records of fishes from Johnston Atoll with notes on biogeography. Pac Sci 45:186–203

Lobel PS (1978) Diel, lunar and seasonal periodicity in the reproductive behavior of the Pomacanthid fish, Centropyge potteri and some other reef fishes in Hawaii. Pac Sci 32:193–207

Lobel PS (1980) Herbivory by damselfishes and their role in coral reef community ecology. Bull Mar Sci 30:273–289

Lobel PS (1989) Ocean current variability and the spawning season of Hawaiian Reef fishes. Environ Biol Fish 24:161–171

Lobel PS (1988) Ocean Current and plume study, Johnston Atoll: Rpt. to U.S. Army Engineers, Honolulu, HI, 31 pp

Lobel PS (1997) Comparative settlement age of damselfish larvae (Plectroglyphidodon imparipennis, Pomacentridae) from Hawaii and Johnston Atoll. Biol Bull 193:281–283

Lobel PS (2003) Marine life of Johnston Atoll, Central Pacific Ocean. Natural World, Vida, OR, 128 pp

Lobel PS, Kerr LM (1996) Johnston Atoll Phase II Marine Sediment and Biota Sampling Field Report.

Tech Rpt to the USAF, US Army, Defense Nuclear Agency, USFWS, USCG, 58 pp

Lobel PS, Kerr LM (1998) Ecological Risk Assessment for the nearshore marine environment of the former HO storage site, Johnston Atoll. Tech Rpt to the USAF, US Army, Defense Nuclear Agency, USFWS, USCG, 128 pp

Lobel PS, Kerr LM (2002) Status of contaminants in Johnston Atoll lagoon sediments after 70 years of U.S. military occupation. Proc 9th Int Coral Reef Symp 2:861–866

Lobel PS, Lobel L (2004) Annotated checklist of the fishes of Wake Atoll. Pac Sci 58(1):65–90

Lobel PS, Robinson AR (1983) Reef fishes at sea: ocean currents and the advection of larvae. Symp Ser Undersea Res 1(1):29–38. NOAA Undersea Res Prog, Rockville, MD

Lobel PS, Robinson AR (1986) Transport and entrapment of fish larvae by ocean mesoscale eddies and currents in Hawaiian waters. Deep-Sea Res 33(4):483–500

Lobel PS, Schrieber BA, McCloskey G, O'Shea L (2003) An ecological assessment of Johnston Atoll. Publ US Army Technical Report, Aberdeen, MD, 49 pp

Maragos JE, Jokiel PL (1986) Reef corals of Johnston Atoll: one of the world's most isolated reefs. Coral Reefs 4:141–150

Mitchum GT (1995) The source of 90-day oscillations at Wake Island. J Geophys Res 100:2459–2475

Myers RF (1999) *Micronesian Reef Fishes*. Third edition. Coral Graphics. Barrigada, GU, 330 pp

National Oceanographic and Atmospheric Administration (1992) Summary of climatology normals, means, and extremes for Wake Island, Pacific

National Oceanographic and Atmospheric Administration (2006a) PIFSC Cruise report CR-06-008, 100 pp

National Oceanographic and Atmospheric Administration (2006b) PIFSC Cruise report CR-06-024, 45 pp

National Oceanographic and Atmospheric Administration (2007) PIFSC Cruise report CR-07-008, 55pp

Ralston S (1981) Aspects of the reproductive biology and feeding ecology of *Chaetodon miliaris*, a Hawaiian endemic butterflyfish. Environ Biol Fish 6:167–176

Randall JE (1955) A revision of the surgeon fish genus *Ctenochaetus*, family Acanthuridae, with descriptions of five new species. Zoologica (NY) 40 (pt 4, no. 15):149–166

Randall JE (1992) Endemism of fishes in Oceania. UNEP. Coastal resources and systems of the Pacific Basin: investigation and steps toward protective management. UNEP Regional Seas Reports and Studies No. 147, pp 55–69

Randall JE (1996) Shore fishes of the Hawaii. Natural World, Vida, OR, 216 pp

Randall JE (1998) Zoogeography of shore fishes of the Indo-Pacific region. Zool Stud 37:227–268

Randall JE (2007) Reef and shore fishes of the Hawaiian Islands. University of Hawaii, Honolulu Sea Grant Program, 546 pp

Randall JE, Carlson BA (1997) *Ammolabrus dicrus*, a new genus and species of labrid fish from the Hawaiian Islands. Pac Sci 51:29–35

Randall JE, Lobel PS, Chave EH (1985) Annotated checklist of the fishes of Johnston Island. Pac Sci 39(1):24–80

Randall JE, Smith CL, Feinberg MN (1990) Report on fish collections from Rapa, French Polynesia. Amer Mus Novitat 296:1–42

Randall JE, Earle JL, Pyle RL, Parrish JD, Hayes T (1993) Annotated checklist of the fishes of Midway Atoll, Northwestern Hawaiian Islands. Pac Sci 47:356–400

Rotondo GM, Springer VG, Scott GAJ, Schlanger SO (1981) Plate movement and island integration – a possible mechanism in the formation of endemic biotas, with special reference to the Hawaiian Islands. Syst Zool 30:12–21

Schultz LP, Herald ES, Lachner EA, Welander AD, Woods LP (1953) Fishes of the Marshall and Marianas Islands. Bull US Nat Mus 202, 1:685

Schreiber E A, Doherty PA, Schenk GA (2001) Effects of a chemical weapons plant on red-tailed tropicbirds. J Wildlife Manag 65:685–695

Smith RM, Swain J (1882) Notes on a collection of fishes from Johnston Island, including descriptions of five new species. Proc US Nat Mus 5:119–143

Springer VG (1982) Pacific Plate Biogeography, with special reference to shorefishes. Smithsonian Contributions to Zoology. Smithsonian Institution Press, Washington, DC, 367, 182 pp

Tsuda RT, Abbott IA, Foster KB (2006) Marine Benthic Algae from Wake Atoll. Micronesica 38(2):207–219

US Army Space and Strategic Defense Command (1992) Real-property master plan (long-range component) Wake Island. TA92-09. Environmental and Engineering Office CSSD-EN, Huntsville, AL

US Department of Interior, Fish and Wildlife Service (1999) Baseline marine biological survey, Peacock Point outfall and other point-source discharges, Wake Atoll, Pacific Islands Ecoregion, Honolulu, HI, 23 pp

US Department of Interior, National Park Service (1984) National Register of Historic Places Inventory-Nomination form: Wake Island, September 16

Wennekens MP (1969) Johnston Island regional oceanography. Forecasting currents eddies, island wake. Office of Naval Research, San Francisco, CA

Wiens HJ (1962) Atoll environment and ecology. Yale University Press, New Haven, CT, 532 pp

Wiltshire JC (1990) Precious Metal Accumulation in Manganese Crusts from the Hawaiian and Johnston Island Exclusive Economic Zones. In: Hunt M, Doenges S, Stubbs GS (eds) Studies related to continental margins, Bureau of Economic Geology University of Texas, Austin, TX, pp 45–53

Wyrtki K (1975) Fluctuations of the dynamic topography in the Pacific Ocean. J Phys Oceanogr 5:450–459

18

Geologic Setting and Geomorphology of Coral Reefs in the Mariana Islands (Guam and Commonwealth of the Northern Mariana Islands)

Bernhard M. Riegl, Samuel J. Purkis, Peter Houk, Genevieve Cabrera, and Richard E. Dodge

18.1 Introduction

These westernmost territories of the United States are where "America's day begins", and contain the most diverse coral reefs under the US flag (Randall 1995; Paulay 2003). The Mariana Islands have a long history of dedicated coral reef investigations by international and US researchers begun by Agassiz (1903). Despite having been discovered by the Spanish in 1521, these islands were rarely visited by Westerners due to their isolation. The Spanish used the Marianas as a trade-route stop-over between the Philippines and South America, a history that later attracted others interested in finding treasure. In 1742, Tinian was visited by the British Commodore Anson who became rich and famous after pirating Spanish treasure, and in 1765 and 1767, Byron and Wallis revisited the island in continued search of treasure and territories. Accounts of Anson's and Byron's voyages were some of the earliest documentations of the Marianas. In Guam, research received a boost when the USA acquired it from Spain following the Spanish American War, and in the CNMI, when Germany bought the islands in 1899. Coral reef investigations began in 1903 when Agassiz reported about the Marianas in the course of his decade of coral reef expeditions around the world from 1893 to 1902. The German explorer Prowazek, reported on the "raised coral limestones" of the CNMI in 1913. Under Japanese mandate, Sugawara (1934) investigated the coral reefs and Holocene limestones on Rota just prior to the Second World War. At a similar time, Stearns (1937, 1940a, b) reported on Guam's geology, hydrogeology and drew attention to the erosional notches, which have remained informative features for many subsequent investigation (Shepard et al. 1967; Easton et al. 1978; Kayanne et al. 1993; Dickinson 2000). From these beginnings, perspectives regarding eustacy in the entire Pacific Ocean were gained. After the war, intensive US-funded mapping and geologic research led to a series of United States Geological Survey publications (Cloud et al. 1956; Doan et al. 1960; Tracey et al. 1964). The seventh International Coral Reef Symposium took place in Guam and highlighted carbonate and reef studies in the Marianas, and Siegrist and Randall (1992) provide an informative overview and guide. Another excellent review of geology and hydrogeology is Mink and Vacher (1997). In addition to the raised limestone islands of the southernmost Mariana islands, the archipelago also includes nine emergent, purely volcanic islands. A rich literature exists with regards to the geological position of the Marianas on the Izu-Bonin-Mariana Arch, and the associated Mariana Trench (Hussong and Uyeda 1981; Clift and Lee 1998; Yamasaki and Murakami 1998; Stern and Smoot 1998; Stern et al. 2004) (Fig 18.1).

B.M. Riegl and R.E. Dodge (eds.), *Coral Reefs of the USA*,
© Springer Science + Business Media B.V. 2008

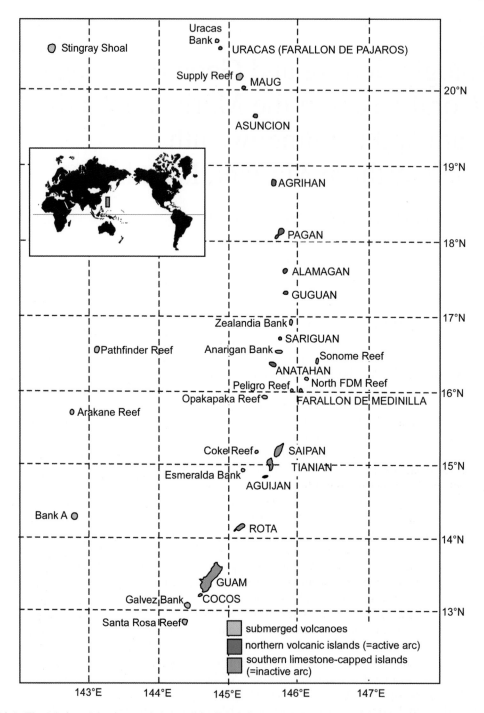

FIG. 18.1. The Mariana Islands, consisting politically of Guam and the Commonwealth of the Northern Mariana Islands, the westernmost territories of the United States of America

18.2 A Brief History of the Mariana Islands as a US Territory

The *Marianas Archipelago* is a group of 15 small raised coral limestone and volcanic islands that are today a part of the US American political family. The chain lies 6,000 miles southwest of San Francisco, 3,700 miles west of Honolulu, 1,500 miles south of Tokyo, and 1,500 miles east of Manila. Guam is the largest and southernmost island and has been an unincorporated territory of the USA since 1898. The 14 other islands situated to the north of Guam became the Commonwealth of the Northern Mariana Islands (CNMI) in political union with the USA in 1976. Saipan is the capitol of the CNMI and it is where approximately 90% of the CNMI's population lives. The other two largely inhabited islands are Tinian and Rota. The USA is a very recent participant, however, in a succession of ruling countries that reflect the archipelago's rich and diverse 500-year political history. This history has not only shaped the islands into the geopolitical entities that they presently are, but more importantly, it has left a lasting mark on the archipelago's indigenous Chamorro people whose collective presence in the islands spans possibly 3,500 years (Rainbird 1994).

Anthropological consensus contends that the ancient Chamorros originated from Island Southeast Asia, migrated through the northern Philippines (Rainbird 1994) before finally settling in the Mariana Islands. As the ancient culture was an oral culture, there were no written records in existence when the Western world stumbled upon the islands early in the sixteenth century; at least not in the sense that the modern world understands written records to be. Documented descriptions of the ancient people therefore come from Western sources and it must be understood that some of these written accounts reflect a purely Western point of interpretation.

One of the earliest surviving descriptions of the ancient Chamorros comes from the chronicler Pigafetta who in 1521 recorded the explorer Ferdinand Magellan's expedition's encounter with the islanders:

[E]ach of these people lives according to his own will, for they are bound to no lord. They are handsome, tawny, and well built and go naked except for the women who have a piece of soft bark covering their private area. Their hair is extraordinarily black. Some men sport beards and their hair is waist-length The women wear their hair exceptionally long, flowing loosely almost to the ground. Their teeth are red and black, which is thought to be beautiful and they cover their hair and body with coconut and sesame seed oil. The women do not engage in fieldwork, but remain in the wood plank and thatched roof houses and weave mats, baskets, and other items with palm leaves. They have not weapons save spears with fishbone tips. For amusement, they sail their boats, which leap from wave to wave like dolphins and can interchange bow and stern at will. …

Although he sailed for Spain, Magellan did not lay claim to the archipelago. It was not until 1565 that the islands were officially declared possessions of the Spanish crown, but yet a little over a century again passed before civil authorities and Catholic missionaries arrived in 1668 and began the colonization of the Marianas.

Over the course of the 50 years that followed, the spread of foreign diseases and the battles that erupted between the Chamorros and the Spanish authorities decimated the population and nearly annihilated the indigenous culture, which may have been populous at the time of contact. Estimates range from around 50,000 (Bratring 1806) to 5,000. The authorities' need for control was so drastic that the Spanish military sailed throughout the entire archipelago and rounded up most of the rebellious factions and along with non-combatants, temporarily placed them in two Spanish mission villages on Saipan before finally transporting them to Guam where they were forced to reside until the mid to late nineteenth century when they again were permitted to return to the northern islands. The Chamorros persisted throughout the 230 years of Spanish rule and continued to use the indigenous language (also called Chamorro but which increasingly incorporated Spanish words) and practiced subsistence fishing and farming. Between the years 1815 and 1818 settlers from the outer Caroline Islands of what is today Yap and Chuuk arrived in the Northern Marianas and were allowed by the Spanish government to settle on Saipan and Tinian. Along with the indigenous Chamorros, the Carolinians consider the Northern Mariana Islands their home.

A big change came about for the archipelago in 1898. Spain and the USA had gone to war and Spain's defeat accorded Guam its new political status as an American territory. The USA were not interested in any of the islands to the north and as a result, Spain sold to Germany what later became the Commonwealth of the Northern Mariana Islands. Those Chamorros that had gained permission to return to Saipan now lived under German rule for a 15-year period (1899–1914). At the outbreak of World War I in 1914, Germany left the Northern Mariana Islands and its other Pacific possessions. Japan immediately assumed control of much of the Pacific for the next 30 years (1914–1944). Although it was never an official member of the League of Nations, the USA continually contested Japan's sanctioned control of most of Micronesia, which included the Northern Mariana Islands. The political scenario changed yet again during World War II. American administration of the Northern Mariana Islands, which continues to this day, began when US military forces wrested control of Saipan from Japan on 9 July 1944. Guam, which was invaded and seized by Japanese forces 8 December 1941, was reclaimed by the US in August 1944.

Guam once again was administered by the US Department of the Navy. The Northern Mariana Islands, Palau, Yap, Ponape, Truk, and the Marshall Islands were awarded to the USA by the United Nations under a trusteeship agreement wherein America was entrusted to develop the socio-economic and political spheres such that the citizens of the these Micronesian Islands would eventually be able to chart their own political course as independent island nations. Before the Trusteeship Agreement expired, negotiations between these Micronesian Islands and the US were being respectively carried out. The ties with the US were not completely severed. The Republic of Belau (Palau), the Federated States of Micronesia (Yap, Pohnpei, Chuuk, and Kosrae), and the Republic of the Marshall Islands are known today as the Freely Associated States of Micronesia in political union with the USA. In the Marianas, a clear distinction exists between Guam and the CNMI with respect to the political union with the US. Guam is a Territory, while the Northern Mariana Islands are a Commonwealth established by Covenant with the US. Both Guam and the CNMI elect a Governor and a Lt. Governor,

respectively, and they each have a bicameral legislature and a judicial branch of government. Guam has a non-voting delegate in Congress. The CNMI currently still does not have such a delegate although it does elect a Resident Representative to Washington, DC with each general election.

18.3 General Geologic Setting of the Mariana Islands

The Mariana Islands are part of the Izu-Bonin-Mariana (IBM) arc system which stretches over about 2,800 km from Japan to south of Guam. This sytem is characterized by subduction of the Mesozoic Pacific plate towards WNW beneath the West Philippine plate. In the process, the Mariana arc has split to form two remnant arcs, the Palau-Kyushu Ridge and the West Mariana Ridge. Between these two ridges exist two extensional basins, the Parece Vela Basin and the Mariana Trough (Karig 1971; Hussong and Uyeda 1981; Clift and Lee 1998) (Fig. 18.2). The Mariana forearc region shows evidence of the extensive vertical movements resulting from seamount collision and fracturing associated with the plate motions constraining arc configuration (Fryer et al. 1992). The forearc basement of the IBM system was formed after the initiation of "infant" arc volcanism and is a consequence of either the trapping of old, most probably Philippine Sea, oceanic crust or by intra-oceanic island arc rifting and volcanism (Stern et al. 2004). The West Philippine Basin is the oldest back-arc basin in the Phillipine Sea and is either this piece of trapped ocean floor or an early back-arc basin. Prior to 45 Ma (million years), the present Palau-Kyushu Ridge was a transform fault and Pacific plate subduction was directly beneath the Asian continent. It is assumed that a change in direction of motion of the Pacific plate caused initial subduction to begin in the early Eocene at about 43–45 Ma, almost simultaneously throughout the IBM system. This timing would coincide with the bend in the Hawaiian–Emperor chain (see Chapter 13, Rooney et al.), originally believed to have formed at 43 Ma (Dalrymple and Clague 1976; Patriat and Achache 1984). However, its age has been under discussion (e.g. Norton 1995) and has recently been suggested to have formed closer to 50 Ma (Sharp and Clague 2006). Also, causative mechanisms for the bend are being discussed, possibilities are changes in plate motion (as required for above interpretation of arc activity in the Marianas) or displacement of the hotspot (see also Chapter 13, Rooney et al. and Chapter 20, Birkeland et al.). Recent results suggest that plate motion is at the very least a component, if not the unique explanation, for the change in direction of some hotspot trails (Steinberger et al. 2004) and preliminary evidence from the southern Pacific between Australia and Antarctica also detected plate motion (Whittaker and Müller 2006; see Chapter 13, Rooney et al.). This evidence supports the above-mentioned change from a transform fault to subduction in the IBM system. Whatever the cause, subduction does happen in the Marianas region and is evidenced by boninite and arc-tholeiite lavas on Chichi Jima and Bonin Island (Bloomer 1983; Clift and Lee 1998; Stern et al. 2004). The Palau-Kyushu Ridge itself was built between 43 and 32 Ma, at which time its rifting formed the Parece Vela basin by back-arc spreading and eastward migration of the fore arc region north of Palau (Reagan and Meijer 1984; Mink and Vacher 1997). Continuous arc volcanism from around 20 Ma to about 9–5 Ma formed the West Mariana Arc region. Back arc spreading set in again and the West Mariana Ridge rifted to form the Mariana Trough. This displaced the forearc region even further east and prior to 1.3 Ma arc volcanism formed the presently active Mariana Islands chain (Meijer et al. 1983; Reagan and Meijer 1984; Randall 1995; Fig. 18.2). At present, the subduction velocities at the Mariana margin are ~40 mm/year towards WNW (Stern et al. 2004). The Pacific plate descends beneath the West Philippine plate at an angle of ~20° to about 60 km and steepens at >100 km to near-vertical (Fryer et al. 1992). Mrozowski et al. (1981) (in the Marianas) and Horine et al. (1990) (in the Izu-Bonin) proposed that only minor sediment accretion has occurred along the entire convergent margin. The lithology of dredged samples from the slopes of the Mariana Trench (island arc tholeiites and boninites) suggests that almost no sediment or oceanic crustal accretion occurs along the Mariana margin (Bloomer 1983; Fryer et al. 1992).

18.4 Geology of Guam and Rota

18.4.1 Formations and Rock Units

Guam, at the southwestern end of the West Mariana Ridge, was volcanically active from 32–20 Ma to 9–5 Ma (Reagan and Meijer 1984). Over time, the West Mariana Ridge has rifted, creating the Mariana Trough and the Mariana Ridge, where the southern Mariana Islands, including Guam, are situated. Renewed volcanic activity began before 1.3 Ma but is associated only with the active ridge, not Guam. The main volcanic basement of Guam resulted from the first volcanic phase and is evidenced in two formations: the Facpi Formation of late Middle Eocene age and the Alutom Formation of late Eocene to early Oligocene age (Fig. 18.3). The Facpi Formation is about 200–150 m thick and consists mainly of submarine boninite lavas occurring as pillow lavas and pillow breccias with tholeiitic basalt dikes. It also has infills of limestones, according to Siegrist and Randall (1992) probably pelagic sediments, consisting of recrystallized micrite. The oldest parts of the Facpi Formation have been dated to 43 Ma with the dikes having Oligocene age (Reagan and Meijer 1984; Siegrist and Randall 1992).

The younger Alutom Formation is about 400 m thick and consists of tuffaceaous shales, sandstones, volcanic breccias, lapilli conglomerates, some pillow lavas, sills and frequent turbidite sequences (Reagan and Meijer 1984; Siegrist and Randall 1992). The stratigraphic position was determined by larger foraminifera (Cole 1963) to upper Eocene-lower Oligocene and trace fossils (worm trails) suggest depositional water-depths of below 500 m (Kilmer 1991). Calcareous shales occur in the Mahlac member in the early Lower Oligocene (Siegrist and Randall 1992). Coral-rich floats also occur on the Alutom Formation (Schlanger 1964), but are completely recrystallized and replaced which results in unavailability of any certain age estimates. Siegrist and Randall (1992) suggest Alutom age but are uncertain whether these floats are *in situ* or allochthonous. In any case, they suggest a significant shallowing of Guam. The upper Eocene Matansa Limestone on Saipan, which may be coeval (Siegrist and Randall 1992), suggests that this is indeed plausible (Cloud et al. 1956).

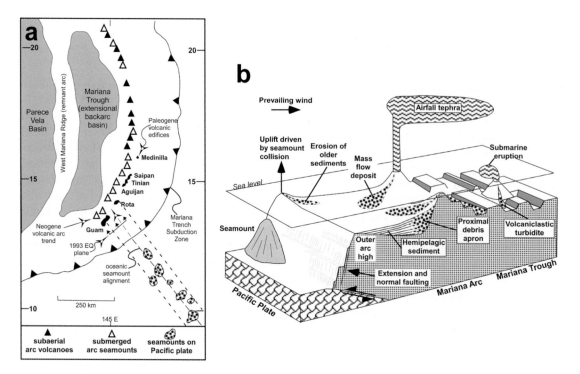

FIG. 18.2. (**a**) The Mariana Islands consist of Neogene island volcanoes and submerged seamounts in the north – with recent coral growth but no well-developed reefs or limestones. In the south, the islands consist of paleogene volcanic bases with thick Paleo- and Neogene limestone cover, as well as well-developed recent reef growth. A seamount chain is being subducted underneath Guam and Rota, and likely is the cause of tectonic uplift experienced by these two islands which is stronger than on the other southern Marianas. (Modified from Dickinson 2000, by permission of Journal of Coastal Research). (**b**) Todays Mariana Islands are situated in the fore-arc region that has separated by back-arc spreading in the Mariana Trough from the West Mariana Ridge, which itself was separated by back-arc spreading in the Parece Vela basin from its original position along the Palau-Kyushu Ridge (not shown). The block diagram shows the position of the islands relative to the the subducting Pacific Plate and the back-arc region, as well as volcaniclastic depositional processes (Modified from Clift and Lee 1998, by permission of Blackwell)

Serious carbonate sedimentation began in the upper Oligocene/Lower Miocene and is encountered in the Maemong Limestone, a member of the Umatac Formation which is of shallow-water origin and consists of mud- to wackestones with some coral (*Porites, Acropora*) rudstones. Although reefal organisms such as corals, *Tridacna, Halimeda* and crustose algae are commonly found, they all occur *ex situ* and no coherent reef edifice has been found to date. Variable fresh and weathered volcanic detritus is intermixed. Schlanger (1964) and Siegrist and Randall (1992) suggest this episode to represent a north-to-south shallowing around an emergent volcanic high. These limestones precipitated around the three volcanic highs on Guam. The foraminiferal assemblage is similar to that of

Saipan's Tagpochau Limestone, which may therefore be older than mentioned in the text further below (Siegrist and Randall 1992).

The Bonya Limestone is spatially restricted to relatively small areas of SE and NE Guam, reaches a thickness of 10–15 m and is of Miocene age. It is primarily a deeper-water facies (deeper in northern than in southern Guam) but carries reef coral and coralline algae detritus. Reefs are not preserved, so this limestone is likely an off-reef unit on the seaward slope deposited between volcanic cones. Another Miocene unit is the Alifan Limestone that conformably overlies the Bonya Limestone or the Talisay Clay, which is the oldest signal for a terrestrial environment on Guam (Siegrist and Randall 1992). It contains mudstones that

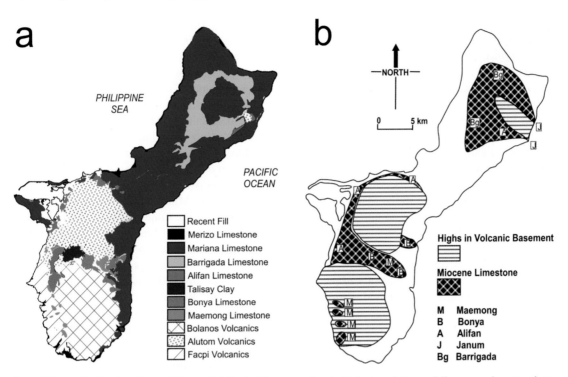

FIG. 18.3. (**a**) Surficial geology of Guam. (**b**) Early Miocene volcanic highs in relation to Miocene carbonates show that the latter formed primarily as ramp facies on and around the volcanic cones (Modified from Siegrist and Randall 1992, by permission of the International Society for Reef Studies)

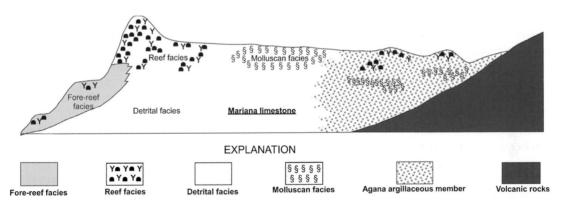

FIG. 18.4. Facies in the Plio–Pleistocene Mariana Limestone (Modified from Tracey et al. 1964, by permission of the United States Geological Survey)

grade to sparry packstones with well-preserved corals and reefal organisms (*Porites, Acropora, Halimeda*). Tayama (1952) interpreted an atoll, while Tracey et al. (1964) and Schlanger (1964) suggest the outcrops to represent a backreef or lagoonal environment. The geographically smallest limestone is the Janum Limestone of Tortinian and Pontian age which is about 20 m thick and consists of three units of deep fore-reef to basinal facies (Siegrist and Randall 1992). The Mio/Pliocene Barrigada Limestone underlies much of the northern plain and crops out in an annular fashion.

It is a chalky foraminiferal-algal wackestone with some larger fossils, in particular corals (*Porites*, *Astraeopora*). All Miocene limestones on Guam are completely recrystallized to calcite and in the Alifan Limestone, individual dolomite rhombs were reported by Schlanger (1964). The Bonia and Alifan Limestones are heavily karstified, and also the Barrigada Limestone shows sinkholes and karren (Siegrist and Randall 1992). While essentially all limestone units show reefal fauna among their components, no Miocene reefs crop out anywhere in Guam or anywhere else in the Mariana Islands. Where the reefs are, remains an unresolved question (Siegrist and Randall 1992).

The most widely distributed limestone in the Mariana Islands is the Plio–Pleistocene Mariana Limestone, which is the dominant carbonate on Guam, Rota, Aguijan, Tinian and Farallon de Medinilla (Siegrist and Randall 1992; Randall 1995) and also of great importance on Saipan. It contains well-developed reefal and peri-reefal facies as well as the Agana argillaceous member (Tracey et al. 1964; Fig. 18.4). On Guam, it can reach a thickness of 175 m. Reef margin facies generally crop out as the most seaward facies and coincide often with the edges of terraces. Rocks are coralline and coralgal boundstones, coarse grainstones as well as packstones and wackestones. They contain micritized reef bioclasts and intraclasts. *In situ* branching and massive corals occur and Siegrist and Randall (1992) report a rich fauna, reminiscent of the situation in modern reefs (*Acropora*, *Favia*, *Goniastrea*, *Leptoria*, *Platygyra*, *Pocillopora*, *Porites*, *Stylophora*, *Symphyllia*, and *Turbinaria*). The fore-reef facies consists of massive to flaggy packstones that dip by 5°–15° seaward and can be cross-bedded. The detrital facies is the most common and contains facies dominated variably by benthic foraminifera, corals, molluscs, oysters, snails, and *Halimeda*. Corals are frequently observed in growth position. The Mariana Limestone has thoroughly altered porosity and mineralogy and displays many karst features (Siegrist and Randall 1992; Fig. 18.5). The youngest limestone is the Merizo Limestone (Tayama 1952), which will be discussed in detail in the section about Holocene reefs.

Rota is situated less than 50 miles north of Guam, and the two islands therefore share much history and many rock units. However, unlike Guam, Rota is primarily (>90%) covered by carbonates with only few areas of outcropping volcanic rocks. Apparently, the majority of the island is covered by the Mariana Limestone (Carruth 2005), which makes up the northern part of Guam, and also covers the majority of Tinian and a large portion of Saipan. Core logs in Carruth (2005) suggest that other limestone units found in Guam may exist underneath the Mariana Limestone in Rota as well. Near the shoreline, excellent exposures of the Milencatan Limestone occur (Sugawara 1934; Kayanne et al. 1993 refer to it as Milencatan while Bell and Siegrist (1988) refer to it as Mirakattan Limestone). The Milencatan limestone is equivalent to the Merizo Limestone on Guam and represents early Holocene reefal limestones of the Flandrian sea level highstand, which peaked about 4 ka (discussed below). The importance of this limestone as sea level indicator is discussed in detail by Easton et al. (1978), Randall and Siegrist (1988), Bell and Siegrist (1988), Siegrist and Randall (1992), Kayanne et al. (1993), Dickinson (2000).

18.4.2 Relevance of Bedrock Geology for Modern Coral Reefs

Guam shows two distinct geological zones: the south is made up by the old igneous rock basement, the north by a geologically young limestone plain. Igneous rocks form volcanoes, extrusive and intrusive bodies that need not respect any pre-existing morphology but, particularly when extrusive, form their own landscape, for example, volcanic cones. Therefore southern Guam is steeper and has the highest hills. Limestones can only form under water and while carbonate platforms aggrade and can form precipitous morphology at their flanks, once caught up to the water surface, carbonate platforms tend to fill up and then primarily prograde towards the basin (Schlager 2005). Thus, the northern limestone province of Guam is largely flat with steep edges and the island's elevation roughly represent the level of the platform's possible aggradation and the island's outline represents the previous edge of the carbonate platform – defined by coral reefs when it grew.

Also, igneous rocks and limestones show markedly different weathering patterns. Igneous rocks rapidly weather to deep soils, so-called laterites. These are the characteristic reddish soils known from tropical volcanic islands. While densely vegetated, these soils assure high plant and associated

FIG. 18.5. Weathered Mariana Limestone (Plio–Pleistocene) at (**a**) the Wedding Cake (Rota) and (**b**) Suicide Cliff (Tinian). (**c**) Bird Island on Saipan is a tuffaceous part of the Sankakuyama Formation. The cliffs behind the island are Mariana and Tagpochau Limestones. (**d**) Brecchia and flows in the Sankakuyama Formation (Saipan)

animal biomass and diversity in the dense forests that cover them in the natural state. The positive morphology of the former volcanic cones attracts orographic rain, maintaining an environment for very active weathering of the rocks, creating soil and maintaining a good environment for vegetation. When this vegetation is removed, however, the laterites are easily eroded and carried downslope, eventually ending up in the coastal zone. Guam has been heavily deforested and soil-erosion with concomitant impacts by sedimentation on the coastal coral reefs is a major problem. Suspended sediment loads in southern Guam can reach over 1,000 mg.l^{-1} and significant impacts on the coral reefs have been demonstrated (Wolanski et al. 2003, 2004; see also Chapter 19, Richmond et al.). Even under natural conditions, on a completely

vegetated island with minimal soil erosion, the strong runoff has clear effects on reef building. Strong pulses of dramatically reduced salinity suppress reef growth near river mouths. Thus, two mechanisms aid in the formation of well-developed embayments along the igneous coast: bedrock erosion by the rivers that forms small bays and suppression of reef growth in the river mouth's immediate vicinity. Since coral reefs tend to form only few dozens of meters away from the river mouths and certainly form along the distal reaches and the outsides of the bays' morpology becomes accentuated by reef growth over time.

Limestones do not weather to soils, but are dissolved under the influence of meteoric water (Esteban and Klappa 1983). The result is a so-called karst landscape which is highly rugose with caves, caverns, and gnarly, sharp surface rocks. Due to the limestone's high porosity, that is further augmented by the dissolution features, water tends to collect less at the surface to run off in the form of creeks and rivers, but mostly percolates into the limestone where it can form important groundwater systems (Mink and Vacher 2004. The result is less riverine input with associated sedimentation into the nearshore marine environment, which is good for coral reef growth. Along the northern limestone coast of Guam, reefs are not as clearly discontinuous along embayments and run-off from land is far less of a problem than along the igneous south.

Similar principles apply to all other islands.

18.5   Geology of Saipan and Tinian

Saipan has an area of 125 km² and is the largest of the 14 islands in the CNMI. It is situated at latitude 15°12′ N and longitude 145°45′ E, about 3,740

miles west–southwest of Honolulu and midway between Japan and New Guinea. Saipan was built by island-arc volcanics which is evidenced in cones of pyroclastics and flows of andesitic to dacitic composition. Erosional processes reworked the volcanic material and produced clastic sediments that later cemented to sandstones, conglomerates, and breccias. Volcanism on Saipan probably ended in the Oligocene, after which marine deposition produced thick limestone sequences. Tectonic uplift has raised these limestones to elevations of 470 m over above sea level (Karig, 1971). Saipan thus consists of an andesitic–dacitic volcanic core overlain by sandstones, conglomerates and breccias which are capped by limestones that cover about 95% of the island's surface (Fig. 18.6).

The oldest rocks (41 Ma) of late Eocene age (Meijer et al. 1983) are found in the Sankakuyama Formation, which consists of dacitic flow and pyroclastic rocks and underlie all other exposed rock units. The formation's overall thickness is unknown; but is at least 600 m based on a section

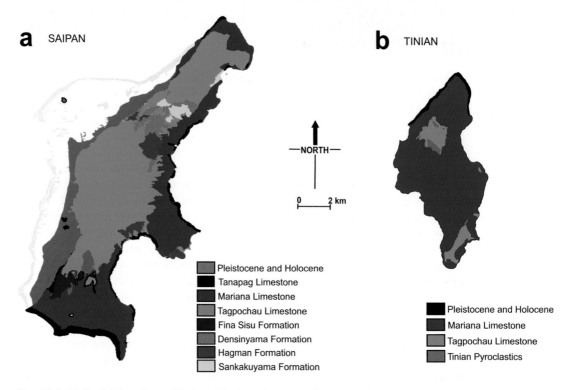

FIG. 18.6. (**a**) Surficial geology of Saipan. The large lagoon to the west is Tanapag Lagoon (Modified after Carruth 2003) (**b**) Surficial geology of Tinian (Modified after Doan et al. 1960, Gingerich 2002 by permission of the United States Geological Survey)

calculated from exposed rocks along the flanks of Mount Achugao, which is interpreted by Cloud et al. (1956) to be a remnant of a stratified Eocene composite volcanic cone. Two other Eocene volcanic formations occur around Mount Achugao. These are the andesitic pyroclastic rocks, lava flows, and water-laid volcanogenic sediments of the Hagman Formation and the marine transitional rocks, volcanogenic sediments, and andesitic breccia of the Densinyama Formation. The estimated thickness of the Hagman Formation is about 350 m that of the Densinyama Formation thins to zero from a maximum of about 230 m. The Fina-Sisu Formation, consisting of calcareous marine tuffs and andesite flows, forms deeply weathered outcrops in southern Saipan, of middle Miocene age (13 Ma; Meijer et al. 1983).

Several limestone units occur (Fig. 18.6). The Matansa Limestone is an Upper Eocene carbonate bank deposit (Cloud et al. 1956) and may be coeval with the coral-rich floats in Guam's Alutom Formation (Siegrist and Randall 1992). Saipan's center is primarily covered by the Tagpochau Limestone (named after the island's highest peak) which is of early Miocene age (Doan et al. 1960). It is a complex of calcareous clastic rocks, composed of fine- to coarse-grained, partially recrystallized broken limestone fragments including reworked volcanic fragments and clays intergrading with one another and distinguished from other fragmental limestones on Saipan mainly by fossil evidence (Cloud et al. 1956). The most widely distributed facies of the Tagpochau Limestone is a compact, mostly pure, pink to white, inequigranular limestone. The formation also includes impure limestone and sedimentary facies of reworked volcanic materials. The unit thickness ranges from near zero at exposures near Mount Achugao, to at least 300 m based on composite sections from the cliffs on the northern end of the island.

Much of Saipan's eastern side is occupied by Pliocene Mariana Limestone, which is equivalent to the same unit found on Guam and Rota and has been discussed in more detail above. In Saipan, it is a reefal limestone with argillaceous rubbly facies (Carruth 2003) and is light colored (dirty white to brownish), coarsely porous, finely to coarsely fragmental with frequent coral remains. The Mariana Limestone differs from the Tagpochau Limestone primarily in the abundance

of corals, and the modern aspect of its fossil assemblage. Thickness of the Mariana Limestone on Saipan ranges from near zero to about 170 m and the largest areas of outcrop are in southern Saipan on the Kagman Peninsula and the Bañaderu area on the northern end of the island. The Mariana Limestone does not occur along the entire western coast that faces the lagoon, but it disappears under Tanapag lagoon (Fig. 18.6). It is therefore likely that the structural control of the lagoon rim (=barrier reef) is determined by the course of the Mariana limestone. The lagoon and the coastal lowlands seaward of the Tagpochau Limestone are probably fill, overlying the Mariana Limestone.

The Tanapag Limestone is a raised reef limestone generally below 30 m elevation and of Pleistocene to Holocene age that includes dirty white to brownish coral–algal reef limestone and bioclastic limestone. The rock is mostly well indurated and porous with well preserved coral heads and mollusc shells. The Tanapag Limestone closely resembles parts of the massive facies of the Mariana Limestone; however, it is restricted to elevations of less than 30 m and its surface ordinarily is constructional rather than erosional (Cloud et al. 1956). The Tanapag Limestone appears to have formed mainly as fringing reefs on an emerging surface and its maximum thickness probably is less than 20 m. In its younger, Holocene, parts it is probably correlative to the Merizo Limestone of Guam and the Milencatan Limestone of Rota (Dickinson 2000). It occurs primarily along the steep eastern shoreline of Saipan and landward of the Pleistocene and Holocene alluvium that forms the western coastal plane and the surficial fill of the wide, barrier-reef protected Tanapag lagoon. Mañagaha Island (Fig. 18.17), at the western tip of Tanapag lagoon consists of the same youngest, Pleistocene/Holocene unit.

In Tinian, which is situated 5 km south of Saipan, the oldest rocks belong to the Eocene Tinian Pyroclastic Rocks, a unit which probably underlies all others and has an approximate thickness of 400 m (Doan et al. 1960). The unit is exposed in the north central highlands and in the island's southern section (Fig. 18.6b). Rocks consist of fine to coarse-grained consolidated ash and angular volcanic fragments. Most outcrops are highly altered and weathered to clay. The rest of Tinian is covered by limestones. The Miocene

Tagpochau Limestone, the same unit as on Saipan is exposed on about 15% of the island and occurs mainly around the exposures of the Tinian pyroclastics. The unit thickens from 0 to almost 250 m thickness (Doan et al. 1960; Gingerich 2002). As on Saipan, it is composed of fine- to coarse-grained, broken limestones with about 5% volcanic fragments and clays. The majority of Tinian is covered by the widely distributed Mariana Limestone of Pliocene age, which is the same unit as on all other islands. It covers roughly 80% of Tinian and thickens from 0 to about 150 m in all directions from near the Tinian pyroclastics (Doan et al. 1960). Like on Saipan, the limestones on Tinian are heavily faulted with subsequent formation of many karst caves (Stafford et al. 2005). Recent alluvium and colluvium reaches thicknesses up to 13 m.

Farallon de Medinilla is the northernmost of the calcareous Marianas and is also mostly covered by Mariana Limestone (Randall 1995; Fig. 18.1)

18.6   Geology of the Volcanic Northern Marianas

North of Farallon de Medinilla, the Mariana Islands bear no limestone caps and are volcanically active. This active island arc stretches from Anatahan to Uracas (Meijer and Reagan 1981). Two islands, Sarigan and Agrihan, are featured more closely to demonstrate the marked geologic differences between the active northern and the now-passive Marianas southern islands. Towards the southern end of the volcanic chain is Sarigan, which is 3.1 km² in area and has a maximum elevation of 538 m. It is the subaerial portion of a cone of 1.8 km height and 15 km basal diameter with an age of the subaerial cone of <1.0 Ma (Meijer and Reagan 1981). The cone is truncated with a well-defined crater. The lower slopes are buttressed by flow sequences near sea level while the upper slopes are covered by pyroclastic sequences (Fig. 18.7a). The tallest, most massive and subaerially exposed volcano is Agrihan (Fig. 18.7b). This

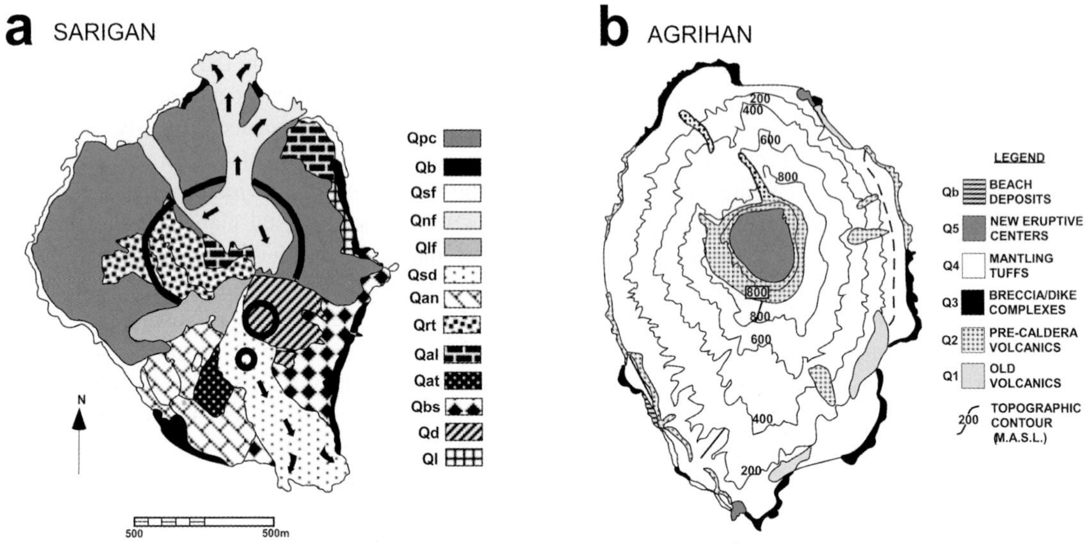

FIG. 18.7. Two examples of the volcanic northern Marianas (**a**) Sarigan. (Modified from Meijer and Reagan 1981, by permission of Springer). Flows are Qaternary, hence the "Q" prefix, Qbs = basal sequence, Qsf = shoreline flow; QPc = pre-collapse series, Qrt = ridge top flows, Qan = andesite series, Qnf = north flow, Qsd = south dome, Qd = north dome, Qlf = lobate flow. (**b**) Agrihan. (Modified from Stern 1978, by permission of Blackwell Publishing). The islands are made up entirely by volcanic rocks and have no limestone cap like the islands south of Farallon de Medinilla. All units are volcanics of different age and composition, but no limestones are present

island is smaller than many of the southern lime-stone islands in the CNMI (6.2×9.7km) and falls into three distinct geomorphological provinces: the shoreline, the slope, and the central caldera. The shoreline is steep and only the more protected SW shore has a sand beach which consists of black sand, in stark contrast to the white, calcareous sands of the southern islands. The flanks of the volcano are covered by andesitic pyroclastics consisting of different age groups. On some of the

volcanic islands, reef frameworks are presently developed (Pagan, Anatahan; Randall 1985).

18.7   The Physical Environment and Reef Growth in the Marianas

A hot and humid tropical climate exists throughout the Marianas with mean annual air temperatures of 27°C and daily fluctuation of ~6°C (Fig. 18.8).

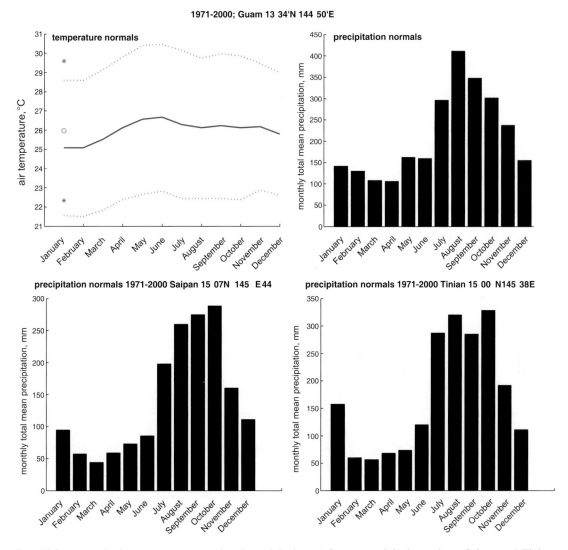

FIG. 18.8. Atmospheric temperature record and precipitation at Guam, precipitation only at Saipan and Tinian. (NOAA, National Climatic Data Center). Normals are uninterrupted measurements for 3 consecutive decades. Original data were translated into metric units. Lines in temperature graph represent monthly means, red circle annual mean, stars annual means of minima and maxima. Well-developed difference between dry and wet season is visible

Relative humidity usually ranges between 65–80% during the day and 85–100% at night (Mink and Vacher 1997). Average Rainfall on Guam is about 216–292 cm/year (85–115 in./year; Gingerich 2003), on Tinian 109–246 cm/year (43–97 in./year; Gingerich 2002) and on Saipan it ranges from 86 to 368 cm/year (34–145 in./year; Carruth 2003; Fig. 18.8). Although the islands are relatively small with a low elevation (400 m or less), significant orographic effects are felt. Guam and the CNMI have well-defined wet and dry seasons. About 70% of the total annual rainfall is recorded during the wet season (July–) with the rest falling during the dry season (January–June). The most intense rainfall events are associated with the passage of typhoons during the wet season. Exceptionally dry years recur about once every 4 years in correlation with episodes of El Niño Southern Oscillation (ENSO) in the Pacific (Lander 1994). Deforestation and repeated burning on the volcanic formations throughout the Marianas has led to sedimentation problems on adjacent coral reefs when seasonally pulsed rainfall events erode lateritic soils from steep slopes (Wolanski et al. 2003, 2004).

Long-term seasonal sea surface temperatures (SST) vary between ~25°C in the wintertime to ~29°C during summer months (Eldredge 1983; Fig. 18.9). A significant positive excursion with several unusually hot years was encountered in the mid 1940s, with 1946 having a higher mean annual temperature than 1998, one of the warmest years on record worldwide. Climate induced coral bleaching such as in 1998 may thus also have occurred in the mid 1940s.

The Mariana Islands are situated within the Northwest Pacific tropical cyclone basin, and are frequently approached by typhoons from the SE direction. Tropical storms are frequent in the western Pacific during the rainy season, with ~50% of all storms recorded between 1949 and 1969 forming in the vicinity of the Marianas (127°–147° E, 10°–23° N). Randall and Eldredge (1977) showed that reef damage was surprisingly small after the passage of typhoon Pamela (1976) over Guam, suggesting adaptation of the reef-building coral community to repeated large-swell events. However, typhoon associated reef damage has been evidenced in Tinian following the heavy 1997 typhoon season and coral cover near the crest diminished to 5% from over 50% in some locations (M. Trianni, personal communication 2006). Extensive damage to *Poecillopora* has also been reported at Farallon de Medinilla, Sarigan and sites around Saipan (M. Trianni, personal communication).

The North Equatorial Current (east to west) represents the predominant oceanic circulation pattern influencing the Marianas, however, the northernmost islands are seasonally bathed by the

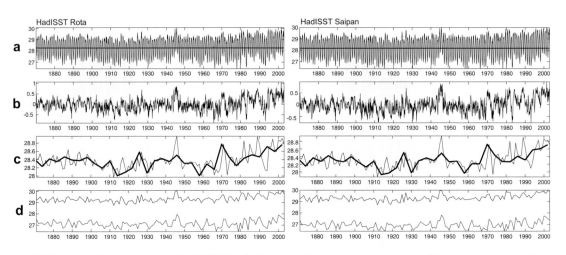

FIG. 18.9. Hadley Center's synthetic HadISST data for Rota and Saipan, as representatives of the southern and northern limestone islands. (**a**) monthly temperature mean (**b**) monthly excursion from 3-month mean (**c**) annual mean and trend line as 3-point cubic spline (**d**) monthly maxima and minima (Data remain Crown copyright, used by permission of the Hadley Center)

Subtropical Counter Current (west to east). During the summer months (June–September) when the northeast tradewinds weaken, the Subtropical Counter Current has a greater influence on the Marianas, as evidenced through recent bleaching events affecting reefs in the northern CNMI (Bonjean and Lagerloef 2002; Eldredge 1983). During the remaining months, strong trade winds and persistent northeast swells, generated by cold fronts moving westward off the Asian continent, prevail. Typically, north–northeast currents flow around each island with the exception of storm events that produce south–southwest currents when tropical storms pass (Eldredge 1983). Contemporary, regional oceanographic patterns explain why modern reefs of the Mariana Islands have more coral species in common with the Marshall Islands compared with Palau, despite being geographically closer to Palau (Randall 1995).

Tides are relatively small, with a mean tidal range of Tinian of 45 cm and a spring tide range of 65 cm, the mean high water interval is approximately 7¾ hours (Doan et al. 1960). Randall (1991) noted a severe low-tide event that exposed many of the shallow water *Acropora* and *Pocillopora* corals on the Holocene reefs throughout the Marianas. It is unknown if, and when, similar events may have occurred in the past, and their frequency.

18.8 Occurrence, Geomorphology and Pattern in Holocene Reefs

In the southern Mariana Islands, two distinct major Holocene episodes of reef building occur: first an early Holocene episode (5.5–3 ka), expressed in the Merizo, Milencatan, and Tanapag limestones, and then the modern reef (2.8 ka to present). The Merizo Limestone (also called "2 m limestone") on Guam, the Milencatan Limestone on Rota and the Tanapag Limestone on Saipan, are emergent reef facies, beautifully preserved, that occur in the low supratidal to shallow subtidal zone (2–4 m above mean sea level – hence the common name on Guam). The emergence of these reefs is due to a combination of global eustatic sea level variability and tectonic factors. The latter is evidenced by the general height differences between Guam and Rota (Kayanne

et al. 1993; Siegrist and Randall 1992). Depositional thickness varies on Guam and Rota from 1 to 5 m (Fig. 18.10). On Guam, thickness is 4–5 m at Aga Point and Ylig Bay, on the south and east coasts respectively, where it rests on Mariana Limestone (Kayanne et al. 1993). On the southwest coast where it rests on Facpi Formation and breccias, it is only one meter thick (Siegrist and Randall 1992). These limestones are clearly coral-dominated and contain similar coral and framebuilder communities to those occurring on the modern reefs (Fig. 18.11). On Rota, Bell and Siegrist (1988) recorded the following growth history:

- Beginning at 5.5 ka sea level had flooded a bioclastic facies of the Milencatan Limestone and corals settled on pre-existing highs about 80 m behind today's reef margin.
- Between 5.5 and 4.7 ka, corals accreted primarily under rising sea level conditions and produced buttresses about 1.5 m high. These were probably wave-energy-related groove-and-spur type structures. The buttresses were dominated by corymbose *Acropora* and thick algal crusts (Fig. 18.11c and d). Large *Porites* mounds occurred in more protected areas.
- Around 4.7 ka sea level rise slowed and the reefs began to catch up to the sea-surface, which was located at ~2.5–3 m above present sea level (Bell and Siegrist 1988). Kayanne et al. (1993) put the date of maximum sea level at 4.2 ka on both Rota and Guam, which is consistent with general predictions of this highstand to have occurred around 4 ka, and its variable persistence between 4 and 2 ka (Pirazzoli and Montaggioni 1988; Dickinson 2000). The tops of the coral frameworks were encrusted by digitate coralline algae, which Bell and Siegrist (1988) and Kayanne et al. (1993) interpret as shallow, relatively high-energy facies. Similar facies are found today in corresponding environments (Fig. 18.11).
- Between 4.7 and 2.9 ka sea level began to fall and the spur-and-groove reefs became exposed. By 2.85 ka corals started to settle on the present reef flat and the present-day algal ridge began to develop.

A mixture of sea level variability and tectonic uplift complicates the sea level signal on Rota and

FIG. 18.10. The Milencatan Limestone in Rota is an early Holocene reef limestone that grew at raised sea level and has also experienced tectonic uplift (**a**) emergence is lowest along the central N coast of Rota, (**b**) close to the tip of the Wedding Cake near SongSong village at Rota, the emergence of the Milencatan Limestone (2.7 m) matches the sea level notch in the Mariana Limestone (**c, d**) the emerged Milencatan Limestone emerges to different heights in Rota; near SongSong it forms a continuous ridge enclosing a lagoon with good Recent coral growth (**c, d**) on the lagoonal edge of framestone outcrop (**e**) more elevated sea level notches than on the shore (**f**) can be found further inland in the Mariana Limestone

FIG. 18.11. Much of the early Holocene Merizo and Milancatan Limestones is made up by coral framestones with corymbose *Acropora* (**a**, **c**) and *Pocillopora* (**b**) the dominant genera. The corymbose Acropora facies is more widely distributed on Guam and Rota (here near Sonsong, Rota). (**e**) A modern equivalent, corymbose *Acropora monticulosa* form the top of a reef buttress that has caught up with the surface (near Chulu Beach, Tinian) (**f**) Corymbose *Acropora* facies on a reef block that was transported above high-water line by a tropical cyclone in 2004 (Tinian)

Guam. Using dated corals and algae in relation to erosional morphology, Kayanne et al. (1993) and Dickinson (2000) developed calibrated sea level curves for the southern Marianas that show an actual rise of 1.8 m between 6 and 4.2 ka, and suggested the rest of the emergence to be resultant from tectonic uplift. The timing of the sea level rise and highstand coincides with predicted melting of North Atlantic and Antarctic ice shields. Dickinson (2000) suggests the hydroisostatic rise and subsequent drawdown to be due to forebulge unflexing and the downflexing of continental shelves due to weight release associated with ice-shield meting. He also suggested the local tectonic movements in Rota and Guam to be associated with seamount chain subduction under the two islands (Fig. 18.2a).

At Ylig Point, Guam, Siegrist et al. (1984a, b) describe two facies from the Merizo Limestone: a coral–algal boundstone facies, and a silty to sandy packstone to wackestone facies, neither of the two heavily altered by diagenesis. Overall, *in situ* corals and coralline algae make up about 60% of the rock, landward, in the sandy facies, sand grades to silt and then to micrite. Dominant cements are high-magnesium calcite, and abundant silty infill occurs in pores. At Rota, Bell and Siegrist (1988) described three facies: firstly a coral framestone facies composed primarily of corymbose *Acropora* (primarily *A. humilis* according to Kayanne et al. (1993)) and in more leeward settings of *Porites* mounds, considered to be the catchup phase of reef growth. Secondly, an algal bindstone facies, which is interpreted as the keep-up algal ridge and as capping buttresses. Thirdly, a detrital facies at the leeward margin of the reef. Also in Rota, high-magnesium cements were common and, to a lesser extent, aragonite cements were found. Bell and Siegrist (1988) suggest two major phases of deposition. First, there was a rapid accretion of the coral framework under rising sea level conditions. The rapid growth of the reef allayed the effects of marine diagenesis, which became stronger in the second phase, when the algal cap was developed. Then, numerous macroborings increased surface area and therefore increased the effects of marine diagenesis (primarily cement precipitation). Reef alteration was primarily surficial and resulted mainly in lithification. However, cement facies were not specific to exposure or windward–leeward position in the reef.

A more exhaustive analysis of variations in coral community structure and dominant coral growth forms in the Merizo Limestone at Ylig Point in Guam can be found in Siegrist et al. (1984a) and Randall et al. (1984). An overview of coral community structure in the framestone facies on Guam in general can be found in Siegrist and Randall (1985). A total of 45 species, which still occur in the area in the Recent, were recorded. At Aga Point and Ylig Point, paleocommunities in different exposure were preserved and studied. Despite a local *Pocillopora*-rich facies at Aga Point, in the subsurface (revealed by cores) and throughout other outcrops, the dominant corals were corymbose *Acropora*. Siegrist and Randall (1985) suggest that the *Pocillopora* facies was a phenomenon of the latest growth-phase in which the reef had caught up to sea level and a new niche had developed. This phase is referred to as senescence. Previously, during the proliferation phase, the corymbose *Acropora* community fuelled the rapid growth and aggradation of the reefs. Prior to that phase a drowning phase had occurred during post-Wisconsin transgression when an older erosion terrace had been flooded and thus the stage set for reef growth. After senescence, reef growth did not cease, but the above described hydro-isostatic sea level drop exposed the first Holocene reef growth phase of the Merizo Limestone.

Early Holocene reef morphology has a strong influence on the morphology of the present coral reefs. Randall et al. (1988) studied the evolution of morphology of fringing reefs on Guam, as did Kayanne et al. (1993) on Rota (Fig. 18.12). It was clearly shown that the position of the early Holocene fringing reef, present as the Merizo Limestone on Guam and the Milancatan Limestone on Rota, displace late Holocene reef growth seaward. The early Holocene limestones are well-eroded, especially the softer, landward sections behind the corymbose coral facies, where a detrital facies is often found (Siegrist et al. 1984a). Thus, lagoons have developed where protection by the raised limestone is evident, that can be several meters deep (for example, on Rota's southern tip near Songsong village, Fig. 18.12 and Fig.18.10 b, d, f) and have good coral growth.

Coral habitat in the Mariana Islands (Fig. 18.13) was mapped by NOAA and areas identified as bearing coral are shown in Fig. 18.14. These are true

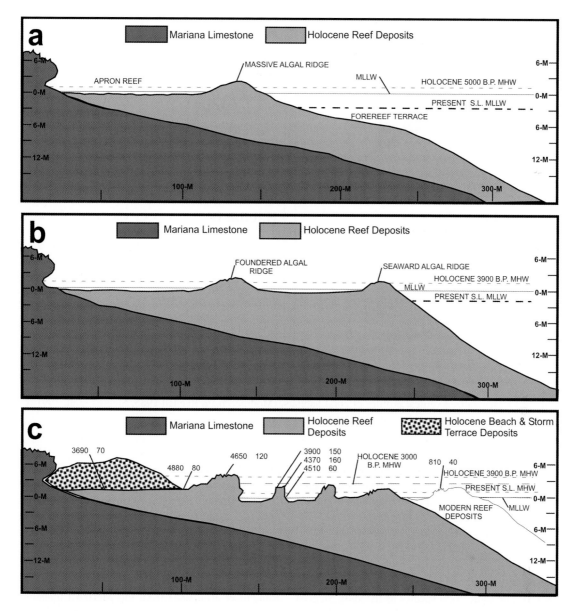

FIG. 18.12. Stages of Holocene reef growth on Guam. (Modified from Randall and Siegrist 1988, by permission of the International Society for Reef Studies). It is clearly visible how strongly the morphology of present-day reef growth is determined by early Holocene reef growth (**a**) early Holocene fringing reefs at about 5 ka (**b**) fully formed early Holocene fringing reef at sea level highstand at about 3.9 ka (**c**) late Holocene fringing reef with its upper portion and eroded as Merizo Limestone

coral reefs in the southern Mariana Islands. In the northern Marianas, the relatively large areas identified as bearing coral are not framework- or biogenic morphology-producing. At basin scale, the spatial arrangement of habitats across the Pacific have been shown to be fractal (Purkis et al. 2007).

The implication is that statistical patterns of size, shape, and complexity of patchiness in these environments is consistent within and between island chains, irrespective of whether an island is dominated by fringing or lagoonal facies. Since these relationships follow power laws, there also exists

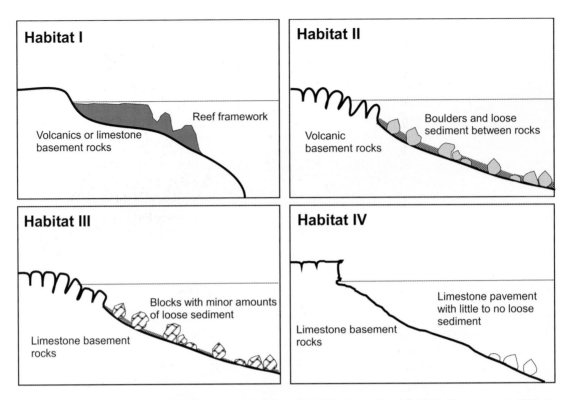

FIG. 18.13. Generalized vertical profiles of coral habitats. (Modified from Randall 1985, Siegrist et al. 1991, by permission of the International Society for Reef Studies)

the possibility that habitat patchiness is predictable across scale. Purkis et al. (2007) showed such scale-invariance to persist from hundreds of metres to many kilometers.

Not only the morphology of present-day coral reefs is similar to that of the early Holocene, but the distribution and composition of benthic communities of the primary calcareous reef-builders is similar too. The same coral species as listed by Siegrist and Randall (1985) and Randall et al. (1984) from the early Holocene occur on modern reefs. For instance, Siegrist et al. (1984a) and Randall et al. (1984) found that the Merizo Limestone on Guam was primarily built by corymbose *Acropora* while reef slopes and more sheltered areas consist of large *Porites*- or *Goniastrea*-built buttresses. This situation is still observed on the modern reef where the corymbose *A. monticulosa* is a particularly striking species on Tinian and Rota with large colonies that can cover and build buttresses entirely by itself (Figs. 18.11e, 18.15a). Other corymbose species common in the shallow habitats belong

primarily to the *A. humilis* group of species. The high energy reef crest zone, where spur-and-groove structures form, are often occupied by calcareous algae if aerial exposure at low tide is evident. Dominance by *Pocillopora* is common elsewhere where lateral accretion occurs. It is, however, unclear whether this *Pocillopora* facies was a distinct, seaward high-energy facies, as described by Randall et al. (1984) from the Merizo Limestone at Ylig Point on Guam. On present-day reefs, the corymbose *Acropora*, *Pocillopora*, and calcareous red algae can switch dominance in the same habitat (Fig. 18.15c, d); perhaps a consequence of mean sea-level or the vagaries of recruitment. The shallowest reef environments in the Merizo Limestone and the present-day reefs are covered by a densely calcified algal ridge, often emergent at low tide and sometimes with dense fleshy algal overgrowth on the calcareous ridge, behind which in more sheltered areas extensive rhodolith beds can be found (Fig. 18.15e, f). Landward of this algal ridge, another zone of corymbose *Acropora* can be found.

FIG. 18.14. Overview of the 15 reefscapes that compose the CNMI-Guam island chain. The figure portrays benthic habitat maps produced by NOAA and areas in orange were identified as containing coral habitat. Note that each scale bar is different

Further landward in lagoons that are eroded into the Merizo or Milencatan Limestone, *Heliopora coerulea* is usually found in well-drained areas mixed together with infrequent but large stands of arborescent *Acropora* and *Porites* micro-atolls.

Present-day coral frameworks are best developed in the southern limestone islands where narrow fringing reefs are prevalent, and influenced by the underlying morphology (see Section 18.4.2). In southern Guam many small bays have discontinuous

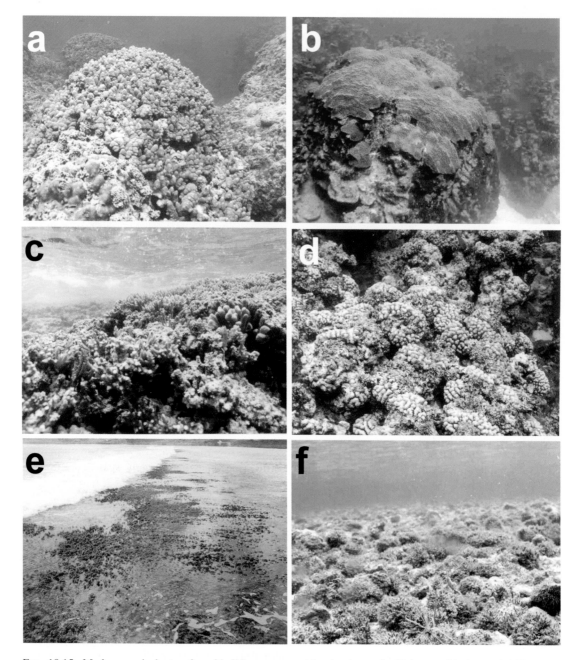

Fig. 18.15. Modern equivalents of reef-building communities in the early Holocene Merizo and Milencatan Limestones of Guam and Rota occur throughout the southern Marianas (**a**) the corymbose *Acropora monticulosa* can build, or at least blanket, entire buttresses; 2 m depth, Chulu Beach, Tinian (**b**) in deeper areas, large *Porites*-built buttresses occur. Clearly visible is the loss of living tissue suffered by this specimen in the *Acanthaster planci* (crown-of-thorns starfish) outbreaks of the 1960s and 1970s (Randall et al. 1988; Randall 1991; Kojis and Quinn 2001). (**c**) The corymbose *Acropora* facies is the widest distributed in the Merizo and Milencatan Limestones (also Fig. 18.11c). Here the windward upper reef slope at Chulu Beach, Tinian is mainly covered by *Acropora digitifera*. (**d**) Reef edges are usually covered predominantly by *Pocillopora*, here *P. meandrina*. (**e**) The reef crests are characterized by a well-cemented algal ridge of encrusting coralline algae, sometimes overgrown by fleshy macroalgae, Tanapag lagoon barrier reef, Saipan (**f**) more sheltered parts of the reef flat can harbor dense beds of geniculate red algae, like here *Lithophyton* Songsong, Rota

reef growth due to rivers that empty into the bay. Reefs increase in thickness and cover of organisms away from the river mouth (Wolanski et al. 2003, 2004). The notable exceptions to the general pattern of narrow fringing reefs hugging the islands' shorelines exist in Guam at Apra Harbour, around Cocos Lagoon, and in Saipan, around Tanapag Lagoon. These are the only instances where barrier reefs exist in the Marianas (Fig.18.16). The development of these lagoons is clearly structurally controlled. Figure 18.6 shows the Mariana Limestone, which forms an outer belt of calcareous sediment around the older Tagpochau Limestones, to disappear underneath Tanapag Lagoon on Saipan. It is likely that the Mariana Limestone was displaced along a fault line, since most major faults on Saipan strike approximately SW–NE. The alignment of the slumped Mariana Limestone would structurally control the shape and position of the Holocene Barrier reef.

The Cocos Block in southern Guam forms a roughly triangular barrier-reef with an enclosed lagoon. The Cocos reef and lagoon are probably growing on a basement of Umatac Formation, a block of which may have dropped along a fault that strikes parallel to the Talafofo fault zone and the Adelup fault (Randall 1979). Modern reef development in the Cocos Lagoon (Guam) and Tanapag Lagoon (Saipan) consists of a barrier reef, which follows the seaward structural controls of the subsurface and a fringing reef along the shorelines inside the lagoon (Emery 1962; Randall 1979). Also Tinian had a small section of barrier reef, but it was transformed into a port by filling and dredging during and immediately after World War II. A well-developed and large patch reef, 245 m across in its longest direction, is found just to the south of Tinian Harbor, the former barrier reef.

Just like in the Merizo and in the Milencatan Limestones, a very clear biotic zonation is found across the reefs and lagoons of Saipan and Guam. Due to the expansive lagoonal setting it is different than on the narrow fringing reefs. The zonation in Saipan was described by Cloud (1959) and is still in place today, despite several biotic disturbances to the corals in the area (crown-of-thorns starfish outbreaks, bleaching events; Richmond et al. 2002; Chapter 19, Richmond et al.). Beginning from seaward, a reef slope is encountered that can vary widely in coral cover. The upper reef slope, which

experiences the highest wave-energy exhibits very low coral cover (Fig. 18.17-1). Groove-and-spur structures are more common in the northern and southern sections of the barrier reef, where it approaches the land, than in the central, widest section. Cloud (1959) considered the groove-and-spur structures primarily erosional. The reef slope abuts a well-cemented reef crest, strongly encrusted by coralline algae and partially with dense brown algae overgrowth. The reef crest slopes lagoonward and towards its edge acquires a more or less dense overgrowth of massive corals and small corymbose *Acropora*. Towards the inner edge, a well-developed zone of *A. palifera* dominance is found (Fig.18.17-3). *A. palifera* is not the only coral occurring here, rather many massive *Porites* and *Heliopora* also occur, but these corals are first subordinate to increase lagoonward where they then displace *A. palifera* dominance and form a mixed-massive coral assemblage (Fig. 18.17-5). Embedded in and surrounded by this zone are areas with large (tens of meters diameter) thickets of *A. formosa*. Many of these colonies suffered mortality during the 1998 and possibly also 2002 bleaching years, but vigorous growth continued afterwards. The branching *Acropora* thickets form well-developed, elongate spurs that stretch lagoonward. Toward the center of the lagoon, coral growth becomes increasingly sparse as cover by sandy sediment increases. Rocky outcroppings with low coral cover and increasingly high algal cover towards the land as well as small patch reefs occur irregularly in the lagoon. Towards the shoreline, dense seagrass beds are encountered.

18.9   Conclusions

The Mariana Islands are home to a rich coral fauna that have been building framework reefs in the older, southern islands, since the Miocene. Thus, the southern islands have important carbonate caps. Where volcanic rocks crop out, higher hills concentrate orographic rain that leads to stronger run-off of lateritic soils due to rampant deforestation. This is a problem particularly on Guam and to a lesser extent on Saipan. Even in the southern islands, framework reefs are not uniformly distributed and non-framebuilding environments occur (Fig. 18.13). The latter dominate the northern, younger,

FIG. 18.16. Views of Marianas coral reefs (**a**) Well developed fringing reef near Agana, Guam (**b**) narrow fringing reef at Chulu Beach in northern Tinian (**c**) clearly visible groove-sand-spurs, reef-slope buttresses and exposed early Holocene reef limestone in the immediate supratidal, Aguian (**d**) raised early Holocene limestone in the immediate supratidal and older Tanapag limestone landward, southern coast, Saipan (**e**) fringing reef inside Tanapag Lagoon, Saipan (**f**) a view across Tanapag Lagoon, Saipan, towards the barrier reef in the distance. This is the widest lagoon in the Marianas

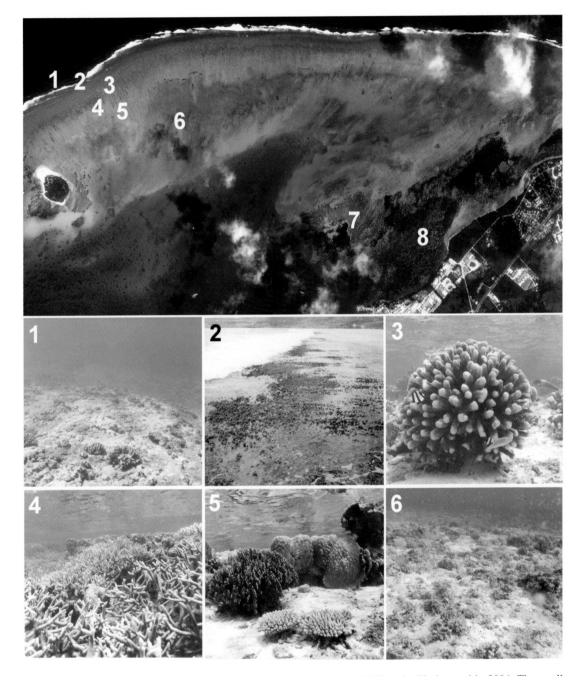

FIG. 18.17. Biotic zonation in Tanapag lagoon as described by Cloud (1959) and still observed in 2004. The small island to the left is Mañagaha Island, the shoreline of Saipan is to the right (**1**) seaward reef slope (**2**) emergent reef flat (**3**) *Acropora palifera* zone (**4**) *Acropora formosa* zone (**5**) massive coral zone (**6**) lagoonal sandy hardgrounds with algae (**7**) are hardgrounds with sparse corals and seagrass (**8**) are dense seagrass beds

islands. On Guam and Saipan large lagoon/barrier reef systems occur, that are structurally controlled by blocks slumped along fault lines. In the southern islands, raised early Holocene reefs occur.

Throughout the entire Holocene, the reefs of the Mariana Islands maintained a clear coral zonation with remarkable constancy. In this respect they differ markedly from the coral reefs in the US Atlantic. In the latter, the composition of the benthic communities and their zonation, in particular with regards to the dominant scleractinia, has changed dramatically. In the Southeast Florida continental reef tract, an early Holocene dominance by *A. palmata* and framebuilding by scleractinian corals (Banks et al. 2007) has been modified to a near-complete absence of *Acropora palmata* (*A. cervicornis* being conspicuously present, but not framebuilding) and framebuilding in general. Southeast Florida relict reefs are dominated today by a sponge and gorgonian-dominated community, not by scleractinia (Moyer et al. 2003; see also Chapter 5, Banks et al.). In the Florida Keys (Chapter 2, Lidz et al. and Chapter 3, Jaap et al.), *A. palmata* was a dominant framebuilder throughout the Holocene, but has suffered spectacular dieback and most present-day framebuilding by this species has all but ceased. This would be comparable to a situation in the Marianas where the entire corymbose *Acropora* and *Pocillopora* assemblages were lost. The concomitantly observed severe restriction of the Atlantic *A. cervicornis* would find an equivalent in the loss of the lagoonal *A. formosa* and other open arborescent *Acropora* beds in Saipan, which has not occurred despite similar stresses (predators, bleaching, nearby human population) as in the Atlantic. Thus, at its eastern and western confines, the USA is home to coral reefs that have widely diverged in their dynamics and further comparisons can only be fruitful.

Acknowledgements Support to NCRI by NOAA grant NA16OA1443. NCRI contribution #90. We thank M. Trianni for support in the field and critical review. The CNMI FWS and DEQ aided with permits and logistics during visits.

References

Agassiz A (1903) The coral reefs of the tropical Pacific. Mem Mus Comp Zool Harvard 28:1–410

Bell SC, Siegrist HG Jr (1988) Patterns in reef diagenesis: Rota, Mariana Islands. Proc 6th Int Coral Reef Sym, Australia 3:547–552

Banks KW, Riegl BM, Shinn EA, Piller WE, Dodge RE (2007) Geomorphology Of the Southeast Florida reef tract (Miami-Dade, Broward, and Palm Beach Counties, USA). Coral Reefs 26

Bloomer SH (1983) Distribution and origin of igneous rocks from the landward slopes of the Mariana Trench: implications for its structure and evolution. J Geophys Res 88:7411–7428

Bonjean F, Lagerloef GSE (2002) Diagnostic model and analysis of the surface currents in the tropical Pacific Ocean. J Phys Oceanogr 32:2938–2954

Bratring FWA (1806) Ueber die Ladronen- oder Marien-Inselgruppe in dem nördlichen Stillen Meere. Allgem Geogr Ephemer (Weimar) 21(3):261–275; and 21(4):369–397

Carruth RL (2003) Ground-water resources of Saipan, Commonwealth of the Northern Mariana Islands. USGS Water Resources Investigations Report 03–4178, 5 pp

Carruth RL (2005) Construction, geologic, and hydrologic data from five exploratory wells on Rota, Commonwealth of the Northern Mariana Islands, 1999. USGS Open-File Report 2005–1042, 40 pp

Clift PD, Lee J (1998) Temporal evolution of the Mariana arc during rifting of the Mariana Trough traced through the volcaniclastic record. Isl Arc 7:496–512

Cloud PE, Schmidt RG, Burke HW (1956) Geology of Saipan, Mariana Islands, Part 1. General Geology: US Geol Surv Prof Pap 280-A, 126 pp

Cloud PE (1959) Geology of Saipan, Mariana Islands, Part 4. Submarine Topography and Shoal Water Ecology. US Geol Surv Prof Pap 280-K, pp 361–445

Cole WS (1963) Tertiary larger foraminifera from Guam. US geol Surv Prof Pap 280-A, 126 pp

Dalrymple GB, Clague DA (1976) Age of the Hawaiian–Emperor bend. Earth Planet Sci Lett 31:313–329

Dickinson WR (2000) Hydro-isostatic and tectonic influences on emergent Holocene paleoshorelines in the Mariana Islands, Western Pacific Ocean. J Coastal Res 16(3):735–746

Doan DB, Burke HW, May HG, Stensland CH (1960) Military geology of Tinian, Mariana Islands. Chief of Engineers, US Army, 149 pp

Easton WH, Ku TL, Randall RH (1978) Recent reefs and shore lines of Guam. Micronesica 14:1–11

Eldredge LG (1983) Summary of environmental and fishing information on Guam and the Commonwealth of the Northern Mariana Islands: Historical background, description of the islands, and review of climate, oceanography, and submarine topography around Guam and the Northern Mariana Islands. NOAA-TM-NMFS- SWFC-40, 181 pp

Emery KO (1962) Marine Geology of Guam. Geol Surv Prof Pap 403-B, 76pp.

Esteban M, Klappa CF (1983) Subaerial exposure environment. In: Scholle PA, Bebout DG, Moore CH (eds) Carbonate depositional environments. Am Assoc Petrol Geol Mem 33: 2–95

Fryer P, Pearce JA, Stokking LB et al. (1992) Proc ODP Sci Results 125. College Station, TX (Ocean Drilling Program)

Gingerich SB (2002) Geohydrology and numerical simulations of alternative pumping distributions and the effects of drought on the ground-water flow system of Tinian, Commonwealth of the Northern Mariana Islands. USGS Water Resources Investigations Report 03–4077, 46pp

Gingerich SB (2003) Hydrologic resources of Guam. USGS Water Resources Investigations Report 03–4126, 3pp

Horine RL, Moore GF, Taylor B (1990) Structure of the outer Izu-Bonin forearc from seismic-reflection profiling and gravity modeling. In: Fryer P, Pearce JA, Stokking LB et al. (eds) Proc ODP Init Repts 125: College Station, TX (Ocean Drilling Program), pp 81–94

Hussong DM, Uyeda S (1981) Tectonic processes and the historyof the Mariana Arc: a synthesis of the results of Deep Sea Drilling Project Leg 60. In: Hussong DM, Uyeda S et al. (eds) Init Repts DSDP, 60: US Government Printing Office, Washington, DC, pp 909–929

Karig DH (1971) Structural history of the Mariana arc. Geol Soc Am Bull 82(2):323–44

Kayanne H, Ishii T, Matsumoto E, Yonekura N (1993) Late Holocene sea-level change on Rota and Guam, Mariana Islands, and its constraints on geophysical predictions. Quat Res 40:189–200

Kilmer F (1991) Microfossils in the Alutom Formation, Guam (abstract) South Pacific Science Congr, Honolulu, May 1991

Kojis BL, Quinn NJ (2001) The importance of regional differences in hard coral recruitment rates for determining the need for coral restoration. Bull Mar Sci 69(2):967–974

Lander MA (1994) Meteorological factors associated with drought on Guam: technical report No. 75, Water and Energy Research Institute of the Western Pacific, University of Guam, Guam, 39 pp

Meijer A, Reagan M (1981) Petrology and geochemistry of the island of Sarigan in the Mariana Arc; Calc-alcaline volcanism in an oceanic setting. Contrib Mineral Petrol 77:337–354

Meijer A, Reagan M, Ellis H, Shafiqullah M, Sutter J, Damon P, Kling S (1983) Chronology of volcanic events in the eastern Philippine Sea. In: Hayes DE (ed) The tectonic and geologic evolution of southeast Asian seas and islands: Part 2: Am Geophys Union, Geophys Monogr 27:349–359

Mink JF, Vacher HL (1997) Hydrogeology of northern Guam. In: Vacher HL, Quinn T (eds) Geology and Hydrogeology of carbonate islands. Develoments in Sedimentology 54, Elsevier, Amsterdam, The Netherlands, pp 743–761

Moyer RP, Riegl B, Banks K, Dodge RE (2003) Spatial patterns and ecology of benthic communities on a high-latitude South Florida (Broward County, USA) reef system. Coral Reefs 22:447–464

Mrozowski CL, Hayes DE, Taylor B (1981) Multichannel seismic reflection surveys of Leg 60 sites, Deep Sea Drilling Project. In: Hussong D, Uyeda S, et al. (eds) Init Repts DSDP 60: US Government Printing Office, Washington, DC, , pp 57–69

Norton IO (1995) Plate motions in the Pacific: the 43 Ma non event. Tectonics 14:1080–1094

Patriat P, Achache J (1984) India-Eurasia collision chronology has implications for crustal shortening and driving mechanism of plate. Nature 311:615–621

Pualay G (2003) Marine biodiversity of Guam and the Marianas: overview. Micronesica 35:3–25

Pirazzoli PA, Montaggioni LF (1988) Holocene sea level changes in French Polynesia. Palaeogeogr, Palaeoclim, Palaeoecol 68:153–175

Prowazek S (1913) Die deutschen Marianen. Ihre Natur und Geschichte. Leipzig. Johann Ambrosius Barth. iv, 126 pp.

Purkis SJ, Kohler KE, Riegl, BM, Rohmann SO (2007) The statistics of natural shapes in modern coral reef landscapes. J Geol 115:493–508

Rainbird P (1994) Prehistory in the northwest tropical Pacific: the Caroline, Mariana, and Marshall Islands. J World Hist 8(3):293–349

Randall RH (1979) Geologic features within the Guam seashore study area. Univ Guam Marine Lab Techn Rep 55, 53 pp

Randall RH (1985) Habitat geomorphology and community structure of corals in the Mariana Islands. Proc 5th Int Coral Reef Congr, Tahiti 6:261–266

Randall RH (1991) Tanguisson-Tumon, Guam, reef corals before, during, and after the crown-of-thorns starfish (Acanthaster planci) predation. Micronesica 24:274–175

Randall RH (1995) Biogeography of reef-building corals in the Mariana and Palau islands in relation to back-arc rifting and the formation of the Eastern Philippine Sea. Nat Hist Res 3(2):193–210

Randall RH, Eldredge LG (1977) Effects of typhoon Pamela on the coral reefs of Guam. Proc 3rd Int Coral Reef Sym, Miami 2:525–531

Randall RH, Siegrist GH Jr (1988) Geomorphology of the fringing reefs of northern Guam in response to

Holocene sea level changes. Proc 6th Int Coral Reef Sym, Australia 3:473–477

Randall RH, Siegrist HG, Siegrist AW (1984) Community structure of reef-building corals on a recently raised Holocene reef on Guam, Mariana Islands. Paleontogr Am 54:394–398

Randall RH, Rogers SD, Irish EE, Wilkins SC, Smith BD, Amesbury SS (1988) A marine survey of the Obyan-Naftan area, Saipan, Mariana Islands. University of Guam Marine Lab Tech Rep 90, 73 pp

Reagan MK, Meijer A (1984) Geology and geochemistry of early arc-volcanic rocks from Guam. Geol Soc Am Bull 95:701–713

Richmond R, Kelty R, Craig P, Emaurois C, Green A, Birkeland C, Davis G, Edward A, Golbuu Y, Gutierrez J, Houk P, Idechong N, Maragos J, Paulay G, Starmer J, Tafileichig A, Trianni M, van der Velde N (2002) Status of the coral reefs in Micronesia and American Samoa: US Affiliated and Freely Associated States in the Pacific. In: Wilkinson C (ed) Status of Coral Reefs of the World 2002. Australian Institute of Marine Science, Australia, pp 217–236

Schlager W (2005) Carbonate sedimentology and sequence stratigraphy. SEPM Concepts in Sedimentology and Paleontology 8, 200 pp

Schlanger SO (1964) Petrology of limeston es of Guam. US Geol Surv Prof Pap 403-D, 52 pp

Sharp WD, Clague DA (2006) 50-Ma initiation of Hawaii-Emperor bend records major change in Pacific plate motion. Science 313:1281–1284

Shepard FP, Curray JR, Newman WA, Bloom AL, Newell ND, Tracey JI, Veeh HH (1967) Holocene changes in sea level: evidence from Micronesia. Science 157:542–544

Siegrist AW, Randall RH, Siegrist HG (1984a) Functional morphological group variation within an emergent Holocene reef, Ylig Point, Guam. Paleontogr Am 54:390–393

Siegrist HG, Randall RH (1985) Community structure and petrography of an emergent Holocene reef limestone on Guam. Proc 5th Int Coral Reef Congr, Tahiti 6:563–568

Siegrist HG, Randall RH (1992) Carbonate geology of Guam. Proc 7th Int Coral Reef Sym 2:1195–1216

Siegrist HG, Randall RH, Siegrist AW (1984b) Petrography of the Merizo Limestone, an emergent Holocene reef, Ylig Point, Guam. Paleontogr Am 54:399–405

Siegrist HG, Randall RH, Edwards CA (1991) Shallow reef-front detrital sediments from the Northern Mariana Islands. Micronesica 24:231–248

Stafford K, Mylroie J, Taborosi D, Jenson J, Mylroie J (2005) Karst development on Tinian, Commonwealth of the Northern Mariana Islands: controls on dissolution in relation to the carbonate island karst model. J Cave Karst Studies 67:14–27

Stearns HT (1937) Geology of the water resources of the island of Guam, Mariana Islands. US Navy Manuscript Report

Stearns HT (1940a) Geologic history of Guam (abstr). Geol Soc Am Bull 52:1948

Stearns HT (1940b) Shore benches on north Pacific islands. Geol Soc Am Bull 52:773–780

Steinberger B, Sutherland R, O'Connell RJ (2004) Prediction of Emperor-Hawai'i seamount locations from a revised model of global plate motion and mantle flow. Nature 430:167–173

Stern RJ (1978) Agrihan: an introduction to the geology of an active volcano in the northern Mariana Island Arc. Bull Volcanol 41:43–55

Stern RJ, Smoot NC (1998) A bathymetric overview of the Marianas forearc. Isl Arc 7:525–540

Stern RJ, Fouch MJ, Klemperer S (2004) An overview of the Izu-Bonin-Mariana subduction factory. In: Eiler J (ed) Inside the Subduction Factory. Geophys Monogr 138:175–223

Sugawara S (1934) Topography, Geology and coral reefs on Rota Island. MSc thesis, Tohuku Imperial University (in Japanese)

Tayama R (1952) Coral reefs of the South Seas. Bull Japan Hydro Office 11: 292

Tracey JI Jr, Schlanger SO, Stark JT, Doan DB, May HD (1964) General geology of Guam. US Geol Surv Prof Pap 403-A, 104 pp

Whittaker JM, Müller RD (2006) New 1-minute satellite altimetry reveals major Australian-Antarctic plate reorganisation at Hawaiian-Emperor bend time. EOS Trans Am Geophys Union, Fall Meeting Supplement 87: Abstract T51F-08

Wolanski E, Richmond RH, Davis G, Bonito V (2003) Water and fine sediment dynamics in transient river plumes in a small, reef-fringed bay, Guam. Estuarine Coastal Shelf Sci 56:1029–1040

Wolanski E, Richmond RH, McCook L (2004) A model of the effects of land-based, human activities on the health of coral reefs in the Great Barrier Reef and in Fouha Bay, Guam, Micronesia. J Mar Syst 46:133–144

Yamasaki T, Murakami F (1998) Asymmetric rifting of the northern Mariana Trough. Isl Arc 7:460–470

19

Aspects of Biology and Ecological Functioning of Coral Reefs in Guam and the Commonwealth of the Northern Mariana Islands

Robert H. Richmond, Peter Houk, Michael Trianni, Eric Wolanski, Gerry Davis, Victor Bonito, and Valerie J. Paul

19.1 Introduction

The Mariana Islands are a chain of 16 volcanic peaks stretching over a distance of approximately 2,500 km from 13° to 21° N latitude and centered at 145° E longitude (Fig. 19.1). Politically, the area is divided into two jurisdictions, Guam and the Commonwealth of the Northern Mariana Islands. Guam is a US territory located at 13°28′ N, 144°45′ E and is the southernmost island in the Mariana Archipelago. It is the largest island in Micronesia, with an area of 560 km² and a maximum elevation of approximately 405 m above sea level. The northern portion of the island is relatively flat and consists primarily of uplifted limestone. The southern half of the island is primarily volcanic, with more topographic relief, and large areas of highly erodible lateritic soils (Siegrist and Randall 1992; Chapter 18, Riegl et al.). The island possesses fringing reefs, patch reefs, submerged reefs, offshore banks, and a barrier reef surrounding the southern shores. The reef margin varies in width, from tens of meters along some of the windward areas, to well over 100 m. The combined area of coral reefs and lagoons is approximately 69 km² in nearshore waters between 0–3 nmi, and an additional 110 km² in federal waters greater than 3 nmi offshore (Hunter 1995).

Guam was ceded to the US in 1898 following the Spanish-American War, and was placed under the administration of the Department of the Navy, with a US appointed governor. It was occupied by the Japanese from 1941 to 1944, after which time it was retaken by US Forces. US President Truman signed the Organic Act in 1949 establishing Guam as an unincorporated territory of the United States and granting citizenship to the island's people. A territorial college was established in 1952, which later became the University of Guam, accredited by the Western Association of Schools and Colleges. The University of Guam Marine Laboratory was established in 1970, and became a center for regional research on coral reefs. The Marine Laboratory has been largely responsible for the wealth of information available on coral reef taxonomy, biology and ecology in the western Pacific, and maintains a database and taxonomic collection as well as access to many regional coral reef-related publications.

The Commonwealth of the Northern Mariana Islands (CNMI), like Guam, was part of the Trust Territory of the Pacific Islands under US administration following World War II. The island chain, named after the widow of Spain's King Phillip IV, Mariana of Austria, became a Commonwealth of the United States in the 1970s, following the approval of a covenant, followed by the ratification of a local constitution. This group of islands extends from the inhabited island of Rota (Luta), just to the north of Guam, to the uninhabited island of Farallon de Parajos. There are three active volcanoes between this northern most point and the

B.M. Riegl and R.E. Dodge (eds.), *Coral Reefs of the USA*,
© Springer Science + Business Media B.V. 2008

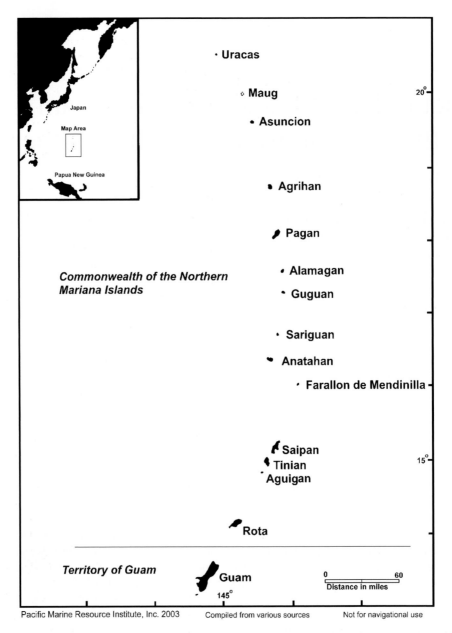

Fig. 19.1. Location of Guam and the Commonwealth of the Northern Mariana Islands in the western Pacific

main island of Saipan, which is the most populated and developed island in the CNMI.

The entire Marianas chain is an island arc system which lies to the west of the Marianas Trench, an active subduction zone where the Pacific Plate is being subducted under the Philippine Plate. Geologically, the group can be divided into the older, inactive volcanic islands of Guam, Rota, Aguijan, Tinian, Saipan and Farallon de Medenilla, which have substantial limestone deposits, and the nine younger, volcanically active northern islands that are offset to the west, including Anatahan, Sarigan, Guguan, Alamagan, Pagan, Agrihan, Asuncion, Maug and Farallon de Parajos (Randall 2003; Chapter 18, Riegl et al.).

19.2   Climate and Oceanographic Conditions

Guam and the CNMI are truly tropical islands, with a typical air temperature of approximately 28°, with a range from 24°C to 30°C, and similar seawater temperatures affecting coastal coral reefs. There is a pronounced dry season from December through June, with a rainy season typically extending from July through November. These islands are regularly hit by typhoons and tropical storms that can occur throughout the year (more often June through November, but major storms have hit during the months of December and January). During these events, winds can exceed 180 miles/h (290 km/h), with associated heavy rains, and large oceanic waves hitting the reefs and shores. Normal rainfall is variable among years, with an average of 218 cm, with the documented record of over 400 cm for Guam in 1976, an El Niño year.

The Mariana Islands are affected by seasonal tradewinds, which normally come out of the northeast or east, averaging 15 knots. As such, the corals and reefs on the windward exposures exhibit signs of wave activity, including more robust forms and skeletal characteristics, wave-mediated orientations and spur-and-groove formations.

The islands are also affected by the near-surface North Pacific Equatorial current that generally flows westward with speeds 0.1–0.2 m s^{-1}, larger in the south than in the north (Fig. 19.2a). On meeting the land mass of Guam, a 35 km long, slab-shaped island, these currents are deflected and generate unsteady eddies in its lee and areas of convergence of smaller-scale, localized currents (Wolanski et al. 2003a). A number of transient eddies off the tips of the island were apparent, the smallest eddies were at the scale of local topographic features such as headlands and embayments, while other eddies were island-size (Fig. 19.2b). In addition, large (200 km in diameter) oceanic eddies occasionally also travel past Guam and can generate for periods

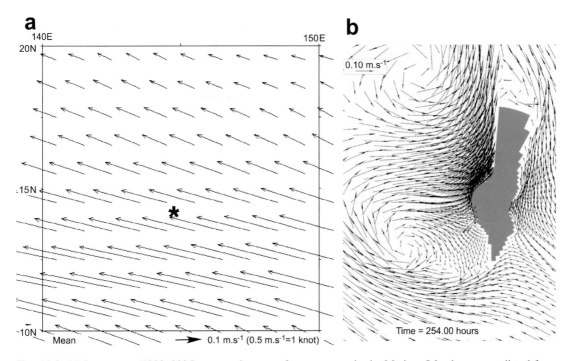

FIG. 19.2. (a) Long-term (1993–2206) averaged near-surface, currents, in the Mariana Islands area, predicted from NOAA's OSCAR satellite data. The area shown is 10–30° N and 140–140° E. The horizontal resolution of the data is 1°. * = Guam Island. (b) An example of the predicted synoptic distribution of the predicted near-surface currents around Guam for a northwestward far-field current impinging on Guam (Reproduced from Wolanski et al. 2003a by permission of Elsevier)

typically 1–2 weeks in duration, currents in the opposite direction to those generated by the North Pacific Equatorial Current. Eddies around Guam are sufficiently energetic to return fish and coral eggs and larvae to their natal reefs in Guam, thereby enabling self-seeding of coral reefs in Guam, while recruitment from far-away reefs is also possible and would vary enormously from year to year dependent on the oceanic currents at the time of spawning. Numerical models also predict that large (up to 30 m amplitude) island-generated internal waves may occur around Guam, however, no observations are presently available to support this prediction.

19.3 Biogeographical Setting of the Area

Guam and the CNMI are oceanic islands, far removed from any continental landmasses. The nearest neighbors are the islands of the Republic of Palau, 1,295 km to the southwest and the Federated States of Micronesia to the southwest (841 km) through the southeast (2,204 km). The Philippines lie to the west (2,300 km), and Japan, ca. 2,500 km to the north.

These islands are located in an area of high marine biodiversity, just to the east of the "Coral Triangle" of highest marine biodiversity (New Guinea, Indonesia and the Philippines). The Marianas fauna includes over 375 species of scleractinian corals, 195 species of echinoderms, 650 crustaceans and 1,000 species of reef and shorefishes (see review issue of Micronesica, Vol. 35–36, 2003; Randall and Myers 1983; Randall 2003; Porter et al. 2005).

The marine flora and fauna are relatively well-studied and documented through a variety of peer-reviewed publications, technical reports, environmental impact statements and surveys (see the University of Guam Marine Laboratory web site, http://www.uog.edu/marinelab, for a list of contributions).

19.4 Biodiversity of Major Organismal Groups

19.4.1 Corals and Invertebrates

Randall (2003) reported a total of 377 scleractinian corals covering 20 families from the Marianas

Islands (Fig. 19.3). Guam's reefs contain all of the major genera of reef-building corals, notably species of *Acropora, Porites, Pocillopora, Favia, Favites, Montipora, Fungia, Pavona, Montastrea, Leptoria, Leptastrea, Psammacora,* and *Galaxea.* Apra Harbor, protected from most oceanic swells a majority of the time, has well-developed shallow patch and fringing reefs, dominated by *Porites* spp., with a degree of vertical zonation from the surface to a depth of 20–30 m. Large anemones hosting different species of clown fish are also found in the Harbor. Within this bay, a few deeper mounds (20–25 m) exist with a good diversity and abundance of large sponges. Sclerosponges have been found in caves just to the south of the harbor. There is a steep drop off to the west of Apra Harbor, over 1,000 ft, which was identified as a possible OTEC (ocean thermal energy conversion) site. There is a pronounced vertical zonation of corals and other species along this slope.

Piti Bay, to the north of Apra Harbor, contains a unique and expansive rubble zone that is home to a number of new invertebrate records and previously undescribed species. There are rich benthic epifauna and infauna that includes a variety of holothurians (*Thelenota ananas, Bohadschia marmorata, B. argus, Holothuria atra, H. edulis, Actinopyga mauritiana,* and *Holothuria nobilis,* Fig. 19.4*),* molluscs and crustaceans. Several "bomb holes" that are actually karst caves with collapsed ceilings, are present within the bay, one of which contains an underwater observatory built in the late 1980s. This controversial project was opposed by many within the research and local communities. Corals (mostly *Porites* spp.) and soft corals (dominated by *Sinularia* and *Sarcophyton* spp.) line the edges of the "bomb holes".

Several outbreaks of *Acanthaster plancii,* the crown-of-thorns starfish occurred the 1970s and 1980s (Colgan 1987; Bonito 2002; Quinn and Kojis 2003). Additionally, the corallivorous starfish *Culcita novaeguinea* and the coral eating mollusc *Drupella* sp. are also found on the reefs of the Mariana Islands.

The molluscan fauna of Guam and the CNMI is rich, with approximately 800 species of prosobranch gastropods identified from shallow water reef habitats (Smith 2003). Giant clams are found on the reefs of Guam and the CNMI, with evidence that populations of *Tridacna gigas* and possibly

Fig. 19.3. Some representative members of the Mariana Islands coral reef fauna (photos by Megan Berkle) (**a**) *Pocillopora eydouxi* (scleractinia) (**b**) *Acropora* cf. *abrolhosensis* (scleractinia) (**c**) *Acropora (Isopora) palifera* (scleractinia) (**d**) *Sinularia* cf. *leptoclada* (alcyonacea) (**e**) *Millepora platyphylla* (milleporina) (**f**) *Heliopora coerulaea* (helioporacea)

T. crocea were collected to the point of local extinction (Smith 2003), while the smaller species including *T. maxima*, and *T. squamosa* are still found. Bailey-Brock (1999) reports 101 polychaete species from coral reefs on Guam and 15 from Saipan.

It can be assumed that these marked differences are more related to sampling than true faunistic differences between the islands. The polychaete fauna on Marianas reef flats is similar to that in comparable habitats of Hawaii, Enewetak, and Indonesia.

Fig. 19.4. Some members of the echinoderm fauna of the Mariana Islands (photos by Megan Berkle) (**a**) *Actinopyga mauritiana*, the surf redfish (**b**) *Holothuria fuscopunctata* (**c**) *Stichopus chloronotus* (**d**) *Linckia laevigata*

19.4.2   Zonation and Community Patterns

There are a variety of coral reef types in the Mariana Islands, including barrier reefs, fringing reefs, patch reefs, and submerged reefs associated with offshore banks (Chapter 18, Riegl et al.). Well developed carbonate platforms are found extending from shore on most of the islands, although these are less pronounced on the northern and geologically younger islands. The Mariana Islands have three species of seagrasses, *Enhalus acoroides*, *Halophila minor*, and *Halodule univervis*, which can be found in well-developed seagrass beds (Tsuda et al. 1977). Also coastal mangroves occur (*Avicennia marina*, *Bruguiera gymnorrhiza*, *Heritiera littoralis*, *Hibiscus tiliaceus*, *Lumnitzera littorea*, *Nypa fruticans*, *Rhizophora mucronata*, and *Xylocarpus moluccensis*) in the Marianas, but

many areas have been impacted throughout the island chain by dredging and coastal development (Mueller-Dombois and Fosberg 1998).

Within the Mariana Archipelago the most notable, broad-scale, reef-community zonation pattern exists between the northern, volcanically active islands and the southern, raised limestone islands (Fig. 19.5). A recent survey of 40 fringing reefs throughout the northern islands found that while coral diversity and colony surface area are significantly lower on the northern islands compared with the southern (mean of 62 species per site and 206 cm^2, mean of 82 species, 312 cm^2, respectively; Houk, 2005), population density is similar (mean of 144 and 139 colonies per site, respectively). This suggests that recruitment is not limiting, rather that harsh environmental conditions select against species settlement and growth (Randall 1985; Houk and van Woesik 2006). In support of this hypothesis, Randall (1985) found

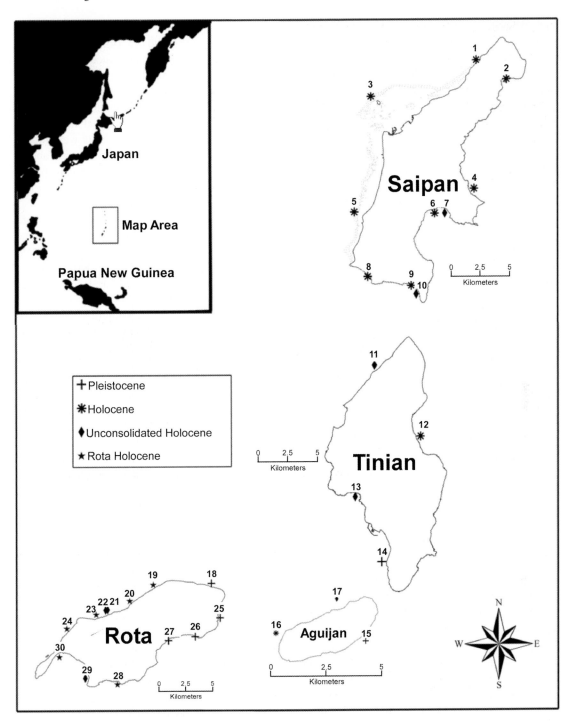

Fig. 19.5. A map of CNMI's southern islands showing long-term monitoring locations and geomorphological settings. The distance between islands is not drawn to scale

a significant relationship between percent coral coverage and geomorphological habitat complexity. The failure of fringing reefs to form on much of the coastline around the northern islands is attributed to: (1) unfavorable bathymetry, (2) a lack of favorable substrate on which corals can settle and grow, (3) a high exposure to wave energy, (4) the re-suspension of volcanic ash, and (5) volcanic eruptions. While it is clear that the failure of reefs to form is ultimately a consequence of the environment (Sheppard 1982; van Woesik and Done 1997; Montaggioni 2005), the proposed mechanism halting growth in the northern islands is a combination of natural disturbances acting against coral colony growth.

Corals reef growth in the southern islands of Saipan, Tinian, Aguijan, Rota, and Guam has not been uniform throughout the Holocene, which has resulted in further community zonation patterns. Typical reef growth, whereby a reef first aggrades to sea-level which is then followed by progradation, is rare throughout these islands. Relatively few late Holocene reefs are found throughout the Marianas, a trend common to high latitude reefs (above 15°) (Harriott and Banks 2002; Nozawa et al. 2006). Most modern reef flats represent flooded and abraded surfaces of early Holocene reefs formed on planed Pleistocene planation. In many locations late Holocene reef growth is absent, which creates a diverse assemblage of shallow-water habitats (Randall and Siegrist 1988). The largest shallow water system is Tanapag Lagoon in Saipan, which is a good example of this heterogeneous reef-building process insofar as its morphology is determined by pre-existing morphology, rather than modern reef building processes alone (Cloud 1959; Chapter 18, Riegl et al.). In some localities, modern coral communities have formed well-developed spur-and-groove reefs, while others remain entirely devoid of coral-mediated carbonate deposition. The 52 km coastline of Tinian Island only exhibits reef growth on 21%, while the 147 km coastline of Guam is 59% reef-fringed.

Houk and van Woesik (P. Houk, personal communication 2003) identified four distinct geomorphological settings in the CNMI with widely differing modern coral assemblages, and linked coral community persistence through time to geomorphology. They defined settings as: (1) typical Holocene, (2) unconsolidated Holocene, (3) Rota Holocene, and (4) Pleistocene reefs (Fig. 19.6). The community setting of typical Holocene reefs is distinct due to the diverse assemblage of framework building corals, dominated by encrusting *Montipora*, massive *Porites*, and large *Acropora* and *Pocillopora*. Unpublished data from in situ conductivity and temperature sensors

suggest that freshwater discharge through the reef matrix on the unconsolidated Holocene reefs after storms causes the notable difference between typical versus unconsolidated Holocene reefs. Thus, the magnitude and nature of freshwater input may be an influential driver of coral communities (Umezawa et al. 2002). In the unconsolidated reef setting more columnar and fewer encrusting corals were found, likely due to a lack of suitable reef substrate and in protected localities, such unconsolidated reefs can be dominated by mono-specific stands of *Porites rus*, *Coscinaraea columna*, or *Pavona* spp.

Rota's and some of Guam's present-day reefs are super-imposed over the remains of early Holocene reef growth exposed by eustatic sea-level drop coupled with geological uplift ~2,000 years ago (Dickinson 2000; Kayanne et al. 1993; Chapter 18, Riegl et al.). Rota's modern spur-and-groove reef slopes contain small to medium-sized massive colonies, with the exception of *Astreopora* corals that attain larger sizes. The fossilized, raised reef-flats along the coast of Rota are a visual reminder that typical Holocene reefs dominated by branching *Acropora* and *Pocillopora* corals once prevailed (see also Chapter 18, Riegl et al.).

In some areas, substantial modern deposition is absent and small massive corals occur directly on the Pleistocene surface. Juvenile, framework building corals are commonly found but the environment limits their size (Harriott and Banks 2002; van Woesik and Done 1997; Nozawa et al. 2006; Houk 2006).

Interestingly, among the common macroinvertebrates that are annually surveyed at 30 sites throughout the CNMI, only the sea-cucumbers show a positive response to modern reef deposits (Fig. 19.7). The genera *Stichopus* and *Actinopyga* are more abundant on Holocene reef crests, where they are commonly harvested, than in other environments. Sea-urchins and sea-stars showed an opposite trend, increasing in abundance in low-relief reef settings (Fig. 19.7). While sea-cucumbers are in many places in the Pacific critically over-exploited, some good populations remain in the Mariana Islands (Kerr et al. 1993; Trianni and Bryan 2004).

A physiographic zonation pattern common to coral reefs also exists in the Marianas (Darwin 1842; Sheppard 1982; Done 1983), and is best delineated into three habitats: (1) the back-reef/

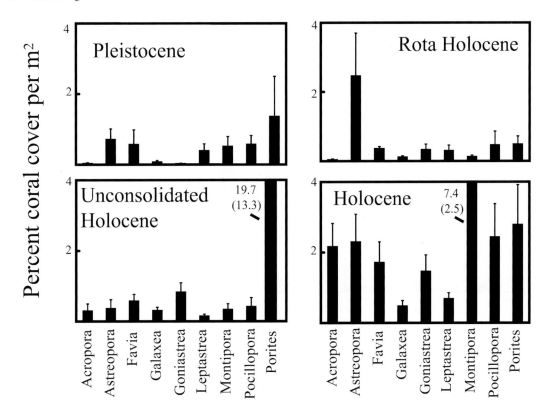

FIG. 19.6. Coral abundance in each geological setting

lagoon, (2) the high-energy reef-crest, and (3) the reef slope. Obvious differences in modern growth occur between these habitats, as dictated by the abiotic environment (light, wave energy, and oxygen levels).

19.5   Fisheries

The harvest of fisheries resources from nearshore coral reefs in the present day CNMI had been occurring since the Prehistoric Phase over 3,000 years before present (Amesbury et al. 1989). Various occurrences of fishing were documented from the logs of different early European explorers, with archaeological and historical evidence and accounts having been summarized by Amesbury and Hunter-Anderson (2003).

No documented information of commercialized harvest from the Spanish Period exists, and prior to the Japanese occupation coral reef resource harvest was primarily conducted for subsistence purposes.

During the German occupancy of the Marianas from 1899 to 1914, it was noted by a German district officer of Guam that Carolinians from Saipan dove for trepang at Aguigan, selling them to Japanese merchants (Amesbury et al. 1989). This activity marks the first documented directed commercial fishing effort by indigenous peoples in the CNMI.

19.5.1   Reef Fisheries Landings

In the Japanese Period (1914–1944) the commercial harvest of fisheries resources was high, as evidenced by over 2,000t landed in 1941 (Smith 1947). Of the 1941 total, about 318t were listed as "other fish", and about 28t as "sea cucumber" and "other shells". Additionally, about 12t of "sharks" were landed. If one-half of the "other fish" and "sharks" categories are considered coral reef species, then about 165t of nearshore reef fish and sharks were also harvested, bringing the total 1941 nearshore coral reef harvest to about 193t. The lack

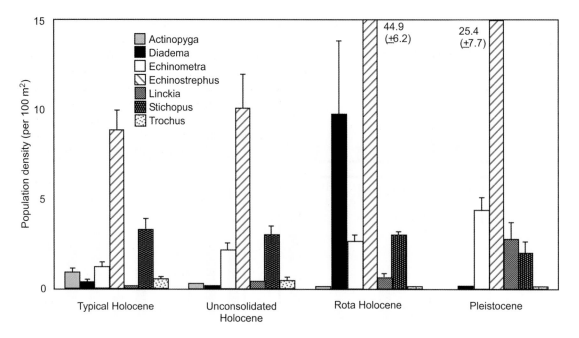

FIG. 19.7. Macroinvertebrate abundance in each geological settings

of available data from the Japanese Period remains a quandary, as it is not known whether the 1941 landings represent a high, low or mean level. The degree of fishing effort during the Japanese Period could have been influential in molding the current nearshore coral reef community structure.

Amesbury and Hunter-Anderson (2003) reviewed archaeological and anthropological data concerning reef fishing in Guam and the Northern Mariana Islands and summarized contemporary landings. They provide a summary of landings from the Trust Territory Administrative Period (1947–1976) in the Northern Mariana Islands from 1948 through 1977, excepting the years 1950–1956 and 1973–1975 when no statistics were available. Although the accuracy of landings is not definable, there appears to be significant variation in landings during this period, and in most years both pelagic and nearshore landings are aggregated. Unlike during the Japanese Period, the majority of fish landed during the Trust Territory Administrative Period (TTAP) were probably nearshore fish, and for non- or small-scale commercial purposes. In the years 1957–1960, the percentage of "tuna" that comprised the total fisheries landings were recorded separately, and ranged from 4% to 33%, averaging

about 17%. As tuna are the primary constituent of contemporary pelagic fisheries landings, it can be assumed that the majority of the total landings consisted of bottom-associated fish and of reef fish since hydraulic and electric reels used in contemporary deeper bottom fisheries were in an incipient stage of development. The long-term average of fisheries landings from this period was about 39 t, and using the 1957–1960 average this figure can be amended to about 33 t.

The primary contemporary fisheries data acquisition method employed in the CNMI is through the collection of commercial sales receipts from fishermen and vendors. The Commercial Purchase Database System (CPDS) is a product of collaboration between the Western Pacific Fishery Information Network (WesPacFIN) and the CNMI Division of Fish and Wildlife (DFW). The CPDS is presumed to capture about 80% of all commercial landings in the CNMI. The documented commercial landings of reef fish indicate a sudden increase from 1981 through 1990, followed by stabilized landings from 1991 through 2002, with a decline from 2003 to 2004 and a rise in 2005–2006 (Fig. 19.8). The CPDS long-term average for commercial landings of reef fish and invertebrates in the CNMI from 1981 to 2006 is about 77 t. Raising this average by the 20% provides a long-term average of about 93 t.

The landings from the three referenced data sources are depicted in Fig. 19.9.

19.5.2 Management of Coral Reef Fisheries Resources in the CNMI

Although historical landings provide some information regarding coral reef fisheries resources, they are insufficient for use as a practical management tool. In addition to the CPDS both shore-based and boat-based creel surveys have been inconsistently implemented since the early 1990s, yielding inadequate trend data for coral reef fisheries. Although creel surveys have been successfully implemented in many jurisdictions and are essential for general

data monitoring and collection, the protocols of data collection as well as the limitations of the data are not always adequate to manage fisheries. All data collection activities in the CNMI are voluntary, as no law exists that requires fishermen to provide the DFW with catch and effort data. As a result, many reef fisheries in the CNMI have been directly monitored and assessed, oftentimes by attaching data reporting conditions to export permits or relying on the goodwill of the fishermen to cooperate. In some cases, management measures have been enacted in agreement with survey and assessment results. Some case examples follow.

19.5.2.1 Scuba Spear Fishery in the CNMI

During the mid 1990s the scuba-spear reef fishery based out of Saipan was monitored and assessed by the Division of Fish and Wildlife (Graham

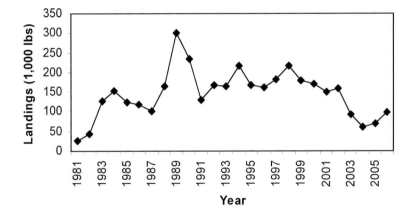

Fig. 19.8. CNMI Division of Fish and Wildlife Commercial Purchase Database Landings, 1981–2006

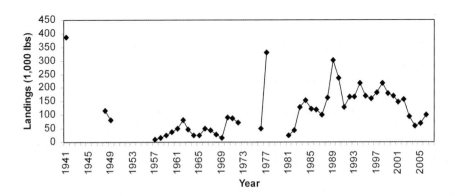

Fig. 19.9. Documented landings of coral reef resources. Japanese Period 1941; Trust Territory Period 1948–1977; CNMI Period 1981–2006

1994; Trianni 1998). This fishery harvested reef fish from nearshore coral reefs around Saipan and Tinian, but also from coral reefs at islands north of Saipan. A number of species harvested in this fishery that represented primary food fish groups were evaluated by comparison of mean fork length (FL) per location (Trianni 1998). Significant differences were observed with mean FL being typically larger for specimens from the Northern Islands, followed by Tinian and Saipan. These results are summarized in Table 19.1. Catch per unit of effort data (CPUE) collected from Saipan and Tinian during two sampling periods, Period 1 (1993–1994) and Period 2 (1995–1996), showed a significant decline from Period 1 to Period 2. CPUE was also found to be lower for Saipan than Tinian, with the lowest CPUE values for each island obtained from leeward aspects. Conclusions drawn from Graham (1994) and Trianni (1998) suggested the ban on the use of scuba-spear fishing as well as indiscriminate

methods such as gill nets. The proposed ban on the use of gill nets was supported by independent reef fish survey in 1979 focusing on fish resources in the Saipan Lagoon (Amesbury et al. 1979), which was repeated in 1997 (Duenas and Associates 1997). Results indicated that primary groups of food fish had decreased in number between surveys (Fig. 19.10).

Fishing effort utilizing scuba-spear fishing eventually became more prevalent on Tinian and Rota, as Saipan-based companies began fishing at those islands when nearshore resources in Saipan became increasingly difficult to harvest. This led to local concerns for reef fish resources on Tinian and Rota, as well as on Saipan. The use of scuba-spear fishing was perceived as being a non-traditional method in direct conflict with the free-diving spear fishermen community. As a result of local community concerns and pressure, the passing of local laws to prohibit the use of scuba-spear

Table 19.1. List of species evaluated for mean fork length by location from the scuba-spear fishery. 1 = largest, 3 = smallest.

Family	Species	Location		
		Saipan	Tinian	Northern Islands
Holocentridae	*Myripristis berndti*	3	2	1
Serranidae	*Epinelophelus fasciatus*	2	2	1
Lutjanidae	*Lutjanus kasmira*	2	2	1
Lethrinidae	*Gnathodentex aurolineatus*	3	2	1
Caesionidae	*Pterocaesio tile*	3	1	2
Scaridae	*Scarus rubrioviolaceus*	2	2	1
	Scarus forsteni	2	3	1
Acanthuridae	*Acanthurus blochii*	2	1	1
	Acanthurus lineatus	2	2	1
	Naso lituratus	3	2	1
	Naso unicornis	2	1	2

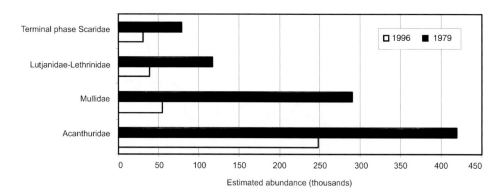

FIG. 19.10. Comparison of primary food fish group abundance from fisheries independent surveys in 1979 and 1997

fishing began in 2000 when Rota local law 12–2 was enacted. Tinian enacted local law 13–01 in 2002, and Saipan and the Northern Islands enacted Saipan and the Northern Islands local law 13–13 in 2003. These laws remain in effect.

19.5.2.2 *Sea Cucumber Fishery in the CNMI*

In October 1995 a sea cucumber fishery that targeted the surf redfish, *Actinopyga mauritiana*, began on the island of Rota (Trianni 2002). The CNMI DFW became aware of the fishery in December 1995 and instituted management measures in January 1996. One problem with the fishery was the lack of baseline abundance data from which to manage harvest levels. As the economic abundance of surf redfish declined on Rota, the fishing company planned to switch harvest to Saipan in mid-1996. Planned pre-harvest surveys on Saipan were not allowed to commence, and harvest began without baseline data to mange from. The DFW created export permit conditions and successfully enforced them, managing the fishery by evaluating CPUE over time (Trianni 2003). As the Saipan CPUE declined with time the fishery was halted and a survey of the resource in harvested areas was commenced in April 1997. The results of that survey, along with results from depletion modeling, indicated from 78% to 90% of initial population sizes in the harvested areas were removed (Trianni 2003). Based on these results, the Saipan sea cucumber fishery was subsequently terminated in May 1997. Although management of sea cucumber fisheries in the CNMI had proven challenging, political pressures in 1997 sought to allow harvest on the island of Tinian. The DFW successfully conducted a pre-harvest survey in fall 1997 and determined that harvest of the surf redfish in Tinian was not sustainable either biologically or economically. Subsequently, no harvest was allowed on Tinian (Trianni and Bryan 2004). It was eventually accepted that sea cucumber fisheries in the CNMI were not sustainable at a commercial level, and any subsequent inquiries for resource harvest be viewed with extreme caution.

During the progression of the sea cucumber fishery from Rota to Saipan, public concerns mounted as to the sustainability of the sea cucumber resource to harvest as well as the impact on the coral reef ecosystem. The CNMI Government eventually recognized the public concern over the harvest of sea cucumbers, an industry that served primarily to satisfy markets in Hong Kong and Taiwan, to the detriment of local coral reef resources. In 1998 House Bill No. 11–144 was introduced, becoming Public Law 11–63 in 1999. This law established a 10-year moratorium on the harvest of sea cucumbers (as well as seaweeds and seagrasses) throughout the CNMI.

19.6 Environmental Factors Influencing Reef Biology

The Mariana Islands are regularly affected by strong typhoons and oceanic swells, with associated high energy waves that can break branching corals, overturn large massive colonies and scour encrusting forms through the abrasive nature of moving sand and rubble. Additionally, the island chain is situated in an area of high tectonic activity, with frequent earthquakes of relatively high magnitudes (in 1993, Guam experienced an event measuring 8.4 on the Richter scale). Faults and slumping have affected reefs throughout the island chain.

Many of the corals studied to date (ca. 40 species) including Acroporids, Poritids, Favids and Pocilloporids, participate in synchronous mass spawning events, 7–10 days following the June and July full moons (Richmond and Hunter 1990; Richmond 1997). Peak reproductive activity corresponds with the rainy season, and periods of reproductive failure due to runoff and coastal pollution have been documented for reefs in the area (Richmond 1996, 1997).

The crown-of-thorns starfish, *Acanthaster planci*, has been responsible for declines in live coral cover resulting from several documented outbreaks during the 1960s, 1970s, 1980s and 1990s. The first documented outbreak occurred in 1967 (Chesher 1969) and Randall (1973) reported a resulting substantial effect on coral cover from approximately 50–60% over a 30 m depth range to less than 20% on the reef front and approximately 1% at deeper depths. Bonito (2002) studied coral community structure at Tanguisson Reef on the leeward side of the island, and demonstrated a significant correlation with outbreaks of *Acanthaster* (Fig. 19.11).

Coral diseases, increases in cyanobacteria, and outbreaks of the encrusting sponge *Terpios* have also affected coral reef community structure on Guam.

19.6.1 Water Quality

Guam underwent a development boom in the late 1980s and early 1990s. Increased stormwater run-off from an airport expansion project, new roads, hotels, shopping centers and golf courses resulted in reduced coastal water quality, especially in bays and areas of restricted water circulation (Richmond

and Davis 2002). The salinity of waters over coastal reefs was found to drop below 28 ppt after periods of rainfall during summer coral spawning events, which was documented to result in reproductive failure (Richmond 1996). Stormwater collection passes into sewer lines and during periods of heavy rain, water is diverted around the sewage treatment plants and discharged directly into the ocean outfall pipes with treatment at best at the primary level. Three of the Island's outfall pipes discharge within 200 m of the shoreline, in depths of 20–25 m and in areas where corals are found. Extension

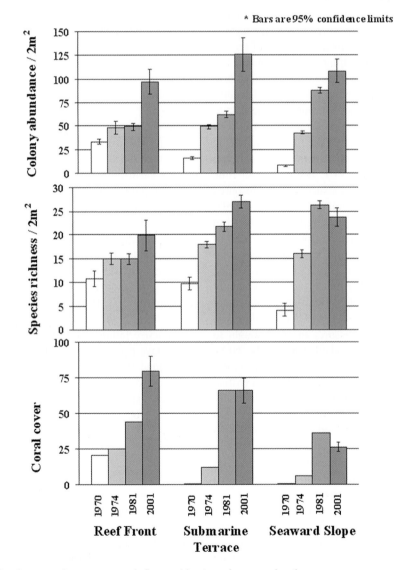

FIG. 19.11. Coral community structure as influenced by *Acanthaster* outbreaks

of the Northern and Central District outfalls into deeper waters further offshore is planned.

A variety of pollutants have been found in the sediments and a variety of marine organisms from Apra Harbor, including PCBs, heavy metals (arsenic, copper, lead, mercury and tin) and PAHs (Denton et al. 1999), which is not surprising based on the high level of military, shipping and related activities occurring within the enclosed bay. The levels reported were considered "mild to moderate," when compared to other harbors around the world, yet are a cause for concern considering re-suspension that could occur as a result of a proposed harbor dredging project. Agrochemicals, both pesticides and fertilizers are also a concern for Guam's coastal waters, resulting from agriculture, golf courses and turf maintenance on the grounds of ocean front hotels.

Discharges into the coastal zone occur via surface runoff on the southern portion of Guam, and from both surface runoff and aquifer discharge on the northern, limestone areas. Due to urbanization, more non-permeable surfaces associated with roads, parking lots and other types of watershed modifications have increased over the past several decades, and are expected to become more of an issue, requiring efforts at integrated watershed management (Wolanski et al. 2004).

19.6.2 Present Status of Reef Health

Guam's reefs range from those in excellent condition to others that have been degraded by anthropogenic stressors. Runoff and sedimentation impacts differ substantially between the northern part of the island which consists of uplifted limestone and the southern part which is primarily volcanic with high topographic relief (see also Chapter 18, Riegl et al.). Assessment and monitoring programs have been ongoing on many of the reefs and associated communities since the 1970s and reveal a variety of human impacts, natural cycles of disturbance, and synergisms between the two.

The studies of Tanguisson Reef show periods of *Acanthaster* predation followed by periods of recovery. The reef front, reef slope and submarine terrace areas are a distance from the Northern District sewer outfall, and have demonstrated repeated periods of coral recruitment and recovery following acute periods of predation. Studies of the reefs adjacent to

the outfall found chronic effects and the sustained loss of species. In contrast to the outer Tanguisson Reef sites, coral communities off Fouha Bay on southern Guam have shown continued losses tied to runoff and sedimentation (Table 19.2; Wolanski et al. 2004).

19.7 Human Impacts and Conservation Issues

Sedimentation is the major anthropogenic problem for the central and southern reefs. For the Ugum River Watershed, soil erosion was estimated at 176,500 t/km^2/year (DeMeo 1995). Forty six percent of this was attributed to roads on slopes, while 34% was contributed from badlands. Ugum Watershed erosion rates doubled from 1975 to 1993 (from 1,547,250 to 3,039,750 t/km^2/year), which was attributed to road construction and development projects. Sediment accumulation on reefs has been documented to substantially reduce both coral diversity and abundance (Randall and Birkeland 1978).

With over one million tourists visiting Guam each year, many from countries where a coral reef conservation ethic is not fully developed, damage to reefs is inevitable. In addition to impacts of SCUBA divers and snorkellers, underwater walking tours using surface-supplied equipment and a large number of personal watercraft (jet skis) have affected reefs and water quality. A coastal use zoning law called the Recreational Water Use Master Plan was passed into law to address these problems but needs enforcement support and to be updated to cover new activities and areas.

Groundings of fishing vessels, recreational watercraft and ships carrying cargo and illegal immigrants have resulted in localized damage to reefs. Guam's main power generation facilities are located on Cabras Island, in the northern portion of Apra Harbor. Elevated temperatures from the dis-

Table 19.2. Coral diversity versus distance from shore in Fouha Bay, Guam in 1978 and in 2003.

	1978	2003
Coral diversity vs distance from shore (cumulative)	0–25 m 3 spp.	0–40 m 0 spp.
	0–75 m 40 spp.	45–90 m 5 spp.
	0–125 89 spp.	100–275 m 41 spp.
	0–200 104 spp.	

charge of seawater used to cool the generators has resulted in coral mortality. The discharge of cleaning chemicals has also occurred, with subsequent impacts on local coral populations.

19.7.1 Fisheries Issues in the CNMI

The nearshore management of coral reef fisheries resources in the CNMI involves long-term low resolution monitoring through the CPDS and both boat-based and shore-based creel surveys, as well as directed monitoring of fisheries on an as-needed basis. These approaches have proved challenging, especially without the legal requirement of fishermen to provide data to the DFW. The result has been the institution of regulations banning the use of certain methods, such as the scuba-spear ban, and legislation that creates harvest moratoria, as with sea cucumbers. Additionally, the CNMI has four no-take Marine Sanctuaries, three on Saipan (Managaha Marine Conservation Area, Bird Island and Forbidden Island Marine Sanctuaries) and one on Rota (Sasanhaya Bay Fish Reserve), with plans to institute a limited take Marine Sanctuary on Tinian. These Marine Sanctuaries are relatively new, and the two oldest no-take reserves, the Managaha Marine Conservation Area (MMCA) and the Sasanhaya Bay Fish Reserve (SBFR) both have demonstrated positive changes over the course of annual assessments from 2000 through 2005/06, with examples provided in Fig. 19.12. In addition to monitoring and assessment of fisheries and Marine Sanctuaries, the Fisheries Research Section (FRS) of the DFW also conducts life history studies of select near shore reef fish. These life history studies provide age and growth parameters required for reliable stock assessments.

The coral reef fisheries resources of the CNMI have become better managed over time, and as research and monitoring activities continue to provide information for use in evaluating resource sustainability, it can be anticipated that coral reef fish resources will approach levels that effectively balance ecological necessities with sustenance needs.

19.7.2 Fisheries Issues on Local Coral Reefs in Guam

Fish populations and catch per unit effort (CPUE) have measurably declined in Guam since data collection began in 1985. Fishing practices, including the use of unattended gill nets, bleach, SCUBA spearfishing and fish traps have contributed to the problem. However, habitat loss due to sedimentation, pollution and physical damage has also been responsible for reduced fish populations.

Guam's Division of Aquatic and Wildlife Resources performed creel censuses that documented a 70% reduction (Fig. 19.13) in coastal fisheries catch from 1985 through 1996 (DAWR Annual Reports), and a comparable drop in catch per unit effort, that led to the passing of Public Law 24-21 in 1997 which established five no-take MPAs around the island and updated the local fishing regulation. These permanent no-take areas cover approximately 18% of Guam's coastal coral reef area and took 14 years to establish legally. This effort was made possible through one of the best coral reef shoreline creel surveys that annually estimates fisheries effort and harvest by method for the entire island. Additionally, the legal changes made through Public Law 24-21 served two primary purposes: to establish management measures to restore sustainable use and to improve enforceability.

A rapid shift in the health of the fisheries was marked by a dramatic move toward commercialization of the reef fisheries in the mid-1980s to support a rapidly growing tourism economy and significant development. Because of the increased demand for fish and a relatively small area of reef, this pressure did not take long to significantly reduce stocks. The rate of decline and the Government's inexperience with sharing fishery information with the public made the installation of effective management approaches publicly contentious. The earliest attempts did not include public input into the development of a strategy and after considerable exchange between government agencies, a series of public hearings were held. While public reactions were negative to the proposed no-take areas and to a number of regulatory changes the positive side was that at least public involvement had begun. Criteria were established to evaluate 60 proposed sites for fisheries management before selecting 9 after significant governmental and public input. A revised package was developed and taken back to public hearing which gained strong support for the changes and the overall need for establishing fisheries management measures. The revised package was then submitted to the legislature and again went through a hearing process after which the regulation package 19-2 was passed.

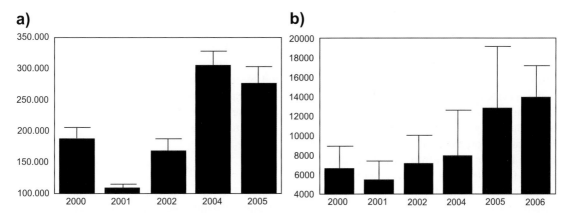

FIG. 19.12. Population estimates of (**a**) sedentary Acanthuridae from the Managaha Marine Conservation Area, and (**b**) terminal phase Scaridae from the Sasanhaya Bay Fish Reserve

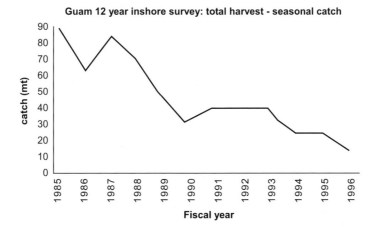

FIG. 19.13. Catch per unit effort (CPUE) for coastal fisheries on Guam from 1985 through 1996

No-take areas were established in 1997 but it was not until January of 2000 that the new regulations were fully enforced and proper signage was installed to inform the public and visitors of those areas that were protected. In total, there are two federal (i.e., War in the Pacific National Historical Park and Guam National Wildlife Refuge) and 11 territorial MPAs, with five of the territorial MPAs no-take marine reserves.

19.8 Case Studies

The Reefs of Guam and the CNMI are affected by a combination of anthropogenic and natural stressors acting synergistically, and there are several case histories worth mentioning.

19.8.1 Typhoons and Tropical Storms

In December 1997, Supertyphoon Paka passed through the Mariana Islands, causing substantial damage to structures, vegetation and the infrastructure of Guam and Rota, with the eyewall passing over both islands. Winds gusting to over 200 mph were recorded, but the record or 236 mph could not be confirmed due to damage to the airfield's recording anemometer. While acute wave damage was observed on coastal reefs, with overturned and fragmented coral colonies, runoff and associated debris became a further confounding issue. Coastal surveys following the storm found a diverse assortment of deposited materials, from sheets of roofing tin causing abrasion to surviving corals to clothing wrapped around coral colonies and causing additional physical damage. Divers recovered and disposed of 40,000 lb of debris during reef cleanups. Six tons were taken from the shallow reef areas and 14 t were recovered in waters from 3 to 20 m in depth. The rapid response was organized and supported through the cooperation and collaboration of local agencies, institutions, dive clubs and community members along with funding and logistical support from NOAA and the Department of the Interior. This effort has been used as a model for developing a rapid response protocol to address reef health following future events and disturbances.

19.8.2 Development and Poor Land Use Practices

Guam and the CNMI, particularly the island of Saipan, underwent a period of rapid economic growth and development during the 1990s, tied to what has been described as the Asian Bubble Economy. Due to the geographic location of the Mariana Islands, the tourism which served to fuel economic activity was primarily from Asia (Japan), and hence the "flavor" of development was also culturally tied to the funding source. Hotels, golf courses and tourist attractions were developed rapidly while the regulatory agencies and infrastructure struggled to keep up. The agency guidance and public testimony presented at public hearings were largely ignored, the environmental impact and assessment process was routinely violated and regulatory enforcement was insufficient to prevent substantial increases in erosion and sedimentation from poor land use practices. Foreign laborers brought in under the H-2 visa program, many of whom did not speak any English, were unaware of local fishing and collecting regulations, and often used destructive fishing practices. The results of these problems included user conflicts, failed infrastructure (power, water and sewer), extensive runoff and sedimentation from land clearing, and substantial impacts to coastal coral reefs and associated resources. While numerous damaging projects and examples of governmental failures to require and enforce adequate mitigation measures abound, the key outcome of value is the lesson learned that these islands need to address carrying capacity issues, develop and implement a more effective planning and review process, and communities need to have the political will to manage

the next episode if there are to be robust coral reefs left as a legacy for future generations. Promises of economic sustainability were made that never occurred and environmental losses were substantial. As these islands prepare for a major increase in military activity due to base closures in Asia, adequate environmental precautions are critical to limit future losses.

19.8.3 Fouha Bay, Guam

Fouha Bay in southern Guam is an excellent example to illustrate the plight of coastal coral reefs throughout the world that are impacted by runoff and sedimentation. Runoff often contains a variety of toxicants, including hydrocarbons, pesticides and heavy metals. Freshwater alone can reduced the success of coral spawning events by osmotically stressing coral gametes and larvae, and the addition of pollutants can affect corals and other coral reef organisms at all life history stages (Richmond et al. 2007). Sediment can affect coral reefs through direct physical abrasion and burial, attenuation of light necessary for zooxanthellae to photosynthesize, through the burial of infaunal bioturbating organisms leading to localized anoxic conditions, and as a carrier of toxicants from land. Development has continued to be a concern in the absence of adequate efforts at integrated watershed management. Fouha Bay, on the southern shore of Guam, was studied in the 1970s and again from 2002 to 2005, and the results provide clear documentation of the impacts of runoff and sedimentation on coastal coral reef ecosystems (Table 19.2).

Fouha Bay receives inputs via a small river from a watershed of approximately 5 km^2. Suspended sediment concentration exceeded 500 mg l^{-1} for a few hours following rain events and can even reach more than 1,000 mg l^{-1}(Wolanski et al. 2003b). River floods are usually very short, typically lasting only a few hours. A near-surface river plume is formed that spreads throughout Fouha Bay (Fig. 19.14). As the tidal currents are weak, the main flushing process for the river plume is the baroclinic current driven by the river discharge. Also this current is weak, typically only a few centimeters per second and only exists during the river flood. Thereafter, the river plume floats passively over the oceanic waters in Fouha Bay. The residence time of the river plume in Fouha Bay – which is only 400 m in size – is large (>1 day). During that time over

75% of the fine riverine sediment flocculates and settles out from the near surface plume, in the process smothering corals and forming a near-bottom nepheloid layer. This sediment is resuspended by swell waves that occur during the passage of a hurricane (typhoon) in the surrounding ocean. During such events the suspended sediment concentration reached 2,000 mg l[-1] for several days (Fig. 19.14). Coral reef ecohydrology models suggest that such events are harmful to corals and prevent coral regeneration (Wolanski et al. 2004).

19.9 Conclusions

The southern Mariana Islands have well-developed coral reefs that are presently subjected to a variety of stresses of natural and anthropogenic nature.

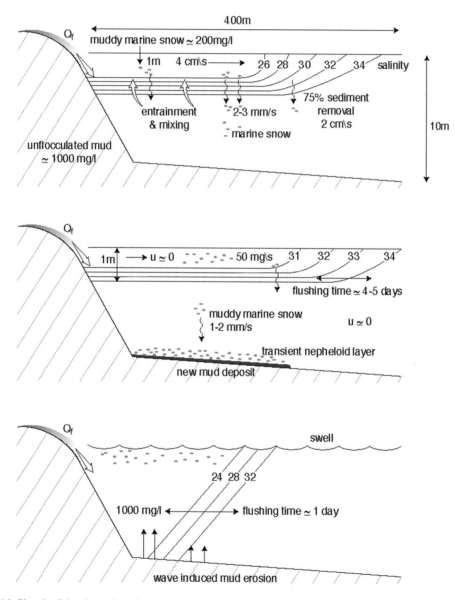

FIG. 19.14. Sketch of the dynamics of river and fine sediment in the transient river plume in Fouha Bay (top) during the river flood in calm weather, (middle) after the river flood in calm weather, and (bottom) during the river flood under a typhoon-driven swell (Reproduced from Wolanski et al. 2003b by permission of Elsevier)

They harbor one of the richest coral reef faunas in the USA. Coral reefs respond in their development to structural constraints by antecedent topography formed by early Holocene and Pleistocene reefs. Nature of bedrock is an important determinant of sedimentation stress, which is higher in areas of igneous rocks that weathers to lateritic soils and is vulnerable to erosion after deforeststaion. Fouha Bay in southern Guam is a case study for reefs under sedimentation stress. Natural impacts that have led to degradation of coral reef resources were outbreaks of crown-of-thorn starfish in the 1970s and 1980s as ell as the passage of typhoons. Increased development pressure has created significant decline of coral reefs and associated resources and forced the regulatory agencies to learn many valuable lessons. Efficiency of coral reef governance has improved but many challenges remain.

Acknowledgements We thank M. Berkle of Rota for allowing use of her pictures.

References

Amesbury JR, Hunter-Anderson RL (2003) Review of archaeological and historical data concerning reef fishing in the U.S. flag islands of Micronesia: Guam and the Northern Mariana Islands. Prepared for Western Pacific Regional Fishery Management Council by Micronesian Archaeological Services, 139 pp

Amesbury JR, Hunter-Anderson RL, Wells EF (1989) Native fishing rights and limited entry in the CNMI. Prepared for Western Pacific Regional Fishery Management Council by Micronesian Archaeological Services, 129 pp

Amesbury SS, Lassuy DR, Myers RF, Tyndzik V (1979) A survey of the fish resources of Saipan Lagoon. University of Guam Marine Laboratory Technical Report No. 52, 58 pp

Bailey-Brock JH (1999) Ecology and biodiversity of coral reef polychaetes of Guam and Saipan, Mariana Islands. Int Rev Hydrobiol 84:181–196

Bonito V (2002) Tanguisson reef: changes in coral community structure driven by *Acanthaster planci* predation? M.S. thesis, University of Guam Marine Laboratory, 56 pp

Chesher RH (1969) Destruction of Pacific corals by the sea star *Acanthaster planci*. Science 165:280–283

Cloud PE (1959) Geology of Saipan, Mariana Islands, Part 4. Submarine topography and shoal water ecology. US Geol Surv Prof Pap 280(K):361–445

Colgan M (1987) Coral Reef recovery on Guam (Micronesia) after catastrophic predation by *Acanthaster planci*. Ecology 68:1592–1605

Darwin C (1842) The structure and distribution of coral reefs. Smith, Elder and Co, London

De Meo (1995) Resource assessment, Ugum watershed, Guam. U.S. Department of Agriculture Natural Resources Conservation Service Report, 102 pp

Denton GL, Conception HR, Wood V, Eflin, V, Pangelinan GT (1999) Heavy metals, PCB's, and PAH's in marine organisms from Four Harbor locations on Guam. A Pilot Study. Water and Environmental Research Institute, University of Guam, 151 pp

Dickinson WR (2000) Hydro-isostatic and tectonic influences on emergent Holocene paleoshorelines in the Mariana Islands, Western Pacific Ocean. J Coastal Res 16(3):735–746

Done TJ (1983) Coral zonation: its nature and significance. In: Barnes DJ (ed) Perspectives on coral reefs. Australian Inst Mar Sci, Townsville:107–147

Duenas & Associates (1997) Saipan Lagoon use management plan, survey of sea cucumber and fish in the Saipan Lagoon, Northern Mariana Islands. Prepared for CNMI Coastal Resources Management Office, 55 pp

Graham T (1994) Biological analysis of the nearshore reef fish fishery of Saipan and Tinian. Commonwealth of the Northern Mariana Islands, Division of Fish and Wildlife Technical Report 94–02, 124 pp

Harriott VJ, Banks SA (2002) Latitudinal variation in coral communities in eastern Australia: a qualitative biophysical model of factors regulating coral reefs. Coral Reefs 21:83–94

Houk P (2006) Spatial Distribution of Coral Reef Communities and Reef Growth in the Commonwealth of the Northern Mariana Islands. Florida Institute of Technololgy, Melbourne, FL, 260 pp

Houk P, vanWoesik R (2006) Coral Reef benthic video surveys facilitate long-term monitoring in the Commonwealth of the Northern Mariana Islands: toward an optimal sampling strategy. Pac Sci 60:177–189

Hunter CL (1995) Review of coral reefs around American flag Pacific islands and assessment of need, value, and feasibility of establishing a coral reef fishery management plan for the western Pacific region. Final Report, Western Pacific Regional Fishery Management Council, 30 pp

Kayanne H, Ishii T, Matsumoto E, Yonekura N (1993) Late Holocene sea-level change on Rota and Guam, Mariana Islands, and its constraints on geophysical predictions. Quat Res 40:189–200

Kerr AM, Stoffel EM, Yoon RL (1993) Abundance distribution of Holothuroids (Echinodermata: Holothuroidea) on a windward and leeward fringing coral reef, Guam, Mariana Islands. Bull Mar Sci 52(2):780–791

Montaggioni L (2005) History of Indo-Pacific coral reef systems since the last glaciation: development patterns and controlling factors. Earth-Sci Rev 71:1–75

Mueller-Dombois D, Fosberg FR (1998) Vegetation of the Tropical Pacific Islands. Springer, New York, 733 pp

Nozawa Y, Tokeshi M, Nojima S (2006) Reproduction and recruitment of scleractinian corals in a high latitude coral community, Amakusa, southwestern Japan. Mar Biol 149:1047–1058

Porter V, Leberer T, Gawel M, Gutierrez J, Burdick D, Torres V, Lujan E (2005) The stats of coral reef ecosystems of Guam. In: Waddell JE (ed) The State of Coral Reef Ecosystems of the United States and Pacific Freely Associated States: 2005. NOAA Technical Memorandum NOS NCCOS 11, 522 pp

Quinn NJ, Kojis BL (2003) The dynamics of coral reef community structure and recruitment patterns around Rota, Saipan, and Tinian, western Pacific. Bull Mar Sci 72:979–996

Randall RH (1973) Distribution of corals after *Acanthaster planci* (L.) infestation at Tanguisson Point, Guam. Micronesica 9:212–222

Randall RH (1985) Habitat geomorphology and community structure of corals in the Mariana Islands. Proc 5th Int Coral Reef Congr, Tahiti 6:261–266

Randall RH (2003) An annotated checklist of hydrozoan and scleractinian corals collected from Guam and other Mariana Islands. Micronesica 35–36:121–137

Randall RH, Birkeland C (1978) Guam's reefs and beaches, Part II. Sedimentation Studies at Fouha Bay and Ylig Bay. University of Guam Marine Laboratory Technical Report No 47, 34 pp

Randall RH, Myers RF (1983) Guide to the coastal resources of Guam, Vol. II. The Corals. University of Guam Marine Laboratory Contribution 189, 128 pp

Randall RH, Siegrist GH (1988) Geomorphology of the fringing reefs of northern Guam in response to Holocene sealevel changes. Proc 6th Int Coral Reef Sym 3:473–477

Richmond RH (1996) Effects of coastal run-off on coral reefs. Biol Conserv 76:211

Richmond RH (1997) Reproduction and recruitment in corals: critical links in the persistence of reefs. In: Birkeland C (ed) Life and Death of Coral Reefs, pp 123–245

Richmond RH, Davis G (2002) Status of the Coral Reefs of Guam. In: Turgeon DD et al. (eds) The State of Coral Reef Ecosystems of the United States and Pacific Freely Associated States: 2002. NOAA/NCCOS, Silver Spring, MD, 265 pp

Richmond RH, Hunter CL (1990) Reproduction and recruitment of corals: comparisons among the Caribbean, the tropical Pacific, and the Red Sea. Mar Ecol Progr Ser 60:185–203

Richmond RH, Rongo T, Golbuu Y, Victor S, Idechong N, Davis G, Kostka W, Neth L, Hamnett M, Wolanski E (2007) Watersheds and coral reefs: conservation science, policy and implementation. BioScience 57:598–607

Sheppard CRC (1982) Coral populations on reef slopes and their major control. Mar Biol 7:83–115

Siegrist HG, Randall RH (1992) Carbonate geology of Guam. Proc 7th Int Coral Reef Sym 2:1195–1216

Smith BD (2003) Prosobranch gastropods of Guam. Micronesica 35–36:244–270

Smith RO (1947) Survey of the fisheries of the former Japanese Mandated Islands. USFWS Fishery Leaflet 273,106 pp

Trianni MS (1998) Summary and further analysis of the Nearshore Reef Fishery of the Northern Mariana Islands. CNMI DFW Technical Report 98–02, 64 pp

Trianni MS (2002) Summary of data collected from the sea cucumber fishery on Rota, Commonwealth of the Northern Mariana Islands. Beche-de-Mer Info Bull 16:5–11

Trianni MS (2003) Evaluation of the resource following the sea cucumber fishery of Saipan, Northern Mariana Islands. Proc 9th Int Coral Reef Symp :829–835

Trianni MS, Bryan PG (2004) Survey and estimates of commercially viable populations of the sea cucumber Actinopyga mauritiana (Echinodermata: Holothuroidea), on Tinian Island, Commonwealth of the Northern Mariana Islands. Pac Sci 58:91–98

Tsuda RT, Fosberg FR, Sachet MH (1977) Distribution of seagrasses in Micronesia, Micronesica 13:191–198

van Woesik R, Done T (1997) Coral communities and reef growth in the southern Great Barrier Reef. Coral Reefs 16:103–115

Umezawa Y, Miyajima T, Kayanne H, Koike I (2002) Significance of groundwater nitrogen discharge into coral reefs at Ishigaki Island, Southwest of Japan. Coral Reefs 21:346–356

Wolanski E, Richmond RH, Davis G, Deleersnijder E, Leben RR (2003a) Eddies around Guam, Mariana Island Group. Cont Shelf Res 23:991–1003

Wolanski E, Richmond RH, Davis G, Bonito V (2003b) Water and fine sediment dynamics in transient river plumes in a small, reef-fringed bay, Guam. Est Coast Shelf Sci 56:1029–1043

Wolanski E, Richmond RH, McCook L (2004) A model of the effects of land-based, human activities on the health of coral reefs in the Great Barrier Reef and in Fouha Bay, Guam, Micronesia. J Mar Syst 46:133–144

20

Geologic Setting and Ecological Functioning of Coral Reefs in American Samoa

Charles Birkeland, Peter Craig, Douglas Fenner, Lance Smith, William E. Kiene, and Bernhard Riegl

20.1 Introduction

American Samoa is rich in coral reefs and all islands are more or less fringed by coral reefs. Although structurally not part of the Samoan chain, political American Samoa includes Rose Atoll, a true atoll, and Swains Island. The coral reefs of American Samoa are integrated into a national protected areas system with the National Park of American Samoa (US Department of Interior) managing some coral reefs on the north coast of Tutuila near Vatia and along the shores of southern Ofu, and southeastern Ta'u, while the National Marine Sanctuary Program (US Department of Commerce) manages Fagatele Bay.

Although debated among historians, many believe that the Samoan Islands were originally inhabited as early as 1000 BC. Thus, the division between American and independent Samoa is very recent and pre-Western history of both Samoan groups is inextricably linked. The Manu'a Islands (Ofu, Olosega, Ta'u) of American Samoa have one of the oldest histories of Polynesia, and the Tuimanu'a title, formerly held by the highest chief of the Manu'a islands, is considered the oldest chiefly title. The title's name is obviously derived from the islands' name and its prestige is because the Manu'a Islands were, at least according to Samoan oral tradition, the first islands settled in Polynesia. During the Tongan occupation of Samoa, Manu'a was the only island group that remained independent because of the familial relationship between the Tuimanu'a and the Tuitonga, who was descended from a former Tuimanu'a. The islands of Tutuila and Aunu'u were culturally connected to Upolu Island in what is now independent Samoa. Still today, all the Samoan Islands are politically connected through the chieftain system and through family connections.

Samoa was not reached by European explorers until the eighteenth century. Early Western contact did not get off to a good start because of a battle in the eighteenth century between French explorers and islanders in Tutuila, which earned the Samoans a fierce reputation. In March 1889, a German naval force shelled a village in Samoa and also destroyed some American property. An impending battle between three American warships, ready to open fire on the three German ships, was made moot by an intervening typhoon – which sank both the German and American fleets. After an outbreak of tribal warfare led the British to lose interest in Samoa, Americans and Germans divided the islands in the Treaty of Berlin in June 1899. Western Samoa went to Germany, and eastern Samoa became American Samoa. The US Navy quickly built a coaling station in Pago Pago harbor for its Pacific Squadron and appointed a local Secretary. The Navy secured a Deed of Cession of Tutuila in 1900 and a Deed of Cession of Manu'a in 1904. In 1914, New Zealand occupied German Samoa, which then became British and, finally, independent in January 1962.

Samoans were at times at odds with the colonial governments, which led to the creation of the Mau movements (Mau meaning "opinion" or "testimony" in Samoan), essentially non-violent resistance movements which were largely suppressed. In

1940, the port of Pago Pago became a training and staging area for the US Marine Corps. During the war years, the United States built roads, airstrips, docks and medical facilities, exposing island residents to the American way of life and causing some major damage to the coral reefs.

After the war, American Samoan chiefs led an attempt to incorporate American Samoa, but this was defeated in Congress. This led to the creation of a local legislature, the American Samoan Fono which meets in the village of Fagatogo, the official capital of the territory. The Navy-appointed governor was replaced by a locally elected one, and American Samoa is now self-governing under a constitution of July 1967.

20.2   Geographic Setting

American Samoa is situated in the south Pacific Ocean on the Pacific Plate, the largest of the world's tectonic plates. Like other archipelagoes on the Pacific Plate (e.g., Hawaiian, Society, Marquesa, Tuamotu, and Caroline Islands), the Samoan Islands are being carried along the path traveled by the plate, moving from southeast towards the northwest at about 7 cm/year. The Samoan Islands are presently about 14° S and fall in the path of the South Equatorial Current, so the broad-scale water movement is from the east towards the west through Samoa. The greatest coral-reef diversity lies far downstream from Samoa, to the west. There are over 500 species of reef-building corals in the southwest Pacific, about 288 in American Samoa, a little over 100 in the Society Islands, and less than 50 on the Pacific coast of the Americas. This pattern of intermediate diversity in American Samoa is also observed in other taxonomic groups such as echinoderms and coral-reef fishes.

The Samoan Islands are not only upstream from the centers of diversity, but they are also relatively isolated and distant from other island groups in comparison with islands of the western Atlantic. This isolation may contribute to the apparently special resilience of the coral communities. The coral populations of American Samoa have been severely affected by large-scale acute disturbances such as outbreaks of the coral-eating crown-of-thorns starfish *Acanthaster planci*, hurricanes, and bleaching in response to seawater warming. When allowed a 15-year interval between disturbances, the coral

communities have recovered (Fig. 20.1). This is in contrast to the western Atlantic where there has been a continual degradation of coral reef systems for a half a century (Gardner et al. 2003). The relatively small area of the tropical western Atlantic allows broadscale events on continents to affect the whole region (Hallock et al. 1993; Garrison et al. 2003). The nutrients (Hallock et al. 1993), pollutants (Garrison et al. 2003), and diseases (Lessios et al. 1984) can disperse across the entire region. American Samoan reefs have managed to maintain resilience by receiving disturbances only as acute events and being largely isolated from nearby big land masses. Overfishing, however, has been chronic and so the fish communities have not been as resilient as the corals (Zeller et al. 2006, 2007).

20.3   Geologic Setting of American Samoan Coral Reefs

The Samoan Islands are part of a roughly 3,000 km long chain of islands, seamounts, shallow banks and atolls (Fig. 20.2). The US territory of American Samoa consists of five high islands, one low island and one atoll. As with other archipelagoes on the Pacific Plate in which the islands are formed over melting anomalies, believed to be caused by mantle plumes also called hot spots (Morgan 1971), island ages generally decrease from oldest in the northwest to youngest in the southeast as the plate moves over the hotspot and continues towards the northwest. The Samoan Islands are situated in the South Pacific Superswell – South Pacific Isotopic and Thermal Anomaly (SOPITA) region (McNutt and Fisher 1987), with an unusually shallow seafloor that is likely caused by anomalously hot and buoyant asthenosphere (Koppers et al. 1998; Dickinson 1998, 2001).

The volcanic features of this chain are oriented in a pattern consistent with an origin caused by movement of the Pacific plate over a hotspot (Natland 1980; Menard 1986). However, superficially observed, the age and size progression appears backward if compared with the Hawaiian, Caroline, and Society Islands. The northwestern end of the Samoan chain had originally been considered to be located near Savai'i, an active volcano, while the southeastern end of the chain is marked by Rose Atoll. In the other mentioned lineaments, submerged volcanic edifices, drowned or capped by atolls, occur on the northwestern

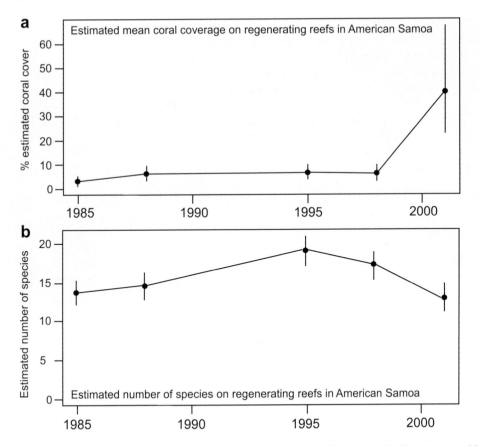

FIG. 20.1. McArdle (2003) analysed the data from most of the reports listed in Section 16.10. The response of Samoan reefs to disturbance is a 15-year recovery from large-scale acute factors (e.g., Colgan 1987 for Guam and Sano 2000 for Iriomote Island). During the first decade or so, numerous small corals recruit to the area but a superficial view of the area looks as if recovery is not occurring and one could fear that there has been a phase shift. However, each of the abundant but tiny recruits start growing and the coral communities spring back remarkably rapidly in the final half a decade (cf. Table 13 in Colgan 1987 and Table 1 in Sano 2000). The decrease of local (within transect) diversity has been speculated to be a result of competition for space

ends of the chains – and not in the southeastern, as Rose atoll does. Since the Pacific Plate is rigid and drifts in a westerly or northwesterly direction, this arrangement appears anomalous (Menard 1986) and the origin of the Samoan chain as a hotspot trace has been disputed. Thus at one end of the chain would be the inactive Rose atoll while the other end would be characterized by the active volcanism at Savai'i – which is exactly opposite to the situation observed in the other lineaments. However, to the west of Rose atoll is situated Vailulu'u submarine volcano (originally called Rockne volcano, then Fa'afafine; Johnson 1984; Hart et al. 1999), which has been identified as the active Samoan hotspot (Hart et al.

2000) and which is the true southeastern extreme of the Samoan chain. Rose atoll does not form part of the Samoan chain as such. It is possible that the plume center that is now apparently situated at Vailulu'u has migrated NE over the past 40 Ma (million years, Fig. 20.3; Hart et al. 2004) but there is controversy over how much of the relative motion of the Pacific Plate is due to movement of the plate versus drift of the hotspot itself and there is a possibility that the apparent motion of the mantle plume may be due to other factors, like true polar wander (P. Wessel, personal communication 2007). Recent paleomagnetic evidence suggests true polar wander, rather than plume migration, to have caused the apparent

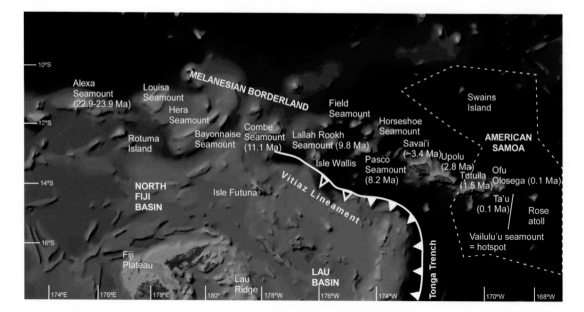

FIG. 20.2. American Samoa is situated in the Samoan Chain, a linear hotspot trace originating at Vailulu'u Seamount (the active hotspot) and extending through the Manu'a group (Ofu, Olosega and Ta'u) to Tutuila. The largest volcanic edifices in the Samoan group, Upolu and Savai'i, also the largest islands, are part of western Samoa. The Samoan Chain then abruptly changes into a series of seamounts extending via Pasco, Lallah Rookh, Waterwich and Combe until Alexa Seamount in the west. The southern portion of the Pacific Plate is subducted into the Tonga Trench, which turns into a transform fault called the Vitiaz Lineament. Stresses associated with flexural warping of the Pacific plate at the plate boundary are generally considered responsible for the rejuvenated volcanism at Savai'i/Upolu and the unusual appearance of the Samoan Chain with the presently active volcanoes in the west, not the east. Rose Atoll is the only seamount with a carbonate cap and is not part of the Samoan hotspot trail. Ages of volcanic edifices based on Keating (1992), Hart et al. (2004), Clouard and Bonneville (2005). EEZ boundary of American Samoa (broken white line) is only approximate and for orientation

displacement of the plume in the Hawaiian–Emperor chain, so it is possible that this is the mechanism observed in Samoa as well (see also Chapter 13, Rooney et al.). In general, a lively scientific debate exists regarding the plume hypothesis and which melting anomalies really represent mantle plumes, whether these really exist or what exact mechanism is involved (e.g., papers in Foulger et al. 2005). Further to the northwest of Vailulu'u are, as would be expected, young islands with tholeiitic lavas of the shield-building phase (like Ta'u, Ofu, Olosega, <1 Ma in age). The larger and older western Samoan Islands are characterized by alcalic basalts of second-ary, younger activity (Fig. 20.4), which is also shown by series of cinder cones along the islands' axis and flows originating from there. In the Hawaiian Islands, this phase of rejuvenated volcanism relative to the shield-building volcanism is seen largely as an effect of crustal loading (see Chapter 11 Fletcher

et al., and Chapter 13, Rooney et al.). In Samoa, however, renewed volcanism has been linked to flexural warping of the Pacific Plate (Menard 1986; Natland 2004). The northwestern Samoa Islands are situated only 100–150 km from a flexural upwarp (the trench forebulge) that is maintained at the bend where the Pacific Plate changes from subduction in the Tonga trench to the transform fault of the Vitiaz Lineamant (Fig. 20.2). This upwarp may be the cause of a 300 km rift that is observed along the Samoan islands (in particular Savai'i and Upolu), manifested by an almost linear series of craters striking parallel to the islands' axes (Fig. 20.4).

While the original explanation for the renewed volcanism in Samoa saw it as a response to Pacific Plate flexure associated with subduction in the Tonga Trench (Menard 1986; Wright and White 1986; Natland 2004), Hart et al. (2004) in contrary believe that volcanism in western Samoa is rejuvenated

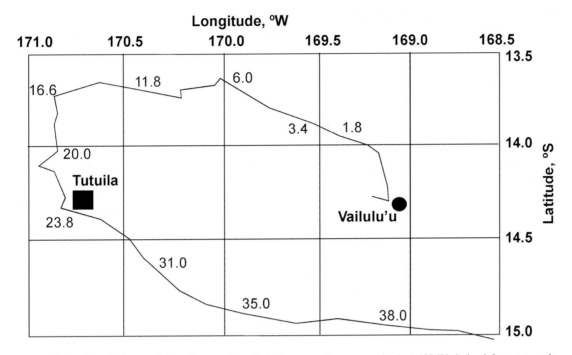

FIG. 20.3. Backtracked potential surface motion of the Samoan plume over the last 40 MY derived from a mantle dynamics model of Steinberger (2000, 2002) and Steinberger et al. (2004), taken from Hart et al. (2004). (By permission of Elsevier Ltd.) However, rather than plume migration, other mechanisms, like polar wander, may have caused the apparent displacement of the plume on the plate (see also Chapter 13, Rooney et al.)

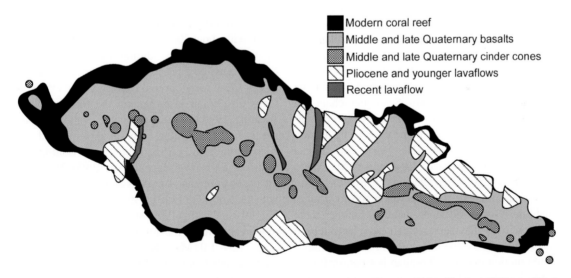

FIG. 20.4. Generalized geologic map of Upolu in Western Samoa. (From Stearns 1944; Keating 1992; modified by permission of the Geological Society of America and Springer) showing the linear arrangement of most recent volcanism (Middle and late Quaternary cinder cones) along the central axis of the island. The situation is similar at Savai'i. The steep side of the island has a fringing reef, the less steeply sloping side of the island has a barrier reef that initiated on the drowned edge of the island as it was subducted. (Modified from Stearns 1944, by permission of the Geological Society of America). The situation in Tutuila was not dissimilar with a drowned barrier reef (due to the island's rapid submergence) along the southern coast – presently only fringing reefs are developed

due to the islands approaching another hotspot that has formed seamounts along other lineaments. From the large western Samoan Islands, a series of seamounts extends further west and follows the general Samoan lineament. Aging and geochemical data suggest that seamounts as far away as Combe and Alexa have Samoan pedigree which suggests that all the islands and seamounts west of and including Vailulu'u were formed from the same single hotspot (Hart et al. 2004; Fig. 20.2).

The Samoan Island chain thus resembles the Hawai'i/Emperor Chain as described in Chapters 11 (Fletcher et al.) and 13 (Rooney et al.) inasmuch as it also ends in a linear seamount chain.

Several significant differences exist between the Samoan and Hawaiian Chains. In Hawai'i, the active mantle plume is situated underneath the emergent island of Hawai'i and has created a much bigger volcanic edifice than the Samoan hotspot at Vailulu'u, which is not emergent. A similarity with the Hawaiian chain is volcanism at the still submerged Lo'ihi (see Chapter 13, Rooney et al.). Hawaiian Islands progressively diminish in size towards the west, and with the exception of subsidence and emergence due to flexural volcano loading (Dickinson 2001) and some rejuvenated volcanism, generally sink progressively deeper. In Samoa, islands get bigger towards the west (Fig. 20.5), with abundant renewed volcanism in particular at Savai'i and Upolu. From then, the Samoan Islands brusquely give way to a seamount chain. No gentle progression of sinking is observed. Between the Hawaiian and the Emperor Seamount chain, an approximately 45 degree offset in direction is indication for either a change in direction of the Pacific Plate at between 43 Ma (Dalrymple and Clague 1976; Patriat and Achache 1984) and 50 Ma (Sharp and Clague (2006); see Chapter 13; Rooney et al.) or a southward drift of the Hawaiian hotspot prior to that time (Norton 1995). As mentioned earlier, there is a possibility that the Samoan hotspot has not been stable (Fig. 20.3), and may have, in a more complicated motion than the Hawaiian hotspot, drifted in a clockwise movement from its initiation at 40 to 16.6 Ma in a northwesterly direction, and then in an easterly, then southeasterly and finally northwesterly direction again (Steinberger 2002). Steinberger et al. (2004) suggest that motion of hotspots is influenced by plume-distortion due to global mantle flow, but controversy remains. Whether due to hotspot

migration or polar wander, in both the Samoan and Hawaiian chains the apparent position of the hotspot has not remained stable which caused a deviation of volcanic edifices from a strictly linear trend.

The islands of American Samoa are therefore mostly basaltic and have virtually no subaerial carbonate cover. Ta'u is a young, shield volcanic island that has been modified by erosion, collapse and renewed volcanism, situated 48 km from Vailulu'u. The main shield volcano that built Ta'u was the Lata Shield, within which a caldera was formed that now forms the southern coastline of the island. Smaller satellite shields were formed later (Stearns 1944; Stice and McCoy 1968; Izuka 2005). About nine miles from Ta'u are Ofu and Olusega, which are also surrounded by coral reefs. These two islands are in reality part of the same volcanic edifice – only separated by a narrow straits – which is also made up by a complex of volcanic cones and is a shield volcano.

Tutuila is the largest (137 km^2) and most populated island (55,000 in 2004) in American Samoa. It was built by several hot-spot shield volcanoes in the Pliocene to Holocene and was at the active center at about 2 Ma (Hart et al. 2004). Much of the island consists of a ridge of steep mountains that rise from sea level to about 710 m. The area of gentlest morphology is the Tafuna–Leone Plain at Tutuila's SW coast, with about 200 m relief. The plain was formed by volcanic eruptions during the Holocene that covered parts of a pre-existing barrier reef (Stearns 1944). Coral reefs covered by very young extrusive rocks also occur elsewhere in the Samoan chain (Keating 1992). While within the Samoan Chain the island elongations, vent alignments and rift zones strike mostly ESE, Tutuila is aberrant in that it strikes ENE (Walker and Eyre 1995). Based on the observation that Tutuila is approximately in line with the North Fiji Fracture Zone (Vitiaz Lineament, Fig. 20.2), Walker (1999) speculates that strike-slip motion occurred at the time of activity in the fracture zone (1.54–1.03 Ma; McDougall 1985) coinciding with the positioning of Tutuila over the hot spot. The strain associated with this motion would have aided in shaping the island, suggested by en-echelon dyke complexes in the Masefau and Fagaitua areas. Modern volcanism in southern Tutuila, associated with the Leone lavas that cover the former barrier reef, is aligned approximately perpendicular to these earlier rift zones. Although Tutuila is a young

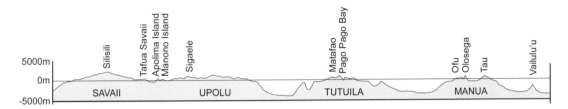

FIG. 20.5. Different to the situation in the Hawaiian Islands, the islands along the Samoan Chain increase in size towards the west, due to rejuvenated volcanism at Upolu and Savai'i in association either with lithospheric flexure or approach to another hotspot. West of Savai'i the chain continues as seamounts, east of the Manu'a group, is the active hotspot (Modified from Stearns 1944, by permission of Geological Society of America)

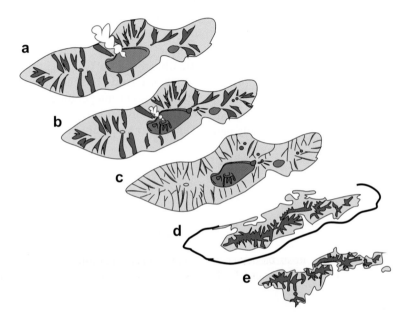

FIG. 20.6. Stages in the evolution of Tutuila as envisaged by Stearns (1944). (**a**) The building phase during the final extrusion of primitive olivine basalts, (**b**) final stage during extrusion of trachyte and differentiated lavas, (**c**) cessation of volcanism and beginning of stream erosion, (**d**) subsidence and growth of a barrier reef on the subsided margin, (**e**) further subsidence and submergence of the barrier reef, extrusion of the Leone volcanics (Modified from Stearns 1944, by permission of Geological Society of America)

island (~1.5 million years), it has already submerged faster than the reefs can grow (Fig. 20.6), leaving former barrier reefs as submerged offshore banks (e.g., Taema Bank off the mouth of Pago Pago Bay, Nafanua Bank as a westward extension of Aunu'u Island, as well as a number of other banks north of Tutuila). These banks are covered by diverse communities of living corals, but the reef formation was slower than sea-level rise. Holocene sea-level history of Tutuila is complex, since it lies on the one

hand near the crest of flexural upwarp of Savai'i, but also within the cone of flexural subsidence of Ta'u (Dickinson 2001).

Rose Atoll (Fig. 20.7) is situated at 14°32′ S 168°08′ W, 240 km ESE of Tutuila at the extreme eastern end of the Samoa Islands. It has a surface area of 640 ha, its lagoon is only 2 km wide with a land area (Rose and Sand Islands) of 0.2 km², and therewith it is one of the smallest atolls of the world (Rodgers et al. 2003). These are pure carbonate islands and have no outcrops of basalts (however, loose pieces of basalt were found on Rose atoll. Whether these really derive from there or not is a point of discussion; Mayor 1924b; Rodgers et al. 2003). Rose Atoll was discovered by Louis de Freycinet on 21 October 1819 on his voyage around the world on the *Uranie* and *Physicienne*. He named it for his wife, who made the voyage with him. Rose Atoll's position outside the Samoan hotspot trail (Hart et al. 2004) and its obvious older age than the hot spot to the west suggest that structurally it is not a part of the Samoan islands. A rich literature exists for this faraway place (Rodgers et al. 1993). Besides being of geologic interest, Rose Atoll is also the home to many giant clams, ~97% of American Samoa's entire population (Green and Craig 1999).

The northernmost island of the Territory of American Samoa is Swains Island (Fig. 20.2), which is situated approximately 320 km north of the Samoan hot spot track and is geologically part of the Tokelau volcanic chain. It is about 2 km long with a central fresh water lake that is cut off from the ocean.

20.4 Climate and Oceanography

In general, the oceanographic conditions are excellent for coral growth which may contribute to the resilience of corals on the forereefs. The average of nearshore Secchi disk readings are 27.4 m (Whaylen and Fenner 2006) and hermatypic corals and green algae can thrive to depths of at least 50 m in Fagatele Bay National Marine Sanctuary. The water temperature is usually around 28°C on the forereef (Fig. 20.8), although locally, in shallow pools on the backreef, the water temperatures can reach 35°C and fluctuate through a range of 6°C in a few hours. The tidal range is about 1 m.

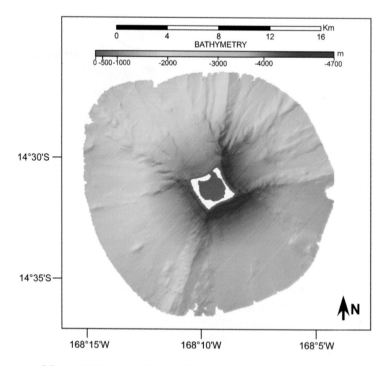

FIG. 20.7. Bathymetry of Rose Atoll as provided by NOAA bathymetry (Image courtesy NOAA Pacific Islands Mapping Center)

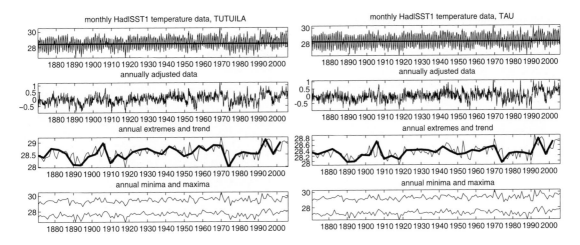

FIG. 20.8. Synthetic Hadley Center HadlSST sea surface temperature for the $1 \times 1°$ tiles centered in Tutuila and Ta'u from 1870 to 2006 (Crown Copyright. Used by permission of Hadley Center, UK)

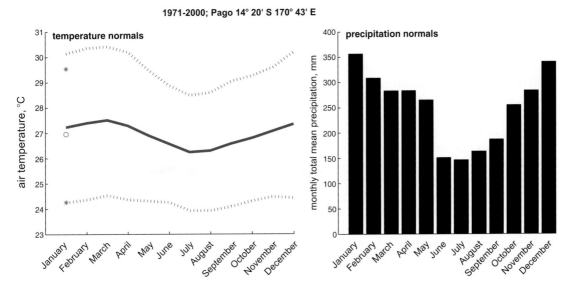

FIG. 20.9. Atmospheric temperature record and precipitation at Pago Pago airport, Tutuila. (NOAA, National Climatic Data Center.) Normals are uninterrupted measurements for three consecutive decades. Original data were translated into metric units. Lines in temperature graph represent monthly means, red circle annual mean, stars annual means of minima and maxima

The climate is characteristic of the tropics with high humidity (averages about 80%) and warm air temperatures (21–32°C). Rainfall averages about 5 m/year (Fig. 20.9) and so careless land management often leads to large-scale run-off of sediment and rubbish onto the reefs after heavy rains.

20.5   History of Biological Research

Quantitative surveys of coral reefs in American Samoa began 90 years ago in Pago Pago Harbor (Mayor 1924a) with transects set up at Aua and

FIG. 20.10. Two scans from Mayor (1924a). Left panel: The map showing Mayor's and Cary's transects. Right panel: *Acropora* exposed on the intertidal near the Aua transect line (Reprinted from Mayor 1924a, with permission of the Carnegie Institution, Washington, DC)

Utelei by Alfred Mayor in 1917 (Fig. 20.10). The transect at Aua was evaluated for ecology, the other primarily for drilling cores. Aua was resurveyed in 1973 (Dahl and Lamberts 1977), 1980 (Dahl 1981), 1995 (Green et al. 1997a), 1999 (Birkeland and Green 1999), 2000 (Birkeland and Belliveau 2000), and 2002 (Green 2002), and is now being resurveyed at every opportunity. Alfred Mayor reported rich coral communities on the reefs around Aua and his photographs show abundant *Acropora* colonies exposed at low tide (Fig. 20.10) at the outer end of the transect (Mayor 1924a). Whereas the reefs on the outer coast are renowned for their resilience, the reefs within Pago Pago Harbor began to succumb to chronic stress with the establishment of two tuna canneries in the inner harbor that began operation in 1956. A chronology of major events affecting coral reefs in American Samoa, and transect surveys of corals and their reports, are given in Section 16.10. Alfred Mayor also studied rates of growth of scleractinians and Lewis Cary studied growth of alcyonaceans between 1917 and 1920.

Lewis Cary (1931) also began surveys in 1917. His transects at Utelei were across the harbor from Mayor's Aua transect (Fig. 20.10). Cornish and DiDonato (2004) repeated Cary's transects to the extent that they still existed in 2002 (Fig. 20.11). Cary also obtained a series of limestone core samples 85, 175, and 280 m from shore along Transect 1 to basalt bedrock at 20,

40 and 40 m, respectively. Alcyonaceans were prevalent along Cary's transects and 75% of the reef flat pavement was composed of compacted spicules deposited by the hermatypic alcyonaceans. *Sinularia polydactyla* (Fig. 20.12a) is a major hermatypic reef-building species of soft coral in some locations in the Pacific, and Cary's reef cores at Utelei showed that the compacted spicules of *S. polydactyla* (presumably Cary's *Sclerophytum densum*) formed a major portion of the 40 m thick reef structure. Cornish and DiDonato (2004) pointed out that the reclamation of the reef for construction and mortality of the alcyonaceans due to pollution from the canneries and other human activities has made the reef framework substantially more vulnerable to erosion.

Surveys of corals and reef fishes around American Samoa became a regular endeavor starting in 1979 with the 1978/79 outbreak of *Acanthaster planci* (Fig. 20.12b). The Government of American Samoa Department of Marine and Wildlife Resources sponsored surveys of corals and fishes around American Samoa in 1979, 1982, 1985, 1988, 1991, 1995, 1996, 1997, 1998, 2001, 2002 (Fig. 20.13) and 2005. A reef at Ofu was surveyed in 1993 and 2000, Swains Island in 1996, Rose Atoll in 1997 (Wegmann and Holzwarth 2007).

In April 1985, upon decision by the Government of American Samoa and the US Department of Commerce Marine Sanctuary Program to estab-

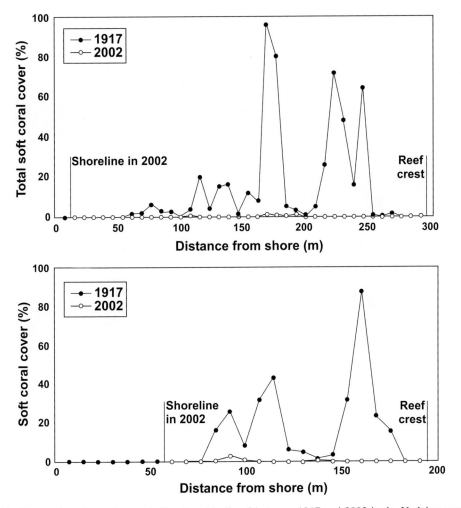

FIG. 20.11. Changes in soft coral cover in line 1 and in line 5 between 1917 and 2002 in the Utelei transect (From Cornish and DiDonato 2004, with permission of Elsevier Science)

lish a National Marine Sanctuary at Fagatele Bay, permanent transect markers were established in Fagatele Bay. Surveys of corals and fishes were performed in 1985, 1988, 1995, 1998, 2001 and 2004. Another survey is planned for August 2007 and it is expected that they be repeated every third year.

To study acclimatization and adaptation of corals to climate change, monitoring of the wide range of water temperatures in small shallow backreef pools on Ofu island was begun in 1999. In this area, temperatures could fluctuate daily by 6°C (Fig. 20.14), yet about 80 species of corals appeared to be in good health (Craig et al. 2001). This diverse community of Ofu lagoon corals appears to be also resilient to wide daily fluctuations in pH (Fig. 20.15) and dissolved oxygen (Fig. 20.16) and thus may offer some insight into mechanisms of corals to respond to ocean acidification and other global changes in the physical environment. The National Park Service, in Cooperation with the USGS, the University of Hawaii, the Rosenstiel School of Marine and Atmospheric Science, and Stanford University, set up an itinerant marine laboratory in the American Samoa National Park on Ofu for long term studies of biochemical, physiological and genetic mechanisms of coral adjustment to climate change.

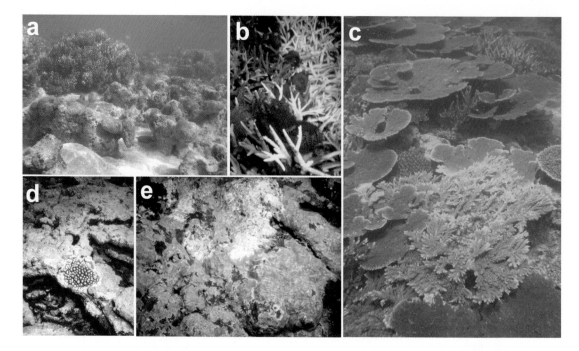

FIG. 20.12. (a) The hermatypic alcyonacean *Sinularia polydactyla* forming solid reef framework. (Picture from the the Piti Bombholes, Guam, since the species is by far not as common now as when encountered by Mayor and Carey.) (b) Phalanx of *Acanthaster planci* thoroughly removing living coral tissue from the coral community on the western coast of Aunu'u island in 1979. (c) A complex system of acroporid corals competing for space on the western forereef slope of Aunu'u Island. (d) Crustose coralline algae in Fagatele Bay, American Samoa, enhances coral recovery and reef community resilience by binding the loose substrata after a hurricane. (e) Coralline Lethal Orange Disease (CLOD), a bacterial film that move across living coralline algae and leaving bare calcium carbonate in its wake

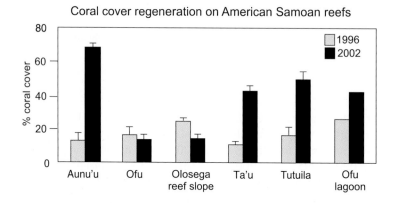

FIG. 20.13. Mean living coral cover (+/− se) on some of the islands of American Samoa during recovery following two hurricanes and a major coral bleaching event in the early 1990s (Data from Birkeland et al. 2004)

20.6 Biodiversity

The total number of scleractinian coral species names in the technical reports of the transect surveys to date is 337, but removal of synonymies and unidentified species reduce the number of names to 329. This number of coral species fits what would be expected along the west to east reduction in diversity across the Pacific. Fagatele Bay National Marine Sanctuary is only one

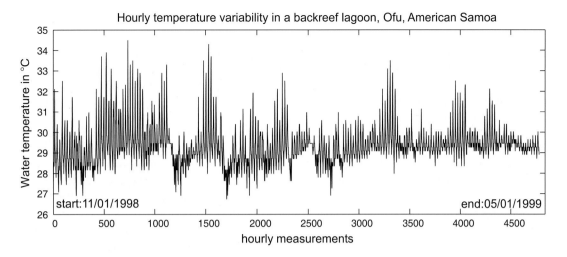

FIG. 20.14. Fluctuating hourly seawater temperatures in the backreef lagoons at Ofu Island, American Samoa (Data communicated by Craig et al. 2001)

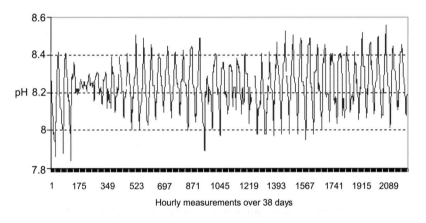

FIG. 20.15. Fluctuating hourly seawater pH measurements in the backreef lagoons of Ofu Island

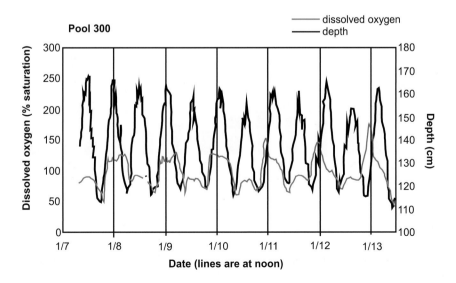

FIG. 20.16. Fluctuating levels of dissolved oxygen in seawater measurements in the backreef lagoons of Ofu Island

quarter square mile (66 ha, or 0.66 km²) in area, but hosts 200 species of coral and 271 species of reef fish.

Stony (scleractinian) corals appear to have spread eastward across the Pacific more effectively than have the soft (octo-) corals. Whaylen and Fenner (2006) report ten genera of octocorals (with *Cladiella* being by far the most prevalent), two genera of antipatharians (black corals), three genera of zoanthids, and three genera of coralliomopharia. Coles et al. (2003) report about 20 species of octocorals, 5 anemones, and 6 zoanthids. Although octocorals are an order of magnitude less diverse than the scleractinians (in striking contrast to the smaller, more continental western Atlantic), *Sinularia polydactyla* nevertheless is (or was half a century ago) a major reef builder. While Cary (1931) found *Nephthya flexile* to be abundant in areas protected from wave action, and particularly well fitted to withstand sedimentation, it now seems to be extinct in American Samoa.

Echinoderm species richness also decreases from west to east across the tropical Pacific. From Indonesia to the Marshall Islands to American Samoa, respectively, the number of coral-reef crinoid species are 91, 6, 6; asteroids 66, 17, 11; ophiuroids 142, 42, 23; echinoids 43, 17, 10; holothuroids 141, 23, 16 (Birkeland 1989).

About 945 species of reef-associated fishes are known from American Samoa (Fig. 20.17). This is consistent with the gradient in decreasing diversity from west to east across the central Pacific (Allen 2003).

Coles et al. (2003) pointed out that the large-scale usage of docking facilities in Pago Pago Harbor for cargo offloading and delivery of tuna to the canneries, plus cleaning of large vessel hulls at the dry-dock facilities, makes reefs in Pago Pago Harbor and American Samoa in general susceptible to introduction of nonindigenous marine species. Although about 17 nonindigenous and 11 cryptogenic (uncertain origin, but with indications of being introduced) species were found in Pago Pago Harbor, only two introduced species, an ectoproct and a polychaete, were found outside the harbor (Coles et al. 2003). This is in striking contrast to Hawaii where 343 marine introduced species (including 287 invertebrates, 20 fishes, 24 algae and 12 flowering plants) have been found (http://www2.bishopmuseum.org/HBS/invertguide/index.htm). On Hawaiian coral reefs, introduced species have been causing substantial ecological damage from the intertidal to depths of over 100 m on the forereef slopes. It is tempting to speculate that the species diversity on coral reefs of American Samoa outside Pago Pago Harbor in comparison to Hawaii somehow leads to resistance against invasion by introduced species.

20.7   Zonation and Community Patterns

On a scale larger than depth zonation, differences exist in distributions of some reef organisms on differently exposed sides of the islands. Whaylen and

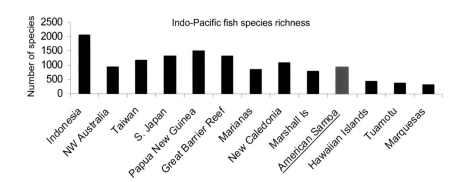

FIG. 20.17. The number of fish species near American Samoa in comparison with the number of fish species near other locations (The data are from Allen 2003)

Fenner (2006) and Sabater and Tofaeono (2006) found significantly more cover of crustose coralline algae on the south coast of Tutuila than on the north coast and conversely, filamentous algae was more prevalent on the north coast. Also Rose Atoll has already been noted by Mayor (1924b) as having an exceptionally well-developed shallow calcareous algae ridge largely made up by *Porolithon* sp. He noted that Rose Atoll had the densest growth of calcareous algae he had encountered anywhere, so much that it could be called a "*Lithothamnion*-atoll rather than a coral atoll" (Mayor 1924b, p. 77). Swains Island, on the other hand, shows the dominance of corals as so often observed elsewhere in American Samoa (Fig. 20.18).

20.8 Effects of Human Activities and Conservation Issues

The American Samoa Environmental Protection Agency ranked the watersheds on Tutuila in terms of influence of human activities. Whaylen and Fenner (2006) assessed the correlations between human activities and living coral cover at 35 sites from the combined data from their own surveys and the surveys of Sabater and Tofaeono (2006) and found no significant correlation. Likewise, they did not find a correlation between human activities and either crustose coralline algal cover or filamentous algal cover. On an island-wide scale, human activities are not correlated with the patterns of distribution of the benthic communities, partially because the effects of human activities are confounded or masked by the major effects of natural disturbances such as hurricanes, bleaching from thermal and UV stress, and predation by *Acanthaster planci*.

20.8.1 Land Management

On a smaller scale, there are very obvious effects of human activities. The reef flats onto which the rivers empty at the base of watersheds can be damaged from chronic sedimentation. About the only coral now residing on the heavily sedimented reef flat at the mouth of Amanave Bay is *Leptastrea purpurea*, although the structure of the reef is made up of skeletons of a diverse array of corals which attest to healthier reef conditions in the past. Fagasa Bay is silty and the coral cover is very low (6.8%). The crustose coralline algal cover is also low (9.5%), but the filamentous algal cover is especially high (45%; Whaylen and Fenner 2006). Luxuriant and diverse coral reefs existed in Pago Pago Harbor a century ago, but reefs of the inner harbor were obliterated in the 1920s to allow construction of a US naval base. About

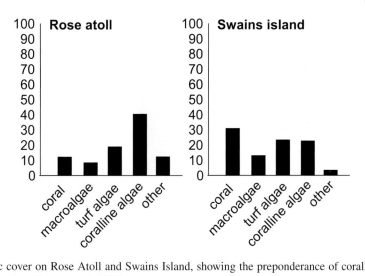

FIG. 20.18. Benthic cover on Rose Atoll and Swains Island, showing the preponderance of coralline algae on Rose Atoll (Data redrawn from Vroom et al. 2006)

95% of the reefs in the inner harbor have been buried in silt (Wass 1983). In the 1940s, further inshore areas were dredged for landfill and the inner Utulei reef was filled for development of a tank farm and to widen the coastal road. In 1956, the first of two tuna canneries began operation on the north shore of the inner harbor. Untreated sewage, polluted streams, and untreated waste from the two canneries caused the death of most corals in the inner harbor and corals in the mid to outer harbor were also not doing well. In 1992, the tuna canneries extended wastewater outflow pipes to the outer harbor where dilution is stronger. In the late 1990s, coral colonies began to be observed growing in the inner harbor and *Acropora hyacinthus* was observed recruiting and growing near the Rainmaker Hotel at the boundary of the inner and outer harbor (Green 2002).

20.8.2   Fishing Pressure

The 945 species of reef fishes in American Samoa are slightly more than would be expected at the longitude of the islands on the west to east gradient in numbers of coral-reef fish species (Fig. 20.17). This could indicate that the full array of habitats necessary to provide the necessary conditions for the life history stages for reef fishes is still in operation. The lack of habitat could lead to local extinction and so the wealth of species indicates that the critical habitats are still represented.

Although the number of species of reef fishes in American Samoa is as great as would be expected, their biomass is remarkably low (Figs. 20.19 and 20.20). Commercial fishing has been in effect for a few decades as the human population grew and residents began to earn a salary and purchase more of their food at the market. Before 1994, most commercial fishing was done by handlines, nets or spearfishing by free diving. From 1994 to 2002, the commercial fishermen used underwater lights and scuba to harvest sleeping fish. The commercial catch increased 15-fold (Fig. 20.21) until eloquent protests at public hearings compelled the Governor to issue an executive order to stop night fishing with scuba until the matter could be discussed and new regulations developed.

The biomass of fish species selected for surveying averaged 56 g/m^2 with a range of 29–114 g/m^2, or 56 mt/km^2 (0.56 mt/ha). The typical biomass of fishes (Fig. 19) on the western tropical Atlantic coral reefs was 160–200 mt/km^2 (Randall 1963; Munro 1983) and on Pacific coral reefs 93–239 mt/km^2 (Goldman and Talbot 1976; Williams and Hatcher 1983). However, this low biomass on American Samoan reefs is at a level characteristic of areas heavily fished (Fig. 20.20).

The number of species and abundance of coral-reef fishes are not noticeably different than what might be expected for central Pacific reefs, but the scarcity of larger fishes in recent years is striking (Figs. 20.22 and 20.23). Large serranids were seen and caught more regularly in the 1980s than they

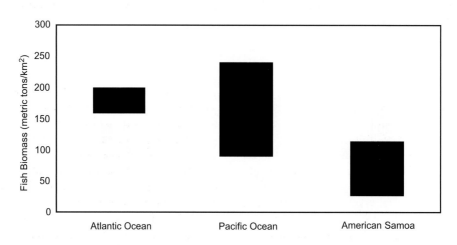

FIG. 20.19. The biomass of fishes on American Samoan reefs in comparison with the biomass of fishes typical of the Atlantic and Pacific reefs from literature from 1983 or prior times

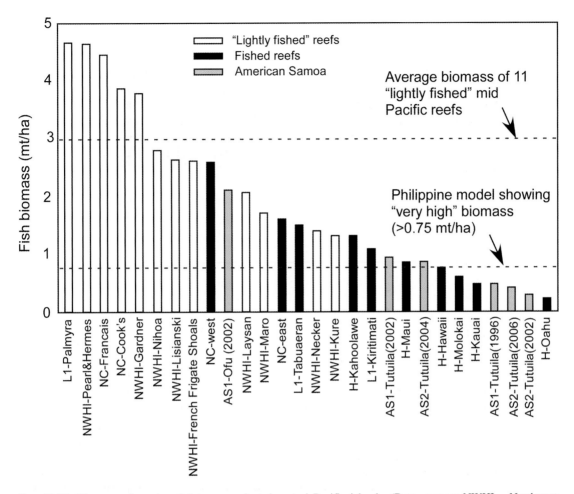

FIG. 20.20. Biomass of coral-reef fishes on selected central Pacific islands. (Data sources: NWHI – Northwest Hawaiian Island and H – main Hawaiian Island (Friedlander and DeMartini 2002; A. Friedlander, personal communication), LI - Line Is. (Stevenson et al. 2007), NC – New Caledonia (Letourneur et al. 2000), AS1 – American Samoa (Green 2002), AS2 – American Samoa (R. Brainard, personal communication, 2007), line for Philippine model (Hilomen et al. 2000)

are now (C. Birkeland personal observation, 1979, early 1980s). In a survey of 11 sites in 2005, only four individual *Cheilinus undulatus* were seen with a mean length of 64 cm, the largest individual being only 110 cm, about half of the maximum attainable size. Only a single shark, a whitetip reef shark *Triaenodon obesus*, was observed.

The extraordinarily low biomass of fishes and scarcity of large fishes might be explained by fishing pressure on the larger fishes. The shallow fringing reefs are all accessible close to shore. Samoans have been fishing the reefs of American Samoa for about

3,300 years. Artisanal fishers can severely exploit and degrade fisheries resources, especially the larger fishes, shortly after prehistoric arrival (Jennings and Polunin 1997, Wing and Wing 2001).

Motivation to manage requires a realistic perception of the state of the resources, and so the shifting baseline is one of the more powerful phenomena undermining responsible programs. A recent (18 June 2007) issue of the Samoan News reported a statement from a scientist that the American Samoan "fisheries are clearly sustainable, and marine protected areas for management purposes are not

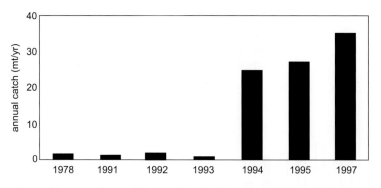

FIG. 20.21. Estimated annual harvest of parrotfishes on Tutuila Island from 1978 to 1997 (Page 1998)

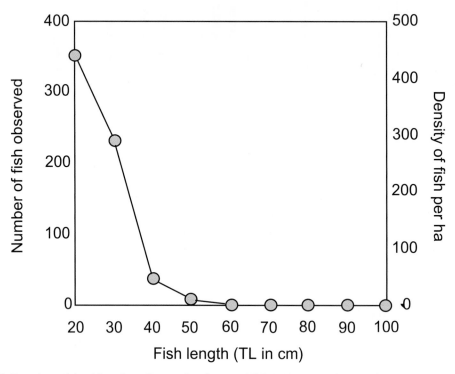

FIG. 20.22. Lengths and densities of standing stocks of targeted fish (>19 cm, species combined) at 17 sites on Tutuila Island in 2002 (Modified from data from Green 2002)

needed as the fisheries are replenishing themselves." This statement must be based on a view that the low biomass and scarcity of large fishes are the natural characteristics of reef communities of American Samoa. When monitoring of resources begins after stocks have been substantially reduced, the perception of what is natural can be lowered several fold. For a heuristic example of how the scale of assessment can potentially alter our perceptions, Hilomen et al. (2000) summarized the findings of fish biomass on 227 transects from throughout the Philippine Archipelago. In this report, Hilomen et al. (2000) strongly emphasized that the condition of the majority of coral-reef fish stocks in the Philippines are very poor. They were aware of the true situation. But in presenting their findings, they

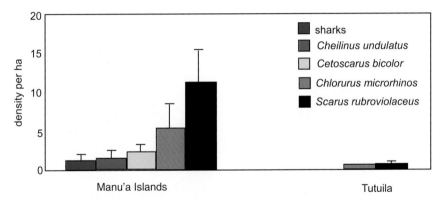

FIG. 20.23. Mean density (+/− se) of large reef fishes on Tutuila and in the Manu'a Islands in 2002 (Modified from data from Green 2002)

categorized the top 11% in the "high" (0.35–0.75 mt/ha) to "very high" (>0.75 mt/ha) relative to the range of conditions within the Philippines. These top 11% were indeed high or very high within the context of present day Philippines reef fish stocks, but the high or very high levels of biomass were one nineth or one fourth the average level for lightly fished areas (Fig. 20.20). One of the very high sites in the Philippines had a stock of 2.20 mt/ha and so the reefs of the Philippines probably have potential for higher standing stocks if the fishing pressure were released. Hilomen et al. (2000) clearly expressed the dire state of overfishing in the Philippines, but we believe that if the levels of biomass characterized as "high" or "very high" in terms of the spread of the data at hand were taken out of context, this could facilitate a shifting baseline in general perception.

20.9 Present Status of Reef Health

The coral communities on American Samoa are a rich mosaic of patches in various stages of recovery from an array of disturbances such as crown-of-thorns predation, hurricanes, blast-fishing, ship groundings, and bleaching from warm-water stress (Section 20.10). On the outer reef slopes, disturbances have generally been acute, and recovery started soon after the event (Green et al. 1999). Some localized and well-defined areas experience chronic stresses, such as sedimentation at the mouths of rivers or in the backs of bays (Houk et al. 2005), and have not

been recovering for decades. If unusual seawater warming becomes more frequent, widespread and of longer duration, and if the increased atmospheric carbon leads to seawater acidification, the coral communities on the forereef might start to lose their resilience.

The capacity of coral communities on forereefs to recover from disturbances is probably partially a result of the ability of crustose coralline algae to bind loose rubble into a stable substratum (Fig. 20.12d). Crustose coralline algae seem especially prevalent on the forereefs of American Samoa. Six years after the 1979 outbreak of *Acanthaster planci*, the coralline algae generally covered about 57% of the forereef slope, with about 65% cover at 3 and 5 m depths (Birkeland et al. 1987). Other algae covered about 21% of the substrata. Living coral cover averaged 12.6%, but was patchily distributed during this period of recovery, with results of 30 m transects within Fagatele Bay at any one time ranging from 0.9% to 64.4%.

Recent surveys have found that the reefs are still in similar condition, but further along in recovery from more recent disturbance events, i.e., damaging hurricanes in 1990 and in 1991, and serious bleaching in 1994. Whaylen and Fenner (2006) and Sabater and Tofaeono (2006) found 28% and 27% coral cover on 11 and 24 sites (for a total of 35 different sites) around Tutuila. Whaylen and Fenner (2006) reported an average 35% cover by crustose coralline algae and concluded that in recent times, the prevalence of living corals and crustose coralline algae, and the scarcity of macroalgae (2% cover at

Fig. 20.24. Coastal community on Olosega. The human populations are concentrated along the coast because of the steep terrain on American Samoa

11 sites) indicated that the coral reef community is presently in good health. In most other areas, especially forereef sites, the reefs of American Samoa are still resilient compared to those of the western Atlantic (Gardner et al. 2003; Pandolfi et al. 2003).

While coral diseases are common on American Samoan reefs, especially neoplasms or "tumors" and white syndromes, they presently do not appear to have a substantial effect on the coral populations to date. However, increasing water temperature, sedimentation and pollution might stress corals and weaken their resistance and coral disease may become a more serious factor in the future if climate change becomes more influential.

Not only corals, but also crustose coralline algae are infected by at least two diseases that are especially common in Fagatele Bay National Marine Sanctuary: a bacterium – Coralline Lethal Disease (CLOD) and

a fungus – "lichen-like" black crust disease. CLOD grabs attention as a bright orange band followed by a white patch (Fig. 20.12e). Despite the fact that CLOD is frequently noticed, the crustose coralline algae appear to grow fast enough to replace any surface killed by CLOD. At the present time, CLOD and the fungal disease do not appear to be a threat.

Uncontrolled human population growth can interact synergistically with the steep topography of the islands (Fig. 20.24) and the rainfall to exacerbate environmental problems (Craig et al. 2001). Since rainfall averages 5 m/year in the mountainous areas in which most of the land has a slope greater than 70%, heavy rains can cause erosion and deliver spectacular amounts of debris, garbage and sediment onto the reefs of Pago Pago and other watersheds. Mayor (1924a) reported on torrential rains that delivered silt that blanketed the reefs and killed a substantial number of corals. Wells (1988) described significant portions of reefs in Pago Pago, Faga'ita, and Leone Bays as having been buried under sediment. Organic pollution and untreated sewage have also locally affected reefs around American Samoa (Wells 1988). The greatest threats, however, may ultimately arise from climate change and acidification of ocean waters, while increased input of sediment and pollution from increased population growth may reduce the vitality and resilience of corals and increase their vulnerability to succumbing to disease.

While corals on the forereef appear resilient, the fish populations do not appear to recover their normal size distributions. Although the diversity and abundance of fishes may be as expected, the biomass is only about a quarter of that on a lightly fished coral reef (Fig. 20.20) and very few large individuals are seen. The apex predators are very rare. With increased human population density and advances in technology, overfishing may have become a chronic problem.

20.10   A Chronology of Major Events and Surveys of Coral Reefs in American Samoa

1917–1920: Alfred A. Mayor established the permanent Aua transect in Pago Pago Harbor and also described the reefs of the other islands, including Rose Atoll (Mayor 1924a, b)

1920: Louis Cary surveyed soft corals and the massive reef they constructed at Utelei in Pago Pago Harbor (Cary 1931)

1920s: Destruction of coral reefs on southern shore of inner Pago Pago Harbor by construction of naval base

1938: Outbreak of *Acanthaster planci*

1940s: Fill over inner Utelei Reef for tank farm

1950s: One third of Mayor's Aua transect destroyed by sand excavation for road construction at Aua

1954: Construction of first tuna cannery

1960s: Fill over start of Cary Transects 1 and 5 for widening of coastal road

1963: Construction of second tuna cannery

1973: First resurvey of Mayor's transect (Dahl and Lamberts 1977)

1977–1979: Outbreak of crown-of-thorns starfish *Acantha-ster planci*

1979: Survey of *Acanthaster* and corals around Tutuila (Birkeland et al. 1985, 1987; Dahl 1981)

1981: Tropical cyclone Esau

1985: Fagatele Bay National Marine Sanctuary officially established by US Department of Commerce and American Samoa Department of Commerce. Quantitave baseline established for coral-reef monitoring (Birkeland et al. 1987)

1987: Tropical cyclone Tusi hits Manu'a Islands hard

1988: Resurvey of Fagatele Bay NMS (Birkeland et al. 1994)

1990: Tropical cyclone Ofa hits Tutuila hard

1991: Tropical cyclone Val hits Tutuila hard – nine ships grounded on Pago Pago reefs

1991, 1992: Coastal resources inventory (Maragos et al. 1994).

1992: Extension of industrial wastewater discharge pipe from canneries from inner Pago Pago Harbor to the outer harbor

1993: Surveys on Ofu for the National Park of American Samoa (Hunter et al. 1993)

1994–2002: Commercial fishing at night with scuba banned by Governor's Executive Order in 2002

1994: Major coral bleaching associated with unusually warm seawater

1995–1997: Surveys in Fagatele Bay, Mayor's transect, and the rest of American Samoa including Swains Island and Rose Atoll (Birkeland et al. 1996; Mundy 1996; Green 1996a, b; Green et al. 1997a, b)

1998: Extreme low tides kill exposed corals

1998: Surveys of Swains Island (Page and Green 1998)

1998–2001: Numerous surveys on Tutuila (Green and Hunter 1998; Birkeland and Green 1999; Birkeland et al. 2004)

2000: Nine ships finally removed from Pago Pago reefs. Prior and after the removal, resurveys of Mayor's Aua transect (Birkeland and Green 1999; Birkeland and Belliveau 2000)

2002: Minor coral bleaching associated with unusually warm seawater

2002: Numerous coral reef surveys (Fisk and Birkeland 2002; Green 2002; Cornish and DiDonato 2004) of Tutuila and Manu'a islands

2002: Surveys for introduced marine species (Coles et al. 2003) at Tutuila and Manu'a islands

2003: Major coral bleaching associated with unusually warm seawater

2004: Tropical cyclone Heta
2004: Resurvey of Fagatele Bay NMS and Mayor's Aua transect (Green et al. 2005)
2005: Extreme low tides kill exposed corals
2005: Tropical cyclone Olaf hits Manu'a Islands hard
2007: Resurvey of Fagatele Bay NMS

References

Allen GR (2003) Reef fishes of Milne Bay Province, Papua New Guinea. Pages 46–55. In: Allen GR, Kinch JP, McKenna SA, Seeto P (eds) A rapid marine biodiversity assessment of Milne Bay Province, Papua New Guinea – Survey II (2000). RAP Bulletin of Biological Assessment 29. Conservation International, Washington, DC, 172 pp

Birkeland C, Randall RH, Amesbury SS (1985) Coral and reef-fish assessment of the Fagatele Bay National Marine Sanctuary. Report to NOAA, 126 pp

Birkeland C, Randall RH, Wass R, Smith BD, Wilkens S (1987) Biological resource assessment of the Fagatele Bay National Marine Sanctuary. NOAA Technical Memorandum NOS MEMD 3, 232 pp

Birkeland C (1989) The influence of echinoderms on coral-reef communities. In: Jangoux M, Lawrence JM (eds) Echinoderm Studies 3. Balkema, Rotterdam, The Netherlands, pp 1–79

Birkeland C, Randall RH, Amesbury, SS (1994) Coral and reef-fish assessment of the Fagatele Bay National Marine Sanctuary. Report to the National Oceanic and Atmospheric Administration US Department of Commerce, 126 pp

Birkeland C, Randall RH, Green AL, Smith BD, Wilkins S (1996) Changes in the coral reef communities of Fagatele Bay National Marine Sanctuary and Tutuila Island (American Samoa) over the last two decades. Report to the National Oceanic and Atmospheric Administration, US Department of Commerce, 225 pp

Birkeland C, Green AL (1999) Resurvey of the Aua Transect prior to the ship removal. Report to the National Oceanic and Atmospheric Administration, 10 pp

Birkeland C, Belliveau SA (2000) Resurvey of the Aua Transect After the Ship Removal. Report to the National Oceanic and Atmospheric Administration US Department of Commerce, 2 pp

Birkeland C, Green AL, Mundy C, Miller K (2004) Long term monitoring of Fagatele Bay National Marine Sanctuary and Tutuila Island (American Samoa) 1985 to 2001: summary of surveys conducted in 1998 and 2001. Report to the National Oceanic and Atmospheric Administration, US Department of Commerce, 158 pp

Cary LR (1931) Studies on the coral reefs of Tutuila, American Samoa, with special reference to the Alcyonaria. Papers from the Tortugas Laboratory, Carnegie Instit Wash 37:53–98

Clouard V, Bonneville A (2005) Ages of seamounts, islands and Plateaus on the Pacific plate. In: Foulger GR, Natland JR, Presnall DC, Anderson DL (eds) Plates, plumes and paradigms. Geol Soc Am Spec Pap 381, pp 71–90

Coles SL, Reath PR, Skelton PA, BonitoV, DeFelice RC, Basch L (2003) Introduced marine species in Pago Pago Harbor, Fagatele Bay and the National Park Coast, American Samoa. Bishop Museum Technical Report No 26, Honolulu, HI, 182 pp

Colgan MW (1987) Coral recovery on Guam (Micronesia) after catastrophic predation by *Acanthaster planci*: a study of community development. Ecology 68:1592–1605

Cornish AS, DiDonato EM (2004) Resurvey of a reef flat in American Samoa after 85 years reveals devastation to a soft coral (Alcyonacea) community. Mar Pollut Bull 48:768–777

Craig PC, Birkeland C, Belliveau S (2001) High temperatures tolerated by a diverse assemblage of shallow-water corals in American Samoa. Coral Reefs 20:185–189

Dahl AL (1981) Monitoring coral reefs for urban impact. B Mar Sci 31:544–551

Dahl AL, Lamberts AE (1977) Environmental impact on a Samoan coral reef: a resurvey of Mayor's 1917 transect. Pac Sci 31:209–319

Dalrymple GB, Clague DA (1976) Age of the Hawaiian–Emperor bend. Earth Planet Sci Lett 31:313–329

Dickinson WR (1998) Geomorphology and geodynamics of the Cook-Austral island-seamount chain in the Saouth Pacific Ocean: implications for hotspots and plumes. Int Geol Rev 40:1039–1075

Dickinson WR (2001) Paleoshoreline record of relative Holocene sea levels on Pacific islands. Earth-Sci Rev 55:191–234

Fisk D, Birkeland C (2002) Status of coral communities in American Samoa. A re-survey of long-term monitoring sites. Report to the Department of Marine and Wildlife Resources, Government of American Samoa, 134 pp

Foulger GR, Natland JR, Presnall DC, Anderson DL (2005) Plates, plumes and paradigms. Geol Soc Am Spec Pap 381, 881 pp

Friedlander AM, DeMartini EE (2002) Contrasts in density, size, and biomass of reef fishes between the northwestern and the Main Hawaiian Islands: the effects of fishing down apex predators. Mar Ecol Prog Ser 230:253–264

Gardner TA, Côté IM, Gill JA, Grant A, Watkinson A (2003) Long-term region-wide declines in Caribbean corals. Science 301:958–960

Garrison VH, Shinn EA, Foreman WT, Griffin DW, Holmes CW, Kellogg CA, Majewski MS, Richardson

LL, Ritchie KB, Smith GW (2003) African and Asian dust: from desert soils to coral reefs. BioScience 53:469–480

Goldman B, Talbot FH (1976) Aspects of the ecology of coral reef fishes. In: Jones OA, Endean R (eds) Biology and Geology of Coral Reefs III. Biology 2. Academic, New York, pp 125–254

Green AL (1996a) Status of the coral reefs of the Samoan Archipelago. Report to the Department of Marine and Wildlife Resources, Government of American Samoa, 120 pp

Green AL (1996b) Coral reefs of Swains Island, American Samoa. Report to the Department of Marine and Wildlife Resources, Government of American Samoa, 2 pp

Green AL, Birkeland CE, Randall RH, Smith BD, Wilkins S (1997) 78 years of coral reef degradation in Pago Pago Harbour: a quantitative record. Proc 8th Int Coral Reef Sym 2:1883–1888

Green AL, Burgett J, Molina M, Palawski D, Gabrielson P (1997) The impact of a ship grounding and associated fuel spill at Rose Atoll National Wildlife Refuge, American Samoa. A Report to the US Fish and Wildlife Service, Honolulu, HI, 60 pp

Green AL, Hunter C (1998) A preliminary survey of the coral reef resources in the Tutuila Unit of the National Park of American Samoa. Final Report to the US National Park Service, American Samoa, 42 pp

Green A, Craig P (1999) Population size and structure of giant clams at Rose Atoll, an important refuge in the Samoan Archipelago. Coral Reefs 18:205–212

Green AL, Birkeland CE, Randall RH (1999) Twenty years of disturbance and change in Fagatele Bay National Marine Sanctuary, American Samoa. Pac Sci 53:376–400

Green AL (2002) Status of coral reefs on the main volcanic islands of American Samoa: a resurvey of long term monitoring sites (benthic communities, fish communities, and key invertebrates). Report to Department of Marine and Wildlife Resources, Pago Pago, American Samoa, 135 pp

Green AL, Miller K, Mundy C (2005) Long term monitoring of Fagatele Bay National Marine Sanctuary. Tutuila Island, American Samoa: results of surveys conducted in 2004, including a re-survey of the historic Aua transect. Report to US Department of Commerce and American Samoa Government, 93 pp

Hallock P, Müller-Karger FE, Halas JC (1993) Coral reef decline. Natl Geogr Res Explor 9:358–378

Hart SR, Staudigel H, Kurz MD, Blusztajn J, Workman R, Saal A, Koppers A, Hauri EH Lyons S (1999) Fa'afafine volcano: the active Samoan hotspot. Trans Am Geophys Union (Eos) 46, F1102, Suppl [abs]

Hart SR, Staudigel H, Koppers AAP, Blustajn J, Baker ET, Workman R, Jackson M, Hauri E, Kurz M, Sims K,

Fornari D, Saal A, Lyons S (2000) Vailulu'u undersea volcano: the new Samoa. Geochem Geophys Geosyst GC0000108: 1–13

Hart SR, Coetzee M, Workman RK, Blustajn J, Johnson KTM, Sinton JM, Steinberger B, Hwakins JW (2004) Genesis of the Western Samoa seamount province: age, geochemical fingerprint and tectonics. Earth Planet Sc Lett 227:37–56

Hilomen VV, Nañola CL, Dantis AL (2000) Status of Philippine reef fish communities. In: Licuanan WY, ED Gomez (eds) Philippine coral reefs, reef fishes, and associated fisheries: status and recommendations to improve their management. GCRMN Report Appendix B, 154 pp

Houk P, Didonato G, Iguel J, Van Woesik R (2005) Assessing the effects of non-point source pollution on American Samoa's coral reef communities. Environ Monitor Assess 107:11–27

Hunter CL, Friedander AM, Magruder WM, Meier KZ (1993) Ofu reef survey: baseline assessment and recommendations for long-term monitoring of the proposed National Park, Ofu, American Samoa. Final report to the US National Park Service, American Samoa, 92 pp

Izuka S (2005) Reconaissance of the hydrogeology of Ta'u, American Samoa. US Geol Surv Invest Rep 2004–5240, 20 pp

Jennings S, Polunin NVC (1997) Impacts of predator depletion by fishing on the biomass and diversity of non-target reef fish communities. Coral Reefs 16:71–82

Johnson RH (1984) Exploration of three submarine volcanoes in the South Pacific. Res Rep Nat Geogr Soc 16:405–420

Keating BH (1992) The geology of the Samoan Islands. In: Keating BH, Bolton BR (eds) Geology and Offshore Mineral Resources of the Central Pacific Basin. Circum-Pacific Council for Energy and Mineral Resources Earth Science Series 14. Springer, New York, pp 127–178

Koppers AAP, Staudigel H, Wijbrans JR, Pringle MS (1998) The Magellan seamount trail: implications for Cretaceous hotspot volcanism and absolute Pacific plate motion. Earth Planet Sc Lett 163:53–68

Lessios HA, Robertson DR, Cubit JD (1984) Spread of *Diadema* mass mortality through the Caribbean. Science 226:335–337

Letourneur Y, Kulbicki M, Labrosse P (2000) Fish stock assessment of the northern New Caledonia lagoons: 1 – Structure and stocks of coral reef fish communities. Aquat Living Res 13:65–76

Maragos JE, Hunter CL, Meier KZ (1994) Reefs and corals observed during the 1991–1992 American Samoa coastal resources inventory. Report to the Department of Marine and Wildlife Resources, American Samoa, 50 pp

Mayor AG (1924a) Structure and ecology of Samoan reefs. Carnegie Inst Pub 340:1–25

Mayor AG (1924b) Rose Atoll, American Samoa. Carnegie Inst Pub 340:73–91

McArdle (2003) Report: Statistical Analyses for Coral Reef Advisory Group, American Samoa, 142 pp

McDougall I (1985) Age and evolution of the volcanoes of Tutuila, American Samoa. Pac Sci 39:311–320

McNutt MK, Fisher KM (1987) The South Pacific Superswell. In: Keating B, Fryer P, Batiza R, Boehlert G (eds) Seamounts, Islands and Atolls, Amer Geophys Union, Washington, DC, pp 25–34

Menard HW (1986) Islands. HW Freeman, New York, 205 pp

Morgan WJ (1971) Convection plumes in the lower mantle. Nature 230:42–43

Mundy C (1996) A quantitative survey of the corals of American Samoa. Report to the Department of Marine and Wildlife resources, Government of American Samoa, 24 pp

Munro JL (ed) (1983) Caribbean coral reef fishery resources. ICLARM Studies and Reviews 7, 276 pp

Natland J (1980) The progression of volcanism in the Samoan linear volcanic chain. Am J Sci 280-A:709–735

Natland JH (2004) The Samoan Chain: a shallow lithospheric fracture system. www.mantleplumes.org

Norton IO (1995) Plate motions in the Pacific: the 43 Ma non event. Tectonics 14:1080–1094

Page M (1998) The biology, community structure, growth and artisanal catch of parrotfishes of American Samoa. Report to the Department of Marine and Wildlife Resources, Government of American Samoa, 78 pp

Page M, Green A (1998) Status of the coral reefs of Swains Island 1998. Report to the Department of Marine and Wildlife Resources, Government of American Samoa, 9 pp

Pandolfi JM, Bradbury RH, Sala E, Hughes TP, Bjorndal KA, Cooke RG, McArdle D, McClenachan L, Newman MJH, Paredes G, Warner RR, Jackson JBC (2003) Global trajectories of the long-term decline of coral reef ecosystems. Science 301:955–958

Patriat P, Achache J (1984) India-Eurasia collision chronology has implications for crustal shortening and driving mechanism of plate. Nature 311:615–621

Randall JE (1963) An analysis of the fish populations of artificial and natural reefs in the Virgin islands. Carib J Sci 3:31–47

Rodgers KA, McAllan IAW, Cantrell C, Ponwith BJ (1993) Rose Atoll: an annotated bibliography. Tech Reps Austral Mus 9, 37 pp

Rodgers KA, Sutherland FL, Hoskin PWO (2003) Basalts from Rose Atoll, American Samoa. Rec Aus Mus 55:141–152

Sabater M, Tofaeono S (2006) Spatial variations in biomass, abundance and species composition of key reef species in American Samoa. Report to the Department of Marine and Wildlife Resources, Government of American Samoa, 54 pp

Sano M (2000) Stability of reef fish assemblages: responses to coral recovery after catastrophic predation by *Acanthaster planci*. Mar Ecol Progr Ser 198:121–130

Sharp WD, Clague DA (2006) 50-Ma initiation of Hawaii-Emperor bend records major change in Pacific plate motion. Science 313:1281–1284

Stearns HT (1944) Geology of the Samoan Islands. Geol Soc Am Bull 55:1279–1332

Steinberger B (2000) Plumes in a convecting mantle; models and observations for individual hotspots. J Geophys Res 105:11127–11152

Steinberger B (2002) Motion of the Easter hot spot relative to Hawaii and Louiseville hotspots. Geochem Geophys Geosyst 3(11): 8503, doi:10.1029/2002GC000334

Steinberger B, Sutherland R, O'Connell (2004) Prediction of Emperor–Hawaii seamount locations from a revised model of plate motion and mantle flow. Nature 430:167–173

Stevenson C, Katz LS, Micheli F, Block B, Heiman KW, Perle C, Weng K, Dunbar R, Witting J (2007) High apex predator biomass on remote Pacific islands. Coral Reefs 26:47–51

Stice GD, McCoy FW Jr (1968) The geology of the Manua Islands, Samoa: Pac Sci 22: 427–457.

Vroom PS, Page KN, Kenyon JC, Brainard RE (2006) Algae-dominated reefs. Am Sci 94:430–437

Walker GPL (1999) Volcanic rift zones and their intrusion swarms. J Volcanol Geotherm Res 94:21–34

Walker GPL, Eyre PR (1995) Dike complexes in American Samoa. J Volcanol Geotherm Res 69:241–254

Wass RC (1983) The shoreline fishery of American Samoa – past and present. In: UNESCO (ed) Marine and coastal processes in the Pacific: ecological aspects of coastal zone management. Papers presented at a UNESCO seminar held at Motupore Island Research Centre, University of Papua New Guinea, 14–17 July 1980, pp 51–83

Wegmann A, Holzwarth S (2007) Rose Atoll National Wildlife Refuge Research Compen dium. Report to the US Fish and Wildlife Service, 93pp

Wells SM (1988) Coral Reefs of the World 3: Central and Western Pacific. UNEP and IUNC, Cambridge, 329 pp

Whaylen L, Fenner D (2006) Report of 2005 American Samoan Coral Reef Monitoring Program (ASCRMP), Expanded Edition. Report for Department of Marine and Wildlife Resources and Coral Reef Advisory Group, 64 pp

Williams DMcB, Hatcher A (1983) Structure of fish communities on outer slopes of inshore, mid-shelf and outer shelf reefs of the Great Barrier Reef. Mar Ecol Progr Ser 10:239–250

Wing SR, Wing ES (2001) Prehistoric fisheries in the Caribbean. Coral Reefs 20:1–8

Wright E, White WM (1986) The origin of Samoa: new evidence from Sr, Nd and Pb isotopes. Earth Planet Sc Lett 81:151–162

Zeller D, Booth S, Craig P, Pauly D (2006) Reconstruction of coral reef fisheries catches in American Samoa, 1950–2002. Coral Reefs 25:144–152

Zeller D, Booth S, Pauly D (2007) Fisheries contributions to GDP: underestimating small-scale fisheries in the Pacific. Mar Res Econom 21: 355–374

21
Deep-Water Coral Reefs of the United States

Charles G. Messing, John K. Reed, Sandra D. Brooke, and Steve W. Ross

21.1 Introduction

Reef-building, or hermatypic, scleractinian corals are widely treated as tropical organisms, confined to warm, sunlit, fully marine habitats by their obligatory endosymbiotic phototrophic algae (zooxanthellae) and stenotopic environmental requirements. However, hermatypic scleractinians lacking zooxanthellae are also the primary architects of extensive, complex, biogenic build-ups at outer shelf to bathyal depths from subarctic (71° N) to subantarctic waters (55° S) at temperatures between 4°C and 12°C (Squires 1965; Stanley and Cairns 1988; Cairns 1995; Fosså et al. 2000; Mortensen et al. 2001; Roberts et al. 2006). Such assemblages, usually called deep-sea or cold-water coral reefs, may cover as much (or more) area of seafloor as the 284,300 km^2 estimated for shallow warm-water reefs (Freiwald and Roberts 2005 cited in T. Williams et al. 2006). Most commonly, they consist of accumulated skeletal debris and captured sediment; living colonies, when present, are restricted to a superficial, often discontinuous, veneer (Teichert 1958; Stetson et al. 1962; Wilson 1979; Mullins et al. 1981; Mortensen et al. 1995; Freiwald et al. 2002; Reed 2002b, Reed et al. 2006). Although Rogers (1999) suggested that such coralligenic topographic features fall within the definition of a coral reef based on their physical and biological characteristics, they lie at depths too great to constitute navigational hazards – the traditional diagnostic character of a reef. As such, they have also been called banks, mounds and bioherms, the latter a geological term reflecting their

specifically biological origin. Off the southeastern United States, limestone ridges called lithoherms consist of layers of coral-containing skeletal debris and sediment lithified by successive episodes of sub-sea cementation, rather than unconsolidated sediments (Neumann et al. 1977; Messing et al. 1990; Paull et al. 2000; Reed et al. 2005c, 2006).

Branching scleractinians may also dominate deep-sea assemblages in which individual colonies or thickets grow on existing, usually elevated, hard substrates but do not form constructional accumulations (Messing 2003). By extension, groves or gardens of individual colonies of octocorals and antipatharians (e.g., Boehlert and Genin 1987; MacIsaac et al. 2001; McAllister and Alfonso 2001; Etnoyer and Morgan 2005; Leverette and Metaxas 2005; Mortensen and Buhl-Mortensen 2005; Watling and Auster 2005; Roberts et al. 2006) have also been treated as deep-sea coral environments. Like scleractinian-dominated communities, they create complex habitat suitable both for their own further settlement and growth, and for the development of diverse, sometimes highly specific, associated faunal assemblages (Koslow and Gowlett-Jones 1998). The great longevity of individual colonies (perhaps > 1,000 years or even >2,000 years) of some primnoid octocorals, antipatharians and the arborescent zoanthid, *Gerardia* sp. (Griffin and Druffel 1989; Druffel et al. 1995; Risk et al. 1998; Witherell and Coon 2001; B. Williams et al. 2006; Roark et al. 2006) offers an ecological time dimension suitable for the evolution of specific assemblages on a par with those developed by biohermal species. That is,

B.M. Riegl and R.E. Dodge (eds.), *Coral Reefs of the USA*,
© Springer Science + Business Media B.V. 2008

deep non-constructional groves also appear to exist for long enough periods for associated organisms to adapt to their particular conditions. However, such non-constructional assemblages lie beyond any usage of the term reef and are not discussed further here. Hexactinellid sponges also produce deep-water "reefs" (Rice et al. 1990; Conway et al. 1991, 2001, 2005a, b), though they appear to be less widespread than those built by corals.

Over a dozen taxa of scleractinian corals may be important contributors to deep-water biogenic build-ups (Stanley and Cairns 1988; Rogers 1999), although not all are principal framework builders. Of the latter, all exhibit branching morphologies. A few have restricted distributions, e.g., *Oculina varicosa* off the southeastern United States, *Goniocorella dumosa* off New Zealand, and *Enallopsammia profunda* (Figs. 21.1b, 21.2a, 21.4d) from New England to the Caribbean, while others such as *Madrepora oculata* (Fig. 21.1a) and *Solenosmilia variabilis* (Fig. 21.2b) are virtually cosmopolitan. Among the most widespread, and certainly the best known, of deep-water constructional taxa is *Lophelia pertusa* (Figs 21.1c, d, 21.2c, 21.4c), which occurs throughout most of the Atlantic from Labrador and Norway to Brazil and Tristan da Cunha, including the Gulf of Mexico and Mediterranean, and in the Indian and eastern Pacific Oceans at depths ranging from ~40 m off Norway to 2,170 m at low latitudes (Cairns 1979, 1994; Smith et al. 1997; Rogers 1999; Cairns and Chapman 2001; Mortensen et al. 2001). The almost continuous band along the northeastern Atlantic margin from Scandinavia to Morocco has been investigated to the greatest extent (e.g., Zibrowius, 1980; Frederiksen et al. 1992; Mortensen et al. 1995, 2001; Kenyon et al. 2003; Freiwald et al. 1999, 2000, 2002; De Mol et al. 2002; Fosså et al. 2002; Masson et al. 2003; Freiwald and Roberts 2005; Taviani et al. 2005; Wheeler et al. 2005a, Meinis et al., 2006; Roberts et al. 2006). In fact, when Forbes (1854, 1859) first outlined the bathymetric distribution of marine life in European seas, he recognized deep-sea corals as defining the deepest faunal zone.

Lophelia pertusa ranges from isolated colonies (Fig. 21.1c) to enormous accumulations kilometers in length and up to 350 m in vertical relief (Stetson et al. 1962; Reed 2002b; Mienis et al. 2006; Wheeler et al. 2005c, 2007; T. Williams et al. 2006). As might be expected from its broad distribution, it occurs in a wide range of geologic settings, including mud mounds and lithoherms in the Strait of Florida and on the Blake Plateau (Neumann and Ball 1970; Neumann et al. 1977; Messing et al. 1990; Paull et al. 2000; Reed 2002b; Reed et al. 2006; Ross and Nizinski, in press), authigenic carbonate outcrops in the Gulf of Mexico (Schroeder 2001, 2005), regional unconformities west of Ireland (Kenyon et al. 1998; De Mol et al. 2002; Huvenne et al. 2007; T. Williams et al. 2006), seamount-crest mega-ripples off Spain (Freiwald 2000), submarine canyons (Mortensen and Buhl-Mortensen 2005), and iceberg plough marks and glacial moraines off Norway (Freiwald et al. 1999, 2002; Mortensen et al. 2001).

Despite the broad spectrum of conditions under which *L. pertusa* occurs, all deep-water coral accumulations appear to be associated with oceanic boundary conditions or with seabed topographic features that may modify prevailing hydrodynamic regimes (e.g., Reed 1980, 1983, 2002b; Messing et al. 1990; Frederiksen et al. 1992; Freiwald 2000; Schroeder 2001; Mortensen et al. 2001; Freiwald et al. 2002; White et al. 2005). Freiwald et al. 2002 (and references therein) outlined five features that deep coral reefs appear to share: (1) attachment to pre-existing elevated topography; (2) growth below local storm wave base; (3) association with at least episodic benthic currents strong enough to prevent burial by sustained deposition of fine sediments; (4) stable, narrowly limited physicochemical conditions (stenohaline, stenothermal), and (5) a preference for settlement within the locally most saline water mass, which often corresponds with a seasonally occurring oxygen minimum layer. Recently, Guinotte et al. (2006) reported that a strong relationship exists between the frequency of cold-water coral records and the depth of the aragonite saturation horizon, with a deep (>2,000 m) horizon perhaps accounting for coral abundance in the northeast Atlantic and a shallow (50–600 m) horizon associated with few scleractinian records in the north Pacific (despite extensive sampling efforts).

Hovland et al. (1998) and Henriet et al. (1998) proposed that hydrocarbon seepage might generate conditions favorable to *L. pertusa* growth. However, no evidence exists for utilization of hydrocarbons by deep-reef corals or associated organisms, nor does any significant correlation exist between *L. pertusa* reef density and pockmarks possibly

FIG. 21.1. Deep-water colonial corals. (**a**) *Madrepora oculata*. (**b**) *Enallopsammia profunda*. (**c**) *Lophelia pertusa*, isolated colony. (**d**) *L. pertusa*, thickets of dead standing coral capped by living colonies. All photographs taken from the submersible *Johnson-Sea-Link I* off the east coast of Florida (Photos: J.K. Reed and S.D. Brooke)

generated by seepage (Mortensen et al. 2001). In coring through the full 155 m vertical relief of the Challenger Mound in the Porcupine Seabight southwest of Ireland, T. Williams et al. (2006) found no evidence that the mound, which consisted entirely of unconsolidated *Lophelia*-bearing sediments, accumulated via microbial activity associated with hydrocarbon seepage.

The progression of growth and development of deep-sea coral bioherms is understood in principle, if not in detail. Essentially, a coral colony forms by settlement of a planula larva and subsequent asexual reproduction. As the colony grows, the interior branches die (likely because of stagnation of water) leaving living exterior branches surrounding a dead core that eventually weakens, collapses and may be buried by sediment (Wilson 1979). Colonies

coalesce and over time the coral debris and sediment form the foundation of the reef, with living branches always at the outermost edges of each colony (Mullins et al. 1981). Coral debris with sediment may be hundreds of feet deep in places (T. Williams et al. 2006). Mounds off Ireland exhibit depositional cycles alternating between coral growth and hemipelagic sedimentation that reflect glacial versus interglacial conditions during the Pleistocene (Roberts et al. 2006; Rüggeberg et al. 2007). The Challenger Mound core mentioned above revealed that mound growth, which began between 2.0 and 2.7 Mya and ended before 0.46 Mya, took place chiefly during interglacial periods and did not require an initial hard substrate. South of the limit of the ice sheets, mounds may have been accumulating continuously

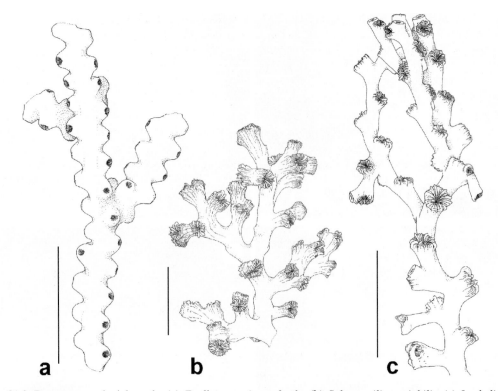

FIG. 21.2. Deep-water colonial corals. (**a**) *Enallopsammia profunda*. (**b**) *Solenosmilia variabilis*. (**c**) *Lophelia pertusa*. Scale bars: a, c = 5 cm; b = 3 cm. Cairns, S.D. 1981. Marine Flora and Fauna of the Northeastern United States. Scleractinia. NOAA Techn. Rep. NMFS Circular 438, US Dept. Commerce. (Scanned from original drawings by C.G. Messing)

over the last 50 ka, although sequences frequently show hiatuses and unconformities (Dorschel et al. 2005; Schröder-Ritzrau et al. 2005; Roberts et al. 2006). However, no information exists about the length of time between a coral colony's death and its disassembly and decomposition into rubble.

The broad geographic and bathymetric distribution of *L. pertusa* guarantees exposure to different conditions of flow, temperature, water chemistry and sedimentation, and, by extension, the development of different associated faunas. In fact, the only features of widely separated *L. pertusa* build-ups to which associated organisms may respond as commonalities are the limestone skeletal substrate, the specific size ranges of branches and inter-branch spaces, and perhaps common flow patterns generated around similarly shaped colonies and build-ups. The possibility also exists, however, that *L. pertusa* actually represents several cryptic species (Rogers 1999).

In the northeastern Atlantic few invertebrate organisms live on or in the living coral itself, e.g., the parasitic foraminifer *Hyrrokin sarcophaga*, the polychaete *Eunice norvegica*, the tanaidacean *Apseudes spinosus*, the bivalve *Delectopecten vitreus*, and the gastropod *Alvania jeffreysi* (Jensen and Frederiksen 1992; Mortensen 2001; Freiwald et al. 2002). Coral associates have been less well studied elsewhere, but on Blake Plateau reefs, such invertebrates as urchins (*Echinus* spp.), brittlestars (*Ophiacantha bidentata*), brisingid sea stars, sponges, galatheoids, polychaetes, and hydroids are abundant on both living and dead corals (S.W. Ross et al., unpublished data). The dead interlocking skeletal branches provide a complex three-dimensional environment for a diverse associated fauna belonging to almost every non-herbivorous trophic guild: sessile and sedentary suspension feeders and predators, mobile grazers (of sessile invertebrate prey), predators and detritivores, and their commensals and parasites. Over 1,300 species have been found associated with *L. pertusa* in the northeastern Atlantic (Roberts et al. 2006). Quantitative records include 297 invertebrate species associated with dredged *L. pertusa* blocks and rubble in 260 m on the Faeroe Shelf (Jensen and Frederiksen 1992), 313 species from the *L. pertusa*-dominated Galway Mound, Porcupine Seabight, west of Ireland (Henry and Roberts 2007), and over 360 recorded by Reed (2002a and references therein) for *Oculina varicosa* reefs off the southeastern US in 60–100 m.

A few taxa span much of the North Atlantic range of *L. pertusa*, e.g., the fan sponge *Phakellia ventilabrum*, the branching coral *Madrepora oculata*, and the solitary corals *Desmophyllum dianthus* (formerly *D. cristagalli*) and *Stenocyathus vermiformis* (Fosshagen and Hoisæter 1992; Mortensen et al. 1995; Freiwald and Mortensen 2000; Freiwald et al. 2002). However, major differences exist among taxa associated with and adjacent to the coral framework even between geographically adjacent areas. For example, Jensen and Frederiksen (1992) found only 61 out of 595 *L. pertusa*-associated invertebrate species shared between their study off the Faeroes and that of Burdon-Jones and Tambs-Lyche (1960) off Norway. Comparisons of the Faeroe and Norway faunas with Danois' (1948) records from the Bay of Biscay halved the number of shared taxa. Although several references to macrofauna associated with *L. pertusa* in deep warm temperate and subtropical waters exist (e.g., Reed, personal communication in Cairns and Stanley 1981; Mullins et al. 1981; Messing 1984; Messing et al. 1990; Paull et al. 2000; Reed 2002b, Reed et al. 2006; Brooks et al. 2007; Ross and Nizinski, 2007), no detailed quantitative analyses have yet been published. As an example of variations among *Lophelia*-associated assemblages at lower latitudes, the large arborescent zoanthid *Gerardia* sp. and four stalked crinoid species occur on or adjacent to lithoherms on the Bahama side of the Strait of Florida (Messing et al. 1990). However, none have ever been recorded from the north central Strait less than 40 km distant despite extensive dredging and submersible observations (Meyer et al. 1978; Reed et al. 2006).

In addition to the relatively recent discovery of the great extent and unexpected richness of deep-sea reefs, the recent downslope movement of commercial fishing and hydrocarbon exploration has brought these habitats into the sphere of economic exploitation and has triggered a major effort toward their conservation and protection (e.g., Rogers 1999; Freiwald 2000; Fosså et al. 2002; Reed 2002b, papers in Willison et al. 2001 and in Freiwald and Roberts 2005). Several studies have also documented significant deterioration as well as outright destruction of these habitats by trawling (Koslow and Gowlett-Jones 1998; Fosså et al. 1999, 2002; Krieger 2001; Hall-Spencer et al. 2002; Reed et al. 2005a; Kitahara 2005; Wheeler et al. 2005b; Reed et al. 2007).

21.2 Deep-Sea Coral Reefs of the United States

Figure 21.3 illustrates the known distribution of deep-sea framework-building coral species in continental US and adjacent waters. Reefs in the form of mounds, banks, ridges, pinnacles and lithoherms occur primarily along the southeastern continental margin from North Carolina southward through the Strait of Florida and into the eastern and northern Gulf of Mexico (Fig. 21.3c). The dominant framework-builders deeper than 200 m are *Lophelia pertusa* and *Enallopsammia profunda*, with lesser contributions by *Madrepora oculata* and *Solenosmilia variabilis* (Reed 2002b; Reed et al. 2005c, 2006; Ross and Nizinski, in press). *Oculina varicosa* dominates one shallower area (60–100 m) off east-central Florida (Reed 2002a; Fig. 21.3c). As mentioned above, *L. pertusa*, *M. oculata* and *S. variabilis* have broad distributions. *E. profunda* is restricted to the western North Atlantic (Georges Bank to Lesser Antilles) in 403–1,748 m (chiefly <1,000 m). Another species, *E. rostrata*, is widely but sparsely recorded in US waters – northern Gulf of Mexico (H. Roberts et al., personal communication, 2006), off Georgia and on Bear Seamount in the New England Seamount chain (Cairns 1979; Moore et al. 2004) – as well as along the insular slopes of the Strait of Florida, but it does not appear to construct mounds. Non-constructional deep-sea coral communities dominated by scleractinians, stylasterid hydrocorals, octocorals and antipatharians also occur along most American continental margins as well as in Alaskan and Hawaiian waters. Although understanding of the

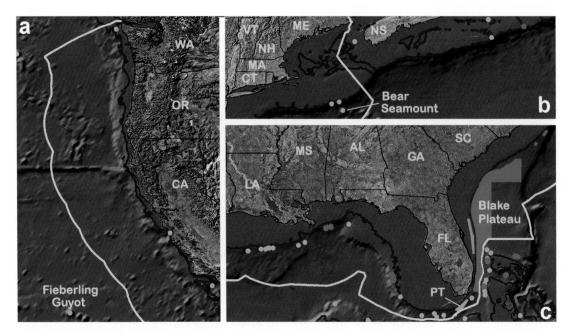

FIG. 21.3. Deep-water coral sites in the United States and adjacent waters. Red dots = records of chiefly *L. pertusa* constructional build-ups. Orange dots = non-constructional records of *L. pertusa*, *E. profunda*, *M. oculata* and *S. variabilis*. (**a**) Pacific coast. (**b**) Northeast including Atlantic Canadian waters; small red square at upper right = *Lophelia* Conservation area. (**c**) Southeast and Gulf of Mexico; large red polygon = Deep-water Coral Habitat of Particular Concern (HAPC) proposed by the South Atlantic Fishery Management Council, including the Stetson Banks, Savannah Banks, East Florida pinnacles and Miami Terrace; two small red polygons at upper right = North Carolina HAPCs; orange polygon = stylasterid-dominated HAPC on the Pourtalès Terrace (PT); green polygon = known extent of *Oculina varicosa* habitat; four red dots at lower right = *L. pertusa* lithoherms and mounds reported by Neumann et al. (1977), Messing et al. (1990), Grasmueck et al. (2006) and A.C. Neumann and C.G. Messing (unpublished data, 1979), and *S. variabilis* mounds (farthest right) by Mullins et al. (1981) in Bahamian waters. See text for other sources. Yellow line = Exclusive Economic Zone (EEZ) boundary. Dark blue line = 200 m contour

distribution of these habitats has advanced substantially in the last several years, only a small percentage of known sites have been investigated in any detail or examined directly. Expeditions continue to discover new sites. Sophisticated mapping technology, such as ship- and AUV-mounted multibeam sonar systems with backscatter data, side-scan sonar systems, and sub-bottom seismic profilers have provided detailed bottom imagery with resolution to 1–3 m. Geographic Information Systems integrate mapping data with other geospatial information (e.g., fishing pressure, management areas, biological, geological, physicochemical observations, geophysical structure, and hydrodynamics) to generate detailed and precise maps and data sets and produce a foundation for robust system analyses, predictions, and management protocols.

21.2.1 Blake Plateau

The continental slope of the southeastern United States is dominated by a unique, wedge-shaped feature known as the Blake Plateau (Fig. 21.3c), a northward extension of the Bahamian carbonate province. From its narrow northern apex off North Carolina, the plateau broadens to the south, flanked to the west in 400–600 m by the Florida–Hatteras slope and gradually sloping eastward to about 1,200 m where the Blake Escarpment, lying roughly along 77° W longitude, drops precipitously to the deep North Atlantic floor (~4,800 m). To the south, the plateau abuts the northern margin of the Bahama Platform, though its precise border is unclear. Seismic profiles and fathometer surveys have revealed hundreds of scarps, mounds, plateaus, pinnacles and depressions across the plateau, many bearing or built by deep-sea corals (Mullins et al. 1981; Paull et al. 2000; Popenoe and Manheim 2001; Reed and Ross 2005; Reed et al. 2006; Ross and Nizinski, in press). A broad convex ramp on the northern Plateau, the Charleston Bump, extends southeastward from the continental shelf off South Carolina in 250–750 m. Its shoaling seafloor deflects the Gulf Stream, intensifying bottom flow and maintaining a largely nondepositional environment (Sedberry et al. 2001; Popenoe and Manheim 2001; Wenner and Barans 2001). The Gulf Stream has also heavily eroded the western plateau margin, chiefly during the Pleistocene, exposing many depressions, Cretaceous- to Miocene-age hard substrates, and mounds (Uchupi 1967). Much of the Plateau is capped by indurated layers of phosphorite, including pavements, loose slabs and rubble, although some areas in >700 m have become impregnated with ferromanganese oxides (Popenoe and Manheim 2001; Wenner and Barans 2001).

Deep-sea corals were first reported from the Blake Plateau in the late 1880s (Agassiz 1888). However, aside from chiefly geological surveys, little research was conducted on them until the last decade (Paull et al. 2000; reviewed by Ross and Nizinski, in press; Partyka et al., in press). Some of the deep-sea coral study areas on the Plateau have been individually named (e.g., Reed and Ross, 2005; Reed et al. 2006), implying separate, discrete areas of coral habitat. Although these coral habitats are actually larger and more continuous than the separate names imply, we use them here for convenience and until the details of mound distributions are more clearly understood.

Ross et al. (upublished data) observed two *Lophelia* morphologies on the Blake Plateau: a slender, branching, upright form and a lower, thick, stubby form. Cairns (1979) also noted the two but stated that they did not rise to subspecific status. The factors that control these growth patterns (e.g., genetic, epigenetic or environmental) remain unknown. Differences in growth among corals on the same mound are likely caused by differences in environmental parameters, such as current-mediated food supply or hydrodynamic forcing. Colonies at and near mound crests tend to be elongated, suggesting a relationship with locally elevated flow regimes. Because branches bearing individual polyps fork alternately in this form, it is relatively easy to follow time-lines from

oldest to youngest. Colonies on mound flanks and bases subject to more resuspended sediment tend to develop a more chaotic growth form in which time-lines are difficult to identify. However, no clear dividing line exists between the two, and the transition from one extreme to the other seems gradual. Nonetheless, the possibility exists that one form might facilitate mound formation more than the other.

21.2.1.1 North Carolina

Following Uchupi's (1967) first record of a coral mound off Cape Lookout, Rowe and Menzies (1969) referred to discontinuous banks of *Lophelia* sp. along the 450-m isobath based on echosounder recordings. A drop camera photograph in 458 m in the same area showed thickets of coral (referred to as *Lophohelia* sp.) on a steep slope (Menzies et al. 1970). Subsequent work in this area has been carried out by an interdisciplinary team headed by one of us (SWR) that described three major coral mound systems (Ross 2006; Ross and Nizinski, in press). These mounds represent the northernmost known coral build ups off the southeastern US. Significant deep-water coral habitats are unknown along the east coast again until southeast of Cape Cod.

From north to south, these three sites are Cape Lookout *Lophelia* Bank A, Cape Lookout *Lophelia* Bank B, and Cape Fear *Lophelia* Bank (Ross and Quattrini, 2007). All form physically similar mounds and ridges in 366–450 m that rise as much as 80–100 m above the seafloor and appear to consist of a sediment/coral rubble matrix topped with almost monotypic stands of *L. pertusa*. Other contributing scleractinians include *Madrepora oculata* and *Enallopsammia profunda*. Solitary corals, sponges, and anemones are also abundant. *In situ* temperatures measured during submersible dives range from 5°C to 12°C (Ross and Quattrini 2007).

Living *L. pertusa* covers extensive areas of mound flanks and crests and is accompanied by abundant dead colonies and coral rubble interspersed with sandy channels. Extensive areas of coral rubble surround the mounds, especially at the bases of the mounds/ridges, and may be quite thick locally. The southernmost mound system, off Cape Fear exhibits some of the most rugged habitat and vertical relief on any of the three sites examined. It is more isolated and lacks the smaller mounds that surround both of the Cape Lookout sites (Ross et al., 2006, unpublished data).

All three sites support similar abundant and diverse macroinvertebrate faunas, notably chirostylids (chiefly *Eumunida picta*) and the brisingid asteroid *Novodinia antillensis* (Ross et al., unpublished data). Several other obvious taxa occur throughout much of the region, e.g., *Echinus gracilis* (Echinoidea) and *Rochinia crassa* (Brachyura). Fish assemblages, so far totaling 33 reef-associated species, also appear similar but differ from deep reef communities to the south (S.W. Ross and A.M. Quattrini, personal communication, 2007). Dominant species at Cape Lookout A were *Laemonema barbatulum*, *Helicolenus dactylopterus*, and *Hoplostethus occidentalis*; at Cape Lookout B: *H. dactylopterus*, *L. barbatulum*, and *Laemonema melanurum*, and at Cape Fear: *Beryx decadactylus*, *L. barbatulum*, and *Polyprion americanus*.

Additional mound systems may also exist off North Carolina. George (2002) found the solitary coral *Bathypsammia tintinnabulum* in an area in 650–750 m that he called the Agassiz Coral Hills. However, multibeam mapping of this area revealed a flat bottom with no hint of corals (Ross et al., unpublished data).

21.2.1.2 South Carolina

From South Carolina southward, deep-sea coral habitats are more variable and occur chiefly in deeper water than those off North Carolina. Most research has concentrated around the Charleston Bump and Stetson Banks (Stetson et al. 1962; Sedberry et al. 2001).

As mentioned above, the Charleston Bump has an important influence on the Gulf Stream (papers in Sedberry et al. 2001). Bottom topography is often quite rugged, with vertical cliffs in places and local vertical relief exceeding 100 m. Bottom currents are generally strong and sediment cover sparse. Exposed phosphoritic pavements, loose slabs and rubble are typical. Coral and sponge communities range from absent to dense. The fish assemblage recorded by Weaver and Sedberry (2001) over chiefly rocky substrates includes a variety of small sharks (e.g., *Centroscyllium fabricii*, *Dalatias*

licha, *Etmopterus* spp.), *Helicolenus dactylopterus*, *Beryx splendens* and *B. decadactylus*, and *Polyprion americanus*, and differs substantially from that associated with the coral-associated assemblage on the Savannah Bank off Georgia at similar depths (Ross and Quattrini 2007) (see below). Ross and Quattrini (2007) suggest that different habitats and sampling methods might account for the faunal differences.

The Stetson Bank (or Stetson Reefs) is a large region of extremely rugged topography and diverse bottom types that lies chiefly east of the outer Charleston Bump near the eastern edge of the plateau, in 640–869 m (~220 km southeast of Charleston, SC) (Stetson et al. 1962; Reed 2002; Reed et al. 2006). It is one of the deeper and more complex of the Blake Plateau deep-sea coral areas. The Bank includes numerous mounds up to 146-m high spread over a 6,174-km² area (Reed and Ross 2005; Reed et al. 2006; Ross and Nizinski, in press). Stetson et al. (1962) first described these features from seismic profiles, dredge samples and drop camera photographs as steep-sloped mounds with active coral growth on top, including live colonies up to 50 cm across. *Enallopsammia profunda* (referred to as *Dendrophyllia profunda*) dominated in all areas with *L. pertusa* (called *L. prolifera*) concentrated on mound crests. Associated organisms included hydroids, octocorals, echinoderms, anemones and ophiuroids.

More recently, fathometer transects (Reed et al. 2006) and multibeam mapping (Ross et al., unpublished data) have revealed dozens of individual pinnacles, mounds and other complex features including ridges, valleys and trenches within only a small portion of the Bank. A diverse assemblage of corals and other sessile invertebrates are most abundant on slopes and crests of peaks and ridges. A particularly deep trench that expands into a plain on the eastern side (Ross et al., unpublished data) supports abundant corals along its western rim (S.W. Ross and A.M. Quattrini, personal communication, 2007). As an example of one feature mapped in detail and examined via submersible (Reed et al. 2006), Stetson's Pinnacle is one of the tallest coral lithoherms known, rising over a distance of ~0.9 km from 780 m at the south base to a peak in 627 m. On a transect up the southern flank, the lower slope (~762–701 m) rises at 10–30° with a series of 3- to 4-m-high ridges and terraces covered with nearly 100% *L. pertusa* rubble, as well as live colonies 15- to 30-cm tall and standing dead colonies 30- to 60-cm tall. Exposed rock, including slumped slabs, is restricted to steeper eroded ridge faces. Slopes increase from ~45° to 60° between 701 and 677 m and become steeper toward the pinnacle crest. From 671 m to the crest, a complex and rugged topography consists of 60–90° rock walls and 3- to 9-m tall rock outcrops, with live *L. pertusa* colonies up to 60-cm tall becoming more common and completely covering some rock ledges. The pinnacle crest is a flat lithified plateau ~190 m across in 625–628 m edged by a 30-cm-thick crust overhanging a 3–6 m vertical escarpment (Reed et al. 2006).

Although *L. pertusa* dominates, *E. profunda* and *S. variabilis* also occur here. Other important major taxa are stylasterids, octocorals, sponges and a variety of solitary scleractinians. Octocoral taxa include members of the Primnoidae (e.g., *Plumarella* sp.), Paramuriceidae, Isididae (e.g., *Keratoisis* sp.), Stolonifera, and Nephtheidae. Antipatharians include *Leiopathes* spp. and *Bathypathes* spp. Eighteen identified sponge taxa include Pachastrellidae (Fig. 21.4a, b), Corallistidae, Hexactinellida (Fig. 21.4b, c), *Geodia* sp., and *Leiodermatium* sp. Sponge density, diversity and size increase on the steeper upper slopes (671–625 m), with abundant Hexactinellida, Pachastrellidae and massive *Spongosorites* sp. Live *L. pertusa* and rubble, *Phakellia* sp. fan sponges up to 50-cm tall, and numerous other demosponges dominate the peak plateau. Other organisms include hydroids, anemones, zoanthids and solitary corals (Reed et al. 2006; Ross et al., unpublished data). Mobile macroinvertebrates include *Eumunida picta*, the crabs *Bathynectes longispina* (Portunidae) and *Chaceon fenneri* (Geryonidae), *Stylocidaris* sp. and *Echinus* sp. and dense populations of ophiuroids. The benthic fish fauna is most similar to that of the Savannah Banks and Jacksonville Banks to the south (S.W. Ross and A.M. Quattrini, personal communication, 2007). Of the 19 species of coral-associated species identified so far, *Nezumia sclerorhynchus* and *Laemonema melanurum* are most abundant (Ross and Quattrini, in review). Others include *Beryx decadactylus*, *Hoplostethus occidentalis* and phycids (Reed et al. 2006; Ross and Quattrini, in review).

21.2.1.3 Georgia–Florida

Deep-water coral habitats have been investigated at three sites here: an area off Georgia called the Savannah Banks, a large area straddling the Georgia–Florida border known as the Jacksonville Banks, and a long series of sites to the south – the East Florida *Lophelia* Pinnacles (Reed et al. 2006; Partyka et al., in press).

The Savannah Banks (or Savannah Lithoherms; Reed et al. 2006) occur on the western end of the Charleston Bump at the foot of the continental shelf, ~167 km east of Savannah, Georgia. A pair of DSRV *Alvin* dives first documented the presence of both *Enallopsammia* (as *Dendrophyllia*) *profunda* and *Lophelia pertusa* (as *Lophelia* sp.) (Milliman et al. 1967). The topography in the area varies from rugged, high-profile hard bottoms and low profile, flat hard bottoms (more offshore) to numerous mounds and ridges with varying topography having a heavier sediment load compared with other sites (Reed et al. 2006; Partyka et al., in press). At the more offshore sites rock types include manganese–phosphorite pavement and nodules, foraminiferan limestone, and calcareous mudstone. Strong bottom currents are often observed, and attached fauna exhibit considerable variation in density and distribution (Milliman et al. 1967; Reed 2002b; Reed et al. 2006; Ross et al., unpublished data). At the more inshore sites, scattered patches of deepsea corals are often smaller than at other sites. The bottom is mostly covered with *Lophelia* rubble and dead standing thickets, with scattered low-profile living corals (e.g., *Stylaster*, *Lophelia*) and sponges (Milliman et al. 1967; Reed 2002b; Reed et al. 2006; S.W. Ross et al. 2006).

Three different reports give an idea of the range of variation. Milliman et al. (1967) reported elongate coral mounds up to 1-km long, dominated by *E. profunda* and with moderate slopes (25–37°) reaching 54-m relief but with no visible rock outcrops. Wenner and Barans (1990, 2001) described a variety of 15- to 23-m-tall mounds in 503–555 m, some thinly veneered with fine sediment and dead coral fragments, and others with dense living *L. pertusa* and *E. profunda*. They also noted "relic coral populations … represented by extensive deposits of dead and broken coral branches with only small, widely spaced colonies of live coral across the mounds" (p 166). Fathometer transects

Fig. 21.4. Organisms associated with *L. pertusa* habitats, South Carolina to Florida. (**a**) Pachastrellid sponge (left) and bamboo coral (*Isidella* sp.) (right) with solitary corals (lower right); Stetson's Reef, 644 m. (**b**) Hexactinellid sponge *Heterotella* sp. (left), pachastrellid sponge (center right) and plexaurid octocoral *Eunicella* sp. (above right); Savannah Lithoherm, 511 m. (**c**) Standing dead *Lophelia pertusa* with some living branches (white), hexactinellid sponge *Aphrocallistes* sp. (lower left), primnoid octocoral *Plumarella* sp. (upper right), numerous small orange solitary corals, and the chyrostylid squat lobster *Eumunida picta* (center); off Cape Canaveral; East Florida *Lophelia* Reef, 785 m. (**d**) Thicket of dead standing *Enallopsammia profunda* with a few live branches and a synaphobranchid eel; off Fort Lauderdale, Florida; Strait of Florida axis, 780 m. (**e**) Slope with numerous stylasterid hydrocorals (white), a large antipatharian (center), antedonid crinoids (far left and right), and orange chyrostylid squat lobsters; Miami Terrace, 283 m. (**f**) Cup-shaped choristid sponge and pink stylasterid hydrocoral; Pourtalès Terrace, 173 m (Photos: J.K. Reed [a–c, e, f] and C.G. Messing [d])

revealed mounds to a depth of 582 m with signatures that suggest hard substrate beneath but not within the mounds.

Reed et al. (2006) described several features based on fathometer transects and submersible observations. One of these (their site 1) is a massive feature in 549 m with an overall vertical relief of 83 m that consists of five major pinnacles rising 9–61 m above the basal mass. The south flank of one pinnacle begins with a 10–20° slope with ~90% cover of coarse sand, coral rubble and some 15-cm rock ledges. Its peak is a sharp NW–SE-oriented ridge, perpendicular to the prevailing 50 cm/s current. By contrast, the north slope is a series of terraces or shallow depressions, each ~9 to 15 m wide, separated by 3-m high 30–45° scarps, including one leading to the ridge crest. Exposed rock surfaces are black phosphoritic pavements. Sessile macrofauna occur on terrace pavements but in greatest abundance along outcrop edges and on the crest. Maximum sponge and octocoral cover reaches 10%, while *L. pertusa* colonies (15 to 30-cm diameter) account for only ~1% of coverage. Dominant taxa include several octocorals 15 to 20 cm tall, e.g., Primnoidae, Paramuriceidae, and *Antipathes* sp. to 1 m tall. The most common sponges include large *Phakellia ventilabrum* (30- to 90-cm diameter), Pachastrellidae (30 cm), Choristida (30 cm) and Hexactinellida. The most obvious and common mobile fauna includes decapod crustaceans (*Chaceon fenneri* and Chirostylidae) and a few mollusks. Fishes include *Xiphias gladius*, *Squalus* sp., Macrouridae and *H. dactylopterus*. Ross and Quattrini (in review) identified 16 species of reef-associated fishes on the inshore Savannah Bank sites, with *N. sclerorhynchus* and *L. melanurum* the two dominant taxa.

Hundreds of high-relief *Lophelia* reefs occur from southern Georgia to Jupiter, Florida, at the base of the Florida–Hatteras slope in 400–866 m. Topographic highs, most with at least some coral development, are abundant and nearly continuous from the Jacksonville area to south of Cape Canaveral (Ayers and Pilkey 1981; Paull et al. 2000; Reed 2002b; Reed et al. 2006; Partyka et al., in press, S.W. Ross et al., unpublished data). Reed (2002b) and Reed et al. (2002, 2006 and unpublished data) mapped almost 300 features from 8 to 168 m in vertical relief along a 222-km stretch from Jacksonville to Jupiter. So far, nearly 10%

have been ground-truthed via submersible or ROV, confirming the presence of *L. pertusa* and/or *E. profunda*. Those in the northern portion of this area off Jacksonville and southern Georgia are primarily lithoherms – rock mounds capped with coral, coral rubble and other attached fauna (Paull et al. 2000), and are discussed here. The features from south of St. Augustine to Jupiter are described below.

Reed et al. (2006) described one feature off Jacksonville, Florida, as a massive lithoherm 5.7-km long (N–S) rising from a base in 701 m and consisting of at least seven individual peaks with heights of 30–60 m and a total relief of 157 m. Submersible dives on the tallest peak found a maximum relief of 107 m with the crest in 544 m. Eastern and western flanks of sand, mud, rock pavement and rubble steepen from 20–30° to 50–80° near the top. The southern 30–40° slope bears a series of lithified terraces and dense thickets of 30- to 60-cm tall dead and live *L. pertusa* mostly restricted to the tops of mounds and ridges and terrace edges. One peak at 565 m has dense thickets of ~20% live and dead standing *L. pertusa* and thick outcrops of coral rubble. Dominant sessile fauna includes *L. pertusa*, abundant octocorals (e.g., *Placogorgia* sp., *Chrysogorgia* sp., Paramuriceidae, *Anthomastus* sp. *Nephthya* sp. and numerous Isididae) and sponges (e.g., *Geodia* sp., *Phakellia* sp., *Spongosorites* sp. Petrosiidae, Pachastrellidae, and Hexactinellida). Ross and Quattrini (2007) described the 15 benthic reef-associated fishes identified from deep reef sites off Jacksonville as most similar to the Savannah and Stetson fish assemblages, and also dominated by *N. sclerorhynchus* and *L. melanurum*.

The features from south of St. Augustine to Jupiter appear to be primarily unconsolidated sediment and coral rubble mounds capped with dense 1-m tall thickets of *L. pertusa* and *E. profunda*, some *M. oculata* and varying amounts of coral debris and dead standing coral. As an example, Reed et al. (2006) described a *L. pertusa* mound off Cape Canaveral with 44 m relief that rises to a series of ridged peaks in 713–722 m. The lower southern flank is a 10–20° slope of fine sand with a series of 1- to 3m-high dunes or ridges supporting ~50% cover of *L. pertusa* thickets consisting chiefly of 1-m tall dead standing colonies topped by ~1–10% cover of living colonies (15–30 cm) on the outer ridge edges. Most of the coral is intact with little broken dead coral rubble in the rippled intervening sand. The upper slope steepens first to 45° and then 70–80° in the uppermost 10 m. The pinnacle crest – a narrow 5- to 10-m-wide ridge – supports up to 100% cover of coral thickets that reach 1.5-m high, with ~10–20% living coral colonies – both *L. pertusa* and *E. profunda* – reaching 90-cm high. The upper 10 m of the northern slope is nearly vertical (70–80°) with a series of coral thickets on terraces or ridges below the crest. Scattered *M. oculata* and some stylasterids are also present. Dominant octocorals include *Plumarella* sp. (Fig. 21.4c), *Isidella* sp. (Fig. 21.4a), *Keratoisis flexibilis*, *Anthomastus* sp. and *Nephthya* sp. Dominant sponges include several species of Hexactinellida, large yellow demosponges (60- to 90-cm diameter), Pachastrellidae, and *Phakellia* sp. Motile fauna are chiefly echinoderms (e.g., a cidaroid and *Hygrosoma*? sp. urchins and comatulid crinoids) and decapod crustaceans (e.g., *Chaceon fenneri* and Chirostylidae). Ross et al. (unpublished) note that these southern sites support a more diverse invertebrate fauna than those further north. Ross and Quattrini (2007) observed 13 deep coral-associated benthic fish species at two sites off Cape Canaveral. They noted that the assemblage, dominated by *Synaphobranchus* spp. and *N. sclerorhynchus*, differs from all areas to the north.

21.2.2 Strait of Florida

The Strait of Florida is a reversed L-shaped trough that separates the Florida Peninsula from the Bahama Platform and Cuba. Its northern arm runs southward from 27°30′ N latitude at the northern margin of the Bahama Platform, while its southern arm opens into the Gulf of Mexico at 83° W longitude. The two arms converge on Cay Sal Bank, a triangular outlier of the Bahama Platform that lies between Great Bahama Bank and Cuba. The Strait's primary hydrographic feature, the Florida Current, is an upstream segment of the Gulf Stream proper that funnels back to the North Atlantic much of the water volume lost to the Southern Hemisphere via the deep thermohaline conveyor. It has been studied in substantial detail via both modeling and synoptic observation (e.g., Düing 1973, 1975; Kielmann and Düing 1974; Lee et al. 1992). Mariners have used the Strait and its powerful current since the early stages of European exploration to carry ships northward from the trade wind belt and Spanish Main to the prevailing westerlies for the return trip to Europe.

Much of the Florida margin of the northern Strait, at least as far south as the Palm Beaches, represents a continuation of the Florida–Hatteras slope environments to the north. *L. pertusa* mud mounds that begin south of St. Augustine continue at least as far south as Miami, Florida. These mounds may extend into the southern Strait off the Florida Keys, but little mapping has occurred here. Corals associated with two geological features of the central and southern Strait – the Miami and Pourtalès Terraces – do not form constructional accumulations and are treated separately below. The Bahamian and Cuban margins of the Strait exhibit insular environments distinctly different from those of the Florida margin and are not discussed further here (see, e.g., Neumann and Ball, 1970; Neumann et al. 1977; Mullins and Neumann 1979; Mullins et al. 1981; Messing et al. 1990; Anselmetti et al. 2000; Grasmueck et al. 2006).

Along the deep axis of the Strait, coral bioherms exist as mounds of unconsolidated muddy sand in a matrix of coral rubble. Neumann and Ball (1970) first observed thickets of unidentified branching coral capping sediment mounds 0.5-m high and 3 to 4 m long east of the base of the Miami Terrace in 719–825 m. Off Fort Lauderdale, Florida, in 773–784 m, thickets and mounds of dead standing *E. profunda* cover low mounds up to 50 m across and 0.3- to 4.5-m high with 0–10% living coral cover (usually at most a few small twigs) (Messing et al. 2006). These features, restricted to a mile-wide band adjacent to the Exclusive Economic Zone boundary with the Bahama Islands (though they continue into Bahamian waters), are separated by areas of dense to sparse coral rubble and flat or rippled sediment. Accompanying macrofauna includes sponges (e.g., *Heterotella* sp. [Fig. 21.4b], *Hyalonema* sp., *Farrea* sp.), octocorals (e.g., *Plumarella* sp., *Eunicella* sp. [Fig. 21.4b], and Isididae) an echinothurid echinoid and a few fishes (Rajidae, Synaphobranchidae [Fig. 21.4d], Chimaeridae) (Brooke et al. 2006; Messing et al. 2006). Submersible observations have also revealed five high-relief coral bioherms regions between the base of the Miami Terrace and Bimini I. on the Great Bahama Bank that were first discovered by multi-beam seismic surveys using an AUV (Grasmueck et al. 2007).

Although substantial trawling and dredging records exist throughout the Strait, most results are scattered throughout the literature for individual taxonomic groups, and neither the invertebrate nor fish fauna of deep coral habitats has been adequately surveyed. Known species richness is almost certainly much greater than currently recognized.

21.2.2.1 East Florida Oculina varicosa Reefs

Although Florida's *Oculina varicosa* reefs (or banks) occur in warmer and shallower water than is characteristic of other deep- and cold-water coral build-ups (Roberts et al. 2006), they are included here because they are similar in structure and physiology to azooxanthellate, deep-water reefs.

Deep-water *Oculina varicosa* reefs occur exclusively along a 167-km region on the eastern Florida shelf edge from ~27°20′ N to ~29°30′ N at depths of 60–100 m (Reed 1980, 2002a; Reed et al. 2005a; Fig. 21.3c). The reef system consists of hundreds of individual coral pinnacles, mounds, and ridges that range from 3 to 35 m in height and up to 100–300 m in width (Fig. 21.5). These mounds appear to be true bioherms, that is, coral-capped mounds of sand, mud, and coral debris that overlay an oolitic limestone base formed during the Holocene transgression nearly 18,000 YBP (Macintyre and Milliman 1970; Reed 1980). Individual *O. varicosa* colonies reach 2 m in diameter and 2 m in height. Colonies less than 25 cm across are often 100% alive. Larger colonies are usually dead in the center, probably from water stagnation due to the dense branching framework, with only the outer 10–30 cm alive.

Moe (1963) first described the shelf-edge region from St. Augustine to Fort Pierce, Florida, from interviews with commercial fishermen. He described "The Peaks" in 35–65 fathoms (64–119 m) as highly irregular coral rock with many peaks and ledges up to 50 fathoms (91 m) high that supported a fishery of red and vermilion snappers, red grouper and triggerfish (see also Koenig et al. 2000). The first detailed submersible surveys began in the 1970s (e.g., Avent et al. 1977; Reed 1980). In 1984, NOAA designated a 315-km² region as the *Oculina* Habitat of Particular Concern (OHAPC), which established the first deep-sea coral marine protected area in the world and prohibited bottom trawling, dredging, bottom longlines and anchoring. However, the protected area covered less than 30% of the reef system known at the time. Reed (1980) had already confirmed that *O. varicosa* bioherms extended at least 55 km north of the OHAPC, and Moe (1963) showed high-relief coral mounds at least to 30° N latitude, about 166 km north of the northern OHAPC boundary.

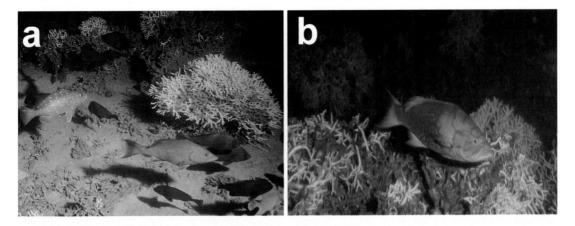

FIG. 21.5. *Oculina varicosa* habitat, central eastern Florida, 80 m. (**a**) Living thicket (right); dead standing coral topped by living colonies (rear), groupers, angelfish and surgeonfishes. (**b**) Terminal supermale phase of scamp, *Mycteroperca phenax*, on *O. varicosa* thickets (Photos: J.K. Reed)

The OHAPC was expanded to 1,029 km² in 2000 by extending the boundary north to Cape Canaveral. Yet, it still did not cover the known extent of the coral habitat. In 2002 and 2003, a multi-beam echo sounder survey mapped the bathymetry of the coral habitat to a resolution of ~3 m and covered ~29% of the OHAPC (295 km²) (Reed et al. 2005a). The survey estimated that more than 100 bioherms may exist in the unprotected area west of the current OHAPC. Current NOAA bathymetric charts show a zone of high-relief features that parallel the 80-m depth contour and extend from the northern boundary of the OHAPC (28°30′ N) off Cape Canaveral to at least 29°30′ N, with the densest zone – as yet unprotected – from Canaveral to 29°15′ N. Unfortunately, many of the *Oculina* reefs in the recently protected northern region of the OHAPC have already been severely damaged from bottom trawling for rock shrimp, in some cases with nearly 100% loss of living coral (Reed et al., 2007).

Scientific divers conducted lockout dives from the *Johnson-Sea-Link* submersibles to study *O. varicosa* growth rates and community ecology on the 80-m reefs (Reed 1981; Reed et al. 1982; Reed and Mikkelsen 1987). Growth rates average 16 mm/year, similar to that of *L. pertusa* (Mortensen and Rapp 1998). The reefs support a diverse community of invertebrates and fishes. Quantitative collections of 42 small *O. varicosa* colonies yielded over 20,000 invertebrate specimens belonging to 230 species of mollusks, 50 decapods, 47 amphipods, 21 echinoderms, 15 pycnogonids, and numerous other as yet unidentified taxa (Reed et al. 1982; Reed and Hoskin 1987; Reed and Mikkelsen 1987; Child 1998). However, larger sessile invertebrates such as the massive sponges and octocorals common on the deeper *L. pertusa* reefs are not common.

At least 70 associated fish species have been recorded (Reed and Gilmore 1982). In the 1970s and 1980s, the deep-water *Oculina* reefs formed impressive breeding grounds for commercially important populations of snapper (*Lutjanus* spp.) and grouper (Serranidae), in particular, dense populations of gag (*Mycteroperca microlepis*) and scamp (*M. phenax*) (Fig. 21.5b) grouper (G. Gilmore, personal observations, 1980; Reed 2002a). Scamp were seasonally abundant, forming dense spawning aggregations of several hundred individuals per hectare (Gilmore and Jones 1992). By the

1990s commercial and recreational overfishing had decimated these dense aggregations, and a ban on bottom hook and line fishing was enacted in 1994 (Koenig et al. 2005).

A hypothetical trophic model of the deep-water *O. varicosa* ecosystem shows that standing coral provides refuge for dense invertebrate communities including polychaete worms, mollusks, crustaceans, sponges and octocorals, which provide food for smaller reef fish and, by extension, larger benthic and pelagic fish (Reed 2002a; George et al. 2007). Economically important grouper species including gag associate closely with the intact coral habitat (Gilmore and Jones 1992; Koenig et al. 2005). Therefore, significant habitat loss, in particular that of intact living and dead standing coral, will dramatically and perhaps catastrophically alter the entire ecosystem. Also, the effects of overfishing here are unknown, e.g., whether the lack of top predators might affect the whole reef system even if the coral framework remains intact and alive. Dramatic shifts in the community structure of smaller prey might ultimately affect the coral itself.

21.2.3. Gulf of Mexico

Schroeder et al. (2005), Schroeder (2007) and Continental Shelf Associates (2007) provide the most recent information on distribution of deep-water hermatypic corals in the Gulf of Mexico. Most of the 24 records listed in Schroeder et al. (2005) were isolated samples collected via dredge or trawl with no information about habitat or substrate. Records of *Madrepora oculata* span much of the Gulf continental margin, including west Florida, Louisiana, Texas, northeastern Mexico, and the northeastern margin of Campeche Bank. In contrast, according to current knowledge, *L. pertusa* appears to be restricted to the west Florida shelf and northern Gulf with only three areas that represent constructional accumulations. Additional records can be found in Cairns (1979, 1993) for *Enallopsammia profunda*, *Solenosmilia variabilis* and *Madrepora carolina*, a colonial but ahermatypic species (not included in Fig. 21.3c). *S. variabilis* is a primary framework-builder just outside US waters on deep mounds north of Little Bahama Bank (Mullins et al. 1981; Fig. 21.3c).

Of the possible constructional sites, Moore and Bullis (1960) recorded *L. pertusa* (as *Lophelia prolifera*) from a trawl collection east of the Mississippi Delta (depth 421–512 m) following recognition of an unusual small hillock on a sonar trace. They estimated the largest feature recorded in their single transect as up to 55 m in vertical relief and over 300-m long, although they suggested that, if several smaller nearby mounds were dead portions of the larger feature, the entire structure might exceed 1,220 m in length. The mounds lay in a depression between two slopes, an arrangement reminiscent of the *L. pertusa* mounds at the base of the lower Miami Terrace adjacent to an elongated sediment drift that rises to the east (Neumann and Ball 1970; Malloy and Hurley 1970). However, subsequent seismic and submarine surveys found no traces of coral or hard substrate at the reported location. Schroeder et al. (2005) suggested that a similar site 7 km to the west might represent the original site. Here, *L. pertusa* was observed in 343 m on a low-relief mound at the northern edge of an exposed 2-km-long carbonate rock complex that abutted a submarine canyon with a similar profile to the fathometer trace in Moore and Bullis (1960).

Schroeder (2002) first reported well-developed *L. pertusa* assemblages on Viosca Knoll, an isolated 90-m-tall knoll on the upper slope of De Soto Canyon east of the Mississippi Delta (29°9.5' N, 88°1.0' W) in 434- to 530-m depth. A more recent report (Schroeder 2007) provided additional details. The flanks range from moderate to steep and consist of lithified carbonate outcrops and terraces (including slabs, boulders and rubble), as well as slopes, hummocks and swales of unconsolidated sediments. This is currently the best known and most studied *L. pertusa* site in the northern Gulf of Mexico (e.g., Schroeder et al. 2005; Continental Shelf Associates 2007). The percent cover of live *L. pertusa* here varies between 2.1% and 45.5% of the substrate, with the percentage of dead coral consistently higher than live coral (Continental Shelf Associates 2007). Coral colonies are not evenly distributed; dense aggregations of large colonies are interspersed with areas of sediment and coral rubble, shell hash, and occasionally chemosynthetic communities. The diversity of coral-associated fauna is low, but some species were consistently observed in high coral areas, including an echinoid (*Echinus tylodes*), comatulid crinoid (*Comatonia cristata*), "venus fly trap" anemone (Hormathiidae), and two common galatheoid crabs (*Eumunida picta*

and *Munidopsis* sp.) (Continental Shelf Associates 2007). Schroeder (2007) also noted the abundance and wide distribution of unidentified antipatharians, the primnoid octocoral *Callogorgia americana delta* and tubeworm colonies. Sulak et al. (in press) reviewed the fish fauna on the coral habitats of Viosca Knoll. Other areas of the northern Gulf, primarily within the Green Canyon area (e.g., 27°44.8' N, 4.81' N, 91°13.44' W and 27°35.89' N, 91°49.60' W), also support *L. pertusa* colonies but not to the same extent as Viosca Knoll (Schroeder et al. 2005; Continental Shelf Associates 2007). Well-developed *L. pertusa* colonies have also been documented from a World War II-era deep-water wreck called the *Gulfpenn* in the northern Gulf of Mexico (Church et al. 2007; Brooke and Schroeder, in press).

The third constructional site in the Gulf of Mexico consists of numerous 5- to 15-m-tall lithoherms, some capped with thickets of live and dead *L. pertusa*, that extend for at least 20 km along the southwestern Florida shelf margin in 428–466 m (Reed et al. 2006). Newton et al. (1987) first described the area from limited dredge and seismic surveys. A small portion of the area was mapped using a Seabeam 2112 swath bathymetric sonar system (Reed et al. 2003, 2006); ROV operations recorded nine lithoherms within a 100-m radius and examined two. The features appear to consist of rugged black phosphorite-coated limestone boulders and outcrops capped with 0.5- to 1.0-m-tall thickets of *L. pertusa*. A transect up a 12-m tall, 60-m wide feature in 466 m revealed a series of terraces on a rugged 45–70° phosphoritic limestone slope that consisted of boulders and outcrops with deep crevices. Thickets of live and dead *L. pertusa* up to almost a meter across occur on some terraces, with more on the northeastern slope than the northwestern face, but they are most abundant on the crest ridge. Reed et al. (2006) estimated coral cover at <5% to >50%, with 1–20% living. Other common sessile benthic taxa include the sponges *Heterotella* sp., Pachastrellidae, Petrosiidae and Choristida; the black corals *Antipathes* sp. and *Cirrhipathes* sp.; bamboo corals (Isididae) and hydrocorals (Stylasteridae). Common mobile invertebrates include Mollusca, Holothuroidea, Crinoidea, and decapod crustaceans (*Chaceon fenneri* and Chirostylidae). Nine species of fish included anthiines, *Chlorophthalmus agassizi*,

Conger oceanicus, *H. dactylopterus*, *L. melanurum*, *Polymixia* sp. and *Urophycis* sp. The numerous lithoherms revealed by the limited mapping effort (3.4 km²) suggest extensive unexplored coral and fish habitat in this region.

21.3 Non-constructional Deep-Water Coral Assemblages

This section includes those areas where deep-water hermatypic corals occur but do not accumulate as topographically recognizable biohermal build-ups.

21.3.1 Miami Terrace

The Miami Terrace is the primary physiographic feature of the southeastern margin of the Florida Peninsula north of the Florida Keys. Its geology has been investigated in substantial detail via high-resolution seismic reflection profiling, rock dredge sampling and submersible observations (e.g., Kofoed and Malloy 1965; Rona and Clay 1966; Malloy and Hurley 1970; Neumann and Ball 1970; Ballard and Uchupi 1971; Mullins and Neumann 1979; Reed et al. 2006). Siegler (1959) first reported the Terrace as "an old coral reef" following a University of Miami echo-sounding survey in 1958. Subsequent soundings and seismic reflection profiles revealed a drowned, narrow carbonate platform that interrupts the smooth profile of the Florida–Hatteras slope for about 120 km between the north end of Key Largo and Fort Lauderdale. The Terrace reaches its greatest width (22.2 km) off Miami (Kofoed and Malloy 1965), and tapers to the north and south where it disappears under prograding sediments.

A distinct upper terrace in ~200–375 m characterized by irregular, sediment-free, karst-like topography is separated from narrower, discontinuous lower terrace in ~600–700 m by a discontinuous sometimes double ridge that may reach 90-m high (Neumann and Ball 1970; Ballard and Uchupi 1971; Mullins and Neumann 1979; Reed et al. 2006). Hard substrates consist of dense conglomeratic phosphorites and phosphatic limestones in the form of pavements, veneers, ledges and loose clasts ranging from enormous slabs and boulders to nodules and gravel (Mullins and Neumann 1979; Reed et al. 2006; Messing et al. 2006).

Direct observations were first made in 1967–1971 using the submersibles *Aluminaut* (Neumann and Ball 1970), *Ben Franklin* (Ballard and Uchupi 1971) and *Alvin* (Mullins and Neumann 1979) and more recently using the *Johnson Sea Link* and an ROV (Reed et al. 2006; Messing et al. 2006). The thickets of branching ahermatypic corals that cap mounds of muddy sand beyond the base of the Terrace (~700–825 m) have been described above. At the base of the narrow northern end of the Terrace off Fort Lauderdale, Messing et al. (2006) observed a few isolated colonies of *Lophelia pertusa* and one of *Madrepora oculata* in 472–488 m with a few standing dead colonies to 509 m. However, extensive areas of coral skeletal debris ranged from carpets that in some places near the Terrace base appeared at least partly cemented, to scattered partly buried twigs found as deep as 526 m. The irregular Terrace face in ~300–350 m supported numerous small stylasterids on pavements and slabs, and large colonies along ledges, but no colonial scleractinians or coral rubble. Other fauna included the sponges *Phakellia* sp., Geodiidae, Pachastrellidae, and Lithistida; the octocorals *Isidella* sp. and *Pseudodrifa nigra*, ophiuroids and comatulid crinoids.

Reed et al. (2006) found that habitats and faunal zonation varied substantially among the several submersible dives made on the broader central portion of the Terrace. However, they found the lower slopes and flat top relatively barren, while the steep slopes and escarpments, especially near the ridge crests, support often dense assemblages of sponges, octocorals and corals. In particular, the margins of the ridge crests often support a continuous luxuriant overhanging battlement of *Lophelia* sometimes with a substantial accumulation of rubble, dead standing coral and additional living colonies just below, although the debris does not appear to significantly alter the Terrace profile. Elsewhere on the Terrace, corals include scattered colonies and thickets of *Lophelia* (Fig. 21.1c), both living and dead, *Madrepora oculata*, *Enallopsammia profunda* and stylasterids. Other fauna include at least 14 taxa of sponges (e.g., *Heterotella* sp., *Spongosorites* sp., *Geodia* sp., *Vetulina* sp., Raspailiidae, Pachastrellidae and Corallistidae), octocorals (e.g., *Paramuricea* sp., *Placogorgia* sp., *Plumarella* sp. and Isididae), antipatharians, the decapods *Eumunida picta* and

Chaceon fenneri, the ophiuroids *Asteroporpa* spp. and *Gorgonocephalus arcticus*, the crinoid *Comatonia cristata* and the echinoid *Stylocidaris* sp. Common fishes include *Chlorophthalmus agassizi*, *Conger oceanicus*, *Cyttopsis rosea*, *Helicolenus dactylopterus*, *Laemonema melanurum*, *Nezumia* sp., *Polyprion americanus* and schools of carangids.

21.3.2 Pourtalès Terrace Lithoherms

The Pourtalès Terrace is a long narrow triangular feature that parallels the Florida Keys for 213 km at depths of 200–450 m and exhibits extensive high-relief hard substrates. The apex of its maximum width extends about half way across the Strait of Florida toward Cay Sal Bank (Jordan 1954; Jordan and Stewart 1961; Jordan et al. 1964; Gomberg 1976; Land and Paull 2000; Reed et al. 2005c). Prograding sediments separate it from the Miami Terrace to the north. Numerous topographic features referred to as bioherms by Reed et al. (2006) occur chiefly on the eastern half of the terrace and range from 12 to 120 m in vertical relief. Their generally steep eastern flanks are broken by series of flat rock terraces separated by pavement or coarse sand and rubble slopes. Terrace edges frequently extend as horizontal crusts that overhang vertical 1- to 2-m scarps. Bioherm crests are broad pavement plateaus with few ledges or outcrops. Although it is unknown if these features are constructed of skeletal material, they are included here because the crests and terraces support dense assemblages of stylasterid hydrocorals (up to 96/m^2) (Fig. 21.4e, f) accompanied by apparently thick rubble deposits. The western flanks generally slope more gradually than the eastern flanks, lack ledges and are covered with more sediment (Reed et al. 2005c). Figure 21.3c indicates the portion of the terrace included within the HAPC proposed by the South Atlantic Fishery Management Council.

Submersible dives recorded a total of 49 sponge (densities to 80/m^2) and 22 cnidarian taxa, the latter including five stylasterids, 11 octocorals (to 48/m^2), three Antipatharia and the colonial scleractinian *Solenosmilia variabilis* (Reed et al. 2005c). No *L. pertusa* was observed, although Gomberg (1976) found *L. pertusa* and *M. oculata* skeletal debris in sediment samples. A total of 30 fish taxa identified during submersible dives include several of potential commercial

importance: *Caulolatilus microps*, *Epinephelus drummondhayi*, *E. flavolimbatus*, *E. nigritus*, *E. niveatus*, *Helicolenus dactylopterus*, *Pagrus pagrus*, *P. iwamotoi*, *Scombrus dumerili*, *Urophycis* sp., Scorpaenidae and Carcharhinidae.

21.3.3 Northwestern Atlantic Ocean

The northwestern Atlantic margin of the United States is divisible into the Mid-Atlantic Bight (shelf and slope waters from Cape Hatteras to Cape Cod), Georges Bank (Nantucket Shoals to Nova Scotia) and the semi-enclosed Gulf of Maine. In the first two areas, deep hard substrates appear to be confined largely to shelf-margin submarine canyons. They are more widespread as ledges and rocky bottoms in the Gulf of Maine (Packer et al., in press). Wherever suitable hard substrates appear, the primary habitat-forming corals are azooxanthellate, ahermatypic octocorals, including *Primnoa resedaeformis*, *Paragorgia arborea*, *Keratoisis ornata*, *Acanella arbuscula* and *Paramuricea grandis*. Off Atlantic Canada, *P. resedaeformis* and *P. arborea* may form dense forests or thickets (Buhl-Mortensen and Mortensen 2005; Butler 2005; Gass and Willison 2005), the latter sometimes exceeding 2 m in height (Strychar et al. 2005), but these organisms do not develop biohermal or reefal build-ups.

L. pertusa (recorded as *L. prolifera*), *E. profunda* and *S. variabilis* are known from one or a few trawl/dredge records each off Georges Bank (S.D. Cairns, January 2007, personal communication; Fig. 21.3b). These include one colony each of *L. pertusa* from Oceanographer Canyon and one of *S. variabilis* from Lydonia Canyon (Packer et al., in press). *L. pertusa* is also known from a single record in the Jordan Basin southeast of Maine but within Canadian waters (Gass and Willison 2005). Further to the east, all three occur on the New England seamount chain (S.D. Cairns 1981 and January 2007, personal communication; Les Watling, January 2007, personal communication), a series of about 30 extinct formerly volcanic hotspot peaks that extends about 1,100 km from Georges Bank to the northeast of Bermuda. Four lie within the US Exclusive Economic Zone (EEZ) (Packer et al., in press). Watling (January 2007, personal communication) found locally abundant *L. pertusa* on Bear Seamount, but only as individual colonies to 30-cm high, never as reefs. Other corals here include large isolated fans of *Enallopsammia* sp. (probably *E. rostrata*), and massive bunches of *Desmophyllum dianthus*, which also occurs as fossils up to 250-KY old on all seamounts sampled. Moore et al. (2003) also recorded *L. pertusa* from Bear Seamount, but noted that the few fragmentary specimens were not all identifiable. They omitted it from a subsequent paper (Moore et al. 2004) and recorded *Enallopsammia rostrata* instead.

Although no evidence of constructional build-ups exists in northeastern US waters, *L. pertusa* may occur as reefs in eastern Canada, or may have until recently. Following discovery of isolated colonies at the Stone Fence – the projecting far eastern end of the Scotia Shelf facing the Laurentian Channel – an additional survey found an area of *L. pertusa* rubble 1-km long by 500-m wide (Gass and Willison 2005). Fishers had reported high concentrations of *L. pertusa* as well as octocorals here, but most were destroyed by commercial fish trawling beginning in the 1950s (Butler 2005). The discovery and the damage led the Department of Fisheries and Oceans to close a 15-km^2 area to fishing in 2003 as the *Lophelia* Conservation Area (Fig. 21.3b). Isolated records also exist from the Gully, a large submarine canyon near the eastern end of the Scotia shelf (Mortensen and Buhl-Mortensen 2005), and edges of holes on Misaine Bank east of Nova Scotia (Gass and Willison 2005).

21.3.4 Northeastern Pacific Ocean

As with the northwestern Atlantic, the dominant habitat-forming corals along the Pacific coast of the United States, including Alaska and Hawai'i, are ahermatypic octocorals accompanied to a lesser extent by antipatharians. Etnoyer and Morgan (2005) listed 99 erect skeletonized taxa belonging to the families Isididae, Primnoidae, Corallidae, Paragorgiidae and Antipathidae. In the Aleutian Islands alone, which support the highest abundances and richness of corals in Alaskan waters, Heifetz et al. (2005) listed 69 taxa, including 10 solitary scleractinians and 25 stylasterid hydrocorals. However, none produce build-ups of any kind. The dominant genera here are *Callogorgia*, *Primnoa*, *Paragorgia*, *Fanellia* (=*Callogorgia*), *Thouarella*, and *Arthrogorgia* (Heifetz 2000).

Primnoa resaediformis and *Paragorgia arborea* reach 3 m in height and create extensive complex habitat occupied by high-diversity assemblages including commercial fish species, e.g., *Sebastes* spp., *Sebastolobus alascanus* and *Pleurogrammus monopterygius* (Risk et al. 1998; Heifetz 2002). As a result, proposals for designating some of these areas as Habitats of Particular Concern that would prohibit bottom-associated commercial fishing are being developed. However, these habitats lie outside the broad definition of coral reefs and are not considered further here.

By contrast, records of deep- or cold-water potentially hermatypic scleractinian corals from the American Pacific are few, scattered and largely limited to individual colonies. Etnoyer and Morgan (2005) list only ten records of caryophyllids (*Lophelia* sp.) and two records of *Madrepora oculata*, although they refer to unidentified recent findings that suggest that "*potentially* [their italics] hermatypic scleractinians, e.g., *Dendrophyllia* spp., may have been overlooked in the Northeast Pacific" (p 340). Collection data in the National Museum of Natural History (Smithsonian Institution) includes six records of *L. pertusa* (including *L. prolifera*) from Washington State to the Channel Islands and Fieberling Guyot (211–488 m; SE of California and outside the US EEZ), and five records of *Madrepora oculata* from the Channel Islands, Fieberling Guyot and northwestern Hawaiian chain (?84 m, 440–832 m) (Fig. 21.3a). Conway et al. (2005a) found *Solenosmilia variabilis* and *Lophelia* sp. in benthic samples and observed localized debris fields off British Columbia. Several dendrophylliids occur in deep Hawaiian waters and *Dendrophyllia oldroydae* is known from off southern California in 99–366 m, but is not known to be hermatypic.

However, *L. pertusa* has recently been observed via ROV in the Olympic Coast National Marine Sanctuary off Washington State at three sites that suggest possible build-ups: in patches on a rocky ledge in 271 m (Hyland et al. 2005); as a large coral rubble mound (perhaps representing decades of accumulation) with living colonies on the rubble as well as on an adjacent large boulder, and a broad field of *L. pertusa* and solitary coral (*Desmophyllum* sp.) rubble and octocoral fragments associated with derelict gear, trawl tracks and tumbled boulders (Bowlby et al. 2005).

Bioherms up to 21-m high chiefly in 164–240 m and up to 600 m across occur in the Queen Charlotte Basin off British Columbia, Canada (Conway et al. 1991, 2001, 2005b). However, these are constructed by three species of hexactinellid sponges – *Farrea occa*, *Aphrocallistes vastus* and *Heterochone calyx*. Build-up occurs via intergrowth of skeletons, secondary holdfast development, larval settlement and sediment entrapment (Krautter et al. 2001; Conway et al. 2005b). Some mounds have existed for up to 9,000 years (Conway et al. 2001). Conway et al. (2005b) note that the association of the sponges with a diverse fish and invertebrate fauna off the Alaskan and Washington State shelves is "indicated by a large sponge bycatch coupled with the rockfish catches in the trawl fishery" (p 606), but provide no direct information about the possibility of sponge bioherms in US waters.

21.4 Summary

Deep-sea coral reefs occur in US waters chiefly along the southeastern continental margin from North Carolina to southeastern Florida with smaller localized build-ups in the eastern and northern Gulf of Mexico. Reefs up to several kilometers long and sometimes exceeding 150 m in vertical relief are primarily biohermal mounds composed of skeletal debris and sediment and dominated by *Lophelia pertusa* with usually secondary contributions by *Enallopsammia profunda*, *Madrepora oculata* and *Solenosmilia variabilis*. Depths range chiefly from 500 to over 800 m, although non-constructional records exist from as little as ~300 m. No evidence exists for constructional build-ups in northeastern US waters, although *L. pertusa* may occur as reefs in eastern Canada. Similarly, the few records of deep-sea hermatypic corals from the American Pacific are limited to individual colonies, although recent observations of *L. pertusa* off Washington State suggest possible build-ups.

References

Agassiz A (1888) *Three cruises of the United States Coast and Geodetic Survey Steamer "Blake."* Houghton Mifflin, Boston.

Anselmetti FS, Eberli GP, Zan-Dong D (2000) From the Great Bahama Bank into the Straits of Florida: a margin architecture controlled by sea level fluctuations and ocean currents. Geol Soc Am Bull 112:829–844

Avent RM, King ME, Gore RH (1977) Topographic and faunal studies of shelf-edge prominences off the central eastern Florida coast. Intl Rev Ges Hydrobiol 62:185–208

Ayers MW, Pilkey OH (1981) Piston cores and surficial sediment investigations of the Florida-Hatteras Slope and inner Blake Plateau. In: P Popenoe (Ed) Environmental geologic studies on the southeastern Atlantic Outer Continental Shelf: US Geol Surv Open-File Rep 81-582A, Reston VA, pp. 1–89

Ballard R, Uchupi E (1971) Geological observations of the Miami Terrace from the submersible *Ben Franklin*. Marine Tech Soc J 5:43–48

Boehlert GW, Genin A (1987) A review of the effects of seamounts on biological processes. In: B Keating, P Fryer, R Batiza GW Boehlert (Eds) Seamounts, Islands and Atolls. Geophysical Monograph 43, Amer Geophys Union, Washington, DC, pp 319–344

Bowlby E, Hyland J, Brancato MS, Cooksey C, Intelmann S (2005) Preliminary discoveries of scleractinian coral *Lophelia pertusa* and other deep-sea coral and sponge communities in the Olympic Coast National Marine Sanctuary 3rd Intl Sympos Deep-Sea Corals, Sci and Mgmt, Programs and Abstracts. University of Miami, Miami, FL, 28 Nov–2 Dec 2005, p 160

Brooke SD, Messing CG, Reed JK, Gilmore RG (2006) Exploration of deep-sea coral ecosystems along the east coast of Florida. 11th Intl Deep-Sea Biol Sympos, Southampton, UK, 10–14 July, Abstracts pp 35–36

Brooke S, Schroeder WW (2007) State of Deep Coral Ecosystems in the Gulf of Mexico region: Texas to the Florida Straits. In: SE Lumsden, Hourigan TF, Bruckner AW, Dorr G (Eds) The State of Deep Coral Ecosystems of the United States. NOAA Technical Memorandum CRCP-3. Silver Spring MD, pp 271–306

Brooks RA, Nizinski MS, Ross SW, Sulak KJ (2007) Sublethal injury rate in a deepwater Ophiuroid, *Ophiacantha bidentata*, an important component of western Atlantic *Lophelia* reef communities. Mar Biol 152:307–314

Buhl-Mortensen L, Mortensen PB (2005) Distribution and diversity of species associated with deep-sea gorgonian corals off Atlantic Canada. In: A Freiwald, JM Roberts (Eds) Cold-water Corals and Ecosystems. Springer, Berlin, pp 849–879

Burdon-Jones C, Tambs-Lyche H (1960) Observations on the fauna of the North Bratthomen stone-coral reef near Bergen. Årbok f Univers Bergen, Mat.-Naturv Ser 1960(4):1–24.

Butler M (2005) Conserving corals in Atlantic Canada: a historical perspective. In: A Freiwald, JM Roberts (Eds) Cold-water Corals and Ecosystems. Springer, Berlin, pp 1199–1209

Cairns SD (1979) The deep-water scleractinia of the Caribbean Sea and adjacent waters. Stud Fauna Curaçao 57(180):1–341

Cairns SD (1981) Marine flora and fauna of the northeastern United States: scleractinia. NOAA Tech Rep NMFS Circ no 438, 14 pp

Cairns SD (1993) A checklist of the ahermatypic scleractinia of the Gulf of Mexico, with the description of a new species. Gulf Res Repts 6:9–15

Cairns SD (1994) Scleractinia of the temperate North Pacific. Smithson Contrib Zool no 557, 150 pp

Cairns SD (1995) The marine fauna of New Zealand: scleractinia (Cnidaria: Anthozoa). New Zealand Oceanogr Inst Mem 103, 210 pp

Cairns SD, Chapman RE (2001) Biogeographic affinities of the North Atlantic deep-water Scleractinia. In: JHM Willison et al. (Eds) Proc 1st Intl Sympos Deep-sea Corals, 30 Jul–2 Aug 2000, Ecology Action Centre, Nova Scotia Museum, Halifax, Canada

Cairns SD, Stanley GD Jr (1981) Ahermatypic coral banks: living and fossil counterparts. Proc 4th Intl Coral Reef Sympos Manila 1:611–618

Child CA (1998) *Nymphon torulum*, new species and other Pycnogonida associated with the coral *Oculina varicosa* on the east coast of Florida. Bull Mar Sci 63:595–604

Church R, Warren D, Cullimore R, Johnston L, Schroeder W, Patterson W, Shirley T, Kilgour M, Morris N, Moore J (2007) The archaeology and biological analysis of World War II shipwrecks in the Gulf of Mexico: the artificial reef effect in deepwater. US Department of the Interior, Minerals Management Service, Gulf of Mexico OCS Region, New Orleans, LA. OCS Study MMS 2007–015, 387 pp

Continental Shelf Associates (2007) Characterization of Northern Gulf of Mexico deepwater hard bottom communities with emphasis on *Lophelia* coral. US Dept Interior, Minerals Management Service, Gulf of Mexico OCS Region, New Orleans, LA. OCS Study MMS 2007–044, 169 pp + appendices

Conway KW, Barrie JV, Austin WC, Luternauer JL (1991) Holocene sponge bioherms on the western Canadian continental shelf. Cont Shelf Res 11:771–790

Conway KW, Krautter M, Barrie JV, Neuweiler M (2001) Hexactinellid sponge reefs on the Canadian continental shelf: a unique "living fossil." Geosci Canada 28:71–78

Conway KW, Barrie JV, Austin WC, Hill PR, Krautter M (2005a) Deep-water sponge and coral habitats in the coastal waters of British Columbia, Canada: multibeam and ROV survey results. 3rd Intl Sympos Deep-Sea Corals, Sci and Mgmt, Programs and Abstracts, University of Miami, Miami, FL, 28 Nov–2 Dec 2005, p 32

Conway KW, Krautter M, Barrie JV, Whitney F, Thomson RE, Reiswig H, Lehnert H, Mungov G, Bertram M (2005b) Sponge reefs in the Queen Charlotte Basin,

Canada: controls on distribution, growth and development. In: A Freiwald, JM Roberts (Eds) Cold-water Corals and Ecosystems. Springer, Berlin, pp 605–621

Danois E Le (1948) Les profondeurs de la mer. Trente ans de recherché sur la faune sous-marine au large des côte des France. Payot, Paris, 303 pp

De Mol B, Van Rensbergen P, Pillen S, Van Herreweghe K, Van Rooij D, McDonnell A, Huvenne V, Ivanov M, Swennen R, Henriet JP (2002) Large deep-water coral banks in the Porcupine Basin, southwest Ireland. Mar Geol 188:193–231

Dorschel B, Hebbeln D, Rüggeberg A, Dullo W-C, Freiwald A (2005) Growth and erosion of a cold-water coral covered carbonate mound in the NE Atlantic during the late Pleistocene. Earth Planet Sci Lett 233:33–44

Druffel ERM, Griffin S, Witter A, Nelson E, Southon J, Kashgarian M, Vogel J (1995) *Gerardia*: bristlecone pine of the deep-sea? Geochim Cosmochim Acta 59:5031–5036

Düing W (1973) Some evidence for long-period barotropic waves in the Florida Current. J Phys Oceanogr 3(3):343–346

Düing W (1975) Synoptic studies of transients in the Florida Current. J Mar Res 33(1):53–73

Etnoyer P, Morgan LE (2005) Habitat-forming deep-sea corals in the Northeast Pacific Ocean. In: A Freiwald, JM Roberts (Eds) Cold-water Corals and Ecosystems. Springer, Berlin, pp 331–343

Forbes E (1854) Distribution of marine life. In: K Johnston (Ed) Physical Atlas of Natural Phenomena, Edinburgh, UK

Forbes E (1859) The Natural History of European Seas. In: RG Austen (Ed) John Van Voorst, London, 306 pp

Fosså JH, Furevik DM, Mortensen PB, Hovland M (1999) Effects of bottom trawling on *Lophelia* deep water coral reefs in Norway (poster). Ecosystem Effects of Fishing, ICES, March 1999, Montpelier, France

Fosså JH, Mortensen PB, Furevik DM (2000) *Lophelia*-korallrev langs norskekysten forekomst og tilstand. Fisken og Havet 2:1–94

Fosså JH, Mortensen PB, Furevik DM (2002) The deep-water coral *Lophelia pertusa* in Norwegian waters: distribution and fishery impacts. Hydrobiologia 417:1–12

Fosshagen A, Hoisæter T (1992) The second Norwegian record of the deep-water coral, *Desmophyllum cristagalli* Milne-Edwards and Haime, 1848 (Cnidaria: Scleractinia). Sarsia 77:291–292

Frederiksen R, Jensen A, Westerberg H (1992) The distribution of the Scleractinian coral *Lophelia pertusa* around the Faeroe Islands and the relation to internal tidal mixing. Sarsia 77:157–171

Freiwald A (2000) The Atlantic Coral Ecosystem Study (ACES): a margin-wide assessment of corals and their environmental sensitivities in Europe's deep waters. EurOCEAN 2000 Project synopses, pp 1–317

Freiwald A, Mortensen PB (2000) The first record of the deep-water coral *Stenocyathus vermiformis* (Pourtalès, 1868) (Scleractinia, Guyniidae) from Norwegian waters. Sarsia 85:275–276

Freiwald A, Roberts JM (Eds) (2005) Cold-water Corals and Ecosystems. Springer, Berlin, 1243 pp

Freiwald A, Wilson JB, Henrich R (1999) Grounding Pleistocene icebergs shape recent deep-water coral reefs. Sed Geol 125:1–8

Freiwald A, Hühnerbach V, Lindberg B, Wilson J, Campbell J (2002) The Sula Reef Complex, Norwegian Shelf. Facies 47:179–200

Gass SE, Willison JHM (2005) An assessment of the distribution of deep-sea corals in Atlantic Canada by using both scientific and local forms of knowledge. In: A Freiwald, JM Roberts (Eds) Cold-water Corals and Ecosystems. Springer, Berlin, pp 223–245

George RY (2002) Ben Franklin temperate reef and deep sea 'AGassiz Coral Hills' in the Blake Plateau off North Carolina. Hydrobiologia 471:71–81

George RY, Okey TA, Reed JK, Stone RP (2007) Ecosystem-based fisheries management of seamount and deep-sea coral reefs in U. S. waters: conceptual models for proactive decisions. In: RY George, SD Cairns (Eds) Conservation and adaptive management of seamounts and deep-sea coral ecosystems. Rosenstiel School of Marine and Atmospheric Science, Univ Miami. Miami, pp 9–30

Gilmore RG, Jones RS (1992) Color variation and associated behavior in the epinepheline groupers, *Mycteroperca microlepis* (Goode and Bean) and *M. phenax* Jordan and Swain. Bull Mar Sci 51(1):83–103

Gomberg D (1976) Geology of the Pourtales Terrace, Straits of Florida. Ph.D. dissertation, University of Miami, Miami, FL

Griffin S, Druffel ERM (1989) Sources of carbon to deep-sea corals. Radiocarbon 31:533–543

Grasmueck M, Eberli GP, Viggiano DA, Correa T, Rathwell G, Luo J (2006) Autonomous underwater vehicle (AUV) mapping reveals coral mound distribution, morphology, and oceanography in deep water of the Straits of Florida. Geophys Res Lett 33, L23616, doi:10.1029/2006GL027734

Grasmueck M, Eberli GP, Correa T, Viggiano DA, Luo J, Wyatt GJ, Wright AE, Reed JK, Pomponi SA (2007) AUV-Based environmental characterization of deep-water coral mounds in the Straits of Florida. 2007 Offshore Technology Conference, 30 Apr–3 May 2007, Houston, TX, OTC 18510

Guinotte JM, Orr J, Cairns S, Freiwald A, Morgan L, George R (2006) Will human-induced changes in seawater chemistry alter the distribution of deep-sea scleractinian corals? Front Ecol Environ 4(3):141–146

Hall-Spencer J, Allain V, Fossa JH (2002) Trawling damage to Northeast Atlantic ancient coral reefs. Proc Roy Acad Sci B 269:507–511

Heifetz J (2000) Coral in Alaska: distribution, abundance and species associations. Proc Nova Scotia Inst Sci Proc 1st Intl Sympos Deep-Sea Corals, 30 Jul–2 Aug 2000, Ecology Action Centre, Nova Scotia Museum, Halifax, Nova Scotia, Canada

Heifetz J (2002) Coral in Alaska: distribution, abundance, and species associations. Hydrobiologia 471(1–3):19–28

Heifetz J, Wing BL, Stone RP, Malecha PW, Courtney DL (2005) Corals of the Aleutian Islands. Fish Oceanogr 14 (Suppl 1):131–138

Henriet JP, De Mol B, Pillen S, Vanneste M, van Rooij D, Versteeg W, Croker PF, Shannon PM, Unnithan V, Bouriak S, Chachkine P (1998) Gas hydrate crystals may help build reefs. Nature 391:64–649

Henry L-A, Roberts JM (2007) Biodiversity and ecological composition of macrobenthos on cold-water coral mounds and adjacent off-mound habitat in the bathyal Porcupine Seabight, NE Atlantic. Deep-Sea Res I 54:654–672

Hovland M, Mortensen PB, Brattegard T, Strass P, Rokoengen K (1998) Ahermatypic coral banks off mid-Norway: evidence for a link with seepage of light hydrocarbons. Palaios 13:198–200

Huvenne VAI, Bailey WR, Shannon PM, Naeth J, Di Primio R, Henriet JP, Horsfield B de Haas H, Wheeler A, Olu-Le Roy K (2007) The Magellan mound province in the Porcupine Basin. Intl J Earth Sci 96: 85–101

Hyland J, Cooksey C, Bowlby E, Brancato MS, Intelmann S (2005) A pilot survey of deepwater coral/sponge assemblages and their susceptibility to fishing/harvest impacts at the Olympic Coast National Marine Sanctuary (OCNMS). Cruise Report for NOAA Ship McARTHUR II Cruise AR-04–04: Leg 2. NOAA Technical Memorandum NOS NCCOS 15. NOAA/NOS Ctr Coastal Environ Health Biomolec Res, Charleston, SC, 13 p

Jensen A, Frederiksen R (1992) The fauna associated with the bank-forming deepwater coral *Lophelia pertusa* (Scleractinaria) on the Faeroe Shelf. Sarsia 77:53–69

Jordan G (1954) Large sink holes in Straits of Florida. Bull Am Assoc Petrol Geol 38:1810–1817

Jordan GF, Stewart HB (1961) Submarine topography of the western Straits of Florida. Geol Soc Amer Bull 72(7):1051–1058

Jordan GF, Malloy RM, Kofoed JW (1964) Bathymetry and geology of Pourtales Terrace, Florida. Mar Geol 1(3):259–287

Kenyon NH, Akhmetzhanov AM, Wheeler AJ, van Weering TCE, de Haas H, Ivanov MK (2003) Giant carbonate mud mounds in the southern Rockall Trough. Mar Geol 195:5–30

Kielmann J, Düing W (1974) Tidal and sub-inertial fluctuations in the Florida Current. J Phys Oceanogr 4(2):227–236

Kitahara MV (2005) Industrial fisheries impact on the deep-sea Scleractinia in southern Brazil. 3rd Intl

Sympos Deep-sea Corals, Sci and Mgmt, Programs and Abstracts. University of Miami, Miami, FL, 28 Nov–2 Dec 2005, p 237

Koenig CC, Coleman FC, Grimes C, Fitzhugh G, Scanlon K, Gledhill C, Grace M (2000) Protection of fish spawning habitat for the conservation of warm-temperate reef-fish fisheries of shelf-edge reefs of Florida. Bull Mar Sci 66:593–616

Koenig CC, Shepard AN, Reed JK, Coleman FC, Brooke SD, Brusher J, Scanlon KM (2005) Habitat and fish populations in the deep-sea *Oculina* coral ecosystem of the western Atlantic. Amer Fish Soc Sympos 41:795–805

Kofoed JW, Malloy RM (1965) Bathymetry of the Miami Terrace. Southeastern Geology 6(3):159–165

Koslow JA, Gowlett-Jones K (1998) The seamount fauna off Southern Tasmania: benthic communities, their conservation and impacts of trawling. Final Report to Environment Australia and Fisheries Research Development Corp., Fisheries Research and Development Corp, Australia, 104 pp (cited in Rogers 1999)

Krautter M, Conway KW, Barrie JV, Neuweiler M (2001) Discovery of a "living dinosaur": globally unique modern hexactinellid sponge reefs off British Columbia, Canada. Facies 44:265–282

Krieger KJ (2001) Coral (*Primnoa*) impacted by fishing gear in the Gulf of Alaska. In: JHM Willison et al. (Eds) Proc 1st Intl Sympos Deep-sea Corals, 30 Jul–2 Aug 2000, Ecology Action Centre, Nova Scotia Museum, Halifax, Nova Scotia, Canada, pp 106–116

Land L, Paull C (2000) Submarine karst belt rimming the continental slope in the Straits of Florida. Geo-Mar Lett 20:23–132

Lee T, Rooth C, Williams E, McGowan M, Szmant AF, Clarke ME (1992) Influence of Florida current, gyres and wind-driven circulation on larvae transport and recruitment in the Florida Keys Coral Reefs. J Cont Shelf Res 12(78):971–1002

Leverette TL, Metaxas A (2005) Predicting habitat for two species of deep-water corals on the Canadian Atlantic continental shelf and slope. In: A Freiwald, JM Roberts (Eds) Cold-water Corals and Ecosystems. Springer, Berlin, pp 467–479

Macintyre IG, Milliman JD (1970) Physiographic features on the outer shelf and upper continental slope, Atlantic continental margin, southeastern United States. Geol Soc Am Bull 81:2577–2598

MacIsaac K, Bourbonnais C, Kenchington E, Gordon D Jr, Gass S (2001) Observations on the occurrence and habitat preference of corals in Atlantic Canada. In: JHM Willison et al. (Eds) Proc 1st Intl Sympos Deep-sea Corals, 30 Jul–2 Aug 2000, Ecology Action Centre, Nova Scotia Museum, Halifax, Nova Scotia, Canada, pp 58–75

Malloy RJ, Hurley RJ (1970) Geomorphology and geological structure: Straits of Florida. Geol Soc Amer Bull 81:1947–1972 + 2 maps

Masson DG, Bett BJ, Billett DSM, Jacobs CL, Wheeler AJ, Wynn RB (2003) The origin of deep-water, coral-topped mounds in the northern Rockall Trough, Northeast Atlantic. Mar Geol 192:215–237

McAllister DE, Alfonso N (2001) The distribution and conservation of deep-water corals on Canada's west coast. In: Willison JHM et al. (Eds) Proc 1st Intl Sympos Deep-sea Corals, 30 Jul–2 Aug 2000, Ecology Action Centre, Nova Scotia Museum, Halifax, Nova Scotia, Canada, pp 126–144

Menzies RJ, George RY, Rowe GT (1970) Abyssal Environment and Ecology of the World Oceans. Wiley, New York

Messing CG (1984) Brooding and paedomorphosis in the deep-water feather star *Comatilia iridometriformis* (Echinodermata: Crinoidea). Mar Biol 80:83–91

Messing CG (2003) Biozonation on deep-water carbonate mounds and associated hardgrounds along the western margin of Little Bahama Bank, with notes on the Caicos Platform Island Slope. 11th Sympos Geol Bahamas, Gerace Res Ctr, San Salvador, Bahamas, June 2002

Messing CG, Neumann AC, Lang JC (1990) Biozonation of deep-water lithoherms and associated hardgrounds in the northeastern Straits of Florida. Palaios 5(1):15–33

Messing CG, Walker BK, Dodge RE, Reed JK (2006) Calypso U.S. Pipeline, LLC, Mile Post (MP) 31-MP 0, deep-water marine benthic video survey final report. Submitted to: Calypso US Pipeline, LLC, 46 pp + appendices

Meyer DL, Messing CG, Macurda DB Jr (1978) Zoogeography of tropical western Atlantic Crinoidea (Echinodermata). Bull Mar Sci 28(3):412–441

Mienis F, van Weering T, de Haas H, de Stigter H, Huvenne V, Wheeler A (2006) Carbonate mound development at the SW Rockall Trough margin based on high resolution TOBI and seismic recording. Mar Geol 233:1–19

Milliman JD, Manheim FT, Pratt RM, Zarudski EF (1967) Alvin dives on the continental margin off the southeastern United States, July 2–13, 1967. Tech Rep Woods Hole Oceanographic Institution 67–80, 48 pp

Moe M (1963) A survey of offshore fishing in Florida. Prof. Paper Series No. 4, Florida State Board of Conservation Marine Laboratory, St. Petersburg, FL, 117 pp

Moore D, Bullis H Jr (1960) A deep water coral reef in the Gulf of Mexico. Bull Mar Sci 10:125–128

Moore JA, Vecchione M, Collette BB, Gibbons R, Hartel KE, Galbraith JK, Turnipseed M, Southworth M, Watkins E (2003) Biodiversity of Bear Seamount, New England Seamount chain: results of exploratory trawling. J Northwest Atl Fish Sci 31:363–372

Moore JA, Vecchione M, Collette BB, Gibbons R, Hartel KE (2004) Selected fauna of Bear Seamount (New England Seamount chain), and the presence of "natural invader" species. Arch Fish Mar Res 51(1–3):241–250

Mortensen PB (2001) Aquarium observations on the deep-water coral *Lophelia pertusa* (L., 1758)

(Scleractinia) and selected associated invertebrates. Ophelia 54(2):83–104

Mortensen PB, Buhl-Mortensen L (2005) Deep-water corals and their habitats in The Gully, a submarine canyon off Atlantic Canada. In: A Freiwald, JM Roberts (Eds) Cold-water Corals and Ecosystems. Springer, Berlin, pp 247–277

Mortensen PB, Rapp HT (1998) Oxygen and carbon isotope ratios related to growth line patterns in skeletons of *Lophelia pertusa* (L.) (Anthozoa, Scleractinia): implications for determination of linear extension rates. Sarsia 83:433–446

Mortensen PB, Hovland M, Brattegard T, Farestveit R (1995) Deep water bioherms of the scleractinian coral *Lophelia pertusa* (L.) at 64°N on the Norwegian shelf: structure and associated megafauna. Sarsia 80:145–158

Mortensen PB, Hovland MT, Fosså JH, Furevik DM (2001) Distribution, abundance and size of *Lophelia pertusa* coral reefs in mid-Norway in relation to seabed characteristics. J Mar Biol Assoc UK 81:581–597

Mullins HT, Neumann AC (1979) Deep carbonate bank margin structure and sedimentation in the northern Bahamas. SEPM Special Publ 27:165–192

Mullins HT, Newton CR, Heath K, Vanburen HM (1981) Modern deep-water coral mounds north of Little Bahama Bank: criteria for recognition of deep-water coral bioherms in the rock record. J Sed Petrol 51(3):999–1013

Neumann AC, Ball MM (1970) Submersible observations in the Straits of Florida: geology and bottom currents. Geol Soc Amer Bull 81:2861–2874

Neumann AC, Kofoed JW, Keller GH (1977) Lithoherms in the Straits of Florida. Geology 5:4–10

Newton CR, Mullins H, Gardulski F, Hine A, Dix G (1987) Coral mounds on the west Florida slope: unanswered questions regarding the development of deepwater banks. Palaios 2:359–367

Packer DB, Boelke D, Guida V, and McGee L (2007) State of Deep Coral Ecosystems in the Northeastern United States Region: Maine to Cape Hatteras. In: SE Lumsden, Hourigan TF, Bruckner AW, Dorr G (Eds) The State of Deep Coral Ecosystems of the United States. NOAA Technical Memorandum CRCP-3. Silver Spring MD, pp 195–232

Partyka ML, Ross SW, Quattrini AM, Sedberry GR, Birdsong TW, Potter J (in press) Southeastern United States Deep Sea Corals (SEADESC) Initiative: A Collaborative Effort to Characterize Areas of Habitat-forming Deep-sea Corals. NOAA Tech. Memo. Silver Spring, MD

Paull CK, Neumann AC, am Ende BA, Ussler W, Rodriguez NM (2000) Lithoherms on the Florida Hatteras Slope. Mar Geol 136:83–101

Popenoe P, Manheim FT (2001) Origin and history of the Charleston Bump-geological formations, currents, bottom conditions, and their relationship to wreckfish habitats on the Blake Plateau. In: GR Sedberry (Ed) Island in the Stream: Oceanography and Fisheries of the Charleston Bump. Amer Fish Soc Sympos 25, Bethesda MD, pp 43–93

Reed JK (1980) Distribution and structure of deep-water *Oculina varicosa* coral reefs off central eastern Florida. Bull Mar Sci 30:667–677

Reed JK (1981) *In situ* growth rates of the scleractinian coral *Oculina varicosa* occurring with zooxanthellae on 6-m reefs and without on 80-m banks. In: IJ Dogma Jr (Ed) Proc 4th Intl Coral Reef Sympos 2:201–206

Reed JK (1983) Nearshore and shelf-edge *Oculina* coral reefs: the effects of upwelling on coral growth and on the associated faunal communities. NOAA Sympos Ser Undersea Res 1:119–124

Reed JK (2002a) Deep-water *Oculina* coral reefs of Florida: biology, impacts, and management. Hydrobiologia 471:43–55

Reed JK (2002b) Comparison of deep-water coral reefs and lithoherms off southeastern USA. Hydrobiologia 471:57–69

Reed JK, Gilmore R (1982) Nomination of a Habitat Area of Particular Concern (HAPC). In: J Brawner (Ed) Fishery Management Plan, Final Environmental Impact Statement for Coral and Coral Reefs, Gulf of Mexico and South Atlantic Fishery Management Councils, pp L-20–L-42

Reed JK, Hoskin CM (1987) Biological and geological processes at the shelf edge investigated with submersibles. In: RA Cooper, AN Shepard (Eds) Scientific Application of Current Diving Technology on the U.S. Continental Shelf, vol. 2, NOAA Sympos Ser Undersea Res, vol 2, Washington, DC, pp 191–199

Reed JK, Mikkelsen PM (1987) The molluscan community associated with the scleractinian coral *Oculina varicosa*. Bull Mar Sci 40:99–131

Reed JK, Ross SW (2005) Deep-water reefs off the southeastern U.S.: recent discoveries and research. Current–J Mar Educ 21(4):33–37

Reed JK, Gore RH, Scotto LE, Wilson KA (1982) Community composition, structure, areal and trophic relationships of decapods associated with shallow- and deep-water *Oculina varicosa* coral reefs. Bull Mar Sci 32:761–786

Reed JK, Pomponi S, Frank T, Widder E (2002) Islands in the Stream 2002: Exploring Underwater Oases. Mission Three Summary: Discovery of New Resources with Pharmaceutical Potential; Vision and Bioluminescence in Deep-sea Benthos. NOAA Ocean Exploration web site: http://oceanexplorer.noaa.gov/explorations/02sab/logs/summary/summary.html, HBOI DBMR Misc Cont No 208, 29 pp

Reed JK, Wright A, Pomponi S (2003) Discovery of new resources with pharmaceutical potential in the Gulf of Mexico. Mission Summary Report, 2003

National Oceanic and Atmospheric Administration Office of Ocean Exploration, HBOI DBMR Misc. Cont. Number 224, 31 pp

Reed JK, Shepard A, Koenig C, Scanlon K, Gilmore G (2005a) Mapping, habitat characterization, and fish surveys of the deep-water *Oculina* coral reef marine protected area: a review of historical and current research. In: A Freiwald, JM Roberts (Eds) Cold-water Corals and Ecosystems. Springer, Berlin, pp 443–465

Reed JK, Koenig C, Shepard A (2005b) Effects of bottom trawling on a deep-water coral reef. 3rd Intl Sympos Deep-sea Corals, Science and Management, Programs and Abstracts, University of Miami, Miami, FL, 28 Nov–2 Dec 2005, p 240

Reed JK, Pomponi SA, Weaver D, Paull CK, Wright AE (2005c) Deep-water sinkholes and bioherms of South Florida and the Pourtalès Terrace – Habitat and Fauna. Bull Mar Sci 77(2):267–296

Reed JK, Weaver DC, Pomponi SA (2006) Habitat and fauna of deep-water *Lophelia pertusa* coral reefs off the southeastern U.S.: Blake Plateau, Straits of Florida, and Gulf of Mexico. Bull Mar Sci 78(2):343–375

Reed JK, Koenig CC, Shepard AN (2007) Impact of bottom trawling on a deep-water *Oculina* coral ecosystem off Florida. Bull Mar Sci 81:481–496.

Rice AL, Thurston MH, New AL (1990) Dense aggregations of a hexactinellid sponge, *Pheronema carpenteri*, in the Porcupine Seabight (northeast Atlantic Ocean), and possible causes. Progr Oceanogr 24:179–196

Risk MJ, McAllister DE, Behnken L (1998) Conservation of cold-and warm-water seafans: threatened ancient gorgonian groves. Sea Wind 10(4):20–22

Roark EB, Guilderson TP, Dunbar RB, Ingram BL (2006) Radiocarbon-based ages and growth rates of Hawaiian deep-sea corals. Mar Ecol Progr Ser 327:1–14

Roberts JM, Wheeler AJ, Freiwald A (2006) Reefs of the deep: the biology and geology of cold-water coral ecosystems. Science 312:543–547

Rogers AD (1999) The biology of *Lophelia pertusa* (Linnaeus 1758) and other deep-water reef-forming corals and impacts from human activities. Intl Rev Hydrobiol 84:315–406

Rona PA, Clay CS (1966) Continuous seismic profiles of the continental terrace off southeastern Florida. Geol Soc Amer Bull 77:31–44

Ross SW (2006) Review of distribution, habitats, and associated fauna of deep water corals reefs on the southeastern United States continental slope (North Carolina to Cape Canaveral, FL). Report to the South Atlantic Fishery Management Council, Charleston, SC, 36 pp

Ross SW and Nizinski MS (2007) State of the U.S. Deep Coral Ecosystems in the Southeastern United States Region: Cape Hatteras to the Florida Straits. In: SE Lumsden, Hourigan TF, Bruckner AW, Dorr G (Eds) The State of Deep Coral Ecosystems of the United States. NOAA Technical Memorandum CRCP-3. Silver Spring MD, pp 233–270

Ross SW, Quattrini AM (2007) The fish fauna associated with deep coral banks off the southeastern United States. Deep-Sea Res I 54(6):975–1007

Rowe GT, Menzies RJ (1969) Zonation of large benthic invertebrates in the deep-sea off the Carolinas. Deep-Sea Res 16:531–537

Rüggeberg A, Dullo C, Dorschel B, Hebbeln D (2007) Environmental changes and growth history of a cold-water carbonate mound (Propeller Mound, Porcupine Seabight). Intl J Earth Sci 96:57–72

Schröder-Ritzrau A, Freiwald A, Mangini A (2005) U/Th-dating of deep-water corals from the eastern North Atlantic and the western Mediterranean Sea. In: A Freiwald, JM Roberts (Eds) Cold-water Corals and Ecosystems. Springer, Berlin, pp 157–172

Schroeder WW (2001) Video documentation of the geology and distribution of *Lophelia prolifera* at a deep-water reef site in the northeastern Gulf of Mexico. In: JHM Willison et al. (Eds) Proc 1st Intl Sympos Deep-sea Corals, 30 Jul–2 Aug 2000, Ecology Action Centre, Nova Scotia Museum, Halifax, Nova Scotia, Canada, pp 224–225

Schroeder WW (2002) Observations of *Lophelia pertusa* and the surficial geology at a deep-water site in the northeastern Gulf of Mexico. Hydrobiologia 471:29–33

Schroeder WW (2005) Seabed characteristics at sites where *Lophelia pertusa* occur in the northern and eastern Gulf of Mexico. 3rd Intl Sympos Deep-sea Corals, Science and Management, Programs and Abstracts. University of Miami, Miami, FL, 28 Nov–2 Dec 2005, p 37

Schroeder WW (2007) Seafloor characteristics and distribution patterns of *Lophelia pertusa* and other sessile megafauna at two upper-slope sites in the northeastern Gulf of Mexico. US Dept Interior, Minerals Management Service, Gulf of Mexico OCS Region, New Orleans, LA. OCS Study MMS 2007–035, 56 pp

Schroeder WW, Brooke SD, Olson JB, Phaneuff B, McDonough III JJ, Etnoyer P (2005) Occurrence of deep-water *Lophelia pertusa* and *Madrepora oculata* in the Gulf of Mexico. In: A Freiwald, JM Roberts (Eds) Cold-water Corals and Ecosystems. Springer, Berlin, pp 297–307

Sedberry GR, McGovern JC, Pashuk O (2001) The Charleston Bump: an island of essential fish habitat in the Gulf Stream. In: GR Sedberry (Ed) Island in the Stream: Oceanography and Fisheries of the Charleston Bump. Amer Fish Soc Sympos 25. Bethesda, MD, pp 3–23

Siegler VB (1959) Reconnaissance survey of the bathymetry of the Straits of Florida. University of Miami Inst Mar Sci, Marine Lab Rept 59–3, Miami, FL, 9 pp

Smith JE, Risk HP, Schwarcz HP, McConnaughey TA (1997) Rapid climate change in the North Atlantic during the Younger Dryas recorded by deep-sea corals. Nature 386:818–820

Squires DF (1965) Neoplasia in a Coral? Science 148(3669):503–505

Stanley GD Jr, Cairns SD (1988) Constructional azooxanthellate coral communities: an overview with implications for the fossil record. Palaios 3:233–242

Stetson TR, Squires DF, Pratt RM (1962) Coral banks occurring in deep water on the Blake Plateau. Amer Mus Novit 2114:1–39

Strychar KB, Hamilton LC, Kenchington EL, Scott DB (2005) Genetic circumscription of deep-water coral species in Canada using 18S rRNA. In: A Freiwald, JM Roberts (Eds) Cold-water Corals and Ecosystems. Springer, Berlin, pp 679–690

Sulak KJ, Brooks RA, Luke KE, Norem AD, Randall M, Quaid AJ, Yeargin GE, Miller JM, Harden WM, Caruso JH, Ross SW (2007) Demersal fishes associated with *Lophelia pertusa* coral and hard-substrate biotopes on the continental slope, northern Gulf of Mexico. In: RY George, SD Cairns (Eds) Conservation and adaptive management of seamounts and deep-sea coral ecosystems. Rosenstiel School of Marine and Atmospheric Science, Univ Miami. Miami, pp 65–92

Taviani M, Freiwald A, Zibrowius H (2005) Deep coral growth in the Mediterranean Sea: an overview. In: A Freiwald, JM Roberts (Eds) Cold-water Corals and Ecosystems. Springer, Berlin, pp 137–156

Teichert, C (1958) Cold- and deep-water coral banks. Bull Amer Assoc petrol Geol 42:1064–1082

Uchupi E (1967) The continental margin south of Cape Hatteras, North Carolina: shallow structure. Southeastern Geol 8:155–177

Watling L, Auster PJ (2005) Distribution of deep-water Alcyonacea off the Northeast Coast of the United States. In: A Freiwald, JM Roberts (Eds) Cold-water Corals and Ecosystems. Springer, Berlin, pp 279–296

Weaver DC, Sedberry GR (2001) Trophic subsidies at the Charleston Bump: food web structure of reef fishes on the continental slope of the southeastern United States. In: GR Sedberry (Ed) Island in the Stream: Oceanography and Fisheries of the Charleston Bump. Amer Fish Soc Sympos 25, Bethesda MD, pp 137–152

Wenner EL, Barans CA (1990) In situ estimates of density of golden crab, *Chaceon fenneri*, from habitats on the continental slope, southeastern U.S. Bull Mar Sci 46:723–734

Wenner EL, Barans CA (2001) Benthic habitats and associated fauna of the upper- and middle-continental slope near the Charleston Bump. In: GR Sedberry (Ed) Island in the Stream: Oceanography and Fisheries of the Charleston Bump. Amer Fish Soc Sympos 25, Bethesda, MD, pp 161–175

Wheeler AJ, Beck T, Thiede J, Klages M, Grehan A, Monteys FX, Polarstern ARK XIX/3a Shipboard Party (2005a) Deep-water cold-water coral carbonate mounds on the Porcupine Bank, Irish margin: preliminary results from Polarstern ARK-XIX/3a ROV cruise. In: A Freiwald, JM Roberts (Eds) Cold-water Corals and Ecosystems. Springer, Berlin, pp 323–333

Wheeler AJ, Bett BJ, Billett DSM, Masson DG, Mayor D (2005b) The impact of demersal trawling on NE Atlantic deep-water coral habitats: the case of the Darwin Mounds, UK. In: J Thomas, P Barnes (Eds) Benthic Habitats and the Effects of Fishing. Am Fish Soc Sympos 41, Bethesda, MD, pp 807–818

Wheeler AJ, Kozachenko M, Beyer A, Foubert A, Huvenne VAI, Klages M, Masson DG, Olu-Le Roy K, Thiede J (2005c) Sedimentary processes and carbonate mounds in the Belgica mound province, Porcupine Seabight, NE Atlantic. In: A Freiwald, JM Roberts (Eds) Cold-water Corals and Ecosystems. Springer, Berlin, pp 533–564

Wheeler AJ, Beyer A, Freiwald A, de Haas H, Huvenne VAI, Kozachenko M, Olu-Le Roy K (2007) Morphology and environment of deep-water coral mounds on the NW European margin. Intl J Earth Sci 96:37–56

White M, Mohn C, de Stigter H, Mottram G (2005) Deep-water coral development as a function of hydrodynamics and surface productivity around submarine banks of the Rockall Trough, NE Atlantic. In: A Freiwald, JM Roberts (Eds) Cold-water Corals and Ecosystems. Springer, Berlin, pp 503–514

Williams B, Risk MJ, Ross SW, Sulak KJ (2006) Deep-water Antipatharians: proxies of environmental change. Geology 34:773–776

Williams T, Kano A, Ferdelman T, Henriet J-P, Abe K, Andres MS, Bjerager M, Browning EL, Cragg BA, De Mol B, Dorschel B, Foubert A, Frannk TD, Fuwa Y, Gaillot P, Gharib JJ, Gregg JM, Huvenne VAI, Léonide P, Li X, Mangelsdorf K, Tanaka A, Monteys X, Novosel I, Sakai S, Samarkin VA, Sasaki K, Spivack AJ, Takashima C, Titschack J (2006) Cold-water coral mounds revealed. EOS 87(47):525–526

Willison JHM, Hall J, Gass SE, Kenchington ELR, Butler M, Doherty P (Eds) (2001) Proc 1st Intl Sympos Deep-sea Corals, 30 Jul–2 Aug 2000, Ecology Action Centre, Nova Scotia Museum, Halifax, Nova Scotia, Canada

Wilson JB (1979) 'Patch' development of the deep-water coral *Lophelia pertusa* (L.) on Rockall Bank. J Mar Biol Assoc UK 59:165–177

Witherell D, Coon C (2001) Protecting corals off Alaska from fishing impacts. Proc 1st Intl Sympos Deep-sea Corals, 30 Jul–2 Aug 2000, Ecology Action Centre, Nova Scotia Museum, Halifax, Nova Scotia, Canada

Zibrowius H (1980) Les scléractiniaires de la Méditerranee et de l'Atlantique nord-oriental. Mém Ínst Oceanogr Monaco 11:1–284

Index

Coral Reefs of the World

1. B.M. Riegl and R.E. Dodge (eds.): *Coral Reefs of the World Vol. I.* Coral Reefs of the USA. 2008.
ISBN 978-1-4020-6846-1